Dieter K. Fütterer

Detlef Damaske

Georg Kleinschmidt

Hubert Miller

Franz Tessensohn

(Editors)

ANTARCTICA

Contributions to Global Earth Sciences

Dieter K. Fütterer
Detlef Damaske
Georg Kleinschmidt
Hubert Miller
Franz Tessensohn
(Editors)

ANTARCTICA

Contributions to Global Earth Sciences

Proceedings of the IX International Symposium of Antarctic Earth Sciences Potsdam, 2003

With 289 Figures, 47 in color

Springer

Editors

Prof. Dr. Dieter Karl Fütterer
Alfred Wegener Institute for Polar and Marine Research
P.O. Box 12 01 61, 27515 Bremerhaven, Germany
E-mail: dfuetterer@awi-bremerhaven.de

Dr. Detlef Damaske
Federal Institute of Geosciences and Natural Resources (BGR)
Stilleweg 2, 30655 Hannover, Germany
E-mail: d.damaske@bgr.de

Prof. Dr. Georg Kleinschmidt
Institute for Geology and Paleontology, J. W.-Goethe-University
Senckenberganlage 32, 60054 Frankfurt a. M., Germany
E-mail: kleinschmidt@em.uni-frankfurt.de

Prof. Dr. Dr. h.c. Hubert Miller
Dept. of Earth and Environmental Sciences, Section Geology, Ludwig-Maximilians-University
Luisenstr. 37, 80333 München, Germany
E-mail: h.miller@lmu.de

Dr. Franz Tessensohn
Lindenring 6, 29352 Adelheidsdorf, Germany
E-mail: franz.tessensohn@tiscali.de

Cover photo: Scenic impression of Marguerite Bay coastline, Antarctic Peninsula, West Antarctica (*photograph:* AWI). *Inset left:* Multispectral satellite image map of Antarctica using Advanced Very High Resolution Radiometer (AVHRR) (*image:* USGS). *Inset right:* Multibeam swath-sonar record of sub-sea volcanic structures in Bransfield Strait, Antarctic Peninsula (*image:* AWI).

Library of Congress Control Number: 2005936395

ISBN-10 3-540-30673-0 **Springer Berlin Heidelberg New York**
ISBN-13 978-3-540-30673-3 **Springer Berlin Heidelberg New York**

Springer is a part of Springer Science+Business Media
springeronline.com

Printed in Germany

Cover design: Erich Kirchner, Heidelberg
Typesetting: Büro Stasch · Bayreuth (stasch@stasch.com)
Production: Almas Schimmel
Printing and binding: Stürtz AG, Würzburg

Printed on acid-free paper 32/3141/as – 5 4 3 2 1 0

Preface

The almost completely ice covered Antarctic continent and the seasonally to permanently ice covered Southern Ocean surrounding it, have in no way lost their fascination and interest for geoscientists. With increasing scientific and public awareness for global change matters, the Antarctic region receives particular interest. Ongoing research on marine sediments, lake sediments and ice cores shed more and more light on climate history not only of the Antarctic region in particular but also on a global scale. For decades, Antarctic hard rock research relied on comparisons of the continent's scarce outcrops with those of the better exposed parts of Gondwana for an understanding of the geology of Antarctica. More recently, the results of broader Antarctic research, including of the deep ocean floor, have been introduced to help understanding Gondwana accretion and dispersal as a whole.

340 participants from 28 countries registered for the IX International Symposium of Antarctic Earth Sciences, held in the magnificent premises of the Potsdam University (see group photo) from 8 to 12 September 2003. The aim of the conference was to focus on the role of the Antarctic continent and the surrounding Southern Ocean in global geodynamics and paleoenvironmental evolution. This was pursued by a number of major themes, each subdivided in a number of sessions. Oral presentations added up to 175 and about 200 posters and maps, in three poster sessions, were contributed. Out of these, about 80 contributions were submitted for publication in the Proceedings Volume and 60 papers were finally accepted after peer-review.

At the beginning of planning for the IX International Symposium of Antarctic Earth Sciences the Steering Committee had discussed if it might be more appropriate to publish all submitted papers in one "big, more multidisciplinary volume" – as was the tradition so far – or better to publish in various more specific journals in order to maximise impact. There was a long controversial discussion which, in the end, reached no conclusion and the subject was kept open for consideration of the National Organising Committee, which finally decided to return to the concept of all earlier Antarctic earth science symposia.

As a consequence, you have this book in your hands, containing some 60 papers covering the traditionally wide field of Antarctic geoscientific research. The volume is organized into eight chapters, which to a large extent, follow the major themes of the symposium.

The contributions to the session "*James Ross and Seymour Islands*" and the results of the workshop on "*Seymour and James Ross Island paleoenvironments across the K/T boundary*", held before the symposium in Potsdam, will be published by the convenors in a separate volume entitled "*Cretaceous–Tertiary high-latitude paleoenvironments, James Ross Basin, Antarctica*".

Another special workshop on "*East–West Antarctic Tectonics and Gondwana Break-up 60° W to 30° E*" was held in the general framework of this ISAES conference. Seven papers, dealing with new geophysical interpretations and regional compilations of the East Antarctic continental margin from the Weddell Sea to the Cosmonaut Sea off Enderby Land, were published separately in a special issue of *Marine Geophysical Researches,* Vol. 25 (2005).

The Editors

9th International Symposium on Antarctic Earth Sciences, Potsdam, Germany. Photograph of participants, 11 September 2003, in front of the magnificent building of the Comuns, opposite the New Palace (Neues Palais), where one of the lecture halls of the symposium was placed. Today, the Comuns, part of the impressive New Palace complex which was built by Frederick the Great as a summer residence between 1763 and 1769, are used by the University of Potsdam. It seems more than strange to think that this imposing building, with its great double flights of steps and its columned halls, housed the domestic quarters and the kitchens, and that servants lived behind these facades

Acknowledgments

Scientific conferences are in no way self-organizing events but depend on great efforts from many people, not all of whom appear visibly on the stage. There is the International Steering Committee, which offered its generous advice, especially during the early phase of planning. There is the National Organizing Committee, which made the symposium happen, with the tremendous support of many local volunteers, students and scientists, again unnamed. Special thanks in this field go to Claudia Kirsch, Gabriela Schlaffer and Renate Wanke at the desk of the Conference Office for their untiring efforts in helping the participants to cope with local problems, informing on changes to the daily programme, and keeping the files in the symposium's background. Many thanks go also to the leaders of the excellent field excursions held before and after the symposium, as well as for various kinds of sightseeing tours in the Potsdam area, sometimes organized *ad hoc* during the symposium.

Last but not least, there is this book, which kept the editors busier than initially expected, so that the dream of publishing the volume within a year could not materialize. The book would never have materialized without the great effort of the numerous colleagues involved in the refereeing process. Many thanks to these colleagues!

Anandakrishnan, Sridhar, University Park
Barrett, Peter, Wellington
Behrend, John, Boulder
Bockheim, James, Madison
Bozzo, Emanuele, Genova
Brancolini, Giuliano, Trieste
Camerlenghi, Angelo, Barcelona
Capra, Alessandro, Taranto
Dalziel, Ian, Austin
Damm, Volkmar, Hannover
Dietrich, Reinhard, Dresden
Eckstaller, Alfons, Bremerhaven
Eisen, Olaf, Bremerhaven
Ferraccioli, Fausto, Cambridge
Fitzsimons, Ian, Perth
Froitzheim, Niko, Bonn
Goodge, John, Duluth
Grobe, Hannes, Bremerhaven
Harley, Simon, Edinburgh
Henjes-Kunst, Friedhelm, Hannover
Hillenbrand, Claus-Dieter, Cambridge
Horn, Peter, München
Jokat, Wilfried, Bremerhaven
Kleinschmidt, Georg, Frankfurt a. M.
Kölbl-Ebert, Martina, Eichstätt
Kraus, Stefan, München
Anderson, John, Houston
Beblo, Martin, Fürstenfeldbruck
Birkenmajer, Krzysztof, Krakow
Boger, Steve, Melbourne
Bradshaw, John, Christchurch
Buggisch, Werner, Erlangen
Capponi, Giovanni, Genova
Cremer, Holger, Utrecht
Damaske, Detlef, Hannover
Diester-Haass, Liselotte, Saarbrücken
Eagles, Graeme, Bremerhaven
Ehrmann, Werner, Leipzig
Fabian, Karl, Bremen
Finn, Carol, Denver
Frimmel, Hartmut, Würzburg
Gohl, Karsten, Bremerhaven
Gore, Damian, Sydney
Haak, Volker, Potsdam
Hegner, Ernst, München
Hervé, Francisco, Santiago de Chile
Hinz, Karl, Hannover
Jacobs, Joachim, Bremen
Kind, Rainer, Potsdam
Klemm, Dietrich, Dießen
Korth, Wilfried, Berlin
Krause, Reinhard, Bremerhaven

Kristoffersen, Yngve, Bergen
Larter, Robert, Cambridge
LeMasurier, Wesley, Denver
Lowry, Roy, Merseyside
Lukas, Sven, St. Andrews
Melles, Martin, Leipzig
Miller, Heinz, Bremerhaven
Müller, Christian, Bremerhaven
Olesch, Martin, Bremen
Pfeiffer, Eva-Maria, Hamburg
Reading, Anya, Canberra
Ricci, Carlo Alberto, Siena
Roland, Norbert, Hannover
Rutgers v. d. Loeff, Michiel, Bremerhaven
Saul, Joachim, Potsdam
Schenke, Hans-Werner, Bremerhaven
Schulz, Hartmut, Tübingen
Skinner, David, Lower Hutt
Steinhage, Daniel, Bremerhaven
Studinger, Michael, Palisades
Ten Brink, Uri, Woods Hole
Uenzelmann-Neben, Gabriele, Bremerhaven
Viereck-Götte, Lothar, Jena
Wiens, Douglas, St. Louis
Wilson, Christopher, Melbourne
Winckler, Stefan, Würzburg
Wörner, Gerhard, Göttingen
Läufer, Andreas, Hannover
Lawson, Jenifer, Chicago
López-Martínez, Jerónimo, Madrid
Lucchi, Renata, Barcelona
Mäusbacher, Roland, Jena
Meschede, Martin, Greifswald
Morelli, Andrea, Rome
Nowaczyk, Norbert, Potsdam
Pankhurst, Robert, Nottingham
Poutanen, Markku, Masala
Reitmayer, Gernot, Hannover
Röser, Hans, Hannover
Rowell, Albert, Lawrence
Salvini, Francesco, Roma
Scheinert, Mirko, Dresden
Schüssler, Ulrich, Würzburg
Shibuya, Kazuo, Tokyo
Sorlien, Christopher, Tappan
Storey, Bryan, Christchurch
Talarico, Franco, Siena
Trouw, Rudolph, Rio de Janeiro
Vaughan, Alan, Cambridge
Vogt, Steffen, Freiburg
Willan, Robert, Aberdeen
Wilson, Gary, Dunedin
Wise, Sherwood, Tallahassee

IX International Symposium on Antarctic Earth Sciences Held in Potsdam (Germany) from 8 to 12 September 2003

International Steering Committee

Fred Davey, Lower Hutt, New Zealand
Dieter K. Fütterer, Bremerhaven, Germany
Garrik Grikurov, St. Petersburg, Russia
Jerónimo López-Martínez, Madrid, Spain
Hubert Miller, München, Germany
Carlo Alberto Ricci, Siena, Italy
Roland Schlich, Strasbourg, France
Michael R.A. Thomson, Cambridge, United Kingdom
Rudolf A.J. Trouw, Rio de Janeiro, Brazil
Peter Noel Webb, Columbus/Ohio, USA

National Organizing Committee

Detlef Damaske, Federal Institute of Geosciences and Resources (BGR), Hannover
Reinhard Dietrich, Technical University of Dresden
Diedrich Fritzsche, Alfred Wegener Institute (AWI), Potsdam
Dieter K. Fütterer, Alfred Wegener Institute (AWI), Bremerhaven
Hans-W. Hubberten, Alfred Wegener Institute (AWI), Potsdam
Georg Kleinschmidt, University of Frankfurt a.M.
Heinz Miller, Alfred Wegener Institute (AWI), Bremerhaven
Hubert Miller, University of München (LMU)
Franz Tessensohn, Federal Institute of Geosciences and Resources (BGR), Hannover

Symposium Sponsors

Deutsche Forschungsgemeinschaft (DFG)
Scientific Committee on Antarctic Research (SCAR)
Alfred-Wegener-Institut für Polar- und Meeresforschung (AWI)
Bundesanstalt für Geowissenschaften und Rohstoffe (BGR)
Cambridge University Press, UK
Helicopter New Zealand LTD, New Zealand
Helicopter Resources PTY LTD, Australia
Kässbohrer Geländefahrzeug AG, Germany
Reederei Ferdinand Laiesz, Germany
Rieber Shipping of Norway

Contents

Theme 1
History of Antarctic Research ... 1

A. B. Ford
1.1 The Road to Gondwana via the Early SCAR Symposia ... 3

C. Lüdecke
1.2 Exploring the Unknown:
History of the First German South Polar Expedition 1901–1903 ... 7

Theme 2
Antarctica – The Old Core ... 13

M. Funaki · P. Dolinsky · N. Ishikawa · A. Yamazaki
2.1 Characteristics of Metamorphosed Banded Iron Formation
and Its Relation to the Magnetic Anomaly in the Mt. Riiser-Larsen Area,
Amundsen Bay, Enderby Land, Antarctica ... 15

T. Kawasaki · Y. Motoyoshi
2.2 Experimental Constraints on the Decompressional *P-T* Paths
of Rundvågshetta Granulites, Lützow-Holm Complex, East Antarctica ... 23

S. Baba · M. Owada · E. S. Grew · K. Shiraishi
2.3 Sapphirine – Orthopyroxene – Garnet Granulite from Schirmacher Hills,
Central Dronning Maud Land ... 37

M. J. D'Souza · A. V. K. Prasad · R. Ravindra
2.4 Genesis of Ferropotassic A-Type Granitoids of Mühlig-Hofmannfjella,
Central Dronning Maud Land, East Antarctica ... 45

A. K. Engvik · S. Elvevold
2.5 Late Pan-African Fluid Infiltration in the Mühlig-Hofmann-
and Filchnerfjella of Central Dronning Maud Land, East Antarctica ... 55

Y. Motoyoshi · T. Hokada · K. Shiraishi
2.6 Electron Microprobe (EMP) Dating on Monazite from Forefinger Point
Granulites, East Antarctica: Implication for Pan-African Overprint ... 63

E. V. Mikhalsky · A. A. Laiba · B. V. Beliatsky
2.7 Tectonic Subdivision of the Prince Charles Mountains:
A Review of Geologic and Isotopic Data ... 69

A. V. Golynsky · V. N. Masolov · V. S. Volnukhin · D. A. Golynsky
2.8 Crustal Provinces of the Prince Charles Mountains Region and Surrounding Areas in the Light of Aeromagnetic Data 83

A. V. Golynsky · D. A. Golynsky · V. N. Masolov · V. S. Volnukhin
2.9 Magnetic Anomalies of the Grove Mountains Region and Their Geological Significance 95

Theme 3
The Continent Beneath the Ice 107

A. V. Golynsky · M. Chiappini · D. Damaske · F. Ferraccioli · C. A. Finn · T. Ishihara H. R. Kim · L. Kovacs · V. N. Masolov · P. Morris · R. von Frese
3.1 ADMAP – A Digital Magnetic Anomaly Map of the Antarctic 109

R. E. Bell · M. Studinger · G. Karner · C. A. Finn · D. D. Blankenship
3.2 Identifying Major Sedimentary Basins Beneath the West Antarctic Ice Sheet from Aeromagnetic Data Analysis 117

D. S. Wilson · B. P. Luyendyk
3.3 Bedrock Plateaus within the Ross Embayment and beneath the West Antarctic Ice Sheet, Formed by Marine Erosion in Late Tertiary Time 123

I. Y. Filina · D. D. Blankenship · L. Roy · M. K. Sen · T. G. Richter · J. W. Holt
3.4 Inversion of Airborne Gravity Data Acquired over Subglacial Lakes in East Antarctica 129

V. N. Masolov · S. V. Popov · V. V. Lukin · A. N. Sheremetyev · A. M. Popkov
3.5 Russian Geophysical Studies of Lake Vostok, Central East Antarctica 135

S. V. Popov · A. N. Lastochkin · V. N. Masolov · A. M. Popkov
3.6 Morphology of the Subglacial Bed Relief of Lake Vostok Basin Area (Central East Antarctica) Based on RES and Seismic Data 141

M. Yamashita · H. Miyamachi · M. Kanao · T. Matsushima · S. Toda · M. Takada · A. Watanabe
3.7 Deep Reflection Imaging beneath the Mizuho Plateau, East Antarctica, by SEAL-2002 Seismic Experiment 147

S. Pondrelli · L. Margheriti · S. Danesi
3.8 Seismic Anisotropy beneath Northern Victoria Land from SKS Splitting Analysis 155

Theme 4
Gondwana Margins in Antarctica 163

C. A. Finn · J. W. Goodge · D. Damaske · C. M. Fanning
4.1 Scouting Craton's Edge in Paleo-Pacific Gondwana 165

G. Kleinschmidt · A. L. Läufer
4.2 The Matusevich Fracture Zone in Oates Land, East Antarctica 175

E. Stump · B. Gootee · F. Talarico
4.3 Tectonic Model for Development of the Byrd Glacier Discontinuity and Surrounding Regions of the Transantarctic Mountains during the Neoproterozoic – Early Paleozoic 181

B. Gootee · E. Stump
4.4 Depositional Environments of the Byrd Group, Byrd Glacier Area: A Cambrian Record of Sedimentation, Tectonism, and Magmatism 191

A. L. Läufer · G. Kleinschmidt · F. Rossetti
4.5 Late-Ross Structures in the Wilson Terrane in the Rennick Glacier Area (Northern Victoria Land, Antarctica) 195

C. J. Adams
4.6 Style of Uplift of Paleozoic Terranes in Northern Victoria Land, Antarctica: Evidence from K-Ar Age Patterns 205

Theme 5
Antarctic Peninsula Active Margin Tectonics 215

F. Hervé · H. Miller · C. Pimpirev
5.1 Patagonia – Antarctica Connections before Gondwana Break-Up 217

T. Janik · P. Środa · M. Grad · A. Guterch
5.2 Moho Depth along the Antarctic Peninsula and Crustal Structure across the Landward Projection of the Hero Fracture Zone 229

J. Galindo-Zaldívar · J. C. Balanyá · F. Bohoyo · A. Jabaloy · A. Maldonado
J. M. Martínez-Martínez · J. Rodríguez-Fernández · E. Suriñach
5.3 Crustal Thinning and the Development of Deep Depressions at the Scotia-Antarctic Plate Boundary (Southern Margin of Discovery Bank, Antarctica) ... 237

J. Galindo-Zaldívar · L. Gamboa · A. Maldonado · S. Nakao · Y. Bochu
5.4 Bransfield Basin Tectonic Evolution 243

C. Pimpirev · K. Stoykova · M. Ivanov · D. Dimov
5.5 The Sedimentary Sequences of Hurd Peninsula, Livingston Island, South Shetland Islands: Part of the Late Jurassic–Cretaceous Depositional History of the Antarctic Peninsula 249

J. F. Dumont · E. Santana · F. Hervé · C. Zapata
5.6 Regional Structures and Geodynamic Evolution of North Greenwich (Fort Williams Point) and Dee Islands, South Shetland Islands 255

S. B. Kim · Y. K. Sohn · M. Y. Choe
5.7 The Eocene Volcaniclastic Sejong Formation, Barton Peninsula, King George Island, Antarctica: Evolving Arc Volcanism from Precursory Fire Fountaining to Vulcanian Eruptions 261

J. Galindo-Zaldívar · A. Maestro · J. López-Martínez · C. S. de Galdeano
5.8 Elephant Island Recent Tectonics in the Framework of the Scotia-Antarctic-South Shetland Block Triple Junction (NE Antarctic Peninsula) ... 271

J. López-Martínez · R. A. J. Trouw · J. Galindo-Zaldívar · A. Maestro · L. S. A. Simões
F. F. Medeiros · C. C. Trouw
5.9 Tectonics and Geomorphology of Elephant Island, South Shetland Islands ... 277

M. Berrocoso · A. García-García · J. Martín-Dávila · M. Catalán-Morollón · M. Astiz
M. E. Ramírez · C. Torrecillas · J. M. E. de Salamanca
5.10 Geodynamical Studies on Deception Island: DECVOL and GEODEC Projects .. 283

Theme 6
Antarctic Rift Tectonics ... 289

D. H. Elliot · E. H. Fortner · C. B. Grimes
6.1 Mawson Breccias Intrude Beacon Strata at Allan Hills, South Victoria Land: Regional Implications ... 291

W. E. LeMasurier
6.2 What Supports the Marie Byrd Land Dome? An Evaluation of Potential Uplift Mechanisms in a Continental Rift System ... 299

F. J. Davey · L. De Santis
6.3 A Multi-Phase Rifting Model for the Victoria Land Basin, Western Ross Sea ... 303

C. R. Fielding · S. A. Henrys · T. J. Wilson
6.4 Rift History of the Western Victoria Land Basin: A new Perspective Based on Integration of Cores with Seismic Reflection Data ... 309

S. C. Cande · J. M. Stock
6.5 Constraints on the Timing of Extension in the Northern Basin, Ross Sea ... 319

J. B. Colwell · H. M. J. Stagg · N. G. Direen · G. Bernardel · I. Borissova
6.6 The Structure of the Continental Margin off Wilkes Land and Terre Adélie Coast, East Antarctica ... 327

P. E. O'Brien · S. Stanley · R. Parums
6.7 Post-Rift Continental Slope and Rise Sediments from 38° E to 164° E, East Antarctica ... 341

Theme 7
Antarctic Neotectonics, Observatories and Data Bases ... 349

A. M. Reading
7.1 On Seismic Strain-Release within the Antarctic Plate ... 351

M. Scheinert · E. Ivins · R. Dietrich · A. Rülke
7.2 Vertical Crustal Deformation in Dronning Maud Land, Antarctica: Observation versus Model Prediction ... 357

M. Kanao · K. Kaminuma
7.3 Seismic Activity Associated with Surface Environmental Changes of the Earth System, around Syowa Station, East Antarctica ... 361

R. Kh. Greku · V. P. Usenko · T. R. Greku
7.4 Geodynamic Features and Density Structure of the Earth's Interior of the Antarctic and Surrounded Regions with the Gravimetric Tomography Method ... 369

A. Meloni · L. R. Gaya-Piqué · P. De Michelis · A. De Santis
7.5 Some Recent Characteristics of Geomagnetic Secular Variations in Antarctica ... 377

R. Greku · G. Milinevsky · Y. Ladanovsky · P. Bahmach · T. Greku
7.6 Topographic and Geodetic Research by GPS, Echosounding and ERS Altimetric, and SAR Interferometric Surveys during Ukrainian Antarctic Expeditions in the West Antarctic ... 383

M. Berrocoso · A. Fernández-Ros · C. Torrecillas · J. M. E. de Salamanca · M. E. Ramírez
A. Pérez-Peña · M. J. González · R. Páez · Y. Jiménez · A. García-García · M. Tárraga
F. García-García
7.7 Geodetic Research on Deception Island 391

C. Torrecillas · M. Berrocoso · A. García-García
7.8 The Multidisciplinary Scientific Information Support System (SIMAC) for Deception Island 397

H. Grobe · M. Diepenbroek · N. Dittert · M. Reinke · R. Sieger
7.9 Archiving and Distributing Earth-Science Data with the PANGAEA Information System 403

Theme 8
Sediments as Indicators for Antarctic Environment and Climate 407

C. Hanfland · W. Geibert · I. Vöge
8.1 Tracing Marine Processes in the Southern Ocean by Means of Naturally Occurring Radionuclides 409

H. Matsuoka · M. Funaki
8.2 Normalized Remanence in Sediments from Offshore Wilkes Land, East Antarctica 415

M. Pistolato · T. Quaia · L. Marinoni · L. M. Vitturi · C. Salvi · G. Salvi · M. Setti · A. Brambati
8.3 Grain Size, Mineralogy and Geochemistry in Late Quaternary Sediments from the Western Ross Sea outer Slope as Proxies for Climate Changes 423

P. Barker · E. Thomas
8.4 Potential of the Scotia Sea Region for Determining the Onset and Development of the Antarctic Circumpolar Current 433

A. Maldonado · A. Barnolas · F. Bohoyo · C. Escutia · J. Galindo-Zaldívar · J. Hernández-Molina
A. Jabaloy · F. J. Lobo · C. H. Nelson · J. Rodríguez-Fernández · L. Somoza · E. Suriñach · J. T. Vázquez
8.5 Seismic Stratigraphy of Miocene to Recent Sedimentary Deposits in the Central Scotia Sea and Northern Weddell Sea: Influence of Bottom Flows (Antarctica) 441

B. Wagner · H. Cremer
8.6 Limnology and Sedimentary Record of Radok Lake, Amery Oasis, East Antarctica 447

J. A. Strelin · T. Sone · J. Mori · C. A. Torielli · T. Nakamura
8.7 New Data Related to Holocene Landform Development and Climatic Change from James Ross Island, Antarctic Peninsula 455

J. Mori · T. Sone · J. A. Strelin · C. A. Torielli
8.8 Surface Movement of Stone-Banked Lobes and Terraces on Rink Crags Plateau, James Ross Island, Antarctic Peninsula 461

A. Navas · J. López-Martínez · J. Casas · J. Machín · J. J. Durán · E. Serrano · J.-A. Cuchi
8.9 Soil Characteristics along a Transect on Raised Marine Surfaces on Byers Peninsula, Livingston Island, South Shetland Islands 467

Index 475

List of Contributors

Adams, Christopher J. · (*205*)
Institute of Geological and Nuclear Sciences, PO Box 30368, Lower Hutt, New Zealand, <argon@gns.cri.nz>

Astiz, Mar · (283)
Departamento de Volcanología, Museo Nacional de Ciencias Naturales, C/ José Gutiérrez Abascal 2, 28006, Madrid, Spain

Baba, Sotaro · (*37*)
Department of Natural Environment, University of the Ryukyus, Senbaru 1, Nishihara, Okinawa 903-0213, Japan

Bakhmach, Pavel · (383)
ECOMM Co, 18/7 Kutuzov St., 01133 Kyiv, Ukraine

Balanyá, Juan Carlos · (237)
Departamento de Ciencias Ambientales, Univ. Pablo de Olavide, Sevilla, Spain;
and: Instituto Andaluz Ciencias de la Tierra, CSIC/Universidad Granada, 18002 Granada, Spain

Barker, Peter · (433)
Threshers Barn, Whitcott Keysett, Clun, Shropshire SY7 8QE, UK, <pfbarker@tiscali.co.uk>

Barnolas, Antonio · (441)
Instituto Geológico y Minero de España, Ríos Rosas, 23, 28003 Madrid, Spain

Beliatsky, Boris V. · (69)
IGGP, Makarova emb. 2, St. Petersburg 199034, Russia

Bell, Robin E. · (*117*)
Lamont-Doherty Earth Observatory of Columbia University, 61 Route 9W, Palisades, NY 10964, USA

Bernardel, George · (327)
Geoscience Australia, GPO Box 378, Canberra, ACT, Australia

Berrocoso, Manuel · (*283*, *391*, 397)
Laboratorio de Astronomía y Geodesia, Departamento de Matemáticas, Facultad de Ciencias, Universidad de Cádiz, Campus Río San Pedro, 11510 Puerto Real, Cádiz, Spain, <manuel.berrocoso@uca.es>

Blankenship, Donald D. · (117, 129)
Institute for Geophysics, John A. and Katherine G. Jackson School of Geosciences, University of Texas at Austin, 4412 Spicewood Springs Rd., Bldg. 600 Austin, TX 78759-8500, USA

Bochu, Yao · (243)
Guangzhou Marine Geological Survey, Guangzhou, PR China

Bohoyo, Fernando · (237, 441)
Instituto Andaluz Ciencias de la Tierra, CSIC/Universidad Granada, 18002 Granada, Spain, <fbohoyo@ugr.es>

Borissova, Irina · (327)
Geoscience Australia, GPO Box 378, Canberra, ACT, Australia

Brambati, Antonio · (423)
Dipartimento di Scienze Geologiche, Ambientali e Marine, University of Trieste, Via E. Weiss 2, 34127 Trieste, Italy

Cande, Steven C. · *(319)*
Scripps Institution of Oceanography, Mail Code 0220, La Jolla, CA 92093-0220, USA, <scande@ucsd.edu>

Casas, José · (467)
Departamento de Química Agrícola, Geología y Geoquímica, Facultad de Ciencias, Universidad Autónoma de Madrid, 28049 Madrid, Spain, <jeronimo.lopez@uam.es>; *and:* Centro de Ciencias Medioambientales, CSIC, Serrano 115, 28006 Madrid, Spain, <j.casassainzdeaja@uam.es>

Catalán-Morollón, Manuel · (283)
Sección de Geofísica, Real Instituto y Observatorio de la Armada, 11110 San Fernando, Cádiz, Spain

Chiappini, Massimo · (109)
INGV, via di Vigna Murata 605, 00143 Roma, Italy

Choe, Moon Young · (261)
Korea Polar Research Institute, Korea Ocean Research and Development Institute, Ansan 426-744, Korea, <mychoe@kopri.re.kr>

Colwell, James B. · *(327)*
Geoscience Australia, GPO Box 378, Canberra, ACT, Australia

Cremer, Holger · (447)
Utrecht University, Department of Palaeoecology, Laboratory of Palaeobotany and Palynology, Budapestlaan 4, 3584 CD Utrecht, The Netherlands;
present address: Netherlands Organization of Applied Scientific Research TNO, Core area Built Environment and Geosciences, Geological Survey of the Netherlands, Princetonlaan 6, 3584 CB Utrecht, The Netherlands, <holger.cremer@tno.nl>

Cuchi, José-Antonio · (467)
Escuela Politécnica Superior de Huesca, Universidad de Zaragoza, 22071 Huesca, Spain, <cuchi@unizar.es>

D'Souza, Mervin J. · *(45)*
Antarctica Division, Geological Survey of India, Faridabad, India, <antgsi@vsnl.net>

Damaske, Detlef · (109, 165)
Bundesanstalt für Geowissenschaften und Rohstoffe (BGR), Stilleweg 2, 30655 Hannover, Germany

Danesi, Stefania · (155)
Instituto Nazionale di Geofisica e Vulcanologia, Via di Vigna Murata 605, 00143 Rome, Italy

Davey, Fred J. · *(303)*
Institute of Geological and Nuclear Sciences, Lower Hutt, New Zealand, <F.davey@gns.cri.nz>

de Galdeano, Carlos Sanz · (271)
Instituto Andaluz de Ciencias de la Tierra, CSIC/Universidad de Granada, 18002 Granada, Spain, <csanz@ugr.es>

De Michelis, Paola · (377)
Instituto Nazionale di Geofisica e Vulcanologia, INGV, Via di Vigna Murata 605, 00143 Rome, Italy

de Salamanca, José Manuel Enríquez · (283, 391)
Laboratorio de Astronomía y Geodesia, Facultad de Ciencias, Universidad de Cádiz, Campus Río San Pedro, 11510 Puerto Real, Cádiz, Spain

De Santis, Angelo · (377)
Instituto Nazionale di Geofisica e Vulcanologia, INGV, Via di Vigna Murata 605, 00143 Rome, Italy

De Santis, Laura · (303)
Instituto Nazionale di Oceanografia e di Geofisica Sperimentale, Trieste, Italy, <ldesantis@OGS.trieste.it>

Diepenbroek, Michael · (403)
Center for Marine Environmental Sciences, Leobener Str. 26, 28359 Bremen, Germany

Dietrich, Reinhard · (357)
Technische Universität Dresden, Institut für Planetare Geodäsie, Helmholtzstraße 10, 01062 Dresden, Germany

Dimov, Dimo · (249)
Sofia University St. Kliment Ohridski, Department of Geology and Paleontology, Tsar Osvoboditel 15, 1000 Sofia, Bulgaria, <polar@gea.uni-sofia.bg>

Direen, Nicholas G. · (327)
Continental Evolution Research Group, University of Adelaide, SA, Australia

Dittert, Nicolas · (403)
Center for Marine Environmental Sciences, Leobener Str. 26, 28359 Bremen, Germany

Dolinsky, Peter · (15)
Geophysical Institute of Slovak Academy of Sciences, Komarno 108, Hurbanovo 947 01, Slovakia

Dumont, Jean Francois · (*255*)
IRD-Geosciences Azur, Observatoire Oceanographique, BP 48 La Darsse, 06235 Villefranche sur Mer, France, <dumon@obs-vlfr.fr>

Durán, Juan José · (467)
Instituto Geológico y Minero de España, Ríos Rosas 23, 28003 Madrid, <jj.duran@igme.es>

Elliot, David H. · (*281*)
Department of Geological Sciences and Byrd Polar Research Center, The Ohio State University, Columbus, Ohio 43210, USA

Elvevold, Synnøve · (55)
Norwegian Polar Institute, N-9296 Tromsø, Norway, <elvevold@npolar.no>

Engvik, Ane K. · (*55*)
Geological Survey of Norway, N-7491 Trondheim, Norway, <ane.engvik@geo.ntnu.no>

Escutia, Carlota · (441)
Instituto Andaluz Ciencias de la Tierra, CSIC/Universidad Granada, 18002 Granada, Spain

Fanning, C. Mark · (165)
Research School of Earth Sciences, The Australian National University, Mills Road, Canberra, ACT 0200, Australia

Fernández-Ros, Alberto · (391)
Laboratorio de Astronomía y Geodesia, Facultad de Ciencias, Universidad de Cádiz, Campus Río San Pedro, 11510 Puerto Real, Cádiz, Spain

Ferraccioli, Fausto · (109)
DIPTERIS Università di Genova, Viale Benedetto XV, 5 16132 Genova, Italy;
now: British Antarctic Survey, BAS, High Cross, Madingley Road, Cambridge CB3 OET, UK

Fielding, Christopher R. · (*309*)
Department of Geosciences, 214 Bessey Hall, University of Nebraska-Lincoln, NE 68588-0340, USA, <cfielding2@unl.edu>

Filina, Irina Y. · (*129*)
Department of Geological Sciences, John A. and Katherine G. Jackson School of Geosciences, University of Texas at Austin, 1 University Station C1100, Austin, TX 78712-0254, USA;
and: Institute for Geophysics, John A. and Katherine G. Jackson School of Geosciences, University of Texas at Austin, 4412 Spicewood Springs Rd. #600 Austin, TX 78759-8500, USA

Finn, Carol A. · (109, 117, *165*)
U.S. Geological Survey, MS 945, Denver Federal Center, Denver, CO 80226, USA

Ford, Arthur B. · (*3*)
Denali Associates, 400 Ringwood Avenue, Menlo Park, CA 94025, USA, <abford@aol.com>

Fortner, Everett H. · (291)
Department of Geological Sciences and Byrd Polar Research Center, The Ohio State University, Columbus, Ohio 43210, USA

Funaki, Minoru · (*15*, 415)
National Institute of Polar Research, 9–10 Kaga 1 Itabashi, Tokyo 173-8515, Japan

Galindo-Zaldívar, Jesús · (*237*, *243*, *271*, 277, 441)
Departamento de Geodinámica, Facultad de Ciencias, Universidad de Granada, 18071 Granada, Spain, <jgalindo@ugr.es>

Gamboa, Luiz · (243)
Petróleo Brasileiro S.A. (PETROBRAS), Rio de Janeiro, Brazil;
and: Universidade Federal Fluminenese, Brazil, <gamboa@petrobras.br>

García-García, Alicia · (283, 391, 397)
Departamento de Volcanología, Museo Nacional de Ciencias Naturales, C/José Gutiérrez Abascal 2, 28006 Madrid, Spain

García-García, Francisco · (391)
Departamento de Ingeniería Cartográfica, Geodesia y Fotogrametría, E.T.S.I. Geodésica, Cartográfica y Topográfica, Universidad Politécnica de Valencia, Camino de Vera, s/n, 46022 Valencia, Spain

Gaya-Piqué, Luis R. · (377)
Instituto Nazionale di Geofisica e Vulcanologia, INGV, Via di Vigna Murata 605, 00143 Rome, Italy;
and: Observatori de l'Ebre, Horta Alta 38, 43520 Roquetes, Spain

Geibert, Walter · (409)
Alfred Wegener Institute for Polar and Marine Research, Am Handelshafen 12, 27570 Bremerhaven, Germany

Golynsky, Alexander V. · (*83*, *95*, *109*)
VNIIOkeangeologia, Angliysky Avenue 1, St. Petersburg 190121, Russia, <sasha@vniio.nw.ru>

Golynsky, Dmitry A. · (83, 95)
SPbSU, 7/9 Universitetskaya nab., St. Petersburg 199034, Russia

González, M. José · (391)
Laboratorio de Astronomía y Geodesia, Facultad de Ciencias, Universidad de Cádiz, Campus Río San Pedro, 11510 Puerto Real, Cádiz, Spain

Goodge, John W. · (165)
Department of Geological Sciences, University of Minnesota, Duluth, MN 55812, USA

Gootee, Brian · (181, *191*)
Department of Geological Sciences, Arizona State University, Tempe, Arizona AZ 85287-1404, USA

Grad, Marek · (229)
Institute of Geophysics, University of Warsaw, Pasteura 7, 02-093 Warsaw, Poland

Greku, Rudolf Kh. · (*369*, *383*)
Institute of Geological Sciences, National Academy of Sciences of Ukraine, 55B Gonchara St., 01054 Kiev, Ukraine, <satmar@svitonline.com>

Greku, Tatyana R. · (369, 383)
Institute of Geological Sciences, National Academy of Sciences of Ukraine, 55B Gonchara St., 01054 Kiev, Ukraine, <satmar@svitonline.com>

Grew, Edward S. · (37)
Department of Earth Sciences, University of Maine, 5790 Bryand Research Center Orono, Maine 04469-5790, USA

Grimes, Craig B. · (291)
Department of Geological Sciences and Byrd Polar Research Center, The Ohio State University, Columbus, Ohio 43210, USA

Grobe, Hannes · (*403*)
Alfred Wegener Institute for Polar and Marine Research, Am Alten Hafen 26, 27568 Bremerhaven, Germany, <hgrobe@awi-bremerhaven.de>

Guterch, Aleksander · (229)
Institute of Geophysics, Polish Academy of Sciences, Ks. Janusza 64, 01-452 Warsaw, Poland

Hanfland, Claudia · (*409*)
Alfred Wegener Institute for Polar and Marine Research, Am Handelshafen 12, 27570 Bremerhaven, Germany, <chanfland@awi-bremerhaven.de>

Henrys, Stuart A. · (309)
Institute of Geological and Nuclear Sciences, PO Box 30-368, Lower Hutt, New Zealand

Hernández-Molina, Javier · (441)
Facultad de Ciencias del Mar, Departamento de Geociencias Marinas, 36200 Vigo, Spain

Hervé, Francisco · (*217*, 255)
Universidad de Chile, Departamento de Geología, Casilla 13518, Correo 21, Santiago, Chile, <fhervé@cec.uchile.cl>

Hokada, Tomokazu · (63)
National Institute of Polar Research, 1-9-10 Kaga, Itabashi-ku, Tokyo 173-8515, Japan

Holt, John W. · (129)
Institute for Geophysics, John A. and Katherine G. Jackson School of Geosciences, University of Texas at Austin, 4412 Spicewood Springs Rd. #600 Austin, TX 78759-8500, USA

Ishihara, Takemi · (109)
Geological Survey of Japan, 1-1-3, Higashi, SUKUBA Ibaraki 305, Japan

Ishikawa, Naoto · (15)
Kyoto University, Nihonmatsu-cho, Yoshida Sakyo-ku Kyoto 606-8501, Japan

Ivanov, Marin · (249)
Sofia University St. Kliment Ohridski, Department of Geology and Paleontology, Tsar Osvoboditel 15, 1000 Sofia, Bulgaria, <mivanov@gea.uni-sofia.bg>

Ivins, Erik · (357)
Jet Propulsion Laboratory, California Institute of Technology, Pasadena CA 91109-8099, USA

Jabaloy, Antonio · (237, 441)
Departamento de Geodinámica, Universidad de Granada, 18071 Granada, Spain, <jabaloy@ugr.es>

Janik, Tomasz · (*229*)
Institute of Geophysics, Polish Academy of Sciences, Ks. Janusza 64, 01-452 Warsaw, Poland, <janik@igf.edu.pl>

Jiménez, Yolanda · (391)
Laboratorio de Astronomía y Geodesia, Facultad de Ciencias, Universidad de Cádiz, Campus Río San Pedro, 11510 Puerto Real, Cádiz, Spain

Kaminuma, Katsutada · (361)
Department of Earth Science, National Institute of Polar Research, 9-10 kaga-1, Itabashi-ku, Tokyo 173-8515, Japan, <kaminuma@nipr.ac.jp>

Kanao, Masaki · (147, *361*)
Department of Earth Science, National Institute of Polar Research, 9-10 kaga-1, Itabashi-ku, Tokyo 173-8515, Japan, <kanao@nipr.ac.jp>

Karner, Garry · (117)
Lamont-Doherty Earth Observatory of Columbia University, 61 Route 9W, Palisades, NY 10964, USA

Kawasaki, Toshisuke · (*23*)
Department of Earth Sciences, Faculty of Science, Ehime University, Bunkyo-cho 2–5,
Matsuyama 790-8577, Japan, <toshkawa@sci.ehime-u.ac.jp>

Kim, Hyung Rae · (109)
Ohio State University, 381, Mendenhall Lab 125 S. Oval Mall, Columbus, OH 43210-1398, USA

Kim, Seung Bum · (*261*)
Korea Polar Research Institute, Korea Ocean Research and Development Institute, Ansan 426-744,
Korea, <sbkim@kopri.re.kr>;
present address: Domestic Exploration Team II, Korea National Oil Corporation, Anyang 431-711,
Korea, <sbkim@knoc.co.kr>

Kleinschmidt, Georg · (*175*, 195)
Geologisch-Paläontologisches Institut der J. W.-Goethe-Universität, Senckenberganlage 32,
60054 Frankfurt am Main, Germany, <kleinschmidt@em.uni-frankfurt.de>

Kovacs, Luis · (109)
NRL, 4555 Overlook Ave. SW, Washington, DC, 20375-5320, USA

Ladanovsky, Yuriy · (383)
ECOMM Co, 18/7 Kutuzov St., 01133 Kyiv, Ukraine, <lada@ecomm.kiev.ua>

Laiba, Anatoly A. · (69)
Polar Marine Geological Research Expedition (PMGRE), Pobeda Street 24, Lomonosov 189510, Russia

Lastochkin, Alexander N. · (141)
St. Petersburg State University (SPSU), 7/9 Universitetskaya nab., 191164 St. Petersburg, Russia

Läufer, Andreas L. · (175, *195*)
Geologisch-Paläontologisches Institut, J. W.-Goethe-Universität, Senckenberganlage 32,
60054 Frankfurt am Main, Germany, <laeufer@em.uni-frankfurt.de>;
present address: Bundesanstalt für Geowissenschaften und Rohstoffe (BGR), Stilleweg 2,
30655 Hannover, Germany

LeMasurier, Wesley E. · (*299*)
Institute of Arctic and Alpine Research (INSTAAR), University of Colorado, Boulder, CO 80309-0450,
USA, <Wesley.LeMasurier@colorado.edu>

Lobo, Francisco J. · (441)
Instituto Andaluz Ciencias de la Tierra, CSIC/Universidad Granada, 18002 Granada, Spain

López-Martínez, Jerónimo · (271, *277*, 467)
Departamento de Química Agrícola, Geología y Geoquímica, Facultad de Ciencias,
Universidad Autónoma de Madrid, 28049 Madrid, Spain, <jeronimo.lopez@uam.es>

Lüdecke, Cornelia · (*7*)
Valleystrasse 40, 81371 München, Germany, <C.Luedecke@lrz.uni-muenchen.de>

Lukin, Valeriy V. · (135)
Russian Antarctic Expedition (RAE), 38 Bering St., 199397 St. Petersburg, Russia

Luyendyk, Bruce P. · (123)
Dept. of Geological Sciences and Inst. for Crustal Studies, University of California,
Santa Barbara, CA 93106, USA, <luyendyk@geol.ucsb.edu>

Machín, Javier · (467)
Estación Experimental de Aula Dei, CSIC, Apartado 202, 50080 Zaragoza, Spain

Maestro, Adolfo · (271, 277)
Servicio de Geología Marina, Instituto Geológico y Minero de España, Ríos Rosas 23, 28003 Madrid,
Spain, <a.maestro@igme.es>

Maldonado, Andrés · (237, 243, *441*)
Instituto Andaluz de Ciencias de la Tierra, CSIC-Universidad de Granada, 18071-Granada, Spain, <amaldona@ugr.es>

Margheriti, Lucia · (155)
Instituto Nazionale di Geofisica e Vulcanologia, Via di Vigna Murata 605, 00143 Rome, Italy

Marinoni, Luigi · (423)
Dipartimento di Scienze della Terra, University of Pavia, Via Ferrata 1, 27100 Pavia, Italy

Martín-Dávila, José · (283)
Sección de Geofísica, Real Instituto y Observatorio de la Armada, 11110 San Fernando, Cádiz, Spain

Martínez-Martínez, José Miguel · (237)
Departamento de Geodinámica, Universidad de Granada, 18071 Granada, Spain, <jmmm@ugr.es>; *and:* Instituto Andaluz Ciencias de la Tierra, CSIC/Universidad Granada, 18002 Granada, Spain

Masolov, Valeriy N. · (83, 95, 109, *135*, 141)
Polar Marine Geological Research Expedition (PMGRE), 24 Pobeda St., 188512 St. Petersburg, Lomonosov, Russia

Matsuoka, Haruka · (*415*)
National Institute of Polar Research, Kaga 1-chome, Itabashi-ku, Tokyo 173-8515, Japan; *present address:* Gakushuin University, Computer Center, 1-5-1 Mejiro, Toshima-ku, Tokyo 171-8588, Japan

Matsushima, Takeshi · (147)
Institute of Seismology and Volcanology, Faculty of Sciences, Kyushu University, Shinyama 2, Shimabara 855-0843, Japan

Medeiros, Felipe F. · (277)
Dpt. Geologia, Universidade Federal do Rio de Janeiro, CEP21949-900 Brasil

Meloni, Antonio · (*377*)
Instituto Nazionale di Geofisica e Vulcanologia, INGV, Via di Vigna Murata 605, 00143 Rome, Italy, <meloni@ingv.it>

Mikhalsky, Evgeny V. · (*69*)
VNIIOkeangeologia, Angliysky Avenue 1, St. Petersburg 190121, Russia

Milinevsky, Gennady · (383)
Ukrainian Antarctic Center, Kyiv/Ukraine, <antarc@carrier.kiev.ua>

Miller, Hubert · (217)
Ludwig-Maximilians-Universität, Department of Earth and Environmental Sciences, Sektion Geologie, Luisenstraße 37, 80333 München, Germany, <h.miller@lmu.de>

Miyamachi, Hiroki · (147)
Faculty of Science, Kagoshima University, 1-21-35 Kourimoto, Kagoshima 890-0065, Japan

Mori, Junko · (455, *461*)
Graduate School of Engineering, Hokkaido University, N13 W8, Kita-ku, Sapporo 060-8628, Japan, <jmori@eng.hokudai.ac.jp>

Morris, Peter · (109)
British Antarctic Survey, BAS, High Cross, Madingley Road, Cambridge CB3 OET, UK

Motoyoshi, Yoichi · (23, *63*)
National Institute of Polar Research, Kaga 1-chome, Itabashi-ku, Tokyo 173-8515, Japan, <motoyosi@nipr.ac.jp>

Nakamura, Toshio · (455)
Dating and Materials Research Center, Nagoya University, Furo-cho, Chikusa-ku Nagoya 464-8602, Japan

Nakao, Seizo · (243)
Marine Geology Department, Geological Survey of Japan

Navas, Ana · (*467*)
Estación Experimental de Aula Dei, CSIC, Apartado 202, 50080 Zaragoza, Spain, <anavas@eead.csic.es>

Nelson, C. Hans · (441)
Instituto Andaluz Ciencias de la Tierra, CSIC/Universidad Granada, 18002 Granada, Spain

O'Brien, Philip E. · (*341*)
Geoscience Australia, GPO Box 378 Canberra, 2601, Australia, <Phil.OBrien@ga.gov.au>

Owada, Masaaki · (37)
Department of Earth Sciences, Yamaguchi University, Yoshida 1677-1, Yamaguchi 753-8512, Japan

Páez, Raúl · (391)
Laboratorio de Astronomía y Geodesia, Facultad de Ciencias, Universidad de Cádiz,
Campus Río San Pedro, 11510 Puerto Real, Cádiz, Spain

Parums, Robert · (341)
Geoscience Australia, GPO Box 378 Canberra, 2601, Australia

Pérez-Peña, Alejandro · (391)
Laboratorio de Astronomía y Geodesia, Facultad de Ciencias, Universidad de Cádiz,
Campus Río San Pedro, 11510 Puerto Real, Cádiz, Spain

Pimpirev, Christo · (217, *249*)
Sofia University St. Kliment Ohridski, Department of Geology and Paleontology, Tsar Osvoboditel 15,
1000 Sofia, Bulgaria, <polar@gea.uni-sofia.bg>

Pistolato, Mario · (*423*)
Dipartimento di Scienze Ambientali, University of Venezia, Dorsoduro 2137, 30123 Venezia, Italy

Pondrelli, Silvia · (*155*)
Instituto Nazionale di Geofisica e Vulcanologia, Via D. Creti 12, 40128 Bologna, Italy, <pondrelli@bo.ingv.it>

Popkov, Anatoly M. · (135, 141)
Polar Marine Geological Research Expedition (PMGRE), 24 Pobeda St., 188512 St. Petersburg,
Lomonosov, Russia

Popov, SergeyV. · (135, *141*)
Polar Marine Geological Research Expedition (PMGRE), 24 Pobeda St., 188512 St. Petersburg,
Lomonosov, Russia

Prasad, A. V. Keshava · (45)
Antarctica Division, Geological Survey of India, Faridabad, India, <antgsi@vsnl.net>

Quaia, Tullio · (423)
Dipartimento di Scienze Geologiche, Ambientali e Marine, University of Trieste, Via E. Weiss 2,
34127 Trieste, Italy

Ramírez, M. Eva · (283, 391)
Laboratorio de Astronomía y Geodesia, Facultad de Ciencias, Universidad de Cádiz,
Campus Río San Pedro, 11510 Puerto Real, Cádiz, Spain, <mariaeva.ramirez@uca.es>

Ravindra, Rasik · (45)
Antarctica Division, Geological Survey of India, Faridabad, India, <antgsi@vsnl.net>

Reading, Anya M. · (*351*)
Research School of Earth Sciences, Australian National University, Canberra, ACT 0200, Australia,
<anya@rses.anu.edu.au>

Reinke, Manfred · (403)
Alfred Wegener Institute for Polar and Marine Research, Am Alten Hafen 26, 27568 Bremerhaven, Germany

Richter, Thomas G. · (129)
Institute for Geophysics, John A. and Katherine G. Jackson School of Geosciences,
University of Texas at Austin, 4412 Spicewood Springs Rd. #600 Austin, TX 78759-8500, USA

Rodríguez-Fernández, José · (237, 441)
Departamento de Geodinámica, Universidad de Granada, 18071 Granada, Spain

Rossetti, Frederico · (195)
Dipartimento di Scienze Geologiche, Università Roma Tre, Largo S.L. Murialdo 1, 00146 Roma, Italy

Roy, Lopamudra · (129)
Institute for Geophysics, John A. and Katherine G. Jackson School of Geosciences, University of Texas at Austin, 4412 Spicewood Springs Rd. #600 Austin, TX 78759-8500, USA; *and:* Department of Applied Geophysics, Indian School of Mines, Dhanbad – 826004, Jharkhand, India

Rülke, Axel · (357)
Technische Universität Dresden, Institut für Planetare Geodäsie, Helmholtzstraße 10, 01062 Dresden, Germany

Salvi, Cristinamaria · (423)
Dipartimento di Scienze Geologiche, Ambientali e Marine, University of Trieste, Via E. Weiss 2, 34127 Trieste, Italy

Salvi, Gianguido · (423)
Dipartimento di Scienze Geologiche, Ambientali e Marine, University of Trieste, Via E. Weiss 2, 34127 Trieste, Italy

Santana, Essy · (255)
INOCAR, Unidad Convemar, Base Naval Sur, Av. 25 de Julio, via Puerto Marítimo, POX 5940, Guayaquil, Ecuador

Scheinert, Mirko · (357)
Technische Universität Dresden, Institut für Planetare Geodäsie, Helmholtzstraße 10, 01062 Dresden, Germany, <mikro@ipg.geo.tu-dresden.de>

Sen, Mrinal K. · (129)
Institute for Geophysics, John A. and Katherine G. Jackson School of Geosciences, University of Texas at Austin, 4412 Spicewood Springs Rd. #600 Austin, TX 78759-8500, USA

Serrano, Enrique · (467)
Departamento de Geografía, Universidad de Valladolid, 47011, Valladolid, Spain, <serranoe@fyl.uva.es>

Setti, Massimo · (423)
Dipartimento di Scienze della Terra, University of Pavia, Via Ferrata 1, 27100 Pavia, Italy

Sheremetyev, Alexander N. · (135)
Polar Marine Geological Research Expedition (PMGRE), 24 Pobeda St., 188512 St. Petersburg, Lomonosov, Russia

Shiraishi, Kazuyuki · (37, 63)
Department of Crustal Studies, National Institute of Polar Research, Kaga 1-9-10, Itabashi, Tokyo 173-8515, Japan

Sieger, Rainer · (403)
Alfred Wegener Institute for Polar and Marine Research, Am Alten Hafen 26, 27568 Bremerhaven, Germany

Simões, Luiz S. A. · (277)
Dpt. Petrologia e Metalogenia, Universidade Estadual Paulista, Brasil, <lsimoes@rc.unesp.br>

Sohn, Young Kwan · (261)
Department of Earth and Environmental Sciences, Gyeongsang National University, Jinju 660-701, Korea, <yksohn@gsnu.ac.kr>

Somoza, Luis · (441)
Instituto Geológico y Minero de España, Ríos Rosas 23, 28003 Madrid, Spain

Sone, Toshio · (455, 461)
Institute of Low Temperature Science, Hokkaido University, N19 W8, Kita-ku, Sapporo 060-0918, Japan

Środa, Piotr · (229)
Institute of Geophysics, Polish Academy of Sciences, Ks. Janusza 64, 01-452 Warsaw, Poland

Stagg, Howard M. J. · (327)
Geoscience Australia, GPO Box 378, Canberra, ACT, Australia

Stanley, Shawn · (341)
Geoscience Australia, GPO Box 378 Canberra, 2601, Australia

Stock, Joann M. · (319)
California Institute of Technology, Mail Stop 252-21, Pasadena, CA 91125, USA,
<jstock@gps.caltech.edu>

Stoykova, Kristalina · (249)
Geological Institute Bulgarian Academy of Sciences, Department of Paleontology and Stratigraphy,
24 Acad. G. Boncev Str., 1113 Sofia, Bulgaria, <stoykova@geology.bas.bg>

Strelin, Jorge A. · (*455*, 461)
Instituto Antártico Argentino, Centro Austral de Investigaciones Científicas,
and Universidad de Córdoba, Argentina;
present address: Departamento de Geología Básica, Universidad Nacional de Córdoba, Ciudad
Universitaria, Avda. Vélez Sársfield 1611, 5000 Córdoba, Argentina

Studinger, Michael · (117)
Lamont-Doherty Earth Observatory of Columbia University, 61 Route 9W, Palisades, NY 10964, USA

Stump, Edmund · (*181*, 191)
Department of Geological Sciences, Arizona State University, Tempe, Arizona AZ 85287-1404, USA

Suriñach, Emma · (237, 441)
Departament de Geodinàmica i Geofísica, Universitat de Barcelona, 08028 Barcelona, Spain,
<emma.surinach@ub.edu>

Takada, Masamitsu · (147)
Institute of Seismology and Volcanology, Graduate School of Science, Hokkaido University,
N10S8 Kita-ku, Sapporo 060-0810, Japan

Talarico, Franco · (181)
Dipartimento di Scienze della Terra, Università di Siena, Via del Laterino 8, 53100 Siena, Italy

Tárraga, Marta · (391)
Departamento de Volcanología, Museo Nacional de Ciencias Naturales, C/José Gutiérrez Abascal 2,
28006, Madrid, Spain

Thomas, Ellen · (433)
Center for the Study of Global Change, Dept. of Geology and Geophysics, Yale University,
New Haven CT 06520-1809, USA

Toda, Shigeru · (147)
Department of Earth Science, Faculty of Education, Aichi University of Education, Hirosawa 1, Igaya,
Kariya, Aichi 448-8542, Japan

Torielli, Cesar A. · (455, 461)
Departamento de Geología Básica, Universidad Nacional de Córdoba, Ciudad Universitaria,
Avda. Vélez Sársfield 1611, 5000 Córdoba, Argentina

Torrecillas, Cristina · (283, 391, *397*)
Laboratorio de Astronomía y Geodesia, Departamento de Matemáticas, Facultad de Ciencias,
Universidad de Cádiz, Campus Río San Pedro, 11510 Puerto Real, Cádiz, Spain,
<cristina.torrecillas@uca.es>

Trouw, Camilo C. · (277)
Dpt. Geologia, Universidade Federal do Rio de Janeiro, CEP21949-900 Brasil

Trouw, Rudolph A. J. · (277)
Dpt. Geologia, Universidade Federal do Rio de Janeiro, CEP21949-900 Brasil, <rajtrouw@hotmail.com>

Usenko, Victor P. · (369)
Institute of Geological Sciences, National Academy of Sciences of Ukraine, 55B Gonchara Str., 01054 Kiev, Ukraine, <satmar@svitonline.com>

Vázquez, Juán Tomás · (441)
Facultad de Ciencias del Mar, Universidad de Cádiz, 11510 Puerto Real, Cádiz, Spain

Vitturi, Laura Menegazzo · (423)
Dipartimento di Scienze Ambientali, University of Venezia, Dorsoduro 2137, 30123 Venezia, Italy

Vöge, Ingrid · (409)
Alfred Wegener Institute for Polar and Marine Research, Am Handelshafen 12, 27570 Bremerhaven, Germany

Volnukhin, Vyacheslav S. · (83, 95)
Polar Marine Geological Research Expedition (PMGRE), 24 Pobeda St., Lomonosov 189510, Russia

von Frese, Ralph · (109)
Ohio State University, 381, Mendenhall Lab 125 S. Oval Mall, Columbus, OH 43210-1398, USA

Wagner, Bernd · (447)
University of Leipzig, Institute for Geophysics and Geology, Talstraße 35, 04103 Leipzig, Germany, <wagnerb@rz.uni-leipzig.de>

Watanabe, Atsushi · (147)
Department of Earth and Planetary Sciences, Graduate School of Sciences, Kyushu University, 6-10-1 Hakozaki, Fukuoka 812-8581, Japan

Wilson, Douglas S. · (123)
Dept. of Geological Sciences and Inst. for Crustal Studies, University of California, Santa Barbara, CA 93106, USA, <dwilson@geol.ucsb.edu>;
and: Marine Science Inst., University of California, Santa Barbara, CA 93106, USA

Wilson, Terry J. · (309)
Department of Geological Sciences, Ohio State University, 155 South Oval Mall, Columbus, OH 43210-1522, USA

Yamashita, Mikiya · (147)
Department of Polar Science, The Graduate University for Advanced Studies, 1-9-10 Kaga, Itabashiku, Tokyo 173-8515, Japan

Yamazaki, Akira · (15)
Meteorological Research Institute, 1-1 Nagamine Tukuba, Ibaraki 305-0052, Japan

Zapata, Carlos · (255)
INOCAR, Unidad Convemar, Base Naval Sur, Av. 25 de Julio, via Puerto Marítimo, POX 5940, Guayaquil, Ecuador

Theme 1
History of Antarctic Research

Chapter 1.1
The Road to Gondwana via the Early SCAR Symposia

Chapter 1.2
Exploring the Unknown: History of the First German South Polar Expedition 1901–1903

Only two contributions have been submitted to this theme which concern various aspects of history, the history of Antarctic earth sciences events and the history of research expeditions into the Antarctic.

A. Ford (Chap. 1.1) presents his personal impression of the Antarctic earth sciences meetings he participated in. He was one out of only 45 participants of the first meeting taking place in Cape Town in 1963, at a time when continental rigidity and fixity still characterized the view of northern hemisphere geologists. The modern view of continental drift and plate tectonics gained more respectability only during the Oslo meeting in 1970.

C. Lüdecke (Chap. 1.2) presents an outline of the First German South Polar Expedition in 1901–1903, describing briefly its organisational background in international cooperation, its planning and realization as well as the unfortunate valuation of its scientific results by Emperor Wilhelm II.

The Road to Gondwana via the Early SCAR Symposia

Arthur B. Ford
Denali Associates, 400 Ringwood Avenue, Menlo Park, CA 94025, USA, <abford@aol.com>

Introduction

Earth-science symposia under the Scientific Committee on Antarctic Research (SCAR) have now spanned forty years. In 1963 an International Symposium on Antarctic Geology was held in Cape Town, South Africa. At that time, before the formulation and acceptance of plate tectonic theory, its predecessor, Alfred Wegener's (1912, 1915, 1924) and Alexander du Toit's (1937) theory of continental drift was widely ridiculed at northern hemisphere major universities. The Cape Town meeting played a significant role in acceptance of the theory of continental drift by Antarctic geologists from the northern hemisphere. Concepts of Antarctica's geology as reported in volumes from the Cape Town meeting and following two symposia, Oslo, Norway (1970), and Madison, Wisconsin, USA (1977), evolved rapidly, synchronously with advance and general acceptance of plate tectonics theory. Only two papers at Cape Town specifically addressed Antarctica related to the continental drift question, both by South Africans (Lester King and Edna Plumstead).

An increasing pace of earth-science research in Antarctica is indicated by the four- or fife-year intervals of succeeding International Symposia on Antarctic Earth Sciences (ISAES; Table 1.1-1). Registrations rose from 45 from nine nations at Cape Town (Table 1.1-2) and 138 (ISAES-2), to more than 200, from 15 nations at ISAES-3. The rising numbers of papers published in the first three symposia, respectively 76, 126, and 151 reflect the surge of Antarctic earth-science research in those early years, in no small way reflecting rapid improvements in field transport technology. Two scientific thresholds were crossed over that time: (1) the recalcitrant general acceptance of continental drift by northern hemisphere geologists; and then (2) the rapid acceptance by all of plate tectonics theory that replaced drift theory in the late 1960s.

This report is largely a personal narrative from my attendance at all of the Antarctic earth sciences symposia through the present one at Potsdam.

Table 1.1-1. International Symposia of Antarctic Earth Sciences (ISAES) of the Scientific Committee of Antarctic Research (SCAR) and resulting proceedings volumes

1. Cape Town	South Africa 1963 Adie RJ (1964) Antarctic geology. North Holland Publication, Amsterdam, pp 758
2. Oslo	Norway 1970 Adie RJ (1972) Antarctic geology and geophysics. Universitetsforlaget, Oslo, pp 876
3. Madison	USA 1977 Craddock C (1982) Antarctic geoscience. University of Wisconsin Press, Madison, pp 1172
4. Adelaide	Australia 1982 Oliver RL, James PR, Jago JB (1983) Antarctic earth sciences. Cambridge University Press, Cambridge, pp 697
5. Cambridge	U.K. 1987 Thomson RMA, Crame, JA, Thomson JW (1991) Geological evolution of Antarctica. Cambridge University Press, Cambridge, pp 722
6. Tokyo	Japan 1991 Yoshida Y, Kaminuma K, Shiraishi K (1992) Recent progress in Antarctic earth science. Terra Scientific Publications, Tokyo, pp 706
7. Siena	Italy 1995 Ricci CA (1997) The Antarctic Region: geological evolution and processes. Terra Antartica Publication, Siena, pp 1206
8. Wellington	New Zealand 1999 Gamble JA, Skinner DNB, Henrys S (2002) Antarctica at the close of a millenium. Royal Society of New Zealand, Bull 35, pp 652
9. Potsdam	Potsdam 2003 Fütterer, DK, Damaske D, Kleinschmidt G, Miller H, Tessensohn F (2005) Antarctica: contributions to global earth science. Springer-Verlag, Berlin Heidelberg New York

From: Fütterer DK, Damaske D, Kleinschmidt G, Miller H, Tessensohn F (eds) (2006) Antarctica: Contributions to global earth sciences. Springer-Verlag, Berlin Heidelberg New York, pp 3–6

The History

The earliest geologists to visit Antarctic regions came in small sailing ships from various nations and were steeped in doctrines of "Catastrophism", "Uniformitarianism", "Neptunism", and "Plutonism" of the late 18th century. Tingey (1996) reviews the history of geological studies in Antarctica up to The Great War of 1914. Earth-science study of Antarctica, however, had barely begun before the 1957–1958 International Geophysical Year (IGY) and was largely based on a few visits from ships or reconnaissance dog-team or over-snow vehicle traverses. Though some geology was undertaken during IGY, the subject was not an official one of that program. Following IGY, investigations were beginning to be carried out on a more systematic basis and particularly in previously unvisited interior regions of the continent, largely by the few nations with logistical capability for such work.

Political impetus was certainly involved. The Antarctic Treaty of 1961 did not contain any restraints on exploration of the continent's potential mineral resources, for which there were many considered at the time. Accordingly, the USA and the Soviet Union - antagonists in the ongoing Cold War of the time - and other nations were generous in support of geological investigations. The late 1950s and 1960s were wonderful times for geologists to obtain research grants for their studies of unvisited lands of the continent. Ongoing interest in Antarctica's potential resources is shown by a special section with seven papers on the subject at ISAES-4 (Adelaide, Australia, 1982). After the 1957–1958 International Geophysical Year the multidisciplinary Scientific Committee on Antarctic Research (SCAR) was established by the International Council of Scientific Unions in order to coordinate research under the Antarctic Treaty signed in 1959.

By 1962 geologists had made inroads into many new areas of the continent and SCAR advisors recommended that a meeting be held for geologists of the twelve Antarctic Treaty nations to meet and discuss their findings. South Africa, and especially Cape Town, was the perfect setting for the First International Symposium on Antarctic Geology (ISAG), to be held at the University of Cape Town 19–21 September 1963. The Antarctic Treaty had been in effect for only two years and was generally considered a highly successful political experiment for the Cold War. Geologically there could not have been a more appropriate site for that meeting. Geology was not included in IGY studies but that oversight was made up for by the rapid expansion of research by the early 1960s. Importantly, the meeting preceded the revolution in earth sciences brought by the new theory of plate tectonics later in the 1960s.

The 1963 Cape Town meeting brought together 45 geologists and a few geophysicists to present reports on their Antarctic research and to debate that notion of continental drift (Fig. 1.1-1). SCAR invited all full-member nations (12 at that time, but 27 by 2002) to participate in this first-of-its-kind symposium, but three did not participate (Belgium, Chile, Soviet Union). More than one-half of the participants were from the southern hemisphere, and though

Table 1.1-2.
The Cape Town meeting 1963, countries and participants

Southern Hemisphere			
Argentina	H. A. Orlando	R. N. M. Panzarini	
Australia	I. R. McLeod		
New Zealand	G. Warren	R. W. Willett	
South Africa	B. B. Brock J. H. Genis L. C. King E. A. K. Middlemost L. O. Nicolaysen I. C. Rust F. C. Truter	V. von Brunn T. W. Gevers M. H. Martin D. C. Neethling J. D. T. Otto E. S. W. Simpson J. E. de Villiers	A. O. Fuller I. W. Hälbich M. Mathias A. R. Newton E. P. Plumstead W. J. Talbot E. Westall
Northern Hemisphere			
France	P. Bellair		
Japan	T. Tatsumi		
Norway	T. F. W. Barth	T. Gjelsvik	
United Kingdom	R. J. Adie D. H. Griffiths	W. Campbell Smith G. de Q. Robin	
United States of America	E. E. Angino R. H. Dott Jr. W. Hamilton D. L. Schmidt	R. F. Black A. B. Ford A. Mirsky D. Stewart	C. Craddock L. M. Gould R. L. Nichols

Fig. 1.1-1.
Participants in the 1963 Cape Town International Symposium on Antarctic Geology (Adie 1994): *Front row:* I. R. McLeod, L. M. Gould, P. Bellair, R. J. Adie, R. W. Willett, R. N. M. Panzarini, F. C. Truter, G. de Q. Robin, T. F. W. Barth, E. S. W. Simpson, T. Tatsumi, T. Gjelsvik. *Second row:* H. A. Orlando, L. C. King, W. Hamilton, D. H. Griffiths, R. H. Dott Jr, C. Craddock, G. Warren, D. C. Neethling, J. E. deVilliers. *Third row:* Edna P. Plumstead, A. B. Ford, D. Schmidt, R. F. Black, V. von Brunn, I. C. Rust, W. Campbell Smith. *Fourth row:* E. Boden, Mrs V. Harvey, Miss R. Abernethy, Mrs W. Kings, M. H. Martin, A. Mirsky, R. L. Nichols, D. Stewart, I. W. Hälbich, A. O. Fuller. *Fifth row:* B. B. Brock, L. O. Nicolaysen, Morna Malthias, T. W. Gevers, Mrs D. F. Murcott, Miss E. Westall, E. E. Angino, H. J. Claassens, W. J. Talbot

predominantly from the Republic of South Africa, South Africans presented only six of the 76 papers published.

That first meeting was dominated by geologists. The ensuing multidisciplinary expansion in studies quickly brought a name change to International Symposium on Antarctic Earth Sciences (ISAES) for the second of these meetings, in Oslo, 1970, a name that continues. South African Alexander du Toit's (1937) bible of continental drift. "Our Wandering Continents", that demonstrated Antarctica's key role in Gondwana was largely unknown to northern geologists invading Antarctica. Du Toit presciently foresaw that geologic events of the South African Cape province would someday be found recorded in mountains just then seen by Lincoln Ellsworth at the head of the Weddell Sea (Sentinel Range, Ellsworth Mountains) on his 1935 transcontinental flight. Surprisingly in South Africa, continental drift was the chief subject of only one of Cape Town's 76 papers – one by Lester King, who himself had not worked in Antarctica. Those who had worked in areas around the Weddell Sea, including myself, never envisioned in their Cape Town reports how their areas might fit into extra-Antarctic settings. That is not surprising. Northern hemisphere geologists like myself were trained in concepts of continental rigidity and fixity. Ideas of movement of continents laterally were ridiculed, though surprisingly, horizontal nappe and thrust movements were in vogue by alpine structural geologists.

Political and societal events often intertwine with scientific ones. 1963 was a time of considerable unrest at the southern tip of Africa. We read news headlines at the time such as: "S.A. protest on apartheid at U.N. fails;" "U.N. told: expel S.A., begin arms blockade" (both Cape Times, 19.09.1963).

I was trained in structural geology and petrology in the times of eugeosynclines, fixity of continents, Barrowian belts and granitization (University of Washington, Seattle, 1954–1958) by alpine geologist Professor Peter Misch, who himself mapped nappes of the Pyrenees that showed great lateral transport. In those days Misch was considered a radical mobilist for his concepts of tens of kilometers of thrusting in mountain belts of western North America. Misch was a product of Germany's Göttingen University, and a member of the tragic 1934 Willy Merkl Nanga Parbat expedition in the Karakoram Himalaya. Of Jewish ancestry, Misch could not return home and spent WW II years teaching in China, during which his wife did not survive the Holocaust. Those were difficult times, as they were in early 1960s at the time of the first International Symposium on Antarctic Geology, during the early Cold War and not long after establishment of official apartheid in South Africa. Most of us had never heard of "apartheid" when we arrived at the Cape Town airport. We would soon find out.

F. C. Truter, former Director of the South African Geological Survey, led an incredible field excursion around all of South Africa for the purpose of showing northern-hemisphere sceptics of continental drift the same rocks and stratigraphy that they had seen in the Ellsworth and Pensacola Mountains and Transantarctic Mountains (Fig. 1.1-2). It was convincing. That 1963 Cape Town field trip was pivotal and changed the way Antarctic geologists from the northern hemisphere looked at southern lands. After that, in the formations of Antarctica's Ellsworth and Pensacola Mountains and elsewhere, we could see the Dwyka, Beaufort and Karroo of South Africa. Du Toit had

Fig. 1.1-2. Field excursion group discusses Dwyka tillite. Profs. Lester King (*center-right*) and Edna Plumstead (*right*) of South Africa discuss origin of diamictite of the Dwyka Tillite and directions of glacier transport with Prof. Bellair of France (*left*)

expected that the Cape Fold belt someday would be found in the mountains of Antarctica around the Weddell Sea. Lincoln Ellsworth's flight discovering the later-named Ellsworth Mountains was later than du Toit's writings. The 1961 first visits here (Webers et al. 1992) discovered the rocks predicted by du Toit, as did Schmidt et al. (1978) in early 1960s mapping of the Neptune Range (Pensacola Mountains). Continental drift gained more respectability at the 1970 Oslo meeting, where one session was dedicated to "Antarctica and Continental Drift", though including only four of the 126 papers. Few papers in other sessions referred to Gondwana relations. "Plate tectonics" had just arrived and is cited only once in the contents of the Oslo volume. Drift and plate tectonics were sufficiently matured by the 1977 Madison meeting that a separate session on the subject was not needed. By then, authors were racing to board the wagon: most papers on topical and regional studies of Antarctica included extra-Antarctic speculations in terms of the new plate tectonics.

Conclusions

Eduard Suess in his "Face of the Earth" (1885) coined the name Gondwána-Land, after the Indian province Gondwana, a name now in common use for that Paleozoic landmass. A variety of explanations were offered for the distribution of the Gondwana floras, among which continental drift was loudly and widely unaccepted until the later 1960s. Arthur Schopenhauer, German philosopher from Danzig (1788–1860), is reported to have once remarked "All truth passes through three stages: First; it is ridiculed; Second, it is violently opposed; and Third, it is accepted as self evident". Wegener's (1915, 1924), du Toit's (1937), and others, even Holmes' (1944) "notions" about the horizontal movements of continents were widely ridiculed by northern hemisphere geological pundits of the 1950s and early 1960s. In my own classes in structural geology I had been deluged with the absolute nonsense of that fantasy of Wegener and du Toit that continents could have drifted to open the Atlantic Ocean. My Prof. Misch at University of Washington was considered a radical mobilist in his day because he believed nappes he mapped to have moved many tens of kilometers – though never continents as a whole. From what I've heard from coeval colleagues the story was the same at most other northern universities before the 1960s.

On the matter of continental drift, the early SCAR earth-science symposia well illustrate Schopenhauer's stages for acceptance of "Truth". Most northern hemisphere participants at the 1963 Cape Town meeting, had been trained in a milieu of general ridicule of Wegener's theory, and their symposium papers largely ignored the subject. Plate-tectonic theory had just arrived by the time of the 1970 ISAES-2 in Oslo, and drift was still under debate though by ever decreasing numbers of skeptics. Debate was essentially over by the time of Madison's 1977 ISAES-3. Continental drift became "self-evident" by way of the overwhelming evidence for plate tectonic mechanism by which it took place.

And now, "Rodinia"? "Antarctica and Rodinia", a separate session of the Potsdam meeting, is for future historians to evaluate.

References

Holmes A (1944) Principles of physical geology. Thomas Nelson & Sons, London

Schmidt DL, Williams PL, Nelson WH (1978) Geologic map of the Schmidt Hills quadrangle and part of the Gambacorta Peak quadrangle, Pensacola Mountains, Antarctica, scale 1 : 250000. Antarctic Geologic Map A-8, U.S. Geol Survey, Washington DC

Suess E (1885) Das Antlitz der Erde 1. Tempsky, Wien

Tingey RJ (1996) How the South was won – A review of the first 150 years of Antarctic geological exploration. Terra Antartica 3:1–10

Toit AL du (1937) Our wandering continents, an hypothesis of continental drifting. Oliver and Boyd, Edinburgh

Webers GF, Craddock C, Splettstoesser JF (eds) (1992) Geology and paleontology of the Ellsworth Mountains, West Antarctica. Geol Soc Amer Mem 170

Wegener A (1912) Die Entstehung der Kontinente. Geol Rundschau 3:276–292

Wegener A (1915) Die Entstehung der Kontinente und Ozeane. Sammlung Vieweg 23, Vieweg, Braunschweig

Wegener A (1924) The origin of continents and oceans. Methuen, London (translation of 3rd German edn 1922)

Exploring the Unknown: History of the First German South Polar Expedition 1901–1903

Cornelia Lüdecke
Valleystrasse 40, 81371 München, Germany, <C.Luedecke@lrz.uni-muenchen.de>

Abstract. At the end of the 19th century the South Pole region was still "terra incognita". Thus a resolution to promote geographical exploration of the Antarctic Regions was passed by VIth International Geographical Congress in London in 1895. Besides political rivalry, a scientific collaboration was adopted during the VIIth International Geographical Congress in Berlin in 1899. The field of work in Antarctica was divided into four quadrants assigning the Weddell and Enderby quadrants to Germany and the Ross and Victoria quadrants to England. Erich von Drygalski (1865–1949), who already had lead two expeditions to the west coast of Greenland (1891, 1892–1893), became leader of the "Deutsche Südpolarexpedition" (1901–1903). The expedition with the newly built first German polar research vessel "Gauss" was financed by the Imperial internal budget. Finally expeditions from Germany, England, Sweden, Scotland, and France took part in the international meteorological and magnetic co-operation. Unfortunately the "Gauss" was beset by ice close to the Antarctic Circle 85 km off the ice-covered coast of Kaiser Wilhelm II. Land, where a winter station was established on sea ice. After 50 weeks of captivity the ship came free and finally sailed home. Emperor Wilhelm II was very disappointed about Drygalski's results, because Robert Falcon Scott (1868–1912) had reached 82° S at the same time. Geographical achievements seemed to be much more valuable than thoroughly measured scientific data, which were to be analyzed and published over three decades.

International Co-Operation in Antarctica

Beginning in 1865, Georg von Neumayer (1826–1909), director of the German Naval Observatory, promoted the dispatch of a German expedition to Antarctica. At the time of the XIth Geographical Conference at Bremen (April 1895), a German Commission for South Polar Exploration was set up to prepare a German South Polar Expedition (Neumayer 1901; Krause 1996). At the turn to the 20th century the South Pole region was still "terra incognita" and an object of speculation. Antarctica – was it a giant atoll or a big continent covered by ice (Lüdecke 1995a)?

During the VIth International Geographical Congress in London (July 1895) a resolution was adopted promoting geographical exploration of the Antarctic Regions (Drygalski 1989). This paved the way for the planning of a German and British expedition. This occurred despite unfavourable political circumstances, which resulted from political rivalry between Germany and Great Britain. Nevertheless both expeditions agreed to engage in scientific collaboration during the VIIth International Geographical Congress in Berlin (1899). There, Clements Markham (1830–1916), president of the Royal Geographical Society, defined the fields of work. He divided Antarctica into four quadrants assigning the Weddell and Enderby quadrants to Germany and the Ross and Victoria quadrants to Britain. International co-operation for meteorological and magnetic observations was arranged to take place from 1 October 1901 until 31 March 1903; this included all merchant and navy ships sailing on a route south of 30° S (Lüdecke 2003a). Besides, the British expedition under the leadership of Robert Falcon Scott (1868–1912), the Swedish expedition under Otto Nordenskjöld (1869–1928) and the Scot expedition under William Speirs Bruce (1867–1921) participated in the collaboration. The agreement was extended until 31 March 1904, when Scott and Bruce wintered a second time and the French expedition (1903–1905) under Jean Charcot (1867–1936) joined. Additional base stations were installed in the sub-Antarctic to compare the meteorological and magnetic influence of the south Polar Regions with data from regions, which were not disturbed by an ice-covered continent (Table 1.2-1).

Planning of the German South Polar Expedition

Erich von Drygalski (1865–1949) from Königsberg (today Kaliningrad) became leader of the first German South Polar Expedition, which was financed by the Imperial internal budget (Drygalski 1989). After his doctorate in geophysics in 1887, he had gathered polar experience as leader of the Greenland expeditions of the Berlin Geographical Society (1891, 1892–1893) (Drygalski 1897). At 1900, Germany had become the second strongest navy power and a serious rival of the British sea power. In this context plans were made for the construction of the first German polar research vessel "Gauss", which was taken over by the Nautical Division of the Admiralty (Lüdecke 2003b). "Gauss" was designed as three-masted auxiliary barquentine, primarily running as a sailing rig, with an auxiliary engine mostly used in ice or when the sailing power was insufficient. The ship was provided with an iron free area for magnetic measurements. Together with the scientific instrumentation it became a flagship of German science.

From: Fütterer DK, Damaske D, Kleinschmidt G, Miller H, Tessensohn F (eds) (2006) Antarctica: Contributions to global earth sciences. Springer-Verlag, Berlin Heidelberg New York, pp 7–12

Table 1.2-1. Base stations and additional stations of the international co-operation

Time	Country	Station	Remark	Observations
1901–1903	Germany	Kerguelen Island	Base station for Drygalski	Magnetism, meteorology
1902–1904	Great Britain	Christchurch (New Zealand)	Base station for Scott	Magnetism
Since 1902	Argentina	Staten Island	Base station for Nordenskjöld	Magnetism, meteorology, hydrography
Since 1902	Germany	Apia (Samoa)	From Academy of Sciences in Göttingen	Magnetism, meteorology, air-electricity, hydrology, seismology
1902–1903	Norway	Iceland, Svalbard, Novaya Zemlya, Bossekop (Norway)	Aurora Borealis Expedition	Magnetism
1903–1904	Scotland	Cape Pembroke (Falkland Islands)	Base station for Bruce	Meteorology

Neumayer defined the sailing route of "Gauss". From the distribution of drift ice and the position of the 0 °C isotherm in winter, he derived the existence of a warm ocean current leading from Kerguelen southward (Lüdecke 1989). Oceanographer Otto Krümmel (1854–1912) considered temporary ice-free conditions of the Weddell Sea. In a confidential letter to Drygalski, he even postulated a large ocean current directing from Kerguelen over the South Pole up to the Weddell Sea (Lüdecke 1995a). Conditions would be similar to those of the Arctic. Starting between 90° E and 100° E, it was expected that the "Gauss" might drift close to the pole.

Heading for the Unknown South

Five officers and 22 seamen were aboard "Gauss". Drygalski and biologist Ernst Vanhöffen (1858–1918) had already wintered together at Greenland (1892–1993) (Lüdecke 2000). Hans Gazert (1870–1861, medical officer), Emil Philippi (1871–1910, geologist), and Friedrich Bidlingmaier (1875–1914, magnetician and meteorologist) completed the group of scientists (Drygalski 1989). The officers participated in various investigations and five seamen assisted in zoological, meteorological and magnetic measurements.

The instructions for the expedition given by Emperor Wilhelm II referred only to a few fundamental principles and thus left Drygalski with a great deal of freedom of action. He followed Alexander von Humboldt's (1769–1859) ideas of comprehensive investigation of an unknown area concerning the three elements earth, water, air, and the living world within (Fig. 1.2-1).

On 11 August 1901 "Gauss" set sail at Kiel for an unknown destination somewhere in the South Indian Ocean at 90° E. Various investigations were made on the cruise. For example, deep-sea soundings verified the extraordinary depth of the Romanche Deep (–7 230 m) close to the equator.

Fig. 1.2-1. Postcard showing Drygalski and the flag of the Ministry of Interior and sketches of Antarctica

Fig. 1.2-2. German base station at Kerguelen in winter 1902 (Lüdecke 1989)

German Base Station at Kerguelen

The German expedition installed an additional base station for meteorological and magnetic measurements at Baie de l'Observatoire on Kerguelen (Fig. 1.2-2). Additionally, a meteorological high station was set up at 160 m on top of a hill and a water gauge was installed at a small inlet close by. With the help of two technicians, biologist Emil Werth (1869–1958, station leader), meteorologist Josef Enzensperger (1873–1903), and earth magnetician Karl Luyken (1874–1947) made their daily observations for a whole year. Although they still could find Kerguelen cabbage and fresh meat, Enzensperger died through Beriberi, a vitamin B_1 deficiency, and was buried close to the station.

Meteorological data covered more than a year representing the cold and stormy climate of Kerguelen very well (Table 1.2-2).

The station was abandoned on 1 April 1903 and all installations were left behind with closed doors and windows. Nevertheless, the buildings were razed to ground by northern gales from in the following decades.

Table 1.2-2. Yearly mean values (January 1902–February 1903) at Kerguelen (49°24' S, 69°53' E)

Parameter	Value
Temperature (°C)	3.2
Air pressurre (hPa)	998
Relative humidity (%)	75
Wind speed (Bft)	4.3
Storm days	66
Cloudiness	7/8
Total precipitation (mm)	851.5
Days with snow cover	115

Overwintering at the Antarctic Circle

Unfortunately, "Gauss" was beset by ice some 85 km off the Antarctic coast at 66°02' S, 89°38' E on 22 February 1902. Here they discovered the ice-capped coast of Kaiser Wilhelm II Land. Luckily the ice was not drifting so they could establish a fixed winter station on sea ice at Posadowsky Bay, 385 m above sea-bottom (Fig. 1.2-3).

Fig. 1.2-3. Meteorological station on the windward side of the "Gauss" (Lüdecke 2003b)

A "variation hut" and an "absolute hut" for the measurements of magnetic parameters were built with wooden boxes and snow blocks. Routine measurements were carried through the whole year.

It was the first time that tides on the high seas far from land were continuously recorded with an improvised device, using the up and down lift of the ship as indicator (Fig. 1.2-4).

Fig. 1.2-4. Observing the tides (Drygalski 1989)

The first scientific ascent with a manned captive balloon in Antarctica took place on 29 March 1902 (Fig. 1.2-5). It revealed a ground inversion over ice with a temperature difference of 3 °C 500 m^{-1}.

Seven sledge trips were undertaken to investigate the surroundings of the "Gauss". During the first trip the ice-free extinct volcano "Gaussberg" with 366 m height was discovered exactly at the border between sea-ice and inland-ice (Fig. 1.2-6). Glaciological investigations were also made. The movement of the inland-ice towards the coast was measured to have a rate up to 11.68 m per month.

Return and Results of the Expedition

After 50 weeks of captivity, the "Gauss" finally broke free on 8 February 1902. Drygalski tried to approach higher latitudes to the west of the wintering place, because he wanted to investigate the coastline or to engage in a drift that would take them into the Weddell Sea. Instead the "Gauss" was beset by ice twice and came into heavy ice pressing during a violent gale. Finally on 8 April Drygalski ordered to sail north again. From the next telegraph office at Cape Town, Drygalski reported to Berlin that the expedition had returned after a successful wintering. Unfortunately, a second attempt to go further south was not allowed by the Ministry of the Interior, because the budget was exhausted. Additional 309 000 M were still outstanding, which Drygalski did not know. Thus the expedition had to sail home ultimately arriving at Kiel on 25 November 1903. At the time of imperialism, when powerful countries competed for the last regions of the earth not yet distributed, the German South Polar Expedition had failed to gain high latitudes (Lüdecke 1992, 1995b). Emperor Wilhelm II was very disappointed with Drygalski's results, because Scott reached

Fig. 1.2-5.
Balloon ascent on 29 March 1902 (priv. possession Mörder, Amberg)

Fig. 1.2-6.
Gaussberg, first climbed on 21 March 1902 (Drygalski 1912)

Table 1.2-3. Yearly mean values (February 1902–February 1903) at the winter Station (66°02' S, 89°38' E)

Parameter	Value
Temperature (°C)	−11.4
Air pressure (hPa)	998
Relative humidity (%)	87
Wind direction	NE to SE
Wind speed (m s^{-1})	7.2
Storm days	69
Cloudiness	7/10
Days with precipitation	251

82° S at the same time, while the "Gauss" had been trapped at the Polar Circle. To avoid costs for further maintenance, the "Gauss" was sold to the Canadian Coast Guard, where the ship sailed under its new name "Arctic".

Over decades, the data were analysed and published in 20 volumes and two atlases (Drygalski 1905–1931). Twelve volumes – instead of the two planned – summarized the biological results. 1 470 new species had been described for the first time. 105 bottom samples had been taken to investigate the sea bottom. The oceanographic results revealed a four-layered stratification of the southern Indian Ocean. Declination at the winter station of "Gauss" amounted to 61° W, inclination to 77.1° S, and horizontal intensity to 0.133 Γ. The magnetic measurements improved the idea of the magnetic field of the earth. Meteorological data described the climate at the Antarctic Circle (Table 1.2-3). Over 600 000 single data of the meteorological co-operation were collected at Berlin and analysed. 913 synoptic charts and 30 charts of different mean values had been constructed and published in 1915.

All investigations indicated that Antarctica is a continent. At the turn to the 20th century, however, reaching high latitudes seemed to be much more valuable than thoroughly measured scientific data, which had to be analysed and published over decades.

References

Drygalski E v (1897) Grönland-Expedition der Gesellschaft für Erdkunde zu Berlin 1891–1893. W. H. Kühl Verlag, Berlin (2 vols)

Drygalski E v (1905–1931) Deutsche Südpolar-Expedition 1901–1903 im Auftrage des Reichsamtes des Innern. Verlag Georg Reimer, Berlin (20 vols, 2 atlases)

Drygalski E v (1989) The southern ice-continent: the German South Polar Expedition aboard the Gauss 1901–1903. Bluntisham Books, Bluntisham and Erskine Press, Norfolk

Krause RA (1996) 1895, Gründerjahr der Deutschen Südpolarforschung. Dt Schiffahrtsarchiv 19:141–162

Lüdecke C (1989) Die Routenfestlegung der ersten deutschen Südpolarexpedition durch Georg von Neumayer und ihre Auswirkung. Polarforschung 59(3):103–111

Lüdecke C (1992) Die erste deutsche Südpolar-Expedition und die Flottenpolitik unter Kaiser Wilhelm II. Hist meereskdl Jb 1:55–75

Lüdecke C (1995a) Ein Meeresstrom über dem Südpol? – Vorstellungen von der Antarktis um die Jahrhundertwende. Hist Meereskdl Jb 3:35–50

Lüdecke C (1995b) Die deutsche Polarforschung seit der Jahrhundertwende und der Einfluß Erich von Drygalskis. Ber Polarforschung 158:1–340, A1–A72

Lüdecke C (2000) Greenland as study area for glaciological theories at the turn of the century. In: Sigurðsson I, Skaptason J (eds) Aspects of Arctic and Sub-Arctic history. University of Iceland Press, Reykjavik, pp 574–582

Lüdecke C (2003a) Scientific collaboration in Antarctica (1901–1903): a challenge in times of political rivalry. Pol Rec 39(208):35–48

Lüdecke C (2003b) "Fest in der See, vortrefflich im Sturm, im Eis stark genug." Der GAUSS – das erste Flaggschiff der deutschen Polarforschung. Mitt Gauß-Ges 40:25–43

Neumayer G v (1901) Auf zum Südpol! 45 Jahre Wirkens zur Förderung der Erforschung der Südpolarregion 1855–1900. Vita Deutsches Verlagshaus, Berlin

Theme 2
Antarctica – The Old Core

Chapter 2.1
Characteristics of Metamorphosed Banded Iron Formation and Its Relation to the Magnetic Anomaly in the Mt. Riiser-Larsen Area, Amundsen Bay, Enderby Land, Antarctica

Chapter 2.2
Experimental Constraints on the Decompressional *P-T* Paths of Rundvågshetta Granulites, Lützow-Holm Complex, East Antarctica

Chapter 2.3
Sapphirine – Orthopyroxene – Garnet Granulite from Schirmacher Hills, Central Dronning Maud Land

Chapter 2.4
Genesis of Ferropotassic A-Type Granitoids of Mühlig-Hofmannfjella, Central Dronning Maud Land, East Antarctica

Chapter 2.5
Late Pan-African Fluid Infiltration in the Mühlig-Hofmann- and Filchnerfjella of Central Dronning Maud Land, East Antarctica

Chapter 2.6
Electron Microprobe (EMP) Dating on Monazite from Forefinger Point Granulites, East Antarctica: Implication for Pan-African Overprint

Chapter 2.7
Tectonic Subdivision of the Prince Charles Mountains: A Review of Geologic and Isotopic Data

Chapter 2.8
Crustal Provinces of the Prince Charles Mountains Region and surrounding Areas in the Light of Aeromagnetic Data

Chapter 2.9
Magnetic Anomalies of the Grove Mountains Region and Their Geological Significance

Traditionally, East Antarctica "east" of the Transantarctic Mountains was regarded as the "East Antarctic Shield" or "East Antarctic Craton" – the oldest part ("core") of the continent. Meanwhile this idea has been revised, and the East Antarctic Craton seems to consist, in reality, of rather a number of smaller cratons (old cores) surrounded by relatively younger mobile belts or orogens. It may be forgiven, that the following papers have been subsumed under the traditional heading "Old Core" with its rather out-dated meaning.

The chapter contains a few papers, which really concern – at least partly – old cores as Funaki et al. (Chap. 2.1), Kawasaki and Motoyoshi (Chap. 2.2), Mikhalsky et al. (Chap. 2.7) and Golynsky et al. (Chap. 2.8). The others concern Grenvillian belts as Baba et al. (Chap. 2.3), Engvik and Elvevold (Chap. 2.5) and Golynsky et al. (Chap. 2.9) and even Pan-African belts as D'Souza et al. (Chap. 2.4), Kawasaki and Motoyoshi (Chap. 2.2), Engvik and Elvevold (Chap. 2.5), Motoyoshi et al. (Chap. 2.6), Mikhalsky et al. (Chap. 2.7) and Golynsky et al. (Chap. 2.9) in East Antarctia. The Pan-African belts are overprinting earlier structures as Grenvillian in Engvik and Elvevold (Chap. 2.5), Motoyoshi et al. (Chap. 2.2) and Golynsky et al. (Chap. 2.9) or cratons as in Kawasaki and Motoyoshi (Chap. 2.2) and Mikhalsky et al. (Chap. 2.7).

Funaki et al. focus on magnetic anomalies due to banded iron formations within the up to >3.9 Ga old Napier Complex of Enderby Land.

Granulite facies metamorphism dominates the three westward following papers: Motoyoshi et al. provide ages by electron microprobe analyses of monazite from the Rayner Complex (Enderby Land). They prove that Pan-African (~500 Ma) granulite facies metamorphism overprinted Grenvillian metamorphism (≤1 000 Ma). (These results should continue to Sri Lanka and India). Kawasaki and Motoyoshi support by experiments that retrograde Pan-African metamorphism in the Lützow-Holm Complex took place within the granulite facies. Presumed Grenville-aged sapphirine granulite is reported from a new locality in the Schirmacheroase by Baba et al. This product of ultrahigh-temperature metamorphism is thought to be suitable to detect the evolution of Rodinia and Gondwana.

In central Dronning Maud Land, Pan-African fluid-infiltration overprinted Grenvillian-aged rocks (Engvik and Elvevold) and geochemical results support the post-collisional intrusion of late Pan-African A-type granitoids (D'Souza et al.). Both processes seem to be of regional importance and occur, or are expected to occur, in other parts of Gondwana.

Mikhalsky et al. outline a terrane model for the Prince Charles Mountains (PCM) of Mac. Robertson Land mainly on the basis of Sm-Nd, Pb-Pb, U-Pb SHRIMP age data. They suggest four ~WSW-ENE running terranes: the Archean Ruker T., the Mesoproterozoic Fisher T., the Meso-Neoproterozoic Beaver T. and the Pan-African reworked Paleoproterozoic Lambert T. Similar models are proposed by Golynsky et al.: They define "magnetic units" on the basis of aeromagnetic data, in Chapter 2.8 five SW-NE to WSW-ENE trending units (16 subunits) for the PCM and for their possible eastern continuation east of the Lambert Glacier in northern Princess Elizabeth Land, and in Chapter 2.9 four W-E to SW-NE trending units (17 subunits) for the Grove Mountains of southern Princess Elizabeth Land. Thus, Mikhalsky et al. and Golynsky et al. infer a preliminary counter model to that of Boger et al. (2001, 2002) and Boger and Miller (2003), who used their structural and age data from PCM for a comprehensive new model for the final Gondwana formation in two steps: *(i)* amalgamation of West Gondwana and Indo-Antarctica and *(ii)* amalgamation of West Gondwana – Indo-Antarctica and the rest of East Gondwana.

Characteristics of Metamorphosed Banded Iron Formation and Its Relation to the Magnetic Anomaly in the Mt. Riiser-Larsen Area, Amundsen Bay, Enderby Land, Antarctica

Minoru Funaki[1] · Peter Dolinsky[2] · Naoto Ishikawa[3] · Akira Yamazaki[4]

[1] National Institute of Polar Research, Tokyo, Japan, 9–10 Kaga 1 Itabashi, Tokyo 173-8515, Japan
[2] Geophysical Institute of Slovak Academy of Sciences, Komarno 108, Hurbanovo 947 01, Slovakia
[3] Kyoto University, Nihonmatsu-cho, Yoshida Sakyo-ku Kyoto 606-8501, Japan
[4] Meteorological Research Institute, 1–1 Nagamine Tukuba Ibaraki 305-0052, Japan

Abstract. The characteristics of the metamorphosed banded iron formation (meta-BIF) in the Mt. Riiser-Larsen area, Amundsen Bay, Enderby Land, Antarctica were investigated in order to understand why a large magnetic anomaly appeared in this region. The study found many meta-BIF layers, consisting of one to nine layers within 30 m thickness in felsic gneiss. The thicknesses of individual layers were usually 0.5–2 m but sometimes varied from zero to 6 m. Almost all strong magnetic anomalies reported by Dolinsky et al. (2002) could be explained by the magnetization of the meta-BIF layers. To explain the source of the magnetic anomaly, the natural remanent magnetization of meta-BIF and its susceptibility are taken in to account. The largest layer including meta-BIF is estimated to be more than $7\,400 \times 213$ m. Magnetite in the layer was examined by thermomagnetic, magnetic hysteresis and microscopic analyses. The results indicated that almost pure magnetite with pseudosingle-domain (PSD) and multi-domain structure is the dominant composition of meta-BIF as well as quartz. The PSD magnetite grains were along the quartz crystal boundaries and formed a network structure. The electric conductivity of the meta-BIF is explained by this network structure rather than being due to the high magnetite concentration.

Introduction

The 42nd Japanese Antarctic Research Expedition (JARE42) conducted a magnetic survey in the Mt. Riiser-Larsen region (66°45' S, 50°45' E), Amundsen Bay, Enderby Land, Antarctica, as denoted by a dotted line in Fig. 2.1-1, in the austral summer season of 2000/2001, using proton mag-

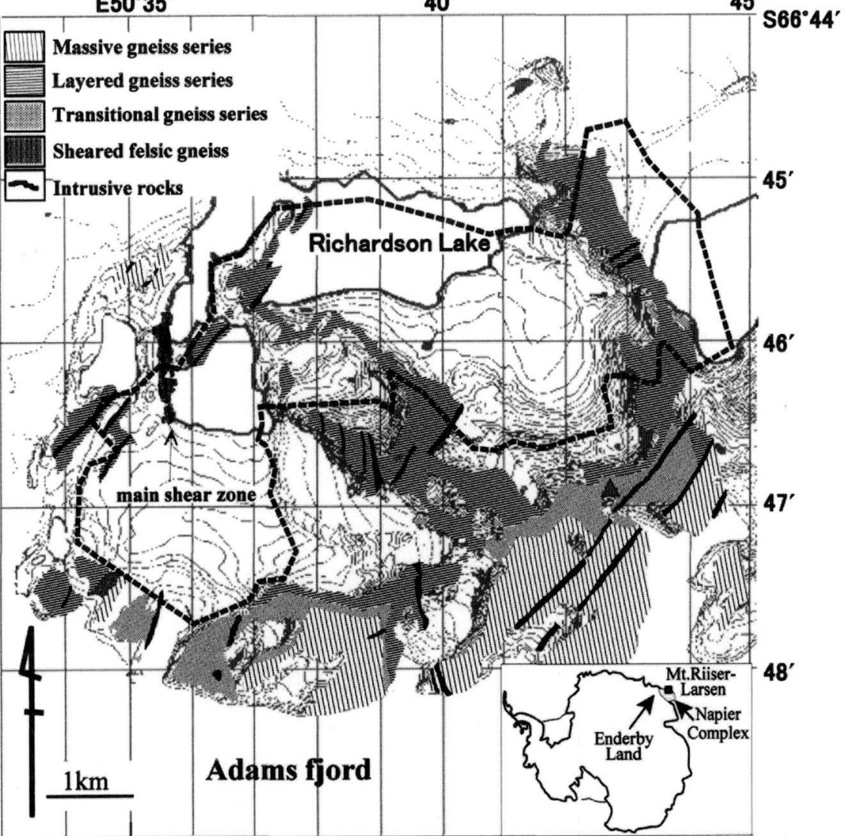

Fig. 2.1-1. Simplified geological and topographical map of the Mt. Riiser-Larsen area (Ishikawa et al. 2000). The area of magnetic survey is denoted by the *dotted line*. *Solid triangle* denotes Mt. Riiser-Larsen

From: Fütterer DK, Damaske D, Kleinschmidt G, Miller H, Tessensohn F (eds) (2006) Antarctica: Contributions to global earth sciences. Springer-Verlag, Berlin Heidelberg New York, pp 15–22

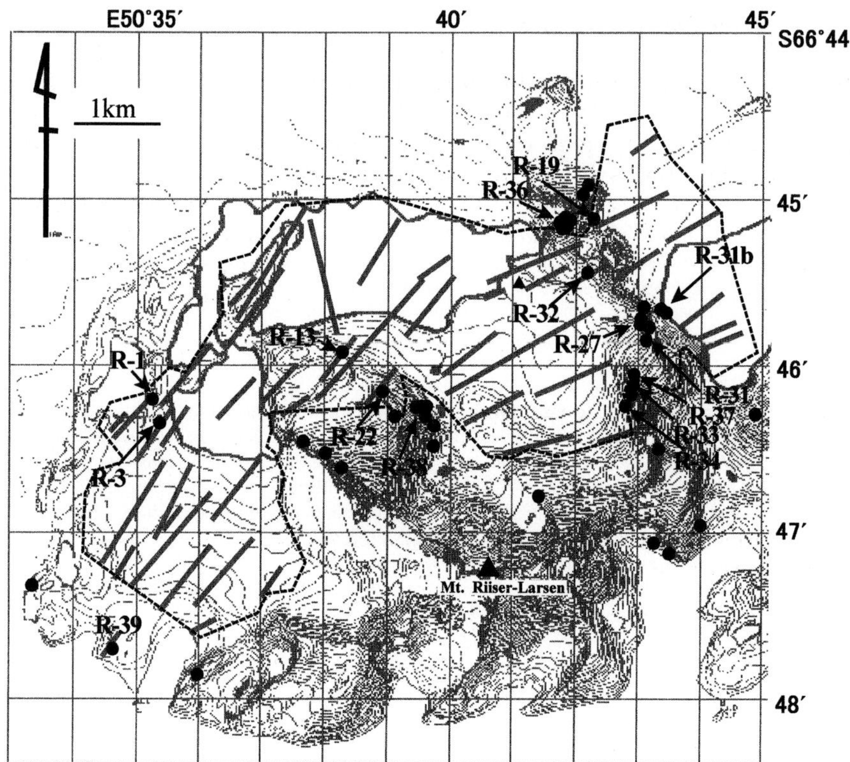

Fig. 2.1-2. Lineation of the magnetic anomalies (*solid lines*) based on Dolinsky et al. (2002) and recognized locations of the meta-BIF layers (*solid circles*). *Solid triangle* denotes Mt. Riiser-Larsen

netometers and global positioning system (GPS), as reported by Dolinsky et al. (2002). The survey area is 3×7 km and within this area the total line of the profile exceeded more than 150 km. The surface in this area consists mainly of moraines, lake and snow field, whereas poor outcrops were distributed along the steep cliffs and mountains ridges. According to the magnetic anomaly pattern (Dolinsky et al. 2002), the strong positive and negative anomalies between −4 000 and 8 000 nT seem to be parallel to the geological structure in this region, as shown in Fig. 2.1-2, which is represented by the extracted data in the range of 0 to −2 000 nT. Ishizuka et al. (1998) reported metamorphosed banded iron formation (meta-BIF) in this region based on the geological surveys by the JARE38. We studied the relation between the magnetic anomalies and meta-BIF and to infer the origin of meta-BIF from the mineralogical evidence.

The Napier complex in Enderby land, East Antarctica, mainly consists of gneissose rocks characterized by metamorphism under high temperature to 1 000 °C (e.g., Sheraton et al. 1987) and high pressure to 11 kPa (Harley and Hensen 1990). The complex has been known as one of the oldest crusts in the world dated to 3 930 ±10 Ma (Black et al. 1986). The JARE38 carried out geological surveys in the Mt. Riiser-Larsen area where the representative outcrop in the Napier complex is situated. The Mt. Riiser-Larsen area consists of metamorphic rocks and unmetamorhosed dolerite intrusions. Ishikawa (2000) summarized geological structures (Fig. 2.1-1) based on their field and laboratory researches as follows. The metamorphic rocks are divided into the layered gneiss series, occurring in the central to northwestern part of this area, and massive gneiss series in the southern to southeastern part. The layered gneiss is characterized by a layered structure composed of garnet pelitic and mafic gneiss, impure quartzite and metamorphosed banded iron formation (meta-BIF). The massive gneiss mainly consists of massive orthopyroxene (OPX) felsic gneiss. Metamorphic foliation strikes NE-SW to E-W and dips at 20–40° to the south or southeast in this area. Doleritic dykes, so-called Amundsen Dykes, intrude throughout the area striking N-S and NE-SW. Large shear zones characterized by mylonite and pseudotachylite appear with the N-S strike at the western part of this area.

Although meta-BIF is minor formation in this area, it is widespread, occurring as layers and sometimes blocks within the garnet felsic gneiss and in rare cases in OPX felsic gneiss. The thickness of the meta-BIF ranges from several centimetres to several tens of centimeters, but it is rarely more than a few meters. Ishizuka et al. (1998) suggested that the origin of meta-BIF is in sedimentary rocks, because the layers were sandwiched between pelitic gneiss. They reported mineral assemblages of meta-BIF to medium-coarse grained magnetite and quartz for the main constituents associated with or without clinopyroxene, OPX and Fe-Ti oxide.

The first magnetic and mineralogical studies were carried out for meta-BIF in the Mt. Riiser-Larsen area (Funaki 1984, 1988). The results indicated that the pseudosingle-domain (PSD)/multi-domain (MD) structures of magnetite carry natural remanent magnetization (NRM) resulting from isothermal remanent magnetization (IRM) and/or viscous remanent magnetization (VRM). Opaque minerals of magnetite (Fe_3O_4), iron sulfide, ilmenite ($FeTiO_3$) and hercynite ($FeAl_2O_4$) were observed by electron probe micro analyzer and optical microscope.

Appearance of Meta-BIF in the Field

To identify the rock type which is causing the large magnetic anomaly, meta-BIF and dolerite dikes were surveyed around the intersections between the lineation of anomaly and outcrops. The meta-BIF layers were found at the outcrops with strong anomalies passing through them. The meta-BIF layers reported by JARE38 (Ishizuka et al. 2000) and found by us (representative sites are numbered R1 to R39) were plotted in Fig. 2.1-2. The common features in outcrops were characteristically a peculiar dark-gray to dark-brown color and were cleaved to rectangular masses which developed cracks along the parallel and vertical gneissosity. The thickness of the layer was usually between 0.5 and 2 m, but sometimes varied suddenly or gradually to more than 6 m (R-36) or to zero. It also disappeared due to faults and dyke intrusions. Sometimes, four to six layers were recognized within a thickness of 30 m. The layers were sandwiched by OPX felsic gneiss in many cases, but they were in direct contact with other felsic gneiss. When the layer shrank gradually, metamorphosed ultramafic orthopyroxenite or OPX-rich OPX felsic gneiss frequently replaced the layer.

The electrical conductivity of the meta-BIF layers was measured in the outcrops by a circuit tester, because magnetite pebbles scattered in this region frequently showed conductivity. This is a qualitative analysis of the resistance of meta-BIF, because the resistance is strongly influenced by a contact resistance between probes and magnetite.

The representative appearances of the meta-BIF layers are as follows (numbers of the layers are denoted in Fig. 2.1-2).

R-13 (66°45.94' S, 50°38.42' E)

A single meta-BIF layer 1 m thick and extending 15 m ran into layered gneiss with strike 46° E and dip 57° SE. It was overlain by the OPX felsic gneiss and underlain by garnet poor felsic gneiss. The NE end of layer was replaced by meta-pyroxenite, but the other one was buried by moraine. It had an electrical resistance of 66 kΩ, when the vertical distance against the meta-BIF strike was 60 cm.

R-19 (66°45.210' S, 50°057' E)

A single meta-BIF layer (maximum thickness 3 m) extending 20 m in a NW-SE direction, as a convex lens, running into garnet-rich felsic gneiss. Both sides of the layer gradually changed to ultramafic meta-orthopyroxenite with internal folding. The layer continued in a NE direction beyond the ridge.

R-32 (66°45.482' S, 50°42.474' E)

Nine meta-BIF layers between 0.05 and 2 m thick extending to less than 10 m with a gentle inclination were sandwiched in garnet-poor garnet felsic gneiss. As the layers were cut vertically by a fault and were sheared by a horizontal fault in this site, some of them were probably originally the same layers. Along the vertical fault, magnetite with 10 cm thickness was formed. In the shear zones, the layer appeared as shiny dark-grey in color with a schistosity like fracture.

R-33 (66°46.18' S, 50°43.09' E; Fig. 2.1-3a)

A single meta-BIF layer was observed from R-34 to R-37 through R-33 sandwiched by OPX felsic gneiss or meta-pyroxenite, and it probably continued to R-31 and R-32b based on its strike and dip (N54E, 35SE). The layer at R-34 was 2 m thick, but it narrowed to 5 cm between R-33 and R-34 and then increased to 3 m between R-33 and R-37. Although the layer was buried by moraine between R37 and R31, it appeared again at R-31. The layer separated into two layers of 0.5 m and 0.15 m thickness at R-31 continuing to R-27 and R-31b. This layer ended in the southern part of R-34 as a dolerite dyke intrusion, but it seems to be extended into the lake with the NE direction at R-31b judging from the magnetic anomaly pattern, as shown in Fig. 2.1-2. Therefore, we estimate that the layer continues more than 1.2 km to the NE-SW direction.

R-36 (66°45.210' S, 50°057' E)

Five meta-BIF layers from 0.6 m to 6 m at R-36-1 to -5 were interlayered within 30 m thick of quartz-rich garnet gneiss or impure quartzite (Fig. 2.1-3b,c). Probably the layers at R-36-1 and -2 were the same layers as deduced from their strike N48W and dip 20SW. The layer at R-36-4 branched out into two layers of 0.6 m in thickness. These layers at R-36 seemed to continue to the north be-

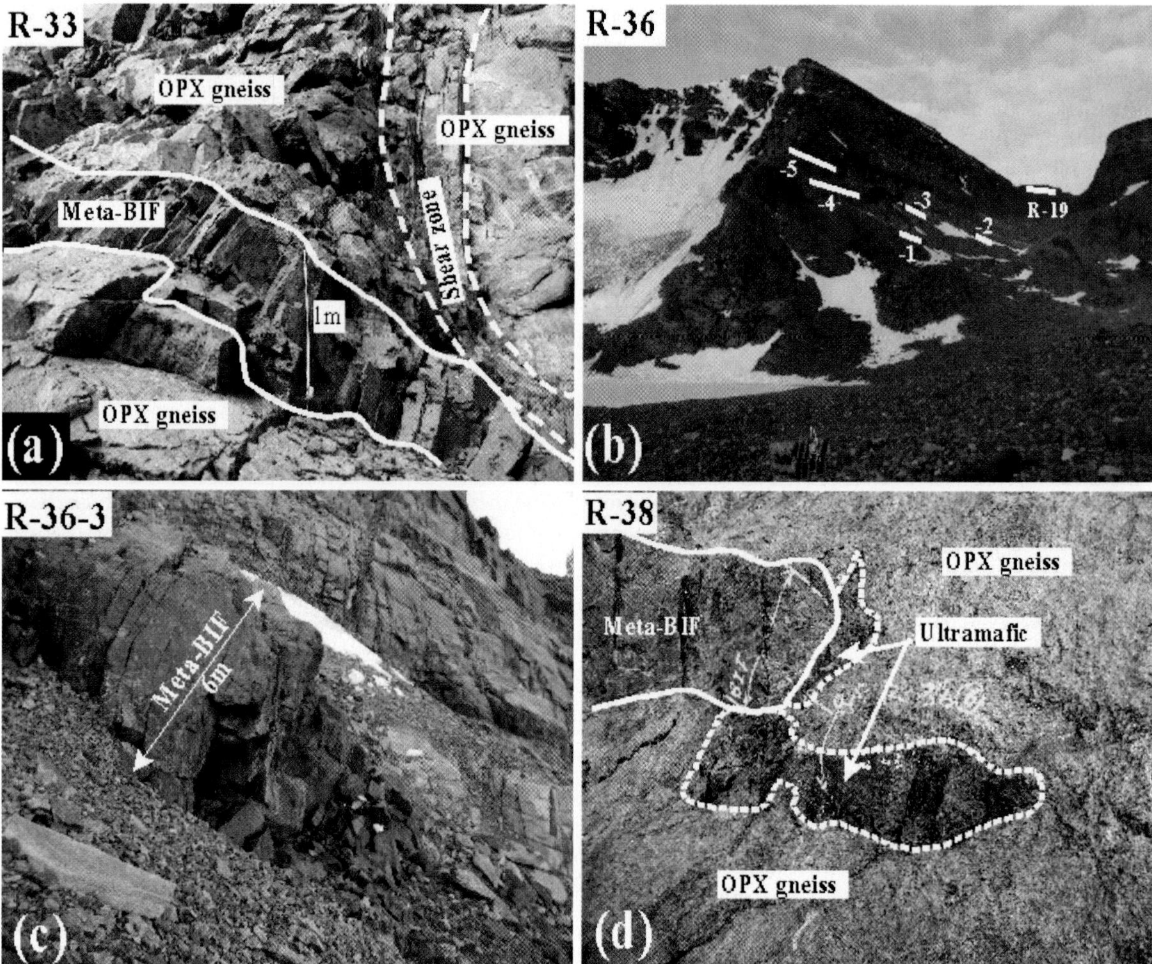

Fig. 2.1-3. Representative outcrops of meta-BIF layers. **a** R-33, the single layer of meta-BIF running in the meta-orthopyroxene (OPX) gneiss. **b** R-36, *white lines* (R-36-1 to -5) denote the meta-BIF layers. The layer of R-19 is seen. **c** R-36-3, the thickest meta-BIF in this area, 6 m in thickness. **d** R-38, metamorphosed ultramafic orthopyroxenite (denoted by '*Ultramafic*') appeared at the end of the meta-BIF layer

yond a ridge, but it was not confirmed due to too steep a cliff. However, any meta-BIF layers were not recognized from the southern end of R-36 to R-32.

R-38 (66°46.34' S, 50°39.62' E)

At least four meta-BIF layers of 20 to 40 m in length and 0.3 to 1.2 m thick were confirmed within 30 m of the layered gneiss. They contacted directly to OPX felsic gneiss in many places, but interfaced with metamorphosed ultramafic orthopyroxenite when they were shrunk and finished (Fig. 2.1-3d). They were folded and cut by a dolerite instruction at the SE end. We estimated these layers on a steep cliff between R-38 and R-22 continuing more than 800 m, because the stratum of OPX felsic gneiss at R-38 could be followed until R-22. Where the colour of the OPX felsic gneiss varied to darker at several places, the meta-BIF layers might be developed.

R-39 (66°47.719' S, 50°34.670' E)

Two meta-BIF layers extending more than 60 m (maximum thickness 4.5 m and 1.5 m) run in parallel with an interval of 12 m in OPX felsic gneiss. These layers narrowed suddenly and disappeared naturally at the NE end of the layers, while the other side was not seen due to burying by snow drift. Any metamorphosed ultramafic rocks as orthopyroxenite were not observed in the boundary between meta-BIF and OPX felsic gneiss.

Mineralogy

The representative meta-BIF samples from each site were polished to observe the structure of opaque minerals under a reflected light microscope. Typical features of the four examples (samples R-1, R-13, R-38 and R-36) are

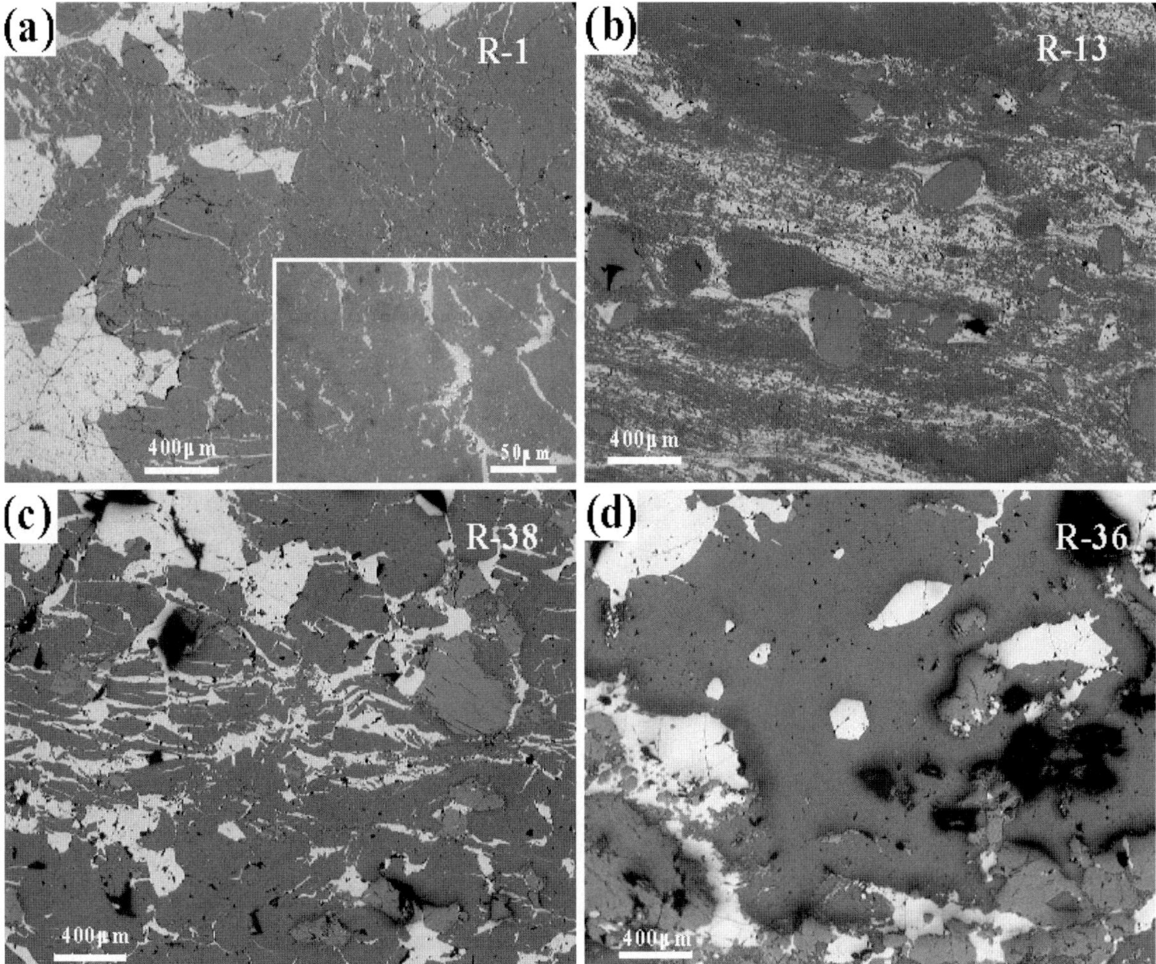

Fig. 2.1-4. Reflected light microscopic photographs of meta-BIF. **a** *R-1:* showing network structure combined with fine-grained magnetite aligned with the silicate crystal boundaries. **b** *R-13:* showing fine-grained magnetite clusters in the matrix but almost no magnetite grains in the larger silicate grains. Mylonite structure can be seen around the larger spherical silicate grains. **c** *R-38:* network structure of magnetite forming along the silicate crystal boundaries. **d** *R-36:* euhedral magnetite grains in the centre, small magnetite grains can be observed in the bottom

shown in Fig. 2.1-4. R-1 (Fig. 2.1-4a) was sampled 30 m east of the main shear zone (Fig. 2.1-1), where several pseudotachylite stringers with 5 to 30 cm width occur in OPX felsic gneiss. In this sample, variable size magnetite grains less than 2 mm were seen. Larger magnetite grains of irregular squarish shape were invaded by silicate stringers and grains. In the silicate field, magnetite stringers, combining fine-grains of magnetite less than 10 μm in diameter, penetrated along the boundaries of silicate crystals from the larger magnetite grains. As the directions of penetration varied from place to places, the deformation effect by shearing does not seem to be too strong. Numerous fine-grained magnetite crystals of a few μm were scattered widely in the quartz crystals (see inset of Fig. 2.1-4a. Electrical conductivity was observed in this layer.

The samples from R-13 (Fig. 2.1-4b) showed that fine-grained magnetite less than 20 μm in diameter was dominant in the matrix associated with fine-grained quartz, but almost no magnetite grains were observed in the larger silicate grains. The magnetite grains were not euhedral and they aligned parallel to the gneissosity in this region. Fine-grained magnetite and silicate drifted characteristically at both sides of silicate nodules. The nodules seem to be down to spherical shapes by shearing transportation. These structures can be explained as a mylonite structure formed by strong kinematic shearing. Electrical conductivity was observed in this layer.

Larger magnetite grains of several mm in diameter and developed stringer characterize sample R-38 (Fig. 2.1-4c). The stringers from a few μm to 200 μm in width appear along the grain boundaries of silicate throughout the visual (3.2 mm) of the microscope. Although the direction of many stringers is almost concordant to the gneissosity in this site, some of them developed vertical and showed a network structure. Almost no fine-grained magnetite was observed in the silicate phase. The sample showed electrical conductivity.

Isolated larger magnetite grains up to several mm were distributed without any structural disturbance by deformation under high pressure and shearing in the sample R-36 (Fig. 2.1-4d). A hexagonal magnetite single crystal 200 µm in diameter was present. Almost no magnetite grains were present in the silicate surrounding the hexagonal magnetite. However, fine-grained magnetites of about 10 µm were scattered in the other area of this sample, as seen at the bottom (Fig. 2.1-4d). The electrical conductivity of this sample was not confirmed.

Magnetic Properties

Representative samples from R-1, R-13, R-22, R-31, R-33, R-36, R-37, R-38 and R-39 were measured for thermomagnetic (J_s-T) curves under an external magnetic field up to 1.0 T in a vacuum of 10^{-3} Pa. Every J_s-T curve, except R-13, was reversible with a Curie point (T_c) at 550–570 °C suggesting almost pure magnetite (Fig. 2.1-5a) and listed in Table 2.1-1. The sample R-13 showed on irreversible curve with a minor T_c = 350 °C and a principal T_c = 550 °C in the heating curve and at T_c = 550 °C in the cooling curve (Fig. 2.1-5b). The magnetization decreased to about 10% after heating in the first-run cycle. As the reversible J_s-T curve with the single Curie point at T_c = 550 °C appeared in the second-run cycle, the T_c = 350 °C is unstable phase as estimating of phase transition from maghemite (γ-phase; weathering product) to hematite (α-phase). Since the J_s value of α-phase is 1/250 versus β-phase (magnetite), the magnetization in J_s-T curve decreased after heating. From these J_s-T curves, the main magnetic minerals in the samples are concluded to be almost pure magnetite.

Magnetic hysteresis curves for these samples were obtained in a magnetic field between +1.0 T and −1.0 T at room temperature. The saturation magnetization (J_s), saturation remanent magnetization (J_R), coercive force (H_c) and remanent coercive force (H_{RC}) were calculated from the curves, as shown in Table 2.1-1, where the curves were compensated by paramagnetic susceptibility between 0.7 and 1.0 T. A magnetite content of 23–68 wt.-% in the samples is estimated from J_s = 21.39–62.27 Am^{2}kg^{-1}, because the J_s value of pure magnetite is 92 Am^{2}kg^{-1}. The magnetic domain state of the representative magnetite grains was inferred based on the ratios of J_R/J_S and H_{RC}/H_C (Day et al. 1977). The values were listed in Table 2.1-1 and were plotted in Fig. 2.1-6, where single-domain (SD), PSD and MD occupy the respective areas as denoted. The samples R-1 and R-13 and R-33 and R38 were plotted on the PSD or MD areas respectively. However, five samples

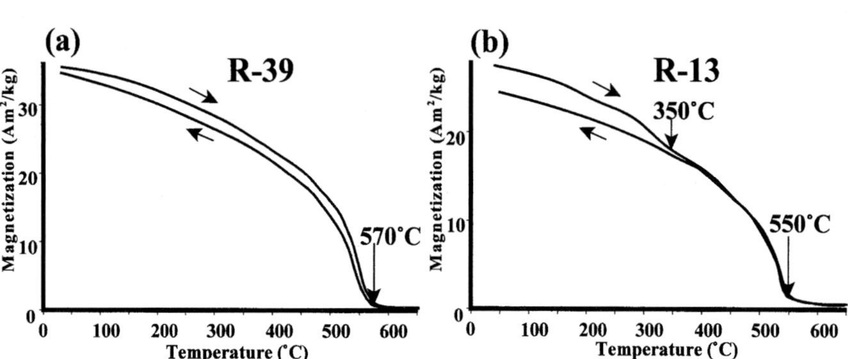

Fig. 2.1-5. First-run thermomagnetic curves (J_s-T curves) of meta-BIF. **a** Typical reversible J_s-T curve (*R-39*); **b** irreversible one (*R-13*). Applied field of 1.0 T and vacuumed to 10^{-3} Pa

Table 2.1-1. Curie point and hysteresis properties

	R-1	R-13	R-22	R-31	R-33	R-36	R-37	R-38	R-39
Main Curie point (°C)	575	550	560	570	540	550	550	570	570
Minor Curie point (°C)		350							
	rev	irrev	rev	rev	rev	rev	rev	rev	rev
NRM (A m^{-1})	27.26	72.88	104.7	64.75	64.10	63.30	40.36	64.81	92.68
χ (SI)	1.005	1.393	1.265	1.539	1.805	1.781	1.806	1.735	1.562
Q ratio (NRM/(χh^a))	0.76	1.48	2.30	1.18	1.01	0.75	0.99	1.04	1.66
J_s (Am^2 kg^{-1})	21.39	26.01	34.84	35.79	62.27	35.01	22.03	30.48	26.69
J_R (Am^2 kg^{-1})	1.854	1.384	1.565	1.510	1.949	1.248	0.925	0.663	1.103
H_C (mT)	10.0	5.1	4.4	4.6	4.6	3.2	4.1	1.7	3.7
H_{RC} (mT)	27.0	10.1	11.6	13.8	19.6	7.1	11.3	18.9	9.4
J_R/J_S	0.087	0.053	0.045	0.042	0.031	0.036	0.042	0.021	0.041
H_{RC}/H_C	2.70	1.98	2.64	3.00	4.26	2.22	2.76	11.18	2.54

[a] h = 45 µT of the geomagnetic field at the Mt. Riiser Larsen area.

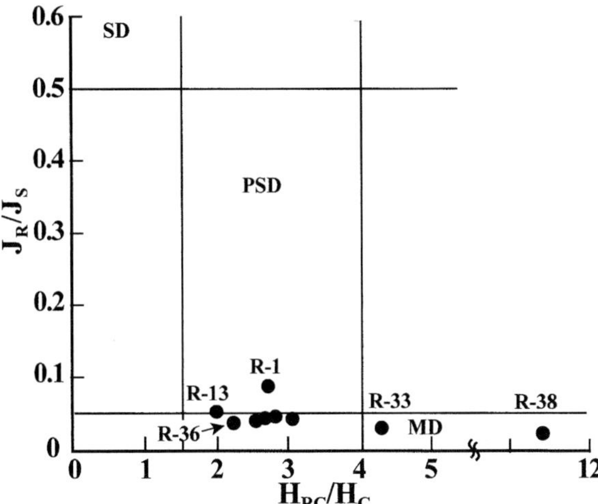

Fig. 2.1-6. Estimation of magnetic domain state using Day plots (Day et al. 1977). *SD:* single domain; *PSD:* pseudosingle domain; *MD:* multi domain

R-22, R-31, R-36, R-37 and R-39 were located to the lower area of PSD. This may be explained by a mixture of PSD and MD grains in the sample; the sample including a small amount of PSD grains versus a large amount of MD ones so as to decrease the J_R/J_S value, while H_{RC}/H_C value is independent of the quantity of PSD grains. The contribution of super-paramagnetic (SP <0.04 µm in diameter) magnetite in decreasing the J_R/J_S value is estimated to be negligibly small because of a larger amount of magnetite grains >1 µm in the samples, as shown in Fig. 2.1-4. If the sample includes a small amount of hematite together with a large amount of magnetite (no evidence of hematite by the microscopic observations), the hysteresis properties of a hematite component cannot be seen due to the strong spontaneous magnetization of magnetite.

NRM intensity (R) and low field susceptibility (χ) were measured for the representative samples from each site, as shown in Table 2.1-1. The Q ratio denotes $R/\chi h$, where h is the geomagnetic field intensity (45 µT in the Mt. Riiser-Larsen area). The Q ratio from 0.76 to 2.30 is derived from the values from R = 27.26 to 104.27 (A m^{-1}) and χ = 1.005 to 1.806 (SI).

Discussion

In the general understanding for dating the Napier Complex, a tonalitic precursor of orthogneiss aged at 3 930 ±10 Ma (U-Pb dating of zircon; Black et al. 1986) was metamorphosed at ca. 2 980, 2 840 and 2 480 Ma (Harley and Black 1997). Probably the ultrahigh temperature metamorphism occurred at about 2 800 Ma or about 2 500 Ma (e.g., Shiraishi et al. 1997; Asami et al. 1998). The metamorphosed rocks of the layered gneiss in the Mt. Riiser-Larsen area have been estimated to originate in sedimentary rocks because of the presence of sedimentary structures in gneiss (e.g., Motoyoshi and Matsueda 1984) and in intrusive rocks for some mafic gneiss (Ishizuka et al. 1998). The precursor of metamorphosed rocks in the layered gneiss accomplished the sedimentation, intrusion and metamorphism between 3 930 and 2 500 Ma.

Field observation in the Mt. Riiser-Larsen area clarified the existence of widely extended meta-BIF layers. These frequently showed natural shrinkage at the end of the layers, but they are cut by faults or dyke intrusions occasionally. One layer through R-31, -27 and -31b extends at least 800 m to the NE SW and 400 m to the NW SE, moreover, it extends to the east as inferred by the magnetic anomaly. Dolinsky et al. (2002) pointed out that the discontinuous lineation more than 7.4 km to the NE-SW direction through R-13 and R-39 (anomaly A, B, C and D in Fig. 2.1-2) is possibly the same layer. These anomalies are also inferred for the wide extension of the meta-BIF layer, because the anomalies appeared at sea-level (R-39), ridge (R-13, 173 m in altitude) and Richardson Lake (13 m in altitude and 53m in depth), where the vertical distance of the anomaly sources is 213 m. From this field evidence, the meta-BIF layer seems to be developed as sheets in the layered gneiss. The layers running in massive gneiss may be same feature estimated from the geological map in this region (Ishizuka et al. 2000). Although Ishizuka et al. (1998) reported that the meta-BIF layers are sandwiched between garnet felsic gneiss and in rare cases between OPX felsic gneiss, our field results showed that the layers run in OPX felsic gneiss as well as the other felsic gneiss. These appearances of meta-BIF layers suggest a sedimentary origin rather than igneous one, although the branched meta-BIF layers from one layer recognized at R-27 and R-36-4 may not be so easy to explain by a simplified sedimentation model. Possibly it is due to the shearing.

The magnetic minerals of meta-BIF layers are almost pure magnetite from the Curie point at 570 °C. The magnetic domain states of magnetite grains were MD (>20 µm) or a mixture of MD and PSD (1 to 20 µm) estimated by Day plots (Fig. 2.1-5). Even if hematite or SP magnetite are included in the sample, it is extremely small compared with MD and PSD magnetite grains by the evidence of the microscopic observations. The result is consistent with that of microscopic observation and Funaki (1988). However, the grains aligned as chains or network structures of fine-grained magnetite (Fig. 2.1-4a–c). A general feature of magnetite in igneous rocks is characterized by isolated euhedral crystals in silicate which resembles the single crystal of magnetite in Fig. 2.1-4d. These complicated features of magnetite distribution can be plausibly explained by metamorphism of banded iron formation (BIF) consisting of fine-grained hematite and quartz as the principal composition. If the precursor hematite grains were transferred to the magnetite grains without melting during metamorphism (reaching to 1 000 °C) (e.g., Sheraton et al. 1987), the fine-grained, larger-grained and networked magnetite structure may be explained; magnetite is driven to the silicate grain boundaries and is grown during the decomposition or growth of silicate grains in the metamorphism. Hydromechanical pressure (P) at 11 kPa (Harley and Hensen 1990) acts to deform and shear the meta-BIF layers, as observed in Fig. 2.1-4b and c. Probably the hexagonal single crystal of magnetite (Fig. 2.1-4d) might be derived from partial melting or recrystallization of meta-BIF after metamorphism at the ultrahigh P stage. The cooling time may have been slow, because a large crystal 200 µm in diameter is included without any deformation.

Magnetite is electrically conductive, but the basalts including numerous magnetite grains are not conductive due to the isolated magnetite grains in silicate. The conductivity of meta-BIF, observed in many layers as represent 66 kΩ · 60 cm^{-1} at R-13, may suggest the magnetite grains were connected with each other through the quartz grain boundary and/or in the quartz grains (Fig. 2.1-4). The magnetite content of 23–68 wt.-% in meta-BIF estimated by hysteresis analyses is extremely high compared with basalt, but it is not responsible for the conductivity. The conductivity may derive from the unique distribution appearing as a network structure of magnetite grains in meta-BIF.

The magnetic anomaly is caused by the NRM and the induced magnetization resulting from the susceptibility of rocks in the ambient magnetic field. Funaki (1984, 1988) reported the strong but unstable NRM due to IRM and VRM for meta-BIF of the Mr. Riiser-Larsen area. It is estimated that the NRM of the layer is parallel to the magnetic field direction. The values of relatively high Q ratios from 0.76 to 2.30, strong susceptibility $\chi = 1.005$ to 1.806 (SI) and strong NRM intensity from $R = 27.26$ to 104.27 (A m^{-1}) of our results are inferred from the magnetic anomaly due to the NRM and induced magnetization. From these viewpoints we concluded that the strong magnetic anomalies result from not only susceptibility but also the NRM of the meta-BIF layers.

Conclusion

The large magnetic anomaly results from the strong NRM and susceptibility of the meta-BIF layers, and the anomaly pattern reported by Dolinsky et al. (2002) is explained by the distribution of meta-BIF layers. One to nine meta-BIF layers run in 30 m thick of felsic gneiss along the strike of the layered gneiss. The thickness is usually 0.5 and 2 m but it varied from zero to 6 m. The same meta-BIF may be observed in the massive gneiss. The largest formation including the meta-BIF layers is estimated to extend more than 7 400 × 213 m, although the layer is frequently discontinuous. Almost pure magnetite of 23–68 wt.-% and quartz are the dominant composition of meta-BIF. Some of the meta-BIF layers preserve traces of metamorphism from hematite to magnetite and shearing history under the ultrahigh T-P condition. Partial melting/recrystallization occurred after the ultrahigh P condition. The combined PSD magnetite grains, sometimes appearing along the quartz crystal boundary as a network structure, is a reason for the electrical conductivity of the meta-BIF layers.

Acknowledgments

The authors wish to express their appreciation to the 42nd Japanese Antarctic Research Expedition for conducting field operations in the Mt. Riiser-Larsen area, and are grateful to Dr. A. Stephenson (University of Newcastle, UK) for correction of English.

References

Asami M, Suzuki K, Grew ES, Adachi M (1998) Chime ages for granulites from the Napier Complex, East Antarctica. Polar Geosci 11:172–199

Black LP, Williams IS, Compston W (1986) Four zircon ages from one rock: the history of a 3 930 Ma old granulite from Mount Sones, Enderby Land, Antarctica. Contrib Mineral Petrol 94:427–437

Day R, Fuller MD, Schmidt VA (1977) Hysteresis properties of titanomagnetites: grain size and composition dependence. Phys Earth Plane Internat 13:260–267

Dolinsky P, Funaki M, Yamazaki A, Ishikawa N, Matsuda T (2002) The results of magnetic survey at Mt. Riiser-Larsen, Amundsen Bay, Enderby land, East Antarctica, by the 42nd Japanese Antarctic Research Expedition. Polar Geosci 15:80–88

Funaki M (1984) Natural remanent magnetization of the Napier Complex in Enderby Land, East Antarctica. Antarct Rec 83:1–10

Funaki M (1988) Paleomagnetic studies of the Archean rocks collected from the Napier Complex in Enderby Land, East Antarctica. Antarctic Rec 32:1–16

Harley SL, Black LP (1997) A revised Archaean chronology for the Napier Complex, Enderby Land, from SHRIMP ion-microprobe studies. Antarct Sci 9:74–91

Harley SL, Hensen BJ (1990) Archean and Proterozoic high-grade terrains of East Antarctic (40–80° E): a case study of diversity in granulite facies metamorphism. In: Ashworth JR, Brown M (eds) High temperature metamorphism and crustal anatexis. Unwin Hyman, London, pp 320–370

Ishizuka H, Ishikawa M, Hokada T, Suzuki S (1998) Geology of the Mt. Riiser-Larsen area of the Napier complex, Enderby Land, East Antarctica. Polar Geosci 11:154–171

Ishikawa M, Hokada T, Ishizuka H, Miura H, Suzuki S, Takada M, Zwarts DP (2000) Mt. Riiser-Larsen. Antarctic Geol Map Ser, Sheet 37

Motoyoshi Y, Matsueda H (1984) Archean granulites from Mt. Riiser-Larsen in Enderby Land, East Antarctica. Mem Nation Inst Polar Res Spec 33:103–125

Sheraton JW, Tingey RJ, Black LP, Offe LA, Ellis DI (1987) Geology of Enderby Land and western Kemp Land, Antarctica. Austral Bur Mineral Resour Bull 223:1–51

Shiraishi K, Ellis DJ, Fanning CM, Hiroi Y, Kagami H, Motoyoshi Y (1997) Reexamination of the metamorphic and protolith ages of the Rayner Complex, Antarctica: evidence for the Cambrian (Pan-African) regional metamorphic event. In: Ricci CA (ed) The Antarctic regions: geological evolution and processes. Terra Antartica Publication, Siena, pp 79–88

Experimental Constraints on the Decompressional *P-T* Paths of Rundvågshetta Granulites, Lützow-Holm Complex, East Antarctica

Toshisuke Kawasaki[1] **· Yoichi Motoyoshi**[2]

[1] Department of Earth Sciences, Faculty of Science, Ehime University, Bunkyo-cho 2–5, Matsuyama 790-8577, Japan, <toshkawa@sci.ehime-u.ac.jp>

[2] National Institute of Polar Research, Kaga 1-chome, Itabashi-ku, Tokyo 173-8515, Japan, <motoyosi@nipr.ac.jp>

Abstract. High-pressure experiments were carried out at pressures 7–15 kbar and temperatures 850–1 150 °C using the piston cylinder apparatus. Experimental data give constraints on the *P-T* path of sillimanite-cordierite-sapphirine granulites from Rundvågshetta, Lützow-Holm Complex, East Antarctica. Combining the previous temperature estimates and the present data, we can infer that the Rundvågshetta granulites experienced the peak metamorphism at 925–1 039 °C and 11.5–15 kbar within the stability field of garnet, orthopyroxene, sapphirine and sillimanite. Subsequent retrograde metamorphism took place at 824–1 010 °C and 6.5–10.8 kbar accompanied by breakdown of garnet and consumption of orthopyroxene and sillimanite within the stability field of orthopyroxene, sapphirine, spinel and cordierite.

Introduction

The metamorphic grade of the Lützow-Holm Complex increases southwestwards from the upper amphibolite facies to granulite facies of the intermediate pressure type (Hiroi et al. 1986). The Rundvågshetta area (Fig. 2.2-1) is located close to the "thermal axis" (Hiroi et al. 1991), and experienced the highest grade of the granulite-facies metamorphism in the Lützow-Holm Complex (Motoyoshi et al. 1989). Shiraishi et al. (2003) reported the SHRIMP age of >2 500 Ma from measurements of cores of zircon crystals in the Rundvågshetta granulites interpreting these as protolith ages (e.g., deterital zircons). They also found the ages of 520–550 Ma for the rims of zircon crystals, interpreted to have grain near the metamorphic peak.

Kawasaki et al. (1993a) estimated the peak metamorphic condition at Rundvågshetta as at least 900 °C or probably higher, based on petrographic study of a sillimanite-cordierite-sapphirine granulite (sample no. SP 92011102A labeled by "I" in Fig. 2.2-1). Motoyoshi and Ishikawa (1997) described two garnet-orthopyroxene-sillimanite granulites (sample nos. SP 93010601-X labeled by "II" and SP RH-111-09 labeled by "Ky"). They found the relict kyanite and sapphirine as inclusions in garnet porphyroblasts in SP RH-111-09. They also found two types of symplectites of sapphirine + orthopyroxene + cordierite + plagioclase and orthopyroxene + cordierite + plagioclase in SP 93010601-X. Harley (1998a,b) estimated the peak temperature of >1 000 °C based on the data of Motoyoshi and Ishikawa (1997). Fraser et al. (2000) obtained the Pan African ages of 517 ±9 Ma and also estimated peak temper-

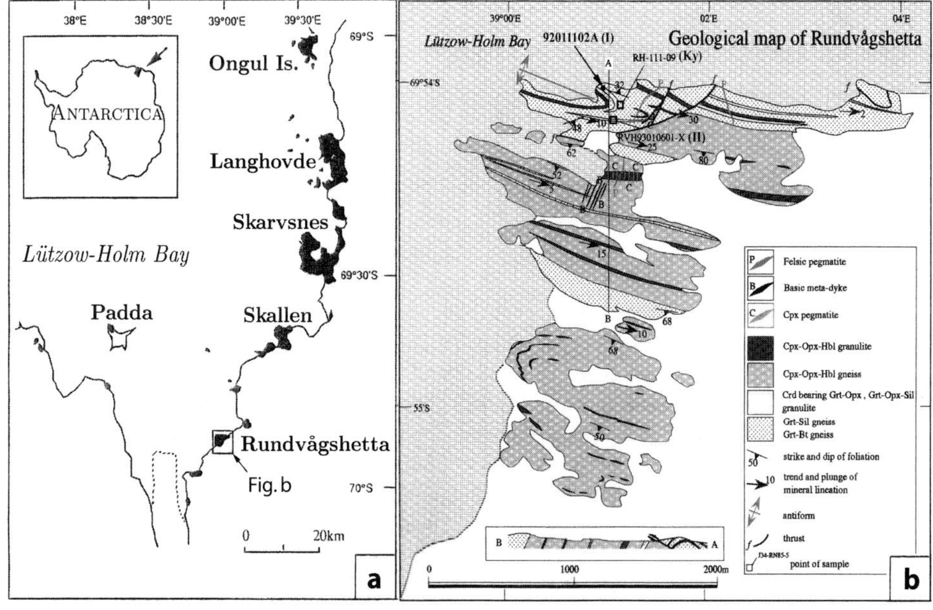

Fig. 2.2-1. Location map and geographical map of Rundvågshetta. **a** Location of Rundvågshetta in Lützow-Holm Bay. **b** Locality of the sample. The geological map is simplified and modified after Motoyoshi and Ishikawa (1997). The sampling point of SP 92011102A is marked by a *filled circle* and labeled by "*I*". Sample nos. SP 93010601-X labeled by "*II*" and SP RH-111-09 labeled by "*Ky*"

From: Fütterer DK, Damaske D, Kleinschmidt G, Miller H, Tessensohn F (eds) (2006) Antarctica: Contributions to global earth sciences. Springer-Verlag, Berlin Heidelberg New York, pp 23–36

Table 2.2-1. Bulk, CIPW norm and modal compositions of a sillimanite-cordierite-sapphirine granulite (Sp 92011102A) from Rundvågshetta, East Antarctica

	Bulk Composition				Mode (20 127 points)	
	wt %	CIPW norm[a]				
SiO_2	44.56	C		10.43	Opx	19.23
TiO_2	1.86	F		32.91	Grt	15.83
Al_2O_3	16.89		or	19.74	Bt	39.49
Fe_2O_3[b]	13.24		ab	11.51	Crd	20.08
MnO	0.09		an	1.66	Spl	0.06
MgO	18.74	Hy		17.40	Spr	2.09
CaO	0.40		en	12.37	Pl	0.50
Na_2O	1.36		fs	5.03	Sil	2.66
K_2O	3.34	Ol		34.82	Qtz	0.02
P_2O_5	0.05		fo	24.04	Total	100.00
Total	100.53		fa	10.78		
	ppm	Z		0.03		
Cr	378.7	il		3.53		
Ni	274.8	cm		0.08		
Rb	197.3	ap		0.12		
Sr	20.2					
V	299.0					
Y	24.6					
Zr	154.3					
Ba	571.4					
Nb	12.5					
Mg / (Mg + Fe)	0.764					

Note: Bulk composition was analyzed by XRF method by Hideo Ishizuka. Volatile components such as H_2O, CO_2 and F are excluded in this analysis. *Bt:* biotite; *Crd:* cordierite; *Grt:* garnet; *Opx:* orthopyroxene; *Pl:* plagioclase; *Qtz:* quartz; *Sil:* sillimanite; *Spl:* spinel; *Spr:* sapphirine. [a] Fe_2O_3 is converted to FeO. [b] Total Fe as Fe_2O_3.

ature of >900 °C at about 11 kbar. Kawasaki and Sato (2002) evaluated the peak metamorphic temperatures of granulites as 925 °C for SP 92011102A and 1 039 °C for SP 93010601-X using the new orthopyroxene-garnet geothermometer (Kawasaki and Motoyoshi 2000). The orthopyroxene-sapphirine thermometer (Kawasaki and Sato 2002) indicated these granulites experienced the subsequent retrograde metamorphism at 824 °C and 1 010 °C, respectively.

In this paper we present the experimental data on the phase relations of the sillimanite-cordierite-sapphirine granulite (SP 92011102A) described by Kawasaki et al. (1993a). These data provide information on the decompressional *P-T* paths of the Rundvågshetta granulites and also give us the important constraints on the *P-T* history of the Lützow-Holm Complex. This kind of approach has previously been carried out with fair success in the evaluation of the *P-T* path of the ultra-high temperature granulites from McIntyre Island, Napier Complex (Kawasaki and Motoyoshi 2000; Kawasaki et al. 2002).

Brief Description of the Rundvågshetta Sillimanite-Cordierite-Sapphirine Granulite

We used a sillimanite-cordierite-sapphirine granulite (sample no. SP 92011102A; Kawasaki et al. 1993a) as a starting material in this study. We collected the granulite during the summer operation (1991–1992) of the 33rd Japanese Antarctic Research Expedition (JARE33) from Rundvågshetta, Lützow-Holm Bay, East Antarctica. The granulite is a melanocratic and coarse-grained rock, and occurs as a concordant thin layer (about 5–10 cm) within the leucocratic quartz-feldspathic gneisses that locally contain large crystals of garnet and sillimanite. Cordierite and sapphirine were easily recognized on the surface of the sample. The granulite consists of garnet, orthopyroxene, cordierite, sapphirine, sillimanite and biotite with small amount of spinel, plagioclase, zircon and quartz. The modal composition of the constituent minerals is given in Table 2.2-1. A very small amount of quartz (0.02 modal-%) is found as inclusions in garnet. Spinel and plagioclase are found in the symplectites with cordierite + sapphirine and plagioclase + sapphirine. The modes of these minerals are less than 0.5%. This granulite contains a large amount of biotite (approximately 39 modal-%). Sapphirine and sillimanite counted as about 2 modal-%.

Kawasaki et al. (1993a) preliminarily reported the chemical compositions of the constituent minerals. The Rundvågshetta granulite SP 92011102A contains garnet with X_{Mg}^{Grt} = 0.49–0.56, orthopyroxene with 5–8 wt.-% Al_2O_3 and X_{Mg}^{Opx} = 0.71–0.78, and biotite with >5 wt.-% of TiO_2 and X_{Mg}^{Bt} = 0.75–0.80. Biotite contains about 1 wt.-% F. This granulite has symplectites and corona textures involving sapphirine (X_{Mg}^{Spr} = 0.78–0.80), orthopyroxene (X_{Mg}^{Opx} = 0.73), cordierite (X_{Mg}^{Crd} = 0.87–0.92), plagioclase (An_{44}–An_{83}) and spinel (X_{Mg}^{Spl} = 0.45–0.49). Mean chemical compositions of these minerals are given in Table 2.2-2 and Fig. 2.2-2.

The decompressional reaction textures are characterized as follows: (1) orthopyroxene and sillimanite are never in direct contact and are separated from each other by a canal of cordierite and/or symplectitic intergrowth of cordierite + sapphirine and sapphirine + plagioclase (Fig. 2.2-3A); and: (2) garnet is locally replaced by a symplectite composed of orthopyroxene + cordierite + sapphirine + spinel + plagioclase (Fig. 2.2-3B). Similar mineral assemblages and reaction textures have been reported from Forefinger Point in Enderby Land (Harley et al. 1990) and the Rauer Group in Prydz Bay (Harley and Fitzsimons 1991).

Experimental Procedures

Starting Material

In the present experiments we used finely powdered granulite as the starting material. Glass starting materials were not employed because the granulite contains a fairly large amount of biotite (about 39 modal-%). In this circumstance, the unquantifiable losses of volatile components including H_2O, F and alkalis during glass preparation would lead to erroneous phase relation that are not relevant to the original bulk composition. Therefore, we employed the dry-powdered rock as starting material to avoid volatile- and alkali-loss. The rock sample was ground to 600 mesh and used for the chemical analysis of the bulk composition (Table 2.2-1). Coarse powder was ground for 1 hour in an agate mortar under ethyl alcohol. Finally rock sample was pulverized to about 10–50 µm in grain size. About 50% of it was less than 30 µm after grinding.

Hideo Ishizuka kindly analyzed the major and trace elements of the Rundvågshetta granulite by means of the XRF technique at the Department of Geology, Kochi University. Data are given in Table 2.2-1. This granulite is characterized by high MgO content (18.74 wt.-%), high Mg/(Fe + Mg) ratio (X_{Mg} = 0.764), extremely low CaO content (0.40 wt.-%), rather high K_2O content (3.34 wt.-%) and high Rb/Sr ratio (≈ 10). The bulk chemistry is reflected in the appearance of normative corundum and olivine,

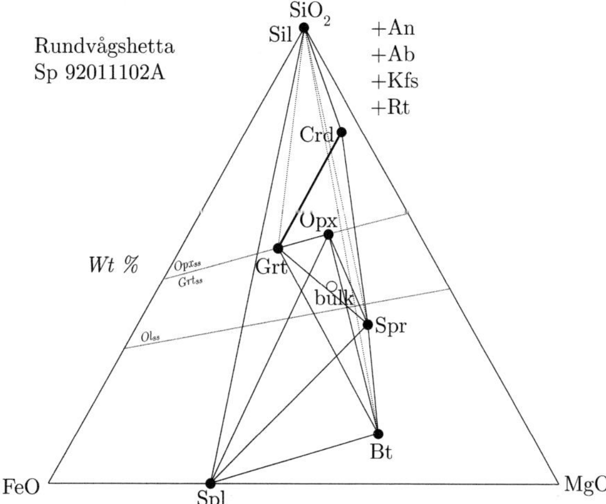

Fig. 2.2-2. Normalized wt.-% SiO_2-FeO-MgO projection of the bulk and minerals of the Rundvågshetta granulite. The bulk composition plots within the silica-undersaturated field far below the garnet-cordierite tie-line defined by McDade and Harley (2001), and near the olivine solid solution line. *Ab:* albite; *An:* anorthite; *Bt:* biotite; *Crd:* cordierite; *Grt:* garnet; *Kfs:* K-feldspar; *Ol:* olivine; *Opx:* orthopyroxene; *Rt:* rutile; *Sil:* sillimanite; *Spl:* spinel; *Spr:* sapphirine

Table 2.2-2. Mean chemical compositions (wt-%) of minerals of a sillimanite-cordierite-sapphirine granulite (Sp 92011102A) from Rundvågshetta, East Antarctica

	Minerals[a]								
	Grt c	Grt r	Opx c	Opx r	Opx s	Spr	Spl	Crd	Bt
SiO_2	39.84	39.57	49.73	49.60	50.44	12.76	0.03	48.38	37.84
TiO_2	0.03	0.02	0.15	0.14	0.16	0.02	n.d.	0.06	5.37
Al_2O_3	22.46	22.79	8.03	6.26	7.04	62.72	60.65	33.55	16.28
Cr_2O_3	0.04	0.05	0.05	0.01	n.d.	0.22	2.28	0.01	0.12
FeO	20.55	21.49	15.10	16.99	16.17	7.05	22.43	2.70	8.80
MnO	0.45	0.54	0.09	0.14	0.10	0.01	n.d.	0.03	0.05
MgO	14.51	13.68	26.06	25.44	24.59	16.23	10.64	12.03	17.65
NiO	n.d.	0.02	0.06	0.06	n.d.	0.10	0.23	0.06	0.10
CaO	0.89	1.16	0.07	0.09	0.07	0.01	n.d.	0.02	0.01
Na_2O	0.04	0.03	0.01	0.02	n.d.	0.03	0.10	0.06	0.26
K_2O	0.03	0.01	0.02	0.03	0.01	0.03	n.d.	0.06	8.64
ZnO	n.d.	0.03	0.10	0.13	0.09	n.d.	3.88	0.05	0.12
F	n.d.	n.d.	n.d.	n.d.	0.09	n.d.	n.d.	n.d.	0.98
–O≡F	–	–	–	–	0.04	–	–	–	0.41
Total	98.84	99.39	99.57	98.90	98.72	99.18	100.23	97.06	95.80
Mg/(Mg+Fe)	0.557	0.532	0.755	0.727	0.731	0.804	0.458	0.888	0.781

[a] Mineral data are given by the mean of 10–20 analyses. Chemical compositions of plagioclase, quartz and sillimanite are omitted in this table. *c:* core; *r:* rim; *s:* symplectite; *Bt:* biotite; *Crd:* cordierite; *Grt:* garnet; *Opx:* orthopyroxene; *Spl:* spinel; *Spr:* sapphirine.

disappearance of normative quartz and diopside and in relatively high ratio of normative orthoclase. The fairly large amounts of normative olivine and corundum indicate that the granulite is chemically saturated in the sapphirine and spinel components, and is highly deficient in the sillimanite component, resulting from the low content of sillimanite (2.66 modal-%, see Table 2.2-1). The bulk of the granulite plots on the silica-undersaturated field (Fig. 2.2-2) below the garnet-cordierite tie-line defined by McDade and Harley (2001). From this chemical feature of the granulite, we could find the sillimanite in only four runs (Table 2.2-3).

Experimental Technique

In the preliminary experiments we adopted the crimping method: one of the edges of noble metal tube was welded

Table 2.2-3. Experimental conditions and phase assemblages

Run No	*P* (kbar)	*T* (°C)	*T* (h)	Capsule	Phases[c]
MS930722	7	850	240	Pt[a]	Opx + Bt + Rt + Spr + Spl + Crd + Pl (Grt,Sil)
MS930819	7	1 050	96	AgPd[a]	Opx + Bt + Rt + Spr + Spl + Gls + V
UNSW960624	7	1 100	266	AgPd[b]	Opx + Bt + Rt + Spr + Spl + Gls + V
02020614B	9	900	952	Au[b]	Opx + Bt + Rt + Crd + Pl (Sil)
MS930917[d]	9	950	232	AgPd[a]	Opx + Bt + Rt + Spr + Spl + Zr + Gls
030623A	9	1 000	598	Mo/Pt	Opx + Bt + Rt + Spr + Spl + Kfs + V
KC940120[d]	9	1 050	100	AgPd[a]	Opx + Bt + Rt + Spr + Spl + Gls
020725A	9	1 050	453	Mo/Pt	Opx + Bt + Rt + Spr + Spl + Kfs + V (Crd)
950721	9	1 100	64	AgPd[a]	Opx + Bt + Rt + Spr + Spl + Gls + V
020902A	9	1 150	176	Mo/Pt	Opx + Bt + Rt + Spl + Gls
KC931222	10	950	220	Pt[a]	Opx + Bt + Rt + Crd + Spr + Spl + Pl + Zr
KC931001	10	1 000	170	Pt[a]	Opx + Bt + Rt + Crd + Spr + Spl + Kfs + Zr (Grt)
KC931129	10	1 050	105	Pt[a]	Opx + Bt + Rt + Spr + Spl + Kfs (Crd)
030402A	10	1 050	340	Mo/Pt	Opx + Bt + Rt + Spr + Spl + Kfs (Crd,Grt)
KC931126	10	1 100	70	AuPd[a]	Opx + Bt + Rt + Spr + Spl + Gls + V
030421A	11	1 000	597	Mo/Pt	Grt + Opx + Bt + Rt + Spr + Spl + Kfs + Pl (Sil,Qtz)
020117A	11	1 050	764	Mo/Pt	Grt + Opx + Bt + Rt + Spr + Spl + Kfs + Zr
030526A	11	1 100	262	Mo/Pt	Opx + Bt + Rt + Spr + Spl + Kfs + Gls
020920A	11	1 150	172	Mo/Pt	Opx + Bt + Rt + Spr + Spl + Kfs + Gls + V
031123A	12	950	770	Mo/Pt	Grt + Opx + Bt + Rt + Spr + Kfs + Zr (Crd)
031030A	12	1 000	575	Mo/Pt	Grt + Opx + Bt + Rt + Spr + Kfs
031020A	12	1 100	230	Mo/Pt	Opx + Bt + Rt + Spr + Spl + Kfs + Gls + V
030613A	12	1 150	172	Mo/Pt	Opx + Bt + Rt + Spr + Spl + Kfs + Zr + Gls + V
021029A	13	1 050	209	Mo/Pt	Grt + Opx + Bt + Rt + Spr + Kfs (Sil)
031001A	13	1 100	119	Mo/Pt	Grt + Opx + Bt + Rt + Spr + Kfs + Zr
021022A	13	1 150	161	Mo/Pt	Grt + Opx + Bt + Rt + Spr + Spl + Zr + Gls + V
980617	15	1 100	266	AuPd[b]	Grt + Opx + Bt + Rt + Spr + Kfs + V

Note: In the experiments using the Mo/Pt double capsule, the inner Mo capsule was put into the outer Pt capsule, two edges of which were welded by carbon arc. [a] An edge of the sample container was welded by carbon arc and the other was crimped twice. [b] Two edges of the noble metal capsule were welded by carbon arc. [c] Relict phases are given in parentheses. Phases are abbreviated after Kretz (1983): *Bt:* biotite; *Crd:* cordierite; *Grt:* garnet; *Gls:* glass; *Kfs:* K-feldspar; *Opx:* orthopyroxene; *Pl:* plagioclase; *Rt:* rutile; *Sil:* sillimanite; *Spl:* spinel; *Spr:* sapphirine; *V:* vapor; *Zr:* Zircon. [d] This run was in failure with crimping. Volatile components from surroundings reduced the solidus of the charge. These data are omitted in Fig. 2.2-5.

by carbon arc and another edge was crimped twice by a fine radio-pliers without welding. Ayers et al. (1992) reported a new technique using a lid on the sample container without welding to seal a capsule for hydrothermal experiments. In cases where we adopted the conventional crimping method, the experiments failed due to addition of the H_2O component from surroundings. We found glass and bubbles in the run no. MS930917, even if the run temperature was about 150 °C below the solidus. We adopted the Mo/Pt double capsule method in the later runs to prevent addition or escape of volatile components. The powdered granulite was packed into an inner Mo foil capsule made by rolling up and crimping two edges with a drill, because the Mo is relatively inert and is useful as a capsule material. The Mo capsule was put into the outer Pt container, whose edge was welded. After dried at 120 °C overnight, another edge of the Pt capsule was welded by carbon arc. Four or five containers were placed in a boron nitride sleeve within a talc + Pyrex glass assembly (Kawasaki and Motoyoshi 2000). We found the low-contrast thin film of MoO_2 with thickness of 2–3 µm coating the inner wall of the molybdenum capsule in all runs. The fO_2 of the charge was kept within the Mo-MoO_2 buffer where the intrinsic fO_2 is 2.9×10^{-16} bar at 1 000 °C (O'Neill 1986).

High-pressure experiments were carried out using a 16.0 mm piston cylinder device at Kochi and Ehime University. We also used ¾-inch piston cylinders at the Institute of the Study of Earth's Interior, Okayama University and the University of New South Wales. The pressure range was between 7 and 15 kbar and the temperature range between 850 and 1 150 °C. Hot piston-in technique was applied for all runs. At about 1 kbar the charge was heated above softening point of Pyrex glass (about 750 °C) of the pressure-transmitting medium and was then compressed to the desired pressure. Finally, sample was heated to the desired temperature. The generated pressures were measured against the oil pressure of the press. Pressures were calibrated by the phase transformations of Bi I–II at room temperature (25.5 kbar: Hall 1971) and by the quartz-coesite transition at 1 000 °C (29.7 kbar: Bohlen and Boettcher 1982). On the basis of these calibrations, we adopted a negative correction of 12.4% to the nominal pressure value in the present experiment. Run pressures in Table 2.2-3 are given as the corrected values.

Electric power was supplied to the graphite heater through tungsten-carbide piston and the stainless steel plug. Temperatures were monitored with a Pt/Pt 13% Rh thermocouple of 0.3 mm in diameter without correction for the pressure effect on emf. We confirm that the temperature gradient as about 1–2 °C mm^{-1} along the axial direction within the sample space at 10.6 kbar and 1 400 °C (Kawasaki et al. 1993b). During each experiment the temperature and pressure were kept constant within ±1% relative. After being kept at a desired pressure and temperature for specified durations, samples were quenched by cutting off the electric supply. The thermocouple reading dropped below 50 °C in a few seconds and to the ambient temperature in several tens of seconds. The pressure was then released to atmospheric pressure over about five minutes, and the run products recovered. The experiments were not reversed, but long durations (64–952 hours), observed systematic changes in mineral assemblage and chemical compositions of the run products suggest runs might be close to equilibrium. All run products were mounted in epoxy resin, and polished for identification of the phases and the examination with electron probe microanalyzer JEOL JXA-8800 at Ehime University. The ZAF correction method was applied. The instrumental conditions were: the accelerating voltage, 15 kV; electron beam current, 5×10^{-9} A; electron beam diameter, 1 µm estimated from the size of contamination spots formed by excitation during analysis.

Run Products

Starting material was the dry powdered rock, which was composed of a mineral aggregate of garnet, orthopyroxene, biotite, cordierite, sapphirine, sillimanite, spinel, plagioclase and quartz. In order to determine the equilibrium phases by the observation of the texture and the chemical composition of run products recrystallized from such reactants, we adopted the following criteria: (1) a newly crystallized mineral is stable; (2) an enlarged mineral is stable; (3) a round-edged mineral is stable; (4) vanished mineral is unstable; and: (5) a corroded mineral is unstable.

Examples of run products are shown in Fig. 2.2-3. We found orthopyroxene, biotite and needle rutile in all runs at pressures from 7 to 15 kbar and temperatures from 850 to 1 150 °C. The survival of the biotite at temperatures up to 1 150 °C is due to its high Ti and F contents (Dymek 1983; Peterson et al. 1991; Hensen and Osanai 1994). The TiO_2 content of biotite in the starting material is about 5 wt.-%, and the F content is about 1 wt.-% (Table 2.2-2). We often found the newly crystallized biotites (abbreviated to Ti-Bt) containing >9 wt.-% TiO_2 and >2 wt.-% F (Table 2.2-4). Figure 2.2-4 shows the examples of the compositional relations among coexisting phases obtained at 10 kbar and 1 100 °C (run no. KCH931126) and at 9 kbar 1 050 °C (run no. 020725A). Orthopyroxene, spinel, sapphirine, biotite and rutile are found in these runs. Glass is found at 10 kbar and 1 100 °C, while K-feldspar at 9 kbar 1 050 °C. These phases plot on the Al_2O_3-TiO_2-(MgO + FeO) and Al_2O_3-SiO_2-(MgO + FeO) planes. The original biotite (abbreviated to Bt) and cordierite are shown by asterisks on these figures. The original biotite plots on the Opx-(Ti-Bt)-Spl, Opx-(Ti-Bt)-Spr and Opx-Rt-Gls (or Kfs) planes. This indicates that the breakdown of the original biotite would derive the combinations of phases ortho-

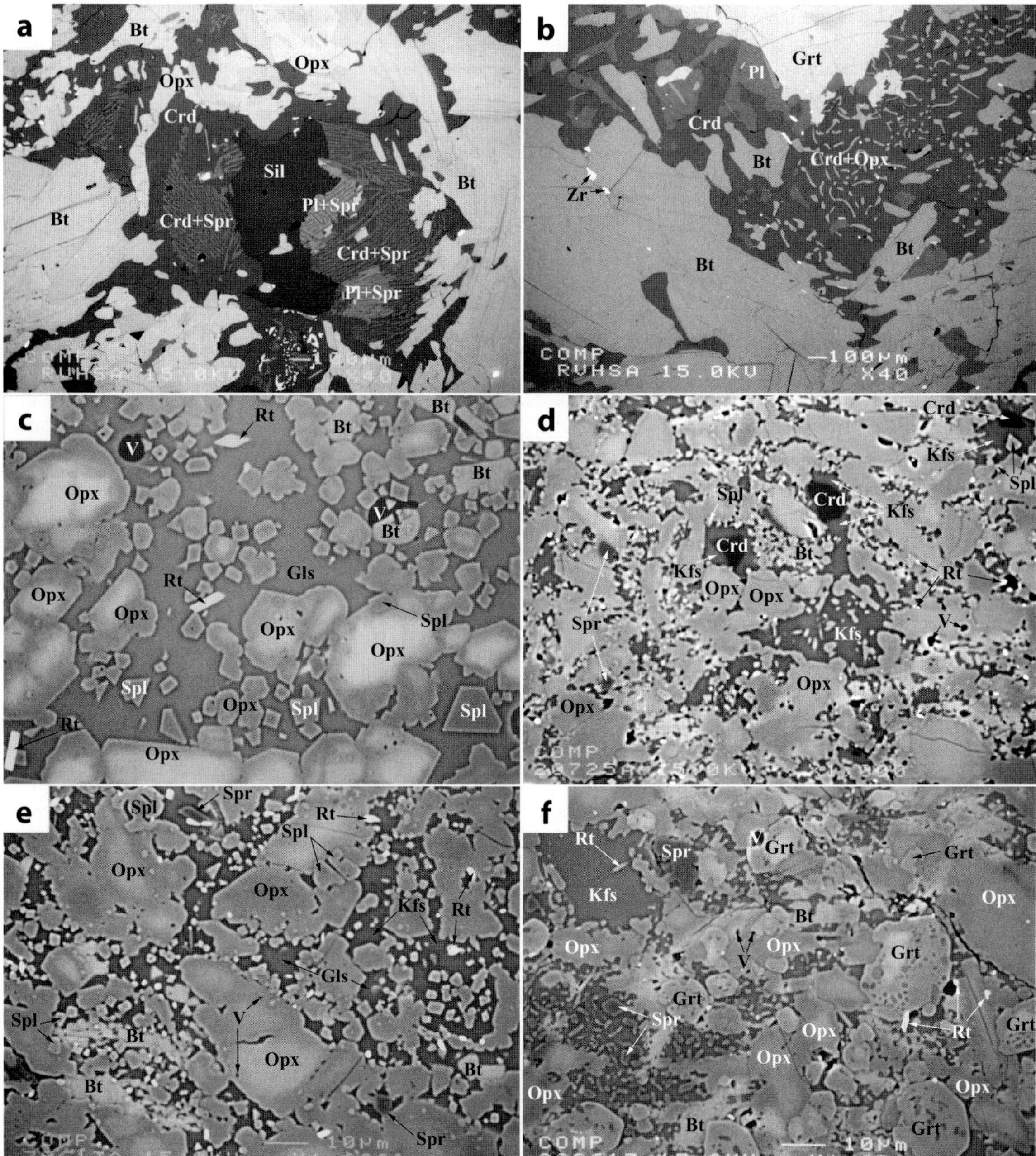

Fig. 2.2-3. Back scattered electron image (BSEI) of the Rundvågshetta granulite and the run products. Photographs **a** and **b** show the symplectitic intergrowths and are in the same scale. The *scale bar* is 100 µm for **a** and **b**. Photographs **c–f** are the run products and are in same scale. The *scale bar* indicates 10 µm. **a** Symplectites of cordierite + sapphirine and plagioclase + sapphirine among the initial orthopyroxene and sillimanite. **b** Symplectitic intergrowth of orthopyroxene and cordierite around garnet. **c** Run no. UNSW960624. Orthopyroxene, biotite, rutile and sapphirine coexist with melt at 7 kbar and 1 100 °C for 266 hours in AgPd capsule. Orthopyroxene shows reverse zoning and has Mg-rich rims. Euhedral biotites and spinels are in direct with orthopyroxene and glass. Cavities of vapor phases are found in the glass. Euhedral fine spinels precipitated at the rims of biotites. **d** Run product of no. 020725A at 9 kbar 1 050 °C for 453 hours in Mo/Pt double capsule. Cordierite is rimed by K-feldspar from the reaction: Bt + Crd ⟶ Rt + Kfs + Opx + Spl + V. Equilibrium assemblage is Opx + Bt + Rt + Spr + Spl + Kfs + V. **e** Run product of no. 030613A at 12 kbar and 1 150 °C for 172 hours in Mo / Pt double capsule. This photograph shows the coexistence of K-feldspar and melt together with orthopyroxene, biotite, rutile, sapphirine, spinel and vapor. Glass is slightly brighter than K-feldspar on this back-scattered electron image. **f** Garnet, orthopyroxene, biotite, rutile, sapphirine and K-feldspar produced at 15 kbar and 1 100 °C for 266 hours in AuPd capsule (run no. 980617). Large garnet includes orthopyroxene, biotite, rutile and sapphirine at Mg-rich rims. Both of large garnet and orthopyroxene show reverse zoning. Fine sapphirine crystallized within K-feldspar accompanying biotite. Spinel disappeared at this condition. *Bt:* Biotite; *Crd:* cordierite; *Gls:* glass; *Grt:* garnet; *Kfs:* K-feldspar; *Opx:* orthopyroxene; *Pl:* plagioclase; *Qtz:* quartz; *Rt:* rutile; *Spl:* spinel; *Spr:* sapphirine; *V:* vapor

pyroxene, spinel, sapphirine, new biotite, rutile, K-feldspar and melt:

$$\text{Bt} \longrightarrow \text{Opx} + \text{Spl} + \text{Ti-Bt} \qquad (2.2\text{-}1)$$

$$\text{Bt} \longrightarrow \text{Opx} + \text{Spr} + \text{Ti-Bt} \qquad (2.2\text{-}2)$$

$$\text{Bt} \longrightarrow \text{Opx} + \text{Rt} + \text{Kfs} \qquad (2.2\text{-}3)$$

$$\text{Bt} \longrightarrow \text{Opx} + \text{Rt} + \text{melt} \qquad (2.2\text{-}4)$$

We found the glass in the run products of run no. UNSW960624 at 7 kbar and 1 100 °C for 266 hours in AgPd capsule (Fig. 2.2-3C). The glass is easily identified as a high-contrast, morphologically clear interstitial phase. Cordierite and plagioclase disappeared, and instead, spinel crystallized as euhedral crystals in the quenched melt. Orthopyroxene reacted with melt and formed euhedral Mg-rich rims.

Fine spinels (<1 µm) also occurred in euhedral biotite. Rare sapphirine was also found in this run product. Rutile and high-Ti biotite (Ti-Bt) were formed by the Reactions 2.2-1, 2.2-2 and 2.2-4. Cavities in the glass were formed by separation of the vapor phase from melt during rapid cooling. At this condition, the equilibrium assemblage is orthopyroxene + biotite + rutile + sapphirine + spinel + melt + vapor.

We found cordierites rimed by K-feldspar in the run products of run no. 020725A at 9 kbar and 1 050 °C for 453 hours in Mo/Pt double capsule (Fig. 2.2-3D). Small rounded crystals of orthopyroxene were found around large and angular ones and in contact with K-feldspar. The Fig. 2.2-3E shows the coexistence of K-feldspar and melt together with orthopyroxene, sapphirine, spinel, bi-

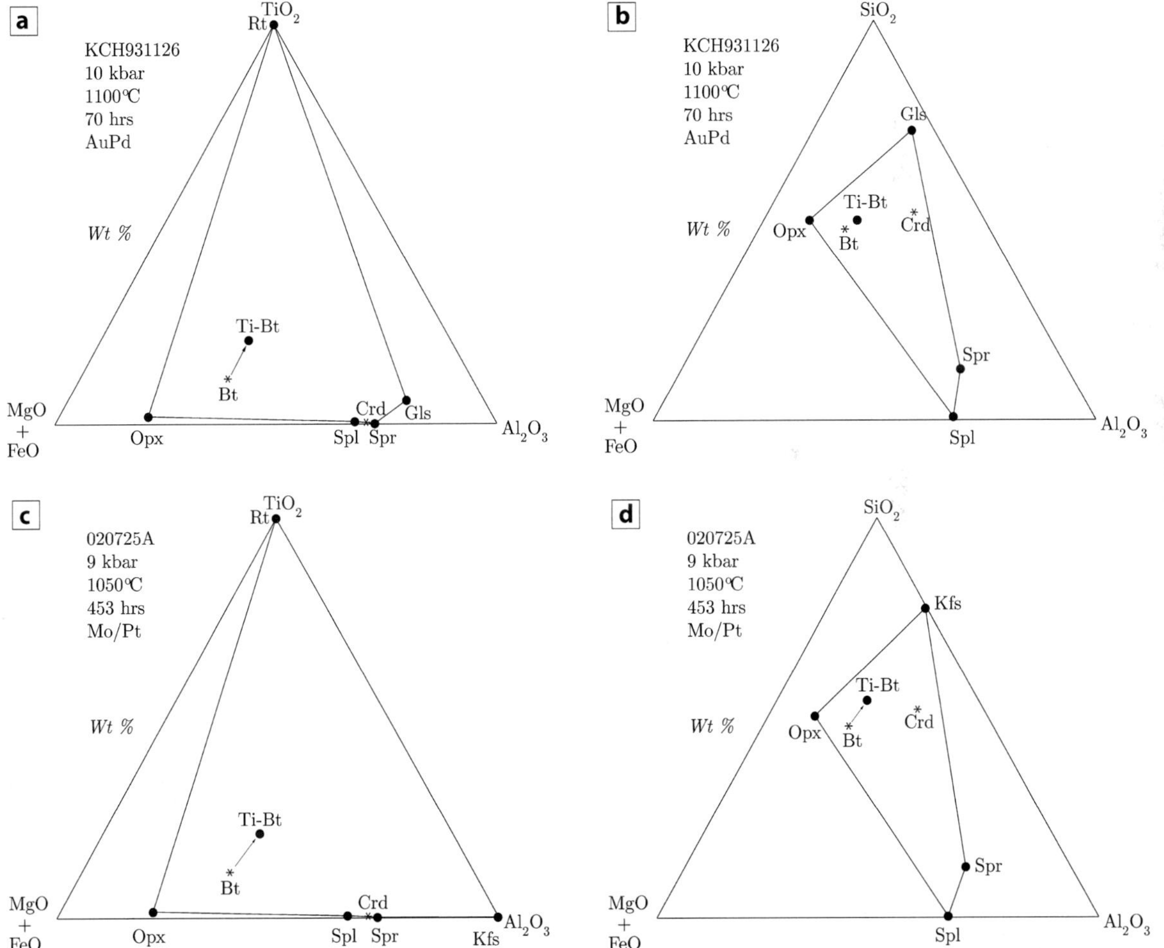

Fig. 2.2-4. Chemical compositions of orthopyroxene, sapphirine, spinel, cordierite, biotite, rutile, K-feldspar and glass on the normalized wt.-% Al_2O_3-TiO_2-(MgO + FeO) and Al_2O_3-SiO_2-(MgO + FeO) planes. The original biotite and cordierite are shown by *asterisks*. The original biotite plots on the Opx-(Ti-Bt)-Spl, Opx-(Ti-Bt)-Spr and Opx-Rt-Gls (or Kfs) planes, while the cordierite on the Spl-Rt-Gls (or Kfs), Spl-(Ti-Bt)-Gls (or Kfs) and Opx-Spr-Gls (or Kfs) planes. The compositional changes in biotites are shown by *arrows*. The Bt-Crd tie line intersects the Spl-Rt and Spl-(Ti-Bt) joins. *Bt:* biotite; *Crd:* cordierite; *Gls:* glass; *Kfs:* K-feldspar; *Opx:* orthopyroxene; *Rt:* rutile; *Spl:* spinel; *Spr:* sapphirine; *Ti-Bt:* newly formed titanian biotite

Table 2.2-4. Representative electron microanalyses of run products

Run No.	MS930722									020725A						
Capsule	Pt[a]									MoPt[b]						
Pressure (kbar)	7									9						
Temperature (°C)	850									1050						
Duration (h)	240									453						
Others	Rt, Sil									Rt						
Phase	**Grt[c]**	**Opx c**	**Opx r**	**Crd c**	**Crd r**	**Spl**	**Spr**	**Bt**	**Pl**	**Opx c**	**Opx r**	**Spl**	**Spr**	**Bt**	**Crd[d]**	**Kfs**
SiO_2	39.94	51.81	50.59	50.88	51.16	0.06	13.06	37.57	45.33	49.84	49.52	0.04	11.95	37.87	49.57	62.86
TiO_2	n.d.	0.14	0.19	n.d.	n.d	0.29	0.08	5.49	n.d.	0.21	0.75	0.36	0.04	10.00	0.04	0.34
Al_2O_3	23.50	6.08	8.21	33.64	33.56	60.68	62.41	17.50	34.77	8.24	10.55	65.87	63.87	15.19	33.13	18.89
Cr_2O_3	0.14	n.d.	0.12	n.d.	n.d.	3.70	0.23	0.05	n.d.	0.08	0.07	0.01	0.66	0.10	0.02	n.d.
FeO	22.01	16.56	17.26	2.43	3.09	23.40	7.24	7.45	0.01	15.94	13.29	16.51	7.40	7.93	3.06	0.51
MnO	0.23	0.17	0.20	0.01	0.04	0.10	n.d.	0.06	n.d.	0.12	0.09	0.07	0.05	0.04	0.03	0.03
MgO	12.99	25.10	23.06	11.87	11.49	10.92	16.15	17.73	n.d.	24.77	25.24	16.00	16.04	14.97	11.36	0.01
NiO	n.d.	n.d.	n.d.	0.16	n.d.	0.10	0.07	0.06	n.d.	0.21	0.06	0.18	0.05	0.02	0.03	n.d.
CaO	1.15	0.06	0.28	n.d.	0.21	0.03	0.09	0.09	18.10	0.04	0.18	0.02	n.d.	0.08	0.16	0.42
Na_2O	n.d.	0.03	0.02	n.d.	n.d.	0.05	n.d.	0.21	1.10	0.01	0.03	0.04	n.d.	0.20	0.10	0.87
K_2O	n.d.	0.01	0.04	0.13	0.07	0.11	0.24	9.43	0.10	0.14	0.09	0.55	n.d.	10.40	0.72	14.73
ZnO	n.a.	0.01	0.06	n.d.	n.d.	0.82	n.d.	n.d.	0.14	0.10	0.11	0.25	0.02	0.14	0.05	0.04
F	n.a.	n.a.	n.a.	n.a.	n.a.	n.a.	n.a	1.34	n.a.	0.03	n.d	n.d.	2.47	2.83	0.07	0.16
−O≡F	–	–	–	–	–	–	–	0.56	–	0.01	–	–	–	1.19	0.03	0.07
Total	99.98	99.97	100.09	99.11	99.71	100.27	99.78	96.42	99.55	99.52	99.96	99.91	100.06	98.75	98.29	98.82
Number of cations for *n* oxygens																
n	**12**	**6**	**6**	**18**	**18**	**4**	**20**	**22**	**8**	**6**	**6**	**4**	**20**	**22**	**18**	**8**
Si	2.980	1.871	1.834	5.055	5.067	0.002	1.558	5.312	2.099	1.809	1.767	0.001	1.424	5.265	5.007	2.940
Ti	–	0.004	0.005	–	–	0.006	0.008	0.583	–	0.006	0.020	0.007	0.004	1.046	0.003	0.012
Al	2.067	0.259	0.351	3.959	3.917	1.920	8.770	2.917	1.898	0.353	0.444	1.992	8.970	2.489	3.944	1.042
Cr	0.008	–	0.003	–	–	0.078	0.022	0.006	–	0.002	0.002	–	0.062	0.011	0.001	–
Fe	1.374	0.500	0.523	0.202	0.256	0.525	0.722	0.881	–	0.484	0.397	0.343	0.737	0.922	0.259	0.020
Mn	0.015	0.005	0.009	0.001	0.012	0.002	–	0.008	–	0.004	0.003	0.001	0.005	0.004	0.002	0.001
Mg	1.445	1.352	1.246	1.758	1.696	0.437	2.870	3.738	–	1.341	1.343	0.632	2.850	3.104	1.711	–
Ni	–	–	–	0.013	–	0.002	0.006	0.007	–	0.001	0.002	0.004	0.004	0.002	0.002	–
Ca	0.092	0.002	0.011	–	0.022	0.001	0.012	0.014	0.898	0.002	0.007	0.001	–	0.012	0.017	0.021
Na	–	0.002	0.001	–	–	0.003	–	0.058	0.099	0.001	0.002	0.002	–	0.053	0.019	0.079
K	–	0.001	0.002	0.016	0.008	0.004	0.036	1.701	0.006	0.006	0.004	0.018	–	1.844	0.092	0.879
Zn	–	–	0.002	–	–	0.016	–	–	0.005	0.003	0.003	0.005	0.002	0.014	0.004	–
F	–	–	–	–	–	–	–	0.597	–	0.003	–	–	–	1.242	0.023	0.023
Σ catio	7.982	3.997	3.986	10.984	10.979	2.995	14.026	15.224	5.004	4.010	3.993	3.006	14.057	14.767	11.061	4.994
Mg/Fe+Mg	0.513	0.730	0.704	0.897	0.869	0.454	0.799	0.809	–	0.735	0.772	0.648	0.795	0.771	0.869	
Ca/(Ca−Na+K)								0.008	0.896					0.006		0.021
Na/(Ca+Na+K)								0.033	0.099					0.028		0.080
K/(Ca+Na+K)								0.960	0.005					0.966		0.898

[a] An edge of the platinum capsule was welded by carbon arc and the other was crimped twice. [b] The inner Mo capsule was put into the outer Pt capsule, two edges of which were welded by carbon arc. [c] Relict garnet is anhedral. [d] Relict cordierite is round-shaped and trapped within K-feldspar. Total Fe as FeO. Glass was analyzed by the defocused X-ray microprobe method (2 μm of beam diameter). *n.a.*: not analyzed; *n.d.*: not detected; *c*: core; *r*: rim; *Bt*: biotite; *Crd*: cordierite; *Grt*: garnet; *Gls*: glass; *Kfs*: K-feldspar; *Opx*: orthopyroxene; *Pl*: plagioclase; *Rt*: rutile; *Sil*: sillimanite; *Spl*: spinel; *Spr*: sapphirine; *V*: vapor; *Zr*: zircon.

Table 2.2-4. *Continued*

Run No.	KC931222					KC931126					
Capsule	Pt[a]					AuPd[a]					
Pressure (kbar)	10					10					
Temperature (°C)	950					1 100					
Duration (h)	220					70					
Others	Rt, Pl, Zr					Rt					
Phase	**Opx**	**Spl**	**Spr**	**Crd**	**Bt**	**Opx c**	**Opx r**	**Spl**	**Spr**	**Bt**	**Gls**
SiO_2	51.58	0.25	12.25	50.16	35.94	50.94	49.15	0.64	13.27	36.31	61.28
TiO_2	0.12	0.07	0.06	n.d.	5.98	0.18	0.88	0.49	0.53	9.69	1.44
Al_2O_3	8.62	64.23	65.44	33.90	16.27	8.61	11.28	65.85	60.97	16.04	19.57
Cr_2O_3	0.07	0.11	0.31	0.04	0.07	0.11	0.07	0.33	0.38	0.05	n.d.
FeO	14.94	23.62	7.81	3.08	7.92	14.90	11.85	13.93	6.58	5.00	2.81
MnO	0.01	0.06	0.05	0.13	0.05	0.09	0.17	0.06	0.05	0.03	n.d.
MgO	25.42	10.82	16.72	11.83	16.00	26.09	26.86	18.78	18.03	16.88	1.75
NiO	0.08	0.22	0.12	0.04	0.10	0.07	0.01	0.10	n.d.	0.04	n.d.
CaO	0.12	0.14	0.18	0.06	0.01	0.08	0.15	0.03	0.05	n.d.	0.75
Na_2O	n.d.	0.12	n.d.	n.d.	0.28	0.01	n.d.	0.03	0.01	0.48	1.01
K_2O	n.d.	0.07	0.02	0.24	10.54	n.d.	n.d.	0.08	0.03	10.63	4.26
ZnO						0.08	0.10	0.09	n.a.	n.a.	0.05
F	n.a.	n.a.	n.a.	n.a.	2.05	n.a.	n.a.	n.a.	n.a.	1.65	n.a.
–O≡F	–	–	–	–	0.86	–	–	–	–	0.70	–
Total	100.97	99.72	101.82	99.49	94.36	101.17	100.53	100.39	99.87	96.10	93.09
Number of cations for *n* oxygens											
n	6	4	20	18	22	6	6	4	20	22	
Si	1.828	0.007	1.437	4.992	5.251	1.807	1.735	0.016	1.580	5.158	
Ti	0.003	0.001	0.006	–	0.657	0.005	0.023	0.009	0.047	1.033	
Al	0.360	2.009	8.907	3.976	2.802	0.360	0.469	1.955	8.551	2.687	
Cr	0.002	0.002	0.029	0.003	0.008	0.003	0.002	0.007	0.036	0.006	
Fe	0.443	0.524	0.766	0.256	0.968	0.442	0.350	0.294	0.655	0.594	
Mn	–	0.001	0.005	0.011	0.006	0.003	0.005	0.001	0.005	0.004	
Mg	1.343	0.428	2.924	1.755	3.485	1.379	1.413	0.705	3.198	3.575	
Ni	0.002	0.005	0.012	0.003	0.012	0.002	–	0.002	–	0.005	
Ca	0.004	0.004	0.002	0.007	0.002	0.003	0.006	0.001	0.006	–	
Na	–	0.006	–	–	0.079	0.001	–	0.001	–	0.132	
K	–	0.002	0.003	0.031	1.965	–	–	0.003	0.004	1.927	
Zn						0.002	0.003	0.002	–	–	
F	–	–	–	–	0.945	–	–	–	–	0.743	
Σ cation	3.987	2.991	14.091	11.034	15.236	4.007	4.006	2.996	14.080	15.121	
Mg/(Fe+Mg)	0.752	0.450	0.792	0.873	0.783	0.757	0.802	0.706	0.830	0.857	0.526
Ca/(Ca+Na+K)					0.001					–	0.097
Na/(Ca+Na+K)					0.039					0.064	0.239
K/(Ca+Na+K)					0.960					0.936	0.663

[a] An edge of the platinum capsule was welded by carbon arc and the other was crimped twice. Total Fe as FeO. Glass was analyzed by the defocused X-ray microprobe method (2 µm of beam diameter). *n.a.*: not analyzed; *n.d.*: not detected; *c*: core; *r*: rim; *Bt*: biotite; *Crd*: cordierite; *Grt*: garnet; *Gls*: glass; *Kfs*: K-feldspar; *Opx*: orthopyroxene; *Pl*: plagioclase; *Rt*: rutile; *Sil*: sillimanite; *Spl*: spinel; *Spr*: sapphirine; *V*: vapor; *Zr*: zircon.

Table 2.2-4. *Continued*

Run No.	020117A							030613A					
Capsule	Mo/Pt[a]							Mo/Pt[a]					
Pressure (kbar)	11							12					
Temperature (°C)	1050							1150					
Duration (h)	764							172					
Others	Rt, Zr							Rt, Zr					
Phase	**Grt**	**Opx c**	**Opx r**	**Spl**	**Spr**	**Bt**	**Kfs**	**Opx**	**Spl**	**Spr**	**Bt**	**Kfs**	**Gls**
SiO_2	39.72	50.38	49.12	0.09	11.99	37.17	61.35	47.82	0.22	15.91	36.99	63.32	67.09
TiO_2	0.56	0.11	0.20	0.29	0.07	8.39	0.18	1.20	0.55	0.68	8.11	0.26	1.20
Al_2O_3	23.32	6.90	9.42	65.38	63.45	16.00	20.73	12.36	64.53	55.75	15.95	18.44	16.31
Cr_2O_3	n.d.	0.14	0.13	0.41	0.27	0.06	0.04	0.11	0.36	0.17	0.08	0.20	n.d.
FeO	17.31	16.57	14.30	17.42	7.19	8.22	0.53	13.15	15.70	8.12	6.65	0.65	2.79
MnO	0.44	0.18	0.03	0.05	0.13	0.01	n.d.	0.20	0.07	0.11	0.02	0.04	0.02
MgO	15.45	24.89	25.40	16.05	16.07	16.08	0.11	24.83	15.84	17.18	16.13	0.22	1.15
NiO	n.d.	0.01	n.d.	0.12	0.05	0.14	n.d.	0.01	0.09	0.02	0.04	n.d.	0.03
CaO	1.76	0.02	0.06	0.01	n.d.	0.05	2.38	0.18	0.06	n.d.	0.04	0.36	0.87
Na_2O	n.d.	n.d.	n.d.	n.d.	n.d.	0.28	1.62	0.01	0.08	0.03	0.20	0.86	0.89
K_2O	0.04	0.04	0.03	0.09	0.01	10.98	13.55	0.13	0.01	0.11	9.64	14.61	7.93
ZnO	0.09	0.10	0.14	0.11	0.11	n.d.	n.d.	0.08	0.25	0.02	0.08	0.05	0.05
F	n.a.	n.a.	n.a.	n.a.	n.a.	1.83	n.d.	n.d.	n.d.	n.d.	3.61	0.13	0.89
–O≡F	–	–	–	–	–	0.77	–	–	–	–	1.52	0.05	0.37
Total	98.69	99.34	98.84	100.01	99.34	98.42	100.48	99.79	97.74	98.10	98.02	98.91	98.85
Number of cations for *n* oxygens													
n	**12**	**6**	**6**	**4**	**20**	**22**	**8**	**6**	**4**	**20**	**22**	**8**	
Si	2.954	1.837	1.782	0.002	1.438	5.226	2.837	1.709	0.006	1.938	5.207	2.956	
Ti	0.031	0.003	0.006	0.006	0.006	0.887	0.006	0.032	0.011	0.062	0.859	0.009	
Al	2.044	0.297	0.403	1.983	8.968	2.651	1.129	0.521	1.989	8.001	2.646	1.015	
Cr	–	0.004	0.004	0.008	0.026	0.006	0.001	0.003	0.007	0.016	0.009	0.001	
Fe	1.077	0.505	0.434	0.375	0.720	0.966	0.020	0.393	0.343	0.827	0.783	0.025	
Mn	0.028	0.005	0.001	0.001	0.012	0.001	–	0.006	0.002	0.012	0.003	0.001	
Mg	1.712	1.353	1.374	0.616	2.872	3.370	0.008	1.323	0.618	3.119	3.384	0.015	
Ni	–	–	–	0.003	0.006	0.016	–	–	0.002	0.002	0.004	–	
Ca	0.140	0.001	0.002	–	–	0.008	0.118	0.007	0.002	–	0.006	0.018	
Na	–	–	–	–	0.134	0.077	0.145	0.001	0.004	0.008	0.054	0.078	
K	0.004	0.002	0.001	0.003	0.002	1.968	0.799	0.002	–	0.018	1.732	0.870	
Zn	0.009	0.003	0.004	0.002	0.010	–	–	0.002	0.005	0.001	0.008	0.002	
F	–	–	–	–	–	0.812	–	–	–	–	1.605	0.019	
Σ cation	7.995	4.011	4.010	2.998	14.060	15.175	5.064	3.998	2.988	14.004	14.697	4.991	
Mg/(Fe+Mg)	0.614	0.728	0.760	0.622	0.799	0.777	–	0.771	0.643	0.790	0.812		0.416
Ca/(Ca+Na+K)						0.004	0.111				0.003	0.019	0.073
Na/(Ca+Na+K)						0.037	0.137				0.030	0.081	0.134
K/(Ca+Na+K)						0.959	0.752				0.967	0.900	0.793

[a] The inner Mo capsule was put into the outer Pt capsule, two edges of which were welded by carbon arc. Total Fe as FeO. Glass was analyzed by the defocused X-ray microprobe method (2 µm of beam diameter). *n.a.*: not analyzed; *n.d.*: not detected; *c*: core; *r*: rim; *Bt*: biotite; *Crd*: cordierite; *Grt*: garnet; *Gls*: glass; *Kfs*: K-feldspar; *Opx*: orthopyroxene; *Pl*: plagioclase; *Rt*: rutile; *Sil*: sillimanite; *Spl*: spinel; *Spr*: sapphirine; *V*: vapor; *Zr*: zircon.

Table 2.2-4. *Continued*

Run No.	980617					
Capsule	AuPd[a]					
Pressure (kbar)	15					
Temperature (°C)	1 100					
Duration (h)	266					
Others	Rt, Kfs					
Phase	**Grt c**	**Grt r**	**Opx c**	**Opx r**	**Spr**	**Bt**
SiO_2	41.32	41.31	50.45	50.44	13.93	37.75
TiO_2	0.15	n.d.	0.24	0.21	0.22	8.18
Al_2O_3	23.56	23.64	7.97	8.66	61.11	16.30
Cr_2O_3	0.09	0.03	0.13	0.07	0.39	0.11
FeO	17.16	14.40	15.14	13.65	7.42	6.12
MnO	0.42	0.27	0.16	0.05	0.05	0.06
MgO	17.26	18.60	24.52	26.73	17.26	16.70
NiO	n.d.	0.02	0.03	0.06	0.09	0.08
CaO	0.98	1.02	0.08	0.08	0.02	0.01
Na_2O	0.01	0.03	0.03	0.01	0.02	0.14
K_2O	0.07	0.01	0.05	0.03	0.04	10.86
ZnO	0.16	0.06	0.10	0.05	0.05	0.03
F	n.a.	n.a.	n.a.	n.a.	n.a.	3.55
–O≡F	–	–	–	–	–	1.49
Total	101.17	99.40	98.91	100.05	100.59	98.40
Number of cations for *n* oxygens						
n	**12**	**12**	**6**	**6**	**20**	**22**
Si	2.982	2.993	1.833	1.800	1.651	5.204
Ti	0.008	–	0.007	0.006	0.020	0.848
Al	2.004	2.018	0.341	0.364	8.531	2.649
Cr	0.005	0.002	0.004	0.002	0.036	0.012
Fe	1.035	0.872	0.460	0.407	0.735	0.705
Mn	0.025	0.017	0.005	0.002	0.005	0.007
Mg	1.857	2.009	1.328	1.422	3.048	3.431
Ni	–	0.001	0.001	0.002	0.009	0.009
Ca	0.076	0.079	0.003	0.003	0.002	0.001
Na	0.001	0.004	0.002	0.001	0.005	0.038
K	0.006	0.001	0.002	0.002	0.005	1.910
Zn	0.009	0.003	0.003	0.001	0.004	0.003
F	–	–	–	–	–	1.547
Σ cations	8.009	8.000	3.990	4.012	14.051	14.818
Mg/(Fe+M)	0.642	0.697	0.743	0.777	0.805	0.830
Ca/(Ca+Na+K)						0.001
Na/(Ca+Na+K)						0.019
K/(Ca+Na+K)						0.980

[a] Two edges of the nobel metal capsul were welded by carbon arc. Total Fe as FeO. *n.a.:* not analyzed; *n.d.:* not detected; *c:* core; *r:* rim; *Bt:* biotite; *Crd:* cordierite; *Grt:* garnet; *Gls:* glass; *Kfs:* K-feldspar; *Opx:* orthopyroxene; *Pl:* plagioclase; *Rt:* rutile; *Sil:* sillimanite; *Spl:* spinel; *Spr:* sapphirine; *V:* vapor; *Zr:* zircon.

otite, rutile and vapor at 12 kbar and 1 150 °C for 172 hours in Mo/Pt double capsule (run no. 030613A). As is seen in Fig. 2.2-4, the original cordierite plots on the Spl-Rt-Gls (or Kfs), Spl-(Ti-Bt)-Gls (or Kfs) and Opx-Spr-Gls (or Kfs) planes. The Bt-Crd tie line intersects the Spl-Rt and Spl-(Ti-Bt) joins. This indicates that the original cordierite and biotite reacted to form the assemblage of orthopyroxene, spinel, sapphirine, high-Ti biotite, rutile, K-feldspar and vapor (or melt) by the Reaction 2.2-5:

$$\text{Bt} + \text{Crd} \longrightarrow \text{Opx} + \text{Spl (or Spr)} + \text{Ti-Bt} + \text{Rt} + \text{Kfs} + \text{V (and/or melt)} \quad (2.2\text{-}5)$$

It is very difficult to distinguish between K-feldspar and glass from the observation of the texture as is seen in Fig. 2.2-3E. Glass is slightly brighter than K-feldspar on the back-scattered electron image. Large crystals of orthopyroxene showed reverse zoning. Aggregations of fine crystals of orthopyroxene in K-feldspar and glass were accompanied by sapphirine, spinel and rutile. Euhedral fine spinels were found on the rims of large orthopyroxene.

At 15 kbar and 1 100 °C for 266 hours, the run product of no. 980617 is garnet, orthopyroxene, biotite, rutile, sapphirine and K-feldspar (Fig. 2.2-3F). Euhedral garnet included orthopyroxene, sapphirine and biotite at the Mg-rich rim grown around the Fe-rich core. Large crystals of orthopyroxene showed reverse zoning. Cavities formed by vapor phase were found in the crystals of orthopyroxene and at grain boundaries between orthopyroxene and garnet. Fine sapphirine crystallized within K-feldspar accompanying biotite. Spinel disappeared at this condition.

Experimental Results

Experimental results are summarized in Table 2.2-3. We found relict phases in some runs; these are parenthesized in this table. We found sillimanite in four runs as the relict phase. Although the granulite contains 2.66 modal.-% of sillimanite, the bulk chemistry is characterized by the disappearance of normative quartz and the appearance of normative olivine, hypersthene, feldspar and corundum (Table 2.2-1). The bulk composition plots far below the garnet-cordierite tie-line, which was defined as the boundary between silica-saturated and silica-undersaturated compositions by McDade and Harley (2001), and near the olivine solid solution line (Fig. 2.2-2). We can find the "bulk" inside the chemographic triangles Opx-Spr-Spl, Opx-Spr-Grt, Opx-Bt-Spl and Opx-Bt-Grt. Any triangles, a side of which is the Opx-Sil tie-line, can not include the "bulk". The triangles Sil-Grt-Spr and Sil-Grt-Bt include both "bulk" and "Opx". The three compositional points of "bulk", "Opx" and "Grt" plot within the Sil-Spl-Spr and Sil-Spl-Bt planes. These chemographic relations

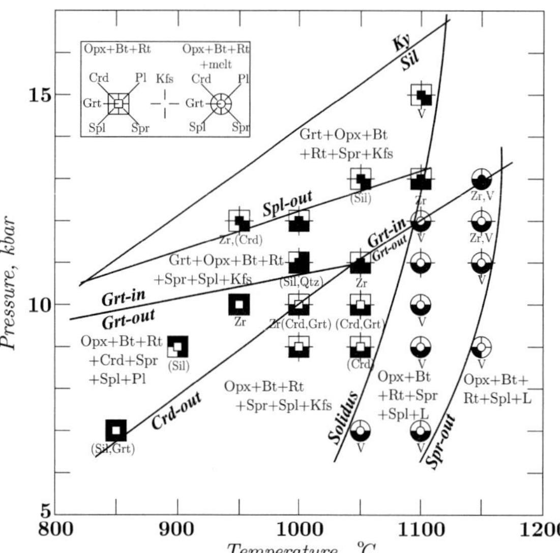

Fig. 2.2-5. Schematic drawing of the phase relation of the sillimanite-cordierite-sapphirine granulite from Rundvågshetta. Note that orthopyroxene, biotite and rutile occurred in all runs. Relict phases, labeled in the *parentheses*, and additional phases are shown at the lower side of run data. The sillimanite-kyanite phase boundary (Holdaway 1971) is shown in this figure. The spinel-out curve intersects with sillimanite-kyanite phase boundary at about 10.6 kbar and 820 °C. These two curves meet with the solidus of 1 120 °C at pressures 13.5 kbar and 16.5 kbar, respectively. The cordierite-out curve intersects with the garnet-in/out curve at about 11 kbar and 1 050 °C

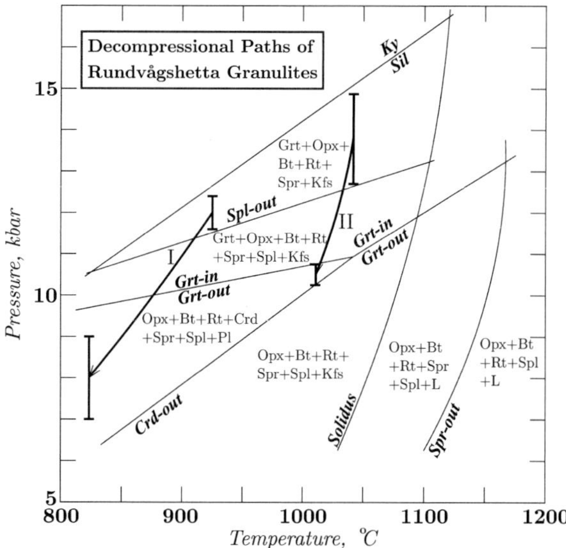

Fig. 2.2-6. Decompressional paths of Rundvågshetta granulites. The new orthopyroxene-garnet geothermometer (Kawasaki and Motoyoshi 2000) indicates the peak metamorphic temperatures of granulites as 925 °C for SP 92011102A labeled by "*I*" and 1 039 °C for SP 93010601-X labeled by "*II*" within the stability field of garnet + orthopyroxene + biotite + rutile + sapphirine + K-feldspar. The subsequent retrograde metamorphism occurred at 824 °C and 1 010 °C, respectively, estimated by the orthopyroxene-sapphirine thermometer (Kawasaki and Sato, 2002) in the stability field of orthopyroxene + biotite + rutile + cordierite + sapphirine + spinel + plagioclase

indicate that the sillimanite would be consumed to produce orthopyroxene and/or garnet during the high-pressure and high-temperature experiments due to its low content.

The phase assemblages are schematically illustrated in Fig. 2.2-5. The sillimanite ⇌ kyanite phase boundary (Holdaway 1971) is given in this figure. The solidus is rather steep with temperature. The stability field of cordierite is restricted within the *P-T* region bounded by the garnet-in/out curve and cordierite-out curve:

$$P\ (\text{kbar}) = 0.0212T\ (°\text{C}) - 11.4 \qquad (2.2\text{-}6)$$

We confirm that the stable phase assemblage is orthopyroxene + biotite + rutile + cordierite + sapphirine + spinel + plagioclase in this field.

The run no. 020117A was carried out at 11 kbar and 1 050 °C for 764 hours. We found the very small amount of garnet coexisting with orthopyroxene, biotite, sapphirine, spinel, K-feldspar and vapor in the inner Mo capsule. We recognized the relict crystals of garnet with the corroded texture in the two runs (KC931001 at 10 kbar and 1 000 °C; and: 030402A at 10 kbar and 1 050 °C), we conclude garnet is unstable below 11 kbar at 1 050 °C. At temperatures above 1 100 °C, garnet is unstable below 13 kbar. The garnet-in/out curve is given in Fig. 2.2-5. Spinel is stable within the stability field of garnet up to the spinel-out curve approximated by

$$P\ (\text{kbar}) = 0.0098T\ (°\text{C}) + 2.5 \qquad (2.2\text{-}7)$$

Spinel is unstable above this curve, and the stable assemblage changes to garnet + orthopyroxene + biotite + rutile + sapphirine + K-feldspar. As has been mentioned before, the Rundvågshetta granulite initially contained garnet, orthopyroxene and sillimanite although spinel is locally associated with the symplectites of sapphirine + cordierite. This implies that the maximum pressure, which the Rundvågshetta granulite experienced during the metamorphism, was below sillimanite ⇌ kyanite phase boundary (Holdaway 1971) and minimum pressure was above spinel-out curve. The spinel-out curve intersects with the sillimanite-kyanite phase boundary at about 10.6 kbar and 820 °C. These two curves meet with the solidus at about 1 120 °C and pressures 13.5 kbar and 16.5 kbar, respectively. This indicates the *P-T* conditions of the peak metamorphism in the Rundvågshetta area are specified in the field bounded by these three curves at pressures from 10.6 kbar to 16.5 kbar and temperatures from 820 °C to 1 120 °C. Then the granulite experienced the retrograde metamorphism into the garnet unstable and cordierite + plagioclase stable fields because we found the symplectite of cordierite and orthopyroxene around garnet, and the sapphirine + cordierite + plagioclase symplectite between orthopyroxene and sillimanite.

Discussion and Conclusion

At 10 kbar and 950 °C (run no. KC931222), mineral compositions are indistinguishable from those in the starting material except appearance of rutile and extinction of garnet (see Table 2.2-4). This suggests that the Rundvågshetta granulite equilibrated at similar temperature conditions at higher pressure conditions or that the starting material did not react at all. Our previous estimations of the peak metamorphic temperatures of 900 °C (Kawasaki et al. 1993a) and 925 °C (Kawasaki and Sato 2002) evaluated by the new garnet-orthopyroxene thermometer (Kawasaki and Motoyoshi 2000) are consistent with the present experimental results. Another granulite (sample no. SP 93010601-X) reported by Motoyoshi and Ishikawa (1997) indicates the temperature of 1 039 °C. The orthopyroxene-sapphirine thermometer (Kawasaki and Sato 2002) applied to the symplectites of these granulites SP 92011102A and SP 93010601-X, resulting the temperatures of 824 and 1 010 °C, respectively. These temperatures are plotted in Fig. 2.2-6. This indicates that the Rundvågshetta granulites experienced the peak metamorphism at 925 °C and 11.5–12.5 kbar for SP 92011102A and at 1 039 °C and 12.6–15 kbar for SP 93010601-X within the stability field of garnet + orthopyroxene + sapphirine bounded by sillimanite ⇌ kyanite reaction and spinel-out curve. Yoshimura (2003) found the coexistence of sapphirine and quartz in the porphyroblasts of garnet in the Rundvågshetta granulite. This supports our idea that garnet, orthopyroxene and sapphirine in the Rundvågshetta granulites were equilibrated during the peak metamorphism. Subsequently the retrograde path intersected the garnet-in/out curve at about 10 kbar and 870 °C for SP 92011102A and at about 10.8 kbar and 1 120 °C for SP 93010601-X, then the original orthopyroxene and sillimanite were consumed to form sapphirine and cordierite by the following reaction:

$$\text{orthopyroxene} + \text{sillimanite} \longrightarrow \text{sapphirine} + \text{cordierite} \qquad (2.2\text{-}8)$$

where garnet is no longer stable and breaks down to orthopyroxene, spinel and cordierite (Kawasaki et al. 1993a). Then these granulites suffered the retrograde metamorphism at 824 °C and 6.5–9.5 kbar for SP 92011102A and at 1 010 °C and 10.2–10.8 kbar for SP 93010601-X within the stability filed of orthopyroxene + cordierite + sapphirine + spinel + plagioclase framed by the garnet-in/out and cordierite-out curves. The possible retrograde paths are schematically illustrated in Fig. 2.2-6. The record of the subsequent retrograde metamorphism is provided by the closure temperatures of 612 °C for SP 92011102A and 521 °C for SP 93010601-X estimated by the Fe-Mg orthopyroxene-cordierite thermometer (Sakai and Kawasaki 1997).

Acknowledgments

We express our hearty thanks to Hideo Ishizuka for his kind help for the chemical analyses of the bulk composition of the Rundvågshetta granulite by the X-ray fluorescence spectroscopy. We often discussed about the high-temperature metamorphism and the anatexis of the lower crustal materials with Kei Sato, Yasutaka Yoshimura, Masaaki Owada, Yasuhito Osanai, Makoto Arima, Yoshikuni Hiroi and Kazuyuki Shiraishi. We obtained the idea of the present study through discussions with them. We also acknowledge Bastiaan Jan Hensen who critically read this manuscript and gave us the constructive comments toward its improvement at the very early stage of this study. He gave us the chance of the research programs of 1996 for sabbatical at the University of New South Wales. Kazuhiro Suzuki, Moonsup Cho and Nobutaka Tsuchiya gave us severe comments. We express our gratitudes to Yoshikuni Hiroi for his criticism at the early stage of this study. Simon Harley and Franco Talarico read this paper and gave us invaluable comments. Part of the expenses of this study was defrayed by the Grant-in-Aid for Scientific Research from the Ministry of Education, Science and Culture of the Japanese Government (nos. 04640740, 09640571, 11640481 and 14654093 to T. Kawasaki).

References

Ayers JC, Brenan JB, Watson EB, Wark DA, Minarik WG (1992) A new capsule technique for hydrothermal experiments using the piston-cylinder apparatus. Amer Mineral 77:1080–1086

Bohlen S, Boettcher AL (1982) The quartz $\rightleftharpoons$ coesite transformation: A precise determination and the effects of other components. J Geophys Res 87:7073–7078

Dymek RF (1983) Titanium, aluminum and interlayer cation substitutions in biotite from high-grade gneisses, West Greenland. Amer Mineral 68:880–899

Fraser G, McDougall L, Ellis DJ, Williams IS (2000) Timing and rate of isothermal decompression in Pan-African granulites from Rundvagshetta, East Antarctica. J Metamorphic Geol 18:441–454

Hall HT (1971) Fixed points near room temperature. In: Lloyd EC (ed) Proceedings of Symposium of Accurate Characterization of High Pressure Environment. NBS Spec Pub 326:313–314

Harley SL (1998a) An appraisal of peak temperatures and thermal histories in ultrahigh-temperature (UHT) crustal metamorphism: The significance of aluminous orthopyroxene. Mem Natl Inst Polar Res Spec Issue 53:49–73

Harley SL (1998b) On the occurrence and characterization of ultrahigh-temperature crustal metamorphism. In: Treloar PJ, O'Brien PJ (eds) What drives metamorphism and metamorphic reaction? Geol Soc London Spec Pub 138:81–107

Harley SL, Hensen BJ, Sheraton JW (1990) Two-stage decompression in orthopyroxene-sillimanite granulites from Forefinger Point, Enderby Land, Antarctica: implications for the evolution of the Archaean Napier Complex. J Metamorph Geol 8:591–613

Harley SL, Fitzsimons ICW (1991) Pressure-temperature evolution of metapelitic granulites in a polymetamorphic terrane: the Rauer Group, East Antarctica. J Metamorph Geol 9:231–243

Hensen BJ, Osanai Y (1994) Experimental study of dehydration melting of F-bearing biotite in model pelitic compositions. Mineral Mag 58A:410–411

Hiroi Y, Shiraishi K, Motoyoshi Y, Kanisawa S, Yanai K, Kizaki K (1986) Mode of occurrence, bulk chemical compositions, and mineral textures of ultramafic rocks in the Lützow-Holm Complex, East Antarctica. Mem National Inst Polar Res Spec Issue 43:62–84

Hiroi Y, Shiraishi K, Motoyoshi Y (1991) Late Proterozoic paired metamorphic complexes in East Antarctica, with special reference to the tectonic significance of ultramafic rocks. In: Thomson MRA, Crame JA, Thomson JW (eds) Geological evolution of Antarctica. Cambridge University Press, Cambridge, pp 83–87

Holdaway MJ (1971) Stability of andalusite and aluminum silicate phase diagram. Amer J Sci 271:97–131

Kawasaki T, Motoyoshi Y (2000) High-pressure and high-temperature phase relations of an orthopyroxene granulite from McIntyre Island, Enderby Land, East Antarctica. Polar Geosci 13:114–133

Kawasaki T, Sato K (2002) Experimental study of Fe-Mg exchange reaction between orthopyroxene and sapphirine and its calibration as a geothermometer. Gondwana Res 4:741–747

Kawasaki T, Ishikawa M, Motoyoshi Y (1993a) A preliminary report on cordierite-bearing assemblages from Rundvågshetta, Lützow-Holm Bay, East Antarctica: evidence for a decompressional *P-T* path? Proc National Inst Polar Res Sympos Antarct Geosci 6:47–56

Kawasaki T, Okusako K, Nishiyama T (1993b) Anhydrous and water-saturated melting experiments of an olivine andesite from Mt. Yakushi-Yama, northeastern Shikoku, Japan. Island Arc 2:228–237

Kawasaki T, Sato K, Motoyoshi Y (2002) Experimental constraints on the thermal peak of a granulite from McIntyre Island, Enderby Land, East Antarctica. Gondwana Res 4:749–756

Kretz R (1983) Symbols for rock-forming minerals. Amer Mineral 68:277–279

McDade P, Harley SL (2001) A petrogenetic grid for aluminous granulite facies metapelites in the KFMASH system. J Metamorphic Geol 19:45–59

Motoyoshi Y, Ishikawa M (1997) Metamorphic and structural evolution of granulites from Rundvågshetta, Lützow-Holm Bay, East Antarctica. In: Ricci CA (ed) The Antarctic region: geological evolution and processes. Terra Antartica Publication, Siena, pp 65–72

Motoyoshi Y, Matsubara S, Matsueda H (1989) *P-T* evolution of the granulite-facies rocks of the Lützow-Holm Bay region, East Antarctica. In: Daly JS, Cliff RA, Yardley BWD (eds) Evolution of metamorphic belts. Geol Soc Spec Pub 43:325–329

O'Neill HStC (1986) No-MoO_2 (MOM) oxygen buffer and the free energy of formation of MoO_2. Amer Mineral 71:1007–1010

Peterson JW, Chacko T, Kuehner SM (1991) The effects of fluorine on the vapor-absent melting of phlogopite + quartz: Implications for deep-crustal processes. Amer Mineral 76:470–476

Sakai S, Kawasaki T (1997) An experimental study of Fe-Mg partitioning between orthopyroxene and cordierite in the Mg-rich portion of the $Mg_3Al_2Si_3O_{12}$-$Fe_3Al_2Si_3O_{12}$ system at atmospheric pressure: its calibration of geothermometry for high-temperature granulites and igneous rocks. Proc National Inst Polar Res Sympos Antarct Geosci 10:150–159

Shiraishi K, Hokada T, Fanning CM, Misawa K, Motoyoshi Y (2003) Timing of thermal events in eastern Dronning Maud Land, East Antarctica. Polar Geosci 16:76–99

Yoshimura Y, Motoyoshi Y, Miyamoto, T (2003) Sapphirine-garnet-prthopyroxene granulite from Rundvågshetta in the Lützow-Holm complex, East Antarctica. The 23rd Sympos Antarct Geosci Abstract, 112:74

Sapphirine – Orthopyroxene – Garnet Granulite from Schirmacher Hills, Central Dronning Maud Land

Sotaro Baba[1] · Masaaki Owada[2] · Edward S. Grew[3] · Kazuyuki Shiraishi[4]

[1] Department of Natural Environment, University of the Ryukyus, Senbaru 1, Nishihara, Okinawa, 903-0213 Japan
[2] Department of Earth Sciences, Yamaguchi University, Yoshida 1677-1, Yamaguchi, 753-8512 Japan
[3] Department of Earth Sciences, University of Maine, 5790 Bryand Research Center Orono, Maine 04469-5790 USA
[4] Department of Crustal Studies, National Institute of Polar Research, Kaga 1-9-10, Itabashi, Tokyo, 173-8515 Japan

Abstract. A third locality of sapphirine granulite was discovered in the Schirmacher Hills, central Dronning Maud Land, East Antarctica; one unique for the association with orthopyroxene and garnet. Sapphirine occurs as inclusions together with orthopyroxene within garnet and as discrete grains together with spinel, secondary orthopyroxene in cordierite coronas. Orthopyroxene porphyroblasts have the highest Al_2O_3 contents (~10 wt.-%) so far reported from Dronning Maud Land. The earlier of two metamorphic stages inferred for the sapphirine-bearing granulite is characterized by breakdown of sapphirine and orthopyroxene to form garnet at peak conditions approaching 950–1 050 °C and 9–10 kbar. The later stage is decompression when spinel-cordierite-orthopyroxene formed at about 950–980 °C and 8 kbar. The age of the ultrahigh-temperature event could be ca. 1.15 Ga, somewhat older than the late Mesoproterozoic granulite-facies event elsewhere in central Dronning Maud Land. Metamorphic temperatures estimated to exceed 900 °C imply that the tectonothermal regime in the Schirmacher Hills was distinct and could be critical in determining relationships between central Dronning Maud Land and metamorphic belts in southern Africa and in understanding the evolution of Rodinia and Gondwana.

Introduction

Sapphirine has been reported widely from the Precambrian rocks of East Antarctica, including four localities in Dronning Maud Land (DML) (Table 2.3-1). However, a sapphirine-orthopyroxene-garnet association has not yet been found in DML. In this paper, we describe a third sapphirine locality in the Schirmacher Hills where orthopyroxene coexisted with garnet, and propose that coexisting sapphirine, garnet and high-Al orthopyroxene are evidence for ultrahigh-temperature (UHT) metamorphism. The mineral abbreviations used in equations, tables and figures are from Kretz (1983).

Geological Outline

The Schirmacher Hills (11°20'–11°55' E, 70°45' S), an isolated exposure on the Princess Astrid coast, is underlain by high-grade polymetamorphic rocks cut by pre- and post-metamorphic dykes (Fig. 2.3-1, e.g., Ravich 1982; Grew 1983; Stackebrandt et al. 1988; Shiraishi et al. 1988; Sengupta 1988, 1993). The metamorphic rocks are divided into following units, listed in order of structural level, lowest first: (1) quartzofeldspathic gneiss, including charnockite, and mafic granulite; (2) augen gneiss; (3) metasediments, including pelitic gneiss and calc-silicate granulite, mafic granulite, and charnockite; (4) garnet-biotite gneiss, biotite gneiss, and biotite-hornblende gneiss (Sengupta 1993). These rocks were affected by granulite-facies metamorphism that was overprinted by at least one amphibolite-facies event, a complex evolution supported by available geochronological data (Grew and Manton 1983; Verma et al. 1987; Mikhalsky et al. 2003). In brief, U-Pb data on allanite, monazite and zircon suggest a granulite-

Table 2.3-1. Previously reported sapphirine-bearing metamorphic rocks in DML

Locality	Sapphirine paragenesis	Inferred *P-T* condition	Reference
Schirmacher Hills			
367	Grt-Spr-Bt-Crn-Pl-Spl[b]-Ky[a]-Sil[a]	$T \approx 750$~800 °C, $P = 5$~8 kbar	Grew (1983)
410A	Sil-Bt-Grt-Spr-Crd-Spl[b]-Crn[b]-Pl[a]-Kfs[a]		Grew (1983)
This study	Opx-Grt-Pl-Crd-Spr-Spl[a]-Ky[a]-Qtz[a]	T 950~1000 °C, $P = 8$~10 kbar	This study
Sør Rondane Mountains			
Auskampane	Spr-Phl-Crd-Pl-Ged-Opx-Spl[b]-Crn	$T = 820$~870 °C	Ishizuka et al. (1995)
Balchenfjella	Spr-Ky-, Spr-Spl and Ged included in Grt; Grt-Bt-Pl-Spl-Crn in matrix	$T \approx 700$ °C, $P \geq 8$ bar	Asami et al. (1994)

[a] As trace. [b] Inclusion.

From: Fütterer DK, Damaske D, Kleinschmidt G, Miller H, Tessensohn F (eds) (2006) Antarctica: Contributions to global earth sciences. Springer-Verlag, Berlin Heidelberg New York, pp 37–44

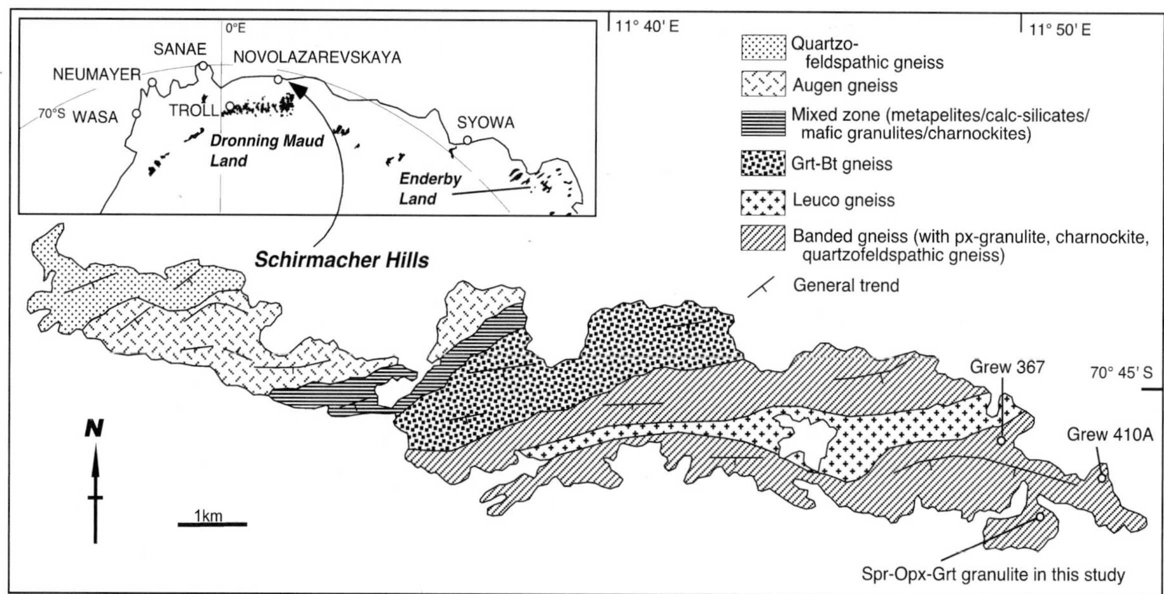

Fig. 2.3-1. Simplified geological map of Schirmacher Hills (modified after Rameshwar Rao et al. 1997) showing localities of sapphirine-bearing granulites reported by Grew (1983, nos. *367* and *410A*) and in the present paper

facies event at ca. 1.15 Ga and amphibolite-facies at ca. 0.625 Ga, as well as several igneous and hydrothermal events. The granulite-facies metamorphism is considered to be related to the 1.0–1.2 Ga Grenvillian event documented in the neighbouring high grade gneiss regions of central DML (e.g., Ravikant 1998).

Occurence of Sapphirine-Bearing Rocks

Sapphirine-orthopyroxene-garnet granulite is exposed in a small outcrop (Fig. 2.3-2) in the eastern part of Schirmacher Hills near where two other sapphirine-bearing granulites have been reported (Fig. 2.3-1). Two domains compose the sapphirine-orthopyroxene-garnet granulite: one coarse-grained, relatively massive, and light-colored consisting of orthopyroxene, garnet and plagioclase; the second fine-grained, weakly banded, and dark-colored consisting of garnet, cordierite, orthopyroxene, and biotite. Surrounding rocks include hornblende-biotite gneiss and biotite gneiss that are locally intercalated with garnet-bearing mafic gneiss and garnet-sillimanite gneiss.

Sapphirine occurs mainly as inclusions within garnet (Fig. 2.3-3a) and rarely orthopyroxene, and as discrete grains together with spinel and secondary orthopyroxene in cordierite coronas (Fig. 2.3-3b). Large orthopyroxene porphyroblasts are dominant in the light-colored domains, and contain tiny rutile needles (Fig. 2.3-3c). Secondary orthopyroxene occurs at the garnet margin together with cordierite, in places as a symplectite, in the light-colored domains, and with cordierite and spinel in the dark-colored domains. Garnet grains are finer than orthopyroxene porphyroblasts, and in places occur along the margins of the porphyroblasts (Fig. 2.3-3c). Rarely, garnet occurs as a reaction rim around sapphirine (Fig. 2.3-3d). Kyanite (identified using laser Raman spectroscopy and showing pale purple pleochroism) is enclosed in plagioclase that is surrounded by a garnet aggregate. Secondary, fine-grained biotite has developed along the grain boundaries between garnet and orthopyroxene, in the cordierite corona, and in the plagioclase matrix. Quartz, zircon, rutile and apatite are accessory minerals.

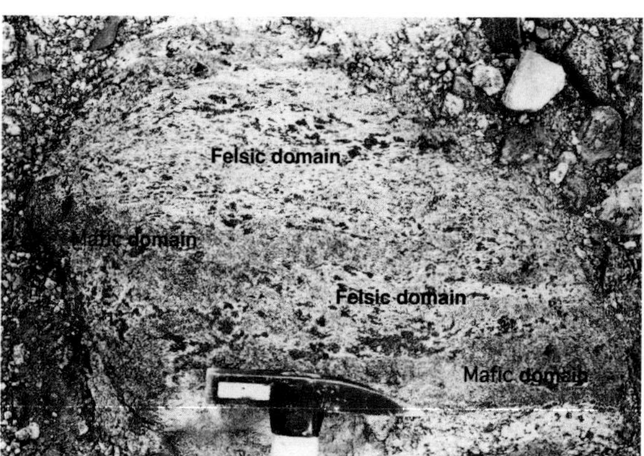

Fig. 2.3-2. Photograph of sapphirine-bearing orthopyroxene-garnet granulite. Orthopyroxene porphyroblasts are prominent in light-colored, relatively massive "felsic" domains, whereas small garnet grains are abundant in dark-colored "mafic" domains showing indistinct banding

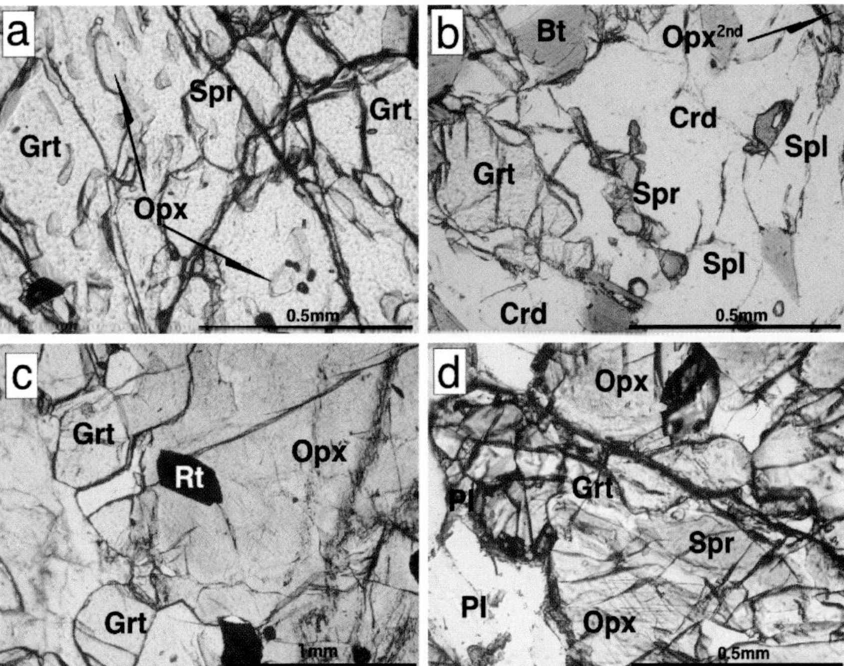

Fig. 2.3-3. Photomicrographs showing mineral relationships in the sapphirine-bearing granulite. **a** Sapphirine (*Spr*) and orthopyroxene (*Opx*) inclusions in a garnet (*Grt*) porphyroblast in the light-colored domain. **b** Discrete sapphirine grain is being replaced by spinel (*Spl*) in cordierite. Secondary orthopyroxene also present, but not in view. *Bt:* biotite. **c** garnet is developed at the marginal part of an orthopyroxene porphyroblast. *Rt:* rutile. **d** Sapphirine grains are rimmed by garnet in the edge of the orthopyroxene porphyroblast. *Pl:* plagioclase

Mineral Composition

The minerals were analyzed with a wavelength-dispersive electron microprobe (JEOL JXA-8800M) at the National Institute of Polar Research (Japan) using natural minerals and synthetic oxides as standards, an accelerating voltage of 15 kV and a specimen current of 12 nA (Table 2.3-2).

Al_2O_3 contents of orthopyroxene porphyroblasts in both the light-colored and dark-colored domains reach 10 wt.-%; X_{Mg} ranges from 0.70 to 0.74. Symplectitic orthopyroxene shows slightly lower Al_2O_3 contents (up to 8.5 wt.-%) and a similar range of X_{Mg} to the porphyroblasts. Orthopyroxene inclusions in garnet both in light-colored and dark-colored domains, have higher X_{Mg} (0.74–0.77) and lower Al_2O_3 (up to 7.5 wt.-%) than those in other textural settings.

In both domains X_{Mg} has a wider range (0.45–0.55) in garnet porphyroblasts than in orthopyroxene porphyroblasts, and garnet X_{Mg} decreases towards the rims. The proportions of grossular and spessartine range from 0.011 to 0.014 and from 0.003 to 0.006, respectively.

Sapphirine composition varies in relation to textural setting. The sapphirine inclusions generally have Al_2O_3-rich composition with Al_2O_3 ranging from 61.8 to 64.3 wt.-% in both domains. On the other hand, discrete sapphirine grains in the dark-colored domain are less aluminous (60.0–62.3 wt.-% Al_2O_3) and richer in SiO_2 (13.4–14.0 wt.-% SiO_2). Fe^{3+} content calculated using charge balance does not exceed 0.14 per 20 oxygens.

Cordierite coronas have fairly similar X_{Mg} values (0.88–0.90) regardless of the coexisting phase. Spinel shows compositional variations of X_{Mg} (0.40–0.46), ZnO (1.4–1.9 wt.-%) and Cr_2O_3 (3.4–6.6 wt.-%). Aluminosilicate is rare; the one analyzed kyanite grain contains 0.2 wt.-% Fe_2O_3. Rare biotite enclosed adjacent to sapphirine in orthopyroxene porphyroblasts contains less TiO_2 (2.8 wt.-%) and has a higher X_{Mg} ratio (0.86) than most biotite flakes along the boundaries between grains of orthopyroxene and garnet. These biotite flakes contain up to 4.0 wt.-% TiO_2 and their X_{Mg} ranges from 0.78 to 0.88. Matrix plagioclase is oligoclase (X_{An} = 0.15–0.18).

Metamorphic Reactions and Conditions

Metamorphic reactions have been deduced from textural relationships. Inclusions of sapphirine and orthopyroxene within garnet are inferred to have been stable before garnet formation both in the light-colored and dark-colored domains. Possible divariant reactions in the FeO-MgO-Al_2O_3-SiO_2 (FMAS) system are

$$\text{Spr} + \text{Opx} + \text{Qtz} = \text{Grt} \qquad (2.3\text{-}1)$$

$$\text{Spr} + \text{Opx} + \text{Ky/Sil} = \text{Grt} \qquad (2.3\text{-}2)$$

In the light-colored domain, quartz is sparingly present in the matrix, whereas in the dark-colored domain, kyanite occurs as a relic phase. Thus, Reactions 2.3-1 and 2.3-2

Table 2.3-2.
Representative chemical composition of the minerals

Mineral	Opx		Grt		Spr		Crd	Spl	Pl
Note	core	sec	core	rim	in Grt	disc		in Crd	
SiO_2	49.11	51.16	40.68	40.53	12.90	14.01	50.03	0.00	63.97
TiO_2	0.15	0.09	0.00	0.09	0.01	0.00	0.01	0.00	0.00
Al_2O_3	10.17	7.17	23.25	22.95	63.29	60.16	33.77	58.68	22.19
Cr_2O_3	0.00	0.16	0.00	0.00	0.42	0.54	0.04	4.55	0.00
FeO*	16.70	16.57	22.95	24.31	7.52	9.21	2.71	24.41	0.14
MnO	0.03	0.05	0.19	0.23	0.05	0.02	0.00	0.01	0.01
MgO	23.76	24.69	14.09	12.94	16.29	15.68	11.81	10.07	0.01
CaO	0.01	0.03	0.41	0.50	0.00	0.00	0.02	0.00	3.40
Na_2O	0.02	0.00	0.04	0.00	0.00	0.00	0.03	0.00	9.71
K_2O	0.01	0.00	0.00	0.00	0.00	0.00	0.00	0.00	0.12
ZnO	–	–	–	–	–	–	–	1.42	–
Total	99.96	99.92	101.61	101.55	100.48	99.62	98.42	99.13	99.55
O	6	6	12	12	20	20	18	4	8
Si	1.775	1.849	2.987	3.000	1.525	1.685	5.010	0.000	2.836
Ti	0.004	0.002	0.000	0.005	0.001	0.000	0.001	0.000	0.000
Al	0.433	0.305	2.012	2.002	8.816	8.525	3.986	1.899	1.159
Cr	0.000	0.005	0.000	0.000	0.039	0.051	0.003	0.099	0.000
Fe^{3+}	0.011	0.000	0.020	–	0.093	0.054	–	0.002	0.005
Fe^{2+}	0.494	0.501	1.389	1.505	0.650	0.872	0.227	0.559	–
Mn	0.001	0.002	0.012	0.014	0.005	0.002	0.000	0.000	0.000
Mg	1.280	1.330	1.542	1.427	2.871	2.811	1.763	0.412	0.001
Ca	0.000	0.001	0.032	0.040	0.000	0.000	0.002	0.000	0.161
Na	0.001	0.000	0.006	0.000	0.000	0.000	0.006	0.000	0.835
K	0.000	0.000	0.000	0.000	0.000	0.000	0.000	0.000	0.007
Zn	–	–	–	–	–	–	–	0.029	0.000
Total cation	4.000	3.994	8.000	7.994	14.000	14.000	10.998	3.000	5.004
X_{Mg}	0.72	0.73	0.53	0.49	0.82	0.76	0.89	0.42	–
$XAl_{(Al/2)}$	0.217	0.153	–	–	–	–	–	–	–
Fe^{3+}/Fe^{total}	0.02	–	0.01	–	0.13	0.06	–	0.00	–

*FeO**: total Fe as FeO; *sec:* secondary mineral; *dis:* as discrete mineral; *Fe^{3+}*: estimated from charge balance.

are plausible as a garnet-forming reaction for the light-colored domain and dark-colored domain, respectively. However, direct evidence of coexistence of quartz and/or sillimanite with orthopyroxene and sapphirine is lacking. A scanning survey with the electron microprobe of micro-inclusions in garnet did not reveal additional phases. One possible interpretation is that modal amounts of quartz and aluminosilicate in this rock were low to start with, and these phases were consumed by the formation of garnet. However, we cannot exclude the possibility of other garnet-forming reactions at the present stage of our investigation.

The discrete sapphirine grains in the dark-colored domain could be relics remaining after Sil/Ky were consumed by Reaction 2.3-2 that subsequently reacted with garnet by the univariant FMAS reaction:

$$\text{Grt} + \text{Spr} = \text{Spl} + \text{Crd} + \text{Opx} \quad (2.3\text{-}3)$$

This reaction implies near isothermal decompression (Fig. 2.3-4). In the light-colored domain, formation of secondary orthopyroxene and cordierite is consistent with the following quartz-bearing FMAS divariant reaction:

$$\text{Grt} + \text{Qtz} = \text{Crd} + \text{Opx} \qquad (2.3\text{-}4)$$

The *P-T* conditions of equilibration of the earlier metamorphic stage have been estimated using experimentally calibrated geothermobarometers based on garnet-orthopyroxene and garnet-orthopyroxene-plagioclase-quartz equilibria and the compositions of the cores of associated orthopyroxene and garnet (Fig. 2.3-5, Table 2.3-3). Temperatures calculated using garnet-orthopyroxene Fe-Mg exchange thermometry (Harley 1984; Lee and Ganguly 1988; Bhattacharya et al. 1991) range from 797 to 953 °C. These temperature estimates are minima because of possible Fe-Mg re-equilibration during cooling (e.g., Fitzsimons and Harley 1994). Pressures calculated using garnet-orthopyroxene barometry (Harley and Green 1982) and garnet-orthopyroxene-plagioclase-quartz barometry (Newton and Perkins 1982) at 900 °C are 5.7 to 6.2 kbar and 7.9 to 8.7 kbar, respectively.

The high Al_2O_3 contents in orthopyroxene porphyroblasts give 1 000–1 050 °C and 9–10 kbar (Fig. 2.3-5, area A) using the calibration by Hensen and Harley (1990) and 960 °C and 8 kbar (Fig. 2.3-5, area B) using the calibration of Harley (1998). However, both calibrations are based on Al isopleths for orthopyroxene associated with garnet, sillimanite and quartz, and thus the *P-T* estimates are approximate. Temperatures calculated using isopleths by Kelsey et al. (2003) at 9–10 kbar ($X_{Mg} = 70$, low Al model) are 975–1 000 °C.

In summary, *P-T* estimates and inferred reactions suggest a clockwise *P-T* path beginning with an isobaric temperature increase from the kyanite field to the shaded area (Fig. 2.3-4) representing peak conditions of ~950–1 050 °C at ~9–10 kbar followed by decompression when garnet-

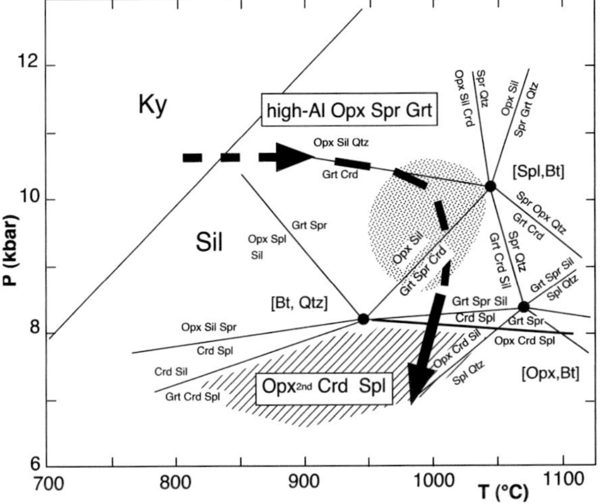

Fig. 2.3-4. *P-T* grid for FMAS system at low oxygen fugacity (Hensen and Harley 1990; Bertrand et al. 1991; Harley 1998) showing univariant reactions and possible *P-T* path for Schirmacher Hills

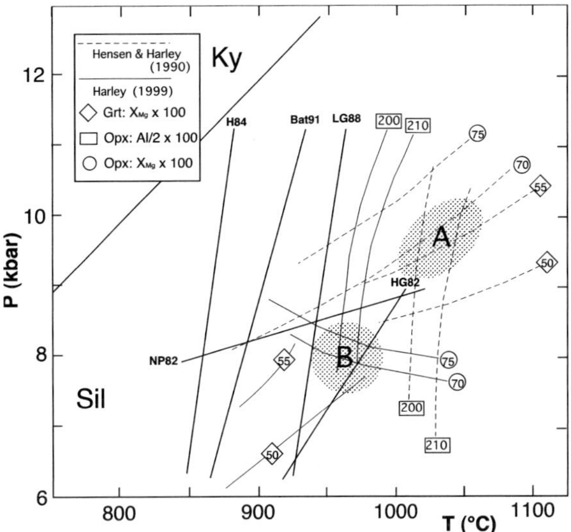

Fig. 2.3-5. *P-T* plot showing geothermometry applied to garnet and orthopyroxene association and partial X_{Al} and X_{Mg} isopleths applied to high-Al_2O_3 orthopyroxene. *Dashed* and *thin solid lines* are after Hensen and Harley (1990) and Harley (1998) respectively. *Shaded fields A* and *B* represent estimated peak conditions after Hensen and Harley (1990) and Harley (1998) respectively

Table 2.3-3. Thermobarometry of the garnet – orthopyroxene association

Analysis No.	Grt				Opx					Temperature (°C)			Pressure (kbar)	
	X_{Mg}	X_{grs}	X_{alm}	X_{pyp}	X_{Mg}	X_{Al}	P_{ref}	T_{ref}	K_D	H84	LG88	Bhat91	HG82	NP82
10-21 core-core	0.529	0.012	0.463	0.520	0.720	0.204	10	900	2.29	851	928	895	6.1	8.7
30-26 core-core	0.527	0.013	0.465	0.518	0.720	0.214	10	900	2.31	874	953	917	6.7	8.3
69-76 core-core	0.514	0.012	0.478	0.505	0.727	0.203	10	900	2.52	797	869	843	6.2	8.1
77-75 core-core	0.518	0.012	0.474	0.510	0.700	0.192	10	900	2.17	867	945	910	6.1	7.9

X_{Mg} = Mg / (Fe + Mg); X_{grs} = Ca / (Fe + Mg + Ca + Mn); X_{alm} = Fe / (Fe + Mg + Ca + Mn); X_{pyp} = Mg / (Fe + Mg + Ca + Mn); X_{Al} = Al / 2 per formula unit of 4 cations; P_{ref} = reference pressure (kbar); T_{ref} = reference temperature (°C); K_D = (Fe / Mg)Grt / (Fe / Mg)Opx; H84: Harley (1984). LG88: Lee and Ganguly (1988); Bhat91: Bhattacharya et al. (1991); HG82: Harley and Green (1982); NP82: Newton and Perkins (1982).

sapphirine broke down to orthopyroxene-spinel-cordierite at around 950–980 °C at 8 kbar (Fig. 2.3-4). The high temperatures during decompression are indicated by the high Al_2O_3 content of secondary orthopyroxene (6.0–8.5 wt.-%), although these estimates may be too high due to the additional components Cr and Zn in spinel (e.g., Hensen and Harley 1990).

Discussion

The sapphirine bearing rocks reported from Dronning Maud Land are texturally complex and the assemblages differ from one to another (Table 2.3-1). Except for the Balchenfjella example, sapphirine is interpreted to have formed at the peak of metamorphism, and much of the textural complexity developed on the retrograde path. The assemblage reported here is distinctive for the association of garnet with orthopyroxene and the very high Al_2O_3 content of orthopyroxene, which is the first evidence for ultrahigh-temperature conditions in metamorphic rocks exposed in the Schirmacher Hills. The other two sapphirine-garnet-bearing rocks from the Schirmacher Hills contain corundum, sillimanite and kyanite, but no orthopyroxene. Grew (1983) attributed kyanite in sample no. 367 to a later event, but textural relations suggest kyanite in our sample could be a relic of an assemblage largely replaced on the prograde path.

Clockwise *P-T* paths involving high pressures and temperatures during an early stage followed by a decrease in both pressure and temperature are characteristic of DML (Fig. 2.3-6). In the Sør Rondane and Gjelsvikfjella regions, re-heating caused by granitic intrusions added a kink to the retrograde path. Because the thermal effect was local, the paths can still be considered as clockwise as a whole. Maximum temperature conditions of other DML regions are estimated to have been about 800 °C. However, metamorphic ages are contradictory. Several researchers have

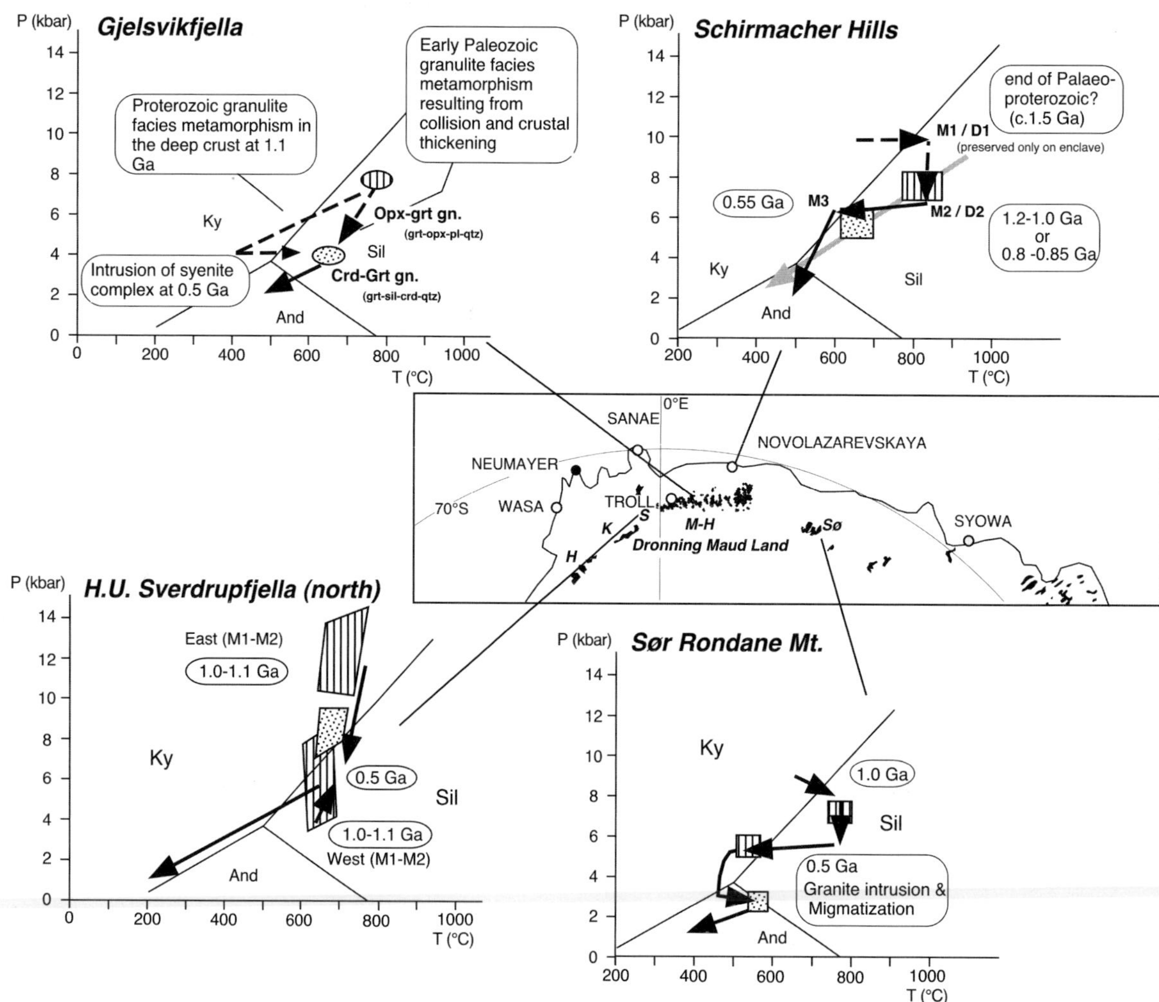

Fig. 2.3-6. Map showing *P-T* paths previously reported for DML. Sources: Gjelsvikfjella (Bucher-Nurminen and Ohta 1993), H. U. Sverdrupfjella (Grantham et al. 1995), Schirmacher Hills (Ravikant and Kundu 1998; Rameshwar Rao et al. 1997), and: Sør Rondane Mountains (Asami et al. 1992)

proposed two stages of metamorphism, i.e., an early stage Grenvillian in age (1.0–1.2 Ga) and a later stage Pan African in age (0.5 Ga) in H. U. Sverdrupfjella and Gjelsvikfjella (e.g., Bucher-Nurminen and Ohta 1993; Grantham et al. 1995). Exposures in western DML, e.g., Kirwanveggen and Heimefrontfjella, preserve 1.1–1.0 Ga ages (U-Pb zircon metamorphic age, Frimmel 2004 and references cited therein) and show no evidence for a Pan-African overprint. In contrast, exposures in central DML, Gjelsvikfjella, show both ca. 1.16 Ga protolith age and ca. 0.5 Ga migmatization age (SIMS dating of zircon from migmatite, Paulsson and Austrheim 2003). Jacobs et al. (2003) also reported SHRIMP zircon data from Gjelsvikfjella and Mühlig-Hofmann-Gebirge in central DML, and recognized a protracted tectono-thermal event involving upper-amphibolite to granulite-facies metamorphism, magmatic intrusion and isothermal decompression between ca. 0.56 and 0.49 Ga. In summary, chronological correlation of the metamorphic *P-T* paths in Fig. 2.3-6 is not possible with available data.

The limited geochronological data from Schirmacher Hills (Grew and Manton 1983; Mikhalsky et al. 2003) can be interpreted to date the ultrahigh-temperature granulite-facies event at ca. 1.15 Ga and the amphibolite-facies overprint at ca. 0.625 Ga; the decompressional stage should have terminated before ca. 0.6 Ga. The earlier event is somewhat older than late Mesoproterozoic granulite facies elsewhere in Dronning Maud Land (e.g., between ~1.08 and ~1.13 Ga, Jacobs et al. 1998). Metamorphic temperatures estimated to exceed 900 °C imply that the tectonothermal regime in the Schirmacher Hills was distinct from those in other parts of DML. Future studies are needed to fully characterize this distinctive regime, which could be critical in determining whether central DML was linked to the Mozambique belt (e.g., Jacobs et al. 1998) or the Namaqua-Natal-Maud belt (e.g., Jacobs et al. 2003); a fuller characterization could also lead to a better understanding of the evolution of Rodinia and Gondwana.

Conclusions

1. Sapphirine granulite from a new locality in the Schirmacher Hills is distinctive for the association of orthopyroxene with garnet. Orthopyroxene porphyroblasts have high Al_2O_3 contents (~10 wt.-%), the highest reported in DML
2. Two stages of metamorphism can be inferred on the basis of mineral textures. The earlier stage is characterized by the breakdown of sapphirine and orthopyroxene to form garnet at near-peak conditions of over 900 °C (950~1 050 °C), and the later stage by post-peak decompression with formation of spinel-cordierite-orthopyroxene at about 950 °C and <8 kbar.
3. Correlation between Schirmacher Hills and other regions in DML is not possible at present. Nonetheless, extremely high temperature metamorphic conditions of Schirmacher Hills suggest the existence of a distinctive tectonothermal event between 1.2 Ga and 0.6 Ga.

Acknowledgments

The present paper is based on fieldwork carried out while at Novolazarevskaya Station during the Japan-Germany-Norway joint expedition in the Mühlig-Hofmanfjella region in 2001/2002. We would like to thank S. Elvevold, A. Läufer and I. Manson for their co-operation in the fieldwork. We acknowledge the logistical support provided by the Norwegian Polar Institute, the German Alfred-Wegener-Institute for Polar and Marine Research and National Institute of Polar Research (Japan). We also thank T. Hokada, Y. Motoyoshi, Y. Osanai and T. Kawasaki for their discussions and analytical support.

This work was partly supported by a Grant-in-Aid for Scientific Research from the Japan Society for Promotion of Sciences to S. B. (No. 16740289) and K. S. (No. 13440151).

References

Asami M, Osanai Y, Shiraishi K, Makimoto H (1992) Metamorphic evolution of the Sør Rondane Mountains, East Antarctica. In: Yoshida Y, Kaminuma K, Shiraishi K (eds) Recent progress in Antarctic earth science. Terra pub, Tokyo, pp 7–15

Asami M, Makimoto H, Grew ES (1994) Relict sapphirine in pyropic garnet from the eastern Sør Rondane mountains, Antarctica (Abstract). Proc NIPR Symp Antarct Geosci 7:179

Bertrand P, Ellis DJ, Green DH (1991) The stability of sapphirine-quartz and hypersthene-sillimanite-quartz assemblage: an experimental investigation in the system FeO-MgO-Al_2O_3-SiO_2 under H_2O and CO_2 conditions. Contrib Mineral Petrol 108:55–71

Bhattacharya A, Krishnakumar KR, Raith M, Sen SK (1991) An improved set of a-X parameters for Fe-Mg-Ca garnets and refinements of the orthopyroxene-garnet thermometer and the orthopyroxene-garnet-plagioclase-quartz barometer. J Petrol 32:629–656

Bucher-Nurminen K, Ohta Y (1993) Granulite and garnet-cordierite gneisses from Dronning Maud Land, Antarctica. J Metamorph Geol 11:691–703

Fitzsimons ICW, Harley SL (1994) The influence of retrograde cation exchange on granulite *P-T* estimates and a convergence technique for the recovery of peak metamorphic conditions. J Petrol 35:543–576

Frimmel HE (2004) Formation of a late Mesoproterozoic supercontinent: the South Africa – East Antarctica connection. In: Eriksson PG, et al. (eds) The Precambrian earth: tempos and events. Devel Precambrian Geol 12:240–255

Grantham GH, Jackson C, Moyes AB, Groenewald PB, Harris PD, Ferrar G, Krynauw JR (1995) The tectonothermal evolution of the Kirwanveggen-H.U. Sverdrupfjella areas, Dronning Maud Land, Antarctica. Precamb Res 75:209–229

Grew ES (1983) Sapphirine-garnet and associated parageneses in Antarctica. In: Oliver RL, James PR, Jago JB (eds) Antarctic earth science. Austral Acad Sci, Canberra, pp 40–43

Grew ES, Manton WI (1983) Geochronologic studies in East Antarctica: reconnaissance uranium/thorium/lead data from rocks in the Schirmacher Hills and Mount Stinear. Antarctic J US 18(5):6–8

Harley SL (1984) An experimental study of the partitioning of Fe and Mg between garnet and orthopyroxene. Contrib Mineral Petrol 86:359–373

Harley SL (1998) On the occurrence and characterization of ultra-high-temperature crustal metamorphism. In: Treloar PJ, O'Brien PJ (eds) What drives metamorphism and metamorphic reactions? Geol Soc Lond Spec Publ 138:81–107

Harley SL, Green DH (1982) Garnet-orthopyroxene barometry for granulites and peridotites. Nature 300:697–701

Hensen BJ, Harley SL (1990) Graphical analysis of *P-T-X* relations in granulite facies metapelite. In: Ashworth JR, Brown M (eds) High-temperature metamorphism and crustal anatexis. Unwin Hyman, London, pp 19–56

Ishizuka H, Suzuki S, Kojima H (1995) Mineral paragenesis of the sapphirine-bearing rock from the Austkampane area of the Sør Rondane Mountains, East Antarctica. Proc NIPR Symp Antarctic Geosci 8:65–74

Jacobs J, Fanning CM, Henjes-Kunst, F, Olesch M, Paech H-J (1998) Continuation of the Mozambique Belt into East Antarctica: Grenville-age metamorphism and polyphase Pan-African high-grade events in central Dronning Maud Land. J Geol 106:385–406

Jacobs J, Fanning CM, Bauer W (2003) Timing of Grenville-age vs. Pan-African medium- to high grade metamorphism in western Dronning Maud Land (East Antarctica) and significance for correlations in Rodinia and Gondwana. Precamb Res 125:1–20

Kelsey DE, White RW, Powell R (2003) Orthopyroxene-sillimanite-quartz assemblages: distribution, petrology, quantitative *P-T-X* constraints and *P-T* paths. J Metamorph Geol 21:439–453

Kretz R (1983) Symbols for rock-forming minerals. Amer Mineral 68:277-279

Lee HY, Ganguly J (1988) Equilibrium compositions of coexisting garnet and orthopyroxene: experimental determinations in the system FeO-MgO-Al_2O_3-SiO_2, and applications. J Petrol 29:93–113

Mikhalsky EV, Hahne K, Wetzel H-U, Henjes-Kunst F, Beliatsky BV (2003) Geological evolution of the Schirmacher Hills from U-Pb zircon dating and a comparison with the Wohlthat Massif, central Dronning Maud Land. Abstr 9th Int Symp Antarctic Earth Science, Terra Nostra 2003(4):229

Newton RC, Perkins D III (1982) Thermodynamic calibration of geobarometers based on the assemblages garnet-plagioclase-orthopyroxene(clinopyroxene)-quartz. Amer Mineral 67: 203–222

Paulsson O, Austrheim H (2003) A geochronological and geochemical study of rocks from Gjelsvikfjella, Dronning Maud Land, Antarctica – implications for Mesoproterozoic correlations and assembly of Gondwana. Precamb Res 125:113–138

Rameshwar Rao D, Sharma R, Gururajan NS (1997) Mafic granulites of Schirmacher region, East Antarctica: fluid inclusion and geothermobarometric studies focusing on the Proterozoic evolution of crust. Trans Royal Soc Edinburgh Earth Sci 88: 1–17

Ravich MG (1982) The Lower Precambrian of Antarctica. In: Craddock C (ed) Antarctic geoscience. Univ of Wisconsin Press, Madison, pp 421–427

Ravikant V (1998) Preliminary thermal modelling of the massif anorthosite-charnockitic gneiss interface from Gruber Mountains, central Dronning Maud Land, East Antarctica. J Geol Soc India 52:287–300

Ravikant V, Kundu A (1998) Reaction textures of retrograde pressure-temperature-deformation paths from granulites of Schirmacher Hills, East Antarctica. J Geol Soc India 51: 305–314

Sengupta S (1988) History of successive deformations in relation to metamorphism-migmatitic events in the Schirmacher Hills, Queen Maud Land, East Antarctica. J Geol Soc India 32:295–319

Sengupta S (1993) Tectonothermal history recorded in mafic dykes and enclaves of gneissic basement in the Schirmacher Hills, East Antarctica. Precamb Res 63:273–291

Shiraishi K, Kanisawa S, Ishikawa K (1988) Geochemistry of post-orogenic mafic dike rocks from the eastern Queen Maud Land, East Antarctica. Proc NIPR Symp Antarctic Geosci 2:117–132

Stackebrandt W, Kaempf H, Wetzel HU (1988) The geological setting of the Schirmacher Oasis, Queen Maud Land, East Antarctica. Z Geol Wiss 16:661–665

Verma SK, Mittal GS, Dayal AM (1987) K-Ar dating of some rocks from Schirmacher Oasis, Dronning Maud Land, East Antarctica. Fourth Indian Expedition to Antarctica, Scientific Rep DOP Tech Publ 4:43–54

Genesis of Ferropotassic A-Type Granitoids of Mühlig-Hofmannfjella, Central Dronning Maud Land, East Antarctica

Mervin J. D'Souza · A. V. Keshava Prasad · Rasik Ravindra
Antarctica Division, Geological Survey of India, Faridabad, India, <antgsi@vsnl.net>

Abstract. Ferrosilite-fayalite bearing charnockite and biotite-hornblende bearing granite are exposed in Mühlig-Hofmannfjella, central Dronning Maud Land of East Antarctica. Both are interpreted as essentially parts of a single pluton in spite of their contrasting mineral assemblages. Based on petrologic and geochemical studies, it is proposed that H_2O-undersaturated parent magma with igneous crustal component that fractionated under different oxygen fugacity conditions resulted in the Mühlig-Hofmannfjella granitoids.

Introduction

Magmatic rocks associated with the Pan-African tectonic event form a large component of the lithological units exposed in central Dronning Maud Land (CDML). These magmatic rocks have been dated to represent an early Pan-African phase and a post-collisional late phase (Jacobs et al. 1998). The early intrusives (600 Ma) are the anorthosite massif exposed in the Grubergebirge and the Schüssel Mountains which was followed by granodiorites in Conradgebirge and charnockite-monzodiorite bodies of Wohlthat Massivet. The voluminous plutonic bodies exposed (Fig. 2.4-1a) in the Petermannketten (Joshi et al. 1991), the Conradgebirge and the Filchnerfjella have been dated at 530 to 510 Ma and they represent post-collisional A-type granitoids (Jacobs et al. 1998; Mikhalsky et al. 1995; Ravindra and Pandit 2000).

In the present study, the coarse undeformed granitoid exposed in Mühlig-Hofmannfjella (MH) forms the base for detailed geochemical and petrochemical studies in an attempt to characterize these rocks within the framework of dominant Pan-African magmatic activity recognized in the CDML.

Geological Outline

The geology of central Dronning Maud Land comprising the Wohlthatmassivet, Orvinfjella and the Mühlig-Hofmannfjella was first described by Ravich and Soloviev (1966) and Ravich and Kamanev (1972) and later by Indian and German geologists (Joshi et al. 1991; D'Souza et al. 1996; Bohrmann and Fritzsche 1995). Geological studies carried out in the area (Jacobs et al. 1998; Jacobs et al. 2003; Ohta et al. 1990; Paulsson and Austrheim 2003) has indicated that the exposed part of CDML has a very large component of Grenvillian (~1 000 Ma) crust which has been extensively modified during the Pan-African orogeny (600–500 Ma).

The oldest recognized rocks in CDML are the thick sequence of metaigneous and sedimentary rocks comprising banded orthogneiss, metapelites, metapsammites, calcsilicates, pyroxene granulites and amphibolites. The banded orthogneiss has been interpreted as representing a bi-modal volcanic sequence. The basement rocks of CDML have indicated an earliest crystallization age of around 1 150–1 100 Ma (Jacobs et al. 2003) and high grade metamorphism between 1 090 and 1 050 Ma. The Grenvillian basement was later re-metamorphosed during the Pan-African orogeny associated with collision of East and West Gondwana around 550 Ma. (Jacobs et al. 1998; Markl et al. 2003). The Pan African orogeny in CDML is mainly represented as a prominent magmatic event which was activated before the collision with the intrusion of massif anorthosite. The magmatic activity apparently continued during the entire period and climaxed with the formation of post-collision extensional regime (530–510 Ma) which produced large volumes of granitic and syenitic rocks (Jacobs et al. 2003).

Field Description and Petrology

The Mühlig-Hofmannfjella granitoid pluton is exposed between. 71°30' to 6°30' S and 5°30' to 72°15' E (Fig. 2.4-1). These rocks are very coarse-grained, porphyritic and undeformed and are composed of K-feldspar (45–55%), quartz (30–35%), plagioclase (5–20%) and ferromagnesian minerals (up to 5%). The K-feldspar megacrysts measure up to 4 cm in length. The matrix is made up of plagioclase, quartz and ferromagnesian minerals. The rock having reddish colour has been distinguished as charnockite whereas the rock with whitish gray colour has been designated as granite. The granite and charnockite are seen to occur as patches on cliffs and slopes at several places and also in distinct zones demarcated by a horizontal line where the lower zone is of charnockite.

From: Fütterer DK, Damaske D, Kleinschmidt G, Miller H, Tessensohn F (eds) (2006) Antarctica: Contributions to global earth sciences. Springer-Verlag, Berlin Heidelberg New York, pp 45–54

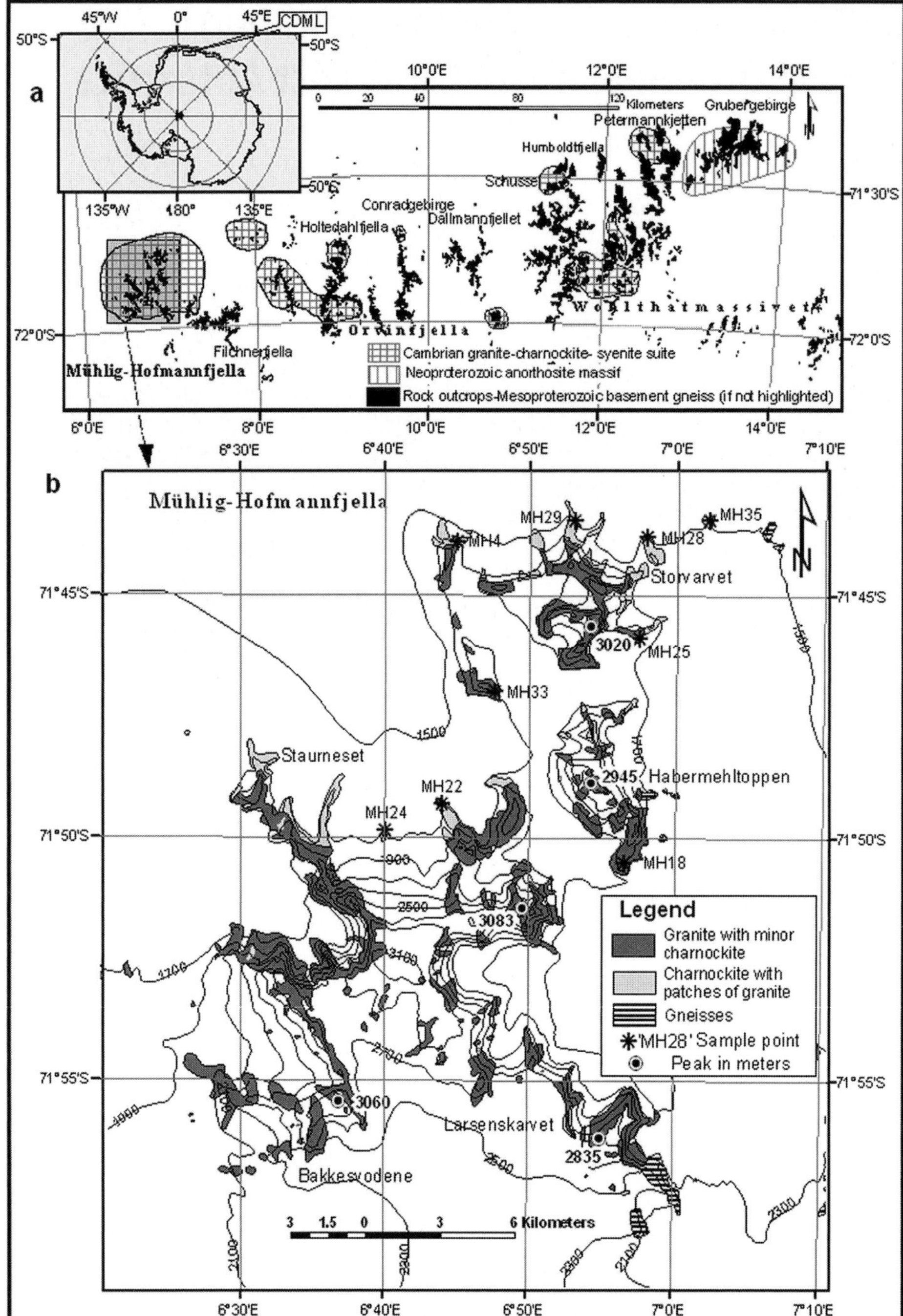

Fig. 2.4-1.
a Lay out of outcrops in central Dronning Maud Land, East Antarctica, showing distribution of granite-charnockite-syenite plutons, anorthosite massif and the basement gneisses. **b** Geological Map of eastern Mühlig-Hofmannfjella with sample locations

Restites of orthogneiss and charnockite (fine-grained, foliated) are very common. The proportion of restite was higher along the NE margin and the southern part of pluton and in the higher reaches of the pluton. The granitoids contain discrete shear zones.

The granitoid under the microscope is a heterogranular rock indicating compositional variation from granite/charnockite to granodiorite. Quartz occupying the interstitial space is comparatively less in the rocks. The reddish coloured charnockite and the granite show textural similarity and differ only by the presence of fayalite and ferrosilite in charnockite. The alkali feldspar is mesoperthitic orthoclase. The proportion of plagioclase in perthite greatly varies. Plagioclase occurs as interstitial mineral. The mesoperthite is often mantled by myrmekite comprising plagioclase and quartz. The mafic silicates and oxides occur as interstitial minerals. Biotite is mostly primary but also occurs as biotite-quartz symplectite rims

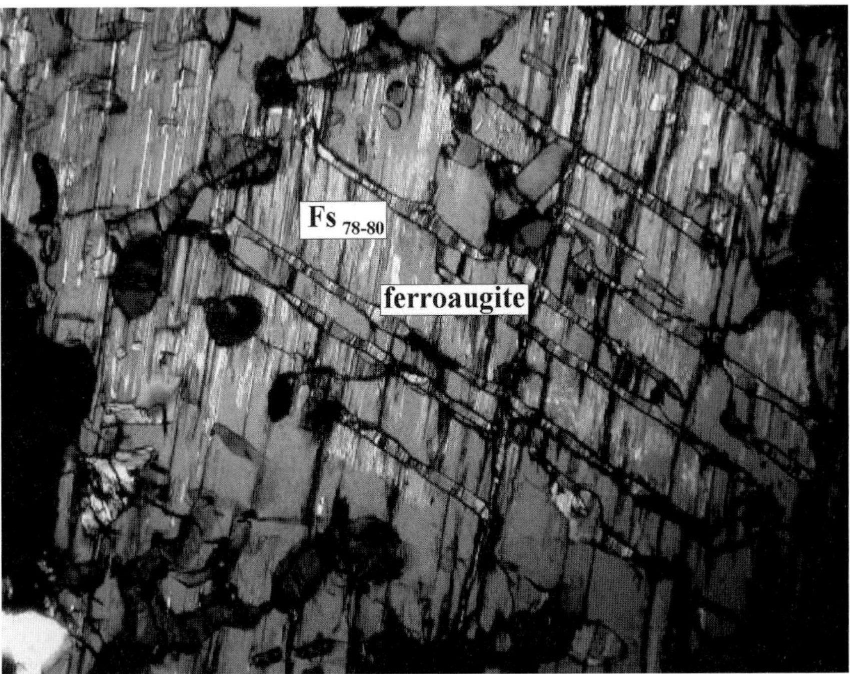

Fig. 2.4-2. Inverted pigeonite with ferroaugite lamellae in a sample of charnockite from Mühlig-Hofmannfjella area (MH 28)

around orthopyroxenes and amphiboles. Amphibole is anhedral, strongly pleochroic and green and brown in colour. In the charnockite, ferrosilite and fayalite are present in several thin sections studied but both these minerals are not observed in contact with each other. Inverted pigeonite is present in three samples. The accessory phases are mainly euhedral zircon, apatite, ilmenite and magnetite. Ilmenite is distinctly more abundant than magnetite. Fluorite specks were detected in two samples. Fayalite and ferrosilite frequently show alteration to amorphous Fe-silicate.

Mineral Chemistry

EPMA analysis were carried out using a CAMAECA-SX51 at the Geological Survey of India (GSI), Faridabad. Fayalite and ferrosilite are restricted to charnockite. The composition of fayalite is homogeneous at Fa_{94} and ferrosilite averages at Fs_{80-81}. In inverted pigeonite (Fig. 2.4-2), the ferrosilite (Fs_{78-80}) contains lamellae of ferroaugite ($Wo_{46}En_{16}Fs_{38}$). Amphibole is uniformly present in granites and charnockites. It is relatively iron-rich and is hastingsite in composition (Fig. 2.4-3).

Biotite is the most common ferromagnesian mineral present in MH granitoids. It is found as subhedral to anhedral grains and as secondary mineral around amphibole and pyroxenes. It is annite in composition. Alkali feldspar occurs as coarse porphyritic megacrysts and also as small grains. The alkali feldspar is orthoclase (Or_{84-90}) while that of the exsolved plagioclase lamellae is andesine (An_{33-39}). Plagioclase is also present in the matrix, which is oligoclase (An_{23-26}) in composition.

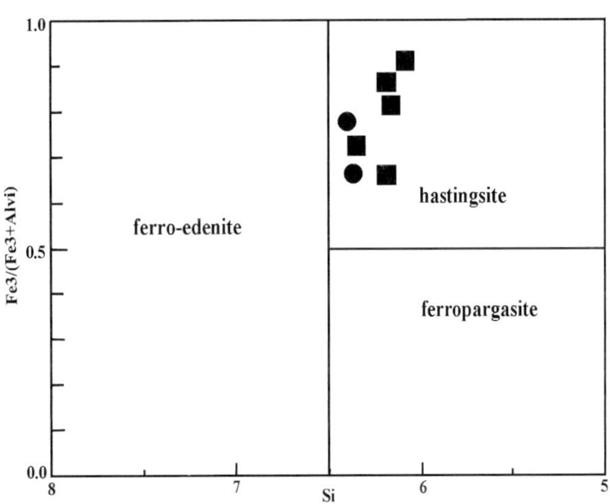

Fig. 2.4-3. $Fe^{3+}/(Fe^{3+} + Al^{vi})$ *vesrus* Si in formula unit diagram for classification of amphibole (Leake 1978). *Symbols* shown are common for all ensuing diagrams in the manuscript

Geochemistry

Whole rock analyses were done using a Phillips XRF spectrometer at the Petrology, Petrochemistry and Ore Dressing (PPOD) laboratory, Airborne Mineral Survey and Exploration (AMSE) Wing, Geological Survey of India, Bangalore, and trace elements were analyzed using ICP-AAS at the Geological Survey of India, Faridabad. REE analysis was done using Instrumental NAA at the Geological Survey of India, Pune. Representative analysis of granitoids is given in Table 2.4-1. The representative analysis of restites occurring within the granitoids (Table 2.4-2)

Table 2.4-1. Whole rock and trace lement analysis of MH granitoids; analysis of major oxides using XRF, analysis of trace elements using ICP

Sample	MH 18A/1	MH 22/1	MH 24/1	MH 25A/1	MH 28/1	MH 28/2	MH 28/4	MH 29/1	MH 33/1	MH 35/1	MH 4/1	MH 4/10	MH 4/11	MH 4/13	MH 4/2	MH 4/3	MH 4/5	MH 4/6	MH 4/7	MH 4/8	MH 4/9
Rock	CH	CH	CH	GR	CH	GR	GR	GR	GR	GR	CH	GR	GR	GR	CH	CG	CH	GR	GR	GR	GR
SiO_2 (%)	66.16	59.03	62.71	70.67	64.71	75.89	70.86	61.22	62.14	73.46	60.95	61.93	63.90	67.58	60.73	62.01	60.03	65.41	61.85	62.32	62.65
TiO_2	0.89	2.25	0.89	0.62	0.67	0.26	0.43	1.19	1.72	0.18	1.04	1.05	0.92	0.55	0.85	0.77	1.05	0.68	1.01	0.82	1.11
Al_2O_3	15.20	14.60	16.65	13.25	16.44	12.00	13.75	16.28	14.45	14.29	15.84	15.72	15.59	15.37	16.97	17.13	16.43	15.38	15.82	16.76	15.91
FeOt	4.40	9.03	5.26	3.57	4.42	1.58	3.27	55.36	6.99	1.51	7.17	5.76	5.02	3.42	6.57	5.31	7.53	4.89	5.41	6.00	4.98
Fe_2O_3t	4.89	10.04	5.85	3.96	4.91	1.76	3.63	5.96	7.76	1.68	7.96	6.41	5.58	3.81	7.31	5.90	8.37	5.44	6.01	6.66	5.53
MnO	0.06	0.13	0.08	0.04	0.05	0.03	0.05	0.09	0.09	0.02	0.09	0.06	0.07	0.04	0.09	0.08	0.09	0.06	0.06	0.07	0.08
MgO	0.87	1.37	0.78	0.48	0.32	0.24	0.33	1.30	0.94	0.46	0.69	1.04	1.13	0.65	0.68	0.52	0.66	n.d.[a]	1.15	0.53	1.23
CaO	2.60	4.93	4.48	1.65	3.33	1.22	1.89	3.40	3.99	1.74	3.68	3.23	2.67	1.55	3.43	3.02	3.78	2.48	3.32	2.98	2.69
Na_2O	3.14	2.61	3.34	2.30	3.11	2.53	2.69	3.42	2.73	3.63	2.94	2.81	3.01	2.50	3.01	3.09	3.00	2.65	2.93	3.04	3.35
K_2O	5.59	3.85	4.14	6.02	5.02	4.84	5.17	5.23	4.36	4.38	5.54	6.42	5.73	7.10	5.82	6.24	5.57	6.29	6.33	5.94	6.06
P_2O_5	0.29	1.02	0.39	0.09	0.15	0.04	0.10	0.45	0.68	0.08	0.29	0.34	0.32	0.14	0.23	0.19	0.29	0.32	0.35	0.19	0.36
BaO	0.22	0.27	0.33	0.33	0.45	0.12	0.30	0.58	0.21	0.09	0.55	0.48	0.43	0.13	0.54	0.57	0.53	0.47	0.48	0.54	0.44
LOI	0.16	n.d.	0.12	n.d.	0.38	0.53	0.30	0.23	0.26	0.20	n.d.	0.25	0.28	0.50	0.29	0.35	0.18	0.02	n.d.	0.07	0.17
Total	100.07	100.10	99.76	99.41	99.57	99.43	99.50	99.35	99.36	99.88	99.57	99.74	99.63	99.92	99.95	99.87	99.98	100.02	99.91	99.93	99.58
Sr (ppm)	402.00	640.00	n.a.[b]	372.00	609.00	121.00	n.a.	1233.00	452.00	232.00	636.00	n.a.	n.a.	n.a.	650.00	n.a.	n.a.	495.00	n.a.	n.a.	n.a.
Y	53.00	65.00	n.a.	43.00	20.00	20.00	n.a.	42.00	72.00	n.a.	35.00	n.a.	n.a.	n.a.	n.a.	n.a.	n.a.	48.00	n.a.	n.a.	n.a.
Zr	305.00	151.00	n.a.	606.00	168.00	226.00	n.a.	155.00	85.00	113.00	565.00	n.a.	n.a.	n.a.	474.00	n.a.	n.a.	507.00	n.a.	n.a.	n.a.
Nb	27.00	255.00	n.a.	23.00	23.00	22.00	n.a.	24.00	31.00	20.00	32.00	n.a.	n.a.	n.a.	n.a.	n.a.	n.a.	30.00	n.a.	n.a.	n.a.
Th	n.a.	n.a.	n.a.	n.a.	n.a.	n.a.	n.a.	n.a.	n.a.	n.a.	7.90	16.00	58.00	100.00	6.40	6.50	8.90	n.a.	20.00	6.90	32.00
Zn	84.00	178.00	n.a.	87.00	75.00	n.a.	n.a.	108.00	151.00	22.00	110.00	n.a.	n.a.	n.a.	n.a.	n.a.	n.a.	85.00	n.a.	n.a.	n.a.
V	n.a.	37.00	n.a.	28.00	34.00	21.00	n.a.	40.00	35.00	29.00	n.a.	n.a.	n.a.	n.a.	n.a.	n.a.	n.a.	18.00	n.a.	n.a.	n.a.
Cr	n.a.	124.00	n.a.	121.00	38.00	78.00	n.a.	86.00	n.a.	81.00	16.00	20.00	30.00	35.00	38.00	28.00	19.00	112.00	36.00	25.00	14.00
Hf	n.a.	n.a.	n.a.	n.a.	n.a.	n.a.	n.a.	n.a.	n.a.	n.a.	22.00	20.00	19.00	17.00	23.00	23.00	24.00	n.a.	26.00	18.00	19.00
Sc	n.a.	n.a.	n.a.	n.a.	n.a.	n.a.	n.a.	n.a.	n.a.	n.a.	17.00	14.00	12.00	6.00	14.00	13.00	18.00	n.a.	18.00	12.00	10.00
Ta	n.a.	n.a.	n.a.	n.a.	n.a.	n.a.	n.a.	n.a.	n.a.	n.a.	0.55	0.56	0.41	0.32	0.32	0.18	0.13	n.a.	0.20	0.52	0.22
Co	n.a.	n.a.	n.a.	n.a.	n.a.	n.a.	n.a.	n.a.	n.a.	n.a.	7.60	5.40	6.00	5.50	7.50	5.60	7.80	n.a.	7.30	5.60	5.60
K_2O/Na_2O	1.78	1.48	1.24	2.62	1.61	1.91	1.92	1.53	1.60	1.21	1.88	2.28	1.90	2.84	1.93	2.02	1.86	2.37	2.16	1.95	1.81
FeOt/ FeOt + MgO	1.87	2.37	1.78	1.48	1.32	1.24	1.33	2.30	1.94	1.46	1.69	2.04	2.13	1.65	1.68	1.52	1.66	n.a.	2.15	1.53	2.23

[a] Not deteccted. [b] Not analyzed.

Table 2.4-2.
Whole rock and trace element analysis of representative enclave rocks occuring within MH granitoids

Sample	MH 15A/2	MH 16/2	MH 16A/2	MH 16A/3	MH 18A/2	MH 19A/5	MH 25A/2
SiO_2 (%)	58.65	74.46	58.99	68.32	61.39	55.16	57.48
TiO_2	1.10	0.18	1.93	0.62	1.59	1.33	2.73
Al_2O_3	15.09	13.4	15.14	12.39	15.24	15.37	13.45
Fe_2O_3t	8.87	1.68	9.31	7.59	7.73	8.98	10.34
MnO	0.12	0.03	0.11	0.077	0.09	0.13	0.13
MgO	3.44	0.26	1.53	0.43	1.48	6.63	1.78
CaO	5.30	1.51	3.88	2.70	3.62	6.50	4.41
Na_2O	3.94	2.24	2.81	2.07	3.21	2.53	2.50
K_2O	2.92	5.97	4.81	5.38	4.62	2.10	4.37
P_2O_5	0.18	0.04	0.65	0.07	0.55	0.52	1.25
BaO	0.04	0.09	0.22	0.12	0.15	0.23	0.33
LOI	0.24	0.21	0.50	0.16	0.13	0.38	0.44
Total	99.93	100.07	99.88	99.92	99.8	99.87	99.21
Ba (ppm)	475.00	853.00	n.a.	1053.00	n.a.	n.a.	n.a.
Sr	210.00	159.00	n.a.	203.00	n.a.	n.a.	n.a.
Y	46.00	n.a.	n.a.	143.00	n.a.	n.a.	n.a.
Zr	233.00	148.00	n.a.	714.00	n.a.	n.a.	n.a.
Nb	20.00	n.a.	n.a.	34	n.a.	n.a.	n.a.
Zn	105.00	30.00	n.a.	n.a.	n.a.	n.a.	n.a.
Ni	49.00	71.00	n.a.	n.a.	n.a.	n.a.	n.a.
V	51.00	25.00	n.a.	13.00	n.a.	n.a.	n.a.
Cr	96.00	152.00	n.a.	105.00	n.a.	n.a.	n.a.
Ce	80.00	31.00	203.00	664.00	418.00	n.a.	n.a.
K_2O/Na_2O	0.74	2.67	1.71	2.60	1.44	0.83	1.75
FeOt/FeOt+MgO	4.44	1.26	2.53	1.43	2.48	7.63	2.78

FeOt recalculated as $Fe_2O_3 \times 0.8998$; *MH15A/2:* diorite; *MH16/2:* fine grained gneissic charrnockite; *MH16A/2* and *MH18A/2:* mafic rich Opx bearing gneissic charnockite; *MH19/5:* diorite; *MH25A/2:* diorite (Bt + Hb + Kfs + Plagg + Opaques; *n.a.:* not analyzed.

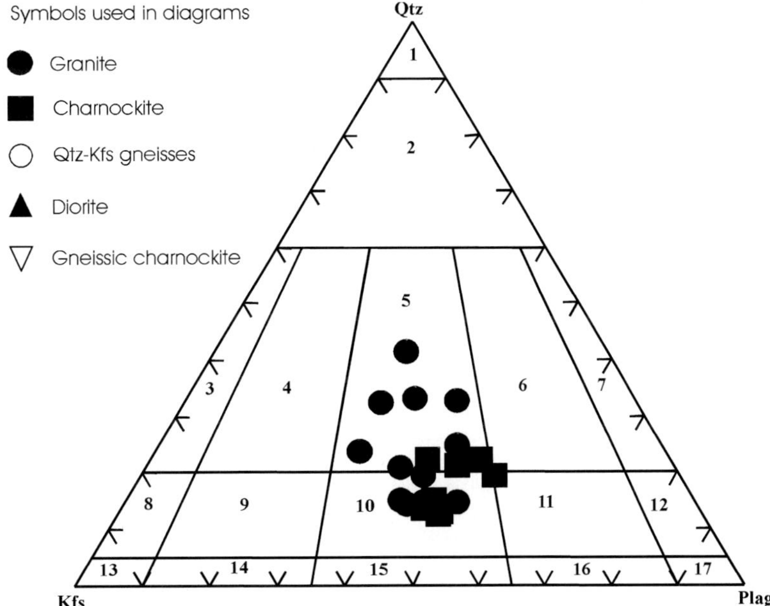

Fig. 2.4-4.
Modal composition *Qtz-Kfs-Plag* ternary diagram of Mühlig-Hofmann (MH) granitoids and enclaves plot in monzo granite and quartz monzonite fields. The fields of plutonic rocks after Le Maitre (1989): *1:* Quartzolite; *2:* Qtz rich granite; *3:* Qtz-Kfs granite; *4:* syeno granite; *5:* monzo granite; *6:* granodiorite; *7:* tonalite; *8:* Qtz-Kfs syenite; *9:* Qtz-syenite; *10:* Qtz-monzonite; *11:* qtz-monzodiorite; *12:* Qtz-diorite; *13:* Kfs-syenite; *14:* syenite; *15:* monzonite

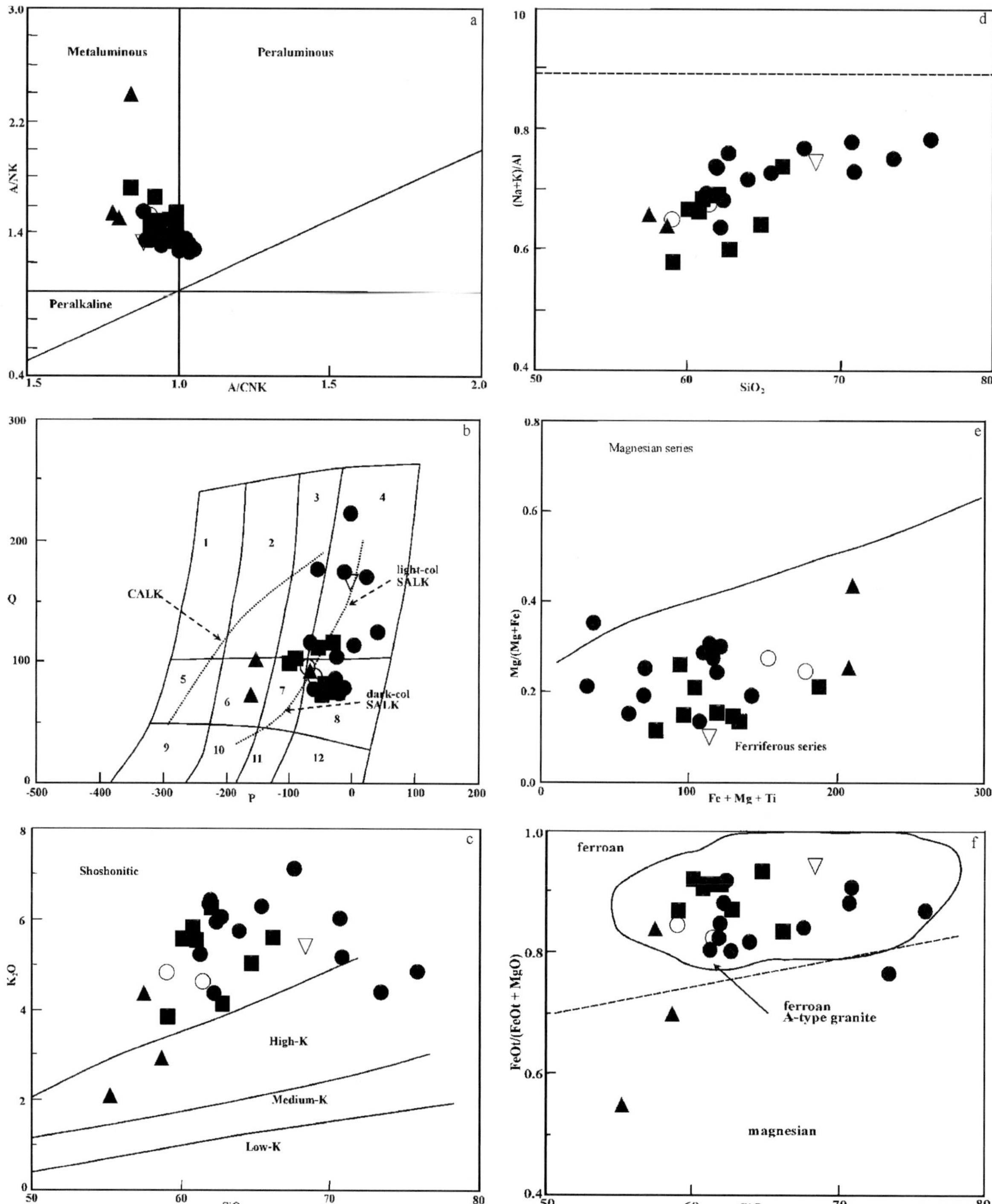

Fig. 2.4-5a–f. **a** Dominantly metaluminous character of Mühlig-Hofmann (MH) granitoids (Shands molar parameters after Maniar and Piccoli (1989). **b** Cationic classification of Debon and Le Fort (1988): $Q = Si/3 - (K + Na + 2Ca/3)$ *versus* $P = K - (Na + Ca)$; shown subalkaline (*SALK*) trend and calc-alkaline (*CALK*) trends: *1:* tonalite; *2:* granodiorite; *3:* adamellite; *4:* granite; *5:* Qtz-diorite; *6:* Qtz-monzodiorite; *7:* Qtz-monzonite; *8:* Qtz-syenite; *9:* gabbro or diorite; *10:* monzodiorite; *11:* monzonite; *12:* syenite. **c** K_2O *versus* SiO_2 (wt.-%) diagram, the limits given are after Rickwood (1989). **d** Agpaitic index (molar (Na/K)/Al *versus* SiO_2. The limit at 0.87 separates subalkaline metaluminous granitoids from alkaline (Liégeois and Black 1987). **e** Ferriferous character in $M = Mg/Mg + Fe$ *versus* $B = Fe + Mg + Ti$ in cationic classification diagram after Debon and Le Fort (1988). **f** FeOt/FeOt + MgO *versus* SiO_2 showing boundary between Ferroan and magnessian granitoids. The A-type granitoid boundary based on 175 samples after Frost et al. (2001)

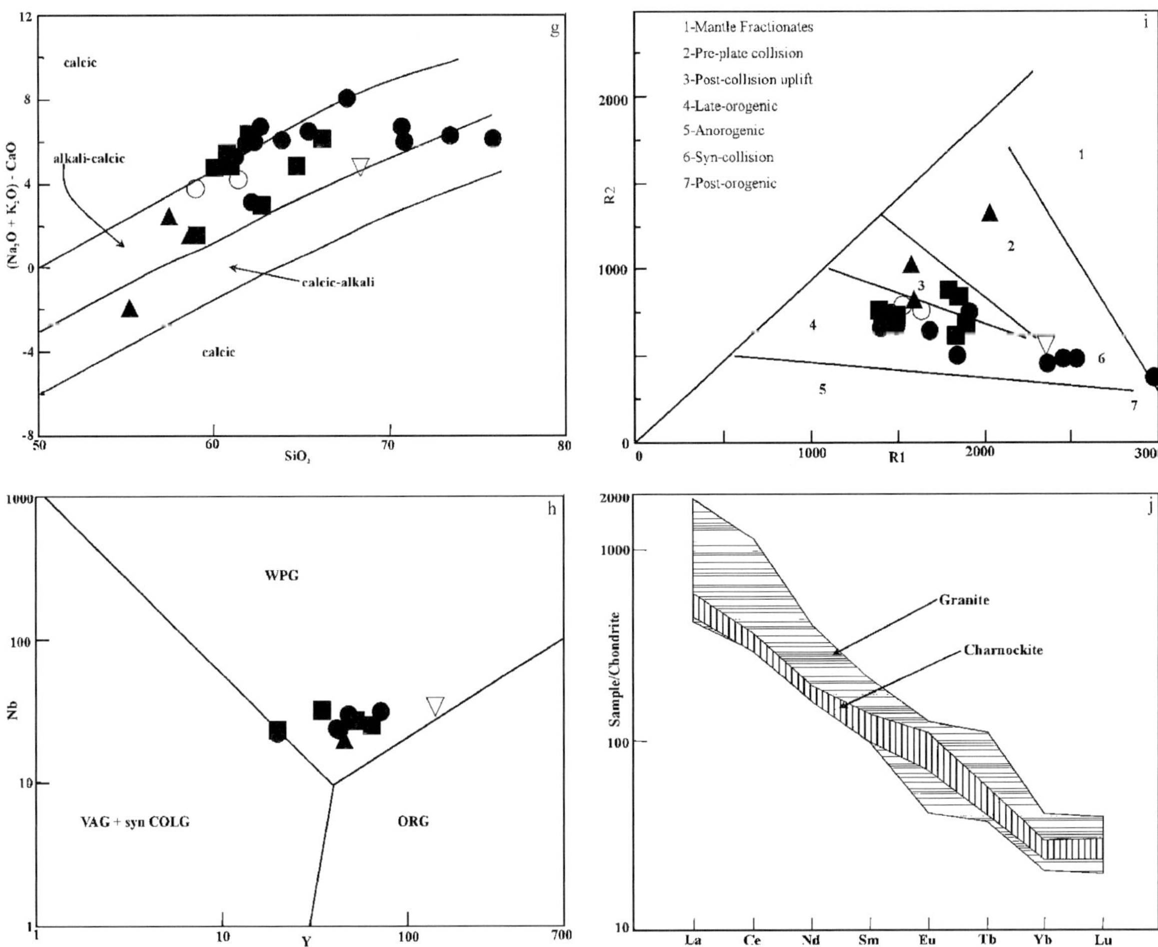

Fig. 2.4-5g–j. g $Na_2O + K_2O - CaO$ *versus* SiO_2 plot after Frost et al. (2001), showing ranges for alkali, alkali-cacic, calcic-alkali and calcic fields. **h** Trace element tectonic discrimination diagram Nb *versus* Y after Pearce et al. (1984). **i** R1-R2 diagram after Bachelor and Bowden (1985). **j** Chondrite normalised REE spidergram. Normalising values after Nakamura (1977)

and the discrimination diagrams that follow suggest that the geochemical composition of enclave rocks compare well with that of host rock.

The MH granitoids are of monzogranite-quartz monzonite composition as indicated by a Qtz-Kfs-Plag plot (Fig. 2.4-4) based on molar norm (not given in the table) and they plot predominantly in metaluminous field in the A/NK versus A/CNK diagram (Fig. 2.4-5a). In TAS diagram of Middlemost (1997) the MH granitoids and associated enclaves show a transalkaline character (not shown). This is corroborated by the cationic classification *P-Q* diagram of Debon and Le Fort (1988). In this *P-Q* diagram, the MH samples show a distinct affinity to dark and light coloured subalkaline trend and plot in quartz syenite to granite field (Fig. 2.4-5b). In the K_2O versus SiO_2 (fields after Rickwood 1989) they plot in the shoshonitic field (Fig. 2.4-5c). The agpatatic index of these rocks is below 0.87 suggesting the rocks lack alkaline character (Fig. 2.4-5d).

On the cationic Mg/Mg + Fe versus Fe + Mg + Ti diagram of Debon and Le Fort (1988) the rocks show a strong ferriferous character (Fig. 2.4-5e). However due to high FeOt/MgO ratios in the AFM diagram they plot in the tholeiitic field. In the FeOt/FeOt + MgO versus SiO_2 and MALI diagram of Frost et al. (2001), the MH granitoids plot in the ferroan A-type granite field and alkali-calcic field respectively (Fig. 2.4-5f and 2.4-5g). The limited trace element data available with author shows a high content of HFSE like Nb, Zr and Ce and also incompatible elements like Ba and Sr. The tectonic discrimination diagrams for the MH granitoids obtained indicate WPG field (Fig. 2.4-5h) in the Nb versus Y diagram of Pearce et al. (1984) and a late orogenic and post collisional tectonic field (Fig. 2.4-5i) in the R1 versus R2 diagram of Bachelor and Bowden (1985). Further these rocks have a high (La/Yb)N ratio (Fig. 2.4-5j) suggesting a fractionated HREE/LREE trend.

Table 2.4-3. REE analysis by instrumental NAA of MH granitoids

Sample	MH 4/1	MH 4/10	MH 4/11	MH 4/13	MH 4/2	MH 4/3	MH 4/5	MH 4/7	MH 4/8	MH 4/9
Rock	CH	GR	GR	GR	CH	CH	CH	GR	GR	GR
La (ppm)	125	185	440	250	115	105	140	245	100	285
Ce	200	285	700	320	185	180	225	320	185	500
Nd	84	105	190	110	78	75	90	106	76	120
Sm	17.0	24.0	32.0	25.0	16.0	15.0	21.0	25.0	15.0	25.0
Eu	4.9	7.2	6.5	2.4	5.2	4.0	6.4	7.3	4.0	7.3
Tb	1.6	2.5	4.1	2.5	1.7	1.5	2.1	3.8	1.4	3.0
Yb	4.5	4.7	5.9	3.7	4.6	4.0	5.0	7.0	3.5	5.2
Lu	0.7	0.7	0.8	0.5	0.67	0.6	0.77	1.00	0.56	0.8
$(La/Yb)_N$	19.9	28.2	53.4	48.4	17.9	18.8	20.0	25.1	20.4	39.3

Discussions

The MH granitoids exhibit a wide variation in the SiO_2 content varying from 59.03% to 75.89%. Based on their geochemical characteristics MH granitoids fit into a distinctive potassic group – A type granitoid (nomenclature after Loiselle and Wones 1979). These rocks have a high FeOt / FeOt + MgO ratio, high K_2O/Na_2O, comparatively low CaO and very low MgO concentrations. Moreover, the trace element data available indicate high contents of HFSE like Nb, Zr and Ce and incompatible elements like Ba and Sr. The presence of inverted pigeonite, high K, Ti P and low Ca, although indicate C-type magma affinity (Kilpatrick and Ellis 1992), very low MgO content and presence of modal fayalite preclude these rocks from the C type charnockites. Further, the MH Pluton is transalkaline ferroan A-type granite, which concurs well with the observation of Frost et al. (2001) that world over the A-type granitoids are overwhelmingly ferroan alkali calcic to calcic-alkali in character.

The SiO_2 content of MH granitoids varies from 59.03% to 75.89%. This variation is reflected by consistent conformable trend in Harker diagram for CaO, MgO, FeOt and TiO_2 as their amount increases with decrease in SiO_2 content (plots not shown). The granite and charnockite plots overlap each other in the Harker diagram suggesting both these rocks to be part of a single pluton and have fractionated from same source. This is also confirmed by the visible textural conformity between granite and charnockites in the outcrops.

Based on the mineral assemblage of charnockite and granite, available geochemical data and a comparison with other ferropotassic A-type granites (Bogaerts et al. 2003; Guimarães et al. 2000; Ferré et al. 1998; Roland 1999) an anhydrous to H_2O-undersaturated parent magma with igneous crustal component is considered as a possible source for the MH granitoids. The involvement of igneous crustal source in the generation of magma is considered to accommodate the transalkaline, potassic and ferriferous character of these rocks. The restitic enclaves like diorite, charnockite (occurring within the granitoids as enclaves) with low SiO_2, high K_2O and FeOt / FeOt + MgO and having composition akin to ferro-monzonite and or ferro-monzodiorite may possibly represent the crustal component in the source magma. Melting of such igneous crustal components must have occurred at high temperature and very low, near-constant oxygen fugacity to crystallize ferrosilite and fayalite in the beginning. As the magma slowly cooled and moved towards shallower depths, it underwent fractional crystallization. During this process, according the Bucher and Frost (1993), the melt can attain water saturation. The resultant increase in the oxygen fugacity facilitated crystallization of ilmenite and magnetite instead of ferrosilite and fayalite. This possibly explains an important observation (made during mineral separation process) that the volume of magnetic minerals (mainly ilmenite and magnetite) present in charnockite is far less compared to that in the granite. The increased H_2O-activity was conducive for the formation of biotite-hornblende bearing granitic assemblage and retrogression of part of the early-crystallized pluton in patches and zones. Thus the charnockite containing fayalite and ferrosilite represents the early-crystallized phase of MH Pluton.

Conclusion

The field disposition of the MH granitoids is very much similar to the Svarthmaren charnockite occurring in western Mühlig-Hoffmanfjella, which has been dated at 500 ±24 Ma by Ohta et al. (1990). Geochemically, the MH granitoids compare very well with similar transalkaline ferro-potassic A-type Pan-African granitoids reported from Nigeria and Brazil (Ferre et al. 1998; Guimaraés et al. 2000) and to some extent with Pan-African A-type granitoid reported from Kerala Khondalite Belt, India, Madagaskar and the Arabian-Nubian Shield (Rajesh 2000; Moghazi 2002; Nédélec et al. 1995). The magmatism associated with Pan-African orogeny in CDML is linked to the continent-continent collision involving east and west Gondwana in the final assembly of Gondwana. This orogenic cycle that started with the intrusion of anorthosite at 600 Ma (Bauer et al. 2002) progressed through continent-continent convergence and metamorphism at 580–560 Ma. The second phase of metamorphism interpreted at 530–510 Ma, was associated with an extensional regime causing intrusion of voluminous granitoids in the area (Jacobs et al. 2003). The MH granitoids, in absence of any deformational or metamorphic feature, is considered to be post-tectonic/post-collisional. The within-plate feature shown by Nb *versus* Y tectonic discrimination diagram and the post-collision/late orogenic feature reflected by the R_1-R_2 diagram (Fig. 2.4-5h and 2.4-5i) conform to this conclusion. The continent - continent convergence led to thickening of the crust followed by collapse of the subducted, predominantly igneous, crust that was involved in the formation of melt giving rise to MH granitoids.

Acknowledgments

This publication is the result of work carried out during the 20th Indian Antarctic Expedition conducted under the aegis of National Center for Antarctic and Ocean Research (NCAOR), Goa, India. The authors wish to acknowledge the co-operation extended by colleagues Jayapaul D. and Dharwadkar A. Critical reviews by Norbert Roland and Hartwig Frimmel have considerably helped to improve this contribution.

References

Bachelor RA, Bowden P (1985) Petrologic interpretation of granitoid rock series using multicationic parameters. Chem Geol 48:43–55

Bauer W, Thomas RJ, Jacobs J (2002) Proterozoic-Cambrian history of Dronning Maud Land in context of Gondwana assembly. In: Windly BF, Dasgupta S (eds) Proterozoic East Gondwana: supercontinent assembly and breakup. Geol Soc London 206:247–269

Bogaerts M, Scaillet B, Liégeois J, Auwera JV (2003) Petrology and geochemistry of the Lyngdal granodiorite (southern Norway) and the role of fractional crystallization in the genesis of Proterozoic ferro-potassic A-type granites. Precamb Res 124:149–184

Bohrmann P, Fritzsche D (1995) The Schirmacher Oasis, Queen Maud Land, East Antarctica and its surroundings. Petermanns Geograph Mitt Ergänzungsheft 289, Justus Perthes, Gotha

Bucher K, Frost BR (1993) Crystallization history of charnockites, Thor range and implication to thermobarometry of cordierite gneisses (abstract). Geol Soc Amer Abst Program 25:448

D'Souza, MJ, Kundu A, Kaul MK (1996) The geology of central Dronning Maud Land (cDML), East Antarctica. Ind Min 50:323–338

Debon F, Le Fort P (1988) A cationic classification of common plutonic rocks and their magmatic associations: principles, method application. Bull de Minéralogie 111:493–510

Ferré EC, Caby R, Peucat JJ, Capdevilla R, Monie P (1998) Pan-African, post-collisional, ferro-potassic granite and quartz-monzonite plutons of eastern Nigeria. Lithos 45:255–279

Frost BR, Barnes CG, Collins WJ, Arculus RJ, Ellis DJ, Frost CD (2001) A geochemical classification for granitic rocks. J Petrol 42:2033–2048

Guimarães IP, Almeida CN, Filho AFD, Araújo JM (2000) Granitoids marking the end of the Brasiliano (Pan-African) orogeny within the central tectonic domain of the Borborema Province. Revista Brasileira Geociencias 30:177–181

Jacobs J, Fanning CM, Henjes-Kunst F, Olesch M, Paech H-J (1998) Continuation of the Mozambique Belt into East Antarctica: Grenville-Age metamorphism and polyphase Pan-African high grade events in central Dronning Maud Land. J Geol 106:385–406

Jacobs J, Baur W, Fanning CM (2003) Late Neoproterozoic/Early Palaeozoic events in central Dronning Maud Land and significance for the southern extension of the East African Orogen into East Antarctica. Precam Res 126:27–53

Joshi A, Pant NC, Parimoo ML (1991) Granites of Petermann ranges, east Antarctica and implication on their genesis. J Geo Soc Ind 38:169–181

Kilpatrick JA, Ellis DJ (1992) C-type magmas: igneous charnockites and their extrusive equivalents. Trans Royal Soc Edinburgh Earth Sci 83:155–164

Leake B (1978) Nomenclature of amphiboles. Can Miner 16:501–520

Le Maitre RW (1989) A classification of igneous rocks and glossary of terms. Recommendations of the International Union of Geological Sciences. Subcommission on the Systematics of Igneous Rocks, 1st edn. Blackwell, Oxford

Liégeois J-P, Black R (1987) Alkaline magmatism subsequent to collision in the Pan-African belt of the Adrar des Iforas. In: Fitton JG, Upton BGJ (eds) Alkaline igneous rocks. Geol Soc, Blackwell Scientific, Oxford, pp 381–401

Loiselle MC, Wones DR (1979) Characteristics and origin of anorogenic granites. Geol Soc Amer Abstr Program 11:468

Maniar PD, Piccoli PM (1989) Tectonic discrimination of granitoids. Geol Soc Amer Bull 101:635–643

Markl G, Abart R, Vennemann T, Sommer H (2003) Mid-crustal metasomatic reaction veins in a spinel peridotite. J Petrol 44:1097–1120

Middlemost EAK (1997) Magmas rocks and planetary development. Longman, Harlow

Mikhalsky EV, Grikurov GE (2003) East Antarctica crust growth from the isotopic data (abstract). 9th Int. Sympos Antarctic. Earth Science, Terra Nostra 2003(4):230–231

Mikhalsky EV, Beliatsky BV, Savva EV, Federov LV, Hahne K (1995) Isotopic systematics and evolution of metamorphic rocks from the northern Humboldt Mountains, the Queen Maud Land, east Antarctica (abstract). VII Internat Sympos Antarctic Earth Sci, Siena

Moghazi AM (2002) Petrology and geochemistry of Pan-African granitoids, Kab Amiri area, Egypt-implications for tectonomagmatic stages in the Nubian Shield evolution. Mineral Petrol 75:41–67

Nakamura N (1977) Determination of REE, Ba, Fe, Mg, Na and K in carbonaceous and ordinary chondrites. Geochem Cosmochim Acta 38:757–775

Nédélec A, Stephens WE, Fallick AE (1995) The Pan-African stratoid granites of Madagascar: alkaline Magmatism in post-collisional extensional setting. J Petrol 36:1367–1391

Ohta Y, Tørudbakken Bo, Shiraishi K (1990) Geology of Gjelsvikfjella and western Mühlig-Hofmannfjella, Dronning Maud Land, East Antarctica. Polar Res 8:538–544

Paulsson O, Austrheim H (2003) A geochronological and geochemical study of rocks from Gjelsvikfjella, Dronning Maud Land, Antarctica – implications for Mesoproterozoic correlations and assembly of Gondwana. Precambrian Res 125:113–138

Pearce JA, Harris NB, Tindle AG (1984) Trace element discrimination diagrams for the tectonic interpretation of granitic rocks. J Petrol 25:956–983

Rajesh HM (2000) Characterization and origin of a compositionally zoned aluminous A-type granite from South India. Geol Mag 137:291–318

Ravich MG, Kamenev EN (1972) Crystalline basement of the Antarctic platform. (1st English edn. 1975, Wiley, New York)

Ravich MG, Soloviev DS (1966) Geology and petrology of the mountains of Central Dronning Maud Land (eastern Antarctica). Trans Sci Res Inst Arctic Geol, Ministry of Geology (transl. in Jerusalem, 1969)

Ravindra R, Pandit MK (2000) Geochemistry and geochronology of A-type granite from northern Humboldt Mountains, East Antarctica: Vestige of Pan-African event. J Geol Soc India 56:253–262

Rickwood PC (1989) Boundary lines within petrologic diagrams, which use oxides of major and minor elements. Lithos 22:247–264

Roland N (1999) Pan-African granitoids in central Dronning Maud Land, East Antarctica. Petrography, geochemistry and their plate tectonic setting. In: Skinner DNB (ed) Programme and abstract 8th Internat Sympos Antarctic Earth Sci

Late Pan-African Fluid Infiltration in the Mühlig-Hofmann- and Filchnerfjella of Central Dronning Maud Land, East Antarctica

Ane K. Engvik[1] · Synnøve Elvevold[2]

[1] Geological Survey of Norway, N-7491 Trondheim, Norway, <ane.engvik@geo.ntnu.no>

[2] Norwegian Polar Institute, N-9296 Tromsø, Norway, <elvevold@npolar.no>

Abstract. The nunataks of Mühlig-Hofmannfjella and Filchnerfjella in central Dronning Maud Land, East Antarctica, comprise a deep-seated metamorphic-plutonic rock complex, dominated by a dark colour due to dark feldspar and containing granulite facies minerals including perthite, plagioclase, orthopyroxene and garnet. The area was affected by a late Pan-African fluid infiltration outcropping as conspicuous light alteration zones restricted to halos around thin granitoid veins. The veins were formed during infiltration of volatile-rich melts, probably originating from underlying magma-chambers. The alteration halos were formed by CO_2-H_2O-volatiles emanating from the veins into the host rock causing hydration of the granulite facies assemblages. The alteration involves a breakdown of orthopyroxene to biotite and sericitisation of plagioclase at crustal conditions around 350–400 °C and 2 kbar. The marked colour change is caused by transformation of feldspars, spread of dusty micas, opaques and fluid inclusions in addition to replacement of coarse to finer grains. The process is locally penetrative indicating that fluid infiltration can affect large rock volumes. The frequent distribution of alteration zones throughout the mountain range independent of lithological variations shows that the fluid infiltration is regionally extensive.

Introduction

In Dronning Maud Land, Mesoproterozoic (ca. 1.1–1.0 Ga) metamorphic rocks were intruded by voluminous intrusions and underwent granulite facies metamorphism and deformation during the Pan-African event (e.g., Mikhalsky et al. 1997; Jacobs et al. 1998; Paulsson and Austrheim 2003). Mühlig-Hofmann- and Filchnerfjella (5–8° E) of Dronning Maud Land consist of series of granitoid igneous rocks emplaced in granulite and upper amphibolite facies metamorphic rocks (Fig. 2.5-1; Bucher-Nurminen and Ohta 1993; Ohta 1999; Engvik and Elvevold 2004), most lithologies typically characterised with a dark colour. The igneous suite includes large intrusions of charnockite, quartz syenite, granite and several generations of dykes.

The area experienced conspicuous fluid infiltration during the later part of the Pan-African event. The fluid-rock interactions are well developed around granitoid veins which cross-cut dark coloured high-grade rocks, clearly visible in outcrop as light coloured altered zones. The alteration zones occur in all the different lithologies in the area, and their abundant distribution throughout a large area suggests that the fluid infiltration is of regional importance. This contribution will describe the mineralogical, textural and mineral chemical changes occurring during the fluid-rock interaction. The data will be used to discuss the cause of the conspicuous colour change, crustal conditions for the infiltration, the origin of the fluids and the implications of the fluid infiltration.

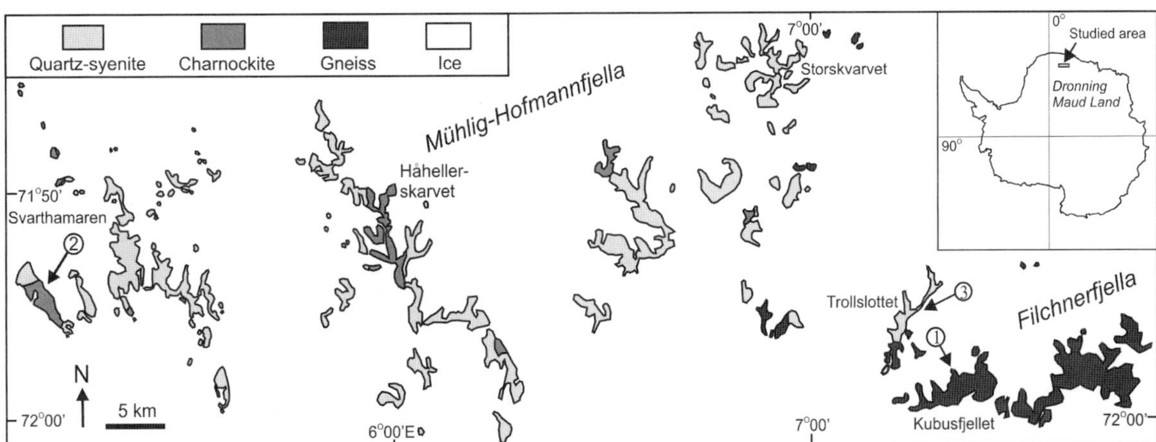

Fig. 2.5-1. Geological map of Mühlig-Hofmann- and Filchnerfjella, central Dronning Maud Land, East Antarctica. The numbers are referring to the type localities of Kubusfjellet (*1*), Svarthamaren (*2*) and Trollslottet (*3*)

From: Fütterer DK, Damaske D, Kleinschmidt G, Miller H, Tessensohn F (eds) (2006) Antarctica: Contributions to global earth sciences. Springer-Verlag, Berlin Heidelberg New York, pp 55–62

Rock Complexes

In the nunataks of Mühlig-Hofmannfjella and Filchnerfjella (5–8° E; Fig. 2.5-1), three main rock complexes of banded gneisses, charnockite and quartz syenite are distinguished. In the following section the lithologies are shortly described, and petrographic and mineral chemical descriptions of three type localities in the respective rock complexes are given.

Banded Gneiss Complex

The probably oldest rock type in the area is a sequence of banded gneisses including brown orthopyroxene-bearing gneiss, leucocratic gneiss, metapelite and garnet amphibolite. The different rock types form layers, which vary in thickness from <1 m up to several tens of meters (Fig. 2.5-2a). Migmatisation has affected large parts of the metamorphic sequence. The leucosomes occur as layers subparallel to the foliation, as nebulous patches and as veins crosscutting the regional fabric. In pelitic rocks, the leucocratic material appears to have segregated out locally from their host rock. The garnet amphibolites are concentrated along foliation-parallel horizons and probably represent disrupted former dykes or sills. The gneiss lithologies and the metamorphic and structural evolution of the rock complex are described by Engvik and Elvevold (2004).

Migmatitic garnet-orthopyroxene gneiss of locality 1 at Kubusfjellet (71°58' S, 07°20' E) is heterogranular and fine- to medium grained. The gneiss reveals the variable grain size on sample scale which is characteristic for different layers defining foliation. Anhedral quartz and feldspars (0.5–2 mm) dominate the gneiss, and occur often with lobate grain boundaries (Fig. 2.5-3a). The feldspars consist of plagioclase (andesine: $An_{35}Ab_{64}Kfs_1$), orthoclase ($An_{<1}Ab_{15}Kfs_{84-85}$), perthite (total chemistry of $An_{<1}Ab_{20}Kfs_{79}$) and minor antiperthite. Myrmekitic textures are common. Biotite (0.5–2 mm: Rich in Ti (0.44–0.58 p.f.u.), F (0.31–0.68 p.f.u.): $Fe^{2+}/(Fe^{2+} + Mg) = 0.66–0.70$) is concentrated in thin layers defining the foliation. Anhedral orthopyroxene (0.5–1 mm: $Fs_{68-74}En_{24-29}Wo_1$) is present in the mafic layers together with biotite (Fig. 2.5-3b). Garnet (0.25–1 mm: $Alm_{78-84}Prp_{5-9}Grs_{9-12}Sps_2$: $Fe^{2+}/(Fe^{2+} + Mg) = 0.90–0.94$) is present as euhedral to subhedral crystals (Fig. 2.5-2b), evenly distributed throughout the rock. Apatite, monazite, zircon and ilmenite are accessory minerals.

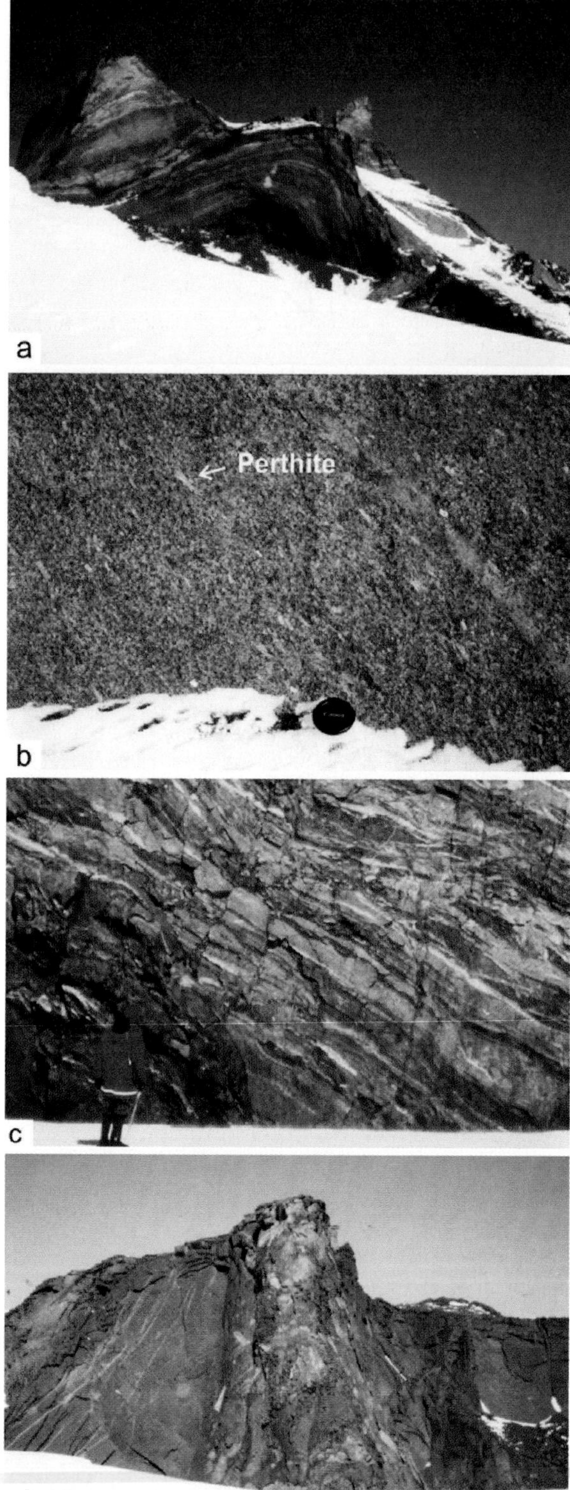

Fig. 2.5-2. a Folded layers of banded gneiss, Kubusfjellet. **b** Massive charnockite laminated by parallel-oriented, euhedral perthite grains, Svarthamaren. **c** Gneissic charnockite banded by granitic, leucocratic and mafic bands, Håhellerskarvet. **d** Zone containing xenoliths through quartz syenite, Håhellerskarvet

Charnockite

A charnockite complex occurs with a distinct reddish to reddish-brown weathering colour, whereas fresh surfaces are greenish grey. The charnockite complex comprises massive, coarse-grained granite with igneous textures and granitic gneiss with well-developed banding and foliation. All gradations between massive granite and gneissic granite are present, and all varieties contain granulite facies mineral assemblages with orthopyroxene + biotite. The massive granite commonly displays a weak planar fabric expressed by parallel-oriented, euhedral, tabular perthite grains (Fig. 2.5-2b) and, locally, diffuse bands containing a larger modal amount of orthopyroxene and biotite. In the banded parts of the charnockite, leucocratic and mafic layers (Fig. 2.5-2c) interlayer granite.

Orthopyroxene-bearing charnockite of locality 2 at Svarthamaren (71°54' S, 05°10' E) displays a heterogranular texture and a fine to medium grain size. Feldspars and quartz dominate the charnockite. The feldspars consist of perthite (total chemistry of $An_{<1}Ab_{13-33}Kfs_{67-87}$) and plagioclase (oligoclase: $An_{23}Ab_{75-76}Kfs_{1-2}$) and constitute commonly phenocrysts with sizes up to 5 mm laminating the granite (Fig. 2.5-2b and 2.5-3c). Quartz shows a rounded grain size with lobate grain boundaries. Minor myrmekite is present. Biotite (0.25–1 mm: Rich in Ti (0.45–0.57 p.f.u.), F (0.40–0.57 p.f.u.): $Fe^{2+}/(Fe^{2+} + Mg)$ = 0.75–0.78) occurs with random orientation. Minor orthopyroxene ($Fs_{76}En_{19}Wo_1$) and hastingsite (rich in F (0.45–0.50 p.f.u.), Ti 0.09 (p.f.u.), Mn (0.10–0.12 p.f.u): $Fe^{2+}/(Fe^{2+} + Mg)$ = 0.73–0.77) are present as anhedral grains (0.25–1 mm). Apatite, zircon, monazite, ilmenite, hematite and Fe-oxides are present as accessories.

Quartz Syenite

Dark brown quartz syenites of different variations are the major rock type in the study area (Fig. 2.5-1). The intrusions belong to a large magmatic complex extending between 06–13° E. The quartz syenites are coarse-grained to pegmatitic, and contain megacrysts of mesoperthitic K-feldspar. Mafic minerals are orthopyroxene, amphibole, biotite and locally fayalite. The quartz syenites intrude the banded gneisses and the charnockite complex, and are thus late in the intrusive history. Brecciated contact relationships along borders to the charnockite complex or zones containing xenoliths are locally observed (Fig. 2.5-2d).

The orthopyroxene-bearing quartz syenite of Trollslottet (locality 3, 71°55' S, 7°15' E) is heterogranular and coarse-grained. The quartz syenite is dominated by perthite (total chemistry of $An_{0-2}Ab_{16-45}Kfs_{53-84}$) and contains in addition quartz, plagioclase (oligoclase: $An_{29}Ab_{69}Kfs_2$), myrmekite, biotite and hastingsite. The perthite form euhedral crystals larger than 1 cm, and plagioclase constitutes euhedral crystals of up to 5 mm. Anhedral quartz occurs as coarse grains of more than 5 mm size. Both plagioclase and quartz constitute in addition anhedral fine-grained (0.5–1 mm) aggregates together with mafic phases surrounding the large grains. Coarse (up to 2.5 mm) biotite (rich in Ti (0.45–0.50 p.f.u.), F (0.60–0.87 p.f.u.): $Fe^{2+}/(Fe^{2+} + Mg)$ = 0.64–0.66) and hastingsite (F-content

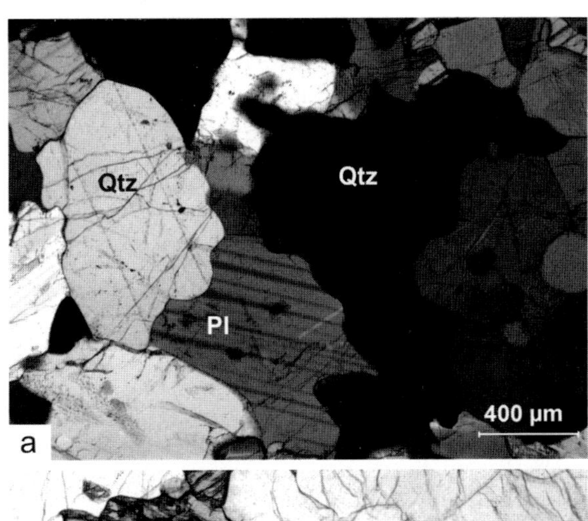

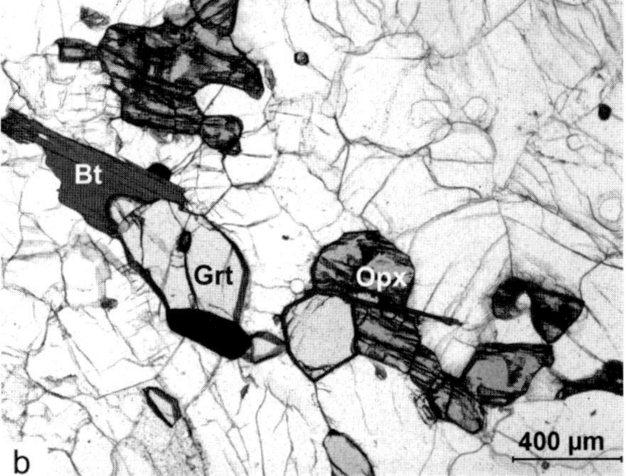

Fig. 2.5-3. Photomicrographs showing microtextures in the unaltered rocks. Abbreviations after Kretz (1983). **a** Lobate grain boundaries of quartz and feldspars in gneiss (crossed nicols, sample AHA205B, locality 1). **b** Subhedral garnet, orthopyroxene and biotite oriented parallel layering in gneiss (sample AHA193A, locality 1). **c** Plagioclase phenocryst laminating the charnockite of Svarthamaren. (crossed nicols, sample AHA13A, locality 2)

0.61–0.62 p.f.u.: Ti-content 0.19–0.20 p.f.u.: $Fe^{2+}/(Fe^{2+} + Mg)$ = 0.64–0.65) dominate the mafic phases. Minor orthopyroxene ($Fs_{69-71}En_{25-26}Wo_2$) and Mn-rich grunerite (0.1–0.25 mm) occur together with quartz and feldspars or are included inside aggregates of biotite and hastingsite. Apatite, ilmenite, zircon, allanite and hematite are accessory minerals.

Light Alteration Zones

Field Occurrence

Most rock types in the area are characterised by a dark brown or reddish brown weathering colour. Alteration of the dark, granulite facies assemblages to whitish or light grey coloured rocks is frequently observed in all three lithologies described above (Fig. 2.5-4). The centre of the light zones is constituted by a granitoid pegmatite or aplitic vein with a thickness ranging between few mm and 15 cm (Fig. 2.5-4a,b). The colour change is associated with a change in mineralogy and microstructure where the dry granulite facies mineral assemblages are hydrated. The widths of the halos range from a dm to several meters and are commonly much wider than the vein-fill. The volumetric proportions of vein-related alteration of the rocks crust is variable, with some areas being barely affected and others criss-crossed by veins (Fig. 2.5-4c). Several nunataks of up to 1 000 m height are composed of mainly altered light rocks which show only small remnants of dark coloured rocks (Fig. 2.5-4d). The alteration phenomenon is observed along the Mühlig-Hofmann- and Filchnerfjella for a minimum length of 150 km, but was also reported in the Wohlthatmassiv by Markl and Piazolo (1998).

Mineralogical, Textural and Mineral Chemical Transformations in the Alteration Zones

The three different lithologies show similar mineralogical and textural transformations related to the alteration. The most important changes are observed in feldspars. The altered rocks are heavily crowded by microcracks compared to unaltered rocks. Within the feldspars, microcracks occur in several sets and are sealed by albite (Fig. 2.5-5a; $An_{1-4}Ab_{96}Kfs_{0-2}$). Plagioclase transforms to albite and white mica, and in samples that show larger degree of alteration, plagioclase grains are partly to completely sericitised (Fig. 2.5-5b). Sericitisation of plagio-

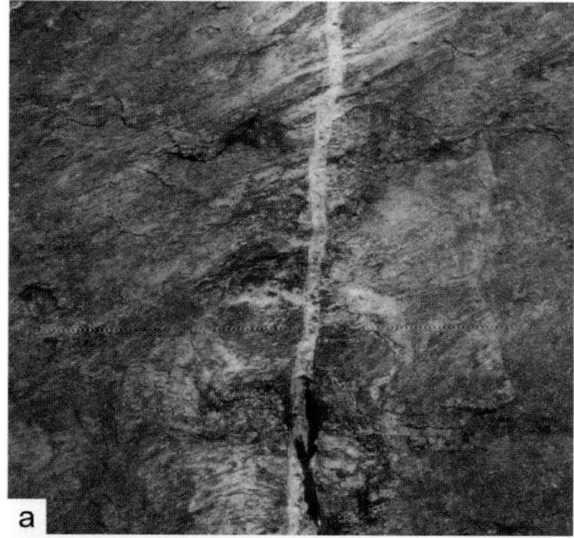

Fig. 2.5-4. a Alteration zone along pegmatite cutting the banded gneisses of Filchnerfjella. **b** Alteration along pegmatite cutting charnockite of Svarthamaren. **c** Two sets of alteration zones cutting the quartz syenite of Trollslottet. The wall is about 300 m high. **d** Nunatak dominated by light-coloured quartz syenite in Storskvarvet, showing only remnants of brown rock. The nunatak is 1 100 m high

clase is more pervasive in the granitoid gneiss, probably due to the higher An-content of the original plagioclase compared to charnockite and quartz syenite. In the gneiss, orthoclase and perthite is generally better preserved than the plagioclase, whereas in the charnockite and quartz syenite, perthite shows a stronger alteration than plagioclase. The transformation of perthite is located along microcracks (Fig. 2.5-5a). The Kfs-component in the

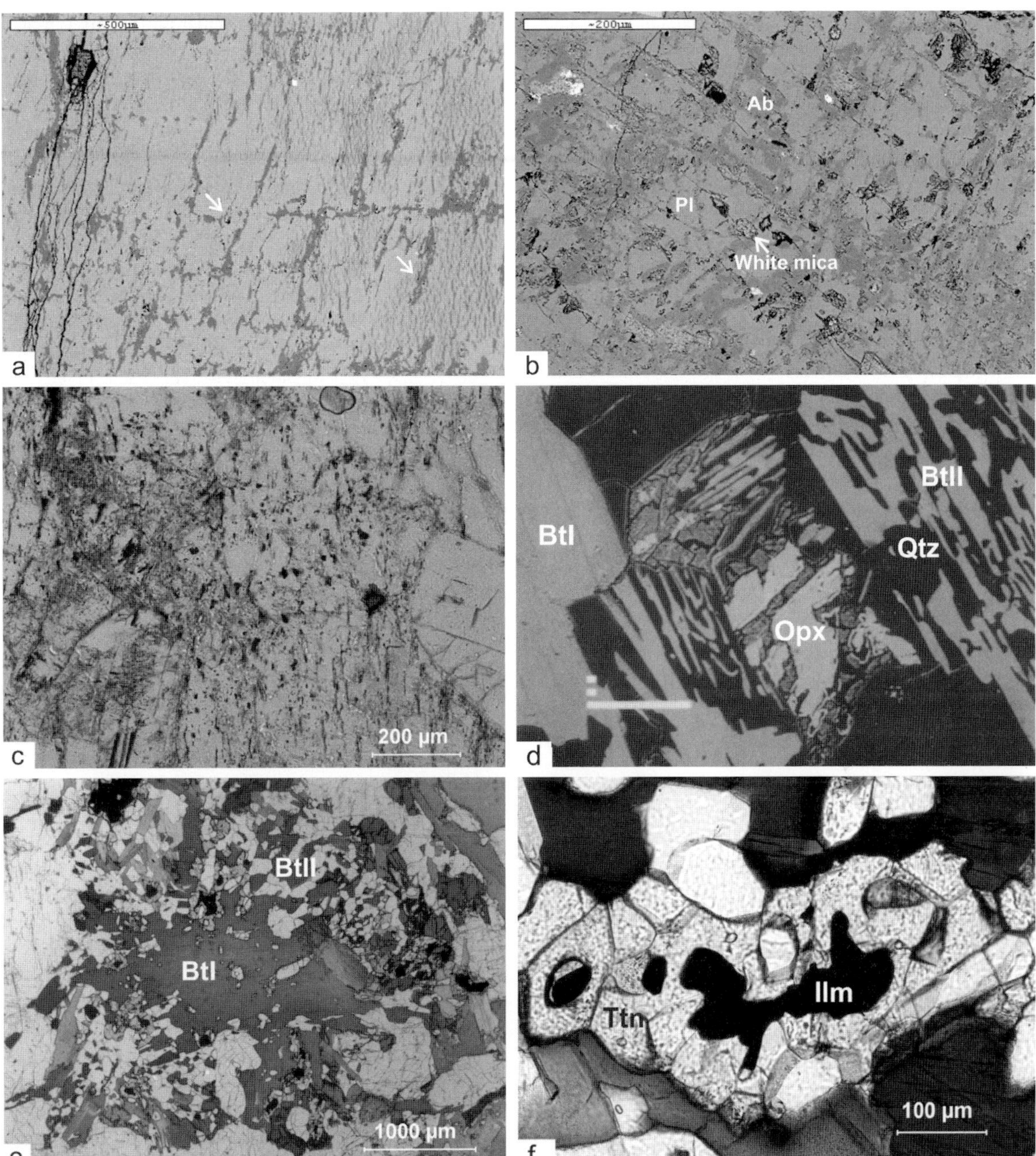

Fig. 2.5-5. BSE-photos and photomicrographs showing mineral and textural transformations in alteration zones. Abbreviations after Kretz (1983). **a** Perthite cut by two sets of albite-filled microcracks (marked by *arrows*). The perthite is partly transformed to microcline, but preserved in the right part of the picture (BSE-photo, sample AHA199, locality 3). **b** Sericitisation of plagioclase in altered gneiss (BSE-photo, sample AHA205G, locality 1). **c** Dusty spread of very fine mineral grains of micas and opaques, fluid inclusions and pores in feldspars of altered charnockite (photomicrograph, sample AHA16B, locality 2). **d** Transformation of orthopyroxene to symplectitic intergrowth biotite (BtII) and quartz in gneiss (BSE-photo, scale bar is 100 µm; sample AHA205B, locality 1). **e** Replacement of original coarse grained biotite (BtI) to finer grains (BtII) in altered quartz syenite (photomicrograph, sample AHA199, locality 3). **f** Corona of titanite surrounding ilmenite in altered quartz syenite (photomicrograph, locality 3, AHA200)

perthite increases along the microcracks, and shows frequently transformation to microcline with composition of $An_{<1}Ab_{7-10}Kfs_{90-93}$. In strongly altered samples, perthite is totally replaced by microcline. The microcracks are crossing grain boundaries between the mineral phases, are healed in quartz where they can be traced as trails of fluid inclusions. A high amount of dusty opaques and biotite are also spread throughout feldspars and quartz in the altered rock (Fig. 2.5-5c). This dusty spread is especially strong in areas with a high microcrack density. In addition, micropores are frequent in the feldspars of the altered rocks, possibly representing former fluid inclusions.

Orthopyroxene is absent in the alteration zones. Symplectites of biotite + quartz (Fig. 2.5-5d) are frequently observed in altered samples, and are interpreted to represent former orthopyroxene. Grunerite, which is present together with orthopyroxene in the quartz syenite, disappears in the alteration zone. Garnet which occurs in the gneiss, is not affected texturally or mineral chemically by the alteration. Coarse biotite, hastingsite, plagioclase and quartz are locally replaced by smaller grains (Fig. 2.5-5e). Ti of the biotites shows a decrease (down to 0.13 p.f.u.; Fig. 2.5-6a), while F and Cl increases (up to 1.08 and 0.18 p.f.u., respectively). $Fe^{2+}/(Fe^{2+} + Mg)$-ratio of the biotite decreases in the gneiss (to 0.63), but shows an increase in the quartz syenite (to 0.69). Similar mineral chemical changes are shown in the hastingsite where Ti decreases (down to 0.10 p.f.u.; Fig. 2.5-6b), while F, Cl and the $Fe^{2+}/(Fe^{2+} + Mg)$-ratio increase (up to 0.37 p.f.u., 0.16 p.f.u. and 0.77, respectively). Ilmenite in the quartz syenite occurs with a corona of titanite in the altered zones (Fig. 2.5-5f) or is completely transformed to titanite. Formation of carbonate, chlorite and green biotite in fine-grained aggregates in the charnockite is also related to the alteration.

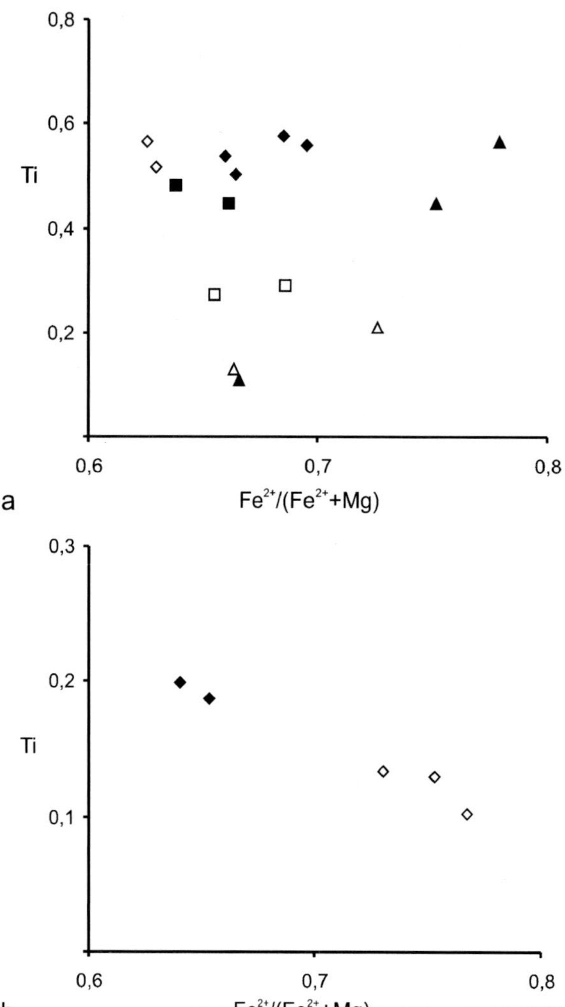

Fig. 2.5-6. Variation in mineral chemical composition of **a** biotites in gneiss (*filled diamonds*) and altered gneiss (*open diamonds*; locality 1), charnockite (*filled triangular*) and altered charnockite (*open triangular*; locality 2), quartz syenite (*filled square*) and altered quartz syenite (*open square*; locality 3). **b** Hastingsite in quartz syenite (*filled diamonds*) and altered quartz syenite (*open diamonds*; locality 3)

Discussion

Fluid-Rock Interaction Causing Colour Change

The colour change in the alteration zones along pegmatitic and aplitic veins is related to infiltration of volatile-rich melts. Engvik et al. (2005) show that the alteration halos represent the wake of the process zone after magma-driven hydraulic fracturing. During cooling of the melt, the volatile emanated into the host rock along microcracks inducing mineral reactions and grain size reduction. The replacement allowed for further fluid transport during breakdown of minerals and production of new grain boundaries. Irregularity of the reaction front in the foliated gneiss compared with the homogenous quartz syenite and charnockite, illustrate that the presence of a foliation plane also enhances infiltration.

The dark granitoid lithologies in Dronning Maud Land show only a minor degree of low-temperature alteration in thin section compared to altered samples. Systematic differences between the dark host rock and the light material of the alteration zones, are correlated with mineral and textural transformations, as observed in thin sections. The mineral transformations are related to the hydration reactions of orthopyroxene to biotite, and sericitisation of plagioclase. The formation of a high density of microcracks, sericitisation, transformation of perthite to microcline, spread of dusty opaques and formation of micropores are pervasive in the light rocks. In addition, replacement of coarse-grained plagioclase, quartz, biotite and amphibole to finer grains are important textural changes. Introduction of fluid inclusions in quartz may also contribute to the bleaching of the rocks. The above described transformations cause a very high density of interfaces in the fine-grained feldspar alteration products compared to the coarse-grained host rock, which cause the macroscopic lightening.

The occurrence of dark feldspars is rare in the exposed crust, most alkali-feldspars in plutonic rocks are light-coloured. Parson and Lee (2000) explain the occurrence of dark feldspars with their escape from reactions with fluids. Fluid-feldspar reactions are the main factor controlling the microtextural evolution of feldspars. Albite that precipitated along microveins through perthite are described by Lee and Parson (1997), who relate their formation to recrystallisation mediated by magmatic fluids and driven by stored elastic energy. Formation of the irregular veins filled by albite, tweed orthoclase and microcline are called unzipping reactions. The formation of the albite-filled microcracks through perthite in this work shows similar textures. However, their continuation across grain boundaries through plagioclase and as fluid-inclusion trails through quartz relates the process to fluid infiltration.

Crustal Level and Temperature Condition for the Fluid Infiltration

The metamorphic gneisses in Filchnerfjella have experienced granulite facies metamorphism with peak conditions of 800–900 °C at intermediate pressures (Engvik and Elvevold 2004), dated between 590–510 Ma in neighbouring areas of central Dronning Maud Land (Mikhalsky et al. 1997; Jacobs et al. 1998). Central Dronning Maud Land underwent isothermal decompression and partial melting under extensional exhumation during the later part of the Pan-African event (Jacobs et al. 2003; Engvik and Elvevold 2004). The Trollslottet quartz syenite, dated to 521 ±4 Ma (Paulsson 2003), was probably intruding during the isothermal decompression.

Replacement reactions and mineral chemical variations in the alteration zones yield indications of crustal *T*-conditions during the fluid infiltration. The lowering of the Ti-content of the replaced hastingsite in the light rocks indicates a decrease in *T* during alteration (Spear 1981). Formation of titanite as observed in the quartz syenite occurs below 550–600 °C at pressures of 2–3 kbar. Textural and mineral chemical evidences show that garnet rim composition does not change by diffusion processes during the alteration, which restricts the temperatures to be below 500 °C (Spear 1993). The increase in Kfs-content in alkali feldspars, as observed along the microcracks in altered rocks, suggest a *T*-decrease (Parson and Lee 2000). The transformation of perthite to microcline as observed in the light rocks happens at *T* below 450 °C. Sericitization of plagioclase is by Que and Allen (1996) observed to occur above 400 °C. Additional information is achieved from fluid inclusion studies in healed microcracks of the altered rocks (Engvik et al. 2003). The two-phase field in the system H_2O-CO_2 was entered during infiltration of the volatiles into the host rock, which for the recorded salinities of 3–6 wt.-% NaCl equiv. means cooling around 400 °C. Combined with the CO_2-densities which values vary between 0.75 and 0.85 g cm^{-3}, a crustal level corresponding to about 2 kbar is suggested for the vein emplacement. Based on the observed mineralogical transformations and the results from fluid inclusion studies, crustal conditions around 350–400 °C and 2 kbar appear feasible for alteration along the pegmatitic and aplitic veins.

Origin of the Fluid

As the alteration halos developed adjacent to the pegmatitic or aplitic veins, crystallising volatile-rich melts are the likely source of the infiltrating fluids. The study of fluid inclusions shows mixed CO_2-H_2O-composition of the volatiles (Engvik et al. 2005). CO_2-H_2O-rich volatiles are usually reported from granitoid magmatism (Roedder 1984) which constitutes the dominant rock type in central Dronning Maud Land. Based on the fluid composition, the regional geology and the appearance of the alteration zones with granitoid veins, a magmatic origin is most likely for the fluids. U-Pb-geochronology on zircons from a granitoid vein in the quartz syenite of Trollslottet (locality 3) gave a crystallisation age of 486 ±6 Ma, while U-Pb-geochronology on titanite grown during the alteration in the host rock yielded an age of 487 ±1 Ma (Paulsson 2003). As the age difference between the crystallisation of the quartz syenite and the cutting vein with related alteration is roughly 35 million years, it is likely that the fluids originated from underlying magma chambers.

Implications

The field relations show that the fluid infiltration along granitoid veins controls the appearance of the rocks in Mühlig-Hofmann- and Filchnerfjella. The marked colour change is conspicuous, but the hydration also causes changes of lustre, grain size and weathering. The dark rocks are easily crumbling, while the altered rocks are more resistant to weathering. The strong weathering of the dark granulite facies rocks is one factor among others controlling the development of morphology in Dronning Maud Land.

The described transformation of dark granulite facies lithologies to light rocks illustrates that metamorphic reactions are controlled by the availability of fluid. The fluid influences rock properties through the metamorphic reactions. Thereby fluid controls directly the strength of rocks and its possibility to deform (Rubie 1990; Carter et al. 1990). Feldspar is a major mineral in the continental crust, and Dimanov et al. (1999) have shown that traces of water will reduce the strength or substantially enhance strain rates of feldspathic rocks. Replacement of phases to finer grains, which is illustrated as a result of fluid infiltration, will lower the rock strength (Fitzgerald and Stünitz 1993). Fluid infiltration affects also geochemistry and mineral chemistry, and should be held in mind while doing thermobarometry and geochronology studies (Erambert and Austrheim 1993; Bomparola et al. 2003).

In the Mühlig-Hofmann- and Filchnerfjella of central Dronning Maud Land, fluid flow can be mapped out by the rock colour. The frequent distribution of alteration zones throughout the mountain range for hundreds of km independent of lithological variation shows that the magma-driven fluid infiltration processes can operate on large scales. The amount of alteration varies, but the fluid infiltration is locally penetrative as whole nunataks of up to 1 000 m high are affected, indicating that fluid-rock interaction can affect large rock volumes. The recorded fluid infiltration in Dronning Maud Land is a phenomenon of regional significance and illustrates that infiltration of fluid must play an important role in geodynamic processes.

Acknowledgment

Fieldwork for this study was carried out during the Norwegian Antarctic Research Expedition 1996/1997 and was financed by the Norwegian Polar Institute and the Norwegian Research Council (NFR). We are thankful to our field partners H. Austrheim and O. Paulsson for their contribution to the fieldwork, discussions and company during our stay in Antarctica. M. Erambert, T. Winje and B. Løken Berg are thanked for assistance during work at the SEM and microprobe. Discussions and comments on an early draft of the paper by H. Austrheim and B. Stöckhert are gratefully acknowledged. We thank M. Olesch and A. Läufer for careful review of the paper. Additional funding from NFR (145569/432 to A. K. E) is gratefully acknowledged.

References

Bomparola RM, Ghezzo C, Belousova E, Dallai L, Griffin WL, O'Reilly SYO (2003) Chemical response of zircon to fluid infiltration and high-*T* deformation: Howard Peaks Intrusive Complex (northern Victoria Land, Antarctica), a case study. Terra Nostra 2003(4):35

Bucher-Nurminen K, Ohta Y (1993) Granulites and garnet-cordierite gneisses from Dronning Maud Land, Antarctica. J Metamorph Geol 11:691–703

Carter NL, Kronenberg AK, Ross JV, Wiltschko DV (1990) Control of fluids on deformation of rocks. In: Knipe RJ, Rutter EH (eds) Deformation mechanisms, rheology and tectonics. Geol Soc London Spec Publ 54:1–13

Dimanov A, Dresen G, Xiao X, Wirth R (1999) Grain boundary diffusion creep of synthetic anorthite aggregates: The effect of water. J Geophys Res 104 B5:0483–10497

Engvik AK, Elvevold S (in press) Pan-African extension and near-isothermal exhumation of a granulite facies terrain, Dronning Maud Land, Antarctica. Geol Mag

Engvik AK, Kalthoff J, Bertram A, Stöckhert B, Austrheim H, Elvevold S (2005) Magma-driven hydraulic fracturing and infiltration of fluids into the damaged host rock, an example from Dronning Maud Land, Antarctica. J Structural Geol 27:839–854

Erambert M, Austrheim H (1993) The effect of fluid and deformation on zoning and inclusion patterns in poly-metamorphic garnets. Contrib Mineral Petrol 115:204–214

Fitzgerald JD, Stünitz H (1993) Deformation of granitoids at low metamorphic grade. I: Reactions and grain size reduction. Tectonophysics 221:269–297

Jacobs J, Fanning CM, Henjes-Kunst F, Olesch M, Paech H-J (1998) Continuation of the Mozambique Belt into East Antarctica: Grenville-Age metamorphism and polyphase Pan-African high-grade events in central Dronning Maud Land. J Geol 106:385–406

Jacobs J, Klemd R, Fanning CM, Bauer W, Colombo F (2003) Extensional collapse of the late Neoproterozoic-Early Paleozoic East African-Antarctic Orogen in central Dronning Maud Land, East Antarctica. In: Yoshida M, Windley BF, Dasgupta S (eds) Proterozoic East Gondwana: supercontinent assembly and breakup. Geol Soc London Spec Publ 206:271–288

Kretz R (1983) Symbols for rock-forming minerals. Amer Mineral 68:277–279

Lee MR, Parson I (1997) Dislocation formation and albitization in alkali feldspars from the Shap granite. Amer Mineral 82:557–570

Markl G, Piazolo S (1998) Halogen-bearing minerals in syenites and high-grade marbles of Dronning Maud Land, Antarctica: monitors of fluid compositional changes during late-magmatic fluid-rock interaction processes. Contrib Mineral Petrol 132:246–268

Mikhalsky EV, Beliatsky EV, Savva EV, Wetzel H-U, Federov LV, Weiser T, Hahne K (1997) Reconnaissance geochronologic data on polymetamorphic and igneous rocks of the Humboldt Mountains, central Queen Maud Land, East Antarctica. In: Ricci CA (ed) The Antarctic region: geological evolution and processes. Terra Antartica Publications, Siena, pp 45–53

Ohta Y (1999) Nature Environment map, Gjelsvikfjella and Western Mühlig-Hofmannfjella, sheets 1 and 2, Dronning Maud Land. Temakart nr. 24, Norsk Polarinstitutt, Tromsø

Parson I, Lee MR (2000) Alkali feldspars as microtextural markers of fluid flow. In: Stober I, Bucher K (eds) Hydrogeology of crystalline rocks. Kluwer Academic Publishers, Amsterdam, pp 27–50

Paulsson O (2003) U-Pb geochronology of tectonothermal events related to the Rodinia and Gondwana supercontinents. Litholund theses no. 2, Univ Lund

Paulsson O, Austrheim H (2003) A geochronological and geochemical study of rocks from Gjelsvikfjella, Dronning Maud Land, Antarctica – implications for Mesoproterozoic correlations and assembly of Gondwana. Precambrian Res 125:113–138

Que M, Allen AR (1996) Sericitization of plagioclase in Rosses Granite Complex, Co. Donegal, Ireland. Mineral Mag 60:927–939

Roedder E (1984) Fluid inclusions. Mineral Soc Amer Rev Mineral 12

Rubie DC (1990) Mechanisms of reaction-enhanced deformability in minerals and rocks. In: Barber DJ Meredith PG (eds) Deformation processes in minerals, ceramics and rocks. Unwin Hyman, London, pp 262–295

Spear FS (1981) An experimental study of hornblende stability and compositional variability in amphibolite. Amer J Sci 281:697–734

Spear FS (1993) Metamorphic phase equilibria and pressure-temperature-time paths. Mineral Soc Amer, Washington

Electron Microprobe (EMP) Dating on Monazite from Forefinger Point Granulites, East Antarctica: Implication for Pan-African Overprint

Yoichi Motoyoshi · Tomokazu Hokada · Kazuyuki Shiraishi
National Institute of Polar Research, 1-9-10 Kaga, Itabashi-ku, Tokyo 173-8515, Japan, <motoyosi@nipr.ac.jp>

Abstract. Electron microprobe (EMP) dating on monazite in granulite-facies rocks from Forefinger Point, East Antarctica, yielded dominant ages of ~500 Ma on matrix monazites. They are associated with secondary cordierite, biotite and sapphirine, formed during nearly isothermal decompression after the high *P-T* assemblages involving garnet, orthopyroxene and sillimanite. Older ages around 750–1 000 Ma are detected in monazite cores and in monazite inclusions in garnet porphyroblast. Combining the available age data and the reaction textures, it becomes evident that the Forefinger Point granulites have been overprinted by a granulite-facies decompressional event of Pan-African age. Moreover, EMP monazite dating imply that the Forefinger Point granulites have experienced at least two stages of metamorphic evolution.

Introduction

Forefinger Point is a small outcrop in Casey Bay, East Antarctica (Fig. 2.6-1), located in the Rayner Complex close to the suspected boundary between the Rayner and Napier complexes (Sheraton et al. 1987). The basement rocks at Forefinger Point are essentially composed of granulite-facies pelitic-psammitic and felsic gneisses. They preserve a variety of marked reaction textures which are suggestive of nearly isothermal decompression (Harley et al. 1990; Motoyoshi et al. 1994, 1995). The Rayner Complex including the Forefinger Point granulites were considered to belong to the Proterozoic mobile belt reworked after the neighboring Archean Napier Complex (e.g., Sheraton et al. 1980, 1987; Harley and Hensen 1990; Harley et al. 1990) at 1 000 ±50 Ma (Grew 1978; Black et al. 1983, 1987). However, recent SHRIMP dating by Shiraishi et al. (1997) yielded abundant concordant U-Pb ages of ~520–540 Ma on zircons in two granulites from Forefinger Point, which imply a metamorphic overprint of late Cambrian (Pan-African) age. Moreover, Asami et al. (1997) reported

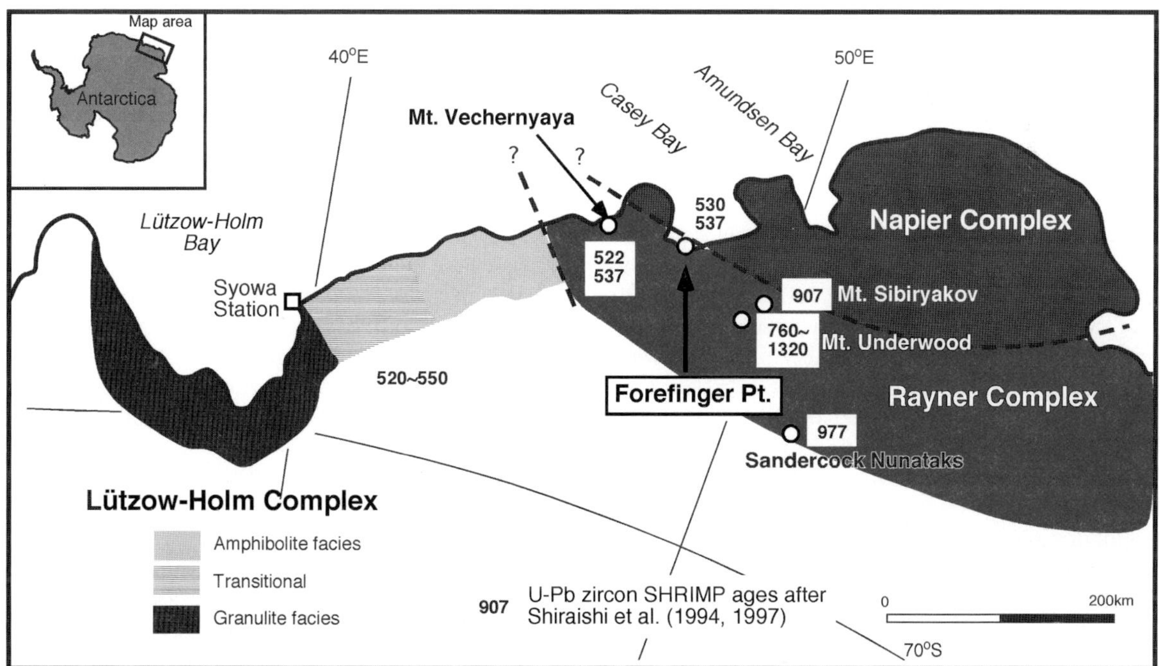

Fig. 2.6-1. Location of Forefinger Point and geological outline of the surrounding areas. U-Pb zircon SHRIMP ages are after Shiraishi et al. (1994, 1997). The estimated boundary between the Rayner Complex and the Napier Complex is after Sheraton et al. (1987). Metamorphic zonation on the Lützow-Holm Complex is after Hiroi et al. (1991)

From: Fütterer DK, Damaske D, Kleinschmidt G, Miller H, Tessensohn F (eds) (2006) Antarctica: Contributions to global earth sciences. Springer-Verlag, Berlin Heidelberg New York, pp 63–68

~530 Ma CHIME (Chemical Th-U-total Pb Isochron Method by Suzuki et al. 1991) ages for monazites in pelitic and charnockitic gneisses from Mt. Vechernyaya which also belongs to the Rayner Complex. In addition, older discordant U-Pb zircon ages of ~760–1 320 Ma, 907 Ma and 977 Ma have been reported from the inland region of the Rayner Complex (Shiraishi et al. 1997). That is to say, the timing of a major orogenic event of the Rayner Complex has not been clearly defined yet.

This paper tries to re-investigate the geochronology of the Forefinger Point granulites by means of electron microprobe (EMP) dating. The main purpose is to confirm the Pan-African overprint in this part of East Antarctica, and to characterize it in relation to the *P-T* evolution.

Samples

Samples used in this study (Sp. 93022206, 93022209, 93022222, 93022223, 93022225 and 93022231; they will be abbreviated as 2209 for 93022209, for example) are all metapelitic granulites with Opx + Sil + Grt, Opx + Grt, and Grt + Sil assemblages formed under the peak metamorphic conditions (Mineral abbreviations are after Kretz 1983). Secondary Crd-bearing assemblages formed at the expense of these peak mineral assemblages, as suggested by the following metamorphic reactions depending on the bulk chemical compositions;

Opx + Sil + Qtz = Crd

Opx + Sil = Crd + Spr

Grt = Opx + Crd ± Spr

Grt + Qtz = Opx + Crd

These reactions invariably suggest nearly isothermal decompression (Harley et al. 1990; Motoyoshi et al. 1994, 1995). On the basis of the model FeO-MgO-Al_2O_3-SiO_2 (FMAS) grid (Hensen and Harley 1990), this decompressional *P-T* path has reasonably passed between two invariant points, namely [Spl] at 11 ±1 kbar and 1 040 °C ±15 °C and [Qtz] 8–9 kbar and 950 °C (Harley et al. 1990). However, there is no evidence that the *P-T* path came down from the Spr + Qtz stability field which characterizes the ultra-high temperature conditions of the neighboring Napier Complex. Harley et al. (1990) proposed that this decompression occurred during Archaean followed by the second stage decompression during Proterozoic.

Monazite in these samples occurs as rounded grains in the matrix in association with secondary cordierite, biotite and sometimes sapphirine. Monazite also occurs as inclusions in garnet. Internal structures of monazite grains were investigated using back scattered electron imaging (BSE) and elemental mapping. The monazites invariably demonstrate very heterogeneous structures due to heterogeneous concentrations of chemical components such as Th.

EMP Dating Method

Quantitative chemical analyses on monazite were performed by means of an electron microprobe analyzer using JEOL JXA-8800 equipped with five wavelength-dispersive spectrometers at the National Institute of Polar Research, Tokyo. The operating conditions were 15 kV accelerating voltage, 0.2 mA beam current and 2 µm beam diameter. In addition to UO_2, ThO_2 and PbO, other REE, SiO_2, CaO, P_2O_5 were also measured, and the intensity data were adjusted using the $\phi\rho(Z)$ correction method (Packwood and Brown 1981). Only analyses higher than 0.02 wt.% PbO were selected for further processing.

CHIME (chemical Th-U-total Pb isochron method) date reduction and calculation of apprent ages were performed by using the age calculation program of Kato et al. (1999).

Results

Pan-African Ages

Monazites which occur in the matrix in Sp. 2206, 2209, 2222, 2223, 2225 and 2231 yielded apparent ages close to 500 Ma (470~550 Ma). Representative results are listed in Table 2.6-1. Among them, examples of apparent ages on some of the grains in Sp. 2209 and 2231, along with the CHIME results (528.2 ±13.6 Ma for Sp. 2209 and 517.3 ±22.1 Ma for Sp. 2231) are presented in Fig. 2.6-2. Because these monazites all occur in the matrix in association with cordierite, sapphirine, biotite, etc., the monazite is inferred to have crystallized simultaneously with these minerals. The CHIME ages of 517–528 Ma are within error identical to SHRIMP U-Pb zircon ages of 530–537 Ma obtained by Shiraishi et al. (1997) from Forefinger Point.

Evidence for Older History

Older ages which predate the Pan-African event were also obtained from the monazites. In Sp. 2222, one of the monazite grains in the matrix (mnz 6) preserved a distinctive euhedral core, which yielded apparent ages of 750–960 Ma (Fig. 2.6-3). Because the rim was dated at around 500 Ma, this core is probably an inherited part crystallized prior to the Pan-African event.

Table 2.6-1. EMP analyses of UO_2, ThO_2, and PbO on monazites from Forfinger Point granulites.

Grain	UO_2	ThO_2	PbO	ThO_2*	Age	Error
Sp. No. 93022206						
Mnz 1	0.07	7.68	0.17	7.91	512	16
Mnz 2	0.08	8.87	0.19	9.12	504	14
Mnz 4	0.11	7.09	0.17	7.47	523	17
Mnz 5	0.16	12.15	0.27	12.67	509	10
Mnz 6-1	0.19	7.11	0.16	7.73	496	17
Mnz 6-2	0.15	7.28	0.17	7.78	502	17
Mnz 7	0.19	7.19	0.17	7.82	499	17
Mnz 8	0.14	6.72	0.16	7.17	531	18
Mnz 9	0.14	6.91	0.16	7.38	514	18
Sp. No. 93022209						
Mnz 17-1	0.20	8.02	0.19	8.70	526	15
Mnz 17-2	0.37	7.57	0.20	8.79	536	15
Mnz 17-3	0.53	7.65	0.21	9.41	522	14
Mnz 18-1	0.99	4.28	0.17	7.58	541	17
Mnz 18-2	0.42	8.78	0.23	10.18	543	13
Mnz 18-3	0.29	15.01	0.35	15.97	524	8
Mnz 19-1	0.31	10.43	0.25	11.44	510	11
Mnz 21-1	0.46	6.76	0.18	8.28	521	16
Mnz 21-2	0.69	6.09	0.20	8.39	558	16
Mnz 21-3	0.65	15.21	0.39	17.37	532	7
Mnz 22-1	0.27	11.17	0.27	12.05	520	11
Mnz 22-2	0.41	10.40	0.26	11.76	519	11
Mnz 22-3	0.21	12.48	0.30	13.18	531	10
Mnz 23-1	0.29	14.47	0.33	15.44	508	8
Mnz 23-2	0.30	12.29	0.29	13.27	521	10
Mnz 24-1	0.22	10.32	0.25	11.05	532	12
Mnz 24-2	0.19	10.31	0.25	10.93	542	12
Mnz 24-3	0.20	10.31	0.24	10.98	525	12
Mnz 24-4	0.14	11.85	0.27	12.31	513	11
Mnz 24-5	0.19	10.14	0.24	10.78	520	12
Sp. No. 93022222						
Mnz 3-1	0.41	7.95	0.20	9.31	506	14
Mnz 3-2	0.14	9.52	0.21	9.98	489	13
Mnz 3-3	0.13	9.38	0.19	9.82	466	13
Mnz 3-4	0.14	8.81	0.20	9.28	503	14
Mnz 3-5	0.14	9.14	0.20	9.60	501	14
Mnz 3-6	0.65	7.95	0.22	10.09	509	13
Mnz 3-7	0.15	9.18	0.21	9.67	507	13
Mnz 3-8	0.44	7.96	0.21	9.41	532	14
Mnz 3-9	0.30	8.82	0.19	9.81	469	13
Mnz 3-10	0.12	8.56	0.19	8.97	505	15
Mnz 6-1	0.26	12.74	0.56	13.63	960	10
Mnz 6-2	0.45	14.39	0.58	15.93	853	8
Mnz 6-5	0.46	14.27	0.58	15.83	862	8
Mnz 6-6	0.48	12.90	0.46	14.53	750	9
Mnz 6-11	0.42	11.20	0.36	12.60	674	10
Sp. No. 93022223						
Mnz 3	0.10	10.33	0.23	10.67	511	12
Mnz 5	0.20	8.81	0.19	9.48	478	14
Mnz 6	0.17	7.45	0.16	8.03	473	16
Mnz 8	0.09	10.19	0.22	10.50	485	12
Mnz 10	0.21	8.22	0.20	8.92	521	15

Grain	UO_2	ThO_2	PbO	ThO_2*	Age	Error
Sp. No. 93022225						
Mnz 11-15	0.20	8.26	0.20	8.92	525	15
Mnz 11-16	1.93	7.31	0.32	13.73	551	9
Mnz 11-17	0.20	12.62	0.29	13.30	513	10
Mnz 11-19	0.21	8.26	0.20	8.96	521	15
Mnz 11-20	0.54	5.57	0.17	7.36	549	18
Mnz 11-21	0.53	5.90	0.18	7.66	542	17
Mnz 11-22	0.53	5.59	0.18	7.36	564	18
Mnz 11-23	2.54	6.26	0.37	14.71	586	9
Mnz 11-24	1.49	4.95	0.23	9.88	551	13
Mnz 12-1	0.79	6.91	0.40	9.62	959	13
Mnz 12-2	1.35	7.16	0.38	11.73	766	11
Mnz 12-3	1.30	6.67	0.37	11.05	780	12
Mnz 12-4	1.51	6.77	0.40	11.86	785	11
Mnz 12-5	1.14	5.54	0.40	9.47	985	14
Mnz 12-6	1.28	6.82	0.41	11.16	850	12
Mnz 12-7	1.38	6.52	0.40	11.20	839	12
Mnz 12-8	1.04	7.26	0.38	10.78	828	12
Mnz 12-9	0.75	7.18	0.46	9.79	1078	13
Mnz 12-10	0.31	7.15	0.40	11.61	713	11
Mnz 12-11	0.77	7.13	0.39	9.78	926	13
Sp. No. 93022231						
Mnz 1	0.67	7.76	0.24	9.99	559	13
Mnz 3-1	0.12	6.56	0.16	6.96	541	19
Mnz 3-2	0.14	6.17	0.15	6.62	541	20
Mnz 3-3	0.09	7.08	0.16	7.38	523	18
Mnz 3-4	0.14	6.22	0.15	6.67	530	20
Mnz 3-5	0.10	5.67	0.12	6.01	477	22
Mnz 3-6	0.13	6.30	0.15	6.72	519	19
Mnz 3-7	0.10	5.32	0.12	5.67	486	23
Mnz 3-8	0.12	6.65	0.15	7.04	518	18
Mnz 3-9	0.11	7.85	0.17	8.20	502	16
Mnz 3-10	0.10	5.71	0.13	6.06	504	22
Mnz 3-11	0.13	7.95	0.19	8.38	521	16
Mnz 3-12	0.09	10.80	0.24	11.11	504	12
Mnz 3-13	0.12	5.86	0.14	6.26	527	21
Mnz 3-14	0.19	7.49	0.18	8.13	523	16
Mnz 3-15	0.12	6.93	0.15	7.33	489	18
Mnz 3-16	0.20	7.71	0.17	8.36	492	16
Mnz 3-17	0.13	6.25	0.15	6.68	522	19
Mnz 3-18	0.10	7.32	0.18	7.65	562	17
Mnz 3-19	0.07	7.17	0.14	7.41	453	18
Mnz 3-20	0.08	7.11	0.15	7.38	473	18
Mnz 3-21	0.11	9.89	0.22	10.27	516	13
Mnz 3-22	0.10	4.89	0.11	5.22	505	25
Mnz 4-1	0.11	6.95	0.16	7.31	521	18
Mnz 4-2	0.12	7.34	0.17	7.73	511	17
Mnz 4-3	0.12	7.65	0.18	8.05	537	16
Mnz 4-4	0.11	7.53	0.16	7.88	490	17
Mnz 4-5	0.14	7.74	0.18	8.19	514	16
Mnz 4-6	0.12	5.92	0.14	6.31	534	21
Mnz 7	0.28	4.45	0.13	5.39	548	24
Mnz 9	0.27	7.31	0.18	8.19	519	16
Mnz 10	0.16	6.33	0.15	6.86	506	19

*ThO_2**: Sum of measured ThO_2 and ThO_2 equivalent of the measured UO_2.

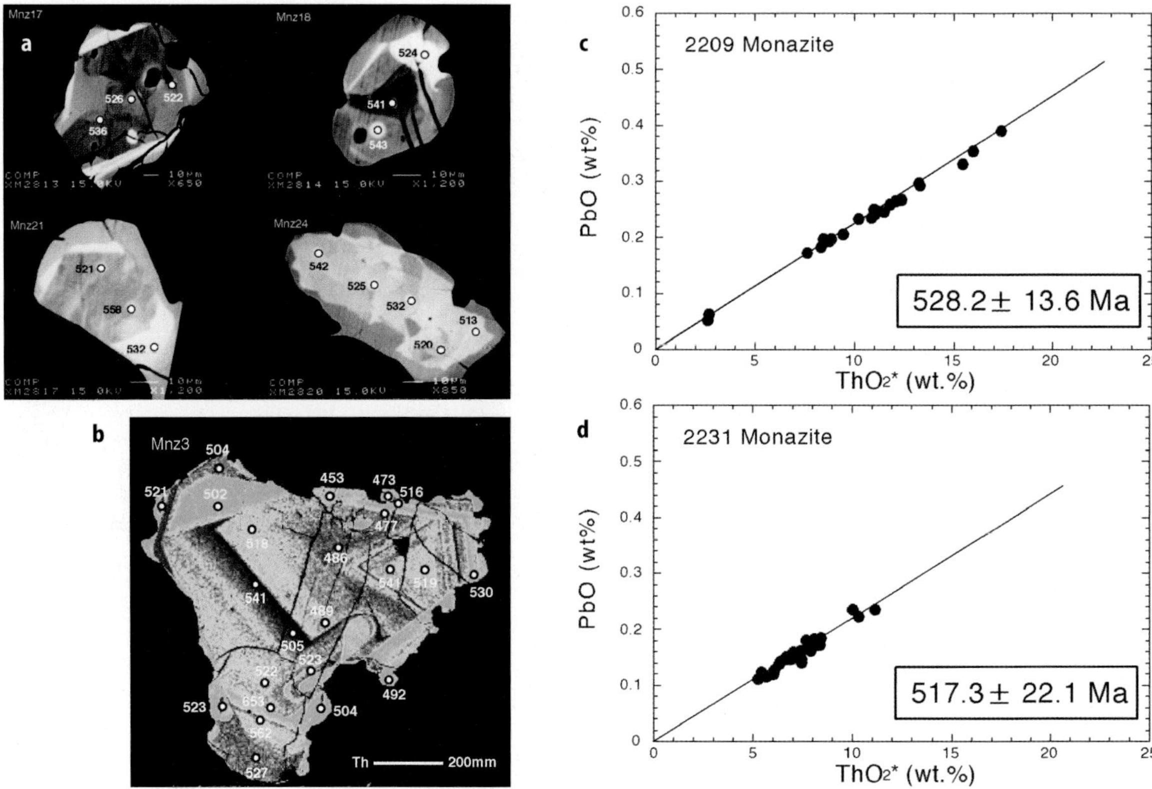

Fig. 2.6-2. a BSE images of representative monazites in Sp. 2209 and apparent ages in Ma. **b** Elemental mapping on Th for monazite (Mnz3) in Sp. 2231 with apparent ages in Ma. **c** CHIME ages on monazites in Sp. 2209. **d** CHIME ages on monazites in Sp. 2231. ThO_2* denotes sum of the measured ThO_2 and ThO_2 equivalent of UO_2 for monazite

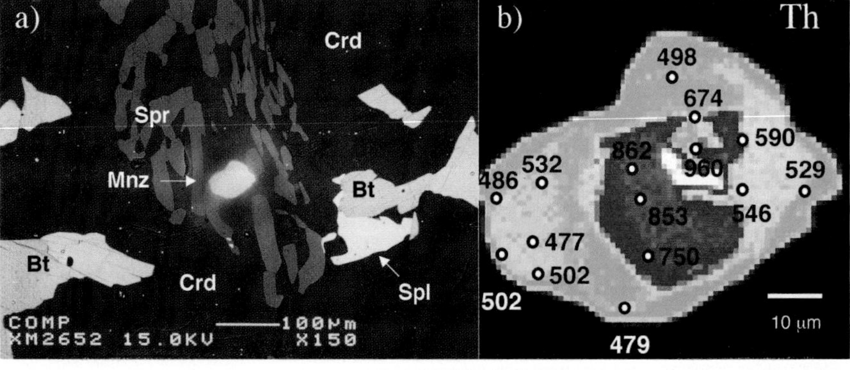

Fig. 2.6-3.
a Mode of occurence of monazite in Sp. 2222 being associated with sapphirine, cordierite and biotite. Back-scattered electron image. **b** Elemental mapping on Th for monazite (Mnz6) in Sp. 2222 with apparent ages in Ma. ThO_2 contents in red core are around 13~14 wt.%, those in rim are around 9~10 wt.%, respectively. Note the euhedral core with older apparent ages

Another example can be seen in Sp. 2225 in which monazite in the matrix yielded Pan-African ages (Fig. 2.6-4a and 2.6-4b), whereas a monazite inclusion in a garnet porphyroblast yielded older ages between 760–1 078 Ma (Fig. 2.6-4c and 2.6-4d).

Discussion

In view of similar ages obtained by SHRIMP dating (Shiraishi et al. 1997) and CHIME dating (present study), it is no doubt that the Forefinger Point granulites have certainly been subjected to the Pan-African overprint at around 520–540 Ma. Although the granulite-facies event which predates the Pan-African overprint has not been well constrained yet, the difference in mode of occurrence and different ages of the monazites clearly suggest that there have been at least two stages of monazite formation, one of Pan-African age and an earlier one of likely 750–1 000 Ma ages, respectively.

Shiraishi et al. (1997) argued on the basis of SHRIMP ages that the western coastal region of the Rayner Com-

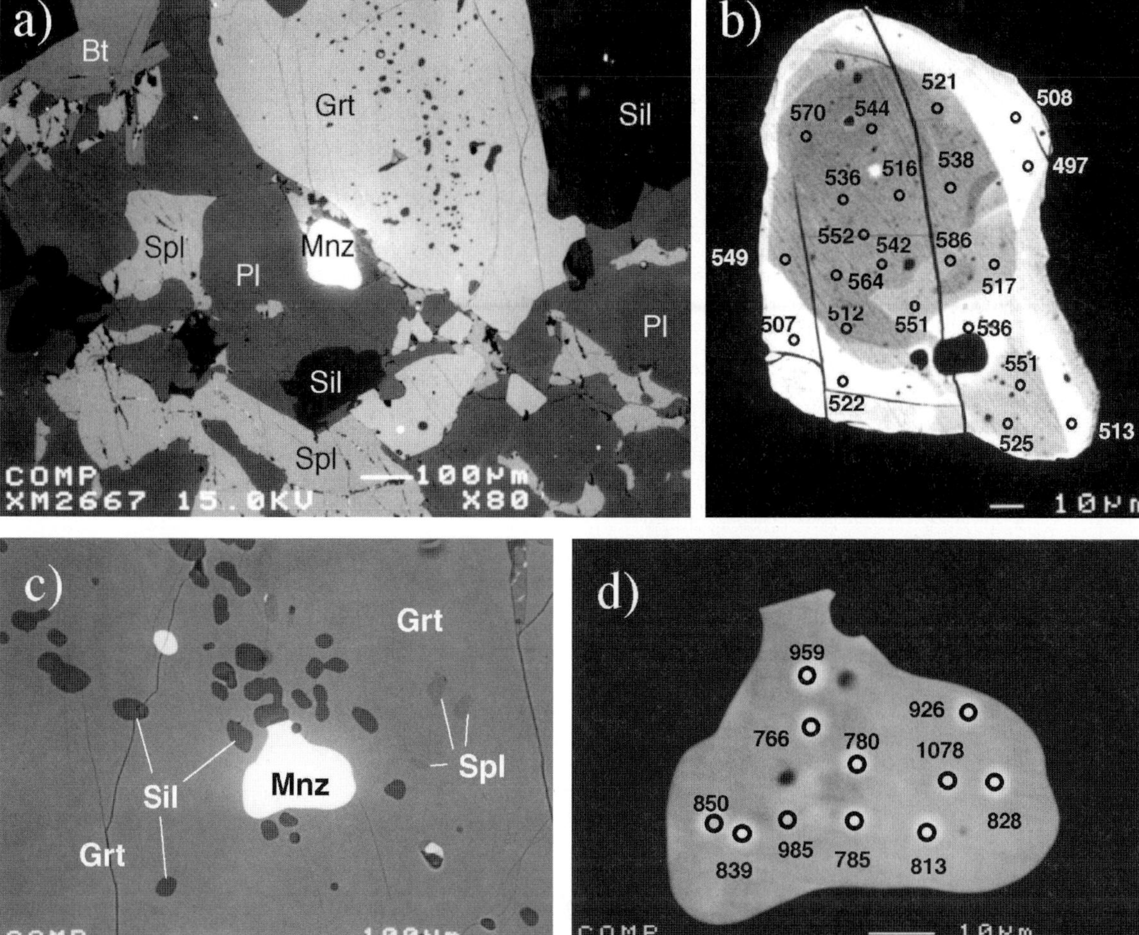

Fig. 2.6-4. Backscattered electron image of monazite in Sp. 2225. **a** Mode of occurrence of monazite in the matrix. **b** Apparent ages in monazite in the matrix. **c** Monazite inclusion (Mnz11) being associated with sillimanite and spinel in garnet. **d** Apparent ages in monazite inclusion

plex including Forefinger Point represents an eastward continuation of the Lützow-Holm Complex which has been also subjected to the regional Pan-African event (Shiraishi et al. 1994, 1997).

On the basis of the monazite ages and reactions textures, the Pan-African event in the Forefinger Point granulites may be characterized by substantial decompression even under high-temperature conditions at granulite grade.

Acknowledgments

We express our thanks to S. Harley and F. Henjes-Kunst for careful and constructive reviews on the manuscript, and to D. Fütterer for editorial handling. This study was supported by a Grant-in-Aid from the Japan Society for the Promotion of Science (JSPS) to K. S. (no. 13440151) and to Yoshifumi Nogi (no. 13640429).

References

Asami M, Suzuki K, Adachi M (1997) Th, U and Pb analytical data and CHIME dating of monazites from metamorphic rocks of the Rayner, Lützow-Holm, Yamato-Belgica and Sør Rondane Complexes, East Antarctica. Proc NIPR Symp Antarct Geosci 10:130–152

Black LP, James PR, Harley SL (1983) Geochronology and geological evolution of metamorphic rocks in the Field Islands area, East Antarctica. J Metam Geol 1:277–303

Black LP, Harley SL, Sun SS, McCulloch MT (1987) The Rayner Complex of East Antarctica: complex isotopic systematics with a Proterozoic mobile belt. J Metam Geol 5:1–26

Grew ES (1978) Precambrian basement at Molodezhnaya station, East Antarctica. Bull Geol Soc Amer 89:801–813

Harley SL, Hensen BJ (1990) Archaean and Proterozoic high-grade terranes of East Antarctica (40–80° E): a case study of diversity in granulite facies metamorphism. In: Ashworth JR, Brown M (eds) High-temperature metamorphism and crustal anatexis. Unwin Hyman, London, pp 320–370

Harley SL, Hensen BJ, Sheraton JW (1990) Two-stage decompression in orthopyroxene-sillimanite granulites from Forefinger Point, Enderby Land, Antarctica: implications for the evolution of the Archaean Napier Complex. J Metam Geol 8:591–613

Hiroi Y, Shiraishi K, Motoyoshi Y (1991) Late Proterozoic paired metamorphic complexes in East Antarctica, with special reference to the tectonic significance of ultramafic rocks. In: Thomson MRA, Crame JA, Thomson JW (eds) Geological evolution of Antarctica. Cambridge University Press, Cambridge, pp 83–87

Kato T, Suzuki K, Adachi M (1999) Computer program for the CHIME age calculation. J Earth Planet Sci 46:49–56

Kretz R (1983) Symbols for rock-forming minerals. Amer Mineral 68:277–279

Motoyoshi Y, Ishikawa M, Fraser GL (1994) Reaction textures in granulites from Forefinger Point, Enderby Land, East Antarctica: an alternative interpretation on the metamorphic evolution of the Rayner Complex. Proc NIPR Symp Antarct Geosci 7:101–114

Motoyoshi Y, Ishikawa M, Fraser GL (1995) Sapphirine-bearing silica-undersaturated granulites from Forefinger Point, Enderby Land, East Antarctica: evidence for a clockwise *P-T* path? Proc NIPR Symp Antarct Geosci 8:121–129

Packwood RH, Brown JD (1981) A Gaussian expression to describe $\varphi(\rho z)$ curves for quantitative electron probe microanalysis. X-ray Spectrom 10:138–146

Sheraton JW, Offe LA, Tingey RJ, Ellis DJ (1980) Enderby Land, Antarctica: an unusual Precambrian high grade metamorphic terrain. J Geol Soc Austral 27:1–18

Sheraton JW, Tingey RJ, Black LP, Offe LA, Ellis DJ (1987) Geology of Enderby Land and western Kemp Land, Antarctica. BMR Bull 223

Shiraishi K, Ellis DJ, Hiroi Y, Fanning CM, Motoyoshi Y, Nakai Y (1994) Cambrian orogenic belt in East Antarctica and Sri Lanka: implications for Gondwana assembly. J Geol 102:47–65

Shiraishi K, Ellis DJ, Fanning CM, Hiroi Y, Kagami H, Motoyoshi Y (1997) Re-examinataion of the metamorphic and protolith ages of the Rayner Complex, Antarctica: evidence for the Cambrian (Pan-African) regional metamorphic event. In: Ricci CA (ed) Antarctic regions: geological evolution and processes. Terra Antartica Publication, Siena, pp 79–88

Suzuki K, Adachi M, Tanaka T (1991) Middle Precambrian provenance of Jurassic sandstone in the Mino Terrane, central Japan: Th-U-total Pb evidence from an electron microprobe monazite study. Sediment Geol 15:141–147

Tectonic Subdivision of the Prince Charles Mountains: A Review of Geologic and Isotopic Data

Evgeny V. Mikhalsky[1] · Anatoly A. Laiba[2] · Boris V. Beliatsky[3]

[1] VNIIOkeangeologia, Angliysky Avenue1, St. Petersburg 190121, Russia
[2] PMGRE, Pobeda Street 24, Lomonosov 189510, Russia
[3] IGGP, Makarova emb. 2, 199034 St. Petersburg 199034, Russia

Abstract. The Prince Charles Mountains have been subject to extensive geological and geophysical investigations by former Soviet, Russian and Australian scientists from the early 1970s. In this paper we summarise, and review available geological and isotopic data, and report results of new isotopic studies (Sm-Nd, Pb-Pb, and U-Pb SHRIMP analyses); field geological data obtained during the PCMEGA 2002/2003 are utilised. The structure of the region is described in terms of four tectonic terranes. Those include Archaean Ruker, Palaeoproterozoic Lambert, Mesoproterozoic Fisher, and Meso- to Neoproterozoic Beaver Terranes. Pan-African activities (granite emplacement and probably tectonics) in the Lambert Terrane are reported. We present a summary of the composition of these terranes, discuss their origin and relationships. We also outline the most striking geological features, and problems, and try to draw attention to those rocks and regional geological features which are important in understanding the composition and evolution of the PCM and might suggest targets for further investigations.

Previous Work

The Prince Charles Mountains (PCM) constitute by far the best exposed cross section through the East Antarctic Shield, extending for over 500 km along the drainage basin of the Lambert Glacier – Amery Ice Shelf system (Fig. 2.7-1). Australian geologists first visited this region between 1955 and 1957. The results of geological observations were presented by Stinear (1956) and Crohn (1959), who recorded in the northern PCM (NPCM) a range of high-grade metamorphic rocks of both sedimentary and igneous origin, as well as orthopyroxene granitoids, and an overlying coal-bearing Permian-Triassic sedimentary sequence. Geological investigations in the southern PCM (SPCM) showed that the rocks there are generally of much lower metamorphic grade (green-schist to amphibolite facies) than those of the NPCM; they include thick metasedimentary sequences and are cut by abundant mafic dykes. The early workers (Tingey 1982, 1991, and references therein), on the basis of numerous Rb-Sr ages, reported the PCM as comprising two major tectonic provinces roughly corresponding to the two parts of the mountain range (NPCM and SPCM). One of the most interesting results of this study (Tingey 1982, 1991) was the recognition that presumed granitic basement and overlying metasediments in the SPCM were of Archaean age (whole-rock Rb-Sr isochrons 2 700 ±90, 2 750 ±400, and 2 760 ±200 Ma), whereas granulite-facies rocks of the NPCM were Meso- to Neoproterozoic (whole-rock Rb-Sr isochrons from 769 ±36 to 1 033 ±85 Ma). This was at variance with earlier ideas that the higher-grade rocks were likely to be older (e.g., Solov'ev 1972). Muscovite-bearing pegmatites cutting the metasediments in the SPCM were dated at 2 589, 2 100, 1 995, and 1 708 Ma, and it was suggested that Palaeoproterozoic, as well as Archaean sequences may be present. A few imprecise Mesoproterozoic isochron ages were also obtained (ca. 1 170 ±230, 1 400 ±150 Ma); these were interpreted as reset ages, and provide some evidence for Mesoproterozoic thermal reworking of the area, possibly in response to a high-grade event in the NPCM.

Soviet geologists first visited the SPCM in 1965, and detailed geological work in the area was carried out in 1971–1974. Deep-seismic soundings, airborne magnetic and gravimetric surveys were carried out, and revealed the main crustal features, with the Lambert Glacier-Amery Ice Shelf rift zone being one of the most important discoveries (Kurinin and Grikurov 1982). Acritarchs were found in the metasediments, and their age was considered to be Mesoproterozoic or Meso-Neoproterozoic (Iltchenko 1972; Ravich et al. 1978). Muscovite-bearing granites were dated at ca. 500 Ma (Halpern and Grikurov 1975). The structure of the SPCM was thought to be dominated by a NE-SW trough filled with Proterozoic sedimentary strata, the Menzies Series and Sodruzhestvo Series of Solov'ev (1972). K-Ar and Rb-Sr dating generally confirmed the results of Tingey (1982), including Meso- to Neoproterozoic thermal reworking (1 040 and 830 Ma; Mikhalsky et al. 2001 and references therein).

Soviet geological investigations between 1982 and 1991 were concentrated in the NPCM, with a few visits to the SPCM. The results of helicopter-supported mapping and geological studies at some mountain massifs were presented by Kamenev et al. (1993). Kamenev et al. (1990, 1993) proposed a threefold tectonic division of the area on structural and lithological grounds. They distinguished two structural zones (provinces or terranes): *(i)* the relatively conservative South Lambert province comprising Archaean granite-greenstone and granite-gneiss-schist

From: Fütterer DK, Damaske D, Kleinschmidt G, Miller H, Tessensohn F (eds) (2006) Antarctica: Contributions to global earth sciences. Springer-Verlag, Berlin Heidelberg New York, pp 69–82

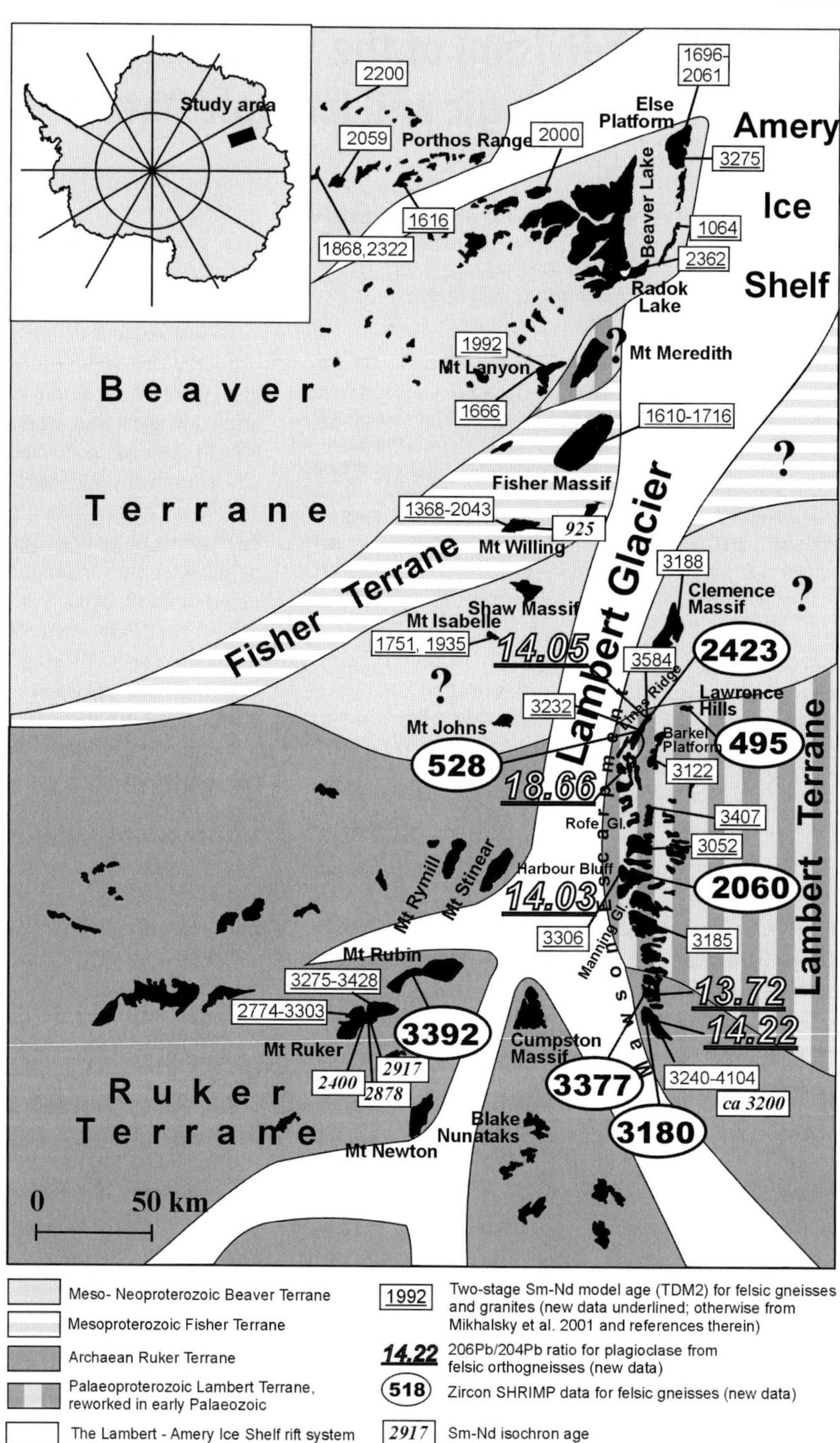

Fig. 2.7-1.
Overview of Prince Charles Mountains tectonic subdivisions, and isotopic data locality map

belts (Ruker Complex), *(ii)* the North Lambert province underlain by a charnockite-granulite belt, which was highly mobile until Cambrian (referred to by Kamenev et al. as the Beaver Complex). These zones were thought to have been brought into contact within the area of supracrustal and granitic rocks in the central part of the PCM (including northern Mawson Escarpment). These rocks were collectively named the Lambert Complex and were considered as either a "buffering" tectonic zone originating by interaction between granite-greenstone and charnockite-granulite belts at mid-crustal levels, or as a higher-temperature equivalents of the granite-greenstone belts themselves (Kamenev et al. 1990). The Lambert Complex was thus supposed to comprise retrogressed Beaver Terrane rocks tectonically intercalated with prograde metamorphosed Ruker Terrane rocks.

Kamenev and Krasnikov (1991) distinguished yet another tectonic province (Fisher Terrane) within the central PCM on the basis of its distinctive lithologies. The Fisher Terrane is characterized by essentially calc-alkaline magmatism, and likely represents an active continental margin or a collage of island arcs and foreland domains (Mikhalsky et al. 1996, 1999).

On this background, based mostly on early Rb-Sr ages and imprecise zircon thermal ionization ages, Mikhalsky et al. (2001) described the structure of the PCM in terms of three terranes: the Ruker Terrane in the SPCM, the Beaver-Lambert Terrane in the central and northern PCM, and the Fisher Terrane in the central PCM, though the Fisher Terrane was considered to be a sub-terrane within the Beaver-Lambert Terrane (Fig. 2.7-1).

Recent ion microprobe studies by Boger et al. (2001, among others) showed distinctive isotopic features of rocks cropping out in the southern and central Mawson Escarpment, and basically confirmed the existence of a separate Lambert Complex as postulated by Kamenev et al. (1993), though they interpreted the origin of the Lambert Complex differently. However, Boger et al. (2001) failed to find any correlation (i.e., a Grenville-age) with the NPCM, the type area for the Beaver-Lambert Terrane of Mikhalsky et al. (2001), but showed for the first time the significance and penetrative character of the 500 Ma Pan-African tectonics in this area.

In 2002/2003 the Prince Charles Mountains Expedition of Germany and Australia (PCMEGA) collected new geological data from an extensive area in the SPCM. In this section we present new isotopic data on rocks collected during PCMEGA and earlier expeditions to the PCM, and we address the so far enigmatic relationships between the different terranes.

Isotopic Studies

More then 60 whole-rock samples were analysed to obtain Sm-Nd model ages (Table 2.7-1). These mostly include felsic orthogneisses and schists from the Mawson Escarpment, Mt. Ruker, and scattered localities in the central and northern PCM. The Sm-Nd studies were conducted at IGGP (St. Petersburg) on a Finnigan MAT-261 solid-source mass spectrometer equipped with eight collectors operating in the mode of the simultaneous determination of the required isotopes. The measured Nd isotopic compositions were adjusted to $^{143}Nd/^{144}Nd = 0.511860$ for the La Jolla standard. The Sm-Nd data presented in Table 2.7-1 include some obtained by the authors earlier, and mentioned by Mikhalsky et al. (2001), but the raw data have never been published. Assuming that most of the area experienced thermal reworking at a subsequent stage of evolution, a two-stage model was applied to calculate the depleted mantle (DM) extraction ages.

The Sm-Nd model ages provide evidence for a clear distinction between the proposed terranes. Typical values are in the range 3.9–3.2 Ga for the Ruker Terrane, 3.4–3.0 Ga for the Lambert Terrane, 2.3–1.6 Ga for the Beaver Terrane, and 1.7–1.4 Ga for the Fisher Terrane. The age of the last significant tectonic activity that might have caused Sm-Nd fractionation (*T* in Table 2.7-1) is defined by zircon U-Pb data available for some localities in the PCM (Boger et al. 2000, 2001; Carson et al. 2000; Mikhalsky et al. 2001).

Three rock groups collected from Mt. Ruker yielded reasonable Sm-Nd isochrons (Fig. 2.7-2). The four green schists (presumably metavolcanics) define an isochron with an age of 2 917 ±82 Ma (MSWD = 0.33, $Nd_i = 0.508446$; Fig. 2.7-2a). Five felsic mica schists from Mt. Ruker, however, show wide variation of isotopic ratios and do not define a reference line. This may be due to sedimentary origin of these rocks, and their isotopic inhomogeneity due to mixing of components derived from various sources. Three metagabbro-dolerite samples and mineral separates (plagioclase, biotite) define an isochron of 2 878 ±65 Ma (MSWD = 0.64, $Nd_i = 0.507985$). Both plagioclase and biotite are thought to be metamorphic minerals, so the age of ca. 2 900 Ma is most likely to reflect that of metamorphism, although the low Nd_i value precludes a long pre-metamorphic crustal residence time. The metavolcanics are likely to be roughly co-eval with the metagabbro-dolerites, but their slightly higher Nd_i values suggest that they originated from a somewhat different mantle source or experienced more pronounced crustal contamination. The granophyric gabbro collected on the southern slopes of Mt. Ruker is one of only a few unaltered rocks. Its whole-rock and mineral compositions (plagioclase, clinopyroxene) define a three-point reference line corresponding to an age of 2 365 ±65 Ma (MSWD = 1.8, $Nd_i = 0.508951$). Two fresh dolerite dyke samples from southeastern Mt. Ruker plot roughly along this line, producing a five-point isochron of 2 400 ±200 Ma (MSWD = 4.33, $Nd_i = 0.508934$). The initial Nd ratio is somewhat higher than in other analysed rock types, providing evidence that these plutonic rocks were derived from a different (less enriched) mantle source.

Hornblende-biotite, biotite granite gneisses and leucogneiss from the southern Mawson Escarpment (Ruker Terrane) have given Sm-Nd whole-rock and mineral-separate isochron age of 3 231 ±130 (10 samples, MSWD = 1.74; Fig. 2.7-2b). Tonalitic orthogneisses from the same area have given very imprecise Sm-Nd whole-rock (four samples), and mineral (orthopyroxene, plagioclase) whole-rock isochron ages of ca. 3 000 and ca. 1 900 Ma, respectively. The latter age presumably reflects a younger thermal overprint. Since orthopyroxene plots on the

Table 2.7-1. Sm-Nd data for rocks from the Prince Charles Mountains

Sample		Rock	Area	[Sm]	[Nd]	$^{147}Sm/^{144}Nd$	$^{143}Nd/^{144}Nd$	2σ	T (Ga)	ε_T	T_{DM}	T_{DM2}
Fisher Terrane												
39115-7	WR	Ms pegmatite	Willing Massif	1.542	8.344	0.11209	0.512242	15	1.3	6.39	1.363	1368
39115-7	Grt	Ms pegmatite	Willing Massif	3.923	6.745	0.35271	0.513685	28				
39115-7	Ms	Ms pegmatite	Willing Massif	1.549	8.853	0.10612	0.512187	24				
39115-7	Mc	Ms pegmatite	Willing Massif	1.421	9.592	0.08986	0.512079	21				
33047-3	WR	Granodiorite	Fisher Massif	4.897	27.34	0.10826	0.512053	11	1,3	3.33	1.586	1621
33047-2	WR	Granodiorite	Fisher Massif	4.229	22.71	0.11256	0.512071	7	1.3	2.96	1.626	1652
34428 g	WR	Granodiorite	Fisher Massif	3.877	21.09	0.11107	0.512030	14	1.3	2.41	1.663	1697
33044-13	WR	Granodiorite	Fisher Massif	4.753	25.49	0.11269	0.512054	5	1.3	2.61	1.653	1681
33044-10	WR	Granodiorite	Fisher Massif	5.243	26.2	0.12094	0.512119	4	1.3	2.50	1.694	1690
34431 g	WR	Granodiorite	Fisher Massif	3.445	15.202	0.13698	0.512305	8	1.3	3.46	1.679	1610
33044-1	WR	Diorite	Fisher Massif	3.758	17.15	0.13244	0.512201	4	1.3	2.19	1.779	1716
34406a	WR	Plagiogranite	Fisher Massif	1.205	6.335	0.11500	0.512066	8	1.3	2.46	1.673	1693
Beaver Terrane												
34230	WR	Felsic orthogneiss	Mt Lanyon	7.815	41.37	0.11456	0.512090	11	1.0	−0.18	1.629	1666
34017	WR	Two-Px granulite	Mt Lanyon	3.512	19.53	0.10905	0.511852	9	1.0	−4.12	1.888	1992
35068	WR	Opx gneiss	Porthos Range	6.354	28.61	0.13468	0.512253	10	1.0	0.43	1.730	1616
34042-1	WR	Opx plagiogneiss	Mt Isabelle	4.216	26.28	0.09728	0.511924	11	1.0	−1.21	1.604	1751
34238	WR	Opx plagiogneiss	Mt Isabelle	3.947	18.21	0.13144	0.512034	12	1.0	−3.44	2.065	1935
35301-2	WR	Grt pegmatite def.	Radock Lake	0.544	3.32	0.09925	0.511558	10	1.0	−8.62	2.115	2362
35025-3	WR	Felsic orthogneiss	Else Platform	6.071	29.55	0.12433	0.511153	20	1.0	−19.8	3.385	3275
33256-1	WR	Gneiss	Clemence Massif	7.789	43.83	0.10762	0.511098	18	1.0	−18.7	2.934	3188
Lambert Terrane												
33512-2	WR	Px plagiogneiss	Barkel Platform	12.96	66.27	0.11822	0.511197	16	2.0	−8.0	3.101	3122
33420-6	WR	Plagiogranite	Rofe Glacier	7.203	44.52	0.09811	0.510754	18	2.0	−11.5	3.141	3407
33526	WR	Grt-Bt plagiogneiss	Rofe Glacier	2.349	11.32	0.12583	0.511341	19	2.0	−7.1	3.121	3052
33504	WR	Bt plagiogneiss	Lines Ridge	8.463	39.96	0.12843	0.511042	12	2.0	−26.80	3.740	3584
33507-1	WR	Granite-gneiss	Lines Ridge	1.948	8.122	0.14544	0.511235	21	0.5	−24.12	4.239	3232
33245-2	WR	Bt orthogneiss	McIntyre Bluff	23.94	129.2	0.11236	0.511080	12	2.0	−8.8	3.096	3185
33536	WR	Bt augen gneiss	Harbour Bluff	12.6	39.78	0.19214	0.512055	19	2.0	−10.25	7.616	3306
Ruker Terrane												
35621-1	WR	Granophyre gabbro	Mt Ruker	0.722	3.141	0.13939	0.511109	18	3.0	−7.73	4.151	3913
35621-1	Px	Granophyre gabbro	Mt Ruker	0.384	1.233	0.18846	0.511897	23				
35621-1	Pl	Granophyre gabbro	Mt Ruker	0.426	2.629	0.09808	0.510488	22				
35634-5	WR	Dolerite	Mt Ruker	6.573	26.72	0.14917	0.511315	18	3.0	−7.49	4.296	3893
35634-3	WR	Dolerite	Mt Ruker	3.981	14.64	0.16495	0.511574	20	3.0	−8.55	4.878	3979
35512-10	WR	Metagabbro-dolerite	Mt Ruker	2.159	8.884	0.14688	0.510755	17	3.0	−17.61	5.395	4711
35512-10	Pl	Metagabbro-dolerite	Mt Ruker	3.101	17.64	0.10658	0.509973	17				

Table 2.7-1. *Continued*

Sample		Rock	Area	[Sm]	[Nd]	$^{147}Sm/^{144}Nd$	$^{143}Nd/^{144}Nd$	2σ	T (Ga)	ε_T	T_{DM}	T_{DM2}
35512-10	Bt	Metagabbro-dolerite	Mt Ruker	2.248	6.058	0.19480	0.511656	21				
35618-8	WR	Metagabbro-dolerite	Mt Ruker	3.147	11.03	0.17246	0.511235	21	3.0	−18.14	6.960	4753
35619-5	WR	Metagabbro-dolerite	Mt Ruker	1.353	5.378	0.15212	0.510859	19	3.0	−17.61	5.597	4711
35841	WR	Green schist	Mt Ruker	6.819	25.35	0.16264	0.511572	15	3.0	−7.69	4.666	3909
35840-2	WR	Green schist	Mt Ruker	6.671	25.93	0.15551	0.511446	15	3.0	−7.39	4.423	3885
35518-1	WR	Green schist	Mt Ruker	3.043	15.73	0.11697	0.510698	16	3.0	−7.08	3.833	3860
35510-15	WR	Green schist	Mt Ruker	2.991	11.78	0.15356	0.511407	16	3,0	−7.40	4.378	3885
35158-17	WR	Felsic schist	Mt Ruker	2.76	15.41	0.10846	0.511104	13	3.0	4.21	2.948	2942
35517-8	WR	Felsic schist	Mt Ruker	2.08	11.95	0.10511	0.511043	9	3.0	4.32	2.943	2933
35517-1	WR	Felsic schist	Mt Ruker	1.15	7.92	0.08768	0.510797	13	3.0	6.27	2.832	2774
35315-2	WR	Felsic schist	Mt Ruker	1.59	6.63	0.14530	0.511041	18	3.0	−11.37	4.652	4207
35158-16	WR	Felsic schist	Mt Ruker	2.77	14.57	0.11475	0.511003	18	3.0	−0.22	3.287	3303
35306-1	WR	Granite	Mt Ruker	5.9	31.01	0.11505	0.510931	8	3.0	−1.75	3.406	3428
35519-1	WR	Granite	Mt Ruker	5.5	31.7	0.10492	0.510746	5	3.0	−1.44	3.347	3403
35305-1	WR	Granite	Mt Ruker	5.36	28.31	0.11438	0.511013	14	3.0	0.12	3.260	3275
35304-2	WR	Granite	Mt Ruker	3.13	16.99	0.11137	0.510874	12	3.0	−1.44	3.368	3402
33013-8	WR	Bt-Hbl gneiss	S. Mawson Escarp.	10.48	38.41	0.16548	0.511610	15	3.0	−8.86	4.820	3938
33014-1	WR	Bt-Hbl gneiss	S. Mawson Escarp.	8.74	42.32	0.12482	0.511154	6	3.0	−1.18	3.401	3381
33014-3	WR	Bt-Hbl gneiss	S. Mawson Escarp.	13.64	61.47	0.13411	0.511369	4	3.0	−0.57	3.390	3331
33014-6	WR	Bt-Hbl gneiss	S. Mawson Escarp.	13.96	59.93	0.14082	0.511486	7	3.0	−0.88	3.459	3357
33018-1	WR	Bt-Hbl gneiss	S. Mawson Escarp.	23.33	94.83	0.14874	0.511655	10	3.0	−0.64	3.487	3337
33024-1	WR	Bt-Hbl gneiss	S. Mawson Escarp.	12.62	62.59	0.12178	0.511078	6	3.0	−1.49	3.414	3406
33025-1	WR	Bt-Hbl gneiss	S. Mawson Escarp.	12.92	59.94	0.13035	0.511283	6	3.0	−0.79	3.393	3350
33025-8k	WR	Bt-Hbl gneiss	S. Mawson Escarp.	7.18	35.12	0.12362	0.511139	8	3.0	−1.00	3.381	3367
33028-1	WR	Bt-Hbl gneiss	S. Mawson Escarp.	7.03	31.33	0.13573	0.511438	7	3.0	0.16	3.327	3272
33025-2	WR	Bt gneiss	S. Mawson Escarp.	6.09	20.73	0.17752	0.512286	8	3.0	0.55	3.623	3240
33027-3	WR	Leucogneiss	S. Mawson Escarp.	2.76	14.21	0.11729	0.511000	6	3.0	−1.27	3.377	3388
33025-4	WR	Pegmatie	S. Mawson Escarp.	5.98	31.83	0.11357	0.511004	11	3.0	0.26	3.247	3264
33027-5	WR	Banded gneiss	S. Mawson Escarp.	5.99	39.72	0.09119	0.510034	4	3.0	−10.09	3.845	4104
33027-6	WR	Banded gneiss	S. Mawson Escarp.	7.67	37.86	0.12244	0.511427	10	3.0	5.12	2.865	2868
33129-6	WR	Bt-Hbl plagiogneiss	S. Mawson Escarp.	3.229	16.3	0.11973	0.510869	12	3.0	−6.81	3.673	3675
33111	WR	Bt plagiogneiss	S. Mawson Escarp.	2.25	12.81	0.10620	0.510714	9	3.0	−2.57	3.431	3494
33111-4	WR	Bt plagiogneiss	S. Mawson Escarp.	1.295	7.483	0.10458	0.510670	8	3.0	−2.81	3.441	3513
33111-1	WR	Opx plagiogneiss	S. Mawson Escarp.	5.12	20.75	0.14930	0.511543	65	3.0	−3.06	3.777	3534
33111-1	Opx	Opx plagiogneiss	S. Mawson Escarp.	9.55	30.07	0.19201	0.512101	13				
33111-1	Pl	Opx plagiogneiss	S. Mawson Escarp.	0.342	2.22	0.09293	0.510871	7				
33111-1	KFsp	Opx plagiogneiss	S. Mawson Escarp.	0.673	3.15	0.12901	0.511480	16				

$T_{DM} = (1/\lambda)\ln\{1 + [(^{143}Nd/^{144}Nd)_{sam.} - (^{143}Nd/^{144}Nd)_{DM}]/[(^{147}Sm/^{144}Nd)_{sam.} - (^{147}Sm/^{144}Nd)_{DM}]\}$, where $^{143}Nd/^{144}Nd_{DM} = 0.513151$, $^{147}Sm/^{144}Nd_{DM} = 0.2136$; $\lambda^{147}Sm = 6\,54 \times 10^{-12}\,y^{-1}$; .
$T_{DM2} = (1/\lambda)\ln\{1 + [(^{143}Nd/^{144}Nd)_{sam.} - (e^{\lambda T} - 1) \times (^{147}Sm/^{144}Nd_{sam.} - {}^{147}Sm/^{144}Nd_C) - (^{143}Nd/^{144}Nd)_{DM}]/[(^{147}Sm/^{144}Nd)_{sam.} - (^{147}Sm/^{144}Nd)_C]\}$, where $^{147}Sm/^{144}Nd_C = 0.12$.

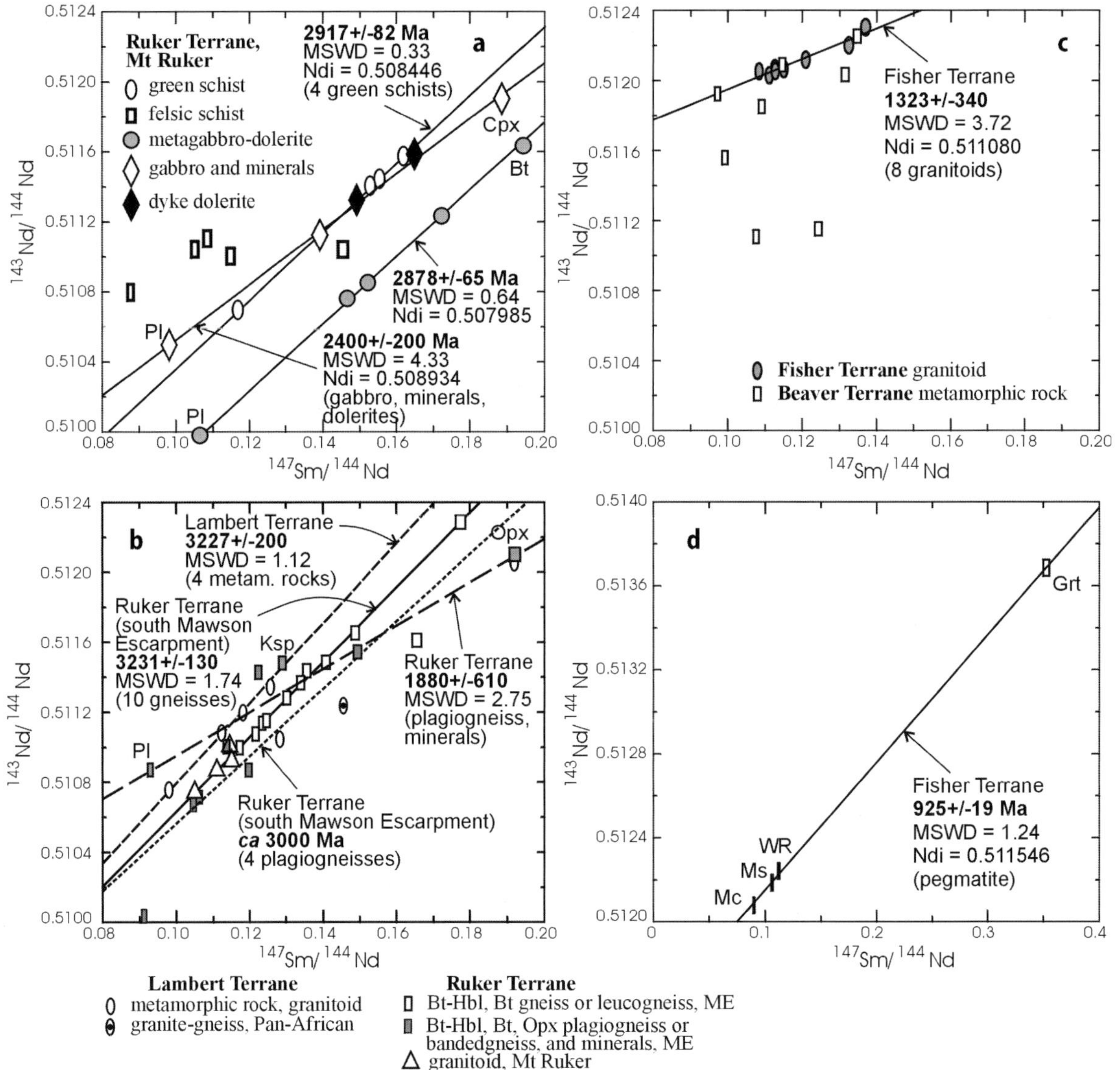

Fig. 2.7-2. Sm-Nd isotope diagrams: **a** metamorphic and mafic plutonic rocks, Mt. Ruker; **b** metamorphic rocks and granitoids from Lambert Terrane and Ruker Terrane (*ME:* Mawson Escarpment); **c** granitoids from Fisher Terrane, and metamorphic rocks from Beaver Terrane; **d** pegmatite from Fisher Terrane (Mt. Willing). Symbols not to uncertainty scale

ca. 1 900 Ma reference line, this thermal event might have caused crystallization of orthopyroxene (incipient charnockitization?). Four (out of six) metamorphic rocks of the Lambert Terrane define a poorly constrained reference line corresponding to the age of ca. 3 200 ±200 Ma (Fig. 2.7-2b), but occurrence of two divergent samples puts serious doubt on the significance of this reference line which may be an artifact. Inclusion of two other low-$^{143}Nd/^{144}Nd$ samples (representing Clemence Massif, and Else Platform) which were collected from the Beaver Terrane but do not follow the relevant isochron (Fig. 2.7-2c) would dramatically change the reference line to ca. 2 000 Ga. Apparently the Sm-Nd data are not sufficient for the Lambert Terrane isochron age constraints.

A post-tectonic Grt-Ms-bearing pegmatite vein from Mt. Willing (Fisher Terrane; sample 39115-7) was dated by the Sm-Nd system. A four-point isochron (whole rock, garnet, muscovite, microcline) gives an age of 925 ±19 Ma (ε_{Nd} = +2.0, MSWD = 1.2; Fig. 2.7-2d).

Whole rock-mineral separate (plagioclase and two clinopyroxene fractions) Sm-Nd dating of a fresh ESE-WNW trending mafic dyke from Mt. Willing (Fisher Terrane) gave an age of 845 ±66 Ma (ε_{Nd} = +2.2, MSWD = 0.66; A. Laiba, unpublished data).

We obtained Pb isotopic data (Table 2.7-2) on plagioclase from orthogneisses from the southern and northern Mawson Escarpment (Fig. 2.7-1), and from a thin pretectonic granitic sheet (granite-gneiss) in the

Table 2.7-2. Lead isotope composition of plagioclase from metamorphic rocks of Mawson Escarpment (ME)

Sample	Rock type	Locality	$^{206}Pb/^{204}Pb$	$^{207}Pb/^{204}Pb$	$^{208}Pb/^{204}Pb$
33504	Bt granite-gneiss	North ME	14.047	13.413	34.423
33507-1	Pre-tectonic granite-gneiss	North ME	18.663	16.189	39.612
33536	Bt granite-gneiss	Central ME	14.033	13.405	34.411
33013-8	Bt-Hbl granite-gneiss	South ME	14.224	13.519	34.589
33129-6	Bt plagiogneiss	South ME	13.725	13.210	34.112

Table 2.7-3. Zircon ion-microprobe ages; for further details see Mikhalsky et al. (in press)

Rock	Age (Ma)	Interpretation
Ruker Terrane		
Granite cobble in Mt. Rubin sediments	3392 ±6 626 ±51	Crystallization Thermal event
Plagiogneiss, south. Mawson Esc.	3377 ±9 3145	Crystallization Thermal event
Hbl-Bt granite, south Mawson Esc.	3180 ±9 3110 587 ±170	Crystallization Metamorphism Thermal event
Lambert Terrane		
Orthogneiss, northern Mawson Esc.	2423 ±18	Crystallization
Orthogneiss, central Mawson Esc.	2060 ±202 330	Metamorphism Inheritance
Pre-tectonic granite-gneiss, northern Mawson Esc.	528 ±6	Crystallization
Syn-tectonic pegmatite, Lawrence Hills	495 ±18 142 ±28	Crystallization Thermal event(?)

northern Mawson Escarpment (Lines Ridge). Plagioclase from all the orthogneisses reveals consistently low lead isotopic ratios ($^{206}Pb/^{204}Pb$ = 13.72–14.22; $^{207}Pb/^{204}Pb$ = 13.21–13.52), while the granite plagioclase has much higher ratios (18.66 and 16.19). These data indicate that the granite originated much later than the granite-gneiss, probably in Pan-African time.

Seven samples were analysed for zircon U-Pb isotope ratios. The measurements were carried out with a SHRIMP-II ion microprobe at the Centre of Isotopic Research (VSEGEI, St. Petersburg, Russia). Five to ten zircon grains from each sample were studied. Each analysis consisted of five scans through the mass range, the spot diameter was about 18 μm and the primary beam intensity about 4 nA. The Pb/U ratios have been normalized relative to a value of 0.0668 for the $^{206}Pb/^{238}U$ ratio of the TEMORA reference zircons, equivalent to an age of 416.75 Ma (Black and Kamo 2003). The results are summarized in Table 2.7-3 (for details see Mikhalsky et al. in press).

Zircon grains in all samples but one have prismatic to long prismatic shape suggestive of magmatic origin, those from the Lambert Terrane are mostly somewhat rounded, and zircons from the Ruker Terrane exhibit thin oscillatory zoning. Zircon grains in sample 33536 are isometric, and most are rounded, which suggests metamorphic origin. Visual cores are rare. Cathodoluminescent images usually show high-U irregular areas which tend to outer grain parts. Nevertheless high-U, and low-U areas, as well as cores, inner and outer parts are indistinguishable in terms of U-Pb isotopic ratios. Most analyses are concordant or nearly concordant.

These dates are in good agreement with those obtained by Boger et al. (2001), who reported granite-gneiss emplacement in Ruker Terrane at 3170 Ma, with an inherited age of 3370 Ma.

In the following sections we outline the main geological features of the PCM from earlier work, as well as considering the new and recently published geochronological data, and the geological observations made by one of the authors (E. V. M.) during the 2002/2003 PCMEGA expedition.

Ruker Terrane

The Ruker Terrane comprises low- to medium-grade metasediments, minor metavolcanics, and granite to granite-gneiss locally cut by abundant variously deformed and metamorphosed mafic dykes and sills of two or three generations.

The rocks from the Ruker Terrane have highly varied $T(DM2)$ values, but most lie between 3 900 and 3 200 Ma, thus reflecting the most ancient protolith ages. However, some felsic schists from Mt. Ruker have lower $T(DM2)$ ages of <3 000 Ma, providing evidence for either source heterogeneity, or addition of juvenile material during the late Archaean. Much more isotopic data are needed to clarify the geological history of the Ruker Terrane.

The rocks of the Ruker Terrane appear to constitute a granite-gneiss basement and one or more supracrustal cover sequences. Presumably cover sequences include the medium-grade Menzies Series of quartzitic to pelitic composition – cut by a 2 580 Ma pegmatite (Tingey 1991) – low-grade metasediments, metavolcanics and BIF of the Ruker Series, and low-grade clastic metasediments of the Sodruzhestvo Series. The latter was long believed to be of Meso- to Neoproterozoic age, based on acritarch studies (Iltchenko 1972) and the apparent lack of presumed Mesoproterozoic mafic dykes, abundant throughout most of the SPCM (Tingey 1991). The chemical composition of these dykes suggested a correlation with Mesoproterozoic mafic dykes in Enderby Land and the Vestfold Hills, which allowed these dykes to be used as a stratigraphic marker. However, during the PCMEGA field work, mafic bodies were found to intrude the Sodruzhestvo Series at Cumpston Massif. But, until the age of these mafic bodies is determined, the age of the Sodruzhestvo Series remains uncertain, and well may be either Neoarchaean or Neoproterozoic. We doubt the validity of using mafic dykes uncorroborated as a stratigraphic marker anyway, and the geological composition and structure rather point to its relatively young age.

The crystalline basement of the Ruker Terrane includes granite and granite-gneiss (Mawson Orthogneiss); it has been suggested that the Menzies Series rocks together with the Mawson Orthogneiss represent a single complex infracrustal assemblage (Kamenev et al. 1990). Two rock associations were distinguished within the Mawson Orthogneiss on compositional grounds (Mikhalsky et al. 2001): *(a)* Y-depleted tonalitic to trondhjemitic orthogneiss similar to the tonalite-trondhjemite-granodiorite (TTG) associations, thought to represent new sialic crust formed by hydrous partial melting of a garnet-and/or amphibole-bearing mafic source, and *(b)* Y-undepleted granite-gneisses enriched in many large-ion and high-field-strength elements, suggesting affinities with A-type granites. TTG-like association seems to be restricted to the south-eastern part of the exposed Ruker Terrane (southern Mawson Escarpment, Cumpston Massif, and Blake Nunataks). The A-type granite-gneisses are more widespread.

Tingey (1982) distinguished three metamorphic events in this area. Granitic basement rocks show evidence for an early (presumably Archaean) high-grade event since they are commonly migmatitic. There is possible petrographic evidence for granulite-facies conditions during early metamorphism: mafic granulite crops out at Mt. Newton, and orthopyroxene-bearing tonalitic gneiss in the southern Mawson Escarpment. It is unclear whether the latter represents an earlier metamorphic or magmatic event, or subsequent charnockitization. These orthopyroxene-bearing rocks do have some primary magmatic textures, but their field appearance clearly indicates subsequent fluid infiltration (charnockitization?).

Later metamorphic events were attributed by Tingey (1982) to various amphibolite to greenschist facies events of Mesoproterozoic to Early Palaeozoic age, but this interpretation is not generally confirmed by recent isotopic studies.

The geological history of the Ruker Terrane may go back to ca. 3 395 Ma (age of a granite cobble in the Sodruzhestvo Series; Mikhalsky et al. in press), i.e., the initial geological processes were Early Archaean, which is also indicated by Sm-Nd model ages (up to 3.9–3.8 Ga). Emplacement of trondhjemite, likely derived by melting of mafic rocks, occurred at ca. 3 380 Ma and granite at ca. 3 180–3 170 Ma (Boger et al. 2001; Mikhalsky et al. in press). The minimum age of subsequent deformation is constrained by an age of ca. 2 645 Ma for an undeformed pegmatite (Boger et al. 2001), and might be approximated by a Sm-Nd isochron age of ca. 3 170 Ma and thermal events recorded by SHRIMP zircon ages of 3 145, and 3 110 Ma.

Biotite granite emplacement at Mt. Ruker was dated at 3 005 ±57 Ma (conventional zircon studies, Mikhalsky et al. 2001). This granite is of near-minimum-melt composition. These granites have quite strongly fractionated REE patterns with negative Eu anomalies (Mikhalsky et al. 2001), consistent with melting of felsic crustal rocks and plagioclase fractionation. The ca. 3 000 Ma granite plutons are more fractionated than most felsic orthogneisses from the Ruker Terrane, but their geochemistry suggests they were derived by partial melting of such rocks (Mikhalsky et al. 2001).

Zircons from Ruker Series metamorphic rocks have a wide range of isotopic ratios indicating a number of thermal events between approximately 3 175 and 2 500 Ma with a major Pb loss at ca. 500 Ma (Mikhalsky et al. 2001).

Metamorphism at Mt. Ruker occurred at ca. 2 900 Ma as evidenced by our Sm-Nd data, which gives a reasonably lower age limit for the Ruker Series, which is probably Mesoarchaean (TDM 3.3–2.9 Ga).

Minor gabbro and dolerite emplacement at Mt. Ruker is dated at 2 365 ±65 Ma (Sm-Nd, this study). The age of these rocks roughly corresponds to the Palaeoproterozoic Rb-Sr ages of high-Mg dykes in the Vestfold Hills and Enderby Land (Collerson and Sheraton 1986; Sheraton and Black 1981). However, none of the analysed Mt. Ruker rocks are of high-Mg composition, although a number of undated high-Mg ultramafic dykes have been reported from this area (Mikhalsky et al. 2001).

Sm-Nd data on Opx-bearing tonalitic gneiss from the southern Mawson Escarpment show that a thermal event occurred at about 1 900 Ma, although the nature of this event is unclear (we believe fluid infiltration was involved).

The earliest mafic dykes in the Mawson Escarpment are Neoarchaean (pre-2 650 Ma, Boger et al. 2001), while many others apparently post-date the ca. 1 900 Ma fluid infiltration event, and thus may well be Mesoproterozoic.

An important feature of the Ruker Terrane is the presence of a tectonic melange of metamorphosed ultramafic rocks, including peridotites, in shear zones in the southern Mawson Escarpment, as well as similar, but much less deformed rocks at Mt. Ruker. These rocks have tholeiite to basaltic komatiite (Mt. Ruker) or peridotitic komatiite (Mawson Escarpment) compositions. The Mawson Escarpment rocks have non-chondritic major and trace element ratios (CaO/TiO_2, CaO/Al_2O_3, Zr/Y, Ti/Zr, etc.), which may be attributed to cumulus processes, and therefore it is possible that they are of intrusive rather than volcanic origin (Mikhalsky et al. 2001). Whatever their origin, these rocks may mark some Archaean sutures. Their age at Mt. Ruker has already been shown to be Archaean (as they are apparently pre-metamorphic, i.e. pre-2 900 Ma).

Beaver and Fisher Terranes

The Beaver Terrane is a long-lived polymetamorphic and polydeformational mobile belt, active between 1 150 and 900 Ma. Peak metamorphic conditions and maximum deformation were attained at ca. 990 Ma, the waning stages lasting until ca. 940 to 900 Ma (Boger et al. 2000). A thick tectonic slice of serpentinite in the Radok Lake area (a protrusion?), probably co-eval with D_4 deformation and thrusting dated at ca. 900 Ma by Boger et al. (2000), gave an age of 1 165 ±13 Ma (conventional U-Pb monazite analysis, Mikhalsky et al. 2001), which reflects a Mesoproterozoic event of unknown nature about 150 Ma before the metamorphic peak in the Beaver Terrane.

Typical Beaver Terrane lithologies in the NPCM are high-grade paragneiss and felsic to mafic orthogneiss, granite gneiss, minor metagabbro, and late-tectonic orthopyroxene granitoids. Igneous protoliths of many granulites may have been produced in a subduction-zone environment (Munksgaard et al. 1992). The rocks of the Beaver Terrane have *T*(DM2) ages mostly in the range 2.3–1.6 Ga, indicating that crust-forming processes occurred in the Palaeo- to Mesoproterozoic. One much older model age (3.2 Ga) for an orthogneiss from Else Platform suggests the involvement of a more ancient source in that area. This area in fact is located within the Lambert Glacier rift system and may represent a boundary between, or comprises a collage of, various crustal blocks or sub-terranes.

The southern, though not marginal, part of the Beaver Terrane consists of a distinctive rock association of low- to medium-grade metamorphic rocks (metavolcanics), minor metasediments, granite and gabbro (the Fisher Terrane). The Fisher Terrane rocks have DM2 model ages of 1.7–1.4 Ga (Mikhalsky et al. 2001, and references therein), which are similar to model ages obtained for the Beaver Terrane, although they tend to be somewhat younger. However, the mantle source of these rocks was probably not typical depleted mantle, having lower $\varepsilon_{Nd} = +2$ to $+4$ at 1.3 Ga (Mikhalsky et al. 1996). The mantle source composition of the Beaver Terrane rocks cannot easily be evaluated at the present stage of this study.

The Fisher Terrane is largely composed of mafic to felsic schist and gneiss of apparent calc-alkaline affinity. Metabasalts plot within the island-arc or volcanic-arc fields in geochemical discrimination diagrams. Most metavolcanics, especially basaltic andesite and andesite show a prominent negative Nb anomaly on a primitive mantle-normalised spidergram. Granitoids also plot in volcanic-arc fields in discrimination diagrams (Mikhalsky et al. 2001, and references therein). These features point to formation of the Fisher Terrane rocks in a subduction-related continental-margin environment. Volcanic activity was dated at ca. 1 300 Ma by both conventional (Beliatsky et al. 1994) and SHRIMP (Kinny et al. 1997) zircon studies. Emplacement of tonalite, granite, and gabbro intrusives ceased by about 1 190 Ma, lower amphibolite-facies metamorphism occurred at ca. 1 100 Ma, and late granitoids were emplaced at ca. 1 020 Ma (Mikhalsky et al. 2001, and references therein). Most of these ages do not correlate with known events in the Beaver Terrane. However, the youngest pegmatite vein in the Fisher Terrane (925 ±19 Ma, this study) is co-eval with granitic dykes in the Beaver Terrane (930 ±14, Boger et al. 2000; 910 ±18, Carson et al. 2000). Such dykes mark the waning stages of deformation in the Beaver Terrane, whereas deformation probably ceased by ca. 1 000 Ma in the Fisher Terrane, although thermal or metamorphic event at ca. 810 Ma is recorded by zircon growth in metagabbro.

The relationship between the Fisher and Beaver Terranes remain the most enigmatic feature of PCM geology. The Fisher Terrane apparently escaped the early Neoproterozoic high-grade metamorphic event which affected areas both north and south of it. However, only Rb-Sr age data are available for the latter area and modern geochronological studies are needed. Until such data are available, a correlation between this area and the rest of the NPCM must remain uncertain. It is possible that this area is allochthonous. On the other hand, a correlation with the Palaeoproterozoic Lambert Terrane (see below) is also possible.

It should be noted that a single zircon fraction from probably tuffaceous felsic schist from the Fisher Massif yielded a discordant $^{207}Pb/^{206}Pb$ age of ca. 2 560 Ma

(Beliatsky et al. 1994), providing evidence for an Archaean component in the sediment source region.

A few occurrences of amphibolite-facies rocks in the NPCM (e.g., Radok Lake area, Boger et al. 2000; Mt. Meredith, Kinny et al. 1997) support the idea of a more widespread occurrence of Fisher Terrane or Lambert Terrane rocks.

A large amount of the Beaver and Fisher Terranes probably originated in the Palaeo(?)- and Mesoproterozoic as accretional complexes formed by collision of magmatic arcs and back-arc basins prior to final collision between the large lithospheric blocks. One of these is represented by the Ruker Terrane and Vestfold Hills and the other comprising Enderby Land and India. The Mesoproterozoic orogen of the NPCM (and probably of a wider areas) is thus probably of accretional, rather than collisional, type.

A notable feature of the Fisher Terrane is the presence of roughly east-west-trending mafic dykes at Mt. Willing (845 ±66 Ma, A. Laiba, unpublished data). These dykes are the only known Neoproterozoic mafic igneous rocks in the region, if not in the whole of East Antarctica, and may be a manifestation of Rodinia break-up. Dyke swarms of broadly similar age (ca. 830 Ma) are known from South China (Li 1999) and central Australia (Wingate 1998), although the Mt. Willing dykes have different chemical composition (A. Laiba, unpublished data).

Lambert Terrane

The area of the central and northern Mawson Escarpment, Clemence Massif, and some outcrops on the western flank of the Lambert Glacier (Mt. Johns, Shaw Massif, and a few smaller nunataks) was distinguished by Kamenev et al. (1993) as the Lambert Terrane. These authors also included Mt. Meredith and nearby outcrops into this terrane which has long been a debatable tectonic province. This area is underlain by high-grade orthogneiss (mostly biotite-quartz-feldspar gneiss) and paragneiss, which differs from the Ruker Terrane rocks in by being generally more strongly layered, containing significant garnet, and generally lacking hornblende, allanite and titanite (Mikhalsky et al. 2001, and references therein). Prominent rock types are metapelite, metapsammite, calc-silicate, and marble. Significant amounts of weakly deformed mafic rocks crop out in the Rofe Glacier area, and ultramafic bodies (probably thrusted tectonic slabs of orthopyroxenite and amphibolite) occur in some localities, suggesting the presence of a suture zone or other major crustal discontinuity. Our U-Pb data argue for a Pan-African age of this tectonic activity, which may have involved folding and thrusting of a mafic to ultramafic underplate or possibly oceanic crust, into granitic basement.

Metamorphic grade in the Lambert Terrane is generally lower than in the Beaver Terrane, but granulite-facies assemblages are known from the very northern part of Mawson Escarpment and Clemence Massif (Mikhalsky et al. 2001). However, the detail geological and geochronological data for Clemence Massif are yet lacking and its tectonic position is not clear. We favour its inclusion into the Beaver Terrane at the present stage of study on the basis of Grenville-age Rb-Sr dating by Tingey (1982), although available single Sm-Nd model *T*(DM) age of 3 188 Ma implies the presence of an ancient source region, which is not typical for the Beaver Terrane.

In spite of its somewhat different lithological composition the whole area was tentatively included in the Beaver-Lambert Terrane (Mikhalsky et al. 2001) on the basis of very similar Rb-Sr ages. However, recent geochronological work by Boger et al. (2001) and our new data demonstrate that this area more likely represents a separate terrane with its own distinct geological history. Apart from containing abundant late- to post-tectonic granites of Pan-African age (550–490 Ma, Boger et al. 2001), the zircon populations in these granites as well as in the basement orthogneisses, reveal signatures of widespread Palaeoproterozoic events (ca. 2 420, 2 330, 2 150, 2 065, 1 850–1 600 Ma), which have not been detected in either Ruker or Beaver Terranes. It is significant that no isotopic evidence for Grenville-age events have yet been found in this area. We conclude from our new data that at least the ca. 2 420, and 2 060 Ma ages reflect geologically meaningful events (orthogneiss emplacement and metamorphism, respectively), but much further work is needed to fully constrain the geological history of this area.

Sm-Nd model ages (3 400–3 000 Ma) of the Lambert Terrane rocks are much higher than those calculated for the Beaver Terrane (2 300–1 600 Ma), but tend to be somewhat lower than those of the Ruker Terrane (>3 200 Ma). These data provide evidence that the Lambert Terrane protoliths have a much longer crustal history than, and may not correlate with, the Beaver Terrane. Nevertheless, some of the Palaeoproterozoic ages from the Lambert Terrane correspond to reported ages of zircon growth at ca. 2 100–1 800 Ma in a paragneiss from Mt. Meredith (Kinny et al. 1997). That suggests that Mt. Meredith may represent a separate block of the Lambert Terrane.[1]

Correlation with Ruker Terrane appears difficult, as the difference between their Sm-Nd model ages is not that striking, although it is just possible that this is due to insufficient sampling of the Ruker Terrane. It is noteworthy that the spectacular mafic dyke swarm in the southern Mawson Escarpment apparently terminates at the boundary of the Ruker and Lambert Terranes, established by abrupt lithological and structural changes there (Kamenev

[1] The U-Pb SHRIMP data quite recently obtained on zirkons from granitic rocks from Mt. Meredith revealed crystallization ages at c. 1 300–1 100 Ma and c. 500 Ma. These data demonstrate the striking age similarity with the Fisher Terrane (Mesoproterozoic), and the Lambert Terrane (Cambrian).

et al. 1990; Boger et al. 2001). The dyke swarm may have its northward continuation in the Manning Glacier area, where undated north-south trending metamorphosed mafic dykes are abundant.

However, further north, in the Rofe Glacier area, the Barkell Platform, Lines Ridge, and Lawrence Hills, mafic dykes are generally lacking, except as rare thin syn-tectonic bodies.

Discussion

The proposed four tectonic divisions (the Ruker, Lambert, Fisher, and Beaver Terranes) of the PCM seem to fit the available isotopic data. At the same time, the Fisher and Beaver Terranes are of similar origin and have similar geochemical features, and perhaps should be considered a single tectonic entity.

The SPCM, which consists of the Ruker and Lambert Terranes, are underlain by material with a long crustal history, starting mostly in the Mesoarchaean. This supports early ideas (Kamenev et al. 1993) that the Lambert Terrane represents a reworked early Precambrian assemblage (but not necessaryly the Ruker Terrane) rather than juvenile Proterozoic crust. However, it is still possible that some Proterozoic mantle-derived material is present in the Lambert Terrane. The mafic-ultramafic bodies in the Rofe Glacier area, and Lawrence Hills, although they have not yet been dated, may be examples of mantle derivates or oceanic crust.

Some rocks cropping out along the eastern flank of the Amery Ice Shelf (Reinbolt Hills, Landing Bluff) also have Archaean Sm-Nd model ages (3.15–3.0 Ga; Beliatsky, unpublished data). In contrast, the rocks of the Beaver and Fisher Terranes on the western flank of the Lambert-Amery Ice Shelf have (with few exceptions in Else Platform, and Clemence Massif) Proterozoic model ages. These data lend credence to a suggestion that the Lambert-Amery rift system may run alongside, rather than at right angles to, a major crustal (lithospheric?) discontinuity marked by abrupt chance in Sm-Nd model ages (>3.0 Ga on the eastern flank and <2.3 Ga on the western Flank of the Lambert-Amery Ice Shelf rift system). However, sampling on the eastern flank of the Amery Ice Shelf is very sporadic, and Mesoproterozoic structures and juvenile rock associations may well run across the Lambert Glacier having not been detected yet on its eastern flank.

In the Lambert Terrane, many zircon growth events (2 400, 2 150, 2 060, 1 800, 1 600 Ma, Boger et al. 2001; Mikhalsky et al. in press) support the suggestion that the Lambert Terrane experienced a long and complex Palaeoproterozoic tectonic history. Granite emplacement and high-grade metamorphism (the latter not yet dated) provide evidence for the orogenic nature of at least some of these events. Thus we propose that the Lambert Terrane is a Palaeoproterozoic orogen, although much more work needs to be done to evaluate its origin and geological history.

Palaeoproterozoic tectonic activity is not well documented in Antarctica, except within the ca. 1.7 Ga Mawson Block (Fanning et al. 1996). This landmass is thought to comprise the Gawler Craton of South Australia, George V Land, Adélie Land, and the central Transantarctic Mountains. Its geological history includes various events dated at ca. 3 150–2 950, 2 700–2 350, 2 000, and 1 850–1 700 Ma (Fitzsimons 2003, and references therein). These ages roughly correspond to those obtained for the Lambert Terrane, thus suggesting that this part of the SPCM can be correlated with the Mawson Block. However, the Ruker Terrane rocks do not seem to correlate well with the Mawson Block, being about 200 Ma older and having older Sm-Nd model ages. Fitzsimons (2003) concluded that one or more Mesoproterozoic sutures may lie between the Mawson Block and areas west of ~100° E (Bunger Hills), thus separating Antarctica into at least two major tectonic domains. Taking into account that the Ruker and Lambert Terranes are bounded to the north by the Mesoproterozoic Fisher Terrane which may partly correlate with the Fraser Complex of the Albany-Fraser Orogen, their correlation with the Mawson Block may be a reasonable suggestion. On the other hand, there is a large sub-glacial gap between the PCM and the Mawson Block, so any correlation can only be speculative. Otherwise a tentative correlation somewhat older West Australia Cratons and (or?) the Palaeoproterozoic Capricorn Orogen with zircon ages of 2 550–2 450, 2 000–1 960, 1 830–1 780, and 1 670–1 620 Ma (Cawood and Tyler 2004) might be proposed.

At the present stage of the study it is hardly possible to determine whether the Lambert Terrane assemblages originated as a distinct complex accreted onto the adjacent Ruker Terrane, and was stabilised by the Mesoproterozoic, or whether it represents a Palaeoproterozoic portion of a long-lived polycyclic mobile belt (Kamenev et al. 1990), serving as a Mesoproterozoic hinterland for the Beaver and Fisher Terranes. Perhaps the Lambert Terrane was bordered by a sinistral transcurrent zone, which was later inherited by the Lambert-Amery rift system and which separated the older crust (Lambert and Ruker Terranes) and younger crust-forming structures (e.g., Fisher Terrane) accreting southwards at a very low-angle to the present Mawson Escarpment, thus leaving it generally unaffected.

The structural position of the Fisher Terrane remains enigmatic due to inadequate knowledge of the geology of the area immediately to the south. It was suggested that the Fisher Terrane accreted onto an older crustal block (Mikhalsky et al. 2001). However, if the proposed correlation between Mt. Meredith and the Lambert Terrane is correct, then the Fisher Terrane would cut across the latter, which might imply an aulocogen origin of the Fisher Terrane. The geographic position of the Fisher Terrane right along the strike of Mesoproterozoic dyke swarms in the Vestfold Hills, if not just coincidence, could be consistent with such speculation. However, rock compositions in the Fisher Terrane are not consistent with this scenario, which would exclude any oceanic opening, and hence Pan-African suturing, since the Neoproterozoic.

Another striking feature of the SPCM is the widespread emplacement of Pan-African granite bodies, as well as prominent deformation of the similar age which caused folding and brought ultramafic material to upper crust. Pan-African processes are widely regarded as the consequence of continental collision (e.g., Boger et al. 2001; Fitzsimons 2003), but we believe that mafic underplate might have been an important tectonic factor as well.

Muscovite-biotite pegmatite and granite veins and dykes show syn- to post-emplacement deformation, providing evidence for extensive tectonic activity during magmatism, which puts doubt on the post-tectonic or anorogenic origin for these rocks advocated by the early researchers (Tingey 1991). It is noteworthy that Pan-African granitic rocks in the SPCM differ in many geochemical respects from otherwise similar rocks in the NPCM, being higher in Al_2O_3 and Na_2O, and lower in light REE, U, and Th (Mikhalsky et al. 2001). These features indicate derivation from different crustal sources.

The east-west-trending Neoproterozoic mafic dykes within the Fisher Terrane may indicate a response to roughly north-south direction of Neoproterozoic extension due to Rodinia breakup. Assuming that the subsequent Pan-African compression inherited this general orientation, we believe that any Pan-African suture zone, if it exists, should trend east-west, rather than north-south.

Acknowledgments

EVM participated in the PCMEGA 2002/2003 as a guest scientist, and very much appreciates an invitation from the scientific leaders N. W. Roland (BGR, Hannover) and C. J. L. Wilson (University of Melbourne), the friendly logistic support by the ANARE team, and invaluable field assistance by the PCMEGA participants. S. A. Sergeev is thanked for carrying out the SHRIMP analyses at the Isotopic Centre (VSEGEI, St Petersburg). We thank J. Sheraton for his valuable comments on the manuscript, and N. W. Roland and an anonymous reviewer for their helpful and constructive review. H. Toms is greatly thanked for improving the language. This contribution was partly made possible by a grant from the Deutsche Forschungsgemeinschaft to EVM.

References

Beliatsky BV, Laiba AA, Mikhalsky EV (1994) U-Pb zircon age of the metavolcanic rocks of Fisher Massif (Prince Charles Mountains, East Antarctica). Antarctic Sci 6:355–358

Beliatsky BV, Mikhalsky EV, Sergeev SA (in prep.) Zircon U-Pb SHRIMP data for the Prince Charles Mountains rocks

Black LP, Kamo SL, et al. (2003) TEMORA 1: A new zircon standard for U-Pb geochronology. Chem Geol 200:155–170

Boger SD, Carson CJ, Wilson CJL, Fanning CM (2000) Neoproterozoic deformation in the Radok Lake region of the NPCM: evidence for a single protracted orogenic event. Precambrian Res 104:1–24

Boger SD, Wilson CJL, Fanning CM (2001) Early Paleozoic tectonism within the East Antarctic craton: the final suture between east and west Gondwana? Geology 29:463–466

Carson CJ, Boger SD Fanning CM, Wilson CJL, Thost D (2000) SHRIMP U-Pb geochronology from Mt. Kirkby, northern Prince Charles Mountains, East Antarctica. Antarctic Sci 12:429–442

Cawood PA, Tyler IM (2004) Assembling and reactivating the Proterozoic Capricorn Orogen: lithotectonic elements, orogenies, and significance. Precambrian Res 128:201–218

Collerson KD, Sheraton JW (1986) Age and geochemical characteristics of a mafic dyke swarm in the Archaean Vestfold Block, Antarctica: inferences about Proterozoic dyke emplacement in Gondwana. J Petrol 27:853–886

Crohn PW (1959) A contribution to the geology and glaciology of the western part of AAT. BMR Bull 52

Fanning CM, Daly SJ, Bennett VC, Menot RP, Peucat JJ, Oliver RL, Monnier O (1995) The "Mawson Block": once contiguous Archean to Proterozoic crust in the East Antarctic shield and Gawler craton, Australia. ISAES VII Abstracts volume, Siena

Fitzsimons ICW (2003) Proterozoic basement provinces of southern and southwestern Australia, and their correlation with Antarctica. In: Yoshida M, et al. (eds) Proterozoic East Gondwana: supercontinent assembly and breakup. Geol Soc London Spec Pub 206:93–130

Halpern M, Grikurov GE (1975) Rubidium-strontium data from the southern Prince Charles Mountains. Antarctic J US 10:9–15

Iltchenko LN (1972) Late Precambrian acritarchs of Antarctica. In: Adie RJ (ed) Antarctic geology and geophysics. Universitetsforlaget, Oslo, pp 599–602

Kamenev EN, Krasnikov NN (1991) The granite-greenstone terrains in the southern Prince Charles Mountains. 6th Internat Sympos Antarctic Earth Sci, Tokyo. Abstract, pp 264–268

Kamenev EN, Kameneva GI, Mikhalsky EV, Andronikov AV (1990) The Prince Charles Mountains and Mawson Escarpment. In: Ivanov VL, Kamenev EN (eds) The geology and mineral resources of Antarctica. Nedra, Moscow, pp 67–113, (in Russian)

Kamenev EN, Andronikov AV, Mikhalsky EV, Krasnikov NN, Stuewe K (1993) Soviet geological maps of the Prince Charles Mountains. Austral J Earth Sci 40:501–517

Kinny PD, Black LP, Sheraton JW (1997) Zircon U-Pb ages and geochemistry of igneous and metamorphic rocks in the northern Prince Charles Mountains. AGSO J Austral Geol Geophys 16:637–654

Kurinin RG, Grikurov GE (1982) Crustal structure of part of East Antarctica from geophysical data. In: Craddock C (ed) Antarctic geoscience. University of Wisconsin Press, Madison, pp 895–901

Li ZX, Li XH, Kinny PD, Wang J (1999) The breakup of Rodinia: did it start with a mantle plume beneath South China? Earth Planet Sci Letters 173:171–181

Mikhalsky EV, Sheraton JW, Laiba AA, Beliatsky BV (1996). Geochemistry and origin of Mesoproterozoic metavolcanic rocks from Fisher Massif. Antarctic Sci 8:85–104

Mikhalsky EV, Laiba AA, Beliatsky BV, Stüwe K (1999) Geological structure of Mount Willing (Prince Charles Mountains, East Antarctica), and some implications for metamorphic rock age and origin. Antarctic Sci 11:338–352

Mikhalsky EV, Sheraton JW, Laiba AA, et al. (2001) Geology of the Prince Charles Mountains, Antarctica. AGSO Bull 247

Mikhalsky EV, Beliatsky BV, Sheraton JW, Roland NW (in press) Two distinct Precambrian terranes in the southern Prince Charles Mountains, East Antarctica: SHRIMP dating and geochemical constraints. Gondwana Res

Munksgaard NC, Thost DE, Hensen BJ (1992) Geochemistry of Proterozoic granulites from northern Prince Charles Mountains, East Antarctica. Antarctic Sci 4:59–69

Ravich MG, Solov'ev DS, Fedorov LV (1978) Geological structure of Mac Robertson Land. Gidrometeoizdat, Leningrad, (in Russian)

Sheraton JW, Black LP (1981) Geochemistry and geochronology of Proterozoic tholeiite dykes of East Antarctica: evidence for mantle metasomatism. Contrib Mineral Petrol 78:305–317

Solov'ev DS (1972) Geological structure of the mountain fringe of the Lambert Glacier and the Amery Ice Shelf. In: Adie RJ (ed) Antarctic geology and geophysics. Universitetsforlaget, Oslo, pp 573–577

Stinear BH (1956) Preliminary report on operations from Mawson base, ANARE 1954–55. BMR Record 1956(44)

Tingey RJ (1982) The geologic evolution of the Prince Charles Mountains – an Antarctic Archean cratonic block. In: Craddock C (ed) Antarctic geoscience. University of Wisconsin Press, Madison, pp 455–464

Tingey RJ (1991) The regional geology of Archaean and Proterozoic rocks in Antarctica. In: Tingey RJ (ed) The geology of Antarctica. Clarendon Press, Oxford, pp 1–58

Wingate MTD, Campbell IH, Compston W, Gibson GM (1998) Ion microprobe U-Pb ages for Neoproterozoic basaltic magmatism in south-central Australia and implications for the breakup of Rodinia. Precambrian Res 87:135–159

Crustal Provinces of the Prince Charles Mountains Region and Surrounding Areas in the Light of Aeromagnetic Data

Alexander V. Golynsky[1] · Valery N. Masolov[2] · Vyacheslav S. Volnukhin[2] · Dmitry A. Golynsky[3]

[1] VNIIOkeangeologia, Angliysky Avenue 1, St. Petersburg 190121 Russia, <sasha@vniio.nw.ru>
[2] PMGRE, Pobeda Street 24, Lomonosov 189510, Russia
[3] SPbSU, 7/9, Universitetskaya nab., St. Petersburg 199034, Russia

Abstract. The aeromagnetic data of the Lambert Glacier – Prince Charles Mountains area provide a rather complex but surprisingly coherent image for studying the geology and tectonic history of this region. Several distinct structural units can be differentiated in the magnetic anomaly data. The aeromagnetic data from the Prince Charles Mountains and surrounding areas reveal the spatial boundaries of the Archaean cratons at the Prince Charles Mountains and Vestfold-Rauer areas and suggest the existence of a previously unknown craton in Princess Elizabeth Land. The magnetic data differentiate the inner structure of the Beaver-Rayner Proterozoic mobile belt and the complex marginal belt of the Archaean cratons reworked by Mesoproterozoic to Neoproterozoic tectonism. The aeromagnetic data clearly indicate no obvious link of the Pan-African mobile belt in Prydz Bay with Lützow-Holm Bay, and provide no evidence that it extends inland towards the Mawson Escarpment or Grove Mountains. Thus, East Gondwana probably was not divided into Indo-Antarctic and Australo-Antarctic sectors as suggested by a number of recent studies.

Introduction

The Prince Charles Mountains-Lambert Glacier-Princess Elizabeth Land region (Fig. 2.8-1) consists of a complex mosaic of metamorphic terranes that range in age from Archaean to early Paleozoic. The southern Prince Charles Mountains preserve Archaean granitic basement rocks and one or two low to medium grade metasedimentary cover sequences. Rocks outcropping in the northern Prince Charles Mountains and in Prydz Bay have been pervasively deformed and metamorphosed under granulite facies conditions by late Mesoproterozoic to early Neoproterozoic and/or late Neoproterozoic to Cambrian orogenic events respectively. The Vestfold Hills Block was formed and stabilized at about 2 500 Ma (Black et al. 1991; Snape et al. 1997). Archaean components in the Rauer Group are 300–700 Ma older than the Vestfold Hills Block indicating a separate Archaean evolution of both terranes (Kinny et al. 1993; Harley et al. 1998).

Intriguing models have been proposed recently for continuation of Pan-African tectonism identified in the Rauer Group and Larsemann Hills through the southern Prince Charles Mountains (Boger et al. 2001; Boger and Miller 2004) or may be linked with equivalent-age tectonism recognized in Lützow-Holm Bay and in the Leewin Complex (Carson et al. 1996; Hensen and Zhou 1997; Fitzsimons 1997). Fitzsimons (2000) suggested that tectonism in Prydz Bay trended inland up to the Gamburtsev Subglacial Mountains and not around the coast as Wilson et al. (1997) inferred. The Prydz Belt of Fitzsimons (2000) is well documented by geochronological and thermobarometric data, but neither the extent of this belt in Antarctica and India nor its nature, i.e. continental collision, intracontinental tectonism or failed rift, has been proved.

In contrast, Boger et al. (2001) have identified Cambrian tectonism along the southeastern edge of the Lambert Terrane in the central part of Mawson Escarpment. Here, syn- and post-tectonic dykes crystallized at ca. 510 and 490 Ma, and garnet-bearing orthogneiss was either emplaced or deformed at ca. 550 Ma. Boger et al. (2001) argued that this 550–500 Ma magmatism and deformation was evidence for the Lambert Terrane being a Cambrian suture, and the continuation of the orogenesis recognized in Prydz Bay (Zhao et al. 1992). Boger et al. (2001) and Boger and Miller (2004) argued that the orogeny continued westwards through the Lambert terrane to intersect with the East African Orogen somewhere in Dronning Maud Land. The proposed Kuunga suture defines the southern margin (in a present-day Antarctic coordinates) of a distinct Indo-Antarctic craton that consisted of part of east Antarctica, including the northern Prince Charles Mountains, the Rayner and Napier Complexes of Enderby Land, as well as potentially most of cratonic India.

Zhao et al. (2000) proposed that the Prydz Bay Complex could continue towards the Grove Mountains, where U-Pb ages of 540–500 Ma for felsic orthogneiss, syntectonic granite, and post-tectonic granodiorite were obtained. They considered that a Pan-African suture runs N–S at Prydz Bay and continues to the central Transantarctic Mountains. However, the extent to which this zone was part of a Pan-African mobile belt (Carson et al. 1996; Boger et al. 2001) is debatable. It seems premature to discard the idea of a continuous Grenvillian terrane in the Prydz Bay area (Hensen and Zhou 1995), because no related suture zone or ophiolite marking the closure of the palaeooceanic basin at that time has been found and the juvenile volcanic rocks are unknown. The extent and signature of ca. 1 000 Ma events in the southern Prydz Bay area remain enigmatic, and Pan-African orogen delineated in this area may be interpreted as intracratonic orogen (Yoshida et al. 2003).

From: Fütterer DK, Damaske D, Kleinschmidt G, Miller H, Tessensohn F (eds) (2006) Antarctica: Contributions to global earth sciences. Springer-Verlag, Berlin Heidelberg New York, pp 83–94

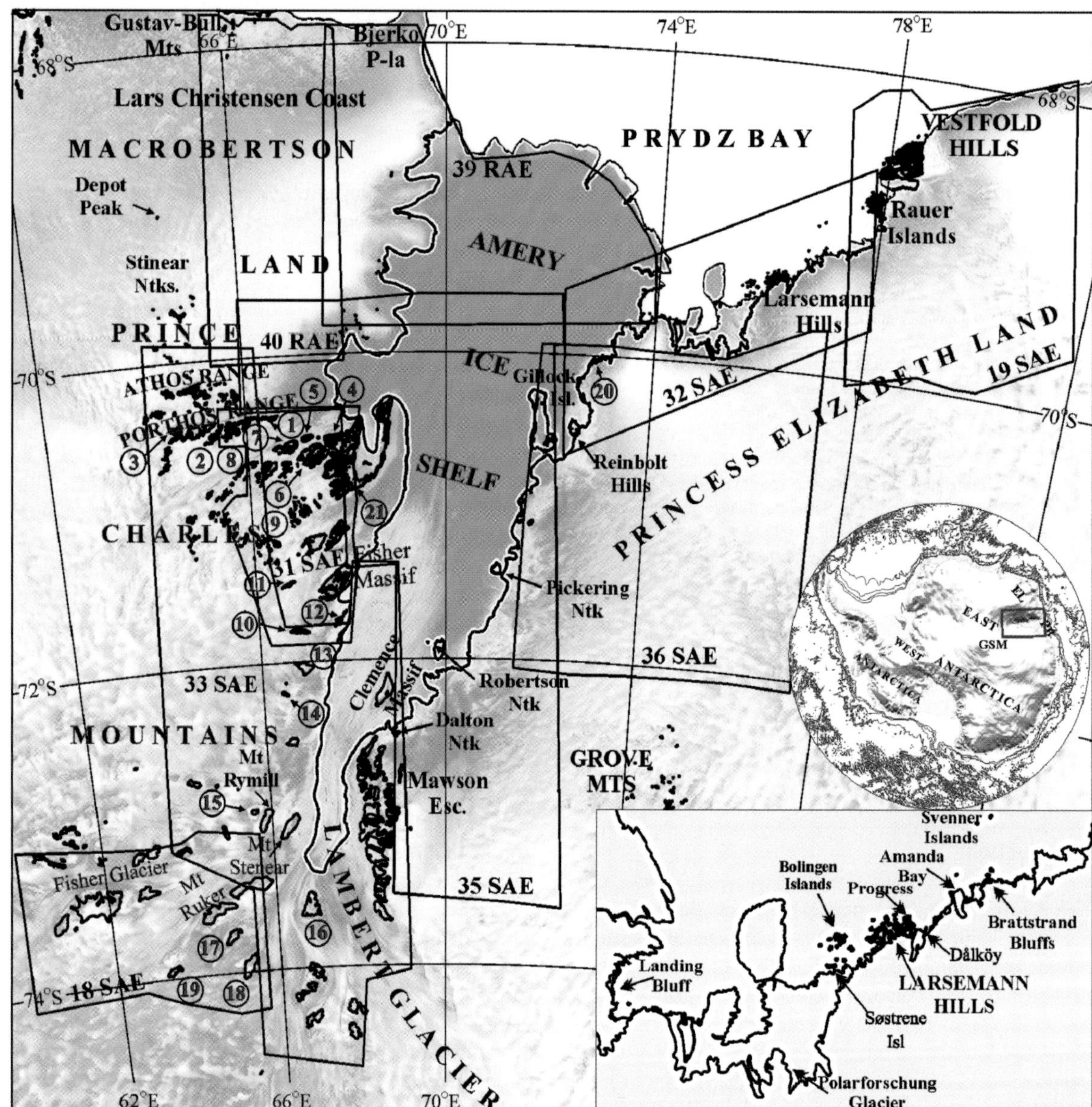

Fig. 2.8-1. Location of aeromagnetic surveys over the Prince Charles Mountains and adjacent areas. *Solid black lines* outline surveys carried out by the PMGRE during the Soviet/Russian expeditions. Basemap is a portion of the RADARSAT composite map. Map projections throughout this paper are polar stereographic. Rock outcrops and coastline are from the Antarctic Digital Database. Insert in the right corner shows the location of place names referred to in the text within the Larsemann Hills area. *GSM:* Gamburtsev Subglacial Mountains; *EL:* Enderby Land; *BP:* Prydz Bay. Location of place names referred to in the text: *1:* Mount McCarty; *2:* Webster Peaks; *3:* Corry Massif; *4:* Loewe Massif; *5:* White Massif; *6:* Amery Peaks; *7:* Thomson Massif; *8:* Charybdis Glacier; *9:* Mount Woinarsky; *10:* Mount Willing; *11:* Mount Collins; *12:* Nilsson Rocks; *13:* Mount Sho; *14:* Mount Izabelle; *15:* Mount Bloomfield; *16:* Cumpston Massif; *17:* Mount Bird;*18:* Mount Newton; *19:* Keyser Ridge; *20:* Jennings Promontory; *21:* McLeod Massif

Magnetic Anomalies

At boundaries between geophysical blocks around the world there are broadly similar types and patterns of magnetic and gravity anomalies (Provodnikov 1975; Gibb and Thomas 1976; Klasner and King 1986; Wellman 1988). The geophysical anomalies and pattern changes at the boundary between older and younger crusts can be one or more of the following three types: *(i)* Changes in trend direction, with trends in the older crust generally truncated by the boundary, and trends in younger crust parallel to the boundary. *(ii)* Major dipole magnetic anomalies along the boundary between the older and the younger crusts, due to an abrupt change at this boundary in average magnetization. *(iii)* Strong, positive magnetic anomalies along the bound-

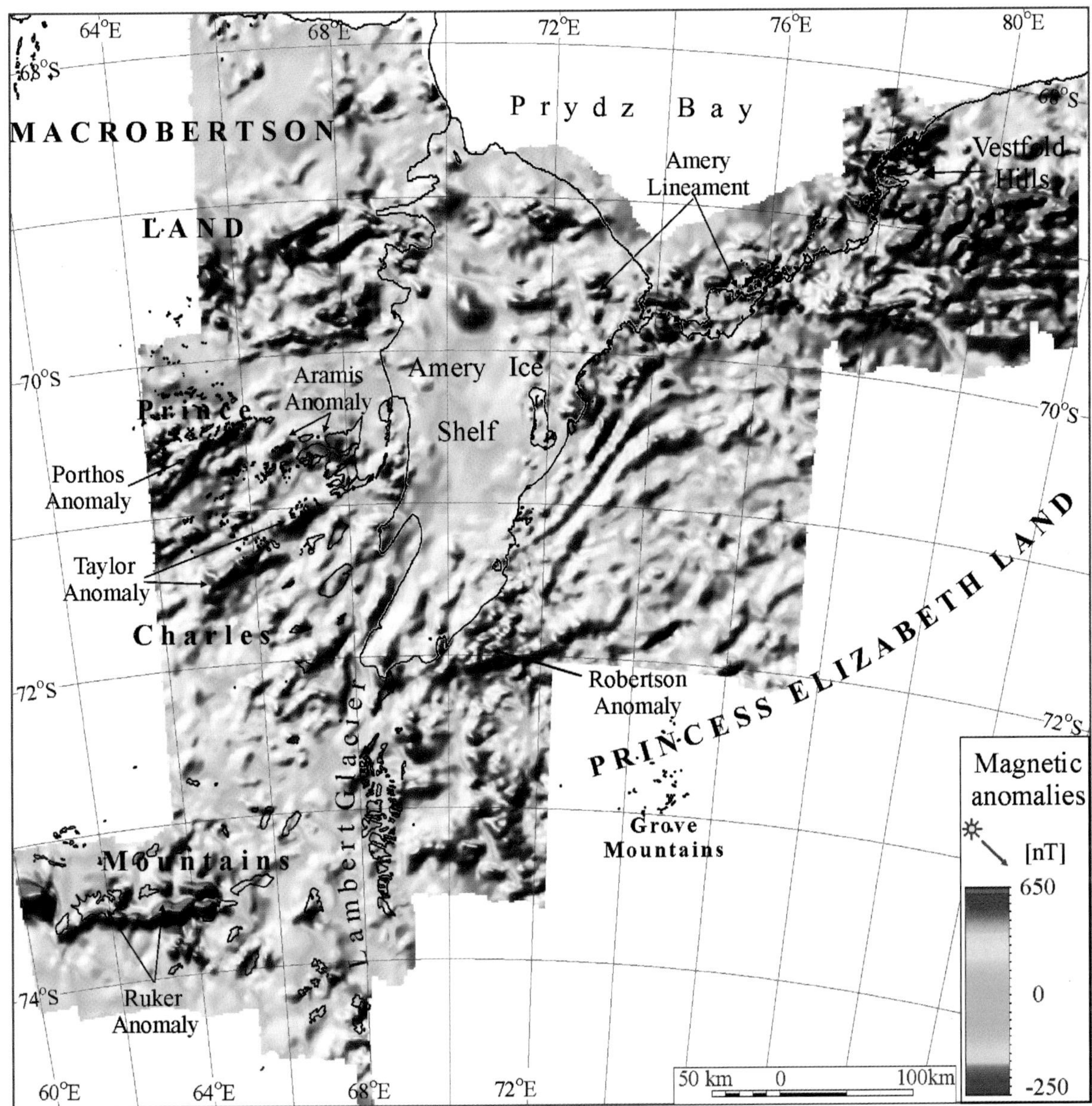

Fig. 2.8-2. Shaded-relief aeromagnetic map of the Prince Charles Mountains and surrounding areas. The map is illuminated from the northwest at an inclination of 45° to reveal smaller relief features

ary due to emplacement of serpentinite and granulites of mid-crustal origin.

Aeromagnetic anomaly maps are extremely useful in tectonic studies and the delineation of the geologic boundaries. Figure 2.8-2 gives a colour shaded-relief map for studying the regional aeromagnetic signatures of Archaen cratons in the Southern Prince Charles Mountains, Vestfold Hills and Rauer Islands that are separated by the extensive late Mesoproterozoic to early Neoproterozoic mobile belt (Mikhalsky et al. 2001). These aeromagnetic data provide a complex, but coherent image of structural features that contribute new understanding of the geology and tectonic history of the study region. The magnetic data assist considerably in correlating known geologic features and trends across ice covered regions. By identifying and correlating features in the magnetic image, we magnetically differentiate the Rayner, Beaver, Ruker, Vestfold and Princess Elizabeth Land structural units (Fig. 2.8-3) that may be further subdivided by conspicuous variations in their magnetic fabric.

Rayner Magnetic Unit

The Rayner magnetic unit occurs in over the Lars Christensen Coast of MacRobertson Land in the northwestern part of the study area (Fig. 2.8-3). Predominantly low-

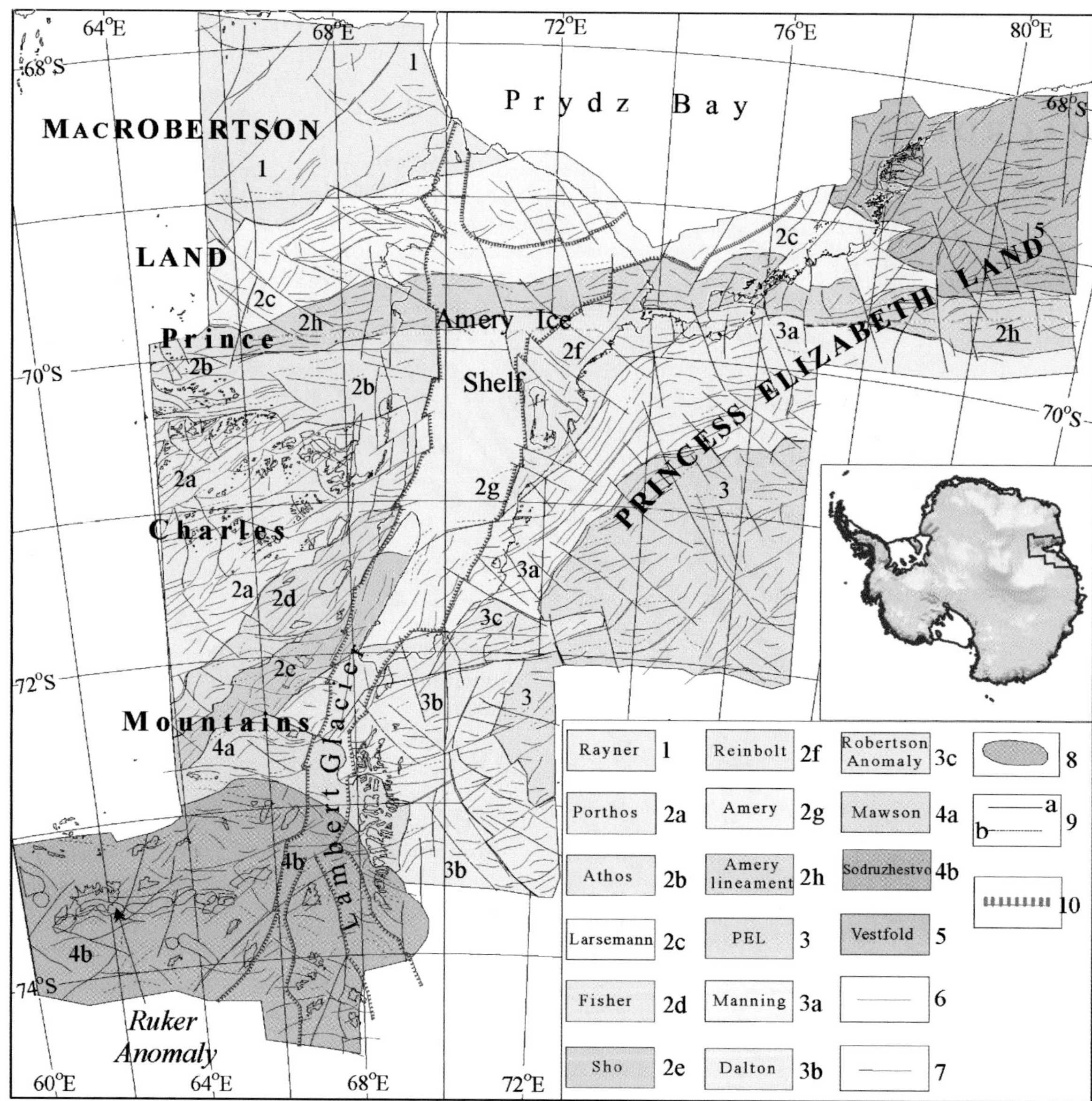

Fig. 2.8-3. Structural map of the Prince Charles Mountains and surrounding areas based on aeromagnetic data. Structural subdivisions of magnetic anomaly data: *1:* Rayner unit; *2:* Beaver unit, *2a:* Porthos sub-unit, *2b:* Athos sub-unit, *2c:* Larsemann sub-unit, *2d:* Fisher sub-unit, *2e:* Sho sub-unit, *2f:* Reinbolt sub-unit, *2g:* Amery sub-unit, *2h:* Amery Lineament; *3:* Princess Elizabeth unit, *3a:* Manning sub-unit, *3b:* Dalton sub-unit, *3c:* Robertson Anomaly; *4:* Ruker unit, *4a:* Mawson sub-unit, *4b:* Sodruzhestvo sub-unit; *5:* Vestfold unit. Magnetic features: *6:* Axis of positive anomalies; *7:* axis of negative anomalies; *8:* high-intensity anomalies associated with iron-banded formation; *9:* transverse discontinuities: *a:* major, *b:* minor; *10:* Lambert rift boundary

amplitude anomalies (150–250 nT) within an area of low gradients characterize the Rayner magnetic unit (Fig. 2.8-2). The flat and almost featureless negative magnetic field generally defines this unit. It is interrupted sporadically by local short-wavelength (10–15 km) circular and elongated anomalies mostly over the Gustav-Bull Mountains and the Bjerkö Peninsula. Its fabric and position suggest that the Mesoproterozoic-Neoproterozoic Rayner Complex (Black et al. 1987; Sheraton et al. 1987; Kelly et al. 2002) may be the source of the Rayner magnetic unit. The Rayner Complex displays a similar magnetic response in Enderby Land (Golynsky et al. 1996) where it is well recognized (Kamenev 1972; Sheraton et al. 1987). The spatial relationships between the characteristic magnetic anomaly patterns and outcropping geological units suggest that short wavelength magnetic anomalies may reflect Neoproterozoic intrusive charnockites which are recognized in the Gustav-Bull Mountains (McLeod 1964) and widely known over Kemp Land and MacRobertson Land (Sheraton 1982; Young et al. 1997).

Beaver Magnetic Unit

The major structural subdivisions of the Beaver magnetic unit (Fig. 2.8-3) are in the northern and central Prince Charles Mountains, and partly within the Amery Ice Shelf and its eastern flank, as well as over both shoulders of the Lambert-Amery rift system. Complex positive bands (50–100 km) of moderate to high amplitude (up to 700 nT) elongated and irregular anomalies characterize the Beaver unit. These continuous bands alternate with negative anomalies that exhibit complex short wavelength positive components. The essential feature of the Beaver unit is a predominantly northeasterly trending magnetic fabric that is clearly recognized over both shoulders of the Lambert-Amery Rift. The predominant trends in the northern part of the unit are roughly latitudinal.

The appearance of high to moderate amplitude linear and/or elongated and sometimes circular magnetic anomalies sharply define the northern and eastern boundaries of the Beaver unit. Prominent alternating linear NE-SW positive and negative low amplitude anomalies over the eastern shoulder of the Lambert-Amery rift delineate the eastern boundary of the Beaver unit (NPCN magnetic unit of Golynsky et al. 2002). The southern boundary of the Beaver unit is not as prominent and no crucial reorganization of the magnetic anomaly patterns is apparent. However, NE-SW trends are dominant in the Beaver unit and E-W trends are common in the Ruker magnetic unit. Thus, these changes in trends of the two neighboring terranes as well as changes of spectral characteristics allow drawing an assumed boundary between two units. It can be located just southward from Sho Massif (Fig. 2.8-2).

Formation of the Lambert Amery rift system led to important modifications of the magnetic anomaly patterns within the Beaver unit. This is manifested basically by a decreasing of magnetic anomaly amplitudes and appearance of quiet and/or negative areas without any visible regularity of trends. Spectral characteristics of anomalies are also changed and become regional. It indicates an increasing depth of causative sources and their disappearance along the axial part of the rift. The analysis of magnetic data can distinguish seven structural sub-units within the Beaver unit that are characterized by individual features. Six sub-units (Larsemann, Athos, Porthos, Fisher, Shaw and Reinbolt sub-units) have a coherent magnetic anomaly pattern, which is common for the whole unit, whereas magnetic fabric within the Amery sub-unit is fairly discordant to general trends of the Beaver unit. The overall pattern of anomalies within the *Amery sub-unit* (Fig. 2.8-3) is due to the depression of the crustal basement within the rift system.

Within the *Athos* and *Porthos sub-units* (Fig. 2.8-3) of the late Mesoproterozoic to early Neoproterozoic Beaver terrane (Mikhalsky et al. 2001), negative and positive anomalies largely reflect the Athos supracrustals and Porthos orthogneiss, respectively (Kamenev et al. 1993; Mikhailov and Sergeyev 1997). The roughly 50–75 km wide Athos sub-unit is clearly differentiated from adjacent terranes by a relatively smooth set of mostly low amplitude (50–200 nT) magnetic anomalies. High-grade metasedimentary paragneiss of the Athos Series (Mikhailov and Sergeyev 1997) with a mean magnetic susceptibility of 47×10^{-5} SI are responsible for the smooth, negative magnetic grain over the Athos Range and surrounding areas. Highly magnetic components within the Athos Paragneiss are negligible and thus they give a negligible response in aeromagnetic data. The highly magnetized rocks of the Porthos Series (Mikhailov and Sergeyev 1997; Mikhalsky et al. 2001) are not known to occur in the Athos Range where short-wavelength, low amplitude magnetic anomalies are clearly visible on the magnetic image. The observed NE-SW spatial trends are consistent with those over the granulite facies rocks of the Porthos Range. Within the Athos sub-unit, these anomalies are probably caused by orthogneiss units or plutonic rocks. It is likely that the predominant sources are charnockites owing to the discrete setting of their magnetic anomalies and their comparatively low magnetic properties relative to other of the igneous rocks (e.g., Mikhailov and Sergeyev (1997) reported mean magnetic susceptibilities of $1\,990\times10^{-5}$ SI, and $1\,750\times10^{-5}$ SI for Porthos Orthogneiss and intrusive charnockite, respectively).

The *Porthos sub-unit* is generally associated with complex anomaly belts of about 50–100 km length formed largely by elongate and partly by irregular moderate to high-intensity amplitudes (up to 700 nT) anomalies (Fig. 2.8-2). These continuous positive anomaly belts alternate with negative low amplitude anomalies of that occasionally include positive short wavelength components. The magnetic grain of the Porthos sub-unit consists of two prominent regional positive anomalies (*Porthos and Taylor anomalies*) separated by an irregular, regional anomaly block (*Aramis block*). The rest of the region is largely dominated by negative long wavelength anomalies that include superimposed short-wavelength magnetic anomalies.

The *Porthos Anomaly* represents a superposition of the different order and predominantly elongated anomalies (Fig. 2.8-2). Their amplitudes varied widely and rich up to 500–600 nT. Sergeyev and Kaulio (1993) and Mikhailov and Sergeyev (1997) showed that the Porthos Range consists mostly of felsic orthogneisses and intrusive charnockites with moderate to high magnetic susceptibilities (1.6–72.1×10^{-3} SI units). While intrusive charnockites within the Porthos Range are observed at several localities such as Mount McCarthy, Webster Peaks, and Corry Massif (Hensen et al. 1997), their broader spatial distribution is poorly known, whereas their spatial dimensions are rather limited. The magnetic properties of these charnockites are not as strong as the charnockites of the

Loewe Massif (Mikhailov and Sergeyev 1997) and thus they do not produce strong magnetic anomalies.

The broadly segmented, linear *Taylor Anomaly* consists of four major segments with moderate wavelength (15–25 km) maxima on the profile of 100–700 nT (Fig. 2.8-2). The amplitude and wavelength variations of the Taylor Anomaly might be explained by variously uplifted crustal blocks along the strike-length of the anomaly. The westernmost segment of the Taylor Anomaly has no geological control. Its magnetic fabric is similar to that associated with the basement gneisses at McLeod Massif and surrounding exposures to the south. The Porthos Orthogneisses are largely the anomaly sources (Mikhailov and Sergeyev 1997). A structural disruption or tectonic fault involving low magnetic Astronomov Paragneiss divides the Taylor Anomaly eastward from Mount Woinarsky. The shaded-relief map shows that the Astronomov Paragneisses are structurally controlled with continuous negative anomalies outlining it to the north and south of the Taylor Anomaly. It is highly likely that the Astronomov Paragneiss is not present in the western and southern parts of the Aramis Range.

The *Aramis Anomaly* (Fig. 2.8-2) exhibits distinctive internal features resulting from the lithological zonation observed in exposed portions of the Aramis Range (Fitzsimons and Thost 1992; Kamenev et al. 1993). Analysis of the published geologic maps (Kamenev et al. 1993; Mikhalsky et al. 2001; Thost et al. 1998) and aeromagnetic data revealed magnetic anomaly correlations with major geological subdivisions that could be traced under the ice cover. It was distinguished that moderate amplitude, broad roughly oval maxima over Loewe Massif, White Massif and the northern Amery Peaks area largely reflect the charnockites of the compositionally varied batholith of Loewe and White Massifs. The distinct quasi-linear 350 nT anomaly corresponds to Thomson Massif composed of Porthos series rocks. It is fairly distinct that the main source of this plutonically derived anomaly shifted distinctly northward from Thomson Massif under the Charybdis Glacier. Magnetic pattern changes to the south of Thomson Massif may reflect the equal mixing of Porthos and Athos Series rocks, as well as the presence of plutonic sequences with lower magnetic properties.

The *Fisher sub-unit* (Fig. 2.8-3) is located close to the southern boundary of the Beaver magnetic unit and characterized largely by low amplitude anomalies. For instance, magnetic anomalies over Fisher Massif, which is composed of Mesoproterozoic rocks (Beliatsky et al. 1994; Kinny et al. 1997), do not exceed 150–200 nT, whereas the gabbroic intrusion of Mount Willing is considered as the main source of a distinctive, roughly 600 nT magnetic anomaly. The form and spatial relationship of the magnetic anomalies are very distinct for this sub-unit. As a rule linear anomalies are not present within it. The sub-unit is further distinguished by an echelon-like pattern of structurally coherent anomalies. The dominant magnetic anomaly trends are NE-SW, although they vary within the sub-unit. Petromagnetic studies show the felsic rocks from Mesoproterozoic Fisher Massif are largely characterized by low magnetic susceptibilities of about $0.05–0.15 \times 10^{-3}$ SI (Sergeyev unpublished data). Thus, they can not be considered as the major source of the aeromagnetic anomalies, whereas the metagabbroic rocks of Fisher Massif that show volume susceptibilities of about 30×10^{-3} SI will be more effective magnetic anomaly sources. Granodiorites, biotite-amphibolite and charnockitic gneisses from Mount Collins yield similarly strong susceptibility values, while values of roughly 26×10^{-3} SI have been noted for the gabbros from Mount Willing and Nilsson Rocks.

The *Sho sub-unit* (Fig. 2.8-3) is related to a category of bordering structures. With high portion of probability it might be considered as a part of the Ruker or Princess Elizabeth Land magnetic unit. The border between these structures cannot be simply defined. As a rule, however, it occupies a broader transition zone where both geologic subdivisions might co-exist. Accordingly, in our concept the Sho sub-unit represents the composite territory between craton and mobile belt. The outstanding feature of this sub-unit is the linear NE-SW trending anomalies with amplitudes of about 300 nT (Fig. 2.8-2). The origin of magnetic anomalies within the sub-unit is not presently known. However, we suggest the biotite-amphibolite gneisses of Shaw Massif with volume susceptibilities of 13.9×10^{-3} SI (Sergeyev unpublished data) may well be considered as a major source of observed magnetic anomalies. Other rocks (e.g., biotite-garnet gneisses; Sergeyev unpublished data) have generally low magnetizations with the average susceptibility value of about 0.78×10^{-3} SI. Biotite-garnet gneisses from Mount Izabelle are characterized by bimodal volume susceptibilities centered on values of roughly 32×10^{-3} and of 0.5×10^{-3} SI.

We interpret the *Reinbolt sub-unit* as the eastward continuation of structural subdivisions within the northern Prince Charles Mountains area (Fig. 2.8-3). However, the structural grain of magnetic anomalies within the Reinbolt sub-unit has no obvious similarity with other subdivisions of the Beaver unit. Especially peculiar is the absence of any regularity in the anomaly pattern of the Reinbolt sub-unit where the anomaly trends are substantially complicated. However, the magnetic data together with available and poorly constrained geologic information clearly show that all orthogneiss outcrops are reflected by positive magnetic anomalies of variable amplitudes.

Positive anomalies do not correspond with all intrusions in the Reinbolt sub-unit that as a rule are located within gradient zones or correspond to negative anomalies. An exception is the charnockite intrusion of Neoproterozoic age (ca. 896 Ma) located in the southern Reinbolt Hills (Grew and Manton 1981) that is associated with local 50 nT anomaly. These anomaly affiliations hold not only for charnockite intrusions of Neoproterozoic age located near the Jennings Promontory (Ravich et al. 1978), but also for the supposedly Cambrian age granites of Gillock Island (Thost et al. 1998). Preliminary evidence indicates that Cambrian granites along the southern coast of Prydz Bay exhibit medium magnetic susceptibilities (Sergeyev 2000, personal communication). However, they are apparently of limited spatial extent because they do not produce any outstanding response in the observed magnetic anomaly pattern.

The *Larsemann sub-unit* occupies the northern part of the Beaver unit over both shoulders of the Amery Ice Shelf and much of the Prydz Bay area (Fig. 2.8-3). The magnetic anomaly fabric within the Larsemann sub-unit is distinctly similar to the pattern of the whole BU. Here, the moderate to high amplitude anomalies form well defined paired anomaly belts. The northern positive belt is poorly constrained due to the lack of outcrops within the area and thus we do not consider it further. The southern belt of positive anomalies is the Amery Lineament (Golynsky et al. 2002). It is broadly segmented and associated with elongated anomalies, although round-shaped and/or circular highs are also widespread. The slightly curvilinear Amery Lineament is up to 35 km wide and approximately 625 km in total length with anomaly amplitudes up to 1 000 nT. This magnetic lineament is one of the most striking features ever recorded by aeromagnetic surveying within the East Antarctic Shield. Hence, understanding of its origin is critical to appreciating the complex geological history of the Prydz Bay coast area and the northern part of Prince Charles Mountains.

The Amery Lineament may continue towards Enderby Land where a roughly 250 km linear anomaly was mapped by the IL-18 reconnaissance aeromagnetic survey flights (Golynsky et al. 2001; Golynsky et al. 2002). However, this linear anomaly has a width of about 75 km that is not consistent with it being the westward continuation of the Amery Lineament, even though its amplitude is fairly similar. Further detailed aeromagnetic surveying over the areas between MacRobertson Land and Enderby Land would clearly increase the geological knowledge of the area. The continuation of the Amery Lineament eastwards is also uncertain owing to the superposition of magnetic anomalies of the Rauer-Vestfold crustal block and the lack of high-quality data. We believe, however, that the Amery Lineament delineates the southern boundary of the Vestfold-Rauer crustal block.

In the northern Prince Charles Mountains late Mesoproterozoic to early Neoproterozoic metasedimentary rocks are most common in the paragneiss sequences (Athos Series) of the Athos Range, Stinear Nunataks, Depot Peak (Stüve and Hand 1992; Scrimgeour and Hand 1997; Mikhalsky et al. 2001) and can not be considered as sources of the intense magnetic anomalies of the Amery Lineament due to their low magnetic properties (Mikhailov and Sergeyev 1997). In our view, the Porthos Orthogneiss with volume susceptibilities of about 30×10^{-3} SI will be more effective magnetic anomaly sources of the Amery Lineament within the northern Prince Charles Mountains, while such rocks not known in the discussed area.

Over the Larsemann Hills, Bolingen Islands and surrounding areas, a fundamentally different, high intensity (up to 1 500 nT) magnetic anomaly pattern is observed. The westernmost part of the Larsemann Hills is outlined by an intensive oval anomaly, whereas the eastern part is characterized by a low amplitude negative anomaly. The northern and southern Bolingen Islands are associated with high-amplitude anomalies separated by a local magnetic minimum. The observed magnetic anomaly pattern is surprisingly coherent with the known geology. For example, the high-intensity anomalies correlate with the Søstrene Orthogneiss and the negative anomalies as a rule correspond to the Brattstrand Paragneiss, which represent basement and cover associations, respectively. However, the relationship between these two interleaved ortho- and paragneiss of the Prydz Complex is uncertain and poorly understood (Sheraton and Collerson 1983; Fitzsimons and Harley 1991; Carson et al. 1995, Fitzsimons 1997). Fitzsimons (1997) regarded most paragneisses to belong to the Neoproterozoic cover unit deposited over the older Mesoproterozoic orthogneiss unit. The age of the Søstrene Orthogneiss is uncertain. It was deformed and metamorphosed during the Pan-African event, but that does not preclude an earlier high-grade history (Carson et al. 1996).

Correlative analysis of the magnetic anomaly data and outcrop geology of the southern Prydz Bay region shows that the Brattstrand Paragneiss at the Brattstrand Bluffs coast-line and the Svenner Islands, considered as a sedimentary cover sequence (Fitzsimons 1997), is associated with the broad negative anomaly of low amplitude (−150 to 200 nT). This magnetic low is punctuated by several local short-wavelength, low amplitude (~50 nT) positive anomalies). Steinnes Peninsula and the northern Svenner Islands are each marked by a local maximum where the Prydz Bay orthogneiss association is recognized (Thost et al. 1998). An additional local maximum (~25 nT) over the Amanda Bay presumably reflects the presence of the kilometer-scale Amanda Bay Granite.

The rock magnetic properties for this area agree with the spatial distribution of magnetic anomalies. For example, low magnetic susceptibility values (0.37×10^{-3} SI) characterize the majority of paragneisses, whereas the basement host rocks yield susceptibility values of the order of $35.3–38.4 \times 10^{-3}$ SI. The intense magnetic anomaly over the western Larsemann Hills probably reflects felsic orthogneiss or the high portion of highly aluminous metapelitic gneisses where susceptibilities reach up to 35.9×10^{-3} SI (Sergeyev unpublished data). Negative low-amplitude anomalies are mainly associated with a number of known post-tectonic A-type granites (Dålköy, Progress, Polarforschung and Landing Bluff) that were intruded at about 500 Ma (Tingey 1991).

It can be concluded that major sources of the positive magnetic anomalies within the Bolingen Islands – Larsemann Hills – Brattstrand Bluff and Svenner Islands area probably are felsic and mafic orthogneisses like those at Søstrene Island that are located at the axis of the Amery Lineament. The Brattstrand Paragneiss, with mineral assemblages and structures developed at ~500 Ma, correlates with negative anomalies of local appearance. The absence of any prominent negative features running towards regions where similar orogenic belts are recognized (e.g., Lützow-Holm Bay area, Mawson Escarpment or Grove Mountains) raises doubts about the continuity of the Pan-African mobile belt.

Ruker Magnetic Unit

The determinant feature of the Ruker magnetic unit (Fig. 2.8-3) is the presence of the high amplitude (up to 2 600 nT), 130 km long Ruker anomaly (Golynsky et al. 2002). The rest of the southern Prince Charles Mountains region is mainly associated with short-wavelength low-amplitude anomalies in a low-gradient area (Fig. 2.8-2). The structural trends of magnetic anomalies are difficult to establish although in the northern part of the unit they stretch W-E. In the eastern part, the trends vary substantially from NE to SE. Differences in the magnetic anomaly pattern allow subdivision of the unit into the *Mawson and Sodruzhestvo sub-units*. The Sodruzhestvo sub-unit corresponds to the central and most stable part of the Archaean to Proterozoic Ruker Terrane that comprises a variety of metasedimentary and metavolcanic rocks (Menzies and Sodruzhestvo supracrustals of Kamenev et al. 1990) which apparently overlie an orthogneiss basement. The Mawson sub-unit is outlined over the outer and reworked part of the craton (the Lambert Series of Kamenev et al. 1993) that includes pronounced components of the Ruker Terrane (Mikhalsky et al. 2001).

The *Mawson sub-unit* includes the central part of the Mawson Escarpment and the western shoulder of the Lambert Glacier (Fig. 2.8-3). Surprisingly smooth and predominantly low amplitude (<100 nT) negative magnetic anomalies are evident. Consideration of the limited number of short-wavelength anomalies of low intensity and extent suggests that the anomaly trends are generally E-W. It is rather obvious from magnetic and geological data (Mikhalsky et al. 2001) that confident correlations between geologic complexes outcropped over the northern and central parts of the Mawson Escarpment and the magnetic anomalies are not feasible at present. The present paucity of geological data for the area makes it very difficult to establish the geological significance of the magnetic anomalies of this unit.

The *Sodruzhestvo sub-unit* is associated with widespread short-wavelength anomalies on a relatively featureless negative background (Fig. 2.8-2). The lack of continuity in the magnetic anomaly pattern is the relevant characteristic of the sub-unit. Magnetic anomalies north of the Ruker Anomaly are less intensive (<150 nT) in comparison with those observed in the southward direction, where amplitudes reach up to 700 nT. This amplitude variation may reflect changes in the subsurface geology, as well as the crustal relief that varies between mountain ranges and outlet glaciers.

The territory northward from Fisher Glacier is largely associated with the metasediments of the greenschist- to amphibolite-facies Menzies Series and younger, lower-grade greenschist-facies Sodruzhestvo Series (Kamenev et al. 1990). Archaean felsic orthogneisses cropping out at Mounts Bloomfield, Rymill and Stinear all correlate with isolated magnetic highs. A remarkable magnetic anomaly pattern overlies Mount Rymill, which at its northern and southern parts is composed of variably deformed granite plutons with supracrustals (Menzies Series) in the middle (Mikhalsky et al. 2001). Distinctive positive anomalies mark both intrusions, whereas a prominent magnetic low outlines the metasedimentary rocks. The northern part of the sub-unit reveals similar relationships between the magnetic features and the causative rocks.

The anomaly fabric of the southern and southeastern parts of the sub-unit is similar to the pattern for the northern area. It clearly can be seen that Archaean granitic rocks for the most part correspond to the local magnetic highs in contrast to the cover sequences that largely correlate with negative anomalies. For instance, one of the largest exposures of the region, Cumpston Massif is composed of felsic orthogneiss to the west and supracrustals in the east. The positive, roughly 200 nT amplitude anomaly located over the western side of the massif clearly seems consistent with this geology.

The relationships between the magnetic anomalies and the tectonically interleaved Archaean to ?Paleoproterozoic metasedimentary rocks and the felsic orthogneiss of the Ruker Terrane (Mikhalsky et al. 2001) are difficult to establish. Many geologic factors are poorly understood and complicate these relationships including the variable lithology of the felsic orthogneiss (granitic to tonalitic and trondhjemitic in composition) and their ratio with supracrustals at exposures, as well as the presence of mafic intrusive rocks (concordant amphibolite bodies) that were widespread in the Late Archaean to ?Paleoproterozoic rocks of the Ruker Terrane. It was distinguished that large outcrops of interlayered sequences at Mount Bird and Mount Newton are associated with positive magnetic anomalies. This is far being the case for the northern and southern parts of Mount Newton which are reflect negative anomalies varying from -100 to -150 nT, or for the outcropping Keyser Ridge where smooth negative anomalies (~ -50 nT) are observed.

The southern Mawson Escarpment comprises mainly of orthogneiss and lithologically distinct suites of strongly folded and sheared metasedimentary and metaigneous rocks that occur as tectonic slabs and nappes (Kamenev et al. 1990). The pronounced layering of tectonically modified Archaean to ?Paleoproterozoic Ruker Terrane rocks produces the outstanding magnetic anomalies characterized by short-linear chains of 2–5 km wavelength, low amplitude positive and negative anomalies. As expected, positive anomalies mark much of the felsic orthogneiss that may occur as tectonic blocks or slabs which sometimes are up to 2–4 km thick (Mikhalsky et al. 2001). Negative anomalies dominate the metasedimentary Menzies Series rocks. However, some of these rocks correlate with the narrow positive anomalies, presumably because of their high portions of metaigneous rocks.

Princess Elizabeth Magnetic Unit

In contrast to the magnetic units described above, the Princess Elizabeth magnetic unit (Grove Mountains magnetic unit, of Golynsky et al. 2002) is situated mostly over the eastern shoulder of the Lambert rift system (Fig. 2.8-3). It occupies a vast region of the Grove Mountains and an area devoid of any exposures to the north, west and northwest. From the anomaly pattern, this area can be further subdivided into five areas (sub-units) with striking magnetic features. These are the Manning, Dalton, Northern, Grove and Gale sub-units (see Golynsky et al. this volume). The magnetic fabric of the Grove Mountains and surrounding areas is quite varied and complex. The spectral composition of magnetic anomalies is compound, the superposition of different order anomalies is of common appearance. Their width varies within the limits of 5–10 km and up to a few tens of kilometres.

One of the crucial features of the Princess Elizabeth unit is the coherent pattern of magnetic anomalies along its margins and lack of intense anomalies within the northern sub-unit. The strongest magnetic anomalies (up to 500–650 nT) are distinguished within the Dalton sub-unit and the Robertson Anomaly. The northern part of the Princess Elizabeth magnetic unit is dominated by short-wavelength, low-intensity anomalies. The bordering magnetic anomalies of the unit trend obliquely to the anomalies of its central part. This behavior is fairly evident over the Manning sub-unit and to a lesser degree over the Dalton sub-unit. These relationships between anomalies of two adjacent terranes strongly suggest that the age of geologic formations responsible for the bordering magnetic anomalies may be fundamentally younger than the age of geologic structures within the central part of the crustal unit. The lack of outcrops within the area greatly limits establishing the origin of these relationships. However, the existence of ancient Archaean nuclei in the surrounding areas suggests that the northern magnetic sub-unit may be a protocraton fragment similar to those observed in the Vestfold Hills/Rauer Islands crustal block and the Ruker Terrane. Verifying this assertion will require additional geophysical and geological information, including conclusive evidence for the presence or absence of Archaean to Paleoproterozoic assemblages. The limited isotopic data available for the Grove Mountains area (Mikhalsky et al. 2001) do not support or refute our assumption, whereas new Sm-Nd data show that central and northern Mawson Escarpment, Clemence Massif and Reinbolt Hills are underlain by Archaean protoliths with TDM Nd model ages of 3.0–3.5, 3.2, 3.0–3.2 Ga, respectively (Mikhalsky et al. this volume). On the other hand, the aeromagnetic data do not support the recently inferred continuity of the Pan-African belt in the southern Prydz Bay coastal area towards the Mawson Escarpment (Boger et al. 2001) and/or Grove Mountains (Zhao et al. 2000).

The Vestfold Magnetic Unit

The Vestfold magnetic unit includes essentially intensely positive (up to 3 000 nT), high-gradient anomalies (Fig. 2.8-2). Rare negative low amplitude anomalies are found mainly at the unit's periphery. An abrupt change in the magnetic anomaly pattern sharply defines the western boundary of the unit along the NW-SE trending gradient zones or lineaments of possibly tectonic origin. There are no conclusive constraints on the location of the southern boundary due to the superposition of the Amery Lineament and anomalies of the unit. The Crooked Lake orthogneiss and possibly the Rauer Group gneiss (Collerson et al. 1983) are responsible for the magnetic anomalies observed over the Vestfold Hills and Rauer Islands. This affiliation is consistent with the occurrence of high-intensity anomalies over a substantial part of the Rauer archipelago excluding it's the central part where the Archaean rocks are mainly exposed (Harley et al. 1998). The Archaean Scherbinina layered complex, interpreted as a polydeformed and metamorphosed Fe-tholeiite intrusive (Harley et al. 1998), is located along the zone of high-gradient magnetic anomalies. Magnetic susceptibility data from the Crooked Lake gneiss of the Vestfold Hills vary from 0.4×10^{-3} to 227×10^{-3} SI.

Tectonic Correlations

Various terranes exposed along the coast and in the interior of study area have been recently correlated based solely on isotopic data. These correlated terranes are separated by wide areas with no exposures and thus may be joined up in numerous ways. However, these correlations must be consistent with the regional structure that aeromagnetic data can perhaps best test at present. Clearly visible on the magnetic image are several paths (e.g., Boger et al. 2001; Boger and Miller 2004; Fitzsimons 2000; Wilson et al. 1997) that may be inferred for the Pan-African tectonism. For instance, the Prydz Bay-Lützow-Holm Bay suture(?) may be correlated with the Amery Lineament. However, in the northern Prince Charles Mountains region Pan-African tectonic units are unknown, whereas in the Prydz Bay coastal area, we find this magnetic feature related with Grenville basement rocks (the Søstrene Orthogneiss) and the Pan-African Brattstrand Paragneiss correlates with negative anomalies. It must be emphasized that the detection of Grenvillian U-Pb zircon ages (Zhao et al. 1995) and Sm-Nd garnet-whole-rock ages (Hensen and Zhou 1995) suggest not only the possible existence of Grenvillian metamorphism in this area, but might also reflect the lower grade, Pan-African metamorphism that did not extensively reset the above ages and magnetic trends have been not pervasively reoriented during a Pan-African tectonism. The magnetic field data lack prominent structural magnetic features (essentially negative continuous belts) that that may help connect regions where similar aged orogenic belts have been recognized. Thus, our analysis raises doubts about continuity of the Pan-African mobile belt. The negative low-amplitude magnetic anomalies over a number of known post-tectonic A-type granites intruded at about 500 Ma (Tingey 1991) are also largely discontinuous.

The aeromagnetic data also extend along the general trends of the Manning and Dalton subunits to accommodate another possible path of the Pan-African suture from Prydz Bay to the southern Prince Charles Mountains. However, there are insufficient geochronology data for any reliable setting for the aeromagnetic subdivisions to be established. For example, the Pb-Pb zircon geochronology of felsic gneiss from Clemence Massif implies an age of about 1 200 Ma for high-grade metamorphism (Belyatsky unpublished data), while the similar rocks from the Pickering Nunatak yield poorly defined Rb-Sr whole-rock ages of 1 042 ±347 Ma (Tingey 1991). The Rb-Sr whole-rock age of 941 ±94 Ma for felsic gneiss from Dalton Nunataks located at the northern extremity of Mawson Escarpment is also relatively poorly constrained (Tingey 1991).

The establishing the path of the Cambrian belt beyond the central Mawson Escarpment where high-resolution isotopic data are obtained (Boger et al. 2001) is also problematical. The aeromagnetic data do not support continuing it towards Prydz Bay or the Grove Mountains region. The prominent Mawson sub-unit belt of magnetic anomalies only provides a sub-latitudinal trace over the western shoulder of the Lambert Glacier. The anomaly belt extends into the area north of Mount Stinear and Mount Rymill that is composed mainly by Archaean to Paleoproterozoic rocks, although Phanerozoic intrusive rocks dominate the northern extremity of Mount Stinear (Mikhalsky et al. 2001).

Finally, the data supporting our arguments are scarce and some may not be conclusive, however the aeromagnetic data from the Prince Charles Mountains and surrounding areas allow to identify the spatial boundaries of the Archaean cratons in the Prince Charles Mountains and Vestfold-Rauer areas and suggest the existence of a previously unknown craton in Princess Elizabeth Land. These cratons potentially belong to a single stable protocraton of Archaean to Paleoproterozoic age. The magnetic data differentiate the inner structures of the Beaver-Rayner Proterozoic mobile belt and the complex features of the margins of the Archaean cratons that were reworked during Mesoproterozoic to Neoproterozoic orogenesis. The aeromagnetic patterns distinctly evident for the Cambrian belts within the Prydz Bay and southern Prince Charles Mountains lack evidence for their possible continuations to support the recent suggestion that East Gondwana may be divided into Indo-Antarctic and Australo-Antarctic sectors (Fitzsimons 2000; Boger et al. 2001; Boger and Miller 2004). The regional tectonic setting of Late Neoproterozoic to Cambrian deformation and metamorphism in the study area remains uncertain, whereas the classical model that the Pan-African events are mostly the result of superposition-reworking of pre-existing Grenvillian belts cannot yet be revised (Yoshida et al. 2003). Clearly, the early models of intracratonic magmatic underplating causing widespread thermal overprinting but limited deformation (Stüwe and Sandiford 1993) need revision and the new models invoking the assembly of East Gondwana during the Pan-African Orogeny should be further examined. Detail geologic studies over the eastern shoulder of the Amery Ice Shelf and the northern Prince Charles Mountains outcrops will help resolve some of the outstanding problems for developing pre-Gondwana continental reconstructions.

Acknowledgements

Without the collaboration of the PMGRE and VNIIOkeangeologia geologists and geophysicists working in the Prince Charles Mountains area for the past several years, the aeromagnetic results would not have been as successful or as useful. We are especially indebted to B. Beliatsky, G. Grikurov, E. Kamenev, A. Kazankov, R. Kurinin, V. Leonov and A. Tscherinov. The authors thank Detlef Damaske and Ian Fitzsimons for their constructive reviews. We are also grateful to Michael Sergeyev for allowing us to cite his magnetic susceptibility determinations. Ralph von Frese is also thanked for his comments that helped shape the published version of this paper.

References

Beliatsky BV, Laiba AA, Mikhalsky EV (1994) U-Pb zircon age of the metavolcanic rocks of Fisher Massif (Prince Charles Mountains, East Antarctica). Antarctic Science 6:355–358

Black LP, Harley SL, Sun S-S, McCulloch MT (1987) The Rayner Complex of East Antarctica: complex isotopic systematics within a Proterozoic mobile belt. J Metamorp Geol 5:1–26

Black LP, Kinny PD, Sheraton JW, Delor CP (1991) Rapid production and evolution of late Archaean felsic crust in the Vestfold Block of East Antarctica. Precambrian Research 50:283–310

Boger SD, Miller JMcL (2004) Terminal suturing of Gondwana and the onset of the Ross-Delamerian Orogeny: the cause and effect of an Early Cambrian reconfiguration of plate motions. Earth Planet Sci Letters 219:35–48

Boger SD, Wilson CJL, Fanning CM (2001) Early Paleozoic tectonism within the east Antarctic craton: the final suture between east and west Gondwana? Geology 29:463–466

Carson CJ, Dirks PGHM, Hand M, Sims JP, Wilson CJL (1995) Compressional and extensional tectonics in low-medium pressure granulites from the Larsemann Hills, East Antarctica. Geol Mag 132:151–170

Carson CJ, Fanning CM, Wilson CJL (1996) Timing of the Progress Granite, Larsemann Hills: additional evidence for early Palaeozoic orogenesis within the east Antarctic Shield and implications for Gondwana assembly. Austral J Earth Sci 43:539–553

Collerson KD, Reid E, Millar D, McCulloch MT (1983) Lithological and Sm-Nd isotopic relationships in the Vestfold Block: implications for Archaen and Proterozoic evolution in East Antarctica. In: Oliver RL, James PR, Jago JB (eds) Antarctic earth science. Australian Academy of Science, Canberra, pp 77–84

Fitzsimons ICW (1997) The Brattstrand Paragneiss and the Søstrene Orthogneiss: A review of Pan-African metamorphism and Grenvillian relics in southern Prydz Bay. In: Ricci CA (ed) The Antarctic region: geological evolution and processes. Terra Antartica Publications, Siena, pp 121–130

Fitzsimons ICW (2000) A review of tectonic events in the East Antarctic Shield, and their implications for Gondwana and earlier supercontinents. J African Earth Sci 31:3–23

Fitzsimons ICW, Harley SL (1991) Geological relationships in high-grade gneiss of the Brattstrand Bluffs coastline, Prydz Bay, East Antarctica. Austral J Earth Sci 38:497–519

Fitzsimons ICW, Thost DE (1992) Geological relationships in high-grade basement gneiss of the northern Prince Charles Mountains, East Antarctica. Austral J Earth Sci 39:173–193

Gibb RA, Thomas MD (1976) Gravity signature of fossil plate boundaries in the Canadian Shield. Nature 262:199–200

Golynsky AV, Masolov VN, Nogi Y, Shibuya K, Tarlowsky C, Wellman P (1996) Magnetic anomalies of Precambrian terranes of the East Antarctic Shield coastal region (20° E–50° E). Proc NIPR Symp Antarct Geosci 9:24–39

Golynsky AV, Alyavdin SV, Masolov VN, Tscherinov AS and Volnukhin VS (2002) The composite magnetic anomaly map of the East Antarctica. Tectonophysics 347:109–120

Golynsky AV, Golynsky DA, Masolov VN, Volnukhin VS (2006). Magnetic anomalies of the Grove Mountains region and their geological significance. In: Fütterer DK, Damaske D, Kleinschmidt G, Miller H, Tessensohn F (eds) Antarctica: Contributions to global earth sciences. Springer, Berlin Heidelberg New York, pp 95–106

Grew ES, Manton WI (1981) Geochronologic studies in East Antarctica: ages of rocks at Reinbolt Hills and Molodezhnaya Station. Antarc J United States 16(5):5–7

Harley SL, Snape I, Black LP (1998) The evolution of a layered meta-igneous complex in the Rauer Group, East Antarctica: evidence for a distinct Archaean terrane. Precambrian Res 89: 175–205

Hensen BJ, Zhou B (1995) A Pan-African granulite facies metamorphic episode in Prydz Bay, Antarctica: evidence from Sm–Nd garnet dating. Austral J Earth Sci 42:249–258

Hensen BJ, Zhou B (1997) East Gondwana amalgamation by Pan-African collision? Evidence from Prydz Bay, East Antarctica. In: Ricci CA (ed) The Antarctic region: geological evolution and progress. Terra Antarctica Publ., Siena, pp 115–119

Hensen BJ, Zhou B, Thost DE (1997) Recognition of multiple high grade metamorphic events with garnet Sm-Nd chronology in the northern Prince Charles Mountains, Antarctica. In: Ricci CA (ed) The Antarctic region: geological evolution and progress. Terra Antarctica Publ., Siena, pp 97–104

Kamenev EN (1972) Geology of Enderby Land. Antarctica 14, M. Nauka, pp 34–58, (in Russian)

Kamenev EN, Kameneva GI, Mikhalsky EV, Andronikov AV (1990) The Prince Charles Mountains and Mawson Escarpment. In: Ivanov VL, Kamenev EN (eds) The geology and mineral resources of Antarctica. Nedra, Moscow, pp 67–113 (in Russian)

Kamenev EN, Andronikov AV, Mikhalsky EV, Krasnikov NN, Stüwe K (1993) Soviet geological maps of the Prince Charles Mountains, East Antarctic Shield. Austral J Earth Sci 40:501–517

Kelly NM, Clarke GL, Fanning CM (2002) A two-stage evolution of the Neoproterozoic Rayner Structural Episode: new U-Pb sensitive high resolution ion microprobe constrains from the Oygarden Group, Kemp Land, East Antarctica. Precambrian Research 116:307–330

Kinny PD, Black LP, Sheraton JW (1993) Zircon ages and the distribution of Archaean and Proterozoic rocks in the Rauer Islands. Antarctic Science 5:193–206

Kinny PD, Black LP, Sheraton JW (1997) Zircon U–Pb ages and geochemistry of igneous and metamorphic rocks in the northern Prince Charles Mountains, Antarctica. AGSO Journal of Australian Geology and Geophysics 16:637–654

Klasner JS, King ER (1986) Precambrian basement geology of North and South Dacota. Can J Earth Sci 23:1083–1102

McLeod IR (1964) An outline of the geology of the sector from longitude 45° E to 80° E, Antarctica. In: Adie RJ (ed), Antarctic geology. North Holland, Amsterdam, pp 237–247

Mikhailov VM, Sergeyev MB (1997) Main lithological formations of high-grade metamorphic terrain of the northern Prince-Charles Mountains, East Antarctica, and their subglacial distribution. In: Ricci CA (ed), The Antarctic region: geological evolution and processes. Terra Antartica Publications, Siena, pp 89–95

Mikhalsky EV, Sheraton JW, Laiba AA, Tingey RJ, Thost DE, Kamenev EN, Fedorov LV (2001) Geology of the Prince Charles Mountains, Antarctica. AGSO Geoscience Australia Bull 247:1–209

Mikhalsky EV, Laiba AA, Beliatsky BV (2006) Main geological features of the Prince Charles Mountains. In: Fütterer DK, Damaske D, Kleinschmidt G, Miller H, Tessensohn F (eds) Antarctica: Contributions to global earth sciences. Springer, Berlin Heidelberg New York, pp 69–82

Provodnikov LY (1975) The basement of platform regions of Siberia. Academy of Science of the USSR, Siberian Branch. Transactions of the Institute of Geology and Geophysics 194:42–57

Ravich MG, Soloviev DS, Fedorov LV (1978) Geological structure of MacRobertson Land, East Antarctica. Gidrometeoizdat, Leningrad, (in Russian)

Scrimgeour I, Hand M (1997) A metamorphic perspective on the Pan African overprint in the Amery area of MacRobertson Land, East Antarctica. Antarctic Science 9:313–335

Sergeyev MB, Kaulio VM (1993) High-grade metamorphic rocks of the northern Prince Charles Mountains, East Antarctica: Their subdivision and history. In: Findlay RH, Unrug R, Banks MR, Veevers JJ (eds) Gondwana eight. Balkema, Rotterdam, pp 153–160

Sheraton JW (1982) Origin of charnockitic rocks of MacRobertson Land. In: Craddock C (ed), Antarctic geoscience. University of Wisconsin Press, Madison, pp 489–497

Sheraton JW, Collerson KD (1983) Archaean and Proterozoic geological relationships in the Vestfold Hills–Prydz Bay area, Antarctica. BMR J Austral Geol Geophys 8:119–128

Sheraton JW, Tingey RJ, Black L, Offe LA, Ellis DJ (1987) Geology of an unusual Precambrian high-grade metamorphic terrane – Enderby and western Kemp Land, Antarctica. Bureau of Mineral Resources, Australia, Bulletin 223

Snape I, Black LP, Harley SL (1997) Refinement of the timing of magmatism, high-grade metamorphism and deformation in the Vestfold Hills, East Antarctica, from new SHRIMP U–Pb zircon geochronology. In: Ricci CA (ed) The Antarctic region: geological evolution and processes. Terra Antarctica Publications, Siena, pp 139–148

Stüwe K, Hand M (1992) Geology and structure of Depot Peak, MacRobertson Land. More evidence for the continuous extent of the 1 000 Ma event of East Antarctica. Austral J Earth Sci 39:211–222

Stüwe K, Sandiford M (1993) A preliminary model for the 500 Ma event in the East Antarctic Shield. In: Findlay RH, Unrug R, Banks MR, Veevers JJ (eds) Gondwana eight. Balkema, Rotterdam, pp 125–130

Thost DE, Leitchenkov GL, O'Brien PE, Tingey RJ, Wellman P, Golynsky AV (1998) Geology of the Lambert Glacier-Prydz Bay region, East Antarctica, 1 : 1 000 000 map. AGSO, Canberra

Tingey RJ (1991) The regional geology of Archaean and Proterozoic rocks in Antarctica. In: Tingey RJ (ed) The geology of Antarctica. Clarendon Press, Oxford, pp 1–58

Wellman P (1988) Development of the Australian Proterozoic crust as inferred from gravity and magnetic anomalies. Precambrian Research 40/41:89–100

Wilson T, Grunow AM, Hanson RE (1997) Gondwana Assembly: The review from southern Africa and East Gondwana. J Geodynamics 23:263–286

Yoshida M, Jacobs J, Santosh M, Rajesh HM (2003) Role of Pan-African events in the Circum-East Antarctic orogen of East Gondwana: a critical overview. In: Yoshida Y, Windley BF, Dasgupta S (eds) Proterozoic of East Gondwana: supercontinent assembly and breakup. Geological Society, London, Special publications 206:57–75

Young DN, Zhao J-X, Ellis DJ, McCulloch MT (1997) Geochemical and Sr-Nd isotopic mapping of source provinces for the Mawson charnockites, East Antarctica: implications for Proterozoic tectonics and Gondwana reconstruction. Precambrian Research 86:1–19

Zhao Y, Song B, Wang Y, Ren L, Li J, Chen T (1992) Geochronology of the late granite in the Larsemann Hills, East Antarctica. In: Yoshida Y, Kaminuma K, Shiraishi K (eds) Recent progress in Antarctic earth science. Terra Scientific Publishing, Tokyo, pp 155–161

Zhao Y, Liu X, Song B, Zhang Z, Li J, Yao Y, Wang Y (1995) Constraints on the stratigraphic age of metasedimentary rocks from the Larsemann Hills, East Antarctica: possible implications for Neoproterozoic tectonics. Precambrian Research 75:175–188

Zhao Y, Liu X, Fanning CM, Liu X (2000) The Grove Mountains, a segment of a Pan-African orogenic belt in East Antarctica. 31st International Geological Congress, Rio de Janeiro, Brazil

Magnetic Anomalies of the Grove Mountains Region and Their Geological Significance

Alexander V. Golynsky[1] · Dmitry A. Golynsky[2] · Valery N. Masolov[3] · Vyacheslav S. Volnukhin[3]

[1] VNIIOkeangeologia, Angliysky Avenue 1, St. Petersburg 190121, Russia, <sasha@vniio.nw.ru>

[2] SPbSU, 7/9, Universitetskaya nab., St. Petersburg, 199034, Russia

[3] PMGRE, Pobeda Street, 24, Lomonosov, 189510, Russia

Abstract. The geology of the Grove Mountains is poorly known. The magnetic anomalies of the Grove Mountains and surrounding areas are characterized by a prominent NE-SW to E-W fabric. Well defined magnetic anomalies along its periphery with the absence of intensive magnetic anomalies in the north are a distinctive feature of this region. The obliqueness of magnetic anomalies along the study area's boundary with respect to its central part suggests that the northern Grove Mountains basement may be much older crust than the neighboring terranes of Meso- to early Neoproterozoic high-grade metamorphic rocks. The existence of two ancient cratonic blocks in the southern Prince Charles Mountains and Vestfold Hills suggests that this region may contain Archaean or Paleoproterozoic crust. The Grove Mountains crustal block is clearly discernible in the aeromagnetic data and can be considered as a region that underwent Grenvillian and/or Pan-African or both tectonism and reworking. The absence of any visible magnetic trends running towards the Prydz Bay coast or central part of the Mawson Escarpment precludes any direct tectonic correlation with these regions.

Introduction

The geologic and tectonic setting of the Grove Mountains is poorly understood. They have been correlated with the southern (e.g., England and Langworthy 1975; Kamenev et al. 1990; Kamenev et al. 1993) and the northern Prince Charles Mountains (e.g., Tingey 1991). They also have been considered as a distinct terrane (e.g., Mikhalsky et al. 2001) because the Pan-African event shows some affinities with the Prydz Bay coast area (Zhao et al. 2000; Mikhalsky et al. 2001; Liu et al. 2002; Fitzsimons 2003). The Grove Mountains and adjacent areas of the Prince Charles Mountain-Prydz Bay region form part of a complex Mesoproterozoic to early Neoproterozoic mobile belt surrounding Archaean cratonic blocks and consisting of three major tectonics provinces (Tingey 1982; Kamenev et al. 1993). They are the late Archaean granulite-gneiss Vestfold Block, the mainly Archaean to Paleoproterozoic Ruker Terrane in the southern Prince Charles Mountains (PCM), and the Mesoproterozoic to early Neoproterozoic mobile belt in the northern and central PCM, eastern Amery Ice Shelf and Prydz Bay coast. Pan-African high-grade metamorphism in the southern Prydz Bay area (Zhao et al. 1992; Dirks and Wilson 1995; Hensen and Zhou 1995; Carson et al. 1996; Hensen et al. 1997; Fitzsimons 1997) is generally agreed.

Isotopic results demonstrate that outcrops of the Mawson Escarpment consist of distinct terranes of two different ages (Boger et al. 2001). The southernmost Ruker Terrane underwent no deformation younger than 2 650 Ma, while the northern Lambert Terrane was deformed during the Cambrian (~550–490 Ma). Boger et al. (2001) argued that this 550–500 Ma magmatism and deformation was evidence for the Lambert Terrane being a Cambrian suture with the continuation of the high-grade belt exposed along the Prydz Bay coast. However, Fitzsimons (2003) suggested that evidence for a Cambrian suture in the Lambert Terrane remains inconclusive unless it can be demonstrated that 550–500 Ma tectonism extends west of the Mawson Escarpment. In general, no conclusive evidence exists that Pan-African tectonism in the Prydz Bay coast area involved ocean closure. There are no exposed ophiolites, high pressure blueschists nor eclogites indicative of subduction and thus no record of convergent margin magmatism preceding the inferred collision (Fitzsimons 2000).

To clarify these speculative relationships better, this paper analyzes the aeromagnetic data collected in the Grove Mountains region during the 1999–2000 field season. Aeromagnetic surveys are powerful for mapping basement structures and unraveling the tectonic history of a region, particularly in areas of poor outcrop. They provide a means to extrapolate known geology exposed in widely separated outcrops into regions covered by younger sediments or ice. They can also help to delineate structural features not clearly recognizable from outcrop mapping. However, correlating the Grove Mountains magnetic anomalies to high-grade metamorphic complexes is rather difficult at present due to the very limited amount of geologic information for the region.

Geology of the Grove Mountains

The Grove Mountains lie to the east of the Prince Charles Mountains and Mawson Escarpment and are about 350 km south of the Prydz Bay coast. These isolated exposures in East Antarctica are largely unstudied despite having been visited by several field parties. Outcrops in

From: Fütterer DK, Damaske D, Kleinschmidt G, Miller H, Tessensohn F (eds) (2006) Antarctica: Contributions to global earth sciences. Springer-Verlag, Berlin Heidelberg New York, pp 95–106

the Grove Mountains comprise layered grey migmatic gneiss, tonalitic to granitic gneiss, quartzite and biotite-garnet paragneiss, with rare mafic schist and calc-silicate rocks. Intrusive granites and granodiorites crop out at few localities. Charnockite was inferred at Mount Harding (Tingey and England 1973), although Liu et al. (2002) found no outcrop of charnockite in the investigated area. Russian geologists who visited the Grove Mountains area during a reconnaissance survey in the 2002–2003 field season also confirmed this finding (D. Vorob'ev and V. Maslov pers. comm. 2003). Thus, it is unclear what kind of a charnockite sample Mikhalsky et al. (2001) analyzed to obtain the Pan-African age of about 504 Ma for the emplacement of the pluton.

A pale felsic gneiss, dark felsic gneiss, mafic granulite, garnet clinopyroxenite, coarse grained two-feldspar granite and fine-grained granodioritic dikes have been identified recently in the Grove Mountains (Liu et al. 2002). Zhao et al. (2000) and Liu et al. (2002) reported ion-microprobe (SHRIMP) U-Pb zircon ages of 529 ±14 Ma for the felsic gneiss from Melvold Nunataks, 534 ±5 Ma for the syntectonic granite, and 501 ±7 Ma for a granodiorite

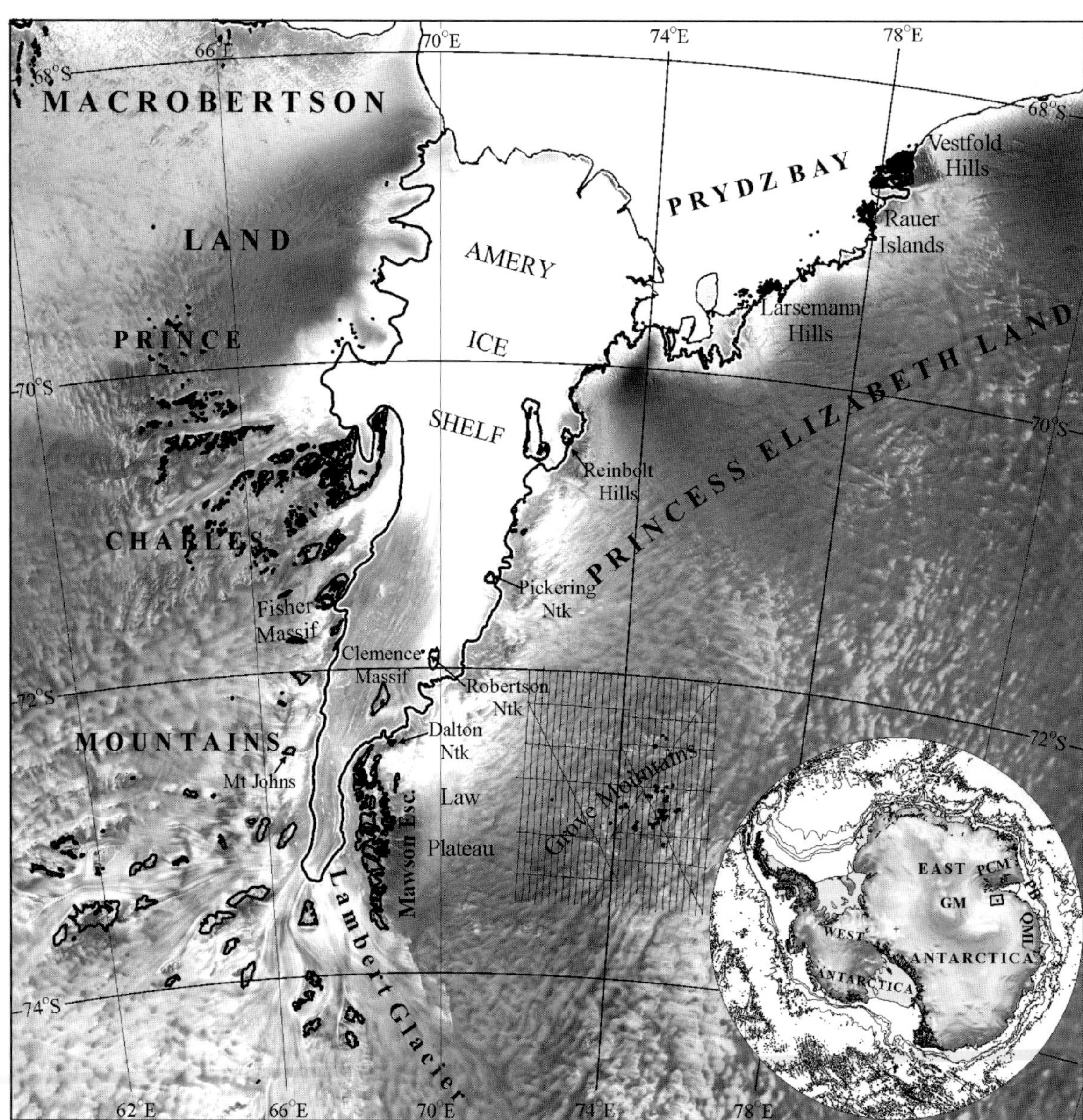

Fig. 2.9-1. Location of aeromagnetic surveys over the Grove Mountains area. Aeromagnetic profiles with the flight-line spacing of about 5 km are shown by *solid black lines*. Basemap is a portion of the RADARSAT composite map. Map projections throughout this paper are polar stereographic. *Insert* in the right corner shows the location of study area within Antarctica. *PCM:* Prince Charles Mountains; *GM:* Gamburtsev Subglacial Mountains; *PB:* Prydz Bay; *QML:* Queen Mary Land

dyke. The older ages (870, 906 and 953 Ma) were interpreted as inherited crystallisation ages. The age data of Zhao et al. (2000) indicate that high-grade metamorphism, deformation, and syntectonic granite emplacement occurred in the Grove Mountains at about 530 Ma. They reported also that the mineralogy of the metamorphic rocks exhibits equilibrium assemblages that suggest a single granulite facies event. This result is very unusual in comparison with other parts of East Antarctica, where typically two periods of granulite facies metamorphism developed widely in the various metamorphic rocks. As suggested by many authors (e.g., Shiraishi et al. 1992; Zhao et al. 1993; Yoshida 1994; Hensen and Zhou 1995; Hensen et al. 1997; Jacobs et al. 1998) all distinguished Pan-African belts carry clear evidence of Grenvillian precursors. In accordance with D. Vorob'ev, V. Maslov and A. Laiba (pers. comm. 2004) the high-metamorphic rocks of the Grove Mountains underwent granulite and amphibolite facies metamorphism. They reported also that their U-Pb SHRIMP zircon analysis identified magmatic zircons in orthogneiss from the southernmost nunatak of the Gale escarpment with an age of about 984 ±9 Ma which they interpreted as the original crystallization age of these meta-igneous units.

Aeromagnetic Survey

The aeromagnetic survey in the Grove Mountains area (71.9–73.6° S; 72–76° E) was carried out by the Polar Marine Geological Research Expedition (PMGRE) in 1999/2000 with 5-km flight-line spacing and 30-km tie-lines. The survey included radio-echo soundings of the sub-ice topography. The aeromagnetic campaign followed closely the layout of the previous PMGRE campaigns for maximum compatibility between neighboring datasets. Nearly 5 450 km of profiles were flown in a N-S direction over an area more than 25 000 km^2. An Antonov-2 fixed-wing aircraft flew the survey at 2 000 m above sea level over ice-covered regions and at 2 500 m over the mountains. A proton magnetometer mounted in a 5 m tail stinger with a sensitivity of 0.01 nT and sample frequency of 1 Hz acquired the total field magnetic data. A base station magnetometer at a field camp located in the northern part of the survey area was used to correct for time variations of the magnetic field. The surveying occurred during the magnetically quietest period of the day and was suspended during magnetic storms. Differential global positioning systems provided navigation accuracy estimated conservatively at about ±50 m.

Each survey was processed separately for base station corrections, removal of the International Geomagnetic Reference Field adjusted for secular variation, and the levelling process were done individually and two maps produced. The field values were adjusted at flight line crossing with tie lines. The root mean squared error after adjustment at the tie lines is less than 5 nT. The lower level survey was upward continued to the 2 500 m level for merging with the higher level section. The upward continued anomaly amplitudes were reduced slightly and did not exceed 15–20 nT of the initial values. The processed magnetic anomaly data were gridded by using a minimum curvature algorithm (Briggs 1974), the grid interval was selected as 1.5 km. The resulting grid is shown in Fig. 2.9-2 as the colour shaded-relief map that was created by using ER Mapper software system (Earth Resource Mapping Pty Ltd. 1995). The obtained data provide the only geophysical information available for the upper crustal structure of this region.

Magnetic Anomaly Map

The magnetic anomaly patterns s of the Grove Mountains and surrounding areas are variable (Fig. 2.9-2). The magnetic anomalies are relatively quiet in the northwestern half of the study area, whereas positive high-frequency anomalies dominate the southeastern half. The magnetic anomalies vary substantially from easterly trends in the northwest and north to NE-SW trends in the southeast. These anomaly trends appear to correspond to the regional strikes of geological structures. The complicated spectral properties of the magnetic anomalies reflect the superposition of commonly occurring anomalies with different spatial scales. The magnetic anomalies range in amplitude from −235 to 400 nT and wavelengths from the 5 km lower limit of survey resolution to tens of kilometers.

To help resolve different geologic features, we applied several map enhancements and analytical techniques including shaded-relief maps of the magnetic anomaly derivatives, and their analytical signal and pseudogravity effects. While we do not show these results in this paper due to space restrictions, these enhancements allowed us to identify important anomaly lineaments, trends and patterns for the geologic synthesis given in Fig. 2.9-3. Further subdivisions of the magnetic anomalies were delineated based on their intensities, spectral properties, and secondary trends.

The distinct magnetic anomaly patterns shown in Fig. 2.9-2 allowed the region to be subdivided into the Grove, Gale, Northern, and Dalton sub-units (Fig. 2.9-3). These sub-units together with the Manning sub-unit belong to the Princess Elizabeth magnetic unit that overlies mainly the eastern shoulder of the Lambert-Amery rift system (Golynsky et al. 2006). The critical attribute of the study area is absence of the intense anomalies observed in the Northern sub-unit together with the coherent anomaly pattern along its periphery, where the magnetic anomalies are oblique to the central anomalies.

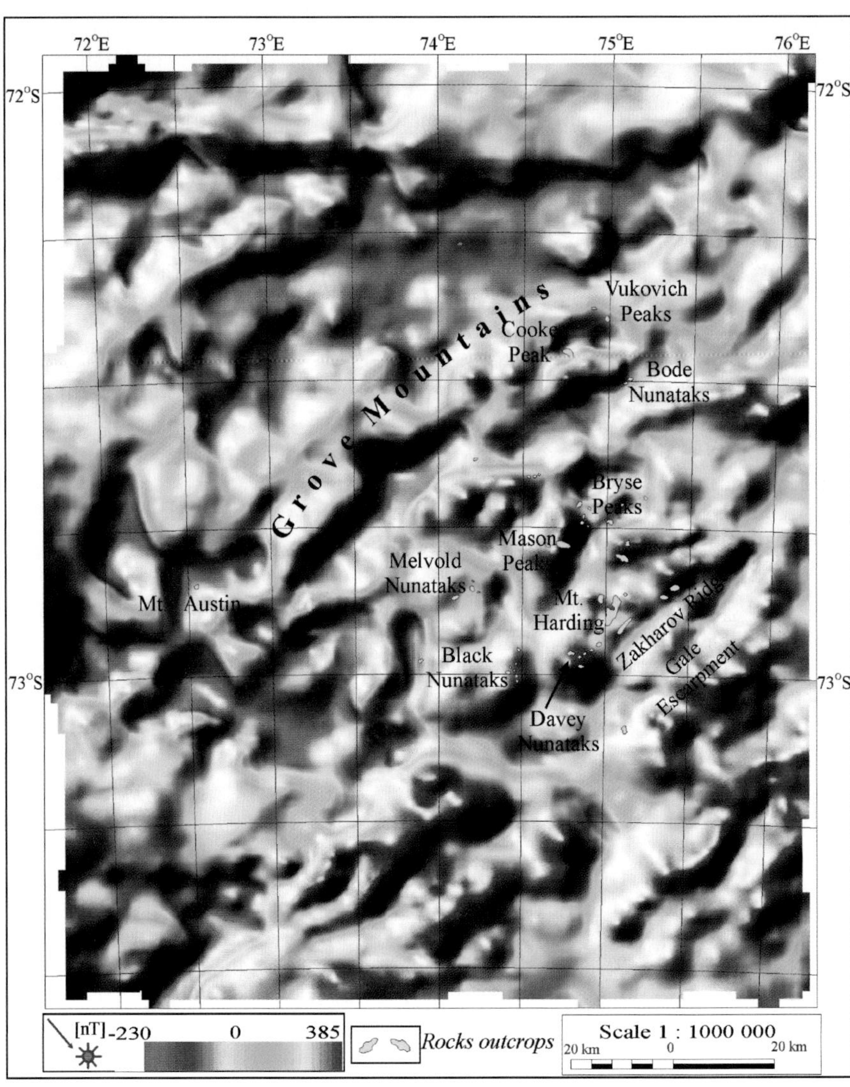

Fig. 2.9-2.
Shaded-relief aeromagnetic map of the Grove Mountains region (illumination is from the northwest at an inclination of 45°)

Grove Sub-Unit

The Grove sub-unit is outlined mainly over the Grove Mountains and is characterized by the magnetic anomalies of moderate amplitude (~200–250 nT; Fig. 2.9-2). The predominant magnetic grain is broken with a number of linear anomalies. The southern boundary of the Grove sub-unit is sharply defined by the curvilinear negative anomaly (*Anomaly N2*, Fig. 2.9-3, 1f) running from the eastern border of the Dalton sub-unit to 75° E (Fig. 2.9-2). It continues further east as a chain of discrete lows. The anomaly pattern the Grove sub-unit can be subdivided into five contrasting crustal blocks with different anomaly forms, sizes, continuities, orientations and crosscutting relationships.

For example, low-to-moderate positive anomalies (50–150 nT) characterize the arcuate *Vukovich* region (Fig. 2.9-3, 1e) in the northern and northwestern parts of the sub-unit (Fig. 2.9-2). NE-SW linear alignments of positive anomalies extend from Mount Austin in the west through Cook and Vukovich Peaks and farther eastwards. The width of the subdivision is constant (~20 km) and does not exceed 25 km. The northern boundary is associated with gentle anomaly gradients, whereas the southern boundary shows steeper, more pronounced gradients. The most significant high-amplitude negative anomaly (~ -325 nT; Fig. 2.9-2) of the Grove Mountains area is located close to the southern boundary of the Vukovich region. The origin of this low is unknown, but its round-shape suggests that it may reflect an intrusion of possibly granitic composition.

The geological origin of the Vukovich region to the Grove sub-unit is relatively uncertain. It is conceivable that it can be considered as a part of the Northern sub-unit due to the many similarities between them. These include similar anomaly intensities and predominant trends, etc. The causative sources of magnetic anomalies are not clari-

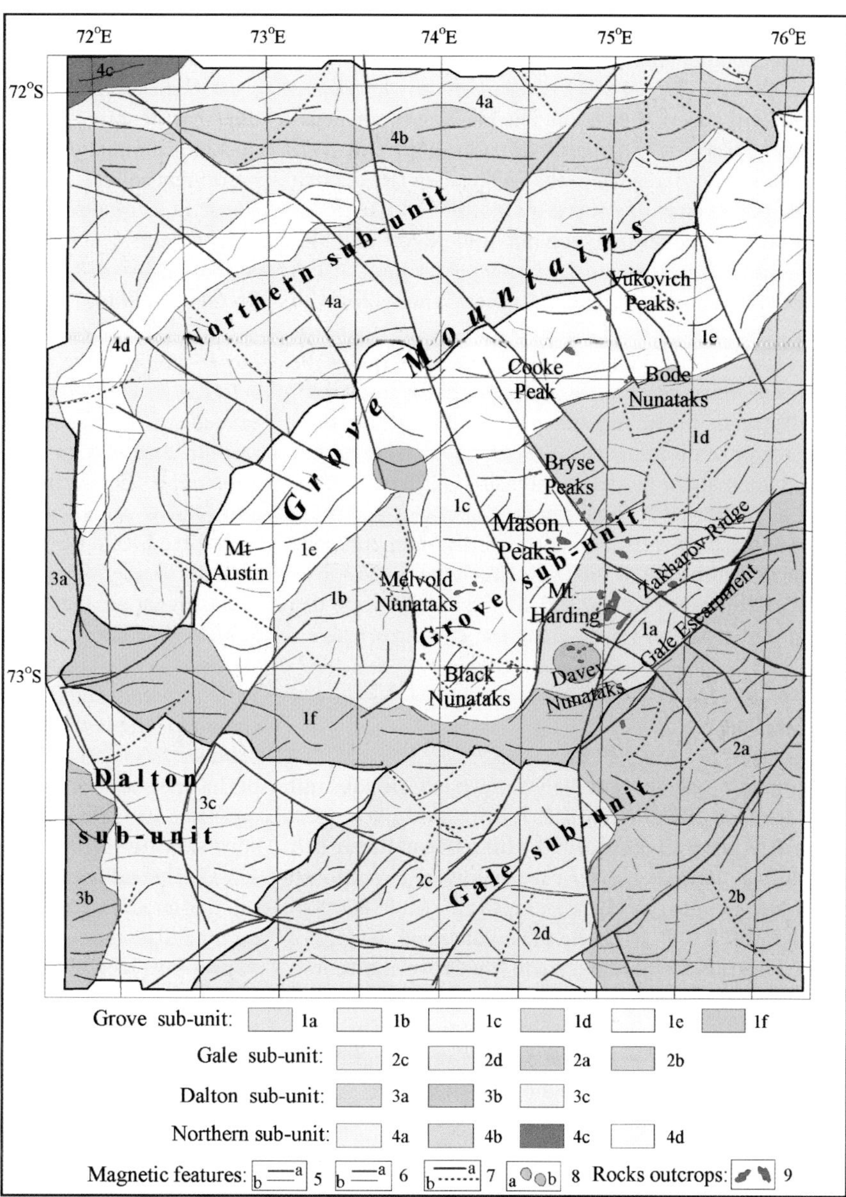

Fig. 2.9-3. Structural map of the Grove Mountains region based on aeromagnetic data. Structural subdivisions of magnetic anomaly data: *1:* Grove Mountains sub-unit: *1a:* Zakharov area, *1b:* Melvold area, *1c:* Black area, *1d:* Bryse area, *1e:* Vukovich area, *1f:* Anomaly N2; *2:* Gale sub-unit: *2a:* North-eastern area, *2b:* South-eastern area, *2c:* Western area, *2d:* Central area; *3:* Dalton sub-unit: *3a:* Northern area, *3b:* Southern area, *3c:* Eastern area; *4:* Northern sub-unit: *4a:* SAE36-RAE45 area, *4b:* anomaly of 72° S, *4c:* Robertson Anomaly, *4d:* Anomaly N1. Magnetic features: *5:* Axis of positive anomalies: *a:* high-moderate intensity, *b:* low intensity; *6:* Axis of negative anomalies: *a:* high-moderate intensity, *b:* low intensity; *7:* Transverse discontinuities: *a:* major, *b:* minor; *8:* Circular magnetic anomalies: *a:* positive, *b:* negative; *9:* Rocks outcrops

fied at present. However, the negative anomalies may largely reflect metasedimentary rocks which are widely distributed in the northern part of the Grove Mountains such as at Cook Peak, Bryse Peaks, Tate Rocks, and the Vukovich Peaks (D. Vorob'ev and V. Maslov pers. comm. 2003).

The *Melvold* region (Fig. 2.9-3, 1b) is found immediately west of the Melvold Nunataks where outcrops are lacking and therefore greatly hinder anomaly interpretation. It is highly likely that metasedimentary rocks similar to those outcropping at Mount Austin (Mikhalsky et al. 2001) are major sources of the negative anomalies, because both regions show distinctive negative anomalies that differentiate these regions from those to the east. The low-amplitude positive anomalies (50–75 nT) do not form clear trends here (Fig. 2.9-2).

Characterized largely by short-wavelength anomalies (5–7 km and 10–15 km) of moderate amplitude (150–200 nT; Fig. 2.9-2), the *Black* Nunataks region (Fig. 2.9-3, 1c) occupies the central part of the Grove sub-unit. The main trend of these anomalies is not possible to determine as they show the distinct superposition of different order anomalies. The positive anomalies align in easterly, northerly, and ENE directions. The weak linearity of magnetic anomalies within the Black Nunataks region is recognized in several places. This anomaly pattern may be explained by the high portion of metaigneous rocks in this region (D. Vorob'ev and V. Maslov pers. comm. 2003).

The magnetic anomaly pattern of the north-eastern part of the Grove sub-unit (*Bryse region*, Fig. 2.9-3, 1d) is associated with linear medium-wavelength anomalies of moderate intensities (150–225 nT). They trend W-E as well as NE-SW. The NE-SW trends are mainly located in the southern part of the region. The low-amplitude negative anomalies (~ -75 nT) are not continuous and typically isometric or irregular in shape. The majority of positive anomalies may be caused by biotite and garnet-biotite granite-gneisses that crop out at the Bryse and Mason Peaks (D. Vorob'ev and V. Maslov pers. comm. 2003).

Moderate, short-wavelength magnetic anomalies with amplitudes ranging from 150 to 250 nT of the *Zakharov region* (Fig. 2.9-3, 1a) form the south-eastern limit of the Grove sub-unit. They are about 20–25 km in length with NE-SW trends prevailing along the southern boundary of the region. A prominent feature of the region is a round-shaped anomaly of moderate intensity (~275 nT) recognized at the Davey Nunataks area. Its circular geometry strongly suggests a relationship with the monzodiorite pluton that crops out at the Davy Nunataks (D. Vorob'ev and V. Maslov pers. comm. 2003). Monzodiorites were also recognized at Mount Harding and the Zakharov Ridge.

Gale Sub-Unit

The complex magnetic anomaly pattern of Gale sub-unit (Fig. 2.9-3) clearly differentiates it from the adjacent terranes. It is generally associated with a broad linear belt of positive anomalies consisting mainly of superposed short-wavelength, moderate amplitude (200–300 nT) anomalies (Fig. 2.9-2). Positive anomalies with lengths of about 50–60 km trend obliquely to the Grove sub-unit in linear bands. The prevalent trend of magnetic anomalies is NE-SW, although a few also show E-W trends. The characteristic anomaly patterns of the sub-unit can be further subdivided into western, north-eastern, and south-eastern components. When compared to other areas, the *western region* (Fig. 2.9-3, 2c) shows a higher level of magnetic field intensity (~300 nT) and steeper gradients. It consists of two composite positive bands, divided by a linear minimum. Sharp changes of the magnetic anomaly fabric along the western border of the region may reflect a tectonic contact of two different terranes or transcurrent lineaments.

Magnetic anomaly amplitudes of the *north-eastern* (Fig. 2.9-3, 2a) region do not exceed 250 nT. It is noteworthy that two complex positive bands are well recognized as was the case for the western region. The boundary between the north-eastern and south-eastern regions is outlined by the distinctive changes in the strike of anomalies.

The intensity of magnetic anomalies in the *south-eastern region* (Fig. 2.9-3, 2b) is notably less (<200 nT) when compared to the neighboring areas. As is usually the case, the diminishing amplitudes leads to decreasing magnetic anomaly gradients. These changes of the magnetic anomaly signature are due to changes in depth to sources and that crustal blocks have been uplifted by varying degrees along the strike-length of the Gale Anomaly belt. The banded magnetic anomalies are striking. They also show a rather prominent coulisse-like pattern with largely by linear and isometric components.

Northern Sub-Unit

Lower-than-average intensity magnetic anomalies characterize the Northern sub-unit, with rare exceptions not exceeding 100 nT. The observed pattern is relatively smooth and regular with linear positive and negative anomalies. They form partial linear chains with easterly and NE-SW trends although single anomalies are also widespread. One of the most prominent features of the area is the slightly curvilinear, eastwardly trending belt of positive anomalies that runs close to 72° S latitude from 72° E to 76° E longitudes that may also extend farther eastwards. Due to its proximity, it was named *the 72° S magnetic anomaly* (Fig. 2.9-3). Structurally it represents a chain of elongated local anomalies divided by regions of subdued or negative anomalies. The five major segments of the anomaly are characterized by distinct wavelength, intensity and fabric properties. The intensities of these segments are surprisingly constant and vary from 150 to 175 nT. The widths of individual segments vary from a maximum of about 15 km to a minimum of about 8–10 km. Segments and distinct offsets of magnetic anomalies axes and appearance of composite anomalies reveal a system of structural displacements which can be interpreted as amagmatic faults located 35–40 km from each other. The origin of the 72° S magnetic anomaly is not immediately evident and it may be caused either by infrastructural variations in the basement or by fault-localized or cut intrusions. We infer that, it is not a suture zone because of the remarkable similarity of the magnetic anomaly patterns on each side of this feature.

The *Robertson Anomaly* (Fig. 2.9-3) occurs at the extreme northwestern corner of the Grove Mountains data set presented here and occupies the central part of the Princess Elizabeth unit (Fig. 2.9-4). It apparently can be outlined over the western flank of the Lambert Glacier bordering between the Beaver and Ruker units (Golynsky et al. 2006). The major characteristics of this subdivision are the oblique pattern of magnetic anomalies along margins of the Princess Elizabeth unit (Fig. 2.9-4) and the absence of the intense anomalies within its inner part. Together with the 72° S magnetic anomaly that might be genetically related to the Robertson Anomaly, it divides the Princess Elizabeth Land magnetic unit into northern and southern components (Fig. 2.9-4). Relative to the 72° S magnetic anomaly, the Robertson Anomaly is characterized by sharp gradients and more complex inner structures. It is about 250 km in length, and consists from three distinct segments that reach a maximum amplitude of more than 600 nT (Fig. 2.9-4). The Robertson Anomaly gradually decreases in amplitude toward the east, where it dies out close to 73.5° E. The westward continuation of the anomaly is problematic and difficult to distinguish northeast of Mount Johns. It appears that the source of the Robertson Anomaly was more homogeneous before the formation of the Lambert rift structure. The development of the Lambert rift led to the appearance of three fragments divided by N-S trending negative lows. Two of these segments are located over the eastern and western shoulders of the rift, whereas the third corresponds to the Clemence Massif area. The linear N-S negative lows are inferred to be related with the Lambert rift bounding faults.

The complex Robertson Anomaly clearly reflects the prolonged geologic evolution of this region. The superposition of effects from processes at different geologic times hampers the interpretation of the anomaly and its relationship with neighboring subdivisions of the magnetic field. The two fragments of the Robertson Anomaly within the Lambert Glacier likely constitute different continuations of the linear magnetic anomalies belonging to the Manning sub-unit (Golynsky et al. 2006). On the other hand, the southern part of the anomaly over the eastern flank of the glacier might be the modified extension of the composite magnetic anomalies from the Dalton sub-unit. If this is the case, then the 72° S magnetic anomaly may well represent the linear continuation of the structurally composite eastern belt recognized in the Dalton sub-unit. In general, the structural relationships between the three subdivisions and the origins of the magnetic sources are not clear at present. Preliminary magnetic property data from the Clemence Massif show that only Bt-Amf gneisses yield the high volume susceptibility of about 43×10^{-3} SI units to account for the observed magnetic anomalies (M. Sergeyev, unpublished data).

Dalton Sub-Unit

This subdivision (Fig. 2.9-3) occurs to the west of the main study area and lies mainly over the Dalton nunataks area and part of the northern Mawson Escarpment (Fig. 2.9-4), as well as over the Law Plateau (Golynsky et al. in press). The major characteristics of the sub-unit are the high-intensity anomalies up to more than 700 nT, the distinct superposition of different order magnetic anomalies, and the predominant E-W trends of the local anomalies that form two composite N-S striking anomaly belts divided by a linear low composed of elongated and circular anomalies.

Both curvilinear anomaly belts are concave to the east and thus are oblique to the structural style of the central Princess Elizabeth Land unit. The eastern belt is characterized by a higher amplitude positive anomalies (>700 nT in places), whereas in the western belt they do not exceed 550 nT. This intensity distribution might be explained by deeper magnetic source depths in the western belt than in the eastern belt. The depth calculations show that the causative sources of the western belt may be located at about 3 km depth. It forms a linear depression running parallel to the eastern flank of Mawson Escarpment. The northern part of this linear depression joins the Lambert rift system close to Clemence Massif, but its southern continuation problematic.

Manning Sub-Unit

While the Manning sub-unit is not present within the study area, it is crucially important for understanding the Prince Charles Mountains region and surrounding area as a whole (Golynsky et al. 2006). The magnetic structure of the Manning sub-unit is extremely straightforward and unique relative to the other regions of the East Antarctic (Golynsky et al. 2001; Golynsky et al. 2002). The major characteristics of the sub-unit are the alternating low amplitude, positive and negative anomalies that are essentially continuously linear and curvilinear throughout the whole magnetic division. Recently acquired, but unpublished PMGRE aeromagnetic data show that it can be traced uninterruptedly from Clemence Massif to the southern margin of the Vestfold Hills crustal block area (Fig. 2.9-4) and further east to 80° E (Leonov pers. comm. 2004).

Tectonic Implications

The origins of the magnetic anomalies within the Grove Mountains area are not well established due to poor outcrops, the absence of geological information and volume susceptibility data for high-grade metamorphic and plutonic rocks. In general, this situation also holds for the whole Princess Elizabeth Land unit. However, the oblique character of the magnetic anomalies observed along the south-eastern, western and north-western boundary of the Princess Elizabeth Land unit relative to its central part (Fig. 2.9-4) suggests that the age of geologic formations responsible for the bordering magnetic anomalies may be fundamentally younger than geologic structures within the central crustal unit. The existence of two ancient cratonic blocks in the southern Prince Charles Mountains and Vestfold Hills suggests that the Northern magnetic sub-unit may reflect a protocraton fragment like those observed in the Vestfold Hills/Rauer Islands crustal block and the Ruker Terrane. The validation of this assumption requires additional geophysical and geological information, as well as geochronology data to differentiate Archaean and

Paleoproterozoic assemblages. The limited isotopic data of the Grove Mountains area (Mikhalsky et al. 2001) neither support nor refute our assumption. These data suggest that the protolith of the Grove Mountains metasediments are either Paleoproterozoic age or derived from source rocks of this age (Mikhalsky et al. 2001).

To test our idea, more extensive geological studies are required of the northern outcrops (Cook Peak and Vukovich Peaks) that do not appear to be affected by younger events. The relative similarity of magnetic patterns within the Northern sub-unit including the area northwards from the 72° S magnetic anomaly and the southern Prince Charles Mountains suggests that both regions may form a terrane of Archean to Paleoproterozoic age that was separated during the Grenvillian extension. However, the magnetic anomaly responses of the Vestfold Hills/Rauer Islands crustal block and the Ruker Terrane are not compatible at all.

The geochronology for reliably matching adjoining aeromagnetic subdivisions of the Manning and Northern subunits is insufficient. Pb-Pb zircon geochronology of felsic gneiss from the Clemence Massif implies an age of about 1 200 Ma for the high-grade metamorphism (Belyatsky unpubl. data), while similar rocks from the Pickering Nunatak about 110 km north-east of the Clemence Massif yielded a poorly defined reconnaissance Rb-Sr whole-rock age of 1 042 ±347 Ma (Tingey 1991). The significance of Rb-Sr whole-rock age of 941 ±94 Ma for the felsic gneiss from the Dalton Nunataks is also poorly constrained (Tingey 1991). While these age data do not correlate conclusively, it seems likely that the Clemence Massif high-grade metamorphic rocks of Mesoproterozoic age responsible for the Robertson Anomaly are somewhat older than in neighboring regions (eastern shoulder of the Lambert rift). Thus, the magnetic anomaly data may indicate an initial stage of tectonism within this region. Clearly, precise age data are needed to test the tectonic correlations more conclusively.

The new Sm-Nd data show that central and northern Mawson Escarpment, Clemence Massif and Reinbolt Hills are underlain by Archaean protoliths with TDM Nd model ages of 3.0–3.5, 3.2, and 3.0–3.2 Ga, respectively (Mikhalsky et al. 2006). These regions are mostly located within the periphery of the Princess Elizabeth magnetic unit that probably is younger than its central part. Therefore, if the bordering anomalies of the unit are associated with the region of reworked early Precambrian assemblages, its central part may contain unexposed protoliths of Princess Elizabeth Land protocraton.

Understanding the occurrence of the Manning subunit linear magnetic anomaly system within the East Antarctic Shield requires additional geological control. However, the anomaly patterns in the neighboring Northern and Manning subunits involve a relatively homogeneous component in the center with obliquely striking anomalies along the periphery. This pattern suggests the presence of older geological assemblages in the central part of the crustal unit with younger units along its margin. Linear stretching of the protocratonic margins by continental rifting may have formed the marginal structures. Thus, igneous activity in the form of linear intrusions or batholiths may have accompanied the formation of the sedimentary basin. The most likely tectonic setting was in an extensional rift environment that may have been behind an active volcanic arc. It is highly likely that the Fisher Terrane composed by metavolcanics and intrusive rocks may well represent the island arc (Mikhalsky et al. 1996).

The Grove Mountains crustal block is clearly discernible in the aeromagnetic data as a region that suffered either Grenvillian and/or Pan-African tectonism and reworking focused within a region of the Gale magnetic subunit. The Grove Mountains and northern Prince Charles Mountains magnetic anomaly patterns are not compatible (Golynsky et al. 2006, Fig. 2.8-2). Therefore, relating the Grove Mountains rocks to the Mesoproterozoic mobile belt of the northern Prince Charles Mountains is problematic, even though the Proterozoic histories of these regions may have some common features.

The continuity of the Pan-African belt distinguished in the southern Prydz Bay coastal area towards the Mawson Escarpment (Boger et al. 2001; Boger et al. 2002), the Gamburtsev Subglacial Mountains (Fitzsimons 2000) and Grove Mountains (Zhao et al. 2000) contradicts the aeromagnetic evidence. The lack of magnetic trends within the Grove Mountains region running towards the Prydz Bay coast area or the central part of the Mawson Escarpment makes the tectonic correlation of these regions very difficult. In the magnetic anomaly map of Fig. 2.9-4, the Gale magnetic sub-unit continues on a southward trajectory apparently to the Dalton sub-unit as a major crustal discontinuity of the Grove Mountains region. Rb-Sr whole-rock ages of 941 ±94 Ma for the Dalton Nunataks felsic gneiss (Tingey 1991) and the U-Pb SHRIMP zircon age of about 984 ±9 Ma for the Gale escarpment orthogneiss (D. Vorob'ev, V. Maslov and A. Laiba pers. comm. 2004) suggest that it was active in Late Proterozoic time. Alternatively, it may extend towards the Gamburtsev Subglacial Mountains, although the aeromagnetic data are insufficient to constrain the lateral extent of the underlying terrane.

The rocks in the southern Prince Charles Mountains (Mawson Escarpment), southern Prydz Bay coastal area and Grove Mountains are widely held to record similar U-Pb ages and probably synchronous deformation during Pan-African orogenesis, even though the regional tectonic setting of this event in these regions is quite uncertain. It is highly likely that the Pan-African belt does not relate to these three areas. The evidence for a Cambrian suture in the Lambert Terrane requires finding the 550–500 Ma tectonically affected rocks at the eastern

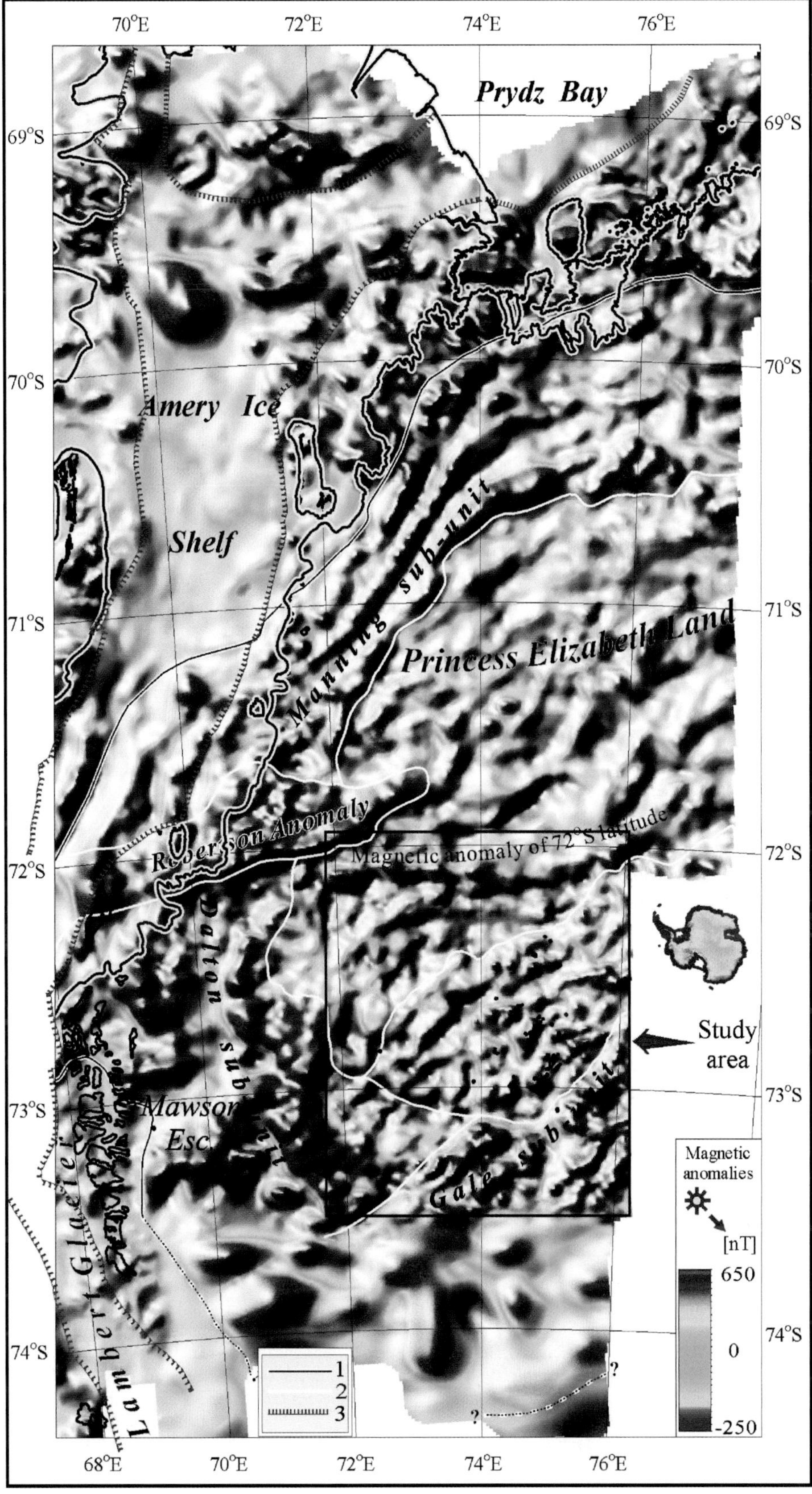

Fig. 2.9-4.
Shaded-relief aeromagnetic map of the eastern shoulder of the Lambert rift system. Image compiled using 5 km dataset north of 73.5° S. South of this area 20 km dataset was used. The boundary of the Princess Elizabeth unit is shown by *heavy white line* (*1*); other subdivisions of magnetic field discussed in the text are shown by thin *white lines* (*2*); the position of the Lambert Rift is marked by *line* (*3*)

shoulder of the Amery Ice Shelf between the Reinbolt Hills and Robertson Nunatak where we have interpreted Grenvillian activity.

The ideas of Carson et al. (1996) and Fitzsimons (2000) that linked Pan-African orogenesis in Prydz Bay with that in the Bunger Hills region and along Australia's western margin cannot be fully tested by the limited magnetic data within Queen Mary Land. However, Golynsky et al. (2002) reported that the Bunger Anomaly over Queen Mary Land may reflect highly deformed rocks of Grenvillian age that outcrop at the Bunger Hills (Black et al. 1992; Sheraton et al. 1993). This anomaly may delineate the extent of the Queen Mary Land craton given the existence of Archean metamorphic rocks in the area between the Mirny Station and the Bunger Hills (Black et al. 1992). Golynsky et al. (2002) argued that the Bunger Hills-Windmill Islands mobile belt may extend into Australia as the Albany-Fraser Orogen (Clarke et al. 1995). Certainly, new aeromagnetic data over Queen Mary Land are necessary to clarify this theory further.

The great uncertainties in the tectonic boundaries beneath the Antarctic ice sheet are due to the general lack of outcrops and limited amount of precise and exhaustive geochronology data. Aeromagnetic data clearly can help to minimize ambiguities of correlating the basement features in East Antarctica for developing pre-Gondwana continental reconstructions.

Acknowledgments

We wish to thank Anatoly Laiba, Vsevolod Maslov and Dmitry Vorob'ev for valuable discussions and allowing us to cite their new age data and geologic information for the Grove Mountains region. We thank the Director of PMGRE for permission to use recently acquired aeromagnetic data. We are grateful to reviewers Steven Boger and Carol Finn for their thoughtful comments and suggestions. Authors are very grateful to Professor Ralph von Frese for the critical reading of the manuscript.

References

Black LP, Sheraton JW, Tingey RJ, McCulloch MT (1992) New U-Pb zircon ages for the Denman Glacier area, East Antarctica, and their significance for Gondwana reconstruction. Antarctic Science 4:447–460

Boger SD, Wilson CJL, Fanning CM (2001) Early Paleozoic tectonism within the east Antarctic craton: the final suture between east and west Gondwana? Geology 29:463–466

Boger SD, Carson CJ, Fanning CM, Hergt JM, Wilson CJL, Woodhead JD (2002) Pan-African intraplate deformation in the northern Prince Charles Mountains, east Antarctica. Earth Planet Sci Letters 195:195–210

Briggs IC (1974) Machine contouring using minimum curvature. Geophysics 39:39–48

Carson CJ, Fanning CM, Wilson CJL (1996) Timing of the Progress Granite, Larsemann Hills, evidence for early Palaeozoic orogenesis within the east Antarctic Shield and implications for Gondwana assembly. Australian J Earth Sci 43:539–553

Clarke GL, Sun SS, White RW (1995) Grenville-age belts and associated older terranes in Australia and Antarctica. AGSO J Austral Geol Geophys 16:25–39

Dirks PHGM, Wilson CJL (1995) Crustal evolution of the East Antarctic mobile belt in Prydz Bay: continental collision at 500 Ma? Precambrian Res 75:189–207

England RN, Langworthy AP (1975) Geological work in Antarctica – 1974. Bureau Min Resour Australia Record 1975(30)

Earth Resource Mapping Pty Ltd. (1995) ER Mapper and ER Storage software and documentation

Fitzsimons ICW (1997) The Brattstrand Paragneiss and the Søstrene Orthogneiss: a review of Pan-African metamorphism and Grenvillian relics in southern Prydz Bay. In: Ricci CA (ed) The Antarctic region: geological evolution and processes. Terra Antartica Publications, Siena, pp 121–130

Fitzsimons ICW (2000) A review of tectonics events in the East Antarctic Shield and their implications for Gondwana and earlier supercontinents. J African Earth Sci 31(1):3–23

Fitzsimons ICW (2003) Proterozoic basement provinces of southwestern Australia and their correlation with Antarctica. In: Yoshida Y, Windley BF, Dasgupta S (eds) Proterozoic of East Gondwana: supercontinent assembly and breakup. Geol Soc London Spec Publ 206:93–130

Golynsky A, Chiappini M, Damaske D, Ferraccioli F, Ferris J, Finn C, Ghidella M, Ishihara T, Johnson A, Kim HR, Kovacs L, LaBrecque J, Masolov V, Nogi Y, Purucker M, Taylor P, Torta M (2001) ADMAP – Magnetic Anomaly Map of Antarctic. In: Morris P, von Frese R (eds) British Antarctic Survey Misc 10, Cambridge

Golynsky AV, Alyavdin SV, Masolov VN, Tscherinov AS, Volnukhin VS (2002) The composite magnetic anomaly map of the East Antarctica. Tectonophysics 347:109–120

Golynsky AV, Masolov VN, Volnukhin VS, Golynsky DA (2006) Crustal provinces of the Prince Charles Mountains region and surrounding areas in a light of aeromagnetic data. In: Fütterer DK, Damaske D, Kleinschmidt G, Miller H, Tessensohn F (eds) Antarctica – Contributions to global earth sciences. Springer, Berlin Heidelberg New York, pp 83–94

Hensen BJ, Zhou B (1995) A Pan-African granulite facies metamorphic episode in Prydz Bay, Antarctica: evidence from Sm-Nd garnet dating. Australian J Earth Sci 42:249–258

Hensen BJ, Zhou B, Thost DE (1997) Recognition of multiple high grade metamorphic events with garnet Sm-Nd chronology in the northern Prince Charles Mountains, Antarctica. In: Ricci CA (ed) The Antarctic region: geological evolution and progress. Terra Antartica Publication, Siena, pp 97–104

Jacobs J, Fanning CM, Henjes-Kunst F, Olesch M, Paech HJ (1998) Continuation of the Mozambique Belt into East Antarctica: Grenville-age metamorphism and polyphase Pan-African high-grade events in Central Dronning Maud Land. J Geol 106: 385–406

Kamenev EN, Kameneva GI, Mikhalsky EV, Andronikov AV (1990) The Prince Charles Mountains and Mawson Escarpment. In: Ivanov VL, Kamenev EN (eds) The geology and mineral resources of Antarctica. Nedra, Moscow, pp 67–113, (in Russian)

Kamenev E, Andronikov AV, Mikhalsky EV, Krasnikov NN, Stüwe K (1993) Soviet geological maps of the Prince Charles Mountains, East Antarctic Shield. Australian J Earth Sci 40:501–517

Liu X, Zhao Y, Liu X (2002) Geological aspects of the Grove Mountains, East Antarctica. In: Gamble JA, Skinner DNB, Henrys S (eds) Antarctica at the close of a millennium. Royal Soc New Zealand Bull 35:161–166

Mikhalsky EV, Sheraton JW, Laiba AA, Beliatsky BV (1996) Geochemistry and origin of Mesoproterozoic metavolcanic rocks from Fisher Massif, Prince Charles Mountains, East Antarctica. Antarctic Sci 8:85–104

Mikhalsky EV, Sheraton JW, Beliatsky BV (2001) Preliminary U-Pb dating of Grove Mountains rocks: Implications for the Proterozoic to early Palaeozoic evolution of the Lambert Glacier-Prydz Bay area (East Antarctica). Terra Antartica 8(1):3–10

Mikhalsky EV, Laiba AA, Beliatsky BV (2006) Main geological features of the Prince Charles Mountains. In: Fütterer DK, Damaske D, Kleinschmidt G, Miller H, Tessensohn F (eds) Antarctica – Contributions to global earth sciences. Springer, Berlin Heidelberg New York, pp 69–82

Sheraton JW, Tingey RJ, Black LP, Oliver RL (1993) Geology of the Bunger Hills area, Antarctica: implications for Gondwana correlations. Antarctic Sci 5:85–102

Shiraishi K, Hiroi Y, Ellis DJ, Fanning CM, Motoyoshi Y, Nakai Y (1992) The first report of a Cambrian orogenic belt in East Antarctica – an ion microprobe study of the Lützow-Holm Complex. In: Yoshida Y (ed) Recent progress in Antarctic earth science. Terra Scientific Publishing, Tokyo, pp 67–73

Tingey RJ (1982) The geologic evolution of the Prince Charles Mountains – an Antarctic Archean cratonic block. In: Craddock C (ed), Antarctic geoscience. Univ Wisconsin Press, Madison, pp 455–464

Tingey RJ (1991) The regional geology of Archaean and Proterozoic rocks in Antarctica. In: Tingey RJ (ed), The geology of Antarctica. Clarendon Press, Oxford, pp 1–58

Tingey RJ, England RN (1973) Geological work in Antarctica – 1972. Bureau Mineral Resources Australia Record 1973(161)

Yoshida M (1994) Tectonothermal history and tectonics of Lützow-Holm Bay area, East Antarctica: a re-interpretation. J Geological Soci India Mem 34:24–45

Zhao Y, Song B, Wang Y, Ren L, Li J, Chen T (1992) Geochronology of the late granite in the Larsemann Hills, East Antarctica. In: Yoshida Y, Kaminuma K, Shiraishi K (eds) Recent progress in Antarctic earth science. Terra Scientific Publishing, Tokyo, pp 155–161

Zhao Y, Song B, Zhang Z, et al. (1993) Early Paleozoic (Pan African) thermal event of the Larsemann Hills and its neighbours, Prydz Bay, East Antarctica. Sci China (Ser B) 38:74–84

Zhao Y, Liu X, Fanning CM, Liu X (2000) The Grove Mountains, a segment of a Pan-African orogenic belt in East Antarctica. Abstracts 31st Internat Geol Congr Rio de Janeiro, Brazil

Theme 3
The Continent Beneath the Ice

Chapter 3.1
ADMAP – A Digital Magnetic Anomaly Map of the Antarctic

Chapter 3.2
Identifying Major Sedimentary Basins Beneath the West Antarctic Ice Sheet from Aeromagnetic Data Analysis

Chapter 3.3
Bedrock Plateaus within the Ross Embayment and beneath the West Antarctic Ice Sheet, Formed by Marine Erosion in Late Tertiary Time

Chapter 3.4
Inversion of Airborne Gravity Data Acquired over Subglacial Lakes in East Antarctica

Chapter 3.5
Russian Geophysical Studies of Lake Vostok, Central East Antarctica

Chapter 3.6
Morphology of the Subglacial Bed Relief of Lake Vostok Basin Area (Central East Antarctica) Based on RES and Seismic Data

Chapter 3.7
Deep Reflection Imaging beneath the Mizuho Plateau, East Antarctica, by SEAL-2002 Seismic Experiment

Chapter 3.8
Seismic Anisotropy beneath Northern Victoria Land from SKS Splitting Analysis

The interior of the Antarctic continent is often simply described as a vast white desert. Indeed, the human eye is only able to observe the "reflected" sunlight of the surface of the ice sheet. However, there is land beneath the ice, which – to a large extent – is still unknown to us. The development of geophysical methods and the advance of logistic technology make it more and more possible to reveal the nature of the rocks as well as the structure and evolution of the earth's crust under the ice.

An important tool is the magnetic field stemming from the rocks of the earth's crust. A compilation of all magnetic data available, mainly acquired by air- or ship-borne surveys, is performed by the ADMAP (Antarctic Digital Magnetic Anomaly Project) group. The map of magnetic anomalies presented by Golynsky et al. (Chap. 3.1) identifies large-scale features at the margins of the continent. The data density is by no way homogeneous and allows detailed interpretations only in a few areas. Large areas in the interior of Antarctica are only crossed by some single traverses and are virtually void of data. Satellite data which appear to fill in these voids cannot replace close-to-surface measurements since the magnetic field is a potential field.

Bell et al. (Chap. 3.2) use aeromagnetic data to identify three major sedimentary basins in the Interior Ross Embayment. Deeper regions in the magnetic basement are interpreted as the floor of sedimentary basins. Since the large Ross Sea basins are all associated with regional Bouguer gravity highs, the association of the magnetic basement lows with positive Bouguer gravity anomalies suggests that the basins under the West Antarctic ice sheet were probably also formed during Late Cretaceous Gondwana break-up.

Ice penetrating radar data, collected in the area of western Marie Byrd Land and Siple Coast in western Antarctica, and seismic reflection data, collected close to Edward VII Peninsula in western Marie Byrd Land in the eastern Ross Sea, show a number of large bedrock plateaus, interpreted by Wilson and Luyendyk (Chap. 3.3) as results of coastal marine erosion at a time when coastal regions of Antarctica were relatively free of ice.

Inversion of gravity data in combination with ice penetrating radar and seismic soundings was used by Filina et al. (Chap. 3.4) to estimate the thicknesses of water and sediment layers of the subglacial Lakes Vostok and Concordia.

Masolov et al. (Chap. 3.5) and Popov et al. (Chap. 3.6) report on the Russian geophysical studies of Lake Vostok station. Radio-Echo-Sounding over the edges of the lake reveals details of the grounding line including a number of small-size subglacial water cavities around the lake. Reflection seismic profiles along the main axis of the lake and across its southern portion yield further details of the geometry of the water body. The subglacial morphology around the Lake Vostok Basin has been identified to consist of different morphological substructures.

Reflection profiles of a deep seismic sounding survey on the Mizuho Plateau in western Enderby Land, presented by Yamashita et al. (Chap. 3.7), reveal prominent reflectors from horizontal and flat planes in depths of 22–23 km and 31–34 km. The depth of Moho reflectors were estimated to 41–42 km.

Seismic anisotropy beneath northern Victoria Land was studied by Pondrelli et al. (Chap. 3.8) using teleseismic events recorded at permanent and temporary broad-band stations in northern Victoria Land. Rather large delay times in a mainly NW-SE fast velocity direction suggest a deep rooted anisotropic layer underneath the Terra Nova Bay area.

ADMAP – A Digital Magnetic Anomaly Map of the Antarctic

Alexander Golynsky[1] · Massimo Chiappini[2] · Detlef Damaske[3] · Fausto Ferraccioli[4] · Carol A. Finn[5] · Takemi Ishihara[6] Hyung Rae Kim[7] · Luis Kovacs[8] · Valery N. Masolov[9] · Peter Morris[10] · Ralph von Frese[7]

[1] VNIIOkeangeologia, Angliysky Avenue 1, St.-Petersburg 190121 Russia, <sasha@vniio.nw.ru>
[2] INGV, via di Vigna Murata, 605, 00143 Roma, Italy
[3] Bundesanstalt für Geowissenschaften und Rohstoffe, BGR, Stilleweg 2, 30655, Hannover, Germany
[4] DIPTERIS Università di Genova, Viale Benedetto XV, 5 16132 Genova, Italy;
Present adress: British Antarctic Survey, BAS, High Cross, Madingley Road, Cambridge CB3 OET, UK
[5] U.S. Geological Survey, Box 25046 Denver, CO 80255, USA
[6] Geological Survey of Japan, 1-1-3, Higashi, SUKUBA Ibaraki 305, Japan
[7] Ohio State University, 381, Mendenhall Lab 125 S. Oval Mall, Columbus, OH 43210-1398, USA
[8] NRL, 4555 Overlook Ave. SW, Washington, DC, 20375-5320, USA
[9] PMGRE, 24 Pobeda St., Lomonosov, 189510, Russia
[10] British Antarctic Survey, BAS, High Cross, Madingley Road, Cambridge CB3 OET, UK

Abstract. For a number of years the multi-national ADMAP working group has been compiling near surface and satellite magnetic data in the region south of 60° S. By the end of 2000, a 5 km grid of magnetic anomalies was produced for the entire region. The map readily portrays the first-order magnetic differences between oceanic and continental regions. The magnetic anomaly pattern over the continent reflects many phases of geological history whilst that over the abyssal plains of the surrounding oceans is dominated mostly by patterns of linear seafloor spreading anomalies and fracture zones. The Antarctic compilation reveals terranes of varying ages, including Proterozoic-Archaean cratons, Proterozoic-Palaeozoic mobile belts, Palaeozoic-Cenozoic magmatic arc systems and other important crustal features. The map delineates intra-continental rifts and major rifts along the Antarctic continental margin, the regional extent of plutons and volcanics, such as the Ferrar dolerites and Kirkpatrick basalts. The magnetic anomaly map of the Antarctic together with other geological and geophysical information provides new perspectives on the break-up of Gondwana and Rodinia evolution.

History

With only a minute part of the Antarctic land surface being free of ice the opportunities for conventional geological mapping are severely restricted and the use of geophysical techniques becomes essential for understanding the regional geology of the continent. Of the various methods available, magnetic surveying typically gives the most immediate information on the rock types beneath the ice. The earliest magnetic surveys were carried out by sledging parties where the slow rate of progress limited the amount and hence the usefulness of the data which could be gathered. Aeromagnetic surveys, by contrast, allow rapid and effective access even to the most inhospitable areas of the Antarctic. For this reason many of the earth science research groups active in the Antarctic have acquired some form of airborne magnetic survey capability using either fixed wing aircraft or helicopters.

By the time airborne surveys in the Antarctic started being used widely, the magnetometer systems available were already quite sophisticated and capable of recording good quality data, although with somewhat lower sensitivity than provided by modern day systems. The main problem with the early flights, indeed one which has persisted until fairly recently, was the navigation. The maps were relatively unreliable since the terrain is often featureless and the electronic navigation systems were quite primitive by today's standards. The use of GPS has greatly improved navigational efforts in the Antarctic.

Offshore, where magnetic surveys were generally carried out using towed proton magnetometers, the problem of navigation was again paramount. In fact data collected before about 1967, when the first satellite navigation systems began to appear on research vessels operating in Antarctic waters, must almost always be viewed as suspect. The distribution of marine tracks shows there were few systematic regional marine magnetic surveys conducted within the area of the new map. Most of the data were mapped along random tracks with rarely the optimum orientations for investigating the magnetic fabrics of any particular area.

Despite all these problems, a considerable amount of data was collected in the Antarctic region during the last 40 years. However due to the fact that acquisition was split between a variety of national agencies whose efforts at best were only loosely coordinated, it has been very difficult to obtain an overview of what had been achieved and what remains to be done. The obvious interest in large magnetic compilations from other parts of the world such as North America (Zietz 1982), the Arctic (Macnab et al. 1995) and Australia (Tarlowski et al. 1996) demonstrated that such a compilation could provide important insights into the tectonic makeup of the region.

From: Fütterer DK, Damaske D, Kleinschmidt G, Miller H, Tessensohn F (eds) (2006) Antarctica: Contributions to global earth sciences. Springer-Verlag, Berlin Heidelberg New York, pp 109–116

Organization

The initiative to produce the present map was supported by IAGA resolutions and SCAR. The ADMAP working group was formally established at a meeting in Cambridge in September 1995. Here a variety of interested parties representing almost all the countries active in gathering magnetic data in the Antarctic came together to consider what data was actually available and where and how the process of compilation could best be achieved. A flight line location map incorporating most of the known surveys was produced for this meeting that was a valuable start for organizing the working group's efforts (Fig. 3.1-1). Further meetings were held in 1997 at Rome, Italy and in 1999 at Columbus, Ohio, USA to review progress and exchange data and interpretations (Johnson et al. 1997; Chiappini et al. 1998).

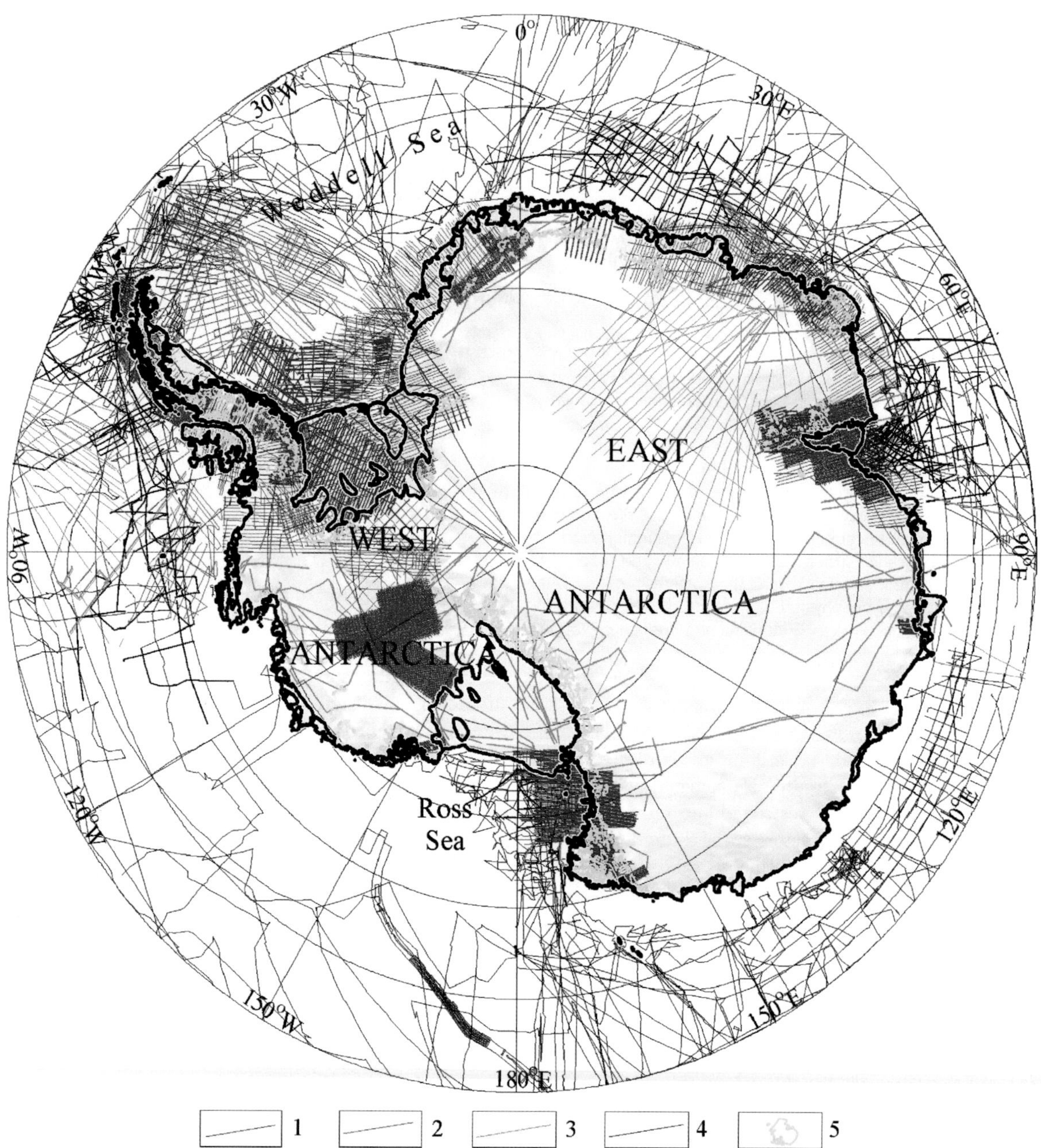

Fig. 3.1-1. Line coverage of near surface magnetic surveys superimposed on a shaded relief map of the surface topography of Antarctica. Map projections throughout this paper are polar stereographic. Rock outcrops and coastline are from the Antarctic Digital Database. *1:* airborne surveys, line spacing ~5 km; *2:* airborne surveys, line spacing ~20 km; *3:* airborne surveys, line spacing ~50 km; *4:* marine surveys, *5:* ice-free ground

The compilation of the near-surface magnetic surveys was divided into three sectors. For the 0°–120° E sector, the data were mostly from Russian surveys (Golynsky et al. 2002a), while data from mainly U.S., German, and Italian surveys were used for the120–240° E sector (Blankenship et al. 1993; Bosum et al. 1989; Bozzo et al. 1999; Chiappini et al. 1999). In the third 240–360° E segment, the data were taken mainly from British, Russian and USAC surveys (Maslanyj et al. 1991; Golynsky et al. 2000a; LaBrecque et al. 1986). Offshore in all sectors, data were available from shipborne and airborne surveys from a wide variety of origins. The magnetic data in each sector were edited for high-frequency errors, levelled and adjusted, and the data quality assessed by statistical analysis of the crossover errors.

In general, conventional aeromagnetic data processing techniques were employed. The more irregular networks of older regional profiles were typically first levelled together to provide a basic framework into which the more recently acquired detailed grids could be fitted. The three regional sector grids were then merged to create the master grid over the Antarctic with minimal mismatch between the data sets along their boundaries.

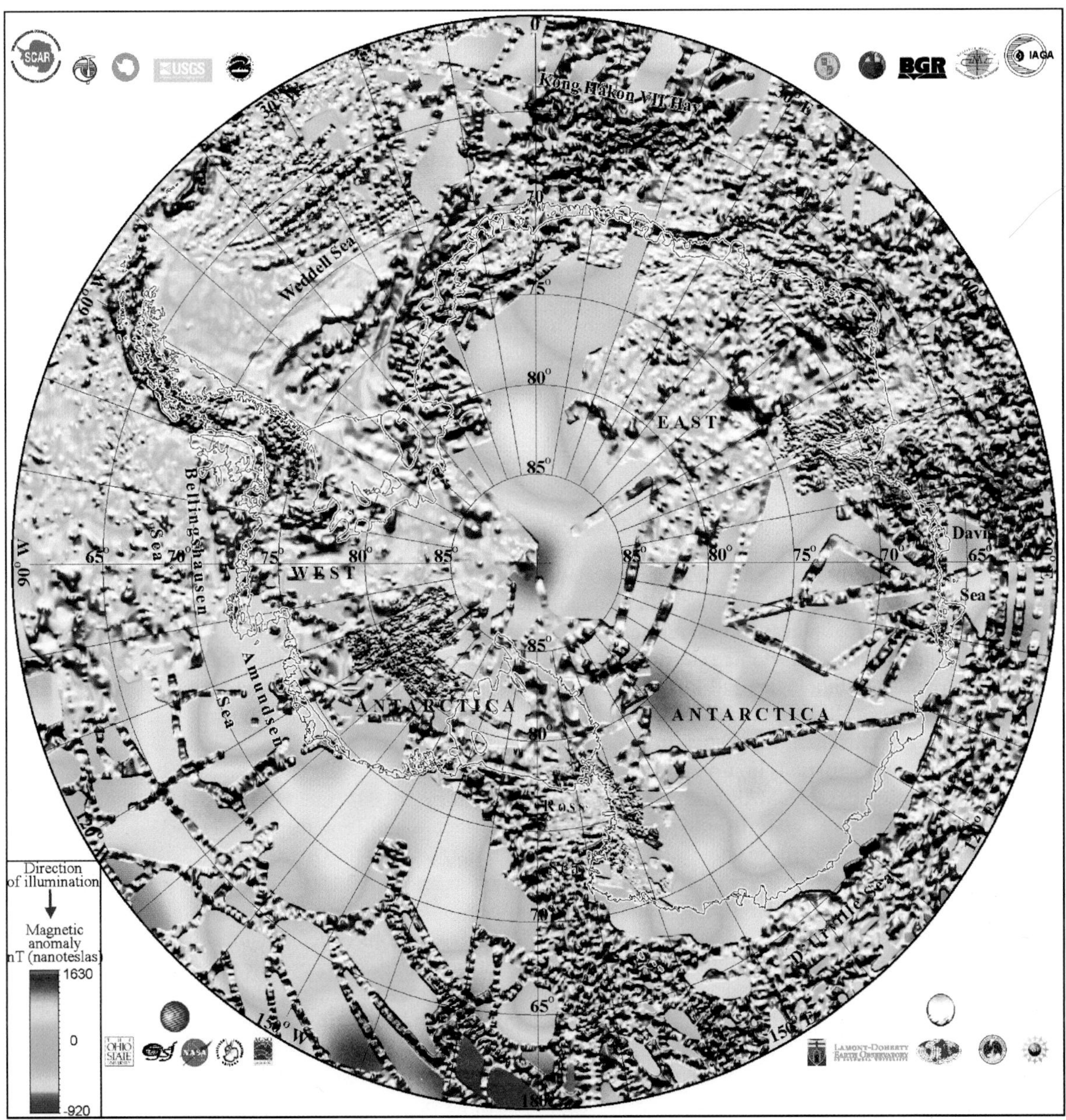

Fig. 3.1-2. Magnetic anomaly map of the Antarctic. The map was produced from near-surface (terrestrial, airborne and shipborne) and satellite (MAGSAT) magnetic observations by the international working group of the Antarctic Digital Anomaly Map Project. More than 0.4 million line-kilometres of near-surface recordings were used in the construction of the map

Significant low frequency components of the magnetic anomaly field were not recovered by the composited near-surface surveys due to their comparatively short profiles and large regional gaps in coverage. To fill in the regional coverage gaps, the satellite magnetic observations from the MAGSAT mission were combined with the low-pass filtered (= 400 km) components of the near-surface anomalies using the inversion procedures of Kim et al. (2003). Adding back the high-cut filtered near-surface anomaly components to the MAGSAT-augmented regional anomalies yielded the comprehensive magnetic anomaly map of the Antarctic shown in Fig. 3.1-2.

The ADMAP anomaly grid was completed in 2000 and presented at a press conference of the spring 2001 AGU meeting. The ADMAP working group organized two AGU conference sessions and another IAGA session featuring Antarctic magnetic anomaly studies. The anomaly map was printed and is being distributed as a British Antarctic Survey publication (Golynsky et al. 2001). Figure 3.1-2 represents the first unified magnetic anomaly map of the whole Antarctic region south of 60° S that integrates the near-surface anomaly surveys acquired by the international community since the International Geophysical Year 1957–1958 through the year 2000 with lithospheric anomaly estimates from the Magsat mission (von Frese et al. 1999).

Geological Implications

The ADMAP compilation shown in shaded relief in Fig. 3.1-2 provides the most complete and coherent view to date of the magnetic properties of the Antarctic crust. The map reveals a wide variation of magnetic anomaly patterns, trends and types reflecting geologic terranes of varying ages and degrees of tectonic reworking, as well as crustal variations in lithologies and metamorphic alterations. However, not all changes of the magnetic fabric signify differences in magnetic properties, in some areas they may be caused by variations in survey specifications. Clearly that at present the ADMAP magnetic anomaly database is largely inadequate to address many of Antarctica's immediate concerns to know more about our planet's crust. Often data coverage is spotty or not available in the geographic area of interest. The acquisition of additional high-resolution magnetic data will enhance the studies of crustal structure and composition, plate tectonics and regional geology.

The Antarctic magnetic anomaly compilation helps to map out various Proterozoic-Archaean cratons; the extents of Proterozoic-Palaeozoic mobile belts; various Palaeozoic-Cenozoic magmatic arc systems and the boundary between East and West Antarctica; the continent-ocean transition, and other regional Antarctic crustal features etc. The map helps to resolve the structural grain in basement, various basement terranes and their suture zones, intra-continental rifts and major rifts along the Antarctic continental margin, and the regional extents of various plutons and volcanics, such as the Ferrar dolerites and Kirkpatrick basalts. For main place names and geographical overview see Fig. 3.1-3.

The first-order difference between oceanic and continental regions is well represented in Fig. 3.1-2 by the complex magnetic anomaly pattern over the continent corresponding to many phases of geological history, whilst over the abyssal plains of the surrounding oceans simpler patterns of linear seafloor spreading anomalies and fracture zones dominate. The pattern of linear magnetic anomalies in the northern Weddell Sea is a remarkable example (Ghidella et al. 2002; Kovacs et al. 2002). The W-E-oriented anomalies suggest that the seafloor of the northern Weddell Sea was created as a direct consequence of South America-Antarctica plate motion. Magnetic anomalies C34 to C6 may be identified with confidence, though the earlier M-series anomalies were not well dated until recently (Livermore and Hunter 1996). New aeromagnetic data collected by the Alfred Wegener Institute along the eastern Weddell Sea and Riiser-Larsen Sea continental margins that have yet to be included into the ADMAP compilation provide new constraints on the timing and geometry of the early Gondwana breakup (Jokat et al. 2003). In the Riiser-Larsen Sea/Mozambique Basin, the first oceanic crust between Africa and Antarctica formed around 155 Ma. In the west the Weddell Rift propagated from west to east with a velocity of about 63 km Myr^{-1} between chrons M19N and M17N (Jokat et al. 2003).

In the southern Indian Ocean, marked disparities in the anomaly pattern of the Riiser-Larsen Sea, Cooperation Sea and Dumont D'Urville Sea strongly suggest different modes of formation for these three segments of seafloor. Magnetic lineations in the Dumont D'Urville Sea were recently revised (Tikku and Cande 1999) by shifting the Magnetic Quiet Zone boundary and anomalies 34y, 33o, 32y, 31o and 27y farther seaward (~100 km). It was assumed that like the Quiet Zone Boundary Anomaly, anomalies 32y, 33o and 34y may not be true isochrons. This fact, together with the oldest sea-floor magnetic anomaly south of the Otway Basin of Australia being identified as anomaly 18, greatly complicates estimating the time when stretching of the continental lithosphere in the conjugate regions ended and breakup occurred.

Strong anomalously positive seafloor magnetic anomalies are observed over the southern margin of the Kerguelen Plateau, the Maud Rise and near the Astrid Ridge in the Riiser-Larsen Sea (Fig. 3.1-2, Fig. 3.1-3). Large anomalies overlie stretched continental fragments capped by basaltic flows that were detached (e.g., Maud Rise and southern Kerguelen Plateau) or extended (e.g., Astrid Ridge) from the Antarctic margin during the early opening of the Indian Ocean. These magnetic anomalies are prob-

Fig. 3.1-3. The location of place names referred to in the text. Basemap is a shaded relief map of the surface topography of Antarctica

ably associated with large submarine igneous provinces. Results of compilation are not expected to change recent models of first-order motion between the Indian and Antarctic Plates, but they should resolve uncertainties about the early histories of opening between these plates.

One of the most conspicuous features of Fig. 3.1-2 is a wide curvilinear belt of positive magnetic anomalies running parallel with the coast of Dronning Maud Land and Enderby Land known as the Antarctic Continental Margin Magnetic Anomaly (Golynsky and Aleshkova 2000). It represents one of the longest continuous tectonic features of Antarctica and presumably marks a continental crustal discontinuity formed during Gondwana breakup.

Another striking feature is the Pacific Margin Anomaly that traverses roughly 3 800 km along the Pacific margins of the Antarctic Peninsula and Ellsworth Land. This anomaly may reflect a complex subduction-related linear batholith that corresponds to the Antarctic Peninsula Mesozoic-Cenozoic magmatic arc (Maslanyj et al. 1991).

On the continent, crustal terranes are characterized by different magnetic signatures that reflect such factors as their ages and lithologies, the degree of reworking, deformations and metamorphic variations. In West Antarctica, the compilation offers unprecedented views of the magnetic anomalies associated with crustal blocks such as the Antarctic Peninsula, Ellsworth-Whitmore Mountains,

Filchner, Haag Nunataks, Marie Byrd Land and Thurston Island (Maslanyj et al. 1991). Thick Paleozoic sedimentary and metasedimentary rocks of the Ellsworth-Whitmore Mountains are associated with a flat and almost featureless magnetic field sporadically interrupted by local short-wavelength anomalies which occur over Middle Jurassic plutonic complexes related to the early stages of Gondwana break-up. The magnetic field in the Haag Nunataks area is characterized by intense short-wavelength positive anomalies with a predominantly NW-SE grain. The aeromagnetic data suggest that the Precambrian basement of the Haag Nunataks may extend beneath the Ronne Ice Shelf, the Ellsworth-Whitmore Mountains and southern Palmer Land (Maslanyj and Storey 1990; Golynsky and Aleshkova 2000).

Well-defined zonations and distinctive patterns of magnetic anomalies are recognized in Western Dronning Maud Land at the edge of the East Antarctic Shield (Golynsky and Aleshkova 2000). Broad featureless magnetic low and weak linear short-wavelength magnetic anomalies delineate the extent of the Archean to Mid-Proterozoic Grunehogna Province. The pronounced Sverdrupfjella-Kirwanveggen Anomaly and alternating bands of elongated positive and negative anomalies define the boundaries of the Middle to Late Proterozoic orogenic belt.

The intricate anomaly pattern over Coats Land defines the extent of the Coats Land crustal block that might be Archean to Early Mesoproterozoic in age (Golynsky et al. 2000b; Jacobs et al. 2003). It delineates various structural elements and characteristic trends of the Shackleton and the Argentina Ranges that are largely composed of high-grade metamorphic basement rocks. The magnetic field of the neighboring Weddell Sea embayment is relatively muted due to the large thickness of sedimentary cover.

In the shield areas of East Antarctica, the aeromagnetic data provide a unique window on basement geology and structural architecture and allow for better defining the boundaries between the Archean stable blocks and the Proterozoic mobile belts, and tracing them beneath the ice sheet (Golynsky et al. 2002). In Enderby Land, a roughly oval belt of mostly positive anomalies differentiates the Archean Napier Terrane from the negative magnetic effects of the Mesoproterozoic Rayner Terrane to the south. The boundary between the Rayner Terrane and the neighboring Lützow-Holm Bay Terrane to the west is clearly identified by a 350 nT anomaly and the difference in the anomaly trends on either side of this anomaly. The elongated, fragmented magnetic highs and intervening lows of the Lützow-Holm Bay Terrane are associated with rocks metamorphosed under granulite facies conditions. The boundary between the Lützow-Holm Bay Terrane and the westward neighboring Yamato-Belgica Terrane is marked by an abrupt change of the magnetic anomaly pattern.

The detailed aeromagnetic data of the Prince Charles Mountains and surrounding areas provide a rather complex but surprisingly coherent image for studying the geology and tectonic history of this region (Golynsky et al. 2002). Several distinct structural units may be differentiated in the magnetic anomaly data. The intense short-wavelength, high-amplitude positive anomalies that extend around the Vestfold Hills are presumably associated with high-grade metamorphic Late Archean craton. The northern Prince Charles Mountains display a predominantly northeasterly trending magnetic fabric that continues to the eastern shoulder of the Lambert Rift. The negative and positive anomalies largely reflect Athos supracrustals and Porthos orthogneisses, respectively, related to the Proterozoic charnockite-granulite terrane of the Beaver Terrane. Elongate and moderate magnetic banding appears to characterize the Late Proterozoic rocks of the Fisher sub-terrane.

The prominent alternating system of linear NE-SW positive and negative anomalies over the eastern shoulder of the Lambert Rift may reflect the western boundary of the newly discovered Princess Elizabeth Land cratonic block like those observed in the Vestfold Hills block and the Ruker Terrane (Golynsky et al. 2002). Metamorphic rocks of the Archean Ruker Terrane are mainly associated with low-amplitude anomalies. The most prominent magnetic lineament within the southern Prince Charles Mountains is the Ruker Anomaly that is associated with a banded iron formation.

Widely spaced aeromagnetic profiles combined with detailed aeromagnetic surveys in the West Antarctic Rift-Transantarctic Mountains system have shown the presence of extensive late Cenozoic volcanic rocks beneath the West Antarctic Ice Sheet, Ross Ice Shelf and Ross Sea continental shelf (Behrendt et al. 1996). In Victoria Land, aeromagnetic images are used to study relationships between tectonic blocks along West Antarctic rift shoulder and prerift features inherited from the Palaeozoic terranes involved in the Ross Orogen, the transition between the Wilson Terrane and the Precambrian East Antarctic Craton, the extent and distribution of Jurassic tholeiitic magmatism (Bosum et al. 1989; Blankenship et al. 1993; Damaske et al. 1994; Ferraccioli and Bozzo 1999).

The compiled map offers a powerful new interpretation tool for extending our knowledge about Gondwana and its evolution. For example, the southern margin of the Dharwar Craton in India makes a convincing fit with the margin of the Napier Complex in Enderby Land (Reeves 1998). Correlative magnetic anomalies are common between southwestern Australian and East Antarctic blocks that were adjacent to each other in conventional Gondwana reconstructions (Tarlowski et al. 1996). The Archean Queen Mary Land craton delineated by the Bunger Anomaly may represent a southern continuation of the Yilgarn Craton, whereas the Bunger Hills-Windmill Islands mobile belt may extend into Australia as the Albany-Fraser Orogen (Golynsky et al. 2002).

Aeromagnetic signatures over northern Victoria Land and southeastern Australia basement have been interpreted to confirm subducted-related Ross-Delamerian orogens at the Pacific margin of early Paleozoic Gondwana (Finn et al. 1999). The plutons in the Glenelg (Australia) and Wilson (Victoria Land) zones may have formed the roots of continental-margin magmatic arcs. Eastward shifting of arc magmatism resulted in the Stavely (Australia) and Bowers (northern Victoria Land) volcanic eruptions onto oceanic forearc crust. The turbidities in the Stawell (Australia) and Robertson Bay (northern Victoria Land zones) shed from the Glenelg and Wilson zones, respectively, were deposited along the trench and onto the subducting oceanic plate. The margin was subsequently truncated by thrust faults and uplifted during the Delamerian and Ross orogenies.

The compiled magnetic anomaly data provide a new window on the poorly known geology of the Antarctic. When interpreted with other geological and geophysical information, they offer considerable promise for developing new insights on the breakup of Gondwana and Rodinia evolution and the creation of the seafloor in the Antarctic.

The Antarctic anomaly map is limited by the highly variable specifications of the surveys and regional gaps in coverage of the near surface surveys. ADMAP is working to improve the compilation for the 50th anniversary of the IGY by incorporating additional high-resolution magnetic data sets and the CHAMP magnetic observations as they become available.

Acknowledgments

We thank Ashley Johnson (Geosoft, UK), Julie Ferris (BAS, UK), John LaBrecque, Michel Purucker and Patrick Taylor (NASA), and Miguel Torta (Observatori de l'Ebre, Spain) for their contributions to ADMAP. Emanuele Bozzo and Hans Roeser are thanked for reviews.

References

Behrendt JC, Saltus R, Damaske D, McCafferty A, Finn C, Blankenship D, Bell RE (1996) Patterns of late Cenozoic volcanic and tectonic activity in the West Antarctic rift system revealed by aeromagnetic surveys. Tectonics 15(2):660–676

Blankenship DD, Bell RE, Hodge SM, Brozena JM, Behrendt JC, Finn CA (1993) Active volcanism beneath the West Antarctic ice sheet. Nature 361:526–529

Bosum W, Damaske D, Roland NW, Behrendt JC, Saltus R (1989) The GANOVEX IV Victoria Land/Ross Sea aeromagnetic survey: interpretation of the anomalies. Geol Jb E38:153–230

Bozzo E, Ferraccioli F, Gambetta M, Caneva G, Spano M, Chiappini M, Damaske D (1999) Recent progress in magnetic anomaly mapping over Victoria Land (Antarctica) and the GITARA 5 survey. Antarctic Science 11(2):209–216

Chiappini M, von Frese R, Ferris J (1998) Effort to develop magnetic anomaly database aids Antarctic research. EOS 23:290

Chiappini M, Ferraccioli F, Bozzo E, Damaske D, Behrendt JC (1999) First stage of INTRAMAP: Integrated Transantarctic Mountains and Ross Sea area magnetic anomaly project. Annali Geofisica 42(2):277–292

Damaske D, Behrendt JC, McCafferty A, Saltus R, Meyer U (1994) Transfer faults in the Western Ross Sea: new evidence from the McMurdo Sound/Ross ice shelf aeromagnetic survey (GANOVEX VI), Antarctic Science 6(3):359–364

Ferraccioli F, Bozzo E (1999) Inherited crustal features and tectonic blocks of Transantarctic Mountains: an aeromagnetic perspective (Victoria Land Antarctica). J Geophys Res 104(11): 25297–25319

Finn C, Moore D, Damaske D, Mackey T (1999) Aeromagnetic legacy of early Paleozoic subduction along the Pacific margin of Gondwana. Geology 27(12):1087–1090

von Frese R, Kim HR, Tan L, Kim JW, Taylor PT, Purucker ME, Alsdorf DE, Raymond CA (1999) Annali di Geofisica 42(2):309–326

Ghidella ME, Yáñez G, LaBrecque JL (2002) Revised tectonic implications for the magnetic anomalies of the western Weddell Sea. Tectonophysics 347:65–86

Golynsky AV, Aleshkova ND (2000) New Aspects of Crustal Structure in the Weddell Sea region from aeromagnetic studies. Polarforschung 67:133–142

Golynsky AV, Masolov VN, Jokat W (2000a) Magnetic anomaly map of the Weddell Sea region: a new compilation of the Russian data. Polarforschung 67:125–132

Golynsky AV, Grikurov GE and Kamenev EN (2000b) Geologic significance of regional magnetic anomalies in Coats Land and western Dronning Maud Land. Polarforschung 67:91–99

Golynsky A, Chiappini M, Damaske D, Ferraccioli F, Ferris J, Finn C, Ghidella M, Ishihara T, Johnson A, Kim HR, Kovacs L, LaBrecque J, Masolov V, Nogi Y, Purucker M, Taylor P, Torta M (2001) ADMAP – Magnetic Anomaly Map of Antarctic. In: Morris P, von Frese R (eds) British Antarctic Survey Misc 10, Cambridge

Golynsky AV, Alyavdin SV, Masolov VN, Tscherinov AS, Volnukhin VS (2002) The composite magnetic anomaly map of the East Antarctica. Tectonophysics 347:109–120

Jacobs J, Fanning CM, Bauer W (2003) Timing of Grenville-age vs. Pan-African medium- to high-grade metamorphism in western Dronning Maud Land (East Antarctica) and significance for correlations in Rodinia and Gondwana. Precambrian Res 125:1–20

Johnson AC, von Frese R, ADMAP Working Group (1997) Magnetic map will define Antarctica's structure. EOS 78(18):185

Jokat W, Boebel T, König M, Meyer U (2003) Timing and geometry of early Gondwana breakup. J Geophys Res 108(9):2428

Kim HR, von Frese R, Taylor PT, Kim JW (2003) CHAMP enhances utility of satellite magnetic observations to augment near-surface magnetic survey coverage. In: Reigber C, Luehr H, Schwintzer P (eds) First CHAMP mission results for gravity, magnetic and atmospheric studies. Springer, Heidelberg, pp 296–301

Kovacs LC, Morris P, Brozena J, Tikku A (2002) Sea floor spreading in the Weddell Sea from magnetic and gravity data. Tectonophysics 347:43–64

LaBrecque J, Cande S, Bell R, Raymond C, Brozena J, Keller M, Parra JC, Vanez G (1986) Aerogeophysical survey yields new data in the Weddell Sea. Antarctic J 21:69–70

Livermore RA, Hunter RJ (1996) Mesozoic seafloor spreading in the southern Weddell Sea. In: Storey BC, King EC, Livermore RA (eds) Weddell Sea tectonics and Gondwana break-up. Geol Soc Spec Publ 108:227–241

Macnab R, Verhoef J, Roest W, Arkani-Hamed J (1995) New database documents the magnetic character of the Arctic and North Atlantic. EOS 76(45):449–458

Maslanyj MP, Storey BC (1990) Regional aeromagnetic anomalies in Ellsworth Land: crustal structure and Mesozoic microplate boundaries within West Antarctica. Tectonics 9(6):1515–1532

Maslanyj MP, Garrett SW, Johnson AC, Renner RGB, Smith AM (1991) Aeromagnetic anomaly map of West Antarctica. BAS GEOMAP series, geophysical map and supplementary text. BAS, Cambridge

Reeves CV (1998) Aeromagnetic and gravity features of continental Gondwana and their relation to continental break-up: more pieces, less puzzle. J Afr Earth Sci 27(1A):153–156

Tarlowski C, Milligan PR, Mackey TE (1996) The magnetic anomaly map of Australia, scale 1:5 000 000. AGSO, Canberra

Tikku AA, Cande S (1999) The oldest magnetic anomalies in the Australian-Antarctic Basin: are they isochrons? J Geophys Res 104:661–677

Zietz I (1982) Composite magnetic anomaly map of the United States, Part A (U.S. Geol Survey Investigations Map GP-954-A)

Identifying Major Sedimentary Basins Beneath the West Antarctic Ice Sheet from Aeromagnetic Data Analysis

Robin E. Bell[1] · Michael Studinger[1] · Garry Karner[1] · Carol A. Finn[2] · Donald D. Blankenship[3]

[1] Lamont-Doherty Earth Observatory of Columbia University, 61 Route 9W, Palisades, NY 10964, USA

[2] U.S. Geological Survey, MS 964, Denver Federal Center, Denver, CO 80225 USA

[3] Institute for Geophysics, John A. and Katherine G. Jackson School of Geosciences, University of Texas at Austin, 4412 Spicewood Springs Rd., Bldg. 600 Austin, TX 78759, USA

Abstract. In the Ross Sea, large sedimentary basins reflect primarily the major extensional event associated with the Late Cretaceous breakup of Gondwana. Within the Interior Ross Embayment, no similar large basins have been identified to date. We have used aerogravity and Werner deconvolution methods applied to aeromagnetic data to map depth to magnetic basement, which helped delineate three major sedimentary basins, the Bentley Subglacial, Onset, and Trunk D Basins.

Introduction

The West Antarctic rift is a region of thinned continental crust, bounded to the south and west by the Transantarctic and Whitmore Mountains and to the north and east by Marie Byrd Land (Tessensohn and Wörner 1991) (Fig. 3.2-1).

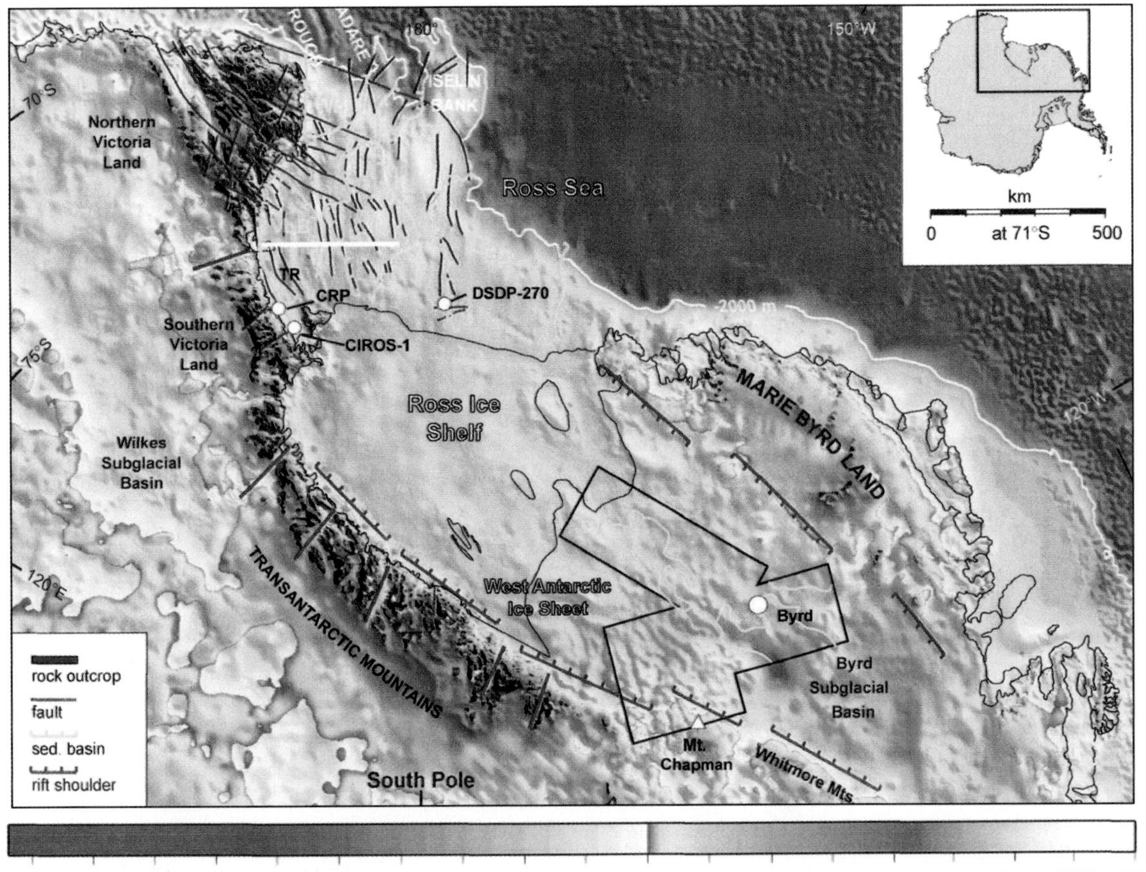

Fig. 3.2-1. Location figure for aerogeophysical data. Aerogeophysical survey area is outlined in *black* and the basins derived from this analysis are shown in *yellow*. The profile shown in Fig. 3.2-2 is illustrated as a *white line* in the western Ross Sea. Image is using data from the BEDMAP compilation of Lytte et al. (2000). Ross Sea basins (*yellow*) are from Busetti et al. (1999) and faults (*red*) from Salvini et al. (1979) and Fitzgerald (2000), and references therein. *CRP:* Cape Roberts Project; *CT:* Central Trough; *EB:* Eastern Basin; *NB:* Northern Basin; *TR:* Terror Rift; *VLB:* Victoria Land Basin

From: Fütterer DK, Damaske D, Kleinschmidt G, Miller H, Tessensohn F (eds) (2006) Antarctica: Contributions to global earth sciences. Springer-Verlag, Berlin Heidelberg New York, pp 117–122

Three distinct phases of tectonic activity have been advanced to explain the formation of this broad low-lying region, an early extensional phase, associated with the Jurassic intrusion of the Ferrar dolerites (Wilson 1992), a Late Cretaceous event linked with the progressive fragmentation of Gondwana (e.g., Cooper et al. 1991; Wilson 1992; Wilson 1995), and a final but minor phase of activity evidenced by Cenozoic faulting in the western Ross Sea as evident in the Terror Rift, dextral offset structures, and the extrusion of bimodal alkalic volcanic rocks (e.g., Behrendt et al. 1991). The amount of extension associated with each of these phases of tectonic activity remains under discussion.

The major basins of the Ross Sea, the Eastern, Central Trough and Victoria Land Basins, are interpreted as primarily the result of Late Cretaceous regional lithospheric stretching and subsidence (Davey and Cooper 1987). The western Ross Sea basins are 100–150 km wide while the Eastern Basin is 300 km wide. These basins parallel the Transantarctic Mountains and are thought to continue beneath the Ross Ice Shelf into the Interior Ross Embayment (Cooper et al. 1991; Munson and Bentley 1992).

Aerogeophysical surveys and seismic studies of the Interior Ross Embayment have imaged several small sedimentary basins but no large basins equivalent in scale to the Ross Sea basins have been recognized. The basins identified to date are relatively narrow (10–40 km wide), with maximum sediment thicknesses of 1–2.5 km (Bell et al. 1998; Studinger et al. 2001; Studinger et al. 2002).

Generally, the sedimentary basins in the ice-covered portions of Antarctica have been identified on the basis of gravity lows and reflection seismic data. However, in the Ross Sea, all the major basins are associated with broad and regional Bouguer gravity highs. While such gravity anomalies can be produced by crustal intrusions, the regional and organized nature of the gravity highs are simply explained in terms of the strengthening of the rifted lithosphere prior to sedimentation (Karner et al. 2005). We use Werner deconvolution of aeromagnetic data to define basement geometry and identify the location of the major sedimentary basins in the Interior Ross Embayment. The gravity sign, positive or negative, helps distinguish between Late Cretaceous (positive) and Cenozoic (negative) extension.

Methodology: Application of Werner Deconvolution to Aeromagnetics

Magnetic data are often used to estimate the depth to magnetic basement in sedimentary basins and passive continental margins (e.g., Klitgord and Behrendt 1979). Magnetic methods can accurately trace the depth of the magnetic basement over sedimentary basins when the sedimentary rocks have weak magnetic susceptibilities relative to the underlying crystalline/volcanic basement. The basic assumption of the Werner method is that all magnetic anomalies are the result of either a sequence of dykes or an interface between juxtaposed half-spaces of

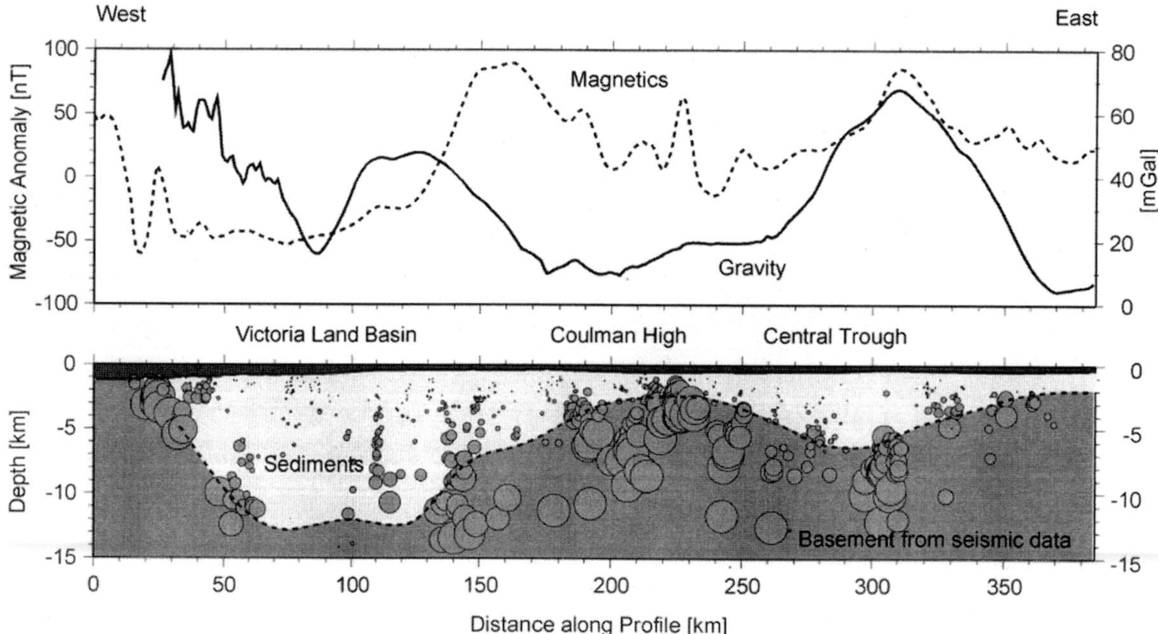

Fig. 3.2-2. Werner solutions along a seismic profile in the Ross Sea located in Fig. 3.2-1. The *circles* are the Werner solutions for depth to magnetic basement with the size of the circles being scaled to the magnetic susceptibility. Magnetic data (Bosum et al. 1989) as well as gravity, bathymetry and basement data are also shown (Childs et al. 1995)

different magnetic susceptibility (Werner 1953). Our Werner deconvolution algorithm uses simple models for the source and a quadratic form for the source/noise interference to determine the magnetization properties of the causative bodies. We use both interface and dike solutions to define the magnetic basement.

To demonstrate the power of the Werner approach, we used Werner deconvolution solutions in the Ross Sea where the basement structure has been determined seismically. The clustering of the high susceptibility Werner solutions along the seismically determined basement (Fig. 3.2-2) demonstrates that the Werner approach can be used to define the basement structure even in the presence of widespread highly magnetic intrusives. The diameter of the plotted solutions relates to susceptibility magnitude.

In order to define the basin structure in the Interior Ross Embayment, we used airborne geophysical data collected onboard a ski-equipped DeHavilland Twin Otter (Bell et al. 1998; Bell et al. 1999; Brozena et al. 1993). The aeromagnetic data were collected with a towed Geometrics 813 proton-precession magnetometer with an estimated precision of about 1 nT (Sweeney et al. 1999). The composite grid was draped from its original flight elevation surface onto a surface 1 500 m above the bedrock elevation surface (Fig. 3.2-3b). Ice-thickness measurements from ice-penetrating radar and laser altimetry of the ice surface topography have been used to derive subglacial topography (Blankenship et al. 2001).

In order to define the basement structures and subsequently the sedimentary basins in the Interior Ross Embayment, we have applied the Werner deconvolution method along 22 profiles (500–800 km long spaced ~25 km apart). Using the upper limit of the Werner solution clusters along each profile, we estimated the depth to magnetic basement and thus regional basin geometry. The resulting magnetic basement depth estimates were gridded to produce a depth to magnetic basement map for the region. (Fig. 3.2-3c).

Definition of Major Interior Ross Embayment Sedimentary Basins

The West Antarctic study area traverses both the Whitmore Mountain block, a region of elevated rugged topography in the southeast, and the generally low-lying thinned crust of the Interior Ross Embayment (Fig. 3.2-3a). To date, no major sedimentary basins equivalent to the large Ross Sea basins have been identified from the subglacial elevation, the Bouguer gravity or the magnetic data over the Interior Ross Embayment. The Werner deconvolution magnetic depth to basement solutions for the entire region varies from 1 800–4 500 m (Fig. 3.2-3c). In the southeast, the Whitmore Block delineated by topography, Bouguer gravity (Fig. 3.2-3d) and seismic constraints on crustal thickness (Clarke et al. 1997), is characterized by relatively deep magnetic basement, on average 3 800 m occasionally disrupted by isolated regions of shallow magnetic depths. The regions of shallow magnetic basement tend to be discrete points or circular structures (82° S 110° W) and are interpreted as Late Cenozoic volcanic edifices (Blankenship et al. 1993).

In contrast to the Whitmore Mountain Block, the Ross Embayment is characterized by generally very shallow magnetic basement depths on average 2 600 m. The northern margin of the study area, adjacent to Marie Byrd Land, is characterized by shallow magnetic basement as is the region to the west of the Whitmore Mountains. We interpret these broad regions of shallow magnetic solutions as basement highs, akin to the Ross Sea basement highs. Between these bordering areas of uniformly shallow basement are regions of much deeper magnetic basement. We have defined regions with magnetic basement depths greater than 3 800 m as sedimentary basins. These regions of deeper magnetic basement are disrupted by small localized, often circular points of shallow magnetic basement, similar to the isolated points of shallow magnetic basement in the Whitmore Block. We interpret the isolated points of shallow magnetic basement as volcanic intrusions. Three large well defined sedimentary basins greater than 100 km in width are delineated in the Werner solutions (Fig. 3.2-3c). The three large basins with average magnetic basement depths of 4 500 m are the Bentley Subglacial Trench north of the Whitmore Mountains, the Onset Basin west of the Whitmore Mountains, and the Trunk D Basin underlying Ice Stream D. The Werner solutions also delineate a smaller basin (at 80.5° S, 127° W) about 40 km in width to the west of Byrd Station and one edge of a sedimentary basin in the northeast.

In the east, the Bentley Subglacial Trench is 100 km wide and over 300 km long and parallels the Whitmore Mountain Front. Situated in the deepest portion of West Antarctica, the topography over the basin varies smoothly from 900–1 700 m below sea level. The Bouguer gravity over this basin is a broad 30 mGal positive anomaly relative to the regional trend (Fig. 3.2-3d). The shape and gravity anomaly of the Bentley Basin is very similar to the Central Basin in the Ross Sea.

In the west, the Trunk D Basin is up to 100 km wide and 250 km long and trends at an angle oblique to the Whitmore Mountain Front. The Trunk D Basin correlates with very rough topography with elevation ranging from 400 m above sea level to 900 m below sea level. The topography over the Trunk D Basin may either reflect ice stream processes, as the seafloor topography in the Ross Sea, or possibly recent tectonics. The Bouguer gravity is not consistently positive. The northern half of the basin is characterized by a positive gravity anomaly while the southern half is associated with a gravity negative. The

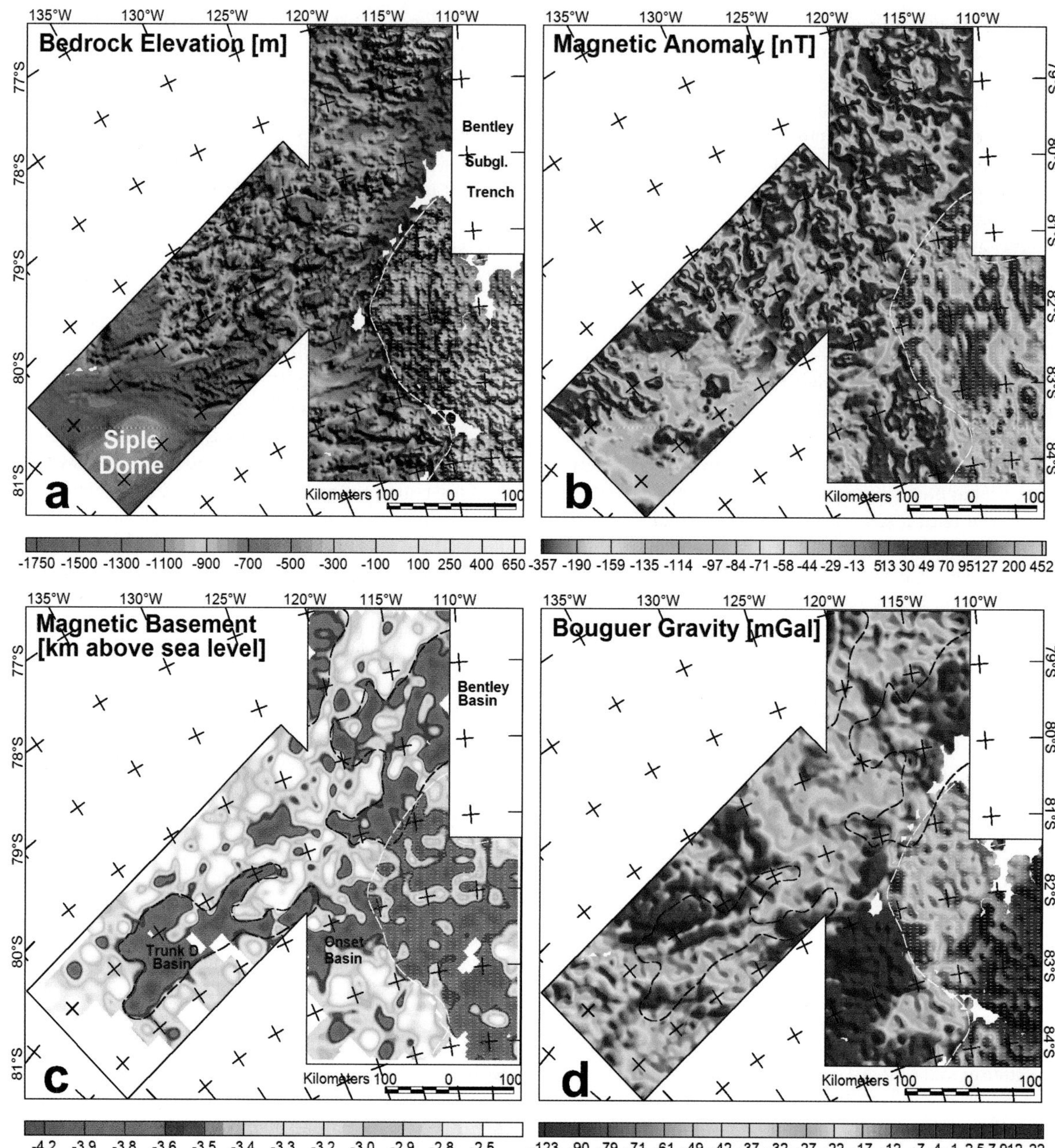

Fig. 3.2-3. Interior Ross Embayment geophysical maps. *White dashed line* shows the boundary between the Whitmore Mountains Block and the Interior Ross Embayment. *Black lines* outline regions of Werner solutions interpreted as sedimentary basins

complex gravity anomaly of this basin resembles the gravity anomaly over the Victoria Land Basin (Late Cretaceous) and the Terror Rift (Cenozoic).

The onset basin lies close to the boundary of the Whitmore Mountains with the Interior Ross Embayment, is 100 km wide and may be the southeastern extension of the Trunk D Basin. The southern margin of this basin correlates well with the basin imaged seismically by Anandakrishnan et al. (1998). The subglacial topography overlying this basin is generally deep and varies smoothly from 900–1 100 m below sea level. This basin is coincident with the northern portion of a major positive Bouguer gravity anomaly.

Conclusions

The unusual positive gravity anomalies of the large Ross Sea basins suggests that identifying the major sedimentary basins within the West Antarctic Rift System cannot be based simply on gravity anomalies. Werner deconvolution of aeromagnetic data can be used to define basement structure and delineate the location and geometry of major sedimentary basins. We identified three basins within the Interior Ross Embayment using magnetic data. These basins are primarily associated with positive Bouguer gravity anomalies and thus we suggest that these basins were probably formed during the Late Cretaceous Gondwana breakup.

Acknowledgments

We thank the Support Office for Aerogeophysical Research for data collection and reduction. Funding for this work was provided by the U.S. National Science Foundation. LDEO contribution 6849.

References

Anandakrishnan S, Blankenship DD, Alley RB, Stoffa PL (1998) Influence of subglacial geology on the position of a West Antarctic ice stream from seismic observations. Nature 394:62–65

Behrendt JC, LeMasurier WE, Cooper AK, Tessensohn F, Trehu A, Damaske D (1991) Geophysical studies of the West Antarctic rift system. Tectonics 10(6):1257–1273

Bell RE, Blankenship DD, Finn CA, Morse DL, Scambos TA, Brozena JM, Hodge SM (1998) Influence of subglacial geology on the onset of a West Antarctic ice stream from aerogeophysical observations. Nature 394:58–62

Bell RE, Childers VA, Arko RA, Blankenship DD, Brozena JM (1999) Airborne gravity and precise positioning for geologic applications. J Geophys Res B, Solid Earth and Planets 104(7):15281–15292

Blankenship DD, Bell RE, Hodge SM, Brozena JM, Behrendt JC, Finn CA (1993) Active volcanism beneath the West Antarctic Ice-Sheet and implications for ice-sheet stability. Nature 361:526–529

Blankenship DD, Morse DL, Finn CA, Bell RE, Peters ME, Kempf SD, Hodge SM, Studinger M, Behrendt MJC, Brozena JM (2001) Geologic controls on the initiation of rapid basal motion for the ice streams of the southeastern Ross Embayment: a geophysical perspective including new airborne radar sounding and laser altimetry results. In: Alley RB, Bindschadler RA (eds) The West Antarctic Ice Sheet: behavior and environment. Antarctic Research Series 77, AGU, Washington DC, pp 283-296

Bosum W, Damaske D, Roland NW, Behrendt JC, Saltus R (1989) The GANOVEX IV Victoria Land/Ross Sea aeromagnetic survey: interpretation of anomalies. Geol Jb E38:153–230

Brozena JM, Jarvis JL, Bell RE, Blankenship DD, Hodge SM, Behrendt JC (1993) CASERTZ 91-92: airborne gravity and surface topography measurements. Antarctic J US 28:1–3

Busetti M, Spadini G, Wateren FM van der, Cloetingh S, Zanolla C (1999) Kinematic modelling of the West Antarctic Rift System, Ross Sea, Antarctica. Global Planet Change 23(14):79–103

Childs JR, Antostrat Ross Sea Working Group (1995) Description of CD-ROM digital data: seismic stratigraphic atlas of the Ross Sea, Antarctica and circum-Antarctic seismic navigation. In: Cooper AK, Barker PF, Brancolini G, Hambrey MJ, Wise SW, Barrett P, Davey FJ, Ehrmann W, Smellie JL, Villa G, Woolfe KJ (eds) Geology and seismic stratigraphy of the Antarctic margin, AGU, Washington DC, pp 287–296

Clarke TS, Burkholder PD, Smithson SB, Bentley CR (1997). Optimum seismic shooting and recording parameters and a preliminary crustal model for the Byrd subglacial basin, Antarctica. In: Ricci CA (ed) The Antarctic region: geological evolution and processes. Terra Antartica Publication, Siena, pp 485–493

Cooper AK, Davey FJ, Hinz K (1991) Crustal extension and origin of sedimentary basins beneath the Ross Sea and Ross Ice Shelf, Antarctica. In: Thomson MRA, JA Crame JA, Thomson JW (eds) Geological evolution of Antarctica. Cambridge University Press, Cambridge, pp 285–291

Davey FJ, Cooper AK (1987). Gravity studies of the Victoria Land Basin and Iselin Bank. In: Cooper AK, Davey FJ (eds) The Antarctic continental margin: geology and geophysics of the western Ross Sea. Circum-Pacific Council Energy Mineral Resources, Houston, pp 119–137

Fitzgerald PG (2002) Tectonics and landscape evolution of the Antarctic plate since the breakup of Gondwana, with an emphasis on the West Antarctic Rift System and the Transantarctic Mountains. In: Gamble JA, Skinner DNB, Henrys SA (eds) Antarctica at the close of a millennium. Royal Soc New Zealand Bull 35:453–469

Karner GD, Studinger M, Bell RE (2005) Gravity anomalies of sedimentary basins and their mechanical implications: application to the Ross Sea basins, west Antarctica. Earth Planet Sci 7558 235:577–596

Klitgord KD, Behrendt JC (1979) Basin structure of the U.S. Atlantic margin. In: Watkins JS, Montadert L, Dickerson PW (eds) Geological and geophysical investigations of continental margins. Amer Assoc Petrol Geol, Tulsa, pp 85–112

Lytte MB, Vaughan DG, BEDMAP-Consortium (2000) BEDMAP – bed topography of the Antarctic. British Antarctic Survey, Cambridge

Munson CG, Bentley CR (eds) (1992) The crustal structure beneath ice stream C and Ridge BC, West Antarctica from seismic refraction and gravity measurements. In: Yoshida Y, Kaminuma K, Shiraishi K (eds) Recent progress in Antarctic earth science. Terra Scientific Publishing, Tokyo, pp 507–514

Salvini F, Storti F (1999) Cenozoic tectonic lineaments of the Terra Nova Bay region, Ross Embayment, Antarctica. In: Wateren FM van der, Cloetingh SAPL (eds) Lithosphere dynamics and environmental change of the Cenozoic West Antarctic Rift system. Elsevier, Amsterdam, pp 129–144

Studinger M, Bell RE, Blankenship DD, Finn CA, Arko RA, Morse DL, Joughin I (2001) Subglacial sediments: a regional geological template for ice flow in West Antarctica. Geophys Res Lett 28(18):3493–3496

Studinger M, Bell RE, Finn CA, Blankenship DD (2002) Mesozoic and Cenozoic extensional tectonics of the West Antarctic. In: Gamble JA, Skinner DNB, Henrys SA (eds) Antarctica at the close of a millennium. Royal Soc New Zealand Bull 35:563–569

Sweeney RE, Finn CA, Blankenship DD, Bell RE, Behrendt JC (1999) Central West Antarctica aeromagnetic data: a web site for distribution of data and maps. U.S. Geol Survey, Reston VA

Tessensohn F, Wörner G (1991) The Ross Sea rift system, Antarctica: structure, evolution and analogues. In: Thomson MRA, Crame JA, Thomson JW (eds) Geological evolution of Antarctica. Cambridge University Press, Cambridge, pp 273–277

Werner S (1953) Interpretation of magnetic anomalies at sheet-like bodies. Sveriges Geologiska Undersøkning Arsbook 43(508)

Wilson TJ (1992) Mesozoic and Cenozoic kinematic evolution of the Transantarctic Mountains. In: Yoshida Y, Kaminuma K, Shiraishi K (eds) Recent progress in Antarctic earth science. Terra Scientific Publishing, Tokyo pp 304–314

Wilson TJ (1995) Cenozoic transtension along the Transantarctic Mountains, West Antarctic rift boundary, South Victoria Land, Antarctica. Tectonics 14:531–545

Bedrock Plateaus within the Ross Embayment and beneath the West Antarctic Ice Sheet, Formed by Marine Erosion in Late Tertiary Time

Douglas S. Wilson[1,2] · Bruce P. Luyendyk[1]

[1] Dept. of Geological Sciences and Inst. for Crustal Studies, University of California, Santa Barbara, CA 93106, USA, <dwilson@geol.ucsb.edu>, <luyendyk@geol.ucsb.edu>

[2] Marine Science Inst., University of California, Santa Barbara, CA 93106, USA

Abstract. Ice penetrating radar (mostly airborne) and marine seismic surveys have revealed plateaus and terraces about 100–350 m below sea level beneath parts of the Ross Embayment including the West Antarctic ice sheet, the Ross Ice Shelf, and the eastern Ross Sea. These surfaces cover many thousands of square kilometers and are separated by bedrock troughs occupied by the West Antarctic ice streams. We interpret these surfaces as remnants of a level surface formed by wave erosion when the coastal regions of Antarctica were relatively free of ice. The flat and level nature of the surfaces that are near the same depth over large distances supports an origin by marine rather than glacial erosion. Marine seismic reflection profiles over one of the plateau remnants show thin, flat-lying glacial marine sediments draped with angular unconformity over gently dipping sediments of early Miocene age. Ice sheet and global sea level histories suggest that the shallower plateaus were last eroded in the middle Miocene, possibly within a warm interval from about 17 to 14 Ma, prior to formation of the modern West Antarctic ice sheet. The plateau surfaces cannot be directly correlated to Ross Sea unconformities but they may be extensions of ~14-Ma unconformity RSU4. The plateaus along the Siple Coast, with depths around 350 m, do not rebound to close to sea level for models of removing past and present ice load. Another possible factor is that western Marie Byrd Land lithosphere was heated in Oligocene time due to substantial extension or intensified mantle plume activity. Subsequent cooling has caused a moderate amount of crustal subsidence since then. These plateaus are similar to the emergent wave cut platforms interrupted by deep fjords, termed "strand" flats that rim the coasts of Norway.

Observations

Background

The Ross Sea and Ross Ice Shelf of Antarctica occupy a region known as the Ross Embayment, a geographic region of sea and ice shelf that separates the subcontinents of East and West Antarctica (Fig. 3.3-1). The Ross Embayment is a low-elevation subset of the geological province of the West Antarctic rift (Behrendt et al. 1991). The continental margin of the Ross Sea includes features unique to the continent, including an over-deepened shelf at a depth of around 500 m, and evidence of multiple advances and retreats of continental glaciations during the Cenozoic (Anderson 1999).

Reconnaissance mapping of the Ross Sea margin has largely been accomplished (Brancolini et al. 1995) except in the far eastern parts of the sea adjacent to western Marie Byrd Land (wMBL). Onshore beyond the grounding line of the ice sheets, little has been known of the

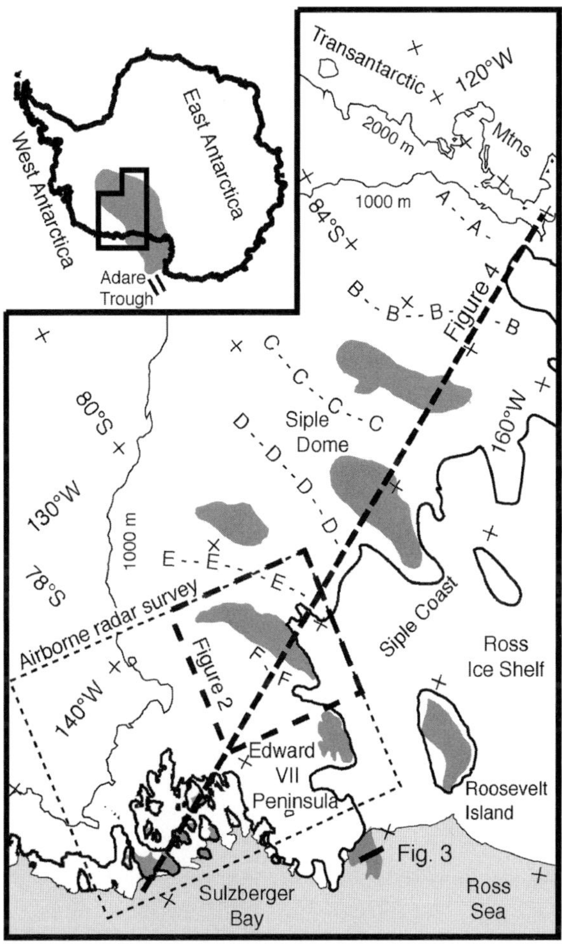

Fig. 3.3-1. Location map with geography, plateau/terrace locations, and locations of Fig. 3.3-2, 3.3-3 and 3.3-4. Flat surfaces mapped under the West Antarctic ice and Ross Sea outlined by *gray shade*. Data are from BEDMAP compilation (Lythe et al. 2001) and from SOAR survey (Luyendyk et al. 2003, *dashed box*). Grounding line in *bold line*, ice shelf edge in *thin line*; Pacific Ocean in *light gray*; surface elevation contour interval 1 000 m. Positions of ice Streams *A* through *F* are indicated by repeated letters. *Grey shade* in inset shows low-elevation continental crust of the Ross Embayment

From: Fütterer DK, Damaske D, Kleinschmidt G, Miller H, Tessensohn F (eds) (2006) Antarctica: Contributions to global earth sciences. Springer-Verlag, Berlin Heidelberg New York, pp 123–128

topography of the bedrock beneath the ice until recently. Radar sounding from the ice surface and from aircraft has mapped sub ice topography in selected locations (BEDMAP http://www.antarctica.ac.uk/bedmap/; (Lythe et al. 2001)). One mapped area of bedrock includes parts surrounding the West Antarctic Ice Streams, fast moving glaciers (km per year) that drain the parts of the West Antarctic Ice Sheet into the Ross Ice Shelf.

A record of West Antarctic glacial history has been interpreted from study of the glacial marine sediments overlying basement in the Ross Sea (Anderson 1999). These studies have revealed eight or more stratigraphic and seismic sequences (Brancolini et al. 1995; Hinz and Block 1983) – the Ross Sea Seismic Sequences (RSS-) numbered from oldest RSS-1 to youngest RSS-8, separated by unconformities (RSU-) numbered from youngest RSU1 to oldest RSU7 (Luyendyk et al. 2001). Sequence RSS-2 (Late Oligocene) and younger are glacial marine or glacial and are separated by unconformities largely believed to be due to the repeated grounding of ice sheets since Oligocene time. It is not generally agreed whether any of these unconformities represent marine transgression or sub aerial erosion. Oligocene ice sheet grounding in shallow water is represented by RSU6 (Bartek et al. 1991). As we discuss below, the plateaus were likely formed in early Middle Miocene time, which encompasses units RSS-4 and RSS-5 with intervening unconformity RSU4.

Airborne and Marine Data

A set of plateaus and terraces are observed at the margins of wMBL and the adjacent Ross Embayment at depths 100–350 m below sea level (m b.s.l.). These surfaces have been mapped by two methods, airborne radar (SOAR: Support Office for Aerogeophysical Surveys at University of Texas, Austin) and marine bathymetric and seismic surveys. Airborne geophysical surveys over a 400 by 400 km area of coastal wMBL mapped bedrock elevations with ice penetrating radar (Luyendyk et al. 2003). This mapping revealed that the basement was faulted into basins and ranges but also found a 200 km by 50 km level surface at 240–270 m b.s.l. at the boundary between the Ross Embayment and coastal wMBL that we named the "Blue Plateau" (Fig. 3.3-2). Aerogeophysical data can be interpreted to show that this plateau is not magnetic as would be expected for a sequence of flat lying lavas. Interpretations of gravity anomalies indicate that the surface is cut into both dense rocks that are probably crystalline and less dense sedimentary rock (Fig. 3d in Luyendyk et al. 2003).

Our airborne radar reveals several discontinuous surfaces on islands and peninsulas inland from Sulzberger Bay (Fig. 3.3-1) that are quite level at depths of 90–120 m b.s.l. Another radar survey (Behrendt et al. 2004) shows that the bed under Siple Dome includes an extensive flat area

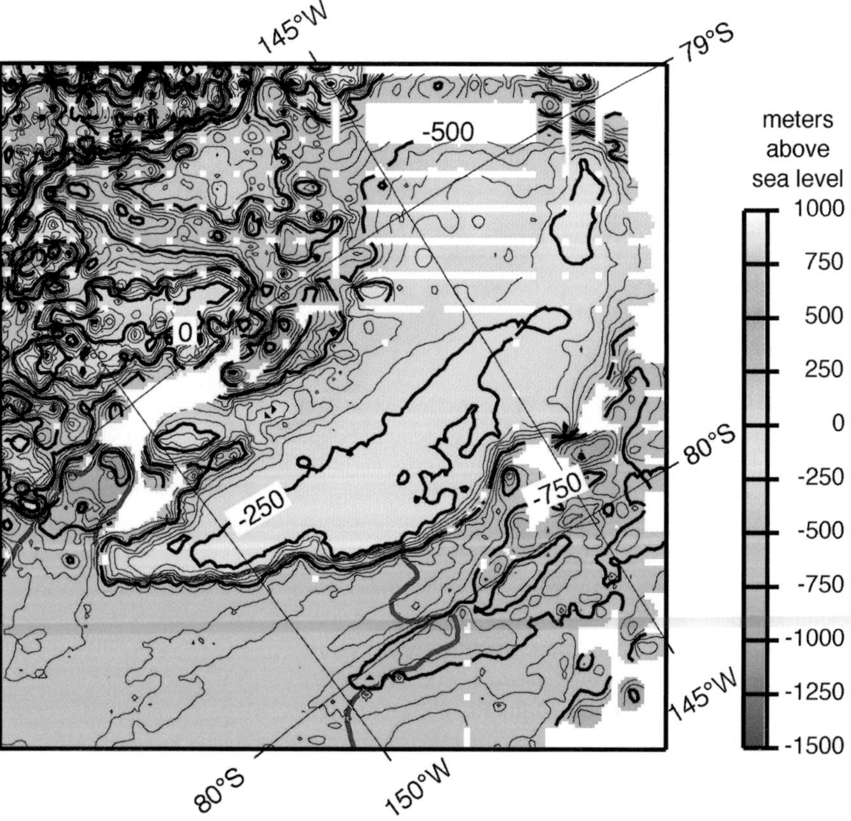

Fig. 3.3-2.
Contour map of Blue Plateau. Location in Fig. 3.3-1. Base of ice elevation map from airborne radar soundings (SOAR project). Grounding line in *magenta*. The depth of the level surface is 240–270 m b.s.l.

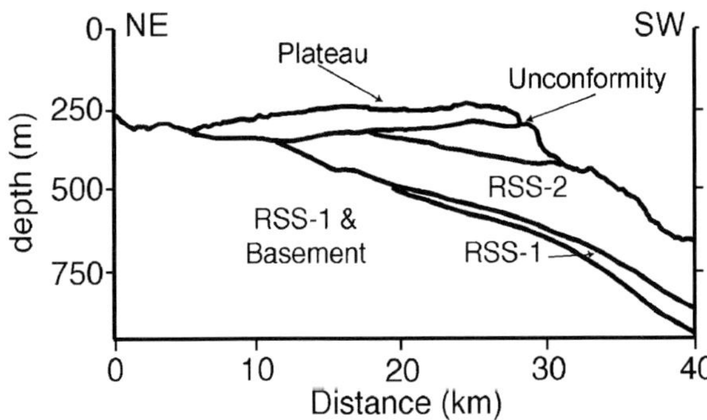

Fig. 3.3-3. Interpreted seismic section NBP-9601-26 across a plateau surface in eastern Ross Sea, showing truncation of ~20–25-Ma strata (RSS-2). Location in Fig. 3.3-1; vertical exaggeration approx. 20×

at 320–350 m b.s.l. Additional plateaus are apparent in the BEDMAP compilation at the divides between Ice Streams B and C and between Streams D and F. The total area of the mapped surfaces is about 50 000 km^2, comparable to the area of West Virginia, or Costa Rica, or the Netherlands.

In the eastern Ross Sea west of the Edward VII Peninsula in wMBL marine geophysical profiles were made across a likely candidate for part of this same surface at 250 m b.s.l. (Fig. 3.3-3). This plateau is formed by flat lying sediments capping a level surface cut into dipping Lower Miocene sediments that are correlated to Ross Sea Stratigraphic Sequence 2 (RSS2 Luyendyk et al. 2001). Therefore the surface was cut during or after the early Miocene.

Discussion

The Surfaces Are Wave Cut Platforms

Our working hypothesis is that these isolated plateaus and terraces were likely once connected as one surface that was formed by marine erosion at wave base during fluctuations in sea level; we refer to it here as the "Blue surface". It has since been dissected by glacial erosion to leave isolated plateaus including the Blue Plateau (Fig. 3.3-2). We assume that the generally flat and level nature of the surfaces and the observation that they are near the same depth over large distances supports an origin by marine rather than glacial erosion. Furthermore, BEDMAP and our bedrock data show that topography under the West Antarctic Ice Sheet is rugged and that the smooth areas we are discussing fringe the seaward borders of the rugged subglacial bed under the ice sheet. A marine erosion origin requires that the Ross Sea and wMBL coastal region was ice-free during the time the surface was cut. These plateaus are younger and lower than the West Antarctic Erosion Surface (WAES) mapped throughout wMBL and correlated to a similar surface in New Zealand (LeMasurier and Landis 1996).

The plateaus we have mapped are interpreted as wave cut platforms similar to those found along many coasts of the world. These submerged Antarctic plateaus bear a resemblance to the strand flats that are characteristic of the coasts of Norway and first noted by Nansen (1922). This type of wave cut platform is found at the mouth of and between fjords. They result from wave action on bedrock that has experienced freeze-thaw cycling. The Norwegian strand flats are exposed up to 50 m above sea level and inland some 50 km (Klemsdal 1982) because of rebound since the Last Glacial Maximum (LGM). We hypothesize that these Ross Embayment plateaus are strand flats formed at wave base that are now below sea level.

Timing

Firm constraints for the age of formation of the level surfaces are generally lacking. Diatoms from upper RSS-2 at DSDP site 270 indicate an age of about 21 Ma for the eroded sediment imaged in Fig. 3.3-3 (Steinhauff et al. 1987). Presumably, several million years elapsed after deposition to form the gentle (~1°) dip of the eroded beds. Our interpretation of wave erosion requires the absence of both grounded ice and stable floating ice, so the surfaces predate the formation of the West Antarctic Ice Sheet. As judged from the presence of ice-rafted debris in the deep ocean east of the Antarctic Peninsula (ODP Leg 113), the ice sheet reached close to its present extent about 8 Ma (Kennett and Barker 1990). More recent drilling west of the Antarctic Peninsula (ODP Leg 178), however, shows abundant ice-rafted debris in the oldest recovered sediments at 10 Ma (Barker and Camerlenghi 2002). Within this broad range of possibilities, we favour an interpretation of erosion during relative lowstands of sea level during the warm interval at 17–14 Ma (Barker and Camerlenghi 2002).

In a study of eastern Ross Sea Cenozoic stratigraphy DeSantis et al. (1999) suggest that unconformity RSU4 was formed at about 14 Ma during a glacial transitional stage in the eastern Ross Sea; this leads to the suggestion that the plateaus are interior remnants of RSU4. In the eastern Ross Sea, RSU4 is preserved west of 164° W, dipping gently north to northwest at depths as shallow as 600 m (Brancolini et al. 1995). The slope of the unconformity surface of about 5 m km^{-1} at shallow depths can reasonably be extrapolated updip to the nearest plateau surfaces on Roosevelt Island and the easternmost Ross Sea, where plateau depths are 200–250 m. Although the lack of continuity precludes a confident age determination, a middle Miocene age for the Blue surface is at least consistent with existing stratigraphic mapping.

Tectonic Subsidence and Post-Glacial Rebound

One obvious test of the hypothesis that the Blue surface formed by wave erosion is whether plausible mechanisms exist to bring the plateaus from near early-middle Miocene sea level to their current depths of 100–350 m b.s.l. Conservative interpretations of absolute sea level histories (e.g., Moore et al. 1987) place 17–14 Ma sea level at about 0–50 m above present sea level, and some interpretations (e.g., Haq et al. 1987) place sea level of that age up to 150 m above present. A mechanism for true subsidence appears necessary, and possible candidates are ice loading, sediment loading, crustal thinning, and lithospheric cooling. Mapping of sediment distribution in the Ross Sea shows that significant accumulations of middle Miocene and younger sediments are restricted to the northern and western Ross Sea, so we do not consider that mechanism further. An interpretation of crustal thinning of Miocene or younger age is difficult to reconcile with the general interpretation of lack of deformation of oceanic lithosphere surrounding the Antarctic continent for that age range, so we also discount that mechanism.

Our interpretation of up to 350 m of subsidence for the Blue surface can be tested quantitatively using simple ice loading and thermal contraction models. The subsidence caused by the increase in ice load is largely determined by the difference in weight between the current ice load and the Miocene ocean load. Additionally, the viscoelastic properties of the Earth's mantle lead to a delay of about 20–30 thousand years between a change in load and achieving the equilibrium vertical displacement due to the new load. Because of this delay, the predicted deflection due to ice load must also account for recent changes in the ice load. We assume that any subsidence that cannot be modelled by ice loading is tectonic, driven primarily by thermal contraction from cooling since the last episode of lithosphere extension with accompanying heating.

Estimating how much subsidence has occurred due to increase in ice load since a warm climate interval in the middle Miocene requires two separate estimates: the equilibrium deflection due to the difference between the Miocene ice load and the present ice load, plus the effect of the load of ice removed over the last several thousand years since the LGM that still depresses the current surface due to viscoelastic memory in the Earth's mantle. An upper bound on the subsidence caused by the present ice load can be safely calculated by assuming that no ice was present in the past, with moderate uncertainties resulting from the poorly known effective elastic thickness of the West Antarctic lithosphere. Uncertainties due to viscoelastic rebound are difficult to evaluate. For simplicity, we assume that these uncertainties can be estimated from differences in the rebound predicted from the Holocene ice-load models summarized by James and Ivins (1998). All models predict maximum current uplift rate in the vicinity of the Siple Coast (Fig. 3.3-4). Extrapolating these rates to equilibrium gives a range of future uplift of about 50–150 m which would bring the surface up to depths of 300–200 m b.s.l. at Siple Dome.

For illustration of restoring the subsidence caused by both present and recently removed ice, we first show the rebound predicted by intermediate ice-load model ICE-4G (Peltier 1994) in Fig. 3.3-4. For the equilibrium ice load, we experimented with different elastic thicknesses and different middle Miocene highstand ice loads on higher-elevation bedrock. These variables contribute several tens of meters to the uncertainty of the subsidence in the area of the Edward VII Peninsula, but much less uncertainty in the area of Siple Dome, which is far from high-elevation bedrock. At Siple Dome the ice load is replaced by a Miocene water load that is only slightly smaller and can be accurately estimated; the subsidence caused by the current ice load is about 100–120 m. Subsidence near Siple Dome from the combined effects of present and recently removed ice is therefore about 150–270 m (current ice load 100–120 m and future rebound 50–150 m). The bedrock now at a depth of about 350 m therefore will not rebound to within 80 m of Miocene sea level (Fig. 3.3-4) and ice loading alone cannot explain why the surface is so far below sea level.

Ice loading can only account for part of the elevation drop of the Blue surface to its present depth, and thermal contraction is the most viable candidate for the remaining part. If the surface were high from lithosphere heating at the time it was cut, then it would subside with time as the lithosphere cooled. Steady tectonic subsidence since the early Miocene could be due to one or more past heating events including Ross Sea rifting in the Late Cretaceous or heating in the Oligocene associated with West Antarctic volcanism and possible hot spot activity (LeMasurier and Rex 1990) or with sea floor spreading in the Adare Trough north of the western Ross Sea (Fig. 3.3-1 Cande et al. 2000). In fact Luyendyk et al. (2001) explain the lack of early Tertiary sediments on the eastern Ross Sea continental shelf by elevation of the region above sea level during that period. Cooling after heating from Cretaceous rifting would not produce an adequate elevation drop because if heating ended at that time, the rate of subsidence would be too slow during Miocene and later time to produce the required subsidence.

If sea floor spreading in the Adare trough was accompanied by distributed continental extension in the western Ross Sea and Siple Coast region, the 43–26 Ma age of the spreading is recent enough to explain the subsidence of the Blue surface. During distributed extension, the average surface elevation subsides due to the isostatic response to crustal thinning. After extension, subsidence continues due to cooling of the hot asthenosphere that wells up during extension. Using McKenzie's (1978) simple model for subsidence resulting from instantaneous, pure-shear extension, an extension age of 35 Ma predicts subsidence of 110 m for 15–0 Ma assuming a stretching factor of 1.25, and 200 m for a stretching factor of 1.5. The current width of the Siple Coast is about 600 km from the

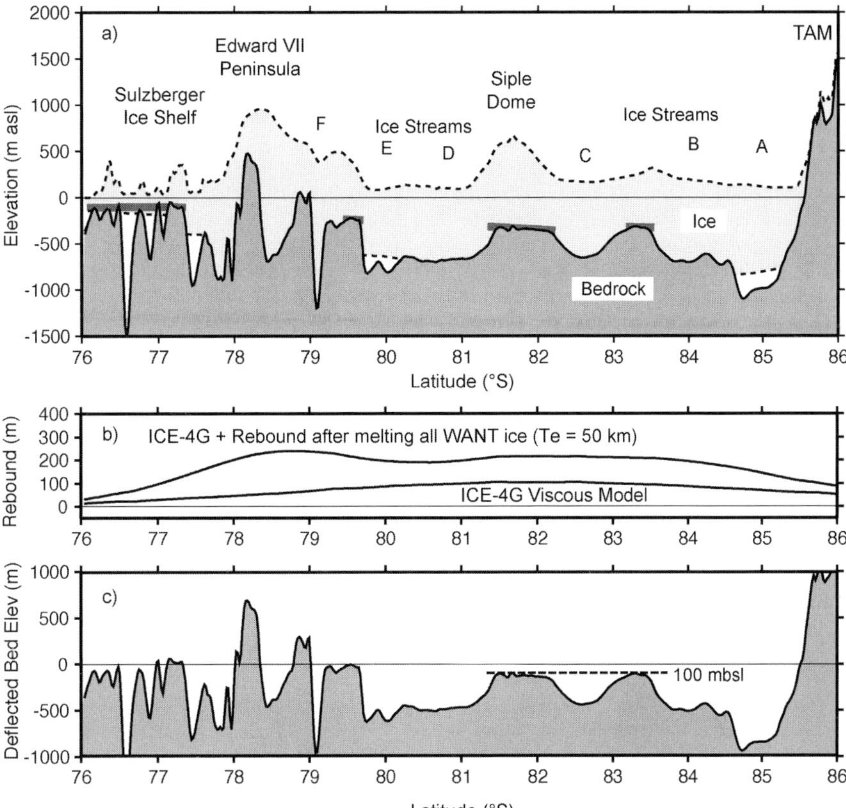

Fig. 3.3-4. Profile following ~149° W (Fig. 3.3-1) showing **a** current ice and bed surfaces, **b** rebound model, and **c** bed surface deflected by the rebound model. *Bold lines* highlight observed flat bed surfaces. Restoring the bed surface based on the rebound model for the effects of past (ICE-4G, Peltier 1994) and present ice load, assuming elastic plate thickness (*Te*) of 50 km, does not bring the Siple Dome Plateau within 100 m of sea level, indicating that an additional mechanism for subsidence is required

Edward VII Peninsula to the TAM, and the Adare Trough extension measured by Cande et al. (2000) is 180 km, implying a stretching factor of 1.4 if comparable extension were distributed across the Siple Coast region. Possible combinations adding up to about 350 m of total subsidence include 100 m from the current ice load, 50 m from the load of recently removed ice in the D91 model (James and Ivins 1998), and 200 m from thermal subsidence at a stretching factor of 1.5, or 100 m from current ice, 100 m from the ICE-4G model, and 150 m from stretching at a factor of about 1.35.

Conclusions

Glacial erosion seems implausible as a cause of the marine and sub ice plateaus we mapped around the eastern and southern borders of the Ross Embayment. Wave erosion, however, can explain the flatness and position of plateau remnants around margins of coastal wMBL. Formation of the Blue surface in the middle Miocene is consistent with gentle tilt of a lower Miocene substrate, allows time for subsidence of the surface after formation, and probably is consistent with ice-free conditions at sea level, permitting wave erosion. Glaciers and ice streams probably have since cut into the surface, dissecting it and causing continuing erosion. The present depths of the plateaus – up to 350 m b.s.l. – requires that they have been subsiding since they were created during middle Miocene. Some subsidence beyond that due to Pleistocene ice loading is needed to explain their present depths. Lithosphere cooling and contraction since late Cretaceous Ross Sea rift extension and heating is not adequate. Tertiary heating and subsequent thermal subsidence within the Ross Embayment including wMBL, is required to explain their present depths. This argument in turn strengthens the role of Oligocene and younger extension as a significant influence on vertical tectonics in this region.

The plateaus we describe are wave cut platforms similar to uplifted strand flats seen fringing the coasts of Norway. If sometime in the future the Ross Embayment plateaus were to emerge above sea level the Siple coast of West Antarctica would comprise strand flats and fjords and resemble that of Scandinavia.

Acknowledgments

We thank the SOAR project of UTIG for collection and processing of the airborne radar data, and Tom James for providing gridded rebound models. We thank Chris Sorlien for interpretation of offshore seismic data. Supported by U.S. NSF grant OPP9615281. Contribution of the Institute for Crustal Studies number 0646.

References

Anderson JB (1999) Antarctic marine geology. Cambridge University Press, New York

Barker PF, Camerlenghi A (2002) Glacial history of the Antarctic Peninsula from Pacific margin sediments. In: Barker PF, Camerlenghi A, Acton GD, Ramsay ATS (eds) Proceedings Ocean Drilling Program, Sci Results 178. Ocean Drilling Program, College Station, pp 1–40

Bartek LR, Vail PR, Anderson JB, Emmet PA, Wu S (1991) Effect of Cenozoic ice sheet fluctuations in Antarctica on the stratigraphic signature of the Neogene. J Geophys Res 96:6753–6778

Behrendt JC, LeMasurier WE, Cooper AK, Tessensohn F, Trehu A, Damaske D (1991) The West Antarctic rift system: a review of geophysical investigations. In: Elliot DH (ed) Contributions to Antarctic research II 53. AGU, Washington DC, pp 67–112

Behrendt JC, Blankenship DD, Morse DL, Bell RE (2004) Shallow-source aeromagnetic anomalies observed over the West Antarctic Ice Sheet compared with coincident bed topography from radar ice sounding – new evidence for glacial "removal" of subglacially erupted late Cenozoic rift-related volcanic edifices. Global Planet Change 42:177–193

Brancolini G, Cooper AK, Coren F (1995) Seismic facies and glacial history in the western Ross Sea (Antarctica). In: Cooper AK, Barker PF Brancolini G (eds) Geology and seismic stratigraphy of the Antarctic margin. AGU, Washington DC, pp 209–234

Cande SC, Stock JM, Müller D, Ishihara T (2000) Cenozoic motion between east and west Antarctica. Nature 404:145–150

DeSantis L, Prato S, Brancolini G, Lovo M, Torelli L (1999) The eastern Ross Sea continental shelf during the Cenozoic: implications for the west Antarctic ice sheet development. Global Plan Change 23:173–196

Haq BU, Hardenbol J, Vail PR (1987) Chronology of fluctuating sea levels since the Triassic. Science 235:1156–1167

Hinz K, Block M (1983) Results of geophysical investigations in the Weddell Sea and in the Ross Sea, Antarctica. Wiley, London, pp 279–291

James TS, Ivins ER (1998) Predictions of Antarctic crustal motions driven by present-day ice sheet evolution and by isostatic memory of the last glacial maximum. J Geophys Res 103:4933–5017

Kennett JP, Barker PF (1990) Latest cretaceous to Cenozoic climate and oceanographic developments in the Weddell Sea, Antarctica: An ocean-drilling perspective. In: Barker PF, Kennett JP (eds) Proceedings Ocean Drilling Program, Sci Results 113. Ocean Drilling Program, College Station, pp 937–960

Klemsdal T (1982) Coastal classification and the coast of Norway. Norsk Geograf Tidskr 36:129–152

LeMasurier WE, Landis CA (1996) Mantle plume activity recorded by low relief erosion surfaces in west Antarctica and New Zealand. Geol Soc Amer Bull 108:1450–1466

LeMasurier WE, Rex DC (1990) Late Cenozoic volcanism on the Antarctic plate: an overview. In: LeMasurier WE, Thompson JW (eds) Volcanoes of the Antarctic plate and southern oceans. Antarct Res Ser 48, AGU, Washington DC, pp 1–17

Luyendyk BP, Sorlien CC, Wilson DS, Bartek LR, Siddoway CH (2001) Structural and tectonic evolution of the Ross Sea rift in the Cape Colbeck region, eastern Ross Sea, Antarctica. Tectonics 20:933–958

Luyendyk BP, Wilson DS, Siddoway CS (2003) Eastern margin of the Ross Sea rift in western Marie Byrd Land, Antarctica: crustal structure and tectonic development. Geochem Geophys Geosyst 4:1090

Lythe MB, Vaughan DG, BEDMAP Consortium (2001) BEDMAP, a new ice thickness and subglacial topographic model of Antarctica. J Geophys Res 106:11335–11351

McKenzie D (1978) Some remarks on the development of sedimentary basins. Earth Planet Sci Letters 40:25–32

Moore Jr. TC, Loutit TS, Greenlee SM (1987) Estimating short-term changes in eustatic sea level. Paleoceanography 3:625–637

Nansen F (1922) The strandflat and isostacy. I kommission hos Jacob Dybwad, Kristiania (Oslo)

Peltier WR (1994) Ice age paleotopography. Science 265:195–201

Steinhauff DM, Renz ME, Harwood DM, Webb P-N (1987) Miocene diatom biostratigraphy of DSDP Hole 272: Stratigraphic relationship to the underlying Miocene of DSDP hole 270, Ross Sea. Antarctic J US 22:123–125

Inversion of Airborne Gravity Data Acquired over Subglacial Lakes in East Antarctica

Irina Y. Filina[1,2] · Donald D. Blankenship[2] · Lopamudra Roy[2,3] · Mrinal K. Sen[2] · Thomas G. Richter[2] · John W. Holt[2]

[1] Department of Geological Sciences, John A. and Katherine G. Jackson School of Geosciences, University of Texas at Austin, 1 University Station C1100, Austin, TX 78712-0254, USA

[2] Institute for Geophysics, John A. and Katherine G. Jackson School of Geosciences, University of Texas at Austin, 4412 Spicewood Springs Rd. #600 Austin, TX 78759-8500, USA

[3] Department of Applied Geophysics, Indian School of Mines, Dhanbad – 826004, Jharkhand, India

Abstract. Airborne gravity data have been acquired over the two largest subglacial lakes in East Antarctica. 2D inversion of these data was performed for several fixed values of density contrast in order to estimate bathymetry and sediment thickness. For Lake Vostok the best agreement between profiles derived from gravity inversion and seismic soundings is achieved for densities 2.55 g cm^{-3} for host rock and 1.85 g cm^{-3} for sediment. The result shows a topographic rise of the lake bottom dividing the lake into two sub-basins. Our inversion results suggest that water thickness in Lake Concordia does not exceed 200 m for all possible density contrasts between ice/water and surrounding rock; the sediment layer cannot be resolved.

Introduction

The majority of subglacial lakes found beneath the Antarctic ice sheet are located in East Antarctica (Dowdeswell and Siegert 1999). This study addresses the two largest subglacial lakes in East Antarctica – Lake Vostok and Lake Concordia. The location of these lakes is shown in Fig. 3.4-1.

The largest known subglacial lake – Lake Vostok – is located beneath the Russian station Vostok, East Antarctica. This lake, covered by 4 km of ice, is about 300 km long and 60 km wide. The thicknesses of the water and sediment layers in Lake Vostok at a few points are known from sparse seismic soundings (Masolov et al. 1999).

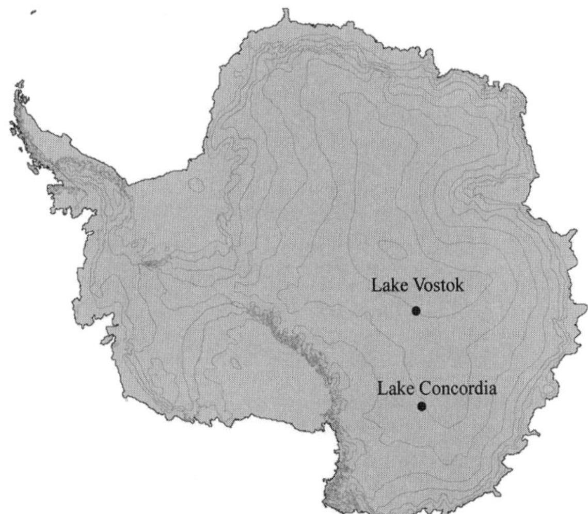

Fig. 3.4-1. The location of Lake Vostok and Lake Concordia

To perform the inversion of the gravity data the densities of subsurface layers must be assumed. There are two bathymetry models of Lake Vostok (Roy et al. 2005; Studinger et al. 2004) derived by inversion of gravity data using an unconstrained bedrock density of 2.67 g cm^{-3}. For this study, several values of bedrock density were tried in order to determine the one which gives the best agreement between both seismic and gravity models. The gravity effect of sediment, which was neglected in Studinger et al. (2004), is included in this study using the assumed density contrast between sediment and bedrock from Roy et al. (2005).

Lake Concordia, the second largest subglacial lake identified in Antarctica, is about 50 km long and 20 km wide. It is located near the Concordia base, Dome C, East Antarctica. The water depth in Lake Concordia was estimated from gravity data to be less then 1 000 m (Tikku et al. 2002).

The main purpose of this study is to estimate the water depth and the sediment thickness in the two largest lakes in East Antarctica by inversion of airborne gravity data.

Gravity Data Acquisition and Reduction

The University of Texas Institute for Geophysics (UTIG) has been performing airborne geophysical surveys in Antarctica since 1991. The UTIG airborne platform is a DeHavilland Twin Otter provided by NSF and instrumented by UTIG with ice-penetrating radar, laser altimeter, magnetometer, a gravimeter, and geodetic GPS receivers. Over 260 000 line-km of simultaneous geophysical measurements, positioning and aircraft altitude measurements have been acquired during nine field seasons.

The gravimeter utilized was a Bell Aerospace BGM-3 on loan from the U.S. Naval Oceanographic Office. The BGM-3 was originally developed for marine gravity work and uses an electromagnetically levitated proof mass to measure vertical acceleration. Verticality of the sensor is maintained by a two-axis gyro-stabilized platform. The platform of the unit used by UTIG has been modified for airborne use through lengthening of its natural oscillation period from the marine default value.

From: Fütterer DK, Damaske D, Kleinschmidt G, Miller H, Tessensohn F (eds) (2006) Antarctica: Contributions to global earth sciences. Springer-Verlag, Berlin Heidelberg New York, pp 129–134

The data was reduced at UTIG (Richter et al. 2001, 2002) by the general method which has become standard in the industry-GPS positions are used to calculate non-gravitational accelerations on the gravimeter which are subtracted from the total signal recorded by the gravimeter. The exact reduction algorithm was tuned to match the characteristics of the BGM-3 and the survey parameters. Notably, the thickness of the ice in the regions surveyed precludes the presence of any measurable gravity field features with wavelengths smaller than several kilometers at the aircraft operating altitude.

Accuracy of the free air gravity values was estimated by several measures related to self consistency of the results for the entire survey grid. These included comparison of results for the lines, which were repeated during the survey and comparison of values at line crossover points within the grid. The Lake Vostok data were the latest and best gravity data set obtained by UTIG. For Lake Vostok the self consistency measures indicate that free air gravity values within the survey grid are repeatable with a statistical distribution of about 1.4 mGal RMS, implying an accuracy on the order of 1 mGal RMS for the free air gravity values, assuming non-correlated random errors. Some correlated errors may have been induced by the noise filtering and other systematic effects. The frequency content of the free air gravity signals calculated is about that expected from the gravity field itself due to the subice topography, so the filters should not have significantly altered the result. No other systematic effects larger than one mGal have been identified.

Method

The forward problem was solved using Talwani's method for calculation of a gravity anomaly due to a 2D body with a polygonal cross section (Grant and West 1965). The gravity response is a function of the geometry of the causative body and its density contrast with the host rock. The density contrast along a profile was assumed to be constant. The densities of ice and water are close, so the ice and water of the lake are considered as one layer. Inversion was performed for a two-layered model, which consisted of ice/water and sediment layers overlying the dense bedrock. The water depth then was found by subtracting the ice thickness, measured by radar sounding, from the total thickness of ice/water layer. The coordinates of these bodies' vertices were chosen as model parameters. The locations of the lakes and the ice thickness/subglacial topography, known from radar sounding, were used to constrain the model. Also, for Lake Vostok the model was constrained by water and sediment thicknesses, known from seismic data (Masolov et al. 1999) with locations shown in Fig. 3.4-2. Inversion was performed using a conjugate gradient algorithm for several fixed values of density contrast between the ice/water body and the surrounding rock. The density contrast between sediment and host rock is chosen to be -0.7 g cm^{-3} as in Roy et al. (2005).

Gravity Inversion over Lake Vostok

The airborne gravity survey over Lake Vostok was performed by UTIG during the 2000–2001 austral summer (Richter et al. 2001, 2002; Studinger et al. 2003). The survey block was 165 by 330 km; each survey line was separated by 7.5 km. Average aircraft height was 3.96 km above mean sea level. The radar sounding bed-echo-strength map is shown in Fig. 3.4-2; Lake Vostok is assumed to be the region of high echo strength.

Four profiles over Lake Vostok were chosen for gravity modeling (Fig. 3.4-2). Before the inversion, the data were reduced to sea level. The gravity anomaly increases

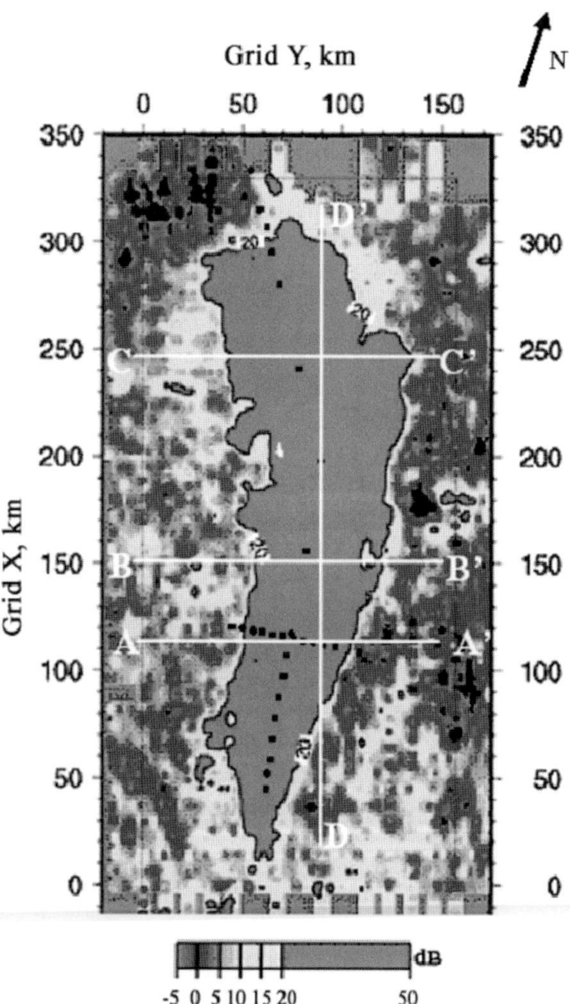

Fig. 3.4-2. Radar sounding bed-echo-strength map of Lake Vostok area (Sasha Carter, personal communication); *white lines* are profiles for inversion, *black dots* are seismic soundings

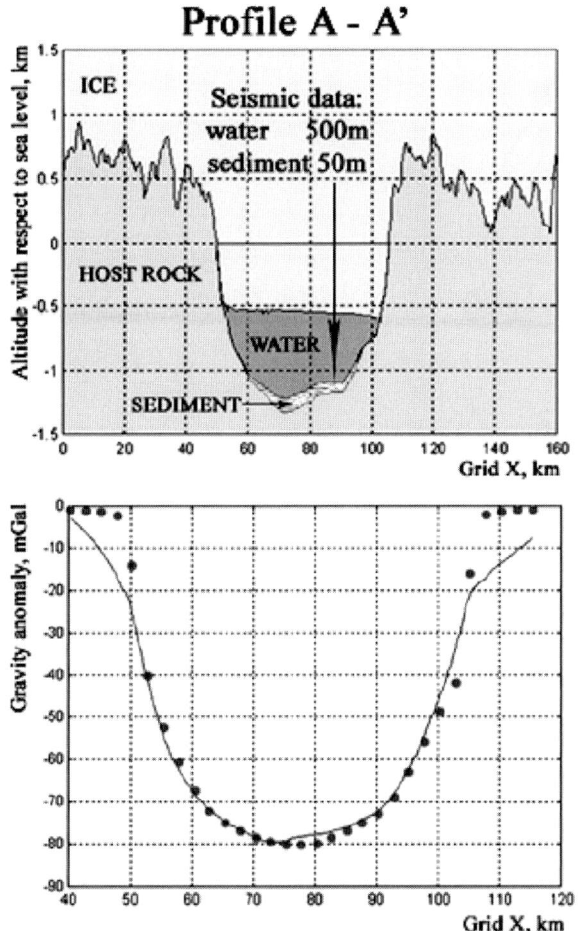

Fig. 3.4-3. Profile *A–A'*, Lake Vostok: *top:* a model for inversion; *bottom:* measured (*solid line*) and calculated (*dots*) gravity anomaly

rapidly by about 70 mGal from the western edge to the eastern edge of the lake basin. Such a sharp increase in gravity indicates significant change in the lower crust from west to east of the lake. To remove the gravity effect from deeper geological structures, a regional trend was found by a cubic spline interpolation, and then subtracted. The best agreement of the gravity model with the results of seismic interpretation at the coincident point on profile A–A' (Fig. 3.4-2 and 3.4-3) was achieved for a density contrast of -1.6 g cm^{-3} between ice/water and host rock, which corresponds to a bedrock density of 2.55 g cm^{-3} (with an ice/water density of 0.95 g cm^{-3}). The calculation over other profiles was performed using this value of density contrast. The result over profile A–A' shows a maximum water thickness to be 750 m; maximum sediment thickness is 120 m. The difference with seismic results at the cross-over point for both water and sediment layers are within 50 m.

The results over other profiles are shown in Fig. 3.4-4. The water layer along profile B–B' is about 600 m thick; the sediment layer becomes thinner (50 m at the cross-

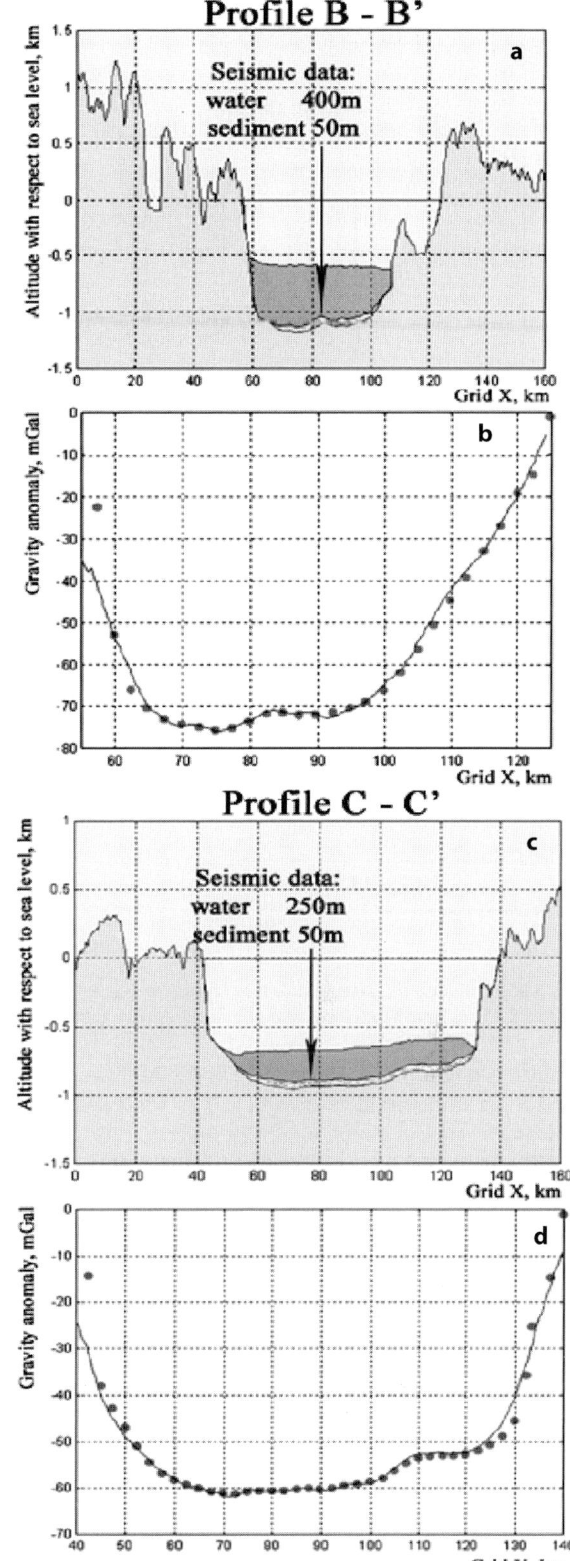

Fig. 3.4-4. Profiles *B–B'* and *C–C'*, Lake Vostok (for legend see Fig. 3.4-3); **a,c** models for inversion for *B–B'* and *C–C'* respectively; **b,d** measured and calculated gravity anomaly respectively

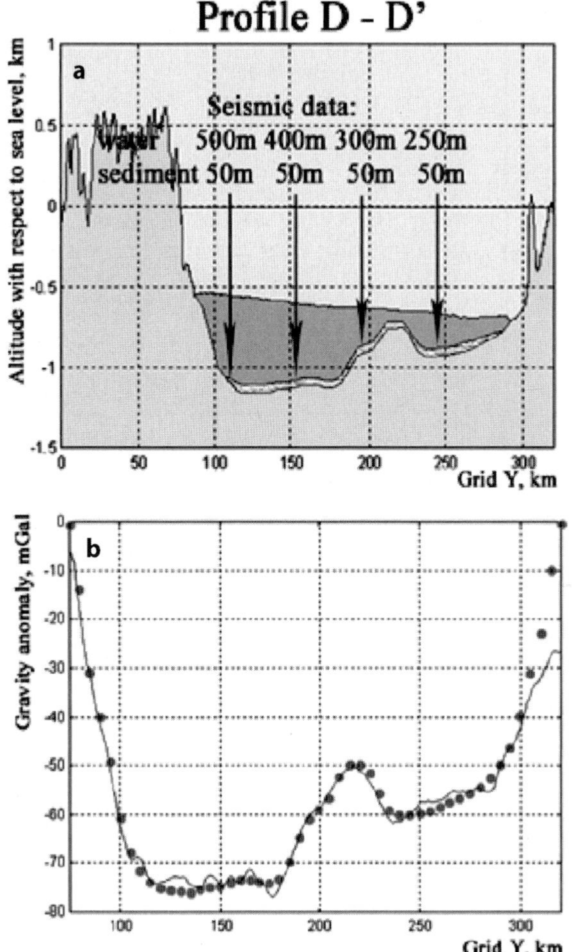

Fig. 3.4-5. Profile *D–D'*, Lake Vostok (see legend in Fig. 3.4-3); **a** a model for inversion; **b** measured and calculated gravity anomaly

point). Along profile C–C' the lake is about 90 km wide and water thickness decreases to 250 m, while sediment thickness remains about 50 m. The inversion for profile D–D' (along the lake) is shown in Fig. 3.4-5. Our results show a significant and sharp rise of the lake bottom in the northern part dividing the lake into two sub-basins. The water thickness over this rise is about 100 m. Since there exist only a few seismic soundings in the northern part of the lake (40 km between shots), this feature in the lake's bottom topography was not recognized in previous seismic surveys.

The largest difference in water thickness derived from gravity and seismics at coincident points is 100 m.

Gravity Inversion over Lake Concordia

The airborne survey over Lake Concordia, East Antarctica, was performed by UTIG during the 1999–2000 austral summer. The radar sounding bed-echo-strength map is shown in Fig. 3.4-6; the dark region in the center, which is about 50 km long and 20 km wide, is assumed to be the lake.

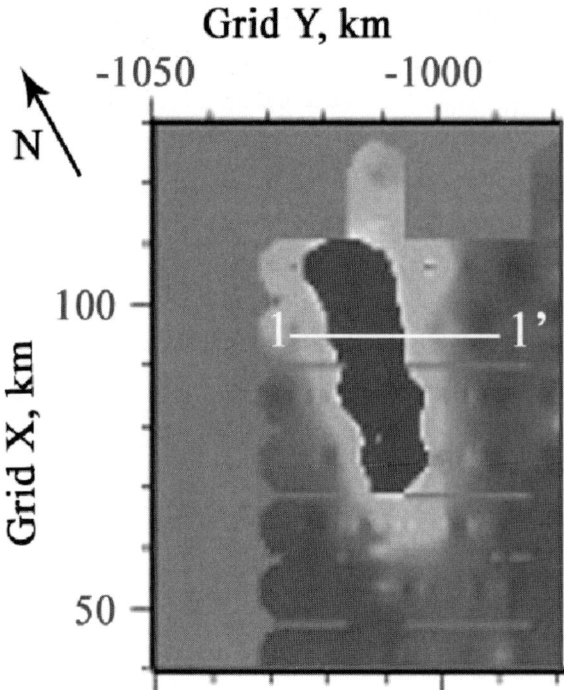

Fig. 3.4-6. Radar sounding bed-echo-strength map over Lake Concordia (Sasha Carter, personal communication); *white line* shows the location of profile in Fig. 3.4-7

The lake is located at the edge of a survey block, creating an uncertainty in the evaluation of the regional trend. Since there are no data for the northern part of the lake, the regional trend was evaluated south of the lake and then extrapolated over the lake. The inversion of the gravity data was performed for three profiles using several fixed values of density contrast (from -1.85 to $-1.55\ \mathrm{g\ cm^{-3}}$ between ice/water and host rock). Figure 3.4-7 shows the result for profile 1–1' (for location see Fig. 3.4-6) using a density contrast of $-1.6\ \mathrm{g\ cm^{-3}}$, which had given the best agreement with seismic results for Lake Vostok. For all density contrasts used, the water thickness in Lake Concordia does not exceed 200 m. The differences in water thickness at cross-over points of inverted profiles for all density contrasts are within 50 m. Since Lake Concordia is relatively shallow, the sediment layer cannot be resolved and it was neglected.

Summary

2D gravity inversion was performed for two Antarctic subglacial lakes using different values of density contrasts between ice/water and the surrounding rock. For Lake Vostok, the best agreement between seismic and gravity

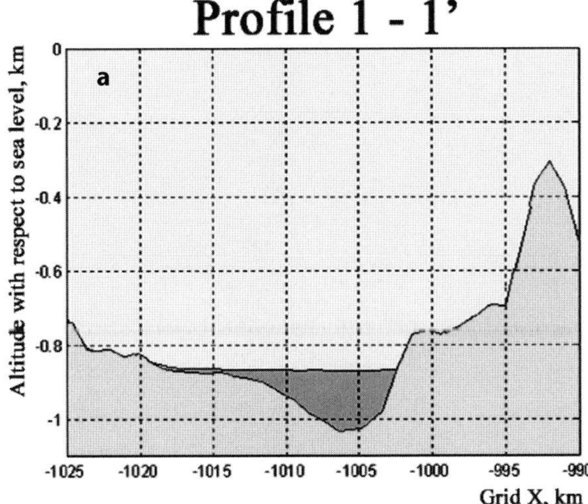

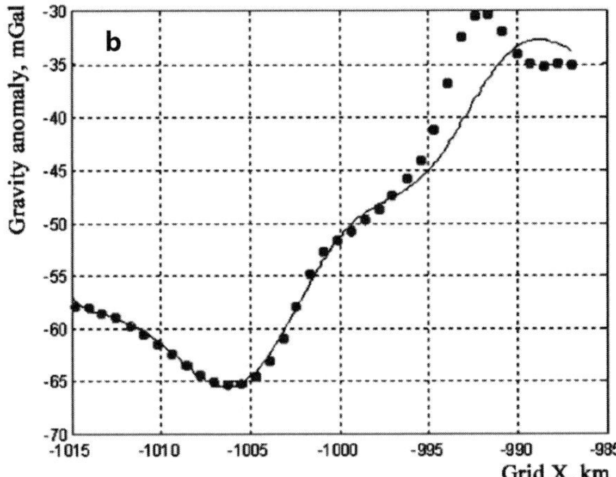

Fig. 3.4-7. Profile *1–1'*, Lake Concordia (for legend see Fig. 3.4-3); **a** a model for inversion; **b** measured and calculated gravity anomaly

models was achieved for density contrasts of -1.6 g cm^{-3} between ice/water and host rock and -0.7 g cm^{-3} for sediment and host rock. The result shows a topographic rise in the northern part of the lake, which divides Lake Vostok into two sub-basins: a large and deep basin in the southern part and a relatively small and shallow one in the north.

The inversion over Lake Concordia using a large range of density contrasts indicates that the water thickness in this lake does not exceed 200 m. Since the lake is relatively shallow, a sediment layer can not be resolved.

Acknowledgments

This work was supported by the John A. and Katherine G. Jackson School of Geosciences, The University of Texas at Austin and Office of Polar Programs of the U.S. National Science Foundation. We would also like to thank the staff of The University of Texas Support Office for Aerogeophysical Research and Sasha Carter for their invaluable contribution. The aerogeophysical field work was proposed collaboratively by Lamont-Doherty Earth Observatory at Columbia University (R. E. Bell and M. Studinger, co-investigators) and the University of Texas at Austin (D. D. Blankenship, D. L. Morse and I. W. Dalziel, co-investigators). The UTIG contribution number is 1714.

References

Grant FS, West GF (1965) Interpretation theory in applied geophysics. McGraw-Hill, New York, pp 583

Dowdeswell JA, Siegert MJ (1999) The dimensions and topographic setting of Antarctic subglacial lakes and implications for large-scale water storage beneath continental ice sheets. Geol Soc Amer Bull 111:254–263

Masolov VN, Kudryavtzev GA, Leitchenkov GL (1999) Earth science studies in the Lake Vostok region: existing data and proposal for future research in subglacial lake exploration. SCAR International Workshop on Subglacial Lake Exploration Report, Cambridge, pp 1–18

Richter TG, Holt JW, Blankenship DD (2001) Airborne gravity over East Antarctica. KIS 2001 Proc Internat Sympos Kinematic Syst Geodesy, Geomatics, Navigation, Banff, pp 576–585

Richter TG, Kempf SD, Holt JW, Morse DL, Blankenship DD, Peters ME (2002) Airborne gravimetry and laser altimetry over Lake Vostok, East Antarctica. EOS 83(19), Abstract B22A-04

Roy L, Sen MK, Blankenship DD, Stoffa PL, Richter TG (2005) Inversion and uncertainty estimation of gravity data using simulated annealing: an application over Lake Vostok, East Antarctica. Geophysics 70:J1–J12

Studinger M, Bell RE, Karner GD, Tikku AA, Holt JW, Morse DL, Richter TG, Kempf SD, Peters ME, Blankenship DD, Sweeney RE, Rystrom V (2003) Ice cover, landscape setting, and geological framework of Lake Vostok, East Antarctica. Earth Planet Sci Letters 205(3/4):195–210

Studinger M, Bell RE, Tikku AA (2004) Estimating the depth and shape of subglacial Lake Vostok's water cavity from aerogravity data. Geophys Res Letters 31

Tikku AA, Bell RE, Studinger M, Tobacco IE (2002) Lake Concordia: a second significant lake in East Antarctica. In: Finn CA, Anandakrishnan S, Goodge JW, Panter KS, Siddoway CS, Wilson TJ (eds) REVEAL workshop report. pp 110

Russian Geophysical Studies of Lake Vostok, Central East Antarctica

Valeriy N. Masolov[1] · Sergey V. Popov[1] · Valeriy V. Lukin[2] · Alexander N. Sheremetyev[1] · Anatoly M. Popkov[1]

[1] Polar Marine Geological Research Expedition (PMGRE), 24 Pobeda st., 188512 St. Petersburg, Lomonosov, Russia

[2] Russian Antarctic Expedition (RAE), 38 Bering st., 199397 St. Petersburg, Russia

Abstract. Since 1995, Polar Marine Geological Research Expedition has performed the geophysical investigations of Lake Vostok, Central East Antarctica. The study of this phenomenon is carried out by means of radio-echo sounding (RES) and reflection seismic. In total, 3 250 km of RES profiles and 194 seismic measurements have been made. These scientific works resulted in mapping the ice thickness, bedrock and sub-ice topography and the Lake Vostok shoreline. We fixed 195 fragments of grounding line, according to RES data, with 169 of them (86%) being reliable while 26 (14%) are questionable. These results are the base for contouring the lake shore. The water table square estimates at some 17 100 km^2. We also detected 22 small-size subglacial water cavities around the Lake Vostok. Subice topography of the Lake Vostok bottom is divided into two main regions: the deep-water and shallow-water basins. The first one with depths from about −1 700 to −800 m is located in the southern part of the lake. We assume its northern part to be shallow-water with the bottom depth of about −940 m.

Introduction

Satellite altimetry (Ridley et al. 1993) revealed a large subglacial lake (named *Lake Vostok*) located northwest from the Russian station Vostok. Geophysical exploration of this phenomenon is important for fundamental scientific and practical problems. Geophysical investigations of the area were started in the middle of the last century. A number of Russian seismic reflections and British radio-echo soundings were carried out in the 1950s to 1970s (Oswald and Robin 1973; Kapitsa et al. 1996). In 1999 Italians produced the airborne geophysical observations to determine the ice thickness, bedrock topography, the lake shape, the amplitude of the bottom reflections and other characteristics (Tabacco et al. 2002). In 2000 Americans carried out the geophysical survey, which completely covered of the lake area (Studinger et al. 2003).

Polar Marine Geological Research Expedition (PMGRE) in framework of the Russian Antarctic Expedition (RAE) initiated and executed a new scientific project dedicated to the study of sub-glacial Lake Vostok region (Masolov et al. 1999, 2001; Popov et al. 2001). The objective of the first phase of this program was to develop the radio-echo sounding and reflection seismic equipment and methods to study the area as a geographical object: the lake geometry and morphology; ice thickness and the lake depth; velocities of the electromagnetic wave propagation in ice and acoustic waves in the Lake Vostok area. The second phase of the Lake Vostok project started in 2002. Its goal is to study Lake Vostok as a geological object: determination of its geological structure and evolution; mapping of ice thickness, bed relief and bathymetry of the Lake Vostok area as well as of its geomorphological features to better understand its nature and evolution.

Data Acquisition

Since 1998, 60-MHz ice radar with repetition frequency of 600 Hz, pulse length of 0.5 μs, pulse power of 60 kW, dynamical range of 180 dB and band of the reception channel of 3 MHz has been used for ice thickness measurements. The reflected signals are digitized by analog-digital transformation device with a sample interval of 50 ns and stacking rate of 256 traces and then registered on PC. The transformer has been developed based on the 12-bit analog-digital converter AD9042AST (Analog Devices Inc.) with SBC-8259 processor (Axiom Technology Co) (Popov et al. 2001, 2003). In total, 3 250 km of the RES data have been collected (Fig. 3.5-1). The first profiles followed the preliminary boundary of Lake Vostok defined by Siegert and Ridley (1998). However, the location of survey profiles was changed during the field work, which allowed us to investigate the lake boundary more effectively.

Reflection seismic investigations began in 1995. Between 1995 and 2002 we obtained 194 seismic measurements to estimate the water depth of the Lake Vostok. During the period of 1995–1999 the Russian seismic station SMOV-0-24 with the analogue registration on the magnetic tape was used. It had 24 channels, recording of 6 and 12 s, noise of the channel of 0.3 μV, cross-talk higher than −36 dB, frequency range of the recording of 10–200 Hz and dynamical range of 80 dB. The acoustic wave registration is performed by the geophone SV-20 with oscillation frequency 20 Hz fixed on the surface and covered by the snow. Starting from the year 2000 we use a digital station with the same technical characteristics. The seismic profiles are located along and across the lake with a dense shot spacing near Vostok Station (Fig. 3.5-1).

From: Fütterer DK, Damaske D, Kleinschmidt G, Miller H, Tessensohn F (eds) (2006) Antarctica: Contributions to global earth sciences. Springer-Verlag, Berlin Heidelberg New York, pp 135–140

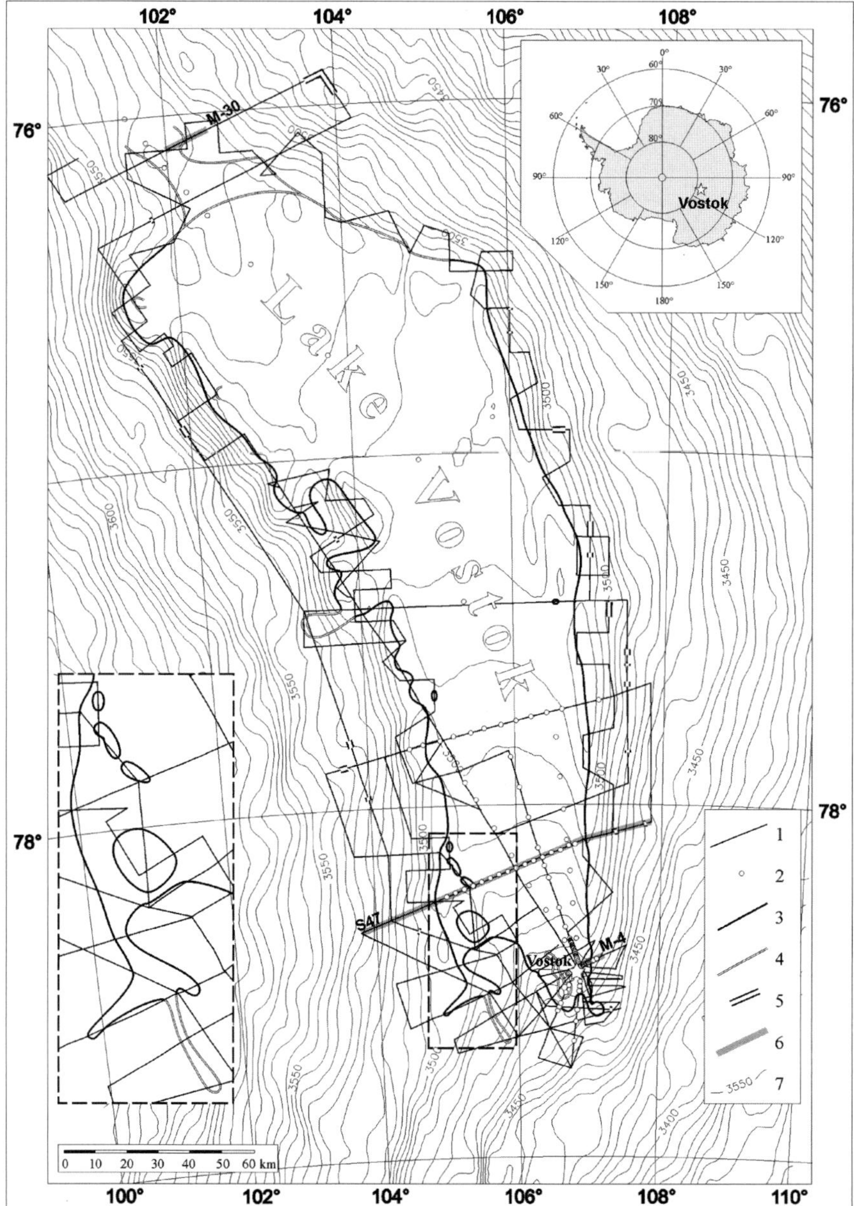

Fig. 3.5-1.
Location map with surface elevation contours. *1:* Russian RES profiles of 1998–2004; *2:* Russian reflection seismic points of 1995–2002; *3:* reliable Lake Vostok boundary; *4:* boundary questionable; *5:* fragments of water cavities; *6:* sections shown in Fig. 3.5-2 and 3.5-3; *7:* surface elevation contours with 5 m spacing

During the first phase we optimized the generation of the seismic waves using detonating cord positioned on the snow surface in contrast to explosive charge put in boreholes, as was practiced before (Popkov et al. 1999). In order to increase the accuracy, we measured the propagation velocities of both seismic and electromagnetic waves. We used the vertical seismic profiling in the 5G-1 borehole located at the Russian station Vostok. Acoustic velocities in ice and water are 3 920 and 1 490 m s^{-1}, respectively. The average velocity within the section (with the snow-firn layer) is 3 810 m s^{-1}. Based on reflection seismic measurements in the borehole we estimated the distance between the borehole bottom (3 623 m) and the water table to be 130 m (Masolov et al. 1999, 2001; Popkov et al. 1999). The velocity of electromagnetic waves in ice, including a snow-firn correction, estimated using a wide angle reflection technique, is 168.4 ±0.5 m µs^{-1} (Popov et al. 2003).

The most typical and interesting profile is the S47 profile across the Lake Vostok (Fig. 3.5-1). The RES and seismic records and the ice sheet section are shown in Fig. 3.5-2. The profile crossed the small island located in south-west part of the lake. It is nicely seen both in the RES (Fig. 3.5-2a) and the seismic records (Fig. 3.5-2b). The lake bottom is flat with about –600 m and –1 300 m height in the western and central parts respectively. The central part is complicated by sub-water valley 400 m deep as shown in Fig. 3.5-2c.

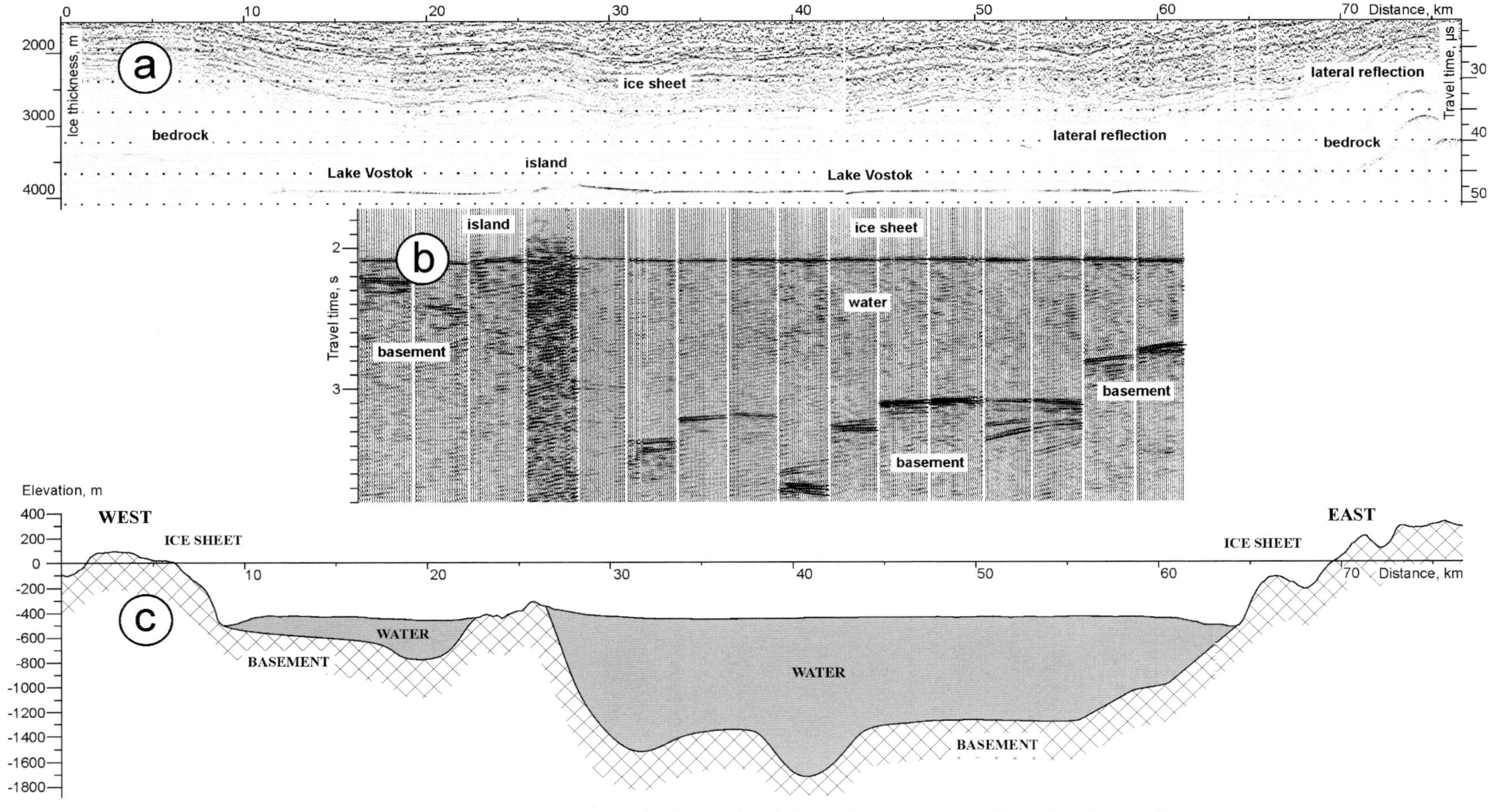

Fig. 3.5-2. Radio-echo (**a**) and seismic (**b**) time sections and interpeted section (**c**) of Lake Vostok and the ice sheet along S47 profile. For location see Fig. 3.5-1

The Features of the Lake Vostok Boundary and Bedrock Topography

One of the main results of the first stage of our investigations was the Lake Vostok boundary charting. Two techniques were used. We analyzed radio-echo sounding reflections from the ice base and morphological features of the subglacial topography. The reflections from ice-water and ice-bedrock are quite different: flat and intensive from the lake water table and weak from the bedrock (Oswald and Robin 1973; Popov et al. 2001). Abrupt slopes, as a rule, mark the boundary of the lake, thus indicating a tectonic nature of Lake Vostok (Masolov et al. 1999, 2001; Popov et al. 2001).

In total we detected 195 fragments of grounding line, according to RES data. 169 points (86%) are reliable while 26 (14%) are questionable. The RES records with a sample of the both types of the grounded zones are shown in Fig. 3.5-3. The points were used to construct a shoreline of Lake Vostok (Fig. 3.5-1). We also identified several islands inside the lake. Some of them are situated near Vostok Station and located on the ice flow line passing through the borehole 5G-1. We suppose that the mineral inclusions detected in the ice core (Jouzel et al. 1999) have been originated from the islands bedrock.

Lake Vostok shore line is oval-shaped complicated by capes, bays and peninsulas located in the west and south. Interestingly, the eastern, northern and northwestern parts of the lake boundary are relatively rectilinear. This suggests the presence of the deep faults along this shoreline that is confirmed by American airborne geophysical data (Studinger et al. 2003). The lake covers about 17 100 km^2.

The northern part of the shoreline of Lake Vostok is still not defined completely. Hypothetically, it could be continued to the north, but most likely the lake is limited as shown in Fig. 3.5-1. The exploration of the northern border of the Lake Vostok will be completed by means of the RES in the nearest future.

Some 30 km to the south-west from Vostok Station three fragments of the water cavities were detected (see the inset in Fig. 3.5-1). The fragments fall in line crossing all the three quasi-parallel RES profiles, where they are observed. The northern fragment joins to the Lake Vostok. The ice base elevations of the mentioned fragments are –270, –116 and +175 m a.s.l. respectively. We assume these water cavities are an appendix of the Lake Vostok (for example, narrow estuary or river). Data to confirm or reject this assumption will be collected during the coming field seasons.

Twenty-two small sized subglacial water cavities with a typical size close to ten kilometers were observed around the Lake Vostok (Fig. 3.5-1). Their elevations range between sea level and +800 m for those located in the east and between –300 and +300 m for the cavities located in the west. An isolated subglacial lake on the cross RES

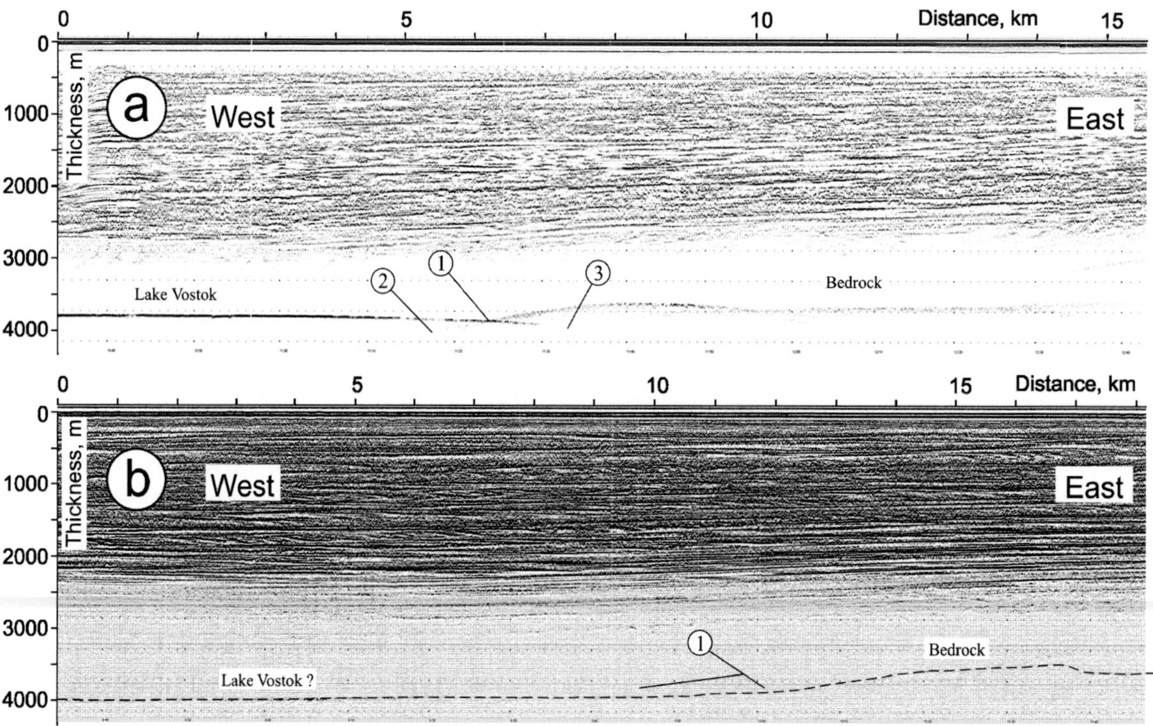

Fig. 3.5-3. RES records with reliable (**a**) and assumed (**b**) grounding zones. *1:* Position of the grounding line; hyperbolic reflections from the lake slope (*2*) and from the Lake Vostok water table (*3*). For location see Fig. 3.5-1; **a** along route M-4; **b** along route M-30

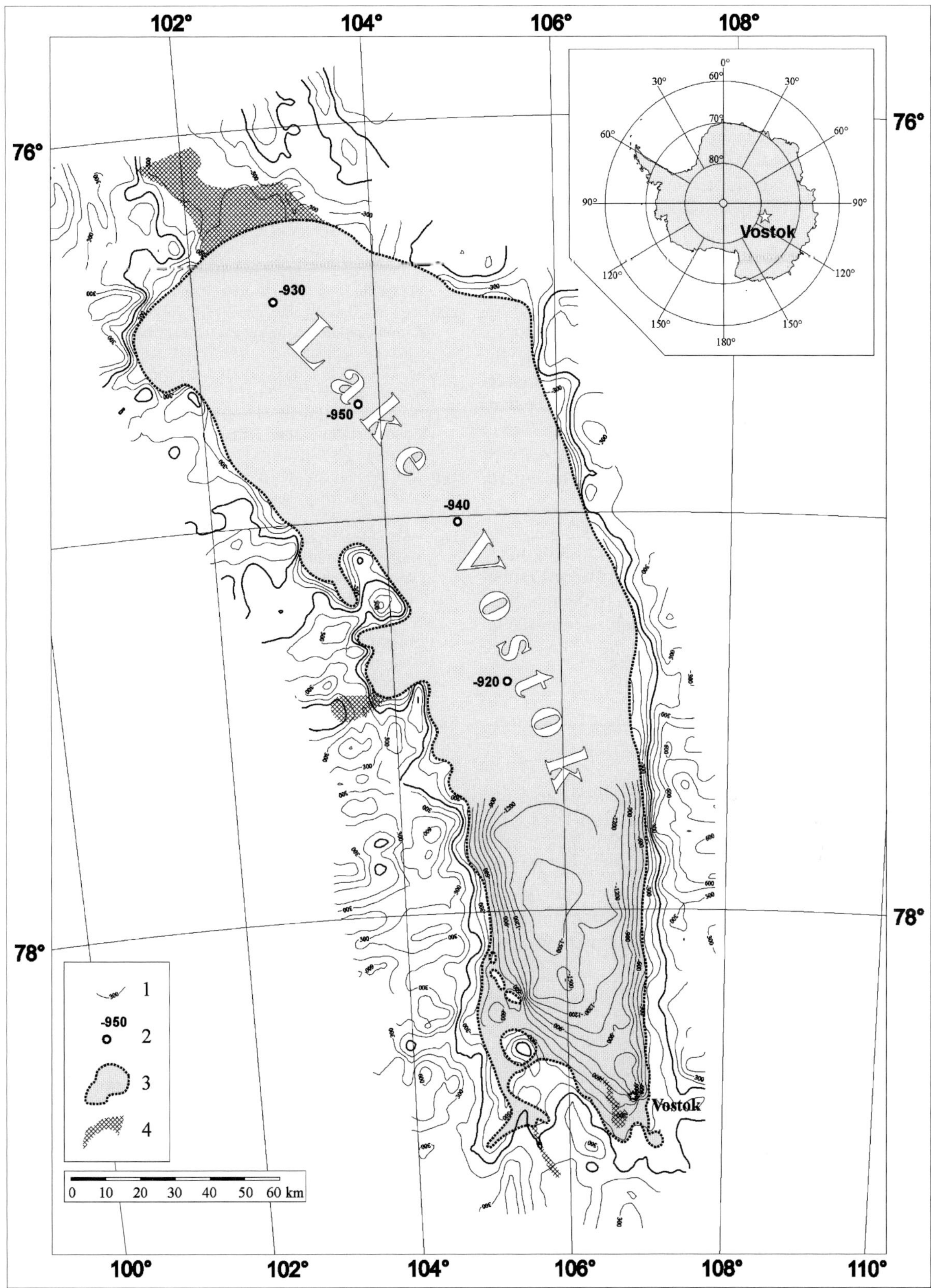

Fig. 3.5-4. Bedrock topography map. *1:* bedrock topography contours with 150 m spacing (*thick line* is the sea level); *2:* reflection seismic measurements with the bedrock elevation; *3:* reliable Lake Vostok water table; *4:* assumed water table

routes is detected to the north-east from the Lake Vostok. The lengths of the fragments of this lake are about 10 km (Fig. 3.5-1).

We used RES and ground reflection seismic data to draw ice thickness, lake depth and bedrock topography maps. In this paper we only present the bedrock topography map (Fig. 3.5-4) as the most interesting. The bedrock morphology is described elsewhere in this volume (Popov et al. 2006). We only focus on the main aspects of the bedrock relief.

Generally, the Lake Vostok bottom is divided into two regions: the deep-water and shallow-water basins. The first one is located in the southern part. It has pear-shaped configuration with depths from about −1 700 to −800 m. The lake depth under the Vostok Station is 680 m and the bottom is about −960 m. Based on four seismic measurements we assume the northern part of the lake is a shallow-water pool with the bottom height of about −940 m (Fig. 3.5-4). The next shallow-water basin was revealed 30 km to the west from the Vostok Station. Its bottom height is about −600 m.

The bedrock relief outside the lake is very different. In general, it is divided into mountain and hilly plane areas. The first one is situated in the north-west (about 800 m in height) and in the central part both to the west and east from the Lake Vostok. Its height is from 400 up to 1 100 m. Other territories around the lake are hilly planes.

Acknowledgments

The authors would like to express their sincere gratitude to Prof. H. Miller (AWI, Germany) for his help with the VSP work and Alexey A. Ekaykin, Igor V. Samsonov and Olga B. Soboleva for their help and comments. The authors are also grateful to Prof. Fütterer and referee Dr. Michael Studinger for their well-wishing comments and suggestions which allowed significant improving the manuscript.

References

Jouzel J, Petit JR, Souchez R, Barkov NI, Lipenkov VYa, Raynaud D, Stievenard M, Vassiliev NI, Verbeke V, Vimeux F (1999) More than 200 meters of lake ice above subglacial Lake Vostok, Antarctica. Science 286:2138–2141

Kapitsa AP, Ridley JK, Robin G de Q, Siegert MJ, Zotikov I (1996) A large deep freshwater lake beneath the ice of central East Antarctica. Nature 381(6584):684–686

Masolov VN, Kudryavtzev GA, Sheremetyev AN, Popkov AM, Popov SV, Lukin VV, Grikurov GE, Leitchenkov GL (1999) Earth science studies in the Lake Vostok Region: existing data and proposals for future research. SCAR International Workshop on Subglacial Lake Exploration Report, Cambridge, pp 1–18

Masolov VN, Lukin VV, Sheremetyev AN, Popov SV (2001) Geophysical investigations of the subglacial lake Vostok in Eastern Antarctica. Doclady Earth Sci 379A(6):734–738

Oswald GKA, Robin G de Q, (1973) Lakes beneath the Antarctic ice sheet. Nature 245:251–254

Popkov AM, Verkulich SR, Masolov VN, Lukin VV (1999) Seismic section in Vostok Station vicinity (Antarctica) as result of investigations in 1997. MGI 86:152–159, (in Russian)

Popov SV, Mironov AV, Sheremetyev AN (2001) Result of ground RES researches of subice lake Vostok in 1998–2000. MGI 89:129–133, (in Russian)

Popov SV, Sheremetyev AN, Masolov VN, Lukin VV, Mironov AV, Luchininov VS (2003) Velocity of radio-wave propagation in ice at Vostok station, Antarctica. J Glaciol 49(165):179–183

Popov SV, Lastochkin AN, Masolov VN, Popkov AM (2006) Morphology of the subglacial bed relief of Lake Vostok basin area (central East Antarctica) based on RES and seismic data. In: Fütterer DK, Damaske D, Kleinschmidt G, Miller H, Tessensohn F (eds) Antarctica – Contributions to global earth sciences. Springer, Berlin Heidelberg New York, pp 141–146

Ridley JK, Cudlip W, Laxon W (1993) Identification of sub-glacial lakes using ERS-1 radar altimeter. J Glaciol 39(133):625–634

Siegert MJ, Ridley JK (1998) An analysis of the ice-sheet surface and subsurface topography above the Vostok Station subglacial lake, central East Antarctica. J Geophys Res 103(5):10195–10207

Studinger M, Bell R, Karner GD, Tikku AA, Holt JW, Morse DL, Richter TG, Kempf SD, Peters ME, Blankenship DD, Sweeney RE, Rystrom VL (2003) Ice cover, landscape setting and geological framework of Lake Vostok, East Antarctica. Earth Planet Sci Letters 205:195–210

Tabacco IE, Bianchi C, Zirizzotti A, Zuccheretti, Forieri A, Della Vedova A (2002) Airborne radar survey above Vostok region, East Central Antarctica: ice thickness and Lake Vostok geometry. J Glaciol 48:62–69

Morphology of the Subglacial Bed Relief of Lake Vostok Basin Area (Central East Antarctica) Based on RES and Seismic Data

Sergey V. Popov[1] · Alexander N. Lastochkin[2] · Valeriy N. Masolov[1] · Anatoly M. Popkov[1]

[1] Polar Marine Geological Research Expedition (PMGRE), 24 Pobeda str., 188512 St. Petersburg, Lomonosov, Russia

[2] St. Petersburg State University (SPSU), 7/9 Universitetskaya nab., 191164 St. Petersburg, Russia

Abstract. During the austral summer field seasons of the 1995–2004 Polar Marine Geological Research Expedition (PMGRE) within the frame of the Russian Antarctic Expedition (RAE) carried out ground-based geophysical investigations in the sub-glacial Lake Vostok area in order to study the ice sheet and bed relief. Geomorphological analysis of the data allowed better understanding of sub-ice and sub-water structures. The most striking structure is the Vostok Basin which subdivides into five main substructures: lake plane, deep-water hollow, sub-water ridges, internal and external slopes. We detected six principal morphological substructures outside the Vostok Basin: lowlands, low hilly planes, high planes, ridged plane, Komsomolskiye Mountains and middle mountain land. A geomorphological chart has been produced.

Introduction

Russian investigations of the Lake Vostok area were started in 1995 by the Polar Marine Geological Research Expedition (PMGRE) within the frame of the Russian Antarctic Expedition (RAE). They were part of the scientific endeavour dedicated to the existence of the large subglacial lake named Vostok (Ridley et al. 1993; Kapitsa et al. 1996) and devoted to ice sheet and bed relief studies. The Russian field work (Fig. 3.6-1) included radio echo sounding (RES) and seismic reflection measurements (Masolov et al. 1999, 2001, 2002; Popov et al. 2001, 2003b, and others).

In 1999, airborne geophysical research (including RES) was carried out over the Lake Vostok region to determine the ice thickness, bedrock topography, lake shape, the amplitude of the bottom reflections and other characteristics (Tabacco et al. 2002). A good coverage of the lake area by the geophysical data was carried out by Americans during the field season of 2000. They collected airborne RES, magnetometric and gravimetric data on the regular network 7.5 × 11.25 km^2 (Studinger et al. 2003; and others). All the data allowed forming the first insights into the subglacial morphology and tectonics of the area (Masolov et al. 2001; Popov et al. 2002; Studinger et al. 2003).

It is necessary to note that the investigations of the Lake Vostok area are crucial for a number of Antarctic scientific fields. Analysis of the ice core data from the unique 5G-1 borehole (drilled in 1991–1998 at Vostok Station) enabled to reconstruct the climate history of our planet over the past 420 ka (Lipenkov et al. 2000; Petit et al. 1999). Besides, comparison of the ice core and RES data, tracing and dating of the radar layers allowed estimation of the features of the ice sheet formation (Mandrikova et al. in press; Siegert et al. 1998; Popov 2003). In this respect the region of Lake Vostok is the most convenient because of the availability of numerous RES lines. Another principle problem is the questions of the formation and existence of the subglacial lakes. Lake Vostok is the biggest and the best studied one by geophysical, glaciological and biological methods. Therefore, understanding of the processes occurring in the lake and in the ice sheet results in understanding the nature of other similar objects (Dowdeswell et al. 2003; Siegert et al. 2001).

The Lake Vostok area is characterized by a major fault system which can be extended for a long distance (Leitchenkov et al. 1998). In this respect, the understanding of the tectonics and geological history of this region is important for studying the East Antarctic deep structure. As a first step, the geomorphological analysis could be used being one of the best ways for bed relief study. The proposed sketch is one of the first of our attempts of geomorphologic analysis of the bed relief which is covered by a thick ice sheet (Popov et al. 2002).

Data and Methods

Russian geophysical investigations were directed to the bed relief and contouring of Lake Vostok. Mapping of the lake bottom was accomplished by seismic reflection work, while RES was applied for mapping of the bedrock topography outside the lake (Masolov et al. 2001, 2002; Popov et al. 2001, 2003b). Between 1995 and 2004, data from 196 seismic reflection shots were collected and about 3 250 km of the RES profiles were obtained. Position of the seismic shots and RES routes are shown in Fig. 3.6-1.

To get accurate seismic data it is necessary to measure the acoustic velocities in ice. Toward this end, during the austral summer field seasons of 1996–1998 we performed vertical seismic profiling in the 5G-1 borehole to define the acoustic velocities. Logging device was kindly provided by Prof. Heinz Miller (AWI, Bremerhaven). The measurements allowed calculating the acoustic velocity in pure

From: Fütterer DK, Damaske D, Kleinschmidt G, Miller H, Tessensohn F (eds) (2006) Antarctica: Contributions to global earth sciences. Springer-Verlag, Berlin Heidelberg New York, pp 141–146

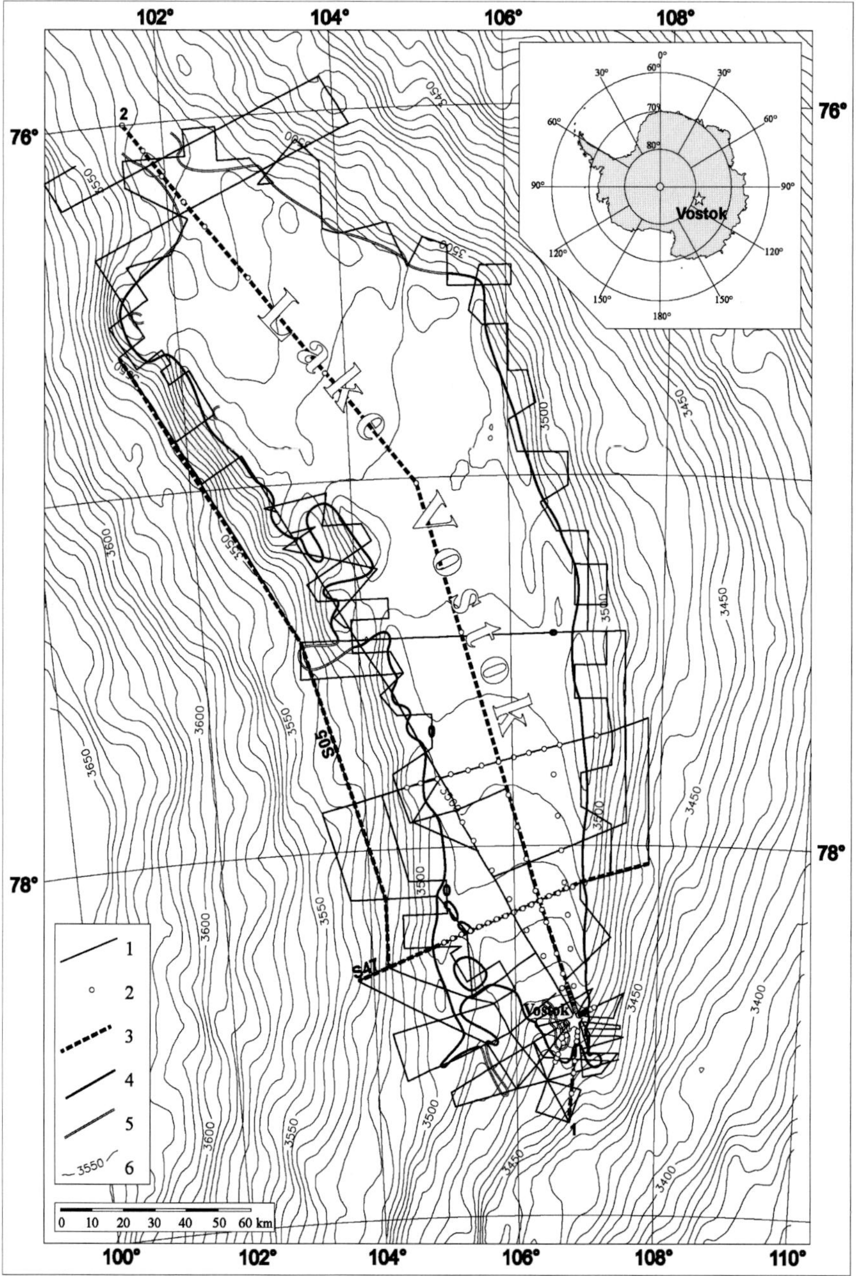

Fig. 3.6-1.
Location map of Lake Vostok area. *1:* Russian RES routes; *2:* sites of Russian reflection seismic soundings; *3:* location of the cross sections shown in Fig. 3.6-2; *4:* confidential Lake Vostok grounding line; *5:* supposed Lake Vostok grounding line; *6:* surface elevation contours (m)

ice using the direct wave. The velocity is 3 920 m s^{-1}. Averaged velocity from the ice surface to the ice base is 3 810 m s^{-1}. This value is used in all our investigations in the Lake Vostok area. Ice thickness in the 5G-1 borehole vicinity, defined using the direct and reflected waves from the ice base (3 750 m), is in good agreement with the RES data (Popov et al. 2003a). During the field season of 1999/2000 wide-angle seismic reflection measurements were performed in the vicinity of Vostok Station to measure the average velocity of the radio wave propagation in ice, which is 168.4 ±0.5 m μs^{-1} (Popov et al. 2003a).

Our geomorphological chart was based on the standard procedure for morphological classification. The boundaries between the regions are the following elements: Lines of maximal steepness of convex or concave forms (Lastochkin 1987, 1991; Spiridonov 1975). The space between the structural elements is genetic homogeneity for scale mapping. Specific features of the regions depend on such morphometric characteristics as relative and absolute height, displacement, gradient, shape and geological information. The last one is inaccessible because the bedrock is covered by ice.

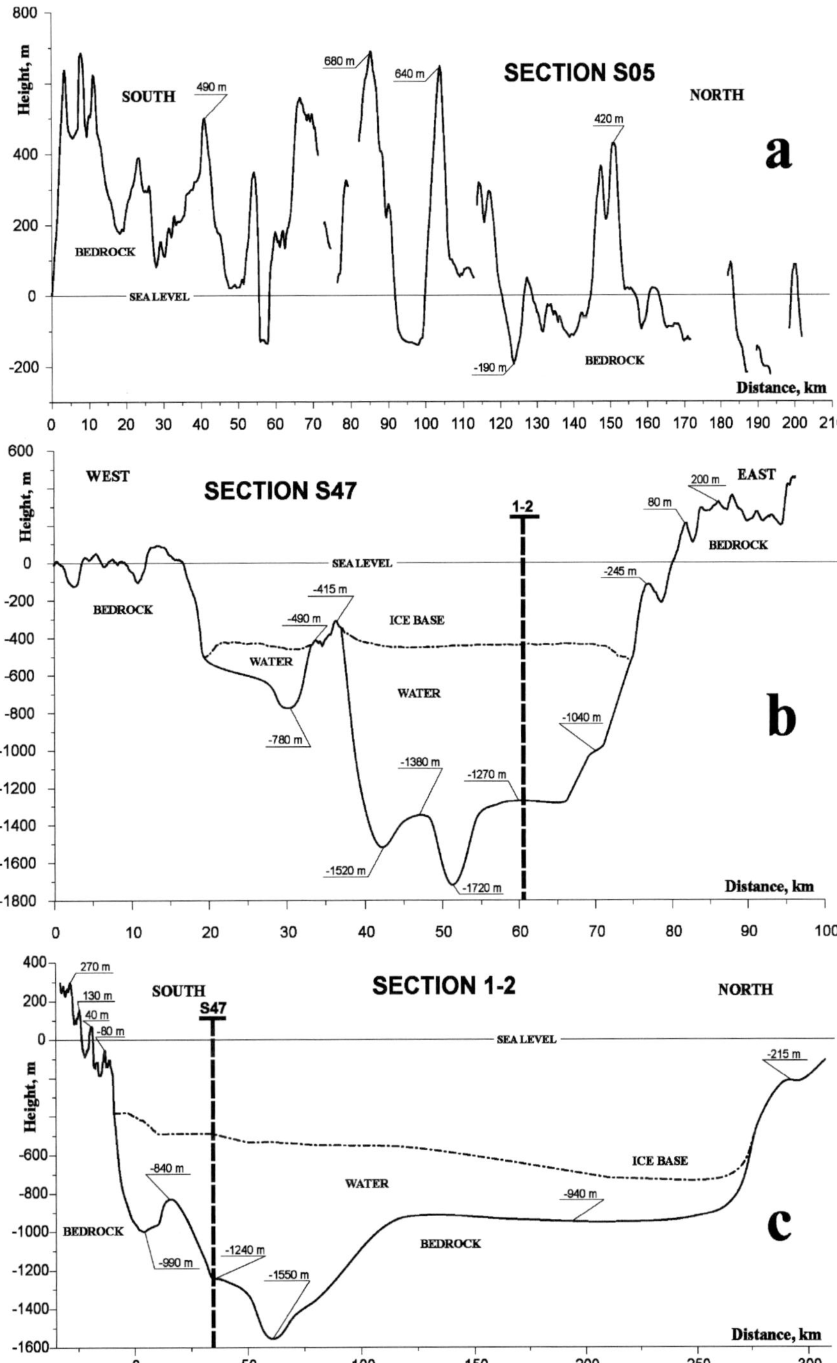

Fig. 3.6-2.
RES and seismic cross sections. The locations of the routes are shown in Fig. 3.6-1

It is necessary to note that the geomorphological chart derived is based on the RES and the seismic cross sections. Some examples are shown in Fig. 3.6-2. The use of the real cross sections is more correct than the use of the contour map since the process of contouring results in smoothing the data. Some smoothing is acceptable for producing and describing the geophysical data but inappropriate for geomorphological analysis.

Morphology

The reflection seismic and RES data show a dominating structure in the region that is the Vostok Basin (Fig. 3.6-3). Its size is approximately 270 × 80 km. The basin configuration is oval-shaped in S-N direction and complicated by a number of small-size structures located along the western side of the basin. The eastern side of the basin is mostly rectilinear south of 77°. The basin boundary is marked by edges of mountain ridges and other enveloped positive forms of the bed relief. The Vostok Basin is subdivided into five main substructures (Fig. 3.6-3): lake plane (LP), deep-water hollow (LD), sub-water ridges (LR), internal (BS) and external slopes (LS).

The lake plain is located in the northern part of the basin. It is possibly represented by sub-horizontal sub-water surface –940 m deep and approximately 150 km long. Our seismic data suggest that the depth variability is about ten meters. This result is based on only four seismic points Fig. 3.6-1, and 3.6-2). It is not enough to fully describe the mentioned large area. However, the randomly located measurements showed practically the same bedrock height. The deep-water hollow has a pear-shaped configuration with a depth from about –1 700 to –800 m. Its relative height is approximately 250 m with a normal slope of about 3° and the size is about 30 × 55 km (Fig. 3.6-2b and 3.6-2c). The relief of the deep-water hollow is complicated by sub-water ridges and valleys with N-W and S-E directions mostly. Their relative heights are from 150 up to 400 m. Sub-water ridges subdivide the southern part of the lake bottom into the deep-water and the shallow-water basins (Fig. 3.6-2b). Its absolute height varies from –780 up to –460 m. The ridge slopes are terraced. The

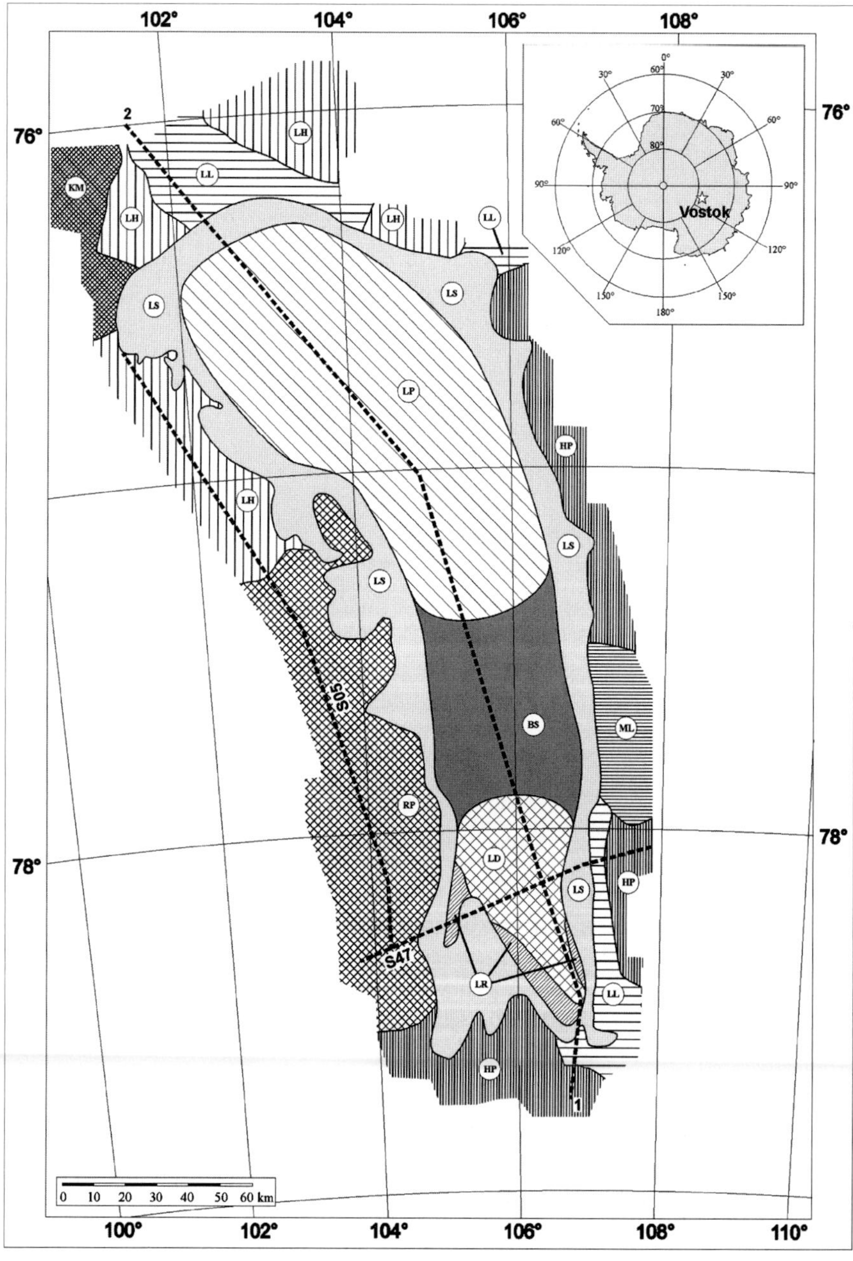

Fig. 3.6-3. Geomorphological pattern of Lake Vostok area. *Dashed lines* show sections of Fig. 3.6-2; *LP:* lake plane; *LD:* deep-water hollow; *LR:* sub-water ridges; *BS:* internal slope; *LS:* external slope; *LL:* lowlands; *LH:* low hilly planes; *HP:* high planes; *RP:* ridged plane, *KM:* Komsomolskiye Mountains, *ML:* middle mountain land

terraces are located at the depths of about –1 380, –1 270, –940, –860, –600, –520, –480 and –460 m. They are characterized by different sizes from 1 up to 7 km long (Fig. 3.6-2b and 3.6-2c). The internal slope is positioned north of the deep basin between the deep-water (in the Vostok Station area) and shallow-water regions. Its depth varies from about –1 400 to –930 m with an angle app. 0.5° (Fig. 3.6-2c). The external slope of the Vostok Basin encircles all the mentioned substructures forming their external boundary. Its width changes from kilometres to several tens of kilometres. Elevation of the external slope varies from –1 300 to about 400 m. The slope angle is approximately 17–22° (Fig. 3.6-2b and 3.6-2c). Its structure is complicated by valleys and canyons app. 1.5 km in width and about 400 m in depth.

Subglacial relief of the area around the Vostok Basin is very different. We identified four principal morphologic substructures there (Fig. 3.6-3): lowlands (LL), low hilly planes (HP), high planes (HP), ridged plane (RP), Komsomolskiye Mountains (KM) and middle mountain land (ML).

The lowlands are located in the north and the south from the Vostok Basin. They are valleys that represent the prolongation of the main Vostok Basin. The absolute height of the lowlands is from about –300 up to sea level in the south and from about –500 m up to –300 m in the north. Its surface is characterized by the slope with the angle of approximately 1° to the Vostok Basin side. The low hilly planes are best expressed in the northern part of the Vostok Basin. Their absolute height is approximately from –100 m up to 100 m and complicated by single hollows and ridges of the W-E directions in the northwestern part and S-N directions in the northern part of the area (Fig. 3.6-2a). Its relative height is about 200 m in general. The ridges width is about 10 km. The high planes are located in the south and east of the Vostok Basin. Their absolute height is about 300 m mainly with several complicated ridges about 100 m in height. The ridged plane is located to the west from the Vostok Basin. Its principal feature is a number of ridges in W-E direction. Having more then 400 m in relative height difference and about 10 km in width, they are divided by a plane which is about 150 m in height. It could be spurs of Sovetskiye Mountains, which are hypothetically located between Lake Vostok and Sovetskaya Station (PMGRE archives, not published). Hypothetical Komsomolskiye Mountains are located between Lake Vostok and Komsomolskaya Station (not published). Their observed small part is about 800 m in height. The last region is the middle mountain land located east from the Vostok Basin. It is probably a fragment of a high ridge. Observed bedrock height is about from 400 up to 1 100 m.

This is one of the first attempts of a morphological analysis of the Lake Vostok bedrock. The authors hope to continue this work using new materials to come from the neighboring areas, which will result in better understanding the Lake Vostok formation.

Acknowledgments

Igor V. Samsonov, Olga B. Soboleva, Daria V. Mandrikova and Dr. Alexey Ekaykin are thanked for their help. The authors express their sincere gratitude to referees Dr. Olaf Eisen, Prof. Heinz Miller and for their well-wishing comments and suggestions which allowed significant improvement of the manuscript.

References

Dowdeswell JA, Siegert MJ (2003) The physiography of modern Antarctic subglacial lakes. In: Fard AM (ed) Subglacial lakes: a planetary perspective. Global Planet Change 35(3/4), pp 221–236

Kapitsa AP, Ridley JK, Robin G de Q, Siegert MJ, Zotikov I (1996) A large deep freshwater lake beneath the ice of central East Antarctica. Nature 381(6584):684–686

Lastochkin AN (1987) Morphodynamical analysis. Nedra, Leningrad, (in Russian)

Lastochkin AN (1991) Relief of the earth surface. Nedra, Leningrad, (in Russian)

Leitchenkov GL, Verkulich SR, Masolov VN (1998) Tectonic setting of lake Vostok and possible information contained in its bottom sediments. Lake Vostok study: Scientific objectives and technological requirements. Internat Workshop March 24–26, 1998, AARI, St. Petersburg, pp 62–65

Lipenkov VYa, Barkov NI, Salamatin AN (2000) History of Antarctic climate and glaciation from the study of the Vostok Station ice core. Probl Arktiki i Antarktiki 72:197–236, (in Russian)

Mandrikova DV, Lipenkov VYa, Popov SV (2003) Ice sheet structure in the subglacial Lake Vostok area (East Antarctic) based on RES data. Mater Glaciol Data, (in Russian)

Masolov VN, Kudryavtzev GA, Sheremetyev AN, et al. (1999) Earth science studies in the Lake Vostok region: existing data and proposals for future research. SCAR Internat Workshop on Subglacial Lake Exploration Report, Cambridge, pp 1–18

Masolov VN, Lukin VV, Sheremetyev AN, Popov SV (2001) Geophysical investigations of the subglacial lake Vostok in Eastern Antarctica. Doclady Earth Sci 379A(6):734–738

Masolov VN, Lukin VV, Popov SV, Popkov AM, Sheremet'ev AN, Kudryavtsev GA (2002) Main results of the seismic and radio echo sounding investigations of the subglacial lake Vostok. Razvedka i ohrana nedr 9:40–44, (in Russian)

Petit JR, Jouzel J, Raynaud D, Barkov NI, et al. (1999) Climate and atmospheric history of the past 420 000 years from the Vostok ice core, Antarctica. Nature 399(6735):429–436

Popov SV (2003) Application the radio echo sounding for the glaciological investigations. Trudy XX–XXI All-Russian sympos Radio echo sounding of the nature environments 3:57–64, (in Russian)

Popov SV, Mironov AV, Sheremetyev AN (2001) Result of ground RES researches of subice lake Vostok in 1998–2000. Mater Glaciol Data 89:129–133, (in Russian)

Popov SV, Lastochkin AN, Popkov AM, Masolov VN, Lukin VV (2002) Results of geomorphologic interpretation of the bed relief in the subglacial Lake Vostok area. EOS 83(19) Suppl., Abstract B22A-02

Popov SV, Sheremetyev AN, Masolov VN, et al. (2003a) Velocity of radio-wave propagation in ice at Vostok station, Antarctica. J Glaciol 49(165):179–183

Popov SV, Sheremetyev AN, Masolov VN, Lukin VV (2003b) Main results of the ground-based RES and glaciological observations in the subglacial Lake Vostok area (East Antarctica) in 1998–2002. Mater Glaciol Data 94:187–193, (in Russian)

Ridley JK, Cudlip W, Laxon W (1993) Identification of subglacial lakes using ERS-1 radar altimeter. J Glaciol 73(133) 625–634

Siegert MJ, Hodgkins R, Dowdeswell JA (1998) Internal radio-echo layering at Vostok station, Antarctica, as an independent stratigraphic control on the ice-core record. Annals Glaciol 27:360–364

Siegert MJ, Ellis-Evans JC, Tranter M, et al. (2001) Physical, chemical and biological processes in Lake Vostok and other Antarctic subglacial lakes. Nature 414:603–609

Spiridonov AI (1975) Geomorphologic mapping. Nedra, Moscow, (in Russian)

Studinger M, Bell R, Karner GD, et al. (2003) Ice cover, landscape setting and geological framework of Lake Vostok, East Antarctica. Earth Planet Sci Letters 205:195–210

Tabacco IE, Bianchi C, Zirizzotti A, Zuccheretti, Forieri A, Della Vedova A (2002) Airborne radar survey above Vostok region, East Central Antarctica: ice thickness and Lake Vostok geometry. J Glaciol 48:62–69

Deep Reflection Imaging beneath the Mizuho Plateau, East Antarctica, by SEAL-2002 Seismic Experiment

Mikiya Yamashita[1] · Hiroki Miyamachi[2] · Masaki Kanao[3] · Takeshi Matsushima[4] · Shigeru Toda[5] · Masamitsu Takada[6] Atsushi Watanabe[7]

[1] Department of Polar Science, The Graduate University for Advanced Studies, 1-9-10 Kaga, Itabashiku, Tokyo 173-8515, Japan
[2] Faculty of Science, Kagoshima University, 1-21-35, Kourimoto, Kagoshima 890-0065, Japan
[3] National Institute of Polar Research, 1-9-10 Kaga, Itabashi-ku, Tokyo 173-8515, Japan
[4] Institute of Seismology and Volcanology, Faculty of Sciences, Kyushu University, Shinyama 2, Shimabara 855-0843, Japan
[5] Department of Earth Science, Faculty of Education, Aichi University of Education, Hirosawa 1, Igaya, Kariya, Aichi 448-8542, Japan
[6] Institute of Seismology and Volcanology, Graduate School of Science, Hokkaido University, N10S8 Kita-ku, Sapporo 060-0810, Japan
[7] Department of Earth and Planetary Sciences, Graduate School of Sciences, Kyushu University, 6-10-1 Hakozaki, Fukuoka 812-8581, Japan

Abstract. A seismic exploration was conducted on the Mizuho Plateau, East Antarctica, during the 2001/2002 austral summer season as the "Structure and Evolution of the East Antarctic Lithosphere (SEAL)" project by the 43rd Japanese Antarctic Research Expedition (JARE-43). The survey line of this exploration (SEAL-2002 profile) was almost perpendicular to the Mizuho inland traverse routes (JARE-41 refraction survey line; SEAL-2000) and was almost parallel to the coastal line along the Lützow-Holm Bay. Several seismic shot records were obtained with clear arrivals of phases until a distance of 150 km in length. We have analyzed two shot data of both ends of the SEAL-2002 profile by using the conventional reflection method. Interval velocities were estimated by applying the normal-move-out (NMO) correction, then the obtained single-fold section obtained explicitly presents the horizontal reflectors originated from the middle crust, the lower crust and the Moho discontinuity. First, the reflector from the top of the middle crust was located at the depth of 23–24 km, which was corresponding to 8–9 s of two way travel time (TWT) in the single-fold section. Next, the reflector from the top of the lower crust was located at a depth of 31–34 km, corresponding to 11–12 s of TWT. The Moho reflector was observed in 13–14 s of TWT and the depth was estimated to be approximately 41–42 km.

Introduction

In order to reveal the lithospheric structure and the evolution process in the areas from western Enderby Land to eastern Dronning Maud Land, East Antarctica, the geoscience program as the "Structure and Evolution of the East Antarctic Lithosphere (SEAL)" has been carried out since 1996–1997 austral summer season in the framework of the Japanese Antarctic Research Expedition (JARE). Several geological and geophysical studies including deep seismic surveys (Kanao 2001) have been conducted from the Napier Complex to the Lützow-Holm Complex (LHC) (e.g., Ishizuka et al. 1998; Tainosho et al. 1997; Tsutsui et al. 2001a,b; Dolinsky et al. 2002) to reveal the difference in several geological terrains from Archean to Paleozoic ages. It is very important for elucidation of the structure of the LHC by carrying out the seismic surveys to evaluate the effects of metamorphic activity at around 500 Ma during the Pan-African orogeny (e.g., Hiroi et al. 1991; Shiraishi et al. 1994).

The seismic refraction surveys in previous times (Pre-SEAL program) in the LHC were conducted by JARE-20, -21 and -22 along the Mizuho traverse routes to investigate a velocity structure of the deeper crust (Ikami et al. 1984; Ito and Ikami 1984; Ikami and Ito 1986). The JARE-41 seismic refraction/wide-angle reflection experiment was conducted along the same Mizuho traverse route as a part of the SEAL project (SEAL-2000) during the austral summer of 1999/2000 (Miyamachi et al. 2001). The purpose of SEAL-2000 experiment was to obtain a more precise structure than that of the Pre-SEAL experiments. Tsutsui et al. (2001a,b) obtained the velocity model of the uppermost crust and the reflection profile down to 20 s in the two-way travel time (TWT). This model exhibited a significant result that the Moho discontinuity was inclined from the inland (42 km depth) to the coastal area (29 km depth) (Fig. 3.7-5a).

In the austral summer season of 2001/2002 (JARE-43), a deep seismic exploration was carried out on the ice sheet of the Mizuho Plateau as a SEAL-2002 profile (Miyamachi et al. 2003a). The objective of this exploration was to estimate the crustal structure perpendicular to the SEAL-2000 profile and to determine the three dimensional crustal model beneath the Mizuho Plateau. Obtained seismic shot records have enough quality for the reflection analyses, so that we were able to investigate the actual architecture of deep crustal reflectors. Spatial distribution of the deep crustal reflectors can also be evaluated by a comparison with the results of SEAL-2000 profile. This paper describes the results of reflection analyses for seismic data obtained by the JARE-43 surveys (SEAL-2002 profile).

Seismic Exploration and Reflection Analysis

JARE-43 Seismic Exploration

The outline of data acquisition by the JARE-43 seismic experiments is as follows. Seismic shot records were obtained with clear arrivals of the later reflected phases by

From: Fütterer DK, Damaske D, Kleinschmidt G, Miller H, Tessensohn F (eds) (2006) Antarctica: Contributions to global earth sciences. Springer-Verlag, Berlin Heidelberg New York, pp 147–154

a total amount of 4 900 kg dynamite charges at the seven explosions along the SEAL-2002 profile, which was almost perpendicular to the SEAL-2000 profile at the H176 cross point (Fig. 3.7-1). The survey line consisted of the 161 temporal seismic stations at an interval of 1 km. A programmable digital data logger was equipped at each station. The details of the seismic operation were described in Miyamachi et al. (2003a). Data acquisition parameters of our analysis are indicated in Table 3.7-1.

Table 3.7-1. Data aquisition parameters of this study

Seismic source	No. of shot points:	2 (SP1 and SP7)
	Amount of dynamite:	700 kg (SP1 and SP7)
	Shot hole depth:	24.5 m (SP1), 28.7 m (SP7)
Receiver	Frequency:	2 Hz
	Station interval:	1 km
	No. of stations:	151–160
Seismic recorder	LS8000SH	
	Sample rate:	5 ms
	Record length:	7 s

Reflection Analyses

Travel time data for the first arrival phases from all explosions were summarized in Miyamachi et al. (2003a). P-wave velocities in the ice sheet and the uppermost crust were obtained by Miyamachi et al. (2003b). Figure 3.7-2 shows the shot records for the shots of SP1 and SP7 after the band-pass filter in frequency range of 4–18 Hz. We can recognize the head waves and the clear later reflected

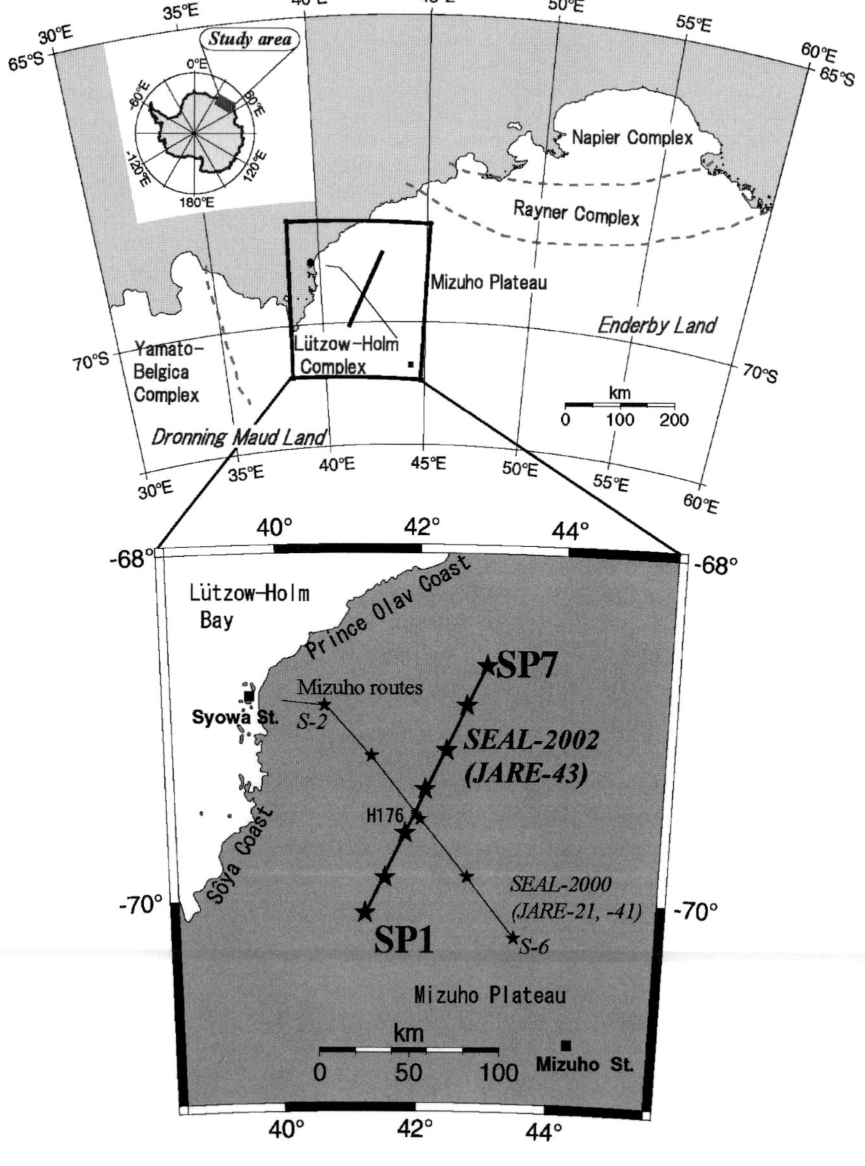

Fig. 3.7-1. Map showing geological setting and SEAL-2002 seismic survey line by JARE-43. *Solid stars* at SP1 and SP7 indicate the two shot points at both ends of the survey line uses in this study; *black line* includes 161 seismic stations

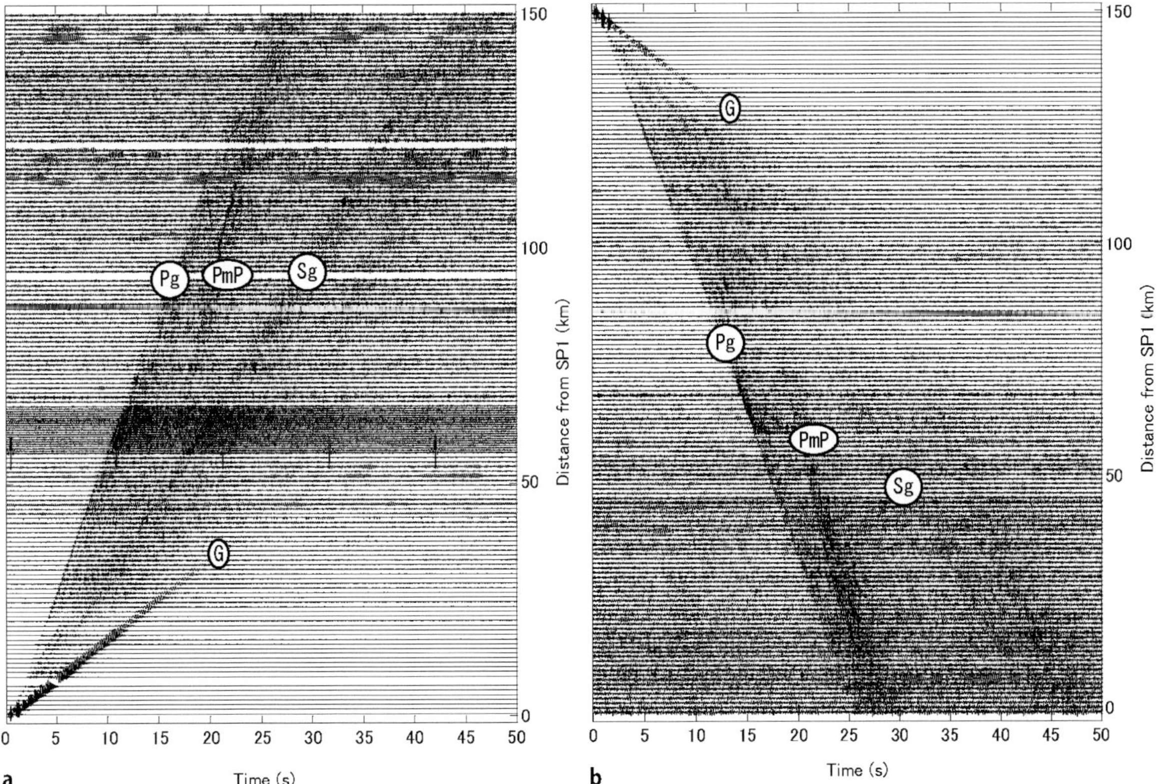

Fig. 3.7-2. Band-pass filtered (4–18 Hz) record section applied for **a** shot SP1 and **b** shot SP7 observed by JARE-43. Each trace is normalized by its maximum amplitude in the trace. *Pg* and *Sg:* P-wave and S-wave refraction wave from upper crust. Clear PmP phases appear in the offset over 100 km. *G:* Surface wave propagated into the ice sheet

phases in a range over 100 km in each shot record, similar to the results by SEAL-2000 profile (Miyamachi et al. 2001 and Tsutsui et al. 2001a,b). Seismic reflection times need a correction to remove a delay due to the thick and low velocity ice-sheet layer. Miyamachi et al. (2003b) obtained the velocity model of the ice-sheet and the uppermost crust by applying the refraction analyses to the SEAL-2002 seismic data. In the model, the thickness and the velocities of the ice-sheet were estimated to be 1.48 km s^{-1}, 3.8 km s^{-1} and 1.24 km, 3.9 km s^{-1} beneath SP1 and SP7, respectively.

A configuration of the stations in the SEAL-2002 profile was planned mainly for the refraction and wide-angle analyses. Thus, the Common-MidPoint (CMP) gather cannot be obtained from the dataset on this JARE-43 experiment. In general, reflection analyses are effective methods in order to investigate the details of the architecture of crustal structure. Tsutsui et al. (2001b) obtained a low-fold section for SEAL-2000 profile by using the wide-angle reflection analyses for all the six shots data by JARE-41 experiment. However, the result had not so much clear reflectors because of the mixture of various kinds of seismic waves.

Therefore, in this study, only two explosion data of SP1 and SP7 at both ends of the survey line were analyzed by simple reflection analyses, namely the Normal MoveOut (NMO) correction. The NMO velocities were assumed to be 6.1 km s^{-1} for the upper crust around 10 s of TWT from typical P-velocities of the Mizuho Plateau (Tsutsui et al. 2001b; Miyamachi et al. 2003) and 6.4 km s^{-1} for the deeper crust at 14 s of TWT from filling the reflection sequence by trial and error, respectively. A data processing flow (Fig. 3.7-3) based on the conventional reflection method (e.g., Yilmaz 1987) was adopted for this study.

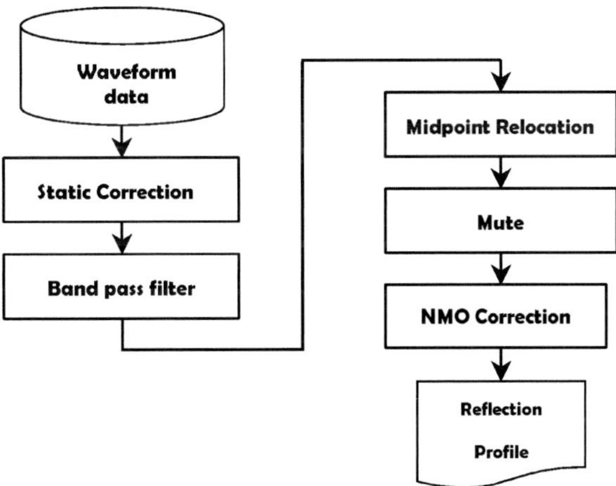

Fig. 3.7-3. Flow chart describing the data processing applied to this study

Correction on Reflection Depths

A two-dimensional velocity structure was obtained from refraction analysis by Ikami et al. (1984) and Ikami and Ito (1986) along the Mizuho Routes (same line of the SEAL-2000 profile). Tsutsui et al. (2001b) estimated the velocity structure by using the NMO correction for the SEAL-2000 large-offset/wide-angle reflection data. Their velocities distribution was lower than that of Ikami and Ito (1986), because the velocity model from the NMO correction reflects the average velocity in the crustal depth.

In this paper, we applied NMO correction for the originally large-offset/wide-angle reflection data by SEAL-2002 with station interval of 1 km, as done by Tsutsui et al. (2001b). Therefore, we cannot discuss the absolute distribution of velocities as determined by the refraction analyses without error estimation of the NMO correction. Kumar et al. (2003) compared the depths of reflections/velocity discontinuities by the large-offset/wide-angle reflection analyses and by the refraction analyses. According to their simulation results, the NMO correction for the large-offset/wide-angle reflection data should overestimate the depths with increasing offset distance.

Therefore, based on the above calculation by Kumar et al. (2003), the depths of reflectors estimated in our study and also for the SEAL-2000 result by Tsutsui et al. (2001b) should presumably be deeper than the actual depths. The overestimated depths could be about 2–5 km at the Moho discontinuity, about 2–4 km at the top of the lower crust and 1–2 km at the top of middle crust, respectively. Furthermore, the amplitude of reflected wave was enhanced by AGC (auto-gain-control) technique, then the detected reflectors were deeper than the actual structure. The true

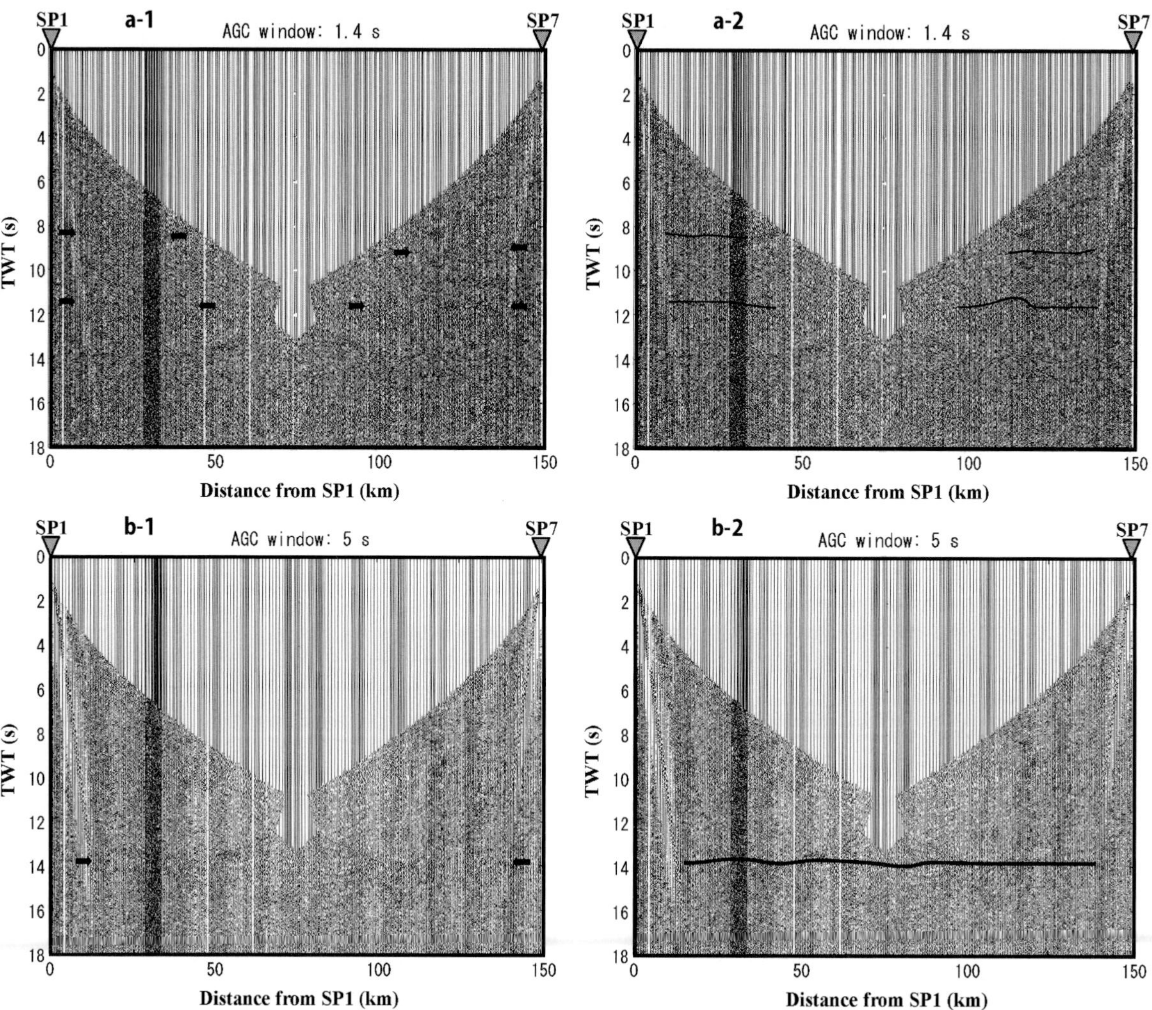

Fig. 3.7-4. Comparison of signal enhancement technique (*AGC*) in CMP based single-fold section compiled for shot *SP1* and *SP7*. Vertical axis indicates two way time (*TWT*) and 0 s corresponds to the surface of the seismic line. Several groups of reflections can be identified as indicated by *black arrows*. **a-1, a-2** *AGC* window 1.4 s; focused on the upper part (<12 s in *TWT*). **b-1, b-2** *AGC* window 5 s; focused on the lower part (>12 s in *TWT*)

dips of these reflectors, moreover, cannot be examined since the horizontal variations in seismic velocities beneath the survey line have not been accurately determined.

Results and Discussion

Figure 3.7-4 shows single-fold reflection profiles obtained by our study. Inspection of the obtained single-fold section reveals some prominent reflectors. Interpretation of reflectors from the single-fold section was illustrated in Fig. 3.7-5. Approximated depths of the crustal reflections were calculated by adopting the simulation results after Kumar et al. (2003), as described the details in the previous chapter. We will, hereafter, discuss the details of the characteristics of obtained reflectors in the following sections.

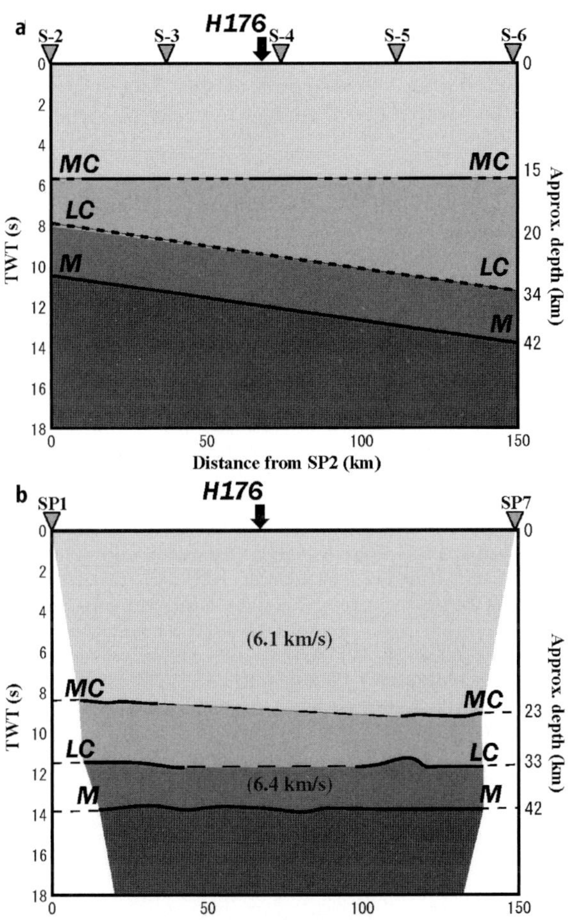

Fig. 3.7-5. Interpretation of reflectors in reflection profile based on single-fold section. *MC:* Reflector from boundary of upper and middle crust; *LC:* boundary of middle and lower crust. *Dashed lines* are the presumed boundary. *M:* Moho reflection. **a** SEAL-2000 profile (Tsutsui et al. 2001b). Right side of vertical axis denotes approximate depth (km) for corresponding with NMO velocities without the correlation of reflection depth. **b** This study. Right side of vertical axis denotes approximate depth (km) for corresponding with NMO velocities

Middle and Lower Crustal Reflections

In Fig. 3.7-4, we found some reflections in the TWT that are corresponding to the depths between the upper crust and the lower crust. Labels of MC (8–9 s of TWT) and LC (11–12 s of TWT) in Fig. 3.7-5b correspond to the reflectors from top of the middle crust and the lower crust, respectively. The depths of the reflectors MC and LC were estimated to approximately 22–23 km and 31–34 km by adopting the correction results after Kumar et al. (2003). A distribution of these reflections, however, has not wide horizontal spacing because the resolved area of our analysis was restricted within approximately 50 km. From single-fold profiles in Fig. 3.7-4, the thickness of lower crust was estimated to be about 8–9 km, which was consistent with the Pre-SEAL results by Ito and Kanao (1995) and the SEAL-2000 results by Tsutsui et al. (2001b) along the Mizuho routes. Thickness of the upper crust was more than 20 km from the results of our reflection study and also from the refraction/wide-angle reflection study for SEAL-2002 profile (Miyamachi et al. 2003). The upper crust has large thickness more than that of the middle and the lower crust.

Miyamachi et al. (2003b), however, have pointed out a seismic velocity discontinuity located at 19 km depth by refraction/wide-angle studies. Moreover, they suggested that that might be a boundary between the upper crust and the middle crust. In addition, they could not recognize the reflections from the boundary between the middle crust and the lower crust. On the contrary, there are no clear reflections observed corresponding to the reflectors MC and LC in the record sections by both the Pre-SEAL (Ikami and Ito 1986) and the SEAL-2000 (Tsutsui et al. 2001b) profiles along the Mizuho routes. These depth variations in the reflectors MC and LC indicate that the thickness of the middle and lower crust change with large variations beneath the Mizuho Plateau. In many cases for refraction analyses, some crustal reflections that were recognized by reflection procedure cannot be exactly identified. Actually, it is fairly difficult to recognize the reflection LC from our individual raw shot records. The signal enhancement technique in reflection analyses, however, can successively detect some weak reflections.

Ishikawa and Kanao (2002) presented the crustal lithologic model of the LHC by a comparison of the metamorphic rock velocities under high-pressure condition with the Pre-SEAL seismic refraction results. They interpreted that the middle and the lower crust are composed of felsic gneiss together with pyroxene granulite and pyroxene granulite with minor amounts of felsic rocks, respectively. An existence of the thin lower part of the crust obtained by this reflection study for SEAL-2002 profile indicates that the lower part of the crust in the LHC had been shaved off and become thin beneath the Mizuho Plateau when the last stage of Gondwana assembly during Pan-African orogenic events. Thick and relatively homogeneous upper crusts in the LHC are assumed to be composed of chiefly felsic gneiss lithology.

Moho and Upper Mantle Reflections

The strong reflection (labeled M) can be recognized from 13 s to 14 s of TWT. This reflection was observed in the offset distance over 100 km for both the shot records of SP1 and SP7. The depth of the reflector M was estimated to be approximately 41–42 km at the both ends of the survey line. If the error of velocities was taken into consideration, the Moho depth from this study was consistent with that from refraction/wide-angle reflection studies for SEAL-2002 dataset by Miyamachi et al. (2003b).

Tsutsui et al. (2001b) reported that the Moho reflector was upward dipping from the inland to the coast along the SEAL-2000 survey line (Fig. 3.7-5a) by reflection analyses without migration. According to their model, the depth of the Moho reflector beneath the H176 site (crossing point between SEAL-2000, and -2002 profiles; Fig. 3.7-1, 3.7-5) was expected to be about 36–38 km (12 s in TWT) by assuming the interval velocities of 6.0–6.2 km s^{-1} in the upper crust and 7.1 km s^{-1} in the lower crust. The difference between their Moho depths and ours might be caused from uncertainty of velocity models from within the crust beneath the survey line. Otherwise, there exist large scale lateral variations in the seismic velocities within the crust beneath the Mizuho Plateau, which presumably involved in the metamorphic grade corresponds to charge of the surface geology from the amphibolite phases (Prince Orav Coast; Fig. 3.7-1) to the granulite phases (Sôya Coast) along the Lützow-Holm Bay.

In addition, Yamashita et al. (2002) applied the mirror image method to the travel times of the Moho reflected waves for the same SEAL-2000 data. They estimated that the depths of the reflected Moho were 42 km at the center of survey line (H176) and the averaged dip angle was 7°, by assuming the uniform velocities of 6.2 km s^{-1} within the crust. Fortunately, their Moho depths were consistent with our reflection result; and their dip angle of the Moho was fairly well coincide with that obtained from reflection analyses by Tsutsui et al. (2001b).

Conclusion

A deep seismic survey was conducted on the Mizuho Plateau by the JARE-43 summer operation as a SEAL-2002 profile. Single-fold reflection profiles were investigated by the reflection analyses for the two shot data at both ends of the survey line. Our reflection profile indicates that reflectors were composed of horizontal and flat planes. The reflections from top of the middle crust was observed in 8–9 s of TWT, and the depths of these reflectors were located at 22–23 km after estimating the error. The reflections from top of the lower crust were identified in 11–12 s of TWT, and the depths of these reflectors were located at 31–34 km. The Moho reflectors were observed in 13–14 s of TWT in the single-fold sections. The depths of the Moho reflectors were estimated to be approximately 41–42 km.

Acknowledgments

The authors sincerely thank to the JARE-42, and -43 related members, particularly for Mrs. Y. Takahashi, D. Kamiya, M. Yanagisawa, N. Ishizaki, K. Nakano, T. Nakamura, N. Yoshida, T. Yasuhara and K. Horiguchi for their great efforts to obtain the seismic data in JARE-43. We also give a special thanks to crews of icebreaker "Shirase".

References

Dolinsky P, Funaki M, Yamazaki A, Ishikawa N, Matsuda T (2002) The results of magnetic surveys at Mt. Riiser-Larsen, Amundsen Bay, Enderby Land, East Antarctica, by 42nd Japanese Antarctic Research Expedition. Polar Geosci 15:80–88

Hiroi Y, Shiraishi K, Motoyoshi Y (1991) Late Proterozoic paired metamorphic complexes in East Antarctica, with special reference to the tectonic significance of ultramafic rocks. In: Thomson MRA, Crame JA, Thomson JW (eds) Geological evolution of Antarctica. Cambridge University Press, Cambridge, pp 83–87

Ikami A, Ito K (1986) Crustal structure in the Mizuho Plateau, East Antarctica, by a two-dimensional ray approximation. J Geod 6:271–283

Ikami A, Ito K, Shibuya K, Kaminuma K (1984) Deep crustal structure along the profile between Syowa and Mizuho Stations, East Antarctica. Mem Nation Inst Polar Res Ser C (Earth Sci) 15:19–28

Ishikawa M, Kanao M (2002) Structure and collision tectonics of Pan-African orogenic belt -Scientific significance of the geotransect for a super continent: Gondwanaland. Bull Earthqu Res Inst Univ Tokyo 77:287–302

Ishizuka H, Ishikawa M, Hokada T, Suzuki S (1998) Geology of the Mt. Riiser-Larsen area of the Napier Complex, Enderby Land, East Antarctica. Polar Geosci 11:157–173

Ito K, Ikami A (1984) Upper crustal structure of the Prince Olav Coast, East Antarctica. Mem Inst Polar Res Ser C 15:13–18

Ito K, Kanao M (1995) Detection of reflected waves from the lower Crust on Mizuho Plateau, East Antarctica. Nankyoku Shiryo (Antarct Rec) 39:233–242

Kanao M (2001) Crustal evolution and deep structure viewed from East Antarctic Shield: Structure and Evolution of the East Antarctic Lithosphere Geotransect Project – Outline and scientific significance. Bull Earthqu Res Inst Univ Tokyo 76:3–12

Kumar P, Sain K, Tewari HC (2003) A direct method of estimating depth to a reflector from seismic wide-angle reflection times. Geophys J Internat 152:740–748

Miyamachi H, Murakami H, Tsutsui T, Toda S, Minta T, Yanagisawa M (2001) A seismic refraction experiment in 2000 on the Mizuho Plateau, East Antarctica (JARE-41) - outline of observations. Nankyoku Shiryo (Antarct Rec) 45:101–147 (in Japanese, English abstract)

Miyamachi H, Toda S, Matsushima T, Takada M, Takahashi Y, Kamiya D, Watanabe A, Yamashita M, Yanagisawa M (2003a) A seismic refraction and wide-angle reflection exploration in 2002 on the Mizuho Plateau, East Antarctica - outline of observations (JARE-43). Nankyoku Shiryo (Antarct Rec), (in Japanese, English abstract)

Miyamachi H, Toda S, Matsushima T, Takada M, Watanabe A, Yamashita M, Kanao M (2003b) A refraction and wide-angle reflection seismic exploration in JARE-43 on the Mizuho Plateau, East Antarctica. Polar Geosci 16:1–21

Shiraishi K, Ellis DJ, Hiroi Y, Fanning CM, Motoyoshi Y, Nakai Y (1994) Cambrian orogenic belt in East Antarctica and Sri Lanka: implications for Gondwana assembly. J Geology 102:47–65

Tainosho Y, Kagami H, Hamamoto T, Takahashi Y (1997) Preliminary result for the Nd and Sr isotope characteristics of the Archaean gneisses from Mount Pardoe, Napier Complex, East Antarctica. Proc NIPR Symp Antarct Geosci 10:92–101

Tsutsui T, Murakami H, Miyamachi H, Toda S, Kanao M (2001a) P-wave velocity structure of the ice sheet and the shallow crust beneath the Mizuho traverse route, East Antarctica, from seismic refraction analysis. Polar Geosci 14:195–211

Tsutsui T, Yamashita M, Murakami H, Miyamachi H, Toda S, Kanao M (2001b) Reflection profiling and velocity structure beneath Mizuho traverse route, East Antarctica. Polar Geosci 14:212–225

Yamashita M, Kanao M, Tsutsui T (2002) Characteristics of the Moho as revealed from explosion seismic reflections beneath the Mizuho Plateau, East Antarctica. Polar Geosci 15:89–103

Yilmaz O (1987) Seismic data processing. In: Doherty SM, Neitzel EB (eds) Investigation in geophysics 2, Soc Explor Geophys

Seismic Anisotropy beneath Northern Victoria Land from SKS Splitting Analysis

Silvia Pondrelli[1] · Lucia Margheriti[2] · Stefania Danesi[2]

[1] Istituto Nazionale di Geofisica e Vulcanologia, Via D. Creti 12, 40128 Bologna, Italy, <pondrelli@bo.ingv.it>

[2] Istituto Nazionale di Geofisica e Vulcanologia, Via di Vigna Murata 605, 00143 Rome, Italy

Abstract. Teleseismic data recorded by temporary and permanent stations located in the Northern Victoria Land region are analysed in order to identify the presence and location of seismic anisotropy. We work on data recorded by 24 temporary seismographic stations deployed between 1993 and 2000 in different zones of the Northern Victoria Land, and by the permanent very broad-band station TNV located near the Italian Base "M. Zucchelli". The temporary networks monitored an area extending from Terra Nova Bay towards the South beyond the David Glacier and up to the Indian Ocean northward. To better constrain our study, we also provide an analysis of data recorded by TNV in the same period of time and we take into account also SKS shear wave splitting measurements performed by Barruol and Hoffman (1999) on data recorded by DRV. This study, to be considered as preliminary, reveals the presence of seismic anisotropy below the study region, with a mainly NW-SE fast velocity direction below the Terra Nova Bay area and rather large delay times, that mean a deep rooted anisotropic layer.

Introduction

The study region, the northern Victoria Land, is mainly characterized by the presence of the Transantarctic Mountains, that border the Ross Sea all along this region. From the tectonic point of view, this high-elevated belt, extending for more than 2 500 km, is considered an asymmetric rift shoulder segmented by several transverse fault systems (Fig. 3.8-1; Behrent et al. 1991; Salvini et al. 1997; Wilson 1999). The origin of the West Antarctic Rift, mainly constitued by the Ross Embayment, is still object of debate. The first hypothesis was to relate it to an active plume centered below Marie Byrd Land (Behrendt et al. 1991), but recently some investigations revealed a complex Cenozoic geodynamic, mainly due to the activation of intraplate right-lateral strike-slip structures, inducing a strong oblique component in the rifting process. The examination of these major tectonic structures may support a transtension-related source for the extension which designed the Ross Sea Embayment and the volcanism related to it (Salvini et al. 1997; Rocchi et al. 2002).

The transition from the Ross Sea extensional basin to the Transantarctic Mountains is abrupt, with a Moho detected at less than 20 km of depth in the Ross Sea and increasing up to more than 40 km of depth beneath the mountain chain (Di Bona et al. 1997 and references therein; Pondrelli et al. 1997; Bannister et al. 2003). This sharp variation marks the presence of a strongly heterogeneous crustal and lithospheric structure. Surface wave tomography as well clearly indicates that the northern Victoria Land region is on the boundary of the East Antarctica craton and the West Antarctic Rift System (Ritzwoller et al. 2003; Danesi and

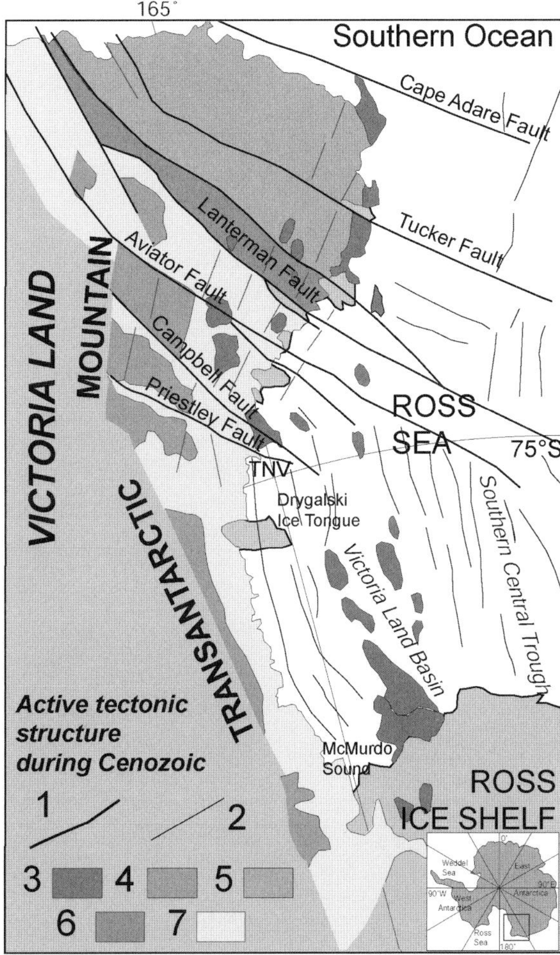

Fig. 3.8-1. Tectonic sketch map of northern Victoria Land (modified from Salvini et al. 1997). *1:* sub-vertical strike-slip fault; *2:* other tectonic lineaments; *3:* Late Cenozoic McMurdo volcanic rocks and intrusive rocks; *4:* Mesozoic Beacon continental deposits, Kirkpatrick Basalt and Ferrar Dolerites; *5:* Robertson Bay Terrane (Cambrian–Devonian); *6:* Bowers Terrane (Cambrian–Devonian); *7:* Wilson Terrane (Precambrian–Ordovician)

From: Fütterer DK, Damaske D, Kleinschmidt G, Miller H, Tessensohn F (eds) (2006) Antarctica: Contributions to global earth sciences. Springer-Verlag, Berlin Heidelberg New York, pp 155–162

Morelli 2001). A dramatic discontinuity in shear wave velocity pattern (see Fig. 4 in Morelli and Danesi, 2004) marks the limit between the Archean shield and the Ross Sea lithosphere, the former showing deep cold continental roots down to 250–300 km (Morelli and Danesi 2004), the latter anomalously warm and stretched. The depth to which seismic tomography can reliably image the Antarctic region is however hampered to 350–400 km, which is not enough to undoubtedly confirm or disprove the presence of a mantle plume head beneath West Antarctica.

Shear wave splitting is generally considered as caused by lattice preferred orientation of anisotropic minerals of the mantle, as olivine (Vinnik et al. 1989; Silver 1996; Savage 1999). Measurements from teleseismic SKS phases are a powerfull instrument to constrain the strain pattern in the mantle, extending the geological analysis at depth and helping in the interpretation of lithosphere kinematics. When shear-waves travel in an anisotropic material, they split into two polarized waves travelling at different velocities. Polarization direction ϕ and time delay δt between these two phases characterise the anisotropy. To detect the presence of seismic anisotropy beneath Northern Victoria Land and to study its relation with the geodynamic of this region, we analyze seismographic data recorded during temporary deployments and by TNV; moreover, we take into account the results of SKS measurements obtained by Barruol and Hoffman (1999) from DRV data records.

Sations and Dataset

Since 1993 several temporary geophysical campaigns have been performed in the Terra Nova Bay region (Cimini et al. 1995; Pondrelli et al. 1997; Della Vedova et al. 1997). On the total we have data to analyze from 24 temporary stations (Table 3.8-1). Most of them were located in the area

Table 3.8-1. Station coordinates and recording times

Station name	Latitude	Longitude	Recording time
ALF9	-75.898	162.58	Dec 1993–Jan 1994
BRA9	-75.83	160.5	Dec 1993–Jan 1994
BT01	-71.11	166.563	Dec 1999
BT02	-71.206	164.512	Dec 1999
BT03	-71.417	162.060	Dec 1999–Jan 2000
BT04	-70.736	159.991	Dec 1999–Jan 2000
BT05	-69.890	158.928	Dec 1999–Jan 2000
BT06	-69.51	157.34	Jan 2000
BT65	-69.354	156.129	Jan 2000
BT07	-69.25	155	Jan 2000
BT08	-68.98	154.19	Jan 2000
CHA9	-75.75	158.3	Dec 1993–Jan 1994
CPW9	-74.63	165.43	1993–1994; 1994–1995; 1997–1998
DEL9	-75.697	157.08	Dec 1993–Jan 1994
ESK9	-74.3	162.65	1995–1996; 1997–1998
INX9	-74.934	163.708	1997–1998
KNT9	-74.53	163.97	1995–1996
MDK9	-74.4	163.97	1993–1994; 1994–1995; 1997–1998
MEL9	-74.32	165.02	1993–1994; 1994–1995; 1997–1998
NAN9	-74.579	162.615	1995–1996; 1997–1998
OAS9	-74.69	164.1	1993–1994; 1994–1995; 1995–1996
OSC9	-74.56	164.85	1997–1998
SKR9	-74.58	161.94	1995–1996; 1997–1998
TNF9	-74.993	162.744	1995–1996; 1997–1998
TNV	-74.7	164.12	Permanent

Table 3.8-2. Station name, event date and location, delta and backazimuth, and fast direction measurements with errors

Station	Event				Ev.-St. distance	Backaz	Fast direct. φ	dφ	Delay time	dt
	Date (yy/mm/dd)	Latitude	Longitude	Depth						
ALF9	94/01/10	−13.33	−69.44	596	85.7	129.7	129.7		null	
BRA9	94/01/10	−13.33	−69.44	596	86.1	131.7	51	2	0.9	0.13
BT06	00/01/21	13.15	125.75	33	85.4	329.2	27	14	0.96	2.05
BT07	00/01/28	26.08	124.5	193	97.6	332.6	332.6		null	
BT08	00/01/28	26.08	124.5	193	97.2	333.3	−7	5	2.16	0.15
CPW9	93/11/29	10.29	126.47	38	88.1	321.8	68	2	2.04	0.26
	97/11/28	−13.74	−68.79	586	85.6	127.8	51	2	1.75	0.22
	98/01/01	23.91	141.91	95	99.5	338.3	23	8	0.55	0.07
ESK9	95/12/03	44.66	149.3	33	119.0	349.1	11	12	1.65	0.57
	97/11/21	22.21	92.7	54	106.0	295.1	59	5	1.7	0.15
	97/12/11	3.93	−75.79	177	102	119.7	−89	9	1.25	0.3
	97/12/18	14.71	144.65	33	89.6	342.6	−78	2	1.35	0.1
	98/01/01	23.91	141.91	95	98.9	340.9	34	4	1.5	0.18
	98/01/10	14.37	−91.47	33	108.0	101.4	−88	6	1.75	0.75
KNT9	95/12/03	44.66	149.3	33	119.3	348.0	33	10	0.7	0.12
MDK9	97/12/18	14.71	144.65	33	89.8	341.3	33	8	0.6	1.22
	98/01/14	−2.11	68.09	10	89.6	263.8	46	6	0.8	0.12
	98/02/11	10.33	124.99	56	88.0	321.7	16	4	2.35	0.23
MEL9	93/12/10	20.91	121.28	12	99.1	319.9	30	6	2.18	0.18
	97/11/28	−13.74	−68.79	586	85.9	128.2	128.2		null	
	98/01/01	23.91	141.91	95	99.2	340.9	58	2	1.9	2.98
	98/01/10	14.37	−91.47	33	107.4	99.1	−53	6	1.4	0.28
NAN9	97/11/15	43.81	145.02	161	118.7	345.6	25	4	1.05	0.12
OAS9	93/11/29	10.29	126.47	38	87.9	323.0	47	18	0.8	0.2
OSC9	97/11/28	−13.74	−68.79	586	85.8	128.3	128.3		null	
	97/12/11	3.93	−75.79	177	101.3	117.5	61	12	0.85	0.23
	97/12/18	14.71	144.65	33	90.0	340.5	54	3	1.95	0.25
	98/01/01	23.91	141.91	95	99.3	338.8	56	2	2.8	0.38
	98/01/10	14.37	−91.47	33	107.4	99.3	31	9	0.8	1.58
SKR9	97/12/11	3.93	−75.79	177	102.0	120.4	42	2	1.7	0.3
TNF9	98/02/03	15.88	−96.3	33	108.1	96.4	96.4		null	
TNV	91/04/22	9.68	−83.08	10	105.2	109.0	42	8	2.25	0.45
	92/01/20	27.96	139.33	512	103.5	337.1	337.1		null	
	93/01/15	43.4	143.26	100	118.4	342.4	−1	12	2.15	0.67
	93/01/19	38.63	133.46	455	114.7	333.4	15	16	1.1	2.23
	95/01/16	34.55	135	16	110.5	334.1	334.1		null	
	95/01/19-1	5.07	−72.92	18	103.2	120.2	47	7	3	0.65
	95/01/19-2	43.33	146.72	63	118.1	345.4	56	6	1.65	0.3
	96/11/21	6.66	126.46	53	85.3	321.8	−69	4	2.15	0.23
	97/11/15	43.81	145.02	161	118.7	343.9	20	16	0.85	0.23
	97/11/21	22.21	92.7	54.4	106.5	293.2	54	14	1.3	0.38
	97/11/28	−13.74	−68.79	586	85.9	128.5	56	6	1.65	0.63
	98/04/03	−8.15	−74.24	164.6	90.1	122.0	67	12	1.55	1.27
	98/05/03	22.31	125.31	33	99.8	323.4	323.4		null	
	98/06/07	15.96	−93.78	86.6	108.4	96.7	45	22	1.6	2.78
	98/06/18	−11.57	−13.89	10	94.2	181.4	65	12	1.55	1.27
	98/08/04	−0.59	−80.39	33	96.0	114.2	42	6	3	0.57
	98/08/23	11.66	−88.04	54.6	105.8	103.6	49	14	2.75	0.88
	98/09/22	11.82	143.15	9.2	87.1	338.9	338.9		null	
	98/10/03	28.5	127.61	226.6	105.5	326.6	326.6		null	

around the Italian Base "M. Zucchelli" and had multiple reoccupations (up to five campaigns, from 1993 to 2000), recording the larger amount of data. On the contrary, northernmost and southernmost sites, respectively related to the last campaign, in 2000, and to the ACRUP1 experiment performed in 1994 (Della Vedova et al. 1997), are those for which only one occupation was done and consequently scarce data were recorded. In the region, the permanent Italian very broad-band seismographic station TNV has been recording continuously since 1989 to present and data availability is rather large.

We selected from NEIS Catalog (http://www.neic.cr.usgs.gov/neis/epic/epic.html) all earthquakes with a magnitude greater than 5.5 occurred during the recording times (from 1993 to 2000), located at a distance between 85° and 120° from our seismographic stations, to collect SKS phases of sufficient energy. The collected dataset includes more than 100 teleseisms, and more than a half of them gave good results (Table 3.8-2). However, only for TNV permanent station the azimuthal coverage was good, while for the other sites the NW and SE quadrants only were well sampled (Table 3.8-2).

Analysis and Discussion

The analysis performed here is done on teleseismic SKS phases, which travel as an S phase in the crust and mantle, as a P phase in the liquid core and convert to an S phase at the core-mantle boundary at the receiver side, polarized in the vertical plane of propagation. If SKS phases travel across an isotropic medium, all the energy propagates on the radial component only. In practice, almost all our data give splitted SKS phases, pointing out that shear waves encountered anisotropic material (Fig. 3.8-2, upper panel). The fast velocity direction ϕ, measured clockwise from the north, corresponds to the direction along which the strain aligns highly anisotropic cristals in the mantle; the delay time δt, measured between the fast and the slow components, is a quantification of the thickness of the anisotropic layer transversed by SKS phases. To determine the fast velocity direction φ and the delay time δt, we use the method of Silver and Chan (1991), that assumes that shear waves traverse a single homogeneous anisotropic layer. The method is

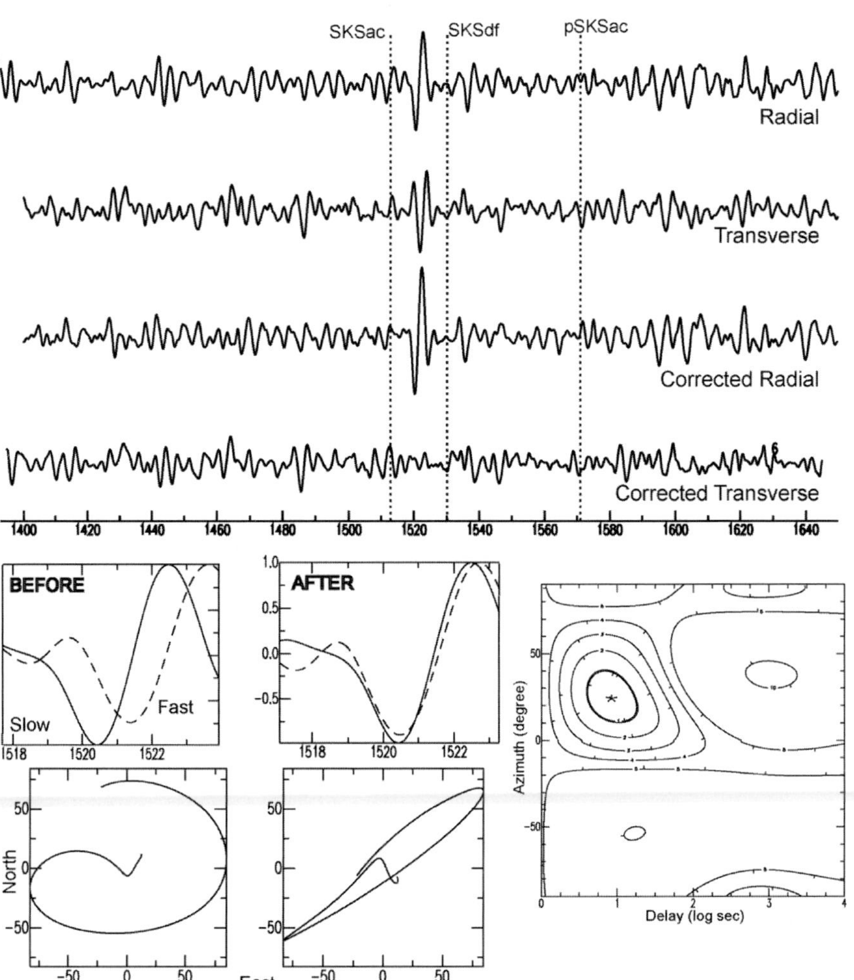

Fig. 3.8-2. Example of performed analysis; event occurred in Japan on November 15, 1997 and was recorded by NAN9 station. *Upper panel:* Radial and transverse components of seismographic recordings before and after the removal of the effect of the anisotropy. *Lower left panels:* Selected SKS phase (above) and its particle motion (below). Fast (*dotted line*) and slow (*continuous line*) phases are superimposed by the removal of the anisotropy effect. *Lower right panel:* Contour plot of the minimized energy on the corrected transverse component; the *star* is the minimum value, *first contour* is the 95% confidence region

based on a grid search over the possible splitting parameters space to find the pair of φ and δt that, when used to remove the anisotropy effect, most successfully removes its (Silver and Chan 1988). This is done by minimizing the energy on the reconstructed transverse component (upper part of Fig. 3.8-2, indicated as Corrected Radial and Transverse components). The error bounds for the estimated splitting parameters are obtained through an F-test analysis, as described by Silver and Chan (1991). When the absence of splitting is detected on the analysed waveform we have a "null" measurement, that doesn't necessarily mean that anisotropy is absent. In fact, a "null" from a single measurement also occur when the initial polarization of the shear wave is parallel to the fast or slow polarization direction of the anisotropic media. For this reason null measurements are characterized by the event back-azimuth (Table 3.8-2).

Results of our analysis reveal the presence of seismic anisotropy below the study region (Fig.3.8-3 and 3.8-4), with different directions moving from the northern to the southern part. Below DRV (Dumont D'Urville, French base) the anisotropy shows an E-W average direction (Barruol and Hoffman 1999). Going eastward, the few data available at stations BT01 to BT08 (deployed only during the 1999/2000 campaign) give a couple of N-S directions. The scarcity of data do not allow any large scale interpretation. Going southward, around the Terra Nova Bay area, our results are better constrained due to the larger amount of data recorded by multiple deployed temporary stations and by the TNV permanent station. In Fig. 3.8-4 data are mapped at the piercing point at 150 km of depth, to identify the presence of different fast velocity direction respect to different back-azimuths. It is evident that the NE-SW fast direction dominates, with also null measurements in agreement. Only few data show a different pattern, as the NW-SE trend obtained for a pair of events recorded at MEL9 or the E-W trend shown by another pair of event recorded at ESK9 (Table 3.8-2). Delay times are 1.6 s on average, and some examples show larger values with respect to world wide evaluation, indicating therefore that the anisotropy layer which splits SKS is deeply rooted. The NE-SW dominant fast velocity direction is in agreement with the Transantarctic Mountains trend, an usual observation along mountain chains, where fast direction is in general parallel to the belt axis (Savage 1999). We therefore consider that this anisotropy is related to the strain induced in the mantle by the formation of the chain; considering the deep rooted structure of the mountains belt, this hypothesis is also supported by the rather large values for the delay times. Concerning data reporting EW and NW-SE fast velocity directions, we can suggest two possible explanations: the pattern could be due to the extensional trend regime that produced the opening of the Ross Sea or, alternatively, the paths could have sampled

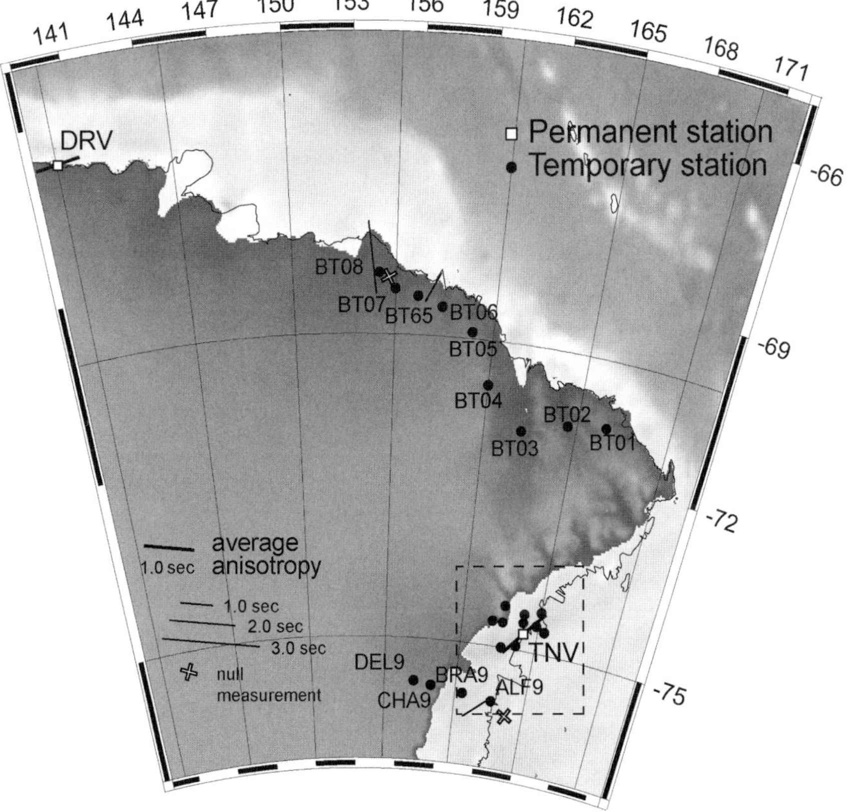

Fig. 3.8-3.
Map of all sites for which data have been analyzed. On *DRV* the average value obtained by Barruol and Hoffman (1999) is mapped. *Thick lines* are average fast direction (used here for *DRV* and *TNV*), *thin lines* are single measurements, mapped at the SKS ray piercing point at 150 km of à depth. Length of lines is proportional to the delay time. *Null measurements* are plotted with two small lines with directions parallel to the two allowed fast direction. *Inset* shows area enlarged in Fig. 3.8-4

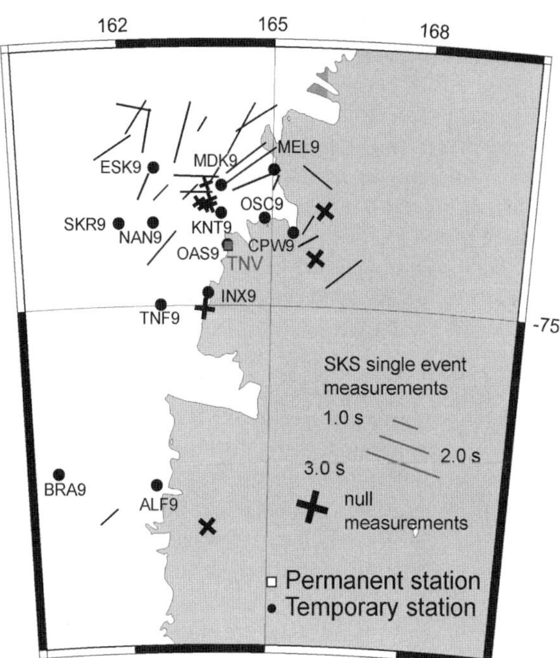

Fig. 3.8-4. Map of fast velocity direction mapped at 150 km depth piercing point for single measurements. For symbols see Fig. 3.8-3

one of the several large strike-slip NW-SE structures mapped in the northern Victoria Land (Fig. 3.8-1; Salvini et al. 1997). However, most of the out-of-trend data are related to events with a SE backazimuth, that therefore along the upward path, sample the warm and stretched lithospehere-astehosphere of the Ross Sea, while on the contrary, NE-SW fast velocity directions seems to be mainly related to data sampling the East Antarctica craton. However, the density and azimuthal distribution of obtained fast direction only allow preliminary hypotheses, that deep structure heterogeneity shown by tomographic images beneath the region (Morelli and Danesi 2004) would authorize. Certainly in the future, with a larger amount of recordings from permanent stations and from recently going on temporary deployments (e.g., Wiens et al. 2003; Bannister et al. 2003), we would better depict the characteristics of seismic anisotropy beneath Northern Victoria Land.

Conclusions

We analyzed more than 100 teleseisms recorded by 24 temporary and one, TNV, permanent stations, deployed in northern Victoria Land region. Our study produced preliminary results, that reveal the presence of seismic anisotropy beneath all of the study region, with different fast velocity direction moving from north to south. However, only around Terra Nova Bay area the data availability was large enough to better depict the anisotropy distribution. Here the dominant fast velocity direction is NE-SW, in agreement with the Transantarctic Mountains trend. Only few examples of E-W and NW-SE directions are recorded, probably due to the extensional trend that allowed the opening of Ross Sea or the presence of large strike-slip structures characterizing the Northern Victoria Land. Delay times show rather large values, on average 1.6 s, indicating that the detected anisotropic layer may be thick and deep rooted. The preliminary character of these results will be solved in the future with a hopefully larger availability of seismographic data.

Acknowledgment

We would like to thank all people involved in installation and maintenance of temporary stations and TNV permanent station. Thanks to D. Wiens and an anonymous reviewer for their useful comments.

Reference

Bannister SyuJ, Leitner B, Kennet BLN (2003) Variations in crustal structure across the transition from West to East Antarctica, Southern Victoria Land. Geophys J Internat 155:870–884

Barruol G, Hoffman R (1999) Upper mantle anisotropy beneath the Geoscope stations. J Geophys Res 104(5):10757–10773

Behrendt JC, LeMasurier WE, Cooper AK, Tessensohn F, Tréhu A, Damaske D (1991) Geophysical studies of the West Antarctic Rift System. Tectonics 10(6):1257–1273

Cimini GB, Amato A, Cerrone M, Chiappini M, Di Bona M,Pondrelli S (1995) Passive seismological studies in the Terra Nova Bay Area (Antarctica): First Results From the 1993–94 Expedition. Terra Antartica 2:81–98

Di Bona M, Amato A, Azzara R, Cimini GB, Colombo D, Pondrelli S (1997) Constraints on the lithospheric structure beneath the Terra Nova Bay area from teleseismic P to S conversion. In: Ricci CA (ed) The Antarctic region: geological evolution and processes. Terra Antartica Publication, Siena, pp 1087–1093

Danesi S, Morelli A (2001) Structure of the upper mantle under the Antarctic Plate from surface wave tomography. Geophys Res Letters 28:4395–4398

Della Vedova, et al. (1997) Crustal structure of the Transantarctic Mountains, western Ross Sea. In: Ricci CA (ed) The Antarctic region: geological evolution and processes. Terra Antartica Publication, Siena

Lawver L, Gahagan L (2003) Evolution of Cenozoic seaways in the Circum-Antarctic Region. Paleogrogr Paleoclimatol Paleoecol (in press)

Morelli A, Danesi S (2004) Seismological imaging of the Antarctic continental lithosphere: a review. Global Planet Change 42:155–165

Pondrelli S, Amato A, Chiappini M, Cimini GB, Colombo D, Di Bona M (1997) ACRUP1 Geotraverse: contribution of teleseismic sata recorded on land. In: Ricci CA (eds) The Antarctic region: geological evolution and processes. Terra Antartica Publication, Siena, pp 632–635

Ritzwoller MH, Shapiro NM, Leahy GM (2003) A resolved mantle anomaly as the cause of the Australian-Antarctic Discordance. J Geophys Re 108(7):2353

Rocchi S, Armienti P, D'Orazio M, Tonarini S, Wijbrans J, Di Vincenzo G (2002) Cenozoic magmatism in the western Ross embayment: role of mantle plume vs. plate dynamics in the development of the West Antarctic Rift System. J Geophys Res 107(9):2195

Silver PG (1996) Seismic anisotropy beneath the continents: probing the depth of geology. Ann Rev Earth Planet Sci 24:385–432

Silver PG, Chan WW (1988) Implication for continental structure and evolution from seismic anisotropy Nature 335:34–39

Silver PG, Chan WW (1991) Shear wave splitting and sub-continental mantle deformation. J Geophys Res 96:16429–16454

Savage MK (1999) Seismic anisotropy and mantle deformation: what have we learned from shear wave splitting? Rev Geophys 37:65–106

Salvini F, Storti F (1999) Cenozoic tectonic lineaments of the Terra Nova Bay region, Ross embayment, Antarctica. Global Planet Change 23:129–144

Salvini F, Brancolini G, Busetti M, Storti F, Mazzarini F, Coren F (1997) Cenozoic geodynamics of the Ross Sea region, Antarctica: crustal extension, intraplate strike-slip faulting, and tectonic inheritance. J Geophys Res 102(11):24669–24696

Vinnik LP, Farra V, Romanowicz B (1989) Azimuthal anisotropy in the Earth from observations of SKS at Geoscope and NARS broadband stations. Bull Seism Soc Amer 79:1542–1558

Wiens DA, Anandakrishnan S, Nyblade A, Fisher JL, Pozgay S, Shore PJ, Voigt D (2003) Preliminary results from the Trans-Antarctic Mountains Seismic Experiment (TAMSEIS). Terra Nostra 2003(4), Abstracts

Wilson TJ (1999) Cenozoic structural segmentation of the Transantarctic Mounatins rift flank in southern Victoria Land. Global Planet Change 23:105–127

Theme 4
Gondwana Margins in Antarctica

Chapter 4.1
Scouting Craton's Edge in Paleo-Pacific Gondwana

Chapter 4.2
The Matusevich Fracture Zone in Oates Land, East Antarctica

Chapter 4.3
Tectonic Model for Development of the Byrd Glacier Discontinuity and Surrounding Regions of the Transantarctic Mountains during the Neoproterozoic – Early Paleozoic

Chapter 4.4
Depositional Environments of the Byrd Group, Byrd Glacier Area: A Cambrian Record of Sedimentation, Tectonism, and Magmatism

Chapter 4.5
Late-Ross Structures in the Wilson Terrane in the Rennick Glacier Area (Northern Victoria Land, Antarctica)

Chapter 4.6
Style of Uplift of Paleozoic Terranes in Northern Victoria Land, Antarctica: Evidence from K-Ar Age Patterns

Between Ross and Scotia Sea, Antarctica comprises one of the best records of the active margin of Gondwana. A major product of the subduction process of the Paleo-Pacific under the East Antarctic craton is the Early Palaeozoic Ross Orogen, a 3 000 km long Andean-type mountain belt with well defined plutonic arc and a subduction-related suture in its northernmost part, North Victoria Land. The six papers in this chapter deal with various aspects of this orogen. Finn et al. (Chap. 4.1) investigate the cratonic interior of the active margin. The authors use data not only from Antarctica, but also from the adjacent Australian Gondwana fragment for their reconstructions. They use two geological fix-points in the Miller Range (Central Transantarctic Mountains) and on the Wilkes Land coast, the better known geology of the Australian cratonic areas as well as satellite and airborne magnetic data to characterize and delineate major cratonic subunits (e.g., Mawson craton). The outer boundary of the thick cratonic crust, just in the location of the linear belt of the Ross Orogen, later becomes the supposed western limit of Palaeozoic and Jurassic volcanism.

The remaining papers deal with Ross orogenic features in the Transantarctic Mountains south of Victoria Land and in northern Victoria Land. Stump et al. (Chap. 4.3) have studied a major boundary across the orogen in the Byrd Glacier area. This boundary between different rock units is interpreted as the result of a left-lateral transpressive terrane accretion of the "Beardmore microcontinent". The interpretation decouples a tectonic Ross Orogeny at 550 Ma from the arc plutonism at 500 Ma.

Gootee and Stump (Chap. 4.4) present a detailed account of the Cambrian, mainly sedimentary units south of the Byrd Glacier discontinuity. The paper thus provides background to the above interpretation of the Stump et al. paper.

In North Victoria Land, Läufer et al. (Chap. 4.5) add new data to the earlier published model of a bilateral thrust system with ocean-ward directed tectonic transport in the east and craton-ward transport in the west. The entire, 500 km wide part of the Wilson terrane west of the Matusevich Glacier is interpreted as a system of thrust sheets detached from the main Wilson terrane after the arc plutonism and therefore of late Ross age.

Adams (Chap. 4.6) takes up an earlier discussion of the K-Ar pattern in North Victoria Land by adding new data from the Terra Nova Bay area and the Oates Coast. The interpretation uses the bilateral thrust set-up discussed above by Läufer and Kleinschmidt to explain the age pattern as a result of a "pop up" structure along the thrust system.

The hypothetic Matusevich Glacier Fault on the Oates Coast has also been taken as part of the above thrust system. Based on a detailed analysis of small scale structures in the area, Kleinschmidt and Läufer (Chap. 4.2) have now shown that there is a younger brittle dextral strike-slip element present. A parallel dextral strike-slip system is present in the Rennick Glacier area. As both faults have the same strike as offshore fracture zones, a possible connection between them is discussed. This leads to the principal question: Can offshore transform faults continue onshore. An additional problem is the fact that the onshore structures are dextral, the offshore transforms, however, sinistral.

Unfortunately, in this chapter there is no contribution on the spectacular high-pressure suture in the Lanterman Range of North Victoria Land.

Scouting Craton's Edge in Paleo-Pacific Gondwana

Carol A. Finn[1] · John W. Goodge[2] · Detlef Damaske[3] · C. Mark Fanning[4]

[1] U.S. Geological Survey, MS 945, Denver Federal Center, Denver, CO 80226, USA
[2] Department of Geological Sciences, University of Minnesota, Duluth, MN 55812, USA
[3] Bundesanstalt für Geowissenschaften und Rohstoffe, Stilleweg 2, Hannover 30655, Germany
[4] Research School of Earth Sciences, The Australian National University, Mills Road, Canberra, ACT 0200, Australia

Abstract. The geology of the ice-covered interior of the East Antarctic shield is completely unknown; inferences about its composition and history are based on extrapolating scant outcrops from the coast inland. Although the shield is clearly composite in nature, a large part of its interior has been represented by a single Precambrian block, termed the Mawson block, that includes the Archean-Mesoproterozoic Gawler and Curnamona cratons of Australia. In Australia, the Mawson block is bounded on the east by Neoproterozoic sedimentary rocks and the superimposed early Paleozoic Delamerian Orogen, marked by curvilinear belts of arc plutons, and on the west by the unexposed Coompana block and Mesoproterozoic Albany-Fraser mobile belt. In Antarctica, these crustal elements are inferred to extend across Wilkes Land and south to the Miller Range region. Aero- and satellite magnetic data provide a means to see through the ice, helping to elucidate the broad composition of the shield. Rocks of the Mawson block in Australia produce distinctive magnetic anomalies; Paleoproterozoic granites and Meso- to Neoproterozoic mafic igneous rocks are associated with high-amplitude, broad-wavelength positive aero- and satellite-magnetic anomalies. The same types of magnetic anomalies can be traced to ice-covered Wilkes Land, Antarctica, and are interpreted to signify similar rocks. However, the diagnostic satellite magnetic high ends ~800 km south of the Antarctic coast, suggesting that the Mawson block is smaller than first proposed and that the remaining East Antarctic shield is composed of several Precambrian crustal blocks of largely undetermined composition and age. Nonetheless, the coincident eastern borders of these magnetic highs and high seismic-velocity anomalies characteristic of the Precambrian shield, together define the edge of thick cratonic lithosphere. East of this boundary, magnetic lows are explained by magnetite-poor upper Neoproterozoic and lower Paleozoic sedimentary rocks, and their metamorphic equivalents, which crop out discontinuously along the Ross margin of Antarctica and in eastern Australia. These rocks are inferred to overlie a Neoproterozoic rift margin, which transects older basement provinces. The coincidence of this cratonic rift boundary with the western limit of Paleozoic and Jurassic magmatism suggests that, although tectonically modified by younger events, the composite Antarctic-Australian shield comprised thick lithosphere that was not penetrated by Paleozoic and younger convergent-margin magmas.

Introduction

The East Antarctic shield is one of Earth's oldest and largest cratonic assemblies, with a long-lived Archean to early Paleozoic history. Based on age and poorly constrained stratigraphic relations, a large part of the shield is proposed to represent a single Precambrian block, the Mawson block comprised of the Archean-Mesoproterozoic Gawler (GC, Fig. 4.1-1) craton in Australia and George V Land (GVL, Fig. 4.1-1), Terra Adélie and the Miller Range (Fig. 4.1-1) (Fanning et al. 1995; Fitzsimons 2000; Fitzsimons 2003; Goodge and Fanning 1999). The earliest events in the Mawson block are recorded in ~3 100 Ma gneisses in the Miller Range with activity continuing intermittently until the amalgamation of the block with the remainder of Gondwana ~1 200–1 000 Ma (Fanning et al. 1995). Because of nearly complete coverage by the East Antarctic ice sheet, however, the crustal architecture of the Wilkes Land sector of East Antarctica is largely unknown. In particular, geological connections between exposures along the coast (Fanning et al. 1995; Fitzsimons 2000; Peucat et al. 2002; Tingey 1991) with those ~1 800 km south in the Miller Range area (MR, Fig. 4.1-1) (Goodge et al. 2001) are largely untested. Recent geological studies in exposed basement of the Transantarctic Mountains (TAM) and Wilkes Land margin suggest tentative correlation of Paleoproterozoic events that may help define the character and extent of the Mawson block. For example, Paleoproterozoic orogenic magmatism, high-grade metamorphism and eclogite formation in the Nimrod Group of the Miller Range at ~1.7 Ga (Goodge et al. 2001) may correspond to igneous and metamorphic events of this age in the Commonwealth Bay region in Antarctica (CB, Fig. 4.1-1) (Oliver and Fanning 1997; Peucat et al. 2002) and in the Gawler craton of Australia (GC, Fig. 4.1-1) (Daly et al. 1998).

Given the extensive ice cover, airborne and satellite geophysical data are a powerful means to characterize broad areas of sub-ice basement and expand our knowledge of the East Antarctic shield interior. Aeromagnetic coverage of the Mawson sector of Antarctica is largely limited to the TAM region, consisting of isolated profiles and transects with widely-spaced flight lines crossing the polar plateau (Fig. 4.1-1). In contrast, Australia is well-covered by low-altitude, high-resolution (flight lines spaced less than ~400 m apart) aeromagnetic data that are routinely used to map buried Precambrian basement (Fig. 4.1-1) (Daly et al. 1998; Myers et al. 1996; Rajagopalan et al. 1993). Satellite magnetic data (CHAllenging Minisatellite Payload, CHAMP) (Fig. 4.1-2), augmenting sparse aeromagnetic data coverage in Antarctica, outline

From: Fütterer DK, Damaske D, Kleinschmidt G, Miller H, Tessensohn F (eds) (2006) Antarctica: Contributions to global earth sciences. Springer-Verlag, Berlin Heidelberg New York, pp 165–174

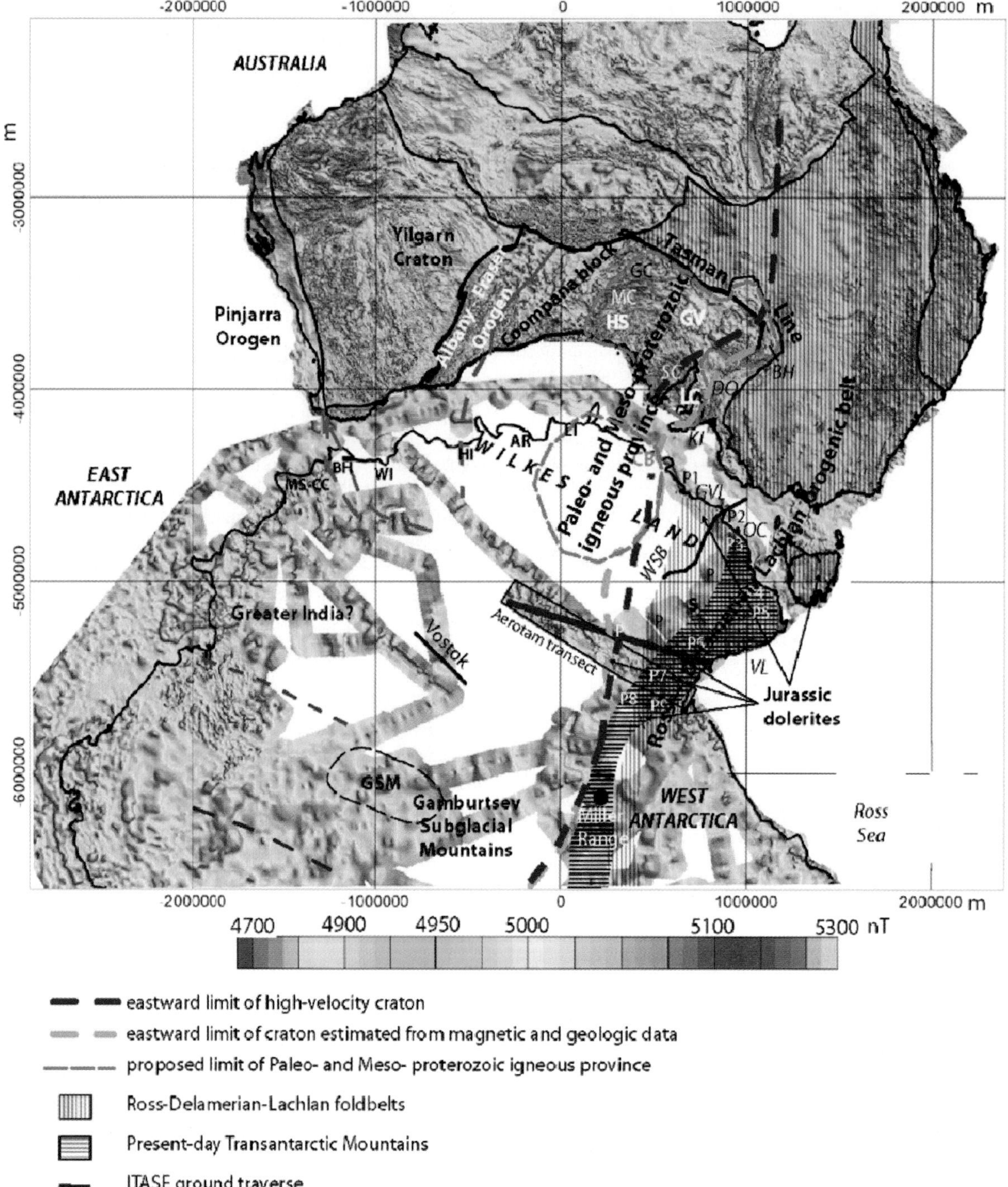

Fig. 4.1-1. Australian (Milligan and Tarlowski 1999; magnetic map of Australia is copyrighted; Australian Geological Survey Organisation 1999) and rotated Antarctic aeromagnetic data (Chiappini et al. 2002; Golynsky et al. 2001; Studinger et al. 2004) with Australia fixed in a Gondwana reconstruction (Lawver et al. 2003). *Grid lines* represent map position in meters in the rotated coordinate system. *Solid black lines* indicate boundaries of major crustal elements in Australia (Shaw et al. 1996), *dashed* where inferred from outcrop (Fitzsimons 2000) and airborne and satellite magnetic data in Antarctica, as follows: Mawson block: *pink dashed line;* and the edges of the Albany-Fraser belt: *brown*. The *thick dashed lines* represent the eastern limit of cratonic blocks in Antarctica inferred from the magnetic (*orange*) and seismic tomography data (*blue*) (Shapiro and Ritzwoller 2002). Australia: *BH:* Broken Hill; *CC:* Curnamona Craton; *DO:* Delamerian Orogen; *GC:* Gawler Craton; *GVL:* George V Land; *HI:* Henry and Chick Islands; *HS:* Hiltaba Suite; *KI:* Kangaroo Island; *LC:* Lincoln Complex; *MC:* Mulgathing Complex; *SB:* Stavely Belt; *SC:* Sleaford Complex. Antarctica: *AR:* Al'bov Rocks; *BH:* Bunger Hills; *CB:* Commonwealth Bay; *LI:* Lewis Island; *MR:* Miller Range; MS-*CC:* Mirny Station to Cape Charcot; *OC:* Oates Coast; *VL:* Victoria Land; *WI:* Windmill Islands. *WSB:* Wilkes Subglacial Basin. Other *letters* refer to anomalies discussed in the text

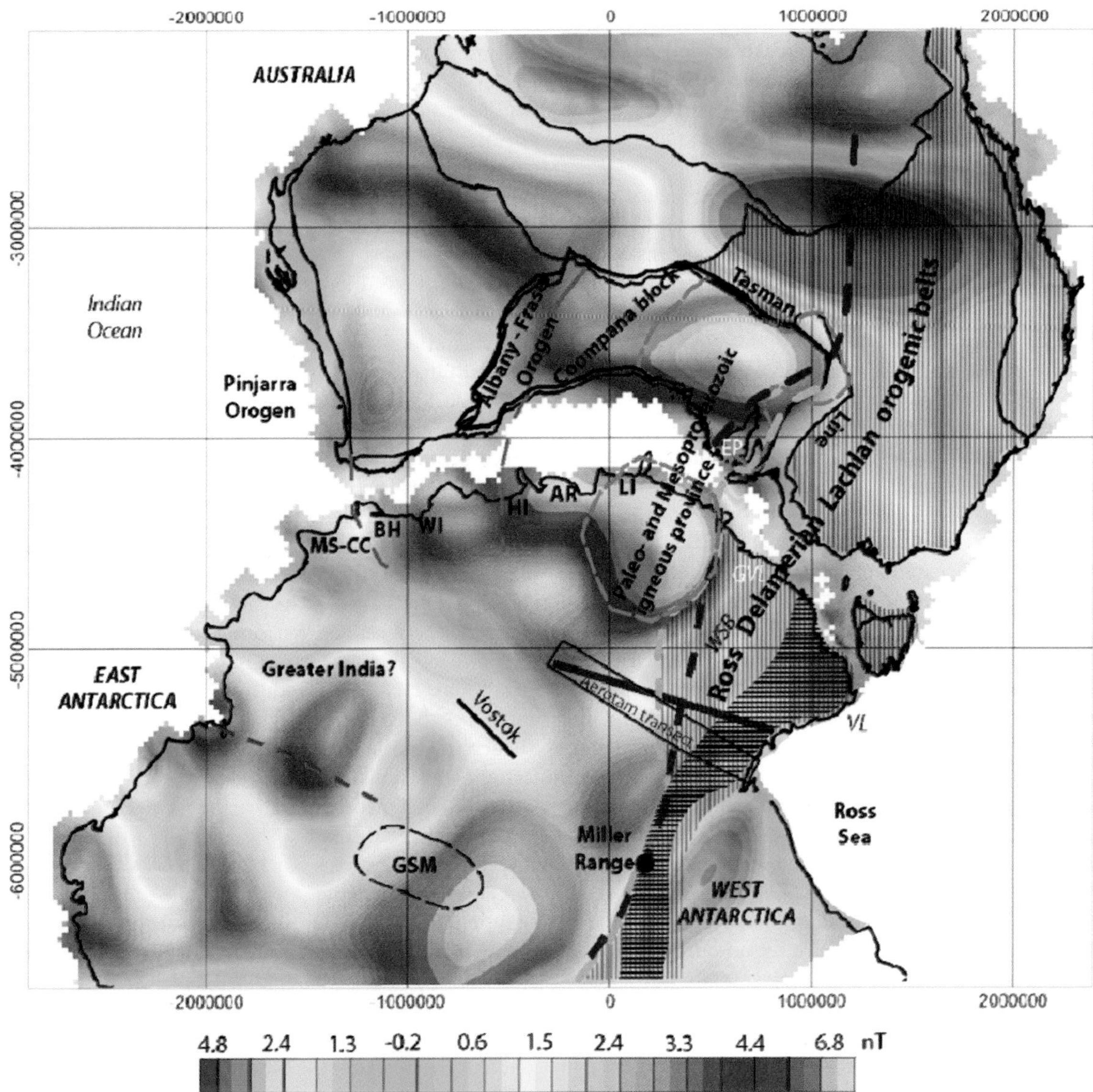

Fig. 4.1-2. Satellite magnetic data (Maus et al. 2002) of Australia and Antarctica rotated in a Gondwana reconstruction (Lawver et al. 2003). Grid coordinate system, legend, lines and place names same as for Fig. 4.1-1

very broad (300–500 km) geologic provinces (Maus et al. 2002). Interpretation of previous satellite data identified a magnetic high over the Gawler craton that extended into Antarctica (Frey et al. 1983; von Frese et al. 1999; von Frese et al. 1986) but did not determine the geologic sources.

To better delineate the Mawson block in Antarctica, we rotated the available airborne (Fig. 4.1-1) and CHAMP satellite (Fig. 4.1-2) magnetic data from Australia and the Wilkes Land sector of Antarctica into their relative Gondwana positions (Dalziel 1991). Comparison of magnetic anomalies from the well-understood Australian portion with those of East Antarctica enables a first attempt to map the extent and composition of the Mawson block, as well as the western limits of the Ross Orogen and Jurassic Ferrar magmatic provinces in Antarctica. Additionally, in conjunction with seismic tomography data (Shapiro and Ritzwoller 2002), the magnetic data help locate the craton edge more broadly through Antarctica.

Interpretation of Magnetic Anomaly Data

Magnetic data reflect variations in the distribution and type of magnetic minerals, primarily magnetite, in the Earth's crust. Magnetic rocks can be mapped from the surface to great depths, depending on their dimensions, shape, and magnetic properties, and on the character of the local geothermal gradient. At regional scales, magnetic highs are often associated with magnetite-bearing batholiths, large volumes of volcanic rocks, and metamorphic rocks derived from mafic igneous protoliths. Magnetic lows can be caused by sedimentary rocks, reversely magnetized volcanic rocks, magnetite-poor plutons, and metamorphic rocks of sedimentary origin.

The extensive coverage and high resolution of aeromagnetic data for Australia (Milligan and Tarlowski 1999) contrast sharply with data from Antarctica (Golynsky et al. 2001), such that only gross comparisons can be made. Despite the variance in survey flight characteristics, broadly similar magnetic anomaly patterns can be identified. Satellite magnetic data collected over ~14 months along ~100° paths in Antarctica at an altitude of 438 km image crustal-scale provinces = 500 km wide (Maus et al. 2002) and, aided by aeromagnetic data, allow for broad geologic correlations.

Aeromagnetic Signatures of the Mawson Block and Adjacent Regions

Aeromagnetic data help delimit the approximate eastern edge of Precambrian rocks in Australia, called the Tasman line (Hill 1951), the precise location, age and tectonic significance of which is debated (Direen and Crawford 2003). Here we use the original definition, the boundary between Precambrian and Phanerozoic rocks (Hill 1951), with no tectonic or age significance implied. West of the line, Precambrian rocks of various ages produce distinctive magnetic anomalies. The Gawler (GC) craton (Fig. 4.1-1 and 4.1-2) comprises Archean to Mesoproterozoic rocks that have remained substantially undeformed since ~1 450 Ma (Parker et al. 1993). Intrinsic to the Gawler craton are ~1.7 Ga metamorphic and igneous events, including bimodal volcanism, granite intrusion, and granulite-facies metamorphism (Daly et al. 1998). These rocks are exposed in places, but much of the Gawler basement remains hidden by younger materials. By imaging the covered basement with aeromagnetic data and obtaining U-Pb ages from limited outcrops, the Precambrian geology and structure of this province were mapped in detail (Daly et al. 1998; also compare (Drexel et al. 1993). Gneisses, iron formations and granites of the Archean Sleaford and Mulgathing complexes, located in the southeast and northwest parts of the craton, respectively (Fig. 4.1-1), can be differentiated with aeromagnetic data. Low-amplitude aeromagnetic lows over Sleaford (SC) and Mulgathing (MC) granites contrast with linear, positive anomalies over iron formations and Paleoproterozoic flows and intrusions (e.g., Lincoln Complex, LC). A satellite magnetic low over the Eyre Peninsula (EP, Fig. 4.1-2) most likely reflects the generally magnetite-poor Archean basement. Broad, high-amplitude positive anomalies relate to the Mesoproterozoic Gawler Ranges bi-modal volcanic suite and comagmatic Hiltaba granites (GV and HS, respectively, Fig. 4.1-1) and granitoids of the Ifould complex (included in HS, Fig. 4.1-1) (Daly et al. 1998; Rajagopalan et al. 1993). The dominant signature in satellite data over this region is a positive anomaly interpreted to be due primarily to magnetite-rich Paleoproterozoic Lincoln Complex intrusions in the southeast, and Mesoproterozoic Gawler volcanics/ Hiltaba granites in the west and central regions (Gawler craton, Fig. 4.1-2).

West of the Gawler craton lies a buried granitic and gneissic continental fragment, the Coompana block (Shaw et al. 1996) (Fig.4.1-1 and 4.1-2), that may have amalgamated with the Gawler craton by at least ~1 450 Ma (Fitzsimons 2003). The sources of the ovoid positive magnetic anomalies are not well determined, but the northeast trends of linear positive anomalies in the east parallel those of the Paleoproterozoic deformation trends in the Gawler craton and might reflect dikes or faults. The western boundary of the Australian Mawson block includes the Albany-Fraser orogenic belt (Fig. 4.1-1 and 4.1-2), a collision zone between the Yilgarn craton, Coompana block and Gawler craton that formed in two stages at about ~1 350–1 260 Ma and 1 210–1 140 Ma (Clark et al. 2000; Fitzsimons 2003) and produces high-amplitude linear anomalies that truncate magnetic fabrics in the adjacent Yilgarn and Gawler cratons. Magnetic highs relate to syntectonic intrusions.

Satellite magnetic highs over the Gawler and Yilgarn cratons reflect thick, magnetic crust, whereas lows characterize the Proterozoic mobile belts consisting of thick sequences of magnetite-poor rocks most likely to be of sedimentary and metasedimentary origin. Northeast trends in the magnetic data reflect deformation associated with the collisions. Compared to the Gawler Craton, the Coompana block is associated with a narrower, lower amplitude east-trending satellite magnetic high, and a gradient characterizes the Albany-Fraser belt (Fig. 4.1-2).

The eastern edge of the Mawson block in Australia is marked in the aeromagnetic data by a broad quiet zone punctuated by magnetic highs related to the early Paleozoic Delamerian Orogen (DO and KI, Fig. 4.1-1). The Delamerian belt is expressed by curvilinear highs (100–500 nT) related to exposed and buried gabbroic and dioritic intrusions representing the roots of a magmatic arc along the early Paleozoic convergent margin of Gondwana (Finn et al. 1999; Flöttmann et al. 1993). These, and mid-Paleozoic igneous rocks associated with ovoid magnetic highs farther east, intrude magnetite-poor greywackes and metamorphosed siliciclastic rocks ranging in age from Neoproterozoic and early Cambrian in the west, possibly deposited along the rifted craton margin (Mancktelow 1990), to Cambrian-Ordovician in the east that produce the magnetic quiet zone.

Aeromagnetic data (flight line spacing ~4 500–10 000 m) from northern Victoria Land, Antarctica (Chiappini et al.

2002; Golynsky et al. 2001; Studinger et al. 2004), display prominent magnetic highs (P, Fig. 4.1-1) averaging 200–500 nT over the Ross Orogen/TAM region, primarily related to exposed and buried gabbroic and dioritic intrusions that form the southern continuation of the Gondwana-margin magmatic arc in Australia (Ferraccioli et al. 2002; Ferraccioli et al. 2003; Finn et al. 1999). Recently collected data over the Oates and George V Coasts of Antarctica (Damaske et al. 2002; Ferraccioli et al. 2003) show similar wavelength (~100–300 km) magnetic highs near scant outcrop of early Paleozoic plutons (M. Fanning, unpublished data) (anomalies P1-2, GVL, Fig. 4.1-1). Although these anomalies have been interpreted as magnetic Precambrian or Jurassic rocks (Damaske et al. 2003; Talarico et al. 2001), their amplitudes, wavelengths and trends are more similar to those associated with Paleozoic plutons in Australia (DO and SB, Fig. 4.1-1) and Antarctica (e.g., anomalies P3–P5) (Ferraccioli et al. 2002; Ferraccioli et al. 2003; Finn et al. 1999) than to Jurassic and Precambrian rocks, suggesting similar Paleozoic sources. If so, these postulated plutons may mark the western edge of the Paleozoic Gondwana margin identified previously by Damaske et al. (2003). Similar wavelength and amplitude highs were interpreted to represent early Paleozoic plutons over the TAM in southern Victoria Land (e.g., P5-6, Fig. 4.1-1) (Bozzo et al. 1997; Ferraccioli and Bozzo 1999; Ferraccioli et al. 2003). However, the Paleozoic plutons do not have a consistent magnetic signature over the entire TAM. Many of the granites are non-magnetic and produce no magnetic anomalies (Damaske and Bosum 1993; Ferraccioli et al. 2002), or magnetic lows as in the Central TAM region (Goodge et al. 2004).

Although arc rocks can produce significant positive anomalies in aeromagnetic data, their thickness is insufficient to produce regional anomalies in satellite data (Fig. 4.1-2). Instead, the satellite data show a broad regional low that reflects the magnetite-poor Neoproterozoic and Cambrian sedimentary basement underlying the early Paleozoic arc, as well as non-magnetic early Paleozoic plutons and metamorphic rocks. The coincidence of the satellite low and local aeromagnetic highs (P1-3) provide further support for the argument that these highs primarily relate to Paleozoic plutons and not magnetic Precambrian crust as suggested previously (Damaske et al. 2003; Talarico et al. 2001). East of the satellite magnetic low are magnetic highs over southeastern Australia (including Tasmania) and Antarctica (Fig. 4.1-2),the origins of which are not clear but may reflect oceanic crust (Finn et al. 1999), abundant magnetite-rich mid-Paleozoic plutons and/or buried Jurassic intrusions.

Superimposed on the Ross Orogen/TAM anomaly pattern are short wavelength (2–10 km), linear 50–300 nT positive anomalies (Fig. 4.1-1) associated with Jurassic dolerites that continue north to Tasmania (Elliot and Gray 1992). The western-most extent of the dolerites is unknown, but along the George V Coast they extend nearly to the western edge of the Paleozoic plutonic belt (Damaske et al. 2003). Farther south, along the Aerotam (Studinger et al. 2004) and ITASE (Ferraccioli et al. 2001) transects (Fig. 4.1-1), models of local high-frequency magnetic highs are consistent with Jurassic dolerite and basalt sources that extend ~400–500 km west of the Transantarctic Mountains.

Based on unique patterns of metamorphism and detrital zircon ages in metapelites from the southern Eyre Peninsula and Commonwealth Bay in Terre Adélie (CB, Fig. 4.1-1), the ~1.7 Ga Gawler signature appears to extend from southern Australia into the Wilkes Land region of East Antarctica (Oliver and Fanning 1997). It is unknown, however, where this basement province extends beyond the narrow coastal exposures of the Commonwealth Bay region (Peucat et al. 2002), due to a lack of aeromagnetic data. Southeast (in rotated coordinates), over the western edge of Victoria Land, long-wavelength (~100–300 km) 200–500 nT highs are linked to Precambrian shield rocks (e.g., S, Fig. 4.1-1) (Bosum et al. 1989; Ferraccioli and Bozzo 1999) or Paleozoic plutons (e.g., P, Fig. 4.1-1) (Bozzo et al. 1997). Scattered magnetic profiles (Golynsky et al. 2001) and the Aerotam (Studinger et al. 2004) and ITASE transects (Ferraccioli et al. 2001) (Fig. 4.1-1) show high-amplitude (700–3 000 nT) anomalies over the polar plateau adjacent to the Transantarctic Mountains. Interpretation of the sub-ice geology in this region is problematic due to the great distance between the data sets, but is aided by satellite observations indicating a continuation of the magnetic high (Frey et al. 1983; von Frese et al. 1986) associated with Paleoproterozoic and Mesoproterozoic igneous provinces of Australia into Wilkes Land (Fig. 4.1-2). The southern terminus of the positive anomaly lies just north of the Aerotam transect (Fig. 4.1-2).

The proposed southern extension of the Mawson block to the area of the Miller Range is based on similar age and thermal histories in rocks with clear evidence of an Archean and Proterozoic ancestry, including: *(a)* juvenile Archean magmatism between 3 150–3 000 Ma; *(b)* crustal stabilization and metamorphism between 2 955–2 900 Ma; *(c)* ultra-metamorphism and magmatism at ~2 500 Ma; *(d)* metamorphism and magmatism between 1 730–1 720 Ma; and *(e)* post-1 700 Ma sedimentation (Fanning et al. 1995; Goodge and Fanning 1999; Goodge et al. 2001). Comparison of satellite magnetic data from the Wilkes Land portion of the Mawson block and Miller Range region show significant differences, however. A magnetic low characterizes the Miller Range region, in contrast to a broad high as observed over Wilkes Land to the north. A magnetic low over the Eyre Peninsula, although similar to that in the Miller Range region is attributed to different geologic sources that is Archean basement in contrast to Proterozoic reworked margin rocks. The lack of evidence for a ~1 600 Ma volcanic event in the Miller Range region that is equivalent to the Gawler and Hiltaba suites in South Australia partially explains the lack of correlated positive magnetic anomalies with the Mawson block. Furthermore, there are no obsereved anomalies we can attribute to igneous rocks equivalent to the Paleoproterozoic Lincoln complex.

Discussion

Distinct high-amplitude positive anomalies in both airborne and satellite magnetic data from Australia image mafic volcanic rocks and intrusions of the Paleoproterozoic Lincoln complex, and Mesoproterozoic Gawler rhyolitic volcanic rocks and Hiltaba granitic intrusions. The extension of the satellite magnetic high south to Wilkes Land, Antarctica, as well as the presence of 1 590 Ma volcanic clasts in glacial moraines on the coast (Peucat et al. 2002), suggest that much of Wilkes Land is underlain by the same Paleo- and Mesoproterozoic igneous rocks. The southern boundary of the satellite magnetic anomaly may indicate their southern limit, but it does not indicate whether older (Archean) Mawson block rocks continue south.

The nature and age of rock units associated with high-amplitude positive aeromagnetic anomalies over the western sector of the Aerotam transect (Fig. 4.1-1) cannot be determined, although the anomalies are characteristic of thick magnetic crust associated with Precambrian cratons. A remaining question is whether this crust belongs to the Mawson block or to some other distinguishable crustal province. Comparison of magnetic anomalies associated with this deeply buried (~3 500 m ice thickness (Studinger et al. 2004)) terrane with the Gawler craton magnetic anomalies shows significant differences (Fig. 4.1-1). Despite the significantly greater depth of burial, the amplitudes of the positive magnetic anomalies (2 000–3 000 peak-to-trough amplitude) are generally higher than in Gawler craton magnetic data (800–1 900 nT) that have been adjusted (by 3 500 m) to match the distance-to-source of the Antarctic data. The northwesterly trends (our rotated coordinate system) of the Aerotam anomalies differ from the east and north trends of positive anomalies over the Paleo- and Mesoproterozoic Gawler rocks (Aertoam and GC, Fig. 4.1-1). In addition, coincident with the magnetic highs over the Paleo- and Mesoproterozoic rocks in Australia are regional Bouguer gravity highs modeled as mid-crustal high-density mafic rocks (Daly et al. 1998; Rajagopalan et al. 1993; Shi 1993). In contrast, the regional magnetic highs over the western portion of the Aerotam and ITASE transects are mostly associated with Bouguer gravity lows inferred to relate to a rift basin (Ferraccioli et al. 2001) or compressional feature (Studinger et al. 2004) not observed in Australia. All of these data sets suggest that the Mawson block does not continue south to the location of the Aerotam transect, indicating a possible terminus to the Mawson block along the southern edge of the satellite magnetic high (pink dashed line, Fig. 4.1-2).

The lack of aeromagnetic data (Fig. 4.1-1) makes delineation of the western edge of the Mawson block difficult. Small outcrops of granodiorite gneiss of unknown age on Lewis and Anton islands (LI, Fig. 4.1-2), >1 090 Ma granitic gneiss at Al'bov Rocks (AR, Fig. 4.1-2), and >750 Ma charnockite and granite gneiss on Chick and Henry islands (HI, Fig. 4.1-2) are comparable to rocks in the Mawson block, Albany-Fraser Orogen, and poorly known Coompana Block to the north (Fig. 4.1-1 and 4.1-2) (Fitzsimons, written comment 2003). The Lewis and Anton islands rocks occur within the region of Mawson block identified by an ovoid satellite magnetic high (Fig. 4.1-2), whereas the exposures to the west are associated with an east-trending (rotated coordinates) magnetic high similar to that over the Coompana block. Therefore, we define the western edge of the Mawson block by the gradient between the prominent ovoid magnetic high corresponding to both the Paleoproterozoic Lincoln complex and Gawler Range Volcanics/Hiltaba Suite granites and the lower amplitude, east-trending high similar to that over the similar-aged Coompana block. Farther west, outcrops in Windmill Islands (WI, Fig. 4.1-2) and Bunger Hills (BH, Fig. 4.1-2) have been equated with the Albany-Fraser Orogen (Fitzsimons 2003; Harris 1995). However, the satellite magnetic signature differs. The inferred southern continuation of the Albany-Fraser Orogen into Antarctica (brown dashed lines, Fig. 4.1-2) is characterized by an east-trending magnetic high, rather than a northwest-trending gradient as seen in Australia, so it is difficult to define the orogen in Antarctica. Outcrops between Mirny Station and Cape Charcot (MS-CC, Fig. 4.1-2) correlate with the Pinjarra Orogen in Australia (Fitzsimons 2003), both of which are associated with low-amplitude satellite magnetic highs (Fig. 4.1-2). The low-amplitude magnetic high over these rocks in Antarctica is truncated on the south by a satellite magnetic low over a province of unknown affinity. A northwest-trending gradient (rotated coordinates) interpreted from aeromagnetic data to relate to a major tectonic boundary at Lake Vostok (Studinger et al. 2003) bounds this low to the southwest (Vostok, Fig. 4.1-2).

In Australia, the eastern edge of the Mawson block is marked by Neoproterozoic-early Cambrian magnetite-poor sedimentary rocks related to rifting along the craton (Mancktelow 1990). These rocks produce lows in both aero- and satellite magnetic data (magnetic quiet zone near DO, Fig. 4.1-1 and east of Tasman line, Fig. 4.1-2). Similarly, the western (rotated coordinate system) end of the George V Land aeromagnetic data (GVL, Fig. 4.1-1), central portions of the Aerotam and ITASE magnetic transects (near orange dashed line, Fig. 4.1-1) and satellite magnetic data (WSB, Fig. 4.1-2) are marked by magnetic lows. These lows are interpreted as magnetite-poor Neoproterozoic and lower Paleozoic sedimentary rocks within the Wilkes Subglacial Basin related to a rift (Ferraccioli et al. 2001) or compressional feature (Studinger et al. 2004). The similarity in satellite magnetic signature for Proterozoic and Paleozoic rocks precludes clear definition of the reworked craton margin, but isotopic data suggest that Proterozoic lower crust does underlie much of the TAM (Borg and DePaolo 1994). Additional evidence for the location of the boundary between the thick, relatively undeformed craton and reworked Neoproterozoic margin comes from seismic tomography data. The eastern edge of high-velocity perturbations interpreted to relate to the Precambrian shield of East Antarctica (Ritzwoller et al. 2001) is approximately coincident with the eastern edge of the satellite magnetic high (Fig. 4.1-2). Given the >500 km resolution of both the satellite magnetic and seismic tomography data sets, this coincidence is strong support for locating the eastern boundary of thick Mawson block lithosphere significantly inboard of the present-day TAM.

South of the proposed Mawson block boundary, the sources of satellite magnetic highs are of unknown nature. The lower amplitude magnetic high northeast of the satellite magnetic high extending from the Miller Range to the Aerotam transect (Fig. 4.1-2) may indicate Precambrian crust. Farther north, the observed satellite magnetic high most likely reflects magnetic oceanic crust inferred to underlie northern Victoria Land (Armadillo et al. 2004; Ferraccioli et al. 2002; Finn et al. 1999). Alternatively, the damped high could relate to large proportions of middle Paleozoic plutons. While the composition of crustal blocks in much of the central East Antarctic shield cannot be determined with the satellite data, the magnetic highs most likely point to the existence of a thick Archean-Mesoproterozoic igneous craton, whereas observed lows probably reflect large regions underlain by Proterozoic and younger sedimentary and volcanic successions, or by Grenville-age or younger mobile belts.

Conclusions

The combination of airborne and satellite magnetic data show that the Mawson block may only extend across part of Wilkes Land, Antarctica (Fig. 4.1-2). The composition of the craton south of the Mawson block is largely unknown. However, the variability of the satellite magnetic anomalies, and proposed tectonic boundary at Vostok based on aeromagnetic data (Studinger et al. 2003), suggest the existence of a heterogeneous, composite province. The coincidence of the eastern limit of high-magnetization and high-velocity crust delimits the edge of thick, relatively undeformed Precambrian (Archean-Paleoproterozoic?) craton. We define the craton edge as the gradient separating magnetic highs related to Precambrian shield basement and lows over magnetite-poor reworked Neoproterozoic-lower Paleozoic craton-margin rocks (Fig. 4.1-2).

The coincidence of the inferred cratonic edge and western limit of Paleozoic and Jurassic magmatism suggests either that the craton margin exerted fundamental control on younger tectonic events or it was significantly modified by these events, in particular the Paleozoic Ross Orogeny. For example, magnetic anomalies associated with Jurassic Ferrar dolerites trend generally northeasterly (rotated coordinates) from the Miller Range (unpublished data) to northern Victoria Land and Tasmania (Fig. 4.1-2), parallel to the craton boundary. The trends of these anomalies change to northwesterly (projected coordinates) along George V Land, corresponding to a similar change in trend of anomalies associated with Paleozoic plutons (GVL, Fig. 4.1-1). These plutons correlate with those on Kangaroo Island, Australia (KI, Fig. 4.1-1), where trends of magnetic highs diverge in an easterly direction from the curvilinear NNE-NNW Delamerian trends and the tomography data show a westward jog in the craton boundary (Fig. 4.1-1). All these observations suggest that the edge of cratonic lithosphere (formed during Rodinia breakup?) could have exerted mechanical control on subsequent volcanism and tectonic events. We further suggest here that the intact craton margin, developed in the Neoproterozoic, lies significantly west of the present Transantarctic Mountains uplift. The Transantarctic Mountains are often considered to mark the eastern edge of the East Antarctic craton, but the edge delineated by magnetic and tomographic data is everywhere to the west and far west along Geroge V Land (Fig. 4.1-1 and 4.1-2). Isotopic data indicate the presence of Paleoproterozoic and younger lower crust beneath much of the presently-exposed Ross Orogen, but there it may be significantly thinned and distended, representing transitional-type basement to the overlying sedimentary cover and granitic intrusions. An example of this type of crust may be found in the Miller Range where non-magnetic lower crust is exposed (Goodge et al. 2004). The combination of extended non-magnetic lower crust with weakly magnetic sedimentary and igneous units could yield the observed satellite magnetic lows marking the proposed craton edge. The locus of Paleozoic and Jurassic Ferrar magmatism consistently extends westward to this craton edge, probably following Paleozoic strata, and may represent a better geologic marker than present-day orographic features.

Success in interpretation of magnetic patterns in the covered Gawler craton gives optimism for identifying similar patterns in the ice-covered East Antarctic shield. Acquisition of new aeromagnetic data over exposures along the margin of the shield and penetrating farther inland will help constrain the age, composition and structure of the shield.

Acknowledgments

We thank Sergei Pisarevsky for rotating the magnetic data into their SWEAT configurations. Helpful reviews by Fausto Ferraccioli, Ian Fitzsimons, Bob Kucks and Anne McCafferty greatly improved the manuscript. CAF's work was funded by National Science Foundation Grant OPP-0232042 and U.S. Geological Survey. JWG's work was funded by National Science Foundation Grant OPP-0232042.

References

Armadillo E, Ferraccioli F, Tabellario G, Bozzo E (2004) Electrical structure across a major ice-covered fault belt in Northern Victoria Land (East Antarctica). Geophys Res Letters 31

Borg SG, DePaolo DJ (1994) Laurentia, Australia, and Antarctica as a Late Proterozoic supercontinent: constraints from isotopic mapping. Geology 22:307–310

Bosum W, Damaske D, Roland NW, Behrendt J, Saltus R (1989) The GANOVEX IV Victoria Land/Ross Sea aeromagnetic survey: interpretation of anomalies. Geol Jb E38:153–230

Bozzo E, Ferraccioli F, Gambetta M, Caneva G, Damaske D, Chiappini M, Meloni A (1997) Aeromagnetic regional setting and some crustal features of central-southern Victoria Land from the GITARA surveys. In: Ricci CA (ed) The Antarctic region: geological evolution and processes. Terra Antartica Publication, Siena, pp 591–596

Chiappini M, Ferraccioli F, Bozzo E, Damaske D (2002) Regional compilation and analysis of aeromagnetic anomalies for the Transantarctic Mountains-Ross Sea sector of the Antarctic. In: von Frese R, Taylor PT, Chiappini M (eds) Tectonophysics 347:121–137

Clark DJ, Hensen BJ, Kinny PD (2000) Geochronological constraints for a two-stage history of the Albany-Fraser Orogen, Western Australia. Precambrian Res 102:155–183

Daly SJ, Fanning CM, Fairclough MC (1998) Tectonic evolution and exploration potential of the Gawler craton, South Australia. J Australian Geol Geophys 17:145–168

Dalziel IWD (1991) Pacific margins of Laurentia and east Antarctica-Australia as a conjugate rift pair: evidence and implications for an Eocambrian supercontinent. Geology 19:598–601

Damaske D, Bosum W (1993) Interpretation of the aeromagnetic anomalies above the Lower Rennick Glacier and the adjacent polar plateau west of the USARP Mountains. Geol Jb E47:139–152

Damaske D, Finn CA, Moeller H-D, Demosthenous C, Anderson ED (2002) Aeromagnetic data centered over Skelton Neve, Antarctica: a web site for distribution of data and maps (on-line edition). U.S. Geol Surv Open-File Rep 02-452, http://pubs.usgs.gov/of/2002/ofr-02-452

Damaske D, Ferraccioli F, Bozzo E (2003) Aeromagnetic anomaly investigations along the Antarctic coast between Yule Bay and Mertz Glacier. Terra Antartica 10:85–96

Direen NG, Crawford AJ (2003) The Tasman line: where is it, what is it, and is it Australia's Rodinia breakup boundary? Austral J Earth Sci 50:491–502

Drexel JF, Preiss WV, Parker AJ (1993) The geology of South Australia. Vol 1, The Precambrian. South Australia Geol Surv pp 242

Elliot CG, Gray DR (1992) Correlations between Tasmania and the Tasman-Transantarctic orogen: evidence for easterly derivation of Tasmania relative to mainland Australia. Geology 20:621–624

Fanning CM, Daly SJ, Bennett VC, Menot RP, Peucat JJ, Oliver RL, Monnier O (1995) The 'Mawson Block': once contiguous Archean to Proterozoic crust in the East Antarctic Shield and the Gawler Craton. In: Ricci CA (ed) Abstracts 7th Annual Symposium on Antarctic Earth Sciences, Siena

Ferraccioli F, Bozzo E (1999) Inherited crustal features and tectonic blocks of the Transantarctic Mountains: an aeromagnetic perspective (Victoria Land, Antarctica). J Geophys Res B 104: 25297–25319

Ferraccioli F, Coren F, Bozzo E, Zanolla C, Gandolfi S, Tabacco IE, Frezzotti M (2001) Rifted crust at the East Antarctic Craton margin: gravity and magnetic interpretation along a traverse across the Wilkes subglacial basin region. Earth Planet Sci Letters 192:407–421

Ferraccioli F, Bozzo E, Capponi G (2002) Aeromagnetic and gravity constraints for an early Paleozoic subduction system of Victoria Land, Antarctica. Geophys Res Letters 29

Ferraccioli F, Damaske D, Bozzo E, Talarico F (2003) The Matusevich aeromagnetic anomaly over Oates Land, East Antarctica. Terra Antartica 10:21–228

Finn CA, Moore D, Damaske D, Mackey T (1999) Aeromagnetic legacy of early Paleozoic subduction along the Pacific margin of Gondwana. Geology 27:1087–1090

Fitzsimons ICW (2000a) Grenville-age basement provinces in East Antarctica: Evidence for three separate collisional orogens. Geology 28:879–882

Fitzsimons ICW (2000b) A review of tectonic events in the East Antarctic Shield and their implications for Gondwana and earlier supercontinents. J African Earth Sci 31:3–23

Fitzsimons ICW (2003) Proterozoic basement provinces of southern and south-western Australia, and their correlation with Antarctica. In: Yoshida M, Windley BF, Dagupta S (eds) Proterozoic East Gondwana: supercontinent assembly and breakup. Spec Publ 206, Geol Soc London, pp 93–130

Flöttmann T, Gibson GM, Kleinshmidt G (1993) Structural continuity of the Ross and Delamerian orogens of Antarctica and Australia along the margin of the paleo-Pacific. Geology 21:319–322

Frey H, Langel R, Mead G, Brown K (1983) Pogo and Pangaea. Tectonophysics 95:181–189

Golynsky A, Chiappini M, Damaske D, Ferraccioli F, Ferris J, Finn C, Ghidella M, Isihara T, Johnson A, Kovacs S, Masolov V, Nogi Y, Purucker M, Taylor P, Torta M (2001) ADMAP: Magnetic Anomaly Map of the Antarctic. In: Morris P, von Frese R (eds) British Antarctic Survey Miscellanous Series, Sheet 10, scale 1:10 000 000

Goodge JW, Fanning CM (1999) 2.5 billion years of punctuated Earth history as recorded in a single rock. Geology 27:1007–1010

Goodge JW, Fanning CM, Bennett VC (2001) U-Pb evidence of ~1.7 Ga crustal tectonism during the Nimrod Orogeny in the Transantarctic Mountains, Antarctica: implications for Proterozoic plate reconstructions. Precambrian Res 112:261–288

Goodge J, Finn C, Damaske D, Abraham J, Moeller H-D, Anderson E, Roland N, Goldmann F, Braddock P, Rieser M (2004) Crustal structure of Ross Orogen revealed by aeromagnetics and gravity data. Geol Soc Amer abstracts programs 36(5):495

Harris LD (1995) Correlation between the Albany, Fraser and Darling mobile belts of Western Australia and Mirny to Windmill Islands in the East Antarctic Shield: implications for Proterozoic Gondwanaland reconstructions. In: Yoshida MS (ed) India and Antarctica during the Precambrian. Mem Geol Soc India 34:47–71

Hill D (1951) Geology. In: Mack G (ed) Handbook of Queensland. Australian Assoc Advanc Sci, Brisbane, pp 13–24

Lawver LA, Dalziel IWD, Gahagan LM, Martin KM, Campbell DA (2003) The Plates 2003 Atlas of Plate Reconstructions (750 Ma to Present Day). Plates Progr Rep 280-0703, UTIG Tech Rep 190

Mancktelow NS (1990) The structure of the southern Adelaide Fold Belt, South Australia In: Jago JB, Moore PS (eds) The evolution of a Late Precambrian-Early Palaeozoic rift complex: the Adelaide Geosyncline. Spec Publ Geol Soc Australia 16:369–395

Maus S, Rother M, Holme R, Luhr H, Olsen N, Haak V (2002) First scalar magnetic anomaly map from CHAMP satellite data indicates weak lithospheric field. Geophys Res Letters 29

Milligan P, Tarlowski C (1999) Magnetic anomaly map of Australia (3rd edn). AGSO, Geoscience Australia, Canberra

Myers JS, Shaw R, Tyler IM (1996) Tectonic evolution of Proterozoic Australia. Tectonics 15:1431–1446

Oliver RL, Fanning CM (1997) Australia and Antarctica: precise correlation of Palaeoproterozoic terrains. In: Ricci CA (ed) The Antarctic region: geological evolution and processes. Terra Antartica Publication, Siena, pp 163–172

Parker AJ, Fanning CM, Flint RB, Martin AR, Rankin LR (1993) Archaean-Early Proterozoic granitoids, metasediments and mylonites of Southern Eyre peninsula, South Australia: field guide 2. Geol Soc Australia Spec Group Tectonics Struct Geol: 1–90

Peucat JJ, Capdevila R, Fanning CM, Menot RP, Pecora L, Testut L (2002) 1.60 Ga felsic volcanic blocks in the moraines of the Terre Adélie craton, Antarctica: comparisons with the Gawler Range volcanics, South Australia. Australian J Earth Sci 49:831–845

Rajagopalan S, Zhiqun S, Major R (1993) Geophysical investigations of volcanic terrains: a case history from the Gawler Range Volcanic Province, South Australia. Explor Geophys 24:69–778

Ritzwoller MH, Shapiro NM, Levshin A, Leahy GM (2001) Crustal and upper mantle structure beneath Antarctica and surrounding oceans. J Geophy Res B 106:30645–30670

Shapiro NM, Ritzwoller M (2002) Monte-Carlo inversion for a global shear velocity model of the crust and upper mantle. Geophys J Internat 151:88–105

Shaw R, Wellman P, Gunn P, Whitaker AJ, Tarlowski C, Morse M (1996) Guide to using the Australian crustal elements map. Australian Geol Surv Org Rec 1996/30

Shi Z (1993) Automatic interpretation of potential field data with applications to estimates of soil thickness and study of deep crustal structures in South Australia. PhD thesis, Adelaide University, Adelaide

Studinger M, Bell RE, Karner GD, Tikku AA, Holt JW, Morse DL, Richter TG, Kempf SD, Peters ME, Blankenship DD, Sweeney R, Rystrom VL (2003) Ice cover, landscape setting, and geological framework of Lake Vostok, East Antarctica. Earth Planet Sci Letters 205:195–210

Studinger M, Bell RE, Buck R, Karner GD, Blankenship DD (2004) Sub-ice geology inland of the Transantarctic Mountains in light of new aerogeophysical data. Earth Planet Sci Letters 220:391–408

Talarico F, Armadillo E, Bozzo E (2001) Antarctic rock magnetic properties: new susceptibility measurements in Oates Land and George V Land. Terra Antartica Rep 5:45–50

Tinge RJ (1991) The regional geology of Archaean and Proterozoic rocks in Antarctica. In: Tingey RJ (ed) The geology of Antarctica. Oxford University Press, Oxford, pp 1–73

von Frese R, Hinze WJ, Olivier R, Bentley CR (1986) Regional magnetic anomaly constraints on continental breakup. Geology 14:68–71

von Frese R, Roman DR, Kim J-H, Kim JW, Anderson AJ (1999) Satellite mapping of the Antarctic gravity field. Annali Geofis 42: 293–308

The Matusevich Fracture Zone in Oates Land, East Antarctica

Georg Kleinschmidt[1] · Andreas L. Läufer[1,2]

[1] Geologisch-Paläontologisches Institut der J. W. Goethe-Universität, Senckenberganlage 32, 60054 Frankfurt am Main, Germany, <kleinschmidt@em.uni-frankfurt.de>

[2] *Present address:* Bundesanstalt für Geowissenschaften und Rohstoffe, Stilleweg 2, 30655 Hannover, Germany

Abstract. The Matusevich Glacier trends 170° totally straight for more than 100 km. For this reason, a major fault was assumed along the glacier formerly. A westward directed ductile thrust system, trending 170°, was subsequently discovered in the upper Matusevich Glacier ("Exiles Thrust"). It formed under amphibolite facies conditions during the Ross Orogeny. Therefore, the course of the Matusevich Glacier was attributed to the Exiles Thrust instead of the postulated simple fault. During GANOVEX VIII/ITALANTARTIDE XV (1999/2000), the small-scale structures at the margins of the Matusevich Glacier were mapped. The most conspicuous and meaningful of these structures are cold, brittle, NW- to N-trending thrusts with slickensides, decorated with quartz fibres and uniformly SW-thrusting (~220°). They occur at the western side of the glacier (Lazarev Mts.). These structures are consistent with strike-slip tectonics along the Matusevich Glacier and could be interpreted as indicators of transpressional tectonics. Unfortunately, corresponding dextral strike-slip faults, which should strike about 170°, could not be observed directly. But 30 km to the west, 165° trending strike-slip structures are exposed at the eastern edge of Outrider Nunatak. Striations on steep fault planes indicate dextral displacement. This strike-slip tectonics produced a flower structure visible in one of the main granite-walls of Outrider Nunatak. Thus the neotectonics of westernmost Oates Land is characterized by brittle dextral strike-slip faulting, following the trend of much older Ross-age ductile thrust tectonics.

A comparable history as at this "Matusevich Fracture Zone" is known from the larger Rennick Glacier. Thus, two brittle dextral strike-slip fault zones are tracing older structures and cross at high angles the coastline of Antarctica. They are co-linear with off-shore fracture zones, the active parts of which are the transform faults between Antarctica and Australia. We discuss, whether the dextral faults on-shore could represent the continuations of the fracture zones off-shore. This idea is supported *(i)* by dextral offsets of the shelf where fracture zones reach Antarctica; *(ii)* possibly by magnetic anomalies along the Matusevich and Rennick Glaciers, which seem to continue off-shore; *(iii)* by several examples of oceanic fracture zones continuing into continental crust.

Introduction

The Matusevich Glacier is located in Oates Land – the Antarctic coastal area between 155° E and 160° E – at about 157.5° E (Fig. 4.2-1). Oates Land's basement belongs to the Ross Orogen and forms the continuation of Australia's Delamerian Orogen. The Matusevich Glacier is about 125 km long and about 10 km wide. Its bed rock morphology forms a classical straight U-shaped valley exceeding 850 m below sea level (Damm 2004). The Matusevich Glacier trends 170° and therefore runs parallel to the Rennick Glacier in northern Victoria Land 120 km to the east of it (Fig. 4.2-1).

The area of the Rennick Glacier is much better known and much better investigated than that of the Matusevich area. The valley of the Rennick Glacier forms a graben (Dow and Neall 1974), defined by downfaulted Jurassic extrusive volcanics and surrounded by uplifted flanks featuring basement rocks. The graben is up to 40 km wide and at least 200 km long (Roland and Tessensohn 1987; Tessensohn 1994). The filling of the graben consists of *(i)* volcanics of the Jurassic Ferrar Group inclusive of superficial volcanics (Kirkpatrick Basalts) as lava flows and pillow basalts (Skinner et al. 1981) and *(ii)* marginally of Permotriassic sediments of the Beacon Supergroup. Small outcrops in the graben indicate that the filling possibly rests on top of low-grade basement rocks. The graben shoulders consist of high grade basement plus sediments of the Beacon Group and sills of the Ferrar Group (GANOVEX Team 1987).

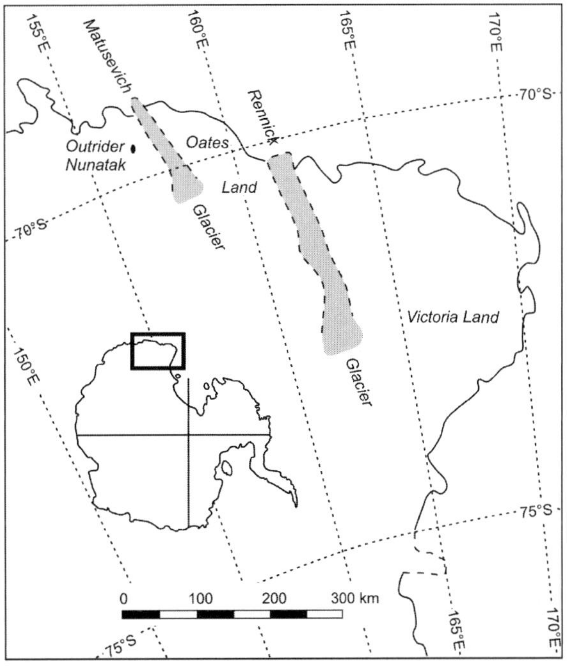

Fig. 4.2-1. The location of the Matusevich Glacier (Oates Land). Note that it lies parallel to the Rennick Glacier (northern Victoria Land)

From: Fütterer DK, Damaske D, Kleinschmidt G, Miller H, Tessensohn F (eds) (2006) Antarctica: Contributions to global earth sciences. Springer-Verlag, Berlin Heidelberg New York, pp 175–180

In detail, the Rennick Graben does not consist of just two parallel straight faults, but its geometry is a rather complicated product of rather complicated processes within its neotectonic framework. (Roland and Tessensohn 1987; Tessensohn 1994).

Recently, Rossetti et al. (2002, 2003a) demonstrated that the present structural architecture of the Rennick structure is dominated by dextral strike-slip tectonics with local transpression and transtension (i.e. both positive and negative flower structures), which overprints an earlier E-W extensional increment. The initiation of these events is estimated at ca. 50–60 Ma and their age is interpreted to be Cenozoic. Evidences are in particular *(i)* a major uplift phase in the Transantarctic Mountains around 55–50 Ma (Lisker 2002), *(ii)* widespread magmatic activity initiating around 48 Ma ago (Armienti and Baroni 1999), *(iii)* a significant change in East-West Antarctic plate configuration and the formation of the Adare Trough as an extinct ridge system mainly between 43 and 28 Ma (Cande et al. 2000), *(iv)* the initiation of large-scale intracontinental right-lateral shear zones in the Ross Sea region around 35 Ma (Salvini et al. 1997), which was recently demonstrated by in-situ Ar-Ar age data of ca. 34 Ma determined on fault-generated pseudotachylytes along the Priestley Fault (Rossetti et al. 2003b). A still older increment involving sinistral motions along the Rennick structure and possibly cogenetic conjugate dextral kinematics is suggested by scarce fault-slip data along the Lanterman Fault system and 3–4 km right-lateral offsets of late-Ross structures, which could relate to the late Mesozoic to early Cenozoic separation of Australia and Antarctica (Damaske and Bosum 1993; Bozzo et al. 1999; Läufer and Rossetti 2003).

But this neotectonic Rennick graben system roughly traces a much older structure: an eastward directed ductile thrust, called "Wilson Thrust" (Flöttmann and Kleinschmidt 1989, 1991a,b, 1993; Kleinschmidt 1992). Its assumed late-Ross age was confirmed by Sm-Nd age determination of 458 ±30 Ma (garnets of undeformed dykes crosscutting deformed granites; Flöttmann pers. comm. 1994). This thrust runs at the western flank of the Rennick Glacier from coast to coast, through entire Victoria Land.

Résumé: The Rennick Glacier follows an old, repeatedly reactivated linear structure. This was demonstrated recently at its Ross Sea-ward prolongation, the Priestley Fault. There, the eastward directed, ductile, Ross-aged reverse fault, outcropping at Black Ridge, was reactivated 34 Ma ago producing the aforementioned pseudotachylytes.

Results from the Matusevich Glacier Area

In contrast to the Rennick Glacier area, the Matusevich Glacier area does not show any evidence for graben formation. It trends 170° totally straight for about 125 km (Fig. 4.2-1). For these reasons, just a "zone of weakness" (McLeod and Gregory 1967) or a major fault (Gair et al. 1969) was initially assumed along the glacier (Fig. 4.2-2a).

About 20 years later, a ductile thrust system was discovered at the uppermost Matusevich Glacier (Flöttmann and Kleinschmidt 1989, 1991a,b, 1993). There, the névé of the glacier is penetrated by the Exiles Nunataks consisting of Granite Harbour Intrusives. They are strongly sheared, their mylonitic shear planes trend 170° (i.e. paralleling the glacier) and dip 30–50° to the NE; shear sense indicatores (S-C structure, σ clasts) clearly show SW-

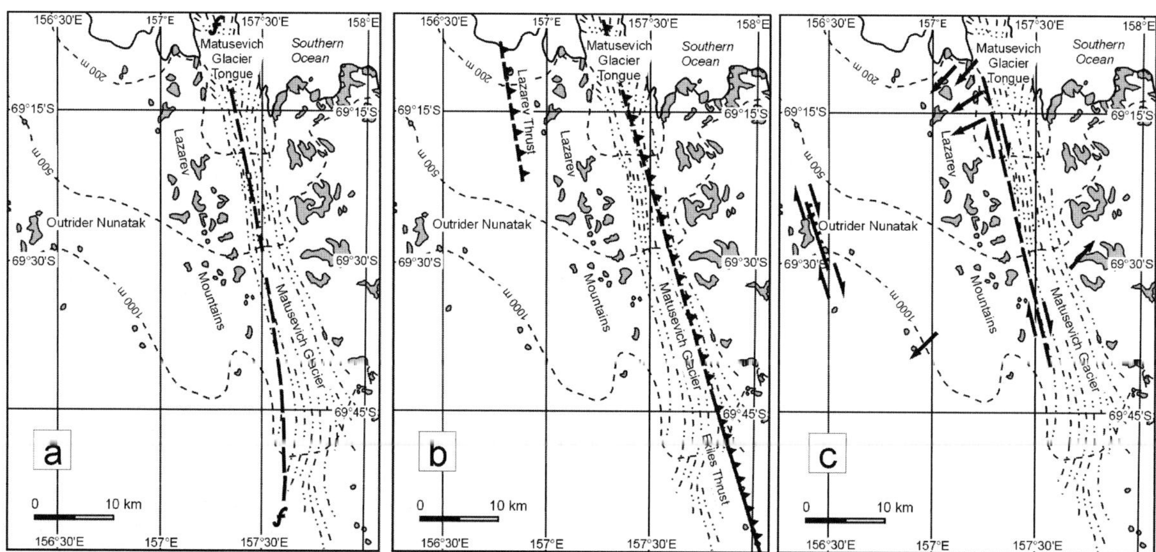

Fig. 4.2-2. Different tectonic interpretations predicting that the Matusevich Glacier is underlain by: **a** simply a fault of undetermined age and kinematics (Gair et al. 1969); **b** a major thrust fault zone of late-Ross age (Flöttmann and Kleinschmidt 1991a,b, 1993); **c** a major strike-slip fault, as supported by our findings including small thrusts (*arrows*: thrust directions paralleling σ_1) and strike-slip faulting at Outrider Nunatak

thrusting (~240°). The thrust system was called "Exiles Thrust". Because of its ductile deformation under amphibolite facies conditions (sillimanite is growing on the shear planes), the thrusting was attributed to the (late) Ross Orogeny. This was confirmed by not well constrained Sm-Nd age determination of 463 ±30 Ma (Flöttmann pers. comm. 1994) and by ^{40}Ar-^{39}Ar plateau ages of 472 ±2 Ma (syntectonic musovite; Roland et al. 2000), i.e. approximately coeval with the Wilson Thrust.

After the discovery of the Exiles Thrust system, we thought, that there is no later brittle faulting and that the course of the Matusevich Glacier is caused by this major thrust system instead of Gair's et al. (1969) postulated fault (Fig. 4.2-2b). The Exiles Thrust of the Matusevich Glacier and the Wilson Thrust of the Rennick Glacier area form a conjugate Ross-aged thrust system. This thurst belt system was used for the reconstruction of the exact relation of Australia and Antarctica within Gondwana (Flöttmann et al. 1993a,b).

During GANOVEX VIII/ITALANTARTIDE XV (1999/2000), the small-scale structures at the margins of the Matusevich Glacier were measured. Strangely enough, only a few ductile structures related to the Ross-aged, ductile Exiles Thrust were recognized, i.e. preferably small thrusts in the area of the upper Matusevich Glacier (Fig. 4.2-3).

Most of the small-scale structures formed at cold and brittle conditions, meaning that they are relatively young. The most conspicuous ones occur in the Lazarev Mountains at the western side of the glacier, especially at their northern part called Burnside Ridge: NW- to N-trending thrusts with slickensides, decorated with quartz fibres. A Schmidt net plot (Fig. 4.2-3) demonstrates, that these structures show uniform thrust-directions of about 220°, forming an acute angle of about 50° to the trend of the glacier. Therefore, these small thrusts and their thrust-directions could be perfectly used as indicators of the σ_1 direction of the regional palaeo-stress field and could be used as indicators for a major strike-slip fault following the Matu-sevich Glacier. The internally consistent overall brittle structure of the Matusevich Glacier area results in dextral strike-slip tectonics similar and parallel to that of the Rennick Glacier (Fig. 4.2-2c). Possibly due to the poor outcrop conditions, the σ_1-parallel directions of thrusting are restricted to the western flank of the glacier. But they should be present as well at its eastern flank in a symmetrical way. But there, only one debatable example was found: a low angle normal fault showing NE-movement (030°; Fig. 4.2-3). This structure would fit to a flower structure due to strike-slip tectonics paralleling the Matusevich Glacier.

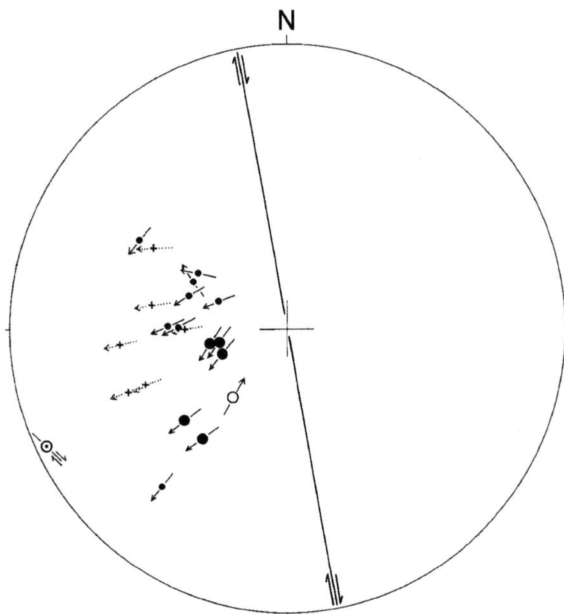

Fig. 4.2-3. Schmidt net plot (method Hoeppener 1955) of cold/brittle structures around Matusevich Glacier. *Dots:* at western side (*large dots:* most reliable data); *circle:* eastern side; *encircled dot:* at Outrider Nunatak, compared with ductile data related to the Exiles Thrust (= ...+...)

No dextral strike-slip faults and no brittle structures following the trend of the glacier (i.e. 170° trending) have been observed close to the Matusevich Glacier. But 30 km to the west, the granite massif of the Archangle Nuntaks is well exposed, the biggest of which is called Outrider Nunatak. The granite at this nunatak shows several of ~170° trending faults. Only one of these structures could have been measured completely: the attitude of fault plane (160/85 E), the lineation (155/60 SE), and the Riedel shears indicating displacement towards SE (Fig. 4.2-3). Thus, these faults show a dextral strike-slip component. This strike-slip tectonics produced a prominent flower structure visible in the main granite wall of Outrider Nunatak (Fig. 4.2-4). Unfortunately, the outcrop is not accessible.

Résumé: Overall, the neotectonics of westernmost Oates Land (around 157° E, Matusevich Glacier and Outrider Nunatak) is characterized by brittle dextral strike-slip faulting, following the trend of much older ductile thrust tectonics.

Fig. 4.2-4. Flower structure in the granite of the Outrider Nunatak trending ca. 165° related to dextral strike-slip tectonics/faulting. View towards SE

Discussion and Conclusions

Although the Matusevich Glacier is much smaller (width of 10 km) than the Rennick Glacier (width up to 25 km) and apart from their parallel trend, a certain analogy regarding their structural evolution is evident. Both the Matusevich and the Rennick structures feature relatively young dextral strike-slip tectonics involving transpression and transtension. Both follow the trace of much older (i.e. late-Ross orogenic) ductile reverse shear zones of roughly the same trend. And both cut the Antarctic coast line of Oates Land and Victoria Land at high angles.

The trends of the two major structures on-shore do not only coincide with, but they also seem to be direct continuations of off-shore fracture zones, the active parts of which represent the transform faults between Antarctica and Australia (Fig. 4.2-5). These fracture zones mark apparent dextral offsets of the Australian-Antarctic spreading ridges, but they of course show sinistral kinematics in their seismically active sections. According to several authors (e.g., Veevers 1987; Sutherland 1995; Cande et al. 2000; Stock and Cande 2002), the fracture zones form a particular dense swarm in front of the Matusevich and Rennick Glaciers, which is called the Tasman Fracture Zone. Therefore, the question may be raised, whether the dextral fault systems on-shore in fact represent the continuations of the intraoceanic fracture zones off-shore. There are several arguments which may support such an unconventional idea:

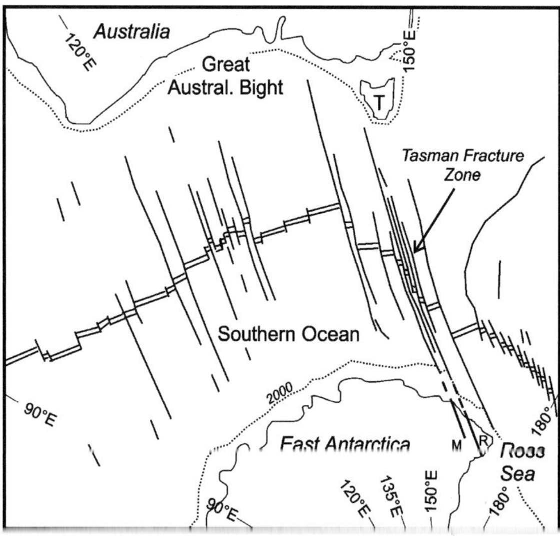

Fig. 4.2-5. The Matusevich Fracture Zone and the parallel running Rennick Graben system regarded as onshore continuations of the offshore fracture zones/transform faults between Australia and Antarctica. Note offsets of the Antarctic shelf! *T:* Tasmania; *M:* Matusevich structure; *R:* Rennick structure. Modified from Veevers (1987), Sutherland (1995) and Cande et al. (2000)

i. The Antarctic continental shelf seems to be dextrally offset, where the intraoceanic fracture zones reach Antarctica, even though the off-shore contour lines in that area are not reliable and require a far better and detailed data base.
ii. Magnetic anomalies along Matusevich and Rennick Glaciers seem to continue off-shore (Finn et al. 1999; Damaske et al. 2003; Ferraccioli et al. 2003). However, the present interpretations of the anomalies differ: The Rennick Anomaly is thought to be related to basic intrusions exposed at the western margin of the glacier and presumably underlying the glacier itself and its eastern margin (Finn et al. 1999). The Matusevich Anomaly is attributed most likely to the high susceptibility of the Exiles granites (Ferraccioli et al. 2003) and at least some of the models strongly support the fundamental importance of the Exiles Thrust system in this area. Perhaps, there will be additional and more conform interpretations in the future than the ones existing at the moment.
iii. There are several examples where oceanic fracture zones do in fact continue into the adjacent continental crust. These are for instance, the Sonne Fault at the southern coast of Iran and Pakistan (Kukowski et al. 2000, 2001) and the St. Paul, Cape Palmas and Grand Cess fracture zones off West Africa (Behrendt et al. 1974).

These arguments lead us to discuss the following four possibilities:

1. Dextral strike-slip tectonics produces dextral offsets of the initial intracontinental rift zones between Antarctica and Australia. The dextral kinematics is preserved in the continuations of the fracture zones on land today. After the initial stage and the formation of the Australian-Antarctic spreading centres, the usual sinistral motion takes place along the active transform portions between the ridge segments.
2. Continuously higher spreading rates east of each transform fracture zone result in excess dextral shear along the Antarctic-ward sections of the entire system. Their passive sections are only apparently passive, which may be provable by different widths of the oceanic anomalies. The suggestions of Salvini et al. (1997) and Tikku and Cande (1999) may at least give some indication of that. On the other hand, however, data by Stock and Cande (2002), do not confirm it.
3. Transform tectonics and ordinary strike-slip tectonics are combined in the way that the rate of dextral shear along the entire fracture zone exceeds the rate of contemporaneous sinistral motion in the active transform portions. Also in this case, the passive sections are only apparently passive, but continue as dextral strike-slip faults into the Antarctic continent.

4. Initial sinistral transtension between Australia and Antarctica is reactivated by dextral strike-slip tectonics in the Cenozoic due to a major change in the regional palaeostress field and the reorientation of the main extension direction as suggested by Cande et al. (2000). The sinistral kinematics is preserved in the active transform faults in the Southern Ocean, whereas dextral kinematics dominates intracontinental shear in Victoria Land and Oates Land.

With the exception of possibility 1, all other scenarios require active tectonics along the Matusevich and Rennick structures, which is supported by recent earth-quakes, for instance in 1974, 1998, and 1999 (Rennick Glacier), and 1952 (Matusevich Glacier) (Reading 2002; Cattaneo et al. 2001).

Anyhow, the existence of the Matusevich fracture zone supports the hypothesis of Salvini and Storti (2003) that transform faults and frature zones off-shore between Australia and Antarctica continue as strike-slip faults on-shore in Victoria Land.

Acknowledgments

The authors thank the Bundesanstalt für Geowissenschaften und Rohstoffe (BGR), Hannover, for the invitation to GANOVEX VIII/ITALANTARTIDE XV; BGR, PNRA and AWI for their logistic support; D. Damaske, V. Damm, F. Henjes-Kunst, N. W. Roland, F. Rossetti and F. Talarico for fruitful discussions during the expedition and afterwards; and last but not least Deutsche Forschungsgemeinschaft, Bonn, for financial support (AZ Kl 429/18-1-3). Additionally *herzlichen Dank* and *mille gracie* to U. Schüßler and F. Ferraccioli for their very helpful reviews.

References

Armienti P, Baroni C (1999) Climatic change in Antarctica recorded by volcanic activity and landscape evolution. Geology 27:617–620

Behrendt JC, Schlee XX, Robb JM (1974) Geophysical evidence for the intersection of the St. Paul, Cape Palmas and Grand Cess fracture zones with the continental margin of Liberia, West Africa. Nature 248:324–326

Bozzo E, Ferraccioli F, Gambetta M, Caneva G, Spano M, Chiappini M, Damaske D (1999) Recent progress in magnetic anomaly mapping over Victoria Land (Antarctica) and the GITARA 5 survey. Antarctic Sci 11:209–216

Cande SC, Stock JM, Müller RD, Ishihara T (2000) Cenozoic motion between East and West Antarctica. Nature 404:145–150

Cattaneo M, Chiappini M, De Gori P (2001) Seismological experiment. Terra Antartica Rep 5:29–43

Damaske D, Bosum W (1993) Interpretation of aeromagnetic anomalies above the lower Rennick Glacier and the adjacent polar plateau west of the USARP Mountains. Geol Jb E47:139–152

Damaske D, Ferraccioli F, Bozzo E (2003) Aeromagnetic investigatons along the Antarctic coast between Mertz Glacier and Yule Bay. Terra Antartica 10:85–96

Damm V (2004) Ice thickness and bed rock map of Matusevich Glacier drainage system (Oates Coast). Terra Antartica 11:85–90

Dow JAS, Neall VE (1974) Geology of the lower Rennick Glacier, northern Victoria Land, Antarctica. New Zealand J Geol Geophys 17:659–714

Ferraccioli F, Damaske D, Bozzo E, Talarico F (2003) The Matusevich aeromagnetic anomaly over Oates Land, East Antarctica. Terra Antartica 10:221–228

Finn C, Moore D, Damaske D, Mackey T (1999) Aeromagnetic legacy of early Paleozoic subduction along the Pacific margin of Gondwana. Geology 27:1087–1090

Flöttmann T, Kleinschmidt G (1989) The structure of northwest Victoria and Oates Lands (Antarctica) and implications to the Gondwana accretion at the Pacific margin of the Antarctic Craton. In: 3rd Meeting Scienze della Terra in Antartide, Siena, pp 52–54

Flöttmann T, Kleinschmidt G (1991a) Opposite thrust systems in northern Victoria Land, Antarctica: Imprints of Gondwana's Paleozoic accretion. Geology 19:45–47

Flöttmann T, Kleinschmidt G (1991b) Kinematics of major structures of North Victoria and Oates Lands, Antarctica. Mem Soc Geol Ital 46:283–289

Flöttmann T, Kleinschmidt G (1993) The structure of Oates Land and implications for the structural style of northern Victoria Land, Antarctica. Geol Jb E47:419–436

Flöttmann T, Gibson GM, Kleinschmidt G (1993a) Structural continuity of the Ross and Delamerian orogens in Antarctica and Australia along the margin of the paleo-Pacific. Geology 21:319–322

Flöttmann T, Kleinschmidt G, Funk T (1993b) Thrust patterns of the Ross/Delamerian orogens in northern Victoria Land and southeastern Australia and their implications for Gondwana reconstructions. In: Findlay RH, Unrug R, Banks MR, Veevers JJ (eds) Gondwana eight. Balkema, Rotterdam, pp 131–322

Gair HS, Sturm A, Carryer SJ, Grindley GW (1969) Geology of northern Victoria Land. Antarctic Map Folio Ser Folio 12: Plate XII, New York

GANOVEX Team (1987) Geological map of north Victoria Land, Antarctica, 1:500 000 – Explanatory notes. Geol Jb B66:7–79

Hoeppener R (1955) Tektonik im Schiefergebirge. Geol Rdsch 44:26–58

Kleinschmidt G (1992) The southern continuation of the Wilson Thrust. Polarforschung 60:117–120

Kukowski N, Schillhorn T, Flueh ER, Huhn K (2000) Newly identified strike-slip plate boundary in the northeastern Arabian Sea. Geology 28:355–358

Kukowski N, Schillhorn T, Huhn K, Rad Uv, Husen S, Flueh ER (2001) Morphotectonics and mechanics of the central Makran accretionary wedge off Pakistan. Marine Geology 173:1–19

Läufer AL, Rossetti F (2003) Late-Ross ductile deformation features in the Wilson Terrane of northern Victoria Land (Antarctica) and their implications for the western front of the Ross Orogen. Terra Antartica 10:141–156

Lisker F (2002) Review of fission track studies in northern Victoria Land, Antarctica – passive margin evolution versus uplift of the Transantarctic Mountains. Tectonophysics 349:57–73

McLeod IR, Gregory CM (1967) Geological investigations along the Antarctic coast between longitudes 108° E and 166° E. Bureau Miner Resourc Geol Geophys Rep 78:1–49

Reading A (2002) Antarctic seismicity and neotectonics. In: Gamble JA, Skinner DNB, Henrys S (eds) Antarctica at the close of a millennium, Royal Soc New Zealand Bull 35:479–484

Roland NW, Tessensohn F (1987) Rennick faulting – an early phase of Ross Sea rifting. Geol Jb B66:203–229

Roland NW, Henjes-Kunst F, Kleinschmidt G, Talarico F (2000) Petrographical, geochemical, and radiometric investigations in northern Victoria Land, Oates Land and George V Coast: towards a better understanding of plate boundary processes in Antarctica. Terra Antartica Rep 5:57–65

Rossetti F, Storti F, Läufer AL (2002) Brittle architecture of the Lanterman Fault and its impact on the final terrane assembly in north Victoria Land, Antarctica. J Geol Soc London 159: 159–173

Rossetti F, Lisker F, Storti F, Läufer AL (2003a) Tectonic and denudational history of the Rennick Graben (North Victoria Land): Implications for the evolution of rifting between East and West Antarctica. Tectonics 22:101b,doi:10.1029/2002TCDO1416

Rossetti F, Di Vincenzo G, Läufer AL, Lisker F, Rocchi S, Storti F (2003b) Cenozoic right-lateral strike-slip faulting in North Victoria Land: an integrated structural, AFT and $^{40}Ar/^{39}Ar$ study. Terra Nostra 2003(4):283–284

Salvini S, Storti F (2003) Do transform faults propagate and terminate in East Antarctica continental lithosphere? Terra Nostra 2003(4):285

Salvini S, Brancolini G, Busetti M, Storti F, Mazzarini F, Coren F (1997) Cenozoic geodynamics of the Ross Sea region, Antarctica: crustal extension, intraplate strike-slip faulting and tectonic inheritance. J Geophys Res 102:24669–24696

Skinner DNB, Tessensohn F, Vetter U (1981) Lavas in the Ferrar Group of Litell Rocks, North Victoria Land, Antarctica. Geol Jb B41:251–259

Stock JM, Cande SC (2002) Tectonic history of Antarctic seafloor in the Australia-New Zealand-South Pacific sector: implications for Antarctic continental tectonics. In: Gamble JA, Skinner DNB, Henrys S (eds) Antarctica at the close of a millennium. Royal Soc New Zealand Bull 35:251–259

Sutherland R (1995) The Australia-Pacific boundary and Cenozoic plate motions in the SW Pacific: Some constraints from Geosat data. Tectonics 14:819–831

Tessensohn F (1994) Structural evolution of the northern end of the Transantarctic Mountains. In: Van der Wateren FM, Verbers ALLM, Tessensohn F (eds) LIRA workshop on landscape evolution. Rijks Geol Dienst, Haarlem, pp 57–61

Tikku AA, Cande SC (1999) The oldest magnetic anomalies in the Australian-Antarctic Basin: Are they isochrons? J Geophys Res 104:661–677

Veevers JJ (1987) Earth history of the southeast Indian Ocean and the conjugate margins of Australia and Antarctica. J Proc Royal Soc New South Wales 120:57–70

Tectonic Model for Development of the Byrd Glacier Discontinuity and Surrounding Regions of the Transantarctic Mountains during the Neoproterozoic – Early Paleozoic

Edmund Stump[1] · Brian Gootee[1] · Franco Talarico[2]

[1] Department of Geological Sciences, Arizona State University, Tempe, Arizona AZ 85287-1404, USA

[2] Dipartimento di Scienze della Terra, Università di Siena, Via del Laterino 8, 53100 Siena, Italy

Abstract. The Byrd Glacier discontinuity is a major tectonic boundary crossing the Ross Orogen, with crystalline rocks to the north and primarily sedimentary rocks to the south. Most models for the tectonic development of the Ross Orogen in the central Transantarctic Mountains consist of two-dimensional transects across the belt, but do not address the major longitudinal contrast at Byrd Glacier. This paper presents a tectonic model centering on the Byrd Glacier discontinuity. Rifting in the Neoproterozoic produced a crustal promontory in the craton margin to the north of Byrd Glacier. Oblique convergence of a terrane (Beardmore microcontinent) during the latest Neoproterozoic and Early Cambrian was accompanied by subduction along the craton margin of East Antarctica. New data presented herein in support of this hypothesis are U-Pb dates of 545.7 ±6.8 Ma and 531.0 ±7.5 Ma on plutonic rocks from the Britannia Range, directly north of Byrd Glacier. After docking of the terrane, subduction stepped out, and Byrd Group was deposited during the Atdabanian-Botomian across the inner margin of the terrane. Beginning in the upper Botomian, reactivation of the sutured boundaries of the terrane resulted in an outpouring of clastic sediment and folding and faulting of the Byrd Group.

Introduction

By the time of publication of the Antarctic Map Folio Series (Anonymous 1969–1970), most of the Transantarctic Mountains (TAM) had been geologically mapped, and a general model had emerged for the formation of the late Proterozoic – early Paleozoic Ross orogenic belt. Following roughly along the trend of the present day TAM (Fig. 4.3-1), a suite of sediments, and in places volcanics (Ross Supergroup), accumulated in a continental margin setting from late Proterozoic to Middle Cambrian time, followed by deformation and metamorphism during the Ross Orogeny, and intrusion of magmas of batholithic proportions, which are found throughout most segments of the TAM. This understanding led eventually to the general consensus that subduction was active outboard of the present day TAM in a Cambro-Ordovician timeframe.

The central Transantarctic Mountains (CTM) figured significantly in interpretations of the evolution of the Ross Orogen. The craton along which the sediments of the Ross Orogen were deposited is exposed in the Miller and Geologists Ranges as outcroppings of amphibolite-grade metamorphics (Nimrod Group) (Grindley 1972; Grindley et al. 1964). The initial passive margin deposits (Beardmore Group) were thought to have accumulated during the Neoproterozoic, throughout a broad region from north

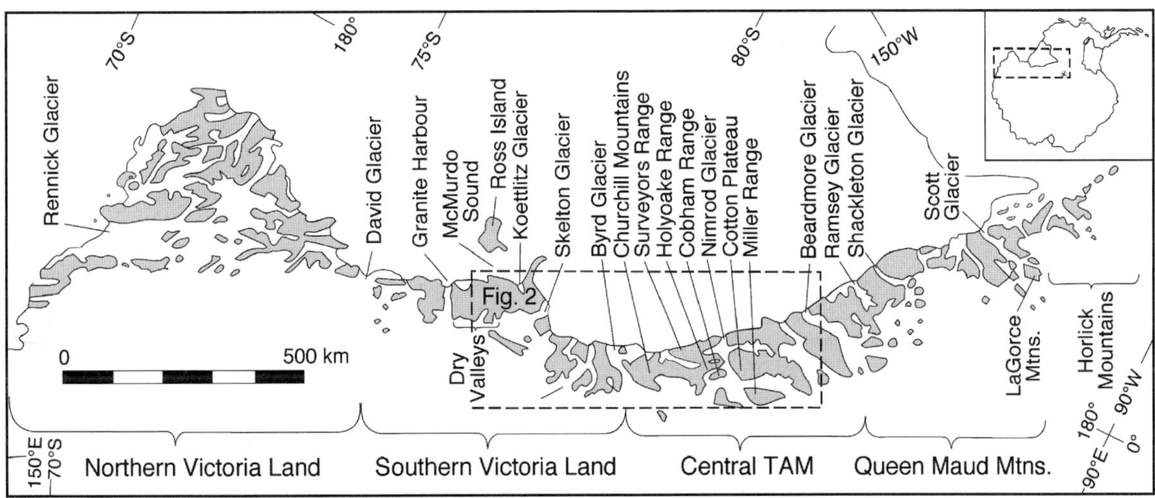

Fig. 4.3-1. Location map of Transantarctic Mountains

From: Fütterer DK, Damaske D, Kleinschmidt G, Miller H, Tessensohn F (eds) (2006) Antarctica: Contributions to global earth sciences. Springer-Verlag, Berlin Heidelberg New York, pp 181–190

of Nimrod Glacier to south of Beardmore Glacier (Gunn and Walcott 1962). Beardmore Group includes Cobham Formation, characterized by greenschist-grade schists, calc-schists, and marbles, conformably overlain by Goldie Formation, a sequence of metamorphosed metagraywacke and argillite (Laird et al. 1971). Folding of the Beardmore Group, recognized at the time by discordant contacts with overlying Lower Cambrian Shackleton Limestone, was named the Beardmore Orogeny (Grindley and McDougall 1969). The Cambrian Byrd Group, composed of Shackleton Limestone and overlying clastic units of the Starshot Formation, Dick Formation, and Douglas Conglomerate (Laird 1963; Skinner 1964), was deformed during the Ross Orogeny sometime during the Middle to Late Cambrian, and then intruded by the Cambro-Ordovician Granite Harbour Intrusives (Gunn and Warren 1962).

The ensuing decades have seen significant additions to and reinterpretations of the bedrock history of the CTM. Borg et al. (1990) presented an important model for the tectonic development of the region. Based on a study of Nd, Sr, and O isotopes in Granite Harbour Intrusives of the CTM, they recognized a major, lower-crustal province boundary beneath Marsh Glacier, with granitic rocks intruding Nimrod Group to the west having model ages (T_{DM}) of ~2.0 Ga and granitic rocks to the east having model ages of ~1.7 Ga. Borg et al. (1990) developed a tectonic model in which a terrane, which they called the "Beardmore microcontinent", converged obliquely on the craton, with Goldie Formation accumulating in the basin between. Suturing of the microcontinent in Neoproterozoic time was accompanied by deformation and erosion of the Goldie Formation, and subsequent deposition of Shackleton Limestone. Further subduction outboard of the Beardmore microcontinent led to volcanism, deformation, and intrusion associated with the Ross Orogeny.

The SWEAT hypothesis, that Laurentia had separated from Antarctica and Australia, provided the rifted cratonic margin along which the Ross Supergroup would be deposited (Moores 1991). It also provided Borg et al. (1990) an opposing margin from which their Beardmore microcontinent was produced.

"Nimrod Orogeny" was the name given to the episode of deformation and metamorphism that affected the Nimrod Group (Grindley and Laird 1969). Dating by Rb-Sr whole-rock analysis estimated this to have occurred during the Paleoproterozoic (Gunner and Faure 1972). Grindley (1972) had mapped a major structure (Endurance thrust) displacing units of the Nimrod Group. Goodge et al. (1991) found that the thrust is a distributed shear zone formed under amphibolite-facies conditions, which dips SW with top-to-the-SSE sense of shear. Dating a succession of intrusions both affected by and post-dating the shearing showed that deformation began before 541 Ma and had waned by 520 Ma (Goodge et al. 1993b). This put in doubt the age of the Nimrod Group, until a more recent study using U-Pb, single-grain zircon analysis showed that deep-crustal metamorphism and magmatism (Nimrod orogeny) occurred 1 730–1 720 Ma, following initial crustal magmagenesis at ca. 3 100–3 000 Ma (Goodge and Fanning 1999; Goodge et al. 2001). The data from the Endurance shear zone were instrumental in development of a model of oblique subduction along the Pacific margin of Antarctic during the Early Cambrian (Goodge et al. 1993a).

The extent of Beardmore Group outcrop, as originally mapped, was considerably reduced when it was found that the age of detrital zircons in a number of samples was as young as Middle(?) Cambrian (Goodge et al. 2002). These sedimentary rocks are now assigned to the Starshot Formation (Myrow et al. 2002a). However, detrital zircons in the suite from Cobham and Goldie formations at Cobham Range and Cotton Plateau are no younger than Grenville age (Goodge et al. 2002). Gabbros and pillow basalts interbedded with Goldie Formation at Cotton Plateau, possibly associated with rifting, have been U-Pb zircon dated at 668 ±1 Ma (Goodge et al. 2002). Thus, the original recognition of Neoproterozoic sedimentation remains, albeit in a restricted area adjacent to the craton margin.

An important detail in the deformational history of Byrd Group was added when Rowell et al. (1988) discovered that Douglas Conglomerate overlies folded Shackleton Limestone unconformably in the northern Holyoake Range. Then Myrow et al. (2002a) showed that the carbonate to clastic transition within the Byrd Group is conformable in the southern Holyoake Range. More recently, we have mapped similar conformable and unconformable relationships in the area to the south of Byrd Glacier (Stump et al. 2004).

Archaeocyathid and trilobite dating of the Shackleton Limestone has shown it to have been deposited during the Atdabanian, Botomian, and possibly Toyonian stages of the Early Cambrian (Hill 1964; Debrenne and Kruse 1986; Palmer and Rowell 1995). Recently, Myrow et al. (2002a) have also found Botomian trilobites in the lower portion of the Starshot Formation where it conformably overlies Shackleton Limestone, constraining the age of the onset of clastic deposition in the CTM.

Since the earliest days of interpreting the evolution of the Ross Orogen, resolving the timing of isotopically dated volcanic and plutonic events with fossil-dated sedimentary episodes has been a fundamental problem. In recent years, with refinements of the Cambrian interval of the Geological Timescale, and with precise U-Pb dates, a clearer picture has emerged (Tucker and McKerrow 1995; Landing et al. 1998; Davidek et al. 1998; Bowring and Erwin 1998). By the timescale of Bowring and Erwin

(1998) the Atdabanian began about 525 Ma and the Botomian/Toyonian ended by about 510 Ma, with the Middle-Upper Cambrian boundary at 500 Ma and the Cambrian-Ordovician boundary at 490 Ma.

The geological patterns that can be followed for several hundred kilometers along the CTM, abruptly end at Byrd Glacier, where folded and thrust-faulted limestones and conglomerates to the south face amphibolite-grade metamorphics and plutonics to the north. In fact, the geology throughout southern Victoria Land (SVL) is distinctly different in detail from the geology of the CTM. In the region from the Dry Valleys to Skelton Glacier, Grindley and Warren (1964) subdivided the metamorphic units into greenschist-grade Skelton Group and amphibolite-grade Koettlitz Group. Recently Cook and Craw (2001, 2002) have lumped all the metamorphic rocks as Skelton Group following the original designation of Gunn and Warren (1962), owing to structural complexity, which has precluded a comprehensive stratigraphic interpretation.

The fundamental, stratigraphic characteristics of the Byrd Group are initial deposition of a thick sequence of nearly pure carbonate sediments followed abruptly by a thick sequence of clastic sediments. This pattern is not seen in any of the occurrences of Skelton Group. In the Skelton Glacier area, the "Marble unit" of Cook and Craw (2002) (Anthill Limestone of Gunn and Warren (1962) and Skinner (1982)) contains centimeter to meter wide layers of quartzite and argillite. Near the base of the marble to the north of Cocks Glacier, polymict, matrix-supported metaconglomerate occurs. The marble overlies a thick succession of clastics, mainly quartzite (Rowell et al. 1993). Above the Marble unit is the "Cocks unit" of Cook and Craw (2002) (Cocks Formation of Skinner (1982)), a schistose metasedimentary unit containing thin pillow basalts, and with conglomeratic lenses near its base. The units of the Skelton Group in the Skelton Glacier area were deformed prior to intrusion of two plutons, each dated at 551 Ma (Rowell et al. 1993; Encarnación and Grunow 1996). This relationship itself precludes correlation of Skelton Group (in the Skelton Glacier area) with Byrd Group.

Although marbles occur at places to the north of Skelton Glacier, again they do not show the thickness and purity of the Shackleton Limestone. For example, the Salmon Marble of Gunn and Warren (1962) and Findlay et al. (1984) in the area between Koettlitz and Blue Glaciers contains thin layers of quartzite, biotite schist, and calc-schist. The earliest plutonism into Skelton Group at the head of Koettlitz Glacier was at 540–530 Ma (Hall et al. 1995; Mellish et al. 2002; Read et al. 2002), again earlier than deposition of Shackleton Limestone.

In the Britannia Range, directly north of Byrd Glacier, poorly studied metasedimentary rocks are assigned to the Horney Formation (Borg et al. 1989). Based on the report

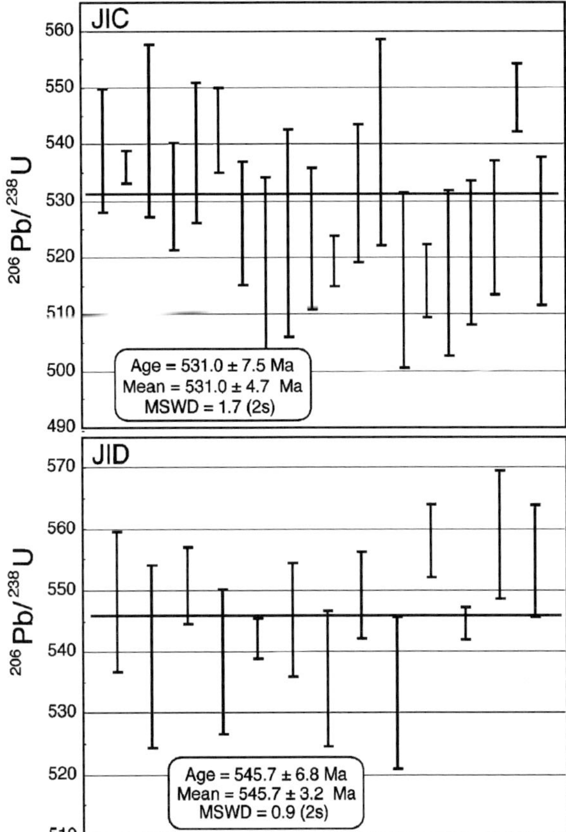

Fig. 4.3-2. Plots of $^{206}Pb/^{238}U$ ages from samples *JIC* and *JID* collected on north side of Byrd Glacier adjacent to Ramseier Glacier

of Borg et al. (1989) and our own reconnaissance observations, there is no marble within the Horney Formation, which is primarily amphibolite-grade micaceous schist and gneiss, with minor calc-silicate layers and pods. Again, these rocks are not a match for the Byrd Group. Until now the oldest reported plutonism in this area is from the Brown Hills, to the north of the Britannia Range, where the Carlyon Granodiorite has been dated at 515 ±8 Ma (Encarnación and Grunow 1996).

Herein we report two new U-Pb dates on plutonic rocks from the western end of the Britannia Range, collected on the buttress to the south of Ramseier Glacier (Fig. 4.3-2, Table 4.3-1). Analyses were made on single grains of zircon using a laser-ablation ICPMS. The older of these intrusions has an age of 545.7 ±6.8 Ma. It is a granite with quartz, plagioclase, microcline, and minor biotite, showing signs of strain and alteration. The quartz is in part polygonized and has undulose extinction. Traces of epidote are associated with the plagioclase, and some of the biotite is altered to chlorite. The younger intrusion, with an age of 531.0 ±7.5 Ma is a diorite with approximately equal proportions of pristine hornblende, biotite, and plagioclase.

Table 4.3-1. U-Pb (zircon) geochronologic analyses by Laser-Ablation Multicollector ICP Mass Spectrometery

Sample-grain	U (ppm)	$^{206}Pb/^{204}Pb$	U/Th	Isotopic ratios							Apparent ages (Ma)					
				$^{207}Pb^*/^{235}U$	± (%)	$^{206}Pb^*/^{238}U$	± (%)	Error corr.	$^{206}Pb/^{207}Pb$	± (%)	$^{206}Pb^*/^{238}U$	± (Ma)	$^{207}Pb^*/^{235}U$	± (Ma)	$^{206}Pb^*/^{207}Pb^*$	± (Ma)
JIC-1	184	3154	2.3	0.67521	4.3	0.08715	2.1	0.49	17.797	3.8	538.7	10.9	524	18	460	84
JIC-2	198	1904	4.0	0.73406	4.5	0.08665	0.6	0.13	16.276	4.5	535.7	2.9	559	19	655	96
JIC-3	210	4513	2.7	0.69199	8.3	0.08773	2.9	0.35	17.480	7.8	542.1	15.2	534	35	500	172
JIC-4	195	2535	2.6	0.69402	5.6	0.08578	1.9	0.34	17.041	5.2	530.5	9.5	535	23	555	114
JIC-5	737	10102	1.3	0.69329	2.9	0.08710	2.4	0.83	17.321	1.6	538.3	12.4	535	12	520	36
JIC-6	320	3799	2.7	0.70028	5.2	0.08775	1.4	0.28	17.278	5.0	542.2	7.5	539	22	525	109
JIC-7	272	2560	2.5	0.65669	6.1	0.08499	2.2	0.36	17.844	5.7	525.8	10.9	513	24	454	125
JIC-8	324	6319	2.1	0.69441	5.9	0.08368	3.2	0.54	16.614	5.0	518.0	16.0	535	25	610	107
JIC-9	1213	5517	1.2	0.66185	4.0	0.08470	3.6	0.92	17.646	1.5	524.1	18.3	516	16	479	34
JIC-10	328	5813	2.1	0.69790	3.6	0.08455	2.5	0.68	16.704	2.6	523.2	12.4	538	15	599	57
JIC-11	222	2451	3.6	0.66531	5.6	0.08387	0.9	0.16	17.381	5.5	519.2	4.4	518	23	512	121
JIC-12	313	7709	1.6	0.67800	3.2	0.08587	2.4	0.74	17.463	2.2	531.1	12.1	526	13	502	47
JIC-13	305	6514	1.6	0.72004	6.7	0.08737	3.5	0.52	16.731	5.7	540.0	18.2	551	29	595	124
JIC-14	267	5530	1.9	0.70525	4.7	0.08329	3.1	0.66	16.283	3.6	515.7	15.4	542	20	654	76
JIC-15	629	7315	1.7	0.66092	2.4	0.08328	1.3	0.55	17.374	2.0	515.7	6.5	515	10	513	43
JIC-16	316	3835	2.5	0.69336	4.2	0.08353	2.9	0.70	16.610	3.0	517.1	14.6	535	18	611	65
JIC-17	354	4226	1.9	0.64985	4.2	0.08412	2.5	0.60	17.847	3.4	520.7	12.7	508	17	454	75
JIC-18	315	4428	3.0	0.68450	8.8	0.08487	2.3	0.27	17.096	8.5	525.1	11.8	530	36	548	186
JIC-19	791	4158	3.3	0.68962	1.9	0.08872	1.2	0.63	17.738	1.5	548.0	6.1	533	8	467	32
JIC-20	630	6492	1.8	0.67676	3.0	0.08475	2.6	0.86	17.267	1.6	524.4	13.1	525	12	527	34
JID-1	1520	2853	6.5	0.71037	4.1	0.08871	2.2	0.52	17.218	3.5	547.9	11.4	545	18	533	77
JID-2	2545	5380	11.1	0.68846	3.5	0.08722	2.9	0.83	17.468	1.9	539.1	14.8	532	14	501	43
JID-3	2488	2613	6.7	0.69960	4.0	0.08915	1.2	0.29	17.570	3.8	550.5	6.2	539	17	488	84
JID-4	6229	3304	5.8	0.67808	3.0	0.08706	2.3	0.77	17.703	1.9	538.1	11.8	526	12	472	42
JID-5	8788	4530	16.0	0.68849	1.2	0.08772	0.6	0.54	17.567	1.0	542.0	3.3	532	5	489	22
JID-6	5884	6617	17.0	0.68582	1.8	0.08822	1.8	0.99	17.735	0.2	545.0	9.2	530	7	468	5
JID-7	4641	9724	19.3	0.67840	2.5	0.08660	2.2	0.85	17.601	1.4	535.4	11.0	526	10	484	30
JID-8	1875	7820	6.7	0.72404	1.7	0.08890	1.3	0.80	16.929	1.0	549.0	7.0	553	7	570	22
JID-9	2696	3563	8.6	0.67881	2.7	0.08622	2.4	0.92	17.513	1.1	533.1	12.4	526	11	495	23
JID-10	1882	1148	8.0	0.73474	11.3	0.09039	1.1	0.10	16.963	11.2	557.9	5.9	559	49	565	245
JID-11	3217	10907	10.3	0.69404	2.3	0.08813	0.5	0.22	17.507	2.3	544.4	2.7	535	10	496	50
JID-12	4242	11191	12.9	0.70864	2.2	0.09056	1.9	0.87	17.619	1.1	558.8	10.4	544	9	482	24
JID-13	4116	7793	18.9	0.70931	2.0	0.08982	1.7	0.84	17.460	1.1	554.5	9.1	544	9	502	24

All errors are reported at the 1-σ level and incorporate only uncertainties from measurement of isotopic ratios. U concentration and U/Th have uncertainty of ~25%. Decay constants: $^{235}U = 9.8485 \times 10^{-10}$, $^{238}U = 1.55125 \times 10^{-10}$, $^{238}U/^{235}U = 137.88$. Isotope ratios are corrected for Pb/U fractionation by comparison with standard zircon with an age of 564 ±4 Ma. Initial Pb composition interpreted from Stacey and Kramers (1975), with uncertainties of 1.0 for $^{206}Pb/^{204}Pb$ and 0.3 for $^{207}Pb/^{204}Pb$.

Tectonic Model

Models that have evolved for the tectonic development of the CTM generally have taken a view in cross section transverse to the orogen (e.g., Borg et al. 1990; Goodge 1997). Such a view is successful for interpretation of the broad region of the CTM; however, it does not take into account the abrupt discontinuity at Byrd Glacier and the geological differences in southern Victoria Land. Our current research is aimed at understanding the tectonic discontinuity followed by Byrd Glacier. Research to the south of Byrd Glacier was begun during the 2000/2001 field season, with results presented elsewhere (Gootee 2002; Stump et al. 2002b; Stump et al. 2004). The following model may serve as a working hypothesis at this stage of our research. It draws on prior research and models of the CTM, in particular Borg et al. (1990) and Goodge et al. (1993a), but goes beyond these in its focus on the Byrd Glacier discontinuity.

Although components of Borg et al.'s (1990) model of the Beardmore microcontinent have been shown to be in error, we feel that the general model of a terrane colliding with the Antarctic continental margin is viable, and so we retain their nomenclature. Their model was based on three sets of isotopic data: (1) the difference in Sm-Nd model ages of Granite Harbour Intrusives across Marsh Glacier (T_{DM} ~ 2.0 Ga in the Miller Range to the west and T_{DM} ~ 1.7 Ga to the east), (2) the presence of basalt and gabbro with oceanic affinity associated with Goldie Formation at Cotton Plateau, adjacent to Marsh Glacier, and (3) T_{DM} of ~1.7 Ga in "Goldie Formation" from outcrops to the east of Cotton Plateau. The first and second of these data sets must be included in any model of the region, the third set not. The T_{DM} data from "Goldie Formation" were emphasized by Borg et al. (1990) to suggest the allochthonous nature of the terrane, with no apparent source on the Antarctic craton to provide detritus of that model age. However, further isotopic investigations on the Granite Harbour Intrusives showed that a crustal province with Sm-Nd model ages in the range 1.6–1.9 Ma underlies most of the Transantarctic Mountains from Shackleton Glacier to northern Victoria Land (Borg and DePaolo 1991; Borg and DePaolo 1994). Furthermore, it has been shown that the "Goldie Formation" rocks on which the analyses were performed belong to the Starshot Formation, and were deposited from local sources along with the associated Douglas Conglomerate in the Early(-Middle) Cambrian (Goodge et al. 2002; Myrow et al. 2002a,b).

Our model encompasses the CTM and part of SVL from Beardmore Glacier to the Dry Valleys. The first panel of Fig. 4.3-3 shows the general geology of this portion of the TAM. The boundaries of the major units on this map are used as geographical reference in the various stages of the model.

Pre-550 Ma

We begin in the Neoproterozoic prior to 550 Ma. The craton had formed in the Archean and Paleoproterozoic. Rifting occurred in the Neoproterozoic, perhaps around 670 Ma (Goodge et al. 2002), with the conjugate continent drifting away. In the area of Byrd Glacier a major offset in the cratonic margin occurred during the rifting. This was first suggested by Stump (1992). Goodge and Dallmeyer (1996) added that this step-like geometry was possibly due to transform faulting in the vicinity of Byrd Glacier, but we postulate that separation occurred along both Byrd Glacier and the craton margin to the west, analogous to the opening of the Red Sea/Gulf of Aden. The Beardmore microcontinent separated during the rifting and lay somewhere to the southeast in present coordinates. It is possible, but not required, that the Beardmore microcontinent was the block that rifted from the reentrant at Byrd Glacier. The rocks immediately to the north of Byrd Glacier and throughout most of Victoria Land are of the same basement-age province as the Beardmore microcontinent (Borg and DePaolo 1994).

Deposition of Beardmore Group occurred in the region between the Beardmore microcontinent and mainland Antarctica. Geochemical and isotopic composition of the gabbro associated with Goldie Formation at Cotton Plateau indicates an oceanic rather than continental setting of magmatism (Borg et al. 1990). The question of how far removed from its present day location this terrane sat is speculative, but the ocean between the terrane and the continent may have been no wider than the Red Sea today. We position the terrane in the second panel of Fig. 4.3-3 in a fairly proximal location.

In SVL clastics, carbonates, and volcanics (Skelton Group) were being deposited during the Neoproterozoic (Skinner 1982; Cook and Craw 2002).

~550 Ma, Ross Orogeny Commences

At some time in the latter part of the Neoproterozoic, subduction commenced along this margin of Antarctica. Indications of convergence are folding of Skelton Group crosscut by plutons dated at 551 ±4 Ma (Rowell et al. 1993; Encarnación and Grunow 1996). An even earlier age for the onset of subduction is suggested by detrital zircons of magmatic origin in Starshot Formation dated at 580–560 Ma (Goodge et al. 2002). The Beardmore microcontinent converged on the continent from the SSE. We postulate that this convergence was oblique in the sector of the CTM (following the direction indicated in the Endurance shear zone (Goodge et al. 1993)), but more orthogonal in the area north of Byrd Glacier due to the geometry established during rifting. If the term Ross orogeny is to be applied to the full cycle of deformation and plutonism affecting the Ross Supergroup, then it began at approximately this time.

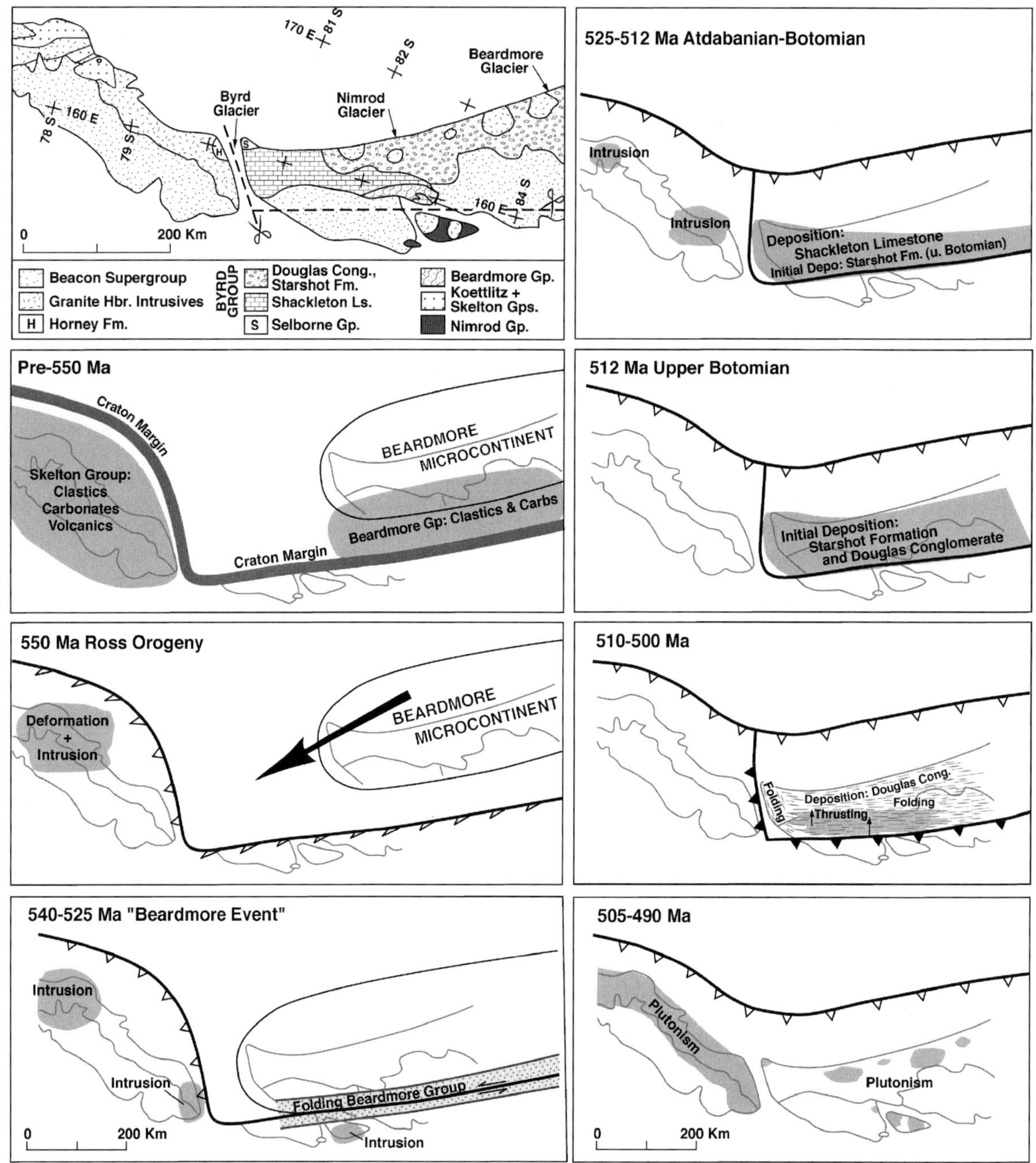

Fig. 4.3-3. Base geological map and tectonic model, southern Victoria Land and central Transantarctic Mountains. First panel is geological basemap. *Lines* between geological units are carried through the model in subsequent panels. The *dashed lines* in first panel represent sutured boundaries of the Beardmore Microcontinent, and the cuts that were made in developing the model. See text for discussion of individual panels of the model

540–525, "Beardmore Event"

By the beginning of the Cambrian the Beardmore microcontinent had begun to collide with the mainland, as indicated by plutonics in the Endurance shear zone dated at 540 Ma (Goodge et al. 1993b). How much earlier this shearing had commenced is unconstrained, but it could have occurred as early as 550 Ma, while plutons were intruding in SVL. Plutonism in the Britannia Range at 545–530 Ma indicates that subduction was continuing in the vicinity of Byrd Glacier during this interval. Beardmore Group was folded in a narrow zone adjacent to the craton, which includes Goldie Formation exposed at Cotton Plateau and

La Gorce Formation in the La Gorce Mountains, 600 km to the south (Stump et al. 2002a). Constraints on folding of La Gorce Formation are detrital zircons as young as 550 Ma (supplied from the magmatism in SVL?) (Vogel et al. 2001), and intrusion of porphyry (Wyatt Formation) dated at 525 ±3 Ma, which cross cuts the folds (Stump et al. 1986; Encarnación and Grunow 1996). A second episode of deformation in the La Gorce Mountains caused high-angle reverse faulting and widespread development of associated cleavage (Stump et al. 1986). At Cotton Plateau, folded Goldie Formation is overlain by basal Shackleton Limestone. Whether this contact is an unconformity or a fault remains controversial (Laird et al. 1971; Myrow et al. 2002a,b). Stump et al. (2002a) argued for two episodes of deformation at Cotton Plateau, the first a folding event recorded only in Goldie Formation, and the second causing folding and major offset of Goldie Formation and Shackleton Limestone along a high-angle shear zone. This is the scenario followed in our model. But even if the unconformity at Cotton Plateau is discounted, folding in the La Gorce Mountains occurred in the same timeframe as movement on the Endurance shear zone.

Identification of the Beardmore orogeny was based primarily on the deformation of Beardmore Group prior to deposition of Shackleton Limestone, but was generalized to encompass any orogenic activity in the Neoproterozoic (Grindley and McDougall 1969). Constraints from detrital zircons in La Gorce Formation now suggest that this folding episode was an Early Cambrian event (perhaps beginning within the last few million years of the Proterozoic) completed by 525 Ma, the beginning of the Atdabanian, which is the age of cross-cutting Wyatt Formation in the La Gorce Mountains and the oldest fossils in the Shackleton Limestone (Palmer and Rowell 1995).

In SVL a series of minor plutons span most of the time period 550–505 Ma when the main episode of Granite Harbour Intrusives began, suggesting subduction to the north of the Byrd Glacier discontinuity throughout that time (Rowell et al. 1993; Cooper et al. 1997; Cox et al. 2000; Allibone and Wysoczanski 2002; Read et al. 2002).

525–512 Ma, Atdabanian-Upper Botomian

We postulate that following docking of the Beardmore microcontinent, the subduction zone stepped out to the outboard side of the terrane. During the Atdabanian-Botomian there is ample evidence that subduction was occurring both south and north of the CTM. In the Queen Maud Mountains, on the outboard side of the docked terrane, volcanism of the Liv Group spanned the time 525–505 Ma (Encarnación and Grunow 1996; Van Schmus et al. 1997; Encarnación et al. 1999; Wareham et al. 2001). In the Darwin Glacier area, the Cooper granodiorite was intruded at 515 ±8 Ma (Simpson and Cooper 2002), and in the Dry Valleys area DV1b suite plutons were intruded at 531 ±10 Ma (Dun Pluton) and 516 ±10 Ma (Calkin Pluton) (Allibone and Wysoczanski 2002). It seems likely that the subduction indicated in the Queen Maud Mountains would have connected along the outboard side of the terrane to SVL.

By the beginning of the Atdabanian (525 Ma), erosion had occurred across the accreted, inboard side of the Beardmore microcontinent and deposition of an east-facing carbonate bank (Shackleton Limestone) had commenced (Rees et al. 1989; Gootee and Stump 2006 this vol.). More than 2 000 meters of carbonate deposition continued through most of the Botomian under quiescent conditions (Stump et al. 2004). That this carbonate basin existed without tectonic effects through this period may be owing to its location at the postulated reentrant to the previously rifted margin.

512 Ma

Before the end of the Botomian, clastic deposition conformably blanketed the carbonate platform, from eroding highlands in the region. In the Holyoake Range, Starshot Formation contains upper Botomian trilobites and is conformably interbedded with Douglas Conglomerate (Myrow et al. 2002a). In the area immediately to the south of Byrd Glacier, clastic and carbonate units interfinger conformably, and the interval is marked by pillow basalts (Stump et al. 2004). A tuff in Shackleton Limestone 50 meters beneath the first, overlying clastic unit has been U-Pb zircon dated as 512 ±5 Ma (Stump et al. 2004). Paleocurrent indicators signal a western source in the Holyoake and Surveyors Ranges (Myrow et al. 2002a,b). A source to the north of Byrd Glacier may also have been supplying sediment to the south at this time.

510–500 Ma

Folding and thrust faulting of Byrd Group began during the cycle of deposition of Starshot Formation and Douglas Conglomerate. Shackleton Limestone was eroded, providing a majority of the clasts for the Douglas, which is seen in unconformable contact with Shackleton at two localities (Rowell et al. 1988; Stump et al. 2004). Throughout the Holyoake Range and Churchill Mountains, the trends of folding and thrust faulting are primarily oriented N-S (Laird et al. 1971; Myrow et al. 2002a,b), whereas in the area immediately south of Byrd Glacier fold trends are ENE parallel to the glacier (Skinner 1964; Stump et al. 2004). We suggest that this pattern reflects a reactivation of the western and northern, sutured boundaries between the Beardmore microcontinent and the cratonic margin at the promontory along Byrd Glacier.

505–490 Ma

Although magmatism began by 550 Ma, as discussed above, it was volumetrically minor. Then between 505–490 Ma voluminous magmatism intruded large regions throughout the TAM (Borg 1983; Borg et al. 1987; Allibone et al. 1993a,b; Allibone and Wysoczanski 2002). The magmatism overlapped the latter stages of deformation and continued post-tectonically. To the south of Byrd Glacier a post-deformation pluton and dikes have been U-Pb dated at 492 ±2 Ma (Stump et al. 2002b). The magmatic flare-up was short-lived, ~15 million years, and was comparable to episodes recorded in the Sierra Nevada batholith of North America, postulated to have been caused by lithospheric-scale underthrusting of the magmatic arc (Ducea 2001). Late-stage dikes and minor plutons were intruded after the flare-up in southern Victoria Land for another 10–15 million years (Allibone and Wysoczanski 2002) and then Ross orogenic activity ended. Erosion cut deeply into the mountain belt and subsequent tectonic activity shifted outward along the continent.

Analytical Methods

Zircons from two samples were analyzed in the Radiogenic Isotope Geochemistry Laboratory of the Department of Geosciences, University of Arizona, with a Micromass Isoprobe multicollector ICPMS equipped with nine faraday collectors, an axial Daly detector, and four ion-counting channels. The Isoprobe is equipped with a New Wave DUV 193 laser ablation system with an emission wavelength of 193 nm. The analyses were conducted on 35–50 micron spots with an output energy of ~40 mJ and a repetition rate of 8 Hz. Each analysis consisted of one 20-s integration on the backgrounds (on peaks with no laser firing) and twelve one-second integrations on peaks with the laser firing. The depth of each ablation pit is ~12 microns. The collector configuration allows simultaneous measurement of ^{204}Pb in a secondary electron multiplier while ^{206}Pb, ^{207}Pb, ^{208}Pb, ^{232}Th, and ^{238}U are measured with Faraday detectors. All analyses were conducted in static mode.

Correction for common Pb was done by measuring ^{206}Pb/^{204}Pb, with the composition of common Pb from Stacey and Kramers (1975) and uncertainties of 1.0 for ^{206}Pb/^{204}Pb and 0.3 for ^{207}Pb/^{204}Pb. Fractionation of ^{206}Pb/^{238}U and ^{206}Pb/^{207}Pb during ablation was monitored by analyzing fragments of a large concordant zircon crystal that has a known (ID-TIMS) age of 564 ±4 Ma (2σ) (G. E. Gehrels, unpublished data). Typically this reference zircon was analyzed once for every four unknowns. The uncertainty arising from this calibration correction, combined with the uncertainty from decay constants and common Pb composition, contributed ~1% systematic error to the ^{206}Pb/^{238}U and ^{206}Pb/^{207}Pb ages (2-sigma level).

The reported ages are based on ^{206}Pb/^{238}U ratios because the errors of the ^{207}Pb/^{235}U and ^{206}Pb/^{207}Pb ratios are significantly greater (Table 4.3-1). This is due in large part to the low intensity (commonly ~1 mv) of the ^{207}Pb signal from these young grains. For each sample, the ^{206}Pb/^{238}U ages are shown on an age plot separate age plot (using plotting program of Ludwig 2001). The final age calculations are based on the weighted mean of the cluster of ^{206}Pb/^{238}U ages, with the error expressed both as the uncertainty of this mean and as the error of the age. The age error is based on the quadratic sum of the weighted mean error and the systematic error. Both are expressed at the 2σ level.

Acknowledgment

We thank Alex Pullen and George Gehrels for the U-Pb analyses. Funding sources include NSF grant OPP-9909463 (E. S.) and PNRA Research Project 4.13 (F. T.).

References

Allibone AH, Wysoczanski R (2002) Initiation of magmatism during the Cambrian-Ordovician Ross orogeny in southern Victoria Land, Antarctica. Geol Soc Amer Bull 114:1007–1018

Allibone AH, Cox SC, Graham IJ, Smillie RW, Johnstone RD, Ellery SG, Palmer K (1993a) Granitoids of the Dry Valleys area, southern Victoria Land, Antarctica: plutons, field relationships, and isotopic dating. New Zealand J Geol Geophys 36:281–297

Allibone AH, Cox, SC, Smillie RW (1993b) Granitoids of the Dry Valleys area, southern Victoria Land: geochemistry and evolution along the early Paleozoic Antarctic craton margin. New Zealand J Geol Geophys 36:299–316

Anonymous (1969–1970) Geologic Maps of Antarctica. American Geographical Society, New York

Borg SG (1983) Petrology and geochemistry of the Queen Maud Batholith, central Transantarctic Mountains, with implications for the Ross Orogeny. In: Oliver RL, James PR, Jago JB (eds) Antarctic earth science. Austral Acad Sci, Canberra, pp 165–169

Borg SG, DePaolo DJ (1991) A tectonic model of the Antarctic Gondwana margin with implications for southeastern Australia: isotopic and geochemical evidence. Tectonophysics 196:339–358

Borg SG, DePaolo DJ (1994) Laurentia, Australia, and Antarctica as a Late Proterozoic supercontinent: constraints from isotopic mapping. Geology 22:307–310

Borg SG, Stump E, Chappell BW, McCulloch MT, Wyborn D, Armstrong RL, Holloway JR (1987) Granitoids of northern Victoria Land, Antarctica: implications of chemical and isotopic variations to regional crustal structure and tectonics. Amer J Sci 287:127–169

Borg SG, DePaolo DJ, Wendlandt ED, Drake TG (1989) Studies of granites and metamorphic rocks, Byrd Glacier area. Antarct J US 24:19–21

Borg SG, DePaolo DJ, Smith BM (1990) Isotopic structure and tectonics of the central Transantarctic Mountains. J Geophys Res 95:6647–6667

Bowring SA, Erwin DH (1998) A new look at evolutionary rates in deep time: uniting paleontology and high-precision geochronology. GSA Today 8:1–8

Cook YA, Craw D (2001) Amalgamation of disparate crustal fragments in the Walcott Bay – Foster Glacier area, South Victoria Land, Antarctica. New Zealand J Geol Geophys 44:403–416

Cook YA, Craw D (2002) Neoproterozoic structural slices in the Ross orogen, Skelton Glacier area, south Victoria Land. New Zealand J Geol Geophys 45:133–143

Cooper AF, Worley BA, Armstrong RA, Price RC (1997) Synorogenic alkaline and carbonatitic magmatism in the Transantarctic Mountains of South Victoria Land, Antarctica. In: Ricci CA (ed) The Antarctic region: geological evolution and processes. Terra Antartica Publication, Siena, pp 245–252

Cox SC, Parkinson DL, Allibone AH, Cooper AF (2000) Isotopic character of Cambro-Ordovician plutonism, southern Victoria Land, Antarctica. New Zealand J Geol Geophys 434:501–520

Davidek K, Landing E, Bowring SA, Westrop SR, Rushton AWA, Fortey RA, Adrian JM (1998) New uppermost Cambrian U-Pb date from Avalonian Wales and age of the Cambrian-Ordovician boundary. Geol Mag 135:305–309

Debrenne F, Kruse PD (1986) Shackleton Limestone archaeocyaths. Alcheringa 10:235–278

Ducea M (2001) The California arc: thick granitic batholiths, eclogitic residues, lithospheric-scale thrusting, and magmatic flare-ups. GSA Today 11: 4–10

Encarnación JP, Grunow AM (1996) Changing magmatic and tectonic styles along the paleo-Pacific margin of Gondwana and the onset of early Paleozoic magmatism in Antarctica. Tectonics 15:1325–1341

Encarnación J, Rowell AJ, Grunow AM (1999) A U-Pb age for the Cambrian Taylor Formation, Antarctica: implications for the Cambrian time scale. J Geol 107:497–504

Findlay RH, Skinner DNB, Craw D (1984) Lithostratigraphy and structure of the Koettlitz Group, McMurdo Sound, Antarctica. New Zealand J Geol Geophys 27:513–536

Goodge JW (1997) Latest Neoproterozoic basin inversion of the Beardmore Group, central Transantarctic Mountains, Antarctica. Tectonics 16:682–701

Goodge JW, Dallmeyer RD (1996) Contrasting thermal evolution within the Ross Orogen, Antarctica Evidence from mineral 40Ar/39Ar Ages. J Geol 104:435–458

Goodge JW, Fanning CM (1999) 2.5 b.y. of punctuated Earth history as recorded in a single rock. Geology 27:1007–1010

Goodge JW, Borg SG, Smith BK, Bennett VC (1991) Tectonic significance of Proterozoic ductile shortening and translation along the Antarctic margin of Gondwana. Earth Planet Sci Letters 102:58–70

Goodge JW, Hansen VL, Peacock SM, Smith BK, Walker NW (1993a) Kinematic evolution of the Miller Range shear zone, central Transantarctic Mountains, Antarctica, and implications for Neoproterozoic to early Paleozoic tectonics of the East Antarctic margin of Gondwana. Tectonics 12:1460–1478

Goodge JW, Walker NW, Hansen VL (1993b) Neoproterozoic-Cambrian basement-involved orogenesis within the Antarctic margin of Gondwana. Geol 21:37–40

Goodge JW, Fanning CM, Bennett VC (2001) U-Pb evidence of ~1.7 Ga crustal tectonism during the Nimrod Orogeny in the Transantarctic Mountains, Antarctica: implications for Proterozoic plate reconstructions. Precambrian Res 112:261–288

Goodge JW, Myrow P, Williams IS, Bowring SA (2002) Age and provenance of the Beardmore Group, Antarctica: constraints on Rodinia supercontinent breakup. J Geol 110:393–406

Gootee B (2002) Geology of the Cambrian Byrd Group, Byrd Glacier area, Antarctica. MSc Thesis, Arizona State University, Tempe

Gootee B, Stump E (2006) Depositional environments of the Byrd Group, Byrd Glacier area: a Cambrian record of sedimentation, tectonism, and magmatism. In: Fütterer DK, Damaske D, Kleinschmidt G, Miller H, Tessensohn F (eds) Antarctica – Contributions to global earth sciences. Springer, Berlin Heidelberg New York, pp 191–194

Grindley GW (1972) Polyphase deformation of the Precambrian Nimrod Group, Central Transantarctic Mountains. In: Adie RJ (ed) Antarctic geology and geophysics. Universitetsforlaget, Oslo, pp 313–318

Grindley GW, Laird MG (1969) Geology of the Shackleton Coast. Antarctic Map Folio Series Folio 12, XIV

Grindley GW, McDougall I (1969) Age and correlation of the Nimrod group and other Precambrian rock units in the Central Transantarctic Mountains, Antarctica. New Zealand J Geol Geophys 12:391–411

Grindley GW, Warren G (1964) Stratigraphic nomenclature and correlation in the western Ross sea region. In: Adie RJ (ed) Antarctic Geology. North-Holland, Amsterdam, pp 314–333

Grindley GW, McGregor VR, Walcott RI (1964) Outline of the geology of the Nimrod-Beardmore-Axel Heiberg glaciers region, Ross Dependency. In: Adie RJ (ed) Antarctic Geology. North-Holland, Amsterdam, pp 206–219

Gunn BM, Walcott RI (1962) The geology of the Mt Markham region, Ross Dependency, Antarctica. New Zealand J Geol Geophys 5:407–426

Gunn BM, Warren G (1962) Geology of Victoria Land between the Mawson and Mulock glaciers, Antarctica. New Zealand Geol Surv Bull 7:1–157

Gunner JD, Faure G (1972) Rubidium-strontium geochronology of the Nimrod Group, central Transantarctic Mountains. In: Adie RJ (ed) Antarctic geology and geophysics. Universitetsforlaget, Oslo, pp 305–311

Hall CE, Cooper AF, Parkinson DL (1995) Early Cambrian carbonatite in Antarctica. J Geol Soc London 152:721–728

Hill D (1964) Archaeocyatha from the Shackleton limestone of the Ross system, Nimrod glacier area, Antarctica. Trans Royal Soc New Zealand Geol 2:137–146

Laird MG (1963) Geomorphology and stratigraphy of the Nimrod glacier-Beaumont bay region, southern Victoria Land, Antarctica. New Zealand J Geol Geophys 6:465–484

Laird MG, Mansergh GD, Chappell JMA (1971) Geology of the central Nimrod Glacier area, Antarctica. New Zealand J Geol Geophys 14:427–468

Landing E, Bowring SA, Davidek KL, Westrop SR, Geyer G, Heldmaier W (1998) Duration of the Early Cambrian: U-Pb ages of volcanic ashes from Avalon and Gondwana. Canadian J Earth Sci 35:329–338

Ludwig KJ (2001) Isoplot/Ex (rev. 2.49). Berkeley Geochronology Center Spec Publ 1a:1–56

Mellish SD, Cooper AF, Walker NW (2002) Panorama Pluton: a composite gabbro-monzodiorite early Ross Orogeny intrusion in southern Victoria Land, Antarctica. In: Gamble JA, Skinner DNB, Henrys S (eds) Antarctica at the close of a millennium. Royal Soc New Zealand Bull 35, Wellington, pp 129–141

Moores EM (1991) Southwest U.S.-East Antarctic (SWEAT) connection: a hypothesis. Geology 19:425–428

Myrow PM, Pope MC, Goodge JW, Fischer W, Palmer AR (2002a) Depositional history of pre-Devonian strata and timing of Ross orogenic tectonism in the central Transantarctic Mountains, Antarctica. Geol Soc Amer Bull 114:1070–1088

Myrow PM, Fischer W, Goodge JW (2002b) Wave-modified turbidites: Combined-flow shoreline and shelf deposits, Cambrian, Antarctica. J Sed Res 72:641–656

Palmer AR, Rowell AJ (1995) Early Cambrian trilobites from the Shackleton Limestone of the central Transantarctic Mountains. Paleontol Soc Mem 45:1–28

Read SE, Cooper AF, Walker NW (2002) Geochemistry and U-Pb geochronology of the Neoproterozoic-Cambrian Koettlitz Glacier alkaline province, Royal Society Range, Transantarctic Mountains, Antarctica. In: Gamble JA, Skinner DNB, Henrys S (eds) Antarctica at the close of a millennium. Royal Soc New Zealand Bull 35, Wellington, pp 143–151

Rees MN, Pratt BR, Rowell AJ (1989) Early Cambrian reefs, reef complexes, and associated lithofacies of the Shackleton Limestone, Transantarctic Mountains. Sedimentology 36:341–361

Rowell AJ, Rees MN, Cooper RA, Pratt BR (1988) Early Paleozoic history of the central Transantarctic Mountains: evidence from the Holyoake Range, Antarctica. New Zealand J Geol Geophys 31:397–404

Rowell AJ, Rees MN, Duebendorfer EM, Wallin ET, Van Schmus WR, Smith EI (1993) An active Neoproterozoic margin: evidence from the Skelton Glacier area, Transantarctic Mountains. J Geol Soc London 150:677–682

Simpson AL, Cooper AF (2002) Geochemistry of the Darwin Glacier region granitoids, southern Victoria Land. Antarctic Sci 14: 425–426

Skinner DNB (1964) A summary of the geology between Byrd and Starshot Glaciers, south Victoria Land. In: Adie RJ (ed) Antarctic geology. North-Holland, Amsterdam, pp 284–292

Skinner DNB (1982) Stratigraphy and structure of low-grade metasedimentary rocks of the Skelton Group, southern Victoria Land: does Teall Greywacke really exist? In: Craddock C (ed) Antarctic geoscience. University of Wisconsin Press, Madison, pp 555–563

Stacey JS, Kramers JD (1975) Approximation of terrestrial lead isotope evolution by a two-stage model. Earth Planet Sci Letters 26:207–221

Stump E (1992) The Ross Orogen of the Transantarctic Mountains in light of the Laurentia-Gondwana split. GSA Today 2:25–31

Stump E, Smit JH, Self S (1986) Timing of events during the late Proterozoic Beardmore Orogeny, Antarctica: geological evidence from the La Gorce Mountains. Geol Soc Amer Bull 97:953–965

Stump E, Edgerton DG, Korsch RJ (2002a) Geological relationships at Cotton Plateau, Nimrod Glacier area, bearing on the tectonic development of the Ross orogen, Transantarctic Mountains, Antarctica. Terra Antartica 9:3–18

Stump E, Foland KA, Van Schmus WR, Brand PK, Dewane TJ, Gootee BF, Talarico F (2002b) Geochronology of deformation, intrusion, and cooling during the Ross orogeny, Byrd Glacier area, Antarctica. Geol Soc Amer, Abstracts Progr 34:560–561

Stump E, Gootee BF, Talarico F, Van Schmus WR, Brand PK, Foland KA, Fanning CM (2004) Correlation of Byrd and Selborne Groups, with implications for the Byrd Glacier discontinuity, central Transantarctic Mountains. New Zealand J Geol Geophys 47:157–171

Tucker RD, McKerrow WS (1995) Early Paleozoic chronology: a review in light of new U-Pb zircon ages from Newfoundland and Britain. Canadian J Earth Sci 32:368–379

Van Schmus WR, McKenna LW, Gonzales DA, Fetter AH, Rowell AJ (1997) U-Pb geochronology of parts of the Pensacola, Thiel, and Queen Maud Mountains, Antarctica. In: Ricci CA (ed) The Antarctic region: geological evolution and processes. Terra Antartica Publication, Siena, pp 187–200

Vogel MB, Wooden JL, Stump E, McWilliams MO (2001) Detrital zircon provenance of Neoproterozoic to Cambrian successions: resolving continental configurations. Gondwana Res 4:809

Wareham CD, Stump E, Storey BC, Millar IL, Riley TR (2001) Petrogenesis of the Cambrian Liv Group, a bimodal volcanic rock suite from the Ross orogen, Transantarctic Mountains. Geol Soc Amer Bull 113:360–372

Depositional Environments of the Byrd Group, Byrd Glacier Area: A Cambrian Record of Sedimentation, Tectonism, and Magmatism

Brian Gootee · Edmund Stump

Department of Geological Sciences, Arizona State University, Tempe, AZ 85287-1404, USA

Abstract. The geology of the Byrd Group immediately south of Byrd Glacier records a major sequence of geologic events, beginning with the development of a carbonate platform (Shackleton Limestone) during the early Atdabanian (approximately 525 Ma), followed by a transitional interval of siliciclastic deposition and volcanism (Starshot Formation) during the late Botomian (approximately 512 Ma), and ending with a coarse cover of clastic molasse deposition (Douglas Conglomerate), no younger than plutonism at 492 ±2 Ma. Thus, the Byrd Group was deposited during a span of less than 33 myr, as a sequence of passive shelf-margin sedimentation through active uplift and erosion related to the Ross Orogeny. The newly subdivided Shackleton Limestone records at least two major depositional cycles, interrupted by a significant karst event. A layer of volcanic ash overlain by a thick layer of argillite in the uppermost Shackleton Limestone records the timing of a conformable carbonate to siliciclastic transition, accompanied by basalt volcanism of the Starshot Formation. Continued clastic deposition of the Starshot Formation coarsens upwards into the Douglas Conglomerate, where the primary sources of clastic detritus are derived from the Shackleton Limestone and possibly much of the Starshot Formation itself.

Introduction

In the central Transantarctic Mountains immediately south of Byrd Glacier (Fig. 4.4-1), the Byrd Group provides a nearly continuous rock sequence from which to study a major transition from Early Cambrian passive carbonate platform sedimentation to uplift and siliciclastic progradation. The results of our field program (NSF Grant no. OPP-9909463) were successful in (1) determining a stratigraphy within the Shackleton Limestone, (2) identifying lithofacies within the Shackleton Limestone, (3) correlating the stratigraphy of the Byrd Group between outcrops, (4) determining the depositional relationship and timing between the Shackleton Limestone, Starshot Formation, and Douglas Conglomerate, and (5) determining the provenance of the Douglas Conglomerate.

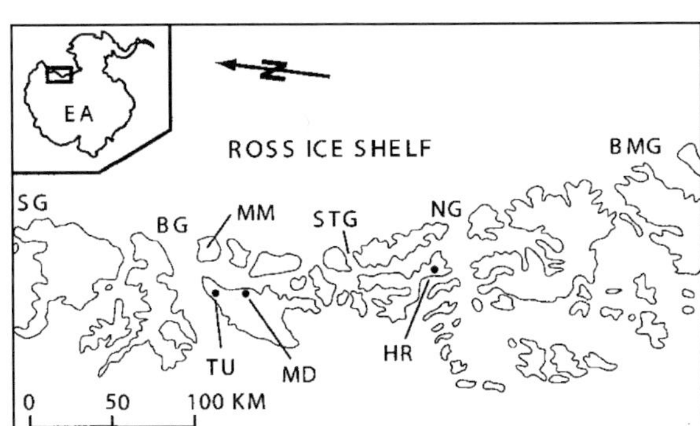

Fig. 4.4-1. Overview and location map of the central Transantarctic Mountains. *EA:* East Antarctica; *SG:* Skelton Glacier; *BG:* Byrd Glacier; *STG:* Starshot Glacier; *NG:* Nimrod Glacier; *BMG:* Beardmore Glacier; *MM:* Mount Madison; *TU:* Mount Tuatara; *MD:* Mount Dick; *HR:* Holyoake Range

Shackleton Limestone

In the course of our mapping, eigth stratigraphic sections of the Shackleton Limestone were measured, with the most complete in excess of 2 200 m. Based on lithologic characteristics recognizable in the field, we have for the first time subdivided the Shackleton Limestone into four informal members (oldest to youngest) (Gootee 2002; Stump et al. 2004): (1) the "Cross-bedded member", (2) the "Cherty member", (3) the "Butterscotch member", and (4) the "Upper member" (Fig. 4.4-2).

Cross-Bedded Member

The Cross-bedded member is best characterized as consisting of predominantly intraclastic, cross-laminated, lime

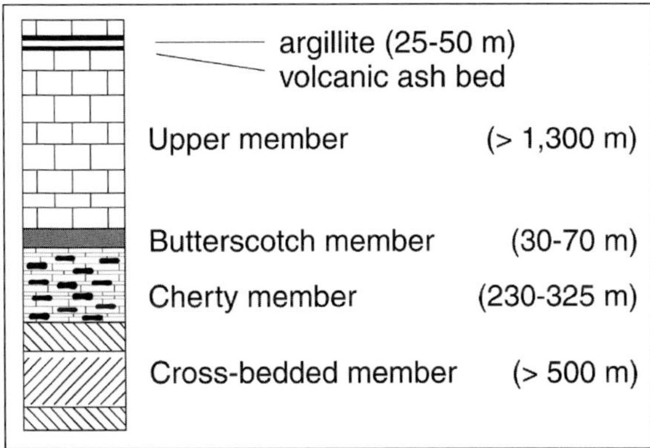

Fig. 4.4-2. Composite stratigraphic section of the Shackleton Limestone, Byrd Glacier area

From: Fütterer DK, Damaske D, Kleinschmidt G, Miller H, Tessensohn F (eds) (2006) Antarctica: Contributions to global earth sciences. Springer-Verlag, Berlin Heidelberg New York, pp 191–194

mudstone to fine-grained limestone and dolostone, with lesser limestone breccia, intraformational flat-pebble conglomerate, and ooid grainstone. Tabular sets of planar crossbeds have cm- to 2-m-bedding. Cross-laminae are enhanced by orange-weathered quartz silicification, which is the facies' most distinguishing characteristic in the field. Thin interbeds of shale and lime mudstone characterize the lowermost section. The limestone-lithoclast breccia is the only marker lithology that appears to correlate between outcrops. The contact between the Cross-bedded and Cherty members appears to be gradational at the horizon where chert begins and cross-bedding ends, up section.

Cherty Member

The Cherty member is characterized by the presence of bedded and nodular chert interspersed with limestone and dolostone. The lower Cherty member consists of intraclastic wackestone, packstone, and shaley limestone. The upper interval consists of alternating light grey and black limestone beds, including bedded and nodular chert, which correlate over several tens of kilometres within the Cherty member. Burrow-mottled lime mudstone, oolitic limestone, rare archaeocyathid bioclasts, and boundstone are also found in the uppermost part, apparently conformable with the overlying Butterscotch member.

Butterscotch Member

The Butterscotch member is a distinctive, 30 to 70 m thick unit in the middle of the Shackleton Limestone that weathers with a distinctive "butterscotch" colour, and provides an excellent marker interval. Throughout its lateral and vertical extent, the member consists of cm-scale, alternating calcareous green argillite and orange-pink limestone. The carbonate layers have a recrystallized lime micrite texture and calcite composition, bioclasts, and trace quartz silt grains. A fenestral texture is common throughout; it is often bladelike in form and filled with coarse, twinned calcite spar and occasionally with geopetal cement. Dark green argillite layers are relatively featureless and commonly sheared along bedding planes.

Upper Member

The Upper member is most extensively exposed at Mt. Tuatara, where it is in excess of 1 200 m. This member is characterized by archaeocyathid bioclasts, algal mats, pisolites, oolites and oncolites, which occur prominently above the Butterscotch member and intermittently throughout the section. In the lower Upper member, a marker interval consists of karst-like features, such as breccia pipes perpendicular to bedding, collapsed cavern roofs, and terra-rossa-stained limestone breccia. Towards the top of the section at Mt. Tuatara are two distinct marker units (Fig. 4.4-2): (1) a yellow-tan volcanic tuff, and (2) a structurally variable, 25 to 50 m thick, dark green-grey, calcareous argillite unit, similar in thickness and stratigraphic position to the argillite bed reported by Burgess and Lammerick (1979) at Mt. Hamilton and the newly designated Holyoake Formation in the Holyoake Range (Myrow et al. 2002a) (Fig. 4.4-1).

Starshot Formation

Skinner (1964) recognized two clastic formations in the area to the south of Byrd Glacier, Douglas Conglomerate, a coarse, polymict conglomerate, and Dick Formation, consisting of sandstone and shale. In the area of Starshot Glacier, Laird (1963) designated similar clastic rocks as the Starshot Formation. Recently, Myrow et al. (2002a) proposed that all clastic rocks of the Byrd Group between the Nimrod and Byrd Glaciers be subdivided into conglomeratic Douglas Conglomerate and finer-grained Starshot Formation. This is the nomenclature that we follow. We have subdivided the Starshot Formation into two units, the Starshot sandstone and Starshot basalts.

Starshot Sandstone

Siliciclastic quartz- and feldspar-arenites, with bedding thicknesses of five to several tens of centimetres, are characteristic of the Starshot sandstone. Both have immature textures consisting of subrounded quartz, feldspar, and limestone, and other lithic grains from distal and proximal sources. Beds containing flute casts and graded calcareous sandstone grade upwards into pebble-polymict conglomerate and coarse sandstone characteristic of the Douglas Conglomerate.

Starshot Basalts

The Starshot basalts consist of pillow basalts and basalt flows, and are inferred to be conformable with the Shackleton Limestone and Starshot sandstone. At Rees et al.'s (1987) locality R17, massive pillow basalts are interbedded with sparse normal-graded dark green sandstone beds. The pillow basalts are bounded sharply up section by black limestone and sandstone, apparently conformably; however, the contact is concealed by a snow patch and fractured rock (Stump et al. 2004). At an unnamed nunatak immediately north of Mt. Dick (spot height of 1 710±), multiple vesicular basalt flows and coarse-grained marbles are interbedded.

Douglas Conglomerate

The Douglas Conglomerate consists of massive bedded, clast-supported, rounded, polymict pebbles and boulders, several hundreds of meters thick (Myrow et al. 2002b). In the area of Mt. Dick, many of the clasts are quartz, but abundant clasts of limestone occur locally near the base of the section, along with individual basalt pillow clasts. Less abundant lithoclasts of second-generation polymict-conglomerate were also observed. The second-generation lithoclasts are composed of siltstone, limestone, fine sandstone, volcanic tuff and metabasite, identical to those lithologies observed in the underlying Starshot Formation and upper Shackleton Limestone. The discovery of an angular unconformity between folded Shackleton Limestone and Douglas Conglomerate, the unconformity itself tilted, is a reinterpretation of Rees et al.'s (1987) thrust fault at Mt. Hamilton (Stump et al. 2004).

Depositional Model of the Byrd Group

The Shackleton Limestone contains first- to possibly second-order sequence stratigraphy, indicative of at least two major cycles. The first cycle grades from platform or shelf-margin deposits at its base (Cross-bedded member) to tidal and supratidal flat deposits at the top (Butterscotch member) (Fig. 4.4-3a). This cycle may represent a long-term offlap sequence caused by aggradation of the carbonate shelf (progradation) and/or a eustatic recession during the early Cambrian, early Atdabanian stage (Algeo and Seslavinsky 1995).

A second depositional cycle is recognized in the upper half of the Shackleton Limestone, a succession, approximately 1 500 m thick, of mainly shallow-water carbonate shelf deposits (Upper member), interpreted to record sedimentation in response to the Cambrian Sauk transgression (Sloss et al. 1960) (Fig. 4.4-3c). However, the lower part of the Upper member records a significant emergence of the carbonate shelf, producing an interval of karst development over tens of kilometres. This event was most likely caused by local tectonism and repeated uplift of the carbonate shelf, and represents a significant disconformity in the Upper member (Fig. 4.4-3b). The existence of an approximately 50 m thick argillite unit at Mt. Tuatara (Fig. 4.4-2) is thought to represent the first significant pulse of detritus in response to crustal contraction and uplift, related to the initiation of the Ross Orogeny in this area (Fig. 4.4-3d). A volcanic ash tuff below the argillite bed provides the first isotopic age of this event at 511.9 ±3.2 Ma (Stump et al. 2004), and an approximate upper bounding depositional age of the Shackleton Limestone. The argillite bed, being the first pulse of clastic material in the area, is at the same stratigraphic position as the Holyoake Formation in the Holyoake Range (Myrow et al. 2002a), which has been fossil and carbon-isotope chemostratigraphically dated to late Botomian (approximately 510–515 Ma) (Myrow et al. 2002a), consistent with our stratigraphic and geochronologic records. These new age data constrain the span of Shackleton deposition to approximately 13 myr, that is, 525 Ma (basal Atdabanian) to 512 Ma (late Botomian).

Basalt flows and pillows of the Starshot Formation that accompany the carbonate to siliciclastic transition in the area south of Byrd Glacier, are coeval with limestone deposition, in an active magmatic and tectonic terrain. The clastic detritus for the Starshot sandstone was likely deposited from proximal Shackleton Limestone and distal basement through fluvial and nearshore marine processes (Fig. 4.4-3e).

The Douglas Conglomerate represents a clastic molasse in response to tectonic uplift. The folded Shackleton Limestone and Starshot Formation were clearly clastic sources for the Douglas Conglomerate (Fig. 4.4-3f). The discovery of an angular unconformity between folded Shackleton Limestone and the Douglas Conglomerate suggests the erosion and removal of both the upper Shackleton Limestone and the Starshot Formation; this is also indicated by the second-generation lithoclasts in the Douglas Conglomerate in close proximity to this unconformity. The extreme thickness of the Douglas Conglomerate suggests a tectonically active source area with frequent input of distal and proximal clastic material into a subsiding basin, or a continually uplifting source area, or both. The abundance of the metabasite lithoclasts at Mt. Dick suggests an extensive source area for this unique lithosome. Additionally, at the Mt. Madison massif, northeast of the study area, metabasites interbedded with marble occur at the transition between extensive marble and extensive schist and meta-conglomerate, that are interpreted as correlatives of the Byrd Group. These lithologies were deformed by an underlying, undeformed granite pluton, with a U/Pb isotopic crystallization age of 492 ±2 Ma (Stump et al. 2002), thus providing an upper bounding depositional and deformational age constraint for the Byrd Group.

References

Algeo TJ, Seslavinsky KB (1995) The Paleozoic world: continental flooding, hypsometry, and sealevel. Amer J Sci 295:787–822

Burgess CJ, Lammerick W (1979) Geology of the Shackleton Limestone (Cambrian) in the Byrd Glacier area. NZ Antarct Rec 2(1):12–16

Gootee BF (2002) Geology of the Cambrian Byrd Group, Byrd Glacier Area, Antarctica. MSc Thesis, Arizona State University, Tempe

Laird MG (1963) Geomorphology and stratigraphy of the Nimrod Glacier, Beaumont Bay region, southern Victoria Land, Antarctica. NZ J Geol Geophys 6:465–484

Myrow PM, Pope MC, Goodge JW, Fischer W, Palmer AR (2002a) Depositional history of pre-Devonian strata and timing of Ross orogenic tectonism in the central Transantarctic Mountains, Antarctica. Geol Soc Amer Bull 114(9):1070–1088

Myrow PM, Fischer W, Goodge JW (2002b) Wave-modified turbidites: combined-flow shoreline and shelf deposits, Cambrian, Antarctica. J Sed Res 72(5):641–656

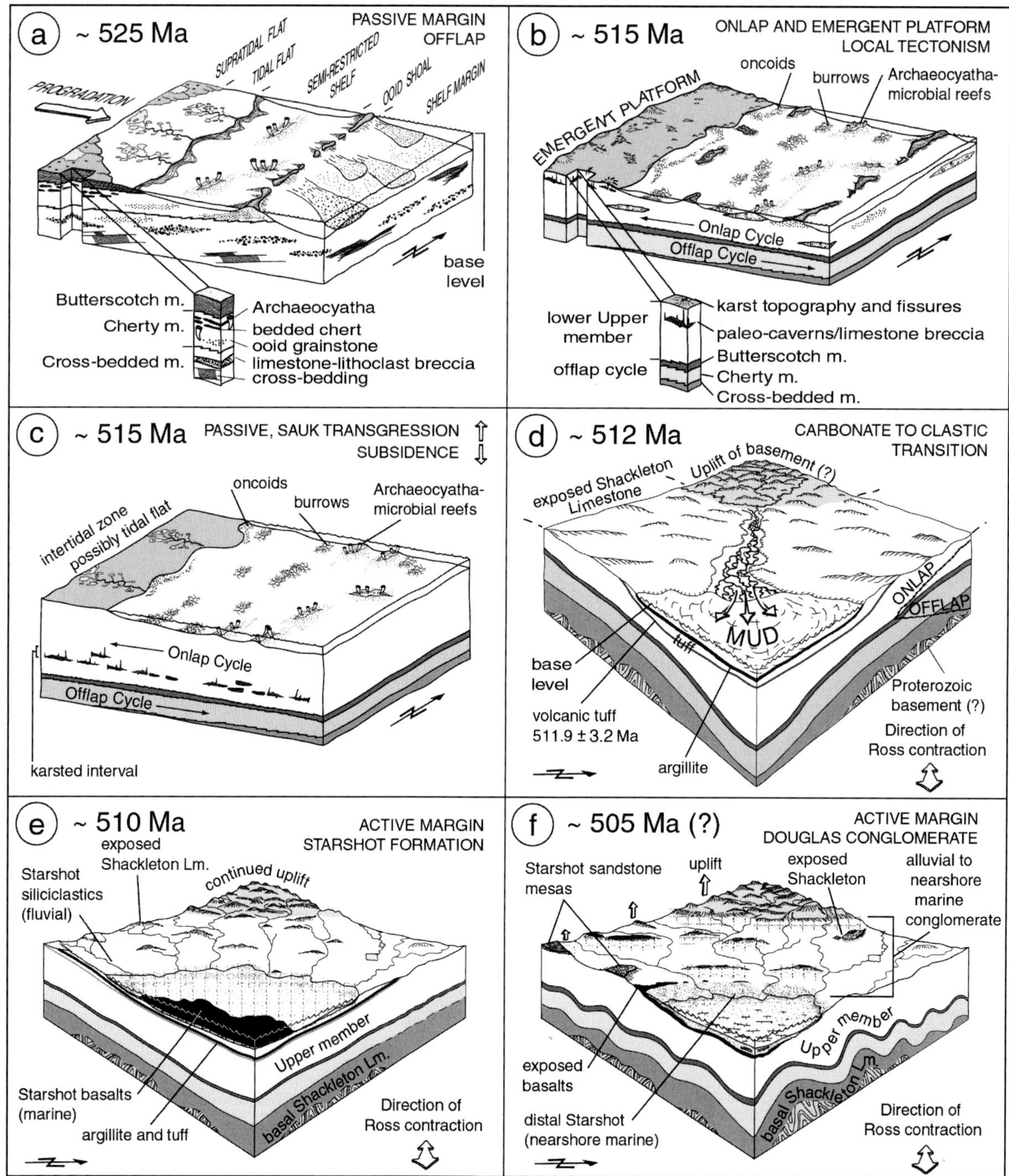

Fig. 4.4-3. Block diagrams illustrating the sequence of geologic events (**a** through **f**) and depositional environments of the Byrd Group (not to scale). Approximate *north arrow* indicated. Depositional and tectonic setting in upper right for each sequence

Rees MN, Girty GH, Panttaja SK, Braddock P (1987) Multiple phases of early Paleozoic deformation in the central Transantarctic Mountains. Antarctic J US 22:33–35

Skinner DNB (1964) A summary of the geology of the region between Byrd and Starshot glaciers, south Victoria Land. In: Adie RJ (ed) Antarctic geology. North-Holland, Amsterdam, pp 284–292

Sloss LL, Dapples EC, Krumbein WC (1960) Lithofacies maps: an atlas of the United States and southern Canada. Wiley, New York

Stump E, Foland KA, Van Schmus WR, Brand PK, Dewane TJ, Gootee BF, Talarico F (2002) Geochronology of deformation, intrusion, and cooling during the Ross Orogeny, Byrd Glacier Area, Antarctica. Abstracts GSA Annual Meeting Program

Stump E, Gootee BF, Talarico F, Van Schmus WR, Brand PK, Foland KA, Fanning CM (2004) Correlation of Byrd and Selborne Groups, with implications for the Byrd Glacier discontinuity, central Transantarctic Mountains, Antarctica. NZ J Geol Geophys 47:157–171

Late-Ross Structures in the Wilson Terrane in the Rennick Glacier Area (Northern Victoria Land, Antarctica)

Andreas L. Läufer[3] · Georg Kleinschmidt[1] · Frederico Rossetti[2]

[1] Johann Wolfgang Goethe Universität, Geologisch-Paläontologisches Institut, Frankfurt a. M., Germany
[2] Dipartimento di Scienze Geologiche, Università Roma Tre, Largo S. L. Murialdo 1, 00146 Roma, Italy
[3] Bundesanstalt für Geowissenschaften und Rohstoffe, Stilleweg 2, 30655 Hannover, Germany

Abstract. Kinematic data from the Rennick Glacier area indicate the presence of two intra-Wilson Terrane late-Ross opposite-directed high-strain reverse shear systems. High-grade rocks are W- and E-ward displaced over low-grade rocks and shallow-level intrusions. The shear zones are offset in a step-like pattern suggesting the presence of ENE trending right-lateral faults. The structural pattern accounts for a relationship between the Exiles and Wilson thrusts in Oates Land, which in our opinion can be traced from the Pacific coast to the Ross Sea. The western front of the Ross Orogen towards the East Antarctic Craton is best interpreted as a broad W-vergent fold-and-thrust belt, along which the intra-Wilson Terrane arc was detached and thrust onto the craton. The shear zones related to the Exiles Thrust system represent the internal, easternmost thrusts of this belt. Based on our data in combination with recent geophysical and geochronological results, the craton-orogen boundary must be located significantly further W than previously inferred. The boundary and hence the inferred termination of the proposed fold-and-trust belt may roughly lie in the area between Mertz and Ninnis Glaciers in George V Land, taking into account a considerable amount of likely post-Ross crustal extension possibly related to the Wilkes Subglacial Basin.

Introduction

The geology of northern Victoria Land (NVL) is characterised by three lithotectonic units, from W to E: the Wilson, Bowers, and Robertson Bay terranes (e.g., GANOVEX Team 1987; Kleinschmidt and Tessensohn 1987). They formed during W-directed subduction associated with magmatic growth and accretion at the palaeo-Pacific margin of Gondwana during the early Palaeozoic Ross Orogeny (Fig. 4.5-1). The Wilson Terrane consists of polyphase metamorphic rocks and Cambro-Ordovician late- to postkinematic Granite Harbour Intrusives. It is unconformably covered by the non-metamorphic Permo-Triassic clastic Beacon Supergroup and Jurassic volcanic Ferrar Supergroup. The Wilson Terrane features high-grade metamorphic and migmatitic rocks that occur close to low-grade metasedimentary units. Detrital zircon dating indicates that the protolith age of the latter is post-Precambrian (Ireland et al. 1999). These neighbouring but contrasting units could be interpreted as *(i)* the low-grade rocks representing the former sedimentary cover of cratonic basement in a pre-Ross passive margin environment (Fanning et al. 1999), or *(ii)* the high- and low-grade units representing different crustal levels of the Ross mobile belt brought together by major crustal shear zones (Flöttmann and Kleinschmidt 1991). A general problem of the early Palaeozoic geology of NVL is the exact location and the character of the western boundary of the Ross Orogen towards the East Antarctic Craton (EAC). It could involve *(i)* W-directed thrusts with or without a molasse basin, *(ii)* a former back-arc basin, or *(iii)* a continuous transition with gradually decreasing Ross-age deformation. In order to gain new hints on the nature and location of the cratonward front of the Ross Orogen, we performed kinematic analyses in ductilely deformed rocks of the Wilson Terrane in the Rennick Glacier area. The data were collected (from E to W) in the Lanterman Range, Morozumi Range, Helliwell Hills, and southern Daniels Range-Emlen Peaks-Outback Nunataks area.

Structural Geology

The polymetamorphic basement of the Lanterman Range (Lanterman metamorphics) shows at least three deformation phases, which is in accordance to the results of Roland et al. (1984), Gibson (1984, 1987), and Capponi et al. (1999, 2002). Strongly isoclinal first-generation folds F_n associated with a first foliation S_n parallel to the axial planes are indicative of a first deformation event D_n. These oldest structures are only locally preserved and are largely overprinted by the main pervasive foliation S_{n+1} associated with tight to slightly isoclinal folds F_{n+1} and a prominent mineral lineation. Both deformation events took place under high-grade metamorphic conditions. The axes of both fold generations F_n and F_{n+1} strike roughly parallel to the mineral lineation. S_{n+1} is refolded by gentle third-generation folds F_{n+2} associated with a weak foliation S_{n+2}. These folds are probably co-genetic to a retrograde deformation event that affected synkinematic, pervasively foliated diorites and tonalites of the Granite Harbour Intrusives. Foliation S_{n+2} is formed by biotite, and quartz grains are stretched defining a co-genetic NE-SW to N-S oriented lineation. Recrystallisation of quartz grains and only weak crystal-plastic deformation of feldspar grains indicate that this late-stage defor-

From: Fütterer DK, Damaske D, Kleinschmidt G, Miller H, Tessensohn F (eds) (2006) Antarctica: Contributions to global earth sciences. Springer-Verlag, Berlin Heidelberg New York, pp 195–204

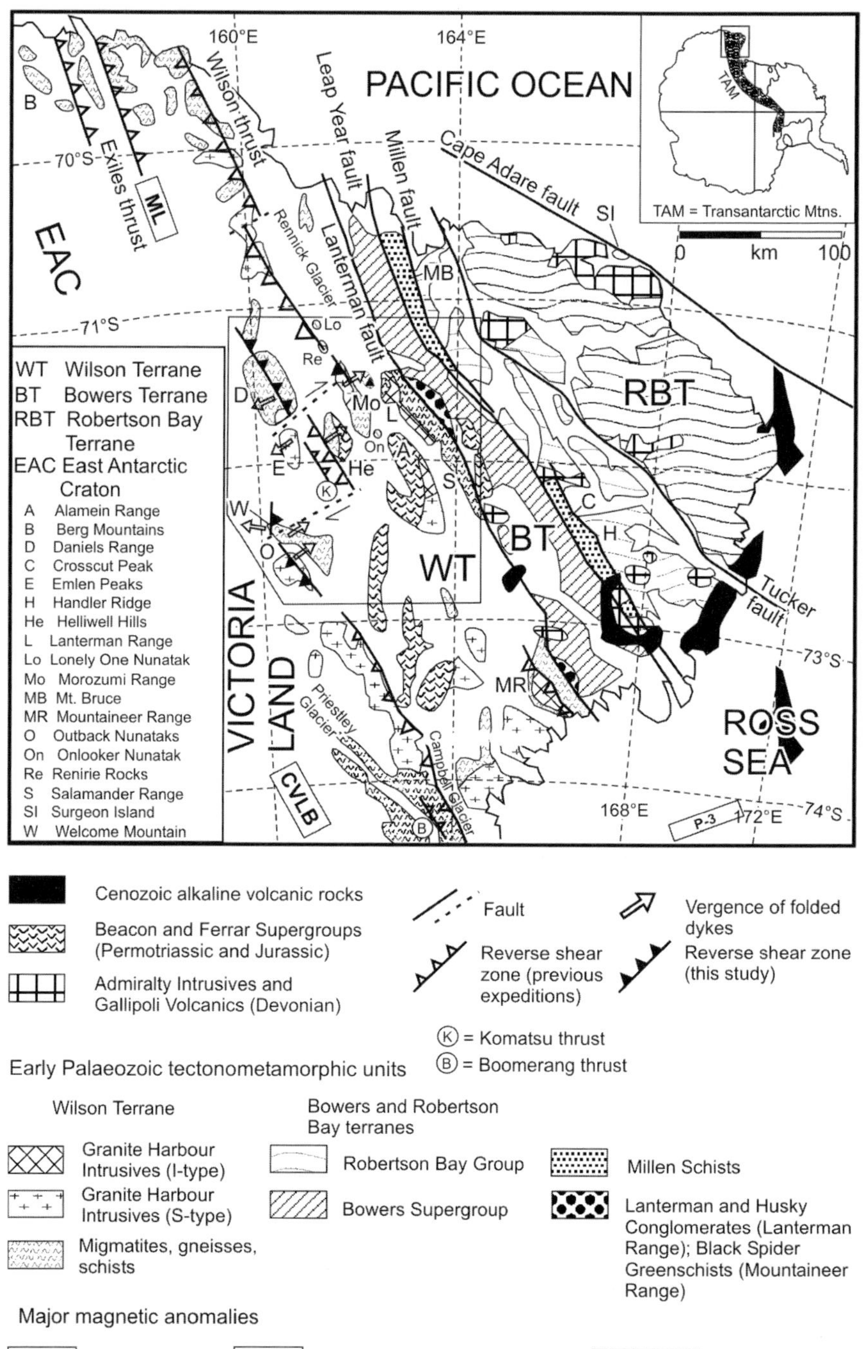

Fig. 4.5-1.
Geological map of Victoria Land with major tectonic and magnetic features. Modified from Ganovex Team (1987), Flöttmann and Kleinschmidt (1991), Kleinschmidt (1992). Magnetic anomalies from Ferraccioli and Bozzo (1999). The study area is shown

mation occurred under low- to medium-grade conditions ($T \sim 400$–500 °C). According to Roland et al. (1984), the intrusion of these plutonic rocks must have occurred post-F_{n+1}, because xenoliths of Lanterman metamorphics already contain the D_{n+1}-structures.

Kinematic analyses of the polymetamorphic rocks of the Lanterman Range and the low-grade overprinted syn-D_{n+2} diorites/tonalites reveal top-to-the E to NE directed sense of shear synkinematic to high-grade peak and subsequent low- to medium-retrograde metamorphism (Fig. 4.5-2). Some data sets (e.g., lan1, lan5) reveal a more or less prominent strike-slip component, which could suggest an oblique collisional setting during the Ross Orogeny (Goodge and Dallmeyer 1996; Capponi et al. 1999, 2002; Musumeci and Pertusati 2001). Younger and non- or only very weakly deformed pegmatites and non-foliated granitoids of the Granite Harbour Intrusives crosscut the whole sequence, but show no high-grade ductile

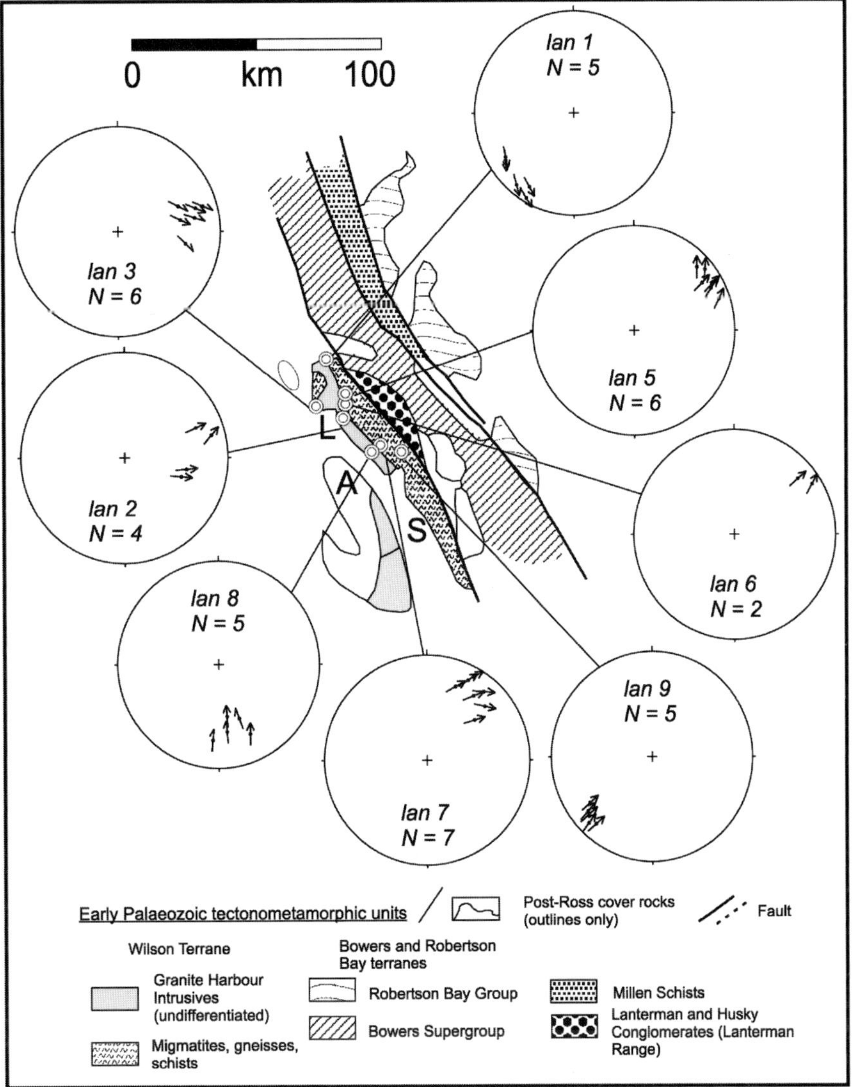

Fig. 4.5-2.
Ductile kinematic data from the Lanterman Range. The data are shown as lower hemisphere projections (foliation planes as *poles*, shear senses as *arrows*). Legend as in Fig. 4.5-1

overprint as in the areas further to the west (i.e. northeastern Morozumi Range, Daniels Range, Outback Nunataks). They are overprinted only by Cenozoic brittle tectonics genetically linked to the Lanterman Fault system (e.g., Rossetti et al. 2002), which causes block tilting and the locally NE dipping foliation and apparent "normal" shear sense of the Lanterman metamorphics (e.g., lan 8, lan 9).

In contrast to the Lanterman Range, the usually non- to weakly metamorphic Granite Harbour Intrusives in the northern Morozumi Range contain higher-shear strain localisation. The magmatic fabric is overprinted by solid-state S-L fabrics involving quartz and subordinate feldspar recrystallisation ($T \sim 400$–$500\,°\mathrm{C}$). This ductile deformation also affected thick quartz veins in the granites, which show strong undulose extinction, grain boundary migration, and recrystallisation. Stretched and recrystallised quartz crystals in the veins are oriented parallel to the stretching lineation in the granites. Up to some tens of metres thick, these shear zones cannot be traced into the central and southern part of the range where the granites are without any pervasive foliation or ductile overprint. Shear senses are reverse and top-to-E to NE directed (Fig. 4.5-3) which is in line with fold vergences of dykes at the north-eastern flank of the Morozumi Range.

A short survey was performed at Komatsu Nunatak W of Helliwell Hills (Fig. 4.5-1 and 4.5-4) where a small ductile reverse shear zone, several tens of metres thick, is characterised by large chunks of metasedimentary rocks within strongly sheared Granite Harbour Intrusives (Kleinschmidt 1992). Shear sense is top-to-W.

Structural data on ductile deformation were furthermore collected in the westernmost nunataks in front of the inland ice sheet. These were the areas between Swanson and Edwards Glaciers in the southern Daniels Range, the Emlen Peaks, and the Outback Nunataks. In

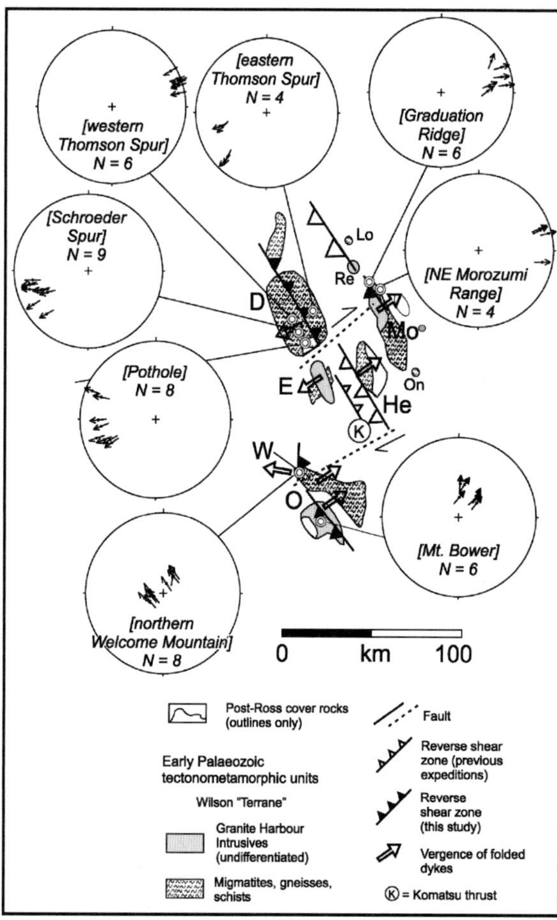

Fig. 4.5-3. Ductile kinematic data from the northern Morozumi Range, southern Daniels Range, and Outback Nunataks. The data are shown as lower hemisphere projections (foliation planes as *poles*, shear senses as *arrows*). Legend as in Fig. 4.5-1

the southern Daniels Range, high-grade, partly migmatitic gneisses in the E (Wilson Gneisses) generally rest tectonically over low-grade metasedimentary rocks (Rennick Schists) and shallow-level Granite Harbour Intrusives in the West. Foliation attitudes are NW-SE to NNW-SSE trending and mostly steeply dipping. Shear senses are consistently top-to-W to NW (Fig. 4.5-3, 4.5-4). At eastern Thomson Spur, in particular, high-grade migmatitic gneisses and equally foliated in-situ granitic melts show asymmetric shear fabrics with top-to-W/SW directed movement. In contrast, low-grade metasedimentary rocks at western Thomson Spur contain a steeply dipping (80–85°) spaced foliation S_2, which overprints an older foliation S_1 associated with a first folding event F_1 indicated by relict tightly to isoclinally folded microlithons in S_2. Similar to the high-grade units further E, these low-grade rocks show shear band fabrics that indicate E-over-W directed displacement co-genetic to S_2. At Schroeder Spur, garnet-bearing leucogranitic pegmatites intruded into cordierite-bearing amphibole-biotite schistose gneisses show ductile deformation recorded by steeply dipping biotite-rich mylonitic foliation planes oriented parallel to the foliation of the country rocks (Fig. 4.5-4). Shear sense indicators in both dykes and schistose gneisses yield top-to-W directed kinematics. Sillimanite in the strongly localised shear zones indicates high-grade deformation temperatures over 500 °C (Holdaway 1971). We observed two types of differently deformed types of Granite Harbour pegmatites: older folded dykes with W-directed vergence of the folds and younger unfolded ones. At "Pothole" at the southern flank of Schroeder Spur, we found an impressive large-scale ductile shear zone reaching a thickness of several tens to some few 100 metres (Fig. 4.5-4). Several with 60–70° steeply E-dipping shear planes with down-dip lineations and top-to-W directed shear sense cut through coarse-grained granites of the Granite Harbour Intrusives. Synkinematically grown sillimanite needles again indicates high deformation temperatures.

The Emlen Peaks are located S of the Daniels Range and consist of medium-grade metamorphics with a NW-SE trending foliation, which are intruded by several granitic rocks (Capponi et al. 1994). During our short reconnaissance, we observed that thick pegmatites and the metamorphic rocks are folded into closed W-vergent folds suggesting W-directed tectonic transport (Fig. 4.5-5), which is in line with our findings from the Daniels Range.

At the northern tip of Welcome Mountain (northernmost Outback Nunataks), metapelitic to metaarenaceous gneissose schists are intruded by several thick Granite Harbour pegmatites. Plagioclase grains in the schists show abundant signs of thermometamorphic overprint (e.g., poikiloblastic textures with inclusions of biotite, quartz,

Fig. 4.5-4. Field examples and thin sections from Schroeder Spur and "Pothole" in the southern Daniels Range and from Komatsu Nunatak W of Helliwell Hills. **a, b** Foliated Granite Harbour leucogranitic dyke with top-to-W shear sense indicated by S-C alignments shown in photo **b. c** Major ductile thrust zone in Granite Harbour Intrusives at "Pothole" with E-over-W displacement. **d** Interpretative field book sketch of the "Pothole" section of photo **c.** Strongly foliated Granite Harbour Intrusives are cross-cut by late-Ross high-strain reverse shear zones. A shallowly dipping brittle thrust displaces the ductile shear zones. Several NNW-SSE striking brittle normal faults reactivate the main foliation and mafik dykes in a step-like pattern towards the Rennick Glacier in the E. Total length of section is approximately 700 m. **e** S-C structure in foliated leucogranitic dyke of photo **a**, W-directed kinematics, crossed polarized light. *B:* biotite; *gt:* garnet; *m:* muscovite. **f** Sillimanite needles in stretched quartz grains. Same sample as photo **e. g** Hornblende-biotite schist. Biotite forms the main foliation of the rock. Asymmetric structures revealed by hornblende and biotite indicate W-directed sense of shear. Same location as photo **e.** **h** Ductile shear zone (Komatsu Thrust) in Granite Harbour rocks at Komatsu Nunatak (for location see Fig. 4.5-1). Shear sense is E-over-W (Kleinschmidt 1992). Flat-lying basalts of te post-Ross (Jurassic) Ferrar Supergroup are visible in the background (Mesa Range). *CPL:* Crossed polarised light; *PPL:* plane polarised light

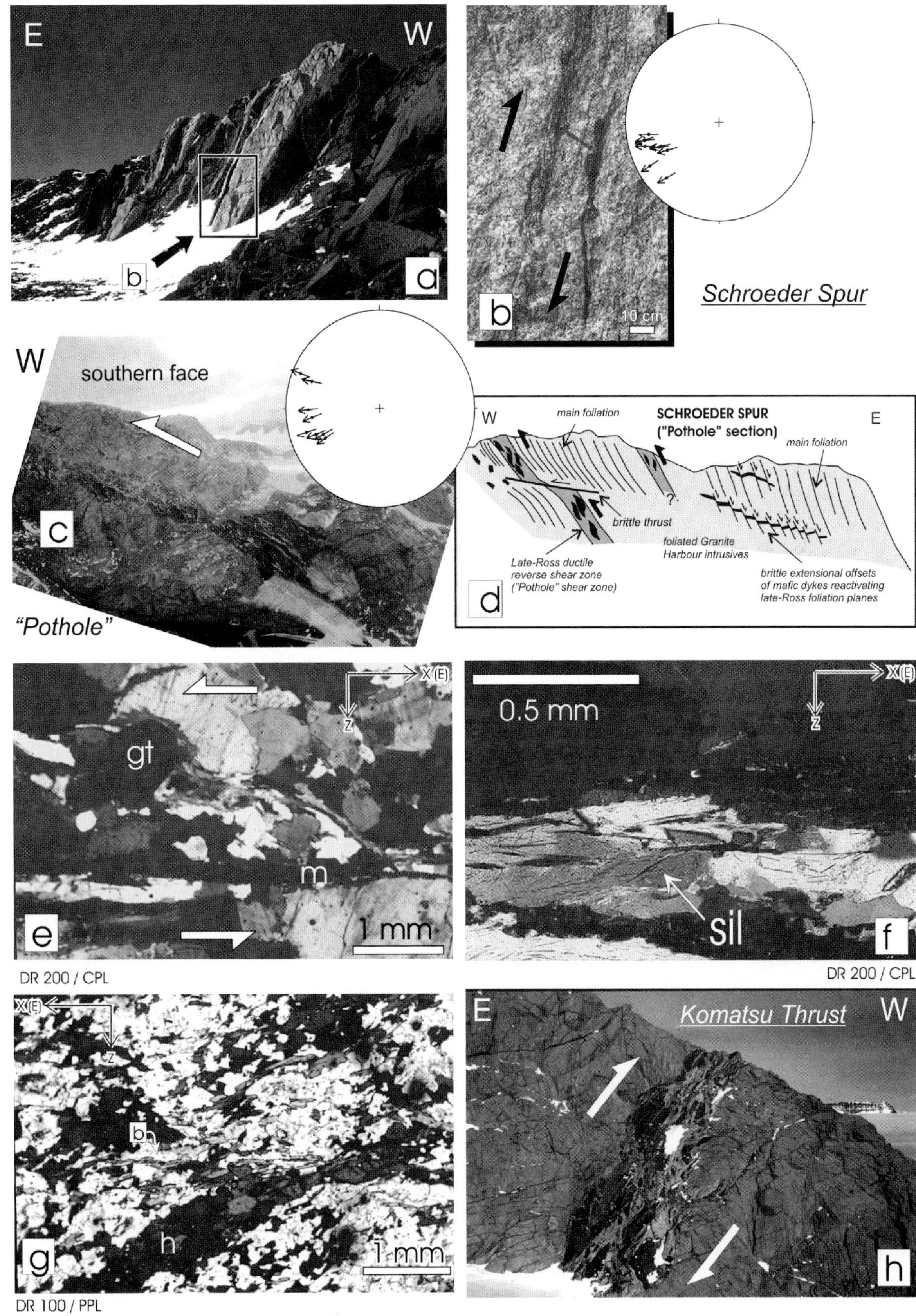
E
W
b
a
b
10 cm
Schroeder Spur
W
southern face
c
"Pothole"
W
main foliation
SCHROEDER SPUR
("Pothole" section)
E
main foliation
brittle thrust
foliated Granite
Harbour intrusives
Late-Ross ductile
reverse shear zone
("Pothole" shear zone)
brittle extensional offsets
of mafic dykes reactivating
late-Ross foliation planes
d
X(E)
Z
gt
m
1 mm
e
DR 200 / CPL
0.5 mm
X(E)
Z
sil
f
DR 200 / CPL
X(E)
Z
b
h
1 mm
g
DR 100 / PPL
E
Komatsu Thrust
W
h

Fig. 4.5-5. Field examples of W-vergent folds of pegmatites in a metasedimentary sequence from **a** Emlen Peaks (*arrow!*) and **b** northern Welcome Mountain

and apatite). Sillimanite needles within a newly formed foliation outlined by biotite flakes indicate temperatures over 500 °C. The foliation is associated with co-genetic crystal-plastic deformation of both quartz and feldspar grains. Top-to-NW tectonic transport, locally containing a certain strike-slip component, is indicated by different shear sense criteria. Again, we can distinguish two pegmatite generations based on their deformation pattern and the different attitude of tourmaline crystals. The older generation of dykes is folded into W-vergent closed folds (Fig. 4.5-5) and contain short and thick tourmalines with well developed crystal faces suggesting sufficient time for crystal growth during magma cooling. The younger dykes are unfolded and contain long and needle-like tourmalines without well-developed crystal faces, which suggests a relatively short time of growth during dyke formation.

Contrasting, E-directed kinematics was found at southern Welcome Mountain and in the central and southern Outback Nunataks (Fig. 4.5-3 and 4.5-6). Roland et al. (1989) report thick pegmatite dykes intruded into the metasedimentary country rocks deformed into SW-plunging and apparently SE-vergent folds at the eastern flank of Welcome Mountain, which contrasts with the opposite-directed shear senses and fold geometries further north. At Mt. Joern, south of Welcome Mountain, E to NE vergent folding affected both metasedimentary rocks and pegmatites. At Mt. Bower still further south, the usually non- to little deformed Granite Harbour Intrusives show strongly localised, narrow high-strain mylonite zones, shallowly dipping to the W (Fig. 4.5-3 and 4.5-6). Shear structures indicate reverse top-to-NE directed displacement. The shear zones show synkinematic growth of garnet-biotite associations. Fibrolite is present as synkinematic needles wrapped around asymmetrically stretched quartz ribbons and as postkinematic needles overgrowing recrystallised quartz (Fig. 4.5-6). These mineral associations and structural patterns are lacking outside the mylonites where the granites are pervasively solid-state foliated as indicated by mica layering. Minimum temperatures of 500 °C within the stability field of sillimanite are indicated by the presence of both syn- and post-kinematically grown sillimanite needles (cf. Holdaway 1971).

Discussion and Conclusions

Top-to-E/NE directed tectonic transport during peak high-grade and subsequent upper low- to medium-retrograde metamorphism in the Lanterman Range is interpreted as relating to early Palaeozoic W-directed subduction at the palaeo-Pacific margin of Gondwana (e.g., Kleinschmidt and Tessensohn 1987; Capponi et al. 1999). Since related structures are crosscut by late- to postkinematic Granite Harbour Intrusives lacking any indication of late-stage, localised high-grade shearing as in the areas further W, the structures are of syn-Ross age. In contrast, kinematic data from the areas further W indicate the presence of two late-Ross opposite-directed reverse shear systems, which reveal a close relationship to the Exiles and Wilson thrust systems (Flöttmann and Kleinschmidt 1991). A central medium- to high-grade metamorphic belt is reversely displaced to the W and E over low-grade metamorphic and shallow-level intrusive rocks. Since the high-strain shear zones overprint the late- to postkinematic Granite Harbour Intrusives, but co-genetic folding has not affected younger late-phase dykes, ductile shearing was contemporaneous with plutonism and a late-Ross age is attested. The data show that the Wilson Thrust can be traced from the known outcrops at Renirie Rocks (Kleinschmidt 1992) into the north-eastern Morozumi Range but not further south. Its southern continuation is located further west on the western flank of Helliwell Hills (Kleinschmidt 1992) and then still further west in the Outback Nunataks, suggesting the presence of right-lateral offsets along ENE trending faults of unknown (late- or post-Ross) age. Similar offsets were suggested for the Exiles-Wilson thrust systems further north (e.g., Kleinschmidt 1992). They indicate a step-like pattern of the whole thrust sys-

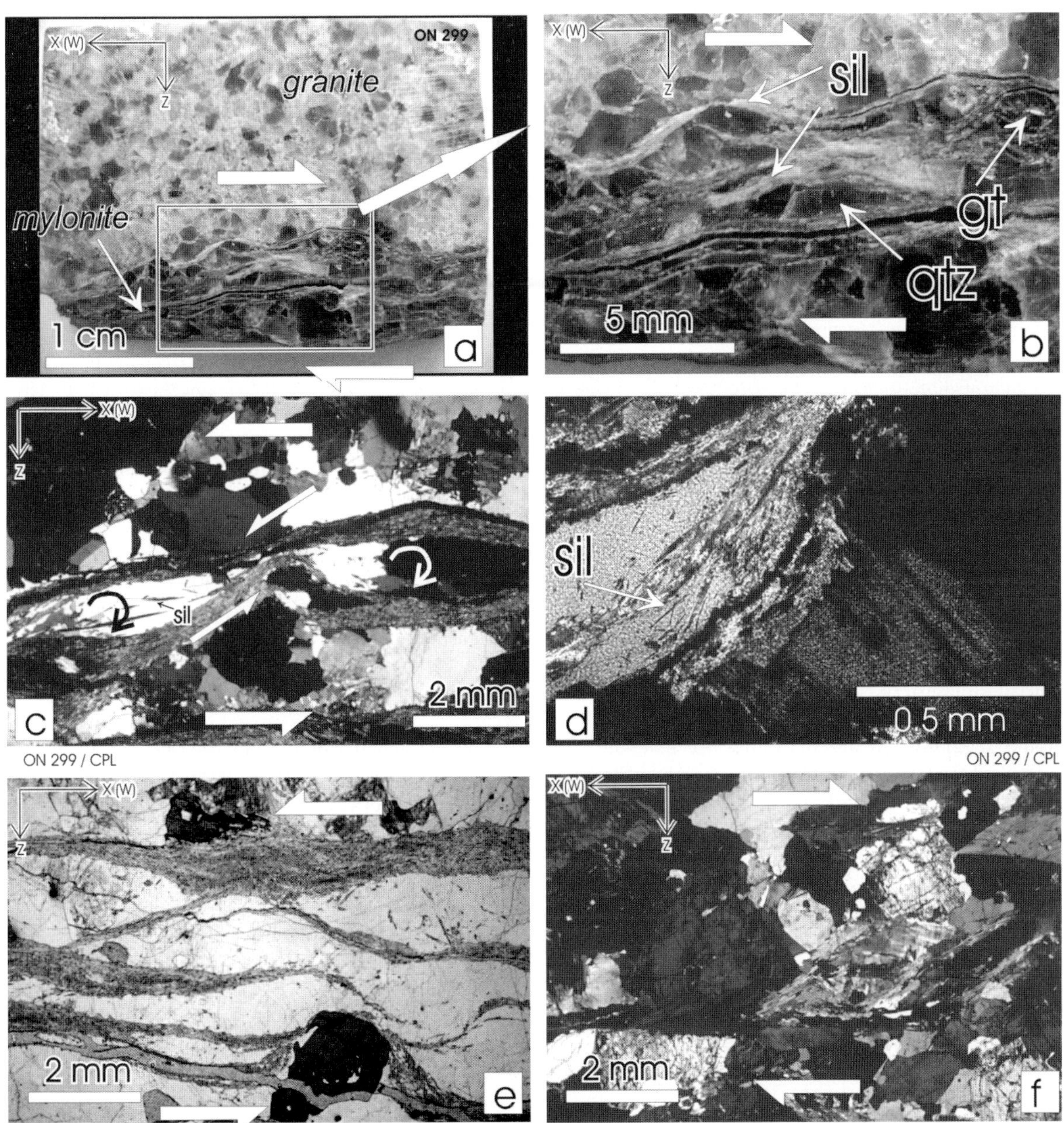

Fig. 4.5-6. Rock slab (**a, b**) and thin sections (**c–f**) of ductilely sheared granitic rocks from Mt. Bower (Outback Nunataks). **a, b** W-dipping, strongly localised mylonite zone in granite consisting of stretched lensoid quartz grains in a sillimanite-rich matrix. Shear sense is towards the E indicated by the asymmetry in the lensoid shape of the grains. **c** Strongly stretched lensoid quartz grains in sillimanite-bearing matrix. Sillimanite (*sil*) is also present inside the quartz grains. Shear sense is top-to-E indicated by asymmetry in the lensoid shape and apparent back-rotation of the quartz grains and the development of shear bands. **d** Enlargement of photo **c** showing sillimanite needles in quartz. **e** Stretched quartz grains in sillimanite-rich matrix. Inclusion trails in the garnet in the lower part of the photo show apparent sinistral rotation in the foliation and top-to-E sense of shear (syntectonic "spiral-Si" garnet: Passchier and Trouw (1996), pp. 162 and 165). **f** S-C alignment of white mica, quartz, and feldspar in a thin shear zone with top-to-E directed shear sense. *CPL:* Crossed polarised light; *PPL:* plane polarised light

tem and might be an explanation for similar segmentations visible in aeromagnetic maps of the Oates Coast region (Damaske et al. 2001). Since no W-directed kinematics as in the areas further north were found in the Outback Nunataks, the southern continuation of the Exiles Thrust should be located still further west hidden under the ice. Aeromagnetic data indicate a major lineament in Oates Land (Matusevich Line: Damaske and Bosum 1993; Matusevich Anomaly: Ferraccioli et al. 2003), which likely coincides with the Exiles Thrust (Flöttmann et al. 1993). A comparable magnetic feature further S (Central Victoria Land Boundary Zone, see Fig. 4.5-1) separates the Prince Albert Mountains from the Deep Freeze Range-Terra Nova Bay area. It is interpreted as the unexposed southern continuation of the Exiles Thrust (Ferraccioli and Bozzo 1999). The trace of the Wilson Thrust from the Rennick Glacier area into central Victoria Land is suggested by structural data from the western margin of Campbell Glacier, particularly from Mount Emison (Castelli et al. 1989), Wishbone Ridge south of Mount Dickson (Palmeri et al. 1989), and Black Ridge-Boomerang Glacier (Musumeci and Pertusati 2000). These data suggest the presence of a large-scale ductile thrust zone paralleling the western margin of Campbell Glacier that is geometrically comparable to the Wilson Thrust (Kleinschmidt 1992). A late-Ross age is indicated by high-strain overprint of the Granite Harbour Intrusives and cross-cutting relationships of undeformed late-phase granitic dykes (Musumeci and Pertusati 2000). Minor opposite-directed reverse shear zones (e.g., Boomerang Thrust at Priestley Glacier: Skinner 1991) are regarded as conjugate faults to the Wilson Thrust as in the case of the Komatsu Thrust west of Helliwell Hills (Kleinschmidt 1992). Further evidence of the continuation of the Wilson Thrust into western Campbell Glacier comes from aeromagnetic and isotopic data showing the presence of two major magnetic lineaments along the Priestley and Campbell faults and a major isotopic break in the basement on both sides of the Campbell Fault (Ferraccioli and Bozzo 1999 and references therein). Storti et al. (2001) attribute the displacement of metamorphic rocks in the Priestley Glacier area to Cenozoic dextral brittle shear along the Priestley Fault rather than to the Ross Orogeny. The impact of Cenozoic brittle tectonics on the inherited Palaeozoic pattern has been attested by e.g., Rossetti et al. (2002, 2003b). Based on Ar-Ar *in situ* dating of fault-generated pseudotachylites, Cenozoic (ca. 34 Ma) activity on the Priestley Fault is documented (Rossetti et al. 2003a). Skinner (1995) reported Palaeozoic K-Ar biotite cooling ages (ca. 486–498 Ma) for a late granitic dyke crosscutting the ductile Boomerang thrust sole. Ductile deformation of the Boomerang Thrust is therefore no younger than early Palaeozoic. Also the high-temperature overprint with synkinematic sillimanite growth in the shear zones in the study area attests to a pre-Cenozoic age. We thus interpret the Priestley Fault as a long-lasted tectonic element with both Palaeozoic and Cenozoic activity.

The presence of a late-Ross bilateral thrust system between the Pacific coast and the Ross Sea allows some speculations on the nature of the boundary of the Ross Orogen towards the East Antarctic Craton (EAC). Based on our data and in comparison to the scenarios given before, the western front of the Ross Orogen is best interpreted as a broad W-vergent fold-and-thrust belt (Fig. 4.5-7), which is in line with the Delamarian Orogen in Australia (Flöttmann et al. 1993). Along this belt, the intra-Wilson Terrane magmatic arc was detached and

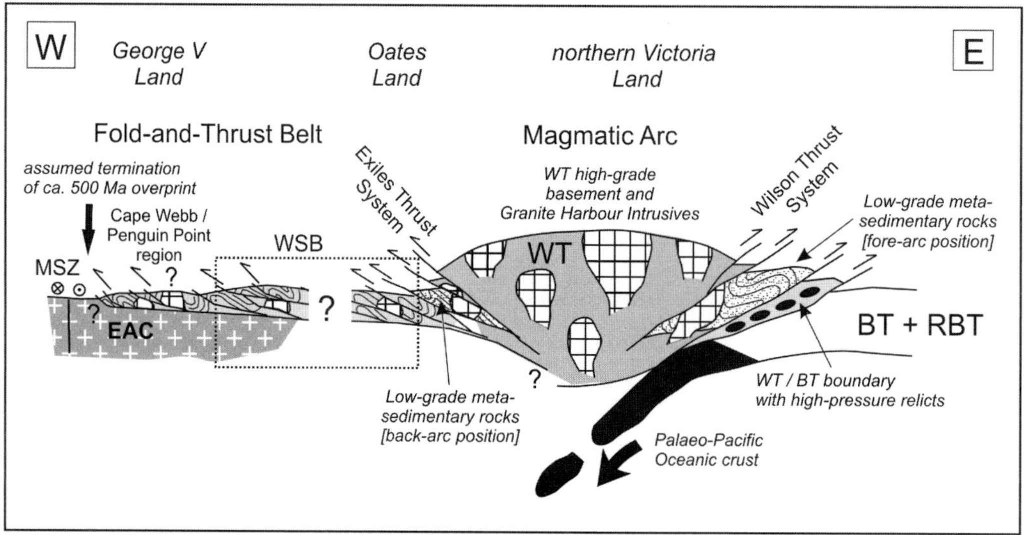

Fig. 4.5-7. Hypothetical tectonic model (not to scale!) of the late-Ross evolution of the Wilson Terrane (*WT*). The high-grade, central Wilson Terrane magmatic arc is detached and thrust W- and E-ward over low-grade metasedimentary units. The western front of the Ross Orogen is interpreted as a broad W-vergent fold-and-thrust belt, which extends possibly into the Mertz and Ninnis Glaciers area based on recent aeromagnetic (Damaske et al. 2003) and geochronological evidence (Fanning et al. 2003). The *stippled rectangle* indicates the rough position of the Wilkes Subglacial Basin (*WSB*), which could involve possible post-Ross large-scale crustal extension overprinting the Ross-orogenic structural edifice. Modified after Flöttmann and Kleinschmidt (1991). *BT:* Bowers Terrane; *RBT:* Robertson Bay Terrane; *MSZ:* Mertz Shear Zone (cf. Talarico and Kleinschmidt 2003)

thrust W-ward onto the craton. The Exiles-related shear systems are interpreted to represent only the internal, easternmost portions of this belt. The actual craton-orogen boundary must be located still further west of the outcropping segments of the Exiles Thrust system and is likely to be hidden under the ice. It was originally interpreted to coincide with a prominent aeromagnetic boundary west of Priestley and Reeves Glaciers located some 350 km south of the study area (Bosum et al. 1989), which was later re-interpreted to reflect Ross-age arc rocks rather than Precambrian shield rocks and to represent the unexposed southern continuation of the Exiles Thrust system (Ferraccioli and Bozzo 1999). However, new aeromagnetic data obtained during the joint German-Italian Ganovex VIII-ItaliAntartide XV programme 1999/2000 (Damaske et al. 2003; Ferraccioli et al. 2003) suggest that the EAC-Ross orogen boundary is located significantly further W than originally proposed, i.e. closer to the western margin of the Wilkes Subglacial Basin rather than close to its eastern margin. This is consistent with recent findings of ca. 500 Ma old granitoids in the Cape Webb-Penguin Point region between Mertz and Ninnis Glaciers in George V Land (Fanning et al. 2003), which contrasts a complex Archean-Palaeoproterozoic record without any ca. 500 Ma old tectonic or thermal overprint in the area west of the Mertz Glacier (Fanning et al. 2002). These new results imply that the Ross-orogenic fold-and-thrust belt proposed in our hypothetical model (Fig. 4.5-7) reaches a total width of several 100 km from the western side of the Transantarctic Mountains to the area between Mertz and Ninnis Glaciers in George V Land. A possible explanation for this rather large extend of the belt may be derived from recent modelling of gravity anomaly data from the unexposed Wilkes Subglacial Basin, which suggests that this basin was formed by broad crustal extension rather than simple flexure beneath at least parts of the basin (Ferraccioli et al. 2001; Damaske et al. 2003 and references therein).

Acknowledgments

This study is part of the Ganovex VIII-ItaliAntartide XV programme in 1999–2000. Sincere thanks to Bundesanstalt für Geowissenschaften und Rohstoffe (BGR) for the inviation and to BGR and the Italian Programma Nazionale die Ricerche in Antartide (PNRA) for logistic support. A. L. and G. K. thank Deutsche Forschungsgemeinschaft for financial support (grants KL 429/18-1 to 3) and Alfred Wegener Institute for Polar and Marine Research for polar equipment. F. R. acknowledges a fellowship of PNRA at Siena University and thanks R. Funiciello and F. Salvini for continuous advice and encouragement.

References

Capponi G, Mecchieri M, Musumeci G, Persutati PC, Ricci CA, Talarico F (1994) A geological transect through the Wilson-Bowers-Robertson Bay terranes junction (northern Victoria Land, Antarctica). IX Italian Antarctic Expedition, Field Data Reports, pp 16–19

Capponi G, Crispini L, Meccheri M (1999) Structural history and tectonic evolution of the boundary between the Wilson and Bowers terranes, Lanterman Range, northern Victoria Land, Antarctica. Tectonophysics 312:249–266

Capponi G, Crispini L, Meccheri M (2002) Tectonic evolution at the boundary between the Wilson and Bowers terranes (northern Victoria Land, Antarctica): structural evidence from the Mountaineer and Lanterman ranges. In: Gamble JA, Skinner DNB, Henrys S (eds) Antarctica at the close of a millennium. Royal Soc New Zealand Bull 35, Wellington, pp 105–112

Castelli D, Lombardo B, Oggiano G, Rossetti P, Talarico, F (1989) Granulite rocks of the Wilson Terrane (North Victoria Land). The Campbell Glacier belt: Field relations, petrography and metamorphic history. 3rd Meeting Sci Terra Antartide, Siena, Riassunti, pp 40–41

Damaske D, Bosum W (1993) Interpretation of the aeromagnetic anomalies above the lower Rennick Glacier and the adjacent polar plateau W of the USARP mountains. Geol Jb E47:139–152

Damaske D, Ferraccioli F, Bozzo E (2001) Aeromagnetic investigations during Ganovex VIII – the BACKTAM transect over the coast of northern Victoria Land, Oates Land, and George V Land. Abstracts Ganovex VIII Workshop Grubenhagen (Germany), pp 24–27

Fanning, CM, Daly, SJ, Bennett VC, Ménot, RP, Peucat, JJ, Oliver, RL, Monnier, O (1995) The "Mawson Block": once contiguous Archean to Proterozoic crust in the East Antarctic Sheld and Gawler Craton. 7th Internat Sympos Antarctic Earth Sci, Abstracts, Siena

Fanning CM, Moore, DH, Bennett VC, Daly SJ, Ménot R-P, Peucat JJ, Oliver RL (1999) The Mawson Continent: The East Antarctic Shield and Gawler Craton, Australia. 8th Internat Sympos Antarctic Earth Sci, Abstracts, Wellington

Fanning CM, Ménot R-P, Pecaut JJ, Pelletier A (2002) A closer examination of the direct links between southern Australia and Terre Adélie and George V Land, Antarctica. In: Preiss VP (ed) Geosciences 2002: expanding horizons, Abstracts 16th Austral Geol Convent, Adelaide, 67:224

Fanning CM, Peucat JJ, Ménot R-P (2003) Wither the Mawson Continent? 9th Intern Sympos Antarctic Earth Sci, Antarctic contributions to global earth science, Potsdam, Germany, Terra Nostra 2003(4):83–84

Ferraccioli F, Bozzo E (1999) Inherited crustal features and tectonic blocks of the Transantarctic Mountains: an aeromagnetic perspective (Victoria Land, Antarctica). J Geophys Res 104: 25297–25319

Ferraccioli F, Coren F, Bozzo E, Zanolla C, Gandolfi S, Tabacco I, Frezzotti M (2001) Rifted (?) crust at the East Antarctic Craton margin: gravity and magnetic interpretation along a traverse across the Wilkes Subglacial Basin region. Earth Planet Sci Letters 192:407–421

Flöttmann T, Kleinschmidt G (1991) Opposite thrust systems in northern Victoria Land, Antarctica: imprints on Gondwana's Palaeozoic accretion. Geology 19:45–47

Flöttmann T, Kleinschmidt G, Funk T (1993) Thrust patterns of the Ross/Delamerian Orogens in northern Victoria Land (Antarctica) and southeastern Australia and their implications for Gondwana reonstructions. In: Findlay RH, Unrug R, Banks MR, Veevers JJ (eds) Gondwana Eight, pp 131–139

GANOVEX Team (1987) Geological map of North Victoria Land, Antarctica, 1:500 000 – Explanatory notes. Geol Jb B66:7–79

Gibson GM (1984) Deformed conglomerates in the eastern Lanterman Range, North Victoria Land, Antarctica. Geol Jb B60:117–141

Gibson GM (1987) Metamorphism and deformation in the Bowers Supergroup: implications for terrane accretion in northern Victoria Land, Antarctica. In: Leitch E, Scheibner E (eds) Terrane accretion and orogenic belts. Amer Geophys Union Geodyn Ser 19:207–219

Goodge JW, Dallmeyer R (1996) Contrasting thermal evolution within the Ross Orogen, Antarctica: evidence from $^{40}Ar/^{39}Ar$ ages. J Geology 104:435–458

Holdaway MJ (1971) Stability of andalusite and the aluminum silicate phase diagram. Amer J Sci 271:97–131

Ireland TR, Weaver SD, Bradshaw JD, Adams C, Gibson GM (1999) Detrital zircon age spectra from the Neoproterozoic – early Palaeozoic Gondwana margin in Antarctica. 8th Internat Sympos Antarctic Earth Sci, Abstracts, Wellington

Kleinschmidt G (1992) The southern continuation of the Wilson Thrust. Polarforschung 60:124–127

Kleinschmidt G, Tessensohn F (1987) Early Palaeozoic W-ward directed subduction at the Pacific margin of Antarctica. In: McKenzie GD (ed) Gondwana Six – structure, tectonics, and geophysics. Geophys Monogr 40, AGU, Washington DC, pp 89–105

Musumeci G, Pertusati P (2000) Structure of the Deep Freeze Range-Eisenhower Range of the Wilson Terrane (North Victoria Land, Antarctica): emplacement of magmatic intrusions in the Early Palaeozoic deformed margin of the East Antarctic Craton. Antarctic Sci 12:89–104

Musumeci G, Pertusati PC (2001) Early Ordovician terrane accretion along the Lanterman Fault Zone (North Victoria Land, Antarctica): new data on correlation between Ross and Delamerian Orogens in the paleo-Pacific margin of Gondwana. J Confer Abstracts 6:379

Palmeri R, Talarico F, Meccheri M, Oggiano G, Pertusati PC, Rastelli N, Ricci CA (1989) Petrographical and structural data along two cross-sections through the Priestley Formation – Priestley Schist boundary in the Boomerang Glacier and Mount Levick area (North Victoria Land), Progress report. 3rd Meeting Sci. della Terra in Antartide, Siena, Riassunti, pp 83–85

Roland NW, Gibson GM, Kleinschmidt G, Schubert W (1984) Metamorphism and structural relations of the Lanterman metamorphics, North Victoria Land, Antarctica. Geol Jb B60: 319–361

Roland NW, Olesch M, Schubert W (1989) Geology and petrology of the western border of the Transantarctic Mountains between the Outback Nunataks and Reeves Glacier, northern Victoria Land, Antarctica. Geol Jb E38:119–141

Rossetti F, Storti F, Läufer AL (2002) Brittle architecture of the Lanterman Fault and its impact on the final terrane amalgamation in North Victoria Land, Antarctica. J Geol Soc London 159: 159–173

Rossetti F, Di Vincenzo G, Läufer AL, Lisker F, Rocchi S, Storti F (2003a) Cenozoic right-lateral faulting in North Victoria Land: an integrated structural, AFT and $^{40}Ar/^{39}Ar$ study. 9th Intern Sympos Antarctic Earth Sci, Antarctic contributions to global earth science, Potsdam, Germany, Terra Nostra 2003(4): 283–284

Rossetti F, Lisker F, Storti F, Läufer AL (2003b) Tectonic and denudational history of the Rennick Graben (North Victoria Land): Implications for the evolution of rifting between East and West Antarctica. Tectonics 22(2)

Skinner DNB (1991) Metamorphic basement contact relations in the southern Wilson Terrane, Terra Nova Bay, Antarctica – The Boomerang thrust. Mem Soc Geol It 46:163–178

Skinner DNB (1995) Metamorphic effects of the Boomerang thrust basement suture in the Wilson Terrane, Deep Freeze Range, northern Victoria Land, Antarctica. Internatt Sympos Antarctic Earth Sci, Siena, Abstracts

Storti F, Rossetti F, Salvini F (2001) Structural architecture and displacement accomodation mechanisms at the termination of the Priestley Fault, northern Victoria Land, Antarctica. Tectonophysics 341:141–161

Talarico F, Kleinschmidt G (2003) Structural and metamorphic evolution of the Mertz Shear Zone (East Antarctic Craton, George V Land): Implications for Australia/Antarctica correlations and East Antarctic Craton/Ross Orogen relationships. Terra Antartica 10:229–248

Style of Uplift of Paleozoic Terranes in Northern Victoria Land, Antarctica: Evidence from K-Ar Age Patterns

Christopher J. Adams

Institute of Geological and Nuclear Sciences, PO Box 30368, Lower Hutt, New Zealand, <argon@gns.cri.nz>

Abstract. K-Ar ages of 82 slate and schist (white-mica-rich whole-rock) samples are reported for Late Precambrian-Early Ordovician metamorphic rocks of the Wilson, Bowers and Robertson Bay terranes of northern Victoria Land. These are amalgamated in two vertical sections along composite NE-SW horizontal profiles across (1) Oates Coast in the north, and (2) Terra Nova Bay area in the south. The ages are in the range 328–517 Ma. Both profiles show some age variation with altitude, but more importantly, they define an inverted wedge shaped pattern, reflecting a "pop-up" structure. This is oriented NW-SE at the eastern margin of the Wilson terrane, and the edges coincide with the Exiles and Wilson Thrusts which cross the region. Ages inside the "pop-up" structure are younger, ca. 460–480 Ma, than those along its eastern and western flanks, ca. 490–520 Ma. The K-Ar age patterns thus demonstrate a late Ross Orogenic age (ca. 460 Ma) for this structure, which may be associated with assembly of the Wilson and Bowers terranes.

Introduction

The Cambrian-Ordovician Ross Orogen forms the axis of the Transantarctic Mountains along the eastern edge of the Antarctic craton (Fig. 4.6-1, inset). At the Ross Sea and Pacific Ocean margins in northern Victoria Land (NVL), it comprises three terranes (Fig. 4.6-1, inset A): (1) a broad, western *Wilson Terrane* of greenschist facies metasediments (Berg Group, Priestley Formation and Priestley Schist), greenschist to amphibolite facies orthogneiss (Wilson Gneiss) and paragneiss, (Rennick Schist), and Late Cambrian-Ordovician granitoid complexes (Granite Harbour Intrusives), (2) a narrow, central *Bowers Terrane* of volcanics (Glasgow Volcanics) and low-grade greenschist facies, Cambrian, metasediments (Sledgers and Mariner Groups) and (3) a broad, eastern *Robertson Bay Terrane*, of greenschist facies, Early Ordovician, turbiditic metasediments (Robertson Bay Group and Millen Schist) (for comprehensive references see Tessensohn et al. 1981). Wilson, Bowers and Robertson Bay terrane rocks are intruded by Devonian-Early Carboniferous Admiralty Intrusives and there are probable scattered, volcanic correlatives (e.g., Gallipoli Volcanics) in all terranes. Major regional metamorphism associated with the Ross Orogeny, occurred in the three terranes in Late Cambrian to Early Ordovician times, now recorded, in part, as regional K-Ar cooling age patterns, 505–455 Ma (Adams et al. 1982; Adams and Kreuzer 1984). However, the terranes are separated by the major Lanterman and Leap Year Faults, between (and close to) which occur anomalously younger (<375 Ma) age patterns, possibly attributable, in part, to terrane assembly (Adams and Kreuzer 1984). In the eastern part of the Wilson Terrane, an east-dipping Exile Thrust, and a west-dipping Wilson Thrust, form a major, inverted wedge-shaped, (or "pop-up") structure of the Ross Orogeny (Flöttman and Kleinschmidt 1991).

Field investigations during German and Italian expeditions of 1989 and 1991 enabled broad rock sampling in profiles along the Oates Coast, Northern Victoria Land, and at Terra Nova Bay, Ross Sea. K-Ar mineral and total rock ages are used here to test the Flöttman and Kleinschmidt (1991) uplift model and, combining with existing K-Ar age data (Adams 1997; Adams and Kreuzer 1984; Borsi et al. 1989; Kreuzer et al. 1987; Tonarini and Rocchi 1994; Vita et al. 1991; Vita-Scaillet 1994), establish a regional Paleozoic uplift history.

K-Ar analytical techniques are given in Adams and Graham (1986). Age errors are given at 95% confidence limits, and are based upon combined percentage reproducibilities of repeated K and Ar analyses of standard slate samples. Sample data are listed in Table 4.6-1, and age data are shown in maps and vertical profiles in Fig. 4.6-1 (Oates Coast sector) and Fig. 4.6-2 (Terra Nova Bay sector).

Age Patterns in the Oates Coast Sector

Along the Oates Coast, the K-Ar dating samples (35 from Table 4.6-1 this study, and 6 from comparable data from Adams 1997) lie within coastal sections straddling the lower Matusevich and Rennick Glaciers (Fig. 4.6-1), and the sampling area crosses important regional terrane structures. To accommodate horizontal and vertical variation within each terrane, the age data are projected normally on to two vertical, ENE-WSW, profiles (A–A' and B–B' in Fig. 4.6-1), and joined to form a composite profile (lower section, Fig. 4.6-1), viewed from the Pacific Ocean, looking inland towards the SSE, as indicated by the open

From: Fütterer DK, Damaske D, Kleinschmidt G, Miller H, Tessensohn F (eds) (2006) Antarctica: Contributions to global earth sciences. Springer-Verlag, Berlin Heidelberg New York, pp 205–214

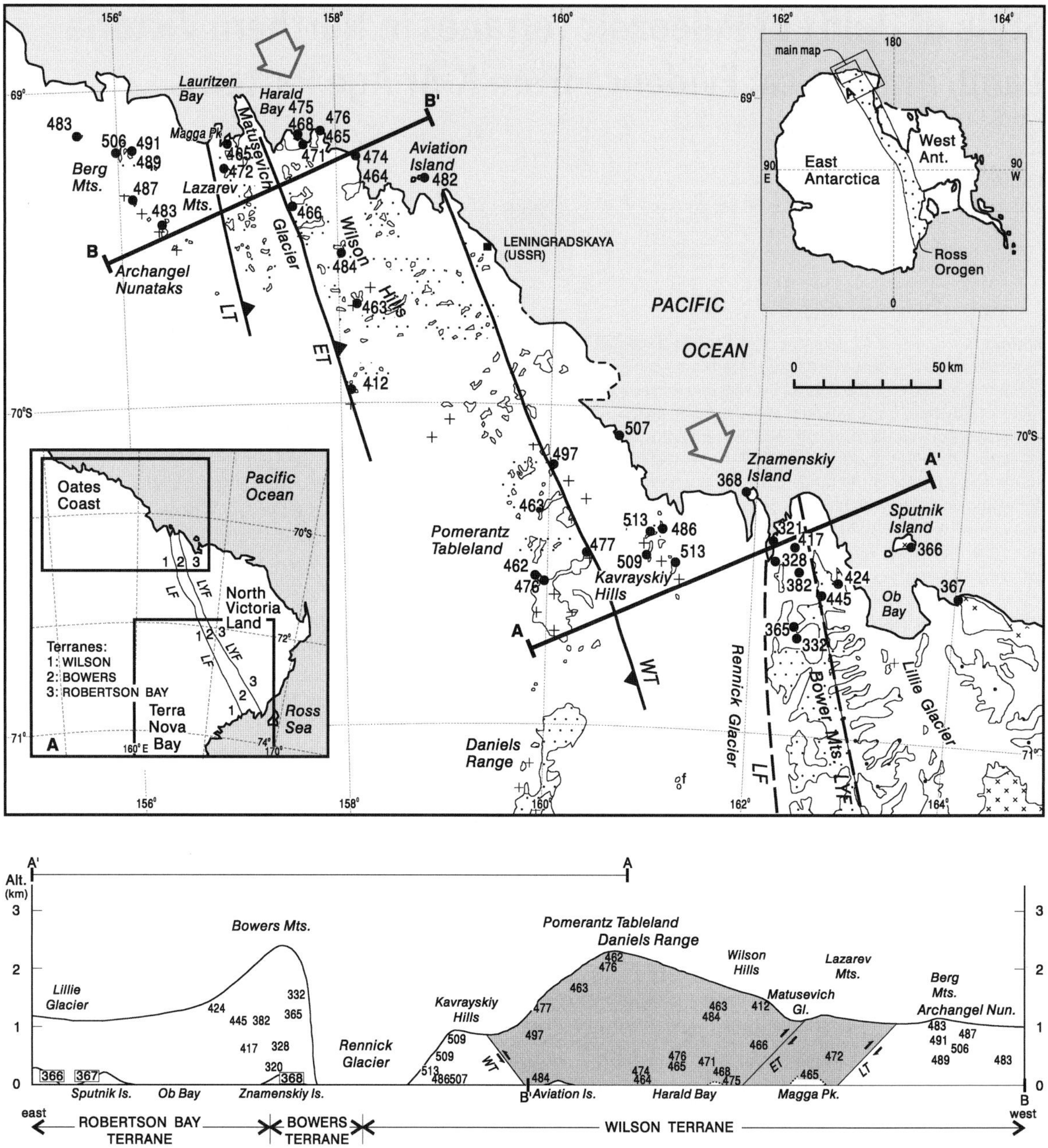

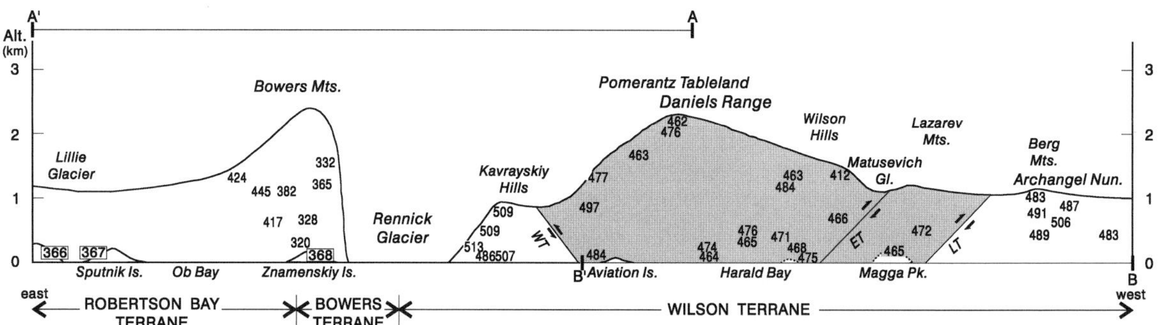

Fig. 4.6-1. Geological map of Oates Coast, and hinterland of the Matusevich and Rennick Glaciers, and K-Ar sample locations. Rock type ornament is as shown in the legend of Fig. 4.6-2. Antarctic *inset diagram* at top right indicates study area with respect to Ross Orogen. *Inset A* at bottom left indicates main tectonostratigraphic terranes of northern Victoria Land, with respect to the two study areas. The terranes are bounded by the Lanterman (*LF*) and Leap Year (*LYF*) faults. The *lines A–A'* and *B–B'* indicate position of vertical profiles used to create the composite K-Ar age profile in the lower part of diagram, and the *open arrows* indicate the viewpoint aspect of this. Age data on the vertical profile are in millions of years (Ma), those in boxes are Admiralty Intrusives. *Grey shaded area* indicates the suggested "pop-up" structure of Flöttman and Kleinschmidt (1991) with its boundary faults (marked on the map) as the Exiles (*ET*) and Wilson (*WT*) Thrusts. *LT* is the Lazarev thrust

arrows. The majority of the ages, 460–513 Ma, are comparable with those of the main 460–524 Ma trend of the Adams and Kreuzer (1984) profile across central NVL (reproduced in Fig. 4.6-3). A minority group of samples, 320–445 Ma, all from within the Bowers Terrane, show the pattern of younger ages decreasing with altitude (Adams and Kreuzer 1984). These main, 460–505 Ma, ages are similar to those of Granite Harbour Intrusives at Terra Nova Bay (Borsi et al. 1989; Vita et al. 1991) which reflect the last regional, thermal (or more strictly speaking, cooling)

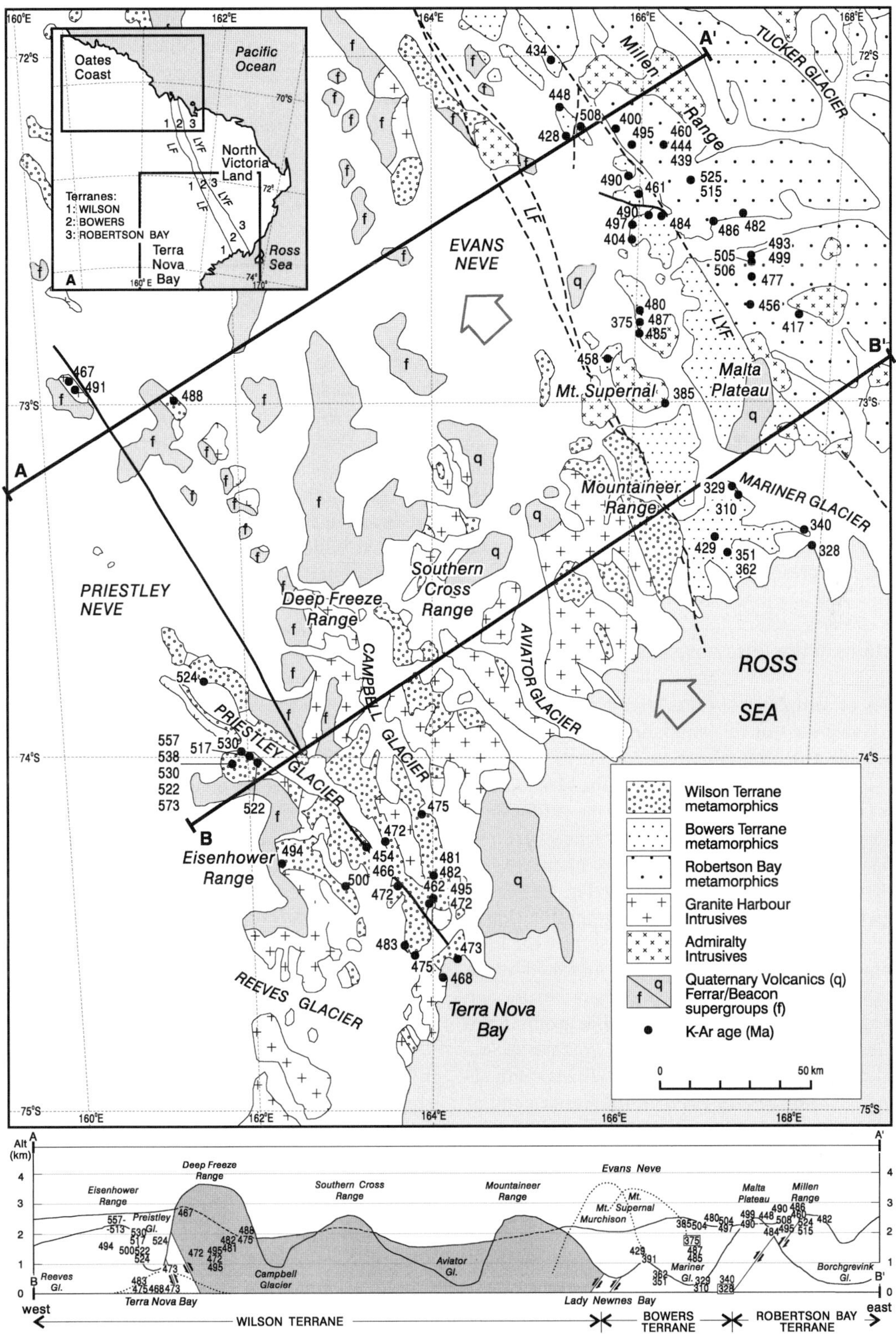

Fig. 4.6-2. Geological map of Terra Nova Bay area, and hinterland of the Priestley and Mariner Glaciers, and K-Ar sample locations. *Inset A:* the principal terranes of northern Victoria Land designated as in Fig. 4.6-1, the terranes are bounded by the Lanterman (*LF*) and Leap Year (*LYF*) faults. The *lines A–A′* and *B–B′* indicate position of vertical profiles used to create the composite K-Ar age profile in the lower part of diagram, and the *open arrows* indicate the viewpoint aspect of this. Age data on the vertical profile are in millions of years (Ma), those in boxes are Admiralty Intrusives. *Grey shaded area* indicates the suggested "pop-up" structure of Flöttman and Kleinschmidt (1991)

event within the Ross Orogen. This thermal effect of Granite Harbour Intrusives is pervasive throughout the Wilson Terrane, and commonly obliterates age evidence for earlier metamorphism. In this respect, the vertical age profile (Fig. 4.6-1) thus reveals an important pattern of K-Ar data in the Wilson Terrane, namely, age groups of low-medium grade metasediment samples at its eastern extremity, 486–513 Ma (Kavrayskiy Hills), and westernmost extremity, 483–506 Ma (Berg Mountains), are older than those of the central portion (Pomerantz Tableland, Daniels Range), 463–484 Ma, entirely from high-grade gneisses and granitoids. These age groups are nicely constrained by the westward-dipping Wilson Thrust in the east, and the eastward-dipping Exiles (but possibly also Lazarev) Thrust in the west (Fig. 4.6-1). The age pattern thus supports the Flöttman and Kleinschmidt (1991) "pop-up" mechanism, with the vertical displacement of a late-Ross cooling age pattern derived from Granite Harbour Intrusives at the Wilson Terrane core (Fig. 4.6-4, shaded area), to the level of early-Ross cooling age patterns derived from metamorphism at higher structural levels, and now preserved only at the structure flanks.

Age Patterns in the Terra Nova Bay Sector

K-Ar ages of samples from the Terra Nova Bay sector (47 from Table 4.6-1 of this study, and augmented with 18 comparable ages from Adams and Kreuzer 1984), fall mainly within two areas which straddle the Mariner and Priestley Glaciers (Fig. 4.6-2), and again the dating sampling area crosses important regional terrane structures. As for Fig. 4.6-1, the age data are projected on to two vertical ENE-WSW profiles (A–A' and B–B' in Fig. 4.6-2), and then superimposed to form a composite profile (lower section, Fig. 4.6-2), but in this case viewed from the Ross Sea looking inland *towards the NNW*, as indicated by the open arrows. In the Wilson Terrane part, the major age group is again 467–495 Ma, whilst a smaller group in the Bowers Terrane, 328–448 Ma (Table 4.6-1), follows the established pattern of ages decreasing (to 310 Ma) with altitude (Adams and Kreuzer 1984). At the western end of the profile, along the western side of the Priestley Glacier, there is again a significant group of older ages, 494–557 Ma, from lower-grade, Priestley Formation metasediments. At these localities Rb-Sr whole-rock isochron ages are 510 ±6 to 570 ±9 Ma (Adams 1997). Clearly both K-Ar and Rb-Sr data sets data reflect earlier events than those recorded by K-Ar cooling age patterns (454–495 Ma) in the extensive Granite Harbour Intrusive complexes to the east, between the Priestley and Aviator Glaciers (Borsi et al. 1989; Vita et al. 1991), and to the north in central NVL (Kreuzer et al. 1987). The early regional metamorphism both here, and in the Oates Coast sector (above), is thus probably Late Neoproterozoic, and certainly predates Bowers and Mariner Group (Cambrian) sedimentation. The division of the cooling age patterns coincides with a major east-dipping fault along the Priestley Glacier (Flöttman and Kleinschmidt 1991). However, from the available age data (this work, and Vita-Scaillet et al. 1994), there appears to be no corresponding older (>500 Ma) age group at the eastern flank of the Wilson Terrane, similar to that in the Oates Coast profile east of the Wilson Thrust, e.g., at Kavrayskiy Hills.

Age Patterns in Bowers and Robertson Bay Terranes

K-Ar and $^{40}Ar/^{39}Ar$ ages (Adams and Kreuzer 1984; and this work Table 4.6-1, Wright and Dallmeyer 1991; Capponi et al. 2002) in Robertson Bay Group slates (and possibly related Millen Schist) at the Bowers/Robertson Bay terrane boundary mostly conform to the broad, 470–490 Ma trend continuing to the east (Fig. 4.6-3). Since Robertson Bay Group includes Tremadocian (490–495 Ma) fossil localities at Handler Ridge (Burrett and Findlay 1984), albeit in unusual limestone enclaves, these ages can be simply interpreted as a cooling pattern following late Early Ordovician metamorphism. However, there are significant exceptions older than this, from 505 ±10 to 525 ±11 Ma, especially immediately west of the Leap Year Fault, possibly the result of incomplete argon degassing during metamorphism or incorporation of excess argon in surviving mineral phases (Wright and Dallmeyer 1991). However, sequences here are tectonically complex and uncertain, so a pre-Ordovician stratigraphic break and older metamorphism remain an alternative possibility. Capponi et al. (2002) report Rb-Sr isochron and $^{40}Ar/^{39}Ar$ actinolite age data for meta-igneous rocks in the northern Robertson Bay terrane and ca. 10 km from the Bowers/Robertson Bay terrane boundary. Their 361 ±70 Ma Rb-Sr age is an errorchron, and its large error does not allow a discrimination of a Ross Orogenic plutonic, or post-Ross Orogenic metamorphic, event. Similarly, the actinolite $^{40}Ar/^{39}Ar$ release pattern has no clear plateau, but shows an increasing age pattern, 345–395 Ma, from low to high temperature, and a secure interpretation is not possible.

Uplift History of the Terranes

The Exiles and Wilson Thrusts define an uplift structure in the Wilson Terrane, which displaces established late-Ross K-Ar cooling age patterns. Since, with one exception (see below), this structure includes ages as young as 462 Ma, then the "pop-up" structure should be younger than this, but still a late-Ross tectonic feature (Kleinschmidt and Tessensohn 1987; Flöttman and Kleinschmidt 1991). However, a single, anomalous, young age, 412 ±8 Ma, from sheared granite near (but above) the Exiles Thrust,

Table 4.6-1. K-Ar ages from Oates Coast sector (Pacific margin) and Terra Nova Bay sector (Ross Sea)

IGNS R. No.	Field No.	Rock type	Geol unit	Location	Latitude °S	Longitude °E	Altitude (m)	K wt.-%	^{40}Ar (rad.)		Age (Ma)
									nl g^{-1}	% tot	
Berg Mountains – Archangel Mountains (data from Adams 1997)											
13752tr	GXV86A	Slate	BG	Berg Mts., E ridge	69°13'	156°05'	720	5.25	115.2	99	491 ±10
13782tr	GXV113	Slate	BG	Archangel Nunatak	69°22'	156°09'	900	5.41	117.5	99	487 ±10
13785tr	GXV116	Slate	BG	West Berg Rocks	69°08'	155°42'	355	5.59	120.2	94	483 ±10
13789tr	GXV120	Slate	BG	Berg Mts. E ridge	69°12'	156°07	330	4.61	100.6	99	489 ±10
13804tr	GXV120	Granite	BAG	Berg Mts. W ridge	69°13'	156°01'	520	7.29	165.3	94	506 ±10
13779rt	GXV110	Granite	BAG	Outrider Nunatak	69°27'	156°25'	1 070	7.31	157.4	96	483 ±10
Lazarev Mountains											
13726bi	GXV60	Granodiorite-gneiss	RS	Coombes Ridge	69°09'	157°05'	230	7.71	159.1	96	465 ±9
13734bi	GXV68	Mica-schist	RS	Drury Nunatak	69°14	156°57'	480	7.47	156.7	98	472 ±9
Matusevich Glacier – Harald Bay											
13687bi	GXV24	Amphibolitic-gneiss	WG	Williamson Head	69°11'	157°59'	400	7.50	158.8	98	476 ±10
13688bi	GXV25	Granodiorite-gneiss	RS	Williamson Head	69°11'	157°59'	380	7.32	151.1	97	465 ±9
13689mu	GVX26	Pegmatite	GHI	Harald Bay	69°14'	157°44'	400	8.40	175.6	93	471 ±9
13690bi	GXV27	Granodiorite	WG	Kartografov Island	69°12'	157°43'	20	7.94	167.5	98	475 ±9
13694bi	GXV31	Granodirorite	GHI	Exiles Nunatak	69°58'	158°04'	1 500	7.61	136.9	98	412 ±8
13695bi	GXV32	Granodiorite	GHI	Mt. Blowaway, NE	69°26'	158°13'	1 300	6.78	139.1	99	463 ±9
13699bi	GXV36	Pegmatite	GHI	Thompson Peak, S	69°29'	157°40'	760	7.20	148.7	84	466 ±9
13771bi	GXV102	Granodiorite	GHI	Harald Bay, W	69°12'	157°43'	50	7.94	165.1	97	468 ±9
13773bi	GXV107	Mica-schist	RS	Celestial Peak	69°31'	158°06'	1 200	7.78	167.9	95	484 ±10
Davis Bay											
13708bi	GXV45	Granodiorite-gneiss	WG	Drake Head	69°14'	158°15'	200	7.92	166.6	97	473 ±9
13713bi	GXV50	Leucogranite	GHI	Drake Head	69°14'	158°15'	120	5.45	112.1	99	464 ±9
13725bi	GXV826A	Granodiorite	GHI	Aviation Island, E	69°17'	158°47'	40	7.40	159.1	98	482 ±10
Pomerantz Tableland											
13675bi	GXV12	Granodiorite	GHI	Armstrong Platform	70°30'	160°09'	1 340	7.30	155.0	98	477 ±10
13677bi	GXV14	Granodiorite	GHI	MacPherson Peak	70°34'	159°51'	2 000	7.24	153.4	94	476 ±10
13678bi	GXV15	Granodiorite	GHI	MacPherson Peak	70°33'	159°44'	2 420	7.64	156.4	98	462 ±9
1368bi	GXV17	Granodiorite	GHI	Mt. Harrison, NE	70°23'	159°40'	1 680	7.46	152.9	98	463 ±9
13682bi	GXV19	Granodiorite	GHI	Robilliard Glacier, SE	70°13'	160°03'	890	7.45	165.03	94	497 ±10
Kavraysky Hills											
13683bi	GXV20	Granite	GHI	Yermak Point	70°07'	160°40'	100	7.10	161.7	87	507 ±10
13813bi	GXV140	Adamellite	GHI	Serrat Glacier	70°28'	161°00'	890	6.98	159.4	95	509 ±10
13814bi	GXV141	Granite	GHI	Kavraysky Hills	70°29'	161°15'	150	5.16	118.9	99	513 ±10
13818bi	GXV1445	Granodiorite-gneiss	GHI	Kavraysky Hills	70°22'	161°05'	90	5.54	119.9	98	485 ±10
13829bi	GXV156	Granodiorite	GHI	Kavraysky Hills	70°23'	161°00'	310	6.91	155.3	92	502 ±10

Table 4.6-1. *Continued*

IGNS R. No.	Field No.	Rock type	Geol unit	Location	Latitude °S	Longitude °E	Altitude (m)	K wt.-%	^{40}Ar (rad.)		Age (Ma)
									nl g^{-1}	% tot	
Bowers Mountains											
13688tr	GXV50	Andesite	GV	Weeder Rock	70°23'	162°02'	160	0.64	8.8	75	320 ±6
13670tr	GXV7	Metavolcanic	GV	Mt. Belolikov, N	70°28'	162°06'	650	1.61	22.5	97	328 ±7
13803tr	MS3509	Slate	SG	Mt. Bruce, NNW	70°26'	162°22'	600	3.10	56.4	96	417 ±8
13809tr	MS3514	Slate	SG	Mt. Bruce, NW	70°31'	162°38'	1 100	2.88	47.6	95	382 ±8
13810tr	MS3519	Slate	RBG	Mt. Sheila, E	70°35'	162°40'	1 600	1.36	26.6	99	445 ±9
13811tr	MS3523	Slate	RBG	Winkler Glacier	70°33'	162°47'	1 290	2.13	39.5	96	424 ±8
13832tr	NS3536	Slate	SG	Arruiz Glacier	70°41'	162°23'	1 200	3.07	48.3	98	365 ±7
13833tr	MS3538	Slate	SG	Arruiz Glacier	70°43'	162°26'	1 500	3.14	44.5	92	332 ±7
13671bi	GXV8	Granite	AI	Znamenskiy Island	70°14'	161°15'	150	7.40	117.4	83	368 ±7
Ob Bay											
13664bi	GXV1	Granite	AI	Cape Williams	70°33'	164°04'	50	7.15	113.0	98	367 ±7
136555bi	GXV2	Granite	AI	Sputnik Island	70°20'	163°31'	50	7.34	115.9	93	366 ±7
Frontier Mountain – Sequence Hills											
14884bi*	FM1	Granite	GHI	Frontier Mountain	72°58'	160°16'	2 800	6.19	128.3	98	467 ±9
14884mu	FM1	Granite	GHI	Frontier Mountain	72°58'	160°16'	2 800	8.45	185.3	98	491 ±10
14891bi	FMB	Mica-schist	RS	Sequence Hills	73°03'	161°19'	2 060	7.21	157.1	94	488 ±10
Foolsmate Glacier – Priestley Glacier											
14667tr	FO8	Slate	PF	Foolsmate Glacier, S	74°02'	161°56'	1 710	4.02	96.3	95	530 ±11
14674tr	FO15	Slate	PF	Foolsmate Glacier, S	74°02'	161°02'	1 830	3.09	71.9	92	517 ±10
14684tr	FO25	Slate	PF	Foolsmate Glacier, W	74°03'	162°01'	1 250	3.89	91.5	96	522 ±10
14695tr	FO36	Slate	PF	Foolsmate Glacier, W	74°04'	162°08'	1 350	3.40	80.3	96	524 ±10
14840tr	FO45	Slate	PF	Foolsmate Glacier, neve	74°05'	161°57'	2 510	1.53	38.8	96	556 ±11
14824tr	FO47	Slate	PF	Foolsmate Glacier, neve	74°05'	161°57'	2 510	3.81	87.9	97	513 ±10
14834tr	FO48	Slate	PF	Foolsmate Glacier, neve	74°05'	161°57'	2 510	2.65	64.6	99	538 ±11
14844tr	FO49	Slate	PF	Foolsmate Glacier, neve	74°05'	161°57'	2 510	3.63	85.9	99	530 ±11
14850tr	FO55	Slate	PF	Fools,ate Glacier, neve	74°05'	161°57'	2 510	1.18	27.8	91	522 ±10
14856tr	FO61	Mica-schist	PS	Wasson Rock	73°51'	161°43'	1 560	4.86	114.9	94	524 ±10
14866tr	FO71	Slate	PF	O'Kane Canyon	74°19'	162°21'	1 600	4.18	92.4	98	494 ±10
14930tr	FO79	Mica-shist	PS	Simpson Crags	74°23'	162°57'	1 370	7.69	172.3	95	500 ±10
Boomerang Glacier – Friestley Glacier											
14709bi	BO9	Mica-schist	PS	Wishbone Ridge	74°27'	163°58'	1 365	7.39	163.5	95	495 ±10
14718bi	BO18	Mica-schist	PS	Wishbone Ridge	74°27'	163°58'	1 350	7.75	162.5	97	472 ±9
14731bi	BO31	PS	PS	Wishbone Ridge	74°28'	163°59'	1 170	7.08	156.8	98	495 ±10
14738bi*	BO37	Mica-schist	PF	Tourmaline Plateau	74°15'	163°15'	1 320	2.77	58.1	97	472 ±9

Table 4.6-1. *Continued*

IGNS R. No.	Field No.	Rock type	Geol unit	Location	Latitude °S	Longitude °E	Altitude (m)	K wt.-%	^{40}Ar (rad.)		Age (Ma)
									nl g^{-1}	% tot	
14744mu	BO42A	Pegmatite	GHI	Black Ridge	74°25'	163°40'	750	8.38	175.9	94	472 ±9
14870bi	BO44	Mica-schist	PS	Mt. Dickason	74°23'	163°59'	1720	7.75	166.4	95	482 ±10
14871bi	BO45	Bi-paragneiss	TNF	Mt. Dickason	74°23'	163°58'	1720	7.04	150.8	98	481 ±10
14878tr	BO52	Mylonite	TNF	Mt. Emison	74°12'	163°47'	1610	2.39	50.4	97	474 ±9
14919bi	BO65	Bi-paragneiss	PS	Cape Sastrugi	74°37'	163°39'	140	6.55	141.1	97	483 ±10
14922bi	BO68	Bi-paragneiss	PS	Snowy Point	74°38'	163°45'	180	7.33	154.8	95	475 ±9
Terra Nova Bay											
14899bi	GO2	Granodiorite-gneiss	TNF	Cape Moebius	74°38'	164°13'	10	7.27	152.9	97	473 ±9
14911bi	ABB5	Mica-schist	PS	Tethys Bay	74°42'	164°04'	25	7.13	148.1	96	468 ±9
Millen Range – Leap Year Glacier											
14747tr	MI2	Slate	MS	Mt. McDonald	72°29'	166°42'	2350	3.01	65.3	86	486 ±10
14757tr	MI12	Slate	RBG	Mt. Hancox	72°25'	167°02'	2360	1.67	36.8	98	493 ±10
14761tr	MI16	Slate	RBG	Mt. Hancox	72°35'	167°02'	2360	1.56	34.9	97	499 ±10
14766tr	MI21	Slate	MS	Mt. Hancox	72°36'	167°02'	2480	3.82	86.6	98	505 ±10
14767tr	MI22	Slate	MS	Mt. Hancox	72°36'	167°92'	2480	1.84	37.3	99	458 ±9
14778tr	MI33	Slate	RBG	Mt. Hussey	72°47'	167°28'	2700	2.09	38.1	94	419 ±8
14785tr	MI40	Slate	MF	Jato Nunatak	72°22'	165°53'	2900	2.81	61.5	98	490 ±10
15110tr	MI48	Slate	MS	Gless Peak	72°13'	165°54'	2480	2.42	53.6	96	495 ±10
15115tr	MI53	Slate	MS	Mueller Glacier	72°16'	165°15'	2310	2.53	51.5	96	469 ±9
15116tr	MI54	Slate	RBG	Mueller Glacier	72°16'	165°15'	2310	3.78	72.9	91	439 ±9
14795tr	MI60	Mica-schist	MS	Mueller glacier	72°16'	165°15'	2310	2.63	51.4	92	444 ±9
14798tr	MI63	Slate	RBG	Lensen Glacier	72°22'	163°30'	2340	2.97	68.8	98	515 ±10
14804tr	MI69	Slate	RBG	Lensen Glacier	72°22'	163°30'	2340	1.84	43.6	95	525 ±11
14952tr	MA37	Slate	LYG	Mt. Hayton	72°03'	165°14'	2180	3.82	72.9	97	434 ±9
14953tr	MA38	Slate	LYG	Pyramid Peak	72°14'	165°30'	2400	4.72	93.4	95	448 ±9
Mariner Glacier – Meander Glacier											
14823tr	MA5	Slate	BSC	Black spider	72°23'	166°51'	1350	2.47	46.5	97	429 ±9
14833tr	MA15	Slate	SG	Index Point	73°23'	166°49'	210	2.93	42.6	98	340 ±7
14938bi	MA23	Granodiorite	AI	Emerging Island	73°24'	168°02'	50	5.73	80.1	97	328 ±7
14940bi	MA25	Granodiorite	AI	Mt. Montreuil	73°03'	166°27'	2420	6.12	102.1	98	385 ±8
14943bi*	MA28	Granodiorite	AI	Lawrence Peaks	72°51'	166°06'	1780	5.57	90.4	94	375 ±8

Notes: IGNS (geochronology archive R) number: *bi:* biotite (* where chloritised); *mu:* muscovite; *tr:* total rock, 200–400 micron size. Geological unit: *AI:* Admiralty Intrusives; *BAG:* Berg and Archangel Granites; *BG:* Berg Group; *BSC:* Black Spider Conglomerate; *GHI:* Granite Harbour Intrusives; *GV:* Glagow Volcanics; *LYG:* Leap Year Group; *MS:* Millen Schist; *PF:* Priestley Formation; *PS:* Priestley Schist; *RS:* Rennick Schist; *SG:* Sledgers Group; *TNF:* Terra Nova Formation; *WG:* Wilson Gneiss. Age (Ma): Decay constants ^{40}K; *beta:* 4.962.10^{-10} yr^{-1}; electron capture: 0.581.10^{-10} yr^{-1}; isotopic abundance ^{40}K/K: 0.01167%; atomic. Age errors: 95% confidence limits.

might indicate uplift as young as mid-Devonian, possibly a younger tectonic event more related to that seen in the adjacent Bowers Terrane. However, the marked decrease in K-Ar ages in the latter, to ca. 300 Ma (Adams and Kreuzer 1984), both with depth (i.e. present-day altitude) and proximity to the Rennick Fault (Fig. 4.6-3), reflects a more localised thermal event at the Wilson/Bowers Terrane boundary, at least in part, younger than ca. 300 Ma. Significantly, the zone of oldest (>500 Ma) K-Ar ages, along the eastern Wilson Terrane margin, intervenes between this young (strike-slip?) fault zone, and the young "pop-up" structure of the Wilson/Exiles Thrusts, suggesting that the two features are probably unconnected.

Capponi et al. (1999) demonstrated a sinistral strike-shearing event in Lanterman Range metamorphic rocks close to the Bowers/Wilson terrane boundary (but not confined solely to that boundary fault) and disturbed $^{40}Ar/^{39}Ar$ age patterns from rocks close to the Lanterman Fault (Capponi et al. 2003) were interpreted by them as probably reflecting a late-Ross thermal event. However, Adams and Kreuzer (1984, Table 2) reported a K-Ar age, 420 ±17 Ma for actinolite on late shear planes in the Lanterman Complex. Capponi et al. (2002) preferred a probable Late Devonian-Early Carboniferous age for an event in the Barber Glacier, northern Robertson Bay terrane, but their dating samples were not taken close to the terrane boundaries themselves, whilst those sample locations of Adams and Kreuzer (1984), Wright and Dallmeyer (1991), and Adams (this work) include many close to (<3 km) terrane boundaries. Also, dated samples of these earlier studies were mostly white micas, which are more likely than amphiboles to record subtle low-temperature, late metamorphic events at the terrane boundary faults. Clearly there is a need for more detailed dating on specific fault fabrics within both zones.

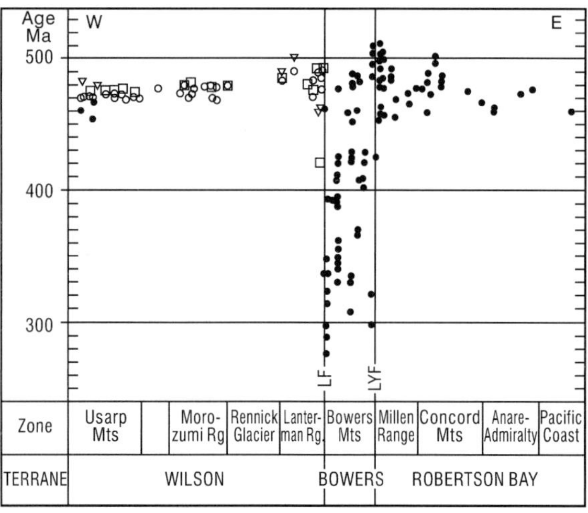

Fig. 4.6-3. Variation of K-Ar ages in North Victoria Land, in relation to Wilson, Bowers and Robertson Bay Terranes (redrawn from data of Adams and Kreuzer 1984). *Solid dots* refer to K-Ar whole-rock ages, all are fine white-mica slates, *open symbols*, various K-Ar mineral (mostly mica) ages. *LF* and *LYF* refer to Lanterman and Leap Year Faults respectively

Fig. 4.6-4. The *Stage II(b)* – Late Ross Orogeny, ca. 480 Ma, in North Victoria Land, Antarctica, of Kleinschmidt and Tessensohn (1987), showing the "pop-up" structure (*shaded area*) defined by Flöttman and Kleinschmidt (1991)

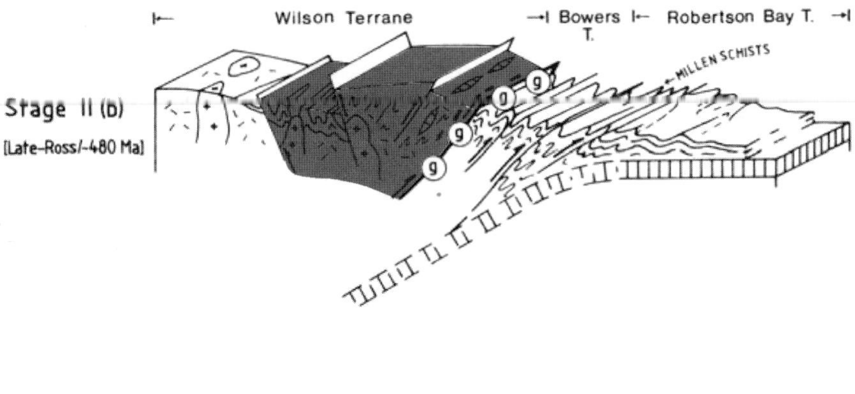

Acknowledgments

I thank Norbert Roland and Carlo Alberto Ricci for invitations to participate in German/Italian Antarctic research activities 1989–1991. Many colleagues on those expeditions are thanked for their ready advice and assistance. The writer also gratefully acknowledges financial support from Alexander von Humboldt Foundation (Germany), and Transantarctic Expedition Association (New Zealand and United Kingdom) to attend ISAES IX in Potsdam and undertake further related studies in Germany. Giovanni Capponi and Peter Horn are thanked for their careful reviews.

References

Adams CJ (1987) Geochronological evolution of the western margin of Northern Victoria Land: Rb-Sr and K-Ar dating of the Berg Group and Berg/Archangel Granites. In: Roland N (ed) German Antarctic North Victoria Land Expedition 1990/91 GANOVEX V, 2. Geol Jb B89:179–104

Adams CJ (1997) Initial strontium isotopic signatures of Late Precambrian-Early Paleozoic metasediments from northern Victoria Land, East Antarctica. In: Ricci C-A (ed) The Antarctic region: geological evolution and processes, Terra Antartica Publication, Siena, pp 227–231

Adams CJ, Graham IJ (1986) Metamorphic and tectonic geochronology of the Torlesse Terrane, Wellington, New Zealand. New Zealand J Geol Geophys 39:157–180

Adams CJ, Kreuzer H (1984) potassium-argon age studies of slates and phyllites from the Bowers and Robertson Bay Terranes, North Victoria Land, Antarctica. In: Roland N (ed) German Antarctic North Victoria Land Expedition 1982/83 GANOVEX III. Geol Jb B60:265–288

Adams CJ, Wodzicki A, Laird MG, Bradshaw JD (1982) Potassium-argon geochronology of the Precambrian-Cambrian Wilson and Robertson Bay Groups and Bowers Supergroup, Northern Victoria Land, Antarctica. In: Craddock C (ed) Antarctic geoscience, Madison, Wisconsin, pp 543–548

Borsi L, Ferrara G, Tonarini S (1989) Geochronological data on Granite Harbour Intrusives from Terra Nova Bay and Priestley Glacier, Victoria Land, Antarctica. Proceedings of the meeting "Earth Science investigations in Antarctica", Siena, 2–3 September 1987. Memoria Societa Italiana 33:161–169

Burrett CF, Findlay RH (1984) Cambrian and Ordovician conodonts from the Robertson Bay Group, Antarctica, and their tectonic significance. Nature 307:723–725

Capponi G, Crispini L, Meccheri M (1999) Structural history and tectonic evolution of the boundary between the Wilson and Bowers terranes, Lanterman Range, northern Victoria Land, Antarctica. Tectonophysics 312:249–266

Capponi G, Castorina F, Di Pisa A, Meccheri M, Petrini R, Villa IM (2002) The metaligeneous rocks of the Barber Glacier area (northern Victoria Land): a clue to the enigmatic Borchgrevink orogeny? In: Gamble J, Skinner D, Henrys S (eds) Antarctica at the close of a millennium, Royal Soc New Zealand Bull 35:99–104

Capponi G, Crispini G, Di Vincenzo G, Palmeri R (2003) Microtextural and petrological insight into shear zones from the Lanterman Range (NVL, Antarctica): contrasting metamorphic evolution at the contact between terranes. Terra Antartica Reports 9:159–162

Flöttman T, Kleinschmidt G (1991) Opposite thrust systems in Northern Victoria Land, Antarctica: Imprints of Gondwana's Paleozoic accretion. Geology 19:45–47

Kleinschmidt G, Tessensohn F (1987) Early Paleozoic westward directed subduction at the Pacific margin of Antarctica. In: McKenzie G (ed) Gondwana Six: structure, tectonics, and geophysics. Geophys Monograph Series 40:89–105

Kreuzer H, Höhndorf A, Lenz H, Müller P, Vetter U (1987) Radiometric ages of Pre-Mesozoic rocks from Northern Victoria Land, Antarctica. In: McKenzie GD (ed) Gondwana Six: structure, tectonics and geophysics, Geophys Monograph Series 40:31–48

Tessensohn F, Duphorn K, Jordan H, Kleinschmidt G, Skinner DNB, Vetter U, Wright TO, Wyborn D (1981) Geological comparison of basement units in North Victoria Land, Antarctica. In: GANOVEX German Antarctic North Victoria Land Expedition 1979/80. Geol Jb B41:31–88

Tonarini S, Rocchi S (1994) Geochronology of Cambro-Ordovician intrusives in Northern Victoria Land: a review. Terra Antartica 1:59–62

Vita G, Lombardo B, Guiliani O (1991) Ordovician uplift pattern in the Wilson Terrane of the Borchgrevink Coast, Victoria Land, Antarctica. Proceedings of the meeting "Earth Science investigations in Antarctica", Siena, 4–6 October 1989. Memoria Societa Italiana 46:257–265

Vita-Scaillet G, Feraud G, Ruffet G, Lombardo B (1994) K/Ar and $^{40}Ar/^{39}Ar$ laser-probe ages of metamorphic micas and amphibole of the Wilson Terrane and Dessent Unit, Northern Victoria Land (Antarctica): their bearing on the regional post-metamorphic cooling history. Terra Antartica 1:59–62

Wright TO, Dallmeyer, RD (1991) The age of cleavage development in the Ross orogen, northern Victoria Land, Antarctica: evidence from $^{40}Ar/^{39}Ar$ whole-rock slate ages. J Structural Geol 13:677–690

Theme 5
Antarctic Peninsula Active Margin Tectonics

Chapter 5.1
Patagonia – Antarctica Connections before Gondwana Break-Up

Chapter 5.2
Moho Depth along the Antarctic Peninsula and Crustal Structure across the Landward Projection of the Hero Fracture Zone

Chapter 5.3
Crustal Thinning and the Development of Deep Depressions at the Scotia-Antarctic Plate Boundary (Southern Margin of Discovery Bank, Antarctica)

Chapter 5.4
Bransfield Basin Tectonic Evolution

Chapter 5.5
The Sedimentary Sequences of Hurd Peninsula, Livingston Island, South Shetland Islands: Part of the Late Jurassic – Cretaceous Depositional History of the Antarctic Peninsula

Chapter 5.6
Regional Structures and Geodynamic Evolution of North Greenwich (Fort Williams Point) and Dee Islands, South Shetland Islands

Chapter 5.7
The Eocene Volcaniclastic Sejong Formation, Barton Peninsula, King George Island, Antarctica: Evolving Arc Volcanism from Precursory Fire Fountaining to Vulcanian Eruptions

Chapter 5.8
Elephant Island Recent Tectonics in the Framework of the Scotia-Antarctic-South Shetland Block Triple Junction (NE Antarctic Peninsula)

Chapter 5.9
Tectonics and Geomorphology of Elephant Island, South Shetland Islands

Chapter 5.10
Geodynamical Studies on Deception Island: DECVOL and GEODEC Projects

The Antarctic Peninsula is the only part of Antarctica which acted as an active margin after Gondwana break-up and dispersal. This active margin is documented by subduction tectonics, magmatism and partly metamorphism since the Jurassic. The situation at this small part of the continent proves to be particularly complicated due to the mixture of active subduction on the one hand, and long-lived passive movement, separating the Peninsula from South America by Gondwana break-up, on the other.

In his chapter, Hervé et al. (Chap. 5.1), following many older attempts, try to find out the way for reconstructing the pre-break-up history of southernmost South America and West Antarctica. Galindo-Zaldívar (Chap. 5.3, 5.4, 5.8), López-Martínez and co-authors (Chap. 5.9) consider the crustal history of the Scotia-Antarctic plate boundary and the evolution of the Bransfield Basin as well as recent tectonics around Elephant Island. The Moho depth alongside the Peninsula is delineated by Janik et al. (Chap. 5.2). Some new data on the Cretaceous and Cenozoic magmatic and structual history of Greenwich Island, Dee Island and King George Island are depicted by Dumont et al. (Chap. 5.6) and Kim et al. (Chap. 5.7) respectively. New and revolutionary Late Cretaceous age data are shown for the Miers Bluff Formation of Hurd Peninsula on Livingston Island by Pimpirev et al. (Chap. 5.5). Berrocoso et al. (Chap. 5.10) deal with the recent geodynamical behaviour of the Deception Island volcano.

The editors regret to say that no contribution on the geological history of the Antarctic Peninsula itself has been sent for publication, so that most part of the Mesozoic history of the Peninsula region is not represented in this context. The results of the workshop "Seymour and James Ross Island Paleoenvironments across the *K*/*T* boundary" will be published in a separate volume elsewhere.

Patagonia – Antarctica Connections before Gondwana Break-Up

Francisco Hervé[1] · Hubert Miller[2] · Christo Pimpirev[3]

[1] Universidad de Chile, Departamento de Geología, Casilla 13518, Correo 21, Santiago, Chile, <fherve@cec.uchile.cl>

[2] Ludwig-Maximilians-Universität, Department of Earth and Environmental Sciences, Sektion Geologie, Luisenstraße 37, 80333 München, Germany, <h.miller@lmu.de>

[3] Sofia University "St. Kl. Ohridski", Department of Geology and Paleontology, 15 Tzar Osvoboditel, 1000 Sofia, Bulgaria, <polar@gea.uni-sofia.bg>

Abstract. The connections between the Antarctic Peninsula and Patagonia, here referred to as South America south of 44° S, are analysed in the light of new geological information and hypotheses. The previously supposed existence of a continuous belt of Late Paleozoic accretionary complexes in the western margin of both Patagonia and the Antarctic Peninsula has been repudiated in recent years, since these complexes are mainly Mesozoic in terms of deposition and metamorphism. This is consistent with paleogeographic models in which the Antarctic Peninsula lays west of Patagonia in pre-break-up times. This disposition is favoured by similarities in provenance between the late Early Permian to ?Late Triassic Duque de York Complex and Trinity Peninsula Group, which share sedimentological characteristics and U-Pb detrital zircon age patterns. After the Early Jurassic Chonide orogeny, the Antarctic Peninsula started to drift southwards, as indicated by paleomagnetic reconstructions of Weddell Sea ocean floor spreading, allowing the present-day margin of Patagonia to become progressively, from north to south, an actively subducting margin. In the Late Jurassic and Late Cretaceous, while the Antarctic Peninsula was drifting to its present position, turbiditic sedimentation took place at Hurd Peninsula, Livingston Island, previously considered to be Triassic in age, which has no equivalent in the Patagonian margin.

The new data presented here, combined with recent data and models from other authors, reveal two major geological problems to be resolved in the future: to identify a late Early Permian magmatic arc which shed detritus to both the Duque de York Complex and the Trinity Peninsula Group, and to resolve the apparent contradiction between left lateral post break-up movements in the Patagonian Andes, and right lateral displacements in the tectonic configuration of the Antarctic Peninsula.

Introduction

The Antarctic Peninsula is one of a group of continental fragments or terranes that constitute West Antarctica (Dalziel and Elliot 1982; Storey 1991; Grunow 1993). It has geologic features that are similar to those of southern South America (Patagonia; Dalziel and Elliot 1973), from which it is separated mainly by Cenozoic oceanic crust (Barker and Burrell 1977) forming the present-day Drake Passage (Fig. 5.1-1). These fragments are considered to have been part of the Gondwana super-continent before its break-up and ensuing dispersal (Dalziel and Elliot 1971, 1973). This paper will focus in the geologic units that predate the formation of the Jurassic Chon Aike large igneous siliceous province which heralded the Gondwana break-up (Pankhurst et al. 2000). Knowledge of the timing of deposition, intrusion and/or metamorphism of the pre-break-up low grade metamorphic units that crop out in the Patagonian Andes and of the metamorphic and igneous units in extra-Andean Patagonia, has increased enormously in the last decade, owing mainly to new field studies in remote areas, new paleontological findings, and the application of modern dating techniques. Similar advances have been made in the Antarctic Peninsula which has been subject to renewed research, leading to new findings and tectonic interpretations. The purpose of this paper is to present an integrated summary of the new data obtained in both areas, both by the authors and by other researchers, and to compare their geologic development in the light of this new information. A hypothesis about the relative paleogeographic positions of the Antarctic Peninsula and Patagonia will be put forward, and geologic problems arising from the new information will be raised.

Geologic Constitution of Patagonia

Patagonia is a broad term used to encompass the southern portion of the South American continent; we will consider here only the area south of 44° S. It consists of the Patagonian and Fuegan segments of the Andes on the Pacific side, and a platform, called extra-Andean Patagonia, extending east of the Andes to the Atlantic coast. In this platform, two main outcrop areas of pre-break-up rocks are termed the North Patagonian Massif to the North and the Deseado Massif to the Southeast. Otherwise, most of extra-Andean Patagonia is covered by the Jurassic volcanic rocks of the Chon Aike – Marifil – El Quemado units, and by the mainly sedimentary Mesozoic and Cenozoic deposits of the Chubut and the Magellan or Austral Basins.

The Patagonian Andes

The Patagonian Andes consist of a topographically rather subdued mountain belt, which bears witness to processes spanning the Late Palaeozoic to the present. The ages and

From: Fütterer DK, Damaske D, Kleinschmidt G, Miller H, Tessensohn F (eds) (2006) Antarctica: Contributions to global earth sciences. Springer-Verlag, Berlin Heidelberg New York, pp 217–228

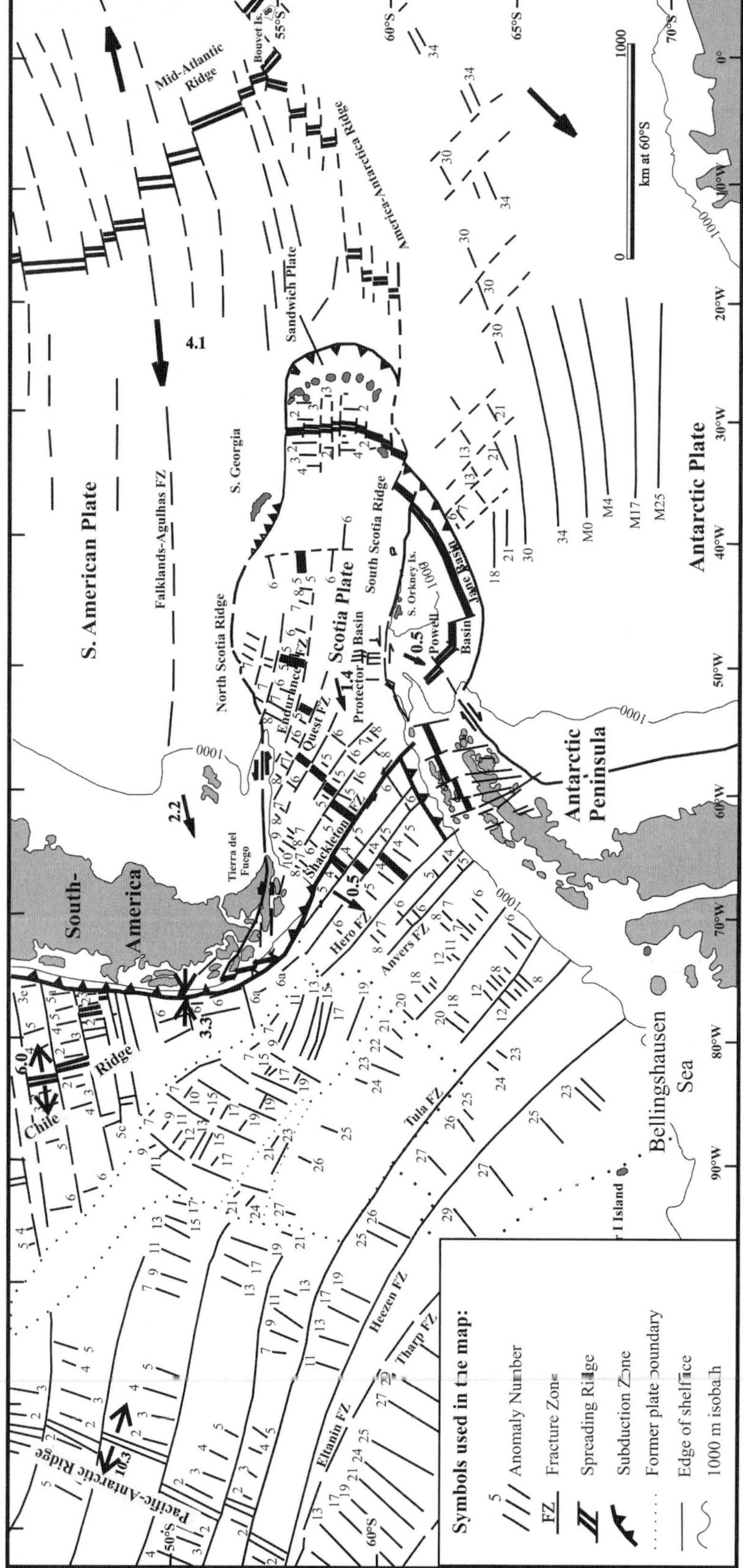

Fig. 5.1-1. Map of the Scotia Arc Region showing the main geographical and structural features. Modified from Veit (2002)

geological characteristics of the pre-break-up units will be briefly summarized below.

North of the Magellan Strait, the most extensive unit of the Patagonian Andes is the Mesozoic to Cenozoic Patagonian Batholith (Bruce et al. 1991). Its earliest (ca. 150 Ma) components (Martin et al. 2001) intrude low grade metamorphic complexes, which at present crop out both west and east of the continuous batholithic belt. These complexes were considered to be time equivalents in the 1:1 000 000 Geologic Map of Chile, sheet 6 (Escobar 1982). Research work in the last decade, however, has modified these views and has led to the subdivision of these units that follows.

Eastern Andes Metamorphic Complex (EAMC)

Hervé et al. (1998, 2003), Faúndez et al. (2002), Augustsson and Bahlburg (2003), Lacassie (2003) and Augustsson (2003) have concluded that this unit has sedimentary components deposited during the late Devonian–Early Carboniferous, as well as younger deposits in their western outcrop areas, ranging into the Permian. They represent deposition in a passive margin environment, and were metamorphosed before the Late Permian (Thomson and Hervé 2002), under lower *P*/*T* conditions than those that are typical of accretionary complexes (Ramírez 2002).

Coastal Accretionary Complexes

The Chonos Metamorphic Complex (CMC), the Madre de Dios Accretionary Complex (MDAC) and the Diego de Almagro Metamorphic Complex (DAMC) crop out west of the Patagonian Batholith (PB) from north to south (Fig. 5.1-2a).

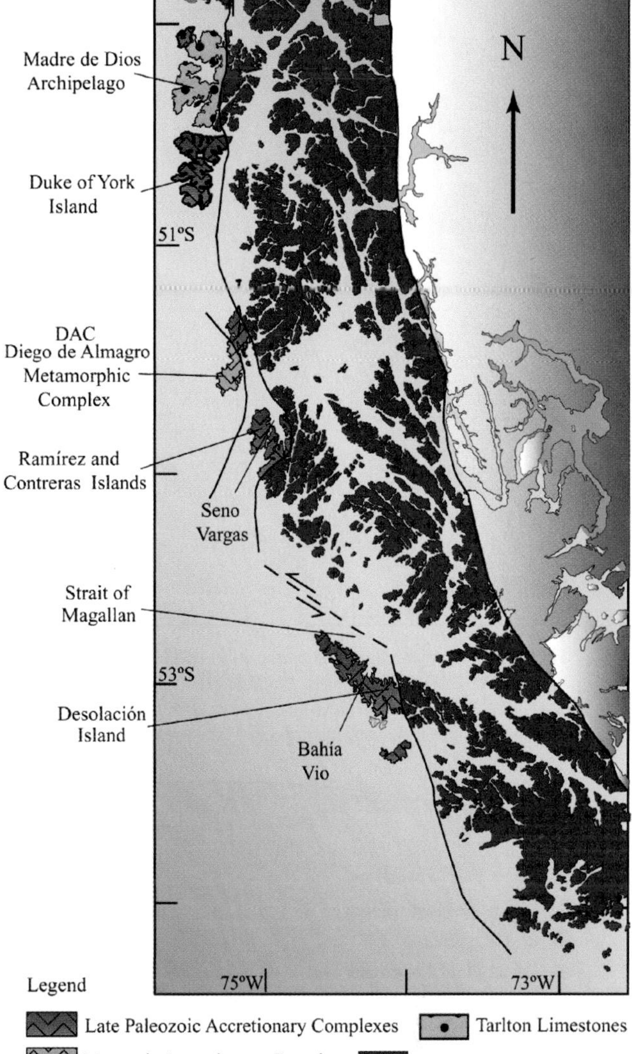

Fig. 5.1-2a. Outcrop areas of the main pre-break-up units in Patagonia. The distribution of the coastal metamorphic complexes south of 50° S

The Chonos Metamorphic Complex has Late Triassic depositional ages, as indicated by fossil fauna (Fang et al. 1998) and by detrital zircon SHRIMP U-Pb age determinations (Hervé and Fanning 2001; Hervé et al. 2003). It consists predominantly of turbidites (Pimpirev et al. 1999), with more restricted occurrences of mafic schists and metacherts. They were metamorphosed under high *P*/*T* metamorphic conditions (Willner et al. 2000) before the Early Jurassic (Thomson and Hervé 2002), a metamorphic event referred to here as the Chonide orogeny.

The Madre de Dios Accretionary Complex is composed of three tectonically interleaved lithostratigraphic units: the Tarlton limestone (TL), the Denaro Complex (DC) and the Duque de York Complex (DYC) (Forsythe and Mpodozis 1979, 1983). The Late Carboniferous-Early Permian (Douglass and Nestell 1976) Tarlton limestone was deposited over a pene-contemporaneous (Ling et al. 1985) oceanic substrate, the Denaro Complex, in an oceanic ridge or sea mount environment and later accreted to the Gondwana continental margin. The DYC represents a turbiditic association, which was deposited uncomformably over the TL, when the latter reached the vicinity of the continental margin (Forsythe and Mpodozis 1983). The DYC has Early Permian radiolarian cherts (Yoshiaki written comm. 2002) and sandstones containing detrital zircons of late Early Permian age (Hervé et al. 2003). The DYC was metamorphosed before the earliest Jurassic (Thomson and Hervé 2002).

The Diego de Almagro Complex is composed of amphibolites, blueschists, greenschists and quartz-mica schists, as well as by an orthogneiss (Hervé et al. 1999). SHRIMP U-Pb ages in zircons from both the orthogneiss and a micaschist (Hervé and Fanning 2003) have yielded Middle Jurassic ages interpreted as the age of crystallization of the igneous forerunners. The high *P*/*T* meta-

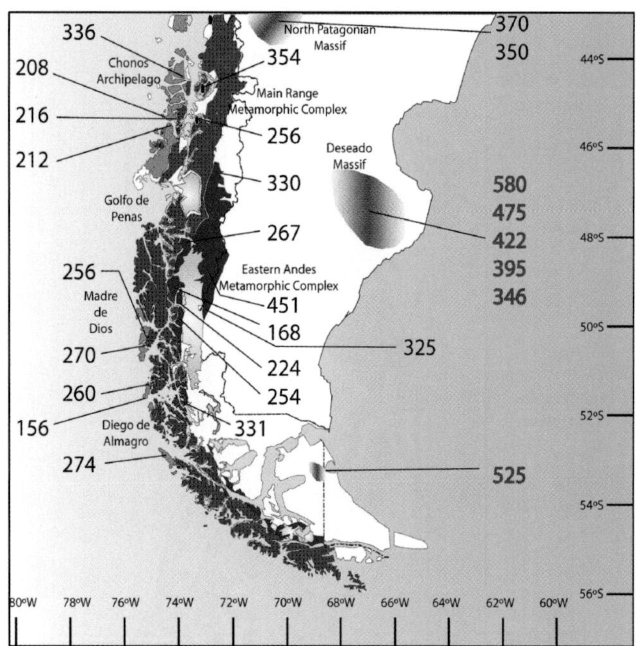

Fig. 5.1-2b. Outcrop areas of the main pre-break-up units in Patagonia with U-Pb SHRIMP zircon ages. The main references are Pankhurst et al. (2003), Söllner et al. (2000) and Hervé et al. (2003). *Black numbers:* younger detrital ages; *red numbers:* igneous crystallization ages

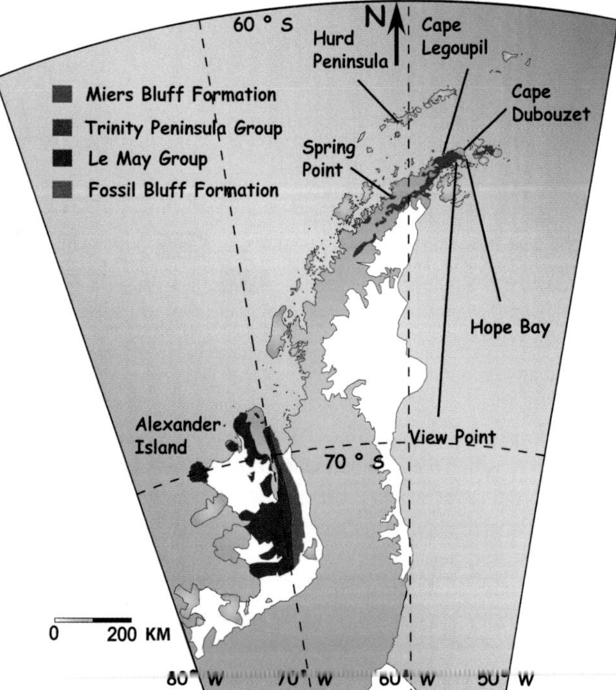

Fig. 5.1-2c. Outcrop areas of the main pre-break-up units in the Antarctic Peninsula. The distribution of rock outcrops constituting the Trinity Peninsula Group of the Antarctic Peninsula, the Miers Bluff Formation of Livingston Island, and the Fossil Bluff Formation and LeMay Group of Alexander Island

morphism which followed developed through the Cretaceous. This unit is in tectonic contact (Forsythe 1981, 1982) with the DYC along the sinistral strike slip Seno Arcabuz shear zone (Olivares et al. 2003).

Extra-Andean Patagonia

The pre-Mesozoic metasedimentary and plutonic units of the extra-Andean Patagonia which crop out in the Deseado massif, have been dated recently by Pankhurst et al. (2003). They include metasedimentary rocks of probable latest Neoproterozoic or Cambrian depositional age, and Cambrian, Ordovician and Late Silurian to early Carboniferous plutonic bodies (Fig. 5.1-2b). Söllner et al. (2000) and Pankhurst et al. (2003) dated ≈ 525 Ma orthogneisses recovered from the bottom of oil wells in Tierra del Fuego, indicating that Early Paleozoic rocks could extend over large tracts of Patagonia under the Mesozoic-Cenozoic cover.

Antarctic Peninsula

Dalziel (1984a) has given full details about the structural geology and tectonic relations of the Southern Scotia Ridge, including the South Shetland Islands and northern Antarctic Peninsula, and its evolution as a fore arc terrane. The tectonic configuration of the whole Antarctic Peninsula region has been recently envisaged as consisting of three domains which could represent the amalgamation of suspect terranes (Vaughan and Storey 2000). A lithostratigraphic chart derived from this model is presented in Fig. 5.1-3. The pre-break-up units in the Antarctic Peninsula (Millar et al. 2002) comprise sparse outcrops of Paleozoic plutonic and metamorphic rocks in the Eastern and Central domains (south-eastern part of the Peninsula), the Trinity Peninsula Group (TPG) which has a wide distribution in the northern part of the Antarctic Peninsula (Fig. 5.1-2c), and the lower part of the LeMay Group in the Western Domain.

Trinity Peninsula Group (TPG)

The TPG consists of a turbidite association, which includes sparse metabasites (Hyden and Tanner 1981). Its outcrops in the Antarctic Peninsula itself have been studied in some detail by Smellie (1991), Trouw et al. (1997) and Birkenmajer (2001). The unit has yielded Triassic fossils (Thomson 1975) from the Legoupil Formation near Base O'Higgins. On the South Orkney Islands late Triassic radiolarian recovered from pelagic cherts (Dalziel et al. 1981) have proven age relations of this oceanic realm to the fore arc subduction complex of the TPG.

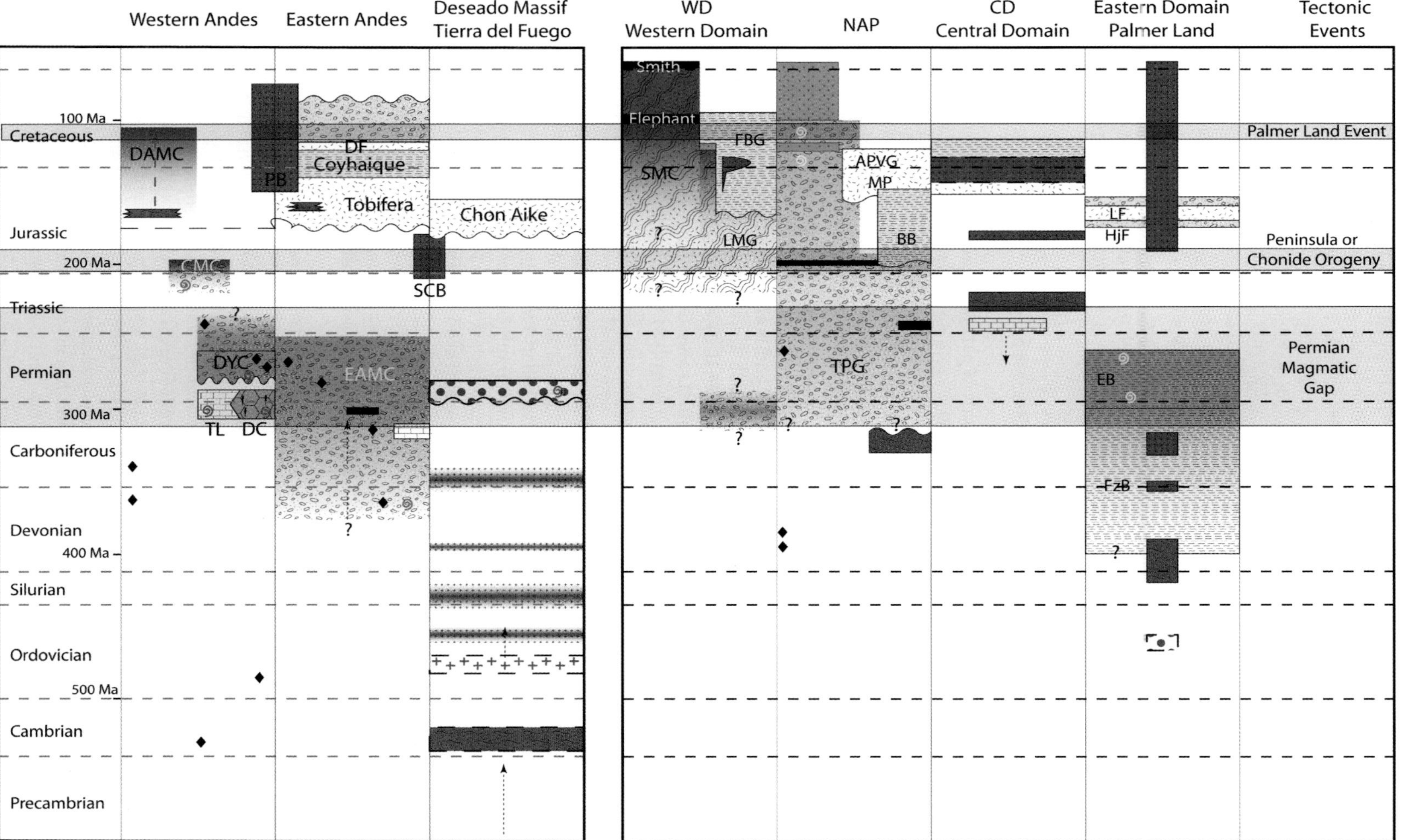

Fig. 5.1-3. Stratigraphic correlation chart between the Antarctic Peninsula and Patagonia pre-break up units. The column headings correspond to a modification of the domains proposed by Vaughan and Storey (2000) for the Antarctic Peninsula. The Graham Land units are presented under the heading for *NAP* (Northern Antarctic Peninsula), separate from the Eastern Domain within which it was included in the original model. *Black diamonds* indicate U-Pb detrital zircon age populations. Abbreviations: Patagonia: *CMC:* Chonos Metamorphic Complex; *DAMC:* Diego de Almagro Metamorphic Complex; *DC:* Denaro Complex; *DF:* Divisadero Formation; *DYC:* Duque de York Complex; *EAMC:* Eastern Andean Metamorphic Complex; *PB:* Patagonian Batholith; *SCB:* Sub-Cordilleran Batholith; *TL:* Tarlton Limestone. Antarctic Peninsula and adjacent islands: *APVG:* Antarctic Peninsula Volcanic Group; *BB:* Botany Bay Group; *EB:* Erehwon Beds; *FBG:* Fossil Bluff Group; *FzB:* Fitzgerald Beds; *HjF:* Hjort Formation; *LF:* Latady Formation; *LMG:* LeMay Group; *MP:* Mount Poster Formation; *NAP:* Northern Antarctic Peninsula; *SMC:* Scotia Metamorphic Complex; *TPG:* Trinity Peninsula Group

In the South Shetland Islands, the Miers Bluff Formation (Fig. 5.1-2c) mainly consisting of turbidites at Hurd Peninsula, Livingston Island, has traditionally been correlated with the TPG and was assigned a Triassic age based on Rb-Sr age determinations (Willan et al. 1994). Also, Deng et al. (2002) describe Late Triassic palinomorphs in the unit. However, Pimpirev et al. (2002) found a late Jurassic ammonite, not in situ, but in a lithology similar to surrounding outcrops near the Bulgarian base. Furthermore, Stoykova et al. (2002) have found calcareous nannoplankton fossils of Late Cretaceous age in rocks of the same area (see also Pimpirev et al. this volume). SHRIMP U-Pb zircon age determinations show a population of Jurassic age (Hervé and Fanning, unpubl. data) thus indicating that at least in part the rocks earlier assigned to the TPG here are younger than the supposed equivalent rocks in the Antarctic Peninsula and in the subsurface rocks of its north-eastern tip (Loske et al. 1990; Miller and Loske 1992).

On Livingston Island, there is a considerable difference in sedimentary facies and structural deformation between the strongly folded, overturned, mostly psammitic turbidites of Hurd Peninsula (Miers Bluff Fm.), on the one hand, and the weakly folded, mostly mudstone, partly volcanic and non-marine series of comparable age on Byers Peninsula (Valenzuela and Hervé 1972; Hathaway and Lomas 1998) on the other. Apparently, the faults separating Hurd Peninsula from the south-eastern and from the north-western parts of Livingston Island (Hobbs 1968) play a much more important role in the tectonics of the South Shetland Islands as considered earlier. The Miers Bluff Fm. of Hurd Peninsula was possibly deposited in a multiple source gravel-rich deep-sea ramp system and was deformed and emplaced along strike-slip faults (Pimpirev et al. 1997). Hence, new aspects are to be added to the terrane concepts presented by Vaughan and Storey (2000).

The Scotia Metamorphic Complex

This complex of metapelites, metabasites, metacherts and marbles was originally considered to be of Precambrian age. Later, it was usually considered to be pre-Upper Jurassic, as it was correlated with the low-grade metamorphic complexes of southern South America, where at Peninsula Staines (Allen 1982) the Upper Jurassic Tobífera volcanic rocks are unconformably deposited over the Staines Complex, now part of the Eastern Andean Metamorphic Complex of Patagonia. This stratigraphic relationship was extrapolated to the Antarctic Peninsula region, on the assumption that the low-grade metamorphic complexes in Patagonia, South Orkney Islands and the Antarctic Peninsula were coeval. Deformation and metamorphism were said to have begun before break-up of Gondwana and to have continued into middle to late Mesozoic or even Cenozoic times (Dalziel 1984a; Trouw et al. 1997).

Rb-Sr, K-Ar (Tanner et al. 1982; Hervé et al. 1984) and Ar-Ar dating (Féraud et al. 2000) have revealed that the metamorphic ages of the Scotia Metamorphic Complex are Mesozoic: Early Cretaceous on Elephant Island, Late Cretaceous on Smith Island. The age of the protolith is still unknown, and some isotopic evidence suggests that part of the Elephant Island group might be Paleozoic in age (e.g., Hervé et al. 1990; Hervé et al. 1991).

The LeMay Group

The western part of Alexander Island consists of a subduction-accretion complex of lithological affinity to the Trinity Peninsula Group and Miers Bluff Fm. (Burn 1984), overlain by Tertiary volcanic rocks. Pillow basalts with oceanic island affinities, arc-derived conglomerates, turbidites and basalt-chert associations are present (Burn 1984). The turbidite lithofacies from the LeMay Group (Fig. 5.1-2c) are similar to those described from the Miers Bluff Fm., and both turbidite sequences may represent deposition in a progradational system (Nell et al. 1989; Pimpirev et al. 1997). Biostratigraphically dated rocks from the LeMay Group range from Lower Jurassic to Albian (Burn 1984; Holdsworth and Nell 1991).

Since the finding of Late Cretaceous nannofossils in the ?uppermost part of the Miers Bluff Fm. on Hurd Peninsula (south-east coast of South Bay, Livingston Island), the resemblance of the Trinity Peninsula Group and Miers Bluff Fm. to the LeMay Group now has also a biostratigraphical aspect. In both regions turbidite rocks are overlain by Jurassic rocks in the east, whereas to the west (Pacific side) they are coeval with them and even younger.

The Antarctic Peninsula Batholith

This batholith extends along the Antarctic Peninsula and the South Shetland Islands. Pre-Jurassic and Jurassic plutons are found chiefly within the central mid Peninsula, whereas Early Cretaceous magmatism occurs very widespread over the whole region. From Late Cretaceous on, granitoids concentrate at the western margin (Pankhurst 1990; Leat et al. 1995). On the contrary, the Patagonian Batholith has a younger middle part flanked by older plutons. As an exception, in the Fuegan Andes the Patagonian Batholith has an age structure which gets younger towards the trench, in a similar way to the northern Antarctic Peninsula Batholith.

Comparison between Patagonian and Antarctic Pre-Break-Up Units

The chronostratigraphic positions of the main pre-break up units in Patagonia and the Antarctic Peninsula are shown in Fig. 5.1-3. Some will be compared in more detail in the following section.

Duque De York Complex (DYC) and Trinity Peninsula Group (TPG)

These two units share a common turbidite lithology, have similar low metamorphic grade, and comparable provenance, as indicated by their U-Pb detrital zircon age patterns. The depositional age is less well established, but in both units it is bracketed between the Early Permian and the Early Jurassic.

They are mainly composed of greywacke-shale alternations, with minor but conspicuous conglomerate and diamictite beds, and scarce metabasic rocks and cherts. Their deformation is variable, and although it is complex and polyphase, primary sedimentary features are preserved in most outcrops. Metamorphic grade, though not thoroughly investigated, reveals a sub-greenschist facies assemblage with predominant white mica, chlorite, albite and quartz in the metapsammopelitic rocks, and pumpellyite-actinolite facies in the mafic rocks. Metamorphism in the DYC took place before the Early Jurassic as indicated by zircon FT data (Thomson and Hervé 2002). The TPG also records a pre-Early Jurassic metamorphic and deformational event, which is registered in Botany Bay, Antarctic Peninsula, where Lower Jurassic sedimentary rocks uncomformably overlie the TPG rocks.

The age patterns of detrital zircons in DYC and TPG rocks are similar, particularly with respect to the younger population of Early Permian zircons present in the rocks analysed so far. The detrital zircon age patterns (Hervé et al. 2003) for rocks from the DYC which crop out at the Diego de Almagro and Madre de Dios archipelagos (Fig. 5.1-4) are characterised by a large peak of Early Permian zircons, with smaller populations of Early Paleozoic, and some Middle Proterozoic grains. Similar patterns are presented by rocks at Isla Desolación (Hervé et al. 2003), which forms the southern shore of the western entrance of the Strait of Magellan in the Patagonian Andes (Fig. 5.1-4). All these patterns are very similar to the one shown by a rock from TPG which crops out at Spring Point, in the western coast of the Antarctic Peninsula (Fig. 5.1-4). A similar pattern is presented by Millar et al. (2002), and taken altogether, they indicate that the main source of detritus for the depositional environment of both TPG and DYC was an Early Permian magmatic province.

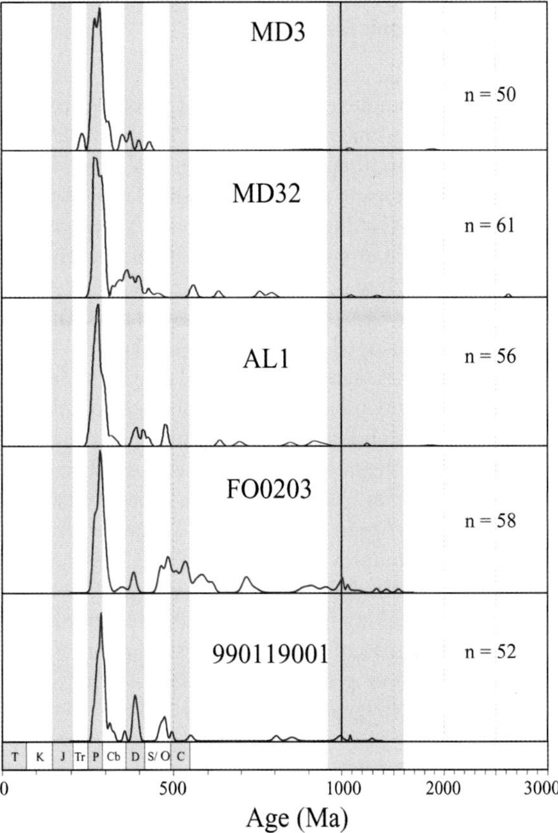

Fig. 5.1-4. U-Pb SHRIMP age – *versus* probability diagrams of detrital zircons of low grade metasandstones of the Duque de York Complex, Patagonia (after Hervé et al. 2003), and of the Trinity Peninsula Group, Antarctica (sample 990119001: Hervé, unpublished data). The *DYC* samples span geographically from the Madre de Dios Archipelago to the northern shore of Isla Desolación. The *TPG* sample is from outcrop at Spring Point, Antarctic Peninsula

Hurd Peninsula

The latest Jurassic (Pimpirev et al. 2002) and late Cretaceous (Stoykova et al. 2002; Pimpirev et al. this volume) sedimentary age of parts of the Miers Bluff Formation at Hurd Peninsula seems now well established (Pimpirev et al. this volume). No rocks of similar age are to be found in the Pacific side of the Patagonian Andes, west of the Patagonian Batholith. However, rocks of similar ages are well known from the Magellan or Austral Basin, east of the Andes, and from the Weddell coast of the Antarctic Peninsula. This outcrop distribution could result from a complete erosion of Jurassic and Cretaceous sedimentary rocks from the western slope of the Andean range in Patagonia, or from a different tectonic environment between the Antarctic Peninsula and the Patagonian Andes during that period.

Jurassic Metamorphic Complexes

Metamorphic complexes with Jurassic protoliths crop out in the Diego de Almagro (Hervé and Fanning 2003) and Diego Ramírez archipelagos (Davidson et al. 1989) in the Patagonian Andes, and in Alexander Island near the Antarctic Peninsula (LeMay Group). They also share in common the presence of metabasites of oceanic provenance, and the high *P*/low *T* nature of their metamorphism (Forsythe 1982; Wilson et al. 1989; Willner et al. 2004; Burn 1984) which developed blueschist assemblages.

In both areas, Patagonian Andes and Alexander Island, tectonic activity compatible with east directed subduction, continued well into the Cretaceous. This is revealed by the dating of the mylonites of the Seno Arcabuz shear zone (Olivares et al. 2003) which constitute the eastern limit of the outcrops in the case of the Diego de Almagro Metamorphic Complex (Hervé et al. 1999). The continuous outcrops of the LeMay Group are cut on the east by the strike-slip LeMay fault, east of which Bathonian to Albian? forearc rocks of the Fossil Bluff Group (Fig. 5.1-2c) are deposited over pre-Middle Jurassic LeMay Group rocks (Burn 1984; Vaughan and Storey 2000).

Paleogeographic Considerations

The lithologic, provenance and time constraints imposed by geological knowledge of the pre-break-up units in Patagonia and the Antarctic Peninsula, offer some possibilities of establishing their relative original positions along the Gondwana margin. Even though the unknowns are manifold, this has been attempted frequently and contradictorily in the past (e.g., Dalziel and Elliot 1971; De Wit 1977; Harrison et al. 1979; Dalziel 1980, 1982, 1983, 1984a,b; Dalziel et al. 1981; Miller 1983a,b; Ghidella and LaBrecque 1997; Lawver et al. 1998; Kovacs et al. 2002; Ghidella et al. 2002). Some new ideas will be presented here.

Paleogeographic Models

Recent paleogeographic reconstructions (Lawver et al. 1998, "tightest-fit model"; Ghidella et al. 2002 and Jokat et al. 2003, "evolution of the Weddell Sea by paleomagnetic research") prefer a relative position in which the Antarctic Peninsula is figured west of and in contact with Patagonia during the Early Jurassic, and reaching as far north as the Golfo de Penas (50° S). This model resulted from paleoplate reconstructions based on the relative positions of New Zealand and West Antarctica (Lawver et al. 1998), but lacked known geological support from Patagonia. However, the concept that the western margin of Patagonia and the Antarctic Peninsula were made up of Upper Paleozoic accretionary complexes (e.g., Hervé et al. 1981; Smellie and Clarkson 1975), thought to be formed at the active continental SW Gondwana margin, was difficult to reconcile with a substantial overlap of the Antarctic Peninsula and Patagonia, favouring instead only a modest latitudinal overlap or a linear continuity between them.

The present knowledge of the depositional and metamorphic ages of the pre-break-up units, is geologically consistent with the Lawver et al. (1998) tight fit model, in

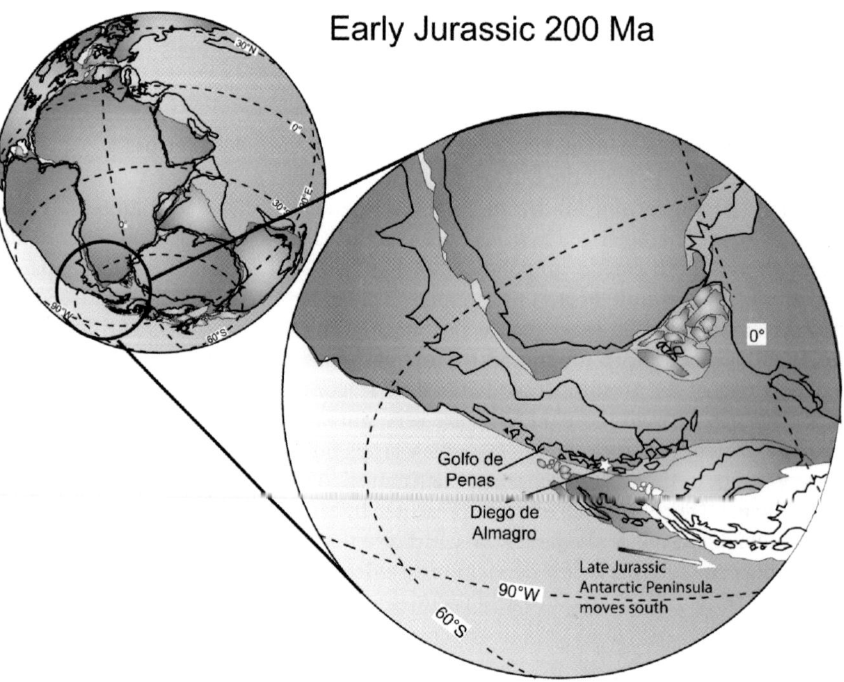

Fig. 5.1-5. Paleogeographic tight-fit model, modified from Lawver et al. (1998) showing the southward displacement of the Antarctic Peninsula which started no later than the Late Jurassic

the sense that the accretionary complexes indicating active subduction are now known to be Middle Jurassic or later. The opening of the Weddell sea since the Jurassic has contributed to the separation of both continental land masses. Recently, Ghidella et al. (2002), analysing the magnetic anomalies of the sea floor in the Scotia Sea-Weddell Sea areas, have produced a series of paleogeographic maps, with the relative positions of the Antarctic Peninsula and Patagonia, from Late Jurassic to the present, which provide new insights. The Ghidella et al. (2002) model implies mainly south directed movement of the Antarctic Peninsula, located west of Patagonia before 100 Ma, which requires left lateral strike-slip movements, along the Patagonia-Antarctic Peninsula boundary, but probably taking place over a wider area. From 100 Ma on, a southeastward shift in the movement of the Antarctic Peninsula becomes predominant. Opening of first the Weddell Sea and later the Scotia Sea may be considered as a sort of mega-pull apart basin, with the Antarctic Peninsula block moving jointly with East Antarctica without being adjacent to it (Grunow 1993).

This model is consistent with the ages of the blueschist complexes in the Patagonian Pacific margin south of 50° S, which are not older than Late Jurassic-Early Cretaceous. It is proposed here that the Duque de York Complex and the Trinity Peninsula Group were once a contiguous rock mass, and that part of the combined complex was displaced south by the movement of the Antarctic Peninsula (Fig. 5.1-5). The Late Jurassic to Late Cretaceous sedimentation on part of Hurd Peninsula took place when the AP was somewhere halfway in its movement to its present position, and has no counterpart in the western margin of Patagonia. The environment could have resembled that of present-day New Zealand, with active strike slip movement occurring inside the Antarctic Peninsula.

New Problems

Mesozoic and Cenozoic strike-slip movements in Patagonia have been considered to be left lateral, on structural (Olivares et al. 2003) and paleomagnetic data (Rapalini et al. 2001). However, they are considered to have been right lateral in the Antarctic Peninsula (Vaughan and Storey 2000; Vaughan et al. 2002). The mode of origin and timing of these movements is not precisely known, so this apparent question remains to be solved.

The existence of abundant Early Permian detrital zircons in both the Duque de York Complex and in the Trinity Peninsula Group suggests that they were deposited in a basin which was fed by detritus from a magmatic arc of late Early Permian age. However, no such magmatic arc is known in the Antarctic Peninsula or in the Patagonian region considered. This raises the question of whether both may have in fact a far more exotic origin along the Gondwana margin, as suggested by Lacassie (2003), considering that voluminous Permian igneous complexes are known in South America north of 42° S (Llambías 1999) and in Australia (e.g., Leitch 1975).

Acknowledgments

This paper is the fruit of a long lasting collaboration between the authors, which have been supported in the field and laboratory work by Instituto Antártico Chileno (FH), the Alfred-Wegener-Institut and Deutsche Forschungsgemeinschaft (HM) and Bulgarian Antarctic Institute (CP). FH acknowledges additional support from Programa Institucional Antártico, Universidad de Chile and FONDECYT project 1010412. R. J. Pankhurst, A. Milne, A. Vaughan and W. Loske have shared field, laboratory work and discussion on these topics during many years. V. Faúndez and J. P. Lacassie prepared some of the figures and helped with the final version of the manuscript. The reviewers R. J. Pankhurst and I. Dalziel have kindly contributed to improve the manuscript.

References

Allen RB (1982) Geología de la Cordillera Sarmiento, Andes Patagónicos entre los 51° y 52° Lat. Sur, Magallanes, Chile. Serv Nac Geol Minería, Bol 38:1–46

Augustsson C (2003) Provenance of Late Paleozoic sediments in the southern Patagonian Andes: age estimates, sources and depositional setting. PhD Thesis, Universität Münster

Augustsson C, Bahlburg H (2003) Active or passive continental margin? Geochemical and Nd isotope constraints of metasediments in the backstop of a pre-Andean accretionary wedge in southernmost Chile (46°30'–48°30' S). In: McCann T, Saintot A (eds) Tracing tectonic deformation using the sedimentary record. Geol Soc London Spec Publ 208:253–268

Barker PF, Burrell J (1977) The opening of the Drake Passage. Marine Geol 25:15–34

Birkenmajer K (2001) Mesozoic and Cenozoic stratigraphic units in parts of the South Shetland Islands and northern Antarctic Peninsula (as used by the Polish Antarctic Programmes). Studia Geol Polonica 118:1–188

Bruce RM, Nelson E, Weaver SG, Lux DR (1991) Temporal and spatial variations in the southern Patagonian batholith: constraints on magmatic arc development. In: Harmon RS, Rapela CW (eds) Andean magmatism and its tectonic setting. Geol Soc Amer Spec Paper 265, pp 1–12

Burn RW (1984) The geology of the LeMay Group, Alexander Island. Brit Antarc Surv Sci Rep 109:1–65

Dalziel IWD (1980) Comment on 'Mesozoic evolution of the Antarctic Peninsula and the southern Andes'. Geology 7:260–261

Dalziel IWD (1982) The early (pre-middle Jurassic) history of the Scotia Arc region. A review and progress report. In: Craddock C (ed) Antarctic Geoscience. University of Wisconsin Press, Madison, pp 111–126

Dalziel IWD (1983) The evolution of the Scotia Arc: a review. In: Oliver RL, James PR, Jago JB (eds) Antarctic Earth Science. Australian Acad Sci, Canberra, pp 283–288

Dalziel IWD (1984a) Tectonic evolution of a forearc terrane, southern Scotia Ridge, Antarctica. Geol Soc Amer Spec Paper 200:1–32

Dalziel IWD (1984b) The Scotia Arc: an international geological laboratory. Episodes 7:8–13

Dalziel IWD, Elliot DH (1971) Evolution of the Scotia Arc. Nature 233:246–252

Dalziel IWD, Elliot DH (1973) The Scotia Arc and Antarctic margin. In: Nairn AFM, Stehli FG (eds) The ocean basins and margins, I. The South Atlantic. Plenum Press, New York, pp 171–246

Dalziel IWD, Elliot DH (1982) West Antarctica, problem child of Gondwanaland. Tectonics 1:3–19

Dalziel IWD, Elliot DH, Jones DL, Thomson JW, Thomson MRA, Wells NA, Zinsmeister WJ (1981) The geological significance of some Triassic microfossils from the South Orkney Islands, Scotia Ridge. Geol Mag 118:15–25

Davidson J, Mpodozis C, Godoy E, Hervé F, Muñoz N (1989) Jurassic accretion of a high buoyancy guyot in southernmost South America: the Diego Ramírez Islands. Rev Geol Chile 16:247–251

De Wit MJ (1977) The evolution of the Scotia Arc as a key to the reconstruction of southwestern Gondwanaland. Tectonophysics 37:53–81

Deng X, Zheng X, Liu X, Ouyang S, Shen Y (2002) Terrestrial palynomorphs from the Miers Bluff Formation of Livingston Island, West Antarctica. In: Gamble JA, Skinner DNB, Henrys S (eds) Antarctica at the close of a millennium. Royal Soc New Zealand Bull 35, pp 269–273

Douglass, RC, Nestell, MK (1976) Late Palaeozoic foraminifera from southern Chile. U.S. Geol Surv Spec Paper 858:1–49

Escobar F (ed) (1982) Mapa Geológico de Chile, escala 1:1 000 000, hoja 6. Servicio Nacional de Geología y Minería, Santiago

Fang Z, Boucot A, Covacevich V, Hervé F (1998) Discovery of Late Triassic fossils in the Chonos Metamorphic Complex, Southern Chile. Rev Geol Chile 25:165–173

Faúndez V, Hervé F, Lacassie JP (2002) Provenance studies of pre-late Jurassic metaturbidite successions of the Patagonian Andes, southern Chile. New Zealand J Geol Geophys 45:411–425

Féraud G, Hervé F, Morata D, Muñoz V, Toloza R (2000) Scotia Metamorphic Complex, Antarctica: evidences for the diachronous build up of a subduction complex. IX Congr Geol Chileno, resum expand 2:374–377

Forsythe RD (1981) Geological investigations of pre-Late Jurassic terranes in the southernmost Andes. PhD thesis, Columbia University, New York

Forsythe RD (1982) The late Palaeozoic and Early Mesozoic evolution of southern South America: A plate tectonic interpretation. J Geol Soc London 139:671–682

Forsythe RD, Mpodozis C (1979) El Archipiélago Madre de Dios, Patagonia Occidental, Magallanes: rasgos generales de la estratigrafía y estructura del basamento pre-Jurásico Superior. Rev Geol Chile 7:13–29

Forsythe RD, Mpodozis C (1983) Geología del Basamento pre-Jurásico Superior en el Archipiélago Madre de Dios, Magallanes, Chile. Serv Nac Geol Minería, Bol 39:1–63

Ghidella ME, LaBrecque JL (1997) The Jurassic conjugate margins of the Weddell Sea: Considerations based on magnetic, gravity and paleobathymetry data. In: Ricci CA (ed) The Antarctic region: geological evolution and processes. Terra Antartica Publication, Siena, pp 441–451

Ghidella ME, Yáñez G, LaBrecque JL (2002) Revised tectonic implications for the magnetic anomalies of the Western Weddell sea. Tectonophysics 347:65–86

Grunow AM (1993) New paleomagnetic data from the Antarctic Peninsula and their tectonic implications. J Geophys Res 98(8): 13815–13833

Harrison CGA, Barron EJ, Hay WW (1979) Mesozoic evolution of the Antarctic Peninsula and the southern Andes. Geology 7:374–378

Hathaway B, Lomas SA (1998) The Upper Jurassic-Lower Cretaceous Byers Group, South Shetland Islands, Antarctica: revised stratigraphy and regional correlations. Cretaceous Res 19:43–67

Hervé F, Fanning CM (2001) Late Triassic zircons in meta-turbidites of the Chonos Metamorphic Complex, southern Chile. Rev Geol Chile 28:91–104

Hervé F, Fanning CM (2003) Early Cretaceous subduction of continental crust at the Diego de Almagro archipelago, southern Chile. Episodes 26:285–289

Hervé F, Nelson E, Kawashita K, Suárez M (1981) New isotopic ages and the timing of orogenic events in the Cordillera Darwin, southernmost Chilean Andes, Earth Planet Sci Letters 55:257–265

Hervé F, Marambio F, Pankhurst R (1984) El Complejo Metamórfico de Scotia en Cabo Lookout, isla Elefante, islas Shetland del Sur, Antártica: evidencias de un metamorfismo cretácico. Serie Científica INACH 31:23–37

Hervé F, Miller H, Loske W, Milne A, Pankhurst RJ (1990) New Rb-Sr age data from the Scotia Metamorphic Complex of Clarence Island, West Antarctica. Zbl Geol Paläont Teil I, (1/2):119–126

Hervé F, Loske W, Miller H, Pankhurst RJ (1991) Chronology of provenance, deposition and metamorphism of deformed fore-arc sequences, southern Scotia Arc. In: Thomson MRA, Crame JA, Thomson JW (eds) Geological evolution of Antarctica. Cambridge University Press, Cambridge, pp 429–435

Hervé F, Aguirre L, Godoy E, Massonne, H-J, Morata D, Pankhurst RJ, Ramírez E, Sepúlveda V, Willner A (1998) Nuevos antecedentes acerca de la edad y las condiciones *P-T* de los Complejos Metamórficos en Aysén, Chile. X Congr Latinoamericano Geología, Buenos Aires, 2, pp 134–137

Hervé F, Prior D, López G, Ramos VA, Rapalini A, Thomson S, Lacassie JP, Fanning M (1999) Mesozoic blueschists from Diego de Almagro, southern Chile. Extended abstract, II South Amer Sympos Isotope Geol, Actas, Córdoba, pp 318–321

Hervé F, Fanning CM, Pankhurst RJ (2003) Detrital zircon age patterns and provenance in the metamorphic complexes of Southern Chile. J South Amer Earth Sci 16:107–123

Hobbs GJ (1968) The geology of the South Shetland Islands: II. The geology and petrology of Livingston Island. Sci Rep Brit Antarc Surv 47:1–34

Holdsworth BK, Nell PAR (1992) Mesozoic radiolarian faunas from the Antarctic Peninsula: age, tectonic and palaeoceanographic significance. J Geol Soc London 149:1003–1020

Hyden G, Tanner PWG (1981) Late Paleozoic-early Mesozoic fore-arc basin sedimentary rocks at the Pacific margin in western Gondwana. Geol Rundschau 70:529–541

Jokat W, Boebel T, König M, Meyer U (2003) Timing and geometry of early Gondwana breakup. J Geophys Res 108(9)

Kovacs LC, Morris P, Brozena J, Tikku A (2002) Seafloor spreading in the Weddell Sea from magnetic and gravity data. Tectonophysics 347:43–64

Lacassie JP (2003) Estudio de la proveniencia sedimentaria de los complejos metamórficos de los Andes Patagónicos (46°–51° Lat. S), mediante la aplicación de redes neuronales e isótopos estables. PhD thesis, Universidad de Chile, Departamento de Geología, Santiago

Lawver LA, Dalziel IWD, Gahagan LM (1998) A tight fit Early Mesozoic Gondwana, a plate reconstruction perspective. Mem Nation Inst Polar Res Spec Issue 53:214–229

Leat PT, Scarrow JH, Millar, IL (1995) On the Antarctic Peninsula batholith. Geol Mag 132(4):399–412

Leitch EC (1975) Plate tectonic interpretation of the Paleozoic history of the New England Fold Belt. Geol Soc Amer Bull 86:141–144

Ling HY, Forsythe RD, Douglass RC (1985) Late Palaeozoic microfaunas from southernmost Chile and their relation to Gondwanaland forearc development. Geology 13:357–360

Llambías EJ (1999) Las rocas ígneas Gondwánicas. 1. El magmatismo Gondwánico durante el Paleozoico superior – Triásico. In: Caminos R (ed) Geología Argentina. Instituto de Geología y Recursos Minerales, Anales 29, pp 349–363

Loske W, Miller H, Milne A, Hervé F (1990) U-Pb zircon ages of xenoliths from Cape Dubouzet, northern Antarctic Peninsula. Zbl Geol Paläont 1990 Teil I:87–95

Martin M, Pankhurst RJ, Fanning DM, Thomson SN, Calderón M, Hervé F (2001) Age distribution on plutons across the southern Patagonian batholith: New U-Pb data on zircons. 3rd South Amer Sympos Isotope Geol, ext abstracts, CD, Soc Geol Chile, Santiago, Chile, pp 585–588

Millar IL, Pankhurst RJ, Fanning CM (2002) Basement chronology of the Antarctic Peninsula: recurrent magmatism and anatexis in the Palaeozoic Gondwana margin. J Geol Soc London 159: 145–157

Miller H (1983a) Gebirgszusammenhänge zwischen Südamerika und der Antarktischen Halbinsel. N Jb Geol Paläont Abh 166: 50–64

Miller H (1983b) The position of Antarctica within Gondwana in the light of Palaeozoic orogenic development. In: Oliver RL, James PR, Jago JB (eds) Antarctic earth science. Australian Acad Sci, Canberra, pp 579–581

Miller H, Loske W (1992) La historia pre-andina de la Peninsula Antártica. In: López-Martínez J (ed) Geología de la Antártica Occidental, Simp T 3. III Congr Geol España y VIII Congr Latinoam Geol, Salamanca, pp 33–42

Nell PAR, Kamenov BK, Pimpirev CT (1989) Field report on combined Anglo-Bulgarian geological studies in northern Alexander Island. Antarctic Sci 1:167–169

Olivares B, Cembrano J, Hervé F, López G, Prior D (2003) Geometría y cinemática de la zona de cizalle Seno Arcabuz, Andes Patagónicos, Chile. Rev Geol Chile 30:39–52

Pankhurst RJ (1990) The Paleozoic and Andean magmatic arcs of West Antarctica and southern South America. In: Kay SM, Rapela C (eds) Plutonism from Antarctica to Alaska. Geol Soc Amer Spec Paper 241:1–8

Pankhurst RJ, Riley TR, Fanning CM, Kelley SP (2000) Episodic silicic volcanism in Patagonia and the Antarctic Peninsula: chronology of magmatism associated with the break-up of Gondwana. J Petrology 41:605–625

Pankhurst RJ, Rapela CW, Loske WP, Márquez M, Fanning CM (2003) Chronological study of the pre-Permian basement rocks of southern Patagonia. J South Amer Earth Sci 16:27–44

Pimpirev C, Vangelov D, Dimov D (1997) Depositional model for the Miers Bluff Formation, Livingston Island, South Shetland Island, Antarctica. C R Acad Bulg Sci 50(11–12):75–78

Pimpirev C, Miller H, Hervé F (1999) Preliminary results on the lithofacies and palaeoenvironmental interpretation of the Palaeozoic turbidite sequence in Chonos Archipielago, Southern Chile. Comunicaciones 48–49:3–12

Pimpirev Ch, Ivanov M, Dimov D, Nikolov T (2002) First find of the Upper Tithonian ammonite genus Blanfordiceras from the Miers Bluff Formation, Livingston Island, South Shetland Islands. N Jb Geol Paläont Mh 2002(6):377–384

Pimpirev C, Stoykova K, Ivanov M, Dimov V (2006) The Miers Bluff Formation, Livingston Island, South Shetland Islands – part of the Late Jurassic–Cretaceous depositional history of the Antarctic Peninsula. In: Fütterer DK, Damaske D, Kleinschmidt G, Miller H, Tessensohn F (eds) Antarctica – Contributions to global earth sciences. Springer, Berlin Heidelberg New York, pp 249–254

Ramírez E (2002) Geotermobarometría en metapelitas de complejos metamórficos de Aysen, Chile. PhD thesis, Departamento de Geología, Universidad de Chile, Santiago, pp 1–143

Rapalini AE, Hervé F, Ramos VA, Singer S (2001) Paleomagnetic evidence for a very large counterclockwise rotation of the Madre de Dios Archipelago, Southern Chile. Earth Planet Sci Letters 184:471–487

Smellie JL (1991) Stratigraphy, provenance and tectonic setting of (?) Late Palaeozoic-Triassic sedimentary sequences in northern Graham Land and South Scotia Ridge. In: Thomson RA, Crame JA, Thomson JW (eds) Geological evolution of Antarctica. Cambridge University Press, Cambridge, pp 411–417

Smellie JL, Clarkson P (1975) Evidence for pre-Jurassic subduction in Western Antarctica. Nature 258:701–702

Söllner F, Miller H, Hervé M (2000) An Early Cambrian granodiorite age from the pre-Andean basement of Tierra del Fuego (Chile): the missing link between South America and Antarctica? J South Amer Earth Sci 13:163–177

Storey BC (1991) The crustal blocks of West Antarctica within Gondwana: reconstruction and break-up model. In: Thomson RA, Crame JA, Thomson JW (eds) Geological evolution of Antarctica. Cambridge University Press, Cambridge, pp 587–592

Stoykova K, Pimpirev CH, Dimov D (2002) Calcareous nannofossils from the Miers Bluff Formation (Livingston Island, South Shetland Islands, Antarctica): first evidence for a late Cretaceous age. Nannoplancton Res 24(2):166–167

Tanner PWG, Pankhurst RJ, Hyden G (1982) Radiometric evidence for the age of the subduction complex in the South Orkney and South Shetland islands, West Antarctica. J Geol Soc London 139:683–690

Thomson MRA (1975) New palaeontological and lithological observations on the Legoupil Formation, North-West Antarctic Peninsula. Brit Antarc Surv Bull 41–42:169–185

Thomson SN, Hervé F (2002) New time constraints for the age of metamorphism at the ancestral Pacific Gondwana margin of southern Chile. Rev Geol Chile 29(2):255–271

Trouw RAJ, Pankhurst RJ, Ribeiro A (1997) On the relation between the Scotia Metamorphic Complex and the Trinity Peninsula Group, Antarctic Peninsula. In: Ricci CA (ed) The Antarctic region: geological evolution and processes. Terra Antartica Publication, Siena, pp 383–389

Valenzuela E, Hervé F (1972) Geology of Byers Peninsula, Livingston Island, South Shetland Islands. In: Adie RJ (ed) Antarctic geology and geophysics. Universitetsforlaget, Oslo

Vaughan APM, Storey BC (2000) The Eastern Palmer Land shear zone: a new terrane accretion model for the Mesozoic development of the Antarctic Peninsula. J Geol Soc London 157:1243–1256

Vaughan APM, Pankhurst RJ, Fanning CM (2002) A mid-Cretaceous age for the Palmer Land event, Antarctic Peninsula: implications for terrane accretion timing and Gondwana palaeolatitudes. J Geol Soc London 159:113–116

Veit A (2002) Vulkanologie und Geochemie pliozäner bis rezenter Vulkanite beiderseits der Bransfield-Straße/West-Antarktis. Rep Polar Marine Res 420:1–177

Willan RCR, Pankhurst RJ, Hervé F (1994) A probable early Triassic age for the Miers Bluff Formation, Livingston Island, South Shetland Islands. Antarctic Sci 6:401–408

Willner A, Hervé F, Massonne H-J (2000) Mineral chemistry and pressure-temperature evolution of two contrasting high-pressure-low-temperature belts in the Chonos Archipelago, southern Chile. J Petrol 41:309–330

Willner AP, Hervé F, Thomson SN, Massonne H-J (2004) Converging *P-T* paths of Mesozoic HP-LT metamorphic units (Diego de Almagro Island, Southern Chile): evidence for juxtaposition during late shortening of an active continental margin. Mineral Petrol 41:43–84

Wilson TJ, Hanson RE, Grunow AM (1989) Multistage melange formation within an accretionary complex, Diego Ramírez Islands, southern Chile. Geology 17:11–14

Moho Depth along the Antarctic Peninsula and Crustal Structure across the Landward Projection of the Hero Fracture Zone

Tomasz Janik[1] · Piotr Środa[1] · Marek Grad[2] · Aleksander Guterch[1]

[1] Institute of Geophysics, Polish Academy of Sciences, Ks. Janusza 64, 01-452 Warsaw, Poland, <janik@igf.edu.pl>

[2] Institute of Geophysics, University of Warsaw, Pasteura 7, 02-093 Warsaw, Poland

Abstract. Results of deep seismic soundings collected during four Polish Geodynamical Expeditions to West Antarctica between 1979 and 1991 were synthesised to produce a map of Moho depth beneath the NW coast of the Antarctic Peninsula. In this paper, we present a new interpretation of the deep landward projection of the Hero Fracture Zone based on two seismic transects. On each transect we found high velocity bodies with $Vp > 7.2$ km s^{-1}, similar to ones we detected previously in Bransfield Strait. However, these bodies are not continuous; they are separated by a zone of lower velocities located SW of Deception Island. In Bransfield Strait an asymmetric "mushroom"-shaped high velocity body was found at a depth interval from 13–18 km down to the Moho boundary at a depth of ca. 30 km. As a summary of results, we present a map of Moho depth along the coast of the Antarctic Peninsula, prepared using previous seismic 2D models. The map shows variations in crustal thickness from 38–42 km along the Antarctic Peninsula shelf in the southern part of the study area to 12–18 km beneath the South Shetland Trench.

Introduction

During four Polish Antarctic geodynamical expeditions between 1979 and 1991, deep seismic sounding (DSS) measurements were performed in the transition zone between the Drake and South Shetland microplates and the Antarctic Plate in West Antarctica. The network of 20 DSS profiles ranging in length from 150 to 320 km covered the western side of the Antarctic Peninsula from Adelaide Island in the south to Elephant Island on the North (Fig. 5.2-1). The seismic results obtained during four expeditions were published in a number of papers (Guterch et al. 1985, 1991, 1998; Grad et al. 1992, 1993, 1997, 2002; Janik 1997a,b; Środa et al. 1997; Środa 2001, 2002). This paper presents new 2D seismic models for a network of six jointly modelled profiles along two seismic transects: Transect I – DSS-10, DSS-7 and DSS-2 (together 660 km long); Transect II – DSS-6, DSS-15 and DSS-5 (460 km). Both transects were sub-parallel to the northwestern coast of the Antarctic Peninsula, crossing the transition zone from a passive to an active margin and main structures of the Bransfield Strait (Fig. 5.2-1). Based on these results and previous models, a map of Moho depth along western coast of the Antarctic Peninsula was prepared.

Tectonic Setting

The tectonic history of the western margin of the Antarctic Peninsula has been reconstructed mainly from surface distribution of marine magnetic anomalies on the neighbouring sea floor by many authors (e.g., Herron and Tucholke 1976; Barker 1982). Ocean crust formed at the Aluk Ridge (Fig. 5.2-1) must have been subducted beneath the peninsula, until segments of the ridge itself arrived at the margin and subduction stopped. Subduction and spreading stopped along well defined segments of the margin at different times, the spreading ridge topography disappeared and the zone where the ridge segment had arrived became a passive margin. These processes repeated in successive segments of the subducting plate, separated by numerous fracture zones. According to Larter and Barker (1991), this process occurred until about 5.5–3.1 million years ago when the last segment of the ridge reached the section of the margin south of the Hero Fracture Zone (HFZ). The date of last ridge segment arrival at margin becomes 6.4–3.3 Ma when the Cande and Kent (1995) magnetic reversal timescale (MRTS) is used (Larter et al. 1997). Between the latter and the Shackleton Fracture Zone (SFZ), there is the last preserved, although now inactive, fragment of the Aluk-Antarctic spreading axis. Spreading has now stopped on this ridge (Larter and Barker 1991; Livemore et al. 2000; Eagles 2004).

The landward projection of HFZ separates the South Shetland Islands and the Bransfield Basin from a continental, passive zone further to the south (Herron and Tucholke 1976). Several tectonic lineaments, discontinuities and geological structures between Deception and Low islands were presented in many papers and interpreted as landward projections of HFZ. Interpretation of detailed long-range side-scan sonar data and multichannel seismic reflection profiles shows that the South Shetland Trench continues 50 km southwest of the HFZ (Tomlinson et al. 1992; Maldonado et al. 1994; Jabaloy et al. 2003). Henriet et al. (1992) presented ~50 km wide HFZ and Smith Island, a blueschist terrane showing evidence of considerable uplift, located exactly in front of the HFZ. They interpreted a trough southwest of HFZ as a fossil rift valley of the Aluk Ridge and suggested

From: Fütterer DK, Damaske D, Kleinschmidt G, Miller H, Tessensohn F (eds) (2006) Antarctica: Contributions to global earth sciences. Springer-Verlag, Berlin Heidelberg New York, pp 229–236

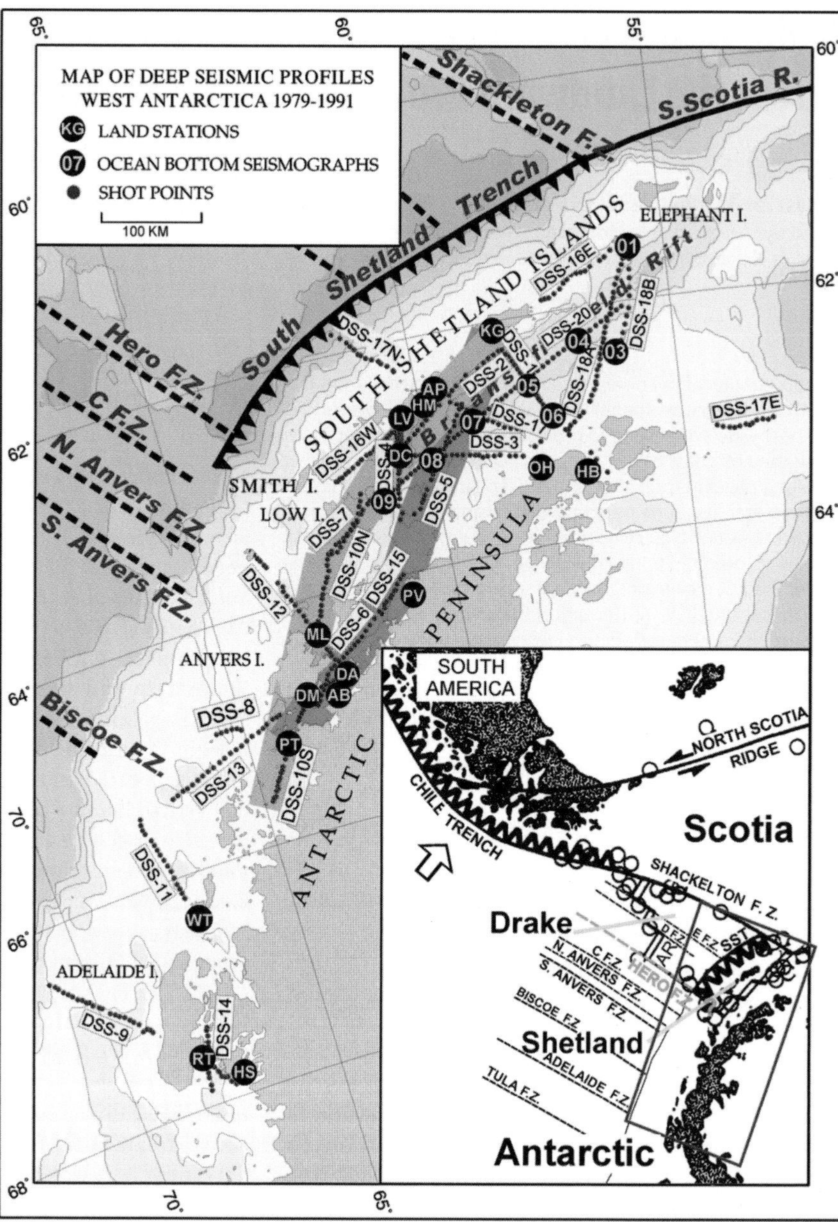

Fig. 5.2-1.
Location of deep seismic sounding (DSS) profiles and transects in West Antarctica. Seismic stations: *AB:* Almirante Brown; *AP:* Arturo Prat; *DA:* Danco; *DC:* at Deception Island; *DM:* Damoy; *HB:* Hope Bay; *HM:* at Half Moon Island; *HS:* Horse Shoe; *KG:* at King George Island; *LV:* at Livingston Island; *ML:* Melchior; *OH:* O'Higgins; *PT:* Petermann; *PV:* Primavera; *RT:* Rothera; *WT:* Watkins; *01–09:* ocean bottom seismographs. Transect I (*pink stripe*): DSS-10 + DSS-7 + DSS-2; Transect II (*green stripe*): DSS-6 + DSS-1 + DSS-5. The positions of fracture zones (F.Z.), the South Scotia Ridge (S. Scotia R.) are taken from Barker (1982), Larter and Barker (1991) with modifications after Tomlinson et al. (1992). Bathymetry contours are plotted at 500 m intervals (ETOPO5 data file, NOAA's NGDC), using GMT software by Wessel and Smith (1995). *Inset:* Plate tectonic elements around the Scotia region from Tectonic Map of the Scotia Arc (1985), and Birkenmajer et al. (1990). Abbreviations: *SST:* South Shetland Trench; *AR:* Aluk Ridge – divergent plate boundary, ridge and transform faults; *open arrows* show direction of subduction; *double, black arrows* present relative motion at the North Scotia Ridge plate boundary; *white circles:* epicentres of the 1963–1985 earthquakes (for earthquakes with mb > 5.0) by Pelayo and Wiens (1989); *dashed lines* show fracture zone trends. The *red box* shows setting of study area

that spreading stopped some 5 Ma ago, shortly before the ridge collided with the trench. The width and position of the ridge associated with HFZ was clear from bathymetry map published by Larter and Barker (1991), and its orientation was confirmed by interpretation of GLORIA data published by Tomlinson et al. (1992). From this, it is clear that any such fossil rift valley would have been oblique to the margin. In fact the trough described by Henriet et al. (1992) is the part of the South Shetland Trench that continues to the south of the HFZ, which is seen very clearly in the GLORIA sidescan sonar data presented by Maldonado et al. (1994). Jabaloy et al. (2003) identified here two active major reverse faults. A seismic reflection profile running along Boyd Strait (Jin et al. 2003), just northwest of the landward projection of the HFZ, shows major structural components similar to those typically observed along the margin to the southwest of the HFZ. The continuation of the post-subduction margin structures to the active margin suggests that the boundary between crust with passive and active margins characteristics is not sharply defined.

Seismic Modelling

Examples of seismic record sections and 2D modelling are presented on Fig. 5.2-2. The travel times of refracted and reflected P waves, correlated in the record sections

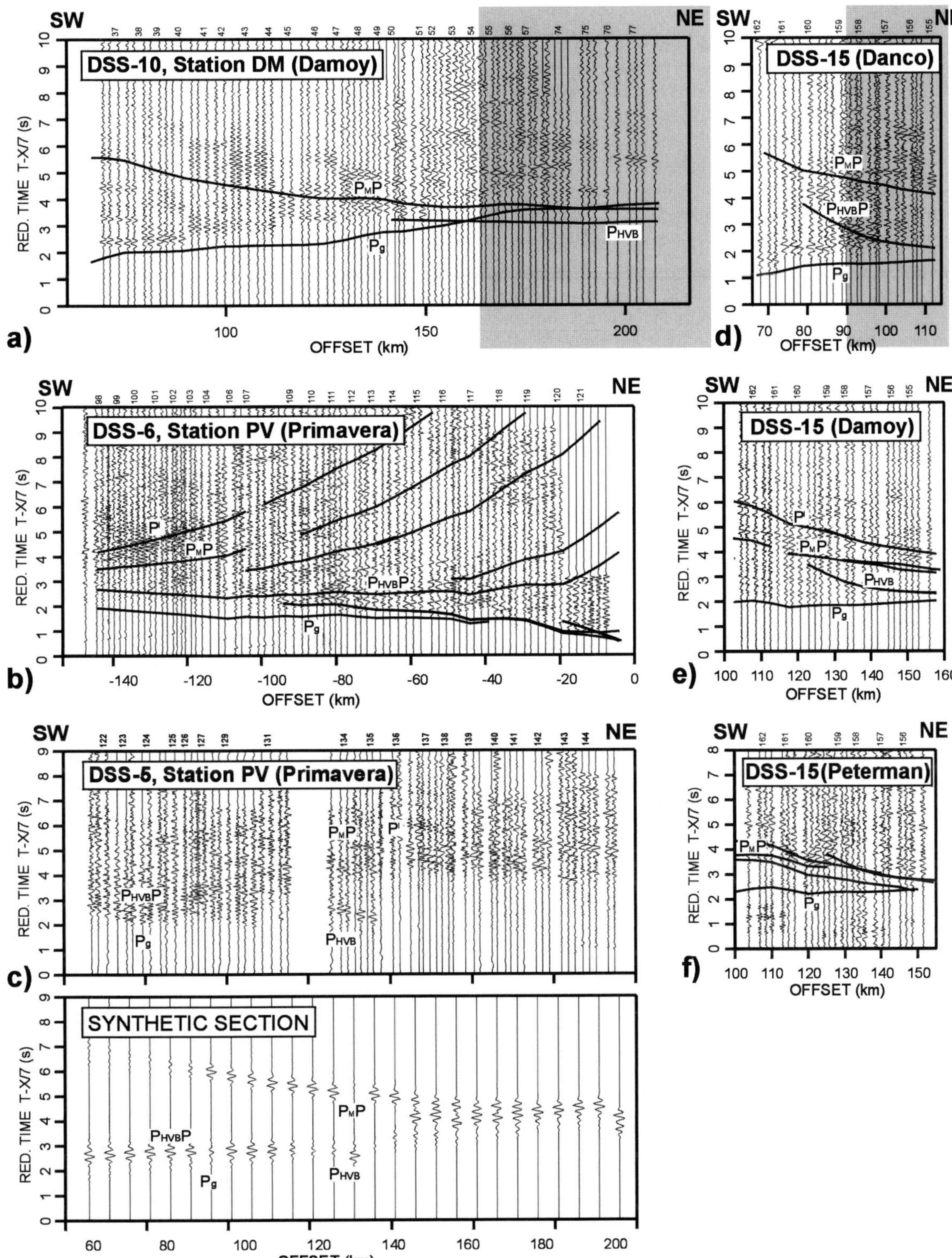

Fig. 5.2-2. Amplitude-normalised seismic record sections and theoretical travel times of P waves calculated for the crustal model for transect I: **a** *DSS-10 profile, station DM* (Damoy I.); and transect II: **b** *DSS-6 profile, station PV* (Primavera); **c** *DSS-5 profile, station PV* (Primavera); and for **d** *DSS-15 profile, station DA* (Danco Island), **e** station *DM* (Damoy), **f** station *PT* (Petermann). Abbreviations: P_g: refracted arrivals from the crust; P_{HVB}: waves penetrated HVB; $P_{HVB}P$: reflection from the top of the HVB; P_MP: reflection from the Moho discontinuity; P^l: lower lithosphere phases. *Bottom diagram* of panel **c** presents synthetic section. Reduction velocity 7.0 km s^{-1}. *Grey shading area* shows range of HVB

along the profiles, provide the basis for the modelling of the velocity distribution and depths to the seismic boundaries in the seismic models of the crust and uppermost mantle (Fig. 5.2-3). For 2D modelling of the velocity structure we applied the ray tracing technique for calculations of traveltimes and synthetic seismograms using the interactive version of package SEIS83 (Červený and Pšenčík 1983) supported by graphical interfaces MODEL (Komminaho 1998) and ZPLOT (Zelt 1994). Results of 2D modelling of the network of linear DSS profiles provide images of the very complicated deep structure of the Earth's crust in the Bransfield Strait and along the margin of the Antarctic Peninsula. Examples of results of the 2D modelling for different parts of the transects are shown in Fig. 5.2-2.

Crustal Models

The seismic models of the crust and uppermost mantle (Fig. 5.2-3) along both transects cross the transition zone between the passive and active margins of the northwestern coast of the Antarctic Peninsula and the main structures of the Bransfield Strait.

Transect I

A ca. 70 km long, high velocity body (HVB), with $Vp > 7.2$ km s^{-1} was modelled at km 360–430 from 12 km down to the Moho boundary located here at a depth of 35 km. The SW part of the model (DSS-10S) is after Środa et al. (1997). In the upper crust, a layer with velocities 6.3–6.4 km s^{-1} down to 12 km depth was found. In the central and NE parts of the transect, layers with $Vp < 5.8$ km s^{-1} were modelled in the upper crust. This difference is also evident in the lower crust where velocities are 7.1–7.25 km s^{-1} in the southwestern part and 7.25–7.4 km s^{-1} in the northeastern part of the transect. The crustal thickness decreases from 36–39 km in the southwestern and central parts of the transect to 31 km in the northeastern part. Sub-Moho velocities of 8.05 km s^{-1} and a deep reflector (55–45 km, at distances 210–600 km) were modelled for the upper mantle.

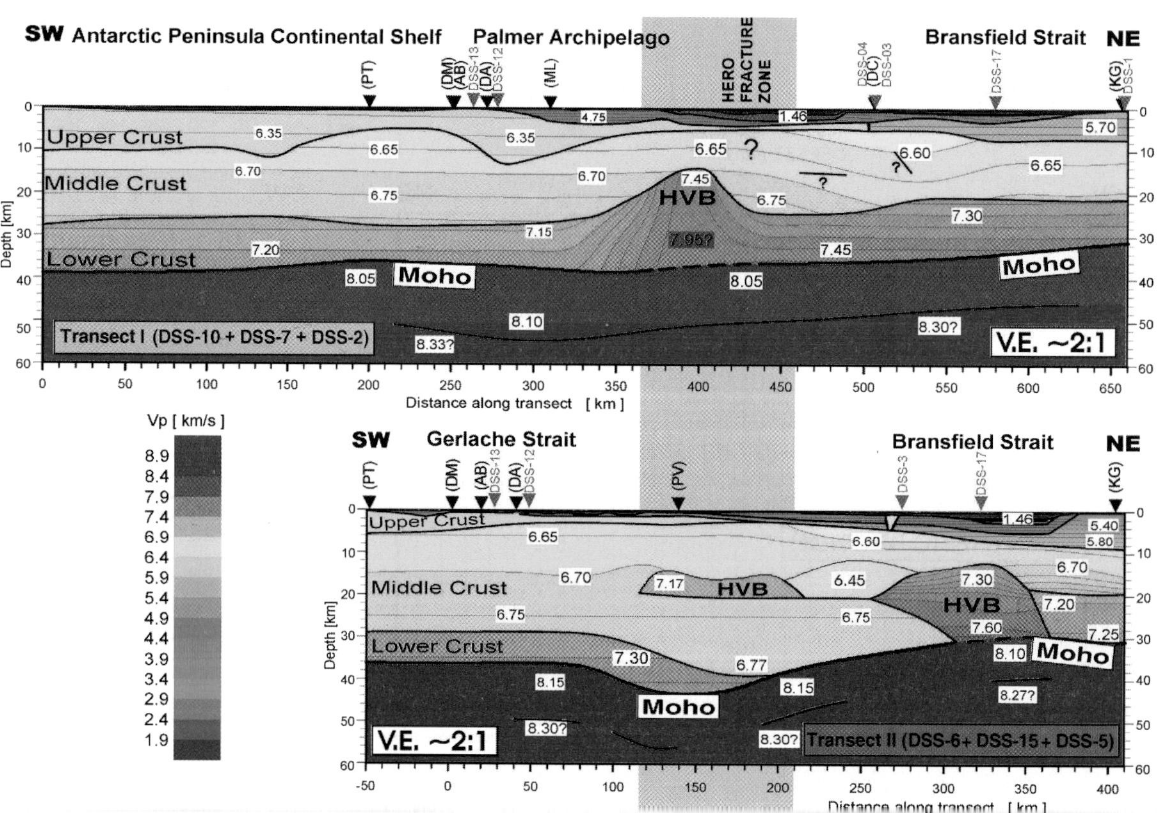

Fig. 5.2-3. Two-dimensional, P-wave velocity models across the contact zone between passive continental margin of the Antarctic Peninsula shelf and the active marginal zone in the area of Bransfield Strait, along two transects between Petermann Island (*PT*) and King George Island, Transect I – profiles: *DSS-10*, *DSS-7* and *DSS-2* (SW part of the model, *DSS-10S*, after Środa et al. (1997); Transect II – profiles: *DSS-6*, *DSS-15* and *DSS-5*. *Thick, black lines* represent major velocity discontinuities (interfaces). *Thin lines* represent velocity isolines with values (km s^{-1}) shown in *white boxes*. The *red box* shows velocity of HVB required by data recorded on Melchior (*ML*) station only. *Black arrows* show positions of stations and *red arrows* crossings with other DSS profiles. For further explanations see Fig. 5.2-1

Transect II

Two high velocity bodies were modelled; the first, ca. 80 km long and 5 km thick, with $Vp > 7.1$ km s^{-1}, was modelled at km 170 at a depth of 15 km; the second one, in the central part of Bransfield Strait, with $Vp = 7.2–7.7$ km s^{-1} at km 250–360 extends from a depth of 13 km down to the Moho boundary. The bodies are separated by a low velocity zone with $Vp < 6.4$ km s^{-1}. In the central part of the transect, velocities $Vp \sim 6.7$ km s^{-1}, typical for middle crust, were modelled down to the Moho boundary which reaches depth of 42 km in this area. The lower crustal velocities are 7.3 km s^{-1} in the 7 km thick southwestern part and 7.2 km s^{-1} in the 12 km thick northeastern part of the model. Sub-Moho velocities 8.10–8.15 km s^{-1} and a deep reflector (55–40 km) at distances 100–400 km were modelled for the upper mantle.

Images of the crustal structure along the transects are different. Along Transect I, located more on the northwest, the HVB ($Vp > 7.2$ km s^{-1}) is very clearly expressed, being about 70 km wide at a depth from 12 km to the Moho boundary. The pattern of HVB on the Transect II, is different, ~100 km wide and only 5 km thick, but the Moho boundary is the deepest beneath HVB (42 km). The existence of a HVB was not confirmed further to the NE. Velocities of a $Vp < 6.5$ km s^{-1} at depths 10–15 km and $Vp = 6.6–6.75$ km s^{-1} at depths 6–20 km were detected on Transects II and I respectively, southwest from the Deception Island. Similar low velocities were detected on DSS-4 at depths 7–15 km (Janik 1997a,b). The seismic results obtained for the Bransfield Strait are summarized in Fig. 5.2-4.

Map of the Moho Depth

Based on the models of crustal structure presented here, as well as on previous results of seismic modelling (Grad et al. 2002; Guterch et al. 1998), a map of isolines of depth to the Moho discontinuity along the Antarctic Peninsula was prepared (Fig. 5.2-5). The depth to the Moho values were taken where available along all profiles. Between the data points the Moho depth was interpolated to cre-

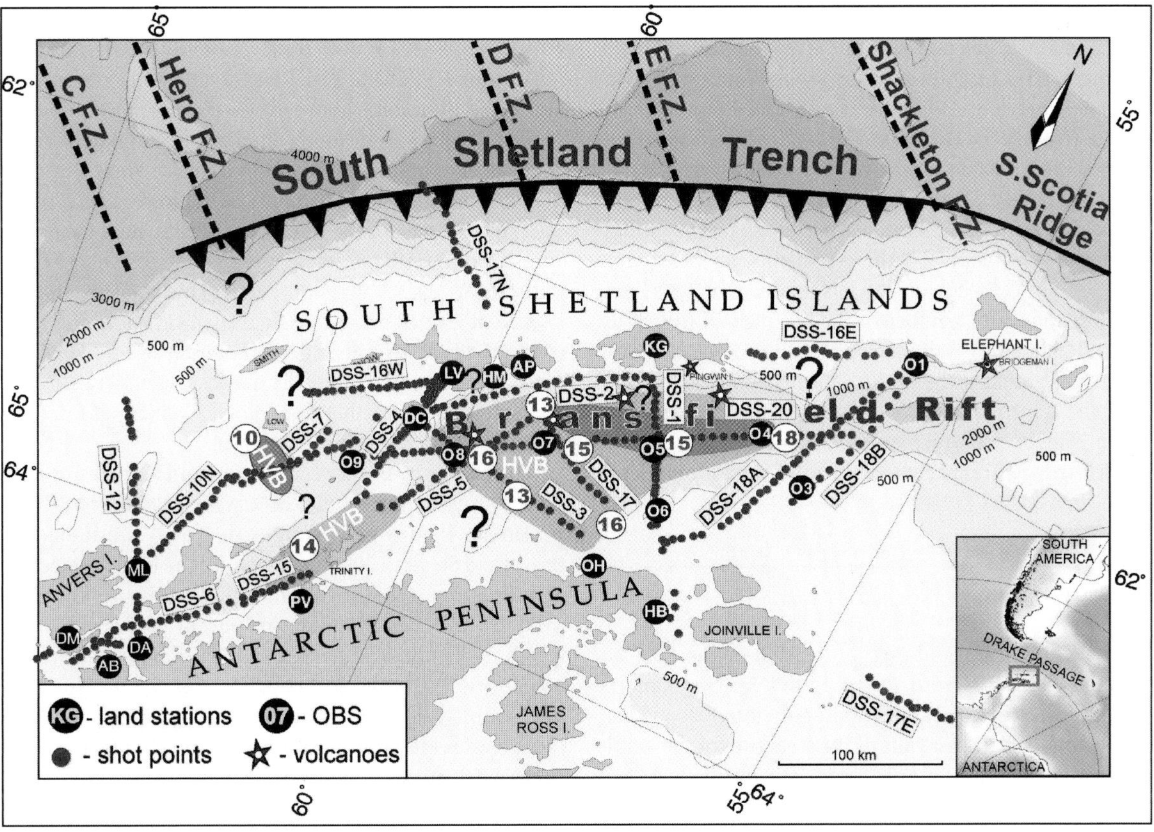

Fig. 5.2-4. P-wave velocity anomalies detected in the crust in the Bransfield Strait marginal basin against of the South Shetland Islands. Range of the previously identified high velocity (mushroom-shaped) body (*HVB*) with $Vp > 7.2$ km s^{-1} in the Bransfield Strait, detected at a depth of 10–18 km, deduced from two-dimensional modelling of the net of intersecting deep seismic profiles. *Violet colour* shows the places where the depth of high velocity body reaches 30 km. At other places (*yellow colour*), the thickness of the body is relatively small (southern part) or unknown (northwest part)

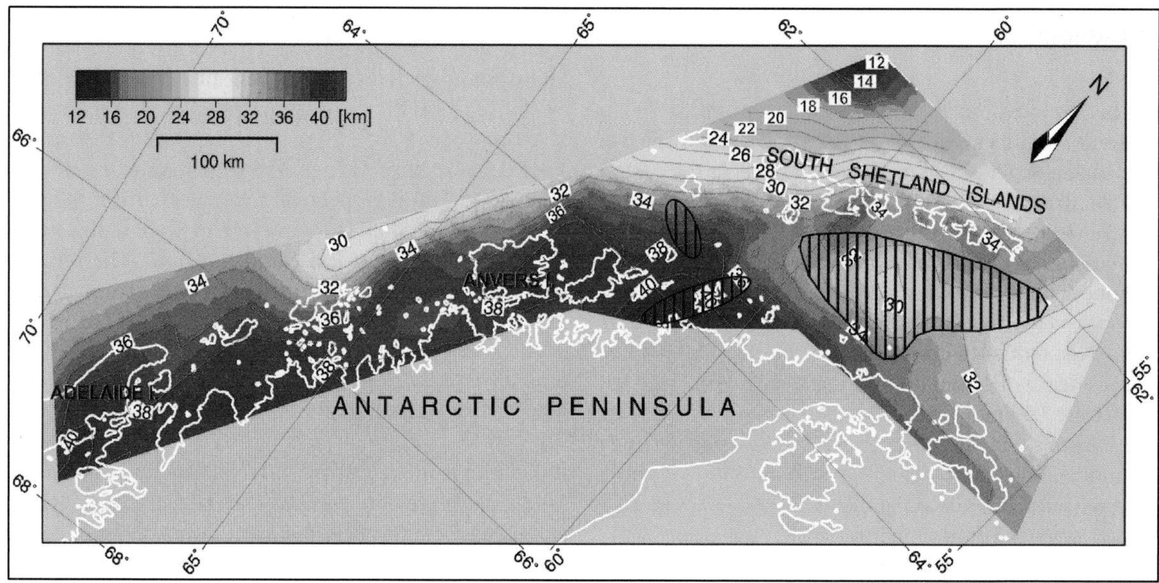

Fig. 5.2-5. Map of the depth to the Moho boundary along the Antarctic Peninsula, based on data from ray-tracing models of the crustal structure. Areas filled with *blue lines* in the central part of the Bransfield Strait mark the extent of the high velocity anomaly, where at a depth of 13–18 km P-wave velocities reach 7.2 km s^{-1}, as well as the extent of two other areas of anomalously high velocity. In the NE part of this area, due to possible existence of a thick crust-mantle transition, the isolines do not reflect the actual Moho depth (see Fig. 5.2-4)

ate a 2D surface. The map shows that the maximum crustal thickness, 38–42 km, occurs along the Antarctic Peninsula shelf between Adelaide Island and Palmer archipelago. Towards the Pacific, the Moho depth decreases and reaches 30–32 km at the edge of study area. In central part of the Bransfield Strait, seismic velocities reach anomalously high value of 7.2 km s^{-1} at a body that is asymmetric "mushroom"-shaped in three-dimensional form (same as in Fig. 5.2-4), the top of which is shaded with blue lines, at depths of 12–18 km. Below the NE part of this body, Moho depth is not well resolved. Along the coast of the Antarctic Peninsula the Moho depth is about 34 km, and along the South Shetland Islands crustal thickness reaches 35 km. The latter area separates Bransfield Strait from the South Shetland Trench, where the minimum Moho depth in the area occurs (12 km).

Discussion

On the SW end of landward projection of the HFZ we detected two HVBs with $Vp > 7.2$ km s^{-1}. The geometry of our network of measurements, for obvious reasons, was not optimal to detect details of the structure of the whole landward projection of HFZ. From our data we can not prove or disprove connection between the HVBs in the south; but we are sure of a their separation from the HVB of Bransfield Strait by a zone of lower velocities. The extend of the HVB in Bransfield Strait is controlled by rifting processes at its NE limit. Can we connect existence of HBVs, expressed on both transects with the landward extension of HFZ? One explanation could be movements along the boundary between the passive and an active margin (?rotation of South Shetland Microplate), which could create conditions for injection of upper mantle material into the crust, or uplift of the lower crust.

In the NE part of the Bransfield Strait where the larger "mushroom" HVB has its root, velocity increases continuously to 7.8 km s^{-1} at 30 km depth. This suggests the existence of a thick crust-mantle transition zone, often encountered in active rift zones, rather than a step-like Moho boundary. Therefore, the area marked by violet colour in Fig. 5.2-4, indicates that the velocity isolines do not reflect the Moho discontinuity depth.

The papers by Barker et al. (2003) and Christeson et al. (2003) present the results of a wide-angle seismic experiment with a net of eight profiles, conducted in Bransfield Strait in 2000. Christeson et al. (2003) detected, similarly to our investigations, velocities >7.25 km s^{-1} at a depths 10–15 km at the central part of Bransfield Strait, which they interpreted as the Moho boundary. In our interpretation it is the top of HVB (Fig. 5.2-4). The well documented upper mantle velocity $Vp > 8.0$ km s^{-1} on the record sections along 300 km of DSS-20 profile (Grad et al. 1997), strong reflections from the Moho boundary observed on DSS-17 profile (Grad et al. 1993) and lower velocities (~6.9 km s^{-1}) detected below HVB along DSS-3 profile (Janik 1997a,b) confirm our interpretation of the Moho boundary at a depth of ~30 km below Bransfield Strait.

Conclusions

- Maximum crustal thickness of 38–42 km occurs along the Antarctic Peninsula shelf between Adelaide Island and Palmer archipelago.
- Towards the Pacific, the Moho depth decreases and reaches 30–32 km at the edge of study area.
- Minimum Moho depth in the study area is beneath the South Shetland Trench (12 km).
- In the central part of the Bransfield Strait seismic velocities reach an anomalously high value of 7.2 km s^{-1} at 13–18 km depth in the "mushroom"-shaped body.
- In the NE part of Bransfield Strait, near the axis of the active rift, velocities increasing from 7.2 to 7.7 km s^{-1} were found.
- The area of shallow Moho depths (ca. 30 km) in central part of the Bransfield Strait extends between the coast of the Antarctic Peninsula, with a Moho depth about 34 km, and South Shetland Islands, with a Moho depth of 35 km.
- On the landward projection of the HFZ, two high velocity bodies with $Vp > 7.2$ km s^{-1} were detected, both separated from the HVB of the Bransfield Strait by a zone of lower velocities with $Vp < 6.6$ km s^{-1}.

Acknowledgments

The authors are grateful to Prof. G. R. Keller from University of Texas at El Paso and to referee, Dr. R. D. Larter from British Antarctic Survey for their constructive criticism and helpful comments.

References

Barker PF (1982) The Cenozoic subduction history of the Pacific margin of the Antarctic Peninsula: ridge-crest interactions. J Geol Soc London 139:787–802

Barker DHN, Christeson GL, Austin JA, Dalziel IWD (2003) Backarc basin evolution and cordilleran orogenesis: insights from new ocean-bottom seismograph refraction profiling in Bransfield Strait, Antarctica. Geology 31(2):107–110

BAS (1985) Tectonic Map of the Scotia Arc, Scale 1:3000000, Sheet 3. Bristish Antarctic Survey Misc 3, Cambridge

Birkenmajer K, Guterch A, Grad M, Janik T, Perchuć E (1990) Lithospheric transect Antarctic Peninsula, South Shetland Islands, West Antarctica. Polish Polar Res 11:241–258

Cande SC, Kent DV (1995) Revised calibration of the geomagnetic polarity timescale for the Late Cretaceous and Cenozoic. J Geophys Res 100:6093–6095

Christeson GL, Barker DHN, Austin JA, Dalziel IWD (2003) Deep crustal strusture of Bransfield Strait: initiation of a backarc basin by rift reactivation and propagation. J Geophys Res 108(10):2492

Červený V, Pšenčík I (1983) Program SEIS83, numerical modelling of seismic wave fields in 2-D laterally varving bedded structures by ray method. Charles University, Praha

Eagles G (2004) Tectonic evolution of the Antarctic Phoenix plate system since 15 Ma. Earth Planet Sci Letters 217:97–109

Gambôa LAP, Maldonado PR (1990) Geophysical investigation in the Bransfield Strait and Bellingshausen Sea, Antarctica. In: John BST (ed) Antarctica as an exploration frontier. Amer Assoc Petrol Geol 31, Tulsa, pp 127–141

Garrett SW, Storey BC (1987) Litospheric extension on the Antarctic Peninsula during Cenozoic subduction. In: Coward MP, Dewey JF, Hancock PL (eds) Continental extension tectonics. Geol Soc London Spec Publ 28:419–431

Grad M, Guterch A, Środa P (1992) Upper crustal structure of Deception Island, the Bransfield Strait, West Antarctica. Antarctic Sci 1:460–476

Grad M, Guterch A, Janik T (1993) Seismic structure of the lithosphere across the zone of subducted Drake plate under the Antarctic plate, West Antarctica. Geophys J Internat 115: 586–600

Grad M, Shiobara H, Janik T, Guterch A, Shimamura H (1997) New seismic crustal model of the Bransfield Rift, West Antarctica from OBS refraction and wide-angle reflection data. Geophys J Internat 130:506–518

Grad M, Guterch A, Janik T, Środa P (2002) In: Gamble JA, Skinner DNB, Henrys S (eds) Antarctica at the close of a millennium. Royal Soc New Zealand Bull 35:493–498

Guterch A, Grad M, Janik T, Perchuć E, Pajchel J (1985) Seismic studies of the crustal structure in West Antarctica 1979–1980 – preliminary results. Tectonophysics 114:411–429

Guterch A, Grad M, Janik T, Perchuć E (1991) Tectonophysical models of the crust between the Antarctic Peninsula and the South Shetland Trench. In: Thomson MRA, Crame JA, Thomson JW (eds) Geological evolution of Antarctica. Cambridge University Press, Cambridge, pp 499–504

Guterch A, Grad M, Janik T, Środa P (1998) Polish Geodynamical Expeditions – seismic structure of West Antarctica. Polish Polar Res 19:113–123

Henriet PJ, Meissner R, Miller H, GRAPE Team (1992) Active margin processes along the Antarctic Peninsula. Tectonophysics 201: 229–253

Herron EM, Tucholke BE (1976) Sea-floor magnetic patterns and basement structure in the souteastern Pacific. In: Hollister CD, Craddock C, et al. (eds) Initial Reports of the Deep Sea Drilling Project. U.S. Government Printing Office, Washington DC, pp 263–278

Jabaloy A, Balanyá J-C, Barnolas A, Galindo-Zaldívar J, Hernández J, Maldonado A, Martínez-Martínez J-M, Rodríguez-Fernánez J, de Galdeano CS, Somoza L, Suriñach E, Vásquez JT (2003) The transition from an active to a passive margin (SW and of the South Shetland Trench, Antarctic Peninsula). Tectonophysics 366:55–81

Janik T (1997a) Seismic crustal structure in the transition zone between Antarctic Peninsula and South Shetland Islands. In: Ricci CA (ed) The Antarctic region: geological evolution and processes. Terra Antartica Publication, Siena, pp 679–684

Janik T (1997b) Seismic crustal structure of the Bransfield Strait, West Antarctica. Polish Polar Res 18:171–225

Jin YK, Larter RD, Kim Y, Nam Sh, Kim KJ (2002) Post-subduction margin structures along Boyd Strait, Antarctic Peninsula. Tectonophysics 346:187–200

Komminaho K (1998) Software manual for programs MODEL and XRAYS – a graphical interface for SEIS83 program package. University of Oulu, Dept Geophys, Rep 20, pp 1–31

Larter RD, Barker PF (1991) Effects of ridge crest-trench interaction on Antarctic-Phoenix spreading: forces on a young subducting plate. J Geophys Res 96(12):19583–19604

Larter RD, Rebesco M, Vanneste LE, Gambôa LAP, Barker PF (1997) Cenozoic tectonic, sedimentary and glacial history of the continental shelf west of Graham Land, Antarctic Peninsula. In: Barker PF, Cooper AK (eds) Geology and seismic stratigraphy of the Antarctic margin, 2. Amer Geophys Union Antarc Res Ser 71:1–27

Livemore RA, Balayá J-C, Maldonado A, Martínez-Martínez J-M, Rodríguez-Fernández J, de Galdeano CS, Galindo-Zaldívar J, Jabaloy A, Somoza L, Hernández J, Molina J, Suriñach E and Viseras C (2000) Autopsy on a dead spreading center: the Phoenix Ridge, Drake Passage, Antarctica. Geology 28:607–610

Maldonado A, Larter R, Aldaya F (1994) Forearc tectonic evolution of the South Shetland margin, Antarctic Peninsula. Tectonics 1:1345–1370

Pelayo AM, Wiens DA (1989) Seismotectonics and relative plate motions in the Scotia Sea region. J Geophys Res 94:7293–7320

Środa P (1999) Modification to software package ZPLOT by CA Zelt. Inst Geophys Polish Acad Sci

Środa P (2001) Three-dimensional modelling of the crustal structure in the contact zone between Antarctic Peninsula and South Pacific from seismic data. Polish Polar Res 22:129–146

Środa P (2002) Three-dimensional seismic modelling of the crustal structure between the South Pacific and the Antarctic Peninsula. In: Gamble JA, Skinner DNB, Henrys S (eds) Antarctica at the close of a millennium. Royal Soc New Zealand Bull 35, pp 555–561

Środa P, Grad M, Guterch A (1997) Seismic models of the Earth's crustal structure between the South Pacific and the Antarctic Peninsula. In: Ricci CA (ed) The Antarctic region: geological evolution and processes. Terra Antartica Publication, Siena, pp 685–689

Tomlinson JS, Pudsey CJ, Livermore RA, Larter RD, Barker PF (1992) Long-range side scan sonar (GLORIA) survey of the Pacific margin of the Antarctic Peninsula. In: Yoshida Y, Kaminuma K, Shiraishi K (eds) Recent progress in Antarctic earth science. Terra Scientific Publishing, Tokyo, pp 423–429

Wessel P, Smith WHF (1995) New version of generic mapping tools released. EOS 76:329

Zelt CA (1994) Software package ZPLOT. Bullard Laboratories, University of Cambridge

Crustal Thinning and the Development of Deep Depressions at the Scotia-Antarctic Plate Boundary (Southern Margin of Discovery Bank, Antarctica)

Jesús Galindo-Zaldívar[1] · Juan Carlos Balanyá[2,3] · Fernando Bohoyo[3] · Antonio Jabaloy[1] · Andrés Maldonado[3]
José Miguel Martínez-Martínez[1,3] · José Rodríguez-Fernández[1] · Emma Suriñach[4]

[1] Departamento de Geodinámica, Universidad de Granada, 18071 Granada, Spain, <jgalindo@ugr.es>, <jabaloy@ugr.es>, <jmmm@ugr.es>

[2] Departamento de Ciencias Ambientales, Univ. Pablo de Olavide, Sevilla, Spain

[3] Instituto Andaluz Ciencias de la Tierra, CSIC/Universidad Granada, 18002 Granada, Spain, <fbohoyo@ugr.es>, <amaldona@ugr.es>, <jrodrig@ugr.es>

[4] Departament de Geodinàmica i Geofísica, Universitat de Barcelona, 08028 Barcelona, Spain, <emma.surinach@ub.edu>

Abstract. Discovery Bank is located at the eastern end of the South Scotia Ridge. The new geophysical data point that this bank is continental in nature and may be a former fragment of the continental bridge that connected South America and the Antarctic Peninsula before the Oligocene and is located at the Scotia-Antarctic plate boundary along the South Scotia Ridge. In this region, continental fragments are bounded by the oceanic crust of the Scotia Sea to the north and of the Weddell Sea to the south. Seismicity indicates that at present active structures related to the plate boundary are located within the continental crust, whereas most of the continental-oceanic crust boundaries seem to be inactive.

The main crustal elements of the Scotia-Antarctic plate boundary at the region of Discovery Bank include, from north to south: the oceanic crust of the Scotia Plate, the Discovery Bank composed of continental crust, a tectonic domain with intermediate features, both in position and nature, between continental and oceanic crusts that includes the Southern Bank, and the oceanic crust of the northern Weddell Sea, which belongs to the Antarctic Plate.

The intermediate domain shows extreme crustal thinning and mantle uplift that are associated to the deep basins in all the MCS profiles, although we do not observe evidence of oceanic spreading. This domain was probably developed during the Late Cenozoic subduction of the Weddell Sea oceanic crust below the Discovery Bank and prior to the recent transcurrent tectonics. The complex bathymetry and structure of the plate boundary are a consequence of the presence of continental and intermediate crustal blocks. Deformation was preferently concentrated here, between the two stable oceanic domains.

Introduction

In the southern Atlantic is located the Scotia Sea that developed since the Oligocene (BAS 1985; Pelayo and Wiens 1989; Barker et al. 1991; Barker 2001). This region accommodate the sinistral transcurrent motion between the major South American and Antarctic Plates that has produced a complex pattern of structures including large continental fragments surrounded by oceanic crust (Fig. 5.3-1). The Scotia Plate, mainly oceanic in nature, was formed by the successive activity of several spreading ridges (BAS 1985; Barker et al. 1991) and constitutes

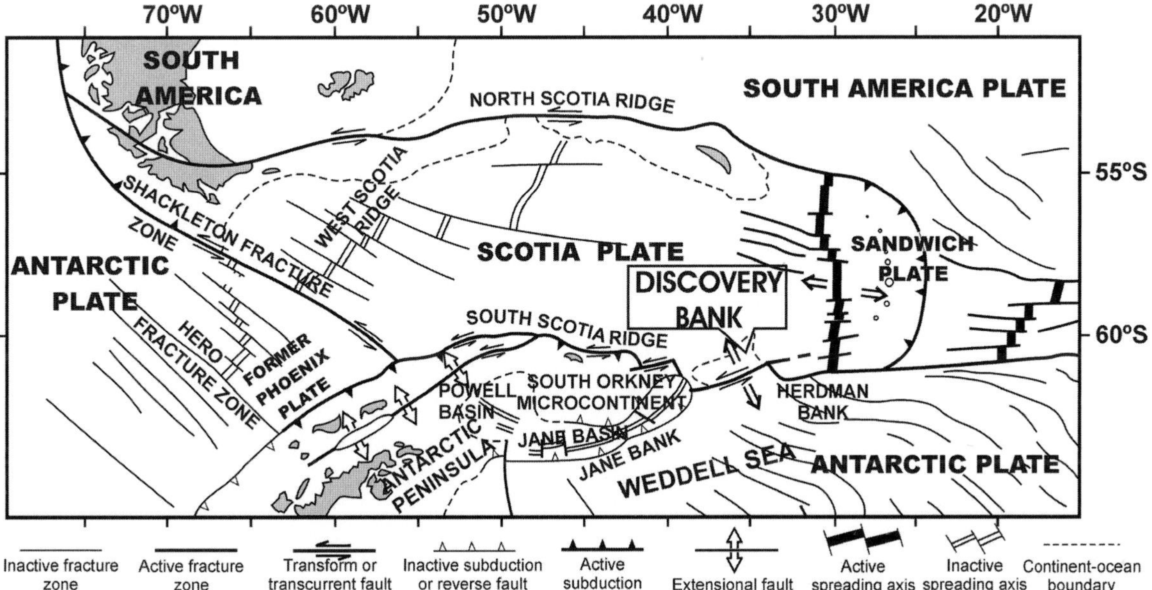

Fig. 5.3-1. Geological setting of the Discovery Bank in the frame of the Scotia Arc

From: Fütterer DK, Damaske D, Kleinschmidt G, Miller H, Tessensohn F (eds) (2006) Antarctica: Contributions to global earth sciences. Springer-Verlag, Berlin Heidelberg New York, pp 237–242

the internal part of the Scotia Arc. The Sandwich Plate overthrusted the South American Plate, developing an arcuate subduction zone at the eastern end of the Scotia Sea. This subduction zone constitutes one of the most active structures of the region with related seismicity (Pelayo and Wiens 1989). Jane Basin developed in Middle Miocene as a back-arc basin related to the subduction of the oceanic crust of the Weddell Sea below the SE margin of the South Orkney Microcontinent (Lawver et al. 1991; Maldonado et al. 1998).

The South Scotia Ridge is located at the boundary of the Scotia and Sandwich Plates with the Antarctic Plate (Fig. 5.3-1). This ridge is formed by a complex array of continental crustal blocks, which includes banks and deep basins. This boundary shows moderate seismicity along a band characterized by tensional, transtensional and transcurrent earthquake focal mechanisms (Pelayo and Wiens 1989; Galindo-Zaldívar et al. 1996). This is an area of intricate relief determined by the presence of adjacent continental and oceanic crusts. The western part of the South Scotia Ridge has been studied in detail by marine surveys that included multichannel seismic reflection, gravity, magnetic and swath bathymetry (Acosta and Uchupi 1996; Galindo-Zaldívar et al. 1996; Klepeis and Lawver 1996; Suriñach et al. 1997; Maldonado et al. 1998). In contrast, in the eastern segment of the plate boundary only general surveys have been previously conducted (Barker 1972, 2001; BAS 1985). Larter et al. (1998) and Bruguier and Livermore (2001) have studied the Scotia-Sandwich spreading ridge and Sandwich trench areas. The seismicity band in the eastern South Scotia Ridge is oblique to the bathymetric features previously recognized in the area (BAS 1985) and do not allow to determine the active structures related to the plate boundary.

The main aim of this contribution is to determine the main tectonic features of the eastern sector of the Scotia-Antarctic plate boundary, in a segment situated between the Discovery and Herdman Banks (Fig. 5.3-1). The study of this region will contribute to understanding the behaviour of different types of crusts involved in complex plate boundaries and the development of tectonic arcs.

Data Set

Ship bathymetric data of the Discovery Bank and surrounding regions are scarce. They were compiled on the Tectonic Map of the Scotia Sea (BAS 1985). There is also a bathymetry data set predicted from the altimetry data of the GEOSAT (Sandwell and Smith 1997), which, while representing a significant aid in the study of this remote region, does not provide detailed resolution for areas of complicated relief. The seismicity database include both the earthquake location and the focal mechanism of some events (BAS 1985; Pelayo and Wiens 1989; USGS 2002). The absence of permanent seismic stations in this area, however, increases the error of location of the hypocenters in respect to other regions, and do not allows to determine with precision the location of active structures.

During the SCAN97 cruise aboard the R/V "Hespérides" (January–February 1997), gravity, magnetic, MCS and swath bathymetry data were recorded along several profiles east of the South Orkney Microcontinent (Fig. 5.3-2 and 5.3-3). Most of the profiles are orthogonal to the topographic features of the plate boundary. Profile SM12 (Fig. 5.3-2 and 5.3-3) shows the main tectonic provinces of this region. Gravity data were obtained with a Bell Aerospace TEXTRON BGM-3 marine gravimeter. The free-air anomaly was determined with Lanzada software (A. Carbó, pers. comm.) and modelled with the GRAVMAG program (Pedley et al. 1993). Total intensity magnetic field data were recorded using a Geometrics G-876 proton precession magnetometer. The IGRF 1995 (I.A.G.A. 1996) was used to calculate the magnetic anomalies. The multichannel seismic reflection profiles were obtained with a tuned array of five BOLT air guns and a streamer with a total length of 2.4 km and 96 channels. The shot interval was 50 m. Data were recorded with a DFS V digital system and a sampling record interval of 2 ms and 10 s record lengths. The data were processed with a sequence, including migration using a DISCO/FOCUS system.

Main Structures of the Plate Boundary

Between the oceanic crusts of the Scotia and Weddell seas, the Discovery and the Herdman Banks and numerous minor elevations and intervening depressions constitute the main physiographic features related to the southern branch of the Scotia Arc. The Discovery Bank has NE-SW elongation and an asymmetrical profile, with a wide irregular slope to the Scotia Sea and a sharp rectilinear SE margin, where an elongated basin over 5 600 m deep is located (Fig. 5.3-2 and 5.3-3). South of the deep basin, several minor highs with a very irregular bathymetry occur in the transition to the Weddell Sea. The Herdman Bank is located in the eastern sector (Fig. 5.3-1 and 5.3-3). This bank shows high seismic activity and it is separated from the Discovery Bank by a deep, long and irregular NE-SW depression. To the west, the seismic activity related to the plate boundary is located along the NE and northern border of the South Orkney Microcontinent.

Most the profiles that cross the region from the Scotia to the Weddell show similar features. Profile SM12 (Fig. 5.3-2 and 5.3-3) is representative of the structures observed in this sector of the Scotia-Antarctic plate boundary. The transition from the oceanic crust of the Scotia Sea to the thinned continental crust of the Discovery Bank is progressive as suggested by the gravity models. In the Scotia Sea, sea bottom is located at about 4 s (twt) and the sedi-

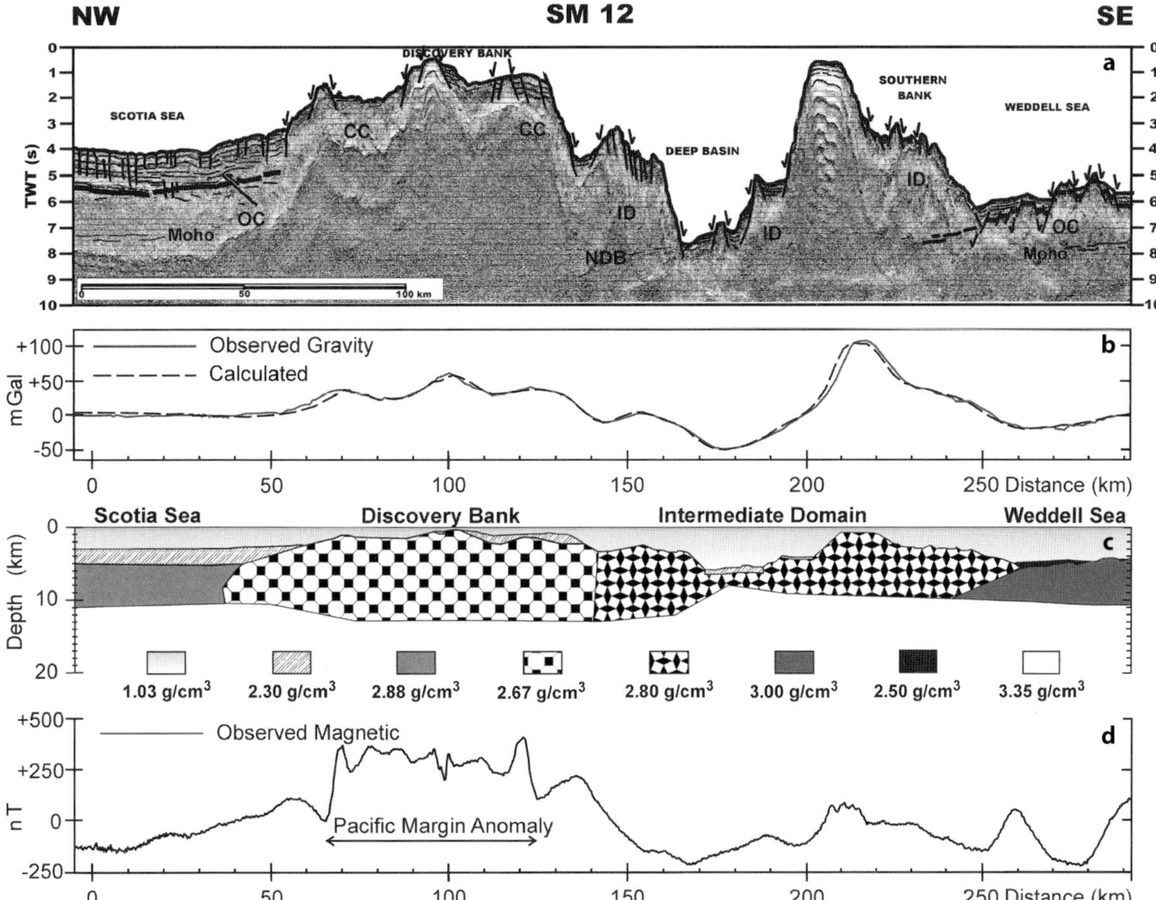

Fig. 5.3-2. Profile *SM12* orthogonal to the Scotia-Antarctic plate boundary (for location see Fig. 5.3-3). **a** MCS profile and interpretation; *OC:* oceanic crust; *CC:* continental crust; *ID:* intermediate domain; *NDB:* north dipping band of reflectors. **b** Observed and calculated free-air gravity anomalies. **c** Free-air gravity model. **d** Total field magnetic anomaly profile

ment layer is up to 1.2 s twt thick. The northern margin of Discovery Bank is affected by reverse faults that are recognized in depth, probably related to an early stage of compression that may have developed during the thrusting or subduction of the Scotia Sea oceanic crust below Discovery Bank. The recent tectonic activity is evidenced by several subvertical and normal faults that affect the sea bottom and are seen in the transition to the Discovery Bank, where perched basins filled by a thin sedimentary cover separate basement highs. This setting is typical of the recent transcurrent or transtensional faulting that has deformed the area.

The southern margin of Discovery Bank is deformed by conjugated normal faults with scarps greater than 3 s (twt), which bound small perched basins and highs (km 130 to 170, profile SM12, Fig. 5.3-2). The faults are related to an elongated depression reaching 7.7 s (twt) depth (Deep Basin) which is filled by a layer of sediments of about 0.5 s (twt). North of the Deep Basin, it is possible to identify a band of northward dipping reflectors that may be related to a former reverse fault, cut by the most recent normal fault, and that do not affect the sedimentary filling of the basin. Southwards, the Southern Bank shows an asymmetric bathymetric profile (Fig. 5.3-2) with a northern margin steeper than the southern margin. The southern segment of the profile corresponds to the Weddell Sea oceanic crust. Reflectors related to the top of the igneous basement seem to extend below the Southern Bank (Fig. 5.3-2), pointing to the location of a former trench related to a subduction zone. Above the igneous basement, a discontinuous depositional sequence with a maximum thickness of 1 s (twt) can be identified. The contact between the oceanic crust and the Southern Bank is sealed by sediments. In the Weddell Sea, some deep reflectors apparently represent the Moho (approximately 8–9 s twt).

The normal and vertical faults that indicate deformation on Discovery Bank, the Deep Basin and the Southern Bank have related scarps and may be active or recently active. One earthquake focal mechanism of transcurrent fault located in the northern margin of Discovery Bank (Pelayo and Wiens 1989), probably related to an E-W oriented fault plane, and two mechanisms of NE-SW oriented normal faults are

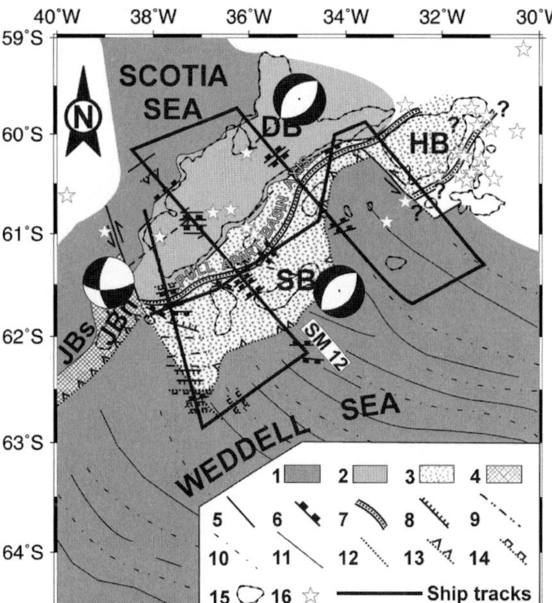

Fig. 5.3-3. Geological sketch of the main tectonic structures in the region of Discovery Bank. *Stars:* earthquake epicentres (1973–2002; USGS 2002). Earthquake focal mechanisms (from Pelayo and Wiens 1989). *1:* Oceanic crust; *2:* continental crust; *3:* intermediate crust; *4:* volcanic arc; *5:* active fracture zone; *6:* active normal fault; *7:* active crustal thinning axis; *8:* scarp; *9:* ridge; *10:* elongated gravity minima. *11:* elongated gravity maxima; *12:* inactive fracture zone; *13:* inactive subduction or reverse fault; *14:* inactive normal fault; *15:* 2 000 m depth bathymetry contour line; *JBs:* Jane Basin; *JBn:* Jane Bank; *DB:* Discovery Bank; *HB:* Herdman Bank; *SB:* Southern Bank

situated in the southern margin, near the Deep Basin, and near the northern border of Discovery Bank (Fig. 5.3-3).

The gravity model of free-air anomalies (Fig. 5.3-2) allows to determine the main features of the deep structure. The geometry of the gravity model is based on the bathymetry of the sea-floor and the sediment thickness derived from profile SM12 (Fig. 5.3-2). In determining depths and thicknesses, we considered velocities of 1 500 m s^{-1}, 2 200 m s^{-1} and 6 340 m s^{-1} respectively for the sea water, the sediments and the igneous oceanic crust, taking into account the velocities of the seismic refraction profiles reported by King et al. (1997) in Powell Basin that constitutes the nearest region with similar features. We considered densities of 3.35 g cm^{-3} for the standard mantle and of 1.03 g cm^{-3} for the sea water. The location of continental, intermediate and oceanic crusts was based on the seismic characteristics of the different units in the profile and the signature of the magnetic anomalies. Taking into account the young age of the oceanic crust of the Scotia Sea (probably about anomaly 5, BAS 1985) and the gravity anomalies, we considered a density of 2.88 g cm^{-3} for the igneous layers of the oceanic crust and a density of 2.3 g cm^{-3} for the sediments. Discovery Bank is made up probably of continental crust (2.67 g cm^{-3}), as pointed by the shallow bathymetry, the seismic character of the basement, with discontinuous reflectors of low amplitude, and the presence of magnetic anomalies related to the Pacific Margin Anomaly (Fig. 5.3-2) (Garrett 1991; Suriñach et al. 1997). Anyway, the presence of magnetic anomalies point that there are basic igneous rocks intruding the thin continental crust. There is a progressive change in the characteristics of the crust southeast of Discovery Bank and the Deep Basin and the Southern Bank are located in a domain of intermediate features (2.8 g cm^{-3}) between the continental crust of Discovery Bank and the oceanic crust of the Weddell Sea. Finally, a typical oceanic crust corresponding to the Weddell Sea is seen in the southern segment of the profile. For the Weddell Sea oceanic crust, a density of 3 g cm^{-3} was assigned by modelling to layers 2 and 3 of the oceanic crust and a density of 2.5 g cm^{-3} to the sediments, considering an age older than this of the Scotia Sea. Those densities fit as best as possible the observed gravity data.

Moho geometry is determined by seismic data and gravity modelling. In the oceanic crust ranges between 10 and 11 km of depth, being slightly deeper in the Scotia Sea than in the Weddell Sea. Discovery Bank is characterized by thin continental crust. In addition, a marked crustal thinning is located in the Deep Basin, southward from Discovery Bank (Fig. 5.3-2). The Southern Bank, which belongs to the Intermediate Domain, reveals a crustal thickness similar to that of the oceanic crust.

Discussion and Conclusions

The eastern segment of the Scotia-Antarctic plate boundary shows a diffuse seismicity and the focal mechanisms of the area point to transcurrent and extensional tectonics (Fig. 5.3-3). The geodynamics of the area also indicate that the plate boundary acts as a sinistral fault zone (BAS 1985). In between the oceanic crusts of the Scotia and Antarctic Plates are located the continental and intermediate blocks that constitute the southern branch of the Scotia Arc. Discovery Bank is formed by thinned continental crust, while southward, around the Deep Basin and the Southern Bank, the change in the seismic character of the crust and in the magnetic and gravity anomaly features suggest that this area probably corresponds to a domain of transitional crust. The contact with the Weddell Sea oceanic crust is attributed to the relict subduction zone probably active during Late Cenozoic times, analogous to the western segments of this contact in the northern Weddell Sea (Maldonado et al. 1998). The intermediate domain is probably formed during the subduction and made up by continental fragments, sediments and magmatic intrusions. In this tectonic setting other contractive structures probably developed, such the reverse faults observed at the northern border of Discovery Bank, and the possible reverse faults related to the northward-dipping band of reflectors observed below the southern border of Discovery Bank (Fig. 5.3-2).

In this sector of the plate boundary the present-day activity is determined by sinistral and transtensional faults. The northern margin of Discovery Bank is deformed by subvertical and high-angle normal faults with associated transcurrent and normal earthquake focal mechanisms (Fig. 5.3-3). The oblique orientation of the southeastern margin of Discovery Bank with respect to the sinistral transcurrent plate boundary permitted the development of normal faults in this margin and also facilitated the growth of the elongated Deep Basin under extensional tectonics (Fig. 5.3-3). Similar deep pull-apart basins occur in the western part of the South Scotia Ridge (Acosta and Uchupi 1996; Galindo-Zaldívar et al. 1996). Gravity model of the region has been constructed taking into account the constraints imposed by the multichannel seismic profile geometry, and taking into account standard density values, that constitute the best possible approach, but not always the unique solution. Although the gravity model points to an extreme crustal thinning, seismic and magnetic profiles show that oceanic spreading has not yet started in this sector in contrasts with other nearby regions (Barker 1972; BAS 1985; Bruguier and Livermore 2001).

The cross section between the Scotia and Weddell seas across the South Orkney Microcontinent, Jane Basin back-arc and Jane Bank island arc (Lawver et al. 1991; Maldonado et al. 1998) has similar features to the studied one. However, the Deep Basin constitute a former back-arc basin, later reactivated as a pull-apart basin. In this case, the Southern Bank would represent an island arc similar to the Jane Bank. These structures end toward the NE in Herdman Bank and toward the SW in Jane Bank-Jane Basin, probably by transform faults parallel to the inactive fracture zones of the Weddell Sea oceanic crust (Fig. 5.3-3). The seismic activity at Herdman Bank may be related to another spreading axis or a crustal thinning axis.

Deformations related to the plate boundary are distributed between the continental crust of the Discovery Bank, and the Intermediate Domain like in the western part of the South Scotia Ridge (Galindo-Zaldívar et al. 1996). Plate boundaries are preferently located in areas with continental crust, which may be more easily deformed than the rigid oceanic crust.

Acknowledgments

The comments of Dr. I. Dalziel and an anonymous referee improved this contribution. Spain's CICYT supported this research through Project ANT99-0817 and REN2001-2143/ANT.

References

Acosta J, Uchupi E (1996) Transtensional tectonics along the South Scotia Ridge, Antarctica. Tectonophysics 267:31–56

Barker PF (1972) A spreading centre in the east Scotia Sea. Earth Planet Sci Letters 15:123–132

Barker PF (2001) Scotia Sea regional tectonic evolution: implications for mantle flow and palaeocirculation. Earth Sci Revs 55:1–39

Barker PF, Dalziel IWD, Storey BC (1991) Tectonic development of the Scotia Arc region. In: Tingey RJ (ed) The geology of Antarctica. Oxford Science Publication, Oxford, pp 215–248

BAS (1985) Tectonic map of the Scotia Arc, Scale 1:3000000. British Antarctic Survey Misc 3, Cambridge

Bruguier NJ, Livermore RA (2001) Enhanced magma supply at the southern East Scotia Ridge: evidence for mantle flow around the subducting slab? Earth Planet Sci Letters 191: 129–144

Galindo-Zaldívar J, Jabaloy A, Maldonado A, Sanz de Galdeano C (1996) Continental fragmentation along the South Scotia Ridge transcurrent plate boundary (NE Antarctic Peninsula). Tectonophysics 242:275–301

Garrett SW (1991) Aeromagnetics studies of crustal blocks and basins in West Antarctica: a review. In: Thomson MRA, Crame JA, Thomson JW (eds) Geological evolution of Antarctica. Cambridge University Press, Cambridge, pp 251–256

IAGA (1996) Division V, Group 8, International Geomagnetic Reference Field, 1995 revision, Geophys J Internat 125:318–321

King E, Leitchenkov G, Galindo-Zaldívar J, Maldonado A, E Lodolo (1997) Crustal structure and sedimentation in Powell Basin. In: Barker PF, Cooper AK (eds) Geology and seismic stratigraphy of the Antarctic margin, 2. Antarctic Res Ser 71, Amer Geophys Union, Washington DC, pp 75–93

Klepeis KA, Lawver LA (1996) Tectonics of the Antarctic-Scotia plate boundary near Elephant and Clarence Islands, west Antarctica. J Geophys Res 101:20211–20231

Larter RD, King EC, Leat PT, Reading AM, Smellie JL, Smythe DK (1998) South Sandwich slices reveal much about arc structure geodynamics and composition. EOS 79:281–285

Lawver LA, Della Vedova B, Von Herzen RP (1991) Heat flow in Jane Basin, northwest Weddell Sea. J Geophys Res 96: 2019–2038

Maldonado A, Zitellini N, Leitchenkov G, Balanyá JC, Coren F, Galindo-Zaldívar J, Jabaloy A, Lodolo E, Rodríguez-Fernández J, Zanolla C (1998) Small ocean basin development along the Scotia-Antarctica plate boundary and in the northern Weddell Sea. Tectonophysics 296:371–401

Pedley RC, Bubsby JP, Dabek ZK (1993) GRAVMAG 5. British Geological Survey

Pelayo AM, Wiens DA (1989) Seismotectonics and relative plate motions in the Scotia Sea Region. J Geophys Res 94:7293–7320

Sandwell DT, Smith WHF (1997) Marine gravity anomaly from Geosat and ERS-1 satellite altimetry. J Geophys Res 102: 10039–10054

Suriñach E, Galindo-Zaldívar J, Maldonado A, Livermore RA (1997) Large amplitude magnetic anomalies in the northern sector of the Powell Basin NE Antarctic Peninsula. Mar Geophys Res 19: 65–80

USGS (2002) National Earthquake Information Center 2002 Web page, http://neic.usgs.gov/neis/epic/epic.html

Bransfield Basin Tectonic Evolution

Jesús Galindo-Zaldívar[1] · Luiz Gamboa[2] · Andrés Maldonado[3] · Seizo Nakao[4] · Yao Bochu[5]

[1] Departamento de Geodinámica, Universidad de Granada, 18071-Granada, Spain, <jgalindo@ugr.es>
[2] Petróleo Brasileiro S.A. (PETROBRAS), Rio de Janeiro, Brazil, and Universidade Federal Fluminenese, Brazil, <gamboa@petrobras.br>
[3] Instituto Andaluz de Ciencias de la Tierra, CSIC-Universidad de Granada, 18071-Granada, Spain, <amaldona@ugr.es>
[4] Marine Geology Department, Geological Survey of Japan
[5] Guangzhou Marine Geological Survey, Guangzhou, PR China

Abstract. Bransfield Basin is located at the Pacific margin of the Antarctic Peninsula and constitutes an incipient oceanic back-arc basin. This basin developed as a consequence of the separation of the South Shetland Block from the Antarctic Peninsula. Analysis of multichannel seismic profiles from Brazilian, Spanish, Japanese and Chinese cruises allows the shallow structure of the Bransfield Basin and its eastward prolongation through the South Scotia Ridge to be studied.

The Bransfield Basin is asymmetrical and taking into account the shallow structures, its opening may be interpreted as related to a low angle normal fault that dips NW, with the South Shetland Block constituting the hanging wall. The margin adjacent to the Antarctic Peninsula exhibits all the features associated with a lower plate passive margin, such as the development of landward tilted half-grabens, the presence of a break-up unconformity, and the deposition of an oceanward dipping "drift" sequence. However, the margin near the South Shetland Islands is typical of an upper-plate margin: poorly nourished, sharp and with high angle faults. Extension is more developed in the Central Bransfield Basin, where a volcanic axis is recognized as the expression of a young spreading center, and there is possibly incipient oceanic crust. In the Bransfield Basin extremities, present-day deposits represent the synrift sequence, and extension continues. The Bransfield Basin probably develops as a consequence of two interacting processes. The main one is rollback of the trench hinge related to the continued sinking of the subducted slab of the former Phoenix Plate. A second process is related to the westwards propagation of the deformations associated with the Scotia-Antarctic plate boundary along the South Scotia Ridge and up to the Bransfield Basin.

Introduction

The structure of the southern part of the Scotia Arc is determined by the left-lateral strike-slip fault zone boundary between the South American and Antarctic Plates along the South Scotia Ridge (Fig. 5.4-1). This boundary extends northwestward across the Drake Passage as a prominent submarine ridge between the South Chile Trench and the South Shetland forearc, called the Shackleton Fracture Zone, with transpressive character. The southern end of the Shackleton Fracture Zone is currently being subducted below the South Shetland forearc (Aldaya and Maldonado 1996). The South Shetland Block may be considered as a crustal element independent from the Antarctic Plate.

Bransfield Strait hosts a NE-SW elongated basin that lies between the NW-facing edge of the Antarctic Peninsula and the South Shetland Block, and is widely interpreted as an extensional marginal sea associated with trench rollback at the South Shetland Trench (Barker 1982; Pelayo and Wiens 1989; Gamboa and Maldonado 1990; Barker and Austin 1998). The South Shetland Trench is the last conspicuous expression of a subduction zone that existed all along the western margin of the Antarctic Peninsula (Barker 1982; Larter and Barker 1991) and that may be active at present (Maldonado et al. 1994). In addition, it is proposed that the rift of the Bransfield Strait merges eastward into the active transcurrent fault zone within the South Scotia Ridge (Galindo-Zaldívar et al. 1996). Meanwhile, González-Casado et al. (2000) propose that the Bransfield Basin develops only as a consequence of the sinistral transcurrence between the Scotia and Antarctic Plates.

The West Scotia and Phoenix-Antarctic Ridges extend across Drake Passage (Fig. 5.4-1). The Phoenix Plate ocean floor was subducted beneath the Antarctic Plate continental lithosphere at the South Shetland Trench, while farther northwest new ocean floor was created at the Phoenix-Antarctic Ridge, which migrated southward (Larter and Barker 1991). Recently, bathymetric and magnetic anomaly data have been used to show that extinction of the last remaining Phoenix Ridge segments of the once-extensive Phoenix-Antarctic spreading centre occurred at magnetic chron C2A (3.3 ±0.2 Ma), synchronous with a ridge-trench collision south of the Hero Fracture Zone (Livermore et al. 2000).

Good markers for understanding the recent tectonic evolution of the area are the elongated bodies of Cretaceous mafic rocks located along the Antarctic Peninsula and that are associated with large magnetic anomaly bands known as the Pacific Margin Anomaly or West Coast Magnetic Anomaly (Garrett et al. 1987). The single elongated anomaly that is recognized in the southern region of the Antarctic Peninsula divides into two branches in the northern Antarctic Peninsula, one each side of the opening of the Bransfield Strait (Garrett et al. 1987), where Bransfield Basin develops.

The aim of this contribution is to analyse the structure of the Bransfield Basin and its prolongation towards the South Scotia Ridge in order to determine the mechanisms involved in its development. A combined set of

From: Fütterer DK, Damaske D, Kleinschmidt G, Miller H, Tessensohn F (eds) (2006) Antarctica: Contributions to global earth sciences. Springer-Verlag, Berlin Heidelberg New York, pp 243–248

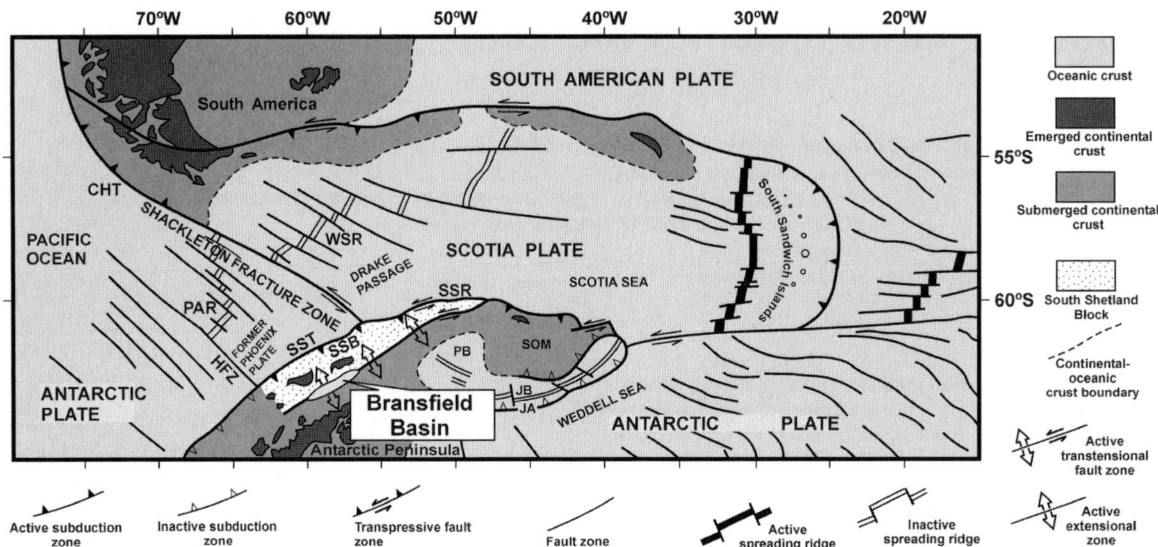

Fig. 5.4-1. Tectonic map of the Scotia Arc. *CHT:* Chile Trench; *HFZ:* Hero Fracture Zone; *JA:* Jane Arc; *JB:* Jane Bank; *PAR:* Phoenix-Antarctic Ridge; *PB:* Powell Basin; *SOM:* South Orkney microcontinent; *SSB:* South Shetland Block; *SSR:* South Scotia Ridge; *SST:* South Shetland Trench; *WSR:* West Scotia Ridge

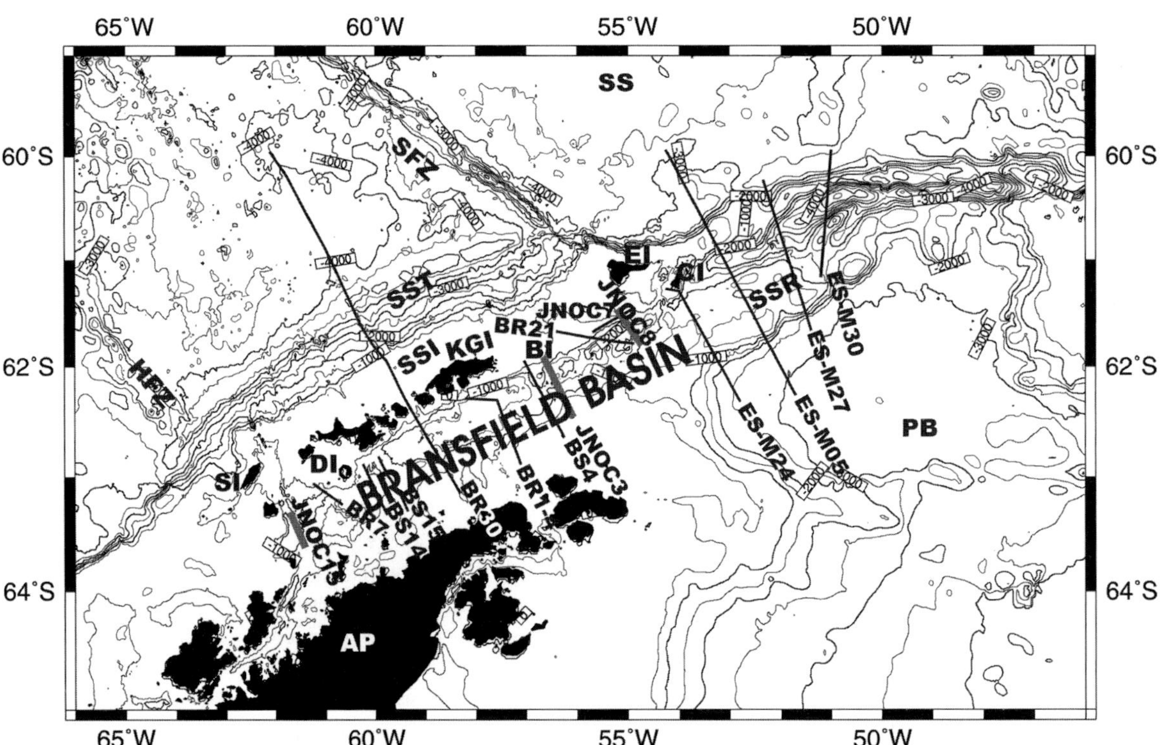

Fig. 5.4-2. Bathymetric map of the study area from satellite and ship tracks data. Isolines are every 500 m. The position of selected MCS profiles from China (*BS*), Brasil (*BR*), Japan (*JNOC*) and Spain (*ES*) considered in this research are indicated. *Thick grey lines:* Profiles illustrated in this contribution. *AP:* Antarctic Peninsula; *BI:* Bridgeman Island; *CI:* Clarence Island; *DI:* Deception Island; *EI:* Elephant Island, *HFZ:* Hero fracture zone; *KGI:* King George Island; *PB:* Powell Basin; *SI:* Smith Island; *SFZ:* Shackleton fracture zone; *SS:* Scotia Sea; *SSI:* South Shetland Islands; *SSR:* South Scotia Ridge; *SST:* South Shetland Trench

multichannel seismic (MCS) profiles from Brazil, Spain, Japan and China have been used to determine the main features of the structure of the Bransfield Strait area (Fig. 5.4-2). Profiles have been processed in different laboratories, with a standard processing sequence including migration.

Bransfield Basin Structure

Bransfield Basin has a lens shape and can be divided into three subbasins: Western, Central and Eastern. Incipient rifting occurs south of Deception Island (Fig. 5.4-3) in the Western Basin. This region is affected, however, by intense normal faulting that created rotated blocks on the continental crust. A fairly thick synrift sequence is observed in half-grabens. The sedimentary fill shows wedges opening towards the NW. Some faulting affects the seafloor, indicating active tectonics and present-day rifting.

The Central Bransfield Basin extends from Deception Island to Bridgeman Island (Fig. 5.4-2), that may constitute outcrops of an incipient spreading center at the axis of the basin, with associated magnetic anomalies (Roach 1978). The southeastern margin presents structures and sedimentary fill that in many ways resemble the features observed along passive margins, with development of syn-rift and post-rift sequences. In this part of the basin, the continental crust has thinned in response to a NW-SE directed extensional tectonic regime, which continues up to the present day as shown by earthquake focal mechanism data (Pelayo and Wiens 1989). Profile JNOC3, located at the eastern extremity of this sub basin (Fig. 5.4-4) clearly illustrates the basin structure. The progressive continental crustal thinning in the Antarctic Peninsula passive margin has isolated a continental crustal block located in the southern margin of the basin, while the northern margin is very steep. The observed structures would be consistent with the presence of a low angle crustal fault located at depth and dipping towards the NW, beneath the South Shetland Islands, and that serves as the detachment plane for the synthetic normal faults that cut a syn-rift sequence deposited along the Antarctic Peninsula side of the basin. The syn-rift sequence can be seen in MCS profiles and the sedimentary fill of the half-grabens has wedge shapes opening towards the SE. In addition, there is a well developed post-rift sequence that is bounded at its base by a break-up unconformity, occurring only at the central part of the Bransfield Basin (Fig. 5.4-4). The stretched continental crust forms a band about 30 km wide along the SE (Antarctic) margin of the basin, whereas a narrow, steep slope topography marks the passage from the South Shetland Islands shelf to the axial area of this part of the Bransfield Basin.

The crustal structure of the axial part of the central Bransfield Basin has been the subject of several studies (Ashcroft 1972; Birkenmajer et al. 1990; Guterch et al. 1990; Grad et al. 1997; Torres et al. 1997). In an earlier investigation using seismic refraction data, Ashcroft (1972) proposed the existence of a stretched continental crust in-

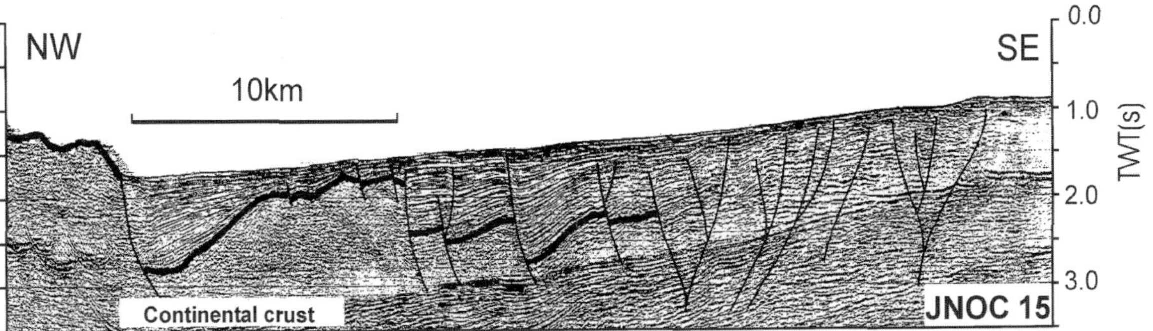

Fig. 5.4-3. Multichannel seismic profile *JNOC15* showing the main structures across the western Bransfield Basin

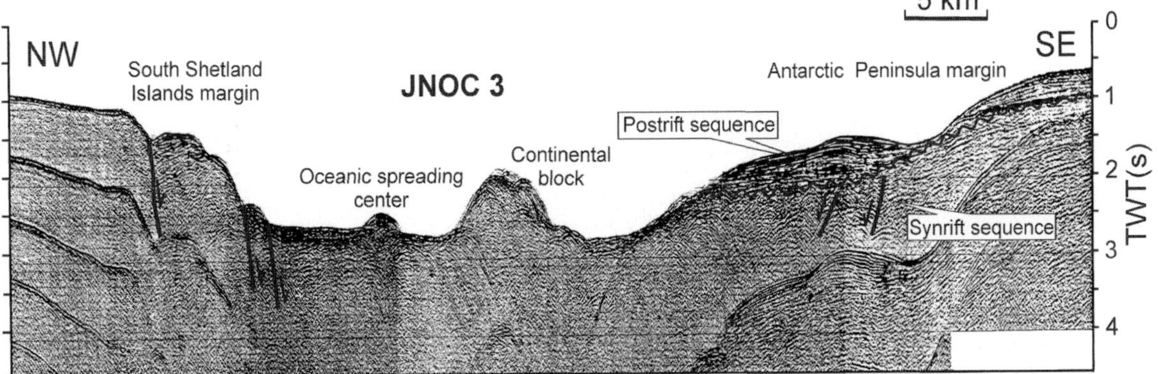

Fig. 5.4-4. Multichannel seismic profile *JNOC3* showing the main structures across the central-eastern Bransfield Basin

truded by an axial volcanic ridge (González-Ferrán 1991), and Moho depths of the order of 14 to 20 km were obtained. Later research obtained much greater Moho depth values of up to 30 km in the same area. Grad et al. (1997), using OBS data, obtained a model of the crustal structure of the Bransfield Basin that, while indicating Moho at depths of 30 km, showed the presence of a high velocity layer ($Vp = 7.3–7.2$ km s^{-1}) in the axial portion of the Bransfield Strait. Torres et al. (1997) modeled the crustal structure at the Central Bransfield Basin using magnetic, gravimetric and seismic data to describe a crustal structure similar to that proposed by Ashcroft (1972), with a shallow Moho below the basement high in the central part of the basin.

The seismicity in the region (Pelayo and Wiens 1989) is concentrated at the ends of the Central Basin at Deception and Bridgeman Islands. The focal earthquake mechanisms determined for shallow and intermediate seismicity (Pelayo and Wiens 1989) indicate NW-SE extensional stresses compatible with the opening of the Bransfield Strait.

The Eastern Bransfield Basin extends northward of Bridgeman Island to a region just south of Elephant Island. The structural style of this sector has features similar to those observed in the Western Bransfield Basin (Fig. 5.4-5): steep normal faults occur along the northern margin. The Eastern Bransfield Basin is characterized by the development of half-graben with sedimentary wedges opening towards the NW, indicating an extensional or transtensional regime. In addition, most of the sediments in the half-grabens show open, large folds indicative of a recent phase of compressional deformation. The deposits representing the synrift phase of the basin are cut by numerous normal faults, some of them affecting the present-day seafloor morphology.

The Boundaries of the South Shetland Block

The South Shetland Block is a tectonic element formed by continental crust. Its northwestern boundary consists of two sections separated by the Shackleton Fracture Zone near its intersection with the Elephant triple junction. The southwestern section is bounded by the South Shetland Trench. The northeastern section is bounded to the N by the sinistral, mainly transpressional fault zone located along the contact between the oceanic crust of the Scotia Plate and the continental blocks of the South Scotia Ridge, presently undergoing moderate or low tectonic activity (Galindo-Zaldívar et al. 1996). The Bransfield Basin and the internal depression of the South Scotia Ridge constitute the southeastern boundary of the block. The southwestern boundary of the South Shetland Block is the most poorly defined one, and may be located in a broad region of crustal deformations related to the transition from active to inactive convergence at the southwestern end of the South Shetland Trench (Maldonado et al. 1994).

Discussion and Conclusions

Bransfield Basin has clearly asymmetrical margins, although the asymmetrical character is variable along the basin. The asymmetry is most developed in the central part that shows two continental margins separated by a volcanic axis that may form incipient oceanic crust. While the northwestern boundary of the basin is sharp, sediment-starved, and deformed by high angle normal faults, generally dipping southeastwards, a thick layer of sediments covers the southeastern margin. In the Central Bransfield Basin, two sedimentary sequences, that may be

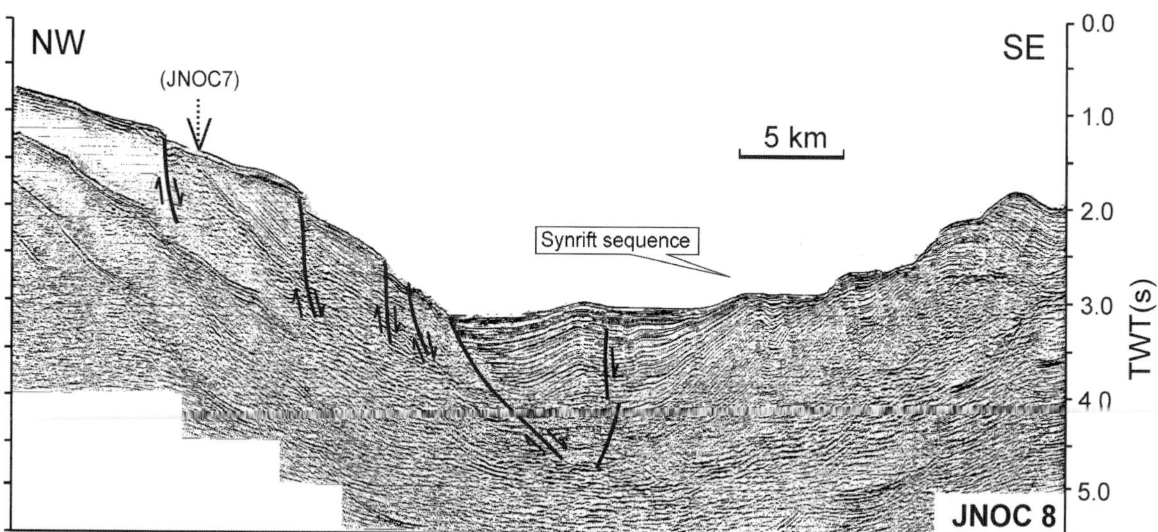

Fig. 5.4-5. Multichannel seismic profile *JNOC8* showing the main structures across the eastern Bransfield Basin

interpreted as synrift and post-rift sequences (Fig. 5.4-4) are separated by a break-up unconformity. The high angle normal faults might be related at depth to a low angle normal fault with a northwestwards dip, and top-to-the-northwest movement (Fig. 5.4-6) similar to the model proposed by Lister et al. (1991) for basin development. In such a model, the South Shetland block is located in the hanging wall of the detachment, while the Antarctic Peninsula constitutes its footwall.

The half-graben ages and asymmetries, meanwhile, vary throughout the Bransfield Basin. In the Western Basin, rifting still occurs over a broad zone of continental crust and a series of half-grabens of differing sizes continues to develop. Despite the fact that a deep depression exists in the Eastern Basin and the continental crust is likewise cut by mafic intrusions at some places in its western end, its features are reminiscent of those of the Western Basin. Barker and Austin (1998) propose a model for the development of Bransfield Basin in which rifting would have progressed from the northeast end of the basin towards the center. However, the maximum opening of the basin, with the development of a break-up unconformity and a volcanic axis, is found in the Central Bransfield Basin (Fig. 5.4-4 and 5.4-6), while rifting is still in its infancy at the Eastern and Western Bransfield subbasins. This suggests that the Bransfield Basin started opening from the Central Basin towards the Eastern and Western Basins or that Eastern and Western Basins are crustal accommodation zones where the basin extensional tectonics are transformed into other movements. Rifting would have initiated in a well-defined zone within the Central Basin, rather than as a propagating rift that started at the end of the Eastern Basin as proposed by Barker and Austin (1998) and González-Casado et al. (2000).

These observations are in agreement with the models that consider the Central Bransfield Basin to have formed by rifting between the Antarctic Peninsula and the South Shetland Block in response to trench roll-back at the South Shetland Trench (Maldonado et al. 1994; Barker and Austin 1998). In addition, the Eastern Bransfield Basin may accommodate westwards propagation of sinistral deformation along the South Scotia Ridge (Barker 1982; Galindo-Zaldívar et al. 1996; Barker and Austin 1998; González-Casado et al. 2000). The variable tectonic features of this southern boundary are determined by the obliquity of the active structures with regard to the stress field: NW-SE extension in the Bransfield Basin, would be consistent with the transtensional sinistral faults in the South Scotia Ridge, evolving eastwards to sinistral faults at the block boundary. Then, Eastern Bransfield Basin opening may be a consequence of the interaction of these two processes.

Acknowledgments

The comments of Dr. G. Eagles and Dr. K. Birkenmajer have improved this contribution. Spain's CICYT supported this research through Project ANT99-0817 and REN2001-2143/ANT and Brazil, through CAPES, Universidade Federal Fluminenese and Petrobras contributions.

References

Aldaya F, Maldonado A (1996) Tectonics of the triple junction at the southern end of the Shackleton Fracture Zone (Antarctic Peninsula). Geo Mar Letters 16:279–286

Ashcroft WA (1972) Crustal structure of the South Shetland Islands and Bransfield Strait. Brit Antarct Surv Sci Rep 66:1–43

Barker PF (1982) The Cenozoic subduction history of the Pacific margin of the Antarctic Peninsula: Ridge crest-trench interactions. J Geol Soc London 139:787–801

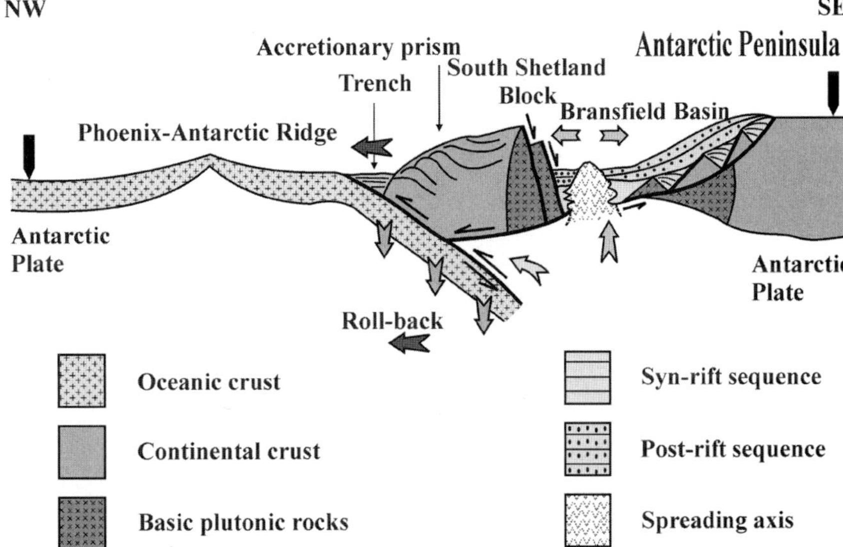

Fig. 5.4-6. Tectonic sketch interpretation of the central sector of Bransfield Basin and the South Shetland Block in the framework of the crustal structure of the Pacific margin of the Antarctic Peninsula. Outcropping gabbros on Elephant and Gibbs islands may be related to an elongated batholith formed by basic plutonic rocks associated to the magnetic anomaly bands (Garrett et al. 1987)

Barker DHN, Austin JA (1998) Rift propagation, detatchment faulting, and associated magmatism in Bransfield Strait, Antarctic Peninsula. J Geophys Res 103:24017–24043

Birkenmajer K (1992) Evolution of the Bransfield basin and rift, West Antarctica. In: Yoshida Y, Kaminuma K, Shirraishi K (eds) Recent progress in Antarctic earth science. Terra Scientific Publications, Tokyo, pp 405–410

Birkenmajer K, Guterch A, Grad M, Janik T, Perchuc E (1990) Lithospheric transect Antarctic Peninsula-South Shetland Islands (West Antarctica). Polish Polar Res 11:241–258

Galindo-Zaldívar J, Jabaloy A, Maldonado A, Sanz de Galdeano, C (1996) Continental fragmentation along the South Scotia Ridge transcurrent plate boundary (NE Antarctic Peninsula). Tectonophysics 242:275–301

Gamboa LAP, Maldonado PR (1990) Geophysical investigations in the Bransfield Strait and in the Bellinghausen Sea – Antarctica. In: St John B (ed) Antarctica as an exploration frontier – hydrocarbon potential, geology and hazards. Amer Assoc Petrol Geol 31:127–141

Garrett SW, Renner RGB, Jones JA, McGibbon KJ (1987) Continental magnetic anomalies and the evolution of the Scotia Arc. Earth Planet Sci Letters 81:273–281

González-Casado JM, Giner-Robles JL, López-Martínez J (2000) Bransfield basin, Antarctic Península: not a normal backarc basin. Geology 28:1043–1046

González-Ferrán O (1991) Volcanic and tectonic evolution of the northern Antarctic Peninsula, Late Cenozoic to Recent. Tectonophysics 114: 89–409

Grad M, Shiobara H, Janik T, Guterch A, Shimamura H (1997) Crustal models of the Bransfield Rift, West Antarctica, from detailed OBS refraction expreriments. Geophys J Internat 130:506–518

Guterch A, Grad M, Janik T, Perchuc E (1991) Tectonophysical models of the crust between the Antarctic Peninsula and the South Shetland trench. In: Thomson MRA, Crame JA, Thomson JW (eds) Geological evolution of Antarctica, Cambridge University Press, Cambridge, pp 499–504

Larter RD, Barker PF (1991) Effects of ridge crest-trench interaction on Antarctic-Phoenix spreading: forces on a young subducting plate. J Geophys Res 96:19586–19607

Lister GS, Etheridge MA, Symonds PA (1991) Detachment models for the formation of passive continental margins. Tectonics 10: 1038–1064

Livermore R, Balanyá JC, Barnolas A, Galindo-Zaldívar J, Hernández J, Jabaloy A, Maldonado A, Martínez JM, Rodríguez-Fernández J, Sanz de Galdeano C, Somoza L, Suriñach E, Viseras C (2000) Autopsy on a dead spreading centre: the Phoenix Ridge. Geology 28:607–610

Maldonado A, Larter RD, Aldaya, F (1994) Forearc tectonic evolution of the South Shetland margin, Antarctic Peninsula. Tectonics 13:1345–1370

Pelayo AM, Wiens DA (1989) Seismotectonics and relative plate motions in the Scotia Sea region. J Geophys Res 94:7293–7320

Roach PJ (1978) The nature of back-arc extension in Bransfield Basin. Geophys J Royal Astron Soc 53:165

Thomas C, Livermore R, Pollitz F (2003) Motion of the Scotia Sea plates. Geophys J Internat 155:789–804

Torres LC, Gomes BS, Gamboa LAG (1997) Determinaçao da Espessura Crustal na Margem Ativa da Regiao Antártica. 5th Internat Congr Brazilian Geophys Soc, Sao Paulo, Brazil, Expanded Abstract 1:67–69

The Sedimentary Sequences of Hurd Peninsula, Livingston Island, South Shetland Islands: Part of the Late Jurassic – Cretaceous Depositional History of the Antarctic Peninsula

Christo Pimpirev[1] · Kristalina Stoykova[2] · Marin Ivanov[1] · Dimo Dimov[1]

[1] Sofia University St. Kliment Ohridski, Department of Geology and Paleontology, Tsar Osvoboditel 15, 1000 Sofia, Bulgaria, <polar@gea.uni-sofia.bg>, <mivanov@gea.uni-sofia.bg>

[2] Geological Institute Bulgarian Academy of Sciences, Department of Paleontology and Stratigraphy, 24 Acad. G. Boncev str., 1113 Sofia, Bulgaria, <stoykova@geology.bas.bg>

Abstract. The age of the sedimentary sequences of Hurd Peninsula (here referred to the Miers Bluff Formation (MBF), has been considered so far as Triassic, coeval of the Trinity Group. Recently, a Tithonian ammonite species was found in a non-*in situ* block, coming from the lowermost unexposed part of the Formation. Our micropaleontological study reveals the occurrence of calcareous nannofossils in six sections. The recorded nannofossil association comprises the following species: *Micula decussata, Calculites obscurus, Arkhangelskiella cymbiformis, Prediscosphaera cretacea, Lucianorhabdus cayeuxii, Cyclagelosphaera reinhardtii, Braarudosphaera bigelowii, Ceratolithoides aculeus, Broinsonia* cf. *parca, Thoracosphaera* sp. indet., *Nephrolithus* sp. indet., *Cretarhabdus* sp. indet., *Watznaueria* sp. indet. It determines a Campanian-Maastrichtian age for the middle and upper part of MBF. Two calcareous nanofossil species, *Prediscosphaera cretacea* and ?*Fasciculithus* sp. indet., found in the Burdick Peak section suggest a Late Maastrichtian to (?) Paleocene age of the uppermost part of the MBF. The sediments of the MBF are possibly coeval of a part of Marambio Group (James Ross Island and Seymour Island) and Williams Point beds (Livingston Island).

Introduction

Livingston Island is the second largest island from the South Shetland archipelago (Fig. 5.5-1). The sedimentary sequences of Hurd Peninsula, which we relate to the Miers Bluff formation (MBF) form an important part of the Late Mesozoic-Cenozoic magmatic arc. The name Miers Bluff Formation (MBF) was introduced by Dalziel (1972). In the last 15 years MBF was extensively studied by Spanish geologists. The sedimentology and lithostratigraphy of the MBF have been described by Arche et al. (1992a,b) and Pallas et al. (1992). The latter divided MBF into the three informal lithostratigraphic units (U1, U2 and U3). The formal lithostratigraphy was introduced by Smellie et al. (1995) who divided MBF into Johnson Dock Member (JDM), Napier Peak Member (NPM) and Moores Peak breccias. Doktor et al. (1994) proposed that the name MBF be restricted to the lowermost unit 1 of Pallas et al. (1992), equivalent to the JDM of Smellie et al. (1995). The removal of the eastern Hurd Peninsula's outcrops from the MBF, however, is not supported by our field observations. The scheme of Smellie et al. (1995) is used in our recent publications (Pimpirev et al. 2000; Dimov and Pimpirev 2002). The latter authors proposed member status for the uppermost unit 3 equivalent to the Moores Peak breccias.

For a long time the depositional age of the MBF was poorly constrained. From poorly preserved, metamorphosed plant debris found in the MBF, Schopf (1973) concluded that its age was younger than Carboniferous, and its Mesozoic age is not excluded. The results of isotope geochronology on detritial zircons from turbiditic sandstones and the clay fraction suggested a wide temporal interval between the Late Carboniferous (?) and the Early

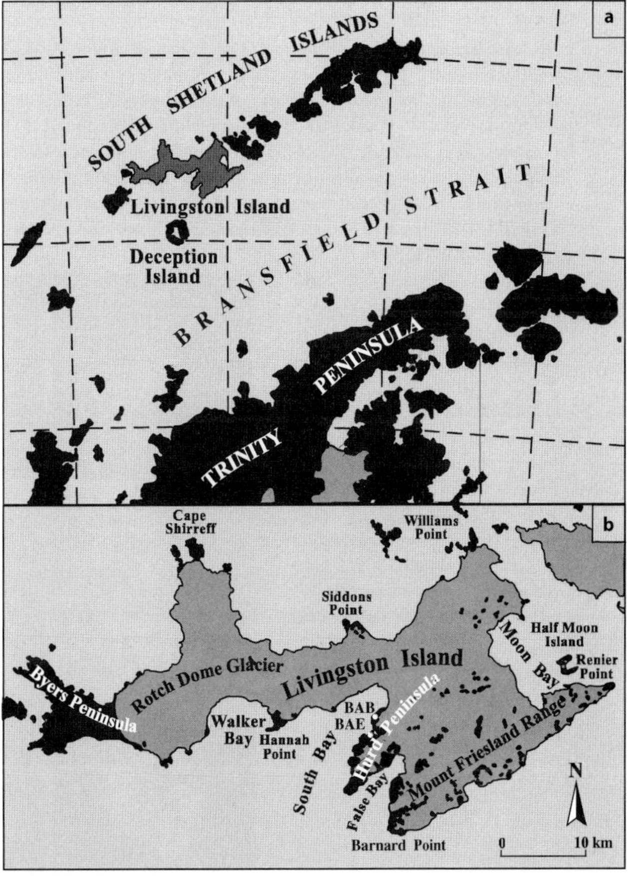

Fig. 5.5-1. a Overview of South Shetland Islands and Trinity Peninsula and **b** location of the studied area

From: Fütterer DK, Damaske D, Kleinschmidt G, Miller H, Tessensohn F (eds) (2006) Antarctica: Contributions to global earth sciences. Springer-Verlag, Berlin Heidelberg New York, pp 249–254

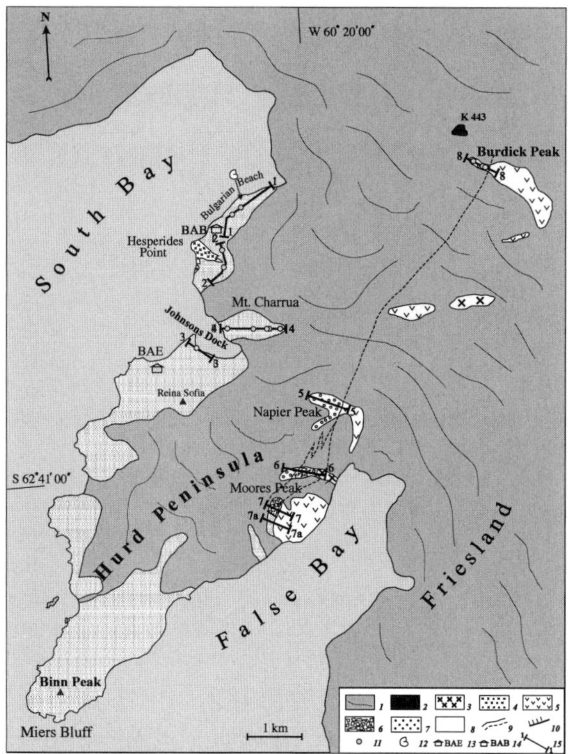

Fig. 5.5-2. Geological map of the region of Bulgarian Antarctic Base and location of studied sections. *1:* Ice; *2:* Inott point Formation; *3:* Friesland tonalite pluton; *4:* Hesperides point pluton; *5:* Antarctic Peninsula Volcanic Group (APVG). Miers Bluff Formation (MBF): *6:* Moores Peak Member; *7:* Pico Napier Member; *8:* Johnsons Dock Member; *9:* boundary of the rock unit; *10:* fault; *11:* location of nannofossil finds; *12:* location of macrofossil finds; *13:* Spanish Antarctic Base (*BAE*); *14:* Bulgarian Antarctic Base (*BAB*); *15:* geological section

Jurassic (Smellie et al. 1984; Miller et al. 1987; Hervé et al. 1991; Hervé 1992; Willan et al. 1994). Recent discovery of a rich palynoflora in the JDM suggests a Late Triassic age (Shu et al. 2000).

The present study, however, summarizes newly obtained macro- and nannofossil data that prove a Tithonian-Maastrichtian age for the MBF (Pimpirev et al. 2002; Stoykova et al. 2002).

Stratigraphy and Age Constrains

The new results for the stratigraphy and age of the MBF are based on detailed fieldwork and geological mapping at a 1:5 000 scale in the areas of the Bulgarian Antarctic base, Charrua Ridge, Johnson Dock Ridge, Moores Peak, Cerro Mirador, Napier Peak, Nunatak de Castillo and Burdick Peak. The MBF sedimentary sequence has been studied for macro- and microfossil content (including calcareous nannofossils) in seven sections (Fig. 5.5-2). A total of 120 samples were analyzed, 16 of them contain calcareous nannofossils. The position of the productive samples is noted on Fig. 5.5-2. Generally, calcareous nannofossils occur rarely, besides their preservation varies from poor to moderate throughout.

We suppose that the lowermost part of the sedidmentary succession is not exposed on the surface. This is indirectly indicated by the presence of scattered blocks of black silty mudstone on the marine terrace of Bulgarian Beach. The blocks often contain poorly preserved belemnites, bivalves and only one age diagnostic ammonite species *Blanfordiceras* sp. aff. *wallichi* (Gray 1832), that indicates a late

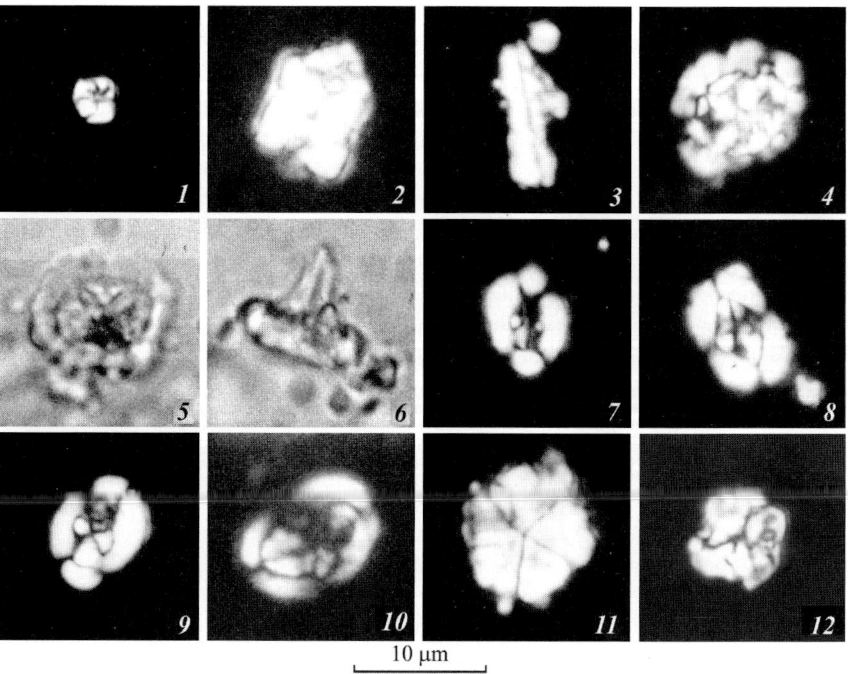

Fig. 5.5-3. Upper Cretaceous nannofossils from the Miers Bluff Formation. *1: Micula* sp., sample P311; *2: Micula decussata*, sample X9; *3: Lucianorhabdus cayeuxii*, sample X9; *4, 5, 6: Prediscosphaera cretacea*, sample X9; *7, 8: Broinsonia* cf. *parca*, sample X9; *9: Arkhangelskiella cymbiformis*, sample X9; *10: Arkhangelskiella cymbiformis*, sample X36; *11: Braarudosphaera bigelowii*, sample X36; *12: Micula decussata*, sample P311

Tithonian age for the lowermost, unexposed part of the formation (Pimpirev et al. 2002).

Sections 1 and 2 on Bulgarian Beach display the lower exposed part of the JDM of the MBF, composed of 320 m thick packages (up to 25 m) medium to coarse grained amalgamated massive sandstones, intercalated with a few meters of thick mudstone/sandstone. These sandstones represent turbidite fan deposits possibly accumulated during a low stand of sea level. Four calcareous nannofossil species were recovered from the mudstone beds of section 2. The samples contain *Micula decussata* Vekshina, *Calculites obscurus* (Deflandre) Prins and Sissingh, *Thoracosphaera* sp. indet. and *Watznaueria* sp. indet., suggesting a Campanian-Maastrichtian age.

The middle and upper parts of the JDM consist of interbedded thick sandstone and mudstone packages and is very well exposed in Johnson Dock Ridge (section 3) and Charrua Ridge (section 4). The mudstone parts of the described sedimentary succession are composed of alternating fine-grained, thin sandstone-mudstone beds and thick massive sandstone strata, possibly related to a sea level rise. In Johnsons Dock section 3 only one poorly preserved nanofossil species, *Calculites obscurus* (Deflandre) Prins and Sissingh is recorded. It indicates a Late Cretaceous age.

From the same outcrop Shu et al. (2000) reported a moderately diverse palynomorph assemblage, comprised of spores, pollen, a few acritarchs and cuticle remains. They suggested a Late Triassic (possibly Norian-Rhaetian) age although preservation is poor and most of identifications are at the generic level. In general, their age determination is based on the rare occurrence of *Aratrisporites*

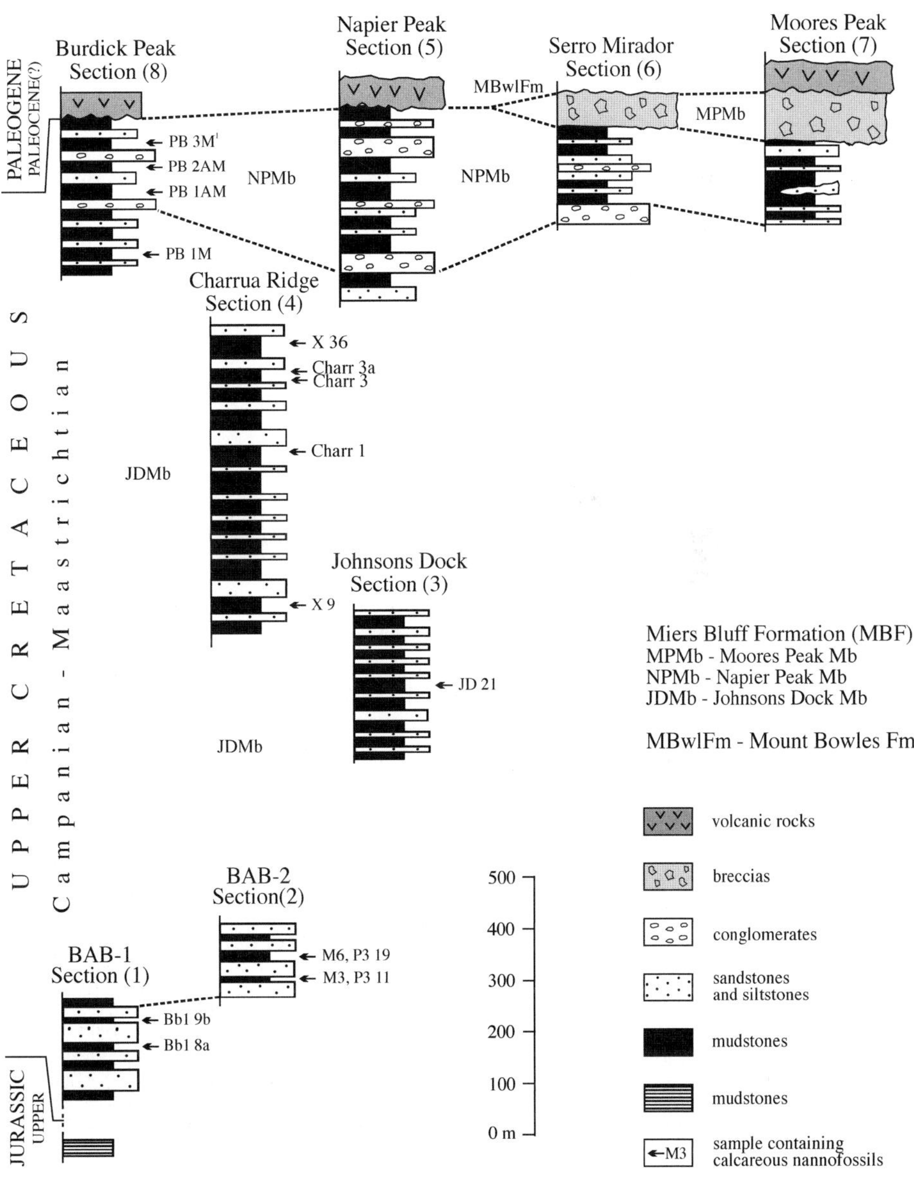

Fig. 5.5-4. Correlation of the studied sections of the Miers Bluff Formation with correct position of fossiliferous samples (calcareous nannofossils)

and *Classopollis* representatives in the assemblages, the dominant non-striate bisaccate pollen of the genus *Alisporites*, as well as by the absence of striate bisaccate pollen grains that dominate Permian flora. We believe this evidence does not contradict an age much younger than Late Triassic. The genus *Classopollis* has an almost worldwide occurrence in Jurassic and Cretaceous strata. In the Southern Hemisphere, its first appearance is considered in the Norian-Rhaetian, but it does not become frequent until the Jurassic. Representatives of the genus *Alisporites* are known in Jurassic and Cretaceous sediments as well. These new and young elements in the assemblage could question the Late Triassic age assessment of the formation. The Triassic taxa could be considered as re-deposited, as supported perhaps by their bad preservation state.

The section exposed in the Charrua Ridge, contains a relatively diverse nannofossil assemblage comprised of *Arkhangelskiella cymbiformis* Vekshina, *Prediscosphaera cretacea* (Arkhangelsky) Gartner, *Prediscosphaera* sp., *Lucianorhabdus cayeuxii* Deflandre, *Broinsonia* cf. *parca* (Stradner) Bukry, *Cyclagelosphaera reinhardtii* (Perch-Nielsen) Romein, *Braarudosphaera bigelowii* (Gran and Braarud) Deflandre, *Nephrolithus corystus* Wind, *Ceratolithoides aculeus* (Stradner) Prins and Sissingh, *Ceratolithoides* cf. *kamptneri* Bramlette and Martini. The specimens are poorly to moderately well preserved and the association undoubtedly proves Campanian-Maastrichtian age of the sediments.

The upper members of the MBF (Napier Peak Mb and Moores Peak Mb) were studied in four sections (Fig. 5.5-4, sections 5–8). They display the final, regressive stage with coarse sedimentation in the Miers Bluff depositional basin. Two calcareous nannofossil species, *Prediscosphaera cretacea* (Arkhangelsky) Gartner and ?*Fasciculithus* sp. indet., found in Burdick Peak section (Fig. 5.5-4, section 8) suggest a Maastrichtian to (?) Paleocene age for the uppermost part of the MBF.

The revision of the age of MBF and its relationships with Mount Bowles Fm. give us ground to assume Early Tertiary age for the volcanic complex. So far its age has been considered as Cretaceous (Smellie 1995). Thereby, the revised age of Mount Bowles Fm. correlates to the well-documented Tertiary volcanic activity in South Shetland Islands at that time.

Correlations and Conclusions

1. The lowermost unexposed part of the MBF of Late Tithonian–Cretaceous (?) age is partly synchronous with the Anchoridge Formation (Byers Peninsula, Livingston Island), The Nordeskjöld Formation (James Ross Island, Graham Land) and the Latady Formation (southeast Palmer Land).
2. The Johnsons Dock Member sediments are possibly age equivalent to parts of the Marambio Group (James Ross Island and Seymour Island) and Williams Point beds (Livingston Island).
3. The depositional sequence of the MBF demonstrates a progradational regressive sedimentary trend with a final coarse impulse during Late Maastrichtian-Paleocene? time. Thus, it can be correlated to the Marambio Group in James Ross and Seymour Islands.

Acknowledgments

The authors are grateful for the critical reviews supplied by Prof. S. W. Wise (Florida State University) and one anonymous referee. Their helpful comments and insights improved the manuscript greatly. Appreciation is also extended to Prof. Hubert Miller for his thoughtful suggestions and support of our research and to Dr. Polina Pavlishina for her consultation on the published Triassic palynomorphs. This study was graciously supported financially by Bulgarian Ministry of the Environment and Water; logistic support came from Bulgarian Antarctic Institute and Spanish Antarctic program.

References

Arche A, López-Martínez J, Martínez de Pison E (1992a) Sedimentology of the Miers Bluff Formation, Livingston Island, South Shetland Islands. In: Yoshida Y, Kaminuma K, Shiraishi K (eds) Recent progress in Antarctic earth science. Terra Scientific Publishing, Tokyo, pp 357–362

Arche A, López-Martínez J, Marfil R (1992b) Petrofacies and provenance of the oldest rocks in Livingston Island, South Shetland Islands. In: López-Martínez J(ed) Geología de Antártida Occidental. Simposios T.3, III Congr Geol España y VIII Congreso Latinoamer Geol, Salamanca, pp 94–103

Dalziel IWD (1972) Large-scale folding in the Scotia arc. In: Adie RD (ed) Antarctic Geology and Geophysics. Universitetsforlaget, Oslo, pp 47–55

Dimov D, Pimpirev C (2002) New data on the field relationships of the Moores Peak breccias, Hurd Peninsula, Livingston Island, South Shetland Islands. Rev Bulg Geol Soc 63(1–3):43–45

Doktor M, Swierczewska A, Tokarski AK (1994) Lithostratigraphy and tectonics of the Miers Bluff Formation at Hurd Peninsula, Livingston Island (West Antarctica). Studia Geol Polonica 104:41–104

Hervé F (1992) Estado actual del conocimiento del metamorfismo y plutonismo en la Península Antárctica al norte de los 65 S y el archipiélago de las Shetland del Sur: revisión y problemas. In: López-Martínez J (ed) Geología de la Antártida Occidental. Simposio T.3, Salamanca, pp 19–30

Hervé F, Loske W, Miller H, Pankhurst RJ (1991) Chronology of provenance, deposition and metamorphism of deformed fore-arc sequence, southern Scotia arc. In: Thomson MRA, Crame JA, Thomson JW (eds) Geological evolution of Antarctica, Cambridge University Press, Cambridge, pp 429–435

Miller H, Loske W, Kramm U (1987) Zircon provenance and Gondvana reconstruction: U-Pb data of detritial zircons from Triassic Trinity Peninsula Group (Antarctica). J South Amer Earth Sci 1:301–307

Pallas R, Muñoz JA, Sabot F (1992) Estratigrafía de la Formación Miers Bluff Isla Livingston, Islas Shetland del Sur. In: López-Martínez H (ed) Geología de la Antártida Occidental. Simposio T.3, Salamanca, pp 105–115

Pimpirev C, Dimov D, Miller H (2000) Field studies of sedimentary sequences in eastern Hurd Peninsula Central Livingston Island, South Shetland Islands. Polarforschung 66(3):1–5

Pimpirev C, Ivanov M, Dimov D, Nikolov T (2002) First find of the Upper Tithonian ammonite genus *Blanfordiceras* from the Miers Bluff Formation, Livingston Island, South Shetland Islands. N Jb Geol Paläont Mh 6:377–384

Schopf JM (1973) Plant material from the Miers Bluff Formation of the South Shetland Islands. Rep Inst Polar Stud Ohio State Univ 45:1–45

Shu O, Xi-Guang D, Yan-Binq S, Xiang-Shenq Z, Xiao-Han L (2000) Late Triassic plant microfossils from the Miers Bluff Formation of Livingston Island, South Shetland Islands, Antarctica. Antarctic Sci 12(2):217–228

Smellie JL, Liesa M, Muñoz JA, Sabot F, Pallas R, Willan RCR (1995) Lithostratigraphy of volcanic and sedimentary sequences in central Livingston Island, South Shetland Islands. Antarctic Sci 7(1):99–113

Smellie JL, Pankhurst RJ, Thomson MRA, Davies RES (1984) The geology of the South Shetland Islands: VI. Stratigraphy, geochemistry and evolution. Brit Ant Surv Sci Rep 87:1–85

Stoykova K, Pimpirev C, Dimov D (2002) Calcareous nannofossils from the Miers Bluff Formation (Livingston Island, South Shetland Islands, Antarctica): first evidence for a late Cretaceous age. J Nannoplankton Res 24(2):166

Willan RCR, Pankhurst RJ, Hervé F (1994) A probable Early Triassic age for the Miers Bluff Formation, Livingston Island, South Shetland Islands. Antarctic Sci 6(3):401–408

Regional Structures and Geodynamic Evolution of North Greenwich (Fort Williams Point) and Dee Islands, South Shetland Islands

Jean Francois Dumont[1] · Essy Santana[2] · Francisco Hervé[3] · Carlos Zapata[2]

[1] IRD-Geosciences Azur, Observatoire Oceanographique, BP 48 La Darsse, 06235 Villefranche sur Mer, France, <dumon@obs-vlfr.fr>

[2] INOCAR, Unidad Convemar, Base Naval Sur, Av. 25 de Julio, via Puerto Marítimo, POX 5940, Guayaquil, Ecuador

[3] Universidad de Chile, Departamento de Geología, Casilla 13518, Correo 21, Santiago, Chile, <fhervé@cec.uchile.cl>

Abstract. Morphological and structural studies of the Fort Williams Point area (Greenwich and Dee Islands) reveal two regional trends, NE-SW and NW-SE. Detailed field observations show that the NW-SE structural direction is associated with dykes and fractures resulting from a NW-SE shortening. This shortening is interpreted as related to active basaltic volcanism during the Mesozoic "Andean" history of the area. The NE-SW structural trend is associated with NE-SW trending sinistral faults post-dating the Cretaceous volcanic units. Kinematic analysis shows a roughly N-S trending shortening. This event is interpreted as related to the opening of the Bransfield Basin located to the south-east.

Introduction

The central position of Greenwich Island (Fig. 5.6-1 and 5.6-2) in the South Shetland archipelago allows discarding structural boundary effects, so that the deformation can be regarded as representative for the regional evolution. The permanent ice cover limits the observations to the littoral margin, but bathymetry from the English Strait between Greenwich and Dee Islands (Fig. 5.6-2) combined with topographic data from the islands provided a continuous database of the area for the morphostructural study. The geologic data presented here cover the areas of Fort Williams (Fig. 5.6-3A) and Ambato points (Fig. 5.6-3B) in north Greenwich Island, and Dee Island (Fig. 5.6-3C) located off Greenwich Island to the north. Correlation and complementary data come from previous work in Livingston and Dee Islands (Smellie et al. 1984, 1996), and from the eastern part of Greenwich Island (Araya and Hervé 1966).

Detailed field analysis was used to define the kinematic characteristics of the main structures. Special effort was dedicated to the identification in the field of fault plane relative motions. The principal objective is an attempt to interpret the regional structures in the framework of the main events that shaped the South Shetland Islands.

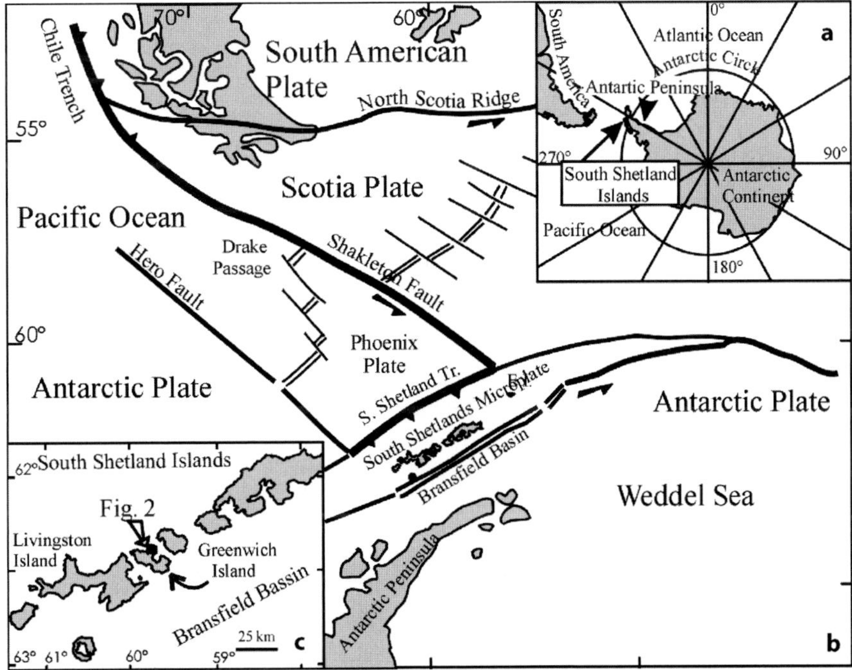

Fig. 5.6-1.
a Location of the South Shetland Islands. **b** Structural sketch of the South Shetland Islands, modified after Prieto et al. (1998) and Galindo-Zaldívar et al. (2004). *E.I.*: Elephant Island. **c** Location of Greenwich Island

From: Fütterer DK, Damaske D, Kleinschmidt G, Miller H, Tessensohn F (eds) (2006) Antarctica: Contributions to global earth sciences. Springer-Verlag, Berlin Heidelberg New York, pp 255–260

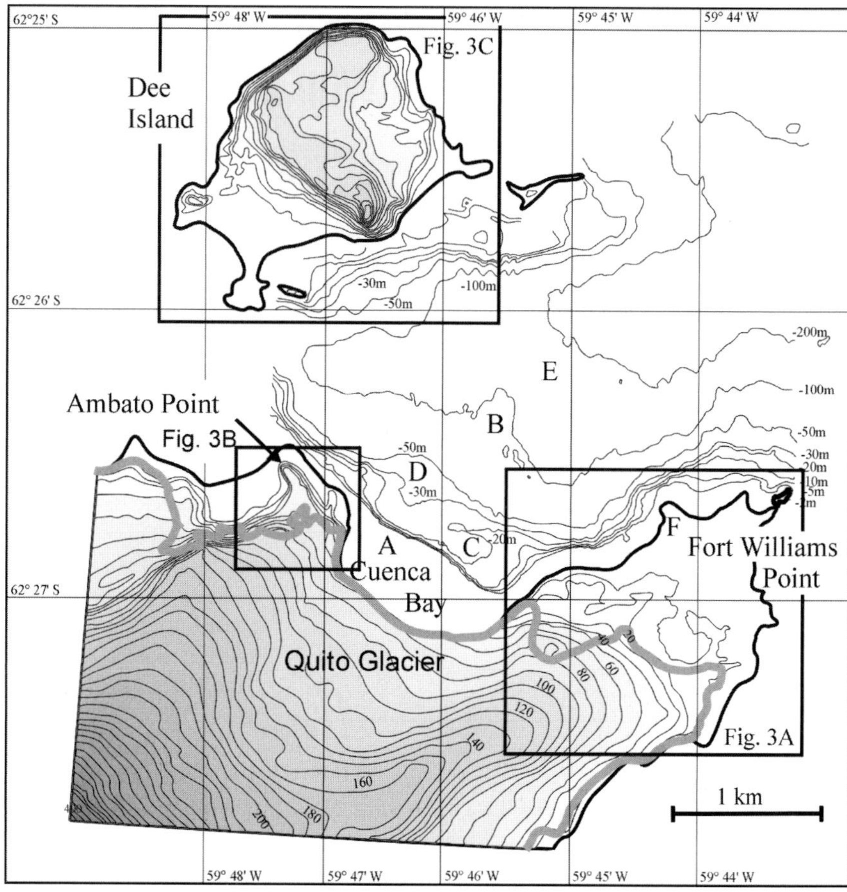

Fig. 5.6-2.
Bathymetry and topography of the study area. The *capital letters* are explained in the text

Geodynamic and Geologic Framework of the Area

The South American and Antarctic Plates remained linked during the Mesozoic along an active Andean margin evidenced by a belt of granite intrusions (Pankhurst 1988). The opening of Drake Passage produced a complex pattern of new wrench faults and microplates (Fig. 5.6-1) (Barker and Burrell 1977; Galindo-Zaldívar et al. 2004; González-Casado et al. 2000; Prieto et al. 1998). Presently the southern Chile Trench extends southward along the sinistral Shackleton Fracture Zone, ending at the triple junction between the Pacific-Antarctic, Weddell-Antarctic and Scotia Plates (Fig. 5.6-1). The structural sketch of the South Shetland Islands shows sinistral motion along the Bransfield Basin (Galindo-Zaldívar et al. 2004; Prieto et al. 1998) (Fig. 5.6-1). The South Shetland volcanic arc is presently inactive due to the lack of convergence along the South Shetland trench (Prieto et al. 1998; Smellie et al. 1984). A Mesozoic to Cenozoic terrain crops out in the central part of the South Shetland arc where Greenwich Island is located. Greenwich Island as well as Dee and Livingston islands are characterised by sub-tabular basalt platforms (Araya and Hervé 1966; Smellie et al. 1984, 1996). Recent dating and correlation indicate that the basalt series observed at Fort Williams Point, Ambato Point and Dee Island belong to the "Older Group" ranging from Coniacian to Maastrichtian in age (Smellie et al. 1996).

The Livingston Island geological map presents NW-SE and NE-SW trending faults (Smellie et al. 1984, 1996). Dyke emplacement is related to several episodes of deformation (Zheng et al. 2003). Late Cretaceous to Paleocene dykes trending N 120° E are related to the main magmatism observed on Livingston Island, and reflect the final stage of the subduction of the Proto-Pacific oceanic crust. The early (N 120° E) and last (N 15–40° E) sets of dykes are associated to the most important faults in Hurd Peninsula, Livingston Island.

Geology of Fort Williams Point Area

Previous descriptions and geological mapping in Greenwich and Dee islands were published by Araya and Hervé (1966) and Smellie et al. (1984). Also unpublished data from Azevedo (1992) and preliminary data collected by S. Benítez and F. Rada during the 1992 Ecuadorian Antarctic Scientific Survey were considered, as well as the petrography from Santana and Dumont (2002).

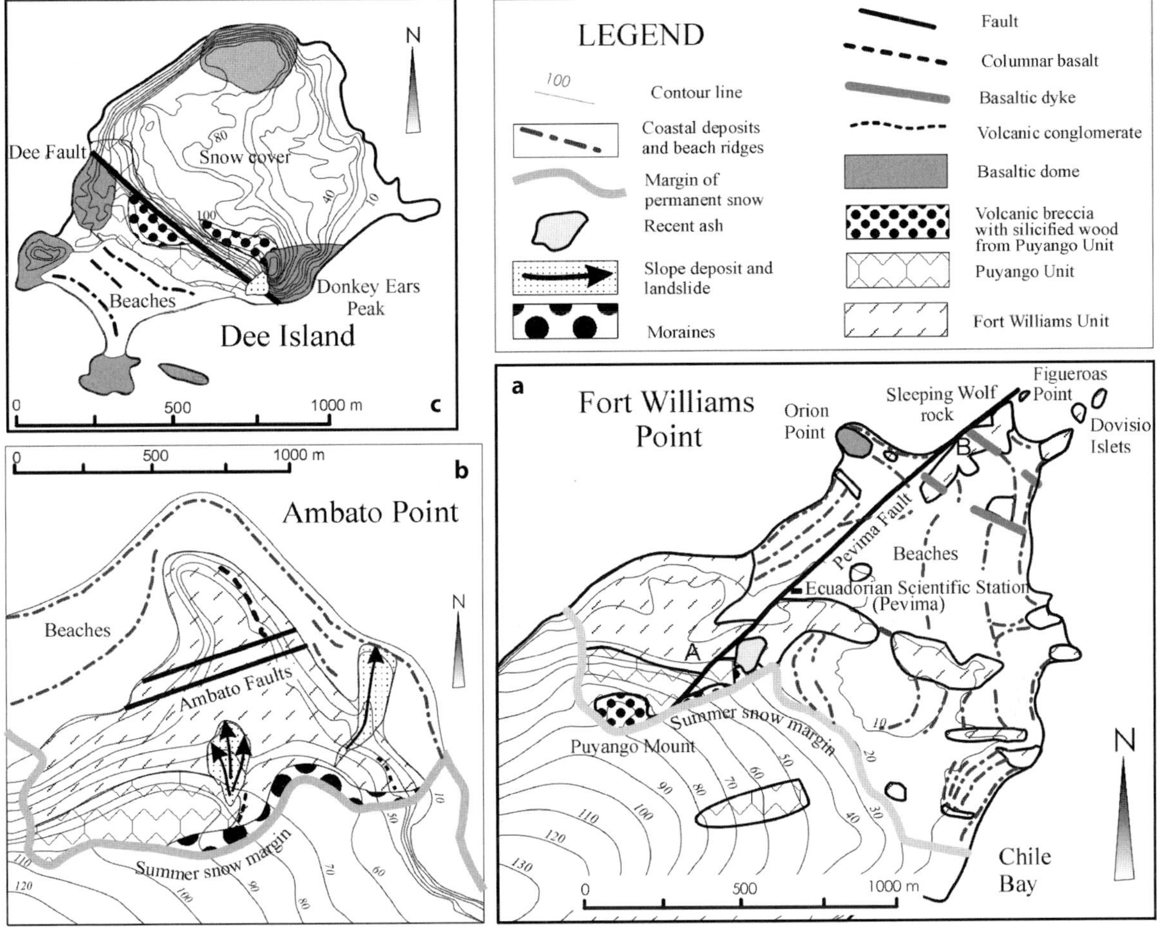

Fig. 5.6-3. Geologic maps of the study area, outlined in Fig. 5.6-2. **a** Fort Williams Point area; **b** Ambato Point area; **c** Dee Island. Location of the three sketches is indicated in Fig. 5.6-2

Two basaltic units were identified. The Fort Williams Unit (Fig. 5.6-3A) consists of a basaltic sill characterized by irregular morphology. Decametre scale circular structures are interpreted as cells of the sill. At least two NW-SE trending dykes cross the sill material at Fort Williams Point. The Puyango Unit is over the Fort Williams Unit in the topography (Fig. 5.6-3A), and crops out in the Puyango Mount, Ambato Point and Dee Island (Fig. 5.6-3b,c). It includes 50–200 m thick lava flows of porphyry to glomeroporphyry basalts with vacuole facies at the top (Santana and Dumont 2002). Thin sedimentary layers and volcanic breccia with silicified wood fragments define the top of the lava flows.

The geological map shows two regional structural trends, respectively NE-SW and NW-SE. The NE-SW trend is represented by the Pevima Fault observed between Pevima Station (name of the Ecuadorian Scientific Station) and Mount Puyango (Fig. 5.6-3A). This fault extends north-eastward to Figueroas Point. Two faults with a similar trend cross Ambato Point (Fig. 5.6-2 and 5.6-3B), determining a morphologic low at the foot of the upper volcanic platform. The Dee Fault illustrates the NW-SE regional trend (Fig. 5.6-3C). Two intrusive domes are located near the fault (Fig. 5.6-3C) but on opposite sides, suggesting a dextral offset.

Analysis of Bathymetric Data

The analysis of bathymetric data is based on a 1:5 000 scale chart of the English Strait between Greenwich and Dee islands (INOCAR 1990), complemented by new data collected during the VIII Proantec Expedition (Fig. 5.6-3). The shallowest contour line is –2 m, giving a good connection with the 1:1 000 and 1:10 000 scale topographic maps of the islands (IGM 1990).

The bottom of the English Strait between Greenwich and Dee islands presents a wide NW-SE trending flat surface sloping gently down to the north-east, from –50 m to about –200 m. This surface is framed within a ENE-WSW border with Dee Island, and sharp NE-SW and NW-SE scarps to the south on the side of Greenwich Island (Fig. 5.6-2). The

NW-SE border running in front of Ambato Point and Cuenca Bay is a straight structural scarp. The upper edge overlooks a 50 m high step presenting different shape and slope along its length: it is steeper in front of Ambato Point and progressively less steep off Cuenca Bay, which is the outlet of the Quito Glacier (Fig. 5.6-2, point A). The inner part of Cuenca Bay is a flat shallow platform (about 2 m deep) interpreted as the erosion bottom of the Quito Glacier before the last glacial retreat. The contorted bathymetry observed between -2 m and -50 m off Cuenca Bay (Fig. 5.6-2, points C and D) is probably due to detritus coming from the lateral moraines of the Quito Glacier. A deviation of the -100 m contour line in B (Fig. 5.6-2) suggests deposits issued from the glacier outlet axial part. The clearness of the NW-SE trending scarp suggests a fault defining a step. This structural trend extends south-eastwards to the Fort Williams Point area along the piedmont of the coastal plain of beach deposits (Fig. 5.6-3A).

The bathymetric data do not show evidence for extension of the Dee Island fault in the English Strait (Fig. 5.6-2, area E). Recent sediments possibly cover and hide the fault scarp, but more probably the flat and framed bottom of the English Strait in this area results from glacier erosion during the glacial periods of the Quaternary, as observed in other parts of the South Shetland Islands (John and Sugden 1971). Basalt platforms favour probably the formation of flat erosion surfaces. To support this hypothesis we mention the box shaped bottom of the English strait in this area, and the fact that the dyke system in Fort Williams Point (Fig. 5.6-2, point F) is parallel to the Dee Fault. The NE-SW trending coast of Fort Williams Point bounds sharply to the south-east the bottom surface of the English Strait (Fig. 5.6-3). This is an important structural trend extending south-westward along the southern border of the Quito Glacier.

Analysis of Faults and Fractures

Tectonic data related to NE-SW trending structures come from the Pevima Fault area (Area A in Fig. 5.6-3A). The evidence of fault motion is observed on calcite coated opened fractures. Slickensides are scarce, but sinistral fault motion is indicated by small pull apart structures (Fig. 5.6-4A1) observed on nearly vertical fault planes cut by approximately horizontal erosion surfaces (Fig. 5.6-4A2). Thick lines on the diagram are open fractures without observed motion; these are interpreted as tension joints indicating a N 10° W to N 30° E trending shortening. The brittle deformation and fault geometry is interpreted as representative of a transtensive tectonic regime at relatively low depth. The results of fault and fracture analysis are coherent with the morphologic characteristics of the Pevima and Ambato faults.

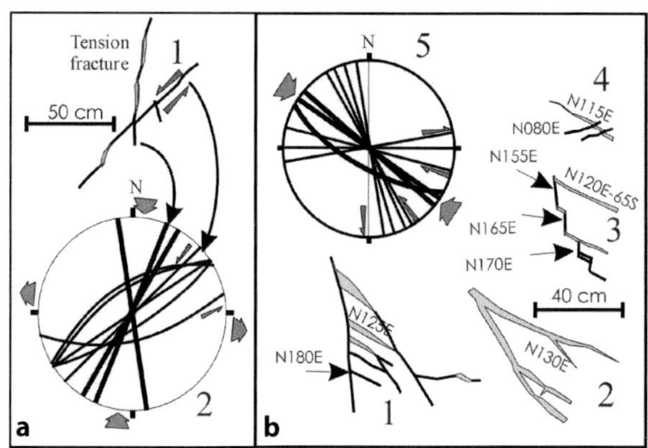

Fig. 5.6-4. a NNE-SSW shortening event at point *A* of Fig. 5.6-3a. *1:* Fractures with pull apart. *2:* Stereographic representation in Wulff net, lower hemisphere projection. *Thin lines* are faults and *thick lines* are tension joints. **b** NW-SE shortening event, located in *B*, Fig. 5.6-3a. *1* to *4* are different patterns of faults; for further explanation see text. *5:* Stereographic representation, *thin lines* represent faults and *thick lines* volcanic dykes

The data related to the NW-SE structural trend come from the area of the Sleeping Wolf Rock at Fort Williams Point (Fig. 5.6-3A, area B). A dense network of fractures is observed close to the NW-SE trending dykes. The fractures show a brittle pattern, with local breccia near the areas of cross-fractures. The lateral offset is identified from pull apart structures and fitting together fault borders, this last characteristic suggesting that the total offset is very limited. Figure 5.6-4B shows the fracture pattern observed on sub-horizontal outcrops. Different styles of fractures are observed (Fig. 5.6-4B, 1 to 4), associating tension joints with small fault offsets. Some of them are almost perfect tension joints (Fig. 5.6-4B2) defining a N 130° E shortening axis. The stereogram determines a shortening angle bracketed between N 105° E and N 150° E (Fig. 5.6-4B5). Most of the tension fractures trend between N 115° E and N 130° E, parallel to the swarm of dykes observed in the same area. The geometrical association of these elements suggests that the area has been submitted to NW-SE shortening probably compressional that is coherent with the trend of intrusive dykes. Our interpretation is that the fracturing is associated with the dyke intrusion that post-dates the sill intrusion but may be contemporaneous with the dome intrusions (Fig. 5.6-3).

Interpretation

The early NW-SE shortening is associated with dykes related to the Cretaceous history. The tension fractures and conjugated faults show N 120° E shortening direction, approximately parallel to the Late Cretaceous dykes observed in Fort Williams Point and Livingston Island (Zheng et al. 2003). It is also consistent with the Late Cretaceous event of Willan and Kelley (1999), and the early event of Santanach et al. (1992).

The recent event is represented by regional faults with striking morphology. The NE-SW trending Pevima Fault cuts all the basalt units, and thus post-dates the Upper Cretaceous. The analysis of fault planes shows sinistral transtension motion. The shortening trend is roughly NNE-SSW (N 10° W to N 30° E), and the strike slip motion suggests ESE-WNW trending extension.

The NW-SE trending Dee Fault shows a normal-dextral motion that cuts the Cretaceous volcanic units and is mechanically coherent with the sinistral motion of the Pevima Fault. However, the Dee Fault is also parallel to basaltic dykes and fractures in Fort Williams Point, and basaltic intrusive domes in Dee Island and Orion Point (Fort Williams area, Fig. 5.6-3A) are located along this structural trend. The petrography of these dykes and domes is similar to that of the lower and upper volcanic units. The structural setting suggests that the present Dee Fault can be interpreted as superimposed on an older fracture zone with associated dykes.

The Pevima, Ambato and Dee faults underline the main morphostructures of the area. These NE-SW and NW-SE trending morphostructures are reported by several authors (Prieto et al. 1998; González-Casado et al. 2000; Galindo-Zaldívar et al. 2004). The Pevima and Dee faults suggest horst style block tectonics, in agreement with the observations of Santanach et al. (1992) and Pallas et al. (1995).

The present deformation of the Bransfield Basin is characterized by NW-SE opening (Prieto et al. 1998; González-Casado et al. 2000; Galindo-Zaldívar et al. 2004). However, Galindo-Zaldivar et al. (2004) point out that the present opening of the Bransfield Basin was preceded by sinistral transtension deformation involving block tectonics, and that sinistral transtension is still active in the Eastern Bransfield Basin. According to Galindo-Zaldívar et al. (2004) this block tectonic event is responsible for the break out of the South Shetland Block since Middle Pliocene. We interpret the Pevima, Ambato and Dee faults as local representatives of this tectonic event, that shape the present morphological appearance of the South Shetland archipelago.

Conclusion

The major NE-SW and NW-SE morphostructural trends observed in NE Greenwich and Dee islands and between the islands are related to two main tectonic events. Brittle fracture zones associated with dyke intrusion characterize the NW-SE trend, interpreted as a NW-SE shortening during the Cretaceous evolution of the area. The NE-SW trending morphostructures are represented by the Pevima Fault, with a sinistral transtensional motion. The NW-SE trending structures were reactivated as dextral faults during this more recent event. The sinistral transtensional motion of the Pevima Fault is parallel to the active Bransfield rift, and interpreted as representative of the break-out of the South Shetland Block during the early stage of the opening of the Bransfield Basin. At a regional scale the present apparent en-echelon alignment of the South Shetland Islands may reflect this structural pattern of dextral and sinistral wrench faults.

Acknowledgment

This study has been made possible thanks to the support of the director of INOCAR CP-NV-EM Fausto López, the Chief of the VIII Proantec (Ecuadorian Antartic Program) Expedition CP-NV-EM Víctor Hugo Yépez, as well as all the Scientific Team of the expedition. We thank Stalin Benítez for unpublished observations, and John Smellie for his comments on the geologic units.

References

Araya R and Hervé F (1966) Estudio geomorfológico y geológico en las Islas Shetland del Sur, Antártica. Public Inst Antártico Chileno 8:76

Azevedo GC (1992) Caracterização geológica, geoquímica e geocronológica da ilha Dee e parte da ilha Greenwich, arquipélago das Shetland do Sul, Antártica. Pós-graduação em Geociências, Thesis, Univ Fed Rio Grande do Sul, Porto Alegre

Barker PF and Burrell J (1977) The opening of the Drake Passage. Marine Geology 25:15–34

Galindo-Zaldívar J, Gambôa L, Maldonado A, Nakao S, Bochu Y (2004) Tectonic development of the Bransfield Basin and its prolongation to the South Scotia Ridge, northern Antartic Peninsula. Marine Geology 206:267–282

González-Casado JM, Giner-Robles JL, López-Martínez J (2000) Bransfield Basin, Antartic Peninsula: not a normal backarc basin. Geology 28(11):1043–1046

IGM (1990) Mapa topográfica de Punta Fort Williams-Isla Dee, Península Antártica. Inst Geográfico Militar, Quito

INOCAR (1990) Mapa IOA-07, Estrecho Ingles, Islas Shetland del Sur, Punta Fort Williams (Isla Greenwich). Inst Oceanográfico de la Armada, Guayaquil

John BS, Sugden DE (1971) Raised marine features and phases of glaciation in the South Shetland Islands. British Antarc Surv Bull 24:45–111

Pallàs R, Vilaplana JM, Sabat F (1995) Geomorphological and neotectonic features of Hurd Peninsula, Livingston Island, South Shetland Islands. Antarctic Science 7(4):395–406

Pankhurst RJ (1988) Los granitos de la Antártica Occidental como una extensión de los arcos magmáticos andinos. Ser Cient INACH 38:107–116

Prieto MJ, Canals M, Ercilla G, Batist M (1998) Structure and geodynamic evolution of the Central Bransfield Basin (NW Antarctica) from seismic reflexion data. Marine Geology 149:17–38

Santana E, Dumont JF (2002) Geología de los alrededores de la Estación Ecuatoriana Pedro Vicente Maldonado (Isla Greenwich) e Isla Dee, Península Antártica. Acta Antártica Ecuatoriana 1:7–32

Santanach P, Pallàs R, Sabat F, Muñoz JA (1992) La fracturación en la Isla Livingston, Islas Shetland del Sur. In: López Martínez J (ed), Geología de la Antártida Occidental, III Cong Geol de España y VIII Cong Latinoamericano de Geología, Salamanca, pp 141–151

Smellie JL, Pankhurst RJ, Thomson MRA, Davies RES (1984) The geology of the South Shetland Islands: VI, stratigraphy, geochemistry and evolution. British Antarc Surv Report 87

Smellie JL, Pallas R, Sabat F, Zheng X (1996) Age and correlation of volcanism in central Livingston Island, South Shetland Islands: K-Ar and geochemical constraints. J South Amer Earth Sci 9(3/4):265–272

Willan RCR, Kelley S (1999) Mafic dyke swarms in the South Shetland Islands volcanic arc: Unravelling multi-episodic magmatism related to subduction and continental rifting. J Geophys Res 104:23051–23068

Zheng X, Kamenov B, Sang H, Monchev P (2003) New radiometric dating of the dykes from the Hurd Peninsula, Livingston Island, South Shetland Islands. J South Amer Earth Sci 15:925–934

The Eocene Volcaniclastic Sejong Formation, Barton Peninsula, King George Island, Antarctica: Evolving Arc Volcanism from Precursory Fire Fountaining to Vulcanian Eruptions

Seung Bum Kim[1] · Young Kwan Sohn[2] · Moon Young Choe[3]

[1] Korea Polar Research Institute, Korea Ocean Research and Development Institute, Ansan 426-744, Korea, <sbkim@kopri.re.kr>
Present address: Domestic Exploration Team II, Korea National Oil Corporation, Anyang 431-711, Korea, <sbkim@knoc.co.kr>

[2] Department of Earth and Environmental Sciences, Gyeongsang National University, Jinju 660-701, Korea, <yksohn@gsnu.ac.kr>

[3] Korea Polar Research Institute, Korea Ocean Research and Development Institute, Ansan 426-744, Korea, <mychoe@kopri.re.kr>

Abstract. The Sejong Formation (100–200 m thick) represents a newly recognized Eocene volcaniclastic unit in Barton Peninsula, King George Island, West Antarctica. Detailed field mapping and lithofacies analysis indicate that the formation can be subdivided into three distinct facies associations (FA): (1) spatter/cinder-cone association (FA I), (2) volcaniclastic-apron association (FA II), and (3) distal-apron association (FA III). FA I, occurring at the base of the formation, comprises massive and jointed basalt lavas, which pass laterally into basaltic agglomerates and agglutinates through a transitional zone of fractured basalt lava flows. These field relations suggest fire-fountaining (Hawaiian) to Strombolian eruptions and subsequent emplacement of "ponded" lavas filling the vents of small-scale spatter/cinder cones at the precursory phase of arc volcanism in Barton Peninsula. FA II, unconformably overlying FA I, is represented by very thick, tabular beds of basaltic to andesitic, welded to non-welded, tuff breccias and lapilli tuffs, emplaced by pyroclastic flows (largely block-and-ash flows), with rare intervening andesite lava flows. FA II indicates onset of the main-phase of explosive and effusive eruptions (Vulcanian), probably associated with repetitive extrusions and collapses of lava domes at the summit crater of a stratovolcano, and thereby formation of large volcaniclastic aprons. The changes in eruption styles probably resulted from generation of more evolved (intermediate) magma, possibly due to compositional differentiation of the parental magma, and interaction of the magma with groundwater. FA III is intercalated with FA II as thin lenses and is characterized by fluvial red sandstone/siltstone couplets, locally alternating with channelized mass-flow conglomerates. FA III represents active hydrologic remobilizations during inter-eruptive periods and thereby development of ephemeral streams and floodplains in lowlands on and beyond the distal volcaniclastic aprons. These eruptive and depositional processes indicate a full emergence (subaerial setting) of the King George Island during the Eocene.

Introduction

The volcanic succession in Barton Peninsula, King George Island, Antarctica (Fig. 5.7-1), has been regarded as an early-stage stratiform complex formed in a volcanic-arc setting (Barton 1965; Birkenmajer 1980; Smellie et al. 1984). It consists of a lower volcaniclastic succession (the Sejong Formation, 100–200 m thick) and an upper succession (ca. 200–300 m thick) of basaltic-andesite lava flows, with an intervening succession of lava/tuff alternations (Fig. 5.7-2; Tokarski 1988; Jin et al. 1991; Birkenmajer 1998; Lee et al. 2002). Plant leaf fossils indicate deposition of the Sejong Formation during the Late Paleocene to Eocene under warm climatic conditions (Tokarski et al. 1987; Chun et al. 1994). Radiometric ages of the lavas indicate more specifically an Eocene age (Jwa et al. 1992; Willan and Armstrong 2002). Numerous previous studies of Barton Peninsula have centered on the hydrothermal history, chemistry, geochronology, palaeofloras and structure (see Willan and Armstrong 2002 for summary), with few studies addressing stratigraphic variations, styles of volcanism, or depositional processes (Yoo et al. 2001). This study focuses on the Sejong Formation and attempts to reconstruct the early volcano-sedimentary evolution, based on detailed lithofacies analysis and field mapping. A revision of the stratigraphy is proposed; the basaltic rocks that were previously attributed to late-stage intrusions (dikes or plugs; Tokarski 1988; Birkenmajer 1998) are reassessed as lava/agglomerate complexes at the base of the Sejong Formation and are designated as the Chottae Member.

Lithofacies and Facies Associations

Based on composition, texture (coherent *versus* clastic) and grain size, ten lithofacies are identified in the Sejong Formation: basaltic lava (lithofacies BL), basaltic agglutinate (BAu), basaltic agglomerate (BAo), basaltic tuff breccia (BTB), basaltic lapilli tuff (BLT), andesitic lava (AL), andesitic tuff breccia (ATB), andesitic lapilli tuff (ALT), reworked conglomerate (C), and sandstone/siltstone couplets (S/Z) (Table 5.7-1). These rocks can be grouped into three facies associations based on the constituent facies, facies sequences, and their field relationships. Each facies association records formation in distinct eruptive and/or depositional environments: (1) spatter/cinder cones (Facies Association (FA) I), (2) volcaniclastic apron (FA II), and (3) distal apron to floodplain (FA III).

Facies Association I

FA I comprises basaltic rocks that form irregular to semicircular patches in map view (Fig. 5.7-2). These units were

From: Fütterer DK, Damaske D, Kleinschmidt G, Miller H, Tessensohn F (eds) (2006) Antarctica: Contributions to global earth sciences. Springer-Verlag, Berlin Heidelberg New York, pp 261–270

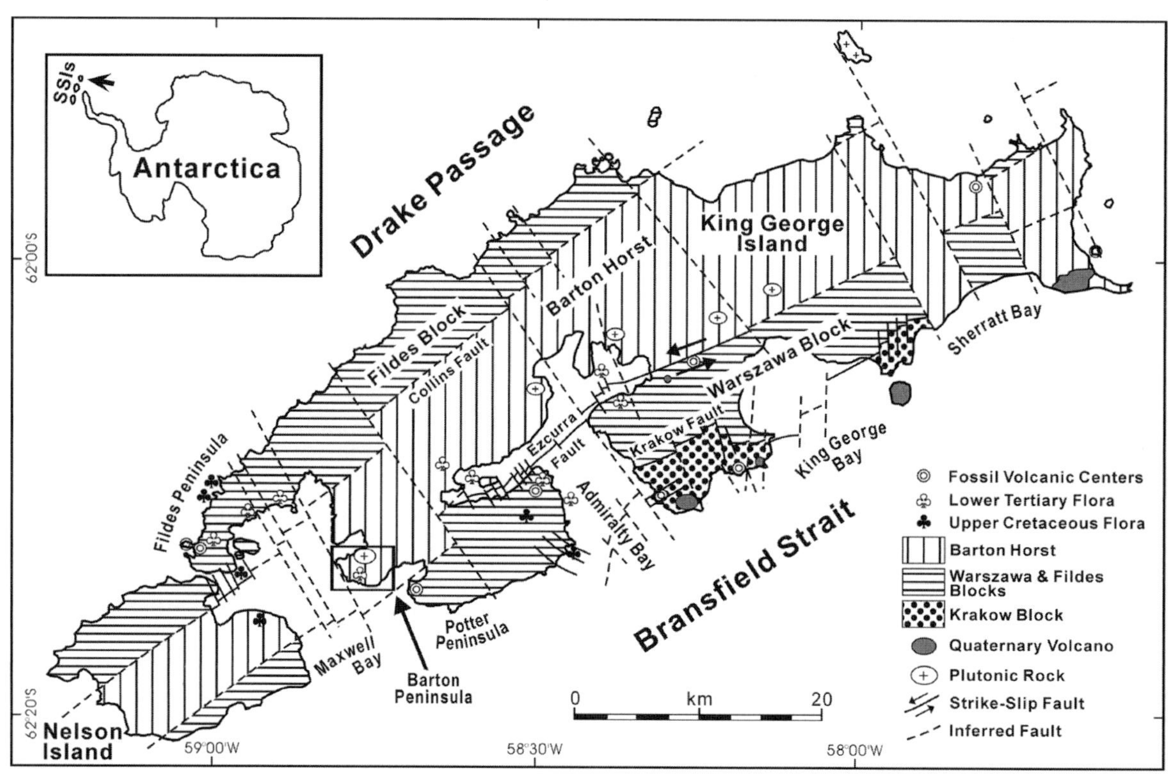

Fig. 5.7-1. Tectonic units of King George Island (redrawn from Birkenmajer 1983). Location of King George Island is given in *inset*. Sites of plutonic rocks and Quaternary volcanoes are based on Barton (1965) and Willan and Armstrong (2002). Study area (Barton Peninsula) is denoted by *arrowed box*. *SSIs:* South Shetland Islands

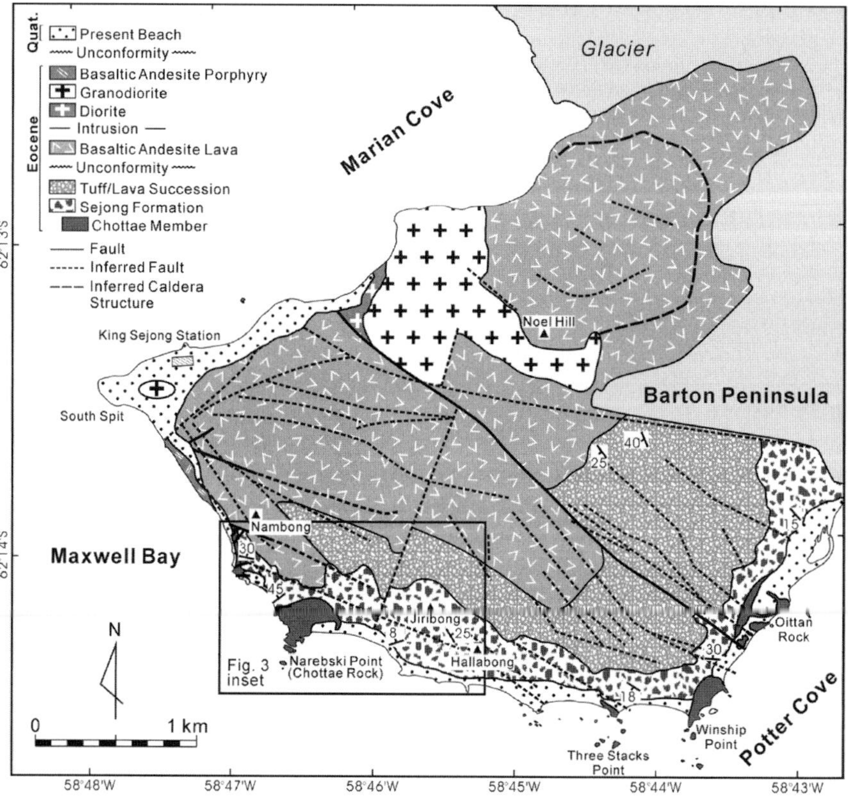

Fig. 5.7-2. Geological map of Barton Peninsula, modified after Lee et al. (2002) and Willan and Armstrong (2002). Subsurface caldera structure is delineated, based on Armstrong (1995)

Table 5.7-1. Description and interpretation of lithofacies of the Sejong Formation

Lithofacies	Description	Interpretation
Basaltic lava (lithofacies BL)	Either massive or compound; massive lava is vertically to subhorizontally jointed; compound lava consists of thin (<1 m) units that are slightly vesicular and locally fragmented, either blocky or fluidal	Massive lava: vent plug. Compound lava: lava flows (Hawaiian effusion) and local brecciation.
Basaltic agglutinate (lithofacies BAu)	Interlayered with basaltic lava flows (lithofacies BL); globular, variably vesiculated bombs and less vesicular, globular to angular lapilli; agglutinated texture with flattened and stretched, highly vesicular basaltic spatters with chilled glassy rims	Transitional between fire-fountain and Strombolian eruptions
Basaltic agglomerate (lithofacies BAo)	Laterally flanking or interlayered with the basaltic lava flows (lithofacies BL); aerodynamically shaped, incipiently to highly vesicular bombs or plastically deformed, fluidal shaped, dense bombs; coarse-grained lapilli tuff matrix	Strombolian eruption
Basaltic tuff breccia (lithofacies BTB)	Interlayered with basaltic or andesitic deposits (lithofacies BLT, ATB, or ALT); ungraded and disorganized; clast- or matrix-supported; boulder-size blocks common; dense, angular blocky clasts dominant with subordinate amounts of globular, variably vesiculated clasts with ragged margins; purple fine-grained ash matrix	Block and ash flow
Basaltic lapilli tuff (lithofacies BLT)	Commonly overlying coarser-grained basaltic deposits (lithofacies BAo or BTB); ungraded (or normally graded) and disorganized; matrix supported; dense, angular blocky clasts with sharp edges predominant; purple fine-grained ash matrix	Scoria and ash flow
Andesitic lava (lithofacies AL)	Porphyritic texture with microcrystalline groundmass; abrupt lobate termination; intense hyaloclastic brecciation	Intermittent effusive eruptions; stubby flows
Andesitic tuff breccia (lithofacies ATB)	Near-top normally graded; disorganized; matrix supported; dense, angular blocky clasts with sharp edges and occasionally rugged margins predominant; pumiceous clasts rare; green fine-grained ash matrix	Block and ash flow
Andesitic lapilli tuff (lithofacies ALT)	Ungraded or normally graded; disorganized; matrix supported; dense, angular blocky clasts dominant; minor subrounded clasts; occasional cobble to boulder-size clasts: commonly welded; occasionally inversely graded strata at the base	Ash flow (partly ground surges)
Reworked conglomerate (lithofacies C)	Interlayered with lithofacies S/Z units; irregularly or sharply based; inverse-to-normally graded; boulder-size clasts common; rounded clasts dominant with common intraclasts; extremely poorly sorted, red sandy siltstone matrix	Reworked debris flow
Sandstone/siltstone couplets (lithofacies S/Z)	Overall fining-upward stacking with lower sandstone-dominant to upper siltstone-dominant divisions; interstratified with distributional or abruptly graded sandstone-siltstone layers; broadly and shallowly channelized geometry; occasional scours at the base; common penecontemporaneous deformation (flames and contorted layers); plant leaf fossils and burrows	Floodplain with ephemeral shallow channels

formerly regarded as late-stage intrusions (dikes or plugs) (Tokarski 1988; Birkenmajer 1998; Lee et al. 2002; Willan and Armstrong 2002). The lack of distinctive discordant contacts and thermal metamorphism affecting the adjacent volcaniclastic rocks however does not support a late-stage origin. Moreover, the outermost basaltic agglutinates and agglomerates are generally draped, either unconformably or locally conformably, by the volcaniclastic rocks of FA II (Fig. 5.7-3). Furthermore, the identification of a gradational transition from central massive/vertically jointed lava (lithofacies BL) to fringing agglutinates and agglomerates (lithofacies BAu and BAo) via a transitional zone of locally brecciated, either blocky or fluidally, compound lava flows (lithofacies BL) in this study suggests intact preserved spatter/cinder cones (Fig. 5.7-3). The basaltic rocks of FA I can therefore be designated as a new stratigraphic unit (Chottae Member) that occupies the base of the Sejong Formation.

The central coherent basalt in the Chottae Member is mostly fresh (moderately altered), forming a distinctive edifice and showing vertical to subhorizontal joints arrayed in a fan-shaped fashion (Fig. 5.7-3, cross sections). This suggests emplacement of the basalt as a ponded lava or a plug, filling the vent (or conduit) of a basaltic volcano. The adjacent piles of compound lava flows dip outward, away from the central massive basalt (Fig. 5.7-3, cross sections). Each flow unit is slightly vesicular, particularly at the top. *In situ* fractured lava occurs locally and comprises either slabby or deformed, flow-foliated clasts, or blocky clasts with curviplanar margins, lacking matrix materials. This suggests autobrecciation of lava flows at transitional rheological conditions between aa and pahoehoe flows. In the outermost part, basaltic agglomerates and agglutinates occur, interlayered with the lava flows. Agglutinates are characterized by flattened and stretched, highly vesicular,

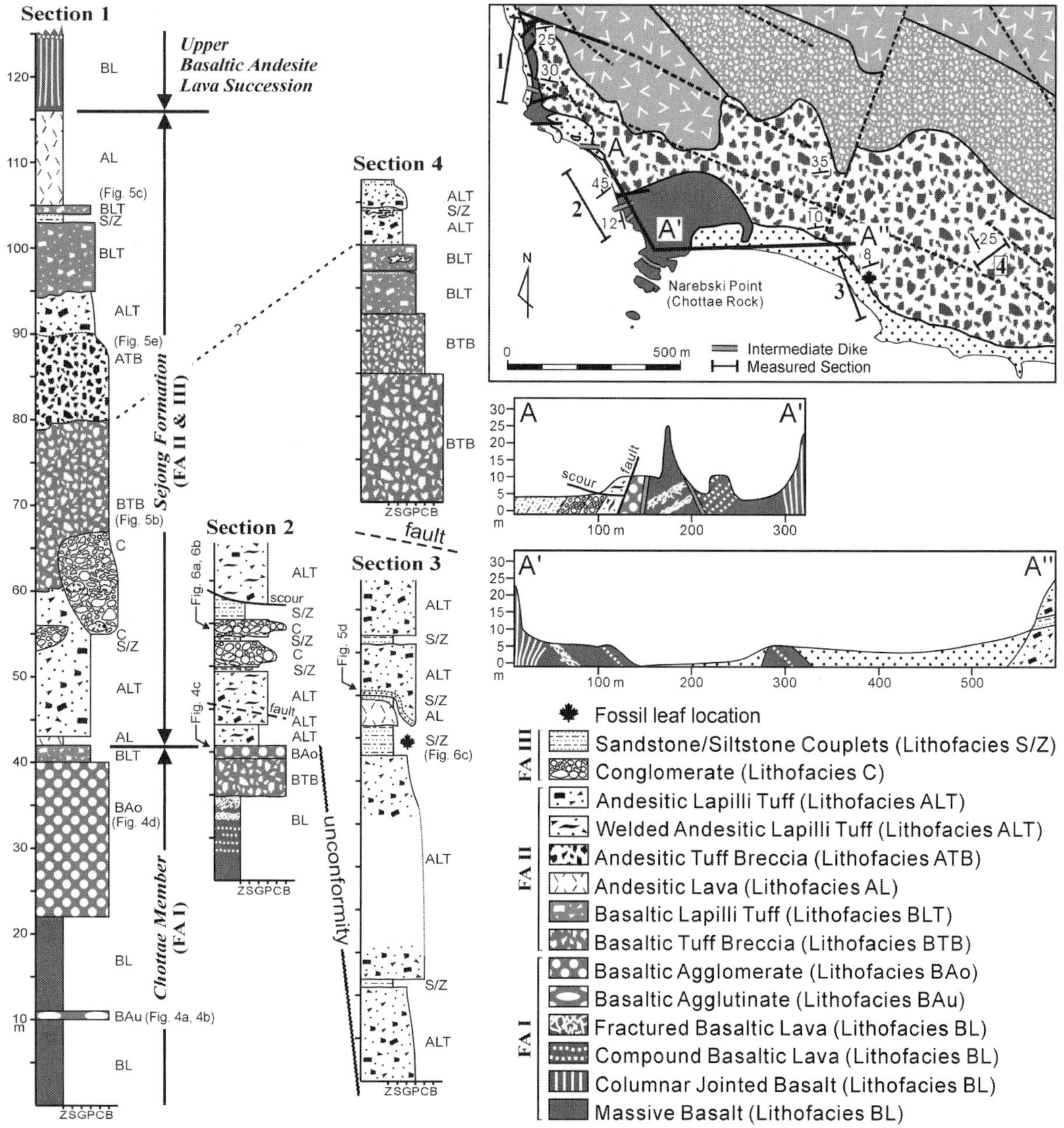

Fig. 5.7-3. Columnar logs and correlation of the Sejong Formation. *Inset map* shows the locations of the measured sections and detailed geological structures in the vicinity (*symbols* as in Fig. 5.7-2)

basaltic spatters with chilled glassy rims between variably vesiculated, either globular or blocky, bombs and lapilli (Fig. 5.7-4a,b), whereas the agglomerates consists mainly of either aerodynamically shaped, vesicular bombs or plastically deformed, fluidal shaped, dense (non- to poorly vesicular) bombs, both set in a coarse-grained lapilli tuff matrix (Fig. 5.7-4c,d). These fabrics indicate ejection of molten lava fragments into the air and subsequent landing in a near-vent area. The sequential transition of lithofacies from ponded lavas, lava flows to agglutinates/agglomerates in FA I suggests Hawaiian fire-fountaining and Strombolian eruptions at basaltic eruptive centers (e.g., spatter or cinder cones) in a subaerial setting and subsequent vent plugging. The lateral persistence of the basaltic agglutinate may indicate a spatter rampart at the initial stage of the eruption. Endogenous dome-like emplacement is unlikely, because there is no evidence of large-scale disruptions or brecciations indicative of mechanical compressions due to succeeding magma pulses.

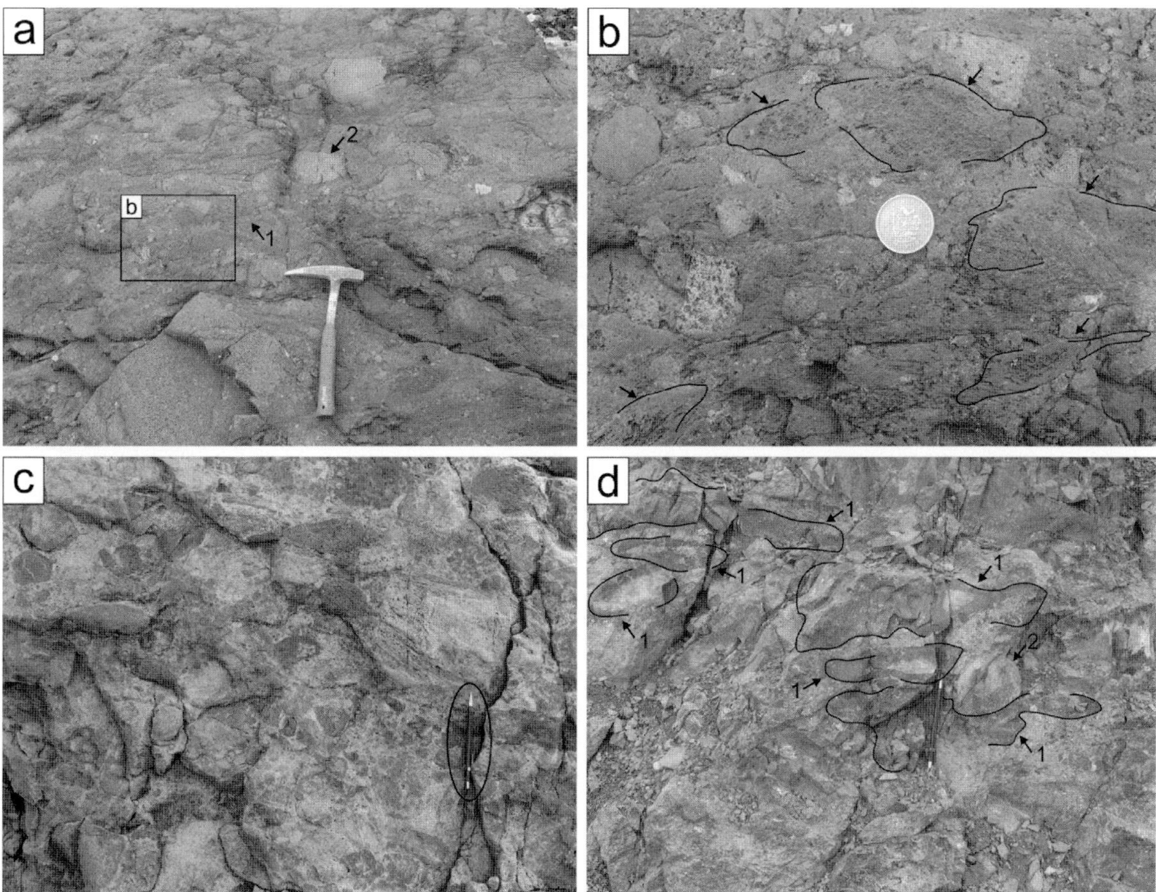

Fig. 5.7-4. Photographs of representative lithofacies of Facies Association I. **a** Basaltic agglutinate (lithofacies BAu) with flattened and stretched spatters (*arrow 1*) aligned subparallel between either globular or blocky, bombs and lapilli (*arrow 2*; *hammer* for scale 32 cm). **b** Close up of **a**. Spatters (partly *outlined* and *arrowed*) are characteristically highly vesicular in the interior with shiny, glassy chilled rims (*arrows*; *coin* for scale 2.6 cm). **c** Basaltic agglomerate (lithofacies BAo) predominated by variably vesiculated, aerodynamically shaped bombs (*pencil* for scale 14.6 cm). **d** Basaltic agglomerate (lithofacies BAo) with cow-dung-like elongated, locally deformed, dense bombs (*arrow 1*) with variably vesiculated, globular bombs (*arrow 2*; *pencil* for scale 14.6 cm)

Facies Association II

FA II comprises the bulk of the Sejong Formation. It lies unconformably or locally conformably above FA I deposits and is interlayered with the deposits of FA III (Fig. 5.7-3). It consists dominantly of very thick, tabular beds of basaltic to andesitic lapilli tuffs and tuff breccias (lithofacies BLT, BTB, ALT and ATB), and rare intervening lava flows (lithofacies AL). Each volcaniclastic unit is characterized by ungraded (or less commonly normally graded near the top) and disorganized fabrics and is either clast or matrix supported (Fig. 5.7-5). Clasts are dominantly blocky shaped with angular corners and sharp (in places ragged) margins, or less commonly with rounded corners and smooth margins (Fig. 5.7-5). Globular or irregularly shaped clasts are also present but to a subordinate amount (Fig. 5.7-5b). Clasts are mostly dense and non- to poorly vesicular, and are generally free of phenocrysts. Highly vesicular (amygdaloidal), scoriaceous or pumiceous particles are rare. Accidental lithic clasts (altered volcanics, sedimentary rocks and quartz-vein fragments) are locally present (Fig. 5.7-5c). The matrix consists predominantly of vitric materials with a subordinate amount of crystal fragments. Welded texture is common in andesitic lapilli tuffs (lithofacies ALT) (Fig. 5.7-5f). At the bases of some andesitic, cobbly lapilli tuff (lithofacies ALT), thin-bedded granule-size andesitic lapilli tuffs occur that show inversely graded strata (Fig. 5.7-5d), which are reminiscent of the layer 1 or the basal layered deposits in ignimbrites (e.g., Sparks et al. 1973; Valentine et al. 1989). An andesitic lava flow (lithofacies AL) shows an abrupt lobate termination (Fig. 5.7-3, section 3).

The thick-bedded nature, lack of stratification and channel incision, and the ungraded and disorganized fabric collectively indicate emplacements by laminar and/or

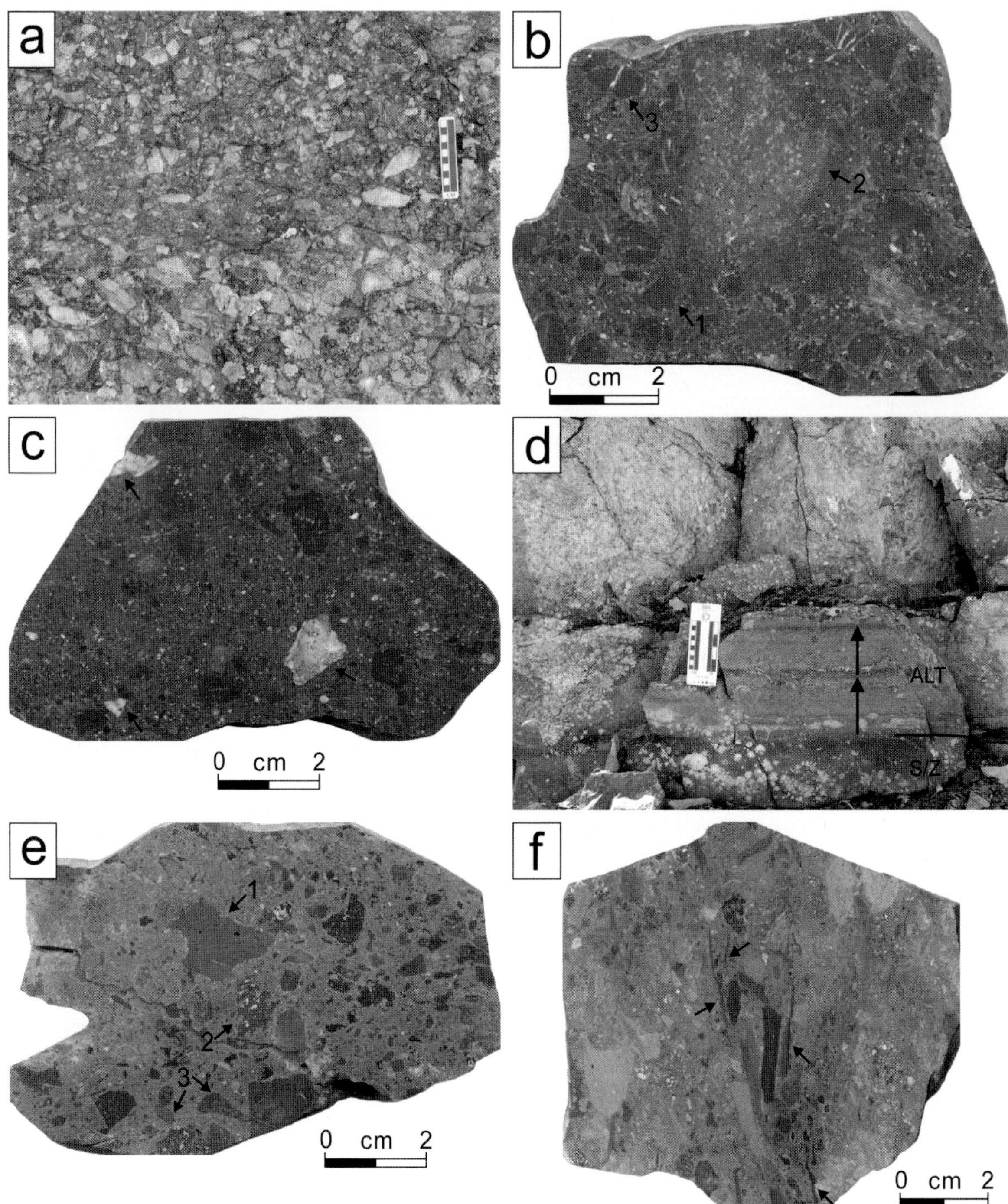

Fig. 5.7-5. Photographs of representative lithofacies of Facies Association II. **a** Basaltic tuff breccia (lithofacies BTB) with clast-supported angular blocks; outcrop near Oittan rock; *scale bar* 10 cm. **b** Slab of basaltic tuff breccia (lithofacies BTB), showing a predominance of dense, non-vesicular, angular blocky clasts occasionally with ragged margins (*arrow 1*); also present vesicular (amygdaloidal), globular shaped clasts with ragged margins (*arrow 2*). *In situ* fragmented clasts (*arrow 3*) separated by tiny quartz veins indicate degassing fracturing or thermal contraction just after deposition. **c** Slab of basaltic lapilli tuff (lithofacies BLT) consisting of dense blocky clasts with minor amounts of accidental lithics of quartz-vein fragments (*arrows*). **d** Andesitic lapilli tuff (lithofacies ALT) with inversely graded strata at the base; *arrows* indicate inverse grading. **e** Slab of andesitic lapilli tuff (lithofacies ALT), showing a predominance of non-vesicular, angular blocky clasts with sharp edges (*arrow 1*); occasional vesicular (amygdaloidal) clasts display highly rugged margins (*arrow 2*); subrounded lapilli (*arrow 3*) suggesting possible "milling" in the eruption centers. **f** Slab of welded andesitic lapilli tuff (lithofacies ALT) with elongate, greenish to dark gray glassy shards; isolated fault block in section 1

plastic flows. The abundance of vitric materials in the matrix, the occasional welded features and the rarity of accidental lithic particles collectively suggest that the deposits are pyroclastic in origin, emplaced most probably by dense pyroclastic flows, rather than being epiclastic or resedimented, as suggested by previous interpretations (Yoo et al. 2001). The predominance of angular blocky, dense clasts, and the large grain size (common boulder-grade clasts) suggest that the pyroclastic flows can be described more appropriately as block-and-ash flows (e.g., Siebe et al. 1993; Carrasco-Nunez 1999), which were generated probably by Vulcanian eruptions (Self et al. 1979; Clarke et al. 2002). The incorporation of globular and/or vesicular clasts as well as dense, angular blocky clasts of diverse lithology suggests an explosive crumbling of lava domes with thick solidified crusts possibly due to a detonation of an ascending magma as a result of interactions with groundwaters (Self et al. 1979; Druitt et al. 2002). The angular blocky clasts can be viewed as cognate lithic particles generated by the fragmentation of the dome crust or chilled margins of ascending magma, whereas the globular clasts were probably formed by disintegration of the hot and plastic interior of the domes or extruding magma. Further gas streaming after the initial explosion may have generated some vesicular juvenile clasts. The appreciable rounding of some blocky clasts with subdued corners and smooth margins can be accounted for by "milling" in the vent by repeated ejection and falling back into the openings.

The virtual lack of ashfall deposits and pedogenic alteration features in FA II suggest rapid accumulation of pyroclastic debris in a near-vent setting. The presence of a probable layer 1 deposit at the bases of some thick ignimbrite units also supports this inference because the runout distance of the most powerful pyroclastic surges hitherto documented only rarely exceeds a few tens of kilometres (e.g., Calder et al. 1999). The FA II deposits are therefore interpreted as slope aprons (volcaniclastic aprons or ring plains) (Palmer and Neal 1991; Siebe et al. 1993), extending from the summit crater of a large stratovolcano.

Facies Association III

FA III occurs as thin lenses or pockets intercalated with FA II deposits (Fig. 5.7-3). It consists mainly of sandstone/siltstone couplets (lithofacies S/Z), locally including resedimented conglomerates (lithofacies C). The sandstone/siltstone couplets are normally graded with a sandstone-dominant base and a mudstone-dominant top, and are commonly inter-stratified with diffuse boundaries (Fig. 5.7-6c). Siltstone layers are mostly homogeneous, red to brown in color and locally bioturbated (vertical burrows). Abundant plant fossils have been found in some

Fig. 5.7-6. Photographs of representative lithofacies of Facies Association III. **a** Reworked conglomerate (lithofacies C), showing an overall inverse grading from a clast-poor lower part to a bouldery, clast-rich upper part; intraclastic siltstone chunks (*arrows*; *hammer* for scale 32 cm). **b** Slab from the upper part of conglomerate in **a.** Note the rounded outlines of each clast, although some with angular corners; common and characteristic incorporation of intraclastic siltstone fragments (*arrows*). **c** Sandstone/siltstone couplets (lithofacies S/Z), showing a thinning- and fining-upward stacking pattern, with scour bases (*arrow 1*) and local incorporation of siltstone rip-up clasts (*arrow 2*; *pencil* for scale 14.6 cm)

layers (Fig. 5.7-3). Sandstone layers are generally thin-bedded with occasional small or broad scours at the base (Fig. 5.7-6c). They are either ungraded or normally graded with an upward increase in silt content. Cross-stratification is relatively rare. The couplets are commonly deformed and distorted with flame structures. Resedimented conglomerates are typically irregularly bedded with protruding clasts. Each unit is inverse-to-normally graded with largest clasts floating in the upper two-thirds. Clasts range in size from pebble to boulder grade; mostly rounded to subangular (Fig. 5.7-6a,b). They consist predominantly of dense basaltic fragments along with some red siltstone chunks. The matrix comprises extremely poorly sorted, red sandy siltstone.

The colour, texture, structure and bedding features of the sandstone/siltstone couplets collectively suggest a predominance of suspension settling from rapidly waning floods. Ephemeral shallow channels are indicated by the scours and rare cross-stratification of sandstone layers. The common deformation structures and the scarce pedogenic features suggest rapid sedimentation due to high flood frequency. The resedimented conglomerates are interpreted as debris-flows, based on the prominent inverse grading, irregular bedding and the presence of protruding clasts. Intimate association of debris flow deposits within the flood deposits is indicative of relative proximity to the source area, i.e., hill slope. FA III is therefore interpreted as floodplains developed in low lands extending from the lower reaches of the volcaniclastic aprons of FA II. The floodplains must have developed during inter-eruptive periods by active hydrologic remobilization processes, i.e., precipitation and flash floods.

Discussion

This study recognizes a new unit of discrete massive basalts (FA I deposits) along the southern coast of Barton Peninsula (e.g., Narebski Point) as fossil vent fillings (plugs) of spatter/cinder cones, contrary to the previous interpretations viewing the basalts as late-stage intrusions. The intrusive origin was suggested, mainly based on the lower degree of alteration and the more evolved (sodic) composition of the basalt compared with those of the upper basalt succession (Birkenmajer 1998; Willan and Armstrong 2002). It should, however, be noted that there lies an intervening, 100–200 m thick, volcaniclastic succession (Sejong Formation) that contains abundant andesitic rocks (tuffs and flows) between the two basalt units. Direct comparison of chemistry of the massive basalts of FA I with that of the upper basalt succession is therefore meaningless or misleading, unless isotope or trace element evidence for derivation from a single parent magma is provided. Lower degree of alteration of the former can be accounted for by less hydrothermal alterations due to low permeability of the massive basalt and the distance from the hydrothermal source (Willan and Armstrong 2002). Furthermore, our new finding of the fringing basaltic agglomerates or agglutinates (lithofacies BAu or BAo) around the massive basalts (Fig. 5.7-3) strongly indicates active eruptions through an open vent. Together with this, the unconformable draping by the volcaniclastic deposits of FA II indicates that the basaltic rocks of FA I were formed at the earliest stage of Barton Peninsula volcanism, negating the former interpretation viewing them as latest-stage intrusions. Suffice it to say that the relative timing of stratigraphic units should be established primarily by the contact relations in the field rather than by other geological or geochemical criteria.

The unconformably overlying FA II deposits show marked differences in composition and lithofacies characteristics compared to the underlying FA I deposits. FA I is characterized by near-vent accumulations of basaltic ejecta (spatters and bombs), lacking appreciable matrix materials, whereas FA II is composed of laterally persistent beds, consisting dominantly of angular-blocky, dense (non- to poorly vesicular), basaltic to andesitic clasts set in a fine-grained ash matrix. These variations can be attributed to the changes in eruption styles from the precursory, localized fire-fountaining (Hawaiian) to Strombolian eruptions at small spatter/cinder cones to extensive Vulcanian eruptions, accompanied by explosive dome collapses at the summit vent of a larger volcanic edifice. The main area of the lava domes could be the inferred subsurface caldera structure to the northeast of Noel Hill (Fig. 5.7-2; Armstrong 1995), although dome extrusions could have also occurred at numerous satellite vents. The change in eruption styles and volcano types can be explained by an increase in explosiveness due to compositional differentiation (evolution) of the parental magma along with possible contribution of external water (e.g., groundwater). Table 5.7-2 summarizes the changes in volcanic phases in chronological order.

Although the present study concentrates on a small well-exposed area (Barton Peninsula), the results can give some insights into the stratigraphy and evolving history of the early-Tertiary volcanic succession of King George Island. Despite the huge efforts to establish a stratigraphy of the island over the last two decades (e.g., Birkenmajer 1983; Smellie et al. 1984), there are still poor correlations among the locally identified rock units (formations or groups). We suggest that better correlations can be achieved through grouping genetically related lithofacies, characterizing their responsible eruption styles and identifying fossil volcanic vents before undertaking geochemical correlations or chronostratigraphic classifications by sparse and equivocal radiometric ages. Our work is put forward to be an example to initiate such an ap-

Table 5.7-2. Summary of volcanic phases of the Sejong Formation

Stage	FA	Eruption style	Volcano form	Sourve vent
Stage 2 (50–44 Ma[a])	FA II	Vulcanian: block-and-ash flows, minor scoria flows and ash flows in relation with dome extrusion and collapse	Stratovolcano	Inferred subsurface caldera structure to the northeast of Noel Hill or indefinable satellite vents
Stage 1 (50–44 Ma)	FA I	Fire fountaining and Strombolian: lava effusions and ballistic ejections	Spatter/cinder cones	Nambong coast, Narebskip Point, Winship Point, Three Stacks Point and Oittan Rock

[a] Hitherto obtained radiometric ages of the rocks of stages 1 and 2 are overlapped in the error ranges.

proach (see also Smellie et al. 1998) and we hope to inspire the workers in adjacent areas. Two points will be particularly helpful regarding regional correlations. One is to identify and isolate basaltic lava/agglutinate/agglomerate complexes and the other is to group block-and-ash flow deposits dominated by blocky, poorly vesiculated clasts. The descriptive details provided here should help in their recognition elsewhere. We have found that there are many volcaniclastic deposits similar to the FA II elsewhere in King George Island, e.g., Fildes Peninsula (our observations), Potter Peninsula (Barton 1965) and Admiralty Bay (Birkenmajer 1980). It is therefore tentatively suggested that Vulcanian eruptions accompanied by lava dome collapses may have played a major role in the formation of stratovolcanoes on King George Island in Eocene times. Given the extensiveness and completeness of the early-Tertiary volcanic sequences on King George Island in the northern Antarctic Peninsula region, establishing a more sophisticated stratigraphy and evolution history for the island can serve as a key to understanding its regional geological and tectonic evolution, particularly in relation to the formation of the South Shetland Islands arc.

Conclusions

Based on detailed lithofacies analysis and field mapping, the volcano-sedimentary evolution of the Sejong Formation is reconstructed. The first stage is marked by the basaltic lava/agglutinate/agglomerate complexes (FA I, Chottae Member) that form discrete semi-circular patches in map view. They represent isolated, small-scale spatter/cinder cones due to fire fountaining (Hawaiian) to Strombolian eruptions. The second stage is represented by repetitive emplacement of basaltic to andesitic block-and-ash flows and attendant andesitic lava flows (FA II), which suggests onset of explosive and effusive eruptions (Vulcanian) of more evolved (intermediate) volcanic centres, probably associated with repeated extrusions and collapses of lava domes at a summit crater of a larger (strato-)volcano. Active hydrologic remobilization processes across the volcaniclastic aprons of FA II in the immediate aftermath of the eruptions resulted in the intercalated floodplain and mass-flow deposits of FA III. The overall characteristics of individual lithofacies and facies associations suggest a deposition of the entire Sejong Formation in a subaerial environment. The changes in eruption style from effusive Hawaiian to explosive Vulcanian types may reflect compositional evolution of arc magmas in association with an increasing contribution of groundwater to the magma.

Acknowledgements

This work was supported by Korea Polar Research Institute (grant PP03103 and PE05001). We are grateful to Mr. Y. M. Jeon for his assistance in the field. We also thank Dr. R. C. R. Willan and an anonymous reviewer for their constructive and critical reviews.

Appendix

Formal Stratigraphic Summary

- Name: Sejong Formation (after Lee et al. 2002).
- Type section: Nambong and Jiribong (sections 1 and 3 in Fig. 5.7-3).
- Thickness/extent: 100–200 m thick; traceable along southern peripheral coast of the Barton Peninsula (Fig. 5.7-2), also occurs in Weaver Peninsula.
- Stratigraphic position: lowest stratigraphic unit in Barton Peninsula, overlain by either basaltic-andesite lava succession or tuff/lava succession.
- Lithology: thick-bedded (>1 m), tabular, basaltic to andesitic tuff breccias and lapilli tuffs with thin lenses (1–5 m thick) of fluvial red sandstone/siltstone couplets and minor reworked debris-flow conglomerates and rare andesitic lava flows.
- Depositional environment: subaerial volcaniclastic aprons with floodplains and ephemeral channels during inter-eruptive periods.
- Age: Eocene (50–44 Ma).
- Regional correlation: Cardozo Cove Group in Admiralty Bay area (Birkenmajer 1998).

- Name: Chottae Member (this study).
- Type sections: Nambong and Narebski Point (sections 1 and 2 in Fig. 5.7-3).
- Thickness/extent: about 30 m thick, discrete semicircular patches with approximate diameters of 400–600 m at Nambong coast, Narebski Point, Three Stacks Point, Winship Point and Oittan Rock (Fig. 5.7-2).
- Stratigraphic position: Basal unit of the Sejong Formation, either conformably or unconformably overlain by the Sejong Formation.
- Lithology: central massive/jointed basalt lavas and fringing basaltic lava flows, agglomerates and agglutinates with minor amounts of basaltic tuff breccias.
- Depositional environment: subaerial spatter/cinder cones.
- Age: Eocene (50–44 Ma).

References

Armstrong DC (1995) Acid sulphate alteration in a magmatic hydrothermal environment, Barton Peninsula, King George Island, Antarctica. Mineral Mag 59:429–441

Barton CM (1965) The Geology of the South Shetland Islands: III. The Stratigraphy of King George Island. Brit Antarct Surv Sci Rept 33:1–44

Birkenmajer K (1980) Geology of Admiralty Bay, King George Island (South Shetland Islands) – an outline. Pol Polar Res 1:29–54

Birkenmajer K (1983) Late Cenozoic phases of block-faulting on King George Island and Nelson Island (South Shetland Islands, Antarctica). Bull Pol Acad Sci Earth Sci 30:21–32

Birkenmajer K (1998) Geological Structure of Barton Peninsula and Weaver Peninsula, Maxwell Bay, King George Island (South Shetland Island, West Antarctica). Bull Pol Acad Sci Earth Sci 46:191–209

Calder ES, Cole PD, Dade WB, Druitt TH, Hoblitt RP, Huppert HE, Ritchie L, Sparks RSJ, Young SR (1999) Mobility of pyroclastic flows and surges at the Soufriere Hills Volcano, Montserrat. Geophys Res Lett 26:537–540

Carrasco-Nunez G (1999) Holocene block-and-ash flows from summit dome activity of Citlaltepetl volcano, Eastern Mexico. J Volcanol Geotherm Res 88:47–66

Chun HY, Chang SG, Lee JI (1994) Biostratigraphic study on the plant fossils from the Barton Peninsula and adjacent areas. J Paleontol Soc Kor 10:69–84

Clarke AB, Voight B, Macedonie G (2002) Transient dynamics of vulcanian explosions and column collapse. Nature 415:897–901

Druitt TH, Young S, Baptie B, Calder E, Clarke A, Cole P, Harford C, Herd R, Luckett R, Ryan G, Sparks RSJ, Voight B (2002) Episodes of cyclic Vulcanian explosive activity with fountain collapse at Soufriere Hills Volcano, Montserrat. In: Druitt TH, Kokelaar BP (eds) The eruption of Soufriere Hills volcano, Montserrat from 1995 to 1999. Geol Soc London Mem 21, pp 281–306

Jin M-S, Lee M-S, Kang P-C, Jwa Y-J (1991) Petrology and geochemistry of the volcano-plutonic rocks in the Barton and the Weaver peninsula, King George Island, Antarctica. Kor J Polar Res 2:107–134

Jwa Y-J, Park B-K, Kim Y (1992) Geochronology and geochemistry of the igneous rocks from Barton and Fildes peninsulas, King George Island: a review. In: Yoshida Y, Kaminuma K, Shiraishi K (eds) Recent progress in Antarctic earth science. Terra Scientific Publication, Tokyo, pp 439–442

Lee JI, Hur SD, Yoo CM, Yeo JP, Kim H, Hwang J, Choe MY, Nam SH, Kim Y, Park B-K, Zheng X, Lopez-Martinez J (2002) Geological map of the Barton and Weaver peninsulas. Kor Ocean Res Devel Inst

Palmer B, Neall V (1991) Contrasting lithofacies architecture in ring-plain deposits related to edifice construction and destruction, the Quaternary Stratford and Opunake Form. Sediment Geol 74:71–88

Self S, Wilson L, Nairn IA (1979) Vulcanian eruption mechanisms. Nature 277:440–443

Siebe C, Abrams M, Sheridan MF (1993) Major Holocene block-and-ash fan at the western slope of ice-capped Pico de Orizaba volcano, Mexico: implications for future hazards. J Volcanol Geotherm Res 59:1–33

Smellie JL, Pankhurst RJ, Thomson MRA, Davies RES (1984) The geology of the South Shetland Islands. VI. stratigraphy, geochemistry and evolution. Brit Antarct Surv Sci Rept 85:1–87

Smellie JL, Millar IL, Rex DC, Butterworth PJ (1998) Subaqueous, basaltic lava dome and carapace breccia on King George Island, South Shetland Islands, Antarctica. Bull Volcanol 59:245–261

Sparks RSJ, Self S, Walker GPL (1973) Products of ignimbrite eruptions. Geology 1:115–118

Tokarski AK (1988) Structural analysis of Barton Peninsula (King George Island, West Antarctica): an example of volcanic arc tectonics. Stud Geol Pol 105:53–63

Tokarski AK, Danowski W, Zastawniak, E (1987) On the age of fossil flora from Barton Peninsula (King George Island, West Antarctica). Pol Polar Res 8:293–302

Yoo CM, Choe MY, Jo HR, Kim Y, Kim KH (2001) Volcaniclastic sedimentation of the Sejong Formation (Late Paleocene–Eocene), Barton Peninsula, King George Island, Antarctica. Ocean Polar Res 23:97–108

Valentine GA, Buesch DC, Fisher RV (1989) Basal layered deposits of the Peach Springs Tuff, northwestern Arizona, USA. Bull Volcanol 51:395–414

Willan RCR, Armstrong DC (2002). Successive geothermal, volcanic-hydrothermal and contact-metasomatic events in Cenozoic volcanic-arc basalts, South Shetland Islands, Antarctica. Geol Mag 139:209–231

Elephant Island Recent Tectonics in the Framework of the Scotia-Antarctic-South Shetland Block Triple Junction (NE Antarctic Peninsula)

Jesús Galindo-Zaldívar[1] · Adolfo Maestro[2] · Jerónimo López-Martínez[3] · Carlos Sanz de Galdeano[4]

[1] Departamento de Geodinámica, Universidad de Granada, 18071 Granada, Spain, <jgalindo@ugr.es>
[2] Instituto Geológico y Minero de España, Ríos Rosas 23, 28003 Madrid, Spain, <a.maestro@igme.es>
[3] Departamento de Química Agrícola, Geología y Geoquímica, Facultad de Ciencias, Universidad Autónoma de Madrid, 28049 Madrid, Spain, <jeronimo.lopez@uam.es>
[4] Instituto Andaluz de Ciencias de la Tierra, CSIC/Universidad de Granada, 18002 Granada, Spain, <csanz@ugr.es>

Abstract. Elephant Island is an outcrop of the South Shetland Block, and is located at the southeastern prolongation of the Shackleton Fracture Zone. The South Shetland Block constitutes a continental fragment of the southern branch of the Scotia Arc and has been separated since the Pliocene from the Antarctic Peninsula by the opening of the Bransfield Strait and the transtensional fault zone that extends along the South Scotia Ridge. The northern boundary of the block is determined by subduction at the South Shetland Trench and its eastwards prolongation. The Shackleton Fracture Zone extends up to the South Shetland Block and constitutes the westernmost segment of the Scotia-Antarctic plate boundary. In this tectonic context, Elephant Island permits study of the rocks and the structures that resulted from deformation of the South Shetland Block near the triple junction with the Antarctic and Scotia Plates.

Several lines of evidence point to important and recent uplift of Elephant Island: presence of HP/LT metamorphic rocks, marine terraces, and high relief. This is the consequence of at least two processes: subduction of the thickened crust related to the Shackleton Fracture Zone that produced local crustal thickening and post glacial isostatic rebound.

This paper reports the results of a study of brittle structures in the western and southern cliffs of Elephant Island. It is shown that recent stresses were variable, in accordance with the tectonic setting. The northwestern sectors are dominated by compression with NE-SW and NW-SE trends respectively related to the Scotia-Antarctic sinistral transcurrence and to subduction of the oceanic lithosphere that was modified in the vicinity of the Shackleton Fracture Zone. In the southern sector, extension predominates with approximately WNW-ESE trend, related to Bransfield Strait opening and transtensional faults located along the South Scotia Ridge. In addition, radial extension has been identified in different sectors that may be related to the uplift of the island.

Introduction

Development of the Scotia Arc since the Oligocene (BAS 1985; Pelayo and Wiens 1989; Barker et al. 1991; Barker 2001) has produced tectonic reorganization of the region situated between South America and the Antarctic Peninsula. The Scotia and Sandwich Plates are located between the much larger South America and Antarctic Plates and accommodate their relative transcurrent and convergent motion (Fig. 5.8-1). This is the main process that causes sinistral motion, both along the northern and southern boundaries of the Scotia Plate. The South Shetland Block is a tectonic element of continental nature between the Scotia and Antarctic Plates at the NE end of the Antarctic Peninsula (Fig. 5.8-1). It is bounded to the north by the South Shetland Trench and its eastward prolongation separates the block from the oceanic crust of the Antarctic and Scotia Plates. The tectonic activity of this boundary varies along strike. Whereas the South Shetland Trench is a marked depression with a depth of more than 4 500 m with active subduction at present according to most authors (Gamboa and Maldonado 1990; Larter and Barker 1991; Maldonado et al. 1994; Aldaya and Maldonado 1996; Jabaloy et al. 2003), east of the Shackleton Fracture Zone the boundary shows low seismicity indicating reduced recent tectonic activity (BAS 1985; Pelayo and Wiens 1989; Galindo-Zaldívar et al. 1996). The southern boundary of the South Shetland Block is located at the Bransfield Strait and the intermediate depressions of the South Scotia Ridge, which is formed by sinistral transtensional faults.

The South Shetland Block was probably isolated by the interaction of two processes:

1. Subduction at the South Shetland Trench may have produced roll-back of the subducting slab since the Pliocene, as a consequence of cessation of spreading of the Phoenix-Antarctic Ridge, leading to the separation of the South Shetland Block, (Barker 1982; Larter and Barker 1991; Maldonado et al. 1994; Barker and Austin 1998).
2. In addition, Bransfield Basin accommodates the westward propagation of sinistral deformation along South Scotia Ridge (Galindo-Zaldívar et al. 1996; Klepeis and Lawver 1996; González-Casado et al. 2000).

The Shackleton Fracture Zone (SFZ) is a NW-SE oriented intra-oceanic transpressive sinistral fault zone with positive relief, forming the Scotia-Antarctic plate boundary. The oceanic lithosphere at the southern end of the Shackleton Fracture Zone is subducted below the South Shetland Block. Fault motion along the Shackleton Fracture Zone has continued to the present as a result of relative movement between the South American and Antarctic Plates (Pelayo and Wiens 1989; Galindo-Zaldívar et al. 1996; Maldonado et al. 2000; González-Casado et al. 2000; Thomas et al. 2003).

From: Fütterer DK, Damaske D, Kleinschmidt G, Miller H, Tessensohn F (eds) (2006) Antarctica: Contributions to global earth sciences. Springer-Verlag, Berlin Heidelberg New York, pp 271–276

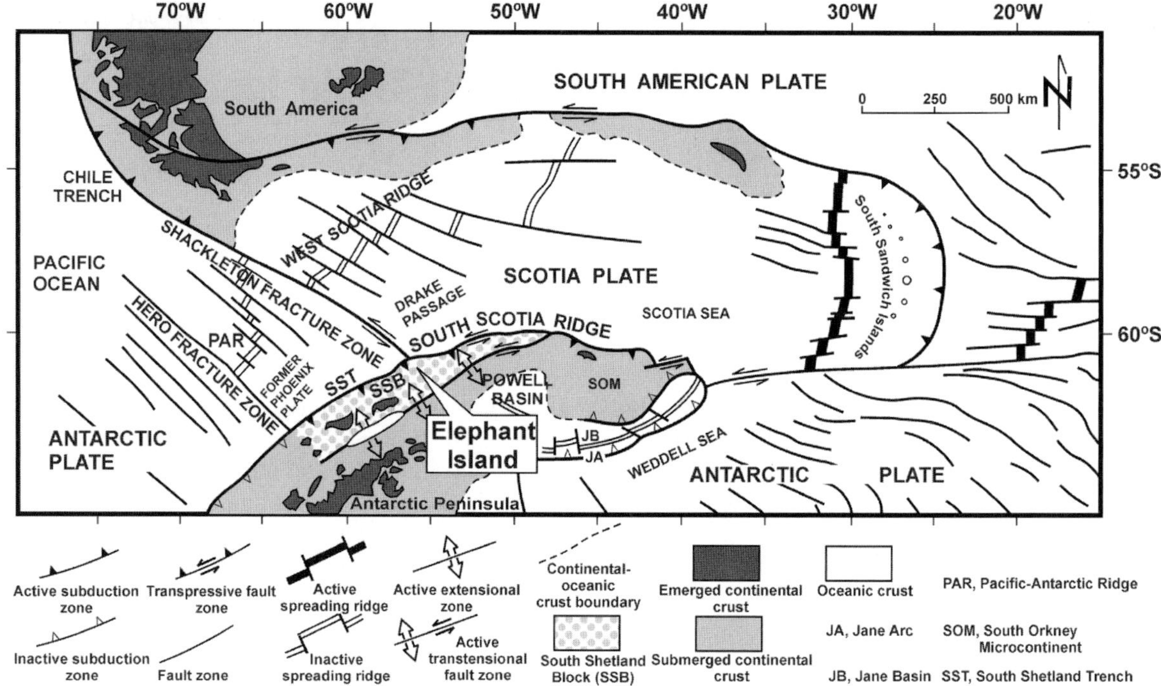

Fig. 5.8-1. Geological setting of Elephant Island in the tectonic framework of the Scotia Arc

Elephant Island is situated southeast of the fracture-fracture-trench triple junction formed by the intersection of the SFZ with the South Shetland Block. It is located very near the present-day active triple junction between South Shetland Block, Scotia and Antarctic Plates.

The main aim of this study is to identify and analyse the recent structures developed on Elephant Island and to contribute to the understanding of the kinematics of this region where most of the active faults are at present below sea level. In addition, this research aims to discuss the different processes involved in the uplift of the island. New field geological observations along the western cliffs of Elephant Island, obtained during November–December 2002 in the framework of a Brazilian-Spanish cooperative research project, constitute the basis of this contribution.

Recent Structures at Elephant Island

Elephant Island is characterized by phyllites, greenschists, blueschists and some metabasites, marbles and quartzites metamorphosed in HP-LT conditions during the Cretaceous (Trouw et al. 1990, 1991, 2000; Grunow et al. 1992). Metamorphic zones have NW-SE orientation, but the main foliation shows E-W to NE-SW trends, generally with steep to subvertical dips. Ductile deformation shows complex overprinting that includes at least three stages of deformation with related foliations, some of them that developed stretching lineations (Trouw et al. 1990, 1991, 2000; Grunow et al. 1992).

Field observations indicate that the ductile structures are affected by later ductile-brittle shear zones, with associated S-C structures indicating mainly reverse faulting. The most recent structures correspond to faults and joints, that were measured along a section located along the steep cliffs of the western coast of the island. The deformation is very intense and the orientation of joints and more than 200 faults indicate the heterogeneous character of brittle structures.

Fault surfaces are generally planar and sharp. They show thin slickensides in amphibolites, quartzites and marbles, whereas in metapelites the development of fault gouges is more common. Pyrite mineralization is present on several fault surfaces. The fault regimes have been determined by different criteria, including sigmoidal structures, trituration tails and displacement of previous markers. However, in most faults only asymmetric steps on fault surfaces could be used as a reliable indicator to determine the fault regime and sense of movement, due to the absence of other markers.

In order to study the variation of the deformation along the section, faults were grouped in six measurement stations, five of them along the northwestern coast between Minstrel Point and Mensa Bay, and one station located at the southern extremity of the island around Cape Lookout (Fig. 5.8-2). The fault sets are variable along the section. The close proximity of normal and reverse faults and the presence of several striae sets on the same fault surface we feel demonstrates reactivation of faults by different stresses. Generally, the northern

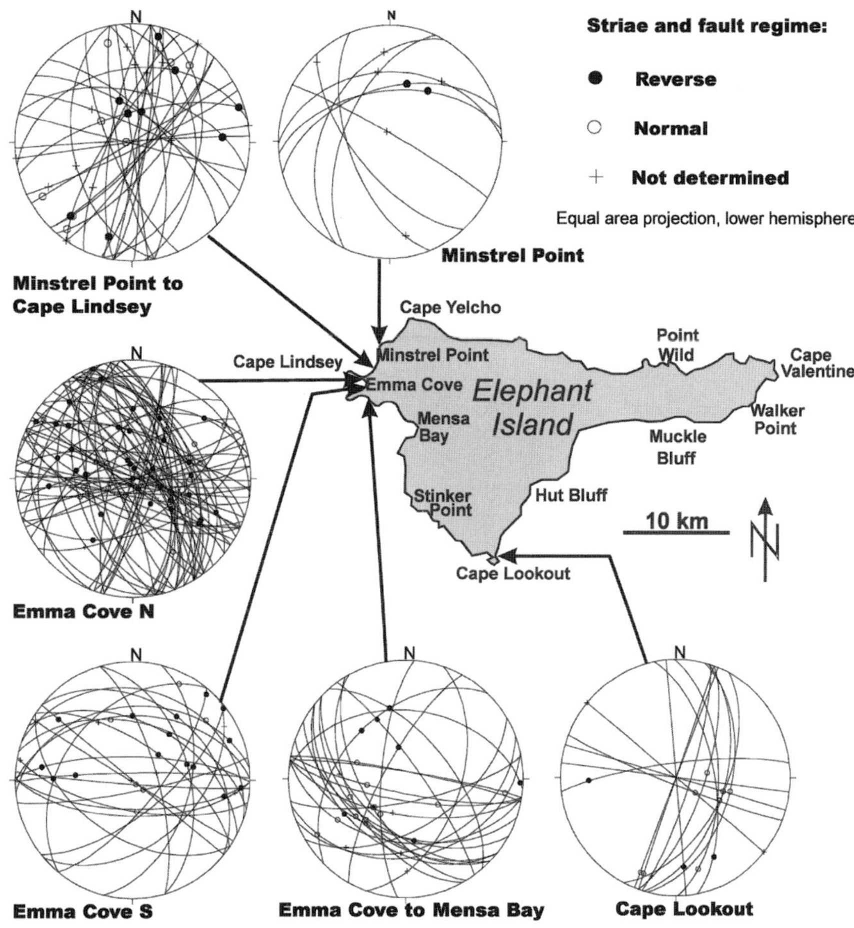

Fig. 5.8-2. Stereographic diagrams of mesoscopic faults. Lower hemisphere, equal area projection

sectors of the section are deformed by reverse faults, with top-to the SE or SW kinematics, whereas the southern sector is dominated by normal faults. At Minstrel Point two main fault sets were identified that include reverse faults and suggest NNE-SSW oriented compression. The faults from Minstrel Point to Emma Cove N show a complex distribution of orientations, with a predominant subvertical set with NNE-SSW trend and a fault set with moderate dip towards the NE. Emma Cove S station shows a complex fault distribution with two dominant sets dipping northwards and southwards and very variable striae orientation. The faults in the Emma Cove to Mensa Bay area show a well developed set with moderate dip towards the SW, with a high dispersion in the striae distribution. Both reverse and normal displacement senses were observed on fault surfaces of similar orientation. At Cape Lookout, two main orthogonal fault sets were recognized, with WNW-ESE subvertical and NNE-SSW steeply dipping ESE orientations, generally with a normal regime component. A dominant dextral regime is determined for the NNE-SSW fault set. Seismicity indicates that some of the faults are still active at least at depth. Pelayo and Wiens (1989) determined an earthquake focal mechanism indicating reverse faulting with NW-SE compression at 10 km depth in the Cape Lindsey area.

Marine terraces with large undulations were recognized along the west coast of Elephant Island. Locally, they reach more than 125 m altitude, indicating considerable recent uplift of the region. Although no granites crop out on Elephant Island, several granite pebbles were found on these marine terraces.

Paleostress Determination from Mesofaults

Faults measured at the different stations show a large dispersion both in fault plane attitudes and striation orientations. Taking into account that overprinted striations on the same fault plane were locally observed, it can be deduced that fault planes were reactivated and paleostresses should not be directly inferred from fault attitudes on diagram observations. In order to study the complex overprinting of brittle deformations, mesofaults were analysed with the Right Dihedra (Angelier and Mechler 1977), Etchecopar et al. (1981) and the Search Grid (Galindo-Zaldívar and González-Lodeiro 1988) paleostress determination methods.

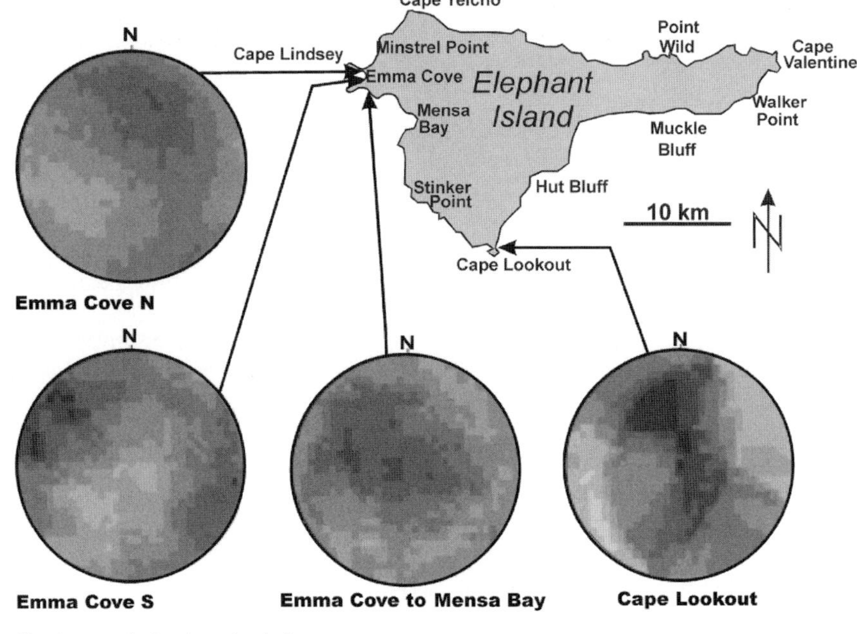

Fig. 5.8-3. Right Dihedra diagrams of mesofaults. Lower hemisphere, equal area projection

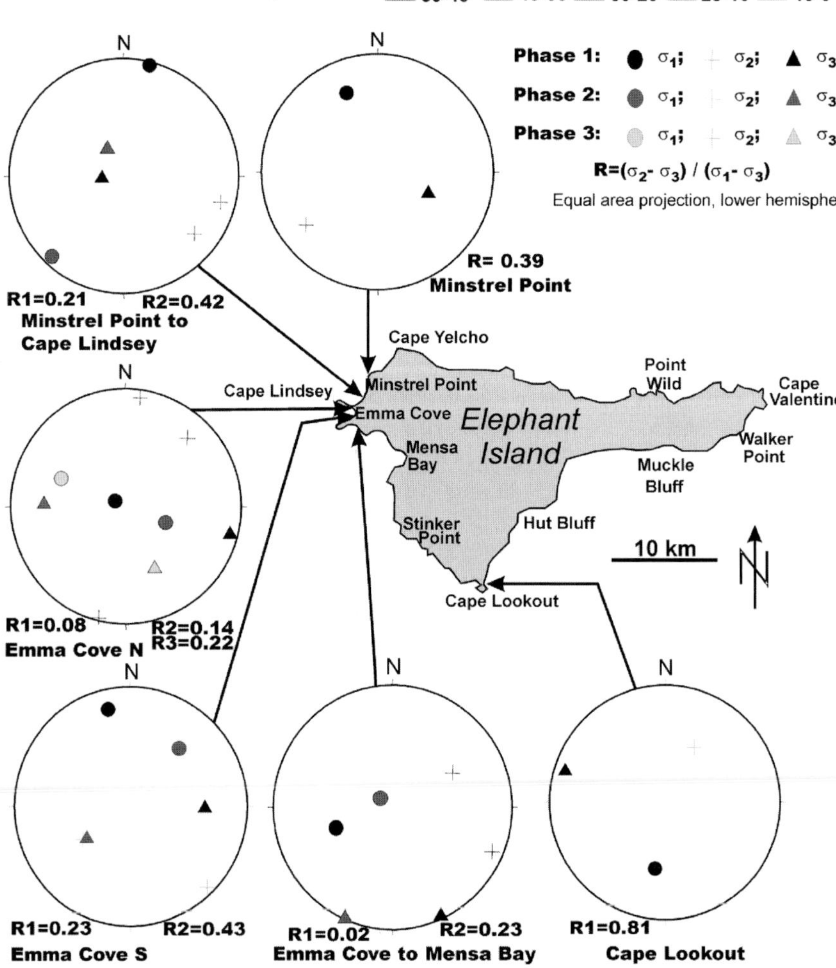

Fig. 5.8-4. Paleostress determined by mesofault analysis with the Search Grid Method. Lower hemisphere, equal area projection

The Right Dihedra diagrams (Fig. 5.8-3) are only shown for those stations with enough fault data, and fault regimes that could be established with confidence. The absence in the diagrams of areas with 0–10% and 90–100% of compression confirms that paleostresses were heterogeneous and faults with incompatible regimes occur in all outcrops. The most representative paleostresses are also variable along the studied section. Whereas in Emma Cove N a roughly oriented E-W extension and northwards inclined compression dominate, in Emma Cove S NNW-inclined compression and SSW-inclined extension are found. The Emma Cove to Mensa Bay area is dominated by poorly defined extension. The Cape Lookout diagram shows well defined ENE-WSW extension. However, although Right Dihedra diagrams constitute a graphical method that indicates fault compatibility and the main features of the dominant stresses, they may produce mistakes in the determination of stresses when overprinted deformation stages are present. In this case do not allow to determine the axial ratio of a single stress ellipsoid.

In order to overcome these problems of the Right Dihedra diagrams in complex regions, we applied the Etchecopar et al. (1981) and the Search Grid (Galindo-Zaldívar and González-Lodeiro 1988) methods. When all the fault regimes are determined, both methods produce similar results, but the latter technique can accommodate, in addition, faults with unknown fault regime. The results obtained using the Search Grid Method (Fig. 5.8-4) also indicate that overprinted deformations are variable on the island. Most of the ellipsoids determined along the northwestern coast are prolate to intermediate, and have a well defined orientation of maximum compression. Mesofaults at Minstrel Point indicate a well defined prolate to intermediate compressive stress, with σ_1 slightly inclined towards the NNW. Between Minstrel Point and Cape Lindsey, overprinted NNE-SSW horizontal compression stresses, with variable axial ratios are well defined. In the Emma Cove area a complex overprinting of stresses was found including compressive stresses with axes inclined towards the N and NE and prolate extensional stresses with subvertical σ_1. The Emma Cove to Mensa Bay area shows radial extension with prolate ellipsoids, while at Cape Lookout a WNW-ESE oblate extensional ellipsoid was determined.

Discussion and Conclusions

Elephant Island ductile deformations and metamorphism were dated as Cretaceous (Trouw et al. 1990, 2000; Grunow et al. 1992). Although the uplift of the HP/LT metamorphic rocks may have occurred at any time since the Cretaceous, several lines of evidence point to a recent age for this uplift: the existence of abrupt relief; the present-day tectonic context that includes the subduction of thickened crust associated with the Shackleton Fracture Zone; the presence of marine terraces at the west coast with exotic granite pebbles, probably emplaced by ice; the existence of seismicity; and finally, the fact that HP/LT metamorphic rock pebbles have not been found in recent sedimentary deposits around the island. At present, there are not enough data to determine the relative contribution to the uplift of the postglacial isostatic recovery and the subduction of lithosphere modified in the vicinity of the Shackleton Fracture Zone below the northern border of the South Shetland Block.

Marine geophysical data indicate that Elephant Island is located in a region of transition between compressive deformations (reverse faults and folds) related to the northern boundary of the South Shetland Block (Barker 1982; Larter and Barker 1991; Maldonado et al. 1994; Barker and Austin 1998) and sinistral transtensional faults related to the South Scotia Ridge internal depressions (Galindo-Zaldívar et al. 1996; Klepeis and Lawver 1996). The mentioned deformation affects rocks as young as Quaternary marine sediments. The recent tectonic activity in Elephant Island is revealed by seismic activity and the existence of a NNW-SSE compressional earthquake focal mechanism at 10 km depth below the west coast (Pelayo and Wiens 1989). Although field observations do not provide evidence of the age of faults, reverse faults similar to those related to the earthquake focal mechanism were observed in the Cape Lindsey area. The existence of several striae orientations on the same fault plane, the irregular distribution of faults and striae and the presence in the same outcrop of faults with incompatible regimes, indicate an overprinting of several brittle deformation stages. The Search Grid Method (Galindo-Zaldívar and González-Lodeiro 1988) that uses all the measured striae, with known and unknown fault regimes, provides a good approach on paleostress evolution.

The northern part of the section (Minstrel Point), was affected by NW-SE compressive stress with prolate ellipsoids that may be related to the subduction of lithosphere modified in the vicinity of the SFZ and the Antarctic Plate below the South Shetland Block. Shallow NW-SE compressive earthquake focal mechanisms below the western part of Elephant Island also support this interpretation. In addition, the undulose shape of marine terraces suggests possible recent folding of the area by compressive deformation. Southwards in the section, NE-SW prolate compressive stress ellipsoids were identified (Cape Lindsey and Emma Cove), that may correspond to propagation of sinistral motion related to the Scotia Plate-South Shetland Block boundary. The southern end of the island, in the Cape Lookout area, is dominated by WNW-ESE extension, with oblate stress ellipsoids indicating a transtensional tectonic regime similar to the one observed along internal depressions of the South Scotia Ridge. Overprinted on these stresses, radial extension has been determined along the whole section and may be related to uplift. The variability of stress along the Elephant Island western coast may be related to the main tectonic processes that determined the boundaries of the island.

Taking into account that the Earth's surface cannot support shear stresses, one main axis of the stress ellipsoid should be vertical and the other two horizontal. Although most of the stress axes determined in Elephant Island show subvertical or subhorizontal orientations, the presence of inclined main axes of paleostress ellipsoids, mainly in the Emma Cove area, suggests that at least part of the brittle deformation developed at depth or has undergone tilting. Anyway, it is not possible to determine the age of the brittle structures; however, the determined stresses are in agreement with the recent tectonic setting of the island, suggesting a probable relation.

The study of the recent tectonic evolution of Elephant Island, including geological field data and marine geophysical data allows a better understanding of the processes related to the evolution of a complex triple junction, where an oceanic transpressive fracture zone intersects with a continental block. The tectonic setting is similar to the one at Smith Island, located at the southwestern extreme of the South Shetland Islands. The data presented here show the sharp transition between compressional and extensional deformations along the boundaries of the narrow and elongated South Shetland continental block.

Acknowledgments

The comments of Dr. A. P. M. Vaughan and Dr. R. A. J. Trouw have improved this contribution. This research has been supported by the project REN2001-0643 of the DGICYT, Spain, and the logistic facilities of the Brazilian Antarctic Program.

References

Aldaya F, Maldonado A (1996) Tectonics of the triple junction at the southern end of the Shackleton Fracture Zone (Antarctic Peninsula). Geo Mar Letters 16:279–286

Angelier J, Mechler P (1977) Sur une méthode graphique de recherche des contraintes principales également utilisable en tectonique et en séismologie: le méthode des dièdres droits. Bull Soc Géol France 7:1309–1318

Barker PF (1982) The Cenozoic subduction history of the Pacific margin of the Antarctic Peninsula: Ridge crest-trench interactions. J Geol Soc London 139:787–801

Barker PF (2001) Scotia Sea regional tectonic evolution: implications for mantle flow and palaeocirculation. Earth Sci Rev 55:1–39

Barker DHN, Austin JA (1998) Rift propagation, detatchment faulting, and associated magmatism in Bransfield Strait, Antarctic Peninsula. J Geophys Res 103:24017–24043

Barker PF, Dalziel IWD, Storey BC (1991) Tectonic development of the Scotia Arc region. In: Tingey RJ (ed) The geology of Antarctica. Oxford Monogr Geol Geophys 17, Oxford Science Publications, Oxford, pp 215–248

BAS (1985) Tectonic map of the Scotia Arc, Sheet (Misc) 3, Scale 1:3.000.000, British Antarctic Survey, Cambridge

Etchecopar A, Vasseur G, Daignieres M (1981) An inverse problem in microtectonics for the determination of stress tensors from fault striation analysis. J Struct Geol 3:5165

Galindo-Zaldívar J, González Lodeiro F (1988) Faulting phase differentiation by means of computer search on a grid pattern. Annales Tectonicae 2:90–97

Galindo-Zaldívar J, Jabaloy A, Maldonado A, Sanz de Galdeano C (1996) Continental fragmentation along the South Scotia Ridge transcurrent plate boundary (NE Antarctic Peninsula). Tectonophysics 242:275–301

Gamboa LAP, Maldonado PR (1990) Geophysical investigations in the Bransfield Strait and in the Bellingshausen Sea – Antarctica. In: St John B (ed) Antarctica as an exploration frontier – hydrocarbon potential, geology and hazards. Amer Assoc Petrol Geol 31:127–141

González-Casado JM, Giner-Robles JL, López-Martínez J (2000) Bransfield Basin, Antarctic Península: not a normal backarc basin. Geology 28:1043–1046

Grunow AM, Dalziel IWD, Harrison TM, Heizler MT (1992) Structural geology and geochronology of subduction complexes along the margin of Gondwanaland: new data from the Antarctic Peninsula and southernmost Andes. Geol Soc Amer Bull 104:1497–1514

Jabaloy A, Balanyá JC, Barnolas A, Galindo-Zaldívar J, Hernández FJ, Maldonado A, Martínez-Martínez JM, Rodríguez-Fernández J, Sanz de Galdeano C, Somoza L, Suriñach E, Vazquez T (2003) The transition from an active margin to a passive one (SW end of the South Shetland Trench, Antarctic Peninsula). Tectonophysics 336:55–81

Klepeis KA, Lawver LA (1996) Tectonics of the Antarctic-Scotia plate boundary near Elephant and Clarence Islands, west Antarctica, J Geophys Res 101:20211–20231

Larter RD, Barker PF (1991) Effects of ridge crest-trench interaction on Antarctic-Phoenix spreading: Forces on a young subducting plate. J Geophys Res 96:19586–19607

Maldonado A, Larter RD, Aldaya F (1994) Forearc tectonic evolution of the South Shetland margin, Antarctic Peninsula. Tectonics 13:1345–1370

Maldonado A, Balanyá JC, Barnolas A, Galindo-Zaldívar J, Hernández J, Jabaloy A, Livermore R, Martínez-Martínez, JM, Rodríguez-Fernández J, Sanz de Galdeano C, Somoza L, Suriñach E, Viseras C (2000) Tectonics of an extinct ridge-transform intersection, Drake Passage (Antarctica). Mar Geophys Res 21:43–68

Pelayo AM, Wiens DA (1989) Seismotectonics and relative plate motions in the Scotia Sea Region. J Geophys Res 94:7293–7320

Thomas C, Livermore R, Pollitz F (2003) Motion of the Scotia Sea plates. Geophys J Internat 155:789–804

Trouw RAJ, Pankhurst RJ, Kawashita K (1990) New radiometric age data from Elephant Island, South Shetland Islands. Zbl Geol Paläont Teil I 1:105–118

Trouw RAJ, Ribeiro A, Paciullo FVP (1991) Structural and metamorphic evolution of the Elephant Island group and Smith Island, South Shetland Islands. In: Thomson MRA, Crame JA, Thomson JW (eds) Geological evolution of Antarctica, Cambridge University Press, Cambridge, pp 423–428

Trouw RAJ, Passchier CW, Valeriano CM, Simoes LSA, Paciullo FVP, Ribeiro A (2000) Deformational evolution of a Cretaceous subduction complex: Elephant Island, South Shetland Islands, Antarctica. Tectonophysics 319:93–110

Tectonics and Geomorphology of Elephant Island, South Shetland Islands

Jerónimo López-Martínez[1] · Rudolph A. J. Trouw[2] · Jesús Galindo-Zaldívar[3] · Adolfo Maestro[4] · Luiz S. A. Simões[5] Felipe F. Medeiros[2] · Camilo C. Trouw[2]

[1] Dpt. de Química Agrícola, Geología y Geoquímica, Facultad de Ciencias, Universidad Autónoma de Madrid, 28049 Madrid, Spain, <jeronimo.lopez@uam.es>

[2] Dpt. Geologia, Universidade Federal do Rio de Janeiro, CEP21949-900 Brasil

[3] Dpt. Geodinámica, Facultad de Ciencias, Universidad de Granada, 18071 Granada, Spain, <jgalindo@ugr.es>

[4] Servicio de Geología Marina, Instituto Geológico y Minero de España, Ríos Rosas, 23, 28003 Madrid, Spain, <a.maestro@igme.es>

[5] Dpt. Petrologia e Metalogenia, Universidade Estadual Paulista, Brasil, <lsimoes@rc.unesp.br>

Abstract. New tectonic and geomorphological data from Elephant Island have been obtained during a cooperative fieldwork campaign carried out in the 2002–2003 season by Brazilian and Spanish groups. The main phases of ductile deformation affecting the high-pressure metamorphic rocks of Elephant Island, D1, D2 and D3, were studied in more detail along a N-S profile in the western sector of the island. D2 was subdivided in two subphases, D2a and D2b. The brittle and brittle-ductile deformation postdating the three main phases was studied systematically for the first time in the island. The measurement of the orientation of more than 200 faults with their kinematics pointed out that normal faults indicating an extensional setting predominate in the southern sector of the island. By contrast, in the northern part of the studied profile, major faults are predominantly reverse, although normal faults are also present, indicating the overprinting of at least two deformation stages. A new 1 : 50 000 topographic map of the island, with 20 m contour interval, was prepared being included in this paper a shadow map from a 3D digital model based on it. Geomorphological features of marine and glacial origin were identified and mapped. The most significant of these are the raised marine platforms up to 150 m a.s.l. at Cape Lindsey area and lower platforms, Holocene raised beaches and moraines at Stinker Point area.

Introduction

Elephant Island forms part of the southern branch of the Scotia Arc, and is located in a complex tectonic setting, near the triple junction of the Scotia, Antarctic and former Phoenix Plates at the intersection of the Shackleton Fracture Zone and the South Shetland Block (Galindo-Zaldívar et al. 1996) (Fig. 5.9-1).

The geological knowledge of Elephant Island is relatively low in comparison to other areas in the region, mainly because of the difficult access. An accurate topographic map of the island does not exist at present and geomorphological and neotectonic studies have not been carried out despite the interest of these topics.

During the 2002–2003 austral summer, cooperative studies between Brazilian and Spanish groups produced new data on the geological evolution of Elephant Island.

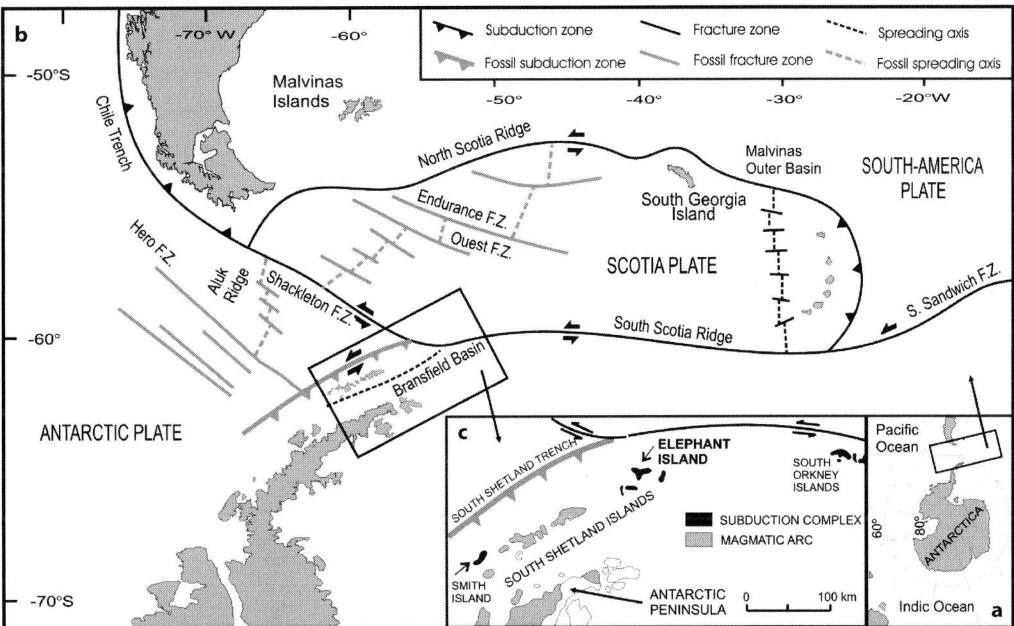

Fig. 5.9-1. a Location of the studied area. **b** Regional tectonic setting of the Scotia Arc and northwest Antarctic Peninsula regions. **c** Geological framework of the Elephant Island area

From: Fütterer DK, Damaske D, Kleinschmidt G, Miller H, Tessensohn F (eds) (2006) Antarctica: Contributions to global earth sciences. Springer-Verlag, Berlin Heidelberg New York, pp 277–282

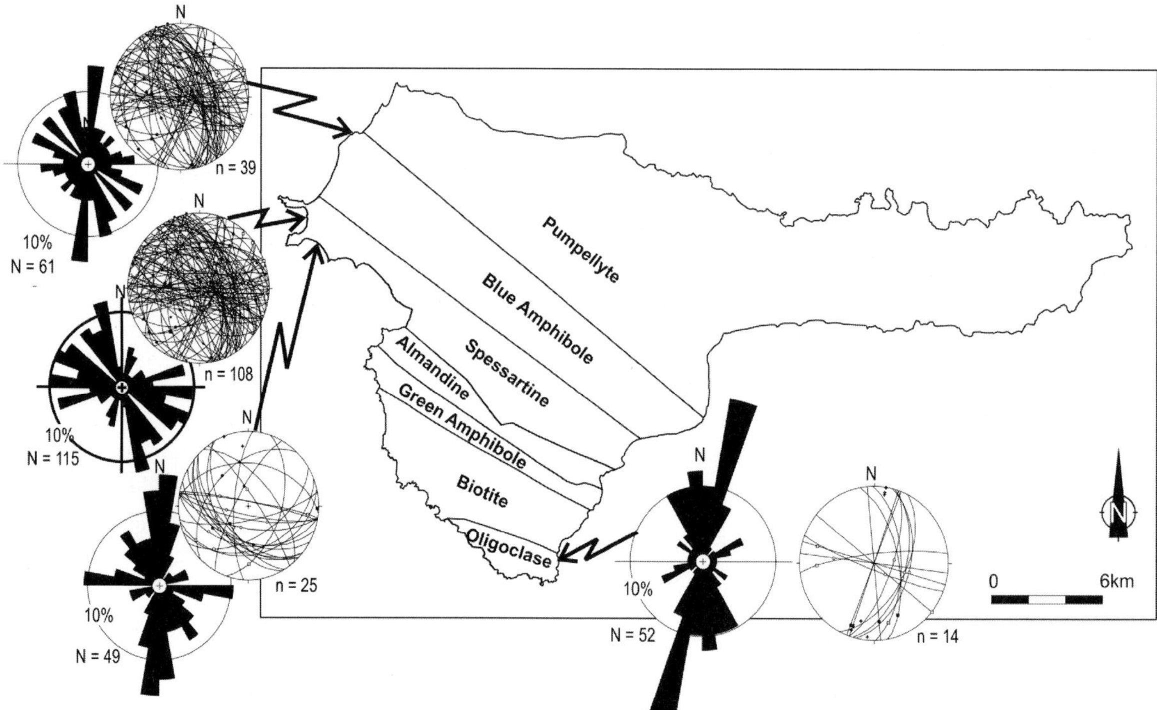

Fig. 5.9-2. Metamorphic map with isogrades defining metamorphic zones. Diagrams show the fracture and fault orientations at outcrop scale in different sectors of the island. *N:* number of fractures; *n:* number of faults

Previous geological studies on Elephant Island are mainly related to ductile deformations and metamorphic petrology (e.g., Dalziel 1984; Grunow et al. 1992; Trouw et al. 1998, 2000). The relatively high-pressure metamorphic rocks correspond to an accretionary prism and yielded ^{39}Ar-^{40}Ar metamorphic ages of about 100 Ma (Grunow et al. 1992). This metamorphism increases from sub-greenschist facies in the northeast to blueschist facies and finally to amphibolite facies in the southwest (Trouw et al. 1998) (Fig. 5.9-2).

Tectonics and Neotectonics

During this study the superposition and the main features of the principal deformation phases, as outlined by Trouw et al. (2000), were studied in more detail, especially along a N-S profile in the western sector of the island. The rocks of the northern part of this profile are affected by a D1 deformation phase that produced a well developed planar-linear fabric, with a sub-vertical E-W striking foliation and steeply westwards plunging stretching lineations. A second deformation phase, D2 (Fig. 5.9-3), that produced crenulation cleavages with related folds, appeared to be more complex than previously thought, and was subdivided into D2a and D2b, according to superposition criteria (Fig. 5.9-4). D2a produced tight to isoclinal folds with axes parallel to the L1 stretching lineation and steep E-W striking axial planes. D2b refolded these structures with steep NE-SW striking axial planes and subvertical axes. Kink bands and open folds with steep axes, indicating E-W shortening constitute the D3 deformation phase. The structures of the southern part of the profile confirmed the scheme outlined in Trouw et al. (2000).

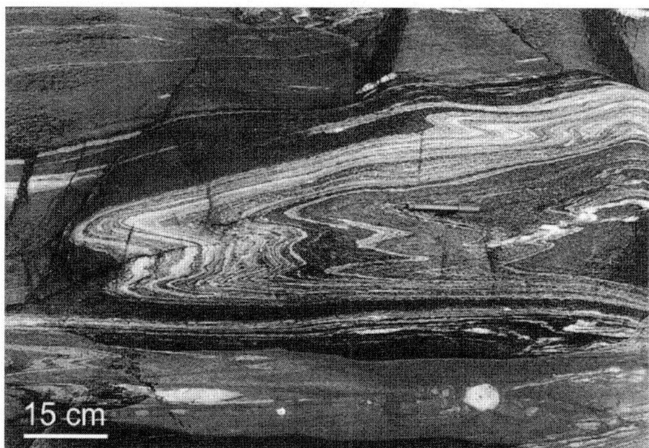

Fig. 5.9-3. Tight D2 folds with subvertical E-W striking axial planes in pink metachert layers interbedded with plagioclase garnet schist; plan view with North upwards, Cape Lookout area

The brittle and brittle-ductile deformation, overprinted on these ductile structures, was analysed systematically for the first time. The orientation of more than 200 faults was measured as well as their kinematic indicators (Fig. 5.9-2). Best indicators are those related to fault gouges that in-

Fig. 5.9-4. Isoclinal D2a folds, refolded by D2b phase with axial planes (S2b) parallel to *pencil*. Quartzite (metachert) and blueschist; plan view with North upwards, Cape Lindsey area

Fig. 5.9-5. Reverse fault with southward vergence in Emma Cove

Fig. 5.9-6. Normal faults trending N-S with dip of about 50° towards the East on the southeastern coast between Cape Lookout and Endurance Glacier

clude cataclastic foliation, crushing and frictional sliding of grains or rock fragments, and structures on fault surfaces such as striations on slickensides and steps, locally with mineral growth of calcite, pyrite and haematite. The major faults observed in the northern part of the studied profile are predominantly reverse with a southward vergence and medium to low dips (Fig. 5.9-5). Normal faults do also occur in this sector, indicating the overprinting of at least two deformation stages, locally revealed by two sets of striations on the same fault plane. In the southern sector of the island normal faults predominate indicating an extensional setting (Fig. 5.9-6). Paleostresses have been identified by means of fault population analyses, using the Right Dihedra (Angelier and Mechler 1977), Etchecopar et al. (1981) and the Search Grid (Galindo-Zaldívar and González-Lodeiro 1988) methods, confirming the overprinting of different deformation stages. The stress ellipsoids in the northern part of the profile have NW-SE and NE-SW subhorizontal main compressional axes, and prolate shape. In addition to these ellipsoids, an extensional stress field was detected in most of the studied area, indicating a radial to E-W oriented extension.

Geomorphology

A new 1:50 000 topographic map of the island, with 20 m contour interval, has been prepared in cooperation with Centro Geográfico del Ejército (Spain) using photogrammetric plotting of aerial photographs and differential GPS field work. The 3D digital model of the island (Fig. 5.9-7), derived from the mentioned map data, allows a more accurate view of the relief and main landforms of Elephant Island.

A geomorphological map of Elephant Island showing the distribution of landforms and surface deposits of different morphogenetic origins is presently being compiled, integrating data of photointerpretation and field observations. It will be published as a separate sheet, together with topographic and geological maps.

The main part of the coast of the island is constituted by very steep cliffs (Fig. 5.9-8). Raised marine erosive platforms at about 130–150 m a.s.l. and about 50–70 m a.s.l. were recognised and mapped at different locations, being especially well developed along the western coast.

There are also less extensive remains of the same origin at intermediate and lower altitudes. A well conserved extensive marine platform at about 140 m a.s.l. has been identified and mapped in detail in the Cape Lindsey area (Fig. 5.9-8).

Striations and moraines on this platform indicate Quaternary glacial fluctuations. Platforms at similar and lower altitudes have been recognised south of Emma Cove and also north of Cape Lindsey up to Minstrel Point. In some cases, as in Cape Lindsey, marine platforms show landward tilting and large undulations.

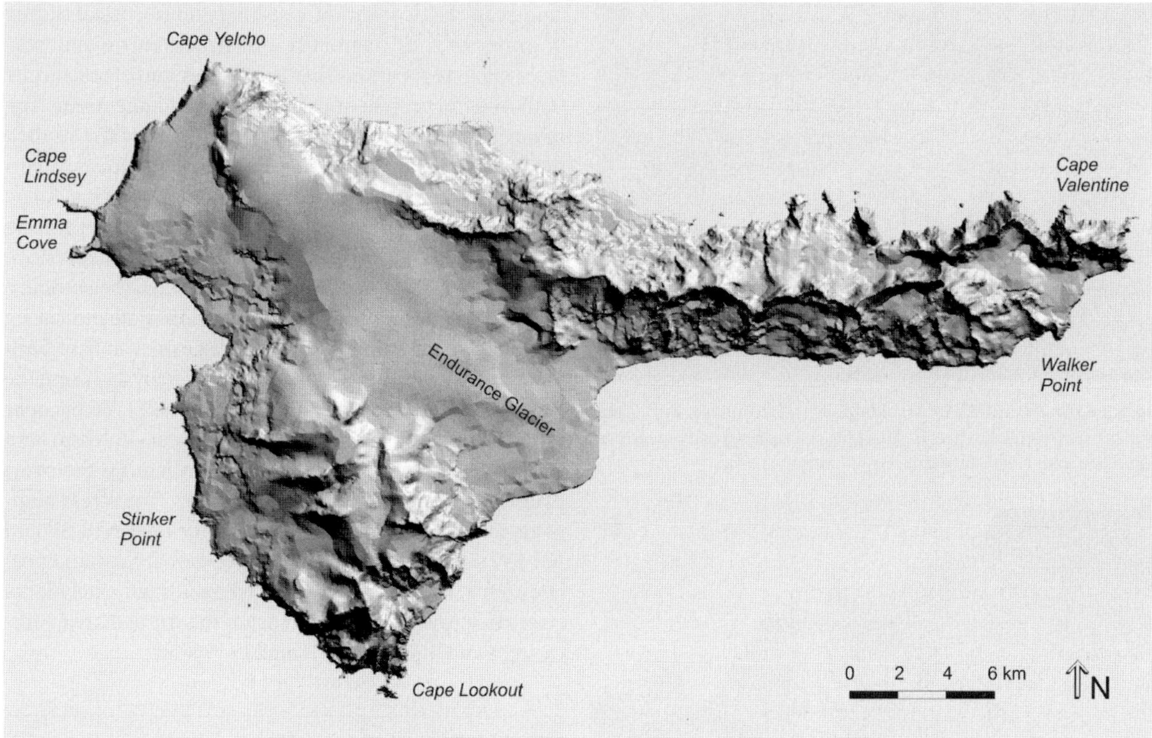

Fig. 5.9-7. Shadow map from 3D digital model of Elephant Island

Fig. 5.9-8. The main part of the coast of Elephant Island is constituted by very steep cliffs. Raised marine erosive platforms around 150 m a.s.l. at Cape Lindsey area. This terrace shows landward tilting. Raised beach deposits are relatively scarce in this area

Fig. 5.9-9. Holocene raised beaches (*HRB*), marine erosive platforms (*MP*), palaeostacks (*PS*) and moraines (*M*) in the Stinker Point area

Glacial erosive and depositional features are also present in the Stinker Point area, where they are associated with intermediate and lower marine platforms and with Holocene raised beaches (Fig. 5.9-9). These younger beaches are mainly restricted to the Stinker Point area.

Discussion and Conclusions

Elephant Island is composed by rocks belonging to a former subduction complex that have undergone a HP-LT Cretaceous metamorphism (e.g., Trouw et al. 1998, 2000). Ductile deformations are evidenced by the planar-linear fabrics that are heterogeneous along the studied transect. While the southern part of the island shows the overprinting of three deformation stages, at the northern part four stages have been recognised by the subdivision of D2 in

D2a and D2b stages. Ductile deformations are probably the result of the subduction tectonic setting, and were active before the development of the Scotia Arc.

The recent setting is determined by a different tectonic scenario that also involves the presence of subduction. The Antarctic-Scotia plate boundary is located at the positive relieves of the Shackleton Fracture Zone, that subducted below the border of the South Shetland Islands continental crust. This active structure is responsible for the elevation of Elephant Island and determines the main features of the brittle deformations. The northwestern part of Elephant Island has undergone recent NW-SE and NE-SW oriented compression evidenced by paleostress determinations from microfaults and field observation of reverse faults. Extensional faults have also been observed and may be related to the northeastward propagation of the Bransfield Basin extension or to the glacio-isostatic uplift.

The presence of raised marine beaches and erosive platforms, common in the South Shetland archipelago, has been previously studied and mapped in different islands (e.g., John and Sugden 1971; Birkenmajer 1981; López-Martínez et al. 1995, 2002). The mentioned marine features have been considered as Holocene up to about 20 m a.s.l. and those at higher altitudes, pre-Holocene (e.g., John and Sugden 1971; Arche et al. 1996). The lack of absolute dating of the pre-Holocene marine features, in Elephant Island as well as in the other islands, hampers a precise reconstruction or correlation of past sea levels.

The marine platforms present on Elephant Island are at similar altitudes as compared to the ones on other islands of the South Shetland archipelago (e.g., John and Sugden 1971; López-Martínez et al. 1995, 2002). However a simple correlation is not possible due to the multiplicity of platforms and the lack of dating. In this context, it is complicated to discriminate the glacio-isostatic and tectonic components in the uplift of the island and to evaluate possible contrasts due to differences in the geotectonic setting between Elephant Island and the other islands to the SW.

According to data compiled on other islands of the archipelago it seems that glacio-isostasy is a major responsible of the recent uplift. However, neotectonics could also have a participation in recent deformation at Elephant Island as revealed by the mentioned undulated morphology of the main raised marine platform at Cape Lindsey.

Holocene raised beaches are scarcer at Elephant Island than at other islands of the archipelago. Reasons for the lack of conservation of beach deposits are the presence of steep relieves close to the coast in the southwest and in the eastern half of the island, the important and active retreat of the cliffs along most of the western coast and the extension down to sea level of the Endurance Glacier.

The new observations obtained increase the knowledge of the geological history of Elephant Island, with more detail on the ductile deformation phases and with new data on neotectonics and relief evolution.

Acknowledgments

This work was supported by the Brazilian Antarctic Programme (PROANTAR), CNPq and Project REN2001-0643 of the Spanish Antarctic Programme. The compilation of the topographic and geodetic data was possible thanks to the collaboration with the Centro Geográfico del Ejército, Spain, and the participation in the fieldwork of C. Vergara and H. Rodríguez. Captain and crew of the vessel Ary Rongel are greatfully acknowledged for the logistic support. The authors thank K. Birkenmajer and F. Hervé for their constructive reviews that improved the manuscript.

References

Angelier J, Mechler P (1977) Sur une méthode graphique de recherche des contraintes principales également utilisable en tectonique et en séismologie: le méthode des dièdres droits. Bull Soc Géol France 7:1309–1318

Arche A, López-Martínez J, Serrano E, Martínez de Pisón E (1996) Marine landforms and deposits. In: López-Martínez J, Thomson MRA, Thomson JW (eds) Geomorphological map of Byers Peninsula, Livingston Island. Brit Ant Surv Geomap Ser 5A, British Antarctic Survey, Cambridge, pp 35–42

Birkenmajer K (1981) Raised marine features and glacial history in the vicinity of H. Arctowski Station (King George Island, South Shetland Islands, West Antarctica). Bull Acad Polonaise Sci, Sér Sci de la Terre 27:59–67

Dalziel IWD (1984) Tectonic evolution of a forearc terrane, southern Scotia Ridge, Antarctica. Geol Soc Amer Spec Pap 200:1–32

Etchecopar A, Vasseur G, Daignieres M (1981) An inverse problem in microtectonics for the determination of stress tensors from fault striation analysis. J Struct Geol 3:51–65

Galindo-Zaldívar J, González-Lodeiro F (1988) Faulting phase differentiation by means of computer search on a grid pattern. Annales Tect 2:90–97

Galindo-Zaldívar J, Jabaloy A, Maldonado A, Sanz de Galdeano C (1996) Continental fragmentation along the South Scotia Ridge transcurrent plate boundary (NE Antarctic Peninsula). Tectonophysics 242:275–301

Grunow AN, Dalziel IWD, Harrison TM, Heitzler MT (1992) Structural geology and geochronology of subduction complexes along the margin of Gondwanaland: new data from the Antarctic Peninsula and southermost Andes. Geol Soc Amer Bull 104: 1497–1514

Jonh BS, Sugden DE (1971) Raised marine features and phases of glaciation in the South Shetland Islands. Brit Ant Surv Bull 24:45–111

López-Martínez J, Martínez de Pisón E, Arche A (1995) Geomorphological Map of Byers Peninsula, scale 1:50000. Brit Ant Surv Geomap Ser 5A, British Antarctic Survey, Cambridge

López-Martínez J, Serrano E, Lee JI (2002) Geomorphological map of Barton and Weaver peninsulas, King George Island, E 1:10000. Korea Ocean Polar Res Inst Seoul

Trouw RAJ, Simões LSA, Valladares C (1998) Metamorphic evolution of a subduction complex, South Shetland Islands, Antarctica. J Metam Geol 16:475–490

Trouw RAJ, Passchier CW, Valeriano CM, Simoes LSA, Paciullo FVP, Ribeiro A (2000) Deformational evolution of a Cretaceous subduction complex: Elephant Island, South Shetland Islands, Antarctica. Tectonophysics 319:93–110

Geodynamical Studies on Deception Island: DECVOL and GEODEC Projects

Manuel Berrocoso[1] · Alicia García-García[2] · José Martín-Dávila[3] · Manuel Catalán-Morollón[3] · Mar Astiz[2] · M. Eva Ramírez[1]
Cristina Torrecillas[1] · José Manuel Enríquez de Salamanca[1]

[1] Laboratorio de Astronomía y Geodesia, Facultad de Ciencias, Universidad de Cádiz, Campus Río San Pedro, 11510 Puerto Real, Cádiz, Spain, <mariaeva.ramirez@uca.es>
[2] Departamento de Volcanología, Museo Nacional de Ciencias Naturales, C/ José Gutiérrez Abascal 2, 28006, Madrid, Spain
[3] Sección de Geofísica, Real Instituto y Observatorio de la Armada, 11110 San Fernando, Cádiz, Spain

Abstract. Deception Island (63° S, 60° W) is situated in the South Shetland Islands, and it is the main active volcano in the Bransfield Strait, with recent eruptions in 1842, 1967, 1969 and 1970. Its volcanic and seismic activity has been monitored from 1986 to study the geodynamic activity on the island. In this paper we present the objectives and some results obtained in DECVOL project, a cross-disciplinary study which was planned to evaluate the volcanic status of the island after the crisis occurred during the 1998/1999 campaign, as well as the goals and activities which have been carried out in the framework of GEODEC project, a multidisciplinary project to continue with the studies on the island.

Introduction

Nowadays, both significant superficial events and considerable seismic activity are monitored on Deception Island. First ones involve areas with fumarolic activity (Fumarole Bay), consisting of gas emissions up to 100 °C, hot soils (Hot Hill), and thermal springs (Pendulum Cove and Whalers' Bay). On the other hand, there are two different sources of seismic activity due to the situation of the island (Fig. 5.10-1): a tectonical event, due to the expansion of the Bransfield Rift and a volcanic event, the subduction process in the North of the South Shetland Islands.

Information resulting from continuous studies since 1987 has shown a scientifically interesting situation in the geodynamical aspect. Antarctic campaigns took place during the austral summer in the Spanish Antarctic bases Gabriel de Castilla, on Deception Island, and Juan Carlos I, on Livingston Island (Fig. 5.10-2). During the 1991/1992 campaign, a series of seismic events was monitored by the Spanish scientific group. It began on January 1, and lasted longer than a month, with almost 900 events and four felt earthquakes. That situation was interpreted as a volcanic reactivation produced by a magma intrusion at 2 km depth in Fumarole Bay (Ortiz et al. 1992a,b). DECVOL campaign, which took place in December 1999, was planned to evaluate the volcanological status of Deception island after the increase of seismic activity in austral summer of 1998–1999 (García et al. 2000). It was a multidisciplinary study which involved geophysical, geochemical and geodetic surveys. After the development of this campaign, and with the aim of continuing the researches on the island, it was necessary to integrate three different subprojects into a big one which covered several investigation in the fields. It was named GEODEC (Geodynamical Researches on Deception Island) and it is an extension of the geodetic and geophysi-

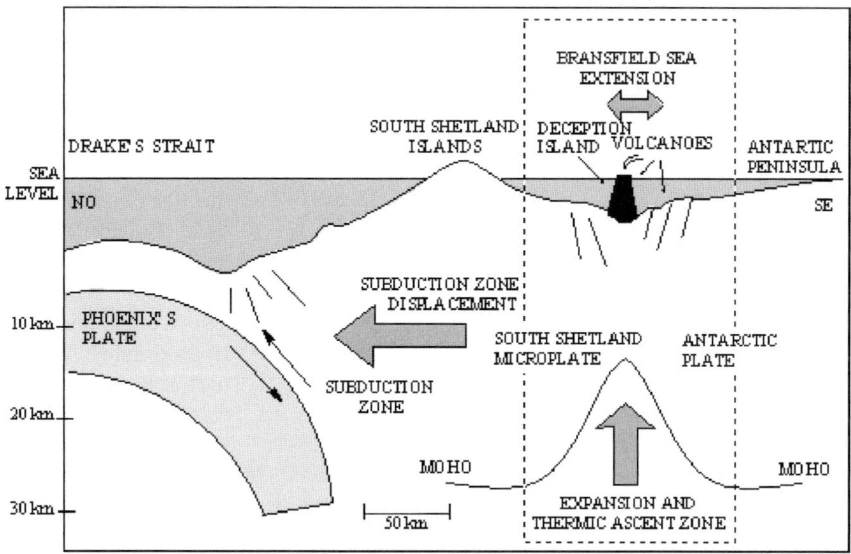

Fig. 5.10-1. Tectonic situation of the South Shetland Islands: geological section from Drake Strait to Antarctic Peninsula

From: Fütterer DK, Damaske D, Kleinschmidt G, Miller H, Tessensohn F (eds) (2006) Antarctica: Contributions to global earth sciences. Springer-Verlag, Berlin Heidelberg New York, pp 283–288

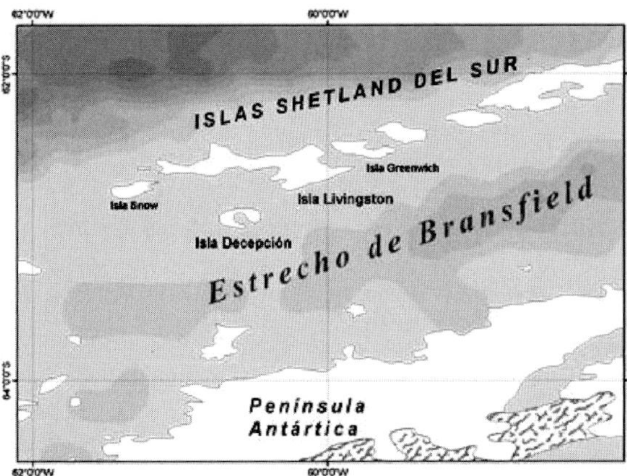

Fig. 5.10-2. Location of the Spanish base Gabriel de Castilla on Deception Island, and the Spanish base Juan Carlos I on Livingston Island

cal investigations into the active Deception volcano which had been carried out thus far.

Decvol Project Objectives

DECVOL multidisciplinary project consisted of the realization of a short campaign set aside for a geodetic and geophysical monitoring and supposed the collaboration among several scientific institutions from different countries. It involved the following fields: ground and marine geomagnetism and gravimetry, hydrography, seismicity, geodesy, geochemistry and thermometry (A. García-García pers. comm.).

For the evaluation of the volcanic status of the island the following objectives were identified:

- Location and quantification of the energy of the volcanotectonic, long-period (LP), and tremor events, by setting a broadband seismic station and a short period seismic network along the coast of the island.
- Resurveying of the local network by GPS receivers. In 1999, the geodetic network consisted of five points on Deception Island and another one 35 km away, at the Spanish Station Juan Carlos I on Livingston Island.
- A Marine Geophysical campaign, which involves bathymetry, gravimetry and magnetometry, covering Deception interior Bay and the marine areas surrounding the island. This campaign was completed with gravimetric and magnetic land measurements in order to link the inner and exterior nets. The gravimetric network was linked to the Argentinian one.
- Realization of a high resolution hydrographic survey.
- Sampling and analysis of fumarolic gasses in Fumarole Bay, Whalers' Bay and Pendulum Cove as well as the measurement of temperatures in anomaly areas.

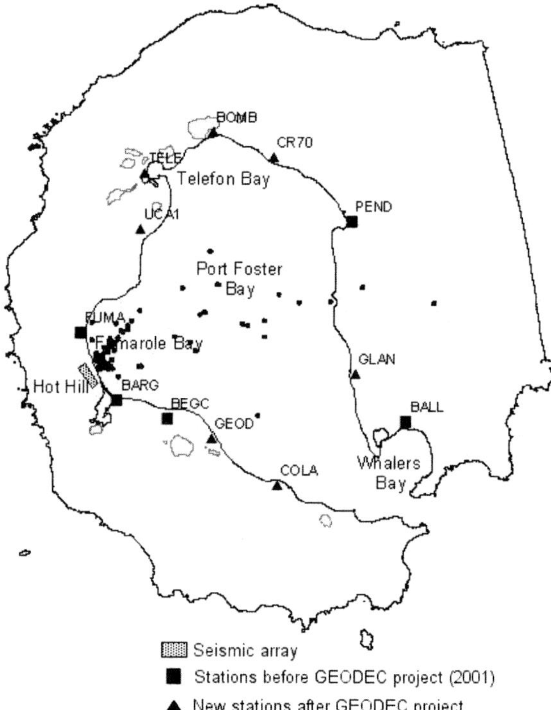

Fig. 5.10-3. Geodetic network on Deception Island and seismic field survey during the 1998/1999 campaign (Ibáñez et al. 2003)

Decvol Results

Although no significant seismic activity was registered during the survey, the network was maintained in operation until the end of the austral summer.

The main results of the different investigations can be summarized as follows:

- Seismicity: A notable increase both in number and energy of seismic activity was detected in the inner part of Deception Island, at a zone where no significant seismic activity was previously recorded. The use of different seismic networks required different technique in the analysis of the data. The results show a high coherence with the main geological structure of the area (Ibáñez et al. 2003).
- Geochemistry: The chemical analysis of the main fumarolic emissions revealed a correlation between the different components and ratios and seismic activity, associated to an effect mainly magmatic.
- Hydrography: The hydrographic survey of the inner part of Port Foster Bay (Fig. 5.10-3) was developed by using a multi echo sounder and a single one, obtaining a 5 m fairsheet in the central part of Port Foster Bay.
- Geodesy: A resurveying of the GPS local network was carried out, getting the horizontal and vertical displace-

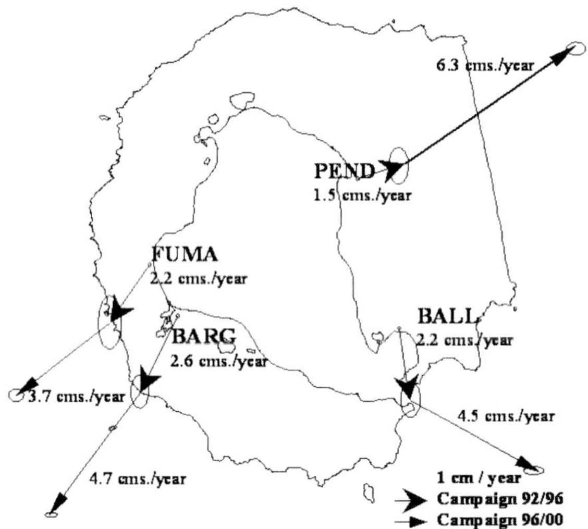

Fig. 5.10-4. Deformation models from 1992 to 1996 campaigns, and from 1996 to 2000 campaign

ment models (Fig. 5.10-3). Data integration of this survey with previous ones proves that the increase in the seismic activity is shown in the values for the velocity in the horizontal and vertical displacements between the 1995/1996 and 1999/2000 campaigns, which are greater than those related to the 1991/1992 and 1995/1996 campaign (Fig. 5.10-4). For the first time the whole network was simultaneously observed, including the data from the Spanish Station Juan Carlos I.

- Marine geomagnetism and gravimetry: Two grids, one of inner Port Foster Bay, the other of the external surrounding seawaters of Deception Island, were covered. Additionally, a recalibration of the fundamental gravimetric network of Deception Island was performed. Measurements were made at the harbour of Ushuaia (Base Ushuaia K), and also at the Spanish Antarctic base Juan Carlos I, on Livingston Island, as well as at fundamental points on Deception Island: Argentinian Antarctic Base (BARG), Spanish Antarctic base Gabriel de Castilla (BEGC), Fumarole Bay (FUMA), Pendulum Cove (PEND) and Whalers' Bay (BALL), some of which had been reobserved in previous campaigns. An additional 31 locations were also considered, covering a large part of the interior perimeter of the island. The obtained anomaly maps were compared to the previous ones and several variations are shown for the two types of anomalies (Blanco 1997).

Analysis of the data led to conclude there was a reactivation of the volcanic activity, which was evident because of the increase in the seismic activity, changes in the composition of the gasses, and the existence of gravimetric and geomagnetic anomalies. It can be interpreted that the reactivation was due to a magma intrusion (at 500 m depth) which produced the volcano-tectonic earthquakes. The long-period events and tremors could be the result of the interaction between the shallow aquifers in the island and the intrusion of incandescent material.

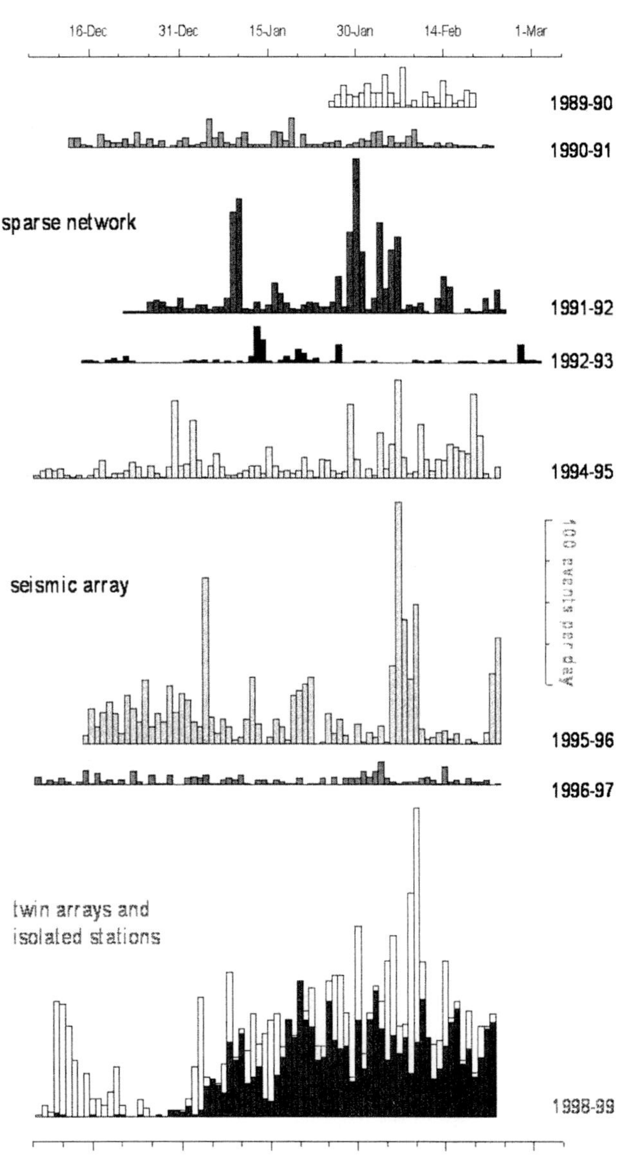

Fig. 5.10-5. Daily number of events on Deception Island (Ibáñez et al. 2003)

These events could be considered as significant evolution of the volcanic activity, indicating a change from stationary to dynamic conditions in the geothermal process present on Deception Island. According to the relation between the frequency of events and the liberated energy, this process was not finished when the 1998/1999 campaign ended (see Fig. 5.10-5). The most affected area has been Fumarole Bay.

Geodec Project and Objectives

With the aim of continuing with the geodynamic studies on the island, a multidisciplinary project named GEODEC began in 2001. It was divided into three subprojects, GEODEC-GEODESIA, GEODEC-MODELOS and GEODEC-GEOMAR (Berrocoso et al. 2002).

In the GEODEC project the geodetic network REGID has been resurveyed, and it has been improved with new points, whose locations were decided according to the recent activity. The extension of the network and the use of a permanent station on Livingston Island will allow a centimetre resolution required for the volcanic monitoring. A marine geophysical campaign following DECVOL campaign will allow the determination of the shallow structure of the Deception volcano. Furthermore, the use of a geographical information system has revealed the importance of the elaboration of risks maps, by integrating all available data (geological, geochemical, geodetic, geophysical, etc.). A detailed map of Deception Island has been produced, contributing to the development of lava flow models, products emission, etc.

Statistical methods have been used to evaluate return periods and deterministic models, by reconstructing past events and forecasting future eruptions and giving rise to a hazard and risk map for the island (Fig. 5.10-6).

The main goals of subproject, GEODEC-GEODESIA (Deformation models, geoid determination, and multidisciplinary scientific information support system) are the establishment of a reference frame to get an accurate position of the scientific data, the establishment of the deformation models of the island, and the detection and monitoring of the volcanic activity.

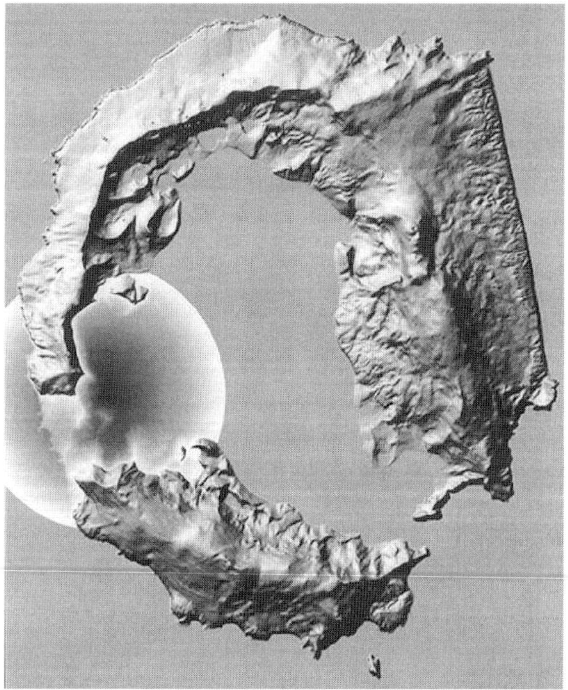

Fig. 5.10-6. Map of natural hazards for Deception Island (Museo Nacional de Ciencias Naturales, Dep. Volcanología)

Resurveying of the existing geodetic network and the addition of new stations will lead us to the establishment of the horizontal deformation models. To get the vertical deformation models, a high precision levelling network has been designed to control the deformation in the most active fractures.

To receive a local geoid for Deception Island, dense gravimetric profiles have been measured. Lastly, the great amount of data available from the first campaigns has shown the need to design an information system for scientific support (SIMAC), which could compile all this information.

A major objective of the subproject GEODEC-MODELOS (Modelling of the geological active processes of the Deception Volcano and its environment from geophysical and geodetic data) is the establishment of a number of superficial deformation models of volcanic areas and the determination of numerical models of caldera behaviour. It is also planned to update the previous models and to design maps of risks and hazards from a digital terrestrial model of high precision.

Subproject GEODEC-GEOMAR (Structure of the volcanic complex on Deception Island and its environment from sea geophysical techniques) will be accomplished by support of the research ship BIO "Hesperides". Main activities will be acquisition of reflection/refraction seismic profiles outside the island as well as within the caldera, the scan of an inner grid to obtain geomagnetic, gravimetric, and echo-sounder data, and the sounding of marine seismometers. Geomagnetic and gravimetric anomalies maps of Deception Island, and a model of the crustal structure as well as of the shallower mantle of the area will be performed.

Acknowledgments

This work has been possible thanks to the projects DECVOL (ref. ANT99.1430.E), and GEODEC (ref. REN2000-0551-C03-01/02/03), funded by the Ministry of Science and Technology of Spain through its National Program of Antarctic Research and Natural Resources.

References

Blanco I (1997) Análisis e interpretación de las anomalías magnéticas de tres calderas volcánicas: Decepción (Shetland del Sur, Antártida), Furnas (San Miguel, Azores) y Las Cañadas del Teide (Tenerife, Canarias). Tesis doctoral, Dpt Física de la Tierra Astronom Astrofís, Univ Complutense Madrid, pp 250

Berrocoso M, Martín-Davila J, Astiz M, Catalán-Morollón M, García A (2002) El Proyecto GEODEC: un estudio multidisciplinar de la actividad geodinámica de la Isla Decepción (Islas Shetland del Sur, Antártida). Asamblea Hispano-Portuguesa Geodes Geofís Valencia, Libro Resúmenes 3ª:767–769

García A, Berrocoso M, Ibáñez J, Martín-Dávila J, Ortiz R (2000) Campaña DECVOL99, Libro de Resúmenes. 2ª Asamblea Hispano-Portuguesa de Geodesia y Geofísica, Lagos pp 359–361

Ibáñez JM, Almendros J, Carmona E (2003) The recent seismo-volcanic activity at Deception Island volcano. Deep Sea Res II 50: 1611–1630

Ortiz R, Vila J, Correig A, García A, Martí J (1992a) Volcanic tremors at Deception Island (South Shetland Islands, Antarctica). JVGR 53:1–4

Ortiz R, Vila J, García A, Camacho A, Díez J, Aparicio A, Soto R, Viramonte J, Risso C, Petrinovic I (1992b) Geophysical features of Deception Island. In: Yoshida Y, Kaminuma K, Shiraischi, K (eds) Recent progress in Antarctic earth sciences. Terra Scientific Publishing Tokyo, pp 443–448

Vila J, Ortiz R, Correig A, García A (1992) Seismic activity on Deception Island. In: Yoshida Y, Kaminuma K, Shiraischi K (eds) Recent progress in Antarctic earth sciences. Terra Scientific Publishing Tokyo, pp 449–517

Theme 6
Antarctic Rift Tectonics

Chapter 6.1
Mawson Breccias Intrude Beacon Strata at Allan Hills, South Victoria Land: Regional Implications

Chapter 6.2
What Supports the Marie Byrd Land Dome? An Evaluation of Potential Uplift Mechanisms in a Continental Rift System

Chapter 6.3
A Multi-Phase Rifting Model for the Victoria Land Basin, Western Ross Sea

Chapter 6.4
Rift History of the Western Victoria Land Basin: A new Perspective Based on Integration of Cores with Seismic Reflection Data

Chapter 6.5
Constraints on the Timing of Extension in the Northern Basin, Ross Sea

Chapter 6.6
The Structure of the Continental Margin off Wilkes Land and Terre Adélie Coast, East Antarctica

Chapter 6.7
Post-Rift Continental Slope and Rise Sediments from 38° E to 164° E, East Antarctica

Rifting marks the beginning of plate fragmentation whereas continental margins present the final product of a "successful" rifting process. In this chapter, aspects of completed rifting, i.e. Antarctica's separation from the surrounding continents during the process of Gondwana fragmentation, are presented. However, Antarctica experiences internal fragmentation expressed by younger rift systems, such as the West Antarctic Rift System (WARS) which is one of the largest on earth. The tectonic development of this rift system, and possible sources of the driving forces behind, are discussed here.

Reflecting the pre-break-up setting of Gondwana in the Transantarctic Mountains, Elliot et al. (Chap. 6.1) focus on the intrusive relationships between pyroclastic rocks of the Mawson Formation and Beacon strata at Alan Hills in southern Victoria Land.

Uplift is a crucial part in active rifting: This topic is the subject of the paper by LeMasurier (Chap. 6.2) who discusses the potential of the Marie Byrd Land Dome within the West Antarctic rift system.

Davey and De Santis (Chap. 6.3) present a model for the development of the 140 km wide Victoria Land Basin (VLB) in the Ross Sea. Angular unconformities of major seismic sequences indicate four rift episodes from 34 Ma to 24 Ma. Fielding et al. (Chap. 6.4) also discuss the rift history of the VLB, but include the early Cenozoic regional uplift (pre 34 Ma) and the early Miocene (post 24 Ma) thermal subsidence and the following Terror Rift formation.

Cande and Stock (Chap. 6.5) look at the northernmost fraction of the WARS, examining the transition from the structures in the Adare Basin to the structures of the Northern Basin of the Ross Sea. Most of the extension in the Adare Basin can be continued into the Northern Basin, thus linking seafloor spreading close to the continental margin to continental extension in the Ross Sea.

The papers by Colwell et al. (Chap. 6.6) and O'Brien et al. (Chap. 6.7) concentrate on the continental margin of large parts of East Antarctica. Newly acquired seismic reflection transects run off the Antarctic coast towards the oceanic crust. While O'Brien et al. classify and describe different sedimentary provinces over the entire margin from Enderby Land to the Mertz Glacier, Colwell et al. investigate the Wilkes-Adélie margin using also refraction, wide angle reflection seismic data together with magnetic and gravity data to delineate the structure of the continental margin underlying the sedimentary sequences.

It is interesting to note that all papers focus on the WARS and the continental margin towards Australia. The major Lambert rift and its possible extension to Antarctica's interior is partially reported on in Theme 2: Antarctica – the Old Core. With the International Polar Year (IPY) coming up in 2007–2009, we are looking forward to new insights into Antarctica's possible interior rift systems, too.

Mawson Breccias Intrude Beacon Strata at Allan Hills, South Victoria Land: Regional Implications

David H. Elliot · Everett H. Fortner · Craig B. Grimes
Department of Geological Sciences and Byrd Polar Research Center, The Ohio State University, Columbus, Ohio 43210, USA

Abstract. At Allan Hills, south Victoria Land, Antarctica, Mawson tuff breccias are intrusive into Permian and Triassic Beacon strata, contrary to previous reports of the relationship. North of Watters Peak, Mawson intrusive and extrusive rocks, a megaclast unit consisting principally of beds of Member C of the Triassic Lashly Formation, and *in situ* brecciated Permian and Triassic country rock, taken together, comprise an assemblage of Jurassic rocks interpreted as a magmatically-driven collapse structure. At adjacent Coombs Hills the contact at one locality is intrusive but at another the relationships are less clear. The intrusive, rather than stratigraphic, relationship between Mawson tuff breccias and Beacon strata leads to reinterpretation of the Prebble Formation field relations at Otway Massif as also intrusive. Previous interpretations, based on a major unconformity at Allan Hills, of an episode of Early Jurassic pre-Ferrar erosion that created significant topography can no longer be supported.

Introduction

Jurassic magmatism recorded by the Ferrar Large Igneous Province, which encompasses Antarctica and southeastern Australasia (Fig. 6.1-1), marks the initial breakup of Gondwana in which Africa and South America separated from Antarctica, Australia and India. Ferrar intrusive rocks in Antarctica include the Dufek intrusion and the Ferrar Dolerite sills and dikes, with extrusive rocks comprising pyroclastic rocks (Mawson, Prebble and Exposure Hill formations) overlain by Kirkpatrick Basalt lava flows. Ferrar magmas were emplaced as sills and dikes principally into the underlying Devonian to Triassic Beacon sequence but locally into basement rocks and into the basaltic pyroclastic deposits. In the Queen Alexandra Range, the contact between the Prebble pyroclastic rocks and older strata is clearly stratigraphic although probably disconformable (Elliot 1996). In north Victoria Land the Exposure Hill Formation is also in stratigraphic contact though in the Mesa Range region continuous sequences are not exposed (Elliot et al. 1986). In south Victoria Land, the contact has been described as an unconformity with as much as 500 m of paleorelief (Ballance and Watters 1971). Recent field work has demonstrated that the contact is intrusive at Coombs Hills (White and McClintock 2001) and this is also the case at Allan Hills. The purpose of this paper is to describe the field relationships at Allan Hills and reconsider the tectonic implications of the geologic relationships between the Beacon and Ferrar extrusive rocks.

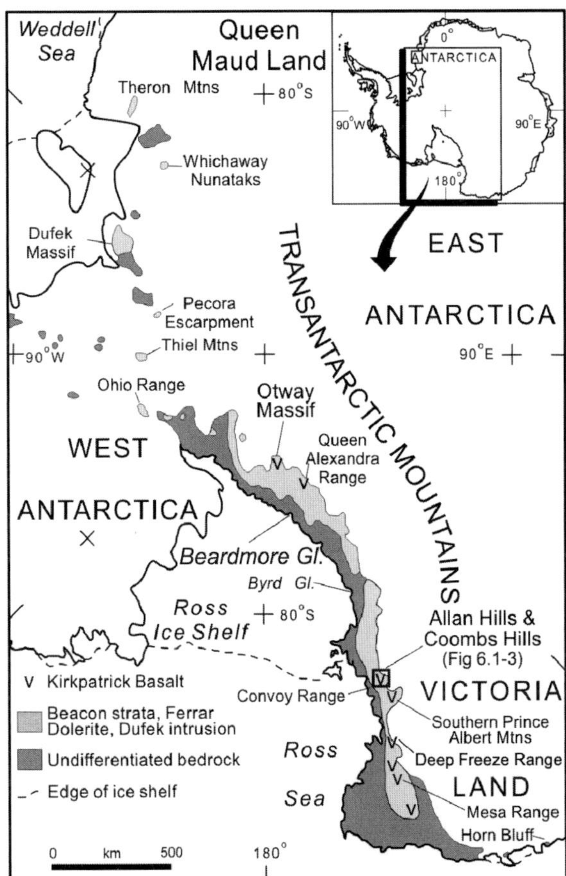

Fig. 6.1-1. Map of Antarctica showing the distribution of the Beacon Supergroup and tholeiitic rocks of the Ferrar Large Igneous Province. Ferrar Dolerite intrusions are co-extensive with the Beacon strata

Regional Geology between the Mackay and Mawson Glacier

The pre-Devonian igneous and metamorphic basement of the Transantarctic Mountains crops out in the foothills of the Convoy Range region (Fig. 6.1-1) and is rep-

From: Fütterer DK, Damaske D, Kleinschmidt G, Miller H, Tessensohn F (eds) (2006) Antarctica: Contributions to global earth sciences. Springer-Verlag, Berlin Heidelberg New York, pp 291–298

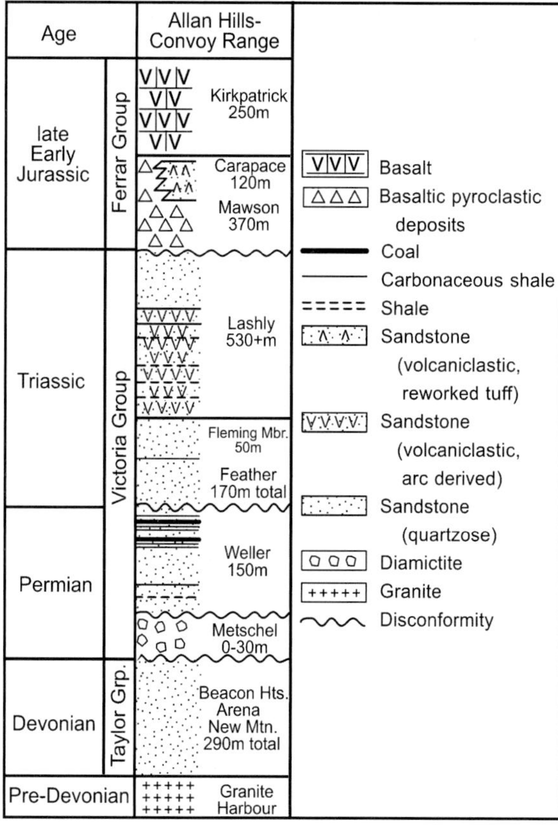

Fig. 6.1-2. Simplified stratigraphic column for the Beacon Supergroup and overlying volcanic strata in the Allan Hills-Coombs Hills-Convoy Range region, south Victoria Land. The Taylor and Victoria groups together comprise the Beacon Supergroup

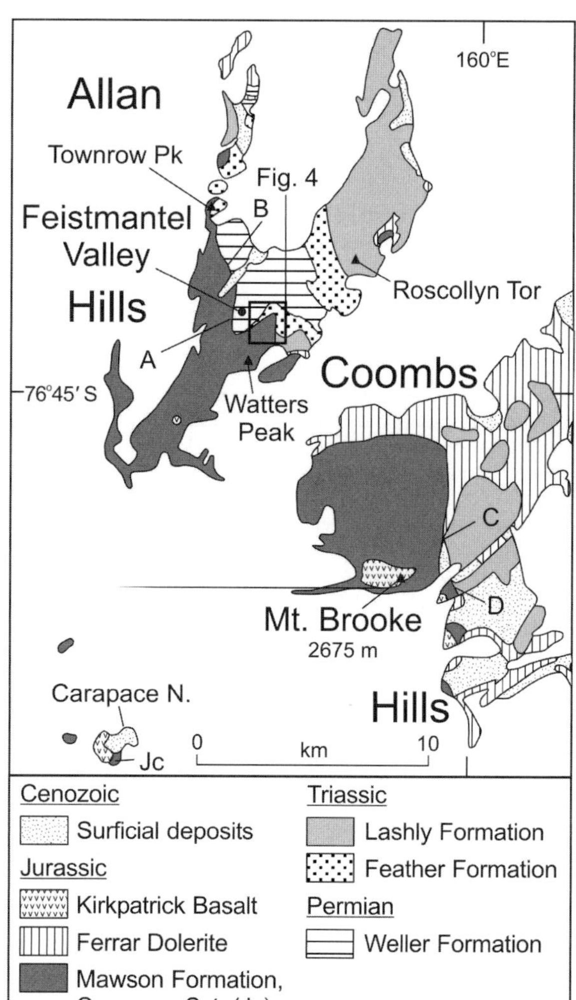

Fig. 6.1-3. Geologic sketch map of the Allan Hills-Coombs Hills region. Ferrar Dolerite dikes and small sheets are widely distributed through the Triassic strata on the northeastern and northwestern arms of Allan Hills. Sites *A*, *B*, *C* and *D* are contacts illustrated in Fig. 6.1-6 to 6.1-10

resented by the Granite Harbour Intrusives (Pocknall et al. 1994). These rocks are overlain by about 300 m of Devonian quartzose strata (Fig. 6.1-2), a thin Permian glacial sequence and about 75 m of Weller Formation coal measures. A further 75 m of the Weller Formation crops out in the Allan Hills (Fig. 6.1-3), but how much overlap of the sections there is, or how much section is missing, is unknown; the coal measures are overlain by about 300 m of Triassic strata (Ballance and Watters 1971; Collinson et al. 1983). Much of Member D of the Lashly Formation is missing at Allan Hills but it is present at Coombs Hills where it is at least 250 m thick, thus making the Lashly Formation in this region about 530 m thick. Ferrar Dolerite sills, dikes, and massive intrusions are particularly prominent in the Convoy Range and Coombs Hills, but relatively minor at Allan Hills. Kirkpatrick Basalt lava flows form a 250 m thick succession at Mt. Brooke, Coombs Hills, a lesser thickness at Carapace Nunatak (Bradshaw 1987), and one small isolated outcrop at the southern end of Allan Hills (P. J. Barrett pers. comm. 2004). Basaltic pyroclastic rocks, however, are thick (as much as 370 m) and well-exposed at Allan and Coombs Hills.

Mawson Formation

The Mawson Formation at Allan Hills and Coombs Hills has been described by Borns and Hall (1969), Ballance and Watters (1971), Hall et al. (1982), Grapes et al. (1974), and Bradshaw (1987). These early studies reported massive and graded tuff-breccia deposits, with stratigraphic thicknesses up to ca. 490 m, and intervals of bedded volcaniclastic sandstone. Explosion breccias, basaltic dikes and cross-cutting intrusion breccias have been described. More recent investigations have shown that the Mawson rocks north of Mt. Brooke, Coombs Hills, form a large vent complex (White and McClintock 2001), although Elliot and Hanson (2001) reported a stratigraphic contact east of Mt. Brooke.

Allan Hills

The contact relations are well exposed both east and west of Feistmantel Valley (Fig. 6.1-3). East of Feistmantel Valley (Fig. 6.1-4), Feather Formation (mainly the uppermost Fleming Member) strata show lateral and vertical transitions into structureless sandstone enclosing disoriented sandstone clasts and elongate micaceous sandstone lenses; the latter are pulled apart into a series of disconnected pieces that overall may exhibit complex deformation. Similar disaggregated and brecciated sandstone is also observed as a vertical pipe-like body a few meters across cutting through *in situ* Feather strata. These disaggregated and/or brecciated sandstones are attributed to phreatic activity. Lack of exposure conceals the transition into a zone of megaclasts, although it has to be steep given the juxtaposition of the rock units. The megaclast unit comprises large rafts, up to tens of meters long, of sedimentary strata. These megaclasts are slightly to strongly deformed, showing gentle to tight folding and fracturing. The megaclasts are dominated by carbonaceous sandstone which is accompanied by lesser amounts of finer grained beds and occasional thin coal seams. The megaclasts are separated by screens, as much as a meter or more wide, of breccia consisting of sandstone clasts in a sand matrix. The megaclast unit grades into an unbedded grey unit comprised of sandstone clasts and sand with increasing proportions of pyroclasts both inward from the contact and vertically. This grey breccia unit is overlain sharply by weakly stratified tuff breccia and lapilli tuff. The basal clast-supported breccia, which can be traced for about 200 m, although becoming less well defined, grades up into matrix-supported tuff breccia. The latter is succeeded by lapilli-tuff and tuff-breccia units. The *in situ* brecciated country rock and the megaclast unit are cut by a Ferrar basaltic dike. All other units are cut by irregular tuff-breccia and lapilli-tuff intrusions, basaltic dikes and pipes, and basaltic diatremes.

The megaclasts have yielded elements of the Triassic Dicroidium flora (plant megafossils and palynomorphs) but not the Permian Glossopteris flora. Within the Triassic sequence, carbonaceous beds and coal (<60 cm thick at Roscollyn Tor, Allan Hills) are confined to Member C of the Lashly Formation. The exposed contact rocks range from the lower Feather Formation, through the Fleming Member and into the basal part of Lashly Member A. Lashly Member C is at least 100 m stratigraphically and paleotopographically higher than any extant contact strata. Thus, the field relations are interpreted to suggest a large-scale collapse structure with formation of the megaclast unit by foundering of large intact blocks of the Lashly Formation (Fig. 6.1-5). The megaclast unit also includes yellow sandstone blocks which could be from either Member B or Member D; if the latter, then the wall of the collapse structure was at least 240 m high.

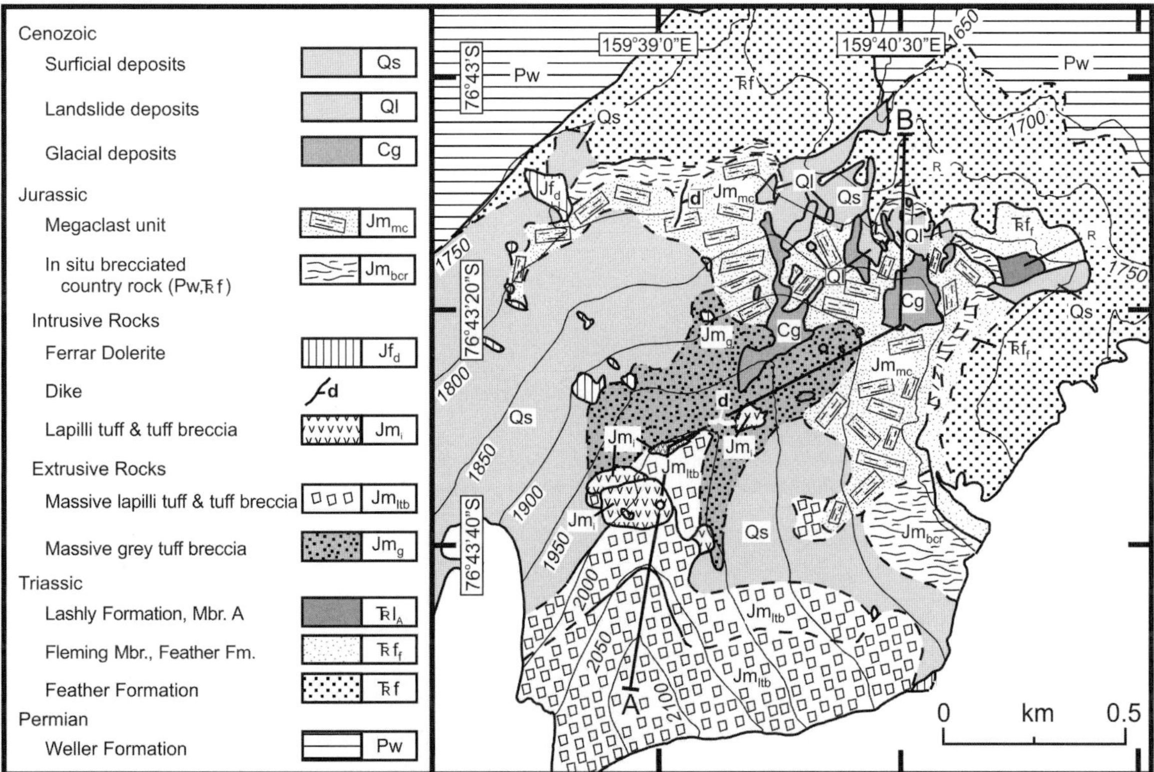

Fig. 6.1-4. Simplified geologic map of the area outlined in Fig. 6.1-3. *A–B:* line of section in Fig. 6.1-4

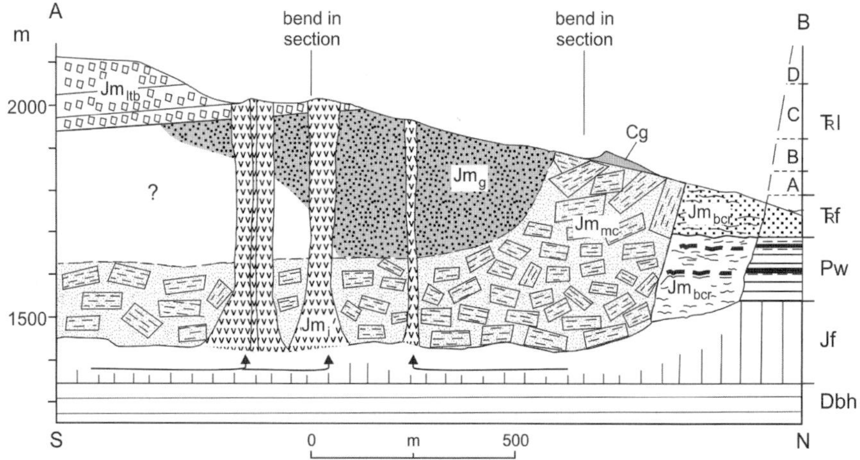

Fig. 6.1-5.
Schematic cross section for the geology outlined in Fig. 6.1-4. Rock symbols as in Fig. 6.1-4; *Dbh:* Beacon Heights Orthoquartzite

Fig. 6.1-6.
Contact relations between the Permian Weller Formation and the Mawson Formation on the west side of Feistmantel Valley (site A, Fig. 6.1-3). A basaltic diatreme cuts through tuff breccia and is in near vertical contact with Permian Weller Formation sandstones. *Figure* for scale

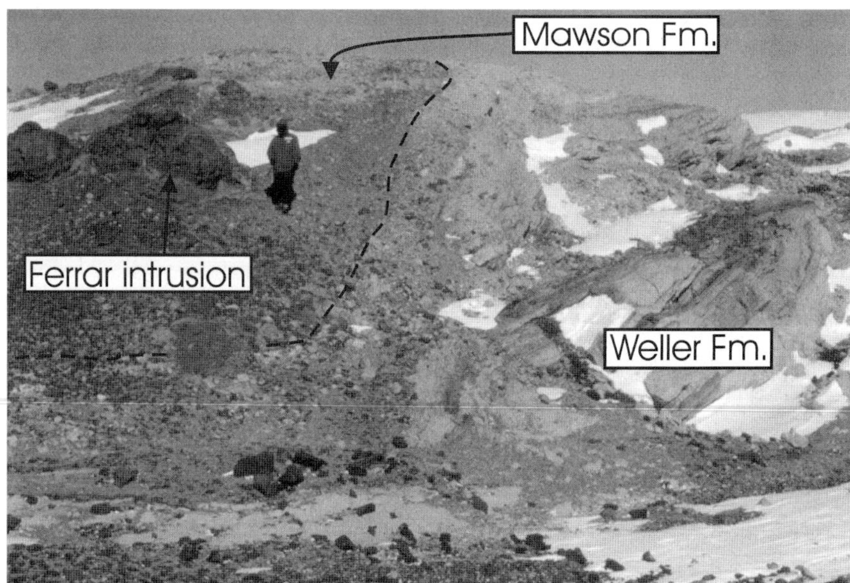

Fig. 6.1-7.
Mawson tuff-breccia intruding the Permian Weller Formation (contact marked by *dashed line*) and cut by small Ferrar dolerite plugs (site B, Fig. 6.1-3). Note the varied attitudes of the disturbed Weller strata and the zone of brecciation adjacent to the contact. *Figure* for scale

West of Feistmantel Valley (site A, Fig. 6.1-3), Mawson rocks are in contact with Permian Weller Formation strata (Fig. 6.1-6). The trace of the contact curves across the terrain; at its southeast end it hooks around to link up with the region just described. To the northwest, and adjacent to the contact (site B, Fig. 6.1-3), Weller strata are cut by intrusive bodies of tuff breccia (Fig. 6.1-7). The actual contact is marked by disturbed Weller strata which are tilted and locally folded, and also display *in situ* disaggregation and stretching of more cohesive beds. Where visible, the actual contact has a dip varying from moderate (ca. 45°) to near vertical. Mawson tuff breccia adjacent to the contact is cut by small (up to 20 m), near circular basaltic diatremes and by tuff-breccia intrusions. Screens of sandstone breccia separate outcrop areas of the tuff breccia. Megaclasts of Beacon strata also occur within the massive tuff breccia.

Coombs Hills

Along the eastern flank of the Mawson outcrop on Coombs Hills, the contact with the Beacon has been observed at only two localities. At one (site C, Fig. 6.1-3), first noted in Elliot and Hanson (2001), the field relations are complex and in part suggest a stratigraphic relationship. Beacon country rocks consist of sandstones that are stratigraphically about 100 m above the base of Member D of the Lashly Formation. The Lashly sandstones at the contact are disturbed, displaying gentle folding and locally disaggregation with loss of bedding, and are zeolitized. In places, the contact is a planar surface, parallel to bedding and dipping at 12–15° westward beneath the Mawson rocks (Fig. 6.1-8). In outcrops along strike to the north of the Fig. 6.1-8 viewpoint, the zeolitized sandstones pass up into a zone consisting of layers and blocks of disaggregated sandstone, zeolitized sandstone, bedded sandstone, basaltic tuff and tuff breccia with a high sand content, and silicic tuff (Fig. 6.1-9). Some sandstones contain sparse basalt pyroclasts, indicating mixing of Mawson material and loose sediment. Other sandstones are grey and carbonaceous, and similar to Lashly Member C. The structureless nature of sandstones at the contact resulted from either surface weathering processes or sub-surface phreatic activity; the latter is supported by the occurrence of phreatic pipes cutting Lashly Member C strata to the east on Coombs Hills. The contact zone is interpreted as the result of juxtaposition of lenses of pyroclastic rocks and sandstone, and intermingling of pyroclastic debris with disaggregated sandstone. The contact zone was formed either by surface processes in which Mawson lahars moved over an erosion surface, entrained clasts, and mixed with loose sediment or by intrusive processes in which mixing occurred in zones parallel to bedding. The silicic tuff clasts, which occur in the contact zone, must have been derived from pre-existing silicic tuff beds, whereas basaltic tuff breccia was contemporaneous. The contact zone is cut by sandstone dikes which also cut the overlying massive tuff breccias. On balance, the variable nature of the contact zone suggests it is more likely to be intrusive and bedding parallel.

At the other locality (site D, Fig. 6.1-3) the trace of the contact on topography indicates it is steep (Fig. 6.1-10). Where the actual contact is exposed, it has been intruded by a Ferrar dolerite dike. The contact rocks belong to Member D of the Lashly Formation. Member D was formally described at Mt. Bastion, south Victoria Land, where it consists of 210+ m of quartzose sandstone but lacks an exposed upper contact (Barrett and Webb 1973). Other described sections in that report have lesser thicknesses of Member D and again no upper contact; at Allan Hills

Fig. 6.1-8. Contact relations between the Mawson Formation and Lashly Formation sandstones on the east flank of Coombs Hills (site C, Fig. 6.1-3). In the foreground, the Mawson tuff breccia overlies a bedding-parallel surface on zeolitized sandstone. In the distance to the right, Mawson tuff breccia overlies a pale-colored contact zone (Fig. 6.1-9) of mixed sandstones, basaltic tuff bodies and silicic tuff clasts. *Figures* for scale

Fig. 6.1-9.
Contact zone between Lashly Formation sandstones and Mawson tuff-breccia (site C, Fig. 6.1-3). The contact zone comprises irregular layers and bodies of sandstone and mixed sand and pyroclastic debris. *1:* Lashly Fm. sandstone with rip-up clasts at base; *2:* sandstone with very sparse pyroclasts, but with a concentration of coarse basalt clasts at the base; *3:* sandstone with basaltic pyroclasts (ash), locally enclosing accretionary lapilli, and wedging out to the south (left in photo); bed cut by small sandstone dikes; *4:* tuff breccia with 90 cm amygdaloidal basalt clast; pinches out to S (left) below a lens of sandstone with basaltic clasts (*5*). *Ice axe* for scale

Fig. 6.1-10.
Intrusive contact between the Mawson Formation and upper beds of Member D of the Lashly Formation on the east flank of Coombs Hills (site D, Fig. 6.1-3). Mawson rocks are massive to weakly stratified. *Ice axe* for scale (5 cm)

only 85 m remain (Collinson et al. 1973). The Lashly Member D at Coombs Hills is 250+ m thick; it is similar in so far as the lower 200 m consist principally of quartzose sandstone, but the poorly exposed upper ca. 50 m consist of finer grained beds which include tuffaceous sandstones with silicic glass shards.

Discussion

The contact between the Mawson rocks and Beacon strata at Allan Hills is demonstrably intrusive, in contrast to previous interpretations of an unconformity with as much as 500 m of paleorelief (Ballance and Watters 1971). This suggested a significant period of erosion, presumably consequent on Early Jurassic regional uplift following the cessation of Beacon deposition (e.g., Elliot 2000). Although the interpretation of the contact as intrusive provides no support for the earlier interpretation of significant paleotopography, evidence from the Beardmore Glacier region has suggested a regional disconformity below the correlative Prebble Formation (Elliot 1996). Further, the petrology of the underlying Lower Jurassic Hanson Formation together with the associated structures (monoclines) suggested vertical tectonism (block faulting) and a rift setting (Elliot and Larsen 1993; Elliot 1996). A rift setting remains the most plausible paleoenvironment for accumulation of the Jurassic extrusive rocks, which include exceptionally thick (more than 100 m) ponded lavas as well as a thick (= 100 m) pillow-palagonite complex (Wörner 1992; Elliot 2002).

The uppermost beds in Member D of the Lashly Formation at Coombs Hills confirm prior inferences, from clasts in the Mawson Formation, of an episode of silicic volcanism that predated basaltic magmatism (Elliot and Hanson 2001). This sequence of shard-bearing sandstones must have originally graded up into silicic tuffs now known only from clasts in the Mawson Formation. Further, if the White and McClintock (2001) model for the Mawson rocks at Coombs Hills is correct, then the thickness of section above the uppermost Lashly D would have been about 150 m, and the strata might have been equivalent to the volcaniclastic Hanson Formation in the Queen Alexandra Range (Elliot 1996). Sandstones carrying silicic shards are also known from the Mesa Range region (Elliot et al. 1986).

Relationships between the Prebble Formation and Beacon rocks in the Queen Alexandra Range (Fig. 6.1-1) are stratigraphic (Elliot 1996; Hanson and Elliot 1996). In the Prince Albert Mountains, although no contact has been found with Beacon strata, the Mawson Formation (Elliot 2002) appears to consist mainly of extra-vent deposits, but with one occurrence of a tuff-breccia body cross cutting another massive tuff breccia. Stratigraphic relations also occur in the Mesa Range, north Victoria Land (Elliot et al. 1986). However, no continuous sequence from Beacon strata to the Kirkpatrick Basalt is exposed, and the only contact observed between sedimentary and pyroclastic rocks is too limited laterally and stratigraphically to be definitive. In the Deep Freeze Range (Fig. 6.1-1) the contact, at least locally, is stratigraphic (Elliot unpublished data). These occurrences in north Victoria Land can be interpreted as extra-vent deposits.

Reinterpretation of the Allan Hills contact, together with the Coombs Hills rocks, provides an alternative explanation for the exceptionally thick tuff breccias of the Prebble Formation at Otway Massif. There, more than 360 m of unbedded tuff breccia underlies Kirkpatrick Basalt lava flows (Elliot and Hanson 2001). Finer grained stratified beds, including base-surge and air fall deposits, are present beneath the basalts, and locally include a paleosol at the contact. On the east flank, Triassic sandstones crop out adjacent to tuff breccia but not in exposed contact. The contact was interpreted as a fault; this field relationship can now be reinterpreted to suggest another large collapse feature. Within this collapse structure, on western Otway Massif, there are disoriented sedimentary megaclasts, as much as 60 m long, enclosed in poorly exposed pyroclastic rocks. Like the megaclast unit north of Watters Peak, they represent down-dropped Triassic strata. On eastern Otway Massif, there is a small isolated outcrop of deformed and folded Dicroidium-bearing Triassic beds, surrounded by surficial debris, which is cut by faults and small lapilli-tuff intrusions. This isolated outcrop may represent either the floor of the postulated collapse structure or a displaced block.

The volcanic breccia deposits at Otway Massif, Coombs Hills and Allan Hills all appear to be the result of collapse of country rock. Collapse features of the dimensions of these, approximately 50–100 km^2 in area each, have not been reported for phreatomagmatic centers of basaltic composition, although they are small compared to silicic calderas. Thus their formation requires some special conditions. The exceptional vertical thicknesses of tuff breccia and lapilli tuff, as much as 370 m, and wide areal extent require large volumes of water for their creation by magma/water interaction. White and McClintock (2001) argued that intra-caldera recycling of earlier formed deposits contributed the water necessary for repeated magma-water interactions, because otherwise seemingly excessive amounts of groundwater would be required, amounts greater than might normally be expected in sedimentary strata without large cavernous voids. Relevant to this is the occurrence of lacustrine beds at the base and within the basalt sequence (Elliot et al. 1991), which indicate periods of internal drainage. Allowing for the uncertainties associated with the volumes of water required, the rift setting inferred for the Hanson Formation in the Beardmore Glacier region (Elliot and Larsen 1993) provides a setting in which abundant groundwater might accumulate, particularly if the rift as a whole was segmented into several topographic lows (Elliot and Fleming 2004), as is the case in other rift valleys such as those in East Africa (Rosendahl 1987).

Conclusions

Intrusive relationships at Allan Hills between pyroclastic rocks of the Mawson Formation and Beacon strata make untenable the inference of a significant period of Early Jurassic erosion and associated development of paleotopography. However, the precise stratigraphic relationships in south Victoria Land between Beacon strata and Ferrar extrusive rocks remain uncertain.

The recognition of the intrusive contacts at Allan and Combs Hills allows reinterpretation of the relationships between the Prebble Formation and Beacon strata at Otway Massif. There, the poorly exposed Prebble rocks also may represent a collapse structure.

Acknowledgments

The authors thank James Elliot and Tim Culley for able assistance in the field. This research has been supported by NSF grant OPP 0087919 to DHE. We thank Peter Barrett and Lothar Viereck-Goette for constructive reviews. Byrd Polar Research Center contribution number 1303.

References

Ballance PF (1977) The Beacon Supergroup in the Allan Hills, central Victoria Land, Antarctica. NZ J Geol Geophys 20:1003–1016

Ballance PF, Watters WA (1971) The Mawson Diamictite and the Carapace Sandstone Formations of the Ferrar Group at Allan Hills and Carapace Nunatak, Victoria Land, Antarctica. NZ J Geol Geophys 14:512–217

Barrett PJ, Webb, PN (1973) Stratigraphic sections of the Beacon Supergroup (Devonian and older (?) To Jurassic) in SVL. Antarctic Data Ser no 3, Victoria University of Wellington, New Zealand

Borns HW, Hall BA (1969) Mawson "Tillite" in Antarctica: Preliminary report of a volcanic deposit of Jurassic Age. Science 166:870–872

Bradshaw MA (1987) Additional field interpretation of the Jurassic sequence at carapace Nunatak and Coombs Hills, south Victoria Land, Antarctica. NZ J Geol Geophys 30:37–49

Collinson JW, Pennington DC, Kemp NR (1983) Sedimentary petrology of Permian-Triassic fluvial rocks in the Allan Hills, central Victoria Land. Antarctic J US 18(5):20–22

Elliot DH (1996) The Hanson Formation: a new stratigraphical unit in Transantarctic Mountains, Antarctica. Antarctic Sci 8:389–394

Elliot DH (2000) Stratigraphy of Jurassic pyroclastic rocks in the Transantarctic Mountains. J African Earth Sci 31:77–89

Elliot DH (2002) Paleovolcanological setting of the Mawson Formation: evidence from the Prince Albert Mountains, Victoria Land. In: Gamble JA, Skinner DNB, Henrys SA (eds) Antarctica at the close of a millenium. Roy Soc NZ Bull 35:185–192

Elliot DH, Fleming TH (2004) Occurrence and dispersal of magmas in the Jurassic Ferrar Large Igneous Province, Antarctica. Gondwana Res 7:225–239

Elliot DH, Hanson RE (2001) Origin of widespread, exceptionally thick basaltic phreatomagmatic tuff breccia in the Middle Jurassic Prebble and Mawson formations, Antarctica. J Volcanol Geotherm Res 110:183–201

Elliot DH, Larsen D (1993) Mesozoic volcanism in the Transantarctic Mountains: depositional environment and tectonic setting. In: Findlay RH, Unrug R, Banks MR, Veevers JJ (eds) Gondwana 8: Assembly, evolution and dispersal. AA Balkema, Rotterdam, pp 397–414

Elliot DH, Haban MA, Siders MA (1986) The Exposure Hill Formation, Mesa Range. In: Stump E (ed) Geological investigations in northern Victoria Land. Amer Geophys Union, Antarct Res Series 46, Washington DC, pp 267–278

Elliot DH, Bigham J, Jones FS (1991) Interbeds and weathering profiles in the Jurassic basalt sequence, Beardmore Glacier region, Antarctica. In: Ulbrich H, Rocha Campos (eds) Proceedings of the Seventh Gondwana Symposium. Univers Sao Paulo, Sao Paulo, pp 289–302

Grapes RH, Reid DL, McPherson JG (1974) Shallow dolerite intrusion and phreatic eruption in the Allan Hills region, Antarctica. NZ J Geol Geophys 17:563–577

Hall BA, Sutter JF, Borns HW (1982) The inception and duration of Mesozoic volcanism in the Allan Hills-Carapace Nunatak area, Victoria Land, Antarctica. In: Craddock C (ed) Antarctic geoscience. Univers Wisconsin Press, Madison, pp 709–713

Hanson RE, Elliot DH (1996) Rift-related Jurassic basaltic phreatomagmatic volcanism in the central Transantarctic Mountains: precursory stage to flood-basalt effusion. Bull Volcanol 58: 327–347

Pocknall DT, Chinn TJ, Sykes T, Skinner DNB (1994) Geology of the Convoy Range area, southern Victoria Land, Antarctica, Scale 1:50000. Inst Geol Nuclear Sci, Geological Map 1

Rosendahl BR (1987) Architecture of continental rifts with special reference to East Africa. Ann Rev Earth Planet Sci 15:445–503

White JDL, McClintock MK (2001) Immense vent complex marks flood-basalt eruption in a wet, failed rift: Coombs Hills Antarctica. Geology 29:935–938

Wörner G (1992) Kirkpatrick lavas, Exposure Hill Formation and Ferrar sills in the Prince Albert Mountains, Victoria Land, Antarctica. Polarforschung 60(2):87–90

What Supports the Marie Byrd Land Dome? An Evaluation of Potential Uplift Mechanisms in a Continental Rift System

Wesley E. LeMasurier
Institute of Arctic and Alpine Research (INSTAAR), University of Colorado, Boulder, CO 80309-0450, USA, <Wesley.LeMasurier@colorado.edu>

Abstract. Volcanism in the Marie Byrd Land (MBL) volcanic province is related to the growth of an 1 200 × 500 km structural dome that lies on the Amundsen Sea coast. Spatial and temporal patterns of volcanic activity suggest that dome uplift began around 29–25 Ma and has continued to the present. Uplift has been accompanied by sometimes voluminous basaltic and felsic volcanism and the development of horst and graben structure, with a maximum of ~3 km of uplift. Estimates of crustal thickness, based on models of gravity data and surface wave dispersion studies, have not resolved questions about the origin of uplift. Mantle plume activity has been proposed; but more detailed tomographic imaging of the mantle, together with seismic determinations of crustal thickness, and the thickness and distribution of sub-ice volcanic rock, are needed to test this, and to answer other petrologic and tectonic questions discussed below.

Introduction

The Marie Byrd Land (MBL) dome is a tectonically elevated portion of the West Antarctic rift system that has been volcanically active since early Oligocene time. The rift system covers an area of ~3 000 × 1 000 km, with an average crustal thickness of ~27–30 km (von Frese et al. 1999; Ritzwoller et al. 2001). With the major exception of the MBL dome, most of the bedrock surface of the rift lies ~500–1 000 m below sea level, after isostatic adjustment to pre-glacial levels (Drewry 1983). It now seems clear that the dome is not a rift shoulder, as originally portrayed (LeMasurier 1990). Its elevation appears to be the result of at least 25 myr of uplift within the region of attenuated crust that defines the rift. In contrast to the transition across the Transantarctic Mountains (TAM), where the crust thickens abruptly to ~40–45 km, the thickness of crust does not appear to change northward from the TAM, across MBL to the Amundsen Sea, implying essentially the same degree of crustal attenuation from the TAM front to the sea. Furthermore, block faulting and alkaline volcanism, both distinguishing features of the rift, are obvious characteristics of the MBL dome. Within the limitations of the available data, uplift of the dome cannot be ascribed to greater crustal thickness, and there is no obvious plate tectonic mechanism for generating uplift and volcanism in this region in Neogene time. On the other hand, the available evidence from seismic tomography seems ambiguous with respect to the existence of a deep mantle plume (cf. Ritzwoller et al. 2001; Sieminski et al. 2003). The purpose of this paper is to provide an overview of geologic and geophysical characteristics of the dome, evaluate mechanisms that might have produced uplift and volcanism, and suggest where more data might provide new insights.

Regional Tectonic Setting

The first-order characteristic of Cenozoic tectonics in the rift system is a stationary plate environment; but the amount of Cenozoic extension has been controversial, and evidence for it is found mainly along the western margin of the Ross Sea. Plate reconstruction studies support plate immobility since ~85 Ma, and suggest <50 km of extension in the Ross Sea embayment since ~85 Ma (Lawver et al. 1991; Lawver and Gahagan 1994). However, Cande et al. (2000) present evidence for ~180 km of extension across the Adare trough (Fig. 6.2-1), between 43–26 Ma, which they suggest represents eastward motion of West Antarctica with respect to East Antarctica. A recent global plate motion study by Steinberger et al. (2004) supports East-West Antarctica motion, but places the time of motion before 47 Ma. Cande et al. (2000) also review evidence for relatively large scale Cenozoic extension in the western Ross Sea and TAM, but there is no evidence as yet for significant Cenozoic extension in the interior of MBL, over 1 000 km to the east. Plate reconstruction studies, marine geophysics, and New Zealand geology all seem to agree that maximum extension and crustal thinning in the rift system preceded and accompanied breakup of the New Zealand-Campbell Plateau block from MBL, around 100–83 Ma (Lawver et al. 1991; LeMasurier and Landis 1996; Larter et al. 2002). In MBL, plate immobility is supported by the absence of hot spot tracks over the past 25 myr, but the amount of Cenozoic extension, while constrained only by plate reconstruction studies, seems much less than in the western Ross Sea region.

From: Fütterer DK, Damaske D, Kleinschmidt G, Miller H, Tessensohn F (eds) (2006) Antarctica: Contributions to global earth sciences. Springer-Verlag, Berlin Heidelberg New York, pp 299–302

Marie Byrd Land Dome

Structure of the Dome

Dome structure is defined by elevations of the West Antarctic erosion surface (WAES) in fault block exposures (LeMasurier and Landis 1996). The WAES has <50 m of relief, where it was surveyed over >30 km of nearly continuous exposure in the Jones Mountains (Craddock et al. 1964). It occurs as flat summits of basement nunataks throughout coastal West Antarctica, commonly truncating ~90 Ma granitoids. Exposures rise from 400–600 m elevations on the coast to 2 700 m at Mt. Petras, on the dome crest (Fig. 6.2-1). Horst and graben structure is manifested by large vertical offsets of the WAES, by basin and range topography, by sharply rectilinear N-S and E-W boundaries of basement nunataks, and by pronounced N-S and E-W alignments of linear volcanic ranges across the entire dome (LeMasurier and Landis 1996). These suggest a simple picture of block-faulted domal uplift over an 1 200 × 500 km area. The magnitude of fault offset cannot be determined precisely because the WAES is the only structural datum in the region, and its position on graben floors is not exposed. A 1 959 seismic traverse across Toney Mountain (Fig. 6.2-1) shows an "unambiguous" contact between basalt and basement rock (presumably the ubiquitous WAES) at ~3 km below sea level (Bentley and Clough 1972). Exposures of the WAES in the Kohler Range (Fig. 6.2-1), ~70 km to the north, are 1 000 m above sea level, suggesting an aggregate offset of ~4 km on intervening faults, and total structural relief from dome crest (2.7 km at Mt. Petras) to graben floors (–3 km at Toney Mountain) on the order of 6 km. There is no perceptible backward rotation of the WAES on any exposures, some having areas of ~30 km^2, suggesting that fault surfaces are planar rather than lystric, and therefore unlikely to represent large amounts of extension. The net surface uplift represented by the MBL dome is hard to estimate without knowing elevations of graben floors near the dome crest. If fault offsets near Mt. Petras are ~4 km (i.e. graben floors ~1 km below sea level), the average elevation of the WAES would be roughly +1 km, representing about 1.5 km surface uplift from the surrounding floor of the rift system. It will obviously require a comprehensive seismic program to produce the data needed for good structural and surface uplift interpretations.

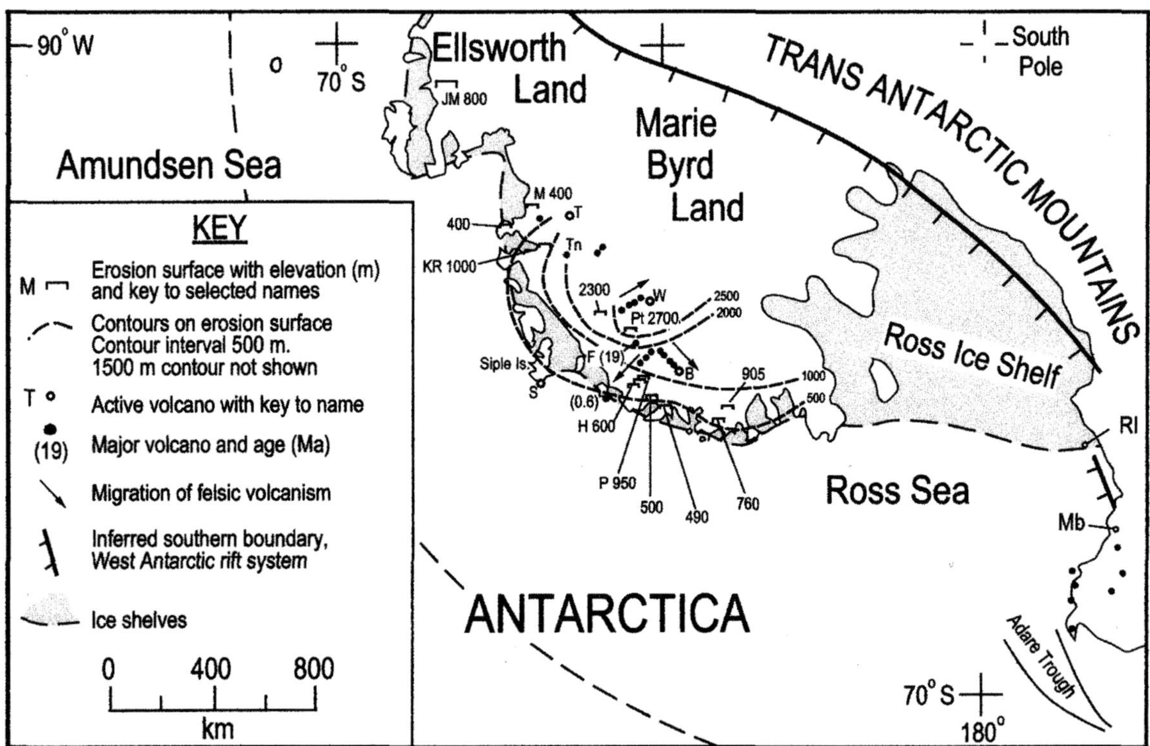

Fig. 6.2-1. Location map showing the Marie Byrd Land dome and volcanic province, and the boundary of the West Antarctic rift system. Elevations of the West Antarctic erosion surface on horst summits define the configuration of the dome, which is shown by structure contours on the erosion surface. The crest of the dome is at Mt. Petras (*Pt*). Felsic rocks in central-type volcanoes (*solid* and *open circles*) are oldest at Mt. Flint (*F*), near the crest, and become systematically younger toward the dome margin, as shown by *arrows*. See text for further explanation. Abbreviations: *B:* Mt. Berlin; *F:* Mt. Flint; *H:* Holmes Bluff; *JM:* Jones Mountains; *KR:* Kohler Range; *M:* Mt. Murphy; *Mb:* Mt. Melbourne; *P:* Patton Bluff; *Pt:* Mt. Petras; *RI:* Ross Island; *S:* Mt. Siple; *T:* Mt. Takahe; *Tn:* Toney Mountain; *W:* Mt. Waesche. Modified from LeMasurier and Landis (1996). Adare Trough location and configuration are from Cande et al. 2000

Timing of Uplift from Patterns of Volcanism

Two patterns strongly suggest that dome uplift and volcanism are interrelated, and provide a means of interpreting timing and rate of uplift.

1. The felsic pattern. Virtually all the felsic rocks in this region occur either as summit sections of large central-type volcanoes (open and closed circles in Fig. 6.2-1), or as the only rocks exposed in these volcanoes. Most of these volcanoes are aligned along N-S and E-W ranges that decrease in age in all directions, away from the crest and toward the flanks of the dome (LeMasurier and Rex 1989). Thus, 19.1 Ma Mt. Flint lies near the dome crest (Fig. 6.2-1), and the active volcanoes lie on its distal flanks. This pattern is believed to represent centrifugal extension of relict NS-EW fracture sets during dome growth, accompanied by the systematic release of felsic magma from crustal magma chambers (LeMasurier and Rex 1989). It appears to record uplift and associated volcanism from ~19 Ma to the present.
2. The basalt pattern. The ages of basaltic rocks that lie on the WAES increase with increasing elevation, e.g., 6.27 Ma basalt overlies the WAES at 600 m at Holmes Bluff, 9.97 Ma basalt overlies the WAES at 950 m at Patton Bluff, 25–29 Ma basaltic tuffs overlie the WAES at 2700 m at Mt. Petras, etc. (Fig. 6.2-1). This pattern is believed to represent extrusion during the initiation of fault offset, when the WAES stood at a low elevation, followed by the rise of each horst block in chronologic succession during dome growth, at a rate of ~105–122 m per myr (LeMasurier and Landis 1996). It appears to record contemporaneous volcanism and fault block uplift from late Oligocene time to ~6 Ma, and the two patterns together suggest contemporaneous volcanism and uplift from 25–29 Ma to the present.

Volumes of Volcanic Products

The volume of rock erupted during dome growth is difficult to estimate because of thick ice cover, and the near absence of geophysical data, but some examples suggest that the volume is large. At Toney Mountain, the top of the basalt section noted above is 2 km above sea level, or ~5 km above its seismic base. At Mt. Murphy, 1400 m of basalt plus ~800 m of trachyte are exposed. Mt. Siple (Fig. 6.2-1), is fully exposed (above ice) and has an estimated volume of ~1800 km^3 (LeMasurier and Thomson 1990), ~5 times the volume of Mt. Shasta, the largest volcano in the Cascade Range (Christiansen 1990).

Another way to view magma flux is to consider the volume of original magma represented by the large volume of felsic rock in MBL. The exposed volume of Mt. Takahe is ~780 km^3 (LeMasurier and Thomson 1990), and appears to be almost entirely trachyte. Petrologic modeling of felsic rocks in MBL indicates they were produced by fractional crystallization of basalt, with virtually no additions from crustal contamination (LeMasurier et al. 2003). Modeling results also suggest that these rocks represent the final 3–6% of the original basalt magma, which implies a roughly 17- to 33-fold enrichment of incompatible trace elements. Enrichments in this range are indeed observed in some comendites and phonolites for the elements Rb and Th, and to a lesser extent Nb and Zr (LeMasurier et al. 2003), suggesting that the 3–6% residual volume firgure is not unreasonable. Thus, just the exposed volume of Mt. Takahe probably represents ~20000 km^3 of basalt. However, the ice thickness around this volcano is >2 km (Drewry 1983), suggesting that the concealed volume may be much greater than the exposed volume. These crude estimates at least provide a basis for suspecting that total magma production over the past ~25–30 myr could have enhanced crustal thickness. Once again, a good seismic program could provide data for much better volume estimates, and for determining (1) if dome uplift is related to crustal thickening due to volcanism and magmatic underplating, or (2) whether extension has compensated for the addition of new material.

Discussion

The fundamental problem of the MBL dome is to explain persistent uplift and voluminous volcanism that began 50–55 myr after the main period of crustal extension, during a time of no perceptible plate motion, in a region with no apparent crustal thickening. The case for buoyant uplift over low density mantle (rather than thickened crust) has been discussed by LeMasurier and Landis (1996). The question may be pursued further by exploring whether this uplift has an upper mantle origin, as recently proposed for the Yellowstone plume (Christiansen et al. 2002), or is the product of deep mantle plume activity. Huismans et al. (2001) present the results of modeling passive and active thinning of the lithosphere in intracontinental rift zones. They discuss examples where asthenospheric upwelling has produced postrift doming and volcanism from 10 myr to as much as 60 myr after rifting. To apply this mechanism to the MBL dome would seem to require an explanation of why uplift was focussed in this portion of the rift system and nowhere else. If additional seismic data could demonstrate that the crust and/or lithosphere was more extended beneath the dome than elsewhere, then asthenospheric upwelling could easily become a preferred model. On the other hand, a deep mantle plume is suggested by the work of Sieminski et al. (2003). They present a 3D SV-wave velocity model of the upper mantle which shows a vertical low velocity structure extending continuously down to transition zone depths beneath the Amundsen Sea and coastal MBL, which they cautiously interpret to be a plume. Certainly this structure extends far below the ~200 km depth identified by Ritzwoller et al. (2001) as the base of the asthenosphere, but the results of the latter study were ambiguous with respect to plume activity. Progress in further defining, or refuting, a deep mantle plume origin for the MBL dome will require more detailed imaging at greater depths, and this in turn will rely on improvements in surface wave tomographic techniques (Sieminski et al. 2003).

Acknowledgments

Field work in Marie Byrd Land and comparative studies have been supported by NSF grants DPP77-27546, DPP80-20836, and OPP9720411. The ideas expressed here have benefited from discussions with John Behrendt, Rob Larter, Anne Sieminski, Joann Stock, and Ruppert Sutherland, and from thoughtful reviews by Bryan Storey and Gerhard Wörner, all of which are gratefully acknowledged.

References

Bentley CR, Clough JW (1972) Antarctic subglacial structure from seismic refraction measurements. In: Adie RJ (ed) Antarctic geology and geophysics. Universitetsforlaget, Oslo, pp 683–691

Cande SC, Stock JM, Müller RD, Ishihara T (2000) Cenozoic motion between East and West Antarctica. Nature 404:145–150

Christiansen RL (1990) Shasta. In: Wood CA, Kienle J (eds) Volcanoes of North America: United States and Canada. University Press, Cambridge, pp 214–216

Christiansen RL, Foulger GR, Evans JR (2002) Upper-mantle origin of the Yellowstone hotspot. Geol Soc Amer Bull 114:1245–1256

Craddock C, Bastien TW, Rutford RH (1964) Geology of the Jones Mountains area. In: Adie RJ (ed) Antarctic geology. North-Holland, Amsterdam, pp 171–187

Drewry DJ (1983) Antarctica: glaciological and geophysical folio. Scott Polar Research Institute, University of Cambridge, 9 sheets

Huismans RS, Podladchikov YY, Cloetingh S (2001) Transition from passive to active rifting: relative importance of asthenospheric doming and passive extension of the lithosphere. J Geophys Res 106:11271–11291

Larter RD, Cunningham AP, Barker PF, Gohl K, Nitsche FO (2002) Tectonic evolution of the Pacific margin of Antarctica, 1. Late Cretaceous tectonic reconstructions. J Geophys Res 107(12):2345

Lawver LA, Gahagan LM (1994) Constraints on timing of extension in the Ross Sea region. Terra Antartica 1:545–552

Lawver LA, Royer J-Y, Sandwell DT, Scotese CR (1991) Evolution of the Antarctic continental margins. In: Thomson MRA, Crame JA, Thomson JW (eds) Geological evolution of Antarctica. Cambridge University Press, Cambridge, pp 533–539

LeMasurier WE (1990) Late Cenozoic volcanism on the Antarctic plate: an overview. In: LeMasurier WE, Thomson JW (eds) Volcanoes of the Antarctic plate and southern oceans. Amer Geophys Union Antarctic Res Ser 48, pp 1–17

LeMasurier WE, Landis CA (1996) Mantle-plume activity recorded by low-relief erosion surfaces in West Antarctica and New Zealand. Geol Soc Amer Bull 108:1450–1466

LeMasurier WE, Futa K, Hole J, Kawachi Y (2003) Polybaric evolution of phonolite, trachyte and rhyolite volcanoes in eastern Marie Byrd Land, Antarctica: controls on peralkalinity and silica saturation. Internat Geol Rev 45:1055–1099

LeMasurier WE, Thomson JW (eds) (1990) Volcanoes of the Antarctic plate and southern oceans. Amer Geophys Union Antarctic Res Ser 48, pp 487

LeMasurier WE, Rex DC (1989) Evolution of linear volcanic ranges in Marie Byrd Land, West Antarctica. J Geophys Res 94:7223–7236

Ritzwoller MH, Shapiro N, Levshin AL, Leahy GM (2001) Crustal and upper mantle structure beneath Antarctica and surrounding oceans. J Geophys Res 106:30645–30670

Sieminski A, Debayle E, Lévêque JJ (2003) Seismic evidence for deep low-velocity anomalies in the transition zone beneath West Antarctica. Earth Planet Sci Letters 216:645–661

Steinberger B, Sutherland R, O'Connell RJ (2004) Prediction of Emperor-Hawaii seamount locations from a revised model of global plate motion and mantle flow. Nature 430:167–173

von Frese R, Tan L, Kim JW, Bentley CR (1999) Antarctic crustal modeling from the spectral correlation of free-air gravity anomalies with the terrain. J Geophys Res 104:25275–2529

A Multi-Phase Rifting Model for the Victoria Land Basin, Western Ross Sea

Fred J. Davey[1] · **Laura De Santis**[2]

[1] Institute of Geological and Nuclear Sciences, Lower Hutt, New Zealand, <F.davey@gns.cri.nz>

[2] Instituto Nazionale di Oceanografia e di Geofisica Sperimentale, Trieste, Italy, <ldesantis@OGS.trieste.it>

Abstract. The Victoria Land Basin is a deep rift margin basin that lies along the Ross Sea margin of the Transantarctic Mountains, Antarctica. Seismic data indicates that the 140 km wide basin contains some 14 km of sediments. Drilling results from the flank of the basin suggests that most of these sediments are late Eocene or younger in age. Four major sequences are defined on the seismic data by angular unconformities indicating several rift episodes. Constraints on the basin formation are derived using flexural cantilever models. The model reproduces the main features (shape, size, timing and broad stratigraphy) of the basin formation moderately well using four rift episodes on five main faults, but does not match the detailed geometry of the seismic stratigraphy.

Introduction

The Transantarctic Mountains (TAM) are a major morphological feature of Antarctica, stretching for about 3 500 km across Antarctica, and with elevations of over 4 500 m (Fig. 6.3-1). They form the major geological boundary in Antarctica, considered to be a rift margin that separates the older, Precambrian craton of East Antarctica from the younger, more geologically active West Antarctica (e.g., Tessensohn and Wörner 1991). This major discontinuity is also reflected by a sharp change in the upper mantle velocity structure, which shows a high seismic velocity upper mantle under East Antarctica and a low velocity, inferred warmer, mantle under West Antarctica (Danesi and Morelli 2000). The Victoria Land Basin (VLB), a deep rift basin, lies along a 300 km long segment of the TAM in southwest Ross Sea (Fig. 6.3-1). The VLB contains up to 14 km of sediments, which were initially considered to comprise two main units approximately equal in thickness, one unit inferred to be of late Cretaceous age and the other of late Cenozoic age (Cooper et al. 1987). Recent drilling on the west flank of the TAM, however, suggests that the age of most of the sediments is late Eocene or younger (Wilson et al. 1998; Florindo et al. 2001). Reflection seismic data across the southern part of the VLB (Fig. 6.3-2a) have been interpreted to show that the section consists of four major sedimentary packages separated by distinct angular unconformities (Fig. 6.3-2b) (De Santis et al. 2001), indicating several episodes of extension.

Several authors have attempted to model the broad features of the rift margin using simple flexural cantilever models (e.g., Stern and ten Brink 1989) and pure shear (e.g., van der Beck et al. 1994; Busetti et al. 1999). However, poor age control has limited the models. In addition, the relatively well defined geometry of the seismic stratigraphy has not been used to constrain rift models. The recent Cape Roberts project (CRST 2000; Florindo et al. 2001) and re-evaluation of the CIROS core in western McMurdo Sound (Wilson et al. 1998) provides significantly more accurate age control for the formation of the basin. In this paper we present initial results of using the flexural cantilever model (Kuznir and Ziegler

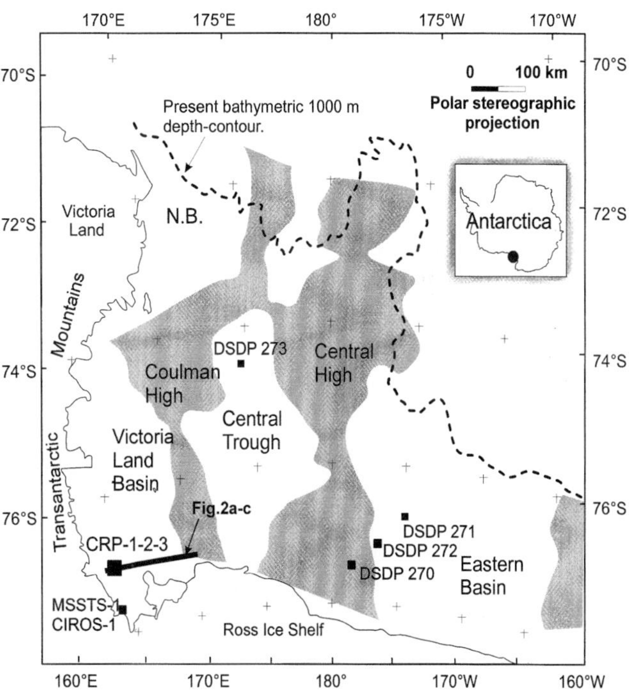

Fig. 6.3-1. Ross Sea, Antarctica, showing the major basins underlying the Ross Sea (*N.B.* is the Northern Basin) and the location of the seismic and modelled profile (*thick line*) across the southern Victoria Land Basin. Deep Sea Drilling Project (*DSDP*) (Hayes and Frakes 1975), Cape Roberts Project (*CRP*) (CRST 2000), *CIROS* (Barrett 1989) and *MSSTS* (Barrett 1986) drill sites marked by *solid squares*. *Inset* shows the location of the figure within Antarctica

From: Fütterer DK, Damaske D, Kleinschmidt G, Miller H, Tessensohn F (eds) (2006) Antarctica: Contributions to global earth sciences. Springer-Verlag, Berlin Heidelberg New York, pp 303–308

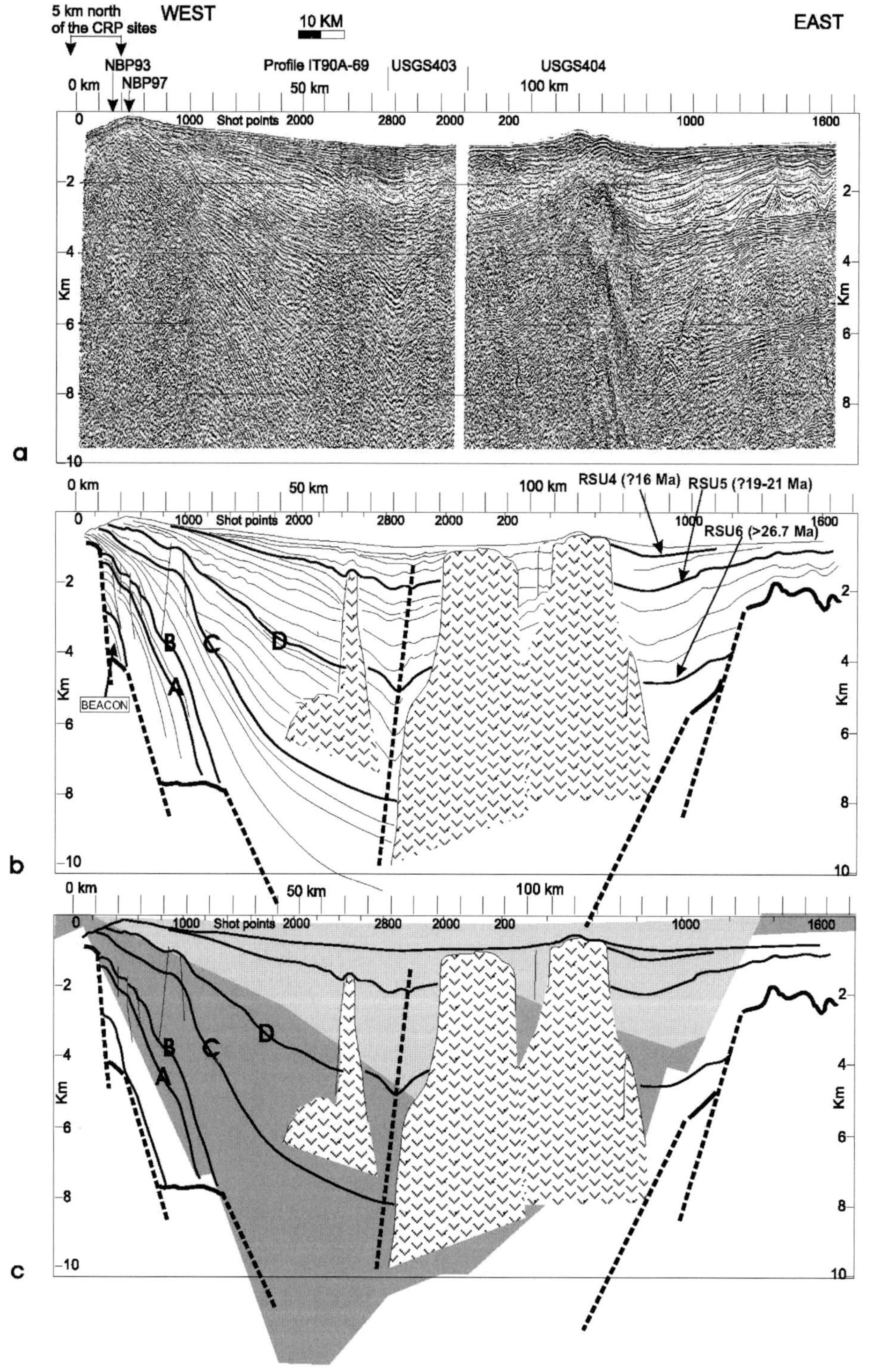

Fig. 6.3-2.
a Depth-converted composite seismic profile (IT69-USGS403/404) across the southern Victoria Land Basin along the line shown in Fig. 6.3-1. Note the profile has been limited to a depth of 9.5 km. **b** Interpretation of the seismic profile. *Solid lines* marked *A–D* are the unconformities related to the proposed rifting. *A:* 34 Ma, *B:* 32.5 Ma, *C:* 29 Ma, *D:* 24 Ma (after De Santis et al. 2001; and based on the Cape Roberts drill-holes). In eastern VLB the *RSU* series of unconformities (Brancolini et al. 1995) are marked. *Dashed lines* mark the major basin forming faults inferred from the seismic data. *V shading* delineates inferred igneous bodies. **c** The final stage (*shaded*) of the rift model for the Victoria Land Basin (Fig. 6.3-3) superimposed on the interpreted seismic profile shown in **b** above. *Light shading* delineates sediments associated with the last rifting event, dark shading delineates sediments from the earlier rifting phases

1992) to reproduce this seismic stratigraphy (Fig. 6.3-2) and provide constraints on the rifting history of the VLB. Simple models of multiple extension rifting are in broad agreement with the basin dimensions, broad stratigraphy and crustal thickness but it has not been possible to reproduce the detailed stratigraphy.

Method

We used the rift modeling program STRETCH (Badley Earth Sciences (Computing) Ltd.) that develops models of a rifted lithosphere for a range of physical variables, e.g., brittle depth, rheology, planar or listric fault, and erosion level. It uses the flexural cantilever model and was developed for rifted continental shelves (Kuznir et al. 1991). However, it does not allow a variable rheology or a variable degree of erosion along a model (as may be expected to occur from East to West Antarctica). Variations in these two parameters would be important for a comprehensive model of the rifting along this boundary since the initial break-up of Gondwana in Mesozoic times, and particularly since uplift of the TAM commenced in the Cretaceous (Fitzgerald 1994). We note however, that the VLB, the major rift margin basin, lies only along a limited part of the TAM rift margin, and that it has a much younger age (late Eocene). These latter two factors can be interpreted as indicating *(i)* an early rift phase, probably associated with the break-up of New Zealand and Australia from Antarctica (100 Ma approx., Davey and Brancolini 1995), forming the Antarctic wide Transantarctic Mountains rift margin and the associated thinning of the West Antarctic-Ross Embayment, and *(ii)* a later, localized, rift that formed the VLB, probably commenced at about 34–37 Ma (Wilson et al. 1998; CRST 2000; Florindo et al. 2001), and formed mainly in the thinned Ross Embayment crust. The pre-existing rift margin along the TAM-Ross Sea boundary may have acted as a zone of weakness for this later rifting, i.e. the initial major east dipping fault on the western margin of VLB. However, the limited along strike extent of the VLB with no major changes in the morphology of the adjacent TAM at its north and south extremities, indicates that for preliminary modeling studies the rifting that formed the VLB can be modeled separately from that forming the TAM uplift. Furthermore, a local (southwest Ross Sea) strike-slip environment has been proposed for causing the VLB (Wilson 1995; Salvini et al. 1997; ten Brink et al. 1997). We have therefore modeled the VLB as a rift occurring in lithosphere of constant rheology since 34 Ma (based on the Cape Roberts drill core results as our profile passes through that location). We use 2D modeling software and do not address the issue of strike-slip tectonics. As noted above, the proposed rifting will have affected the adjacent TAM, but a more comprehensive model is needed to quantify this. However, the simple modeling shown here provides constraints on the amount and timing of rift faulting that formed the VLB and its associated crustal thinning.

Input Parameters and Assumptions

A lithosphere of constant thickness and rheology is assumed as a starting condition for the model. The initial thickness and rheology are parameters that were tested during the modeling. Pre-rift topography was assumed to be at sea level with a preexisting sedimentary section 1 km thick. Syn-rift and post-rift sediments were assumed to be a 40:30:30% sand-silt-clay mix (e.g., Barrett and Anderson 2000). Four separate phases of extension, on one or more of five main faults, were inferred to occur instantaneously, at 34 Ma, 32.5 Ma, 29 Ma, and 24 Ma based on CRP drilling results and the seismic stratigraphy. The seismic stratigraphy indicates that a younger (less than 17 Ma) extensional event may have occurred but we have no age control for it and it has not been considered. Erosion and thermal subsidence were modeled after each rift episode. The Central Trough was modeled as a single rift episode at 34 Ma to allow for some influence of this extension on the adjacent VLB model – it is not very critical since its effect is largely on the Moho morphology. Prograding sedimentation was assumed for the early sedimentation associated with the western brittle faulting when extension and subsidence was greatest, and horizontal basin in-fill for subsequent basin development related to the listric eastern faults.

Output Model

The final model, showing the stages of development of the VLB, is shown in Fig. 6.3-3a–e. A comparison of the final stage with the depth converted seismic section is shown in Fig. 6.3-2c.

The following variables were tested to provide the best fit (by inspection) of the output model (Table 6.3-1) to the interpreted seismic section:

1. location, dip and extension of faults (as derived from interpretation of the seismic section in Fig. 6.3-2b).
2. lithosphere effective elastic thickness (*Te*). A range from $Te = 3$ km to $Te = 50$ km was tested, and the best value to get the observed flexure of the broad (140 km) VLB was $Te = 10$ km. Assuming (1) above, $Te = 15$ km gave too broad a basin and Moho morphology, $Te = 5$ km increased the convex up geometry of the unconformities and would not fit the Moho morphology under the VLB.
3. brittle lithosphere layer. A range from 10 km to 35 km was tested. The preferred value was 15 km to get the closest Moho morphology to that derived from gravity modeling (Davey and Cooper 1987). For planar faults this value defines the depth to which brittle faulting is carried in the model. The software default value of 15 km is a common hypocentral depth for continental earthquakes. For listric faults this defines the depth at which the fault system detaches. Below this depth all deformation is plastic pure-shear for both fault styles. A value of <15 km made it difficult to fit the basin depth, a value = 20 km gave a more asymmetric morphology then that computed from gravity data for the Moho under the VLB.

4. pre-rift crustal thickness. A range from 40 km to 20 km was tested. The best fit value was 33 km. This was thicker than expected. Values <30 km were difficult to model to give the depth of the basin without making the depth to Moho much smaller than the gravity model, whereas depth values >35 km gave depths to Moho that were too deep.
5. amount and timing of erosion after each rift phase. Any topography above sealevel was assumed to erode to 500 m above sealevel after the first three extensional events, and to 200 m above sealevel after the last extension.

The four major rifting phases are as follows:

Phase A (Fig. 6.3-3a) 34 Ma

Initial major rifting phase, corresponding to the rapid subsidence observed at Cape Roberts drill site (CRST 2000; De Santis et al. 2001). A graben structure is needed to get the final basin depth. Planar faulting is required to produce a steep western dip of Moho. The rifting that formed the Central Trough is assumed to occur at this time. Erosion of all the crustal rocks higher than 0.5 km above the sea level is assumed to occur after this extensional phase. Thermal subsidence occurred until 32.5 Ma.

Phase B (Fig. 6.3-3b) 32.5 Ma

East dipping extension occurred on a single planar fault to give nearly the maximum depth of the basin. The total extension (phase A and B) is limited by minimum Moho depth modelled using gravity data. Erosion of all the crustal rocks higher than 0.5 km above the sea level is assumed to occur after this extensional phase. Thermal subsidence occurred until 29 Ma.

Phase C (Fig. 6.3-3c) 29 Ma

West dipping listric extension occurred along two faults on eastern margin of basin. The mid-basin fault is required to emulate the stratigraphy in western part of the basin and depress the sediments in the central part of the basin. A large fault is not detected in this location on the seismic section. However, small offset faulting of strata and volcanic intrusion in the seismic section suggest the occurrence of vertical discontinuities (Fig. 6.3-2). The volcanic intrusions imaged on the seismic data would have provided an additional load on the crust and an associated downwarp. However, the size of the intrusive bodies is not well constrained and only minor disturbance of sediments is apparent on the seismic data, and they have not been included in the modelling. The intrusive bodies may be associated with the proposed mid-basin fault. Thermal subsidence until 24 Ma and erosion of all the crustal rocks higher than 0.5 km above the sea level occurred.

Phase D (Fig. 6.3-3d) 24 Ma

Additional west dipping listric extension occurred along the two faults on eastern margin of basin, with erosion to 0.2 km above sea-level, and thermal post rift subsidence with sediment infill to 0.3 km water depth, from 24 Ma to present.

The modelling software does not retain the detailed geometry of the stratigraphy from earlier rift episodes in its output. We have therefore sketched in the major unconformities on Fig. 6.3-3. To model the stratigraphy imaged on the seismic data, we assumed prograding sedimentation for the basin in-fill associated with the west faults and flat-lying sedimentation for sedimentation associated with the eastern faults. This gives the basement parallel bedding of the sediments laid down after the east dipping faulting occurred and the subsequent stratigraphy of sediments that onlap to the west and were deposited after the west dipping faulting occurred.

Discussion

A major problem with the derived model is that the nature of the flexural cantilever model gives rise to the major unconformities between sediment packages being convex upwards, in contrast to that seen on the stratigra-

Table 6.3-1. Fault parameters used for modeling the VLB

Fault no.	Present X-coord (km)	Present horizontal extension (km)	Present dip (deg)	Dip direct. right = west	Fault type	Time of faulting Ma
1	0	25	20	Right	Planar	34
2	10	15	20	Right	Planar	32.5
3	90	10	20	Left	Planar	34
4	90	10	40	Left	Listric	29.24
5	130	20	20	Left	Listric	29
6	155	15	20	Left	Listric	24
7	210	20	8	Right	Listric	34
8	280	20	8	Left	Listric	34

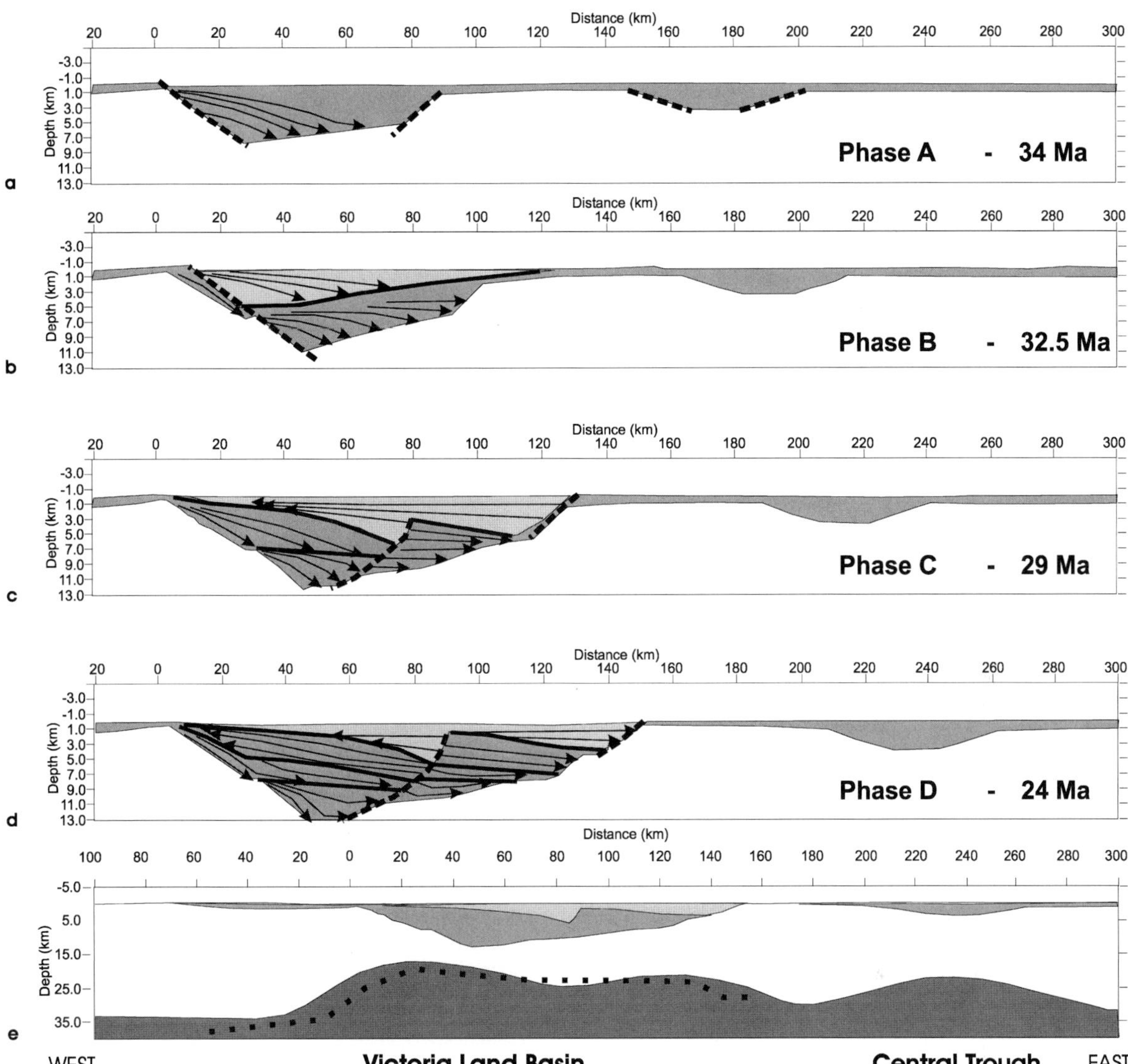

Fig. 6.3-3. The four phases of rifting of the Victoria Land Basin showing the faults and detailed basin structure. The depositional stratigraphy is indicated by the light lines with arrowheads at the onlap margin. Unconformities formed by the four rift phases are shown by *thick lines*. Faults active in each phase are shown by the *heavy dashed line*. For the fourth phase, the upper part is the detailed basin structure, and the lower part the crustal section. The *dotted line* on the crustal structure is the Moho modelled from gravity data (after Davey and Cooper 1987). *Light shading* (new rift phase) and *medium shading* (previous rift phases) are sediments, the crust has *no shading*, *dark shading* is upper mantle

phy. Thermal subsidence is not sufficient to reverse the curvature and give the concave unconformities imaged and we would speculate that a more complex fault model, with blind faults under the sediments in the centre of the basin, where crustal stretching is greatest, is needed. Alternatively, some pure shear model could be proposed, but the large degree of crustal thinning was not obtainable over a relatively small basin width (150 km approx.) with any reasonable pure shear model.

The seismic stratigraphy of the Victoria Land Basin (Fig. 6.3-2) can be matched moderately well by an extensional multiple rifting of a slightly thinned (33 km) continental crust. The model replicates the size and shape of the basin but does not reproduce the detailed stratigraphy well, particularly the concave basins in the western part of the basin, as noted earlier. Our study shows that the development of the Victoria Land Basin since 34 Ma can be modelled by a minimum of five major faults. Rifting commenced on the western margin of VLB and subsequently moved to the eastern margin (a similar conclusion to De Santis et al. 1994). Early, east dipping, brittle planar faulting, with about 50 km of extension, offshore from the TAM rift margin, formed an initial rift graben that reached a depth of about 10 km. Subsequently, major west dipping listric faulting at 29 Ma (30 km extension) and 24 Ma (15 km extension) further extended the rift and resulted in a more symmetric feature. Post rift subsidence of about 2 km has occurred since 24 Ma, giving a final basin depth of about 14 km. The amount of extension is 95 km, and has resulted in a thinning of the continental crust to 6 km and a maximum local stretching factor (beta) of 3. The causes of changes in the locus of faulting from west to east and the style of faulting from planar to listric, and in the location of the eastern faults, has not been addressed. We would speculate that they are probably related to the extreme thinning of the crust by the initial western faulting at 34 Ma and the subsequent internal crustal stresses that this generated.

Acknowledgements

We thank G. Brancolini for useful comment, and C. De Cillia for preparation of the seismic data. Thoughtful reviews by F. Salvini and G. Wilson greatly improved the manuscript. We acknowledge PNRA for support under the Cape Roberts Project (LDS), and NZGSF for support (FJD).

References

Barrett PJ (ed) (1986) Antarctic Cenozoic history from the MSSTS-1 drillhole, McMurdo Sound. DSIR Bull NZ, 237

Barrett PJ (ed) (1989) Antarctic Cenozoic history from the CIROS-1 drillhole, McMurdo Sound. DSIR Bull NZ 245

Busetti M, Spadini G, Van der Wateren F M, Cloetingh S, Zanolla C (1999) Kinematic Modeling of the West Antarctic Rift System, Ross Sea, Antarctica. In: Van der Wateren FM, Cloetingh SAPL (eds) Lithosphere dynamics and environmental change of the Cenozoic West Antarctic Rift System. Global and Planetary Change 23, Special Issue, pp 79–103

CRST (Cape Roberts Science Team) (2000) Summary of results. In: Barrett PJ, Sarti M, Wise S (eds) Studies from the Cape Roberts Project, Ross Sea, Antarctica – Initial report on CRP-3. Terra Antartica 7(1/2):185–209

Cooper AK, Davey FJ, Behrendt JC (1987) Seismic stratigraphy and structure of the Victoria Land Basin, western Ross Sea, Antarctica. In: Cooper AK, Davey FJ (eds) The Antarctic continental margin: geology and geophysics of the western Ross Sea, Circum-Pacific Council Energy and Mineral Resources, Houston, pp 27–65

Davey FJ, Brancolini G (1995) The Late Mesozoic and Cenozoic structural setting of the Ross Sea region. In: Cooper AK, Barker PF, Brancolini G (eds) Geology and seismic stratigraphy of the Antarctic margin. Antarct Res Ser 68, Amer Geophys Union, Washington DC, pp 167–182

Davey FJ, Cooper AK (1987) Gravity studies of the Victoria Land Basin and Iselin Bank. In: Cooper AK, Davey FJ (eds) The Antarctic continental margin: geology and geophysics of the western Ross Sea, Circum-Pacific Council Energy and Mineral Resources, Houston, pp 119–137

Danesi S, Morelli A (2000) Group velocity of Rayleigh waves in the Antarctic region. Phys Earth planet Inter 122:55–66

De Santis L, Brancolini G, Busetti M (1994) Structural evolution of the Victoria Land Basin south of the Drygalski Ice Tongue (western Ross Sea). Terra Antartica 1:107–110

De Santis L, Davey FJ, Prato S, Brancolini G (2001) Subsidence at the Cape Roberts (CRP) drillsites from backstripping techniques. Terra Antartica 8(3):1–5

Florindo F, Wilson GS, Roberts AP, Sagnotti L, Verosub K L (2001) Magnetostratigraphy of Late Eocene–Early Oligocene strata from the CRP-3 core, Victoria Land Basin, Antarctica. Terra Antartica 8(3):599–613

Fitzgerald P G (1994) Thermochronologic constraints on post-Paleozoic tectonic evolution of the central Transantarctic Mountains. Tectonics 13:818–836

Hayes DE, Frakes LA, et al. (1975) Initial Reports Deep Sea Drilling Project Vol 28. U.S. Government Printing Office, Washington DC

Kusznir NJ, Ziegler PA (1992) The mechanics of continental extension and sedimentary basin formation: a simple-shear/pure-shear flexural cantilever model. Tectonophysics 215:117–131

Kusznir NJ, Marsden G, Egan SS (1991) A flexural-cantilever simple-shear/pure-shear model of continental lithosphere extension: applications to the Jeanne d'Arc Basin, Grand Banks and Viking Graben, North Sea. In: Roberts, Yielding, Freeman (eds) The Geometry of Normal Faults. Geol Soc Spec Publ 56:41–60

Salvini F, Brancolini G, Busetti M, Storti F, Mazzarini F, Coren F (1997) Cenozoic geodynamics of the Ross Sea region, Antarctica: crustal extension, intraplate strike-slip faulting, and tectonic inheritance. J Geophys Res 102(11):24669–24696

Stern TA, Ten Brink US (1989) Flexural uplift of the Transantarctic Mountains. J Geophys Res 94:10315–10330

Tessensohn F, Wörner G (1991) The Ross Sea Rift System (Antarctica): structure, evolution and analogues. In: Thompson MRA, Crame JA, Thompson JW (eds) (1997) Geological Evolution of Antarctica. Cambridge University Press, Cambridge, pp 273–277

Ten Brink US, Hackney RI, Bannister S, Stern TA, Makovsky Y (1997) Uplift of the Transantarctic Mountains and the bedrock beneath the East Antarctic ice sheet. J Geophys Res 102:27603–27621

Van der Beek P, Cloetingh S, Andriessen P (1994) Mechanisms of extensional basin formation and vertical motion at rift flanks: Constraints from tectonic modeling and fission track thermochronology. Earth Planet Sci Letters 121:417–433

Wilson T (1995) Cenozoic transtension along the Transantarctic Mountains-West Antarctic rift boundary, southern Victoria Land, Antarctica. Tectonics 14:531–545

Wilson GS, Roberts AP, Verosub KL, Florindo F, Sagnotti L (1998) Magnetobiostratigraphic chronology of the Eocene-Oligocene transition in the CIROS-1 core, Victoria Land margin, Antarctica: Implications for Antarctic glacial history. Geol Soc Amer Bull 110:35–47

Rift History of the Western Victoria Land Basin: A new Perspective Based on Integration of Cores with Seismic Reflection Data

Christopher R. Fielding[1] · Stuart A. Henrys[2] · Terry J. Wilson[3]

[1] Department of Geosciences, 214 Bessey Hall, University of Nebraska-Lincoln, NE 68588-0340, USA, <cfielding2@unl.edu>
[2] Institute of Geological and Nuclear Sciences, PO Box 30-368, Lower Hutt, New Zealand
[3] Department of Geological Sciences, Ohio State University, 155 South Oval Mall, Columbus, OH 43210-1522, USA

Abstract. The results of a stratigraphic study of the western Victoria Land Basin, Antarctica, are summarized. This analysis is based on all existing seismic reflection data integrated with lithological information from fully cored drillholes in the Cape Roberts area of western McMurdo Sound. A number of subsurface seismic reflectors were recognized in the Cape Roberts area and correlated to stratal interfaces previously recognized in the cores. These events were then traced regionally throughout the southern McMurdo Sound, and form the basis for a new seismic stratigraphic subdivision of the Cenozoic section. Key reflectors define boundaries of seismic stratigraphic units, each of which shows distinctive overall cross-sectional geometry and internal reflection character/facies. On this basis, we propose a new model for the evolution of the Victoria Land Basin, invoking five phases of tectonic activity and associated sediment accumulation patterns. Phase 1 (pre-latest Eocene) involved regional uplift and erosion of the Transantarctic Mountains to the immediate west of the basin. Phase 2 (latest Eocene to Early Oligocene) was an Early Rift stage characterized by sediment accumulation in laterally restricted grabens. Phase 3 (Early Oligocene to Early Miocene) was the Main Rift stage, in which sediment accumulation was no longer confined to grabens in the west of the basin, but rather formed an eastward-thickening wedge into the centre of the basin. Phase 4 (Early Miocene) was a consequence of passive thermal subsidence, producing a relatively even blanket of sediment across the entire basin. Phase 5 (post-Early Miocene) was associated with the "Terror Rift" and gave rise to a succession containing both young magmatic rocks and young faults and which thickens markedly into a central depocentre. The new framework allows recognition of thick, post-Early Miocene stratigraphic intervals as yet unsampled by stratigraphic drilling in McMurdo Sound.

Introduction

Despite a considerable history of geological and geophysical exploration, the tectonic history and basin evolution of the Victoria Land Basin (VLB) in the Ross Sea region of Antarctica are poorly understood. Various seismic reflection surveys have been carried out, over more than thirty years (Fig. 6.4-1), leading to several seismic stratigraphic frameworks and consequent interpretations of basin history, but until recently such models have been largely unconstrained by lithological data. This is because sedimentary strata of the VLB are only preserved below the water and ice of McMurdo Sound (Fig. 6.4-1) and do not crop out on land. In the late 1990s, core drilling by the Cape Roberts Project (Cape Roberts Science Team 1998, 1999, 2000) provided a complete lithological transect through the western VLB, allowing firm correlations between the extensive seismic data set and geological reality for the first time. Lithostratigraphic boundaries in Cape Roberts Project (CRP) cores have been correlated to events imaged by seismic reflection data in the immediate vicinity of the CRP drillholes by both CRP scientists (Henrys et al. 2000, 2001; Fielding et al. 2000, 2001) and by Hamilton et al. (2001). However, to date, no-one has attempted to correlate these events regionally throughout the seismic data set for the McMurdo Sound region (Fig. 6.4-1). Furthermore, problems with seismic data quality (in particular, the prominence of sea-floor multiples on seismic sections) have to date hampered a complete understanding of local cross-sectional basin geometry in the vicinity of Cape Roberts.

In this paper, we summarize the results of a comprehensive review of available seismic data. We utilize scientific results from CRP drillholes to constrain the timing and geological meaning of events recognized on seismic data, and use the resulting framework as the basis for a new geological model for the western VLB. We compare our framework with those published previously by others.

Methods

Seismic reflection data from all surveys carried out in McMurdo Sound (Fig. 6.4-1) were collated at the Institute of Geological and Nuclear Sciences (GNS) at Lower Hutt, New Zealand, and loaded onto a Unix workstation running Schlumberger Geoquest (©) interpretation software. Navigational data were used to correct some older lines to a linear distance scale. Seismic lines were then interpreted on a work-station, using cross-line functions to provide realistic four way and other stratigraphic ties. Paper prints were also interpreted independently to serve as a quality control mechanism. Seismic data quality is compromised in the area by the presence of a strong sea-floor multiple due to ice cover and associated "ringing" effects. Since the first multiple lies closer to the sea floor in shallower water, and since all holes drilled to date have been sited on shallow submerged ridges (notably CRP and CIROS holes), the data quality issue is (confoundingly) most acute over the ar-

From: Fütterer DK, Damaske D, Kleinschmidt G, Miller H, Tessensohn F (eds) (2006) Antarctica: Contributions to global earth sciences. Springer-Verlag, Berlin Heidelberg New York, pp 309–318

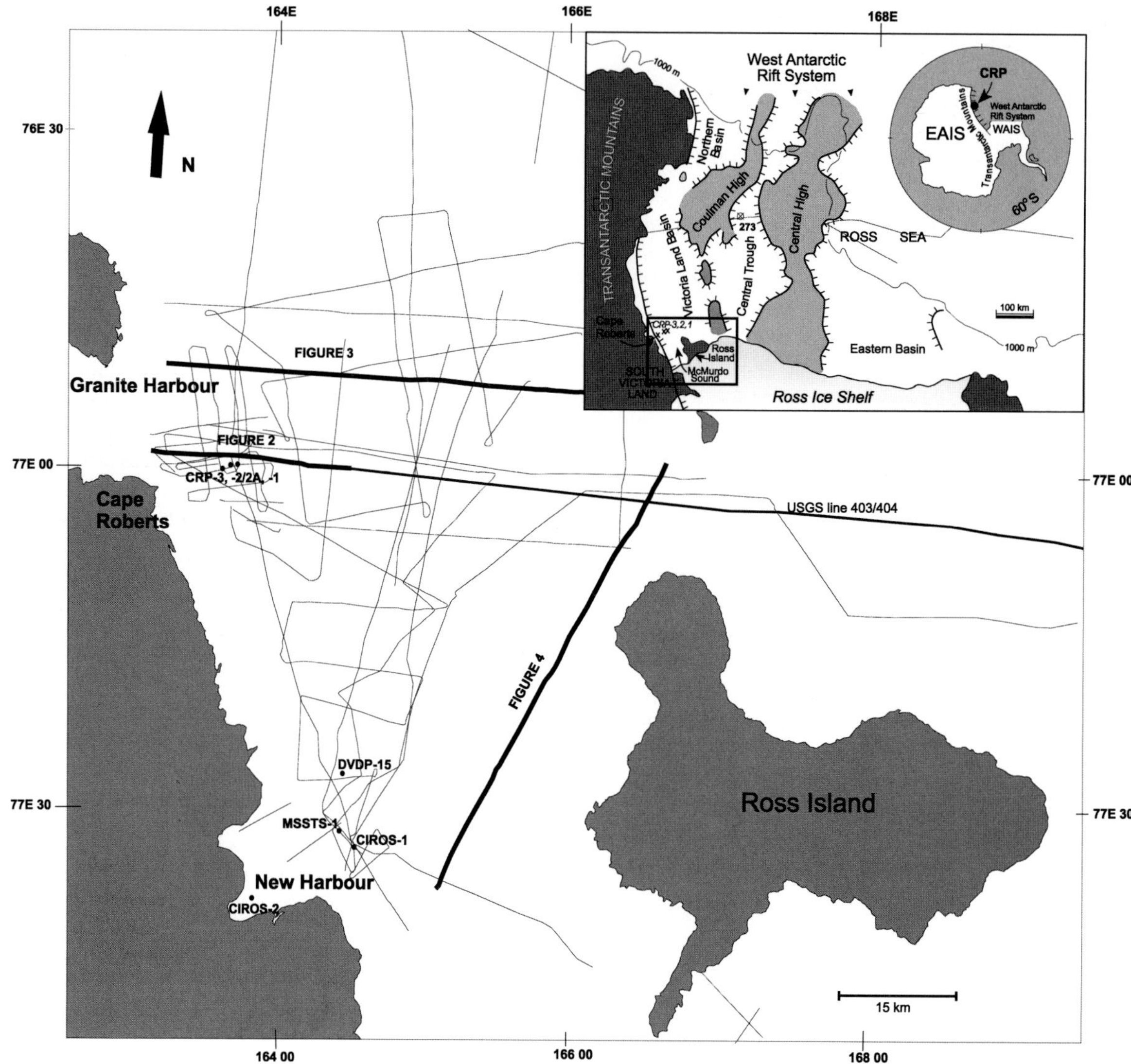

Fig. 6.4-1. Maps showing the regional context and structure of the Victoria Land Basin, the track lines of seismic reflection surveys and relevant drillhole locations. Note the positions of USGS seismic reflection line 403/404 (*bold section* illustrated as Fig. 6.4-2), line IT90a-75 (Fig. 6.4-3) and line IT90a-70 (Fig. 6.4-4)

eas of greatest interest. In order to circumvent this as far as possible, correlations from the Cape Roberts area to the New Harbor (CIROS) area were achieved by tracing events in one area eastward into deeper water, north-south along depositional strike and then westward up into shallow water again. Additionally, important new data on the deeper structure of the basin in the vicinity of Cape Roberts were generated by re-processing of USGS Line 403/404 (Cooper et al. 1987) at GNS to mitigate multiple effects. This line, and the tectonic interpretations that follow from it, are treated separately by Wilson et al. (submitted). In the present paper, a new seismic stratigraphic framework is presented, prominent seismic events are linked to lithostratigraphic horizons recognised in the CRP cores, and this framework is interpreted in terms of basin history.

Regional Geological Considerations

The VLB is broadly north-northwest-elongated, as defined by major bounding faults and horsts (Fig. 6.4-1). Depositional dip is east-northeastward in the western part of the basin, and has an opposite sense in the east. Thus, regional seismic lines such as USGS Line 403/404 (Cooper et al. 1987) show stratal thickening from both west and east into a central depocentre, analogous to the "steer's head" rift geometry of White and McKenzie (1988). The central depocentre is also a zone where the amplitudes of reflectors are attenuated and scattered, interpreted by various workers (e.g., Cooper et al. 1987) as concentrations of magmatic (intrusive and volcanic) rocks. Near the western margin of the basin (CIROS-1, CRP-1, -2/2A, -3), the VLB succession comprises a mostly homoclinal section ranging in age from latest Eocene (ca. 34 Ma) through to late Early Miocene (ca. 17 Ma), unconformably overlain by a generally thin Plio-Pleistocene section. The major angular discordance that separates these two packages has been referred to by many previous workers as the "Ross Sea Unconformity", although it is clear now that more than one major unconformity punctuates the section in some areas.

Seismic Data and Interpretation

The regional seismic reflection line USGS 403/404 passes within 2 km of the Cape Roberts Project drillholes (Fig. 6.4-1). Re-processing has revealed the deeper structure in this region necessary to allow interpretation of early Cenozoic basin formation. The westernmost portion of the re-processed line in the vicinity of Cape Roberts is shown in Fig. 6.4-2 (see Fig. 6.4-1 for location). The major features of the present sea-floor bathymetry are, from the coastline eastward, a narrow shelf adjacent to the coast, a north-south-linear trough with steep margins ("Cape Roberts Rift Basin" of Hamilton et al. 2001), a linear topographic high named the Roberts Ridge (on the western flank of which the CRP holes were sited: Fig. 6.4-2), and a ramp slope that descends from the crest of Roberts Ridge at ca. –50 m to the deepest part of McMurdo Sound (–700 to –800 m). Figure 6.4-2 shows a thick succession beneath Roberts Ridge that thickens and dips eastward and is apparently truncated westward against the margin of the "Cape Roberts Rift Basin" (CRRB). This trough is interpreted as being fault-bounded, and its fill is essentially flat-lying in contrast to the Roberts Ridge section. A further important structural feature evident from the re-processed data is a basement horst downdip from the drilling sites that appears to truncate the lower part of the coherent stratigraphy (below the yellow event: Fig. 6.4-2). A variety of shallow faults have also been interpreted, some directly above the basement horst and others closer to the CRRB.

The most consistently recognizable and persistent reflections in this and other lines were chosen as the basis for a seismic stratigraphic framework. Most of these are major seismic sequence boundaries, and correlate to important stratal interfaces identified independently in the CRP cores (Table 6.4-1). The basal navy blue event is interpreted as the basement unconformity. It is intersected in CRP-3, where it separates probable Devonian quartzose sandstones below from probable latest Eocene, lithic sandstones, conglomerates and breccias above. This event is truncated updip against the margin of the CRRB, can be tentatively interpreted at greater depth within that feature, and also terminates downdip against the afore-mentioned basement horst (Fig. 6.4-2). The overlying section, dominated in CRP-3 by conglomerates passing upward into predominantly sandstones with minor conglomerates, thickens into the area immediately updip of the basement horst and is confined by that feature. This interval corresponds to much of the 800+ m thick, sandstone-dominated latest Eocene to earliest Oligocene section penetrated in CRP-3, which becomes gradually more organized into depositional sequences upward, and becomes lithologically more diverse, with increasing proportions of mudrock and diamictite upward.

The next major seismic marker is the yellow event (Fig. 6.4-2, Table 6.4-1), which corresponds to a major stratal interface in CRP-2/2A (at 443 m below sea floor: m b.s.f.). This horizon separates sections confined by the early

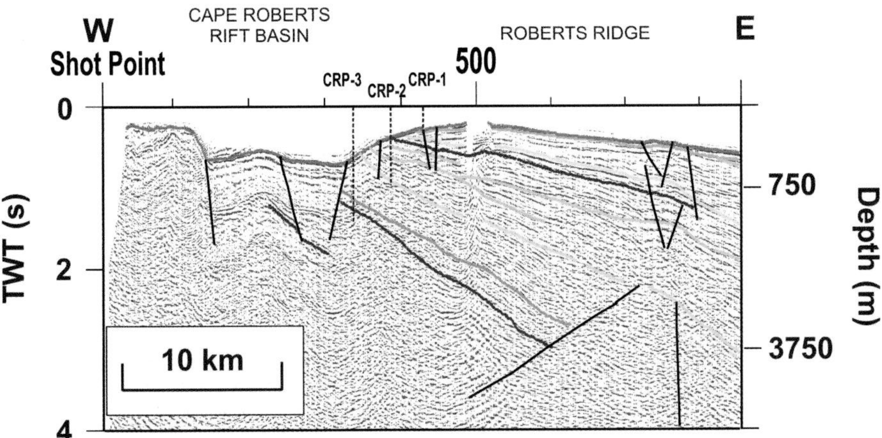

Fig. 6.4-2. Westernmost portion of seismic reflection line USGS 403/404 (re-processed to mitigate the masking effects of the sea-floor multiples), showing the projected positions of Cape Roberts Project drillholes (*CRP-1, -2/2A* and -3), and interpreted to show major faults and seismic events described in the text. *TWT:* Two Way Travel Time. See Table 6.4-1 for key to seismic reflector coding scheme. The tectonic phase 2 (early rift) spans the interval between the *navy blue* and *yellow* reflectors, Phase 3 (main rift) between the *yellow* and *purple-grey* reflectors, phase 4 (thermal subsidence) between the *purple-grey* and *green* reflectors and phase 5 (Terror Rift formation) is the thin interval above the *green* reflectors. The phase 5 section thickens markedly into the central depocentre of the VLB (see Fig. 6.4-3 and 6.4-4)

Table 6.4-1. Key to seismic reflector coding scheme

Cooper et al. 1987	Brancolini et al. 1995	Bartek et al. 1996	Hamilton et al. 2001	Henrys et al. 2000, 2001; Fielding et al. 2000, 2001	This paper	Rift phase
V1	RSU 1-3 RSS 5-8 (Middle Miocene in DSDP273) RSU 4 (Lower Miocene in DSDP 273) RSS 4	A-H	?	Thin intervals only sampled	(Pliocene or Pleistocene in MSSTS/CIROS/CRP)	5
	RSU 4a				Refl. h (dark green)	
V2	RSS 3	I J K	? ? ?	Not sampled		
	RSU 5	L-M	?		Refl. h1 (light green)	
V3	RSS 2	N O P	? ? ?	Not sampled		4
	~RSU 6	P/Q			w1 (beige) 17 Ma ~ top CRP section	
	RSS 1	R	V3			
			90 mbsf CRP-2/2A	?Refl. "a" 52 mbsf CRP-2/2A	w1b (crimson) 19 Ma	
			V4a	Q		
			307 mbsf CRP-2/2A	Refl. "b" 91 mbsf CRP-2/2A	w1a (pink) 21 Ma	
V4		S	V4b	R	(d) = top thick sequences 23.7 Ma	
				Refl. "e" ("d") 186 (130) mbsf CRP-2/2A	w2 (e) (brown + purple-grey)	
				S		3
			443 mbsf CRP-2/2A	Refl. "i" 307 mbsf CRP-2/2A	Base thick sequences 24.1 Ma	
		T	V5a ?	T		
V5			330 mbsf CRP-3	Refl. "l" 443 mbsf CRP-2/2A	w3 (yellow) 29 Ma	
			V5b+V5c ?	U		2
				Refl. "p" 95 mbsf CRP-3		
				V		
				Refl. "r" 225 mbsf CRP-3		
				W		
				Refl. "u" 444 mbsf CRP-3		
				X	w4 (orange) w5 (red) 34 Ma	
				Basement 823 mbsf CRP-3	Basement (navy blue)	
V6		("volcanics")	V6			1
V7		("basement")	V7			

horst-graben topography from overlying strata that are not confined by this structure. Above the yellow event, the section thickens downdip into a more distal depocentre to the east (Fig. 6.4-2). In the CRP cores, this section is sandstone-dominated but lithologically more heterogeneous, including numerous diamictite intervals, and can be divided into depositional sequences bounded by erosion surfaces that are interpreted to record glacial advance/retreat cycles with associated changes in relative sea-level (Fielding et al. 1998, 2000, 2001; Naish et al. 2001). This interval is of Lower Oligocene to basal Lower Miocene age. In addition to the numerous sequence-bounding unconformities, many of which can be recognized as persistent seismic reflectors (Fielding et al. 2000), a major, angular unconformity is mapped at the purple-grey reflector, which truncates the brown reflector (Fig. 6.4-2). This unconformity forms the top of the eastward-thickening stratal wedge mentioned above.

Above this unconformity, the section is more tabular in cross-sectional geometry. The overlying interval corresponds to the upper part of the Lower Miocene section in CRP-2/2A and the section penetrated by CRP-1, which lithologically is composed of highly top-truncated and incomplete depositional sequences otherwise similar to those below with abundant diamictites. The uppermost Miocene horizon penetrated by CRP-1 corresponds to the beige (light brown) event (Fig. 6.4-2), dated at ca. 17 Ma from microfossils. However, up to several hundred metres of additional (as yet unsampled) section clearly concordant with this event is evident from the seismic data. Above this is a prominent unconformity surface coded light green (Fig. 6.4-2). This surface separates the more or less tabular-stratified section below from a section above that thickens more prominently into VLB depocentres (Fig. 6.4-2). A further major stratal interface (coded dark green: Fig. 6.4-2, Table 6.4-1) occurs near the top of the CRP section, and is a prominent, angular unconformity overlain across the Roberts Ridge by only a thin (1–2 cycles thick) section which in CRP is of Pleistocene and Pliocene age. This section also thickens spectacularly eastward. The major (dark green) unconformity in CRP separates lithified strata of Lower Miocene (17 Ma) age from largely unconsolidated, diamictite-dominated Pliocene and Pleistocene strata.

The post-CRP section is better illustrated by seismic profiles that extend into the deep water depocentre of the present McMurdo Sound. Figure 6.4-3 shows part of seismic profile IT90a-75 (near-trace, single channel section), which runs parallel to, and north of the USGS 403/404 line (Fig. 6.4-1). The northward extension of the CRRB and Roberts Ridge are evident near the west end of the transect, as is the Eocene-Miocene section described above (note the location of the beige reflector denoting the top-CRP Miocene horizon). Unlike Fig. 6.4-2, however, Fig. 6.4-3 shows a massive post-17 Ma section (up to 1.5 seconds Two-Way Travel Time [TWT], conservatively

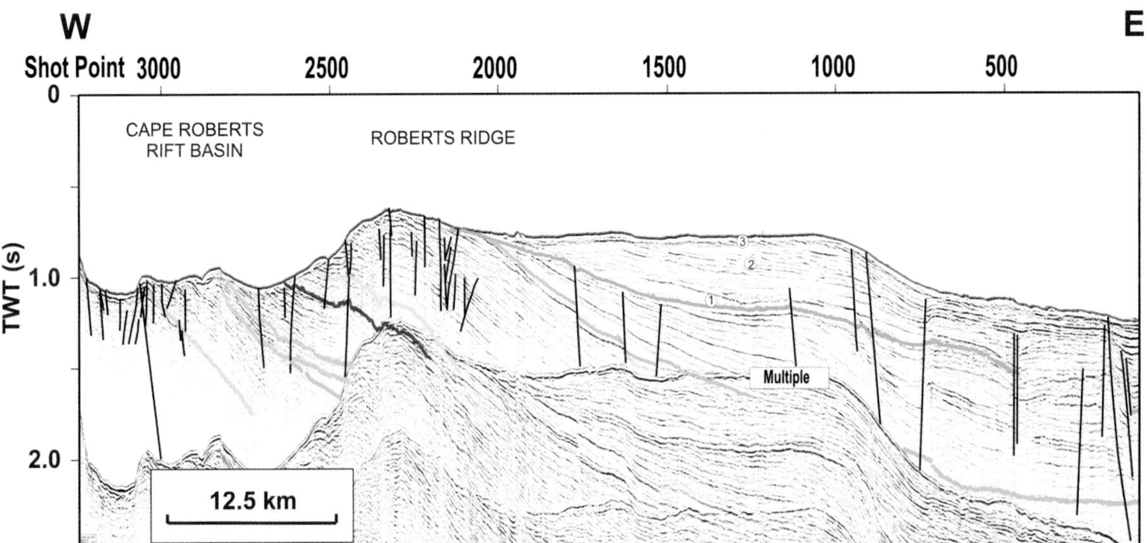

Fig. 6.4-3. Seismic reflection line IT90a-75, interpreted to show major faults and seismic events described in the text. *TWT:* Two Way Tavel Time. See Table 6.4-1 for key to seismic reflector coding scheme. Tectonic phase 2 (early rift) is that interval below the *yellow* reflector (not very clearly imaged), phase 3 (main rift) between the *yellow* and *purple-grey* reflectors, phase 4 (thermal subsidence) between the *purple-grey* and *light green* reflectors, and phase 5 (Terror Rift formation is above the *light green* reflector. The top of the Oligocene-Miocene section cored in the Cape Roberts holes is approximated by the *beige* (*light brown*) reflector, which intersects the sea floor at shot point ca. 2600. Note the dramatic thickening of the phase 5 section eastward into the VLB depocentre, the major angular unconformity represented by the *dark green* reflector, and the development of various seismic facies in the section above the unconformity (including 1. a reflection-free lens-shaped interval directly overlying the unconformity, interpreted as a lowstand slope wedge; 2. shelf-edge clinoforms overlying this wedge that form a pronounced progradational sequence set, and 3. a near-surface interval of parallel, concordant reflectors)

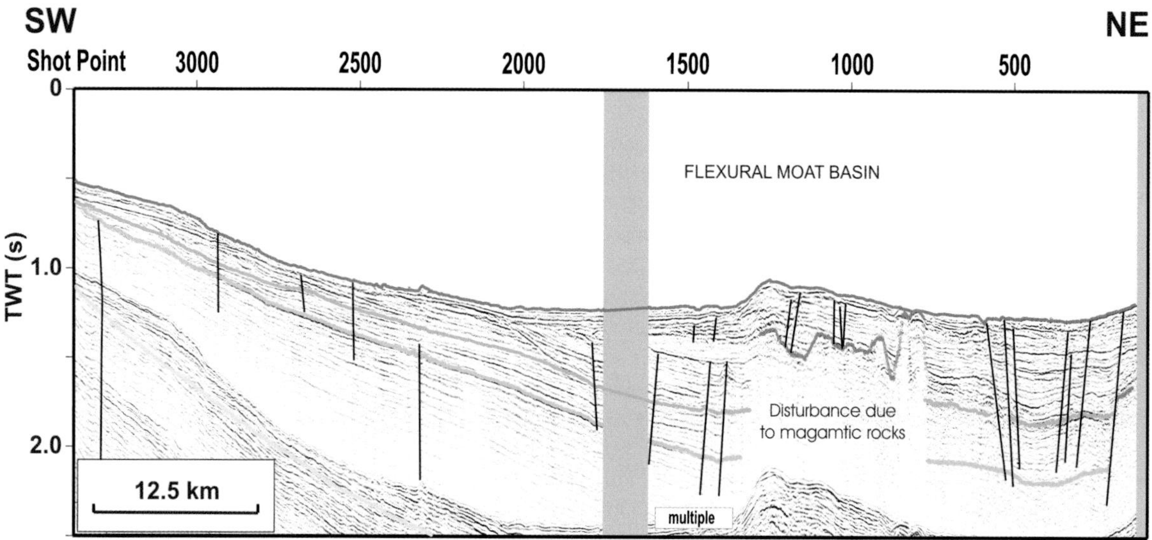

Fig. 6.4-4. Seismic reflection line IT90a-70, interpreted to show major faults and seismic events described in the text. See Table 6.4-1 for key to seismic reflector coding scheme. The *beige reflector*, representing the top of the Oligocene-Miocene section in CRP-1 to -3, lies somewhat below the first sea-floor multiple, the boundary between phase 4 (thermal subsidence) and phase 5 (Terror Rift formation lies at the *light green* reflector, and the major angular unconformity is represented by the *dark green* reflector. The line passes into a deep-water, flexural moat basin associated with the construction of the Ross Island volcanic edifices towards the northeast, and the top of an interpreted zone of magmatic rocks is denoted by the *purple* reflector. A series of further, post-*dark green* event unconformities likely caused by flexural loading is evident in the shallow subsurface between shot points 1 800–2 000

at least 1 500 m of section above the beige reflector: Fig. 6.4-3) preserved east of Roberts Ridge. The lower part of this section (at least 500 ms TWT) is concordant with the beige reflector and is of tabular cross-sectional geometry, suggesting that it forms part of the same succession as the upper CRP interval. Overlying this interval is an eastward-thickening wedge defined at its base by the light green event, and punctuated in the middle by the dark green angular unconformity. A further 300–400 ms of post-unconformity section is evident from this transect, suggesting significant post-Early Miocene sediment accumulation in the depocentre. This interval preserves spectacular examples of various seismic facies, including acoustically transparent wedges of strata on palaeo-slopes (lowstand wedges), and large-scale clinoform sets some of which define distinct shelf-slope breaks (Fig. 6.4-3).

The Neogene section is also clearly imaged by the northeast-trending line IT90a-70 (near-trace, single channel section: Fig. 6.4-4), which shows thickening of the post-beige interval into the deep water axial zone of the VLB, post-dark green unconformities that appear to be related to the formation of a flexural moat basin around Ross Island, and disruption of the seismic data by subsurface magmatic rocks in this region. The major Neogene reflectors were traced from the southwest end of Line IT90a-70 updip via line PD90-46 (see Barrett et al. 1995; Bartek et al. 1996) for illustrations and interpretations of this line) into the vicinity of CIROS-1 and MSSTS-1 (Fig. 6.4-1).

Integrated Stratigraphic Framework

The seismic stratigraphy described above and illustrated in Fig. 6.4-2 through 6.4-4 forms the basis for an interpretation of the evolution of the western VLB in the vicinity of Cape Roberts. Intervals between successive key horizons define seismic stratigraphic units, and changes in the cross-sectional geometry of these units are used to define Basin-Forming Phases 1 to 5 (Table 6.4-1). The following is a summary of the five interpreted phases of tectonic activity together with relevant geological information obtained from drillholes.

1. Regional Uplift and Erosion: Early Cenozoic (Pre-34 Ma)

Evidence from Apatite Fission Track Analysis (Fitzgerald 1992; Fitzgerald and Baldwin 1997) suggests that the Transantarctic Mountains underwent a period of major uplift during the early Cenozoic (55–45 Ma). Furthermore, clasts of granite and metamorphic rocks characteristic of the basement complex in the region west of McMurdo Sound are conspicuous throughout the Cenozoic section of CRP-3, down to immediately above the basal unconformity. The implications of these data are that substantial topography must have existed in the hinterland to the immediate west of CRP holes prior to any sediment delivery into the VLB, and that the topography had by this time excised at least down to the nonconformity with Lower Palaeozoic basement. The only stratigraphic record of this phase of activity is a thin breccia of ?Devonian sandstone clasts directly overlying the basement unconformity in CRP-3. The basement unconformity is defined by the navy blue seismic reflector in this study (Table 6.4-1). The eastward dip of the overlying VLB section in seismic lines also suggests that there was little if any down-to-the-basin normal faulting along the Transantarctic Mountain front during accumulation of the latest Eocene to Early Miocene rift succession.

2. Early Rift: Latest Eocene to Early Oligocene (34–29 Ma)

This package is defined in seismic data as the interval between the navy blue (basement) and yellow reflectors. Recognition of this basal, early rift package was facilitated by the reprocessing of USGS Line 403/404. Downdip of the Roberts Ridge, the interval overlying the basement unconformity is confined to structural lows (grabens) bounded by upstanding blocks (horsts), in a manner typical of early rift topography. Within the westernmost infra-basin, the cross-sectional geometry is an eastward-thickening wedge hinged on the western side. While such a structural and stratigraphic architecture is only visible in updip portions of this and some other lines, we suggest that the same architecture characterizes much if not all of the basin. The yellow event that defines the top of the package is the first reflector that is not confined by the early rift topography but rather is continuous over the tops of horsts into the deeper parts of the basin (eastward).

This graben-confined interval comprises the entire Cenozoic section of CRP-3 (823 m b.s.f. upward) and up to 443 m b.s.f. in CRP-2/2A, an aggregated thickness of ca. 1 000 m on Roberts Ridge. A subsidence model for the CRP cores based on a variety of chronostratigraphic data shows that this interval coincides with a period of very rapid subsidence (Wilson et al. submitted). Lithologically, the lower part of the interval comprises conglomerates and breccias up to boulder grade, progressively becoming interbedded with higher proportions of sandstone, but only minor mudrocks. This basal section has been interpreted as the product of subaerial to ultimately subaqueous base-of-slope fans or aprons, passing upward into a shallow marine environment characterised by oversupply of coarse sediment (Cape Roberts Science Team 2000). This interval, up to ca. 300 m b.s.f. in CRP-3, shows little or no cyclical vertical stacking of lithofacies that might be interpreted in terms of sequence stratigraphy. The upper half of the interval, spanning 300–0 m b.s.f. in CRP-3 and 623–443 m b.s.f. in CRP-2/2A, shows a progressive decrease in gravel grade facies and a concomitant increase in mudrocks (Cape Roberts Science Team 1999). The stratigraphically lowest diamictites also occur in this interval, increasing in abundance upward. These patterns are interpreted to record a progressive decline in the rate of sediment supply to the still rapidly subsiding graben, with sediment accumulation in a variety of shelfal water depths. The development of cyclical facies stacking patterns has been interpreted in terms of cycles of relative sea-level change under conditions of varying sediment supply and accommodation, and under the increasing influence of glacial advance and retreat across the shelf (Fielding et al. 2000).

3. Main Rift: Oligocene (29–24 Ma)

This package is defined from seismic data in the western VLB as an eastward-thickening wedge unconfined by the early horst-and-graben topography (Fielding et al. 2000; Wilson et al. submitted). The interval is bounded at its base by the yellow reflector and at its top by the purple-grey event (Table 6.4-1). The cross-sectional architecture is interpreted to reflect a lateral shift in the locus of subsidence eastward towards a more basin-central depocentre, and thus to record the main phase of rifting in the VLB.

This interval corresponds to 443–186 m b.s.f. in CRP-2/2A. Although according to the most recent seismic interpretation, the purple-grey event corresponds to the base of sequence 9 in CRP-2/2A at 186 m b.s.f. (Henrys et al. 2000), the top of the eastward-thickening stratal wedge may in fact correspond to the top of sequence 9 at 130 m b.s.f., which marks the top of the package of three, thick sequences and an upward change to much thinner, more truncated sequences (Fielding et al. 2000). The corresponding "reflector d", however, cannot be traced regionally and so in this analysis, the upper boundary of the Main Rift Phase is taken at the purple-grey reflector. Subsidence modeling (Wilson et al. submitted) shows that in CRP holes this interval was accumulated during a regime of variable subsidence, generally less rapid than the underlying Early Rift. This interval as a whole is lithologically diverse, with abundant mudrocks, sandstones and diamictites. Lithologies are cyclically stacked, allowing interpretation in terms of sequence stratigraphy (Fielding et al. 2000). The package of three unusually thick and complete sequences at the top of the interval, which Naish et al. (2001) have correlated to Milankovitch band cycles, evidently reflects a discrete (<450 ka) interval of accelerated subsidence. These patterns are interpreted to record sediment accumulation under varying subsidence and sedimentation rates in shallow glacimarine environments influenced by repeated glacial advance-retreat cycles and associated changes in relative sea-level.

4. Thermal Subsidence: Early Miocene (24–Younger Than 17 Ma)

This package is defined from seismic data as ranging from the purple-grey to the light green reflectors (Table 6.4-1). In cross-section, it has a more sheet-like tabular geometry than the underlying unit, and can be traced across the VLB with only a modest thickening into the central depocentre. This geometry is interpreted to reflect a period of passive, thermal subsidence that is a natural consequence of crustal extension. The angular unconformity noted at the purple-grey reflector could record a "rift-drift" unconformity such as is commonly found in other rift fill successions.

The thermal subsidence phase spans the upper part of the CRP section and a significant thickness of overlying, as yet undrilled, section. This interval coincides with a markedly slower subsidence rate, consistent with the above interpretation. Lithologically, the sampled part of the interval comprises generally thin, condensed glacimarine sequences dominated by diamictites formed during the retreat phase of glacial advance-retreat cycles. The Early Miocene section of CRP-2/2A also contains a composite basaltic fallout tephra interval deposited in a shallow marine setting at 21.44 Ma (Cape Roberts Science Team 1999).

5. Terror Rift Formation: ?Late Miocene to ?Present (Younger Than 17 Ma–?)

This package is defined on seismic data as the interval from the light green reflector to the sea floor (Table 6.4-1). In cross-section, the interval thickens from both east and west into the central depocentre. Strong interference patterns in the depocentre region (e.g., Fig. 6.4-4) indicate the presence of significant magmatic rocks in this interval, coinciding with the McMurdo Volcanic Group. The package can be divided into two sub-sections: *(a)* an interval characterized by broadly homoclinal reflectors that thicken passively into the depocentre, and *(b)* a diverse array of seismic stratigraphic units that overlie a major, regional, angular unconformity (dark green event: Table 6.4-1). This upper unit also contains several additional, less prominent unconformities. This interval is largely unsampled by drilling, but near-surface intersections in CIROS-1, -2, MSSTS-1 and CRP holes indicate Pliocene and Pleistocene ages for at least the upper parts. In addition to the evidence of young magmatic activity, a large number of shallow faults and basins are evident from seismic data. This succession as a whole is correlated to the formation of the Terror Rift (Cooper et al. 1987) by transtensional deformation (Wilson 1995; Salvini et al. 1997), over a period that is poorly constrained. Included in this period of activity is the "Cape Roberts Rift Basin" of Hamilton et al. (2001), which from its location landward (i.e., updip) of the Cape Roberts rift succession and its expression as a basin on the present sea-floor, is here considered likely to have formed subsequent to accumulation of the CRP section.

The older, largely homoclinal package suggests that Terror Rift formation developed initially via passive subsidence in the north-south axial zone of the Victoria Land Basin (Cooper et al. 1987). This phase was terminated by a structural upheaval that formed the major angular unconformity noted above. The degree of angularity on the unconformity increases dramatically westward towards the western margin of the VLB, suggesting that it may be associated with a young phase of uplift in the Transantarctic Mountains. The dramatic thickening of post-unconformity section basinward (eastward) in seismic data may also record sediment derivation and dispersal from the rejuvenated Transantarctic Mountains into the Terror Rift depocentre. We have traced the dark green (unconformity) event updip into the site of drillhole MSSTS-1 at ca. 35 m b.s.f., close to a horizon that separates Lower Miocene from Lower Pliocene strata (D. Harwood, pers. comm. 2004). Three seismic stratigraphic units have been recognized in the post-unconformity section, the boundaries between them not necessarily being single horizons. The lowest of these units is a package containing a variety of seismic facies, notably concordant reflections, acoustically bland wedges (?lowstand wedges: Fig. 6.4-3) and clinoform sets, all of which dip towards the depocentre. The second package is dominated by large-scale clinoform sets (Fig. 6.4-3), the flat tops of some of which appear to define palaeo-shelf edges. The upper package comprises mainly tabular, concordant reflections that are dominantly flat-lying (Fig. 6.4-3). This facies (though not necessarily this seismic stratigraphic unit) is most prevalent in the condensed section of shallow-water areas adjacent to drillholes, where the Plio-Pleistocene section has been drilled. Further unconformities are also evident within the post-dark green interval in some seismic lines. This is particularly the case in the deep-water "moat" around the modern volcanic construct of Ross Island. Here, post-dark green unconformities may be the product of flexural loading imposed by the construction of the Ross Island volcanic edifices (Fig. 6.4-4). If this interpretation is correct, then the <4.6 Ma ages from Ross Island magmatic rocks (Kyle 1990) may provide further constraints on these uppermost units.

Discussion and Conclusions

Since the original study of Cooper et al. (1987), subsequent seismic stratigraphic analyses of the McMurdo sound region have divided the section in much greater detail (e.g., Brancolini et al. 1995; Bartek et al. 1996; Henrys et al. 2000, 2001; Hamilton et al. 2001). The present analysis is based on the premise that a detailed (tens of meters vertical scale) seismic stratigraphic framework cannot be sustained across the large area of McMurdo Sound with the quality and quantity of data available. Rather, we have picked distinct, regionally traceable horizons that correspond to significant and recognizable changes in cross-sectional geometry, and have used these as the basis for our framework. Thus, each of the key horizons nominated in this study can be interpreted as recording some change in the tectonic development of the basin. Having completed this analysis, we subsequently discovered that our seismic stratigraphic units, and consequent basin-forming phases, correlate quite well with the original seismic stratigraphic analysis of Cooper et al. (1987). We also note substantial differences in the correlations between seismic reflectors and CRP made by Hamilton et al. (2001) and those made in this study (see Table 6.4-1). By comparing our correlations between the lithostratigraphy in CRP holes and the seismic stratigraphic scheme of Bartek et al. (1996) with the correlations proposed in Hamilton et al. (2001), we believe the correlation of Hamilton et al. (2001) of seismic unit boundaries with CRP cores to be too deep in all cases. It is also worthy of note that the entire pre-Pliocene section drilled in the Cape Roberts Project is contained within a single seismic stratigraphic unit (RSS 1) in the scheme of Brancolini et al. (1995).

This analysis has also shown that significant stratigraphic intervals remain largely or entirely unsampled by drilling (e.g., Fig. 6.4-3 and 6.4-4). Notable among these are the upper part of the homoclinal section overlying CRP holes (Lower Miocene and younger in age) and much of the Terror Rift succession, which in depocentres must reach 1–2 km in thickness (Table 6.4-1). These as yet unknown successions represent potential targets for future scientific drilling endeavours in the region.

Acknowledgements

We thank the Cape Roberts Drillsite and Science Teams for their sterling efforts in acquiring and characterizing the CRP cores. CRF's participation in the Cape Roberts Project was supported by two out-of-cycle research grants from the Australian Research Council. SAH acknowledges support from the Royal Society of New Zealand Marsden Fund. TJW's participation in the Cape Roberts Project was supported by the National Science Foundation under Grant no. OPP-9527394. Continued work by TJW on the McMurdo Sound Basin framework has been supported by NSF-OPP-0313974. Karl Hinz and an anonymous individual are thanked for their reviews of the submitted manuscripts. Giuliano Brancolini and Instituto Nazionale di Oceanografia e Geofisica Sperimentale, Trieste, Italy, are thanked for providing access to seismic lines IT90a-70 and -75 for our analysis.

References

Barrett PJ, Henrys SA, Bartek LR, Brancolini G, Busetti M, Davey FJ, Hannah MJ, Pyne AR (1995) Geology of the margin of the Victoria Land Basin off Cape Roberts, southwest Ross Sea. In: Cooper AK, Barker PF, Brancolini G (eds) Geology and seismic stratigraphy of the Antarctic margin. Amer Geophys Union Antarctic Res Ser 68, Washington DC, pp 183–207

Bartek LR, Henrys SA, Anderson JB, Barrett PJ (1996) Seismic stratigraphy of McMurdo Sound, Antarctica: implications for glacially influenced early Cenozoic eustatic change? Marine Geology 130:79–98

Brancolini G, Cooper AK and Coren F (1995) Seismic facies and glacial history in the western Ross Sea (Antarctica). In: Cooper AK, Barker PF, Brancolini G (eds) Geology and seismic stratigraphy of the Antarctic margin. Amer Geophys Union Antarctic Res Ser 68, Washington DC, pp 209–233

Cape Roberts Science Team (1998) Initial Report on CRP-1, Cape Roberts Project, Antarctica. Terra Antartica 5:1–187

Cape Roberts Science Team (1999) Studies from the Cape Roberts Project, Ross Sea, Antarctica: Initial Report on CRP-2/2A. Terra Antartica 6:1–173

Cape Roberts Science Team (2000) Studies from the Cape Roberts Project, Ross Sea, Antarctica: Initial Report on CRP-3. Terra Antartica 7:1–209

Cooper AK, Davey FJ, Behrendt JC (1987) Seismic stratigraphy and structure of the Victoria Land Basin, western Ross Sea, Antarctica. In: Cooper AK, Davey FJ (eds) The Antarctic continental margin: geology and geophysics of the western Ross Sea. Circum-Pacific Council Energy Mineral Resources, Earth Sci Ser 5B, Houston, pp 27–65

Fielding CR, Woolfe KJ, Howe JA, Lavelle MA (1998) Sequence stratigraphic analysis of CRP-1, Cape Roberts Project, McMurdo Sound, Antarctica. Terra Antartica 5:353–361

Fielding CR, Naish TR, Woolfe KJ, Lavelle MA (2000) Facies analysis and sequence stratigraphy of CRP-2/2A, Victoria Land Basin, Antarctica. Terra Antartica 7:323–338

Fielding CR, Naish TR, Woolfe KJ (2001) Facies architecture of the CRP-3 drillhole, Victoria Land Basin, Antarctica. Terra Antartica 8:217–224

Fitzgerald PG (1992) The Transantarctic Mountains in southern Victoria Land: the application of apatite fission track analysis to a rift-shoulder uplift. Tectonics 11:634–662

Fitzgerald PG, Baldwin S (1997) Detachment fault model for the evolution of the Ross embayment. In: Ricci CA (ed) The Antarctic region – geological evolution and processes. Terra Antartcia Publication, Siena, pp 555–564

Hamilton RJ, Luyendyk BP, Sorlien CC, Bartek LR (2001) Cenozoic tectonics of the Cape Roberts Rift Basin and Transantarctic Mountain Front, southwestern Ross Sea, Antarctica. Tectonics 20:325–342

Henrys SA, Bucker CJ, Bartek LR, Bannister S, Niessen F, Wonik T (2000) Correlation of seismic reflectors with CRP-2/2A, Victoria Land Basin, Antarctica. Terra Antartica 7:221–230

Henrys SA, Bucker CJ, Niessen F, Bartek LR (2001) Correlation of seismic reflectors with the CRP-3 drillhole, Victoria Land Basin, Antarctica. Terra Antartica 8:127–136

Kyle PR (1990) McMurdo Volcanic Group – western Ross embayment: introduction. In: LeMasurier WE, Thomson JW (eds) Volcanoes of the Antarctic Plate and Southern Oceans. Amer Geophys Union Antarctic Res Ser 48, Washington DC, pp 19–25

Naish TR, Woolfe KJ, Wilson GS, Atkins C, Barrett PJ, Bohaty SM, Bucker C, Claps M, Davey F, Dunbar G, Dunn AG, Fielding CR, Florindo F, Hannah M, Harwood DM, Watkins D, Henrys S, Krissek L, Lavelle MA, van der Meer J, McIntosh WC, Niessen F, Passchier S, Powell RD, Roberts AP, Sagnotti AP, Scherer RP, Strong CP, Talarico F, Verosub KL, Webb P-N, Wonik T (2001) Orbitally induced oscillations in the East Antarctic Ice Sheet: direct evidence from the Cape Roberts Drilling Project. Nature 413:719–723

Salvini F, Brancolini G, Busetti M, Storti F, Mazzarini F, Coren F (1997) Cenozoic dynamics of the Ross Sea region, Antarctica: crustal extension, intraplate strike-slip faulting and tectonic inheritance. J Geophys Res 102:24669–24696

White N, McKenzie D (1988) Formation of the "steer's head" geometry of sedimentary basins by differential stretching of the crust and mantle. Geology 16:250–253

Wilson TJ (1995) Cenozoic transtension along the Transantarctic Mountains-West Antarctic Rift boundary, southern Victoria Land, Antarctica. Tectonics 14:531–545

Wilson TJ, Henrys SA, Hannah MJ, Jarrard RD, Fielding CR, Barrett PJ, Paulsen TS (submitted) New rift history for West Antarctica. Geology

Constraints on the Timing of Extension in the Northern Basin, Ross Sea

Steven C. Cande[1] · **Joann M. Stock**[2]

[1] Scripps Institution of Oceanography, Mail Code 0220, La Jolla, CA 92093-0220, USA, <scande@ucsd.edu>

[2] California Institute of Technology, Mail Stop 252-21, Pasadena, CA 91125, USA, <jstock@gps.caltech.edu>

Abstract. Recent kinematic constraints for the region north of the western Ross Sea suggest that there was approximately 150 km of seafloor spreading in the Adare Basin, northeast of Cape Adare, between Chrons 20 and 8 (43 to 26 Ma). This kinematic history has important implications since the 150 km of extension in the Adare Basin occurred immediately north along strike from the Northern Basin of the Ross Sea, whose extensional history is not well known. This paper examines the transition from the structures in the Adare Basin to the structures of the Northern Basin and speculates on the manner in which the extension was accommodated in the Ross Sea. Magnetic anomaly data in the Adare Basin document a sequence of anomalies 18 to 12 formed during a period of very slow spreading. The easternmost part of this sequence, anomalies 16 to 18, coalesces into a single positive anomaly near 72° S, forming a distinct anomaly that can be traced southward from the Adare Basin across the continental margin and down the east side of the Northern Basin to a latitude of roughly 73° S. This observation has important implications for the tectonic history of the Ross Sea since it suggests that most of the extension in the Adare Basin continued into the Northern Basin. This, in turn, suggests that the Northern Basin was formed by a combination of crustal thinning and massive, narrowly focused intrusions.

Introduction

The Cenozoic tectonic evolution of the West Antarctic Rift System has been a major focus of geological research in Antarctica (e.g., Behrendt et al. 1991; Behrendt et al. 1993) including several MCS seismic reflection programs (see in Cooper et al. 1995) and a recent drilling program near Cape Roberts (Barrett et al. 2000). Cenozoic extension between East and West Antarctica produced the deep sedimentary basins in the Ross Sea (Cooper et al. 1987) and the uplift of the Transantarctic Mountains (Fitzgerald 1992). The rift system has clearly experienced several distinct episodes of extension with changing stress directions and strain rates (Salvini et al. 1997). However, because of the complexity of the plate motion history, the connection between the geological structures imaged in the Rift System and the plate motions responsible for them has been poorly known.

The West Antarctic Rift System is the result of the complex interaction of the Pacific, Australia, Lord Howe Rise (prior to 53 Ma), West Antarctic and East Antarctic Plates. Although the major aspects of these plate motions have been known for 30 years (e.g., Molnar et al. 1975), important new details have been revealed by recent work. In particular, Cande et al. (2000a) used magnetic lineations to constrain at least one and perhaps two episodes of major East-West Antarctic separation in Cenozoic time. The younger episode of separation, between Chrons 20 and 8 (43 to 26 Ma), involved about 150 km seafloor spreading in the Adare Basin (Fig. 6.5-1), between Cape Adare and the Hallett Ridge, north of the Ross Sea embayment (Cande et al. 2000a). The older episode is more poorly constrained, but it may have involved up to 100 km of E-W directed extension in the Central Basin,

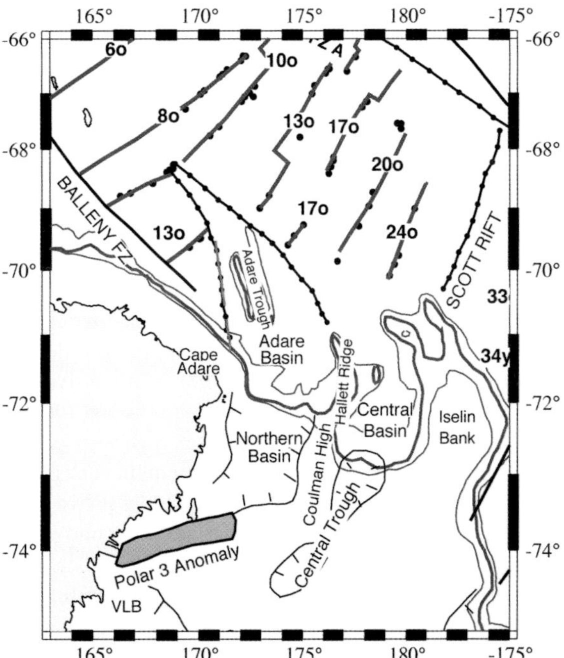

Fig. 6.5-1. Location of basins and structural highs in the western Ross Sea referred to in the text. Boundaries of sedimentary basins are from Davey and Brancolini (1995). The 1000, 1500 (*heavy*) and 2000 m bathymetric contours (*red lines*) are shown. *Blue lines* show magnetic isochrons associated with spreading on the Southeast Indian Ridge. These anomalies constrain the amount and timing of extension in the northern Ross Sea. The Polar 3 Anomaly (*shaded region*) is a prominent magnetic anomaly between the Victoria Land Basin (*VLB*) and the Northern Basin

From: Fütterer DK, Damaske D, Kleinschmidt G, Miller H, Tessensohn F (eds) (2006) Antarctica: Contributions to global earth sciences. Springer-Verlag, Berlin Heidelberg New York, pp 319–326

between the Hallett Ridge and Iselin Bank, between roughly Chrons 27 and 24 (61 to 53 Ma; Cande et al. 2000b). These results have fundamental implications for the timing and amount of Cenozoic extension in the Ross Sea region but they remain to be reconciled with existing geological and geophysical data from the Ross Embayment and the West Antarctic Rift System.

Here we discuss magnetic and gravity data that constrain a small but critical aspect of the tectonic development of the sedimentary basins in the northern Ross Sea (Northern Basin, Central Trough, Fig. 6.5-1). The conventional wisdom is that these basins formed by stretching and thinning of continental crust. However, the striking alignment of shelf and deepwater structures in the northern Ross Sea (e.g., the Coulman High aligns with the Hallett Ridge), combined with the constraints of our kinematic model, suggests that the Northern Basin may have formed by extensive crustal thinning and very focused, massive intrusions synchronous with, and kinematically linked to, seafloor spreading in the Adare Basin. This has important implications for the tectonics of other sedimentary basins in the Ross Sea.

Constraints on Cenozoic Extension in the West Antarctic Rift System

The tectonic history of the Ross Sea region has been difficult to unravel in large part because of the lack of a well-constrained plate kinematic framework. There has clearly been a considerable amount of extension in the Ross Sea region as evidenced by the anomalously thin crust (~20 km) and shallow Moho across the embayment (e.g., Cooper et al. 1987; Davey and Cooper 1987); the total amount of extension has been estimated to be on the order of 400 km (Behrendt et al. 1991). Paleomagnetic studies (DiVenere et al. 1994; Luyendyk et al. 1996) independently indicate between 440 and 1 820 km of relative motion between East and West Antarctica in the last 100 Ma. The timing of the extension, however, has been poorly constrained. Fission track dates on basement rocks from DSDP site 270 in the Ross Sea indicate at least two periods of extension, one in the Jurassic and a second in the late Cretaceous (Fitzgerald and Baldwin 1997). The Late Cretaceous extension, documented in West Antarctica and Marie Byrd Land, on the eastern side of the Ross Sea, is coeval with the separation of the surrounding fragments of Gondwana (New Zealand and West Antarctica) (e.g., Siddoway et al. 2004). Based on regional plate reconstructions, Lawver and Gahagan (1994) argued that the bulk of the extension within the Ross Sea must have taken place in Late Cretaceous time.

Cande et al. (2000a) showed that, contrary to earlier indications, there has probably been a considerable amount of extension between East and West Antarctica in the western Ross Sea region in mid-Cenozoic time. The critical constraint on this motion is the misfit of magnetic anomalies 10 and older on the Antarctic Plate northeast of the Adare Basin (blue lines in Fig. 6.5-1) when rotated back to their conjugate location on the Australian Plate using rotation parameters for Australia-East Antarctica. Specifically, there is a progressively larger misfit for increasingly older anomalies, starting with a 10 km misfit for anomaly 10 and increasing to a roughly 170 km misfit for anomaly 20. Cande et al. (2000a) showed that the systematic misfit of these Southeast Indian Ridge (SEIR) anomalies indicated that there was an extensional boundary between East and West Antarctica that ran south from the SEIR and into the western Ross Sea starting around Chron 20 (43 Ma) and ending around Chron 8 (26 Ma).

Further evidence for the timing and the amount of extension in the region of the northern Ross Sea comes from the presence of linear magnetic anomalies straddling the Adare Trough, a large NNW-trending graben located along the west side of the Adare Basin (Fig. 6.5-2). Cande et al. (2000a) showed that these anomalies were consistent with a period of slow seafloor spreading (12 mm yr^{-1} full rate) between Chrons 20 and 8 (43 to 26 Ma) (see projected profiles and model in Fig. 6.5-3; see Fig. 6.5-2 for location of profiles). The anomaly pattern indicates that the spreading rate was asymmetrical, with the west flank (5.5 mm yr^{-1}) slower than the east flank (7 mm yr^{-1}). Although the anomalies are difficult to recognize because of the slow spreading rate, their identification is consistent with the motion required by the systematic misfit in SEIR anomalies 10 and older from northeast of the Adare Basin. Cande et al. (2000a) concluded that the Adare Basin anomalies were caused by roughly 150 km of seafloor spreading between East and West Antarctica north of the Ross Sea, and that the pole of rotation for this motion was sufficiently far away that there should have been a similar amount of extension further south in the northern Ross Sea.

Our proposed identification of the Adare Basin anomalies is supported by magnetic anomaly data collected on R/VIB Nathaniel Palmer in February 2005 (NBP0501). These data, shown in red in Fig. 6.5-2, demonstrate the existence of a classic magnetic bight between the NNW-SSE striking Adare Basin anomalies and the NE-SW striking SEIR anomalies between 69° S and 70° S. In particular, anomalies 17 and 18 clearly wrap around in a bight between 174° W and 175° W near 70° S. The continuity of the negative anomaly between anomalies 17 and 18 as it wraps around the bight is marked by the dashed white line in Fig. 6.5-2.

The implications of these kinematic constraints for the tectonic evolution of the western Ross Sea basins since the late Eocene have not been developed. Ideally, they would provide a check on a tectonic history independently constrained by drilling and seismic stratigraphy. However, because the sedimentary sequences drilled at Cape Rob-

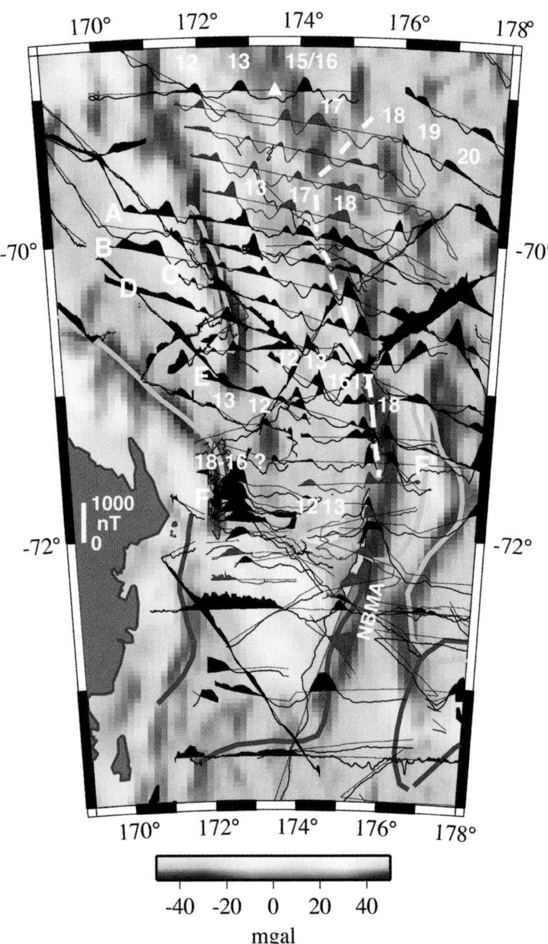

Fig. 6.5-2. Magnetic anomaly data in the region of the Adare Trough shown superimposed on the satellite-derived free-air gravity field of McAdoo and Laxon (1997). Note the apparent continuity of anomalies 16–18 on the east flank of the Adare Basin with the Northern Basin Magnetic Anomaly (*NBMA*). The continuity of this anomaly across the margin suggests that most of the spreading in the Adare Basin was accommodated by extension in the Northern Basin. The *dashed white line* follows the negative anomaly between anomalies 17 and 18 as it wraps around the magnetic bight between the NNW-SSE striking Adare Basin anomalies and the NE-SW striking SEIR anomalies. Positive anomalies are *shaded black* except for the north-south profile near 175° E which is *shaded in gray*. Magnetic data collected on the R/VIB "Nathanial Palmer" in February 2005 are *shaded red*. The 1 500 m bathymetric contour is shown by the *heavy gold line*. The boundaries of the sedimentary basins are shown by the *heavy blue line*. *White triangle* shows location of DSDP site 274. The labels *A* through *F* refer to profiles shown in Fig. 6.5-3

erts proved to be younger than anticipated (Barrett et al. 2000), the previous interpretations of the stratigraphic constraints on the early Cenozoic tectonic evolution of the Ross Sea basins are now in question. At the latitude of Cape Roberts (77° S) there was a major basin-forming event in Eocene-Oligocene time, synchronous with the seafloor spreading in the Adare Basin (Hamilton et al. 2001); however the net amount of extension during this time interval, and how it connects to structures farther north in the Ross Sea, is still not understood. Thus the kinematic constraints of the Adare Basin seafloor spreading anomalies take on added importance. To understand their significance requires tracing the Adare Basin structures southward.

The focus of this paper is the transition from the structures in the Adare Basin to the structures of the northern Ross Sea basins. The Adare Basin structures are best defined in their gravity and magnetic signatures. Although the topographic signature of the Adare Trough disappears around 71° S, a seismic profile at 71°30' S (TH91 Line 28; although not the part of Line 28 shown in Fig. 6.5-4) indicates that a graben is still present in the subsurface at that latitude. This is consistent with the continuity of the gravity low associated with the axis of the trough, which can be traced in the ERS-1 derived free-air gravity field (McAdoo and Laxon 1997) to nearly 72° S, essentially disappearing into the continental slope adjacent to the Northern Basin (Fig. 6.5-2). Magnetic data show that lineations from anomaly 12 to 18 can be clearly traced on the east flank of the trough, just west of the Hallett Ridge, from 70° S to 71°30' S (Fig. 6.5-2). This observation suggests that the Hallett Ridge (and the rest of West Antarctica) was rifted from Cape Adare around Chron 20 (43 Ma) when spreading initiated in the Adare Basin.

The magnetic anomaly pattern on the west side of the Adare Basin, in the conjugate position to anomalies 16–18 in the eastern Adare Basin, is more complicated. First, we note that the anomalies on the west side of the Adare Basin cannot be clearly traced between 71°00' and 71°30' S (Fig. 6.5-2). This might reflect later deformation or strike slip faulting in the region. However, there is a conjugate anomaly to the anomaly 16–18 sequence in the eastern Adare Basin. R/VIB "Nathaniel Palmer" cruise 0209 recorded a prominent linear N-trending positive magnetic anomaly east of the north-south trending margin of Cape Adare. This anomaly is labeled "18–16 ?" in Fig. 6.5-2 and is also shown on the left side of profile F in Fig. 6.5-3. (Note that profile F–F' in Fig. 6.5-3 is a composite profile; no single magnetic profile crosses the Adare Basin at this latitude.) Although this anomaly may be related, in part, to structures associated with the development of the continental margin, it is also consistent with the shape of a coalesced (due to slow spreading) sequence of anomalies 16 to 18, which could have formed by seafloor spreading as the conjugate to the faster spreading (and more easily identified) anomalies 16 to 18 in the eastern Adare Basin. Thus, even at this latitude, there is evidence for a linear pair of conjugate magnetic anomaly sequences generated by seafloor spreading about the axis of the Adare Trough.

The most important new observation is that the southern end of the oldest anomaly sequence on the east side of the Adare Basin, anomalies 16–18, appears to coalesce into a single positive anomaly around 175°30' E, 71°40' S.

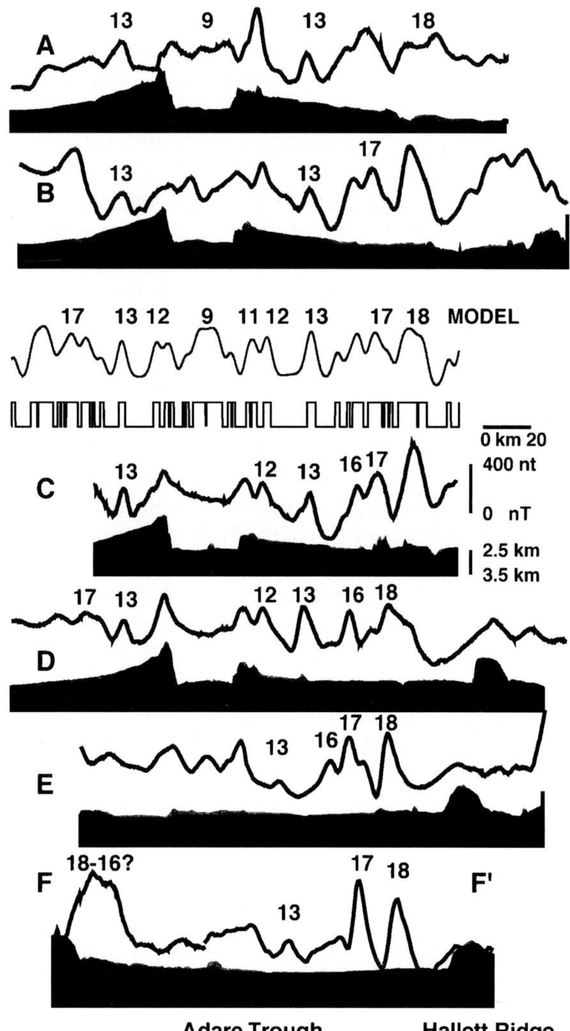

Fig. 6.5-3. Magnetic and topographic profiles across the Adare Trough. Profile *F–F'* is a composite of two profiles (NBP0209 on the left and TH91 on the right) crossing the southern Adare Basin. A magnetic model is shown between profiles *B* and *C* (see Fig. 6.5-2 for locations). Polarity widths based on timescale of Cande and Kent (1995)

From there it can be traced onto the shelf and along the east side of the Northern Basin south to about 73° S (Fig. 6.5-2). The continuity of this anomaly, which we call the Northern Basin Magnetic Anomaly, was first noted in archival magnetics data (shaded black in Fig. 6.5-2), including a line which fortuitously ran along the axis of the anomaly (shaded gray in Fig. 6.5-2), and was confirmed by several additional crossings which were made by the R/VIB "Nathaniel Palmer" in February 2005 (shaded red in Fig. 6.5-2). The Northern Basin Magnetic Anomaly (NBMA), which is continuous from 71°40' to 73°00' S, is very significant since it strongly suggests that there is no offset between the structures of the Adare Basin and the Northern Basin. This, in turn, suggests that the Coulman High was rifted as part of a continuous feature with the Hallett Ridge from the Cape Adare margin at Chron 20. Although there are fewer data on the west side of the basin, the existing lines suggest that the positive anomaly corresponding to anomalies 18–16? on the west flank of the Adare Basin (abutting the Cape Adare margin) might also continue southward into the Northern Basin.

The seismic data in the southern Adare Basin supports the continuity of structures between the Adare Basin and the Northern Basin. TH91 Line 28 (middle line, Fig. 6.5-4) shows that the basement beneath anomalies 16–18 on the east side of the Adare Basin, just west of the Hallett Ridge, is relatively shallow, in fact shallower than basement beneath the younger, central part of the basin. The average depth to the top of the basement, corrected for the sediment cover, is only about 2 300 m, which is about 1 500 m shallower than the expected basement depth for 40 Ma oceanic crust (Müller et al. in press). A north-south crossing seismic line (TH91 Line 17; top line, Fig. 6.5-4) shows that the basement beneath anomalies 16–18 shallows gradually but continuously to the south as the Adare Basin merges into the Northern Basin, without any evidence of offsetting structures. The gradually shallowing basement implies a gradual thickening of the oceanic crust as the Northern Basin is approached. Unfortunately, basement cannot be imaged beneath the multiple in the Northern Basin on this line. Basement on IT88 Line 6 (bottom line, Fig. 6.5-4) beneath magnetic anomaly 16–18 ? on the western side of the Adare Basin, has a similar character to basement on the two TH91 lines in the eastern Adare Basin, which we interpret as supporting the oceanic character of the basement in this region. Unfortunately, basement on this line cannot be followed southward into the Northern Basin.

Where the sedimentary packages can be seen above the basement on these three lines, there are also general similarities in the lower part of the section, as would be expected if the seafloor in these two regions were formerly adjacent. On the TH91 Line 28, a seismically transparent package of sediment directly overlies acoustic basement in the vicinity of magnetic anomaly 16–18 (eastern side). On IT88 Line 6, which is interpreted to run right along the 16–18? anomaly on the (conjugate) western side, the lowest sedimentary package, immediately overlying the deepest basement, is similarly quite transparent. This package lies beneath the early Miocene RSU4a unconformity as defined by Brancolini et al. (1995).

The Bouguer gravity anomaly (Fig. 6.5-5) also supports the continuity of structures from the Adare Basin southward into the Northern Basin. The Bouguer anomaly was calculated for the free-air gravity field of McAdoo and Laxon (1997) and includes corrections for both the sediment water interface, based on a revised topographic map of the Ross Sea (Davey 2004) and the sediment basement interface based on the depth to basement map from the Antostrat volume (Brancolini et al. 1995). The Bouguer anomaly shows that the large free air gravity high at the

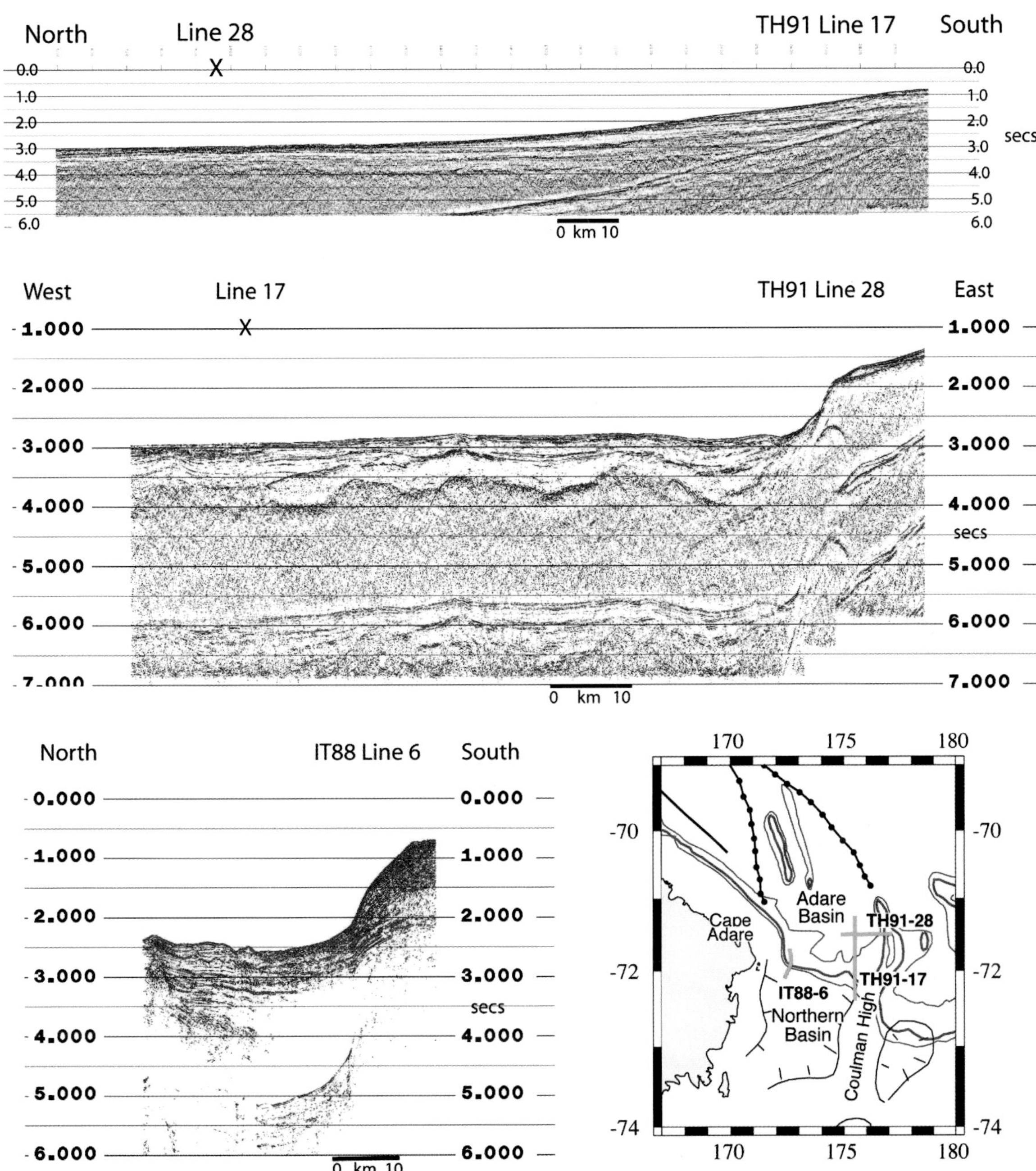

Fig. 6.5-4. Seismic reflection data supporting the continuity of structures between the Adare Basin and Northern Basin. *TH91 Line 28* (*middle profile*) shows that the acoustic basement (which we hypothesize to be oceanic crust) on the east side of the Adare Basin is relatively shallow for its age (40 Ma). *TH91 Line 17* (*top*) shows that this basement shallows gradually but continuously to the south without evidence of any off-setting structures beneath the lower continental slope. *IT88 Line 6* (*bottom*) shows that the basement reflector on the west side of the Adare Basin has a similar character to the basement on the east side of the basin. In the *center* of the *middle frame*, and in the *bottom frame*, which are from approximately conjugate locations on the margins of the Adare Basin rift system, a similarly transparent character of the lowest sediments above basement is apparent. The location of the profiles is shown in the *insert* on the *lower right* and in Fig. 6.5-5. Note difference in scale among the profiles. Seismic lines are available from the SCAR Seismic Data Library System

shelf edge is primarily due to the changing water depth and is not a reflection of a structural feature near the shelf edge. More importantly, Fig. 6.5-5 shows that the Bouguer gravity high over the eastern Adare Basin continues southward into the Northern Basin without an offset at the continental slope. This is contrary to what would be expected if there was a major fault along this margin transforming motion into the Central Basin.

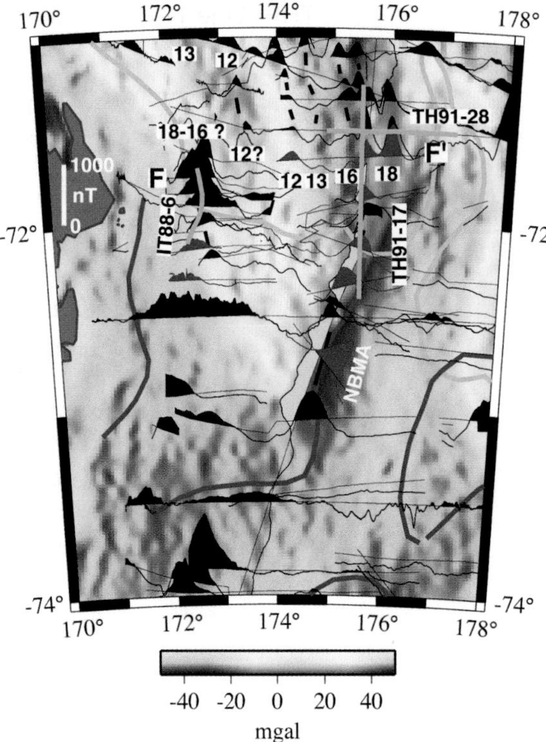

Fig. 6.5-5. The Bouguer gravity anomaly over the southern Adare Basin and Northern Basin. Note that the Bouguer gravity high over the eastern Adare Basin continues into the Northern Basin without any obvious offset near the continental slope (corresponding roughly to the region near the 1 500 m isobath). The Bouguer anomaly includes corrections for water depth and the sediment basement interface. Magnetic anomaly data and other symbols are plotted as in Fig. 6.5-2. The *heavy green lines* show the location of the seismic profiles in Fig. 6.5-4

Implications

The Adare Basin kinematic history places tight constraints on the development of the Northern Basin. The kinematic constraints require that the Hallett Ridge, at least that part adjacent to where we can positively identify anomaly 18 (i.e. north of 71°30' S), was abutting Cape Adare prior to Chron 20. This in turn implies that there was an equivalent amount of extension, about 170 km, across the northern Ross Sea sedimentary basins between Chrons 20 and 8. There are two alternative models by which this extension could have been accommodated.

Model 1

In the first model, the extension is taken up primarily by continental stretching and thinning. If this is the case, it must have been taken up by stretching and thinning over both the Northern Basin and Central Trough since the crustal thickness (~12 km) and inferred β (thinning) factor in the Northern Basin (Davey and Brancolini 1995) preclude it from taking up more than about 50% of the extension. This, in turn, requires that there was a major transform boundary between the Adare Basin and the Northern Basin. Salvini et al. (1997) analyzed all the existing seismic data and concluded that a NW-SE striking fracture zone, which they called the Adare Fracture Zone, separates the shallow water and deep water basins. They asserted that this fracture zone has been active in the last 30 Ma. We have examined the existing seismic data to look for evidence of such a fracture zone and conclude that the data are too sparse to draw this conclusion. Indeed, the apparent continuity of magnetic anomalies 16–18 along the east side of the Adare Basin with the Northern Basin Magnetic Anomaly suggests that the offset across any such fault must be small. Alternatively, the alignment of these magnetic and gravity anomalies is a coincidence.

Model 2

The alternative model is that the Northern Basin took up all of the extension by major localized crustal thinning and massive intrusion into the lower continental crust at the same time as the Adare Basin was spreading. Although this mechanism has been invoked in other basins of the Ross Sea (e.g., Cooper et al. 1987), the process by which the Northern Basin alone accommodated the 150 km of extension would have required extreme thinning and considerable addition of new crustal material from elsewhere, since, if the Coulman High is reconstructed to its "pre-breakup" position at Chron 20, there would have been only a small amount of pre-existing continental crust between the Coulman High and the Cape Adare margin available to intrude and extend. If the Northern Basin Magnetic Anomaly is really continuous with anomalies 16–18 in the Adare Basin, it suggests that the intrusions were very narrowly focused at the time of anomalies 16–18. The argument in favor of this alternative is the continuity of structures across the margin and the lack of evidence for a major transform offset between the Adare Basin and the Northern Basin.

Note that if Model 2 is correct, the percentage of extension inferred by Davey and Brancolini (1995) (with a β factor of ~3) must be underestimated. Calculations of β factor are based on the assumption that continental crust has thinned in place without any lateral flow in the lower crust different from that of the upper crust. These calculations frequently underestimate the amount of extension in continental rift systems with large magnitudes of extension, where lateral flow of the lower or middle crust has occurred to fill in the gap as the two sides of the extensional system move apart (e.g., Block and Royden 1990; Wernicke1992). Other rift systems that are transitional from continental to oceanic have extensive regions of new area that are not fully oceanic crust but instead inferred to be transitional crust (Oskin et al. 2001). If Model 2 is correct, scenarios similar to these will need to be explored for the Northern Basin.

Both of these models also have important implications for the seismic stratigraphic history of the Northern Basin. If there were a major extensional event in the Northern Basin at the same time as spreading in the Adare Trough (40 Ma to 28 Ma), then any stratigraphic packages deposited during this time would be highly faulted and disrupted. The lowest seismic stratigraphic package in the Western Ross Sea that is discussed by Brancolini et al. (1995), RSS-1, is inferred to have an age of early Oligocene and older. RSS-1 appears to be absent in many places, and it is therefore difficult to correlate it across basement highs. However, where it is identified in the Northern Basin, it overlies faulted blocks of acoustic basement and has mostly sub-parallel reflectors (Brancolini et al. 1995). This suggests that it post-dates any major extensional event that would have opened the basin. We therefore suggest that in the Northern Basin, the strata that have been interpreted as RSS-1 are most likely younger than the major extension in the Adare Basin; i.e., no older than latest Oligocene in age. Any units that were deformed or intruded during the major rifting event between 40 and 28 Ma should therefore lie below RSS-1, within what is interpreted as acoustic basement in the seismic sections.

Summary and Conclusions

Recent studies have led to the development of a detailed kinematic model for the region north of the Western Ross Sea. The aspect of this model that has the most important ramifications for the tectonics of the region is the demonstration that there was an episode of extension between East and West Antarctica in mid-Cenozoic time. This period of extension was constrained both by broad scale regional studies that determined the motion of East and West Antarctica using a plate circuit (i.e. by the summation of finite rotations constraining Australia and East Antarctic motion and Australia and West Antarctica motion) and by the analysis of magnetic anomalies straddling the Adare Trough. These results demonstrate that in the region of the Adare Trough there was roughly 150 km of approximately East-West extension between Chrons 20 and 8.

The important question now is what happened to this motion farther south in the western Ross Sea. In particular, it is important to establish whether *(a)* the motion is taken up over a broad area of the northern Ross Sea, i.e. between both the Northern Basin and Central Trough or *(b)* the motion continued straight south into the Northern Basin. Our analysis of magnetic anomalies and gravity data to the south of the Adare Trough suggest the latter, i.e. that the motion continued directly into the Northern Basin. The observations that support this conclusion are as follows:

Magnetic lineations on the east side of the Adare Basin continue southwards for 50 km after the Adare Trough ends at 71° S.

The sequence of positive magnetic anomalies along the east side of the Adare Basin (anomalies 16–18) coalesces into a single positive magnetic anomaly (the Northern Basin Magnetic Anomaly) around 71°40' S that continues along the eastern side of the Northern Basin to 73° S.

The Bouguer gravity high on the east side of the Adare Basin continues across the continental margin and into the Northern Basin without a major offset.

These observations imply structural continuity between the Adare and Northern Basins and, therefore, suggest that the Northern Basin should have a similar history of extension (~150 km of extension between 43 and 26 Ma) as the Adare Basin. This, in turn, suggests that the Northern Basin formed in Cenozoic time, most likely by a combination of major localized crustal thinning and massive, narrowly focused intrusions into the lower continental crust. The style of extension in the Northern Basin may be analogous to that seen in the Northern Gulf of California.

Acknowledgements

We thank the captains, officers, crew and scientific staff of the R/VIB "Nathaniel Palmer" for their dedicated efforts. Dr. Takemi Ishihara generously provided us with seismic reflection and magnetics data from the RV "Hakurei Maru". Two anonymous reviewers provided helpful comments. This work was supported by NSF grants OPP-03-38346 (Cande) and OPP-03-38317 (Stock).

Reference

Barrett PJ, Sarti M, Wise S (ed) (2000) Studies from the Cape Roberts Project, Ross Sea, Antarctica, Initial Report on CRP-3. Terra Antartica 7(1/2):1–209

Behrendt JC, LeMasurier WE, Cooper AK, Tessensohn F, Trehu A, Damaske D (1991) Geophysical studies of the West Antarctic rift system. Tectonics 10:1257–1273

Behrendt JC, Damaske D, Juergen F (1993) Geophysical characteristics of the Western Antarctic rift system. Geol Jb E47:49–101

Brancolini G, Cooper AK, Coren F (1995) Seismic facies and glacial history in the Western Ross Sea (Antarctica). In: Cooper AK, Barker PF, Brancolini G (eds) Geology and seismic stratigraphy of the Antarctic margin. Antarct Res Ser 68:209–233

Block L, Royden LH (1990) Core complex geometries and regional scale flow in the lower crust. Tectonics 9:557–567

Cande SC, Kent DV (1995) Revised calibration of the geomagnetic polarity timescale for the Late Cretaceous and Cenozoic. J Geophys Res 100:6093–6095

Cande SC, Stock J, Müller RD, Ishihara T (2000a) Cenozoic motion between East and West Antarctica. Nature 404:145–150

Cande SC, Stock JM, Müller RD, Ishihara T (2000b) Two stages of Cenozoic separation in the western Ross Sea embayment. EOS 48:F1131

Cooper AK, Davey FJ, Behrendt JC (1987) Seismic stratigraphy and structure of the Victoria Land Basin, western Ross Sea, Antarctica. In: Cooper AK, Davey FJ (eds) The Antarctic continental margin: geology and geophysics of the western Ross Sea. Earth Sci Ser Vol 5B, Circum-Pacific Res. Council, Houston, pp 27–76

Cooper AK, Barker PF, Brancolini G (eds) (1995) Geology and seismic stratigraphy of the Antarctic margin. Antarct Res Ser 68, Amer Geophys Union, Washington DC

Davey FJ (2004) Ross Sea bathymetry, 1:2000000, version 1.0. Institute of Geological & Nuclear Sciences geophys map 16. Inst Geol & Nuclear Sci Ltd., Lower Hutt

Davey FJ, Brancolini G (1995) The late Mesozoic and Cenozoic structural setting of the Ross Sea region. In: Cooper AK, Barker PF, Brancolini G (eds) Geology and seismic stratigraphy of the Antarctic nargin, Antarctic Res Ser 68, Amer Geophys Union, Washington DC, pp 167–182

Davey FJ, Cooper AK (1987) Gravity studies of the Victoria Land Basin and Iselin Bank. In: Cooper AK, Davey FJ (eds) The Antarctic continental margin: geology and geophysics of the western Ross Sea, Earth Sci Ser Vol 5B, Circum-Pacific Res. Council, Houston, pp 27–76

DiVenere VJ, Kent DV, Dalziel I (1994) Mid-Cretaceous paleomagnetic results from Marie Byrd Land, West Antarctica: A test of post-100 Ma relative motion between East and West Antarctica. J Geophys Res 99:15115–15139

Fitzgerald PG (1992) The Transantarctic Mountains in Southern Victoria Land: the application of apatite fission track analysis to a rift-shoulder uplift. Tectonics 11:634–662

Fitzgerald P, Baldwin S (1997) Detachment fault model for the evolution of the Ross Embayment. In: Ricci CA (ed) The Antarctic region: geological evolution and processes, Terra Antartica Publication, Siena, pp 555–564

Hamilton RJ, Luyendyk BP, Sorlien CS (2001) Cenozoic tectonics of the Cape Roberts Rift Basin and Transantarctic Mountains front, southwestern Ross Sea, Antarctica. Tectonics 20:325–342

Lawver LA, Gahagan LM (1994) Constraints on the timing of extension in the Ross Sea region. Terra Antartica 1:545–552

Luyendyk B, Cisowski S, Smith C, Richard S, Kimbrough D (1996) Paleomagnetic study of the northern Ford Ranges, western Marie Byrd Land, West Antarctica: motion between West and East Antarctica. Tectonics 15:122–141

McAdoo D, Laxon S (1997) Antarctic tectonics: constraints from an ERS-1 satellite marine gravity field. Science 276:556–560

Molnar P, Atwater T, Mammerickx J, Smith SM (1975) Magnetic anomalies, bathymetry, and the tectonic evolution of the South Pacific since the late Cretaceous. Geophys J Royal Astronom Soc 40:383–420

Müller RD, Cande SC, Stock JM, Keller WR (in press) Crustal structure and rift flank uplift of the Adare Trough, Antarctica. Geochem Geophys Geosys

Oskin M, Stock J, Martin-Barajas A (2001) Rapid localization of Pacific-North America plate motion in the Gulf of California. Geology 29:459–462

Salvini F, Brancolini G, Busetti M, Storti F, Mazzarini F, et al. (1997) Cenozoic geodynamics of the Ross Sea region, Antarctica – crustal extension, intraplate strike-slip faulting, and tectonic inheritance. J Geophys Res 102:24669–64696

Sandwell DT, Smith W (1997) Marine gravity-anomaly from Geosat and ERS-1 satellite altimetry. J Geophys Res 102:10039–10054

Siddoway CS, Baldwin SL, Fitzgerald PG, Fanning CM, Luyendyk BP (2004) Ross Sea mylonites and the timing of intracontinental extension within the West Antarctic rift system. Geology 32:57–60

Wernicke B (1992) Cenozoic extensional tectonics of the U.S. Cordillera. In: Burchfiel BC, Lipman PW, Zoback M (eds) The Cordilleran Orogen. Geol Soc Amer, Boulder, pp 553–581

The Structure of the Continental Margin off Wilkes Land and Terre Adélie Coast, East Antarctica

James B. Colwell[1] · Howard M. J. Stagg[1] · Nicholas G. Direen[2] · George Bernardel[1] · Irina Borissova[1]

[1] Geoscience Australia, GPO Box 378, Canberra, ACT, Australia
[2] Continental Evolution Research Group, University of Adelaide, SA, Australia

Abstract. In 2001 and 2002, the Australian Government acquired approximately 9000 km of high-quality geophysical data over the margin of East Antarctica between 110–142° E that provide a sound framework for understanding the geology of the region. The data comprise 36-fold deep-seismic, gravity and magnetic data and non-reversed refraction/wide-angle reflection sonobuoys recorded along transects that extend from the lower continental slope out to oceanic crust at a spacing along the margin of approximately 90 km. The continental slope is underlain by a major rift basin beneath which the crust thins oceanwards through extensive faulting of the rift and pre-rift sedimentary section and by mainly ductile deformation of the crystalline crust. Outboard of the margin rift basin, the 90 to 180 km wide continent-ocean transition zone is interpreted to consist primarily of continental crust with magmatic components that can account for the lineated magnetic anomalies that have been interpreted in this zone. The thick sedimentary section in the COT zone is floored by dense lower crustal or mantle rocks indicating massive (>10 km) thinning of the lower and middle crust in this zone. The boundary between the margin rift basin and the COT is marked by a basement ridge which potential field modelling indicates is probably composed of altered/serpentinised peridotite. This ridge is similar in form and interpreted composition to a basement ridge located in a similar structural position at the inboard edge of the COT on the conjugate margin of the Great Australian Bight. On both margins, the ridge is probably the product of mantle up-welling and partial melting focussed at the point of maximum change/necking of crustal thickness. Integrated deep-seismic and potential field interpretations point very strongly to the boundary between unequivocal oceanic crust and largely continental crust of the continent-ocean transition as lying in very deep water, and considerably seaward of most previous interpretations (often based on inadequate seismic data or magnetic data only). We consider the continent-ocean boundary to be well-constrained from 124–131° E and unequivocal from 131-140° E, but open to debate in the sector from 110-124° E. There is a strong degree of pre-breakup symmetry between the conjugate margins of southern Australia and East Antarctica east of about 120° E. In addition to the crustal symmetry, there is also a strong correlation in seismic character between the margins, which allows us to date the major unconformities as probably of base Turonian, Maastrichtian and early Middle Eocene age.

Introduction and Background

The continental margin and the adjacent deep ocean basins off Wilkes Land and Terre Adélie, East Antarctica, occupy a key position for understanding the breakup of eastern Gondwana and the separation of Australia and Antarctica in the Late Cretaceous. However, while the geology of the conjugate southern Australian continental margin is reasonably well documented through seismic and well data, particularly between the Great Australian Bight and Tasmania (e.g., Sayers et al. 2001; Totterdell et al. 2000; Moore et al. 2000; Finlayson et al. 1998), the geology of the East Antarctic margin is still poorly understood.

Multichannel seismic surveys of the Wilkes-Adélie margin were carried out by France in 1982 (ATC-82 survey; Wannesson et al. 1985), the USA in 1984 (L1-84-AN survey; Eittreim and Smith 1987) and Japan in 1982, 1983 and 1994 (TH-82, TH-83, TH-94 and TH-95 surveys; Sato et al. 1984; Tsumuraya et al. 1985; Tanahashi et al. 1987; Ishihara et al. 1996; Tanahashi et al. 1997). With the exception of the TH-94 survey and a long margin strike-line recorded on the TH83 survey, the previous work has been generally concentrated in the area east of 130° E (i.e. offshore eastern Wilkes Land and Terre Adélie). The quality of the data recorded varied from fair to moderate and most lines were located on the continental slope with rare lines in the deep-ocean basin.

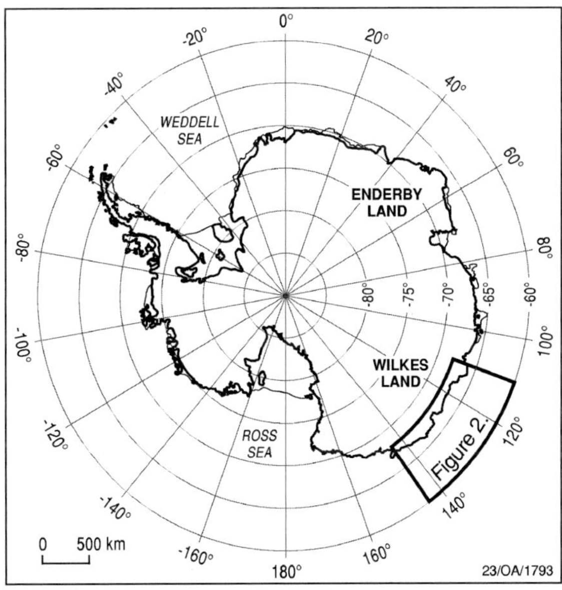

Fig. 6.6-1. Antarctica, showing the study area for this paper

From: Fütterer DK, Damaske D, Kleinschmidt G, Miller H, Tessensohn F (eds) (2006) Antarctica: Contributions to global earth sciences. Springer-Verlag, Berlin Heidelberg New York, pp 327–340

By the end of the 1990s, the total multichannel seismic coverage of the Wilkes-Adélie margin was approximately 11 000 km mainly along irregularly spaced and oriented lines. Published interpretations of these datasets are limited, with the most comprehensive work being that of Eittreim and Hampton (1987) who focussed on the L1-84-AN survey but also included an interpretation of the TH-82 survey. These interpretations were limited by the quality of the available data and the processing techniques available at the time, as well as the irregular line coverage.

In 2001 and 2002, the Australian Government acquired a major new reflection and refraction seismic and potential field dataset along the margin of East Antarctica from 38° to 150° E, the Australian Antarctic and Southern Ocean Profiling Project (Stagg and Schiwy 2002; Surveys GA-228 and GA-229). Approximately 9 000 km of these data were recorded along the margins of Wilkes Land, Terre Adélie and George V Land (Fig. 6.6-1, 6.6-2). Reflection seismic data were recorded 36-fold from a 288 channel, 3 600 m streamer and a 60 litre tuned airgun array, with a record length of 16 s. The processed data are of very high quality and provide extremely clear imaging down to the lower crust and upper mantle as well as clear definition of the post-rift sedimentary packages. The survey lines extend from the mid-continental slope to the deep ocean basin and were regularly spaced at approximately 90 km along the margin. Individual line lengths range from 300 to 450 km.

This paper presents an integrated seismic and potential field interpretation of two representative deep-seismic profiles, concentrating on the transition from thinned continental crust to the adjacent oceanic crust and on the along-margin variations in this transition. We also summarise the interpretation of the entire dataset by reference to a map of the tectonic elements of the margin from 110–144° E.

Interpretation

The interpretation presented here is based on the integration of reflection and refraction seismic data and potential field modelling and is illustrated by reference to interpreted seismic profiles and potential field models (Fig. 6.6-3 to 6.6-8) and a map of the interpreted tectonic elements (Fig. 6.6-9). The interpreted profiles were located off central Wilkes Land (line GA-228/24) and Terre Adélie (line GA-229/06; locations in Fig. 6.6-2).

The reflection seismic data are the primary dataset for the interpretation as they provide clear definition of the seismic stratigraphy and structural geometries down to the middle crust and, in places, down to the deep crust and upper mantle. The potential field modelling (described in detail in Stagg et al. 2005) was designed to quantitatively validate the time-based interpretations of seismic profiles against coincident potential field data, with the objective of enhancing the understanding of the margin crustal framework, and particularly the transition from extended continental to oceanic crust. This was done by combining migrated depth-converted seismic reflection images and their interpretations with density data derived from numeric conversion of seismic refraction and stacking velocities. The gravity and magnetic fields of these petrophysically-attributed geometric models were then concurrently forward-modelled in 3-dimensions, by giving the sections limited strike extents. Limited information on magnetic physical properties was derived from analysis of dredge samples at analogue sites in the Southern Ocean; however, the magnetic properties remain relatively unconstrained. As the quality and definition of the reflection seismic data deteriorates with

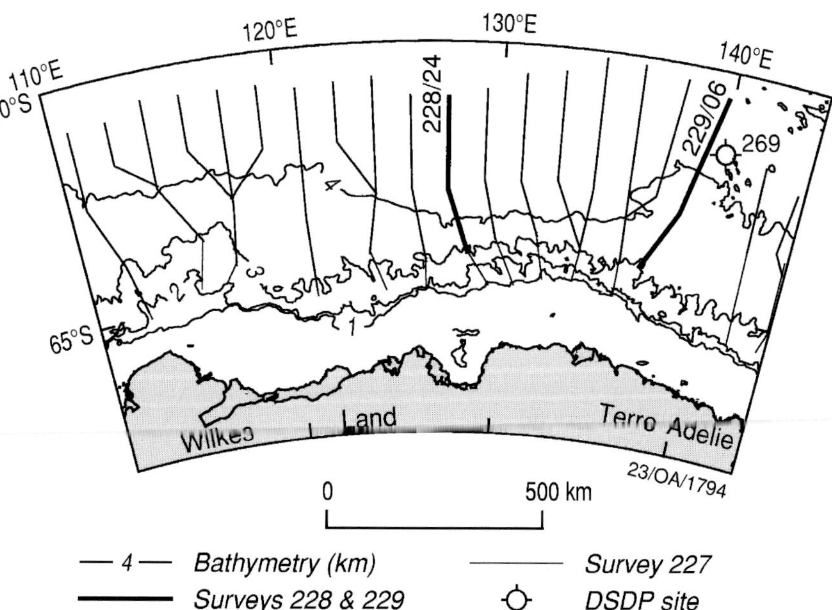

Fig. 6.6-2. Location of deep-seismic profiles recorded by Australia off Wilkes Land and Terre Adélie. Also shown are: bathymetry (isobaths in km); location of DSDP site *269*; and locations of profiles (*GA-228/24* and *GA-229/06*) illustrated in this paper

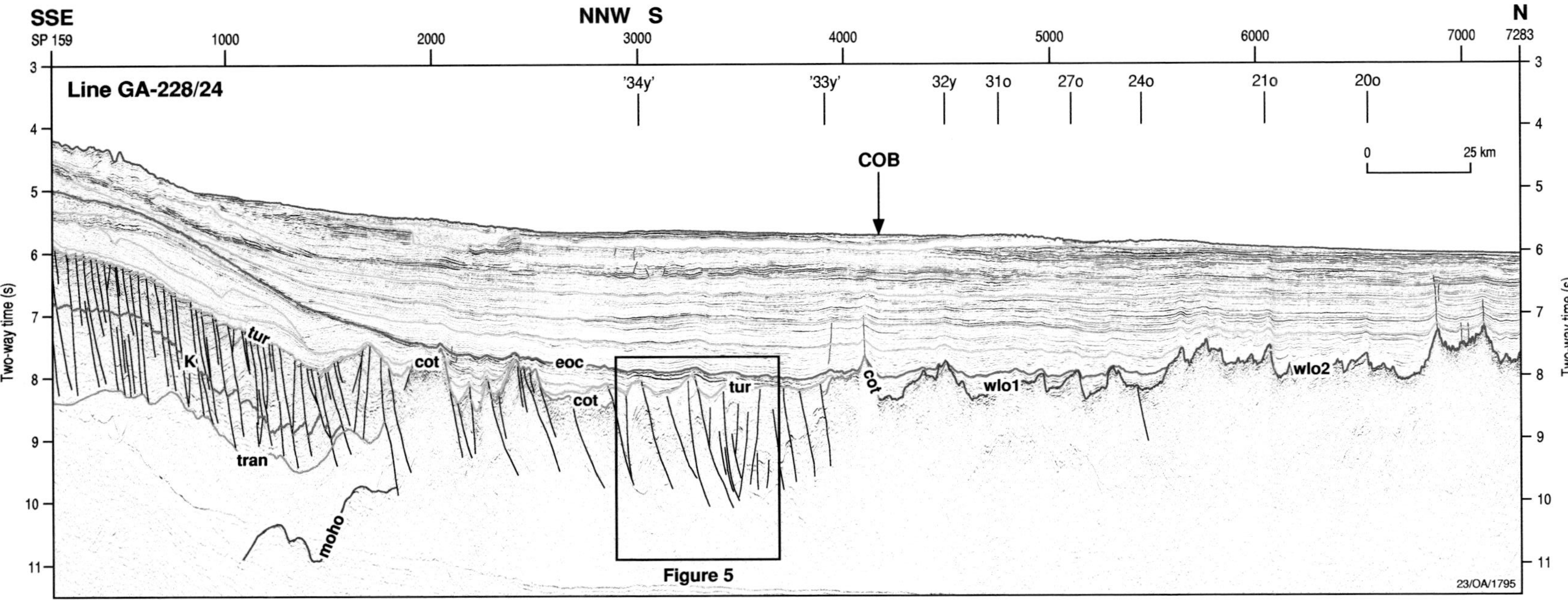

Fig. 6.6-3. Seismic profile *GA-228/24* from offshore central Wilkes Land; location shown in Fig. 6.6-2. Abbreviations: *COB:* approximate location of the continent-ocean boundary; *moho:* reflection Moho; *tran:* top of transparent lower crust; *cot:* continent-ocean transitional crust; *wlo1* and *wlo2:* Wilkes Land oceanic crust; *K:* undefined Cretaceous horizon; *tur:* Turonian unconformity; *eoc:* early Middle Eocene unconformity. Magnetic anomaly pick numbers shown above the section; "*34y*" and "*33y*" are not considered valid seafloor spreading lineation picks as they are located in the interpreted COT zone. *Box near the centre* of the line shows the location of the seismic profile in Fig. 6.6-5 that shows details of the basin located within the *COT* zone

depth in the crystalline crust, it follows that the geometries of deep geological bodies are also quite unconstrained. Therefore, at deep levels, the potential field models should be considered a guide to the gross geology but not to the fine-scale detail. Although only two models from the Wilkes-Adélie margin are illustrated in this paper, a total of six lines were modelled to a consistent level of detail (Stagg et al. 2005). All models showed a strong structural consistency along the margin, and the lines that are illustrated here are considered to be representative.

In the following interpretation, we use two terms that are often used confusingly in studies of rifted continental margins; we therefore define our usage of those terms on the Wilkes-Adélie margin here.

- Continent-ocean transition (COT): a region on the continental margin that lies between the outboard edge of highly-attenuated, but predominantly continental crust and the inboard edge of unequivocal oceanic crust. The COT includes both sedimentary and magmatic components in proportions that vary both along and across the margin.
- Continent-ocean boundary (COB): this marks the inboard edge of unequivocal oceanic crust.

It is important to note that these definitions do not preclude the existence of some fragments of proto-oceanic crust emplaced within the COT. Also, the nature of incipient breakup may lead to the igneous material being intruded/extruded along relatively linear rift segments; some of these may become the sites of oceanic crust injection, whereas others will fail. In some cases, the igneous crust may be sufficiently pervasive to give rise to linear magnetic anomalies that could be interpreted as seafloor-spreading anomalies lying inboard of the COB.

Central Wilkes Land (Line GA-228/24)

The interpretation of this line is shown in Fig. 6.6-3 (seismic reflection data) and 6.6-4 (potential field model). This line clearly illustrates the oceanward transition from highly-extended continental crust at the landward end of the line (inboard of ~SP 1800) through the interpreted COT comprising a mixture of sedimentary and igneous components, to oceanic crust lying outboard of the COB at ~SP 4000.

The landward end of the line is characterised by a rapid thinning of the crystalline continental crust (seismically transparent section below horizon *tran* in Fig. 6.6-3; bodies I1 and I2 in Fig. 6.6-4) from about 10 km to <4 km over a distance of about 80 km (Fig. 6.6-4); the Moho (top of body J, Fig. 6.6-4) shallows from about 18 km to about 12 km over the same distance. This thinning is accompanied by extensive faulting of the overlying pre- and syn-rift section, and erosion and onlap of the post-rift section (Fig. 6.6-3). The unconformity at the top of the rift section (top of major structuring; horizon *tur*, equivalent to horizon K1 of Eittreim and Smith 1987) is postulated to be of Turonian age on the basis of seismic character correlation with a similar, dated unconformity on the conjugate Australian continental margin in the Great Australian Bight (base Tiger unconformity of Totterdell et al. 2000). Much of the crustal thinning appears to be due to ductile deformation and thinning of the transparent lower crustal layer. The outboard edge of the margin rift basin is marked by a basement ridge which the potential field modelling indicates is composed of material with physical properties consistent with those of altered/serpentinised peridotite (normally magnetised, magnetic susceptibility (k) 0.02 SI, densities 2.80 and 2.87×10^3 kg m^{-3}). This ridge (bodies K1 and K2 in Fig. 6.6-4) is similar in form and interpreted composition to a basement ridge located in an equivalent structural position at the inboard edge of the COT on the conjugate margin of the Great Australian Bight (Sayers et al. 2001). The upper parts of these bodies on both margins are marked by mantling caps of noisy, discontinuous, high-amplitude reflections interpreted as basaltic volcanics. Seismic reflections in the inner parts of the ridges beneath the caps show little coherence, although some deep-biting planar faults can be detected offsetting reflectors. Some of these steep structures are wider with high-amplitude reflections, perhaps indicating intrusion of dykes within fault planes. Both ridges are probably very complex bodies, likely comprising magma chambers, sills and dykes that are the product of mantle up-welling and partial melting focussed at the point of maximum change/necking of crustal thickness. These rocks are also likely to have been affected by infusing seawater which would result in them being variably serpentinised. Inboard of the ridges on both margins, altered mantle peridotites (body J in Fig. 6.6-4: 3.05×10^3 kg m^{-3}, k 0.01 SI) are interpreted as underlying the now-brittle continental crust.

The broad COT zone on the Wilkes Land margin is dominated by a 50 km wide, sedimentary basin that contains approximately 2 s TWT of deeply-faulted and highly-rotated syn-rift rocks of probable Cretaceous age (Fig. 6.6-5; body G in Fig. 6.6-4: v_p 4 km s^{-1}; 2.75×10^3 kg m^{-3}; 0 SI). These sediments have a similar seismic character (well-developed stratification with some high-amplitude events that probably represent sills, dykes and volcanics) and structuring to the pre-rift section lying inboard of the peridotite ridge (Fig. 6.6-3 and 6.6-5; body F in Fig. 6.6-4, v_p 3400 m s^{-1}; 2.30×10^3 kg m^{-3}; 0.03 SI). The higher velocity and density, but lower magnetisation of the outer set of fault blocks probably reflects greater degrees of metamorphism/metasomatism during rifting. Faulting of both the sedimentary section and the peridotite ridge largely terminates by Turonian time. The potential field model shows that this depocentre is underlain by three blocks (M1 to M3 in Fig. 6.6-4). These have "noisy" seismic character with few coherent reflections, and an upper surface often bounded by a single, strong continuous high amplitude event which may be a detachment fault similar to those imaged by Reston et al. (2001) and Perez-Gussinyé and Reston (2001) in North Atlantic basins. The modelled properties of the blocks underlying the strong reflector (3.03×10^3 kg m^{-3}; 0.03 SI; 3.10×10^3 kg m^{-3}; 0.01 SI; 3.00×10^3 kg m^{-3}; 0.02 SI) are consistent with abandoned blocks of serpentinised continental lithospheric mantle. This scenario is similar to that modelled by Whitmarsh et al. (2001) on the comparable Iberian non-volcanic COT zone, and mapped by Muentener and Herrmann (2001) in exhumed COT successions in the Swiss Alps.

The potential field model indicates that there has been massive (>10 km) thinning of the lower and middle crust beneath the tilted blocks, with little or no indication of the lower crustal petrophysical properties interpreted in the inboard section of the line beneath the continental slope.

The outer edge of the COT is marked by a basement high (~SP 4000 in Fig. 6.6-3; body O in Fig. 6.6-4) which

Table 6.6-1. Interpretation and body parameters for potential field model for line GA-228/24

Body	Interpretation	Density ($\times10^3$ kg m^{-3})	Magnetisation (A m^{-1})
A-C, E	Post-rift oozes, turbidites and contourites	1.76–2.14	0
D	Altered basaltic flows	2.65	10000
F, G	Siliciclastic tilt-blocks intruded by sills/dykes	2.30, 2.75	5400, 0
H	Siliciclastic middle crust intruded by sills/dykes	2.53	9300
I, N	Crystalline continental crust	2.80, 2.85	0
J	Partially serpentinised upper mantle peridotite	3.05	670
K, O	Mafic igneous complexes and exhumed, altered mantle	2.86, 2.65	1300, 1400
L	Stacked basaltic flows	2.90	6300
M	Variably serpentinised, sheared continental mantle	3.00–3.10	0–2000
P	Layer 2A oceanic crust	2.55	2660–14000
Q	Metasomatised Layer 2B oceanic crust	2.70	0–12800
R	Metasomatised Layer 3 oceanic crust	2.75–3.05	670
Mantle	Unaltered mantle peridotite	3.30	0

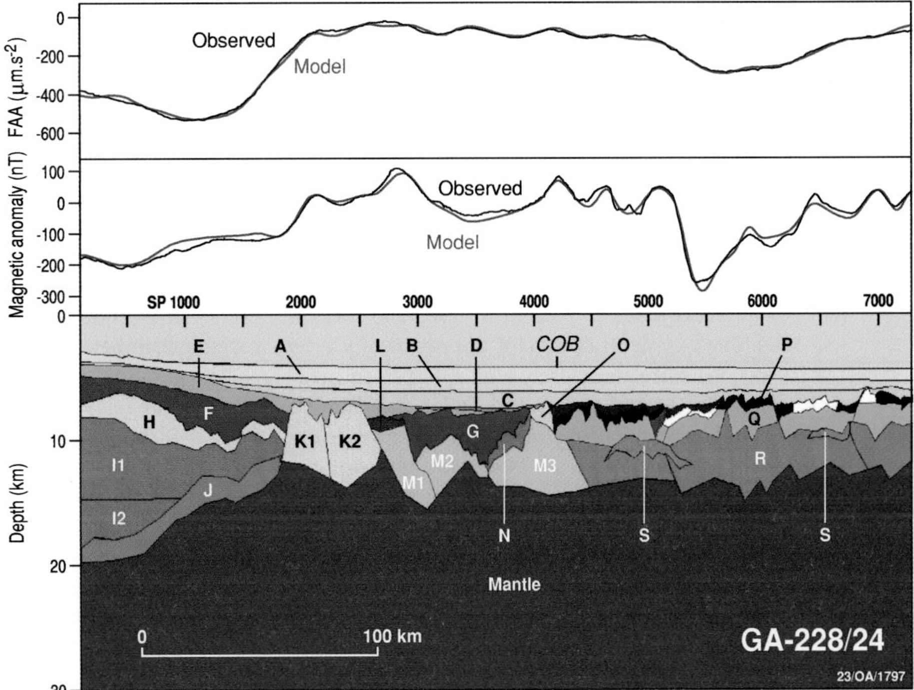

Fig. 6.6-4. Potential field model for line *GA-228/24*, offshore central Wilkes Land; location shown in Fig. 6.6-2. Body interpretation and parameters in Table 6.6-1. *COB:* continent-ocean boundary

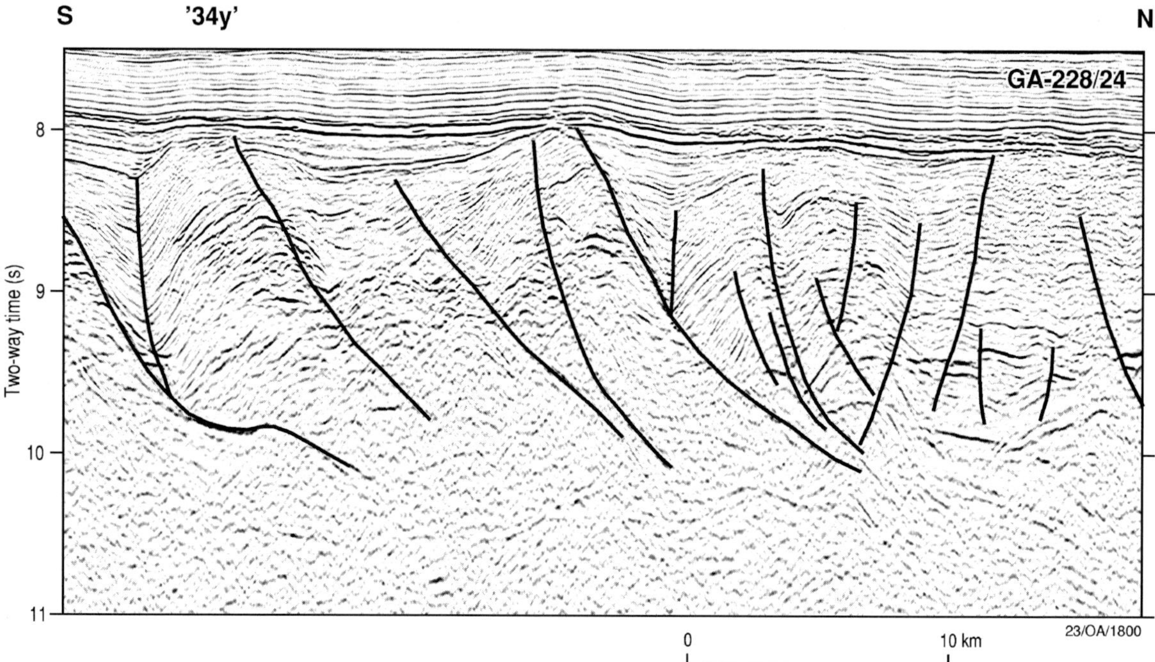

Fig. 6.6-5. Seismic detail of part of line *GA-228/24*, showing the faulted and folded sedimentary basin located within the continent-ocean transition, between the basement (?peridotite) ridge and the COB. The magnetic anomaly pick "*34y*" is clearly not associated with oceanic crust. Location shown in Fig. 6.6-3

is interpreted as comprising mixed intrusions and flows at the locus of breakup. In the seismic reflection data (Fig. 6.6-3) this body is characterised by variable-amplitude, discontinuous reflectors with faint, low-continuity reflectors at depth, particularly at its inboard edge. A thin, non-magnetic, landward-dipping wedge of relatively-continuous reflectors on the landward flank of the outer basement high (body N in Fig. 6.6-4) may be a stranded fragment of lower continental crust.

Immediately outboard of the outer basement high lies the COB and unequivocal oceanic crust. In the reflection seismic data, the COB is characterised by a change from the diverse but sediment-dominated seismic character of the COT (Fig. 6.6-5) to the low-reflectivity seismic character of the oceanic regime. Two types of oceanic basement are recognised (*wlo1* and *wlo2* in Fig. 6.6-3) separated by an oceanwards step-up of in basement level at ~SP 5500. The base level change of approximately 0.4 s TWT (~600 m) correlates closely with interpreted magnetic chron 24o (Early Eocene) and Tikku and Cande's (1999) change in spreading rate from an ultra-slow 1.5 mm a^{-1} to a slow 6.5 mm a^{-1}. Both wlo1 and wlo2 are characterised by a rugged upper surface, interpreted to be partly the result of mechanical extension in a low magma-input environment. Potential field modelling indicates a typical three-layer structure for oceanic crust: layer 2A (body P in Fig. 6.6-4) consists of normally and reversely magnetised blocks of basalt; layer 2B (body Q) comprises sheeted dykes; and layer 3 (body R) is interpreted as gabbro/peridotite. All three layers appear to be cut by faults, which in some cases appear to offset Moho and sole out in the upper mantle. These faults may have assisted the alteration and serpentinisation of much of the crust, producing relatively low densities, particularly in parts of the peridotite layer (altered peridotite layer densities of 2.75×10^3 kg m^{-3} vs unaltered densities of ca. 3.30×10^3 kg m^{-3}). Deep faulting and fracturing of brittle upper crust during early rifting has been interpreted as the mechanism for serpentinisation on other rifted margins, for example west Iberia (Perez-Gussinyé and Reston 2001) and Ireland (Reston et al. 2001). Ingress of water along faults through brittle crust leads to serpentinisation of the uppermost mantle (Hopkinson et al. 2004). High-density bodies (bodies S in Fig. 6.6-4; density $2.8–2.9 \times 10^3$ kg m^{-3}) correlate with bands of higher-amplitude seismic reflectors and may be fossil magma chambers.

The post-rift section on the line (i.e. the section above horizons *tur*, *cot*, *wlo1* and *wlo2* in Fig. 6.6-3; bodies A–E in Fig. 6.6-4) consists of a series of initially downlapping and later onlapping sedimentary packages of Late Cretaceous and Cainozoic age. A prominent margin-wide unconformity is dated as probable base Middle Eocene, based on seismic comparison with a similar unconformity in the Great Australian Bight (the base Dugong horizon of Totterdell et al. 2000). Sedimentary packages above this unconformity (horizon *eoc* in Fig. 6.6-3; top of body E in Fig. 6.6-4) lap onto the surface and have velocities of 1.8–3.4 km s^{-1}, and modelled densities of $1.76–2.10 \times 10^3$ kg m^{-3}. None of these successions is magnetised. None of the shallow sedimentary packages have been dated, due to a lack of core samples other than from a few metres below the seafloor. The total post-rift sedimentary package on this line attains a maximum thickness of ~4 km in the central part of the COT zone (Fig. 6.6-4). As noted by O'Brien and Stanley (2003), the seismic character of the Cainozoic section suggests that it is dominated by turbidite fan deposits with a large slump deposit located at about SP 2000 in the uppermost part of the section (Fig. 6.6-3).

Terre Adélie (Line GA-229/06)

The interpretation of this line is shown in Fig. 6.6-6 (seismic reflection data) and Fig. 6.6-7 (potential field model). The line is located southwest of the major fracture zones which separate the Otway Basin margin of Australia from Antarctica (Miller et al. 2002). The strike-slip movement on these fracture zones has induced structural complexities that are not seen on lines further to the west (e.g., GA-228/24) where extension was essentially normal to the margin.

The most prominent features of this line are the presence of pinch-and-swell structures (boudins) in the lower crust and the associated major deformation of the thick Cretaceous (?and Jurassic) mid-crustal rocks (Fig. 6.6-6 and 6.6-8). The deformation is largely confined to a crustal block, referred to here as the "Adélie Rift Block" (ARB in Fig. 6.6-6), bounded on its landward side by a major landward-dipping, deep-cutting fault system with an associated sediment trough, and oceanward by igneous rocks of the continent-ocean transition zone. The pinch and swell structures of the lower crust and the major fault system at the inner edge of the Adélie Rift Block are the major controls on the distribution of the predominantly post-rift section above the unconformity maas.

Potential field modelling of this line (Fig. 6.6-7) indicates that the landward edge of the Adélie Rift Block is underlain by a dense, magnetised body interpreted as altered/serpentinised mantle (body K in Fig. 6.6-7; density 3.26×10^3 kg m^{-3}, 0.03 SI) which separates blocks of magnetic, crystalline lower crust (body J: 2.95×10^3 kg m^{-3}, 0.04 SI) with probable mafic intrusions (bodies M and L; density 2.85×10^3 kg m^{-3}, 0.04 SI) from thick, pre-rift to syn-rift, strongly layered, faulted and folded sedimentary rocks (bodies H, G and F; densities $2.48–2.53 \times 10^3$ kg m^{-3}, Fig. 6.6-7, 6.6-8). At about SP 1500, the mid-crustal rocks are pierced by a body that is interpreted as a serpentinised gabbro/peridotite ridge (body I; density 2.68×10^3 kg m^{-3}) analogous to the peridotite ridge interpreted on line GA-228/24. As on line GA-228/24, this body is located at the point of maximum necking of the crust. Parts of F, G and H around the basement ridge structure are strongly magnetised (0.03–0.06 SI) with reverse polarity, probably indicating significant volumes of mafic intrusions intruding the sedimentary successions over the ridge. Elsewhere outboard, the overlying sedimentary rocks are non-magnetised. Outboard of the ridge, the thickening of the syn- to post-rift section (Fig. 6.6-6) against a major south-dipping fault (SP 2200) marks the inner edge of the Adélie Rift Block.

The crust of the Adélie Rift Block and the outboard COT zone is complex and distinctive. The basal layer (body N in Fig. 6.6-7) is interpreted from its density (2.87×10^3 kg m^{-3}), geometry, non-magnetic properties and seismic character to be ductile crystalline crust which has formed a series of large-scale (4–5 km amplitude, 20–40 km wavelength) boudin structures at the base of the crust.

As off central Wilkes Land, the COB off Terre Adélie is marked by a change in the seismic character from structurally complex and diverse seismic character of the COT zone to the low-reflectivity seismic character of the oceanic crust. The upper surface of oceanic crust (horizon *wlo2* in Fig. 6.6-6) has a rugged topography, probably at least partly due to faulting. Potential field modelling again indicates a three-layer division of the oceanic crust: an upper layer of normally and reversely magnetised basalt (oceanic crust layer 2A; body P); a middle layer (layer 2B, body Q) comprising dolerite dyke swarms; and a lower layer of gabbro-peridotite (layer 3, body R).

The Adélie transect is blanketed by 1–3 km of Cainozoic sediments, likely to be a mixture of turbidites and pelagic oozes. DSDP site 269 located to the east of the line (Fig. 6.6-2) recovered predominantly silts and clays from up to 958 m below the seafloor. The oldest material recovered was of Middle Oligocene or older age (Kanep 1975).

Discussion

Previous Interpretations

Several interpretations have previously been published of geophysical data acquired off Wilkes Land and Terre Adélie, mainly in the 1980s. The principal findings of these papers, as they relate to the crustal structure, are summarised here.

The interpretation of Wannesson et al. (1985) is based on a single French MCS line of moderate quality acquired off Terre Adélie in 1982. In addition to a thick sedimentary basin beneath the outer shelf and slope, they noted the presence of a structurally high zone in deep water (corresponding to the "Adélie Rift Block" of this paper) which contained a faulted sequence of reflectors estimated to have a thickness of 3 000–4 000 m. They referred to this zone as being an "anomalous oceanic zone", apparently

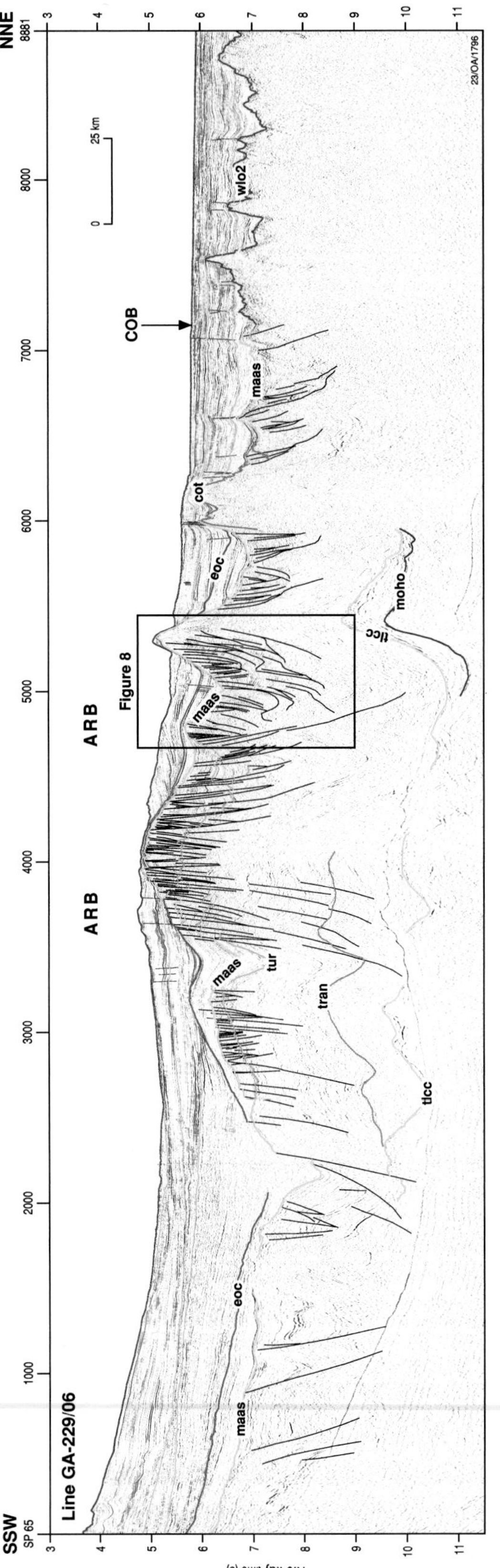

Fig. 6.6-6. Seismic profile *GA-229/06*, offshore Terre Adélie; location shown in Fig. 6.6-2. Abbreviations as for Fig. 6.6-3, plus: *ARB:* Adélie Rift Block; *maas:* top Maastrichtian; *tlcc:* top laminated continental crust. *Box* shows the location of the seismic detail in Fig. 6.6-8

on the basis that previous workers (e.g., Cande and Mutter 1982) had identified seafloor spreading magnetic anomalies in this area and therefore interpreted a continent-ocean boundary further inboard on the margin. Also prominent in these data is a persistent reflection Moho beneath the outer part of the rift basin and beneath the "anomalous oceanic zone".

Eittreim and Smith (1987) interpreted the MCS data set acquired by the USGS L1-84-AN survey, which was located between 130–146° E (offshore eastern Wilkes

Table 6.6-2.
Interpretation and body parameters for potential field model for line GA-229/06

Body	Interpretation	Density ($\times 10^3$ kg m^{-3})	Magnetisation (A m^{-1})
A-E	Post-rift oozes, turbidites and contourites. Marginal marine chalks, coals	1.82–2.41	0
F	Terrigeneous rift fill with interlayered volcanics	2.43	variable: 0–10000
G	Siliciclastic tilt blocks intruded by sills/dykes	2.48–2.53	variable: 0–12000
H	Siliciclastic middle crust	2.55–2.60	0
I	Igneous complexes and highly serpentinised exhumed mantle	2.50–2.68	4000–10000
J	Variable crystalline lower continental crust	2.95, 2.85	0, 8000
K, O	Partially to totally serpentinised upper mantle peridotite	3.26, 2.67	5200, 23500
L	?Mafic–intermediate intrusions	2.85	8000–10000
M	?Intermediate intrusions	2.85	1300–2700
N	Sheared crystalline continental crust	2.87	0
P	Layer 2A oceanic crust	2.50	4000–17400
Q	Layer 2B oceanic crust	2.60	0–12800
R	Layer 3 oceanic crust	2.70	0
Mantle	Unaltered mantle peridotite	3.30	0

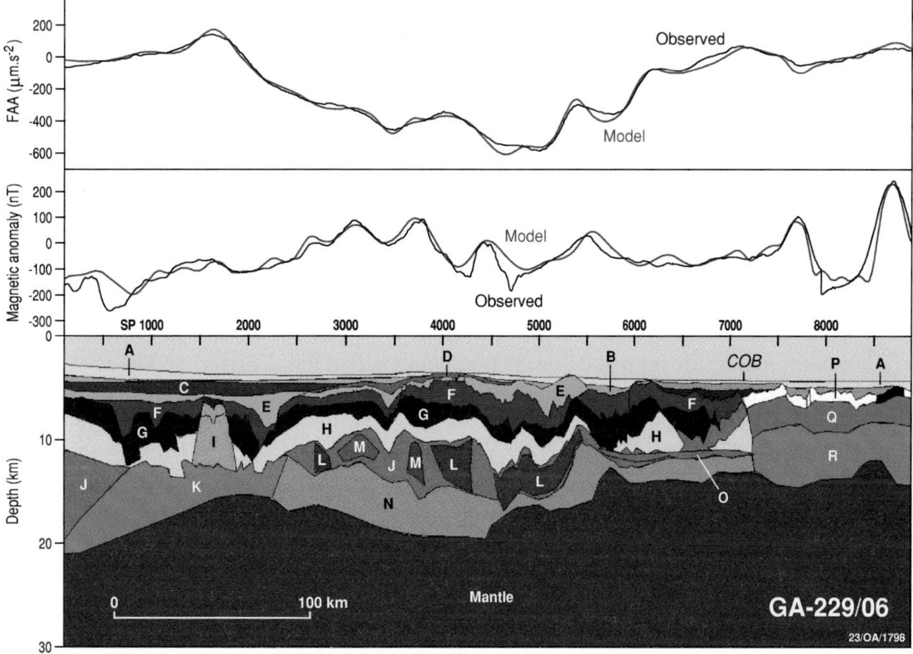

Fig. 6.6-7.
Potential field model for line *GA-229/06*, offshore Terre Adélie; location shown in Fig. 6.6-2. Body interpretation and parameters in Table 6.6-2. *COB:* continent-ocean boundary

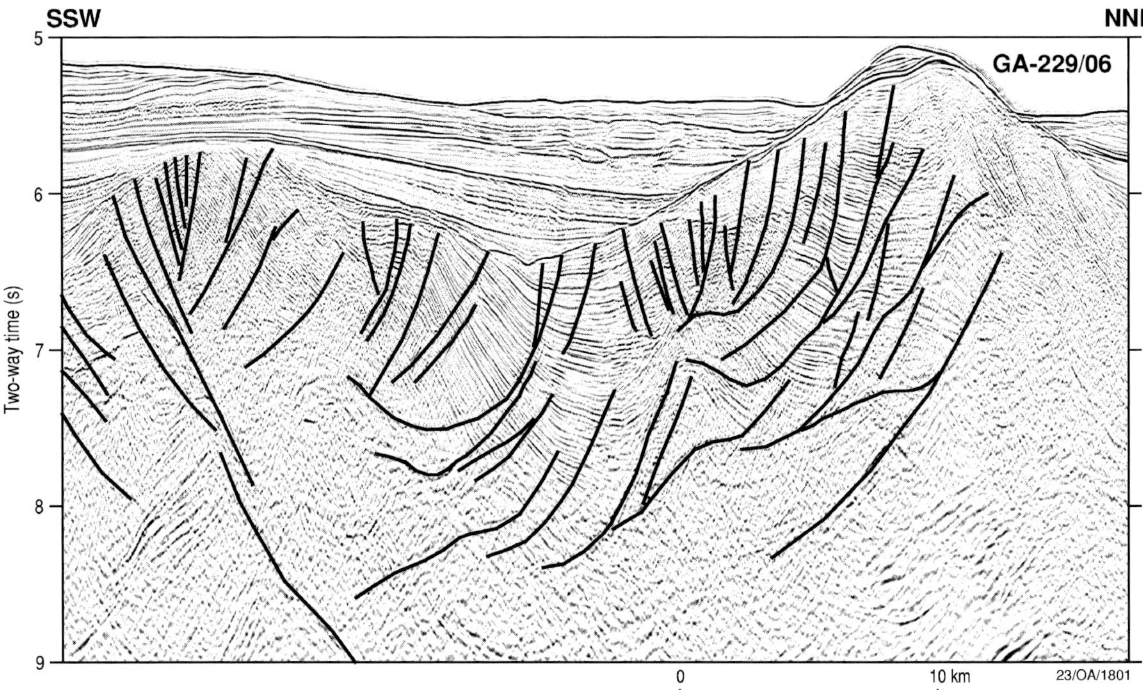

Fig. 6.6-8. Seismic detail, line *GA-229/06*, showing the thick, folded and faulted ?Cretaceous sedimentary section of the Adélie Rift Block, immediately inboard of the continent-ocean transition. Continental rocks (Tanahashi et al. 1997; Yuasa et al. 1997) have been dredged from seamounts to the east of this profile at locations that are outboard of the example section shown here. Location shown in Fig. 6.6-6

Land, Terre Adélie and western George V Land). This data set extends from the shelf edge to the inner edge of the deep-ocean basin, overlapping with the inner ends of the GA-228 and GA-229 lines. Beneath the continental slope, they identified rift-onset (K2) and breakup (K1) unconformities in a broad, margin-parallel sedimentary basin. They also identified a deep-water marginal rift basin, about 40 km in width, adjacent to the inboard edge of elevated basement that they interpreted as oceanic crust. In the vicinity of their continent-ocean boundary, they identified three volcanic sequences, interpreted as flood basalts above thinned continental crust (sequence V1), stratoid or Icelandic basalts of earliest oceanic crust (V2), and normal oceanic crust (V3). Consistent with the interpretation of Wannesson et al. (1985), they identified a persistent reflection Moho beneath the outer part of the rift basin and the oldest oceanic crust and interpreted the COB well inboard on the margin. The trace of the COB interpreted by Eittreim and Smith (1987) is shown in the map of tectonic elements that have been interpreted from the new data set (Fig. 6.6-9).

Interpretation of two long lines acquired on the Japanese TH-82 and TH-95 surveys by Tanahashi et al. (1987, 1997) shows marked differences with the interpretations of both Wannesson et al. (1985) and Eittreim and Smith (1987). The TH-82 and TH-95 data sets extend more than 350 km seaward of the L1-84-AN data, into the deep ocean basin. Tanahashi et al. (1987, 1997) confirm the identification of the Adélie Rift Block (the "marginal high" of Tanahashi et al. 1987). While Tanahashi et al. (1987) did not explicitly identify a COB, they did suggest that the "marginal high" was composed of "anticlinally folded continental and shallow-marine sequences" and they also noted that seafloor spreading probably started to the north of this high in the Middle Eocene. Tanahashi et al. (1997) interpreted the higher-quality TH-95 seismic profile in a similar way and also reported continental rocks (including granite, gneiss, slate and diorite) dredged from seamounts that are outboard of the "marginal high". While it is possible that these rocks were ice-rafted debris, the dredge stations were located on the flanks of seamounts, and we consider this scenario to be unlikely. Yuasa et al. (1997) also reported in situ peridotite blocks dredged from the same area which were interpreted as having "fertile subcontinental characteristics". Based on this seismic and geological evidence, Tanahashi et al. (1997) interpreted a COB at 61.5–62.3° S off Terre Adélie, a location that is approximately 300 km seaward of the location interpreted by Eittreim and Smith (1987).

Discussion of New Data

The two lines described in this paper illustrate most of the characteristic features of the continental margin off Wilkes Land and Terre Adélie. When the interpretation of these lines is combined with the other Geoscience Australia data collected on the margin it is possible to map the major tectonic provinces of this part of the East Antarctic margin (Fig. 6.6-9). This map is consistent with the interpretation of Tanahashi et al. (1987, 1997), but varies from Wannesson et al. (1985) and Eittreim and Smith (1987) in the identification of the continent-ocean boundary. From this interpretation, a number of issues emerge that are discussed here.

A broad (90–180 km wide) COT zone, in which highly-extended continental crustal fragments are heavily intruded and overprinted by volcanics, is interpreted in deep water along the Wilkes-Adélie continental margin (Fig. 6.6-9). The presence of this zone clearly illustrates that Antarctic continental crust extends into very deep water, analogous to the situation for the conjugate Great Australian Bight margin described by Sayers et al. (2001). Inboard of the COT, the geology of the margin is dominated by a major rift and post-rift basin that is bound to the south by probable Archaean-Proterozoic basement (as shown by the onshore geology). It is likely that this basement underlies much of the continental shelf and that the shelf edge off Wilkes Land approximately coincides with a major basement fault zone, as is generally the case on the conjugate Australian margin. The COB (here defined as the inner edge of unequivocal oceanic crust) is confidently located east of about 124° E, but becomes more difficult to define in the west where the distinction between oceanic crust and igneous components of the COT zone is blurred.

The continental margin off Wilkes Land and Terre Adélie can be broadly divided into three along-strike sectors on the basis of the gross structure and geology: the Sabrina Coast Sector (110–124° E); the Banzare Coast sector (124–131° E); and the Adélie sector (131–140° E) (Fig. 6.6-9).

The Sabrina Coast sector, which is not illustrated in detail in this paper, is located approximately opposite the Australia-Antarctic Discordance, and it is possible that this regional-scale lithospheric anomaly has also impacted on structuring on the Antarctic margin. In this sector, the COT is more diffuse than to the east and the COB is difficult to define. While we currently interpret the COB to step out to the north, we note that this disagrees with the interpreted seafloor spreading anomaly identifications of Tikku and Cande (1999) and B. J. Brown (pers. comm.), as are shown in Fig. 6.6-9. There are two obvious possibilities to explain this disagreement:

- The seismic interpretation of the COB presented here is incorrect and it actually lies inboard of the position shown in Fig. 6.6-9.
- The lineated magnetic anomalies shown here as being in the COT were not the product of normal (albeit slow) seafloor spreading.

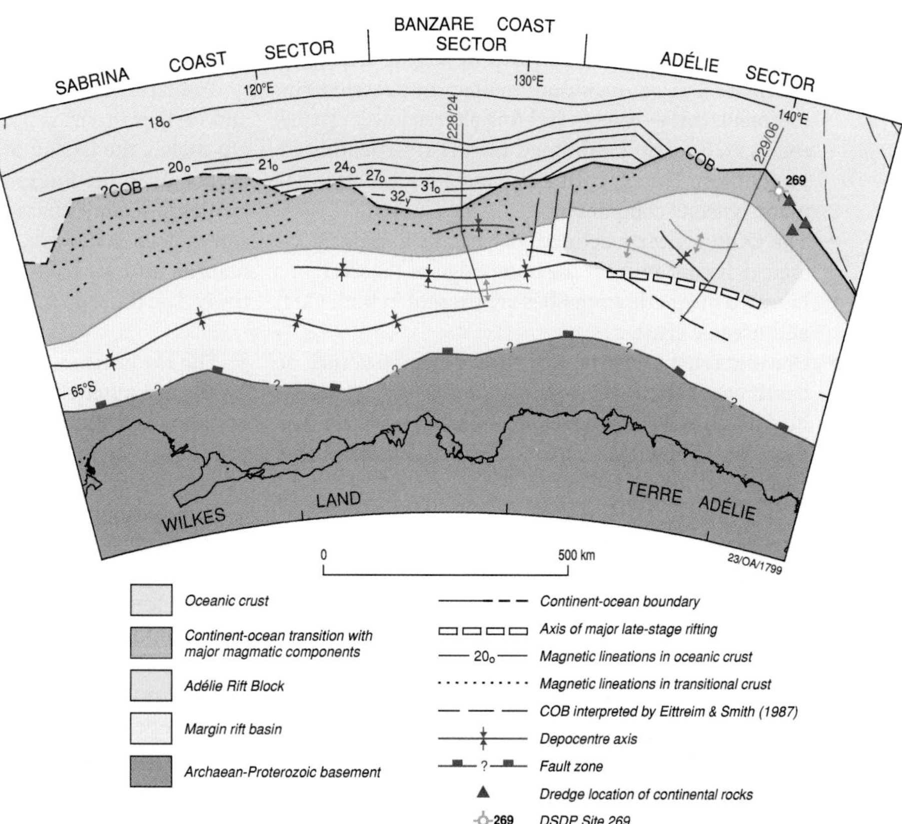

Fig. 6.6-9.
Tectonic elements of the Wilkes-Adélie margin, based on the interpretation of the deep-seismic lines shown in Fig. 6.6-2

On the basis of the reflection seismic character and the potential field modelling undertaken in this sector (Stagg et al. 2005), we prefer the second of these possibilities. In further support of this, we note that Tikku and Cande (1999) found that their reconstructions were characterised by a "large" (40–50 km) amount of continental overlap involving the South Tasman Rise, Tasmania and Victoria Land, Antarctica. One of their suggested reasons for this overlap was that the magnetic anomalies older than 31 may not be true isochrons and it was therefore possible that during the "long period of ultra-slow spreading there was considerable diffuse extension in the AAB which would have distorted the shape of the anomalies".

By contrast, the COB and COT in the adjacent Banzare Coast sector to the east are more clearly defined (see Fig. 6.6-3 and 6.6-4). This sector has marked similarities and symmetry with the conjugate margin in the Great Australian Bight (GAB; Totterdell et al. 2000; Sayers et al. 2001). In particular:

- The inner part of the continental margin is dominated by a thick, faulted Cretaceous–?Jurassic syn-rift section which lies outboard of a major basement fault system and inboard of a prominent basement (?peridotite) ridge marking the inner flank of the COT zone.
- The margin rift basin is underlain by transparent lower crustal material which has undergone ductile deformation and thinning.
- The COT zone is characterised by highly-extended continental crust, including faulted and rotated rift sediments immediately overlying altered lower crustal rocks, and possibly exhumed mantle (?serpentinised peridotites). The COT zone also includes a substantial mafic igneous component.
- The COB is clearly defined and marks a major geological discontinuity at the outer edge of the COT.
- Lineated magnetic anomalies are present in both COT and oceanic crust.
- Oceanic crust has a typical three-layer structure of basalt underlain by sheeted dykes and gabbro/peridotite. This crust is dissected by major faults or shears that span the full thickness of the crust (e.g., Fig. 6.6-4 and Sayers et al. 2001).

The inboard flank of the peridotite ridge was identified as the COB by Eittreim and Smith (1987). We believe that this identification was in error, mainly due to their data not extending sufficiently far into the ocean basin to image the outboard basin shown in Fig. 6.6-5.

Magnetic anomalies, combined with the interpreted seismic data and potential-field modelling on the GAB margin indicate that emplacement of oceanic crust between central Wilkes Land and the GAB commenced at about chron 33o time (i.e. at ~83 Ma, early Campanian; Sayers et al. 2001) and about 12 Ma later that previously interpreted (Tikku and Cande 1999). On the Antarctic margin, the relationship between the COB and the previously identified magnetic anomalies is less clear, although spreading also appears to have commenced in the Banzare Coast sector at about chron 33o time (see Fig. 6.6-9).

To the east, the structure of the Adélie sector is strongly influenced by its proximity to the major strike-slip fracture zones which separate the Otway Basin/west coast of Tasmania margin of Australia from Antarctica. The Adélie Rift Block coincides with a major northward salient of the COB and here is interpreted unequivocally as a continental fragment, both on the basis of its seismic character (e.g., Fig. 6.6-8) and because of the continental rocks dredged by Japanese workers from its outer flank. The Adélie Rift Block has the appearance of a marginal plateau that almost detached from the continent and has since subsided to abyssal depths. The block is bounded on its landward flank by an interpreted failed rift and associated volcanics and on its outer flank by a COT zone that is narrower than that to the west. The COB interpreted in Fig. 6.6-6 is located at the boundary between an apparent sedimentary basin beneath the inner flank of the deep ocean basin and unequivocal oceanic crust. This location is in close agreement with that interpreted by Tanahashi et al. (1997). An alternative location for the COB is the inboard flank of the basement ridge at approximately SP 6000; however, this location would require the development of a deep extensional basin on oceanic crust further outboard, which we consider to be less likely.

The lack of any stratigraphic control on the syn-rift and early post-rift section along the East Antarctic margin makes the dating of the major tectonic/structuring events along the margin difficult. However, strong similarities in seismic character and structure allow comparison to deep-water sequences on the southern Australian margin, and we have used these comparisons to tentatively date the sequences in Fig. 6.6-3 and 6.6-6 as follows.

- The *tur* horizon, which marks the top of major faulting and other structuring in the Sabrina and Banzare Coast sectors on line GA-228/24 (light green unconformity in Fig. 6.6-3; equivalent to horizon K1 of Eittreim and Smith 1987) is dated as base Turonian by comparison of structural style and seismic reflection character with the unconformity at the base of the Tiger sequence on the conjugate GAB continental margin (Totterdell et al. 2000). On the GAB margin this unconformity pre-dates the age of breakup (early chron 33 i.e. ~83 Ma) by about 7–10 Ma. This unconformity is traceable for 1 500 km beneath the continental slope along the Wilkes-Adélie margin and is obviously of regional significance.
- The highly folded and faulted sedimentary section of the Adélie Rift Block has strong character and struc-

tural similarities with the Late Cretaceous Sherbrook Group and the upper part of the Early Cretaceous Otway Supergroup of the Otway Basin of southeast Australia (Moore et al. 2000), and we therefore date the unconformity at the top of this section (horizon *maas*) as Maastrichtian.

- The prominent onlap surface (dark green unconformity in Fig. 6.6-3 and 6.6-6) is dated as early Middle Eocene on the basis of its correlation with similar onlap surfaces in the Great Australian Bight (base Dugong sequence; Totterdell et al. 2000), in the Otway Basin (base Nirranda Group; Moore et al. 2000) and in the Sorell Basin on the west coast of Tasmania (unconformity U5; Hinz et al. 1986).

Conclusions

A number of significant conclusions arise from the interpretation presented in this paper:

1. The Wilkes-Adélie margin is underlain by a major rift basin that extends for at least 1 500 km along the margin. Beneath this basin, crystalline crust thins oceanwards through extensive faulting of the rift and pre-rift sedimentary section and by mainly ductile deformation of the crystalline crust. The total thickness of pre-rift, rift and post-rift sedimentary rocks in this basin is probably at least 7 km.
2. Outboard of the margin rift basin, the 90–180 km wide continent-ocean transition zone is interpreted to consist primarily of continental crust, albeit with strongly magnetised magmatic components that can account for the lineated magnetic anomalies that have been interpreted in this zone. The thick sedimentary section in the COT zone is floored by dense lower crustal or mantle rocks indicating massive (>10 km) thinning of the lower and middle crust. As with the conjugate Australian margin, the first "true" seafloor spreading magnetic anomalies place the start of oceanic spreading at about chron 33o time.
3. The boundary between the margin rift basin and the COT is marked by a basement ridge which potential field modelling indicates is probably composed of altered/serpentinised peridotite. This ridge is similar in form and interpreted composition to a basement ridge located in a similar structural position at the inboard edge of the COT on the conjugate margin of the Great Australian Bight. Both ridges are probably the product of mantle up-welling and partial melting focussed at the point of maximum change/necking of crustal thickness.
4. Integrated deep-seismic and potential field interpretations point very strongly to the boundary between unequivocal oceanic crust and the largely continental crust of the continent-ocean transition as lying in very deep water, and considerably further outboard of earlier interpretations (often based on inadequate seismic data or magnetic data only). We consider the COB to be well-constrained in the Banzare Coast sector and unequivocal in the Adélie sector, but less well-defined in the Sabrina Coast sector. Plate reconstructions that do not take account of a continent-ocean boundary, where it is based on high-quality geophysical data and sample data, can have significant shortcomings.
5. There is a high degree of symmetry between the formerly conjugate margins of southern Australia and East Antarctica east of about 120° E. This symmetry is evident both in the crystalline crust and in the overlying rift section. In contrast, the post-rift sections are markedly different in thickness, reflecting the different depositional histories during much of the Cainozoic. In addition to there being a high degree of symmetry between the margins, there is also a strong correlation in the seismic characters, which gives us some confidence in dating the major unconformities as being of base Turonian, Maastrichtian and early Middle Eocene age.

Acknowledgments

The interpretation underpinning this paper was carried out by the authors with valuable contributions from L. Carson, D. Close and B. Brown. The seismic and potential field data were acquired by Fugro Geoteam AS, under contract to the Commonwealth of Australia, and their professionalism in survey operations is acknowledged. We are also appreciative of the contributions of Andrew Krassay and Tony Stephenson of Geoscience Australia in reviewing this manuscript. The comments supplied by the reviewers, Giuliano Brancolini and Karl Hinz, were also appreciated. The figures were drafted by Silvio Mezzomo. This paper is published with the permission of the Chief Executive Officer, Geoscience Australia.

References

Eittreim SL, Hampton MA (eds) (1987) The Antarctic continental margin: geology and geophysics of offshore Wilkes Land. Earth Science Series, Circum-Pacific Council for Energy and Resources, 5A

Eittreim SL, Smith GL (1987) Seismic sequences and their distribution on the Wilkes Land margin. In: Eittreim SL, Hampton MA (eds) The Antarctic continental margin: geology and geophysics of offshore Wilkes Land. Earth Science Series, Circum-Pacific Council for Energy and Resources, 5A, pp 15–43

Finlayson DM, Lukaszyk I, Collins CDN, Chudyk EC (1998) Otway Continental Margin Transect: crustal architecture from wide-angle seismic profiling across Australia's southern margin. Australian J Earth Sci 45:717–732

Hinz K, Willcox JB, Whiticar M, Kudrass HR, Exon NF, Feary DA (1986) The west Tasmanian margin: an underrated petroleum province? In: Glenie RC (ed) Second south-eastern Australia oil exploration symposium 1985. Petrol Explor Soc Australia, Melbourne, pp 395–410

Hopkinson L, Beard JS, Boulter CA (2004) The hydrothermal plumbing of a serpentinite-hosted detachment: evidence from the West Iberia non-volcanic rifted continental margin. Marine Geology 204:301–315

Ishihara T, Tanahashi M, Sato M, Okuda Y (1996) Preliminary report of geophysical and geological surveys of the west Wilkes Land margin. Proc NIPR Symp Antarc Geosci 9: 91–108

Kanep AG (1975) Cenozoic planktonic foraminifera from Antarctic deep-sea sediments, Leg 28, DSDP. In: Hayes DE, Frakes LA, et al. (eds) Initial Rep Deep Sea Drilling Project, U.S. Government Printing Office, Washington DC, 28:573–584

Miller JMcL, Norvick MS, Wilson CJL (2002) Basement controls on rifting and the associated formation of ocean transform faults – Cretaceous continental extension of the southern margin of Australia. Tectonophysics 359:131–155

Moore AMG, Stagg HMJ, Norvick MS (2000) Deep-water Otway basin: a new assessment of the tectonics and hydrocarbon prospectivity. APPEA J 40:66–84

Muentener O, Herrmann J (2001) The role of lower crust and continental upper mantle during formation of non-volcanic passive margins: evidence from the Alps. In: Wilson RCL, Whitmarsh RB, Froitzheim N (eds) Non-volcanic rifting of continental margins: a comparison of evidence from land and sea. Geol Soc London, Spec Publ 187:267–288

O'Brien PE, Stanley S (2003) Cainozoic continental slope and rise sediments from 38° E to 164° E, East Antarctica. In: Fütterer DK (ed) Antarctica: contributions to global earth science. Abstracts 9th ISAES Potsdam, Terra Nostra 246

Perez-Gussinyé M, Reston TJ (2001) Rheological evolution during extension at nonvolcanic rifted margins: onset of serpentinization and development of detachments leading to continental breakup. J Geophys Res 106(3):961–3975

Reston TJ, Pennell J, Stubenrauch A, Walker I, Perez-Gussinyé M (2001) Detachment faulting, mantle serpentinzation, and serpentinite-mud volcanism beneath the Porcupine Basin, southwest Ireland. Geology 29:587–590

Sato S, Asakura N, Saki T, Oikawa N, Kaneda Y (1984) Preliminary results of geologiacal and geophysical surveys in the Ross Sea and in the Dumont d'Durville Sea, off Antarctica. Mem Nat Inst Polar Res Spec Issue 33:66–92

Sayers J, Symonds PA, Direen NG, Bernardel G (2001) Nature of the continent-ocean transition on the non-volcanic rifted margin of the central Great Australian Bight. In: Wilson RCL, Whitmarsh RB, Froitzheim N (eds) Non-volcanic rifting of continental margins: a comparison of evidence from land and sea. Geol Soc London Spec Publ 187:51–76

Stagg HMJ, Colwell JB, Direen NG, O'Brien PE, Brown BJ, Bernardel G, Borissova I, Carson L, Close DB (2005) Geological framework of the continental margin in the region of the Australian Antarctic Territory. Geosci Australia Record 2004/25

Stagg H, Schiwy S (2002) Marine geophysical surveys completed off Antarctica. Aus Geo News 66:18–19

Tanahashi M, Saki T, Oikawa N, Sato S (1987) An interpretation of the multichannel seismic reflection profiles across the continental margin of the Dumont d'Urville Sea, off Wilkes Land, East Antarctica. In: Eittreim SL, Hampton MA (eds) The Antarctic continental margin: geology and geophysics of offshore Wilkes Land. Earth Science Series, Circum-Pacific Council for Energy and Resources, 5A, pp 1–13

Tanahashi M, Ishihara T, Yuasa M, Murakami M, Nishimura A (1997) Preliminary report of the TH95 geological and geophysical survey results in the Ross Sea and Dumont d'Urville Sea. Proc NIPR Symp Antarc Geosci 10:36–58

Tikku AA, Cande SC (1999) The oldest magnetic anomalies in the Australian-Antarctic Basin: are they isochrons? J Geophys Res 101(1):661–677

Totterdell JM, Blevin JE, Struckmeyer HIM, Bradshaw BE, Colwell JB, Kennard JM (2000) A new sequence framework for the Great Australian Bight: starting with a clean slate. APPEA J 40(1):95–116

Tsumuraya Y, Tanahashi M, Saki T, Machihara T, Asakura N (1985) Preliminary report of the marine geophysical and geological surveys off Wilkes Land, Antarctica in 1983–1984. Mem Nat Inst Polar Res Spec Issue 37:48–62

Wannesson J, Pelras M, Petitperrin B, Perret M, Segoufin J (1985) A geophysical transect of the Adélie margin, East Antarctica. Marine Petrol Geol 2:192–201

Whitmarsh RB, Manatschal G, Minshull TA (2001) Evolution of magma-poor continental margins from rifting to seafloor spreading. Nature 413:150–154

Post-Rift Continental Slope and Rise Sediments from 38° E to 164° E, East Antarctica

Philip E. O'Brien · Shawn Stanley · Robert Parums
Geoscience Australia, GPO Box 378 Canberra, 2601, Australia, <Phil.OBrien@ga.gov.au>

Abstract. The Australian Antarctic and Southern Ocean Profiling Project has acquired more than 20 000 km of north-south seismic reflection transects every 90 km along the East Antarctic continental margin between 38° E to 164° E. These data provide a unique overview of the broad scale depositional patterns around a large part of the Antarctic margin. Each line was examined and the post-rift section classified according to depositional environment.

The depositional environments recognised are:

1. Submarine fans.
2. Contourite drift and canyon complexes.
3. Mixed contourite-turbidite drift sediments.
4. Thin separated drifts.
5. Non-deposition and erosion of older sediments.
6. Prograding upper slope wedges.
7. Distal abyssal plain deposits.

We have recognised nine sedimentary provinces on the continental slope and rise, based on the relative dominance of these environments. The distribution of contourite deposits is controlled by sediment input from the continent and by the shape of the margin. Prydz Bay has provided a large amount of sediment over a long period, producing the thickest post-rift sediment pile on the margin. It has been suggested that major sediment inputs have taken place via the Wilkes sub-glacial Basin, the Aurora Basin and Prydz Bay through the Lambert Graben. Our examination of the data implies that only Prydz Bay, western Enderby Land and the area at about Latitude 120° E have received large influxes of sediment, with western Enderby Land being relatively inactive during the Neogene. Sediment thicknesses are large when compared to the conjugate margin of Australia.

Introduction

The marine sedimentary environments around Antarctica are very poorly known because only a small proportion of the margin has been surveyed systematically with reflection seismic data (ANTOSTRAT 2003). Effort has tended to concentrate in areas that are relatively easy to access or where initial surveys took place in the 1980s. In these areas, there is enough data to allow development of hypotheses for testing by new surveys and scientific drilling. Along the East Antarctic margin, surveys have concentrated in Prydz Bay and on the George V-Terra Adélie-Wilkes Land margin. Here we report on the results of three surveys that collected regularly-distributed seismic lines across the Antarctic margin, often in areas where there were little pre-existing data. The wide distribution of the surveys and the high quality of the data provide an unparalled opportunity to examine sedimentary patterns along a large part of the Antarctic margin.

Data Set

The Australian Antarctic and Southern Ocean Profiling Project (AASOPP) collected geophysical data along the East Antarctic continental margin from 38° E to 164° E in three surveys in the Austral summers of 2000/2001 and 2001/2002 (Fig. 6.7-1). About 20 000 line-km comprised 36-fold deep-seismic data (60 l airgun array source; 3 600 m streamer; 288 channels; 16 s record length), collected with coincident gravity and magnetics data and refraction/wide-angle sonobuoys (Geoscience Australia Surveys GA-228 and GA-229, Fig. 6.7-1). A further 3 425 km of high-speed seismic data of limited penetration (6 l airgun array, 300 m streamer, 24 channels, 8 s record length) were also recorded during Geoscience Australia Survey GA227. All data were stacked and migrated. The deep-seismic lines extend from the mid to lower continental slope out to oceanic crust at an average separation of 90 km along the margin and have an average length of about 320 km. As well as providing excellent definition of the sedimentary section, the data also show structure as deep as the base of the crust, particularly over the outermost part of the continental margin. The shallow-penetration lines concentrating on the upper continental rise and slope.

Methods

Each line was examined and the top of the rift phase of margin development was identified as the surface marking the end of major fault deformation (Stagg et al., submitted). The post-rift section was then classified according to sedimentary environment, based on reflector geometries. This classification was not intended as a detailed

From: Fütterer DK, Damaske D, Kleinschmidt G, Miller H, Tessensohn F (eds) (2006) Antarctica: Contributions to global earth sciences. Springer-Verlag, Berlin Heidelberg New York, pp 341–348

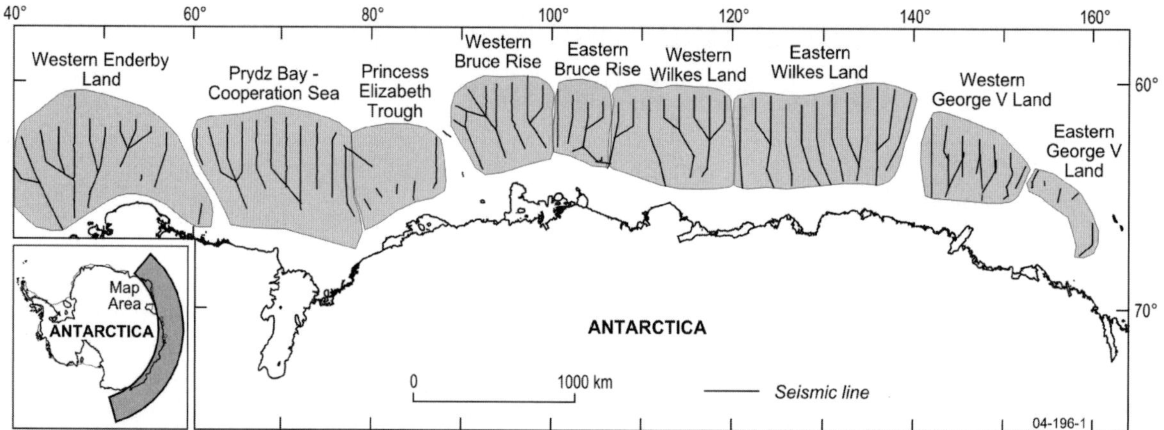

Fig. 6.7-1. Location of AASOPP seismic lines and areas of differing facies assemblage. *Prydz Bay* slope and rise = mixed contourite-turbidite drift deposits. *Wilkes Land* = predominantly turbidite fans. *Western Enderby Basin* = dissected older sediments onlapped by thin contourites. *Eastern George V*, *Eastern Bruce Rise* and *Princess Elizabeth Trough* = thin contourite and basin floor deposits. *Western Bruce Rise*, *Western George V* and *Western Wilkes Land* margins = dissected contourite drift deposits passing out into submarine fan and abyssal plain deposits

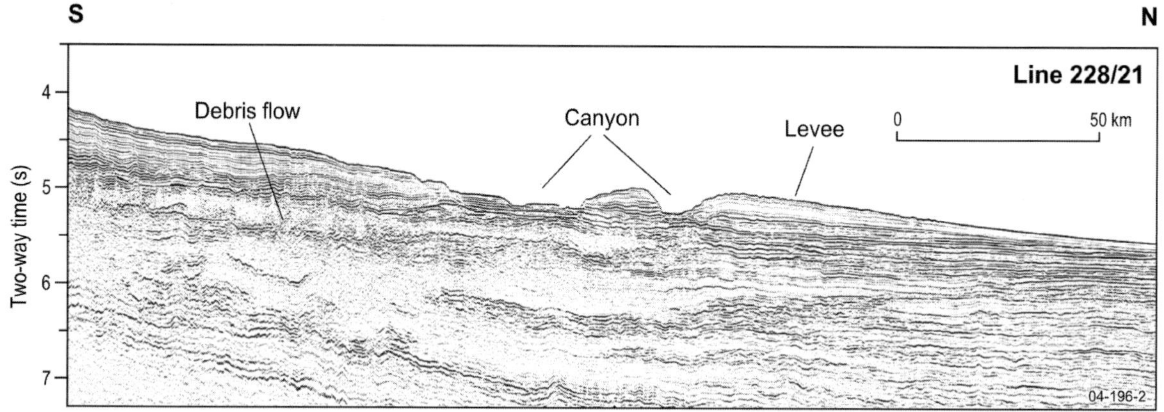

Fig. 6.7-2. Submarine fan sediments showing channel-levee deposits, Line GA *228/21*, Wilkes Land

assessment, but rather to define the broad distribution of sediment type and depositional environment. Naturally there is some gradation between the different environments recognised, especially between turbidite fans and contourite deposits. Previous authors have recognised mixed turbidite-contourite complexes from the Antarctic margin (e.g., Kuvaas and Leitchenkov 1992; Rebesco et al. 1997). A map of the post-rift sediment thickness in kilometers was also produced using seismic reflection two-way time data and converted array stacking velocities.

Classification

Submarine fans. Defined by deep sea channels flanked by levees that are recognised because they are thickest adjacent to the channel and thinning away from the channel. Fan lobes appear as broad mounds with small channels at their crests (Fig. 6.7-2).

Contourite drift and canyon complexes. Large, well-stratified sediment bodies that may be contourites that are cut by canyons (Fig. 6.7-3). A contourite origin is suggested by mounded geometry, high reflector continuity, the absence of channel structures and sediment wave fields deposited over a long period of time (Faugère and Stow 1993; Marani et al. 1993; Rebesco et al. 1997). The proximal parts of these deposits are now strongly dissected by canyons and gullies (Fig. 6.7-3). These deposits may have formed as contourite drifts that are now being eroded by sediment gravity flow proccesses.

Mixed contourite-turbidite drift sediment. Sedimentary sequences several thousand meters thick that show a range of features, including sediment waves and mounded geometry, that suggest contourite deposition, and submarine channels with levees characteristic of turbidite fans (Fig. 6.7-4; Kuvaas and Leitchenkov 1992).

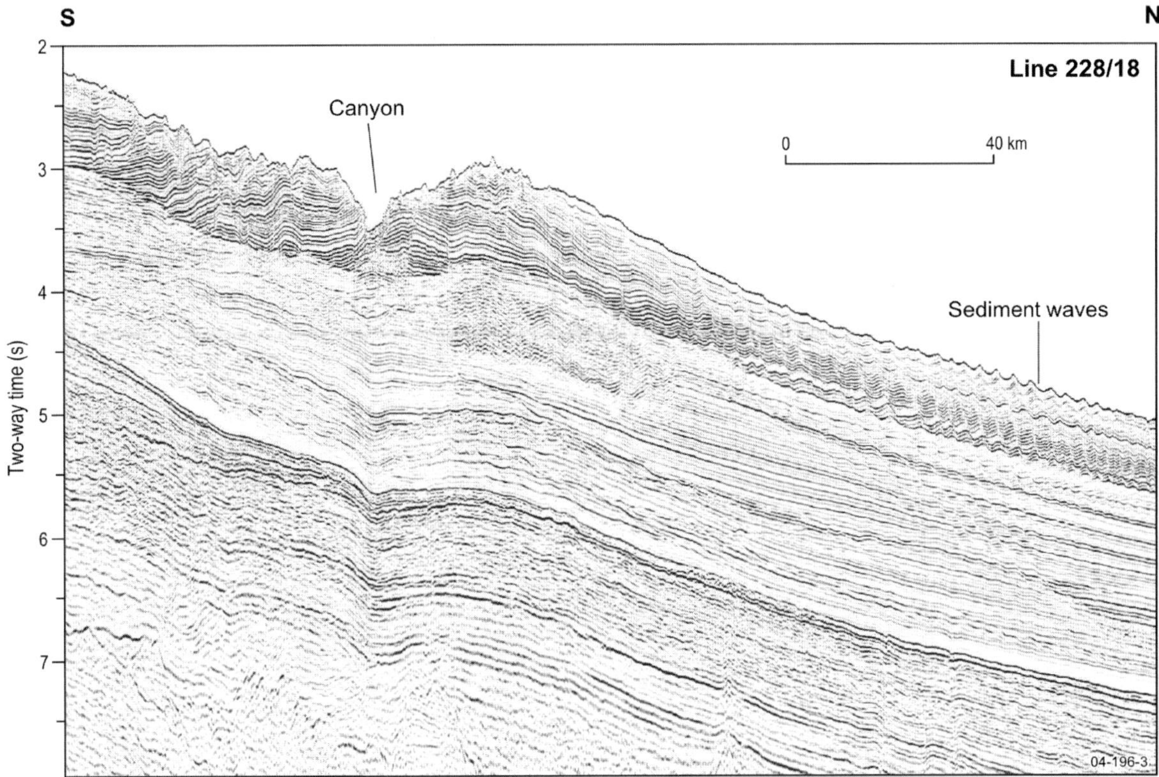

Fig. 6.7-3. Large mound of probable contourite deposits now being dissected by submarine canyons

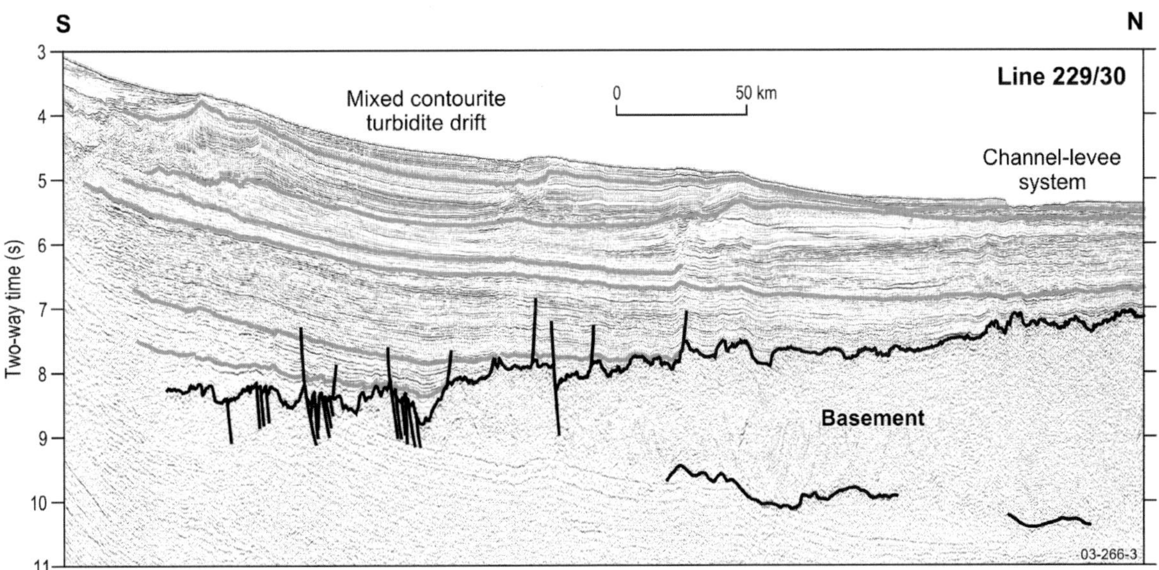

Fig. 6.7-4. Mixed-contourite-turbidite drift sediments, Line GA *229/30*, seaward of Prydz Bay

Thin contourite drifts. Areas of thin, well-bedded sediments forming ridge and moat topography typical of separated drifts (Faugères et al. 1999). Other areas show thinly-bedded, high-continuity reflectors suggesting sheeted drifts (Fig. 6.7-5).

Erosion of older sediments. Sedimentary sections several thousand meters thick of variable reflection geometry and continuity that thin rapidly seaward, that are cut by deep canyons and onlapped by deep-sea sediments (Fig. 6.7-6). Reflectors onlap oceanic crust quite close to the conti-

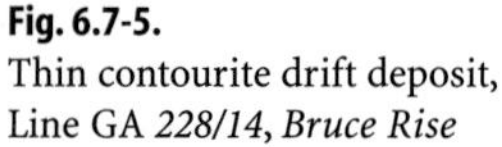

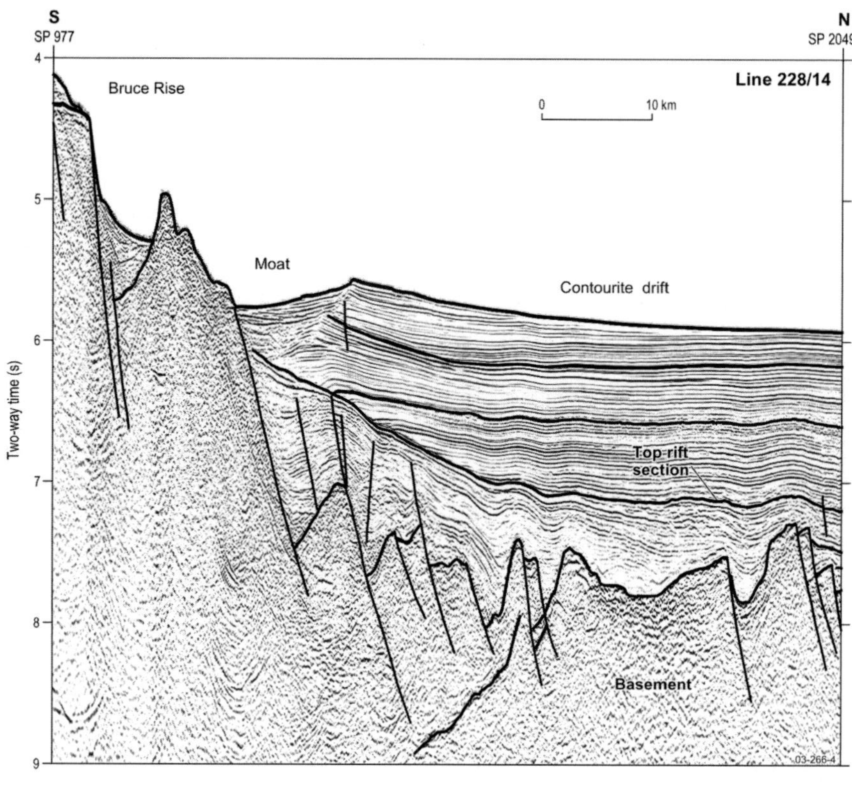

Fig. 6.7-5. Thin contourite drift deposit, Line GA *228/14*, *Bruce Rise*

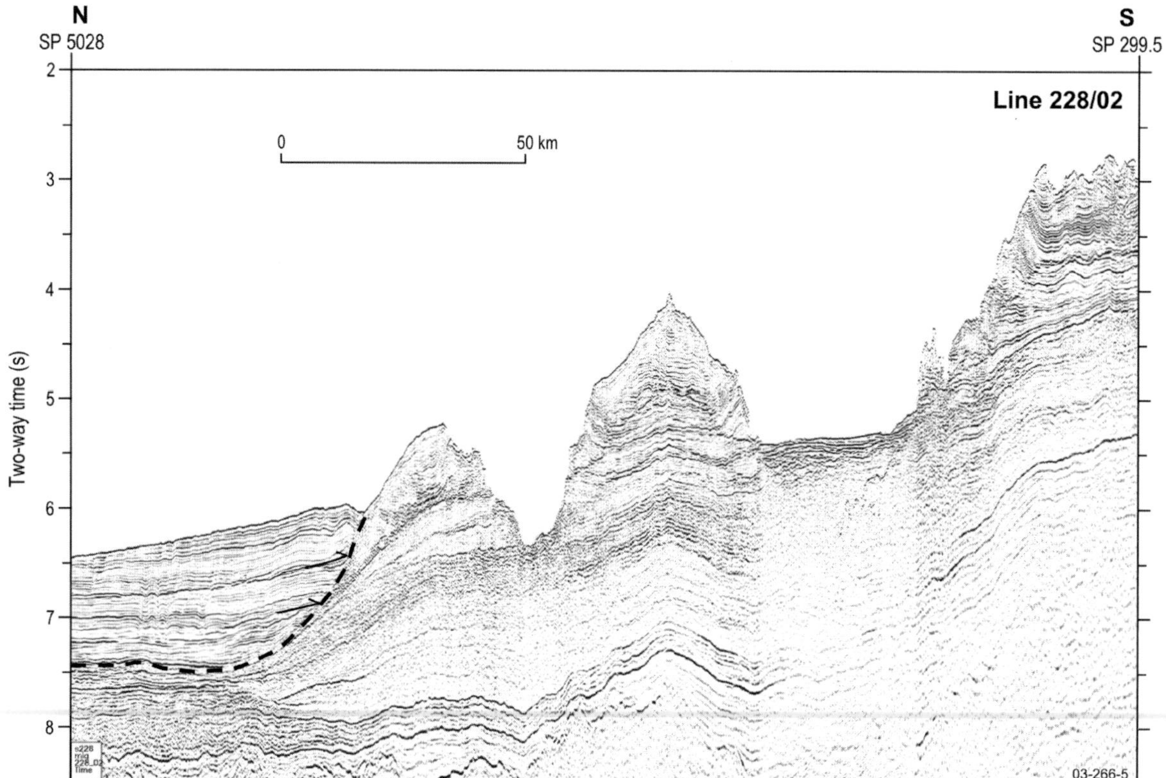

Fig. 6.7-6. Outcropping older sediment onlapped by thin contourites, Line GA *228/02*, western Enderby Land. Contourites onlap along the surface shown

nent, suggesting that these sediments are probably Cretaceous in age.

Distal abyssal plain deposits. Distal abyssal plain sediments are typically flat-lying with parallel, continuous reflectors.

Upper slope deposits. A few survey lines reached the upper slope and shelf edge. They imaged offlapping wedges of sediment that downlap onto lower slope/rise sediments. On some lines, acquired along spurs extending from the shelf edge, reflectors parallel the sea floor from the shelf to the slope.

Distribution of Facies Assemblages

Prydz Bay Slope and Rise

The area seaward of Prydz Bay and the Mac. Robertson Shelf shows the greatest accumulation of sediments, comprising up to 10 km of sediment including 6 km of mixed contourite-turbidite drift deposits. Kuvaas and Leitchenkov (1992) discussed the seismic facies in the area and ODP Leg 188 drilled one of the large sediment drifts (Fig. 6.7-1; O'Brien, Cooper et al. 2001; Cooper and O'Brien 2004). The major features of the area are large sediment mounds separated by modern canyons. The mounds contain broad channels with levees on both sides showing a predominance of vertical accretion.

Highly continuous reflectors and common fields of sediment waves suggest the strong influence of contour currents during deposition. ODP site 1165 predominantly encountered mudstone with silt laminae of probable contourite origin, passing up into massive hemipelagic muds and oozes (Shipboard Party 2001). Of the 999.1 m drilled, only the top 80 m is younger than Miocene with most rapid sedimentation during the Lower Miocene (Shipboard Party 2001; Florindo et al. 2003).

Wilkes Land

The Wilkes Land margin from 120° E and 140° E exhibits predominantly turbidite fan deposits with closely-spaced gullies on the upper slope and broad channel-levee deposits. Reflection continuity is poor to moderate, and large sediment mounds are not present, suggesting turbidite deposition rather than contourite sedimentation. Two very large debris flow deposits extend up to 270 km from the base of the continental slope (Fig. 6.7-2).

Western Enderby Basin (038° E – 055° E)

The western Enderby Land margin features a sedimentary sequence greater than 8 km thick that thins rapidly seaward and is cut by large canyons (Fig. 6.7-6). Well-stratified mounded drift deposits onlap the seaward edge of the sequence (Fig. 6.7-6). The thick succession onlaps oceanic crust close to the margin that probably formed early in the post-rift evolution of the Enderby Basin. This would indicate a Cretaceous age for the west Enderby margin sedimentary succession.

Eastern George V Land, Eastern Bruce Rise and Princess Elizabeth Trough

These areas exhibit thin contourite and basin floor deposits abutting steep, faulted scarps composed of continental basement and older sediments.

Western Bruce Rise, Western George V and Western Wilkes Land

These areas display well-stratified deposits on the lower slope and upper rise with some of the characteristics of contourites (Faugères et al. 1999). They are dissected by submarine canyons and grade seaward into submarine fan and abyssal plain deposits.

Sediment Thickness

Previous surveys along the East Antarctic margin have commented on the presence of a thick sedimentary section. A total thickness in the order of 6 km has been observed inboard and outboard of the shelf edge in a number of places (e.g., Stagg 1985; Eittreim and Smith 1987; Wanneson 1990; Ishihara et al. 1996). We have calculated the thickness of the post-rift section and the lines used tend to be north of the rift-related depocentre (Stagg et al. 2005), so this map reflects the degree to which sediments have spread beyond the rift-related depocentres (Fig. 6.7-7).

The areas of thickest post-rift sediments are western Wilkes Land and the area seaward of Prydz Bay. Both areas include large sediment mounds of contourite or mixed contourite-turbidite origin. Sediments in the eastern Wilkes Land-western George V Land area exceed 6 km in thickness beneath the continental shelf edge but much of this is in rift basins (Wannesson 1990). Post-rift sediment mounds of mixed or tubidite fan origin are also present (De Santis et al. 2003; Escutia et al. 2000). Our new reflection data do not extend into the area they describe.

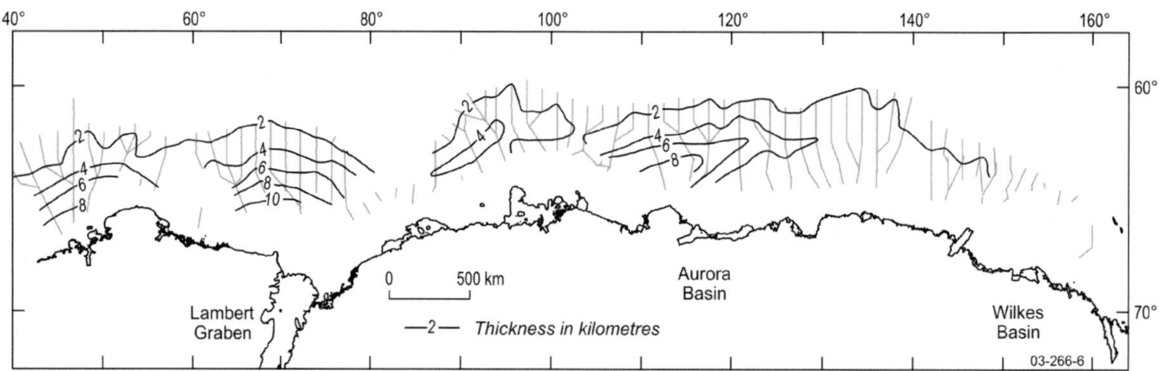

Fig. 6.7-7. Thickness in kilometres of post-rift sediments from 38° E to 164° E, East Antarctic continental margin based on sonobuoy seismic refraction velocities and reflection two-way times. Subglacial topographic and geological features that may have controlled sediment input are, from west to east: *Lambert Graben*, *Aurora Subglacial Basin* and *Wilkes Subglacial Basin*

Discussion

The thickest post-rift sediments are composed of mounded, mixed turbidite-contourite origin. The thickest sediment section is seaward and slightly west of Prydz Bay, suggesting westward advection of sediment supplied from the continent via the Lambert Graben drainage system. The Lambert Graben has probably controlled drainage since the Palaeozoic (Arne 1994). The thickest section off western Wilkes Land is seaward and westward of the Aurora Subglacial Basin and its major outlet glaciers such as the Totten Glacier. Pollard and De Conto (2003) modelled high sediment fluxes in this area, based on subglacial topography, and Bamber et al. (2000) recognised it as a zone of high ice flux in the modern ice sheet. The western Wilkes Land contourite drift is not as clearly connected to a large drainage outlet as those off Prydz Bay, suggesting that it may have formed from fine sediment advected from the whole Wilkes Land coast rather than from a single discharge point.

A surprising observation is the relatively thin post-rift section in the George V-eastern Wilkes Land region. The presence of the Wilkes Subglacial topographic basin caused Pollard and De Conto's (2003) model to predict high sediment flux through that part of the coast. The relatively thin (<4 km) section may indicate that the ice fluxes through this region have not been as high as predicted by the model and may be more in line with the present situation where ice fluxes are not as great as further west (Bamber et al. 2000). The thick west Enderby Land margin sequence, as noted above, probably formed early in the history of margin development and so probably does not relate to ice sheet drainage. Rather, it seems to be an early post-rift basin fill, and therefore may have been deposited soon after separation of Antarctica and India in the Valanginian (Stagg et al. 2005) and be Jurassic to Cretaceous in age.

Age control on the East Antarctic margin is very sparse, with three ODP holes and two DSDP holes providing age data on the slope and rise. Off Prydz Bay, ODP site 1165 suggests that most of the large drifts there formed during the Miocene, and particularly during the Early Miocene (Cooper and O'Brien 2004). De Santis et al. (2003) concluded that most sediment ridge growth on the George V Land margin also took place during the Miocene. Cooper and O'Brien (in press) suggest that the Miocene may have been the time of maximum sedimentation on the slope and rise, because the continent was covered by extensive polythermal glaciers that provided abundant meltwater and fine sediment to the margin. This proposed timing of sedimentation will need to be tested by further mapping of seismic sequences through these new data sets.

Conclusions

New seismic reflection data from the East Antarctic margin between Enderby Land and George V Land allow the broad classification of sediments on the slope and rise. Prograding upper slope wedges form the slope in many places, whereas distal abyssal plain sediments are present on the northern ends of the longest seismic lines. Submarine fans form most of the Wilkes Land slope and rise and thick accumulations of mixed contourite-turbidite deposits are present offshore from Prydz Bay. Contourite sediments cut by canyons are present offshore west of Bruce Rise, western George V and western Wilkes Land. Thin contourite drifts abut steep upper slopes along the eastern George V Land margin, east of Bruce Rise and in the Princess Elizabeth Trough. Thick sediments along the western Enderby Land margin probably pre-date glaciation of the continent.

The thickest post-rift sediments are offshore and to the west of major ice drainage systems with the thickest section seaward of the Prydz Bay-Lambert Graben system. The Wilkes Subglacial Basin does not seem to have been a major sediment input point.

Acknowledgments

We would like to thank Howard Stagg, Karl Hinz and Angelo Carmerlenghi for comments on the manuscript and Joe Mifsud for drafting. The authors publish with the permission of the Chief Executive Officer, Geoscience Australia.

References

ANTOSTRAT (2003) SCAR Antarctic Seismic Data Library (SDLS) web site, http://walrus.wr.usgs.gov/sdls/sdlsp5.html (accessed 19/11/2003)

Arne DC (1994) Phanerozoic exhumation history of northern Prince Charles Mountains (East Antarctica). Antarctic Science 6:69–84

Bamber JL, Vaughan DG, Joughin I, (2000) Widespread complex flow in the interior of the Antarctic Ice Sheet. Science 287:1248–1250

Cooper AK O'Brien PE (2004) Transitions in East Antarctic glacial history - from ODP drilling in the Prydz Bay region. In: Cooper AK, O'Brien PE Pro Ocean Drilling Program, Sci Results 188 (CD-ROM), Texas A&M University, College Station, pp 1–42

De Santis L, Brancolini G, Donda F, (2003) Seismo-stratigraphic analysis of the Wilkes Land continental margin (East Antarctica): influence of glacially driven processes on the Cenozoic deposition. Deep-Sea Res II 50:1563–1594

Escutia C, Eittreim SL, Cooper AK, Nelson CH, (2000) Morphology and acoustic character of the Antarctic Wilkes Land turbidite systems: ice-sheet versus river-sourced fans. J Sediment Res 70:84–93

Eittreim SL, Smith GL (1987) Seismic sequences and their distribution on the Wilkes Land margin. In: Eittreim SL, Hampton MA (eds) The Antarctic continental margin: geology and geophysics of offshore Wilkes Land. Earth Science Series, Circum-Pacific Council for Energy and Resources, 5A, pp 15–43

Faugères JC, Stow DAV, (1993) Bottom-current-controlled sedimentation: a synthesis of the contourite problem. Sediment Geol 82:287–297

Faugères JC, Stow DAV, Imbert P, Viana A (1999) Seismic features diagnostic of contourite drifts. Marine Geology 162:1–38

Florindo F, Bohaty SM, Erwin PS, Richter C, Roberts AP, Whalen PA, Whitehead JM (2003) Magnetobiostratigraphic chronology and palaeoenvironmental history of Cenozoic sequences from ODP sites 1165 and 1166, Prydz Bay, Antarctica. Palaeogeogr Palaeoclimatol Palaeoecol 198:69–100

Ishihara T, Tanahashi M, Sato M, Okuda Y, (1996) Preliminary report of geophysical and geological surveys on the west Wilkes Land margin. Proc NIPR Symp Antarctic Geosci 9:91–108

Kuvvas B, Leitchenkov G, (1992) Glaciomarine turbidite and current controlled deposits in Prydz Bay, Antarctica. Marine Geology 108:365–381

Marani M, Argnani M, Roveri M, Trincardi F (1993) Sediment drifts and erosion surfaces in the central Mediterranean: seismic evidence of bottom current activity. Sediment Geol 82: 207–220

O'Brien PE, Cooper AK, Richter C, et al. (2001) Initial Reports, Prydz Bay - Cooperation Sea, Antarctica: glacial history and paleoceanography. Proc Ocean Drilling Program, Initial Rep 188, Texas A&M Univ, College Station

Pollard D, DeConto RM (2003) Antarctic ice and sediment flux in the Oligocene simulated by a climate-ice-sheet-sediment model. Palaeogeogr Palaeoclimatol Palaeoecol 198:53–68

Rebesco M, Larter RD, Barker PF, Camerlenghi A, Vanneste LE (1997) The history of sedimentation on the continental rise west of the Antarctic Peninsula. In: Barker PF, Cooper AK (eds) Geology and Seismic Stratigraphy of the Antarctic Margin (Pt. 2). Amer Geophys Union Antarctic Res Ser 71:29–50

Shipboard Scientific Party (2001b) Site 1165. In: O'Brien PE, Cooper AK, Richter C, et al. (eds) Proc Ocean Drilling Program, Initial Rep 188. Texas A&M Univ, College Station, pp1–191

Stagg HMJ (1985) The structure and origin of Prydz Bay and MacRobertson Shelf, East Antarctica. Tectonophysics 114: 315–340

Stagg HMJ, Colwell JB, Direen NG, O'Brien PE, Bernadel G, Borissova I, Brown BJ, Ishihara T (2005) Geology of the continental margin of Enderby and Mac. Robertson Lands, East Antarctica: Insights from a regional data set. Mar Geophys Researches 25:183–219, DOI:10.1007/s11001-005-1316-1

Wanneson J (1990) Geology and petroleum potential of the Adélie Coast, East Antarctica. In: St John W (ed) Antarctica as an exploration frontier - hydrocarbon potential, geology and hazards. AAPG Studies in Geology 31:77–88

Theme 7
Antarctic Neotectonics, Observatories and Data Bases

Chapter 7.1
On Seismic Strain-Release within the Antarctic Plate

Chapter 7.2
Vertical Crustal Deformation in Dronning Maud Land, Antarctica: Observation versus Model Prediction

Chapter 7.3
Seismic Activity Associated with Surface Environmental Changes of the Earth System, around Syowa Station, East Antarctica

Chapter 7.4
Geodynamic Features and Density Structure of the Earth's Interior of the Antarctic and Surrounded Regions with the Gravimetric Tomography Method

Chapter 7.5
Some Recent Characteristics of Geomagnetic Secular Variations in Antarctica

Chapter 7.6
Topographic and Geodetic Research by GPS, Echosounding and ERS Altimetric, and SAR Interferometric Surveys during Ukrainian Antarctic Expeditions in the West Antarctic

Chapter 7.7
Geodetic Research on Deception Island

Chapter 7.8
The Multidisciplinary Scientific Information Support System (SIMAC) for Deception Island

Chapter 7.9
Archiving and Distributing Earth-Science Data with the PANGAEA Information System

Antarctica has been said to be an aseismic continent. However, continuous observation with sensitive equipment allows for recognising local seismicity. High precision GPS measurements help to define a new geodetic network and to supply exact data on tiny horizontal and vertical movements. Such and other geological and geophysical data are worthy of being collected in data bases to serve the whole geoscientific community.

Firstly, in this chapter, a valuable comprehensive summary of Antarctic seismicity is given by Reading (Chap. 7.1). In Dronning Maud Land, Scheinert et al. (Chap. 7.2) measured and modelled vertical crustal deformation produced by glacial isostatic adjustment. Local seismic activity around Syowa Station is depicted by Kanao and Kaminuma (Chap. 7.3). Greku et al. (Chap. 7.4) describe the density structure of the Earth's interior in Antarctica. The characteristics of the recent geomagnetic secular variation in Antarctica are shown and developed to a regional magnetic reference model by Meloni et al. (Chap. 7.5). West Antarctica, particularly Argentine Island and Deception Island were targets of geodetic work done by Greku et al. (Chap. 7.6) and Berrocoso et al. (Chap. 7.7) respectively. Torrecillas et al. (Chap. 7.8) developed a specific information support system for Deception Island. Finally Grobe et al. (Chap. 7.9) present PANGAEA, a major comprehensive information system for earth science data.

On Seismic Strain-Release within the Antarctic Plate

Anya M. Reading
Research School of Earth Sciences, Australian National University, Canberra, ACT 0200, Australia, <anya@rses.anu.edu.au>

Abstract. Seismic strain release on the Antarctic continent takes place at a much lower rate than in other continental, intraplate areas. Tectonic and glaciogenic forces controlling this observed distribution have previously been discussed in terms of the Antarctic continent only. Improved locations of large earthquakes in the surrounding, oceanic, Antarctic Plate show that a number of these events, including the great 25 March 1998 earthquake occurring between New Zealand and Antarctica, have intraplate settings. Such large episodes of strain-release suggest that it is more appropriate to address controls on Antarctic seismicity by considering the entire plate, including both the central region of continental crust and the surrounding oceanic crust. The catalogued seismicity for 20 years, between 1981 and 2000, is presented for the whole Antarctic Plate together with a discussion of the tectonic settings of the largest continental events and the oceanic events occurring during the same period. The forces acting on the oceanic and continental regions are discussed in the context of the unique tectonic setting of the Antarctic Plate, which is surrounded almost entirely by ocean-ridges. A correspondance exists between regions of low continental seismicity and the most extensive regions of surrounding oceanic Antarctic Plate. This suggests that the oceanic region, acting as a buffer in some places against plate boundary influences on the Antarctic continent, is an additional factor in controlling the distribution of seismicity.

Introduction

Investigations of Antarctic seismicity (Adams et al. 1985; Kaminuma 1994; Rouland et al. 1992; Reading 2002) have necessarily been dominated by addressing deficiencies in the catalogued record of Antarctic and southern hemisphere earthquakes. It is of immense value to lower the detection threshold of such events, ensuring the most comprehensive appraisal of brittle deformation in the Antarctic crust. Catalogue bias still remains due to the sparse and uneven distribution of Antarctic stations and the urgent need to deploy more stations in the Antarctic interior remains as strong as ever. This contribution, however, presents a wider perspective. Using the higher magnitude earthquakes that are found in the global catalogues, patterns of seismic strain-release throughout the entire Antarctic Plate are investigated including both oceanic and continental lithosphere.

Strain and Earthquake Magnitude

The strain accumulated in the crust over time may be approximated by that released during an earthquake. Radiated seismic waves therefore contain the required information to estimate the strain-release/stress drop (Stein and Wysession 2003). The key quantity is the seismic moment, which is proportional to the stress drop. The seismic moment is related to the moment magnitude as follows:

$$M_W = (\log M_0 / 1.5) - 10.73$$

The magnitude of an earthquake is an observable quantity, hence the spatial distribution of earthquakes of differing magnitudes reflects the way in which deformation is partitioned across the plate. The logarithmic relationship between magnitude and moment shows that a first order approximation of the strain-release/stress drop may be obtained by considering only the larger magnitude earthquakes of a given region. Earthquakes of lesser magnitude by 3 or 4 scale points are 1 000s of times less significant than the large earthquakes in terms of strain release. In this broad-scale survey, body-wave magnitudes, m_b, are accepted as an approximation for the moment magnitude of the small earthquakes of the Antarctic continent. For the large earthquakes of the surrounding oceans, we use M_S (surface wave magnitude) as a working approximation for M_W since M_S is much more widely reported. This approximation is not valid when considering the relative magnitudes of (e.g.) two large earthquakes owing to the saturation of the M_S scale at high values but it is adequate in comparing the distribution of large and small earthquakes as is carried out here.

Seismicity

A near comprehensive appraisal of Antarctic seismicity, including many smaller events not in the global catalogues, was presented by Reading (2002). Small, locally observed earthquakes and those located by the prelimi-

From: Fütterer DK, Damaske D, Kleinschmidt G, Miller H, Tessensohn F (eds) (2006) Antarctica: Contributions to global earth sciences. Springer-Verlag, Berlin Heidelberg New York, pp 351–356

nary International Data Centre (associated with monitoring the Comprehensive Test Ban Treaty) were added to events in the standard catalogue. At the time of writing, no earthquake locations were available from Marie Byrd Land, central West Antarctica. It is now clear beyond doubt that Marie Byrd Land does show low-levels of active seismicity, possibly associated with a largely dormant rift system (Winberry and Anandakrishnan 2003). Every part of the Antarctic continent shows seismic activity if seismic recorders are deployed locally. Such deployments also dramatically lower the detection threshold for small Antarctic earthquakes and confirm that the level of seismicity on the continent is extremely low. One of the oceanic provinces of the Antarctic Plate, in contrast, gave rise to one of the largest earthquakes ever recorded (Antonioli et al. 2002). This occurred on 25 March 1998 close to the Balleny Islands and was initially thought to be a plate boundary earthquake. Closer analysis (e.g., Antonioli et al. 2002) revealed that it consisted of near simultaneous failure on two huge planes in a region of ocean floor not associated with the plate margin. Remarkably, one of the greatest recorded earthquakes of all time has taken place in the plate which is regarded as being the least seismically active, the Antarctic Plate. Choy and McGarr (2002) examined oceanic earthquakes with high apparent stress and suggest that the Balleny Island region earthquake shows exceptional horizontal stresses associated with unusually buoyant lithosphere.

Measured and estimated sea-floor topography (Smith and Sandwell 1997) reveals numerous oceanic lithosphere provinces around the Antarctic Plate that may be similarly capable of sustaining high stress deformation.

Motion and Uplift of the Antarctic Plate

The motion and deformation of the Antarctic Plate was investigated using GPS data by Bouin and Vigny (2000). They found that East Antarctica is moving as a rigid plate, and also observed 1–2 cm of co-seismic displacement associated with the 25 March 1998 great earthquake. Lateral tectonic forces are dominated by spreading at mid-ocean ridges surrounding the Antarctic Plate. Rundquist and Sobolev (2002) review the seismicity and geodynamics of mid-ocean ridges and include a summary of the spreading rates of different ridge segments. Spreading rates surrounding the Antarctic Plate (in cm yr^{-1}) are quite variable as follows: Ant-Pacific = 7.27; Ant-Nazca = 5.93; Ant-South America = 1.75; Ant-African = 1.42; Ant-Australian = 6.77.

James and Ivins (1998) address the likely deformation of the Antarctic Plate due to ice sheet evolution. They calculate the likely uplift rates associated with ice-sheet retreat. Observing this uplift is significant as a constraint on continent-wide glaciological processes but for the purposes of this contribution, it is noted that uplift could be sufficiently fast to generate earthquakes, especially around the periphery of East Antarctica. The challenge of separating tectonic and glaciological controls on the deformation and seismicity of the Antarctic Plate remains.

Data

The catalogued seismicity of the Antarctic Plate over a twenty year interval, 1981–2000, is shown (Fig. 7.1-1). Hypocentre locations are those of the Bulletin of the International Seismological Centre (ISC). Events occurring on the Antarctic continent (m_b greater than 3.5) are indicated by the star symbols. The larger events occurring in oceanic regions (M_S greater than 6.5) are shown by open circles. Implicit in the catalogue, and hence the mapped seismicity, are variations in detection threshold. This is difficult to quantify in practical terms since most of the Antarctic stations that report to the ISC show very large variations in noise levels. During the Antarctic winter, such stations may either be very quiet or subject to considerable wind noise and during the summer, after the sea-ice melts, they may be affected badly by ocean-wave noise. It is unlikely that many events on the Antarctic continent with $m_b > 3.5$ will have been missed during the twenty year interval and therefore the distribution shown is representative of the patterns of seismicity under investigation.

Earthquakes of the Antarctic Continent

i. Victoria Land

Victoria Land, together with the volcanic and other current/recent activity in the Ross Sea (reviewed by Fitzgerald 2002) is the region of the Antarctic continent with the highest seismicity (Reading 2002; excluding the active subduction of the Bransfield Strait). Victoria Land has a concentration of long-standing permanent seismic observatories (near McMurdo, two stations near Scott Base, and at Terra Nova Bay), hence, the detection threshold is sufficient to state with confidence that the coastal region of Victoria Land shows genuinely high levels of seismic deformation. The interior of Victoria Land shows an incorrectly high seismicity in the early catalogues (identified as "poorly located" by Reading 2002). These "hypocentres" are actually later phases of earthquakes from the South Sandwich Islands which have been misidentified as first arrivals. The "well located" events in the same paper show the same distribution as the catalogued hypocentres in this catalogue and are confined to the coastal region.

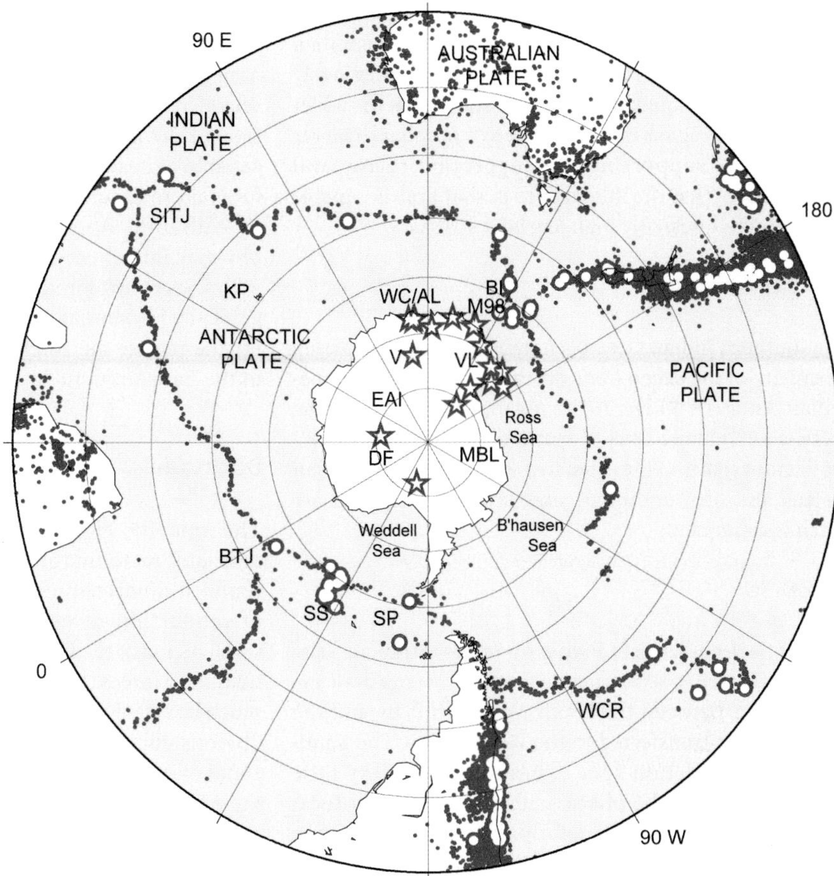

Fig. 7.1-1. The seismicity of the Antarctic Plate and surrounding regions (1981–2000). *Small dots* indicate seismicity catalogued in the Bulletin of the International Seismological Centre; events in the Antarctic continent greater than $m_b = 3.5$ are shown with *stars* and those in the surrounding ocean greater than $M_S = 6.5$ are shown by *open circles*; *M98:* the location of the great Antarctic Plate earthquake of 25 March 1998; *SITJ:* South Indian Triple Junction; *BTJ:* Bouver Triple Junction; *KP:* Kerguelan Plateau; *SS:* South Sandwich; *SP:* Scotia Plate; *MBL:* Marie Byrd Land; *EAI:* East Antarctic Interior; *V:* Vostock; *DF:* Dome Fuji; *VL:* Victoria Land; *WC/AL:* Wilkes Coast/Adélie Land; *BI:* Balleny Islands; *WCR:* West Chile Rise

ii. Terre Adélie/Wilkes Land Coast

The Terre Adélie/Wilkes Land coast continues the coastal distribution of the Victoria Land events. The hypocentres extend westward, until 105° when they abruptly cease. There is no seismicity above the catalogue threshold right around the East Antarctic Coast between 105° and the Weddell Sea. This striking gap coincides with the widest region of oceanic crust surrounding Antarctica (enclosing the thickened Kerguelan Plateau). It extends up to nearly 15° S at the mid-Indian Ocean triple junction.

iii. East Antarctic Interior

A small number of earthquakes have been recorded deep within the East Antarctic interior. Two such events are indicated in Fig. 7.1-1. They occur in the Vostok region and close to Dome Fuji. The Vostok region has been subject to relatively intense geophysical scrutiny since the discovery of the large sub-glacial lake (e.g., Studinger et al. 2003) and is characterised by subglacial topography indicating a major geological boundary. Data are more sparse beneath the Dome Fuji region.

iv. Weddell Sea

Seismic events in the Weddell Sea have a recently improved detection threshold due to the seismic array located near Neumayer Station and in this case the ISC catalogue is less than complete. The Weddell Sea shows a low to moderate level of seismic activity (C. Müller, personal communication). The ice-conditions in the southern part of the embayment limit extensive marine geophysical research in the area. It appears to be a region of thinned continental or thickened oceanic crust, outside of the rigid East Antarctic Plate. The trend of the West Antarctic Rift system between West and East Antarctica, possibly into the Weddell Sea or, alternatively, through Ellsworth Land and into the Bellingshausen Sea, is not known at present.

Earthquakes of the Southern Ocean

i. Balleny Island Region

The Balleny Island region is very active, located at the triple junction between the Pacific, Australian and Antarctic Plates. The region is subject to intense deformation owing to the subduction of the Australian Plate beneath the Pacific Plate. The epicentre of the great earthquake of 25 March 1998 was 250 km from the Antarctic-Australian plate boundary and 100 km from any other previously well-located earthquake. Figure 7.1-1 shows that the wider Balleny Island region has given rise to a number of larger earthquakes, supporting the suggestion (Choy and McGarr 2002) that the lithosphere in that area is capable of sustaining unusually high levels of stress.

ii. Southern Indian Ocean

The southern Indian Ocean is notable for the contrasting seismicity of the ocean floor north and south of the Australian-Antarctic Ridge. To the north, in the Indian Plate, there is a moderate level of scattered intraplate seismicity. Further south, in the extensive region of the Kerguelan Plateau and surrounding oceanic crust, earthquakes are much less frequent.

iii. Scotia Sea

The Scotia Sea is a misfit in this survey of intraplate seismicity, however, it is instructive to note how much of the convergence between the South American Plate and the Scotia Plate is transferred to the Weddell Sea. The South Sandwich subduction zone appears to show very little coupling between the plates, with no thrust fault focal mechanisms and very little deformation in the over-riding plate (Larter et al. 1998). The lack of seismicity south of the strike-slip margin between the Scotia Plate and the Weddell Sea indicates that there is little transfer of stress across the plate boundary.

iv. Bellingshausen Sea

A few earthquakes are observed (Fig. 7.1-1) in the Bellingshausen Sea. An earthquake fault mechanism from an event in 1977 indicates a compressive stress regime (Okal 1980). Marine geophysical studies indicate older compressive structures (Gohl et al. 1997) which may be reactivated by similar or ongoing tectonic forces. The quiet seismic zone along the coast of the Bellingshausen Sea also corresponds to the very wide region of oceanic crust between the Antarctic Continent and the West Chile Rise. Like the region of low seismic activity west of 105° E, low coastal activity and a broad buffer of ocean crust seem to be associated.

Distribution of Strain Release

The pattern of seismic strain-release emerging from the catalogued seismicity shows a low level of continental seismicity, especially around coastal regions closer to the plate margin.

Coastal regions having a very broad region of oceanic or ocean-plateau material between the Antarctic continent and the plate margin seem to show particularly low seismicity. The Balleny Island region shows both a high level of deformation, and that the oceanic lithosphere is capable of sustaining especially high levels of stress before failure occurs.

Within the Antarctic continent, the sub-ice topography may influence the location of seismic strain-release, with scarcely explored tectonic province boundaries (suggested by Fitzsimons 2003) providing possible weaknesses for the release the small amount of seismicity observed in the East Antarctic interior.

Discussion

The Antarctic Plate is subject to more uniform and less dramatic tectonic forces than other plates, being surrounded almost entirely by spreading centres. The absence of subduction-related forces (other that the relatively small region of the Bransfield Strait) implies that the plate-boundary forces that might propagate across the plate are much less than those across other continents. There is no obvious correlation between ridge spreading rate and zones of enhanced seismicity, e.g., fast spreading between the Antarctic and Australian Plates corresponds to high seismicity in that sector of the Antarctic continent, yet fast spreading on the Antarctic-Pacific Ridge has low seismicity in the adjacent Antarctic continent.

Antarctica is subjected to a uniquely high component of glaciogenic lithospheric stress. The coastal distribution of larger continental earthquakes seems to indicate that ice unloading forces are sufficient, in addition to tectonic forces, to cause brittle failure of the lithosphere. Since the ice-related forces are likely to act all around East Antarctica, it appears that the ice-component is not a sufficient addition to the tectonic component to cause failure where there is a broad surrounding region of oceanic Antarctic Plate. Sub-ice topography and relict tectonic province boundaries are likely to control the small amount of seismicity observed within the Antarctic Interior. It seems unlikely that seismicity above the $m_b > 3.5$ threshold is due to neotectonic forces but microseismicity below this level and not discussed in this paper may well be caused by current low-levels of active processes (e.g., Winberry and Anandakrishnan 2003).

It may be that the twenty year time interval chosen for the seismicity plot in Fig. 7.1-1 is substantially shorter than the repeat time of great oceanic lithosphere earthquakes such as those investigated by Choy and McGarr (2002). It may be some decades before it is appropriate to address the above discussion with a quantitative analysis, however, it does seem certain that oceanic intraplate earthquakes are an important part of the seismotectonics of the Antarctic Plate. It is even proposed that the 28 March 1998 quake was caused by post-glacial rebound (Tsuboi et al. 2000). This possibility confirms that then the Antarctic continent and oceanic lithosphere must henceforth be investigated together.

Conclusions

1. The low seismicity of the Antarctic continent must be considered together with the high seismicity of the surrounding intraplate oceanic lithosphere.
2. The wide regions of oceanic lithosphere that surround Antarctica are capable of supporting high levels of stress and may buffer the Antarctic continent from lateral forces due to the spreading ridges.
3. Larger earthquakes in the Antarctic continent are more likely at coastal regions and may be influenced by crustal uplift due to glacial unloading.
4. Earthquake distribution in the continental interior may be influenced by sub-ice topography and tectonic province boundaries.

Acknowledgment

Earthquake hypocentre information was sourced from the International Seismological Centre, UK. Alfons Eckstaller and Sridar Anandakrishnan are thanked for their reviews and consequent improvements to the paper.

References

Adams RD, Hughes AA, Zhang BM (1985) A confirmed earthquake in continental Antarctica. Geophy J Royal Astron Soc 81:489–492

Antonioli A, Cocco M, Das S, Henry C (2002) Dynamic stress triggering during the great 25 March 1998 Antarctic Plate earthquake. Bull Seismol Soc Amer 92:896–903

Bouin M, Vigny C (2000) New constraints on Antarctic plate motion and deformation from GPS data. J Geophys Res 105:28279–28293

Choy GL, McGarr A (2002) Strike-slip earthquakes in the oceanic lithosphere: observations of exceptionally high apparent stress. Geophys J Internat 150:506–523

Fitzgerald P (2002) Tectonics and Landscape evolution of the Antarctic plate since the breakup of Gondwana, with an emphasis on the West Antarctic Rift System and the TransAntarctic Mountains. In: Gamble JA, Skinner DNB, Henrys S (eds) Antarctica at the close of a millennium. Royal Soc New Zealand Bull 35, Wellington, pp 453–469

Fitzsimons ICW (2003) Proterozoic basement provinces of southern and south-western Australia and their correlation with Antarctica. In: Yoshida M, Windely BF, Dasgupta S (eds) Proterozoic East Gondwana: supercontinent assembly and breakup. Geol Soc London Spec Publ 206, London, pp 93–130

Gohl K, Nitsche F, Miller H (1997) Seismic and gravity data reveal Tertiary interplate subduction in the Bellingshausen Sea, southeast Pacific. Geology 25:371–374

James TS, Ivins ER (1998) Predictions of Antarctic crustal motions driven by present-day ice sheet evolution and by isostatic memory of the Last Glacial Maximum. J Geophys Res 103:4993–5017

Kaminuma K (1994) Seismic activity in and around the Antarctic Continent. Terra Antartica Special Issue 1:423–426

Larter RD, King EC, Leat PT, Reading AM (1998) South Sandwich slices reveal much about arc structure, geodynamics and composition. EOS 79(24):281–285

Okal EA (1980) The Bellingshausen Sea earthquake of February 5, 1977: Evidence for ridge-generated compression in the Antarctic plate. Earth Planet Sci Letters 46:306–310

Reading AM (2002) Antarctic seismicity and neotectonics. In: Gamble JA, Skinner DNB, Henrys S (eds) Antarctica at the close of a millennium. Royal Soc New Zealand Bull 35, Wellington, pp 479–484

Rouland G, Condis C, Parmentier C, Souriau A (1992) Previously undetected earthquakes in the southern hemisphere located using long-period geoscope data. Bull Seismol Soc Amer 82: 2448–2463

Rundquist DV, Sobolev PO (2002) Seismicity of mid-oceanic ridges and its geodynamic implications: a review. Earth Sci Rev 58: 143–161

Smith W, Sandwell D (1997) Measured and estimated seafloor topography (version 4.2). World Data Center for Marine Geology and Geophysics, Res Publ RP-1 (poster)

Stein S, Wysession M (2003) An introduction to seismology, earthquakes, and earth structure. Blackwell Publishing, Oxford

Studinger M, Bell RE, Karner GD, Tikku AA, Holt JW, Morse DL, Richter TG, Kempf SD, Peters, ME, Blankenship DD, Sweeney RE, Rystrom VL (2003) Ice cover, landscape setting and geological framework of Lake Vostok, East Antarctica. Earth Planet Sci Letters 205:195–210

Tsuboi S, Kikuchi M, Yamanaka Y, Kanao M (2000) The March 25, 1998 Antarctic earthquake: great earthquake caused by post-glacial rebound. Earth Planets Space 52:133–136

Winberry JP, Anandakrishnan S (2003) Seismicity and neotectonics of West Antarctica. Geophys Res Letters 30(18):1931

Vertical Crustal Deformation in Dronning Maud Land, Antarctica: Observation *versus* Model Prediction

Mirko Scheinert[1] · Erik Ivins[2] · Reinhard Dietrich[1] · Axel Rülke[1]

[1] Technische Universität Dresden, Institut für Planetare Geodäsie, Helmholtzstraße 10, 01062 Dresden, Germany, <mikro@ipg.geo.tu-dresden.de>

[2] Jet Propulsion Laboratory, California Institute of Technology, Pasadena CA 91109-8099, USA

Abstract. During the last decade a variety of geodetic observations have been carried out in central Dronning Maud Land, East Antarctica, in order to investigate geodynamic and glaciologic phenomena. Of special interest is the interaction of recent and historic ice mass changes and the vertical crustal deformation, which is characterized by the rheological properties of the Earth, especially of the crust and the upper mantle. The geodetic information may help to constrain the recent isostatic uplift pattern, the recent and – using additional age information – past ice sheet configuration in central Dronning Maud Land. Coupling the geodetic data with these additional constraints on recent and past ice mass changes allows a self-consistent glacial load history to be investigated. A spectrum of viable load histories will be examined and the respective isostatic deformation signature will be computed, using a flat earth approach. First model computations will be presented and discussed, aiming to reconcile the modelled vertical uplift signature and the observations.

Introduction

During the past two decades, international efforts have been focused on determining the dynamics of the Antarctic ice sheet and its link to past and present climate change. The weight of a more massive past ice sheet pushed downward on solid bedrock. As this weight is removed during the past millennia, the deep solid Earth suffers gravitational viscoelastic strain. Although straight forward conceptually the knowledge of the ice load history and glacial isostatic adjustment (GIA) pattern in Antarctica is quite limited. There is a growing need to combine observations of different sources with modelling on a regional scale (Huybrechts and LeMeur 1999; James and Ivins 1998). The purpose of this brief report is to employ a multidisciplinary data set that constrains the ice load history, to construct a simple GIA model and to compare the pattern and amplitude of these predictions to new GPS uplift data which we have collected and analyzed in Dronning Maud Land.

Field Observations

Over the last decade, the Institut für Planetare Geodäsie of the Technical University of Dresden has been working in Central Dronning Maud Land, East Antarctica, in the

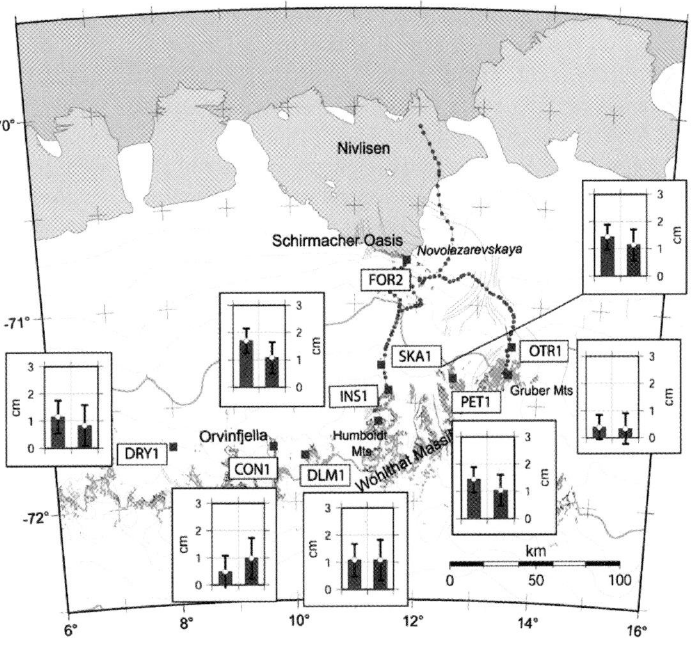

Fig. 7.2-1.
Area of investigation (central Dronning Maud Land, East Antarctica). *Red dots* show location of geodetic-glaciological traverses: Insel traverse: south of Schirmacher Oasis to Humboldt Mountains; Untersee traverse: south-east of Schirmacher Oasis to Gruber Mountains; North traverse: north of Schirmacher Oasis to Nivlisen. *Blue rectangles* show the location of GPS markers on bedrock. The respective subpanels give the height change of GPS sites relative to the reference station FOR2 for the time span of five years: *red bar:* estimation of troposphere parameters at both end points of the respective baseline; *blue bar:* estimation of troposphere parameters at the mountain site relative to the reference station FOR2; *black bars:* errors at same scale

From: Fütterer DK, Damaske D, Kleinschmidt G, Miller H, Tessensohn F (eds) (2006) Antarctica: Contributions to global earth sciences. Springer-Verlag, Berlin Heidelberg New York, pp 357–360

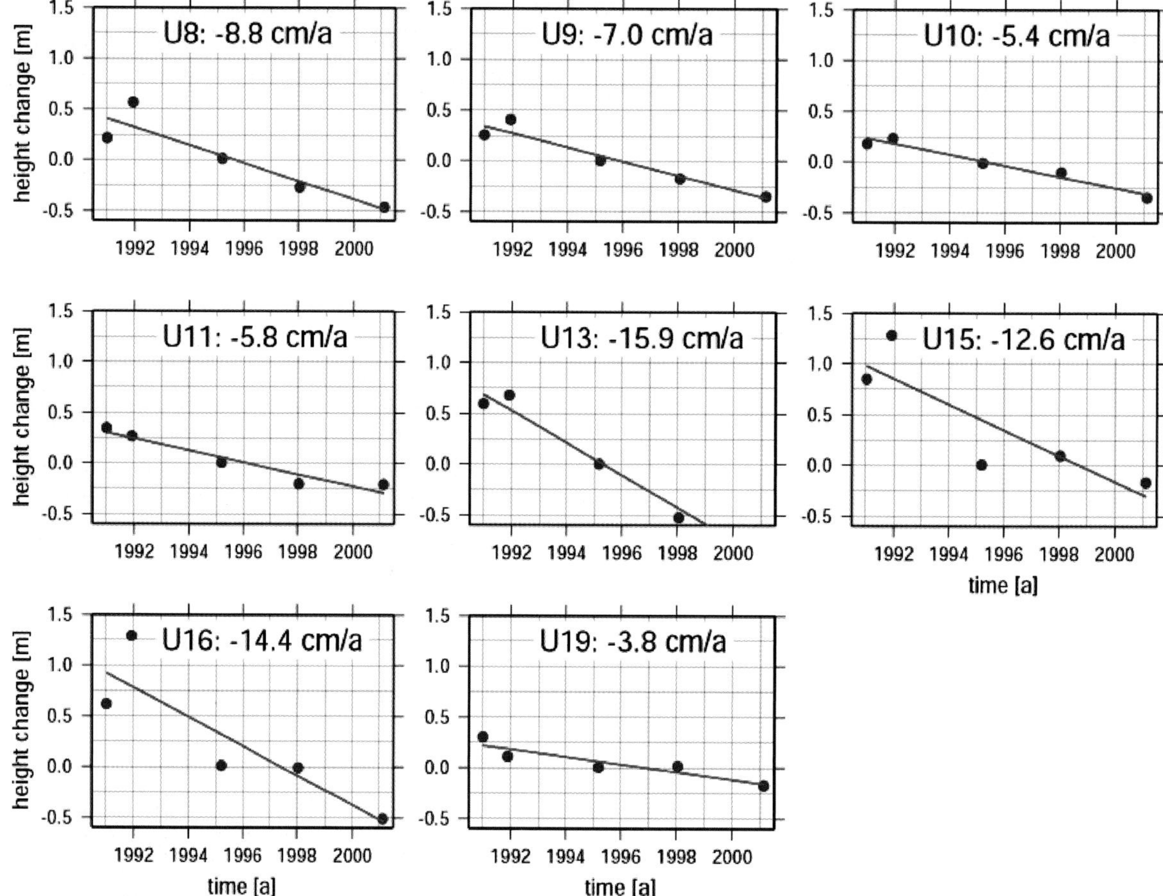

Fig. 7.2-2. Height change at signals of the Untersee traverse (locations are given by *blue dots* in Fig. 7.2-1; with signal numbers increasing from south-east to north-west)

region of the Schirmacher Oasis. A variety of measurements have determined the flow velocity and height change of the ice sheet, as well as the vertical deformation of the crust. There are two geodetic-glaciological traverses running south from the Schirmacher Oasis to the northern edge of Wohlthat Massif (Fig. 7.2-1).

Repeated GPS observation campaigns in 1996 and 2001 allowed determination of the vertical motion of bedrock sites. The analysis yielded a height change of the mountain sites of about 2 mm yr^{-1}, relative to the reference site in the Schirmacher Oasis (Fig. 7.2-2). This result – still preliminary and in need of a third observation campaign – could serve as a constraint on the ongoing viscoelastic GIA and therefore, on models of recent ice mass changes.

Along the geodetic-glaciological traverses height changes at signals were measured. The result for the Untersee traverse, which crosses a large blue ice region characterized by strong ablation, yielded a decrease of the ice surface of about 5 to 15 cm yr^{-1}. Such measurements are indicative of the state of mass balance over a large area between Schirmacher Oasis and Wohlthat Massif.

A minimum age estimate of regional ice retreat was given by Wand and Hermichen (2003), who investigated the occurrence and basal age of mumiyo deposits. (Mumiyo is a deposit of stomach oil of snow petrels, who settle in ice-free areas across the mountains.) They concluded that in the Gruber Mountains (eastern Wohlthat Massif) the ice surface could have been about 900 m higher during the past, whereas in Orvinfjella the ice surface was only about 130 m higher than at present.

Model Results

In order to roughly model the ice load history, which in turn drives present day vertical deformation as observed by GPS, a simple viscoelastic model was incorporated (Wolf 1985; James and Ivins 1999). The lithosphere is treated as an elastic layer, whereas the model upper mantle is a homogeneous incompressible Maxwell viscoelastic halfspace. The perturbation equations for the deformation field are solved, applying a finite axisymmetric stress

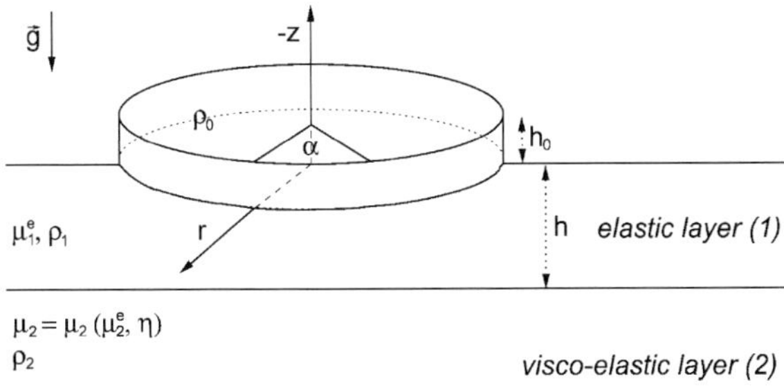

Fig. 7.2-3. Principle of the construction of the flat-earth model and a disk load. Symbols: *h:* lithospheric thickness; *μ:* shear modulus; *ρ:* density; *η:* viscosity; *g:* gravity; *r, α, z:* cylindric coordinates; h_0: thickness of disk

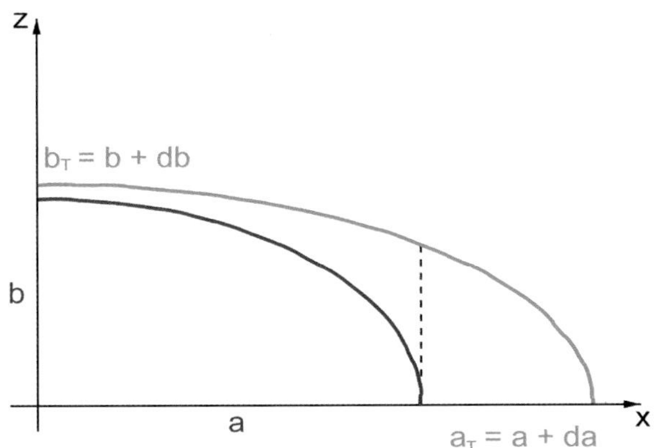

Fig. 7.2-4. Elliptic profile of an ice sheet. The *blue profile* represents the present state, whereas the *green profile* represents the state at the glacial maximum

load (Fig. 7.2-3). The vertical rate of deformation can then be compared to the GPS uplift data collected at the different bedrock sites.

To construct the ice load model, the modified Antarctic ice load history from James and Ivins (1998) was used. For the area of investigation, the ice load model was modified applying an elliptic cross section ice sheet model (Fig. 7.2-4). Two different assumptions were made to construct the ice sheet at glacial maximum: First, it was assumed that the maximum ice extent reached the position of the present ice shelf edge. Second, it was assumed that the ice sheet at glacial maximum ended at the present grounding zone. From the difference of the elliptic profiles, and taking into account the local discretization, the ice load history was refined for the area of investigation. This allows the individual disks to have an appropriate disk height at glacial maximum. A host of model runs were carried out and then compared with the vertical deformation given by the GPS observations. Figure 7.2-5 shows a model run for the first assumption of ice sheet extent at glacial maximum, while Fig. 7.2-6 shows it for the second assumption. We also considered the loading of the adjacent ocean over the past 100 ka.

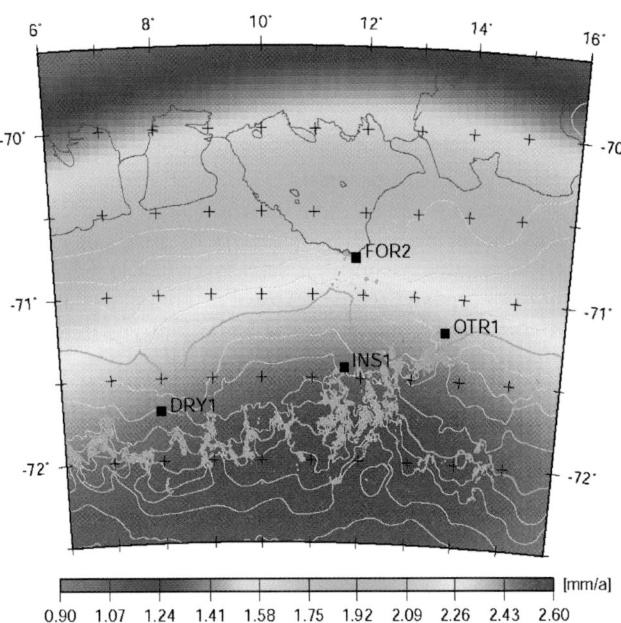

Fig. 7.2-5. Model run, assuming maximum ice extension at present ice shelf edge. Lithospheric thickness = 100 km; viscosity of upper mantle = 10^{21} Pa s; relative height change of mountain sites with respect to the reference site FOR2 = 0.2 to 0.3 mm yr^{-1}

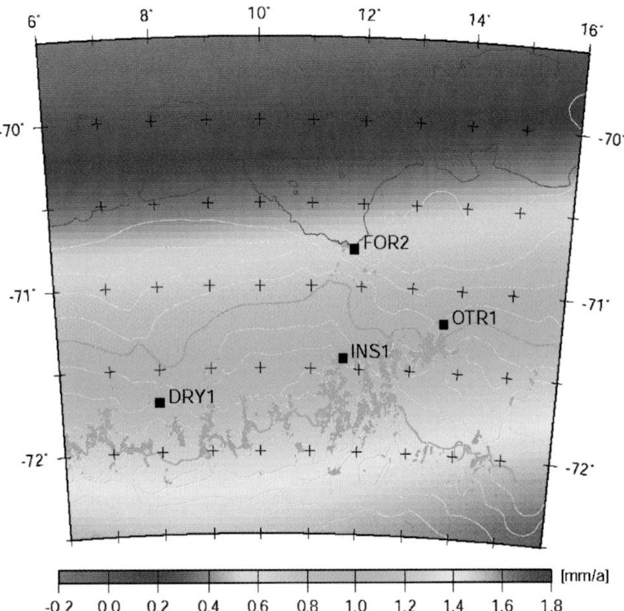

Fig. 7.2-6. Model run, assuming maximum ice extension at present grounding zone. Lithospheric thickness = 100 km; viscosity of upper mantle = 10^{21} Pa s; relative height change of mountain sites with respect to the reference site FOR2 = 0.4 to 0.5 mm yr^{-1}

The effective earth model parameters were chosen to be 100 km for lithospheric thickness and 10^{21} Pa s for viscosity of the mantle. The first model run (Fig. 7.2-5) yielded vertical deformation rates of 0.2 to 0.3 mm yr^{-1}

for the mountain sites relative to the Schirmacher Oasis, the second run 0.4 to 0.5 mm yr^{-1}. Ocean loading accounts for approximately 0.2 mm yr^{-1} 100 km^{-1}. Hence, there is coherence in the sign of the vertical deformation rate, but only 15% and 30% of the GPS relative uplift can be so far explained by GIA processes.

In order to obtain more reliable data to constrain the GIA modelling it will be necessary to carry out a third GPS observation campaign. Such data will also reduce the uncertainty imposed by seasonal or annual signals due to different accumulation – ablation patterns in the region of interest.

References

Hermichen W (1995) The continental ice cover in the surroundings of the Schirmacher Oasis. In: Bormann P, Fritzsche D (eds) The Schirmacher Oasis, Queen Maud Land, East Antarctica, and its surroundings. Justus Perthes, Gotha, pp 221–242

Huybrechts P, Le Meur E (1999) Predicted present-day evolution patterns of ice thickness and bedrock elevation over Greenland and Antarctica. Polar Res 18(2):299–306

Ivins ER, James TS (1999) Simple models for late Holocene and present-day Patagonian glacier fluctuations and predictions of a geodetically detectable isostatic response. Geophys J Internat 138:601–624

James TS, Ivins ER (1997) Global geodetic signature of the Antarctic ice sheet. J Geophys Res 102(1):605–633

James TS, Ivins ER (1998) Predictions of Antarctic crustal motions driven by present-day ice sheet evolution and by isostatic memory of the Last Glacial Maximum. J Geophys Res 103(3):4993–5017

Korth W, Dietrich R, Perlt J, Metzig R (1999) Geodetic indications on recent ice mass changes in an East Antarctic region. Abstract, XXII General Assembly IUGG (July 18–30, 1999), Birmingham

Peltier WR (1998) Postglacial variations in the level of the sea: implications for climate dynamics and solid earth geophysics. Rev Geophys 36(4):603–689

Scheinert M, Dietrich R, Rülke A, Perlt J, Malaimani EC (2003) Precise GPS measurements to infer vertical crustal deformations in central Dronning Maud Land, East Antarctica. In: Paech H (ed) Geol Jb B96, Bundesanstalt Geowiss Rohstoffe, Hannover

Wand U, Hermichen WD (2003) Late Quaternary ice level changes in central Dronning Maud Land, East Antarctica, as inferred from C14-ages of organic deposits ("mumiyo") in snow petrel colonies. In: Paech H (ed) Geol Jb B96, Bundesanstalt Geowiss Rohstoffe, Hannover

Wolf D (1985) The normal modes of a layered, incompressible Maxwell half space. J Geophys Res 57:106–117

Seismic Activity Associated with Surface Environmental Changes of the Earth System, around Syowa Station, East Antarctica

Masaki Kanao · Katsutada Kaminuma
Department of Earth Science, National Institute of Polar Research, 9-10 kaga-1, Itabashi-ku, Tokyo 173-8515, Japan, <kanao@nipr.ac.jp>, <kaminuma@nipr.ac.jp>

Abstract. The Japanese Antarctic Station, Syowa (69° S, 39° E; SYO), is located on Lützow-Holm Bay of western Enderby Land, East Antarctica. Seismic observations at SYO started in 1959, and the arrival-times of the major phases for teleseismic events have been reported from the National Institute of Polar Research every year since 1968. Here, we summarize records from local earthquakes around SYO in the last three decades. In particular, the fifteen years since 1987 divided into three periods are examined in detail, with respect to the location of epicenters and estimation of magnitudes. A three-station seismic array was deployed around SYO in 1987–1989. By using these data, epicenters of local earthquakes were determined for the first time. Many different types of earthquakes, such as a mainshock-aftershock sequence, twin earthquake, and earthquake swarms were detected and clearly identified. The seismic activity during this period was higher than that of the following decade. Earthquake location was concentrated along the coast and central Lützow-Holm Bay.

In the next period between 1990–1996, nine local earthquakes were classified in many different types. The seismicity during this period was very low and magnitudes ranged from 0.1 to 1.4. Hypocenters of four earthquakes out of nine were localized in Lützow-Holm Bay and its northeastern coastal area by a single station method using SYO data. One local event was detected in 1997, two events in 1998 and one event in 2001 and 2003, respectively. The low seismic activity has continued to date (December 2003). The observed offshore location and low level off seismic activity is consistent with a glaciogenic component to the stress field causing the earthquakes.

Introduction

The Japanese Antarctic Syowa Station (69° S, 39° E; SYO), located at the continental margin of western Enderby Land, was established as a contribution to the scientific program of the International Geophysical Year (IGY) in January 1957. Seismological observations at SYO started with a vertical component short period seismograph in 1959. Phase readings of the teleseismic events have been reported to the United States Coastal Geological Survey (USCGS) from 1967 in order to help determine earthquake locations using the global seismic network (Kaminuma 1969, 1976). Following that initial stage, seismological instrumentation at SYO was improved with three-component long and short period seismographs.

In 1990, a three-component digital broadband seismograph (STS-1V, -1H) was installed at SYO (Nagasaka et al. 1992) as one of the permanent stations of the Federation of Digital Seismographic Networks (FDSN). The digital waveform data obtained are published on the web-site of the National Institute of Polar Research (NIPR) and are also available from the PACIFIC21 data center at the Earthquake Research Institute (ERI) of the University of Tokyo. The waveform data are then sent to the Incorporated Research Institutions for Seismology (IRIS)/Data Management Center (DMC) for public use by Internet for global researchers.

During these three decades, phase data of the teleseismic events and the detected earthquake lists at SYO have been sent to the National Earthquake Information Center (NEIC) of the United States Geological Survey (USGS) and also to the International Seismological Center (ISC). The above two data sets have also been published by NIPR once a year, as the "Data Report Series (Seismology)". The Data Reports represent scientific activities of the Japanese Antarctic Research Expedition (JARE, e.g., Kanao 1999). It was generally understood by a majority of seismologists in the IGY era that no extreme earthquakes occurred in the Antarctic area (Gutenberg and Richter 1954). Though the Antarctic region was known as an aseismic region, some significant earthquakes occurred in the Antarctic continent and the surrounding oceans. Earthquakes with magnitude larger than 4.0 and some local events with magnitude less than 3.0 were identified in the first stage of Antarctic research (Kaminuma and Ishida 1971; Adams 1969, 1972, 1982; Adams et al. 1985; Kaminuma 1976, 1990). Although the seismic stations in Antarctica have been operated as part of the global seismic network, no detailed studies of local events have been made until recent publications (e.g., Kaminuma 2000; Bannister and Kennett 2002; Reading 2002).

A temporary seismic array with a three-station network was established around SYO in 1987 and its observations continued until 1989 (Akamatsu et al. 1988, 1989; Kaminuma and Akamatsu 1992). The local seismic activity around SYO was at first analyzed using seismic data from the local network. Micro and/or ultra-micro earthquake activities were recognized during the three years 1987–1989. In this paper, we summarize

From: Fütterer DK, Damaske D, Kleinschmidt G, Miller H, Tessensohn F (eds) (2006) Antarctica: Contributions to global earth sciences. Springer-Verlag, Berlin Heidelberg New York, pp 361–368

the characteristics of local earthquakes around the Lützow-Holm Bay area in the last three decades. Particularly, local seismicity from 1987 until the present is described in detail.

Detection of Local Earthquakes

Several kinds of seismic events, distinct from natural earthquakes, have been recorded on seismograms at SYO. These are sea icequakes and seismic tremors with duration of some ten seconds caused by ice calving at the edge of the continental ice sheet. We can detect local earthquakes on the seismograms at SYO by their waveform patterns. Based on data from the tripartite array in 1987–1989, local events are characterized by the clear onset of P and S phases with S-P times mostly less than 30 s, and the epicenters of those events have been localized in Lützow-Holm Bay and along the coast. In this paper, these events are defined as "local". After 1990, local events were mainly detected by three-component short period seismometers (Fig. 7.3-1; a natural period of 1 Hz; HES seismograph) with maximum magnification of 10 000 at 1 Hz frequency (Kanao 1999).

Fig. 7.3-1. Array of three-component short period seismometers (HES; *front*), and broadband seismometers (STS-1; *behind*) installed at the seismographic hut at Syowa Station (SYO)

To verify local earthquake occurrence, we have examined seismograms and detected local events by individual authors. There is a possibility that other events were hidden on the seismograms during bad weather. Fourteen events detected as local earthquakes are listed in Table 7.3-1 and described with S-P times. The seismograms of the three components of the most recent four events are shown in Fig. 7.3-2. The event on May 03, 1994, for instance, was not counted as a local event in the previous study by Kaminuma and Kanao (1997) because the event had long duration, of more than one minute, with high frequency dominant waveforms. However, it is clear from the three component digital seismograms recorded by the STS-1 seismometer that the event was a local one having clear P and S phases and an S-P time of 16.2 s. This earthquake was the strongest between 1990 and 2001, with magnitude 1.4. The relatively large magnitude corresponds to the longer duration of this event. The magnitudes of the earthquakes, listed in Table 7.3-1, were determined by using the Japanese Meteorological Agency (JMA) method.

Annual Variations in Seismicity

Figure 7.3-3 shows the annual frequency of local earthquakes detected at SYO from 1972 to 2001. The annual frequencies before 1990 were reported by Kaminuma and Akamatsu (1992). The numbers corresponding to 1972, 1973 and 1987 may be larger than those counted in the figure because the actual observation periods during these three years were less than 12 months each. Magnifications of seismographs changed about 10% through the observation period; moreover, this figure does not reflect exact numbers of local earthquakes which occurred in each year. However, the figure at least provides information about local seismic activity around SYO since the 1970s.

Kaminuma and Akamatsu (1992) reported six local earthquakes in 1990. In this paper, however, as the result of detailed evaluation of the six events by a few persons including the present authors as described in the previous section, we find that only one occurred in 1990. In 1990–1996, nine local earthquakes of many different types were recorded including the 1990 event. The seismicity during the period was very low and the magnitudes ranged from 0.1 to 1.4. The seismic activity during 1990–1996 was lower than that in 1987–1989 when the three-station network was in operation (Kaminuma and Akamatsu 1992). In 1997–2003, one local event was detected in 1997, two events in 1998 and one in 2001 and 2003, respectively. Low seismic activity has continued until the present day.

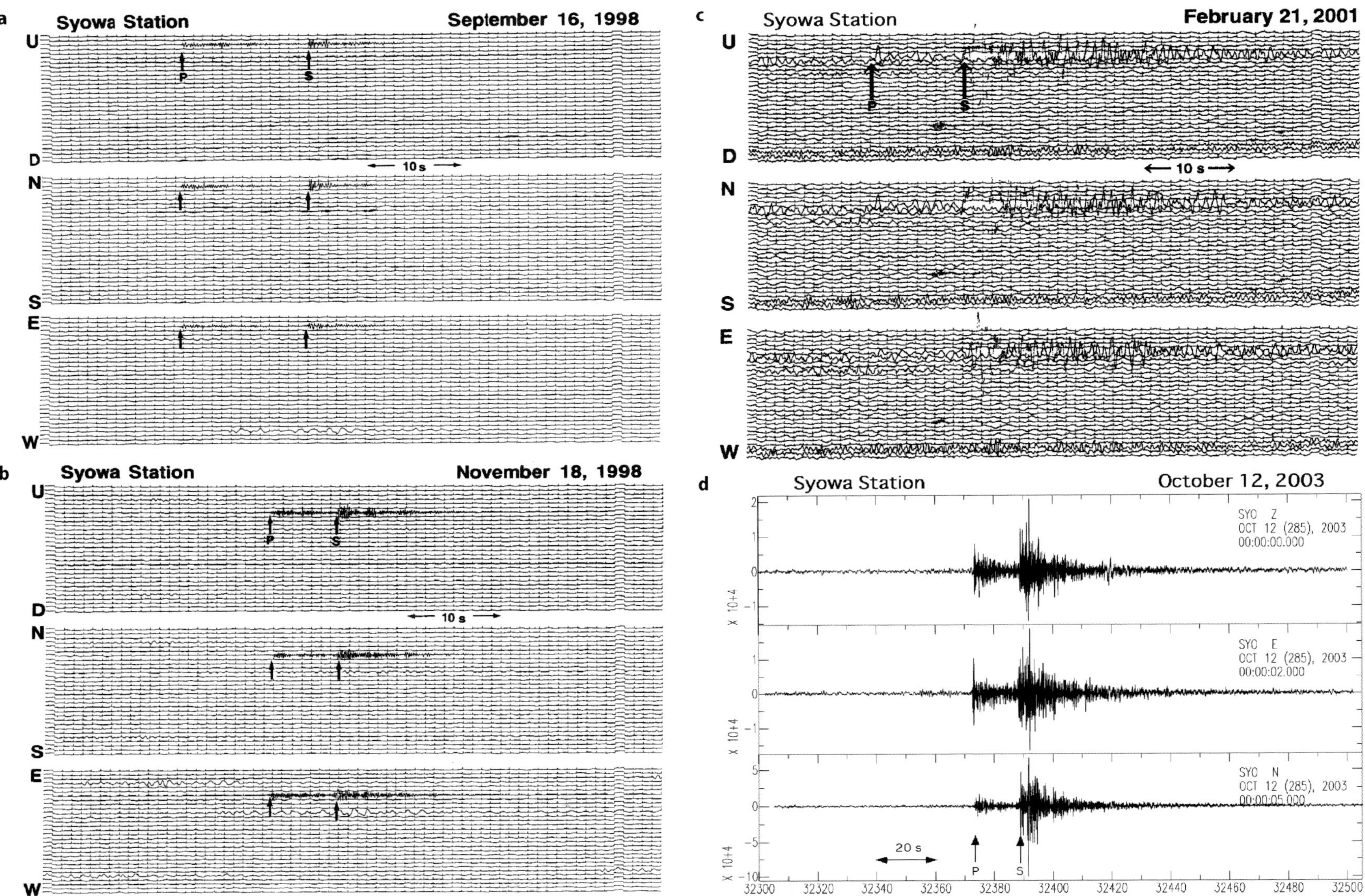

Fig. 7.3-2. Three component analog seismograms of the short period seismometer (HES) for local events on: **a** September 16, 1998; **b** November 18, 1998; **c** February 21, 2001. For the event on **d** October 12, 2003, the digital seismographs for the HES are shown

Table 7.3-1.
List of local earthquakes detected at Syowa Station from 1990. *From left:* occurrence date of the event, kind of first arrival phases, arrival-time of first phases (UTC), P-S residual times, propagation angle calculated from the method described in text, and body wave magnitude estimated by the JMA method

Date	Phase	Arrival time	P-S time (s)	Propagating angle (°)	Magnitude North to East
1990 May 29	iP eS	07 h 36 m 33.5 s 07 h 36 m 47.9 s	14.4	–	–
1991 Jan. 12	iP eS	01 h 51 m 12.4 s 01 h 51 m 27.6 s	15.2	–	–
1991 May 29	iP iS	01 h 17 m 30.8 s 01 h 17 m 40.7 s	9.9	–	–
1992 Jan. 11	iP iS	12 h 49 m 07.1 s 12 h 49 m 10.8 s	3.7	–	–
1992 Sep. 21	iP iS	16 h 14 m 51.8 s 16 h 14 m 59.0 s	7.2	–	–
1993 Dec. 15	iP iS	04 h 13 m 16.5 s 04 h 13 m 28.8 s	12.3	303	0.6
1994 May 03	iP iS	10 h 10 m 13.3 s 10 h 10 m 29.5 s	16.2	60	1.4
1995 Sep. 28	eP eS	08 h 06 m 04.0 s 08 h 06 m 17.0 s	13.0	309	-0.4
1996 Aug. 03	iP iS	14 h 32 m 24.2 s 14 h 32 m 37.4 s	13.2	52	1.1
1997 Sep. 18	iP iS	03 h 11 m 27.7 s 03 h 11 m 32.3 s	4.6	358	-0.5
1998 Sep. 16	eS iS	10 h 32 m 14.8 s 10 h 32 m 28.7 s	13.9	36	0.3
1998 Nov. 18	eP eS	04 h 36 m 24.3 s 04 h 36 m 31.3 s	7.0	–	-0.1
2001 Feb. 21	iP iS	04 h 35 m 13.2 s 04 h 35 m 22.9 s	9.7	186	0.9
2003 Oct. 12	iP eS	08 h 59 m 33.4 s 08 h 59 m 48.8 s	14.4	255	1.2

Epicenters of Local Earthquakes

Earthquake locations were obtained by using the three-station observation network data recorded during 1987–1989. Many different types of earthquakes, such as a mainshock-aftershock, twin earthquakes, earthquake swarms, etc., were detected and identified during that time. The epicentral locations obtained were concentrated on the two restricted areas around SYO. That is, one group was localized along the coast from north to south around SYO, and the other was in the central part of Lützow-Holm Bay, respectively (Fig. 7.3-5a; Kaminuma and Akamatsu 1992).

For the period after 1990, epicenters for local events were determined by using the first particle motion of the initial phase and S-P time detected at SYO. As the first arrival phases of the local events were assumed to be the direct P phases, the propagation directions and incidence angles of the seismic waves were estimated by considering particle motions of the three orthogonal components

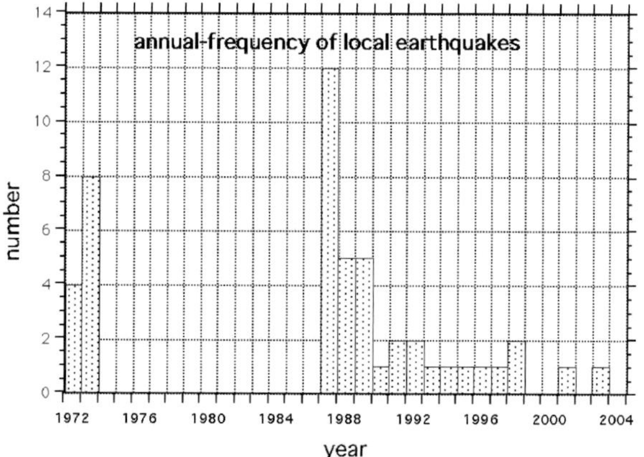

Fig. 7.3-3. Annual frequency of local earthquakes observed at Syowa Station since 1972

from digital seismograms. Figure 7.3-4a,b shows four examples of the horizontal and the vertical trajectories of initial motion, respectively. The propagation directions are determined by horizontal trajectories (Fig. 7.3-4a).

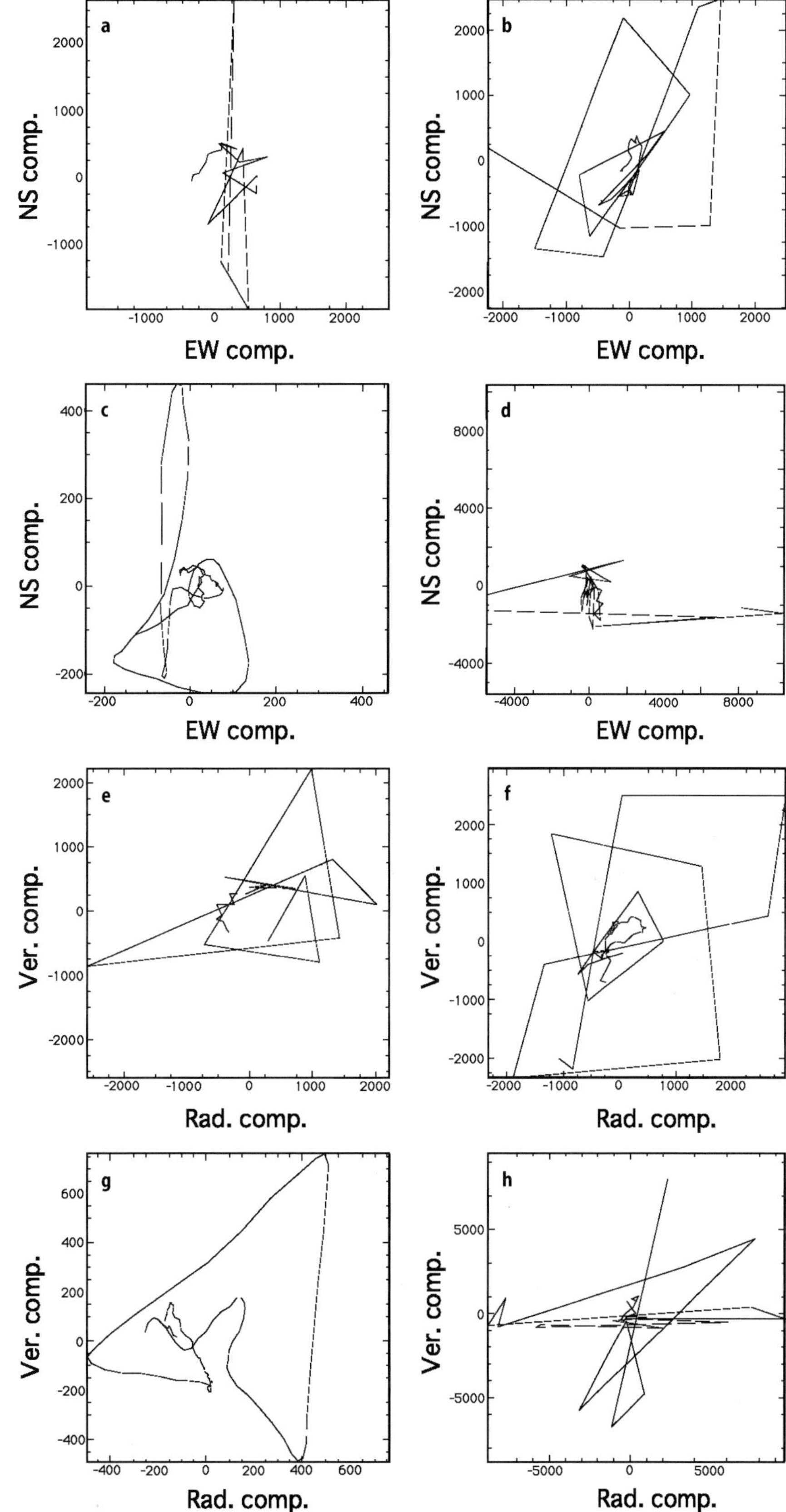

Fig. 7.3-4.
a–d Horizontal trajectories of the initial motion of seismic events. **a** September 18, 1997; **b** September 16, 1998; **c** February 21, 2001; **d** October 12, 2003. **e–h** Vertical trajectories of the initial motion were projected on the vertical plane including the epicenter and SYO. **e** September 18, 1997; **f** September 16, 1998; **g** February 21, 2001; **h** October 12, 2003

The incidence angles were determined by vertical trajectories (Fig. 7.3-4b), which were projected on a vertical plane including the epicenter and the station. The propagating directions to SYO azimuths are also listed in Table 7.3-1. Epicentral distance was calculated by "Oomori's formula" (Oomori 1918) using 6.8 km s^{-1} as "Oomori's constant". This constant was estimated from the local crustal structure (Vp = 6.2 km s^{-1}, Vp / Vs = 1.95) after Ikami et al. (1983).

For the period 1990–1996, Kaminuma et al. (1998) had already reported the epicenters of the four earthquakes (Table 7.3-1) by using the above particle motion method. Figure 7.3-5b shows the epicenter locations of the four local earthquakes determined by this single station method. Locations of the four earthquakes seem to be reasonable, estimated from the previous earthquake locations in 1987–1989. Hypoceters of the local four events for 1997–2003 periods were also determined in almost the same location as in 1990–1996 by using the same methods of particle motions at SYO (Fig. 7.3-5c).

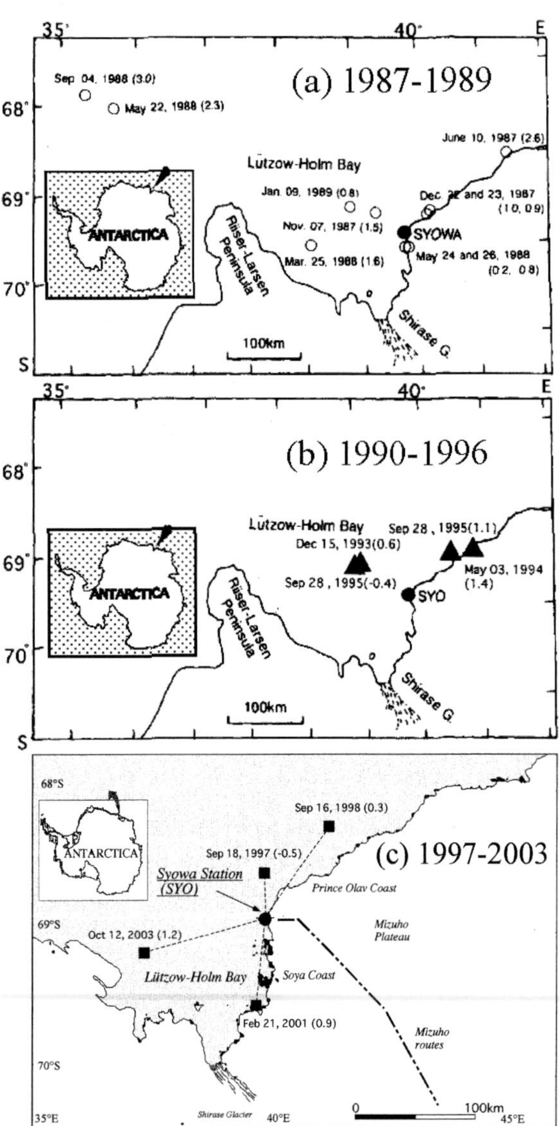

Fig. 7.3-5. Local earthquake locations around Syowa Station with magnitude. **a** 1987–1989; **b** 1990–1996; **c** 1997–2003

Discussion

Some natural earthquakes, moreover, have very similar waveforms to those of icequakes, which were frequently generated by several glacial related ice-mass movements such as of ice-sheets, sea-ice, tide-cracks and icebergs, etc. In some times, we can hardly distinguish the waveforms between the tectonic local earthquakes and those of ice related phenomena. In this study, however, we have chosen the local events by assuming strict detected conditions as indicated in section above. Then the physical phenomenon of local events represents the actual tectonic signature around the SYO area, as part of the margin of the East Antarctic shield.

The influence of glaciogenic forces on the margins of East Antarctica have been investigated by Ivins et al. (2003). They find, that, depending on lithospheric thickness, stress fracturing in offshore locations could be caused by the response of the crust to the meelting East Antarctic ice mass.

Local seismicity around SYO, during the twelve years 1990–2001, was very low compared with those in 1987–1989 when the three-station seismic network was in operation. The capability to detect teleseismic events at SYO in 1987–1993 was studied by Kanao and Kaminuma (1995). Although the detection capability was not so high in 1987–1989 compared with the other periods, high seismicity was recorded during that period. Therefore, there seems to be no relation between high and low detection capability and high local seismicity. It is clear that the number of local seismic events was larger in 1987–1989 and smaller in later years. The present seismicity represents one of the on-going crustal movements, associated with some kinds of tectonic process. One of the candidates to cause the tectonic stress would be related with the post-glacial crustal uplift process, as already pointed out by Kaminuma and others (Kaminuma and Akamatsu 1992; Kaminuma 1996; Kaminuma and Kimura 1997).

In the last few years, there have been several reports presenting the local seismicity around Antarctica by temporary/permanent seismic networks (Bannister and Kennett 2002; Müller and Echstaller 2003; etc). Bannister and Kennett (2002) studied the seismicity around the SBA and VNDA area by a temporary broadband seismic network. They found several interesting features about the location of local events; that is, a majority of the events were located along the coast, particularly in the vicinity of large glaciers. They have suggested several reasons for the occurrence of these events: basal sliding of the continental ice sheet, movement of ice streams associated with several scales of glaciers, movement of sea-ice, and tectonic earthquakes. In order to distinguish the actual origins of these events, they mentioned the importance of determining the earthquake mechanism together with the depths of hypocenters. Müller and Eckstaller (2003) have begun to operate the local seismic network around Neumayer Station. They also determined the hypocenters of local tectonic events in the same locality as by our study; that is, along the coast and the mid area of the surrounding Atka bay. The tectonic mechanism is likely to be similar in Lützow-Holm Bay and around Neumayer Station: post-glacial rebound movements.

A compilation of seismicity around the whole Antarctic continent has recently been reported by Kaminuma (2000) and Reading (2002). Kaminuma (2000) classified the signature of seismicity for tectonic earthquakes into the following three groups: (1) the great earthquakes in the intraplate low seismic regions, (2) microseismic activity at the edge of the continent, and (3) relatively high seismic activity in Wilkes Land. Local seismicity around SYO as revealed by our study was classified into category 2. Local earthquakes in these regions were presumably caused by tectonic stress accumulated with crustal deformation after deglaciation. The effect of ice sheet changes may have caused phenomena such as crustal deformation, earthquake occurrence, faulting system, etc. in the shallow part of the lithosphere beneath Antarctica (e.g., Bannister and Kennett 2002).

As more data accumulate, we will be able to estimate the stress concentration relating to glacial rebound around the Lützow-Holm Bay region.

From 2002 at SYO, we have started continuous digital recording of the HES seismograph, which will enable us to determine the source mechanism of these local events in the near future. This will help to reveal the seismicity around Antarctica and deglacial phenomena such as crustal uplift and sea level change within the Earth environmental system.

Concluding Remarks

In this paper, we have described the characteristics of the time records for local earthquakes around SYO, eastern Droning Maud Land, in the last twelve years. The important results obtained here are summarized as follows:

Seismicity in 1987–1989. A three-station seismic network was operated around SYO in 1987–1989. Epicenters of local earthquakes were determined for the first time during these three years.

Many different types of earthquakes, such as a mainshock-aftershock, twin earthquake, earthquake swarms, etc., were detected and identified at that time.

The seismic activity during this period was higher than that of the following decade. Earthquake locations were concentrated along the coast and the central part of Lützow-Holm Bay.

Seismicity in 1990–1996 and 1997–2003. In 1990–1996, nine local earthquakes were recorded with many different types of events. The seismic activity during this period was very low and the magnitudes ranged from –0.5 to 1.4.

One local event was detected in 1997, two events in 1998 and one event in 2001 and 2003, respectively. The low seismic activity continues to the present day. The ongoing crustal movements are likely to have a component of glaciogenic stress which may be the controlling factor in the location of the microseismicity along the continental margin offshore.

Acknowledgments

We would like to express our sincere thanks to Ms. A. Ibaraki and M. Minegishi of NIPR for their help in scaling the seismic phases of the local events. We would also like to express sincere thanks to Dr. A. Reading of the Australian National University and one anonymous referee for their critical reading of the manuscript and giving useful suggestions for modification. Finally, we express our thanks to Dr. S. Bannister of the Institute of Geological and Nuclear Sciences, NZ, and Dr. C. Müller of the Alfred Wegener Institute for Polar and Marine Research, Germany, for their useful discussions about regional seismicity and tectonics in Antarctica.

References

Adams RD (1969) Small earthquakes in Victoria Land, Antarctica. Nature 224:255–256

Adams RD (1972) Local earthquakes in Victoria Land. In: Adie RJ (ed) Antarctic geology and geophysics. Universitetsforlaget, Oslo, pp 495–499

Adams RD (1982) Source properties of the Oates Land Earthquake. In: Craddock C (ed) Antarctic geosciences. University Wisconsin Press, Madison, pp 955–958

Adams RD, Hughes AA, Zhang BM (1985) A confirmed earthquake in continental Antarctica. Geophys J Royal Astron Soc 81:489–492

Akamatsu J, Ichikawa N, Kaminuma K (1989) Seismic observation with local telemetry network around Syowa Station, East Antarctica. Polar Geosci 3:1–12

Akamatsu J, Yoshikawa S, Kaminuma K (1988) Preliminary report of local seismic activity around Syowa Station, East Antarctica. Polar Geosci 2:1–6

Bannister S, Kennett BLN (2002) Seismic activity in the Transantarctic Mountains – Results from a broadband array deployment. Terra Antartica 9:41–46

Gutenberg B, Richter CF (1954) Seismicity of the Earth and associated phenomena. Princeton University Press, Princeton

Ikami A, Ito K, Shibuya K, Kaminuma K (1983) Crustal structure of the Mizuho Plateau, Antarctica revealed by explosion seismic experiments. In: Oliver RL, James PR, Jago JB (eds) Antarctic earth science. Australian Acad Sci, Canberra, pp 671–674

Ivins ER, James TS, Kleman V (2003) Glacial isostatic stress shadowing by the Antarctic ice sheet. J Geophys Res 108(12):2560

Kaminuma K (1969) The seismological observation and the earthquake detection capability of Syowa Station, Antarctica. Bull Earthquake Res Inst 47:453–466

Kaminuma K (1976) Seismicity in Antarctica. J Phys Earth 24:381–395

Kaminuma K (1990) Local earthquake activities around Syowa Station, East Antarctica. Tohoku Geophys J 35:127–136

Kaminuma K (1996) On the possibility of detecting absolute crustal uplift at Syowa Station, Antarctica. Polar Geosci 9:l6–23

Kaminuma K (2000) A revaluation of the seismicity in the Antarctic. Polar Geosci 13:145–157

Kaminuma K, Akamatsu J (1992) Intermittent micro-seismic activity in the vicinity of Syowa Station, East Antarctica. In: Yoshida Y, Kaminuma K, Shiraishi K (eds) Recent progress in Antarctic earth science. Terra Scientific Publishing, Tokyo, pp 493–497

Kaminuma K, Ishida M (1971) Earthquake activity in Antarctica. Antarct Rec 42:53–60

Kaminuma K, Kanao M (1997) Local earthquakes recorded at Syowa Station, Antarctica. Antarct Rec 41:643–651 (Japanese with Engl. abstract)

Kaminuma K, Kimura I (1997) Leveling survey on East Ongul Island, Antarctica and its implication. Polar Geosci 10:19–25

Kaminuma K, Kanao M, Kubo A (1998) Local earthquake activity around Syowa Station, Antarctica. Polar Geosci 11:23–31

Kanao M (1999) Seismological bulletin of Syowa Station, Antarctica, 1997. JARE Data Rep 236 (Seismology 33), pp 1–65

Kanao M, Kaminuma K (1995) Detection capability of earthquakes recorded at Syowa Station, Antarctica, from 1987 to 1993. Antarct Rec 39:156–169

Müller C, Eckstaller A (2003): Local seismicity detected by the Neumayer seismological network, Dronning Maud Land, Antarctica: tectonic earthquakes and ice-related seismic phenomena. Terra Nostra 2003(4):236–237

Nagasaka K, Kaminuma K, Shibuya K (1992) Seismological observations by a three-component broadband digital seismograph at Syowa Station, Antarctica. In: Yoshida Y, Kaminuma K, Shiraishi K (eds) Recent progress in Antarctic earth science. Terra Scientific Publishing, Tokyo, pp 595–601

Oomori F (1918) Epicentral distance calculated from S-P travel-times. Bull Earthquake Inv Comm 9:33

Reading AM (2002) Antarctic seismicity and neotectonics. In: Gamble JA, Skinner DNB, Henrys S (eds) Antarctica at the close of a millennium, Royal Soc New Zealand Bull 35, Wellington, pp 485–491

Geodynamic Features and Density Structure of the Earth's Interior of the Antarctic and Surrounded Regions with the Gravimetric Tomography Method

Rudolf Kh. Greku · Victor P. Usenko · Tatyana R. Greku
Institute of Geological Sciences, National Academy of Sciences of Ukraine, 55B Gonchara Str., 01054 Kiev, Ukraine, <satmar@svitonline.com>

Abstract. This paper presents the Earth's interior data of the Antarctic and surrounding regions obtained through the gravimetric tomography method developed by the authors and destined for reconstruction and displaying the structural geological inhomogeneities in different layers. Characteristics of the global geoid height model's spherical harmonics are the input data for the method. They are both used for determination of the layers' depths disturbing geopotential and for computing of harmonic dense anomalies. Vertical and lateral sections calculated show a distribution of masses in all range of depths up to 5 300 km, geometry and sizes of density inhomogeneities, their displacement in depth under the impact of geodynamic processes, and also correlation of subsurface bodies with the known topographic features. A significant difference between geodynamic processes in geospheres of the core, mantle and crust is emphasized.

Introduction

Geotectonic researches are constrained by the shortage of the deep Earth's interior data in most cases. This problem relates especially to such key remote and inaccessible areas as the Antarctic and Arctic. A known informative source of such data is the seismic tomography technology (Bijwaard et al. 1998) in which signals of earthquakes and explosions are used. Our gravimetric tomography technique is based on realization of the theoretical approach by Moritz (1990) that the Earth's equipotential surfaces coincide with surfaces of the constant density and on usage of his algorithm of determination of the harmonic dense anomalies through the spherical harmonics of the gravity potential.

Moreover, we used well-known extracting procedure of certain harmonics for determination of a residual (differential) geoid and for mapping of structures in different layers, which disturb the geopotential.

Experienced geophysicists have made approximate evaluations of disturbing layers depths using the geoid harmonics. So, it was noted by Gainanov (1981) that the density inhomogeneities at the center of the Earth are responsible for harmonics of about $2 \le n \le 5$, the lower mantle is responsible for the range of $2 \le n \le 20$, and the upper mantle for the range of $2 \le n \le 100$. It is supposed in another reference work (Allan 1975), that harmonics of up to 8 degrees are probably caused by the impact of masses located in depths of more than 1 000 km, and from 8 to 22 in depths of 50–300 km.

These assessments of a logical conformity between number of a harmonics and depth of a disturbing layer motivated us to determine a numeric dependence in this relationship. Thus, the gravimetric tomography method includes a solution of the following tasks:

1. Determination of harmonic density of anomalous disturbing masses by the geoid spherical functions.
2. Determination of a relationship between degrees of the harmonic series expansion of the geoid topography and depths of disturbing layers of the Earth.
3. Creation and displaying of the structured model of dense inhomogeneities for the studied region.

The Algorithm for Calculation of Anomalous Harmonic Densities

The solution of the inverse gravity problem for determination of density anomalies from given potential anomalies is considered by Moritz (1990) under some conditions. In particular, the distribution of density is taken as a continuous function, which can be approximated uniformly by means of a system of polynomials. Then density, like the potential, can be expanded in series of spherical harmonics by polynomials (harmonic density). The other boundary condition is that a general solution, which corresponds to the zero-potential densities, is determined. That is a distribution of positive and negative densities, which do not change the external gravity potential, as their total mass should be equal to zero. Anomalous harmonic densities are obtained by excluding the general solution from the harmonic densities.

Practically, it is considered that not the general geopotential (normal plus anomalous), but the disturbing (anomalous) potential, an expansion of which is well known in the global scale by a conventional gravity model, such as EGM96. Therefore, the anomalous harmonic densities can be calculated using spherical harmonics of the

From: Fütterer DK, Damaske D, Kleinschmidt G, Miller H, Tessensohn F (eds) (2006) Antarctica: Contributions to global earth sciences. Springer-Verlag, Berlin Heidelberg New York, pp 369–376

disturbing potential. The following final algorithm from Moritz (1990) was used:

$$\rho_h = \sum_{n=2}^{z}\sum_{m=0}^{n}\left(\frac{M(2n+1)(2n+3)}{4\pi R^{n+3}} \times r^n(c_{nm}\cos m\lambda + s_{nm}\sin m\lambda)P_{nm}(\cos\theta)\right)$$

where

- ρ_h = anomalous harmonic density;
- M = mass of the Earth;
- R = radius of the Earth (in a point, to which the value of a geopotential is referred);
- r = radius-vector of an internal point, in which a density disturbing the geopotential is determined,
- $P_{nm}(\cos\theta)$ = Legendre polynomial of nth degree and m^{th} order;
- θ = central angle or spherical distance between R and r;
- c_{nm} and s_{nm} = coefficients of the surface geopotential spherical harmonics.

Transformation of the external spherical harmonics to the internal spherical harmonics is implemented by this algorithm.

Estimation of the Depth of Disturbing Layer by the Number of Harmonics

The density anomalies obtained above are obviously situated in different depths. Therefore a following question remains: which layers of the Earth are responsible for disturbance of the one geopotential anomaly or another? In our method an assessment of the disturbing layer's depth is computed by a known harmonic function in the geoid theory (Moritz 1990) for the case when the external potential of the internal masses confined by a sphere is determined

$$\frac{1}{\ell} = \sum_{n=0}^{\infty}\frac{r^n}{R^{n+1}}P_n(\cos\theta)$$

where ℓ = distance between point, to which the value of a geopotential is referred and point, to which the density disturbing the geopotential is referred.

Calculation was carried out with $n_{\min} = 2$. At the right part of the expression the normalizing coefficient of spherical functions $(2n+1)^{1/2}$ was used (Shimbirev 1975). If $\theta = 0$, then $P_n(\cos\theta) = 1$ for any n, and $\ell = R - r$. Under these conditions and for given values of ℓ, r and R the corresponding values of n were calculated.

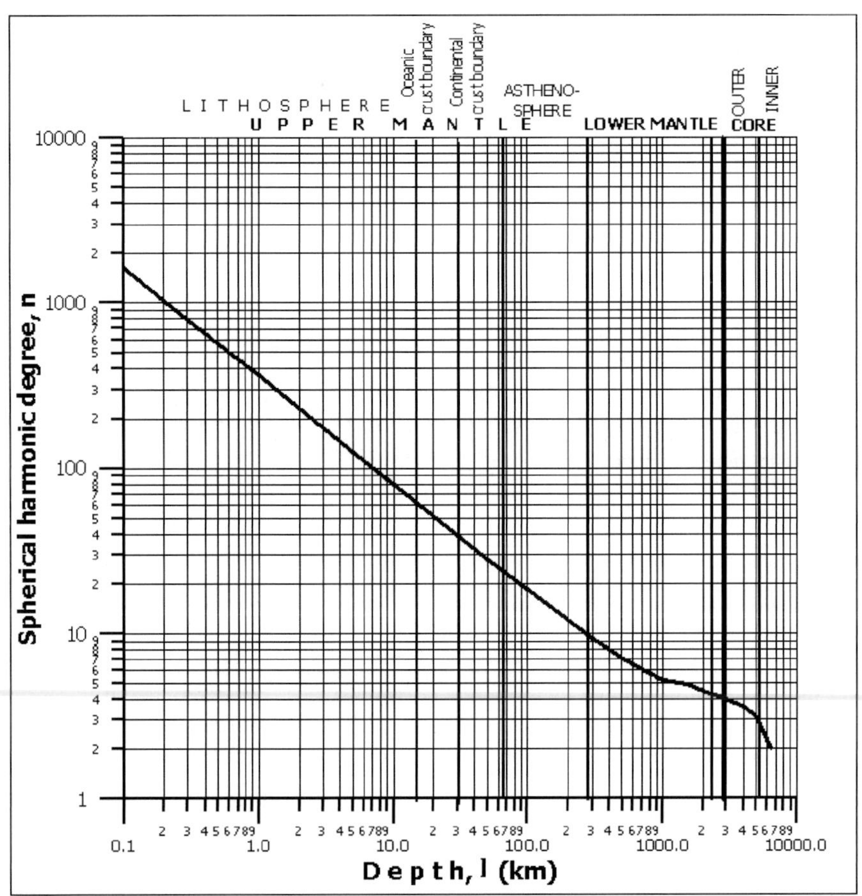

Fig. 7.4-1. Relationship between harmonic degrees n and depths l disturbing layers of the Earth. Value n is a sum of harmonics in a range from degree 2 up to n. Depth l corresponds to upper cover of the disturbing layer; its thickness is considered from the center of the Earth

Obviously, for large values of n the series of r^n/R^{n+1} converges weakly. Therefore a restriction is included into the procedure of calculation. The calculation was ended when at a maximal degree n_{max} the condition was reached

$$\ell - \frac{R}{\sum_{n=0}^{n_{max}} \left(\frac{r}{R}\right)^n} \leq 0.1\ell$$

where coefficient 0.1 means 10% error from ℓ.

Relationship between harmonic degrees n and depths ℓ of disturbing layers is shown in the bilogarithmic diagram (Fig. 7.4-1). Main boundaries of the lithosphere are shown in accordance with the Bullard's (1954) density model of the Earth.

The software developed within the gravimetric tomography method allows us to compute values of heights of both the full geoid (all harmonics of a model used) and differential geoid, values of the anomalous harmomic densities in units of g cm^{-3} and values of upper cover depths of disturbing layers of the Earth. The spherical coefficients of the EGM96 global geopotential geoid model was used.

Some Results from the Antarctic Continent

Structural modeling of the Antarctica continent and separated regions (West Antarctic, West Antarctic Rift System, and East Antarctic) is implemented on the basis of the above technology. Processes at the core-mantle boundary and the thermal situation of the mantle originate dynamics and structure of the Earth's upper layers. We display the structure of the lower and upper layers separately to identify their features in detail.

Figure 7.4-2 shows the differential geoid topography, where heights were computed by harmonics in the range of 2–28. An interval between latitudes and between points along the latitudes is 0.25°. The thickness of the disturbing column at each point is 6 321 km in accordance with the diagram in Fig. 7.4-1. More dense structures have a brighter tint. Only two main structures have been discovered in the region. Epicenters of the structures correspond to known geoid undulations. One of them, the Antarctic undulation is a depression (–64 m) located near the Ross Sea and another, the Conrad undulation is an elevation (+50 m) placed towards the Conrad Rise in the Indian Ocean (Rapp et al. 1991). More detailed level-by-level analysis of the locations of the epicenters' depths shows

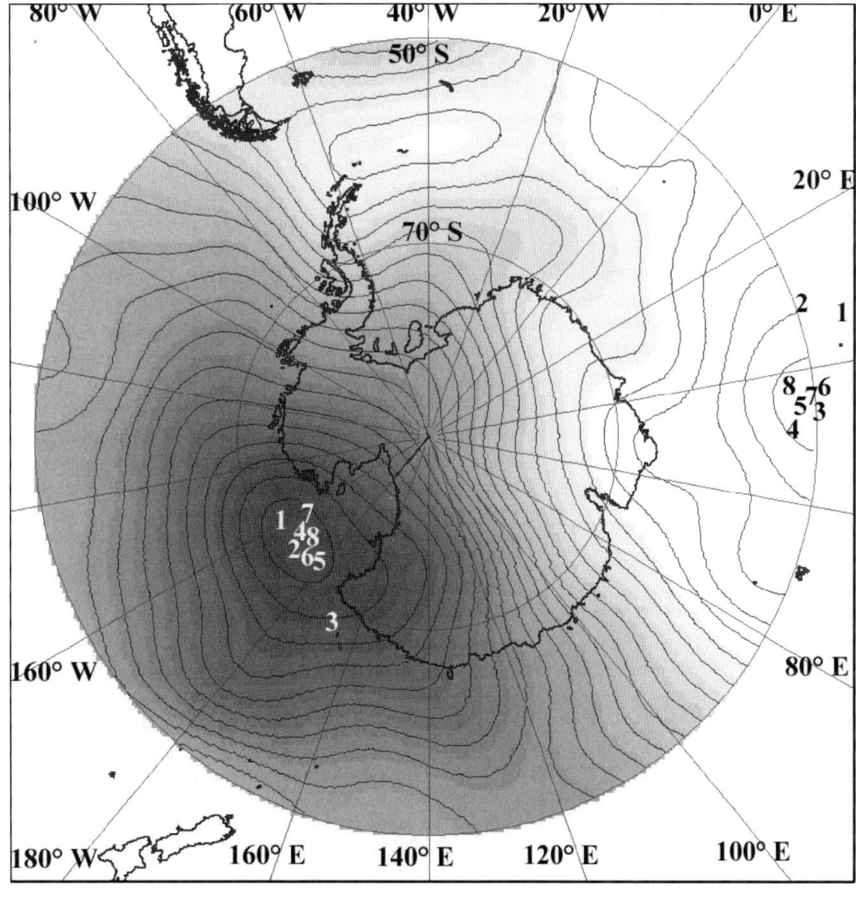

Fig. 7.4-2. Map of the differential geoid topography is in the Polar Stereographic projection with 71° S as the latitude of true scale and 40° W as the central meridian. A distribution of two main dense inhomogeneities in the Antarctic region delimited by 50°S is shown. A range of used harmonics is 2–28. The thickness of the disturbing layer is estimated from the depth of 50 km down to the center of the Earth. Contour interval is 5.0 m. More dense structures have a *brighter tint*. The numbers *1–8* show positions of epicenters of the Antarctic (negative) undulation and Conrad (positive) undulation at different depths. Ranges of harmonics and depths of the upper covers of disturbing layers in accordance with Fig. 7.4-1 are shown in the Table 7.4-1

their coordinates changing (numbers of 1–8 in Fig. 7.4-2). Ranges of harmonics and depths of the upper covers of disturbing layers in accordance with Fig. 7.4-1 are shown in the Table 7.4-1.

Note, the distances between points in the geographic coordinate system on the Earth surface have to be reduced on the corresponding sphere with the radius of $R-\ell$. Maximum amplitudes of displacements for the Antarctic undulation are in a depth range of 5 300–1 500 km and can reach up to 240 km and for the Conrad undulation to 650 km. It is apparent that such significant displacements can influence the orbital characteristics of the Earth. As we can see in Fig. 7.4-2 contours of the Scotia Sea structure is becoming apparent at the depth of 50 km already. This feature is not discovered in greater depths.

Table 7.4-1. Ranges of harmonics and depths of the upper covers of disturbing layers in accordance with numbers of epicenter undulation in Fig. 7.4-2

	Range of harmonics (n)	Depth, l (km)
1	2 – 3	5 300
2	2 – 4	2 800
3	2 – 5	1 500
4	2 – 10	290
5	2 – 12	200
6	2 – 19	100
7	2 – 28	50
8	2 – 360	Surface

Structure of the upper layers can be computed by removing of the lower frequency effect. The structural composition of the Antarctic and surrounding regions in the harmonic range of 28–360 is shown in Fig. 7.4-3. The corresponding disturbing layers have a thickness of 50 km from the terrestrial surface and ocean bottom. This topography is like the subglacial bed elevation from the BEDMAP Project data in the scale of 1 : 1 000 000 (Lythe and Vaughan 2001). Most of the general structures of the Antarctica (Fifield 1987) are visible quite well on the differential geoid map.

Trans Antarctic Vertical Section

A deep structure of the region is shown at the vertical two-dimensional cross-section along meridians 190° E and 44° E between latitudes 30–90° S (Fig. 7.4-4). An ex-

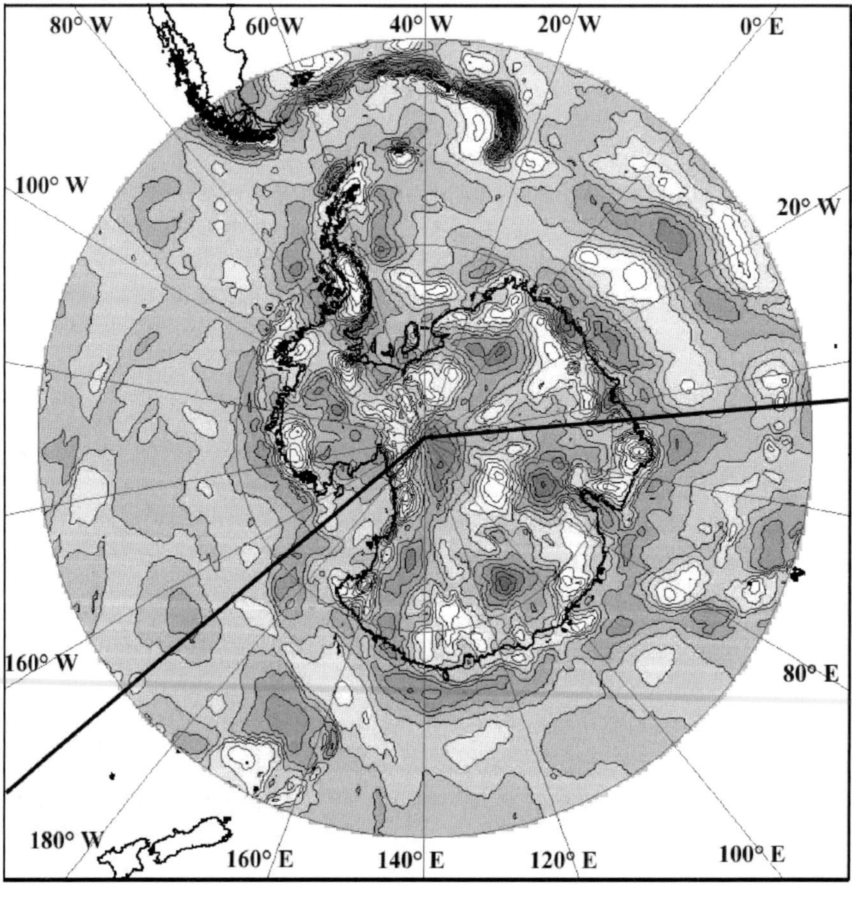

Fig. 7.4-3. Map of the differential geoid topography is in the Polar Stereographic projection with 71° S as the latitude of true scale and 40° W as the central meridian. A distribution of dense inhomogeneities in the Antarctic region delimited by 50° S is shown. A range of harmonics used is 28–360. The thickness of the disturbing layer is 50 km from the terrestrial surface and ocean bottom. Contour interval is 1.0 m. *Solid lines* show locations of the cross-section along directions of 190° E–Pole–44° E between latitude 30° S

tension of the section is 13 365 km. An interval between calculated points is 30 km. This section crosses the whole Antarctic Plate between the Pacific-Antarctic Ridge (PAR), South Pole and South-West Indian Ridge (SWIR). It includes such geographical structures as the Pacific Plate, PAR interplate boundary, epicenter of the Antarctic geoid negative undulation, West Antarctic Rift System (WARS), Polar zone, East Antarctica, Indian Ocean part of the Antarctic Plate, epicenter of the Conrad geoid positive undulation, SWIR and African Plate.

Figure 7.4-4 shows a distribution of the harmonic dense anomalies (section 2) of deep masses from 140 km down to 2 800 km (outer core). Three large-scale zones with a different density are distinguished clearly. The central zone I is related to the Antarctic negative undulation with the lower dense masses. A range of the harmonic dense anomalies is 0 ÷ –0.002 g cm^{-3}. Geographically this zone in depths more than 140 km, includes a part of the Pacific Plate, the West Antarctic between PAR and TAM. Zones of higher density are situated on the right in the figure (II, East Antarctic between TAM and SWIR, 0 ÷ +0.0015 g cm^{-3}) and on the left (III, Pacific Plate, 0 ÷ +0.0006 g cm^{-3}).

Another structural scheme with the suppressed lower layers and displaying of the upper layer higher than 140 km is shown in Fig. 7.4-5. It is impossible to create one picture to distinguish equally the details of all layers. Therefore vertical scales for the blocks in depths 1–15 km,

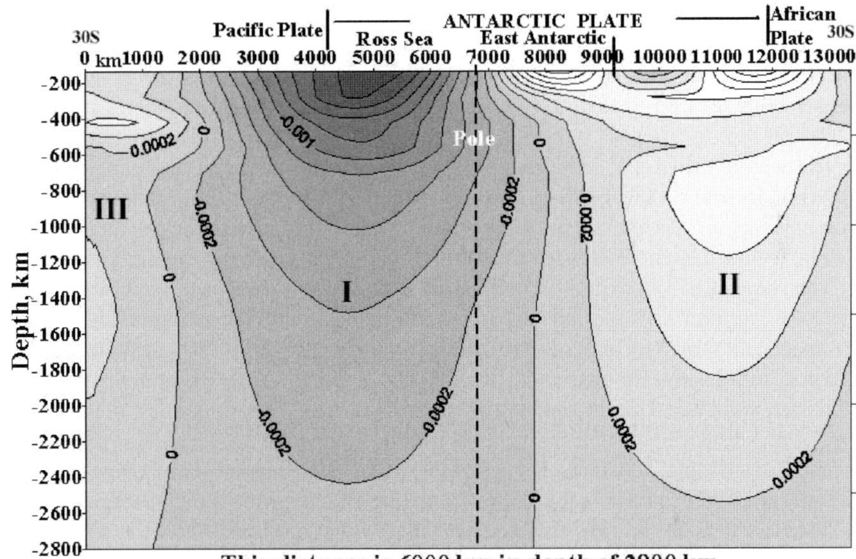

Fig. 7.4-4. Vertical cross-section of density inhomogeneities with the harmonic density anomalies in g cm^{-3} is located along meridians 190° E and 44° E through the South Pole between 30° S. Range of depths is 140–2 800 km (for location of section see Fig. 7.4-3). Contour interval is 0.0002 g cm^{-3}. *Brighter tint* corresponds to more dense structure. *I:* zone of lower dense anomalies confined by the Antarctic undulation; *II:* zone of higher dense anomalies confined by the Conrad undulation; *III:* zone of the Pacific Plate

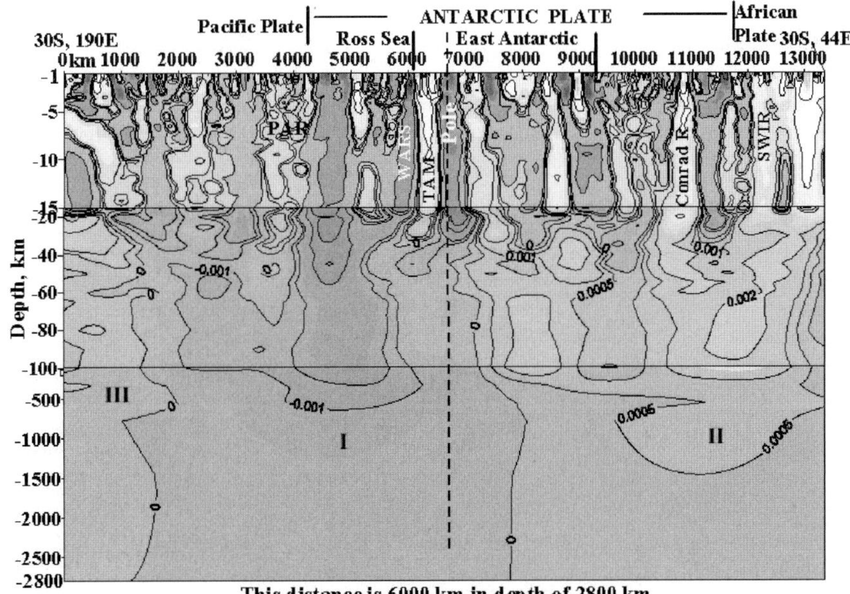

Fig. 7.4-5. Vertical cross-section of density inhomogeneities with the harmonic density anomalies in g cm^{-3} is located along meridians 190° E and 44° E through the South Pole between 30° S. Range of depths is 1–2 800 km. *PAR:* Pacific-Antarctic Ridge; *WARS:* West Antarctic Rift System; *TAM:* Transantarctic Mountains; *Conrad R.:* Conrad Rise; *SWIR:* South-West Indian Ridge. Boundaries of the Plates (*vertical lines above*) are shown from GEBCO (1984). *Vertical scales* for the blocks in depths 1–15 km, 15–100 km and 100–2 800 km are different. For further details see legend of Fig. 7.4-4

15–100 km and 100–2 800 km are different. In Fig. 7.4-4 and 5 we can see a tectonic interaction between zones I, II and III and corresponding displacements of the contour lines in depths 400–500 km, 600–650 km (upper and lower segments of the Benioff zone) and 1 500 km. Gradients and a spatial differentiation of the anomalies increase closer to the surface.

Features of the West Antarctic Rift System Structure

Among many structural features in the upper layers we shall note a structure of the WARS-TAM system. In Fig. 7.4-5 we can see two main channels of the lifting up of a material of the zone I from depths. One of them is located at a distance of 4 500 km in a depth of 60 km. The second is at a distance 6 100 km and in a depth of 20 km, the last is the actual rift. This channel can be blocked by a denser body located to the west at a distance mark of 5 700 km. Besides there is one more channel, which is clearly seen to the east of TAM on a mark of distance of 7 000 km. Its root on the depth of 50 km is turned to the west and supports TAM from below. Apparently, it should also be included to the WARS system.

The complex of WARS and TAM in detail is shown in Fig. 7.4-6. It is possible to see that the erupted cooled blocks are floating off the rift on both sides and dipping with inclination up to 3–4.5 km. The blocks with numbers of 1, 2 and 3 on the left and 4 and 5 on the right of the rift are shown in Fig. 7.4-6. A distance of the blocks from the rift is not symmetric. For example, a distance between blocks of 1–3 is 200–400 km. At the same time the blocks 4 and 5 are united (welded) in to a single mountain system, which root is marked in depths more than 40 km.

Conclusions

Our tomographic data is coordinated and confirms the interpretation of the zone (I) from Grushinsky et al. (2004) and the zone (II) from Storey (1995) as plumes arising from a thermal disturbance at the core-mantle boundary. A nature of the Antarctic plume (I) and Conrad plume (II), apparently, is different. The Antarctic plume has a direct open passage to the top layers and it is characterized by the present-day rifting. The condensed kernel of the Conrad plume at the depths of 650–1 200 km (Fig. 7.4-4, and 7.4-5) is separated almost from the upper horizons by a layer in depths of 400–580 km in the Benioffzone. Apparently, the Antarctic plume near the Ross Sea remained at the same place historically, like the eastern geoblock of Antarctica. At the same time, the Conrad plume was displaced up to 5 000 km (Storey 1995) during the breakup of Gondwanaland and the sea-floor spreading. Moreover, a dynamics of displacement was unequal at different levels. An asymmetrical situation of the upper layers within depths of 400 km in relation to the lower layers of the condensed kernel in depths up to 1 200 km is seen clearly in Fig. 7.4-4. The modeling with the gravimetric tomography method can add the field seismic and geophysical observation by the data of internal topological parameters, by the information on a continuation of surface structures in deep layers and also allows the expansion of the geodynamic and tectonic analysis.

The examples of the lateral maps and vertical cross-sections show the appropriateness of such information to promote the understanding of problems of the neotectonics, geodynamics, boundary and deformation zones. Altimeter data is a base information for determination of the geoid in the ocean. Using the local detail altimetric geoid can provide a higher degree of the harmonic expansion than the global model. It will allow the increase of a resolution by depth in highest sedimentary layers.

Acknowledgments

The authors would like to thank Dr. M. Scheinert for his remarks and valuable comments to the paper.

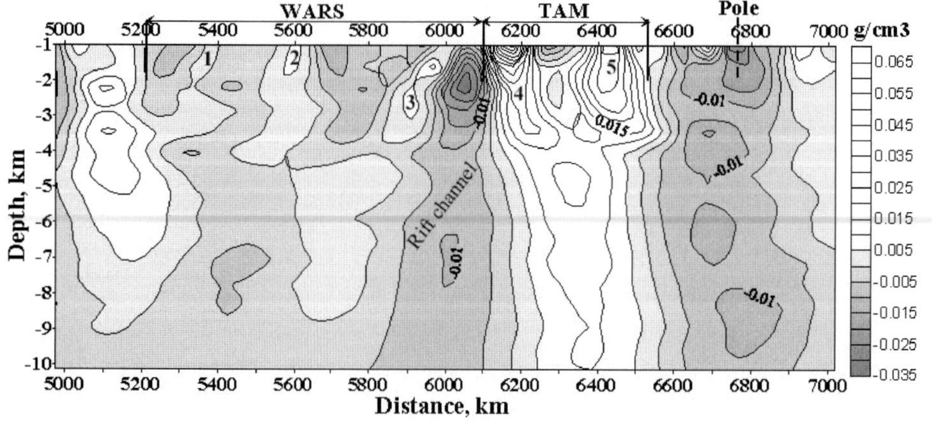

Fig. 7.4-6. Fragment from the cross-section in Fig. 7.4-5 including the West Antarctic Rift System. Range of depths is 1–10 km. *1–3:* spreading blocks, which float off the rift to the west; *4–5:* TAM massif to the east from the rift. Contour interval is 0.005 g cm^{-3}. For further details see legend of Fig. 7.4-5

References

Allan RR (1975) Depth of sources of gravity anomalies. Nature Phys Sci 236(63):22–23

Bijwaard H, Spakman W, Engdahl ER (1998) Closing the gap between regional and global travel time tomography. J Geophys Res 103(12):30055–30078

Bullard EC (1954) The interior of the Earth. In: Kuiper GP (ed) The Earth as a planet. Univ. of Chicago Press, Chicago, pp 57–137

Fifield R (1987) International research in the Antarctic. Oxford Univ Press, Oxford

Gainanov AG (1981) Geology and geophysics of the eastern Indian Ocean floor (by the 54 and 58 cruises data of the RV Vityaz). Nauka, Moscow, (in Russian)

GEBCO (General Bathymetric Chart of the Ocean) (1984) Mercator Projection, Scale 1:35000000. Canadian Hydrographic Service, Ottawa

Grushinsky AN, Stroev PA, Koryakin ED (2004) Structure of the Antarctic lithosphere and its isostasy. Otechestvennaya Geologiya 2:30–36 (in Russian)

Lythe M, Vaughan D, BEDMAP Consortium (2001) BEDMAP: A new ice thickness and subglacial topographic model of Antarctica. J Geophy Res 106(6):11335–11351

Moritz H (1990) The figure of the Earth. Theoretical geodesy and the Earth's interior. Wichmann, Karlsruhe

Rapp H, Wang Y, Pavlis NK (1991) The Ohio State 1991 geopotential and sea surface harmonic coefficient models. Dept Geodetic Sci and Surv, Ohio State Univ, 410

Shimbirev BP (1975) Theory of the Earth's figure. Nedra, Moscow, (in Russian)

Storey BC (1995) The role of mantle plumes in continental breakup: case histories from Gondwanaland. Nature 377:301–308

Some Recent Characteristics of Geomagnetic Secular Variations in Antarctica

Antonio Meloni[1] · **Luis R. Gaya-Piqué**[1,2] · **Paola De Michelis**[1] · **Angelo De Santis**[1]

[1] Instituto Nazionale di Geofisica e Vulcanologia, INGV, Via di Vigna Murata 605, 00143 Rome, Italy, <meloni@ingv.it>

[2] *Also at:* Observatori de l'Ebre, Horta Alta 38, 43520 Roquetes, Spain

Abstract. Some of the most interesting features of the Earth's magnetic field and of its time variations are displayed in polar areas, where the geomagnetic field dipole poles are located. Space time models of the geomagnetic field give a mathematical description that allows generally to undertake a common epoch time reduction of magnetic surveys and to extract magnetic anomaly maps after removing the main part of the geomagnetic field; in addition in polar regions geomagnetic field models allow to follow the location of the geomagnetic dip poles in their time wandering. In this work the development of a dedicated regional magnetic reference model for Antarctica (Antarctic Reference Model, ARM) is presented and compared to the well known IGRF (International Geomagnetic Reference Field) model and it is shown that the first is more appropriate to better study the behaviour of secular variation and its unusual characteristics as observed in Antarctica.

Moreover single and multi-station analyses have been applied to the longest available time series of geomagnetic data for Antarctica in order to investigate the most interesting behaviour of secular variation: the geomagnetic jerks. It was found that geomagnetic jerks are also detectable in Antarctica and that they show a peculiar space time structure.

Introduction

It is known from over 300 years that the geomagnetic field changes with time. Variations in the geomagnetic field observed at the Earth's surface occur on time scales ranging from milliseconds to millions of years. The short-term variations are believed to be primarily of external origin. They arise mostly from electric currents flowing in the ionosphere and magnetosphere and are responsible of the magnetic storms and of the more or less regular daily variation. These rapid variations of the magnetic field are superimposed upon much slower changes occurring over time scales of a few to thousand years. The slower changes, which are related to the dynamo processes acting within the Earth, are generally referred to as geomagnetic secular variation. In detail, the geomagnetic secular variation is the result of inductive effects of fluid motion at the top of the Earth's liquid iron outer core. This means that, at a single epoch, the morphology of the field at the Earth's surface can be used to infer the configuration of flux at the interface between the core and the overlying solid mantle and when the temporal evolution of the field is taken into account in addition to knowledge of the field itself, more can be achieved. Indeed, the secular variation holds information on the flow into the core, and on diffusion of the magnetic field due to the effects of finite electrical conductivity of the core.

One of the most unusual features of the secular variation, for a given field element, is the secular variation impulse commonly named "jerk". A geomagnetic jerk is defined as a rapid change, taking place in a year or two, in the slope of the secular variation or as a step function in the secular acceleration. This means that we can think of the secular variation as a series of straight-line segments separated by geomagnetic jerks. Consequently, the geomagnetic jerks represent a reorganization of the secular variation and are of internal origin, as it has been established through spherical harmonic (Malin and Hodder 1982) and wavelet (Alexandrescu et al. 1996) analysis. Their short time scale implies that they could be due to a change in the fluid flow at the surface of the Earth's core (Waddington et al. 1995) or could result from abrupt magnetic changes generated at the core-mantle boundary. However, little is still understood about their physical origin (Bloxham et al. 2002) and as a consequence they are still largely debated.

Examination of geomagnetic data from worldwide observatories has revealed that this phenomenon is not apparent at all observatories, nor in all components and that it is particularly spectacular in the east (Y) component of European observatories. However, it is usually a worldwide event in extent being present in different components at different points of the Earth's surface and at different times. In particular, it has been noticed that, in the case of the last three jerks of 20th century, for 1969, 1978 and 1991 (the worldwide character of the 2000 jerk is still debated; Mandea et al. 2000), the occurrence of the event in the Southern Hemisphere is a few years after that of the same event in the northern hemisphere (Alexandrescu et al. 1996; De Michelis et al. 2000). These commonly observed time lags for the occurrence epochs are very important because they may imply the existence of higher electrical conductivities in the lower mantle beneath these regions.

We can conclude that the geomagnetic field and its time variation are of geophysical interest, not only for their own sake but also for the knowledge that can be acquired about

From: Fütterer DK, Damaske D, Kleinschmidt G, Miller H, Tessensohn F (eds) (2006) Antarctica: Contributions to global earth sciences. Springer-Verlag, Berlin Heidelberg New York, pp 377–382

some of the Earth's physical properties, such as electrical conductivity of the mantle and motion of fluid core. The structure of the geomagnetic field on the Earth is such that a dipolar field can approximate it, with poles relatively close to geographical poles. The polar regions of the Earth are then a very peculiar area from a geomagnetic point of view. However only at the end of the fifties, we had the first real opportunity for extensive and continuous geophysical investigation in the polar areas, and especially in Antarctica. Before this period less was known about the Antarctic regions than about any other area of comparable size in the world.

In this paper we have focalized our attention on the study of the secular variation and on its impulse (i.e. jerk) in the Antarctic region. In detail, we have reported the development of a regional magnetic reference model to study the behaviour of the secular variation and we have reconstructed the intensity distribution of the last jerks of 20th century.

Models of the Geomagnetic Field

The most important and commonly used method to describe quantitatively the Earth's magnetic field is through spherical harmonic analysis. Various combinations of global observatory, satellite, and magnetic field survey data sets are used to obtain the coefficients of the truncated spherical harmonic series, in order to quantitatively describe the Earth's magnetic field and its secular variation. Models of the field itself are determined essentially from land, sea, airborne, and satellite data while the secular variation models are based mainly on data from continuously recording observatories and to some extent on repeat-station data. The International Geomagnetic Reference Field (IGRF) is the standard mathematical description of the Earth's magnetic field, undertaken by a spherical harmonic analysis implemented by means of Legendre polynomials and Fourier series (e.g., Mandea and Macmillan 2000). Torta et al. (2002) and De Santis et al. (2002) used the Spherical Cap Harmonic Analysis (SCHA; Haines 1985) to represent in a more accurate way the field in the cap delimited by latitude 60° South, that encircles all the Antarctic continent and large part of the Southern Ocean. SCHA is a regionalization of the global model for a spherical cap by means of non integer Legendre polynomials and Fourier series. The Antarctic reference model based on the use of this technique was called Antarctic Reference Model (ARM).

In the most recent ARM version (Gaya-Piqué et al. 2004), annual means of X, Y and Z components registered at Antarctic observatories, as well as a selected subset of satellite total field values data, have been used to develop a model, which is formed by 123 coefficients, that permits the computation of the main field and its secular varia-

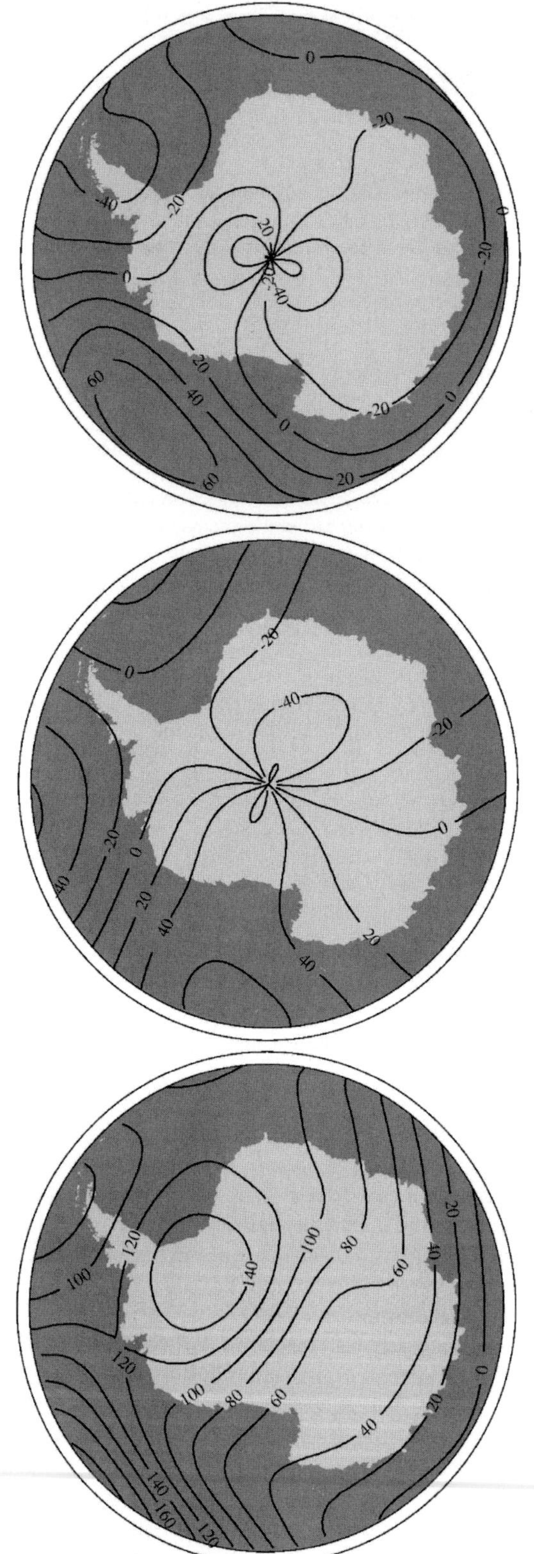

Fig. 7.5-1. Secular variation for X (*top*), Y (*centre*) and Z (*bottom*) components of the magnetic field at sea level for the epoch 1995.0 computed from ARM. Contour lines every 20 nT

tion over the Antarctic regions from 1960 to 2005. The method of the Main Field Differences, proposed by Haines (1993), was applied for the time variation.

The comparison of rms fits to real data (main field values and observatory differences) with those of IGRF allows us to confirm that ARM represents a clear improvement with respect to the global model when representing the field and, especially, its secular variation. Indeed, the comparison of ARM versus IGRF clearly shows that the former fits better than IGRF the observatory differences with regard to the respective mean values for each magnetic element: the rms for *X*, *Y*, and *Z* magnetic elements are respectively 42.6 nT, 55.3 nT, and 121.8 nT for the IGRF, and 23.9 nT, 28.5 nT, and 35.2 nT for the ARM. The clear improvement of ARM with respect to IGRF, almost greater than 50% for all three elements, is probably due to the fact that the latter model presents a constant secular variation pattern at 5-year intervals.

Once the validity of ARM has been checked in terms of rms fits to real input data sets, geomagnetic charts can be depicted for each of the magnetic elements, for a given epoch and elevation (a.s.l.). As an example, we present sea-level maps of the secular variation for *X*, *Y*, and *Z* components at epoch 1995.0 in Fig. 7.5-1.

Study of Geomagnetic Jerks

The internal Earth origin of the geomagnetic jerks has made this event a key factor in understanding the Earth's internal dynamics. The comprehension of this phenomenon and of its space-time distributions is, moreover, directly associated with the solution of the main problem of geomagnetism, i.e. the origin of the magnetic field and its variations, with the evaluation of the electrical conductivity of the lower mantle and with the verification of some hypothesis regarding the internal structure of the Earth. For this reason the knowledge of the spatial and temporal distribution of a given jerk in any part of the world, included the Antarctic region, is particular important. It is known that the secular change is a global phenomenon but with local characteristics. Indeed, the secular change does not proceed in a regular way all over the Earth, giving rise to regions where the field changes more rapidly than elsewhere. As a consequence of this behaviour, the sudden change in the trend of the secular variation that can be found in localized regions does not always reveals a worldwide character. For this reason the investigation of the characteristics of the secular variation trend in the Antarctic region is very important.

Figure 7.5-2 shows the trend of the secular variation relative to different magnetic field components in the case of five selected Antarctic geomagnetic observatories (Argentine Island (AIA), Mawson (MAW), Dumont D'Urville (DRV), Casey (CSY) and Syowa Station (SYO)). The an-

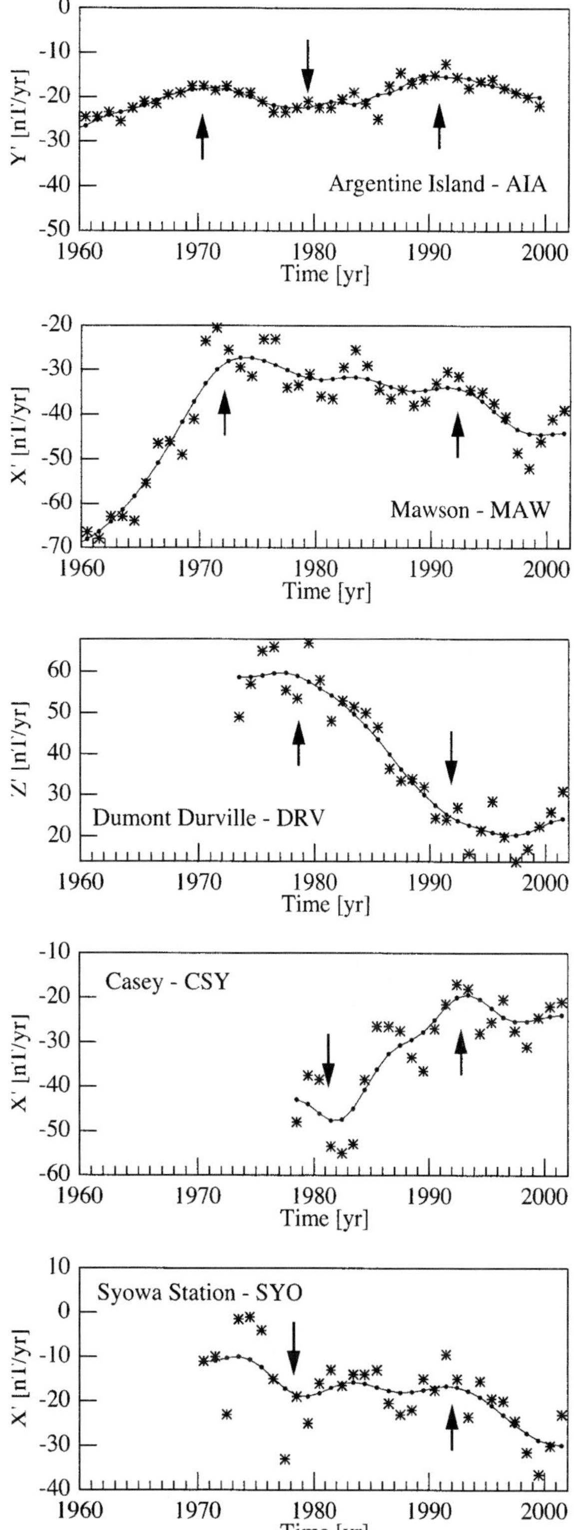

Fig. 7.5-2. Secular variation trend in the case of five different geomagnetic observatories localized in Antarctic region. The *circles* represent the trend of the secular variation after a smoothing operation, the *stars* are the original data and the *arrows* identify jerk events

nual mean values of the geomagnetic field used to evaluate the secular variation come from the British Geological Survey website. The stars represent the secular variation obtained from the original data while the circles represent the trend of the secular variation after a smoothing operation (moving window on three points iterated twice). The arrows show the existence of jerks around: 1969, 1978 and 1991. Actually, these events are temporally shifted with regards to these typical years. As well known, the observatories localized in the Southern Hemisphere are indeed characterized by a time delay in the record of the jerk events. The figure shows immediately the intrinsic difficulties of the jerk study in Antarctica: a limited temporal duration of available data, perturbed data from external magnetic disturbances and, also for this reason, an objective higher difficulty in obtaining high quality data.

Taking into account these problems it is clear that the classical methods of analysis of the geomagnetic jerks (i.e. traditional straight-line fit or wavelet analysis) may be problematic especially if the analysis is limited to Antarctic region without the help of other observatories localized in the Southern Hemisphere. In this paper we reconstructed the intensity maps relative to the last three jerks in Antarctic region by merging the analysis of the trend of the secular variation recorded in Antarctic observatories with that coming from other Southern Hemisphere observatories. In order to evaluate the intensity distribution of a jerk, the secular variation time series must be fitted using a bilinear functional form. The point where the two straight lines intersect defines the time of the jerk event while the difference between the angular coefficients of the two straight lines before and after the jerk event defines the intensity of the phenomenon. Figure 7.5-3 shows intensity maps expressed in [nT yr^{-2}] for the last three jerks of 20th century. The three events appear similar in the intensity distribution structure and all three show a focus with a maximum intensity. A peculiar feature seems to appear in these maps, indeed it looks like the focus of the maximum of the jerk intensity rotates longitudinally during the years, from 120° E to 20° W and finally to 20° E. However, we cannot exclude that this result can be influenced by data non-uniformly distributed or by the data noise, where the noise is defined as error due to instrument drift and malfunction, calibration errors, man-made magnetic signals, and possible errors in the data processing that produces annual means.

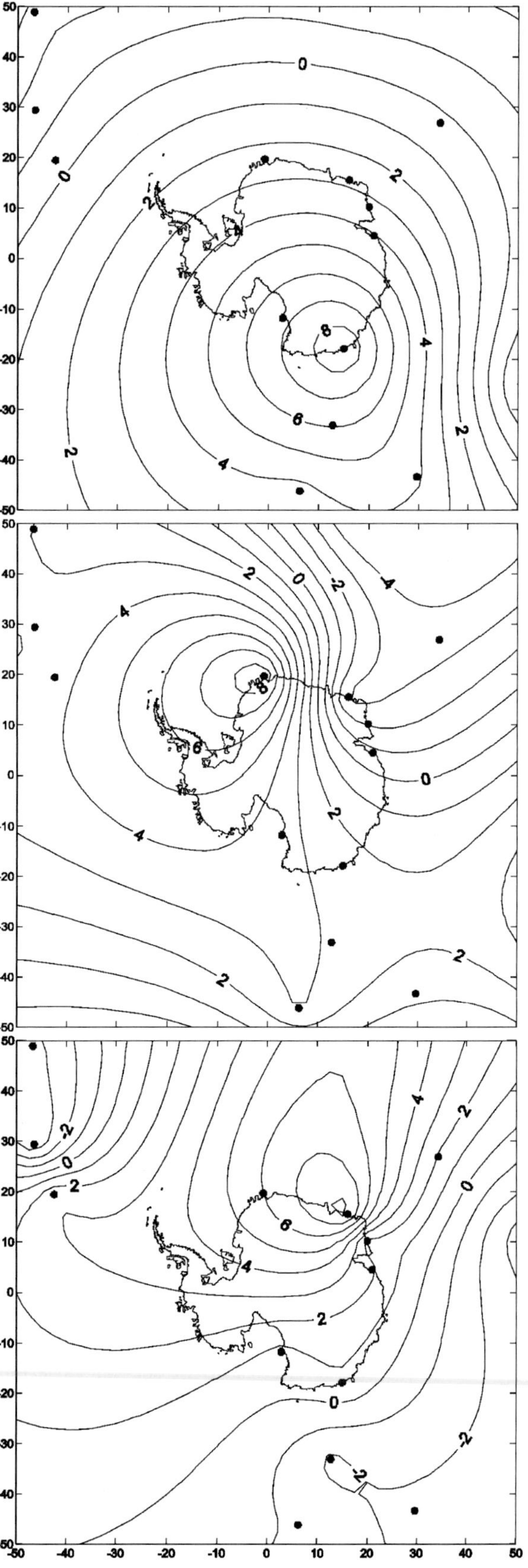

Fig. 7.5-3. Azimuthal equidistant projection of the spatial distribution of the jerk intensity relative to the last three jerks of 20th century: 1969 (*top*), 1978 (*centre*), and 1991 (*bottom*). These distributions, relative to the *Y*-component, have been obtained using observatories located both in Antarctic and in Southern Hemisphere (*full circles*)

Conclusions

Spherical Cap Harmonic Analysis has been applied to obtain a reference model of geomagnetic secular change for Antarctica (ARM) valid for the last forty years. This model improves the fit to the secular variation deduced from observatory data by about 60% relative to IGRF and allows merging data sets taken at different altitudes and epochs in Antarctica. Using this method we have had the opportunity to depict geomagnetic charts for each of the magnetic elements for a given epoch and elevation (a.s.l.).

We have also investigated the secular variation trend in Antarctic region focusing on its particular behaviour: geomagnetic jerks. In detail, using the annual mean values of the geomagnetic field components available for this region we evaluated the intensity distribution of the last three jerks of 20th century (1969, 1978, 1991). A rotation of the jerk focus has been found from a multi-station analysis even if we cannot exclude that this feature can also be influenced by the low quality of the used Antarctic observatory data and/or to the bad distribution coverage of the data. Naturally, we expect that more studies are still needed to better uncover the above found aspects. A better spatial and temporal distribution of the Antarctic observations will be of great importance to improve future investigations on SV characteristics.

References

Alexandrescu M, Gibert D, Hulot G, Le Mouël, Saracco G (1996) Worldwide wavelet analysis of geomagnetic jerks. J Geophys Res 101:21975–21994

Bloxham J, Zatman S, Dumberry M (2002) The origin of geomagnetic jerks. Nature 420:65–68

De Michelis P, Cafarella L, Meloni A (2000) A global analysis of the 1991 geomagnetic jerk. Geophys J Internat 143:545–556

De Santis A, Torta JM, Gaya-Piqué LR (2002) The first Antarctic geomagnetic Reference Model (ARM). Geophys Res Letters 29(8):1129

Gaya-Piqué LR, De Santis A, Torta JM (2004) Use of Champ magnetic data to improve the Antarctic geomagnetic Reference Model. In: Schwintzer P, Wickert J (eds) Earth observation with Champ results from three years in orbit. Springer, Berlin Heidelberg New York, pp 317–322

Haines GV (1985) Spherical Cap Harmonic Analysis. J Geophys Res 90(3):2583–2591

Haines GV (1993) Modelling geomagnetic secular variation by main-field differences. Geophys J Internat 114:490–500

Malin SRC, Hodder BM (1982) Was the 1970 geomagnetic jerk of internal or external origin? Nature 296:726–729

Mandea M, Macmillan S (2000) International Geomagnetic Reference Field – The eighth generation. Earth Planet Space 52:1119–1124

Mandea M, Bellanger E, Le Mouel JL (2000) A geomagnetic jerk for the end of the 20th century? Earth Planet Sci Letters 183:369–373

Torta JM, De Santis A, Chiappini M, von Frese R (2002) A model of the secular change of the Geomagnetic Field for Antarctica. Tectonophysics 347:179–187

Waddington R, Gubbins D, Barber N (1995) Geomagnetic field analysis – V. Determining steady core-surface flows directly from geomagnetic observations. Geophys J Internat 122:326–350

Topographic and Geodetic Research by GPS, Echosounding and ERS Altimetric, and SAR Interferometric Surveys during Ukrainian Antarctic Expeditions in the West Antarctic

Rudolf Greku[1] · Gennady Milinevsky[2] · Yuriy Ladanovsky[3] · Pavel Bakhmach[3] · Tatyana Greku[1]

[1] Institute of Geological Sciences, National Academy of Sciences of Ukraine, 55B Gonchara st., 01054 Kyiv, Ukraine, <satmar@svitonline.com>

[2] Ukrainian Antarctic Center, Kyiv/Ukraine, <antarc@carrier.kiev.ua>

[3] ECOMM Co, 18/7 Kutuzov st., 01133 Kyiv, Ukraine, <lada@ecomm.kiev.ua>

Abstract. The region of the Ukrainian Antarctic geodetic and topographic surveys is the transition zone, which includes the Antarctic Peninsula with ice streams; large longitudinal fractures known as the Grandidier, Penola and Bransfield straits, separating the island part of the wide shelf from the Peninsula; small archipelagos appear as a result of the ancient ice shelf movement overlapping tectonic processes; and the open sea of the Pacific Ocean's south-eastern margin. Such physiographic diversity of the region causes a necessity to use different geodetic research methods. The Ukrainian Vernadsky Station (former British Faraday Station) is located at the middle part of the region on the Galindez Island of the Argentine archipelago. Following investigations in this area were carried out under the auspices of the Ukrainian Antarctic Centre within the Ukrainian Antarctic Research State Program for the SCAR GIANT (Geodetic Infrastructure of Antarctic), ANTEC (Neotectonics of Antarctic) and IBCSO (International Bathymetric Chart of the Southern Ocean) Projects:

- Several days duration seasonal GPS observations at the "SCAR GPS 2002" site on Galindez Island;
- Repeated GPS observations on the British triangulation markers and extension of the geodetic network on the islands;
- Large-scale mapping of islands' topography and terrestrial photogrammetry survey of the island ice cliffs;
- Echosounding of the Argentine archipelago's seabed in unsurveyed shallow water;
- Mapping of the Flask Glacier ice stream of the Antarctic Peninsula using ERS Synthetic Aperture Radar (SAR) interferometry data;
- Geoid determination of the Bellingshausen Sea based on altimeter data for geological purposes.

As a result of the research different maps of the bottom topography by echosoundings, of the island relief by GPS observations, of topographic features of the moving glacier by the satellite data were created. The mass balance of the island's ice cap by unconventional terrestrial photogrammetric technique was also studied.

Seasonal GPS Observations

Initial seasonal GPS observations at the Vernadsky Station (VER1 reference point) for studying the regional tectonic stability were carried out by the Geodetic Institute of the University of Karlsruhe (GIK) in 1998 and 2002. Since 2002 Ukrainian observers have carried out repeated seasonal GPS observations at the "SCAR GPS 2002" reference point. The central marker of the new site was installed by GIK. The distance between these sites is 7.94 m as a difference of the observed coordinates in 1998 and 2002. However, synchronous observations on both points were not fulfilled during the period 2002–2004.

Duration of observations was 15 days in each season of 2002 trough 2004. A dual frequency receiver Trimble GPS Total Station 4700 was used in 2002 and 2004; a Trimble 4800 was used in 2003. The antennas L1/L2 with ground microcentered plate were used in each epoch. The Australian AUSPOS Online GPS Processing Service (Minerals and Geohazards Division, Geoscience Australia) carried out the processing of the RINEX format data. The AUSPOS Service uses International GPS Service (IGS) software products including Precise Orbits, Earth Orientation, Coordinate Solutions to compute precise coordinates in the ITRF2000 system anywhere on Earth. All computations were undertaken using the MicroCosm Software System, which is fully IGS compatible (see Table 7.6-1; http://www.vmsi-microcosm.com).

Computed horizontal components of displacements for periods 2002–2003 and 2003–2004 are 7.2 mm along

Table 7.6-1. IGS stations used for computing; root mean square (RMS) errors of the "SCAR GPS 2002" observed point and annual differences between coordinates measured on the same site

Year	IGS station	RMS of observed point (mm)			Annual differences (mm)		
		Lat	Long	Height	Lat	Long	Height
2002	Ipgs, riog, vesl, sant	7	5	12	2002 minus 2003		
2003	Ipgs, riog, vesl, sant	4	3	19	3.1	6.4	5.0
					2003 minus 2004		
2004	Ipgs, riog, vesl	4	7	15	18.5	25.8	3.0

lpgs: La Plata; *riog:* Rio Grande; *vesl:* Vesleskarvet; *sant:* Santiago.

From: Fütterer DK, Damaske D, Kleinschmidt G, Miller H, Tessensohn F (eds) (2006) Antarctica: Contributions to global earth sciences. Springer-Verlag, Berlin Heidelberg New York, pp 383–390

azimuth 64.47° and 31.8 mm along azimuth 54.38° correspondingly. Obtained computations for the "SCAR GPS 2002" can be compared with the closest PAL1 station data from (Dietrich et al. 2001). Namely, the velocity solution in the north and east components in the ITRF96 are 12 mm yr^{-1} and 15 mm yr^{-1} correspondingly for the PAL1. The similar data show the other solutions with the geophysical model NNR NUVEL-1A are based on the concept of no net rotation (19.17 mm yr^{-1} and 60.62° azimuth) and the actual plate kinematics model APKIM2000.0 (18.49 mm yr^{-1} and 53.78° azimuth).

These results are preliminary since the network of IGS reference stations used for this solution is not optimal due to its geometry, selected sites and base line lengths. A re-processing with other and more IGS reference stations is planned.

Coastal GPS Survey

A local network of GPS sites was established within 10–15 km radius relatively to the "SCAR GPS 2002" geodetic reference point. The GPS seasonal observations dura-

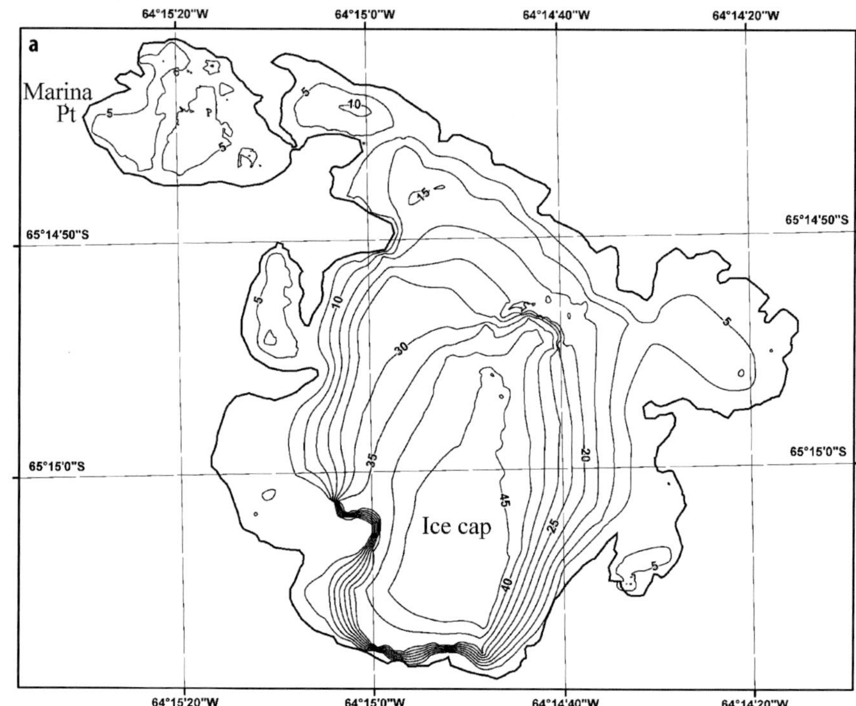

Fig. 7.6-1. Ukrainian Antarctic station on Marina Peninsula, Galindez Island. **a** Topographic map of the Galindez Island from GPS surveys. UTM map projection, Vertical Datum is EGM96 geoid, contour line interval 5 m. **b** 3D image of the Galindez Island draped in the Quick Bird satellite image of 24.01. 2004. View from the west

tion of 15–20 days in period February–March were realized with the one-frequency receivers 4600LS Surveyor, Pathfinder ProXRS and Geoexplorer. Differential postprocessing mode relatively to the "SCAR GPS 2002" reference station with the Trimble Geomatics Office software and Pathfinder Office software was used for positioning with an accuracy of 1–2 cm. Coordinates for more than 3 000 GPS points were determined during seasonal periods of 2002–2004 for different geophysical measurements on the archipelago's islands. A number of different objects in the vicinity of the Vernadsky Station (meteorological and geophysical laboratories, masts and antennas), the British triangulation stations, the tide gauge, ground control points for the stereophotogrammetric survey, and points of geomagnetic, geological and biological sampling were positioned using GPS. 30 points are on a rock outcrop measured with an accuracy of 0.5–0.7 cm and can be used for the repeated observations and the expansion of the local geodetic network. One of the results is (Fig. 7.6-1A) a topographic map of the Galindez Island by more than 1 000 GPS measurements. The accuracy of height measurements is double to the horizontal coordinates' 2–3 cm accuracy. An orthometric system was introduced for the topography, the ellipsoid heights were reduced by the EGM96 geoid. The coastal line and landscape features were drawn by using the Quick Bird satellite image from 24 January 2004 with a spatial resolution of 70 cm. The georeferencing of the image was carried out using more than 100 ground control points observed by GPS with high accuracy. It allowed us to correct the image geometric distortion to match the real GPS coordinates. DEM and contour lines were computed by the GPS observations (ArcGIS v. 8.3 software; Fig. 7.6-1B).

A large part of Galindez Island is occupied by a small ice cap, which degrades due to an increase of the mean annual temperature. A significant change of the ice cap is observed in the south-west in the region of the 50 m high cliffs. Terrestrial photogrammetry technique is used to obtain a detailed morphology, to provide annual monitoring and to create an evolution model of the cap (Glotov et al. 2004, in press). This method is usually applied in surveying (mining) works to determine the volume of excavation. The technology uses stereophotographs of the ice cap cliff on a vertical plane, which is normal to the optical axis of the photocamera. Photoexposures were carried out from GPS fixed geodetic points.

Changes of the vertical frontal topography of the cliff and of the ice volume between two surveys (as a difference between topographies) depend beside others on the instrumental errors in photogrammetry, namely: measurement of a basis between the observation points, focal distance, parallax and integration (smoothing) of the relief. The RMS of the measurement of the decrease of ice was determined with 0.12% of the volume. Observations of 2002–2004 show a reduction in volume of 23 000 m^3 during 2002–2003 and 28 000 m^3 during 2003–2004. This is 2–3% of the whole volume of the Galindez Island ice cap. These results supposes that this ice cap will probably disappear within the century (Glotov et al. 2004, in press).

Sea Floor Topography of the Argentine Islands Archipelago

The archipelago of the Argentine Islands is located on the western shelf of the Antarctic Peninsula. It is separated from the Peninsula by the more than 300 m deep and 7 km wide Penola Strait. The archipelago is a tectonic mesoblock, which is broken into smaller fragments by a system of fractures.

Echosoundings and geological samplings were carried out on an adequately equipped boat in the internal waters between the islands of the archipelago during March and April 1998, and were later continued in 2002 and 2003. The total length of tracks is approx 400 km. Depths and coordinates were recorded at an interval of two seconds, which corresponds to a distance of 5 m. The positioning RMS error of ±23.5 m was determined (1998) by two second interval observations during 24 hours at a fixed point using one single frequency GPS receiver. D-GPS kinematic positioning using a one-frequency receiver with phase-tracking was applied in 2002 and 2003. The phase-tracking software provides an accuracy of 2 cm. The instrumental accuracy of measured depths is not worse than 0.1% of the water depth taking into account that depths are within 70 m in the archipelago. The depths have been corrected for the tidal effects using continuous tide gauge records at Vernadsky Station. The Working Datum level was used as a reference level for the sounding surveys (Hydrogr. Dict. 1994; IOC 2002) and determined as lowest water during observation from 01 January 1996 to 02 April 1998. There is an elevation mark of 0.73 m at the tide gauge.

A digital elevation model (DEM) with 20 m resolution has been computed by Triangulation with Linear Interpolation method and used for the construction of an electronic map of the seabed topography (Fig. 7.6-2; Gozik et al. 2002). The outer areas of the archipelago shown on the map without our soundings have been filled up by sparsely populated depths from the British Nautical Charts 3572 and 3575.

Depths within the archipelago are not more than 70 m. A well visible system of the fracture zones spreads out in NE and NW directions. Comparison of the seabed topography images between the Ukrainian and the British Royal Navy soundings (1964–1965) was carried out. The Original/Master K4655/1 British chart (by the Royal Navy, Royal Naval Antarctic Survey Party, Natural Scale 1 : 12 500, February 1964 to April 1965) with manually pictured depths along regular tracks was used after having been kindly

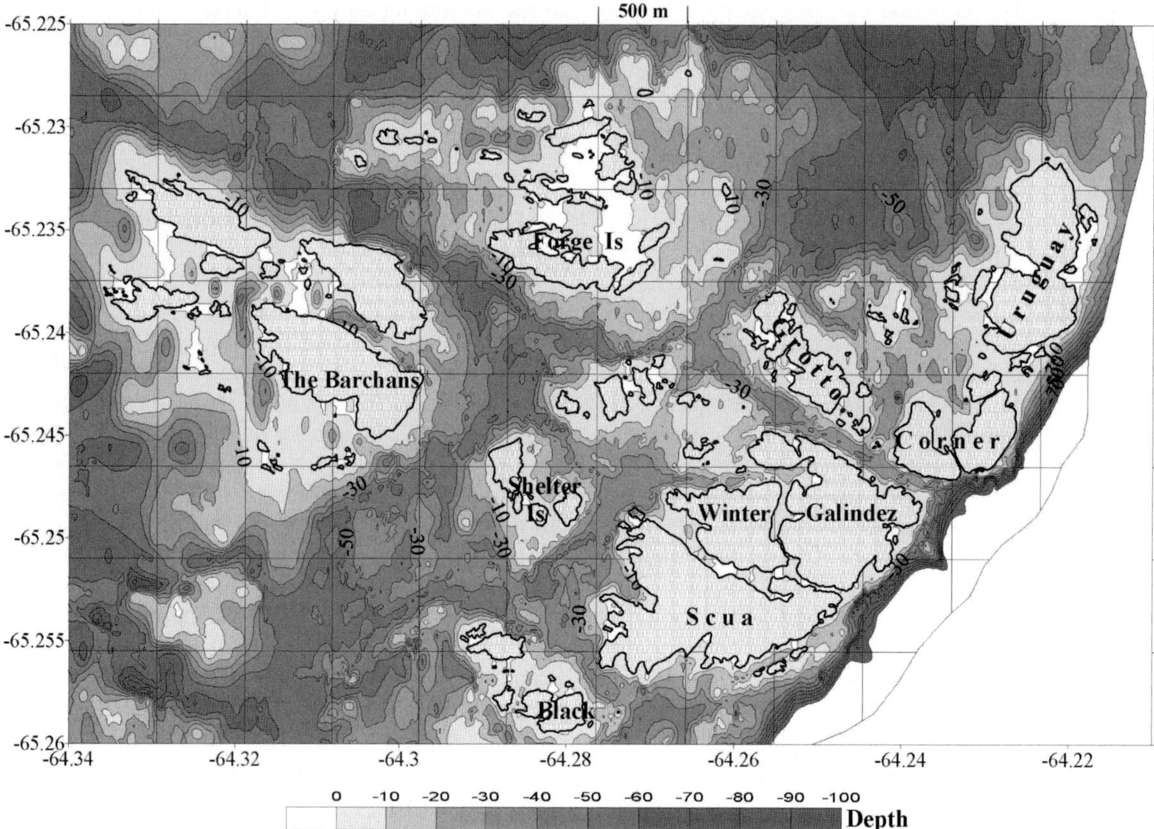

Fig. 7.6-2. Bathymetric map of the Argentine archipelago observed during seasonal Ukrainian expeditions (1998–2003); contour interval is 10 m

provided by the British Antarctic Survey. These data were digitized and used for DEM computing. In general, the images of the seabed features on the maps are similar. Four segments from different morphological areas of the archipelago were used for comparison by 1 430 node pairs. An average difference between depths is 0.2 m and the standard deviation (STD) is 2.8 m. A number of differences between depths less than ±5 m are more than 80%. Both data sets can be combined after some mutual adjustment.

Geoid Determination by ERS-1 Altimeter Data in the Bellingshausen Sea

One purpose of the work is to determine a detailed local geoid surface and to provide new information for Geodesy, Geology (Earth's interior) and Oceanography (reference surface for the sea surface topography) of the region where Ukrainian institutes have carried out marine surveys and satellite monitoring for several years.

ERS-1 global altimeter data for the period of 1992–1998 have been received from the European Space Agency (ESA) and then extracted for the local region within boundaries: latitudes 61–67° S, longitudes 55–73° W. Altimeter measurements (Altimeter Ocean Product, ERS.ALT.OPR02) during several phases of the ERS-1 mission have been used, namely: 3-day repeat period, 35-day period and 168-day period.

The 838 tracks (430 ascending and 408 descending) with a mean spacing of 6.6 km including 46 021 measurements were selected primarily. The preprocessing of the selected data included a statistical assessment and removing of random noise by smoothing along each track, elimination of land and ice located measurements, elimination of abrupt spikes and overly short segments (less than 7 points) which are not effective for crossover analysis. The following physical corrections have been applied: dry and wet tropospheric, ionospheric and electromagnetic biases, as well as tide and waves.

Crossover analysis has been performed to reduce systematical orbit errors using the modified ALTCROSS-ALTCOR software by P. Knudsen (Coren et al. 1997). In total, 836 tracks with 45 994 measurements were used. 34 598 crossovers were determined for adjustment. The mean height difference of all crossing points before correction was 0.282 m and RMS 3.279 m. The mean of the remaining height differences in crossovers become −0.005 m and RMS 0.821 m after applying a least squares adjustment method.

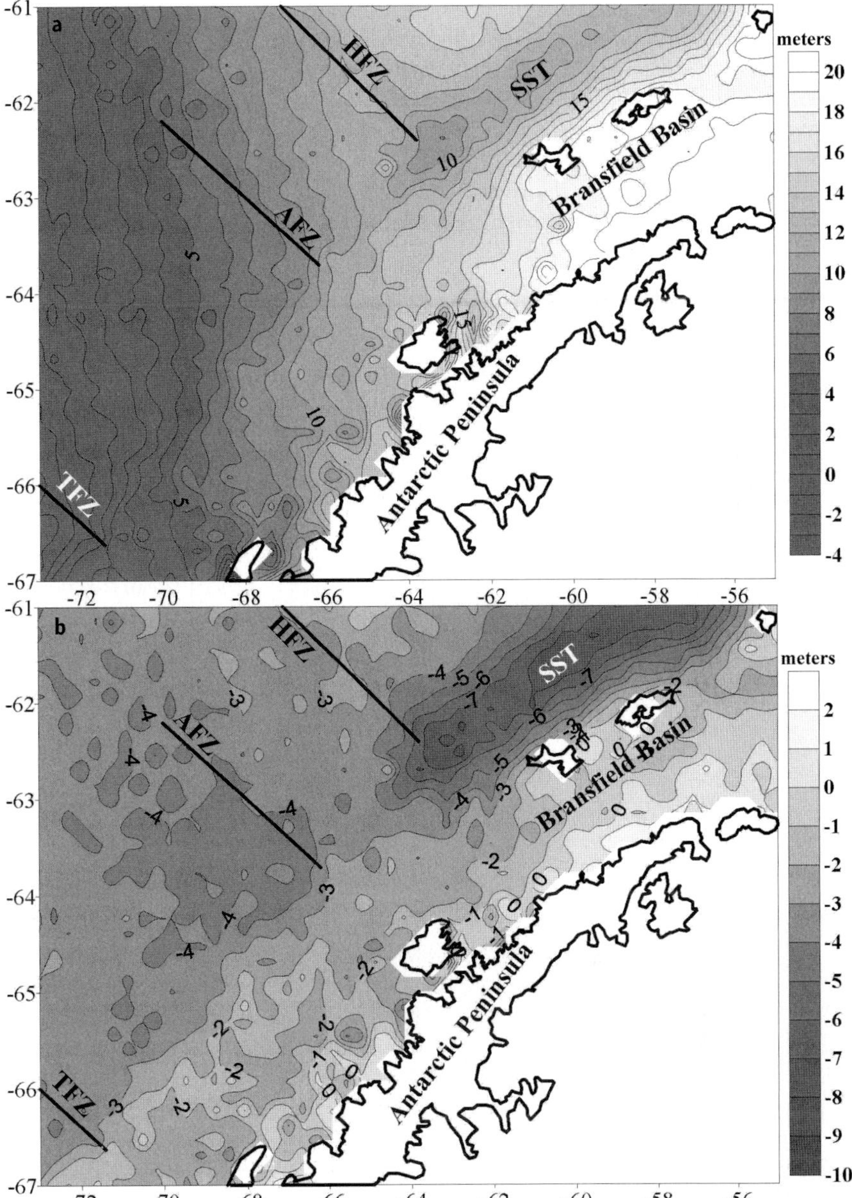

Fig. 7.6-3. Geoid maps in Mercator projection with the central latitude 64° S. **a** Geoid topography of the Bellingshausen Sea by the ERS-1 altimeter data. **b** Residual geoid as a difference between altimetric data and heights from the EGM96 model computed with spherical harmonics at the 2–50 range. Values are in meters; *SST:* South Shetland Trench; *HFZ:* Hero Fracture Zone; *AFZ:* Anvers Fracture Zone; *TFZ:* Tula Fracture Zone

Gridding procedure with cell size of 17×17 km was calculated to create a DEM. Most of the cells include 2–3 altimetric measurements. Such cell size was a compromise between the details of the DEM and adequate density of measurements. Many near coastal measurements are incorrect because of ice. Therefore, these areas are blanked. A contour map of the geoid is shown in Fig. 7.6-3a. The general features of the geoid are well in agreement with the global EGM96 model after the removal of a systematic bias of 4 m. The absolute differences between the altimetric and EGM96 geoid heights have not exceeded a value of 3 m even at the steep continental slope and it is near one meter for other regions.

Data of the geoid and undulations can be used for tectonic interpretation of the Earth's interior. For example, there is an inclined trend of the geoid in our region with height increasing from the south-west to the north-east from 0 up to 15 m. It is caused mainly by distribution of large scale deep geological masses of the Earth. One of them is a well known Antarctic undulation, the epicenter of which is located near the Ross Sea. It is a thinning body which gives falling of the geoid topography up to –65 m (Coren et al. 1997). Another one is the elevated 30 m height of the South America undulation.

It can be seen that the local fracture zones and structure of upper lithosphere blocks are hidden under the trend background. They are displayed (Fig. 7.6-3b) once a long wavelength component has been removed from the full altimetric geoid (Fig. 7.6-3a). The EGM96 global geoid model was used to compute this component from harmonics in range of 2–50 which corresponds to a disturbing layer of the Earth from the core to a depth of 20 km (Greku et al. 2006). It is shown that the Hero Fracture Zone (HFZ) and Anvers FZ divide blocks of different densities where the brighter tint has a more dense structure.

Thus, the geoid data are informative not only for geodesy but for geophysical research, also as it is shown soundly in work (Moritz 1990).

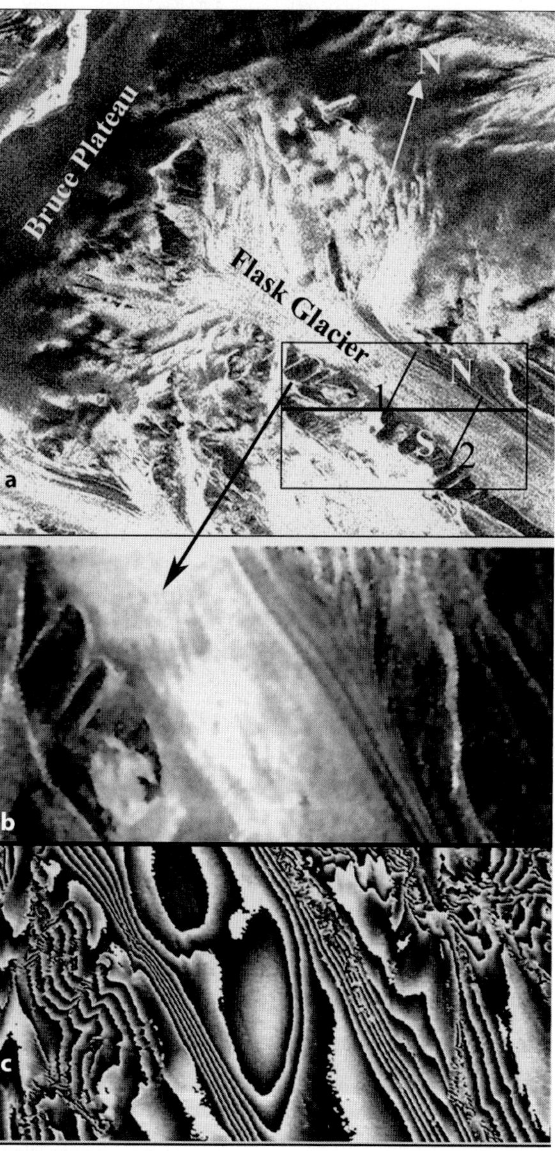

Fig. 7.6-4. Interferometric processing of ERS-1/2 images of 26–27 February 1996 for the Flask Glacier on the Antarctic Peninsula. **a** Fragment of the ERS-2 image of 27 February 1996; area is 36 × 32 km; *S* and *N:* south and north sides of the glacier; *1* and *2:* locations of topographic profiles in Fig. 7.6-6. **b** Amplitude image of a part of the extracted fragment, area of 4 × 15 km. **c** Phase interferogram by two tandem images for the same area as in **b**

Topography of the Flask Glacier from SAR Interferometry

Tandem ERS-1/2 Synthetic Aperture Radar Single Look Complex Images (SAR SLCI) along ascending tracks were acquired by the German tracking station O'Higgins on 26–27 February 1996 and then processed by the D-PAF (German Processing and Archiving Facility). These data were released by the ESA within the Ukrainian project AO358. The area of 100 × 100 km, centre coordinates 65°20' S, 64°15' W is covered by frame 5 835 of track 104. A perpendicular baseline between the images is 87 m and the parallel one is 40 m.

The frame includes the following main provinces: the western shelf of the Antarctic Peninsula, Graham Land, Bruce Plateau and the eastern slope of the Antarctic Peninsula with glaciers flowing to the Weddell Sea. A sector of the frame of 36 × 32 km is shown in Fig. 7.6-4a. The Bruce Plateau is a source of a great volume of ice and snow. The width of the plateau is 5–15 km and the height is 1 700–1 800 m. Separated small flows from Bruce Plateau form the connection to the common Flask Glacier, which has an extension lengthways of 30 km. Extracted from the sector, a fragment of the glacier flow (inset in Fig. 7.6-4a) covers an area of 7 × 15 km. It was processed with the ERDAS Radar Module software for geometric transformation of the SLCI data, decreasing of speckle noise and rectification of the amplitude images (Fig. 7.6-4a). A large scale amplitude image and phase interferogram of the same area for a part (4 × 15 km) of the extracted fragment are shown in Fig. 7.6-4b,c.

Georeferencing of the images was carried out using the corner and central coordinates of the SLCI scenes. Determination of shifts in range and azimuth directions and registration between two SAR images were performed in two stages: *(i)* by cross-correlating the amplitudes of the images (coarse registration) and *(ii)* by fine coherence registration for the subpixel shifts. The coherence by the pair of the images within the test site 64 × 64 pixels in size was 0.65, that leads to an accuracy of the registration as good as 1/125 of a pixel (Prati et al. 1994). A constant elevation increment, corresponding to a 2π phase drop of the interferogram is 106 m. An output space resolution of a pixel is 25 m. An accuracy assessment of DEM has not been computed due to the lack of a reference elevation model for this region.

A coherence and interferogram were computed with the POLIMI INSAR software (Prati et al. 1994; Prati and Rocca 1993). The interferogram shows main topographic features of the glacier. The central elevated part of the main glacier flow is 500–700 m wide. Steep slopes are shown on the interferogram as narrow stripes parallel to the main flow. The unwrapping processing of the phase

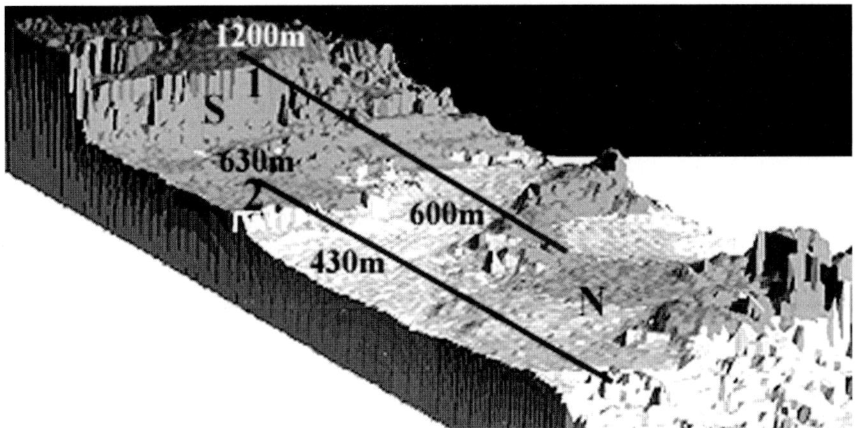

Fig. 7.6-5. Flask Glacier 3D topographic model for the extracted area of 36×32 km. Extreme heights are shown along profiles (see Fig. 7.6-6). For further details see Fig. 7.6-4

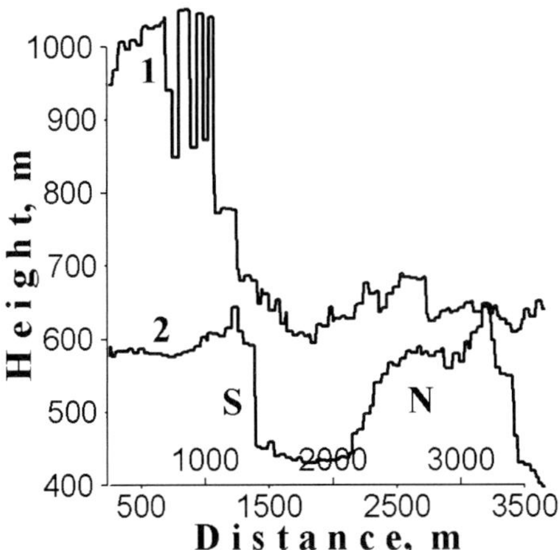

Fig. 7.6-6. Cross profiles of Flask Glacier topography by DEM from the ERS-1/2 SAR interferometry

interferogram for generation of topographic heights, computing of DEM and 3D model (Fig. 7.6-5) were carried out using the ERDAS IFSAR module.

Two cross profiles of the glacier topography by DEM are shown in Fig. 7.6-6. The width of the flow between bordering peaks is 1 300–1 600 m. A depth of the valley is 100–150 m. Minimum heights of the glacier topography vary from 600 m to 430 m for a distance of 4 km between profiles. Heights of the left bordering ridge decrease from 1 200 m to 700 m. The right ridge is not so high, 600–700 m approximately. There is one small glacier flow also farther behind this ridge. Width of this flow is up to 200 m.

Conclusions

As adduced in the paper, examples of geodetic research are apparently the most detailed material for the local and characteristic areas of the West Antarctic. These results can be applied as a background to investigate other phenomena and processes in Antarctica. Thus, a number of thematic, morphometric and geomorphological maps (slope, aspect, curvature, ridge and channel directions) have been drawn up till now using bathymetric and topographic DEMs with the program LandSerf (Wood 1996) and the ENVI/IDL software. The geoid data were used for modeling the deep structure of the Scotia Sea, Antarctic Peninsula and the Antarctic continent with our "gravimetric tomography" technique (Greku et al. 2006).

An extension of the geodetic methods is undertaken for such researches as the determination of the total electron content (TEC) fluctuations in the ionosphere from continuous GPS observations (geophysical aspect), monitoring of kinematics and mass balance of the ice sheet using the SAR differential interferometry (glaciological aspect), investigation of the surrounding archipelagos' seabed, where the depositional history of the sediments is not so well known (geological aspect).

An integrating approach to interpretation and visualization of a wide complex of data are provided within the creation of The National Geographic Information System "Antarctica-Vernadsky" on the base of the ArcGIS software in the Ukrainian Antarctic Centre.

Acknowledgments

Authors would like to thank the ESA for releasing the ERS-1 altimeter data and ERS-1/2 SAR SLCI images and the British Antarctic Survey and the U.K. Hydrographic Office whose sounding data were used for investigation of the archipelago's seabed. We thank Claudia Depenthal from the Geodetic Institute of the University of Karlsruhe who installed the antenna marker of the "SCAR GPS 2002" geodetic station. The authors would also like to thank two anonymous reviewers for theirs remarks and valuable comments. These researches were supported by the Ukrainian Antarctic Centre.

References

Coren F, Marson I, Stoka M, Zanolla C (1997) Computation of the Moho depths and geoid undulation from gravity data in the Ross Sea (Antarctica). In: Ricci CA (ed) The Antarctic region: geological evolution and processes, Terra Antartica Publications, Siena, pp 603–608

Dietrich R, Dach R, Engelhardt G, Ihde J, Korth W, Kutterer HJ, Lindner K, Mayer M, Menge F, Miller H, Müller C, Niemeier W, Perlt J, Pohl M, Salbach H, Schenke HW, Schöne T, Seeber G, Veit A, Völksen C (2001) ITRF coordinates and plate velocities from repeated GPS campaigns in Antarctica – an analysis based on different individual solutions. J Geodesy 74:756–766

Glotov VN, Kovalonok SB, Chizevskiy VV (2004, in press) Determination of quantities of island glaciers using results of digital stereophotogrammetric surveying. Ukrainian Antarctic J 2 (in Ukranian)

Gozik PF, Greku RKh, Usenko VP, et al. (2002) Map of the Argentine Islands archipelago sea-bed relief in the region of the Ukrainian Antarctic Station. Geol J 1:128–13, (in Russian)

Hydrographic Dictionary (1994) Part 1, 5th ed, Spec Publ 32, Monaco

IOC (Intergovernmental Oceanographic Commission) (2002) Manual on sea level measurement and interpretation – vol III. Manuals and Guides 14, UNESCO

GRAVSOFT (1996) A system for geodetic gravity field modeling. Geophys Dept Univ Copenhagen Sympos

Greku RKh, Usenko VP, Greku TR (2006) Geodynamic features and density structure of the Earth's interior of the Antarctic and surrounding regions with the gravimetric tomography method. In: Fütterer DK, Damaske D, Kleinschmidt G, Miller H, Tessensohn F (eds) Antarctica – Contributions to global earth sciences. Springer, Berlin Heidelberg New York, pp 369–376

Moritz H (1990) The figure of the Earth – theoretical geodesy and the Earth's interior. Wichmann, Karlsruhe

Prati C, Rocca F (1993) Use of the spectral shift in SAR interferometry. Proceedings Second ERS-1 Sympos Hamburg, pp 691–696

Prati C, Rocca F, Guarnieri AM, Pasquali P (1994) Report on ERS-1 SAR interferometric techniques and applications. Dipart Elettronica, Politecnico Milano

Wood J (1996) The geomorphological characterization of digital elevation models. Unpubl PhD Thesis, Univ Leicester, Dept Geography, Leicester

Geodetic Research on Deception Island

Manuel Berrocoso[1] · Alberto Fernández-Ros[1] · Cristina Torrecillas[1] · José Manuel Enríquez de Salamanca[1] · M. Eva Ramírez[1] Alejandro Pérez-Peña[1] · M. José González[1] · Raúl Páez[1] · Yolanda Jiménez[1] · Alicia García-García[2] · Marta Tárraga[2] Francisco García-García[3]

[1] Laboratorio de Astronomía y Geodesia, Facultad de Ciencias, Universidad de Cádiz, Campus Río San Pedro, 11510 Puerto Real, Cádiz, Spain, <manuel.berrocoso@uca.es>

[2] Departamento de Volcanología, Museo Nacional de Ciencias Naturales, C/José Gutiérrez Abascal 2, 28006, Madrid, Spain

[3] Departamento de Ingeniería Cartográfica, Geodesia y Fotogrametría, E.T.S.I. Geodésica, Cartográfica y Topográfica, Universidad Politécnica de Valencia, Camino de Vera, s/n, 46022 Valencia, Spain

Abstract. Deception Island (62.93° S, 60.57° W) is one of the few active volcanoes in the Antarctica, whose most recent eruptions took place in 1842, 1967, 1969 and 1970. In the following paper geodetic investigations carried out in this area during the last years are described. During the continuous Spanish campaigns in Antarctica, several scientific groups have developed different projects in order to control deformation the island suffers as a result of its volcanic activity. With this purpose, a geodetic network has been designed and improved. Nowadays, the network consists of twelve stations around Port Foster which are provided with WGS-84 geodetic coordinates with respect to the ITRF2000, and another station at the Spanish Base Juan Carlos I on Livingston Island. Time analysis of these coordinates will lead us to get the horizontal deformation model. On the other hand, a levelling network has been designed to obtain the vertical deformation model. This network is denser in those areas where the volcanic activity is stronger, as at Fumarole Bay and the Hill of Obsidians. GPS, levelling and gravimetric measurements have also been collected in secondary points to obtain an experimental geoid model which makes possible an adequate reference frame for physical applications.

Deception Island: Situation and Characteristics

The South Shetland Islands and the Antarctic Peninsula are two of the few active areas of volcanic and seismic activity of Antarctica. Deception Island is situated at the spreading centre of the Bransfield Strait back-arc on the south western side of the Scotia Sea region (Fig. 7.7-1). It is a young stratovolcano 25 km large in its submerged basal diameter and 15 km of diameter in its emerged part (Ortiz and Vila 1992).

From 1987–1988 on, several Spanish groups have carried out different research projects to study the volcanic activity on the island and its geodynamic consequences. After increase of seismic activity detected during 1998–1999 austral summer, with more than 2000 earthquakes in two months, a cross-disciplinary study was carried out to evaluate the volcanological status of the island (García et al. 2002).

This study was part of the DECVOL project, which took place in December 1999. The main activities carried out were a reobservation of the geodetic network; the seismic register; the analysis of fumarolic gasses; the acquisition of geomagnetic and gravimetric data, and thermometric measurements in hot areas. The volcanic events were modelled and monitored by means of geodetic and geophysical techniques.

The main goals of the geodetic researches which have been done on the island are the establishment of a local reference frame to get an accurate position of scientific data, the determination of a deformation model for the island and the detection and monitoring of the volcanic activity (Berrocoso et al. 2002). In order to achieve these objectives a geodetic network (REGID) and a levelling network (RENID) have been designed and established, and the determination of a local geoid as well as the elaboration of a Multidisciplinary System for Scientific Support (SIMAC) have been carried out.

Regid Network from Its Beginning until GEODEC Project

In the 1987/1988 campaign, the geodetic network consisted of five stations, four of them on Deception Island and one more at the Spanish Base Juan Carlos I on Liv-

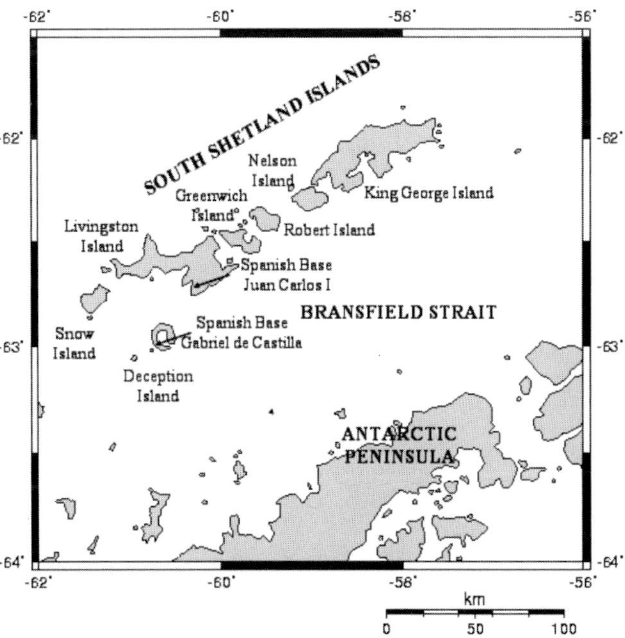

Fig. 7.7-1. Geographic situation of Deception Island

From: Fütterer DK, Damaske D, Kleinschmidt G, Miller H, Tessensohn F (eds) (2006) Antarctica: Contributions to global earth sciences. Springer-Verlag, Berlin Heidelberg New York, pp 391–396

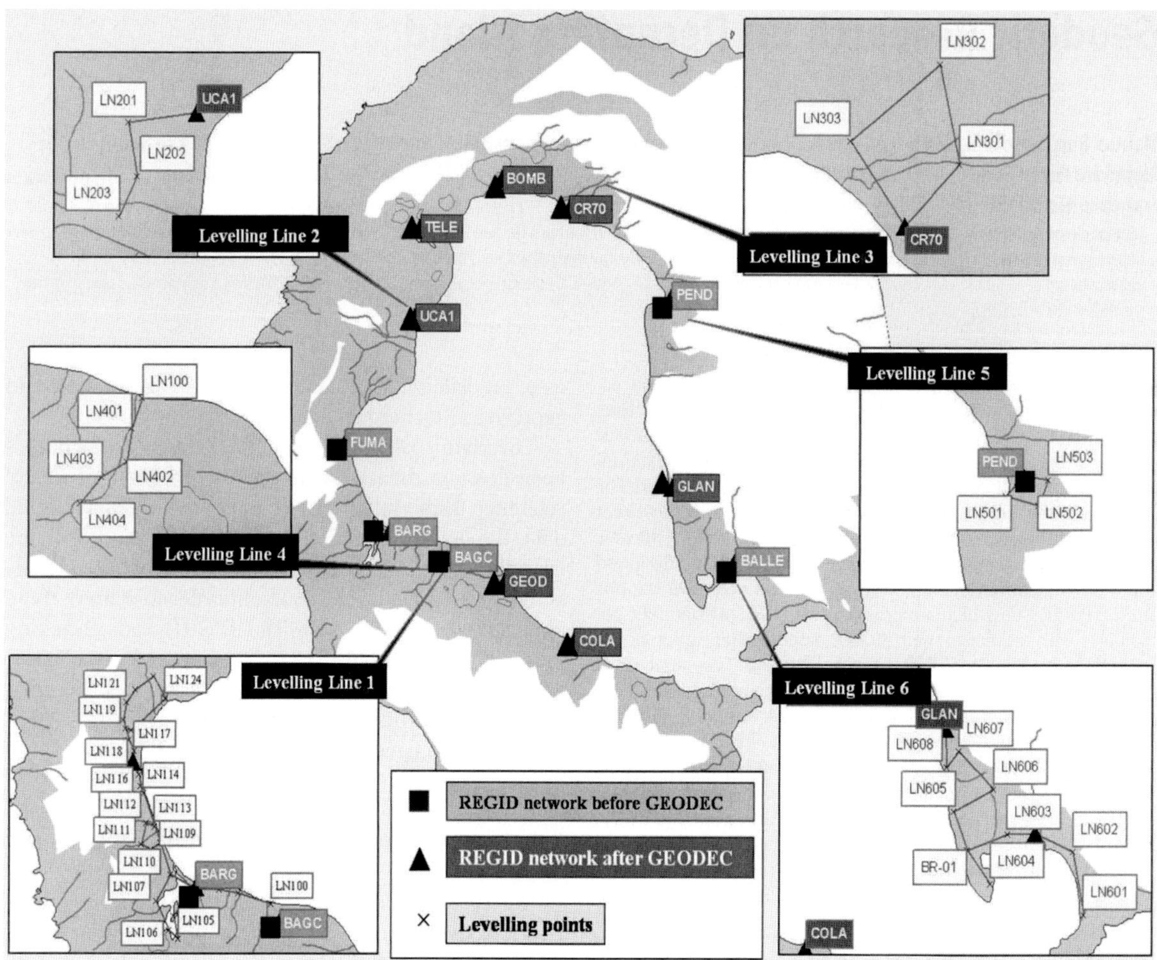

Fig. 7.7-2. REGID Geodetic Network and RENID levelling network on Deception Island

Table 7.7-1. Coordinates of the REGID network stations

Station name	WGS-84		
	Latitude South	Longitude West	Ellipsodical height (m)
BEGC	62°58'43"658	60°40'27"534	82.06
FUMA	62°57'41"022	60°42'59"3496	22.99
BARG	62°58'30"27	60°41'53"3652	22.35
PEND	62°56'09"8592	60°35'34"3752	28.85
BALL	62°58'38"5608	60°33'52'5276	26.08
COLA	62°59'27"9999	60°37'32"0000	48.00
GLAN	62°57'58"3599	60°35'23"8499	27.50
GEOD	62°58'56"4099	60°39'11"729	42.15
UCA1	62°56'28"4100	60°41'28"0899	28.65
CR70	62°55'23"6700	60°38'00"9999	23.63
TELE	62°55'37"9899	60°41'25"5500	23.79
BOMB	62°55'08"4200	60°39'33"8500	23.82
BEJC	62°39'46"492	60°23'20"005	31.49

ingston Island. The stations On Deception Island were placed at the Argentine Base (BARG), at Pendulum Cove (PEND), near the Chilean Base, at Fumarole Bay (FUMA) and at Whalers' Bay (BALL). Some of these vertices were built using existing structures, like the ones in Pendulum Cove and in Whalers' Bay. The main point of the network was at the Argentine Base (BARG). During the 1987/1988 campaign the first GPS observations were acquired on the island. In the 1995/1996 campaign a new station was built near the Spanish base Gabriel de Castilla. This station (BEGC) was set as the main station of the network with the aim of substituting the previous one, sited at the Argentine Base, with a station located in a less active area. Thus, in 1995 the network was formed by five stations on Deception Island (Fig. 7.7-2) and another one on Livingston Island. However, one of the main deficiencies of the network was the lack of stations in the North-South area of the island.

In order to continue the study and control of the geodynamic activity on Deception Island a project called GEODEC was purposed, formed by three subprojects, GEODEC-GEODESIA, GEODEC-MODELOS AND GEODEC-GEOMAR. During the 2001/2002 campaign seven more stations, provided with WGS-84 coordinates, were built: COLA, GEODEC, UCA1, TELE, BOMB, CR70, and GLAN.

The new geodetic marks were well-anchored in the permafrost, with steel bars to fix them and a low height from the ground, making them more stable and involving a poor environmental impact. A standard screw was fixed in one of the corners of the post to allow the use of geodetic instrumentation.

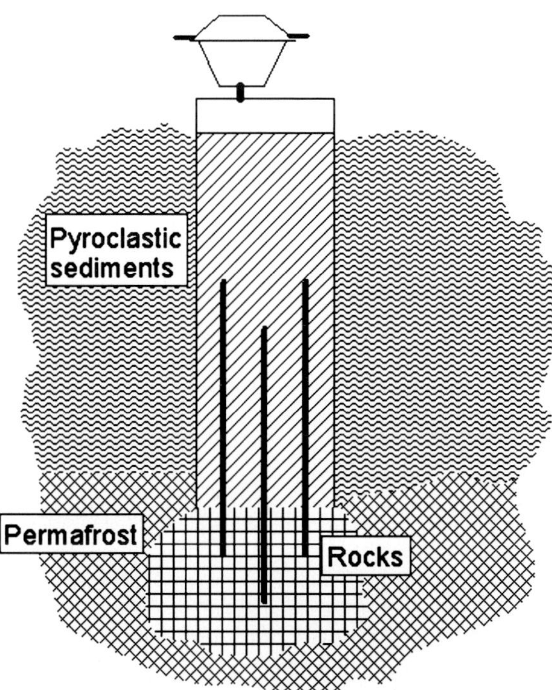

Fig. 7.7-3. Mounting of GPS stations in permafrost ground

The most commonly used instrumentation for the GPS surveying is constituted by TRIMBLE receivers, from its series 4000 and 5000, although ASHTECH products have also been used. Data were processed with the GPSurvey software in a first stage although they have been reprocessed using the BERNESE software, with an accuracy in the order of mm. Preliminary deformation models are presented in Fig. 7.7-4. The increase in the volcanic activity during the 1998/1999 campaign can be interpreted from the deformation model (Fig. 7.7-5) and its comparison to the deformation model of the 1991/1992 to 1995/1996 campaign (Fig. 7.7-4).

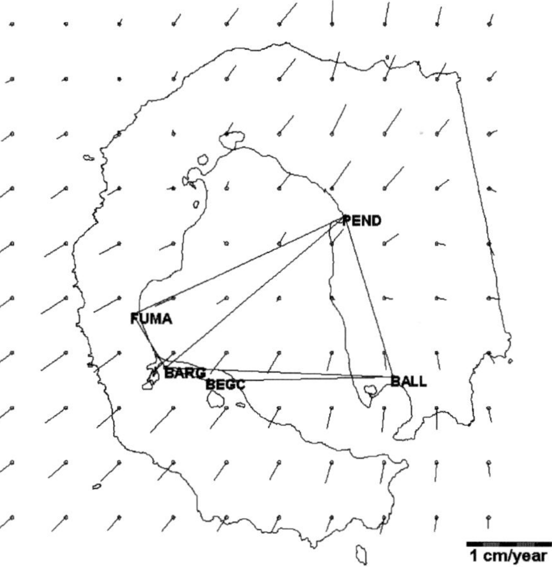

Fig. 7.7-4. Preliminary deformation models of the horizontal displacement on Deception Island between 1991/1992 and 1995/1996 campaigns

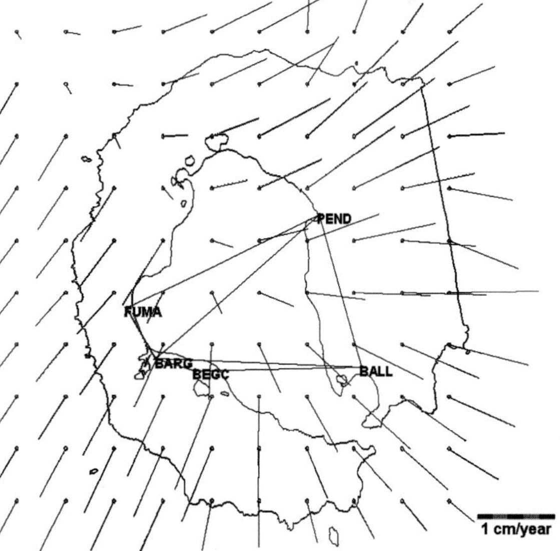

Fig. 7.7-5. Preliminary deformation models of the horizontal displacement in Deception Island between 1995/1996 and 1999/2000 campaigns

The Levelling Network RENID

One of the levelling activities carried out during the 2001/2002 campaign was the establishment of the main bench mark LN000, substituting the BARG bench mark as the main point of the levelling network.

The zero of the vertical datum is set at LN000 benchmark. It was measured with respect to the BARG bench mark, whose height with respect to the sea level was previously attained. The height of LN000 using a tide gauge will be done in a near future. The levelling network consists of six levelling lines distributed around the island. Points of every line are connected to one of the vertices of the geodetic network, except those in line 4, which are connected to the levelling benchmark LN000. This division in six independent lines makes possible the observation of just one of the lines, that one close to where the volcanic activity is concentrated. Every bench mark in every line was positioned using a differential GPS. Precision levelling of the whole network was carried out during the 2001/2002 campaign by means of a Leica NA2. The connections among lines were done during the 2002/2003 campaign, using a Wild T2000 theodolite and a DI5000 range finder. Lines 1, 2, 3 and 5 were connected among them, and levelling line 6 was linked to the geodetic station COLA, also connected to the LN000 bench in line 1 along the coastline. The connection between line 4 and line 1 was done during the last stage of that campaign.

Determination of Local Geoid

For the determination of the geoid undulation it was necessary to determine the ellipsodical height h and the orthometric height H of several points. The ellipsodical heights are obtained from the GPS observations and the second ones from absolute gravimetric measurements and levelling heights. The considered points for the calculus of a local geoid are the stations of the geodetic network, the bench marks of the levelling network and some other specific points (Fig. 7.7-5).

During the 2001/2002 campaign the fundamental gravimetric point in Deception Island was built in the surroundings of the Spanish base Gabriel de Castilla. During the 2002/2003 campaign, the gravimetric measurements were made using a Lacoste-Romberg gravimeter. The reference absolute point was at Punta Arenas (Chile), the control point on the island is at the Spanish base Juan Carlos I.

Design of a Geographic Information System

Since the Spanish Antarctic campaigns began on Deception Island in 1987 a large amount of cartographic data and graphic output on numerous projections has been obtained referenced to several geodetic systems. There is now a need to compile these multidisciplinary data and integrate them into a geographic information system. This would make the existing information about Deception Island more accessible to the scientific world, avoiding the duplication of information and enhancing the collaboration among the different research groups through the exchange of data and the transference of results. The main parts of the SIMAC are the following: (1) a metadata catalogue to store and distribute information about geospatial resources, grouping specific information of a data set so that users can look for wished information; (2) a geodatabase, which is an information system to work with georeferenced data, and (3) a bibliography.

Fig. 7.7-6.
Situation of the main gravimetric point GBEGC in the surroundings of the Spanish base Gabriel Castilla

Preliminary Results

Throughout the last campaigns the following objectives have been achieved:

- The geodetic network REGID was established and continuously improved. The processing of resulting data has given rise to the establishment of preliminary horizontal deformation models (Fig. 7.7-4).
- The design and establishment of a levelling network RENID as a reference to detect vertical displacements in case of a volcanic crisis.
- The determination of an experimental local geoid (Fig. 7.7-7) was obtained from levelling and gravimetric measurements (Berrocoso et al. 2004).
- In relation to the elaboration of the SIMAC, a digital terrain model has been produced, the design of the basic structure, the georeference of the data and the input of cartographic and geological information, among others. The elaboration of this Geographic Information System is in progress now.

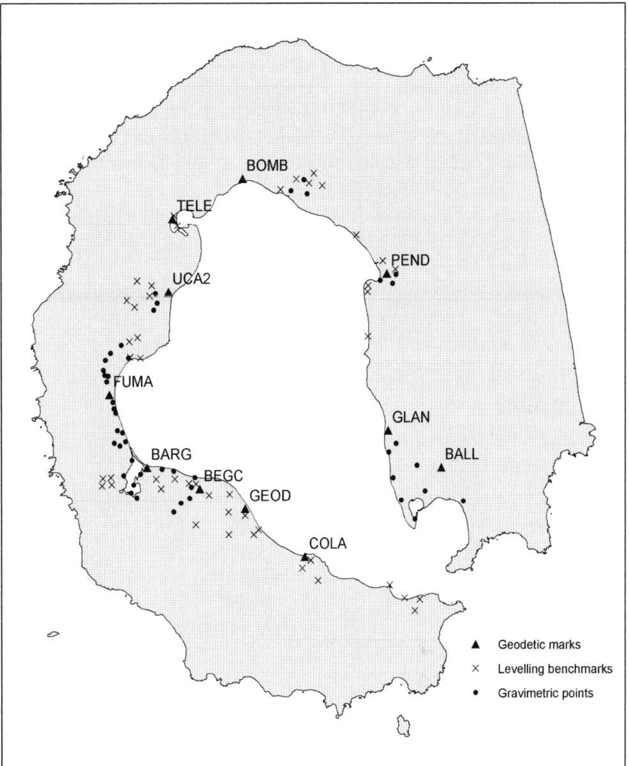

Fig. 7.7-7. Location of the gravimetric measurements

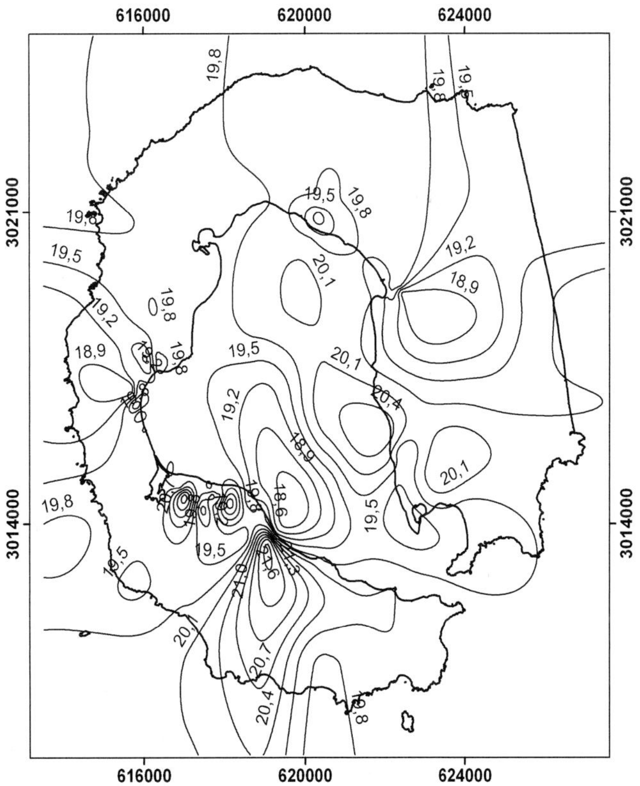

Fig. 7.7-8. Experimental geoid for Deception Island

Acknowledgment

The realization of this work was possible thanks to the subproject "Geodetic Studies on Deception Island: deformation models, geoid determination and SIMAC"-REN 2000-0551-C03-01, as part of GEODEC Project (Geodynamic Activity on Deception Island), funded by the Ministry of Science and Technology of Spain through the National Program of Antarctic Research of Natural Resources. We would also like to thank the "Las Palmas" ship crew and the members of the Spanish base Gabriel de Castilla for their collaboration during the last Antarctic campaigns.

References

Berrocoso M, Martín-Dávila J, Astiz M, Catalán-Morollón M, García A (2002) El proyecto GEODEC: un estudio multidisciplinar de la actividad geodinámica de la Isla Decepción (Islas Shetalnd del Sur, Antártida). Proc 3ª Asamblea Luso-española de Geodesia y Geofisica 2:766–769

Berrocoso M, Enríquez de Salamanca J M, Jiménez Y, Fernández-Ros A, Torrecillas C, Ramírez ME, González Fuentes MJ, Pérez-Peña A, Páez R, Tárraga M, García García A, García García F, Soto R (2004) Determination of an experimental geoid for Deception Island (Antarctica). IAG Internat Sympos GGSM 2004, Porto

Caselli AM, Dos Santos A, Risso C, García A, Ortiz R (2002) Caracterización geoquímica de los gases volcánicos de Isla Decepción (Shetland del Sur, Antártida). Revista de la Asociación Geológica Argentina 57(1)

García A, DECVOL Working Group (submittted) A cross-disciplinary study at Deception Island (South Shetland Islands, Antarctica). Evaluation of the recent volcanological status. J Volcanol Geotherm Res

Ibáñez J, del Pezzo E, Almendros J, La Rocca M, Alguacil G, Ortiz R, García A (2000) Seismovolcanic signals at Deception Island volcano, Antarctica: wave field analysis and source modelling. J Gephys Res 105:13905–13931

Martini M, Giannini L (1988) Deception Island (South Shetland Island): an area of active volcanism in Antarctica. Mem Soc Geol Italia 43:117–122

Ortiz R, Vila J, García A, Camacho AG, Diez JL, Aparicio A, Soto R, Viramonte JG, Risso C, Menegatti N, Petrinovic I (1992) Geophysical Features of Deception Island. In: Yoshida Y, Kåminuma K, Shiraishi K (eds) Recent progress in Antarctic earth science, Terra Scientific Publishing, Tokyo, pp 443–448

Ortiz R, García A, Aparicio A, Blanco I, Felpeto A, Del Rey R, Villegas MT, Ibañez J, Morales J, Del Pezzo E, Olmedillas JC, Astiz M, Vila J, Ramos M, Viramonte JG, del Risso C, Caselli A (1997) Monitoring of the volcanic activity of Deception Island, South Shetland Islands, Antarctica (1986–1995). In: Ricci CA (ed) The Antarctic region: geological evolution and processes. Terra Antartica Publication, Siena, pp 1071–1076

The Multidisciplinary Scientific Information Support System (SIMAC) for Deception Island

Cristina Torrecillas[1] · Manuel Berrocoso[1] · Alicia García-García[2]

[1] Laboratorio de Astronomía y Geodesia, Departamento de Matemáticas, Facultad de Ciencias, Universidad de Cádiz, Campus Río San Pedro, 11510 Puerto Real, Cádiz, Spain, <cristina.torrecillas@uca.es>, <manuel.berrocoso@uca.es>

[2] Departamento de Volcanología, Museo Nacional de Ciencias Naturales, C/José Gutiérrez Abascal 2, 28006 Madrid, Spain

Abstract. Spain has been taking part in Antarctic research through annual austral summer campaigns since 1987. The volcanic Deception Island in the South Shetland Islands is one of the main working areas. The information collected is numerous and diverse so in many cases is the same one. Also, this information is stored in several alphanumeric and graphic digital formats, and the maps are made in different geodetic and cartographic representation systems. The fundamental data are unified and integrated in SIMAC, a unique information system that can to be used as a "gate" through which the various scientific groups working on Deception Island may exchange data. SIMAC is an example of an Information System applied to Earth Sciences.

Background

The Spanish National Antarctic Investigation Programme sponsors numerous research groups on Deception Island, South Shetland Islands (Fig. 7.8-1). Deception Island is an active volcano (R. Ortiz et al. 1992, 1997), its most recent dynamic volcanic activities were in 1967, 1969 and 1970, when eruptions demolished the Chilean station in Pendulum Cove and the British station in Whalers Bay. In 1992, 1995 and 1998 several seismic crisis were detected, producing an increase in volcanic activity and in surface deformation. During the 1998/1999 season, a significant increase in seismic activity occurred due to a small young magma intrusion. This situation made it necessary to plan and carry out a campaign to evaluate the present conditions on the island and to establish the hazard level from a volcanological viewpoint (Ibañez et al. 2002).

This multidisciplinary campaign required the collaboration of different groups investigating applied geodetic, geochemical and geophysical techniques such as Global Positioning System (GPS) satellite measurements, short period and wide band seismic stations, measurements of temperature and fumarolic gas. Also geomagnetic, gravimetric, and bathymetric data were acquired by systematic sweep of the island and of its internal and surrounding waters (García et al. 2000).

Apart from volcano monitoring, the environmental characteristics of the island make it a natural laboratory,

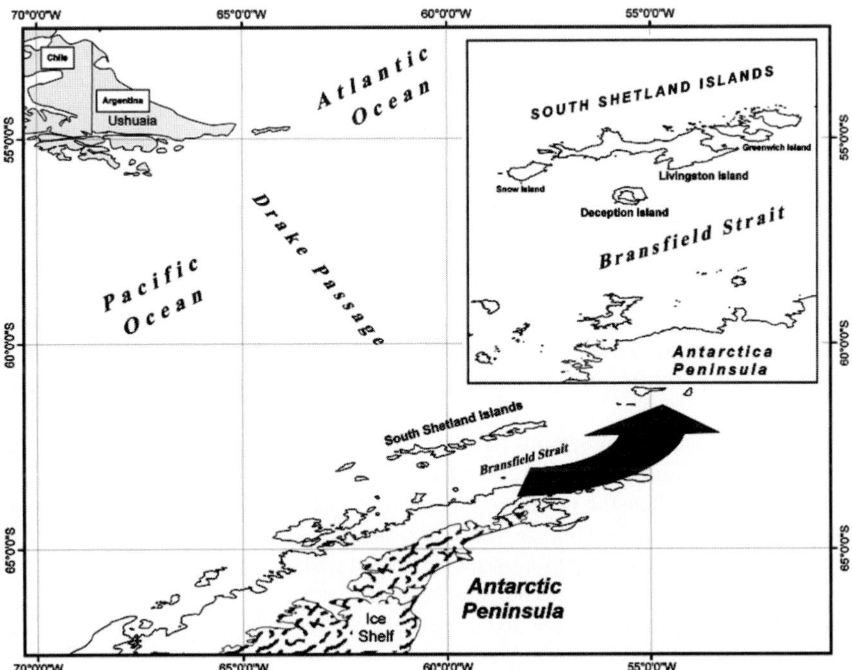

Fig. 7.8-1. Deception Island location map

From: Fütterer DK, Damaske D, Kleinschmidt G, Miller H, Tessensohn F (eds) (2006) Antarctica: Contributions to global earth sciences. Springer-Verlag, Berlin Heidelberg New York, pp 397–402

Fig. 7.8-2.
Diagram of SIMAC process

so several biologic, botanic, limnologic, meteorological or glaciological projects were also developed.

Until 1994, the collected geospatial data were represented above the only reference available: The British Antarctic Survey cartography, of 1:250 000 scale and Lambert conformal conic projection. In 1994, the Geographical Service of the Spanish Army published the Deception Island Topographic Map, with a 1:25 000 scale and UTM projection (SGE 1994) referenced to the 1984 World Geodetic System (WGS84). With this cartography the representation of results improved, but the next disappointment was the non-existence of an unique digital cartography on the island. Finally, every research team collected information according to their data models, so that the same element could be described in several graphic ways (point, line or polygon), with different associated information and several draw scale and precision (mm, m). Now, there is a need to compile these multidisciplinary data and integrate them into a unique information system. This would make the information on Deception Island more accessible to the scientific world, avoid duplicating information and enhance collaboration among research workgroups with multidisciplinary projects through the exchange of data and transfer of results.

The Multidisciplinary Scientific Information Support System (SIMAC) for Deception Island

Geographical Information Systems (GIS) are Information Systems designed for working with georeferenced data. GIS are of great value where the variable or attribute has great geographical influence. Among the advantages offered by GIS, it is an environment for the analysis, creation and study of topological relations, as well as comparative spatial, proximity or adjacency studies (Davis 2001). All these properties and the diversity of the research performed in Deception Island makes GIS the appropriate support device for storing all the scientific information collected throughout the Spanish Antarctic campaigns and it provides a framework for a new multidisciplinary information system.

The Multidisciplinary Scientific Information Support System (SIMAC) is an information system whose main purpose is to handle all the scientific information that exists in a local area under study, in order to facilitate access, data processing and analysis of the scientific objective (see Fig. 7.8-2). The SIMAC information is diverse so its data model is opened to new features don't enclosed in the first version.

SIMAC operates on the basis of several methods to obtain information about the area being studied. First, the SIMAC integrates a metadata catalogue containing and supplying a description of resources. With this catalogue it is possible to consult all the data that exists in SIMAC about Deception Island, whether in graphic or alphanumeric form, as maps or as papers. Secondly, the SIMAC incorporates a geodatabase comprising a graphic dataset with associated information according to the usual GIS structure. This geodatabase constitutes the core of the SIMAC, creating a digital cartographic base, which includes all the existing information (raster and vectorial) and the necessary data to analyze any partial or full scientific aspect of the area under study. Finally, the SIMAC includes a georeferenced bibliography of Deception Island. In addition to this, SIMAC includes the most often highly demanded basic data (mainly topographic, geomorphologic and geological information), establishes a data model and contributes to the spread of information and co-operation among different scientific groups with Deception Island projects.

The SIMAC for Deception Island is developed using ArcGIS[1] 8.2 software.

[1] "ArcGIS" is a trademark of Environmental System Research Institute, INC.

The Metadata Catalogue

A metadata catalogue is a system prepared to store and distribute information about geospatial resources and to organize specific information into a set of data.

The objective of the catalogue is to allow to discover information resources from different sources. It simplifies access to the information and use of the geospatial service program and it supplies the tools to find and obtain data that may be required by future users (Nebert 2001). In preparing the metadata catalogue, a series fields to describe the dataset are defined and completed. These fields provide the information on certain geospatial data such as lineage, precision, spatial extension, dates, or the contact person.

The use of metadata standards for geographic information was considered in order to facilitate input in a future geospatial data structure or geospatial service. In the beginning, the Federal Geographic Data Committee metadata (FGDC 1998) were used but now, the SIMAC metadata catalogue uses the ISO/TC211 19115 international standards (Fig. 7.8-3). This catalogue provides a scientific thesaurus, key words, in Spanish and English to improve efficiency in the search for information on general or specific aspects.

The main problems encountered in preparing the metadata catalogue for Deception Island were the lack of documentation, the difficulty in locating existing data, the diversity of formats and structures of the available information, descriptions and quality information indicators, or the fact that updates were not dated. Neither do we always know the date when data were collected or digitalized. Almost all these problems can be solved by contacting directly with the different research groups on the Deception Island.

The SIMAC Geodatabase

The SIMAC geodatabase is a Geographic Information System. To develop the geodatabase, a data model was designed for different sorts of information. This data model ensures the incorporation into the geospatial data structure without redundancy, avoiding premature data introduction. The organizational structure of the data model is prepared for different users, bearing in mind diverse areas of knowledge and the multidisciplinary character of SIMAC. Also, our method of reality modelling was to classify by subject and common spatial extent, ignoring topologic association, since it would be too complex and confusing for the final results. The final classification is made hierarchically according to the division of information into areas and sub-areas contained in the UNESCO list of knowledge (Table 7.8-1).

Also, the Scientific Committee on Antarctic Research supports the Antarctic Digital Database (ADD Consortium 2000), defined over Polar Stereographic projection on WGS84, integrating a substantial amount of digitised cartography from several countries. In the ADD the only reference to Deception Island is the digitised BAS cartography. This geodatabase is not targeting at scales below about 1:100 000 but SIMAC incorporates an item named ADD_CODE as a link to the ADD data model in all-common topics, to favour compatibility between systems.

The definition of entities is classified considering unit elements, and data relationships are described using Unified Modelling Language diagrams (Fig. 7.8-4). The method for adding input to the SIMAC uses the scan process, vectorisation or introduction by an ASCII file. At this point, each element is assigned to the following digital formats: raster, vectorial or alphanumeric. At last, we define attributes (Table 7.8-2, Fig. 7.8-4), associated information or topological rules (MacDonald 2001). This stage

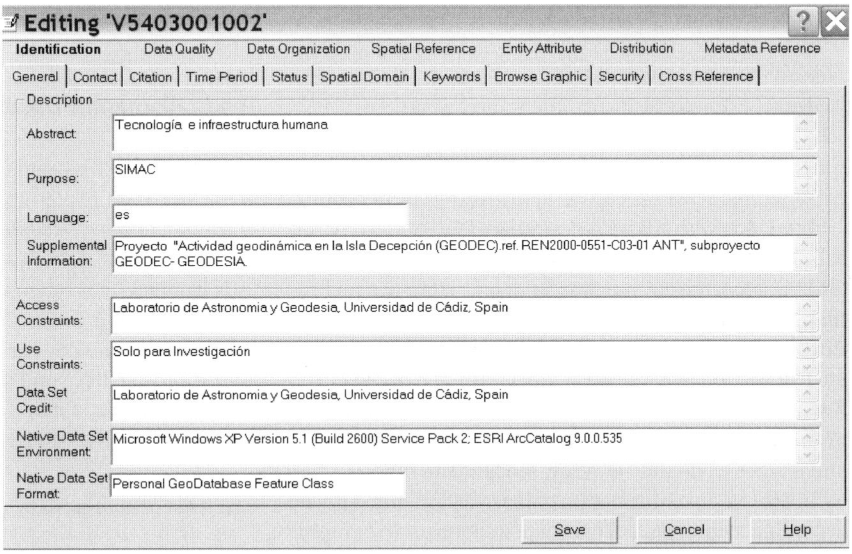

Fig. 7.8-3. An example of metadata editing in ARCGIS

Table 7.8-1. UNESCO List of knowledge selected for SIMAC in first version

UNESCO Code	Description
2400	Life sciences
2401	Zoology
2417	Botany
2500	Earth and space sciences
2502	Climatology
2503	Geochemistry
2503-99	Volcanic geochemistry
2504	Geodesy
2504-02	Geodetic cartography
2504-05	Surveying
2504-06	Physical geodesy
2504-07	Satellite geodesy
2504-90	Geodetic networks and deformation
2505	Geography
2505-07	Physical geography
5402	Historical geography
5403	Human geography
2506	Geology
2506-21	Volcanology
2506-13	Igneous and metamorphic petrology
2506-14	Sedimentary petrology
2506-07	Geomorphology
2506-20	Structural geology
2507	Geophysics
2507-07	Tectonics
2507-05	Seismology and seismic prospecting
2507-01	Geomagnetism and geomagnetic prospecting
2507-02	Gravimetry and gravimetric prospecting
2508	Hydrology
2508-03	Glaciology
2508-06	Hydrography
2510	Oceanography

concludes by implementing the metadata and defining a cartographic scale for each graphic information group.

The Deception Island Topographic Map of Geographical Service of the Spanish Army is the skeleton of the SIMAC geodatabase, but was updated using a panchromatic satellite image from the recently launched Quick Bird satellite. In addition, the geodetic network has been enlarged to improve the georefenciation of new data from five to twelve geodetic vertices, a new geodetic levelling network with nearly a hundred points (Berrocoso 1997) and a 3-meter digital elevation model were built.

In SIMAC, the information gathered was homogenised according to its digital format, data structure, cartographic

Table 7.8-2. Example of some attributes of seismic stations for SIMAC

Item name	Alias	Technical description	Value
Id_esta	Array or station name	A,40	
Id_sensor	Sensor name	A,40	
X	X UTM	F,7.3	Meters
Y	Y UTM	F,7.3	Meters
h	Ellipsoidal height	F,7.3	Meters
Error_HZ	XY accuracy	F,5.3	Meters
Error_V	Z accuracy	F,5.3	Meters
Component	No. components	A,2	1C 3C
Tipo	Station type	A,3	ARR EST OBS
Tipo_sens	Sensor	A,4	BA CP
Modo	Recording mode	A,1	C D
F_ini	Sampling initial frequency	F,5.3	Hz
F_fin	Sampling final frequency	F,5.3	Hz
Fecha_ini	Initial date	I,10	dd/mm/yy
Fecha_fin	Final date	I,10	dd/mm/yy
Proyecto	Investigation project	A,150	
Responsabl	Responsible autority	A,150	
ADD_CODE	ADD code	I,5	
Observacio	Comments	A,150	

projections and datum, after passing a cartographic quality control that determined a scale interval adapted to the accuracy of the data.

The Bibliographic Database

The third SIMAC component is a set of georeferenced bibliographic data related to the studied area. The location/ place field adds a spatial search to the traditional alphanumeric search. A better bibliographic search engine is thus attained. This bibliographic database is implemented with fast-answer forms designed in Visual Basic for Applications in ArcGIS. For creation of the referenced bibliographic database on Deception Island and its environment, MS Access[2] software was chosen because of its versatility and ArcGIS compatibility.

[2] "Access" is a trademark of Microsoft.

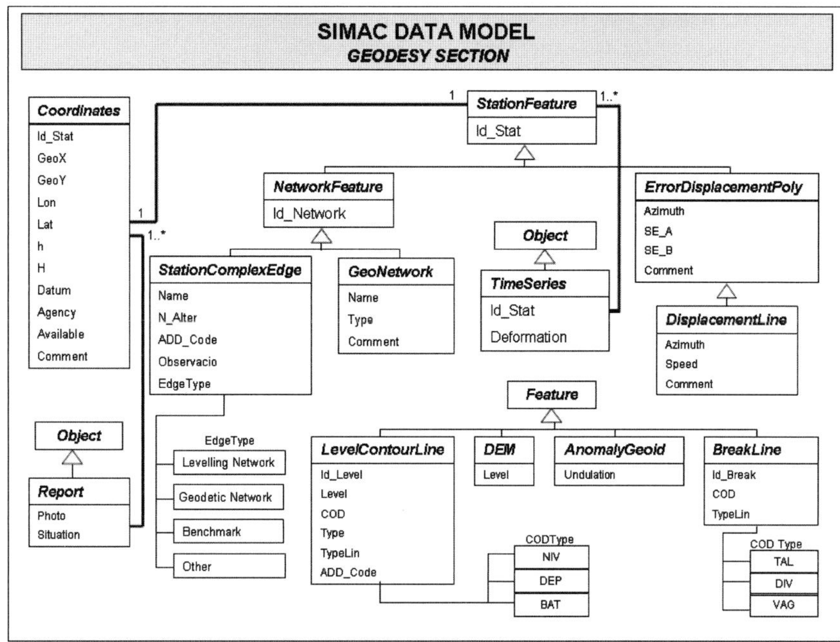

Fig. 7.8-4.
UML graph of the Geodesy section of SIMAC

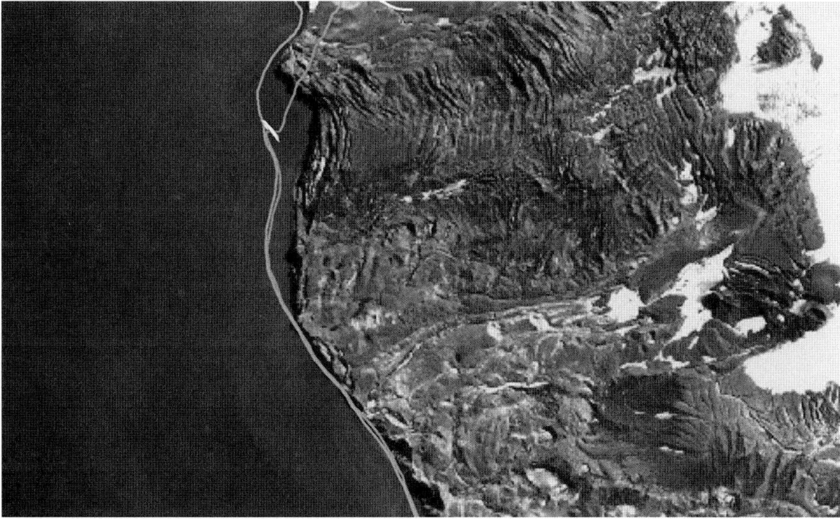

Fig. 7.8-5.
Studying the glacier backward movement

Diffusion and Programme Updates

To bring SIMAC information down to the final user and to his software workspace, several output formats (DXF, E00 and SHP) will be established for vector data as well as for raster (TIFF) and database information (ASCII and Access). The SIMAC Manual gives the appropriate indications to incorporate their data as soon as possible. Meanwhile, the metadata catalogue will be transmitted over the Internet through an on-line service will be available the next summer with a map server.

It is simultaneously updated and maintained so the SIMAC remains current. The collaboration of all the Deception Island research groups is essential for this duty so SIMAC is free. SIMAC diffusion, updates and maintenance will be performed in the Astronomy and Geodesy Laboratory, Department of Mathematics, Faculty of Science (Cádiz University, Spain).

Conclusions

SIMAC is a new scientific Information System with GIS core. Its development has required new information as a digital cartographic base, a 3-meter digital elevation model, the enlargement of the existing geodetic network or the construction of a new geodetic levelling network with nearly a hundred points. The first applications are in progress. A new experimental geoid of the island or a new volcanic risk map are some of them, although several studies are possible just using the SIMAC information such as the studying of the glacier front backward movement from the new QuickBird Image and the older geomorphological information, included in SIMAC both (Fig. 7.8-5).

Acknowledgments

This work has been made possible under the subproject Geodetic Studies On Deception Island: Deformation Models, Geoid Determination And SIMAC (Geodec-Geodesy), referenced as REN 2000-0551-c03-01. It is included in the project Geodynamic Activity On Deception Island (Geodec), financed by the Science and Technology Ministry of Spain through the Antarctic Research sub-programme of the National Natural Resources Programme. We are also grateful for the collaboration and assistance offered during the 2001/2003 seasons by the staff of the "Gabriel de Castilla" Spanish Antarctic base and of the Spanish RV "Las Palmas".

References

ADD consortium (2000) Antarctic Digital Database, Version 3.0. Database, manual and bibliography. Scientific Committee of Antarctic Research, Cambridge, http://www.nerc-bas.ac.uk/public/magic/add_home.html

Berrocoso M (1997) Modelos y formalismos para el tratamiento de observaciones GPS. Aplicación al establecimiento de redes geodésicas y geodinámicas en la Antártida. Spanish Department of Defence, Royal Observatory of the Spanish Army, ROA Bulletin n°1/97

Davis BE (2001) GIS: A visual approach. Onword Press, New York

FGDC Federal Geographic Data Committee (1998) Metadata ad hoc working group. Document FGDC-STD-001-198

García A, Martín Dávila J, Ibañez J, Berrocoso M, Ortiz R (2000) Informe campaña DECVOL. Actividad Científica en la Antártida Española. Spanish National Antarctic Investigation Programme

Ibáñez J, Carmona E, Almendros J, Saccorotti G, Del Pezzo E, Abril M, Ortiz R (2002) The 1998–1999 seismic series at Deception Island volcano, Antarctica. J Vulcanol Geotherm Res 128:65-68

MacDonald A (2001) Building a geodatabase. ESRI Press, Redlands

Nebert DD (2001) The SDI Cookbook. Technical working group chair, GSDI, http://www.nsif.org.za/news/cookbook.pdf

Ortiz R, Vila J, García A, Camacho AG, Diez JL, Aparicio A, Soto R, Viramonte JG, Risso C, Menegatti N, Petrinovic I (1992) Geophysical features of Deception Island. In: Yoshida Y, Kaminuma K, Shiraishi K (eds) Recent progress in Antarctic earth science. Terra Scientific Publishing, Tokyo, pp 443–448

Ortiz R, García A, Aparicio A, Blanco I, Felpeto A, Del Rey R, Villegas MT, Ibanez JM, Morales J, Del Pezzo E, Olmedillas JO, Astiz M, Vila J, Ramos M, Viramonte JG, Risso C, Caselli A (1997) Monitoring of the volcanic activity of Deception Island, South Shetland Islands, Antarctica (1986–1995). In: Ricci CA (ed) The Antarctic region: geological evolution and processes. Terra Antartica Publication, Siena, pp 1071–1076

SGE Servicio Geográfico del Ejercito (1994) Mapa topográfico de la isla Decepción 1:25000, Proyección UTM

UNESCO. http://www.mec.es/consejou/unesco

Archiving and Distributing Earth-Science Data with the PANGAEA Information System

Hannes Grobe[1] · Michael Diepenbroek[2] · Nicolas Dittert[2] · Manfred Reinke[1] · Rainer Sieger[1]

[1] Alfred Wegener Institute for Polar and Marine Research, Am Alten Hafen 26, 27568 Bremerhaven, Germany, <hgrobe@awi-bremerhaven.de>

[2] Center for Marine Environmental Sciences, Leobener Str. 26, 28359 Bremen, Germany

Abstract. PANGAEA Publishing Network for Geoscientific and Environmental Data (http://www.pangaea.de) is an information system aimed at archiving, publishing, and distributing data related to climate variability, the marine environment, and the solid earth. The system is a public "data library" distributing any kind of data to the scientific community through the Internet. Data are stored in a relational database in a consistent format with related meta-information following international standards. Data are georeferenced in space and/ or time, individually configured subsets may be extracted. Any type of information, data and documents may be served (profiles, maps, photos, graphics, text and numbers). Operation by Alfred Wegener Institute for Polar and Marine Research (AWI) and Center for Marine Environmental Sciences (MARUM) is assured in the long-term. Both institutions provide the technical infrastructure, system management and support for data management of projects as well as for individual scientists. Most important collections from Antarctic research archived in PANGAEA so far are the data of the Cape Roberts Project, geological maps and age determinations of rock outcrops, a complete set of JGOFS, WOCE, DSDP and ODP data including those from the Southern Ocean, any marine sediment cores, documentation and analytical data from German expeditions and an increasing inventory of data published by the running EPICA project.

Introduction

In 1990 SEPAN, the predecessor of the PANGAEA information system, was established at AWI for the administration of the geological samples archived in the "Polarstern" Core Repository. The system was reorganized to archive data from paleoclimate research in 1994. After the inclusion of a full spatial and temporal georeference for each single value, the system was able to handle any kind of geodata. Since 1998 the system has been accessible on the Internet via the domain www.pangaea.de. During the last six years, PANGAEA has been used by 23 major projects and many individual scientists for the archiving of scientific primary data (Diepenbroek et al. 1998, 2002). By the end of 2005, the system had an inventory of about 250 000 data sets. Major international projects served with PANGAEA are IMAGES (International Marine Global Change Study) and several EU projects (see http://www.pangaea.de/Projects/). Of major importance for the Antarctic are CRP (Cape Roberts Project) and EPICA (European Project for Ice Coring in Antarctica). PANGAEA has also archived all marine geological data from samples and sediment cores taken by the research vessel "Polarstern" in the Southern Ocean as well as in the Arctic Ocean. The World Data Center for Marine Environmental Sciences (http://www.wdc-mare.org), which was founded in 2001 in Germany as a member of the ICSU World Data Center System (WDC 1996), is using PANGAEA as its central archiving system. PANGAEA operates the European web and publication mirror for the Ocean Drilling Program (ODP). The institutional framework for PANGAEA, including the World Data Center, is supplied by the Alfred Wegener Institute for Polar and Marine Research (AWI) in Bremerhaven and the Center for Marine Environmental Sciences (MARUM) at the University of Bremen. PANGAEA is structure and function unique to any other geoscientific database available to date on the Internet.

The Data Model

The challenge of managing any kind of georeferenced data was met in PANGAEA through a flexible data model. This was implemented by a combination of a simple fully normalized relational database frontended by middleware components and various clients for upload and download.

The model reflects the standard activities for data collection in the geosciences (Fig. 7.9-1). Collaborative activities in a Project carry out expeditions (Campaign) for sampling. During an expedition samples may be taken or measurements are made (Event) at a number of locations (Site). The medium to be investigated (e.g., sediment, rock, water or ice) is subsampled or measured for different analytical procedures (Sample). Finally from each sample or measurement analytical Data result, organized in Data Sets. These main levels are supplemented by related tables comprising information about items as personal, references, parameters or methods.

The essential part of the model is the combination of the "Data", "Parameter" and "Method" tables, which allows the definition and storage of new, unique parameters at any time. Up to a maximum of four different geocodes can be used simultaneously for the description of data points in space and time. These are selected from latitude, longitude, elevation, altitude, date/time, geologi-

From: Fütterer DK, Damaske D, Kleinschmidt G, Miller H, Tessensohn F (eds) (2006) Antarctica: Contributions to global earth sciences. Springer-Verlag, Berlin Heidelberg New York, pp 403–406

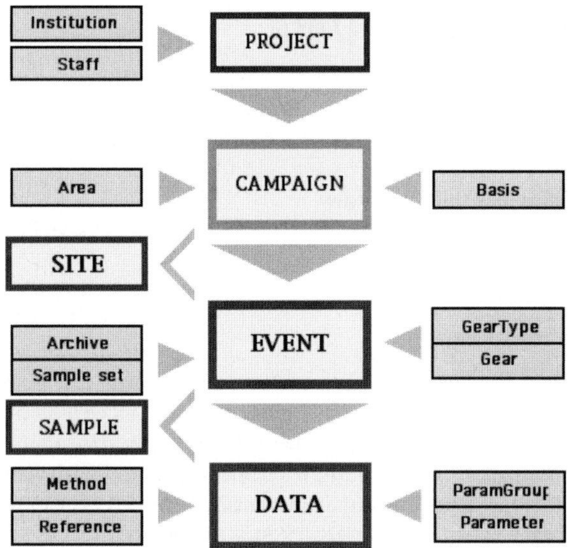

Fig. 7.9-1. The simplified data model of PANGAEA is used as the graphical user interface (GUI) for data import and mining on the Internet. Each *box* represents a table in the relational database. The user can use a uniform retrieval tool to find information in each of the tables

cal age, and depth in different media like water, sediment, rock or ice. Due to the complete georeferencing of each single value, the system allows the combination of any data types and the extraction of individually defined quantities of data. It is therefore a useful prerequisite for the analysis of complex data inventories or data mining (Han and Kamber 2001).

How to Find Data in PANGAEA

A number of clients allow the user to access metainformation and analytical data from the system on different levels of information and technical complexity. The data import client is the central management interface for the data curators and written in a proprietary software; all other clients for export are web-based.

PangaVista

This is a simple web-based search engine which allows the retrieval of predefined datasets in the relational system, referenced by web links. PangaVista makes use of a thesaurus, comprising all the metainformation related to the data, thus allowing a retrieval for any given keyword like an author's name, a parameter, a project or a sample label. Keywords may be combined to create bolean expressions, the syntax of which is similar to the one used by other search engines on the Internet. A map server allows the user to set geographical constraints and shows the locations of those datasets found by a retrieval. The result of a query is a list of short descriptions of the datasets found, with a link to the complete set at the end of each header. Data can be downloaded to a dynamically produced web page in html-format or as a tab-delimited text file. A login is required if data are unpublished. Data sets are typically composed of a number of data series, accompanied by a meta-data description conforming to the ISO19115 standard. The ability to download all data sets found by a retrieval in one step is given through an external program, a web based version is in preparation.

Advanced Retrieval Tool (ART)

This tool provides full access to all tables of the relational system and enables the user to retrieve individually configured subsets of data from the inventory. This provides functionality such as the ability to compare several paleoclimatic records from different archives versus time. The simplified data model is used as the graphical user interface (Fig. 7.9-1). ART is designed as a "data mining tool" to support the production, use and interpretation of comprehensive data collections. As a Java application it runs on any platform and with the most common web browsers. Users are advised to study the "Help" provided or to contact <info@pangaea.de> in case specific data mining requests are needed.

Direct Download Interface (DDI)

This tool provides the functionality to easily distribute and publish data. Each data set in the system has a unique identifier that may be obtained as a URL such as http://doi.pangaea.de/10.1594/PANGAEA.132796. This link can be used in publications as a precise reference to a data set. This technology was used for the first time in a publication edited by Fischer and Wefer (1999) where each publication refers to its primary data through a given link. Links can be defined for any query or retrieval on data as well as to metadata and may also be distributed via email or placed on web pages.

PanCore

This is a web-based interface to search for locations and metadata of sampling sites. Geographical constraints can be set within the included map, which is also used for the display of the resulting list of sites; any result set can be downloaded as a text file. For geological samples, the curator, responsible for the repository where the samples are archived, is given. Thus PanCore enables the user easily to search for samples in a certain area and submit a sample request to the appropriate curator.

4th-Dimension-Client (4D-Client)

This is the administrative tool for the processing and maintenance of any information stored in PANGAEA. It supplies routines through a graphical user interface for the import and editing of analytical data and the definition of all types of meta-information with its relations to the data. The 4D-client is mostly used by the data curators and librarians of PANGAEA and other projects and institutes.

Examples from Geoscientific Investigations in Antarctica

The following examples are given as a short tutorial to show how geoscientific data can be retrieved and how the data are downloaded to the users computer in a consistent format with metaheader. The PANGAEA search engine can be accessed at http://www.pangaea.de/PangaVista. Any expression included in the description of data sets can be used for a search, e.g., names, parameters or labels. The search is not case sensitive; results of a search example are shown in Fig. 7.9-2.

Example 1

The aim of CRP was to drill three stratigraphically overlapping cores in the Ross Sea to provide full coverage of Antarctic glaciation history. The cores were labeled CRP-1, CRP-2/CRP-2A and CRP-3. To retrieve all data from the project simply type in "crp". To obtain data from one hole type in its identifier e.g., "crp-1". If you are looking for a certain parameter in one of the cores, a combination of two expressions with a blank in between can be used (The syntax is the same as used in common search engines). For example, if looking for pollen of the plant family Caryopyllaceae a retrieval in PangaVista can be made with "crp-1 caryopyllaceae". If looking for data from an author, the name and the label can be combined, e.g., "crp-2 kettler". Download in text or html-format starts by clicking on the links given at the end of each metaheader.

Example 2

The EPICA project was running two drill sites on the East Antarctic ice sheet to recover the Pleistocene climate history of Antarctica. Any data related to EPICA can easily be found by typing in "epica". Looking for the age models of the EPICA cores would produce a query like "epica age model". The main cores are accompanied by several short firn cores. To see its distribution ask for "epica firn core" and click on "Show map". Switch to stereographic (S) projection and zoom into the map by using the magnifying glass. If looking for data of a specific site, use the "?" button, click on the dot of interest and find related data sets listed in a new window.

Example 3

To search for geological ages measured by the fission-track method in Antarctica, a retrieval may be started using the map provided with PangaVista. Switch to stereographic

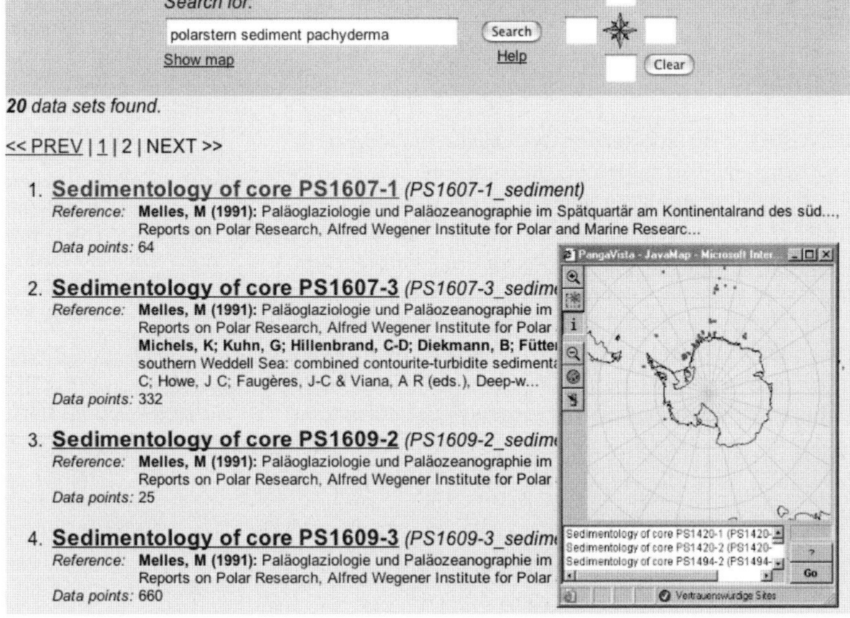

Fig. 7.9-2. When using the search engine PangaVista any combination of keywords is possible. The results are listed and the location of data sets found are plotted on a map. Using the "?" button by pointing on one of the sites, a list with related data sets will be given for download

south projection, choose the button with the arrow/rectangle and drag a rectangle above the Antarctic to set the geographical constraints. The geographical limits will be included in the four fields besides the PangaVista search line. Type in "age fission-track" and press "Search". The retrieval should list the data set of Meier (1999).

Example 4

PANGAEA has archived most of the marine geological data of AWI resulting from the analysis of sediment cores taken by its research vessel "Polarstern" in the Atlantic and Pacific parts of the Southern Ocean. A search for "polarstern sediment" will list more than 2 500 data sets with only the first 200 shown. In such a case, the search has to be defined more restrictive. A search on "polarstern sediment pachyderma" (Fig. 7.9-2) will prompt the user with a list of all published data sets containing data about e.g., the planktonic foraminifera *Neogloboquadrina pachyderma*. The user may scan through the list of sets by using the, «prev and next» buttons.

Example 5

PANGAEA is also used to archive georeferenced graphics or images. A retrieval on "ant-VIII documentation" will show a list of data sets, each containing the metadata of a sediment core and including just a "georeferenced" link to a directory. This directory contains all photos and descriptions of the sediment core in standard file formats (txt, pdf, jpg), presented in an overview with thumbnails. The grafic/photo is available for download in full resolution by a click on a thumbnail.

Conclusion

PANGAEA is an information system for the long-term archiving and distribution of georeferenced data. Due to the flexibility of the data model analytical data from many fields of basic research in natural science can be stored consistently together with the related metainformation necessary for their understanding and usage. With its comprehensive graphical user interfaces and the built-in functionality for upload and download, PANGAEA is an efficient system for scientific data management and data publication. Web-based interfaces to retrieve information from the system range from a simple search engine to a sophisticated data mining tool, the latter allowing the retrieval and combination of any subquantity of analytical data from the full inventory. For the visualization of data, software tools are distributed as freeware from the PANGAEA web site (Schlitzer 1997; Grobe et al. 2000; Sieger et al. 2001; see http://www.pangaea.de/Software). The internal consistency in combination with the use of clients and tools optimized for the users' needs, give data an added value if they are archived in PANGAEA. The major advantage of PANGAEA is its easy accessibility on the Internet providing the scientific community with a library of thousands of valuable data sets even from remote areas like the Antarctic. Any scientist, project or institute is encouraged to contribute to and make use of the PANGAEA library to establish a long-term archive and publication system for earth science data.

Acknowledgments

We are grateful to Roy Lowry and Hartmut Schulz for review and comments. Technical infrastructure and operation of the system is ensured by the computer centre of the Alfred Wegener Institute for Polar and Marine Research (AWI). AWI is a Helmholtz Research Centre, funded by the German Ministry for Education and Research and the State of Bremen.

References

Diepenbroek M, Fütterer DK, Grobe H, Miller H, Reinke M, Sieger R (1998) PANGAEA information system for glaciological data management. Annals Glaciol 27:655–660

Diepenbroek M, Grobe H, Reinke M, Schindler U, Schlitzer R, Sieger R, Wefer G (2002) PANGAEA an information system for environmental sciences. Computer Geosci 28:1201–1210

Fischer G, Wefer G (eds) (1999) Use of proxies in paleoceanography: examples from the South Atlantic. Springer, Berlin Heidelberg New York, http://www.pangaea.de/Projects/SFB261/Use_of_Proxies

Grobe H, Sieger R, Diepenbroek M (2000) PanMap. http://www.pangaea.de/Software/PanMap

Han J, Kamber M (2001) Data mining, concepts and techniques. Morgan Kaufmann Publishers

Schlitzer R (1997) Ocean-Data-View. http://odv.awi-bremerhaven.de

Sieger R, Grobe H, Diepenbroek M (2001) PanPlot. http://www.pangaea.de/Software/PanPlot

WDC (1996) Guide to the World Data Center System. Secretariat of the ICSU Panel on World Data Centers, http://www.ngdc.noaa.gov/wdc/guide/wdcguide.html

Theme 8
Sediments as Indicators for Antarctic Environment and Climate

Chapter 8.1
Tracing Marine Processes in the Southern Ocean by Means of Naturally Occurring Radionuclides

Chapter 8.2
Normalized Remanence in Sediments from Offshore Wilkes Land, East Antarctica

Chapter 8.3
Grain Size, Mineralogy and Geochemistry in Late Quaternary Sediments from the Western Ross Sea outer Slope as Proxies for Climate Changes

Chapter 8.4
Potential of the Scotia Sea Region for Determining the Onset and Development of the Antarctic Circumpolar Current

Chapter 8.5
Seismic Stratigraphy of Miocene to Recent Sedimentary Deposits in the Central Scotia Sea and Northern Weddell Sea: Influence of Bottom Flows (Antarctica)

Chapter 8.6
Limnology and Sedimentary Record of Radok Lake, Amery Oasis, East Antarctica

Chapter 8.7
New Data Related to Holocene Landform Development and Climatic Change from James Ross Island, Antarctic Peninsula

Chapter 8.8
Surface Movement of Stone-Banked Lobes and Terraces on Rink Crags Plateau, James Ross Island, Antarctic Peninsula

Chapter 8.9
Soil Characteristics along a Transect on Raised Marine Surfaces on Byers Peninsula, Livingston Island, South Shetland Islands

The nine papers of this chapter cover an extremely wide topical field of sedimentary research, reaching from oceanic processes via investigating marine sediment cores and interpreting seismic stratigraphy to paleo-lake studies and to the investigation of Holocene landforms and soil formation processes.

The importance of the Southern Ocean area in today's climate system as a whole has become widely accepted. However, details of the system and its feed-back mechanisms are far from being completely understood. Hanfland et al. (Chap. 8.1) use naturally occurring radionuclides as tracers to study marine processes on a scale of a few days to several ten thousand years. They present examples from the Atlantic sector of the Southern Ocean regarding particle transport and large scale movement of water masses.

Deep-sea sediments provide the continuous record of the past geomagnetic field. To investigate the long-term changes of the relative geomagnetic field intensity in the Antarctic Matsuoka and Funaki (Chap. 8.2) analysed in detail a sediment core from the continental margin of Wilkes Land covering the last about 1.1 Ma and compare it with previously published records.

Sediment cores from the northern Ross Sea off Cape Adare were investigated by Pistolato et al. (Chap. 8.3) with respect to the identification of changes in the late Quaternary environment. They have identified cyclical changes in several proxy parameters which are in phase with glacial – interglacial changes of marine isotope stages one to eight (MIS 1–8).

The onset and development of the Circum Antarctic Current (ACC) is considered one – if not the – crucial event of glacial onset in the Eocene – Oligocene period and the later glacial development of the Antarctic. Barker and Thomas (Chap. 8.4) discuss and assess the parameters that might be capable of determining ACC onset, and show that suitable sedimentary records for sampling should be available in the Scotia Sea.

The presence of high-energy contourite sediments in the northern Weddell Sea and Scotia Sea is described by Maldonado et al. (Chap. 8.5) from patterns of high-resolution seismics and sub-bottom echosounding with a tentative age assignment of Early Miocene to Quaternary. They interprete the contourite deposits to be formed under the complex interplay of the ACC and Weddell Sea Bottom Water (WSBW).

A detailed limnological and sedimentary study on Radok Lake, the deepest non-subglacial lake in Antarctica, is presented by Wagner and Cremer (Chap. 8.6). Despite the promising location in the large ice-free area of Amery Oasis, Radok Lake by poor age control and extensive sediment redeposition has turned out to be unsuited for advanced paleoenvironmental studies.

Aspects of Holocene landform development on NW James Ross Island are described in the papers of Strelin et al. (Chap. 8.7) and Mori et al. (Chap. 8.8). The former dealing with relict and active landforms delimited by various Holocene glacial advances. The latter dealing specifically with the formation and structure of stone-banked lobes and terraces.

The paper of Navas et al. (Chap. 8.9) describes a survey on soils carried out on Byers Peninsula, the largest ice-free area in the South Shetland Islands to characterize the main features of soils and formation processes along a transect from raised Holocene beaches to pre-Holocene upper marine platforms.

Tracing Marine Processes in the Southern Ocean by Means of Naturally Occurring Radionuclides

Claudia Hanfland · Walter Geibert · Ingrid Vöge
Alfred Wegener Institute for Polar and Marine Research, Am Handelshafen 12, 27570 Bremerhaven, Germany, <chanfland@awi-bremerhaven.de>

Abstract. Members of the naturally occurring decay series are found throughout the world's oceans, though in activities that vary by several orders of magnitude within the same decay series. They can be distinguished on the basis of their overall reactivity – e.g., adsorption or incorporation – with particles. Physical and biogeochemical processes in the water column, and close to the sediment-water interface, lead to fractionation of mother and daughter nuclides and hence create disequilibria in the decay series. These disequilibria have become powerful tools in the study of marine processes. In order to illustrate their use in marine sciences, three examples are presented for the Atlantic sector of the Southern Ocean. For the group of particle-reactive radionuclides, the distribution of the short-lived isotope ^{234}Th (half-life 24.1 days) is used as a measure of export fluxes from the photic layer. ^{228}Ra and ^{227}Ac (half-lives 5.8 and 21.7 years, respectively), belong to the more soluble nuclides. In contrast to ^{234}Th, they are hardly found in the particulate fraction of a sea water sample but instead exist in the dissolved state. ^{228}Ra is indicative of shelf water input while ^{227}Ac is a tracer for upwelling deep water masses.

Introduction

Since the discovery of radioactivity, unstable isotopes have become the basis of geochronology, a tool critical to modern geology. While common rock dating methods mostly rely on the long-lived members of the naturally occurring decay series ^{238}U–^{206}Pb, ^{235}U–^{207}Pb and ^{232}Th–^{208}Pb (half-lives 4.47×10^9, 7×10^8 and 14×10^9 years, respectively), marine processes take place on much shorter time scales. In view of the overturning rate of the ocean in the order of 1 000–2 000 years (Broecker and Peng 1982), the short-lived intermediate daughter products are of great importance.

Besides the half-life, the chemical and biochemical reactivity of a given element in sea water – its tendency to get adsorbed onto particle surfaces or its participation in biological processes – is important in marine applications. While Uranium or Radium exist mostly in the dissolved state, other elements like Thorium, Polonium or Protactinium are more adsorption-prone and can hence effectively be removed from the water column by attaching to particles that subsequently sink. Polonium is also known to be incorporated into organic tissue (Fisher et al. 1983) and can be removed from the marine environment via this mechanism. From the viewpoint of application, the difference in particle-reactivity can be used as criteria for the distinction between a group of radionuclides used for water mass studies and those more applicable to particle studies. While rocks unaffected by weathering represent a closed system with regard to the short-lived radionuclides – which are consequently found to be in secular equilibrium with their precursor isotopes – this is not the case for the marine system. Instead, strong partitioning of the isotopes can occur within a single decay series between the dissolved and the particulate phase. This results in conspicuous radioactive disequilibria in any given parent-daughter pair.

Uranium, Radium isotopes as well as ^{232}Th enter the ocean mostly by fluvial input. In general, concentrations of radionuclides in the oceans are considerably lower than in both their source rocks on land and in river water (Table 8.1-1). This finding is attributable to various processes taking place during the mobilization and the transport from the source area to the ocean. Estuaries are known to act as filters for radioisotopes (Moore 1992) and have a considerable influence on the budget of an element in ocean water. While Uranium is about a thousand times more abundant in sea water than gold, the majority of the intermediate decay products are only present in trace amounts. For example, the total amount of ^{227}Ac, one of the rarest isotopes in the world's oceans, is estimated to be less than 20 kg.

Table 8.1-1. Average concentrations of selected radionuclides in rocks, river and sea water. Data from Faure (1986), The Open University (1989), Krishnaswami (2001) and Geibert et al. (2002)

Element (ppm)	Basalt	Granitic rocks	River water	Sea water
U	0.43	4.8	0.1–1	0.0032
Th	1.6	21.5	0.01	1×10^{-5}
Pb	3.7	23	Highly variable	5×10^{-7}
^{226}Ra	[a]	[a]	Highly variable	7×10^{-11}
^{227}Ac	[a]	[a]	No data	<20 kg

[a] Isotopes are in radioactive equilibrium with their precursor isotopes in rocks unaffected by weathering processes.

From: Fütterer DK, Damaske D, Kleinschmidt G, Miller H, Tessensohn F (eds) (2006) Antarctica: Contributions to global earth sciences. Springer-Verlag, Berlin Heidelberg New York, pp 409–414

Owing to their low activities, determination of radionuclide concentrations in the marine environment is commonly performed with α-, β- and γ-counting in combination with large volume sampling. However, the ongoing improvement of measurement techniques, especially in the field of mass spectrometry, constantly brings more radionuclides into a new or wider field of application.

The Southern Ocean represents a key area in today's climate system and is often regarded as an early-warning system for the world's climate as a whole. However, the polar environment and its feed-back mechanisms are far from being completely understood. Any precise reconstruction of the palaeoclimatic conditions as well as reliable predictions of future trends involve both a close investigation of the processes and interactions that govern the modern climate, and with it the Southern Ocean, today. In this respect, radionuclide studies can yield valuable additional insights as their specific half-lives provide valuable time-relevant information.

In the following, three examples from the Southern Ocean are given to illustrate the application of radionuclides in the study of marine processes. We show their use regarding particle transport and movement of water masses.

Methods

For the analysis of ^{234}Th, 20 l samples of sea water were filtered (1 µm), followed by MnO_2-precipitation in the supernatant to remove Thorium from solution. The precipitate was filtered, too, and both the particulate and the dissolved fraction were analyzed for their ^{234}Th activity via β-counting.

^{228}Ra and ^{227}Ac were collected by pumping several hundred liters of sea water through two MnO_2-coated filter cartridges put in series. These were either connected to the ship's sea water supply or loaded on time-programmed filtration units. For the determination of the ^{228}Ra activity, the filter cartridges were combusted in a muffle furnace and the ^{228}Ra content of the ash either determined directly by γ-spectrometry or via a modified version of the ^{228}Th-ingrowth method (Moore 1972; Hanfland 2002). ^{227}Ac was brought into solution through leaching of the filter cartridges. After purification by ion column chemistry, its activity was plated on silver planchets and determined via its short-lived daughter ^{227}Th by α-spectrometry. A full description of all the methods used is given in Rutgers van der Loeff and Moore (1999).

^{234}Th as an Indicator of Export Production

^{234}Th (half-life 24.1 days; Table 8.1-2) is selected as representative of a highly particle-reactive element. It is produced in the water column through the decay of ^{238}U, an isotope with conservative behaviour that is evenly distributed throughout the water column. Its activity is related to salinity by ^{238}U [dpm l^{-1}] = 0.0704 Salinity (Rutgers van der Loeff et al. 1997). In an ideal ocean without any particles, the activity between parent and daughter nuclide can be taken as unity, a scenario that is best matched in the open ocean during early spring before the onset of plankton blooms (Fig. 8.1-1a). With only few particles present or at the beginning of a bloom, the ^{234}Th inventory is partitioned between the dissolved and the particulate phase but still sums up to unity (Fig. 8.1-1b). Subsequent sinking of particles leads to export not only of organic matter but also of a fraction of ^{234}Th (Fig. 8.1-1c). Hence, the depletion of ^{234}Th in the upper water column with respect to its parent nuclide ^{238}U can be taken as a measure of export production from the euphotic zone (Coale and Bruland 1985). From this depletion, the export of other constituents like organic carbon or biogenic silica can be calculated assuming the ratio of this material to particulate ^{234}Th has been determined. Once the particle production stops and the remaining particulate fraction sinks to deeper water layers, ^{234}Th can grow back into equilibrium with its precursor.

Table 8.1-2. The ^{238}U-^{206}Pb, ^{235}U-^{207}Pb and ^{232}Th-^{208}Pb decay series with decay schemes of the daughter nuclides presented in this study and their corresponding half-lives

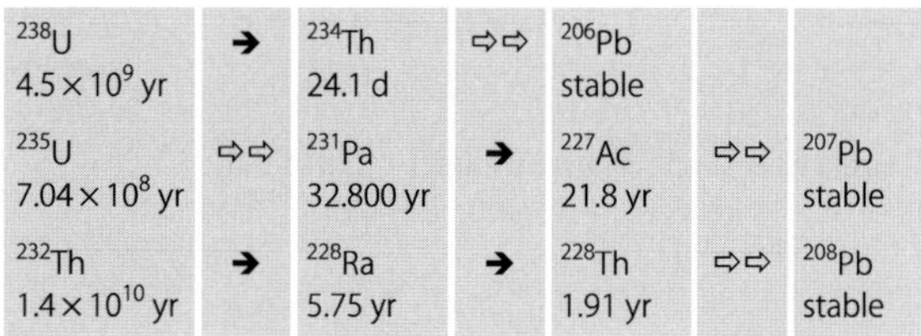

^{238}U 4.5×10^9 yr	➔	^{234}Th 24.1 d	⇨⇨	^{206}Pb stable		
^{235}U 7.04×10^8 yr	⇨⇨	^{231}Pa 32.800 yr	➔	^{227}Ac 21.8 yr	⇨⇨	^{207}Pb stable
^{232}Th 1.4×10^{10} yr	➔	^{228}Ra 5.75 yr	➔	^{228}Th 1.91 yr	⇨⇨	^{208}Pb stable

➔ decay step to next daughter nuclide; ⇨⇨ one or more decay steps have been omitted.

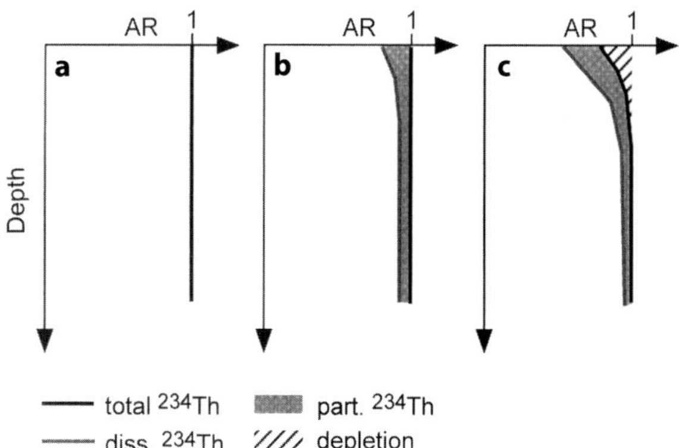

Fig. 8.1-1. Depth profiles showing the partitioning of ^{234}Th between the dissolved and particulate phase. Situation **a** in an ideal ocean without particles; **b** in early spring before the onset of primary production; **c** after development of plankton blooms. Profiles **b** and **c** stylized from data after Rutgers van der Loeff and Berger (1991). *AR:* ^{234}Th/^{238}U activity ratio

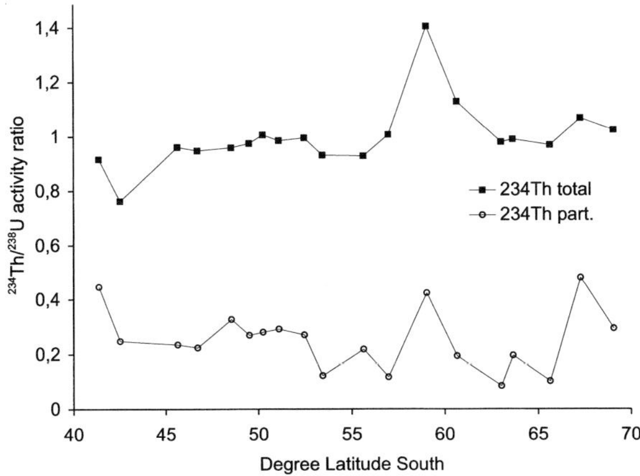

Fig. 8.1-2. ^{234}Th/^{238}U activity ratios during austral spring (RV "Polarstern" expedition ANT-XV/2). For sample processing and analysis of ^{234}Th see Rutgers van der Loeff and Moore (1999)

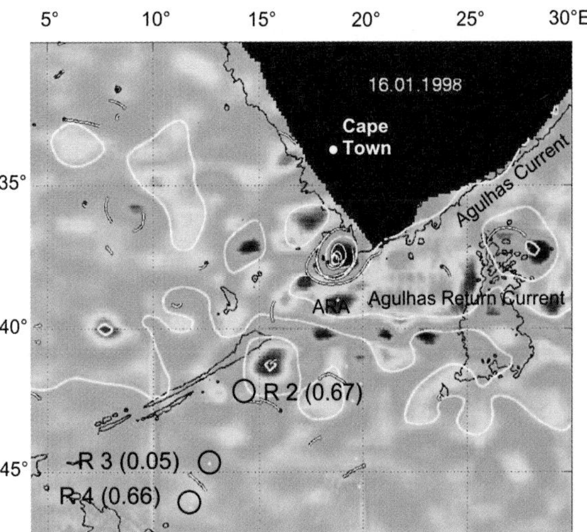

Fig. 8.1-3. Sea surface steric heights with superimposed tracks of KAPEX RAFOS floats; situation shown for 16 January 1998, coinciding with radium sampling during expedition ANT-XV/3 with RV "Polarstern". *Blue colours:* negative anomalies (= cyclones); *yellow-red colours:* positive anomalies (= anticyclones, Agulhas Ring); *white lines:* drifter tracks; *ARA:* Agulhas Retroflection Area. Satellite altimetry data kindly provided and interpreted by O. Boebel, AWI. Drifter tracks are used in an animated time-series version of the figure (not shown) to follow the path of cyclones and anticyclones over time. ^{228}Ra surface water activities determined on the 15/16 January 1998 are given in brackets in dpm/100 kg. For ^{228}Ra sampling and methods see Hanfland (2002)

Figure 8.1-2 shows the particulate and total ^{234}Th content of surface water samples collected during austral spring (November 1997) on a N-S transect across the Antarctic Circumpolar Current (ACC) between Cape Town and Neumayer Base at Atka Bay. Originally, ^{234}Th was determined on this transect as natural yield tracer for the analysis of the ^{230}Th and ^{231}Pa activities (Walter et al. 2001).

The data correspond well with the situation given in Fig. 8.1-1b: ^{234}Th is mostly in secular equilibrium with ^{238}U. Considerable export can only be observed at the approximate position of the Subtropical Front, where primary production seemed to start earlier than further south. The sample at 59° S contained a lot of krill. This is probably the reason for the high total ^{234}Th activity caused by its accumulation in the food chain.

With the development of strong blooms, clear export signals could be observed on the same transect later in season in a subsequent year (March 1999) with highest depletion coinciding with the ACC frontal systems (Usbeck et al. 2002). The repeated determination of ^{234}Th on vertical profiles yielded evidence for the productive nature of the ACC fronts and the related export production.

^{228}Ra as a Tracer for Shelfwater Advection

^{228}Ra (half-life 5.75 years; Table 8.1-2) has been widely used as a tracer for water masses originating from the continental shelves (e.g., Kaufmann et al. 1973; Reid et al. 1979; Rutgers van der Loeff et al. 1995). It is a daughter product of ^{232}Th, which is common in most sediment types but nearly absent in sea water due to its particle-reactive behaviour. Furthermore, ^{232}Th is the starting member of the decay series and can hence not grow in from a soluble parent in the water column, as is the case for ^{234}Th. Part of the ^{228}Ra that is produced through decay of ^{232}Th in the sediment diffuses back into the water column where it can accumulate to high activities when in prolonged contact with sediments. This is the case above the ocean bottom or, even more pronounced, on the continental shelves. According to its relatively short half-life, the activity of ^{228}Ra decreases with distance from the source and is extremely low in the ocean interior and in open ocean surface waters.

The concept of decreasing ^{228}Ra activities in surface waters with distance to the shore has been applied in the region south of South Africa where intense mixing of subtropical and subantarctic water masses takes place: Occlusion of the retroflecting Agulhas Current generates so-called Agulhas Rings that move northwestwards into the Atlantic while perturbations in the flow of the Agulhas Current lead to the spawning of both cyclonic and anticyclonic eddies (Lutjeharms 1996; Boebel et al. 2003). In a sea surface steric height depiction of the area (Fig. 8.1-3), cyclones are represented by negative anomalies (blue colours) while anticyclones and rings are associated with positive anomalies (yellow/red colours).

The cyclones can have different origins. They either move northwards from the subantarctic region and represent Atlantic water, or originate in the course of a Natal Pulse off the coast of South Africa, and are comprised of water from the Indian Ocean. As the whole region is of great importance for interocean water exchange and represents one of the return paths of the thermohaline circulation into the Atlantic Ocean, a better distinction of the origin of the cyclones will contribute to the understanding of water exchange in that region.

Previous studies have shown that waters coming from the Antarctic zone have ^{228}Ra activities close to the detection limit (Hanfland 2002). In contrast, the Agulhas Current can be expected to pick up a strong ^{228}Ra signal while following the continental shelf edge of southern Africa.

Figure 8.1-3 displays the hydrographic situation on 16 January 1998 together with the samples taken for ^{228}Ra surface water activities within 48 hours of this snapshot. The picture reveals an anticyclonic Agulhas Ring centred around 41° S, 16° E. Sample R2 probably gained its activity from this ring. Sample R4 coincides with a subtropical anticyclone coming from further west. This can be deduced by backtracking the position of the associated KAPEX drifter, represented as a white line in Fig. 8.1-3, in a time-sequence of this figure (not shown).

Located between R2 and R4 is a negative sea surface anomaly where sample R3 yields a very low ^{228}Ra activity. It displays the effect of subantarctic water being entrained northwards into subtropical waters.

Based on these encouraging results, specific sampling of cyclones in the Agulhas Retroflection Area was conducted during expedition ANT-XX/2. Preliminary results indeed reveal very high ^{228}Ra activities in the surface water of one of the cyclones, suggesting an origin off the coast of South Africa. More evidence is expected to come from the analysis of complete vertical profiles that are currently being processed.

^{227}Ac as a Tracer for Upwelling of Deep Waters

Analogous to ^{228}Ra, ^{227}Ac (half-life 21.8 years) is produced in sediments through decay of a particle-reactive mother nuclide (^{231}Pa). In contrast to ^{232}Th, the activity of ^{231}Pa in the surface sediment is related to the depth of the overlying water column. Consequently, ^{231}Pa activities are highest in deep-sea sediments which in turn represent the major source for ^{227}Ac released into the sea water. Excess activity of ^{227}Ac (expressed as $^{227}Ac_{ex}$) over its parent nuclide ^{231}Pa in surface waters may therefore be interpreted as evidence for rapid upwelling of deeper water masses. Figure 8.1-4 gives two examples of $^{227}Ac_{ex}$ profiles from the Atlantic and Pacific Ocean. These show the influence of upwelling on the dissolved $^{227}Ac_{ex}$ activities in the water column (modified after Geibert et al. 2002). The location in the Southeast Pacific is characterized by little advection and does not show significant excess of ^{227}Ac. Here, vertical movement of water masses is too slow to allow any $^{227}Ac_{ex}$ signal to reach the surface. In contrast, the sampling station in the Weddell Gyre, situated in the inflow region of North Atlantic Deep water, displays high

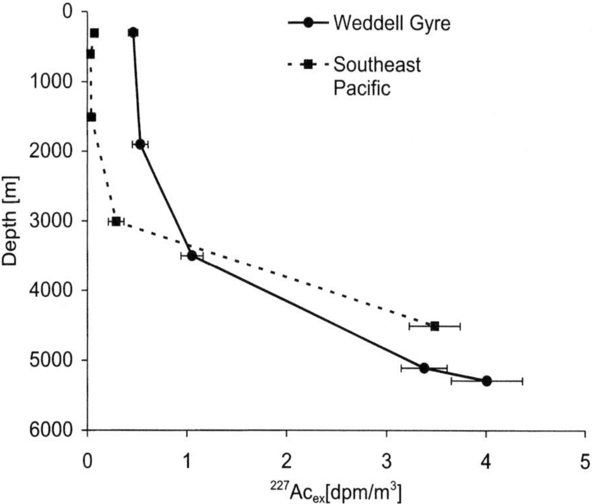

Fig. 8.1-4. $^{227}Ac_{ex}$ profiles from two locations: Weddell Gyre (PS2579-2; 62°59' S, 07°45' E) and Southeast Pacific (PS2660-7; 50°09' S, 89°13' E). For ^{227}Ac sampling and methods see Geibert et al. (2002)

$^{227}Ac_{ex}$ activities in its upper part, thereby indicating a rapid upwelling of deeper water masses. Associated rates of upwelling calculated by means of $^{227}Ac_{ex}$ were in the order of 55 m yr^{-1}. A more detailed discussion of the $^{227}Ac_{ex}$ distribution in different hydrographic regimes can be found in Geibert et al. (2002).

Conclusions

The three examples presented were chosen to illustrate the use and applicability of naturally occurring radionuclides to process studies in the Southern Ocean. When studying particle fluxes, ^{234}Th provides a useful tool for the determination of the export of carbon to deeper water layers. This provides important constraints on whether a region serves as a sink or source for carbon dioxide. The combined analysis of ^{228}Ra and ^{227}Ac allows for a better distinction between lateral advection *versus* deep upwelling of water masses. Due to their specific source regions (continental shelf areas and deep-sea sediments), they provide a powerful pair of tracers to distinguish between different origins of water masses.

To summarise, the decay series are unique in that they offer a large variety of parent-daughter pairs of different reactivities, in combination with half-lives between a few days to several ten thousand years. These features make them a valuable tool in marine sciences. Specific trends or irregularities in their distribution not only confirm expectations derived from other variables, but often give an indication of hitherto unknown or unconsidered processes in a given context. With steadily improving analyzing techniques, the potential of these marine tracers can be expected to further increase in future studies.

Acknowledgments

The authors gratefully acknowledge the efficient technical assistance of the crew of RV "Polarstern" during the sampling program. This work was supported by Deutsche Forschungsgemeinschaft (DFG grants no. RU 712/1 and GE 1118/2-2). Thanks are due to Adrian Webb and Steve Boger for proof-reading of the manuscript. The authors are grateful to M. Rutgers van der Loeff and one anonymous reviewer for helpful comments on the manuscript.

References

Boebel O, Lutjeharms J, Schmid C, Zenk W, Rossby T, Barron C (2003) The Cape Cauldron: a regime of turbulent inter-ocean exchange. Deep Sea Res II 50:57–86

Broecker WS, Peng T-H (1982) Tracers in the sea. Lamont-Doherty Geological Observatory, Palisades, New York

Coale KH, Bruland KW (1985) ^{234}Th:^{238}U disequilibria within the California Current. Limnol Oceanogr 30(1):22–33

Faure G (1986) Principles of isotope geology. Wiley and Sons, New York

Fisher NS, Burns KA, Cherry RD, Heyraud M (1983) Accumulation and cellular distribution of ^{241}Am, ^{210}Po and ^{210}Pb in two marine algae. Mar Ecol Prog Ser 11:233–237

Geibert W, Rutgers van der Loeff MM, Hanfland C, Dauelsberg HJ (2002) Actinium-227 as a deep-sea tracer: sources, distribution and applications. Eart Planet Sci Letters 198:147–165

Hanfland C (2002) Radium-226 and Radium-228 in the Atlantic Sector of the Southern Ocean. Rep Polar Mar Res 431:1–135

Kaufman A, Trier RM, Broecker WS (1973) Distribution of ^{228}Ra in the World Ocean. J Geophys Res 78:8827–8848

Krishnaswami S (2001) Uranium-Thorium series isotopes in ocean profiles. In: Steele JH, Thorpe SA, Turekian KK (eds) Encyclopedia of ocean sciences. Academic Press, San Diego, pp 3146–3156

Lutjeharms JRE (1996) The exchange of water between the South Indian and South Atlantic Oceans. In: Wefer G, Berger WH, Sielder G, Webb DJ (eds) The South Atlantic: present and past circulation. Springer, Berlin Heidelberg New York, pp 125–162

Moore WS (1972) Radium-228: application to thermocline mixing studies. Eart Planet Sci Letters 16:421–422

Moore WS (1992) Radionuclides of the uranium and thorium decay series in the estuarine environment. In: Ivanovitch M, Harmon RS (eds) Uranium series disequilibrium. Applications to earth, marine and environmental science. Clarendon Press, Oxford, pp 334–395

Reid DF, Moore WS, Sackett WM (1979) Temporal variation of ^{228}Ra in the near-surface gulf of Mexico. Earth Planet Sci Letters 43:227–236

Rutgers van der Loeff MM, Berger GW (1991) Scavenging and particle flux seasonal and regional variations in the Southern Ocean (Atlantic sector). Mar Chem 35:553–567

Rutgers van der Loeff MM, Moore WS (1999) Determination of natural radioactive tracers. In: Grasshoff K, Kremling K, Ehrhardt M (eds) Methods of seawater analysis. Wiley-VCH, Weinheim, pp 365–397

Rutgers van der Loeff MM, Key RM, Scholten J, Bauch D, Michel A (1995) ^{228}Ra as a tracer for shelf water in the Arctic Ocean. Deep Sea Res II 42:1533–1553

Rutgers van der Loeff MM, Friedrich J, Bathmann U (1997) Carbon export during the spring bloom at the Antarctic Polar Front, determined with the natural tracer ^{234}Th. Deep Sea Res II 44:457–478

The Open University (1989) Seawater: Its composition, properties and behaviour. Pergamon Press, Oxford

Usbeck R, Rutgers van der Loeff MM, Hoppema M, Schlitzer R (2002) Shallow remineralization in the Weddell Gyre. Geochemistry geophysics geosystems 3(1)

Walter HJ, Geibert W, Rutgers van der Loeff MM, Fischer G, Bathmann U (2001) Shallow vs. deep-water scavenging of ^{231}Pa and ^{230}Th in radionuclide enriched waters of the Atlantic sector of the Southern Ocean. Deep Sea Res I 48:471–493

Normalized Remanence in Sediments from Offshore Wilkes Land, East Antarctica

Haruka Matsuoka[1,2] · **Minoru Funaki**[1]

[1] National Institute of Polar Research, Kaga 1-chome, Itabashi-ku, Tokyo 173-8515, Japan

[2] *Present address:* Gakushuin University, Computer Center, 1-5-1 Mejiro, Toshima-ku, Tokyo 171-8588, Japan

Abstract. In order to investigate the long-term changes of the geomagnetic field intensity in the Antarctic region, paleomagnetic and rock-magnetic studies have been conducted on a deep-sea sediment core obtained from offshore Wilkes Land, East Antarctica. The core covers the last about 1.1 Ma. Stepwise alternating-field (AF) demagnetization of natural remanent magnetization (NRM) revealed that a great majority of samples are characterized by a single stable component of magnetization, sometimes associated with a secondary component completely demagnetized by a 30 mT AF field. Downcore changes of magnetic concentration represented by magnetic susceptibility (χ) and anhysteretic remanent magnetization (ARM) are a factor of five or less. Variations in magnetic grain size and coercivity are estimated to be small from ratio of ARM to χ and median destructive fields of ARM respectively. These results demonstrate that the core is rock-magnetically homogeneous, and thus could be considered to yield relative paleointensity record. The ratio of the NRM demagnetized at 30 mT (NRM_{30mT}) versus the ARM demagnetized at 30 mT (ARM_{30mT}), which is the reasonable parameter to eliminate the effects of the secondary remanence, is interpreted as our best approximation for paleointensity estimation. Absence of correlation between the normalized intensity (NRM_{30mT}/ARM_{30mT}) and the normalizer (ARM_{30mT}) shows the appropriateness of the normalization. The obtained record is similar in general to other worldwide marine records. Such a global synchronicity might be attributed to dipole intensity changes.

Introduction

Deep-sea sediments provide the continuous trace of the past geomagnetic field. Until the pioneer works of establishing sedimentary paleointensity records by Johnson et al. (1975) and Levi and Banerjee (1976), the great majority of paleomagnetic studies were concerned solely with the directional behavior of the field (see Tauxe 1993). Over the last decade, in contrast, a number of relative paleointensity records of the geomagnetic field were obtained from sediments for the past few hundred thousand years, which allowed the construction of global reference paleointensity curves (Guyodo and Valet 1996; Guyodo and Valet 1999) using technical and methodological developments in paleointensity determination. However, the number of long-period paleointensity data spanning more than the last 0.8 Ma is still insufficient to make a standard curve. Moreover, only a few relative paleointensity records derived from sediments have been reported in the Antarctic regions (e.g., Sagnotti et al. 2001; Guyodo et al. 2001). It is therefore important to obtain continuous paleomagnetic records involving both direction and intensity of the field from the Antarctica for understanding of the evolution of the global scale geomagnetic field.

Brachfeld et al. (2000) did a paleomagnetic study on five sedimentary sequences cored in the western margin of the Antarctic Peninsula. They showed a single continuous record of relative paleointensity over the past 9 000 yr as the first high-resolution record of paleosecular variation for the Antarctic Peninsula region. Guyodo et al. (2001) provided a paleomagnetic record for a part of the Matuyama chron, derived from sediments collected from the drift deposits off the western Antarctic Peninsula continental margin, at Ocean Drilling Program (ODP) site 1101 (Leg 178). A relative paleointensity record over the 0.7–1.1 Ma interval and a comparison with eight other paleointensity records over the 0.95–1.1 Ma interval were reported. They constructed a composite record of relative paleointensity for the Jaramillo subchron with nine records from different oceans in the study. Sagnotti et al. (2001) showed relative paleointensity records spanning the last 160 kyr with using three cores obtained from a hemipelagic sediment drift on the continental rise of the Pacific margin of the Antarctic Peninsula during the Sediment Drifts of the Antarctic Offshore (SEDANO) project (Lucchi et al. 2002). The relative paleointensity records were used for improving their age model by comparison with the reference Sint-200 curve (Guyodo and Valet 1996) and the relative paleointensity stack from the North Atlantic, NAPIS-75 (Laj et al. 2000).

The core used in this study was collected on the TH94 cruise (1994–1995) of RV "Hakurei-Maru", carried out by the Technology Research Center, Japan National Oil Corporation (Ishihara et al. 1996). A brief relative paleointensity estimation using NRM intensity normalized by magnetic susceptibility was performed by Sakai et al. (1998) using one side of the core divided in half. For detailed discussion, however, the stability of remanence and the rock-magnetic homogeneity should be proved for paleointensity estimates. More reliable paleointensity estimate using other normalizer such as ARM or isothermal remanent magnetization (IRM) is also necessary. We, therefore, per-

From: Fütterer DK, Damaske D, Kleinschmidt G, Miller H, Tessensohn F (eds) (2006) Antarctica: Contributions to global earth sciences. Springer-Verlag, Berlin Heidelberg New York, pp 415–422

formed careful magnetic measurements and experiments on all specimens resampled from the other side of the core (archived halves of cores). In this article we examine the stability of remanence, the homogeneity of the rock-magnetic properties and the normalized intensity record as one of the relative paleointensity records of the geomagnetic field.

Materials

The core used in this study was obtained from a continental rise site 3 060 m deep at the western part of the Antarctic Wilkes Land margin. The position of the coring site is 63°43.13' S, 112°20.06' E (Fig. 8.2-1). In the vicinity of our study area, Hayes et al. (1975) reported the lithology of the sequence cored from about 3 500 meters deep at site 268 (63°56.99' S, 105°09.34' E) of the Deep Sea Drilling Project (DSPS). The DSDP site 268 depicted in Fig. 8.2-1, has experienced dominantly terrigenous sedimentation and the sediments are chiefly silty clays with the transition from diatom-bearing or diatom-rich above, to nanofossil-rich below (Hayes et al. 1975; Kemp et al. 1975). Other geological and geophysical data from the western part of Wilkes Land margin were obtained from Tsumuraya et al. (1985) and Ishihara et al. (1996).

Our core was collected with a gravity corer. The total length of the core is 540 cm. The surface sediments, at least 20 cm, were lost while recovering. The core was estimated to have been collected perpendicularly, while the north direction of the core is unknown. Subsequently the core was separated into six sequences (numbers: 1 to 5 and C) of 1 m length keeping their orientations and was split in half lengthwise. They were then sealed in split core shaped plastic cases and stored at +5 °C to minimize dehydration of the sediments. The 236 discrete samples of 7 cm^3 (edge length is 2.3 cm) were sequentially taken from the archive halves of sub-cores in August 2001. All the samples were then sealed to minimize dehydration of the sediments.

The sediment material is siliceous silt of brownish gray color. Abundant foraminiferal skeletons in good preservation are observed throughout the core. They clearly reflect the oxidized condition of the sediments. Paleoclimatically induced lithological variations were not observed. This implies that the sediment was deposited in an environmentally stable condition.

Remanence

In this study, detailed demagnetization experiments were conducted on all samples using a SQUID magnetometer with a static three-axis alternating AF demagnetizer and an ARM acquisition coil, which was produced by 2G Enterprises. The magnetometer is installed in a low magnetic field room in the National Institute of Polar Research in Tokyo, Japan. NRM was measured before demagnetization and during stepwise AF demagnetization using a peak field increment of 5 mT in the 0–100 mT range.

The representative AF demagnetization curves described by Zijderveld projections are shown in Fig. 8.2-2. In general, almost all samples from the surface to 460 cm depth have very stable NRM (Fig. 8.2-2a–d). The

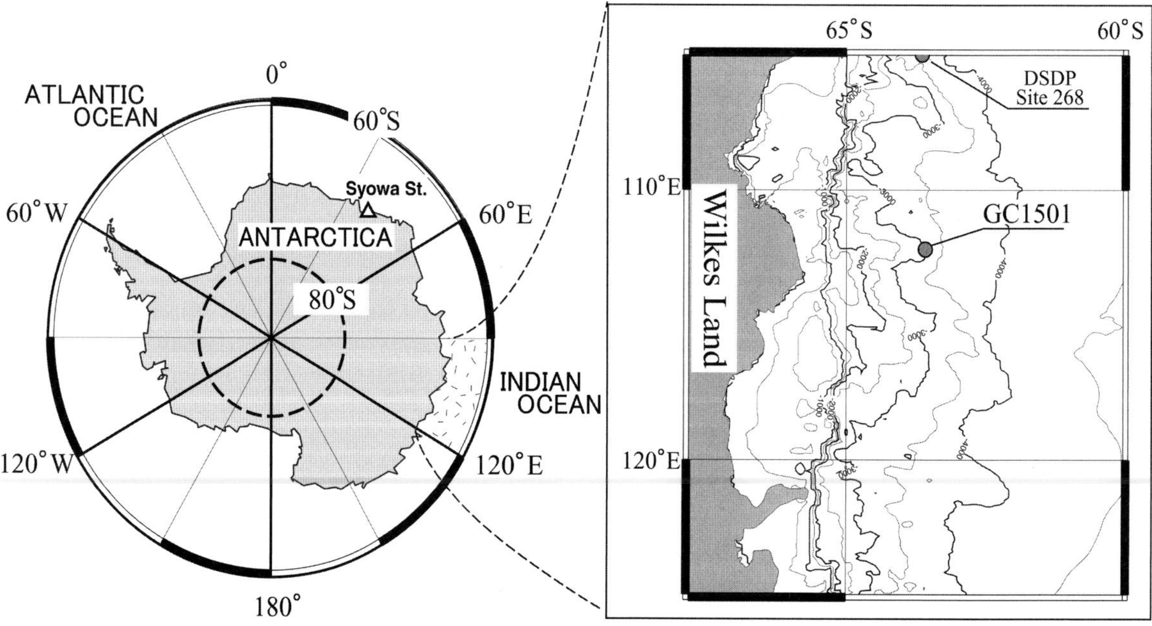

Fig. 8.2-1. Location map showing the coring site (63°43.13' S, 112°20.06' E)

NRM intensities of these samples were on the order of 10^{-1}–10^{-2} A m^{-1} in intensity, which are 10–100 times higher than those commonly obtained from different localities, though such high values are sometimes reported from mid or high latitudes (e.g., Channel et al. 1997; Guyodo et al. 2001). In the lower section of the core, in contrast, many samples showed more unstable (zigzag) demagnetization curves. The demagnetization curve of sample 5-M28 (462.0 cm) and 5-M39 (486.8 cm) showed at least two component magnetizations (Fig. 8.2-2e,f). The soft component of the normal polarity was demagnetized to 20 mT, while the hard component of the reversed polarity seemed to survive up to 100 mT. The intensities were more than 10 times weaker than those of the samples obtained from the upper section. A similar demagnetization curve and intensities appeared in sample C-M07 (Fig. 8.2-2g), while the polarities were opposite for the hard and soft components.

However, a few samples showed only soft NRM components (Fig. 8.2-2h,i). Their NRM were demagnetized

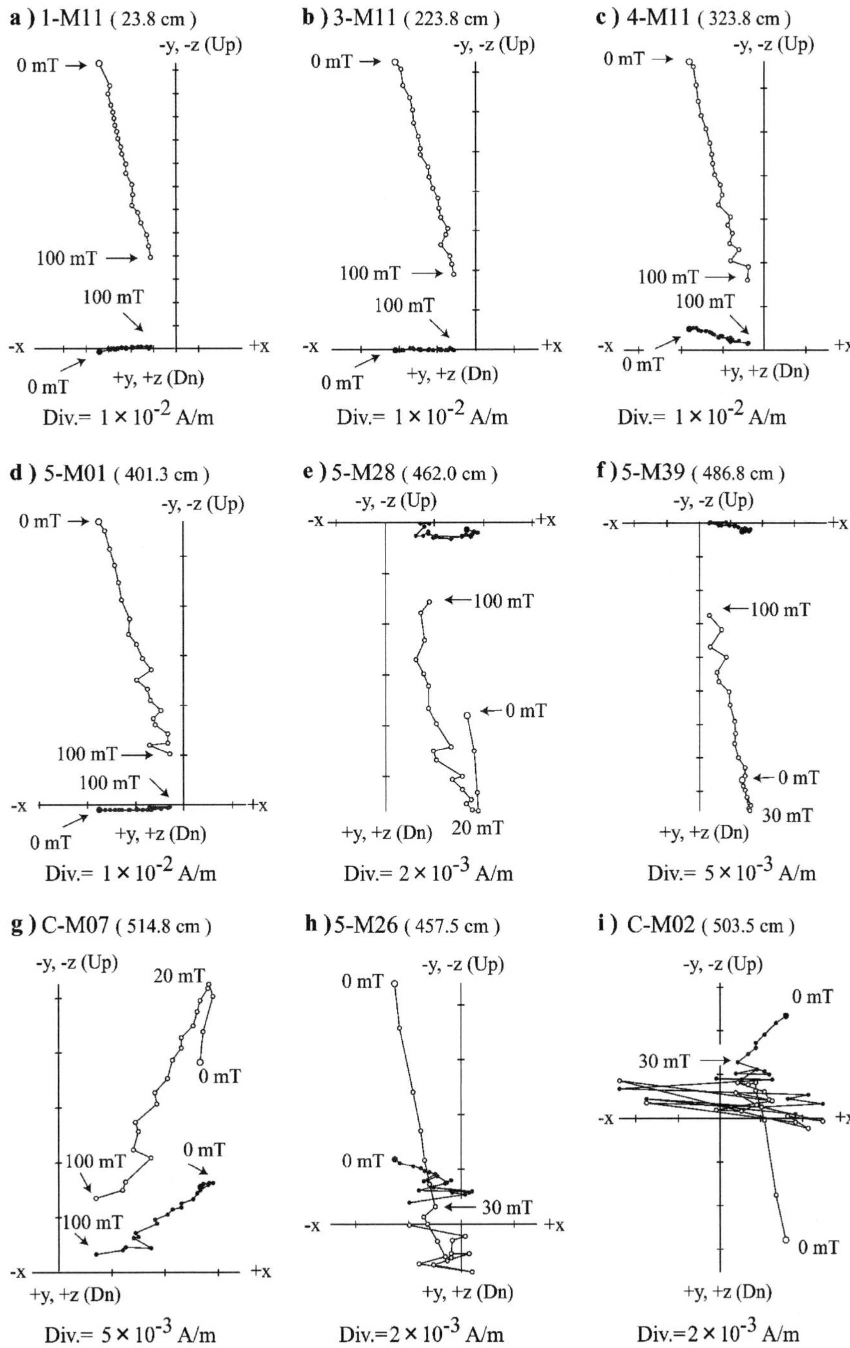

Fig. 8.2-2. Representative results of stepwise AF demagnetization of specimens. *Solid/open circles* show projections on the horizontal/vertical plane

linearly up to 30 mT, but they underwent drastic zigzag variations. The linear demagnetizations between 0 and 30 mT are due to only the soft components. The optimum AF demagnetization field intensity was inferred to be 30 mT, because the soft NRM components of every sample seemed to be demagnetized completely before 30 mT. The high coercive components of NRM have never been demagnetized completely as the residual NRM of 12–33% can survive, even if the samples are demagnetized up to 100 mT. The median destructive field (MDF) in the upper 460 cm was in the range of 37.8–78.0 mT (average 58.0 mT). Such a high stability of NRM is likely to contribute to the establishment of the reliability of our paleomagnetic record.

Downcore variations of the NRM directions after the AF demagnetization at 30 mT (Fig. 8.2-3) show that the core is divided into at least three polarity zones. The normal and reversed polarities of NRM can be compared with the magnetostratigraphy of Cande and Kent (1995). Even though the surface sediments were lost while recovering, the correlation of susceptibility variations with the other cores recovered perfectly from the nearby sites indicates that the uppermost magnetic boundary is presumably assumed to be the Brunhes-Matuyama (B/M) geomagnetic boundary (Sakai et al. 1998). The most plausible fitting of the inclination and declination profiles is represented in Fig. 8.2-3 from the surface to 460 cm depth for Brunhes Chron, from lower than 460 cm for Matuyama Chron and 504 to 526 cm for Jaramillo Subchron.

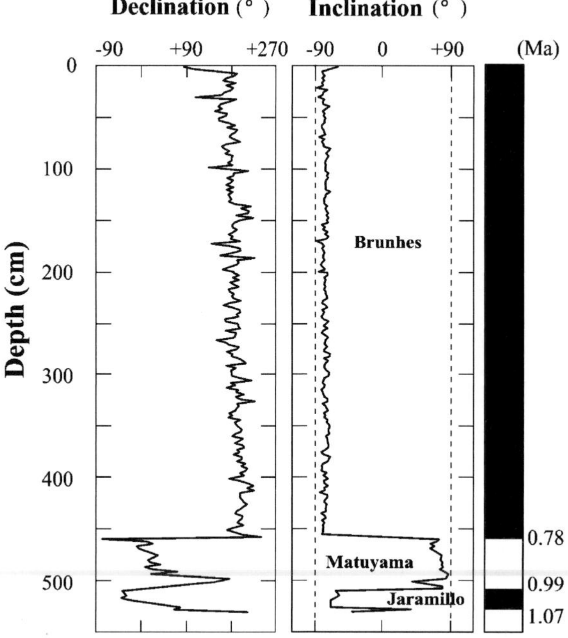

Fig. 8.2-3. Downcore variations in declination and inclination of the NRM after AF demagnetization at 30 mT with the magnetostratigraphy of Cande and Kent (1995). The declinations are relative because the core was not azimuthally oriented

Rock-Magnetic Homogeneity

Downcore changes of magnetic concentration, coercivity, and variations in magnetic grain-size were investigated to determine the rock-magnetic homogeneity and the suitability of core for relative paleointensity estimates.

Magnetic susceptibility (χ) is one of the parameters believed to primarily reflect down-core changes in concentration of magnetic minerals in the sediments (e.g., Tauxe 1993). We measured χ at low frequencies (0.47 kHz) of applied field using a Bartington susceptometer Ms-2 Susceptibility System. The down-core variation of χ is shown in Fig. 8.2-4. In the upper part above 250 cm depth, the down-core curve of χ shows three large peaks with 1.31×10^{-3} SI (30.7 cm depth) for the maximum value and three discrete peaks at the depth of 128 cm (6.83×10^{-4} SI), 137 cm (1.03×10^{-3} SI), 201 cm (1.06×10^{-3} SI). The minimum value of 2.68×10^{-4} SI appears at 244 cm depth. From the depth of 250 cm, χ gradually increases downward up to 6.55×10^{-4} SI (373 cm). Below the depth of 373 cm, χ gradually decreases to 3.05×10^{-4} SI (406 cm). In the interval between 406 and 440 cm, χ is almost stabilized at lower value of about 3.40×10^{-4} SI. In the lower most part below 440 cm depth, the down-core curve shows a gradual increase downward. The overall average value is 5.78×10^{-4} SI.

Anhysteretic remanent magnetization (ARM) is also believed to represent down-core changes of magnetic concentration (e.g., Tauxe 1993). ARM was imparted on the every sample by superimposing a DC biasing field of 0.1 mT on a smoothly decreasing AF field with a peak of 100 mT, and then the ARM was AF demagnetized up to 100 mT. Since our core was obtained from the Antarctic polar region, the dominant NRM component is vertical, as supported by the AF demagnetization results (Fig. 8.2-2), and it was not demagnetized completely even if the demagnetization field was 100 mT. Therefore, ARM was imparted toward the horizontal direction of the core to minimize the effect of the residual NRM after the demagnetization. The downcore curve of ARM intensity after demagnetization by the optimum field (30 mT) is shown in Fig. 8.2-4. It exhibits humps at about 30 cm and 90 cm, and depicts depression in the lower section between 405 and 460 cm. The ARM curve varies from a minimum of 0.07 A m^{-1} to a maximum of 0.21 A m^{-1}. The average value of ARM is 0.12 A m^{-1}.

Median destructive fields of ARM (MDF_{ARM}) were calculated as a crude guide of coercivity. The down-core variation of MDF_{ARM} is shown in Fig. 8.2-4. The MDF_{ARM} varies within remarkable small amplitudes of several mT and is almost stabilized at around 22 mT throughout the core. In the upper part above 228 cm, the down-core curve shows several general peaks and abrupt depressions with relatively large amplitudes of several mT. The minimum value of 19.9 mT appears at 26 cm depth. The MDF_{ARM} value increases from 20.7 mT to 22.9 mT downward in the

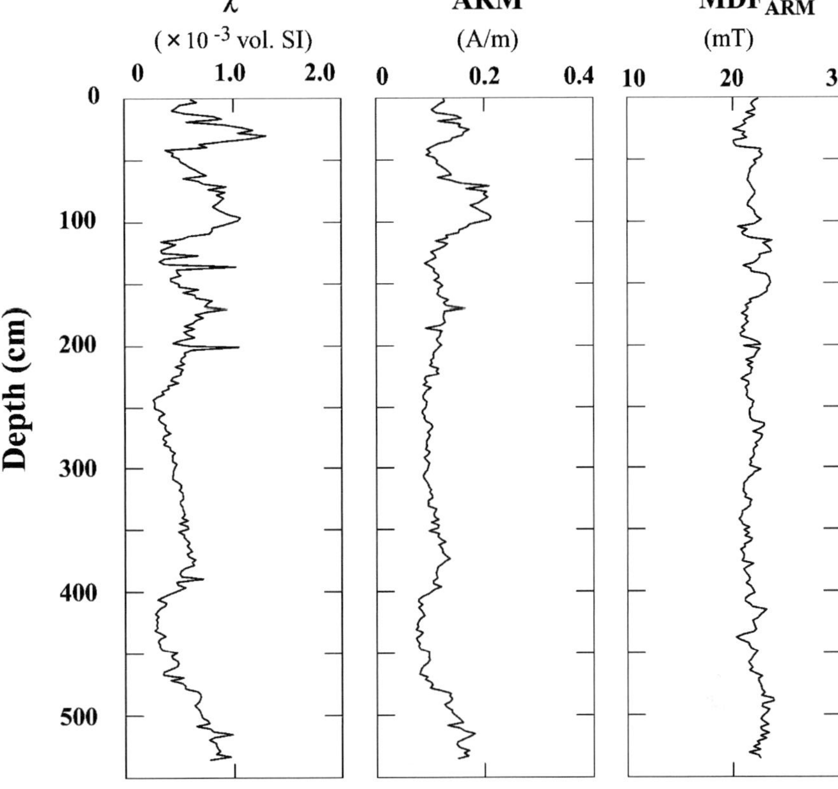

Fig. 8.2-4. Downcore variations in the magnetic susceptibility (χ), *ARM* intensity demagnetized at 30 mT AF field and MDF_{ARM}

interval between 231 cm and 264 cm. From 264 cm depth, the value starts decreasing gradually to a small value of 20.9 mT (313 cm). In the interval between 315 cm and 382 cm, the down-core curve is almost stabilized at around 21 mT. Below that interval, the down-core curve shows a hump between 406 cm and 437 cm, while does a gradual increase downward in the interval between 440 cm and 485 cm. In the lower most part, the baseline value lies at high value of 23 mT. The maximum value of 23.7 mT appears at 489 cm depth. The overall average value is 21.8 mT.

The ratio of ARM or anhysteretic susceptibility (χ_{ARM}) *versus* χ, which has been used for estimating grain-size variations (e.g., Banerjee et al. 1981; Channell et al. 1998), was determined. We made a plot of χ_{ARM} *versus* χ (a Banerjee plot) to test the magnetic grain size distribution of the samples (Fig. 8.2-5). In the plot, the ratio χ_{ARM}/χ is related to grain size, and the distance from the origin is related to concentration of magnetic fractions. Though the distances of each datum from the origin have a certain range, most of these data fall along one clearly defined track. The ratio χ_{ARM}/χ varies from a minimum of 4.0 to a maximum of 13.6. The average value of the ratio is 8.1.

The amplitudes of these measured variations are significantly small. Only the small changes (a factor of 4.9 for χ, 3.0 for ARM, 1.2 for MDF_{ARM}, and 3.4 for χ_{ARM}/χ) were observed in their overall variations. It strongly demonstrated that the sediment core is rock-magnetically homogeneous.

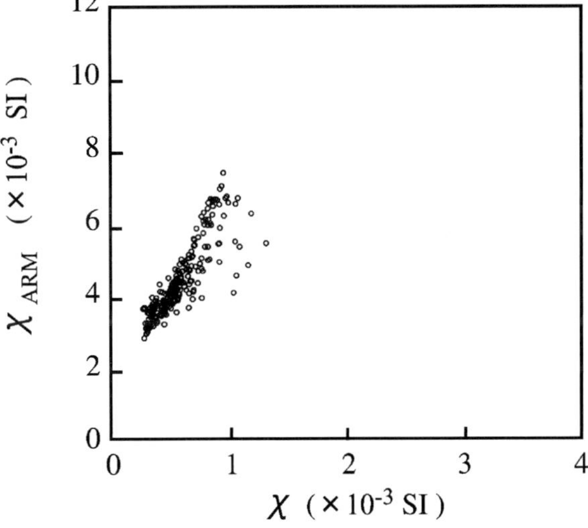

Fig. 8.2-5. Plot of anhysteretic susceptibility (χ_{ARM}) versus susceptibility (χ)

Paleointensity Normalization

NRM is profoundly influenced by the rock-magnetically condition, such as concentration of magnetic minerals, grain size distribution, etc., of sediments and is therefore not a direct indicator of geomagnetic intensity. Thus the normalization of NRM with using an appropriate proxy is essential for paleointensity estimates. Sakai et al. (1998) investigated geomagnetic field intensity with using χ as a normalizer on the same core with ours. However, the parameters measured in the presence of a field, such as χ and saturation magnetization, are hard to relate to remanence measured in an absence of a field, and are disproportionately influenced by superparamagnetic and multi-domain grains which yield minor contribution to the stable remanence (Levi and Banerjee 1976). Thus the parameters such as ARM and IRM should be selected as a normalizer for paleointensity estimates as noted by Levi and Banerjee (1976). In many previous studies, ARM has been most commonly preferred as a normalizer for relative paleointensity estimates (e.g., Stoner et al. 2002; Dinarès-Turell et al. 2002). We thus selected ARM for normalizer. Since ARM acquisition is strongly affected by the amount of particle interaction (Sugiura 1979), King et al. (1983) hence put a limit of a factor of 20–30 on the variation in magnetic concentration and Tauxe (1993) suggested a limit of a factor of 10 for reliable ARM normalization. In our core, the amplitudes of variations of χ and ARM both indicating the variation in magnetic concentration are perfectly within those limits (a factor of 4.9 for χ and 3.0 for ARM), and the amplitude of χ_{ARM}/χ variation is also sufficiently small throughout the core (Fig. 8.2-4; Fig. 8.2-5). It means that the core has a sufficient condition for paleointensity estimates.

The ratio of the NRM demagnetized at 30 mT (NRM_{30mT}) *versus* the ARM demagnetized at 30 mT (ARM_{30mT}), which is the reasonable parameter to eliminate the effects of the secondary remanence, is interpreted as our best approximation for paleointensity estimation. The appropriateness of the normalization was checked by absence of correlation between the normalized intensity (NRM_{30mT}/ARM_{30mT}) and the normalizer (ARM_{30mT}) in Fig. 8.2-6.

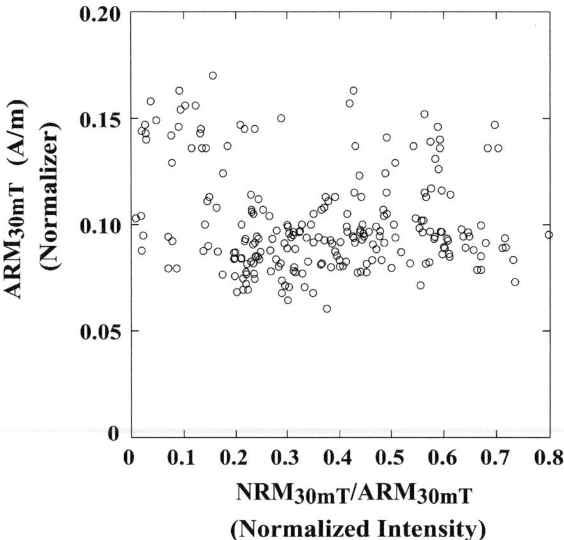

Fig. 8.2-6. Relationship between the normalized intensity (NRM_{30mT}/ARM_{30mT}) and the normalizer (ARM_{30mT})

Results and Discussion

The normalized NRM intensity record of this study is shown in Fig. 8.2-7. For assessing the reliability of the record as a relative paleointensity record, it is important to check a broad scale agreement between records from around the world. In Fig. 8.2-7, our record is compared with two composite paleointensity records and other previously published records, which were well-dated and derived from geographically spread core sites: North Atlantic (Channell and Kleiven 2000), Indian Ocean (Meynadier et al. 1994), the western equatorial Pacific Ocean (Sato and Kobayashi 1998), the western margin of Antarctic Peninsula (Guyodo et al. 2001). For comparison with other records, we briefly defined the initial time scale of our record by the diatom data of Ishihara et al. (1996) obtained from our core and by the magnetostratigraphy in Fig. 8.2-3.

Despite of the long geographic distances, all previously published records contain common features, including, broad peaks centered at around 0.15, 0.35 and 0.65 Ma, and lows around 0.04, 0.11, 0.26, 0.54 and 0.68 Ma (Fig. 8.2-7). Such global synchronicity might be induced from dipole intensity changes and our record seems to involve the possibility to be consistent with those features. Some differences that appear among those published records could be due to non-geomagnetic signal or inaccuracies in dating, as pointed out in previous studies (e.g., Guyodo and Valet 1996; Roberts et al. 1997). In our record, the normalized intensity in the Brunhes chron is much higher than in the Matuyama chron, which is consistent with the result of Sakai et al. (1998) but differs from the other records shown in Fig. 8.2-7. Our record differs from the others also around the Brunhes/Matuyama (B/M) polarity boundary. The other records show sharper increase than ours at the onset of the Matuyama chron. Though the two records, MD90-0940 and KH73-4-7, show relatively slower rate of field recovery following the B/M transition, both of their recovery rates to their large peaks seem much faster than ours. Detailed analyses of the magnetic minerals in our core will be necessary for ascertaining the presence of alterations, which might affect the magnetization and induce discrepancies with other intensity records. A few of the largest decreases, which are acknowledged as excursional events, are not observed in our record. Because excursions are relatively short-term events and can be undetected in the sediments deposited at a low sedimentation rate. In our case, the presumed sedimentation rate is remarkably low (about 2–7 mm kyr^{-1}), hence the thickness of specimens strongly constrain the time resolution of the record by smoothing the intensity variation over a long period of 3–11 kyr. For further study, the improvement of time resolution of our paleointensity record is essential. Thus the magnetic measurement using smaller samples such as plastic cubes of 1 cm^3 is likely to enable us to derive more detailed paleointensity record of the Antarctic region.

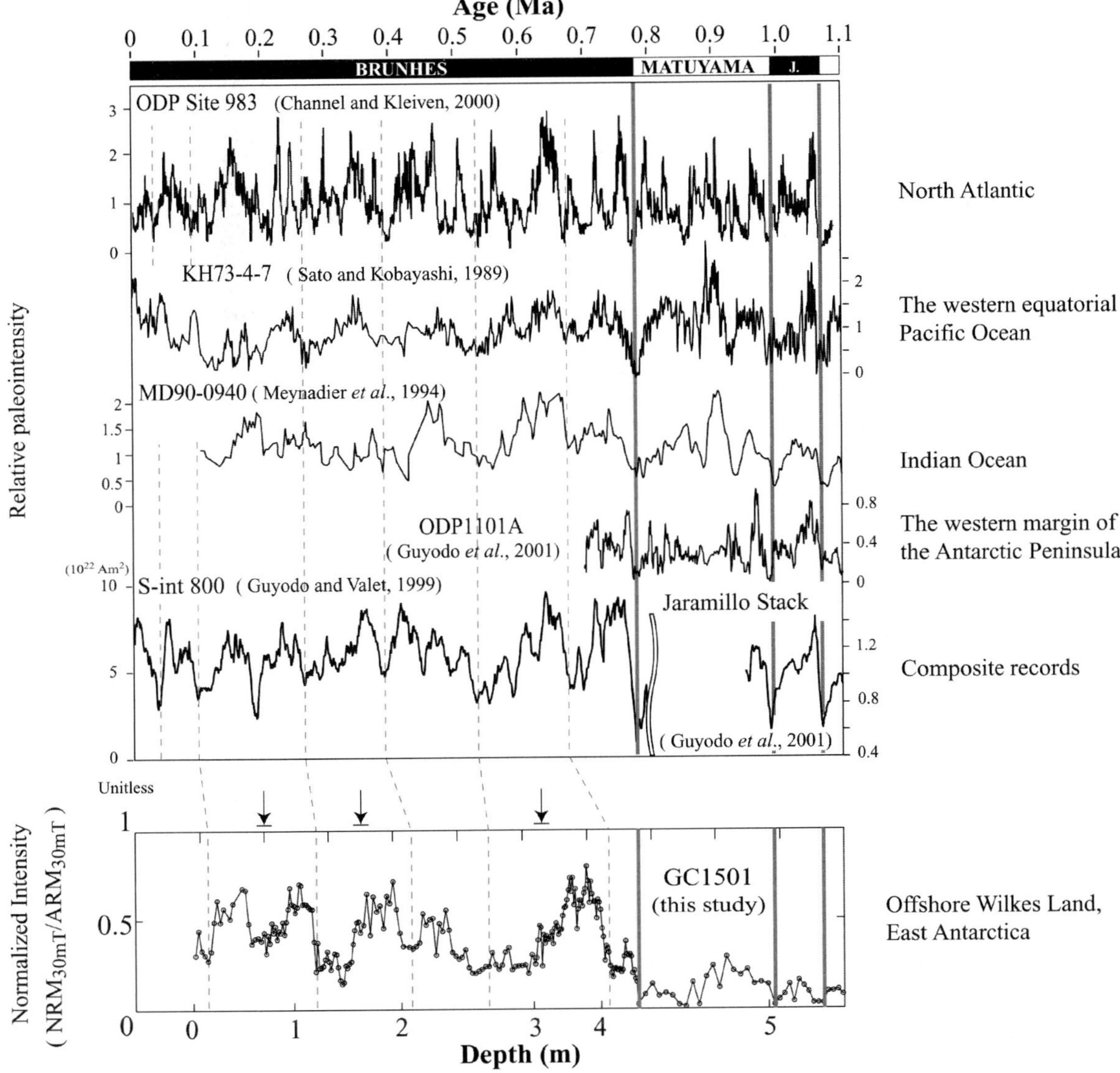

Fig. 8.2-7. Normalized NRM intensity record in comparison with other previously published paleointensity records spanning the last 1.1 Ma. *Arrows* show age control points by the diatom data of Ishihara et al. (1996) obtained from Core GC1501

Acknowledgements

We are grateful to the scientists, officers, and crew on the TH94 cruise of RV "Hakurei-Maru" for obtaining the core used in this study, and the Japan National Oil Corporation for providing the core. We also thank D. Takaharu Sato for his useful suggestions and discussions.

References

Banerjee SK, King JW, Marvin J (1981) A rapid method for magnetic granulometry with application to environmental studies. Geophys Res Letters 8:333–336

Brachfeld S, Acton GD, Guyodo Y, Banerjee SK (2000) High-resolution paleomagnetic records from Holocene sediments from the Palmer Deep, western Antarctic Peninsula. Earth Planet Sci Letters 181:429–441

Cande SC, Kent DV (1995) Revised calibration of the geomagnetic polarity timescale for the Late Cretaceous and Cenozoic. J Geophys Res 100:6093–6095

Channell JET, Kleiven HF (2000) Geomagnetic palaeointensities and astro-chronological ages for the Matuyama-Brunhes boundary and the boundaries of the Jaramillo Subchron: palaeomagnetic and oxygen isotope records from ODP Site 983. Phil Trans Royal Soc Lond 358:1027–1047

Channell JET, Hodell DA, Lehman B (1997) Relative geomagnetic paleointensity and $\delta^{18}O$ at ODP Site 983 (Gardar Drift, North Atlantic) since 350 ka. Earth Planet Sci Letters 153:103–118

Channell JET, Hodell DA, McManus J, Lehman B (1998) Orbital modulation of the Earth's magnetic field intensity. Nature 394:464–468

Dinarès-Turell J, Sagnotti L, Roberts AP (2002) Relative geomagnetic paleointensity from the Jaramillo Subchron to the Matuyama/Brunhes boundary as recorded in a Mediterranean piston core. Earth Planet Sci Letters 194:327–341

Guyodo Y, Valet JP (1996) Relative variations in geomagnetic intensity from sedimentary records: the past 200 000 years. Earth Planet Sci Letters 143:23–36

Guyodo Y, Valet JP (1999) Global changes in intensity of the Earth's magnetic field during the past 800 kyr. Nature 399:249–252

Guyodo Y, Acton GD, Brachfeld S, Channell JET (2001) A sedimentary paleomagnetic record of the Matuyama chron from the Western Antarctic Margin (ODP Site 1101). Earth Planet Sci Letters 191:61–74

Hayes DE, Frakes LA, et al. (1975) Initial Report Deep Sea Drilling Project 28. U.S. Governmental Printing Office, Washington DC, pp 153–177

Ishihara T, Tanahashi M, Sato M, Okuda Y (1996) Preliminary report of geophysical and geological surveys of the west Wilkes Land margin. Proc NIPR Symp Antarct Geosci 9:91–108

Kemp EM, Frakes LA, Hayes DA (1975) Paleoclimatic significance of diachronous biogenic facies, Leg 28, Deep Sea Drilling Project. Initial Report Deep Sea Drilling Project 28, U.S. Governmental Printing Office, Washinton DC, pp 909–917

Johnson HP, Kinoshita H, Merrill RT (1975) Rock magnetism and palaeomagnetism of some North Pacific Deep-Sea sediments. Geol Soc Amer Bull 86:412–420

King JW, Banerjee SK, Marvin J (1983) A new rock magnetic approach to selecting samples for geomagnetic paleointensity studies: Application to paleointensity for the last 4 000 years. J Geophys Res 88:5911–5921

Laj C, Kissel C, Mazaud A, Channell JET, Beer J (2000) North Atlantic paleointensity stack since 75 ka (NAPIS-75) and the duration of the Laschamp event. Phil Trans Royal Soc London 358:1009–1025

Levi S, Banerjee SK (1976) On the possibility of obtaining relative paleointensities from lake sediments. Earth Planet Sci Letters 29:219–226

Lucchi RG, Rebesco M, Busetti M, Caburlotto A, Colizza E, Fontolan G (2002) Sedimentary processes and glacial cycles on the sediment drifts of the Antarctic Peninsula Pacific margin: preliminary results of SEDANO-II project. In: Gamble JA, Skinner DNB, Henrys S (eds) Antarctica at the close of the millennium. Royal Soc New Zealand Bull 35, Wellington, pp 275–280

Meynadier L, Valet JP, Bassinot F, Shackleton N, Guyodo Y (1994) Asymmetrical saw-tooth pattern of the geomagnetic field intensity from equatorial sediments in the Pacific and Indian Oceans. Earth Planet Sci Letters 126:109–127

Roberts AP, Lehman B, Weeks RJ, Verosub KL, Laj C (1997) Relative paleointensity of the geomagnetic field over the last 200 000 years from ODP sites 883 and 884, North Pacific Ocean. Earth Planet Sci Letters 152:11–23

Sagnotti L, Macr P, Camerlenghi A, Rebesco M (2001) Environmental magnetism of Antarctic Late Pleistocene sediments and interhemispheric correlation of climatic events. Earth Planet Sci Letters 192:65–80

Sakai H, Kikawa E, Ishihara T, Kobayashi H, Komori K, Sunagawa A (1998) Paleomagnetic study of marine sediments from Antarctic Sea: Central Wilkes Land margin, Dumont d'Urville Sea and Victoria Land Basin. Polar Geosci 11:222–238

Sato T, Kobayashi K (1989) Long-period secular variations of the Earth's magnetic field revealed by Pacific deep-sea sediment cores. J Geomag Geoelectr 41:147–159

Stoner JS, Laj C, Channell JET, Kissel C (2002) South Atlantic and North Atlantic geomagnetic paleointensity stacks (0–80 ka): implications for inter-hemispheric correlation. Quat Sci Rev 21: 1141–1151

Sugiura N (1979) ARM, TRM and magnetic interactions: concentration dependence. Earth Planet Sci Letters 42:451–455

Tauxe L (1993) Sedimentary records of relative paleointensity of the geomagnetic field: theory and practice. Rev Geophys 31:319–354

Tsumuraya Y, Tanahashi M, Saki T, Machihara T, Asakura N (1985) Preliminary report of the marine geophysical and geological surveys off Wilkes Land, Antarctica in 1983–1984. Mem Nation Inst Polar Res Spec Issue 37:48–62

Grain Size, Mineralogy and Geochemistry in Late Quaternary Sediments from the Western Ross Sea outer Slope as Proxies for Climate Changes

Mario Pistolato[1] · Tullio Quaia[2] · Luigi Marinoni[3] · Laura Menegazzo Vitturi[1] · Cristinamaria Salvi[2] · Gianguido Salvi[2] Massimo Setti[3] · Antonio Brambati[2]

[1] Dipartimento di Scienze Ambientali, University of Venezia, Dorsoduro 2137, I-30123 Venezia, Italy
[2] Dipartimento di Scienze Geologiche, Ambientali e Marine, University of Trieste, Via E. Weiss 2, I-34127 Trieste, Italy
[3] Dipartimento di Scienze della Terra, University of Pavia, Via Ferrata 1, I-27100 Pavia, Italy

Abstract. Textural, mineralogical and geochemical investigations of three sedimentary sequences from the Ross Sea continental slope allow to give some important indications on climatic and environmental changes occurred during the Late Quaternary. The cores show cyclical changes in several proxies (grain size, mineralogical and geochemical parameters) which are in phase with glacial/interglacial changes (MIS 1-8). Such fluctuations are supposed to be driven by changes in transport mechanisms, reworking and provenance of the material, as well as by changes in direction and strength of marine currents induced by variations in the ice coverage.

Introduction

Glacimarine sedimentary sequences from the Ross Sea continental margin provide detailed indications on the Late Quaternary environmental conditions and climate changes. More specifically, grain size, mineralogical and geochemical data obtained from sediment cores recovered from the outer slope can give useful information on long-term climate cycles in the time window of the past few hundred thousand years. Such sedimentary sequences, typically, exhibit low sedimentation rates (Anderson et al. 1979) and do not include hiatuses due to glacial exaration. Conversely, hiatuses may be very common in the sequences from the continental shelf, because of the subglacial erosion exerted by the ice sheet during glacial advances. Cores collected from the outer continental slope frequently show cyclical fluctuations in several proxies (grain size parameters, clay minerals contents, distribution patterns of detrital elements and redox-sensitive elements or palaeoproductivity biogenic tracers), which are in good phase with glacial/interglacial changes (marine isotope stages (MIS) 1–8) (Quaia and Brambati 1997; Ceccaroni et al. 1998; Quaia and Cespuglio 2000; Marinoni et al. 2000; Menegazzo Vitturi et al. 2000). These fluctuations are supposed to be driven by bottom current changes induced by variations in the ice coverage over the continental margin and by changes in sediment transport mechanisms, reworking and provenance, which are, in turn, connected to glacial/interglacial cycles.

According to the Late Quaternary glacimarine sedimentation models proposed by Grobe and Mackensen (1992) and Pudsey (1992) for the Weddell Sea, and by Quaia and Brambati (1997) for the Ross Sea, interglacial periods are characterised by high biological productivity and bottom current intensification due to sea-ice retreat, which lead to an increased transport capacity and a consequent accumulation of better sorted sediment, with more abundant silt and sand fractions. Conversely, during glacial periods, the ice sheet expansion markedly reduces the biological productivity and the bottom currents strength, giving rise to the accumulation of more clayey and less sorted sediment; slow contour currents are the main sedimentary factor which governs the sediment transport and deposition onto the continental slope.

This paper tries to contribute to a better understanding of the sedimentary processes and facies which occur in different depositional environments of the Antarctic continental slope and the relationships existing between these different processes and facies and climate changes.

Geographical Setting

The cores ANTA91-8, ANTA99-c24 and ANTA91-2 were collected from the Ross Sea continental slope north of Cape Adare (Fig. 8.3-1 and Table 8.3-1). This part of the slope is very steep and sinuous. It is dissected by large submarine canyons, which are supposed to have been very active during glacial periods, collecting and discharging large sedimentary inputs from the continental shelf bulldozed by the advancing ice. Current circulation and sedimentary processes in this area are strongly controlled by bottom morphology, which is responsible for distinct depositional patterns. The core site ANTA91-8 lies on a

Table 8.3-1. Sampling sites location and core lengths

Core	Latitude S	Longitude E	Water depth (m)	Core length (cm)
ANTA 91-8	70°46.99'	172°50.39'	2383	511
ANTA 99-c24	70°28.95'	175°57.11'	2750	811
ANTA 91-2	69°59.34'	177°46.85'	3360	613

From: Fütterer DK, Damaske D, Kleinschmidt G, Miller H, Tessensohn F (eds) (2006) Antarctica: Contributions to global earth sciences. Springer-Verlag, Berlin Heidelberg New York, pp 423–432

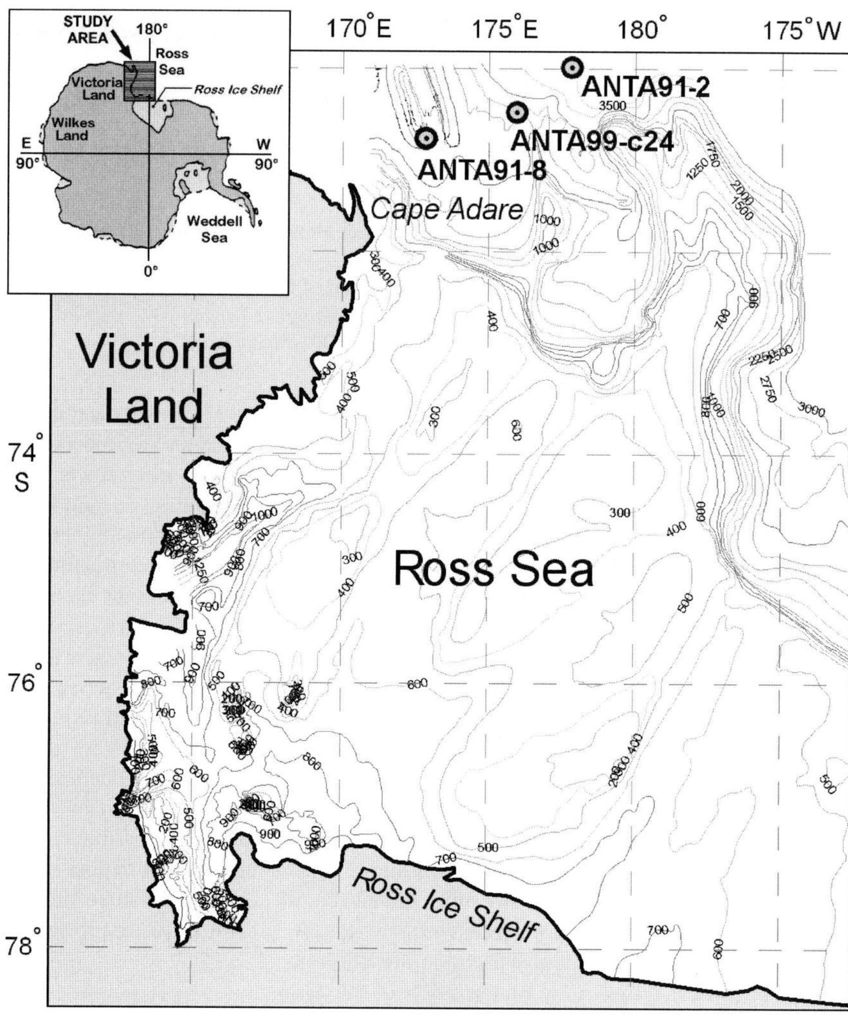

Fig. 8.3-1. Location map for cores *ANTA91-8*, *ANTA91-2* and *ANTA99-c24*

flank of a small topographic high, far from the influence of the above mentioned sedimentary inputs. The core site ANTA91-2 is located off the outlet of a submarine canyon, which cuts the continental margin east of Iselin Bank (along 177° E). The core site ANTA99-c24 lies very close to the canyon outlet.

Materials and Methods

The cores were subsampled at 2–3 cm intervals and textural, mineralogical and geochemical analyses were conducted on several selected samples.

More specifically, 100 samples from core ANTA91-8, 78 samples from core ANTA91-2 and 98 samples from core ANTA99-c24 were chosen for grain size analyses, dried, treated with 10% H_2O_2 and wet-sieved to separate the gravel (>2 mm), sand (2 000–50 μm) and mud (<50 μm) fractions (Nota 1958). The sand fraction was analysed on a Macrogranometer settling tube and the mud fraction on a Micromeritics Sedigraph 5000ET particle size analyser.

Mineralogical analyses were carried out through X-ray powder diffractometry (XRD) on the bulk sediment of all the three cores. On the bulk sediment, the relative abundances of the minerals were estimated from the heights of the main reflections. An evaluation of the amorphous content (biogenic silica) in the bulk sediment was performed by measuring the areas of the diffuse scatter in each sample (Cook et al. 1975). Even if the values reported are not the absolute concentration of biogenic silica in the sediments, they allow the identification of abundance changes of this phase along the core. We also report analyses on the <2 μm fraction (oriented preparation) of core ANTA91-8 (see Marinoni et al. 2000) and some preliminary analyses on core ANTA99-c24. On the clay fraction, the X-ray traces were recorded in natural conditions (air-drying) and after ethylene glycol solvatation. Semiquantitative percentages of each identified mineral phase were then calculated using methodologies reported in Biscaye (1965). TEM (transmission electron microscopy) observations and microanalyses were carried out on the clay fraction of 14 selected samples of core ANTA91-8 by means of a PHILIPS CM 20 coupled with the EDS X-ray spectrometry.

Samples for geochemical analyses were chosen as to define the various units and stages formerly evidenced by grain size analysis and visual description. A total of 86 samples were examined: 40 samples from core ANTA91-8, 16 samples from core ANTA91-2 (0–240 cm, upper part of the core) and 30 samples from core ANTA99-c24. The bulk sediment samples were dissolved and analysed for major, minor and trace elements: Si and K were detected by Atomic Absorption Spectrometry AAS (Perkin Elmer 3100); Cd by AAS coupled with a graphite furnace HGA 500; Al, Ti, Fe, Mg, Mn, Ca, Na, P, Cr, Ni, Co, V, Zn, Pb, Cu were determined by an emission spectrometer equipped with an inductively coupled plasma source ICP (Jobin Yvon H24R). Loss on ignition (L.O.I.) was measured by differential thermogravimetry DTG (Netzsch STA429) at 1 000 °C after pre-heating at 110 °C. AMS-^{14}C datings were performed on the organic fraction of selected levels from the cores. Age determinations were carried out by Geochron Laboratories (Cambridge, Mass., USA) with an accelerator mass spectrometer AMS.

Further specifications regarding analytical procedures are reported in Quaia and Brambati (1997), Marinoni et al. (2000), Menegazzo Vitturi et al. (2000), and Quaia and Cespuglio (2000).

Grain Size and Chronostratigraphy

In all the cores intervals of clayey silts to silty clays alternate down-core, with variable but generally low amounts of sand and gravel clasts, mainly of ice-rafted and, sometimes, of turbiditic origin.

Core ANTA91-8 records (Fig. 8.3-2) the last two main climatic cycles (last 300 ka). A statistical approach to grain size data (factor analysis on frequency percentages at 0.5 phi intervals of the <50 μm fraction) extracted two significant factors (Quaia and Brambati 1997). Factor 1 represents coarser sediment, indicative of sedimentation controlled by stronger bottom currents during warmer periods, whereas factor 2 corresponds to finer sediment, indicative of weaker currents during cooler periods. Textural changes thus reflect past velocity fluctuations of downslope currents (Antarctic Bottom Water (AABW), coming from the mixing of High Salinity Shelf Water (HSSW)), which in turn are related to the different extensions of the ice shelf and the sea ice upon the continental margin during the glacial-interglacial periods: during glacials sluggish contour currents deposited clayey material, whilst during interglacials strong downslope currents gave rise to the accumulation of silty sediments (Quaia and Brambati 1997; Brambati et al. 2002). The factor loadings log is coherent

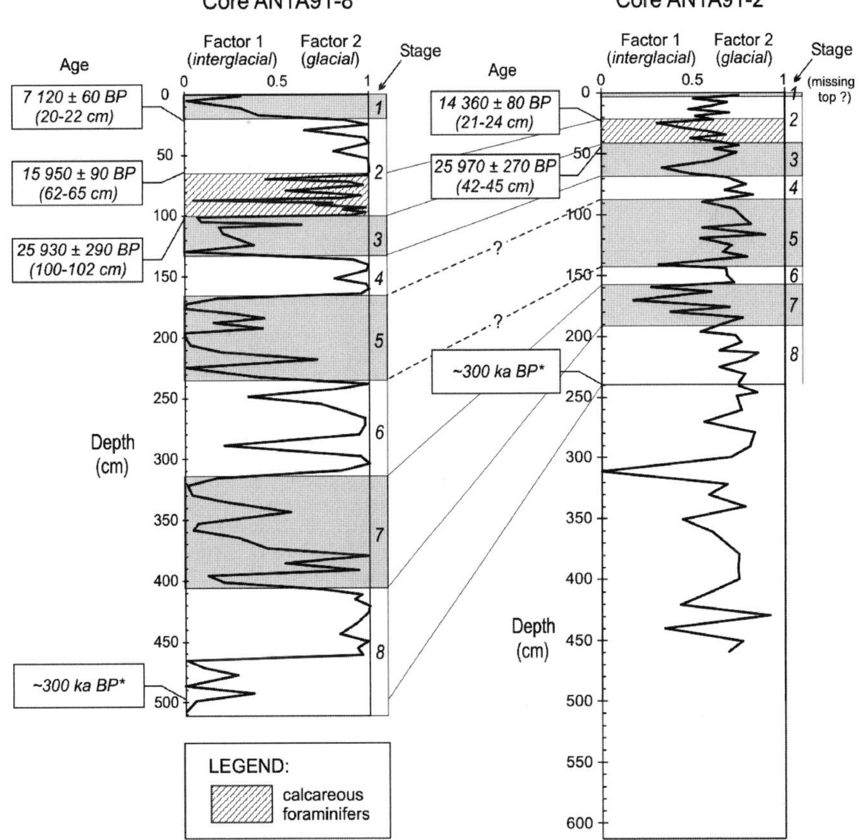

Fig. 8.3-2. Grain size factor loading log and chronostratigraphy of cores *ANTA91-8, ANTA91-2* (after Brambati et al. 2002). *Factor 1* represents interglacial sedimentation, whereas *factor 2* indicates glacial deposition. Stratigraphic correlation is indicated by *solid lines*; *dotted lines* indicate uncertain correlation. Ages from AMS-^{14}C radiometric dating and from the ^{230}Th excess data (marked by an *asterisk*) by Ceccaroni et al. (1998)

with climate changes, and its peaks are well correlated with MIS 1–8 by Martinson et al. (1987). Such subdivision is also confirmed by ^{230}Th excess data (Ceccaroni et al. 1998) and AMS radiocarbon dates (Quaia and Cespuglio 2000).

Core ANTA91-2 is a less expanded sequence: sedimentation rates were found to be almost two times lower than in core ANTA91-8 (Ceccaroni et al. 1998). Several thin laminated layers from very distal turbidites were recognised in this core. Its factor loadings log is, therefore, less clearly correlated with the palaeoclimatic curve (Fig. 8.3-2). Factor loadings correlation between cores ANTA91-8 and 2 is evident for MIS 2, 3, 7 and 8, whereas in core ANTA91-2 the difficult identification of stage five boundaries leads to ambiguous correlations among MIS 4, 5 and 6 (Brambati et al. 2002). Moreover, MIS 1 has reduced thickness, probably due to the loss of top material during the core recovery.

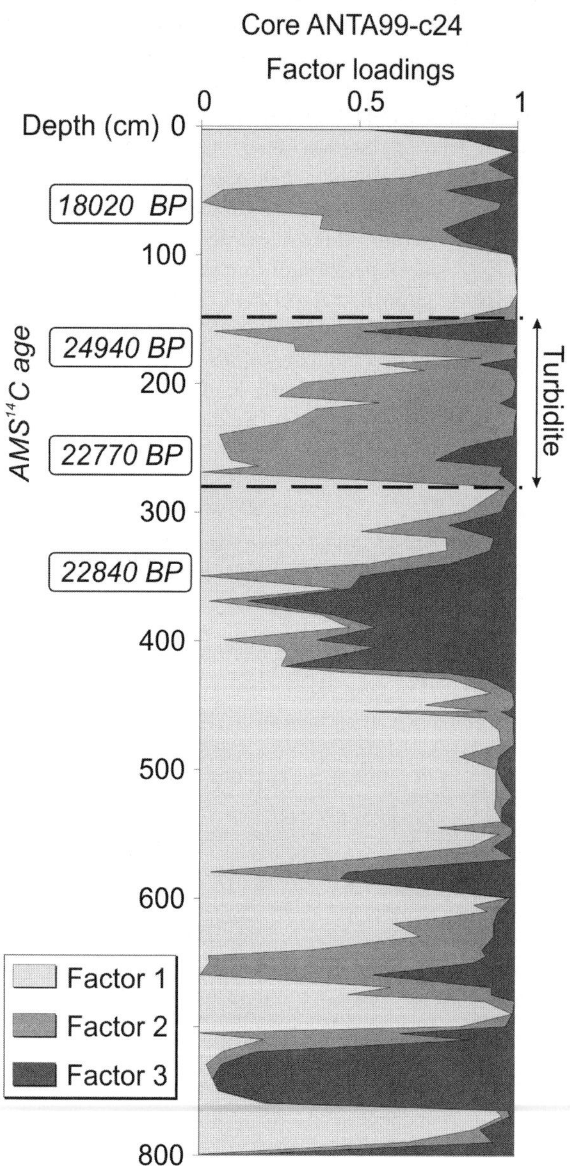

Fig. 8.3-3. Grain size factor loading log and chronostratigraphy of core *ANTA99-c24. Factor 1* represents glacial deposition, *factor 2* a turbiditic mud fraction and *factor 3* interglacial sedimentation. Ages from AMS-^{14}C radiometric dating

Palaeoclimatic signals in core ANTA99-c24 are even less evident. A turbidite interval in this core shows the classical Bouma sequence: a coarse lowermost layer overlain by a fining-upward sequence and a laminated silty-clayey interval. The topmost part of this core is probably lacking, due to mudflow events and/or strong bottom currents which might have been so efficient to result in an erosional event removing the upper part of the Holocene. Furthermore, coring problems are not to be excluded, which might have led to the loss of the core top.

Age reversals support resedimentation events in the upper portion of the core, at least in the interval between 58 and 339 cm (Fig. 8.3-3).

A factor analysis of textural data from core ANTA99-c24 (same variables used for cores ANTA91-8 and ANTA91-2) extracted three factors which account for 93.89% of variance. Grain size frequency distributions of each endmember are reported in Fig. 8.3-4. The endmember of factor 1 is a clayey mud. It corresponds to endmember 1 of core ANTA91-8 and represents a glacial mud fraction deposited by slow bottom currents. The endmember of factor 2 represents a turbiditic mud fraction. The grain size distribution is almost flat (except for the two tails of the curve) thus seeming indicative of mass transport processes with no or very little grain size sorting by marine currents. The endmember of factor 3 is a coarse silt. It corresponds to endmember 2 of core ANTA91-8 and can be interpreted as an interglacial mud fraction, which represents transport processes in a high energy depositional environment (stronger bottom currents, typical of interglacial stages).

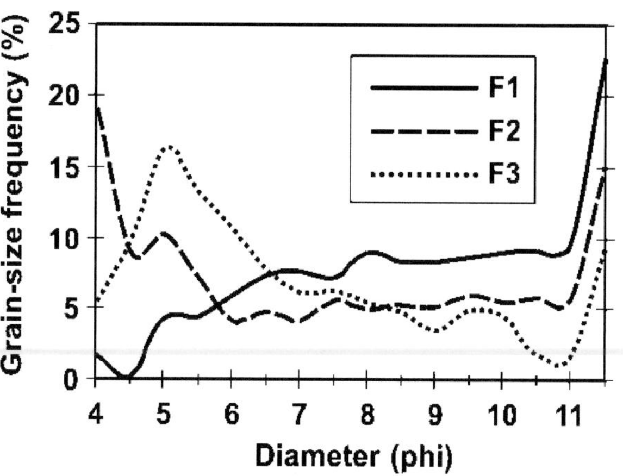

Fig. 8.3-4. Grain size frequency distribution of the three endmembers of core *ANTA99-c24* extracted from the factor analysis

Mineralogy

Bulk sediment mineralogy of the three sequences is composed by large amount of amorphous biogenic silica, related to the occurrence of siliceous organisms. Concerning silicate minerals, quartz and feldspars are the most abundant, mica occurs in relevant amounts, chlorite is subordinate. This mineral assemblage, which mostly represents the product of physical weathering processes, is typical of high-latitude environment, as the composition of detrital mineral suites in polar regions depends mainly on petrographic nature of source rocks (Chamley 1989). Down-core changes in the distribution of the minerals are partly controlled by the grain size of the sediments, as the quartz is higher in coarser fraction, whereas phyllosilicates like chlorite and mica increase in finer fractions.

In core ANTA91-8 the variations of the biogenic silica content reflect the climatic stages; biogenic silica is higher during interglacials and lower during glacials (Fig. 8.3-5). In particular, glacial MIS 6 seems well characterised by constant and lower content of biogenic silica. The variations of biogenic silica are related to variations of productivity in the water (Hillenbrand and Fütterer 2001).

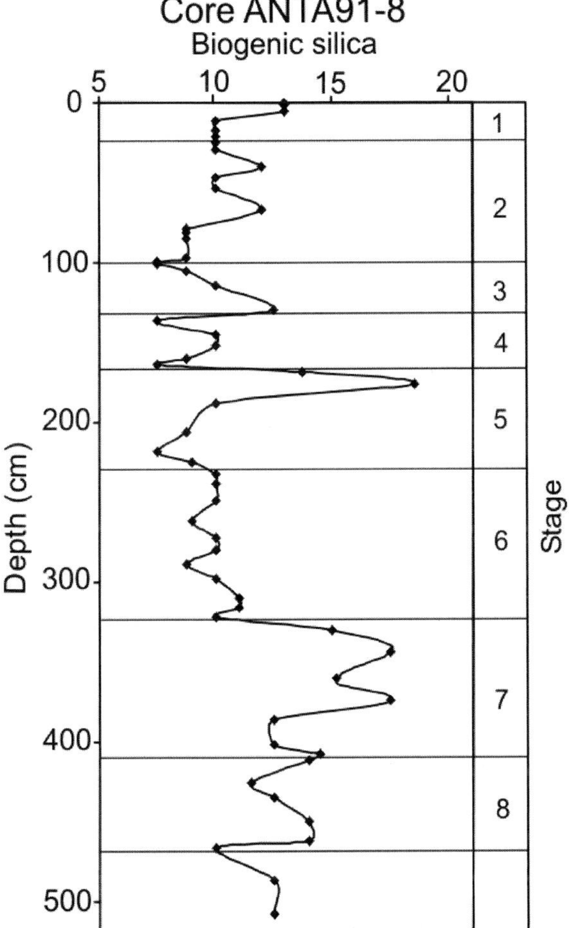

Fig. 8.3-5. Biogenic silica variations in the bulk sediment of core *ANTA91-8*, calculated by measuring the area of the diffuse scatter (Cook et al. 1975). Values are expressed as cm^2 of the diffuse scatter area. On the *right* the isotopic stages by Martinson et al. (1987) are reported

The mineralogical assemblage of the clay fraction of cores ANTA91-8 and ANTA99-c24 is similar, and mainly composed by mica/illite with lower content of smectite and chlorite and traces of kaolinite. Basing on the 5/10 Å peaks intensity ratio (Esquevin 1969), mica in the sequences is largely composed by biotite.

Mica/illite and chlorite in marine sediments are of detrital origin as these minerals are usually the product of physical weathering on different kind of lithologies (Biscaye 1965). Chlorite generally occurs in low-grade metamorphic and basic lithologies while illite tends to derive from more acid ones (Biscaye 1965; Ehrmann et al. 1992). Concerning smectite, the interpretation of its genesis can be controversial, as this mineral may be detrital or authigenic. Authigenic smectites may form in the sea floor or in the sedimentary sequences from processes like hydrothermalism or submarine alteration of volcanic glass and rock fragments (Chamley 1989; Setti et al. 2004). Detrital smectites, in contrast, derive from continental soils; they generally form by hydrolysis under humid climatic conditions, in environments characterised by very slow movement of water (Chamley 1989). In general, smectite formation in the recent Antarctic environment is only a subordinate process, reported from a few soils and tills (Campbell and Claridge 1987; Chamley 1989), even if large smectite concentrations have been reported from glaci-marine sediments in areas with basalt in the hinterland (Ehrmann et al. 1992). It is important to differentiate between detrital and authigenic smectites as only detrital phase can be indicative of paleoclimate and provenance.

TEM investigations performed on the sediments of sequence ANTA91-8 indicates that smectites are dioctahedral terms and can be classified as Al-Fe beidellites. The mean calculated structural formula of the most abundant smectites is as follows $(Si_{3.66}Al_{0.34})O_{10}(Al_{0.98}Mg_{0.44}Fe_{0.66}Ti_{0.03})(OH)_2(K_{0.26}Na_{0.01}Mg_{0.06})$. Smectite microparticles generally show plate shapes like those common in soils. These features indicate that smectites are mostly detrital and formed through chemical weathering processes on parent rocks containing abundant volcanic materials. No marked compositional difference was observed between smectites deposited during glacial stages and those deposited during interglacials.

It is possible to hypothesize the different provenance of clays on the basis of the composition of parent rocks and the characteristics of the currents. The importance of clay minerals distribution to differentiate between glacial and interglacial intervals in sediments from Antarctica has been demonstrated also by other authors (Lucchi et al. 2002).

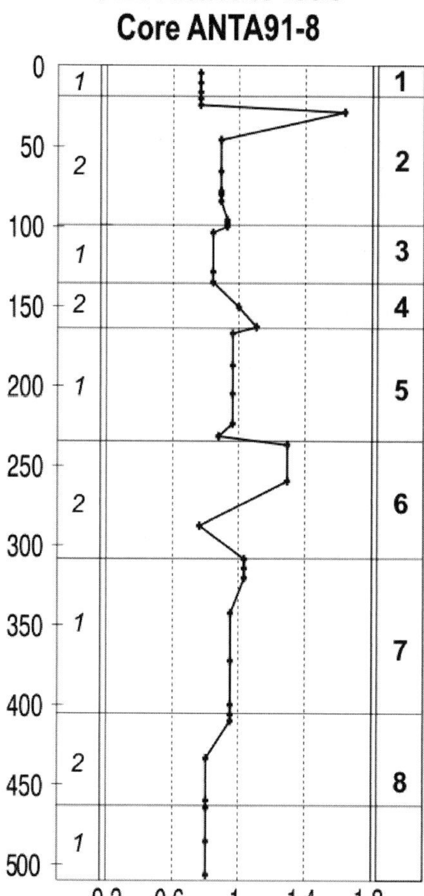

Fig. 8.3-6. Smectite/illite variation in the <2 µm fraction of the sediments of the core *ANTA91-8*. Grain-size cluster as previously defined by Quaia and Brambati (1997) are reported on the *left*; the corresponding isotopic stages by Martinson et al. (1987) are indicated on the *right*

The variations in clay mineral assemblages in core ANTA91-8 appear to reflect the climatic cycles (Fig. 8.3-6) and the trend of the smectite/illite ratio is the most indicative marker. Smectite is higher during interglacials, while chlorite and, in particular, illite increase during glacial phases. The compositional variations are supposed to reflect the different roles played by downslope currents and contour currents on sedimentation. Smectite is more abundant during interglacials, as strong bottom currents were active during warmer phases, because of reduced ice coverage and enhanced high-density water production. Bottom currents resuspended and transported materials enriched in smectite from the inner part of the continental shelf (McMurdo Volcanics, see Fig. 8.3-1) onto the outer slope giving rise to the accumulation of more silty, smectite-rich sediments. Conversely, chlorite and mica/illite, which can derive from the physical weathering on crystalline and/or metamorphic rocks, are more abundant during colder phases, as these minerals were transported by sluggish contour currents that were stronger during glacial periods. Contour currents resuspended and transported fine-grained materials, enriched in illite and chlorite, from the East Antarctic craton (Wilkes Land) to the site investigated. Changes in the weathering regime did not play an important role. Therefore, the trend of clay mineral variations in core ANTA91-8 (Marinoni et al. 2000) supports well the subdivision obtained by the grain size statistical analysis.

Geochemistry

The abundance of major and trace elements shows different ranges in the three cores and the presence of more or less marked fluctuations. In cores ANTA91-8 and ANTA99-c24 the Al_2O_3, Fe_2O_3 and K_2O contents have similar trends characterised by significant vertical variations (Fig. 8.3-7). The depth profiles of trace elements Cr, Ni, Co and V are quite analogous. Only the element ranges are different in the two sequences. Conversely, a quite different behaviour can be observed in core ANTA91-2, where less marked down-core changes are detected. In general, the distribution of major and minor elements is largely controlled by the mineralogical composition of the bulk sediment: Fe_2O_3 and MgO are mostly related to the content of chlorite and mica/illite, while K_2O to mica/illite. SiO_2 can be related to several mineral phases, but the increasing, especially in the core ANTA91-8, is related to higher content of biogenic silica. Positive correlations are also observed between the content of mica plus chlorite and trace elements like V, Zn, Cr, Co, Cu. The content of trace elements is higher in the phyllosilicates because these elements are usually incorporated, or adsorbed, in clay mineral structures.

A statistical approach (cluster analysis) was applied to the compositional data of each core.

Samples from core ANTA91-8 were associated into four clusters. The cluster assignments of samples were plotted with respect to the chronostratigraphy of the core. A very good correlation is evident between the discriminated groups and climatic conditions (cluster 1: cold events; cluster 2 and 3: transition glacial–interglacial; cluster 4: warmer events of interglacial stages) (Menegazzo Vitturi et al. 2000). The geochemical characteristics of each cluster are reported in Table 8.3-2. The sediments deposited during glacial stages (cluster 1) are characterised by low SiO_2 and higher contents of Al_2O_3, TiO_2, Fe_2O_3, MgO, K_2O, L.O.I., Cr, Ni, V, Zn. Conversely, sediments of the interglacial stages (cluster 4) show higher values of SiO_2, Mn, Cu and Cd, and lower contents of Fe_2O_3, MgO, L.O.I., Cr, V and Zn.

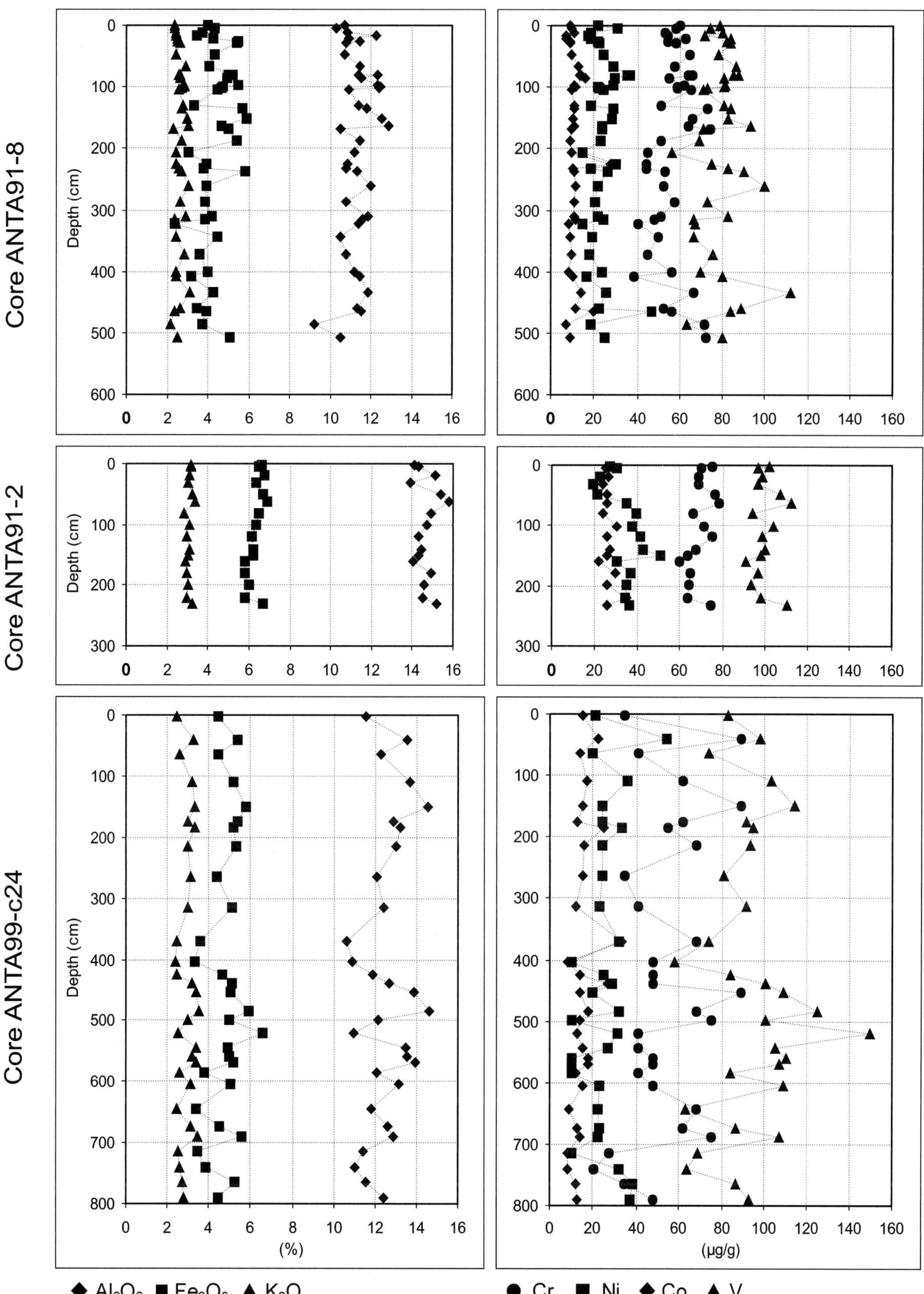

Fig. 8.3-7. Down-core elements distributions of cores *ANTA91-8*, *ANTA91-2* and *ANTA99-c24*

Table 8.3-2. Geochemical composition (average values) of cluster from cores ANTA91-8, ANTA91-2 and ANTA99-c24. Major elements as percent oxides, L.O.I. as percent and trace elements in $\mu g\ g^{-1}$

Core ANTA91-8					Core ANTA91-2				Core ANTA99-c24				
Cluster	1	2	3	4	Cluster	1	2	3	Cluster	1	2	3	4
SiO_2	67.1	69.9	71.4	69.9	SiO_2	62.0	63.1	64.0	SiO_2	69.8	66.3	66.2	67.7
TiO_2	0.8	0.8	0.7	0.7	TiO_2	0.8	0.8	0.8	TiO_2	0.7	0.8	0.8	0.8
Al_2O_3	11.8	10.5	11.3	11.3	Al_2O_3	14.6	14.8	14.6	Al_2O_3	11.8	12.8	12.7	12.6
Fe_2O_3	5.1	4.5	3.4	4.3	Fe_2O_3	6.6	6.4	6.0	Fe_2O_3	4.0	4.9	5.2	4.5
MgO	2.0	1.7	1.6	1.7	MgO	2.8	2.6	2.5	MgO	1.8	2.1	2.2	2.0
CaO	2.1	2.1	1.7	1.8	CaO	2.2	1.4	1.4	CaO	1.8	1.5	1.6	1.7
Na_2O	2.2	2.3	2.4	2.3	K_2O	3.1	3.1	3.0	Na_2O	2.0	1.8	1.8	1.9
K_2O	2.8	2.4	2.6	2.5	Na_2O	2.2	2.2	2.2	K_2O	2.7	3.1	3.0	2.9
P_2O_5	0.2	0.1	0.1	0.1	P_2O_5	0.1	0.1	0.1	P_2O_5	0.1	0.1	0.1	0.1
L.O.I.	5.5	5.2	4.3	4.7	L.O.I.	5.6	5.2	4.5	L.O.I.	5.1	6.1	6.2	5.6
Mn	471	385	419	996	Mn	1077	1103	1149	Mn	480	1766	599	620
Cr	61	68	48	54	Cr	72	68	68	Cr	52	64	53	48
Ni	27	22	18	25	Ni	24	39	37	Ni	21	37	26	10
Co	12	9	10	12	Co	25	255	29	Co	11	25	14	14
V	88	75	75	75	V	100	101	98	V	75	94	100	94
Zn	121	101	100	102	Zn	129	116	99	Zn	75	102	99	92
Pb	12	17	9	18	Pb	29	22	34	Pb	17	18	26	27
Cu	37	35	36	37	Cu	54	55	445	Cu	34	55	47	30
Cd	0.3	0.2	0.1	0.3	Cd	0.2	0.1	0.1	Cd	0.1	0.6	0.7	0.7

Samples from core ANTA91-2 were associated in three clusters. Cluster 1 is well discriminated with respect to the other two and groups sediments of the upper part of the core (from 50 cm to the top). These sediments are characterised by lower contents of SiO_2 and Mn and higher contents of TiO_2, Fe_2O_3, MgO, CaO, K_2O, L.O.I., Cr, Zn and Cd (Table 8.3-2). Conversely, sediments of the clusters 2 and 3 show higher SiO_2 and Mn contents and lower contents of Fe_2O_3, MgO, CaO, K_2O, L.O.I., Cr, Zn and Cd.

Samples from core ANTA99-c24 were associated into four clusters. Cluster 1 is well discriminated with respect to the other three groups. Cluster 4 samples are sited in to the lower part of the core. The sediments of cluster 1 are characterised by high contents of SiO_2 and low of TiO_2, Al_2O_3, Fe_2O_3, MgO, K_2O, L.O.I., Mn, V, Zn and Cd (Table 8.3-2). Conversely, sediments of the clusters 3 and 4 show lower contents of SiO_2 and higher of Al_2O_3, Fe_2O_3, MgO, K_2O, L.O.I., V, Zn, Pb and Cd. Sediments of cluster 2 are characterised by similar abundance of major elements as in cluster 3 and 4, but highest values of Mn, Cr, Ni, Co, Zn, and Cu.

Distribution patterns of major and trace elements highlighted by the peculiar geochemistry of each cluster identified in each core support the cyclicity defined by the grain size distribution and by clay mineral variations. The element fluctuations are controlled by different contents of detrital elements (as Al, Ti, Fe, Mg and K), redox-sensitive elements (Mn, V, Cr, Zn) and biogenic proxies of productivity (Si, Cu and Cd). In each sedimentary sequence was identified one cluster which groups sediments enriched in major detrital elements (as Al, Ti, Fe, Mg and K), and in trace elements as V, Zn, Cr and Ni, with lower Si and Mn contents (cluster 1 in core ANTA91-8, cluster 1 in core ANTA91-2, cluster 4 in core ANTA99-c24). These sediments correspond to clayey muds typical of glacial stages. In each core was also extracted a second cluster which groups samples enriched in Si and Mn, with lower percentages of Al, Ti, Fe, Mg, K and L.O.I. and lower contents of trace elements except for Cu and Cd (cluster 4 in core ANTA91-8, cluster 3 in core ANTA91-2, cluster 1 in core ANTA99-c24). This group corresponds to coarser sediments (silt or clayey silt) typical of interglacial stages.

Conclusions

Statistical analysis on grain size, bulk sediment and clay fraction mineralogy and geochemical data from sediment cores collected in the Antarctic continental slope allows the recognition of glacial/interglacial cycles recorded in the sedimentary sequences. This multidisciplinary statistical approach can give important indications on the provenance of the materials, on transport processes and on changes of palaeoenvironmental conditions. Different geomorphological settings strongly influence the sedimentary response to climatic changes and, therefore, the facies recovered in cores collected from different sampling sites can be very different as well. Glacial sediments from the Ross Sea continental slope are generally characterised by higher clay and lower silt contents, worse sorting and negative skewness, higher contents of major detrital elements (Al, Ti, Fe, Mg, K) and of trace elements (V, Zn, Cr, Ni) and by lower contents of Si and Mn. In the clay fraction, illite and chlorite are higher while smectite is lower. Interglacial sediments are characterised by higher silt and lower clay contents, better sorting and positive skewness, higher values of Si and Mn, lower contents of Al, Ti, Fe, Mg, K, L.O.I and trace elements except for Cu and Cd. In the clay fraction, smectite is higher while illite and chlorite are lower.

The grain size characteristics, clay minerals associations and geochemical composition in the late Quaternary sediments from the northwestern Ross Sea continental slope seem to reflect the changes in the influence of the source areas. Compositional and textural variations are supposed to be the product of the different impact played by the downslope currents and contour currents onto the sediment deposition. During interglacial stages, due to reduced ice coverage and enhanced high-density water production, strong downslope currents resuspend and transport materials from the inner part of the continental shelf onto the outer slope. During glacial stages, sluggish contour currents continuously resuspend and transport fine grained materials from the East Antarctic craton (Wilkes Land) to the site investigated.

During glacials, turbidites are not infrequent, due to the bulldozing effect of the advancing ice sheet onto the continental shelf, which, during glacial maxima, pushes the sediment beyond the shelf break. The instability of the accumulated sediment often originates slumpings, which lead to the presence of layers of proximal to distal turbidite material within the sequences investigated. The thickness and texture of these turbidite layers, which partially mask the paleoclimatic rythmic signal detected by statistical analysis on textural and compositional data, vary according to the morphological features of the coring site area and to its distance from the turbidite source.

Acknowledgments

This work was carried out with the financial support of the Italian Programma Nazionale di Ricerche in Antartide (PNRA). We are grateful to Renata G. Lucchi and to the anonymous referee who greatly improved our manuscript.

References

Anderson JB, Kurtz DD, Weaver FM (1979) Sedimentation on the Antarctic continental slope. In: Doyle LJ, Pilkey O (eds) Geology of continental slope. Soc Econ Paleontol Mineral Spec Publ 27:265–283

Biscaye PE (1965) Mineralogy and sedimentation of recent deep sea clay in the Atlantic Ocean and adjacent seas and oceans. Geol Soc Amer Bull 76:803–832

Brambati A, Melis R, Quaia T, Salvi G (2002) Late Quaternary climatic changes in the Ross Sea area, Antarctica. Royal Soc. In: Gamble JA, Skinner DNB, Henrys S (eds) Antarctica at the close of the millennium. Royal Soc New Zealand Bull 35, Wellington, pp 359–364

Campbell JB, Claridge GGC (1987) Antarctica: soils, weathering processes and environment. Elsevier Publishing, Amsterdam

Ceccaroni L, Frank M, Frignani M, Langone L, Ravaioli M, Mangini A (1998) Late Quaternary fluctuations of biogenic components fluxes on the continental slope of the Ross Sea, Antarctica. J Mar Systems 17:515–525

Chamley H (1989) Clay sedimentology. Springer, Berlin Heidelberg New York

Cook HE, Johnson PD, Matti JC, Zemmels I (1975) Methods of sample preparation and X-Ray diffraction data analysis, X-Ray mineralogy laboratory. Initial Report Deep Sea Drilling Project 28, U.S. Governmental Printing Office, Washington DC, pp 999–1007

Ehrmann WU, Melles M, Kuhn G, Grobe H (1992) Significance of clay mineral assemblages in the Antarctic Ocean. Mar Geol 107:249–273

Esquevin J (1969) Influence de la composition chimique des illites sur la cristallinité. Bull Cent Rech Pau SNPA 3:147–154

Grobe H, Mackensen A (1992) Late Quaternary climatic cycles as recorded in sediments from the Antarctic continental margin in the Antarctic paleoenvironment: a perspective on global change. Antarct Res Ser 56:349–376

Hillenbrand C-D, Fütterer DK (2001) Neogene to Quaternary deposition of opal on the continental rise west of the Antarctic Peninsula, ODP Leg 178, Sites 1095, 1096, and 1101. In: Barker PF, Camerlenghi A, Acton GD, Ramsya ATS (eds) Proc ODP, Sci Results 178: 1-33, <http://www-odp.tamu.edu/publications/178_SR/VOLUME/CHAPTERS/SR178_023.PDF>

Lucchi RG, Rebesco M, Camerlenghi A, Busetti M, Tomadin L, Villa G, Persico D, Morigi C, Bonci MC, Giorgetti G (2002) Mid-late Pleistocene glacimarine sedimentary processes of a high-latitude, deep-sea sediment drift (Antarctic Peninsula Pacific margin). Mar Geol 189:343–370

Marinoni L, Quaia T, Setti M, Lopez-Galindo A, Brambati A (2000) Mineralogy and crystal-chemistry of the clay fraction in core ANTA91-8 (Ross Sea, Antarctica): palaeoclimatic and palaeoenvrionmental implications. Terra Antartica Rep 4:211–216

Martinson DG, Pisias NG, Hays JD, Imbrie J, Moore TC, Shakleton NJ (1987) Age dating and the orbital theory of Ice Ages: development of a high resolution 0 to 300000-year chronostratigraphy. Quat Res 27:1–29

Menegazzo Vitturi L, Pistolato M, Rampazzo G, Visin F (2000) Inorganic geochemical indicators in glacial marine sediments: applications to the palaeoclimatic fluctuations recorded by Late Quaternary sequences in Antarctica. Terra Antartica Rep 4:217–228

Nota DJG (1958) Sediments of the western Guiana shelf. Reports of the Orinoco Shelf Expedition. Vol. II, Mededel Landbouwhogeschool Wageningen, Nederland 58:1–98

Pudsey CJ (1992) Late Quaternary changes in Antarctic bottom water velocity inferred from sediment grain-size in the Northern Weddell Sea. Mar Geol 107:9–33

Quaia T, Brambati A (1997) Climatic stages control on grain-size clusters in core ANTA91-8 (Ross Sea). Geografia Fisica Geodinam Quatern 20:279–282

Quaia T, Cespuglio G (2000) Stable isotope records from the western Ross Sea continental slope (Antarctica): considerations on carbonate preservation. Terra Antartica Rep 4:199–210

Setti M, Marinoni L, López-Galindo A (2004) Mineralogical and geochemical characteristics (major, minor, trace elements and REE) of detrital and authigenic clay minerals in as Cenozoic sequence from Ross Sea, Antarctica. Clay Minerals 39:4055-421

Potential of the Scotia Sea Region for Determining the Onset and Development of the Antarctic Circumpolar Current

Peter Barker[1] · Ellen Thomas[2]

[1] Threshers Barn, Whitcott Keysett, Clun, Shropshire SY7 8QE, UK, <pfbarker@tiscali.co.uk>

[2] Center for the Study of Global Change, Dept. of Geology and Geophysics, Yale University, New Haven CT 06520-1809, USA

Abstract. The strength of interaction between tectonics, ocean circulation and climate is a major concern of palaeoclimate research. To evaluate the strength, we must assess the time of onset and development of the Antarctic Circumpolar Current (ACC) and its likely effects on climate, particularly Antarctic glaciation. Developments in numerical climate modelling, marine geology, tectonics and physical oceanography have cast doubt on widely held assumptions of a causal relationship between the ACC and glacial onset, in the Eocene-Oligocene boundary interval. Here we argue that our best chance to determine ACC onset and development is in the Scotia Sea region ("Drake Passage"), south of South America. There lies the greatest tectonic uncertainty, concerning when a complete deep-water circumpolar pathway was created, and (thus) when the ACC developed as we know it today. There also, the ACC is topographically constrained, and key factors (water mass and sediment distributions, sea-floor spreading history) are sufficiently well known. Determination of the time of onset would enable solution of other questions, such as the nature of Southern Ocean circulation and primary productivity in any period (possibly Oligocene and early Miocene) when Antarctica was glaciated but before a complete circumpolar deep-water pathway existed, and the extent to which ocean circulation changes affected palaeoclimate, particularly Antarctic glaciation. We assess the parameters that might be capable of determining ACC onset, and show that suitable sedimentary records are available in the Scotia Sea region.

Introduction

The Antarctic Circumpolar Current (ACC) is highly influential in the modern climate system. The largest ocean current in the world, it averages about 130 Sv ($10^6\ m^3\ s^{-1}$) volume transport through Drake Passage. It is mainly or entirely wind-driven, but extends to the seabed in most places. It is closely associated with one or more deep-reaching oceanic fronts, and its mean "axial" position is approximated by the locus of the Polar Front (PF in Fig. 8.4-1). The sea surface expression of most of the fronts associated with it (PF, Sub-Antarctic Front SAF, Southern ACC Front SACCF – Fig. 8.4-1) is a sharp southward temperature drop. This association of the ACC with changes in sea surface temperature and thus planktonic biotic assemblage, and its extension to the seabed, have led to its past identification and location using proxies within the geologic record, and to speculation about its effects on palaeoclimate. The ACC developed at some time during the Cenozoic, as deep-water gaps opened around Antarctica, and has been widely viewed as having reduced meridional heat transport, isolating the continent within an annulus of cold water and thus being at least partly responsible for Antarctic glaciation (e.g., Kennett 1977). It is today the major means of exchange of water between oceans, and its onset may have significantly modified Northern Hemisphere climate (Toggweiler and Bjornsson 1999; Sijp and England 2004). Some consider the atmospheric concentration of greenhouse gases (principally CO_2) to have been the prime influence on Antarctic palaeoclimate (e.g., De Conto and Pollard 2003), but the relative contributions of ocean circulation and greenhouse gases to global climate change are not yet understood. In this paper, we assess the potential of the geological record in the Scotia Sea region to reveal the time of onset and subsequent development, and thence probably the climatic influence, of the ACC.

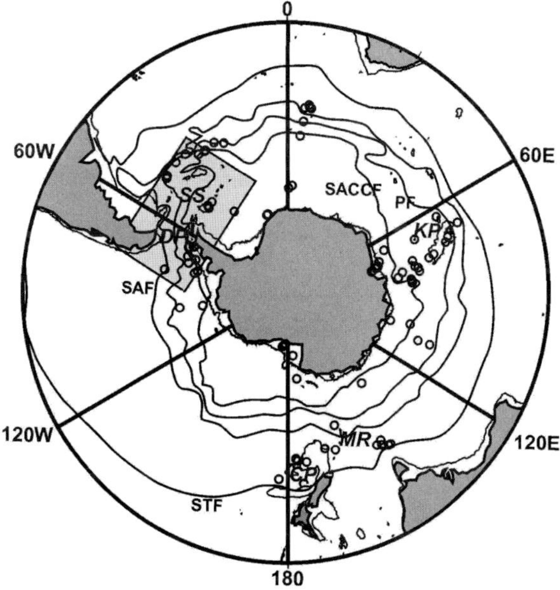

Fig. 8.4-1. ACC fronts (*SAF:* Sub-Antarctic Front; *PF:* Polar Front; *SACCF:* southern ACC front), and Subtropical Front (*STF*) which is not continuous, located by Orsi et al. (1995) based on ca. 100 ship transects. Also existing Southern Ocean DSDP and ODP sites (*open circles*). Figure 8.4-3 shows detail in Scotia Sea region (outline of Fig. 8.4-3 is *lightly shaded box*)

From: Fütterer DK, Damaske D, Kleinschmidt G, Miller H, Tessensohn F (eds) (2006) Antarctica: Contributions to global earth sciences. Springer-Verlag, Berlin Heidelberg New York, pp 433–440

Nature and History of the ACC

The time of onset of the ACC is uncertain, and abundant speculation exists about its onset and effects, despite improvements in our understanding of the geological record and an expanded palaeoceanographic data base. The traditional view is that the ACC began coevally with substantial Antarctic glaciation, in the Eocene-Oligocene boundary interval (the isotope peak Oi-1 is actually shortly after the boundary, e.g., Zachos et al. 2001), and probably caused this glaciation by isolating the continent in a cold-water annulus. In the traditional view also, ACC signature is a southern biosiliceous facies, indicating cold surface waters, with a coeval biocalcareous facies north of an ACC axis, the two separated by a broad zone of non-deposition or erosion corresponding to rapid bottom-water flow at the axis itself.

Recent ODP drilling on Leg 189 (Exon et al. 2001) confirmed that a deep-water gap opened south of Tasmania close to the Eocene-Oligocene boundary, supporting the view of a link between glacial onset and ocean circulation. However, the final barrier in an otherwise-continuous deep-water pathway, essential for continuous, deep-reaching current jets similar to the modern ACC, is generally accepted as lying south of South America, at Drake Passage and other obstructions around the developing Scotia Sea. Estimates of the removal time of this final barrier, based on regional and local tectonics, range from 16–22 Ma to 31–34 Ma (Barker and Burrell 1977, 1982; Barker 2001; Lawver et al. 1992; Lawver and Gahagan 1998, 2003; Livermore et al. 2004. See also Fig. 8.4-2).

Three developments have thrown doubt on the validity of past palaeoceanographic conclusions. First, modern physical oceanography (e.g., Nowlin and Klinck 1986; Gille 1994; Heywood and King 2002) recognises that virtually all ACC transport occurs in narrow jets within deep-reaching fronts that are continuous around Antarctica (the SAF, PF and a less energetic SACCF, Fig. 8.4-1). Second,

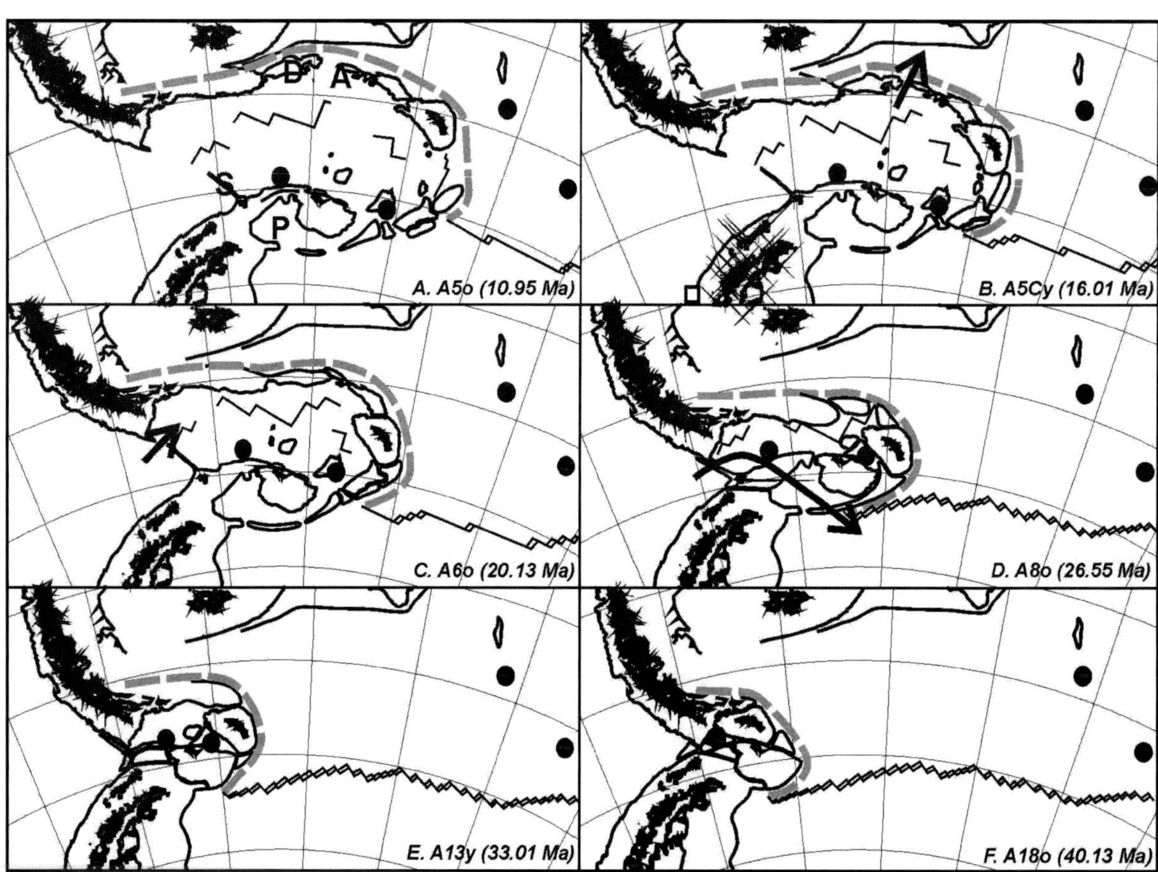

Fig. 8.4-2. Six stages in Scotia Sea evolution from Barker (2001), showing present 2000 m contour, with locations of possible sample sites (Fig. 8.4-3 and 8.4-4) to show that sites may be found on basement sufficiently old to test all existing models of ACC onset and development. In speculating on the time of ACC onset, there is little dispute concerning ocean floor ages, more concerning palaeo-elevations of parts of the Scotia Ridge and Shackleton Fracture Zone (*top left: S:* Shackleton Fracture Zone; *A:* Aurora Bank; *D:* Davis Bank; and *P:* Powell Basin respectively, are such parts; *arrows* show pathways suggested by Barker (2001) and Lawver and Gahagan (2003), at approximate times of opening)

numerical modelling (DeConto and Pollard 2003) suggests that substantial Antarctic glaciation could have developed rapidly as a result of gradual reduction of global atmospheric pCO_2, with or without an ACC. Third, recent results from ODP Leg 189 (e.g., Stickley et al. in press), together with additional numerical modelling (Huber et al. in press), suggest that creation of a deep-water gap south of Tasmania did not significantly change the level of thermal isolation of Antarctica, because of a pre-existing clockwise South Pacific oceanic circulation. In passing, the numerical model underlines the importance of a future focus on the region south of South America, which opened last. It suggests (e.g., Fig. 8.4-3 of Huber et al.) that, with a similar pre-existing ocean circulation, creation of a gap in that region could have resulted in a dramatic change in thermal isolation.

The first two of the above developments gave rise to the following conclusions and possibilities (Barker and Thomas 2004):

1. Existing DSDP and ODP drill sites are too sparse to allow the firm conclusion that sediment biofacies reflect the past existence of large areas of near-constant sea-surface temperature separated by fronts. Fronts are a relatively common feature of the modern ocean, but the conditions that develop and sustain them are not understood, so their past existence as interpreted is in question.
2. If such fronts did exist, it cannot be concluded that they were deep-reaching, or the loci of strong, along-front current jets, or (particularly) that they were continuous. Very few modern frontal zones show these features.
3. The observed sea-surface cooling, with or without intervening fronts, may be a simple effect of continental glaciation (and associated sea-ice formation) rather than an indicator of its cause. Independently of the ACC, much of the Southern Ocean has sea-ice cover for part of the year at present, and at parts of the Antarctic continental margin sea-ice formation is sufficiently persistent and systematic for it to be involved also in the production of cold, salty bottom water. In the past, similar cooling processes may have acted, even with water much warmer than today's arriving at a glaciated Antarctic margin. Cold surface water off Antarctica need not indicate an ACC.
4. Enhanced biogenic production, often taken as a proxy for invigorated circulation, may not have been caused by the ACC. Upwelling, carrying nutrients into the photic zone, is a feature of the present-day Southern Ocean south of the Polar Front but, since cold water sinks at parts of the glaciated Antarctic margin as well as at the fronts, the compensatory upwelling cannot be taken as a certain indicator of an ACC.

Useful Sedimentary Parameters

Future studies in physical oceanography and palaeoclimate modelling may remove some of the uncertainty described above. Given that tectonics cannot easily provide precise and unambiguous answers, direct sampling of sediments may be the only way of determining the onset of the ACC. However, much of the evidence of ACC existence and location adduced in the past from the sedimentary record may not be diagnostic. What parameters remain valid?

1. Grain-size studies of bottom current strength. Modern ACC transport occurs within narrow jets extending to the seabed, rather than within a broad zone, but the jets meander and in places generate detached current rings. Thus, the seabed is swept at intervals by strong bottom currents within a zone much broader than the jet width, but experiences much lower bottom current speeds otherwise (e.g., data in Pudsey and Howe 2002). The ACC, unlike the far steadier and better-known western boundary currents common within the world's oceans, is a high eddy-kinetic-energy (K_e) regime. Grain-size studies cannot distinguish between different bottom current regimes, nor provide useful estimates of mean current flow (mean kinetic energy K_m) in high K_e regimes (McCave et al. 1995). Sedimentation may reflect quieter periods, in between times of more rapid flow, and a greater energy is required for resuspension than to prevent original deposition. Nevertheless, the geological record will reflect in some way the existence of faster currents over an area broader than the jet width, because the periodicity of variation remains shorter than the time resolution of most sediment sections. In summary, grain-size studies on properly-sited samples *will* be capable of indicating an onset and subsequent development of the ACC. Care must be taken, of course, to choose sites beneath the ACC, rather than beneath a bottom current, such as Weddell Sea-produced bottom water, that is clearly related to continental climate, and to choose hemipelagic rather than turbiditic sediments so as to determine local conditions.
2. Studies of current flow direction. Estimates of flow direction from magnetic susceptibility anisotropy of the deposited sediments may be useful, but could be degraded by uncertain directions of fast flow within eddies and rings. Geochemistry and mineralogy, on both clays and ice-rafted detritus (IRD), may provide information on provenance and thus the continuity or otherwise of deep and shallow pathways. Clays would have been transported mainly by deep pathways. The age uncertainties considered here lie entirely within the period of known Antarctic glaciation, so IRD would be available also, to provide information on shallow-

water pathways and the places of origin of icebergs, its coarser-grained component being incapable of bottom-current transport. In addition, studies on radiogenic isotopes (e.g., Nd isotopes in fish teeth) may help establish the overall directions of deep-ocean circulation (e.g., Frank 2002).

Very few measurements of current strength and direction have been made in the past, and the unfavourable distribution of drill sites provides little hope that measurements on existing samples will solve the problem of onset. However, such measurements on new, optimally-sited samples could be complemented by detailed work on biofacies, with the aims of fruitful reinterpretation of the comparatively large number of biofacies studies that already exist, and the separate determination of deep- and shallow-water conditions. The sea-surface palaeotemperature determinations from biofacies studies remain valid and of palaeoceanographic significance, with or without an ACC. For example, the PF probably migrated north of the mid-ocean ridge south of Australia, and of the Kerguelen Plateau, as recently as the latest Miocene or early Pliocene (Kemp et al. 1975; Wei and Wise 1992). Also, recent benthic oxygen isotopic curves (e.g., Zachos et al. 2001) show major changes (in water temperature or ice volume) at several times since glacial onset, that are not satisfactorily explained. A choice of continuous sections for future sampling would ensure that ACC variation since onset could also be examined.

Potential of the Scotia Sea Region

As a region where ACC onset may be determined, the Scotia Sea has several distinct advantages. First, it is generally acknowledged as the region where the final barrier in an otherwise-complete deep-water circumpolar path was removed, this proximity reducing the chance of confusion of an ACC onset-related signal with any other. Second, the confining western North Scotia Ridge prevents northward migration of ACC fronts in response to (for example) externally-induced climate change, making its investigation there potentially less time-consuming. Third, the region is relatively well-known in several respects (see below), so that selection of ACC-related sedimentary sections there may be made more confidently.

In the Scotia Sea region, the SAF and PF cross northern Drake Passage and pass northward through deep gaps in the North Scotia Ridge, to cross the Falkland Plateau. The less prominent SACCF crosses southern Drake Passage and stays within the Central Scotia Sea until east of South Georgia (Fig. 8.4-1 and 8.4-3). To ensure the preservation of a continuous geological record, target sedimentary sections that recorded ACC onset and development should be located away from both the mean frontal positions, which are erosional or non-depositional, and the inner flanks, close to mean positions, where eddies are so common as to induce sediment reworking and hiatuses (see unsuitable sections in, e.g., Fig. 7 of Barker and

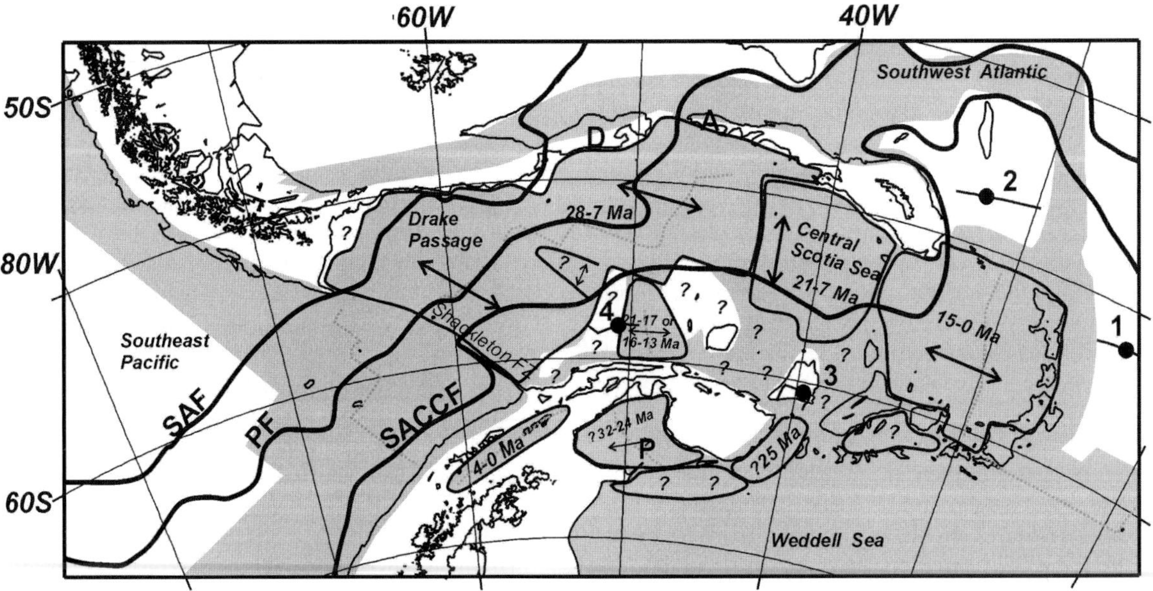

Fig. 8.4-3. Area in Scotia Sea region *shaded*, where sediment sampling would be unwise because of glacial (including glacial turbiditic) sedimentation, Weddell Sea-origin bottom water flow, too-young basement, deformed sediments. Scotia Sea basement ages and bottom-water pathways from Tectonic Map (1985) and Barker (2001). Suitable sections (ship tracks and arrowed sites from Fig. 8.4-4 marked and numbered) avoid these areas. Modern mean loci of component oceanic fronts of the ACC (*SAF*, *PF* and *SACCF*) and key Scotia Ridge components (*A*, *D*, *P*: Aurora Bank, Davis Bank and Powell Basin respectively) are marked, but areas of strong ACC-related bottom currents (along mean loci and inner flanks of meander zones) are not. Continuous but ACC-influenced sedimentation occurs on outer flanks

Burrell 1977 and Fig. 8.4-6 of Howe and Pudsey 1999). The outer flanks are ideal. In general around Antarctica, sections are better selected in the south (closest to the present SACCF, for example) than in the north, because a wide range of proxies that might be associated with an ACC migrated northward with time (the biocalcareous/biosiliceous transition, zones of restricted biological diversity and of non-deposition or erosion, e.g., Kemp et al. 1975; Wei and Wise 1992). Further, sedimentary sections are better located on minor structural elevations, to avoid possible glacial-origin bottom water and unwanted thickening of sections by nepheloid-transported clays.

There are two other main constraints. First, the sediment sections must extend back far enough in time (i.e. latest Eocene) to include the oldest alternative age of ACC onset: most of the floor of the Scotia Sea (and of other nearby areas) is too young. Older basement occurs outside the Scotia Sea, but also inside in a few places (e.g., Toker et al. 1991). The exact age of this older basement is uncertain, but it is generally acknowledged as older than that in dated regions (e.g., Barker 2001; Lawver and Gahagan 2003, and Fig. 8.4-2). Second, ACC onset must be recognisable. Towards this, the best sampling sites would before onset have been quiet – well away from a gyre edge or large anomalous elevation – and after onset would have been within the zone of ACC influence (best defined as the farthest reach of eddies and rings away from the frontal zones that spawned them). Other sections to avoid are deformed sediments (fore-arc accretionary prisms of Pacific South America, the North Scotia Ridge, South

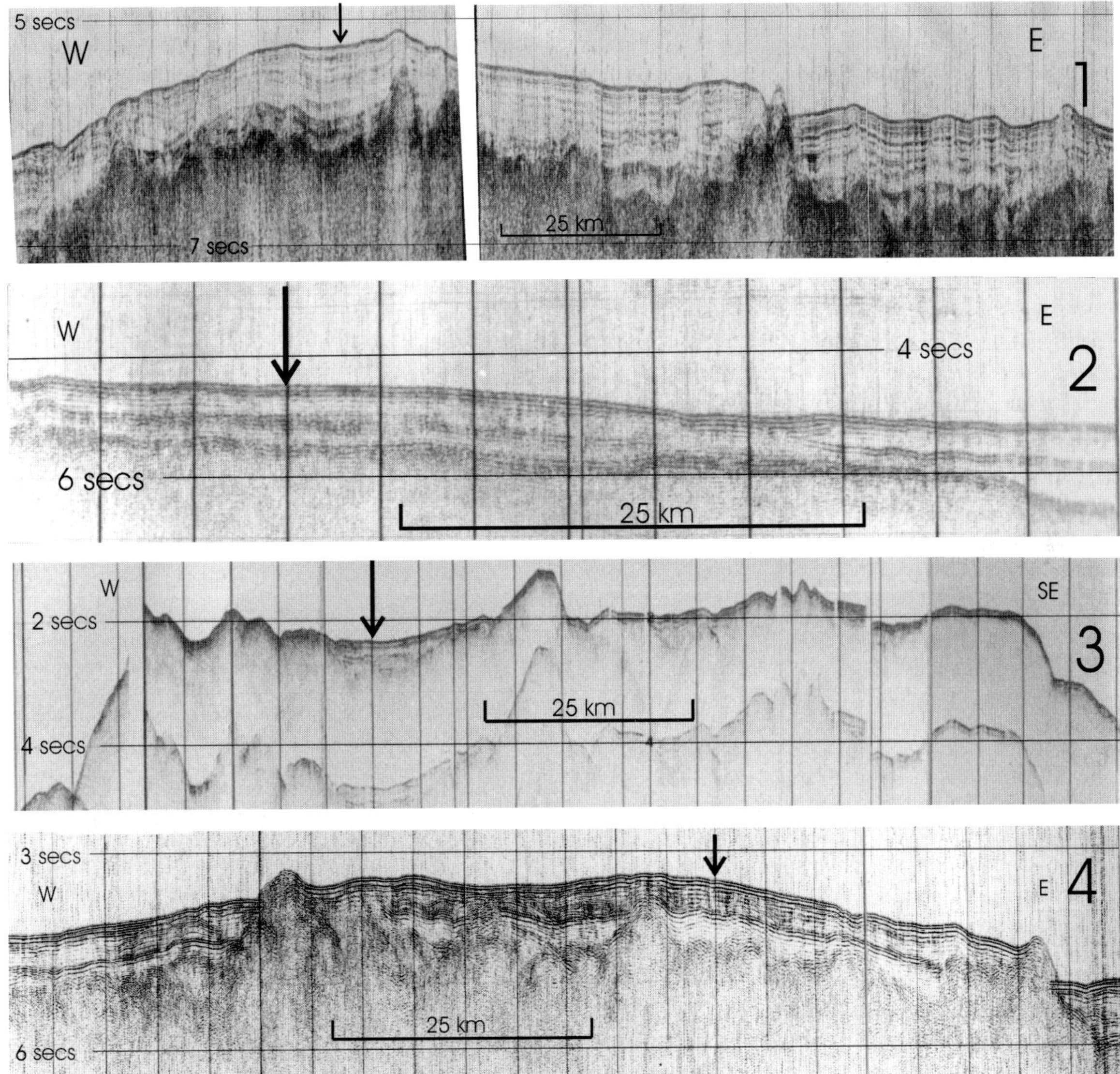

Fig. 8.4-4. Selected seismic profiles showing suitable sedimentary sections (arrowed sections located in Fig. 8.4-2 and 8.4-3), from University of Birmingham data

Sandwich and South Shetland Islands), and glacial and other sediments related primarily to Antarctic climate, such as those deposited beneath glacial-origin bottom water (Weddell Sea, southernmost and eastern Scotia Sea, South Sandwich Trench, western boundaries and Antarctic Peninsula margin), and glacially-initiated turbidites (as off the Antarctic Peninsula: Hollister and Craddock 1976). In the Scotia Sea region, all these provinces are very well-defined. A composite of "where NOT to sample" forms Fig. 8.4-3, which also shows four suggested sections (seismic profiles crossing those sites form Fig. 8.4-4) and the mean loci of the three circum-Antarctic fronts that comprise the ACC. Two suitable areas (sections 1 and 2) lie outside the Scotia Sea, on oceanic basement ca. 70 Ma old now on the outer rise of the South Sandwich Trench (Barker and Lawver 1988) but previously more remote from the growing Scotia Sea, and on ca. 100 Ma old basement on the SE flank of the NE Georgia Rise (Kristoffersen and LaBrecque 1991) respectively. The other two lie on a likely stretched and subsided continental fragment (3) and on old ocean floor (4) in the southern west/central part of the Scotia Sea, generally considered to have been formed during the earliest stages of Scotia Sea development (see Fig. 8.4-2). All show sections that lack unconformities and are parallel-bedded, compatible with mainly pelagic/hemipelagic sedimentation and relatively minor current control. We have not included the more elevated continental regions around the Scotia Sea, which may have risen above sea level (suffering erosion) during break-up, or large parts of the SE Pacific where terrigenous turbidites dominated deposition (Hollister and Craddock 1976), and pelagic/hemipelagic interbeds, representative of local conditions, may or may not exist. The preferred strategy, in view of the importance of determining ACC onset, would be to examine several sections for essentially simultaneous effects.

Conclusions

Re-assessment of ACC onset and development is important for the understanding of global palaeoclimate, despite (or even because of) uncertainties over its role, but should be undertaken with care because some previously-accepted indicators of ACC existence may have other origins. Changes in bottom current strength and deep-water pathways remain valid indicators.

We suggest that the Scotia Sea region is an appropriate location for obtaining sediments that would make such a reassessment possible. It is generally acknowledged as the location of removal of the final barrier in a complete circumpolar deep-water path (this proximity reducing the chance of confusion of an ACC onset signal with others), confines the ACC topographically (reducing the range and complexity of any variation in path) and is relatively well-known (permitting a more discriminating choice of sediments). We identify four suitable sedimentary environments, and show a typical seismic profile from each (Fig. 8.4-4):

- ocean floor of Mesozoic age east of the South Sandwich island arc and trench;
- the southeast flank of the Northeast Georgia Rise (also Mesozoic);
- two areas within the Scotia Sea, as old as possible; one probably on ocean floor, the other probably on a subsided continental fragment.

Acknowledgments

We are grateful for the comments of Dieter Fütterer and anonymous reviewers. PFB thanks the Trans-Antarctic Association for funding his attendance at 9th ISAES. ET thanks NSF for partial funding for her research by Grant EAR-0120727.

References

Barker PF (2001) Scotia Sea regional tectonic evolution: implications for mantle flow and palaeocirculation. Earth Sci Rev 55:1–39

Barker PF, Burrell J (1977) The opening of Drake Passage. Mar Geol 25:15–34

Barker PF, Burrell J (1982) The influence on Southern Ocean circulation, sedimentation and climate of the opening of Drake Passage. In: Craddock C (ed) Antarctic geoscience. Univ Wisconsin Press, Madison, pp 377–385

Barker PF, Lawver LA (1988) South American-Antarctic plate motion over the past 50 Myr, and the evolution of the South American–Antarctic Ridge. Geophys J R Astr Soc 94:377–386

Barker PF, Thomas E (2004) Origin, signature and palaeoclimatic influence of the Antarctic Circumpolar Current. Earth Sci Rev 66:143–162

De Conto R, Pollard D (2003) Rapid Cenozoic glaciation of Antarctica induced by declining atmospheric CO_2. Nature 421:245–249

Exon NF, Kennett JP, Malone MJ, Shipboard Scientific Party (2001) Proc ODP Init Repts 189. Ocean Drilling Program, College Station (CD-ROM)

Frank M (2002) Radiogenic isotopes: tracers of past ocean circulation and erosional input. Rev Geophys 40(1):1001

Gille ST (1994) Mean sea surface height of the Antarctic Circumpolar Current from Geosat data: method and application. J geophys Res 99:18255–18273

Heywood KJ, King BA (2002) Water masses and baroclinic transports in the South Atlantic and Southern oceans. J Mar Res 60:639–676

Hollister CD, Craddock C, et al. (1976) Initial Reports Deep Sea Drilling Project 35

Howe JA, Pudsey CJ (1999) Antarctic Circumpolar deep water: a Quaternary paleoflow record from the northern Scotia Sea, South Atlantic Ocean. J Sed Res 69:847–861

Huber M, Brinkhuis H, Stickley CE, Doos K, Sluijs A, Warnaar J, Schellenberg SA, Williams GL (in press) Eocene circulation in the Southern Ocean: was Antarctica kept warm by subtropical waters? Paleoceanography

Kemp EM, Frakes LE, Hayes DE (1975) Paleoclimatic significance of diachronous biogenic facies, Leg 28, Deep Sea Drilling Project. DSDP Init Repts 28:909–918

Kennett JP (1977) Cenozoic evolution of Antarctic glaciation, the Circum-Antarctic ocean, and their impact on global paleoceanography. J geophys Res 82:3843–3860

Kristoffersen Y, LaBrecque J (1991) On the tectonic history of the Northeast Georgia Rise. In: Ciesielski PF, Kristoffersen Y (eds) Proc ODP Sci Res 114:23–38

Lawver LA, Gahagan LM (1998) Opening of Drake Passage and its impact on Cenozoic ocean circulation. In: Crowley TJ, Burke KC (eds) Tectonic boundary conditions for climate reconstructions. Oxford Univ Press, Oxford, pp 212–223

Lawver LA, Gahagan LM (2003) Evolution of Cenozoic seaways in the circum-Antarctic region. Palaeogeogr Palaeoclimat Palaeoecol 198:11–38

Lawver LA, Gahagan LM, Coffin MF (1992) The development of paleoseaways around Antarctica. In: Kennett JP, Warnke DA (eds) The Antarctic paleoenvironment: a perspective on global change. AGU Ant Res Ser 56:7–30

Livermore R, Eagles G, Morris P, Maldonado A (2004) Shackleton Fracture Zone: no barrier to early circumpolar circulation. Geology 32:797–800

McCave IN, Manighetti B, Robinson SG (1995) Sortable silt and fine sediment size/composition slicing: parameters for palaeocurrent speed and palaeoceanography. Paleoceanography 10:593–610

Nowlin WD Jr, Klinck JM (1986) The physics of the Antarctic Circumpolar Current. Rev Geophys 24:469–491

Orsi AH, Whitworth T III, Nowlin WD Jr (1995) On the meridional extent and fronts of the Antarctic Circumpolar Current. Deep Sea Res I 42:641–473

Pudsey CJ, Howe JA (2002) Mixed biosiliceous-terrigenous sedimentation under the Antarctic Circumpolar Current, Scotia Sea. In: Stow DAV, Pudsey CJ, Howe JA, Faugeres J-C, Viana AR (eds) Deep-water contourites: modern drifts and ancient series, seismic and sedimentary characteristics. Geol Soc Lond Mem 22:325–336

Sijp WP, England MH (2004) Effect of the Drake Passage throughflow on global climate. J Phys Oceanogr 34:1254–1266

Stickley CE, Brinkhuis H, Schellenberg S, Sluijs A, Roehl U, Fuller M, Grauert M, Huber M, Warnaar J, Williams G (in press) Timing and nature of the deepening of the Tasmanian Gateway. Paleoceanography

Toggweiler JR, Bjornsson H (1999) Drake Passage and paleoclimate. J Quat Sci 15:319–328

Toker V, Barker PF, Wise SW Jr (1991) Middle Eocene carbonate-bearing sediment off the northern Antarctic Peninsula. In: Thomson MRA, Crame JA, Thomson JW (eds) Geological evolution of Antarctica. Cambridge University Press, Cambridge, pp 639–644

Wei W, Wise SW Jr (1992) Selected Neogene calcareous nannofossil index taxa of the Southern Ocean: biochronology, biometrics and paleoceanography. In: Wise SW Jr, Schlich R, et al. (eds) Proc ODP Sci Results 120. Ocean Drilling Program, College Station, pp 523–537

Zachos J, Pagani M, Sloan L, Thomas E, Billups K (2001) Trends, rhythms and aberrations in global climate, 65 Ma to present. Science 292:686–693

Seismic Stratigraphy of Miocene to Recent Sedimentary Deposits in the Central Scotia Sea and Northern Weddell Sea: Influence of Bottom Flows (Antarctica)

Andrés Maldonado[1] · Antonio Barnolas[2] · Fernando Bohoyo[1] · Carlota Escutia[1] · Jesús Galindo-Zaldívar[3]
Javier Hernández-Molina[4] · Antonio Jabaloy[3] · Francisco J. Lobo[1] · C. Hans Nelson[1] · José Rodríguez-Fernández[1]
Luis Somoza[2] · Emma Suriñach[5] · Juán Tomás Vázquez[6]

[1] Instituto Andaluz Ciencias de la Tierra, CSIC/Universidad Granada, 18002 Granada, Spain, <amaldona@ugr.es>
[2] Instituto Geológico y Minero de España, Ríos Rosas, 23, 28003 Madrid, Spain
[3] Departamento de Geodinámica, Universidad de Granada, 18071 Granada, Spain
[4] Facultad de Ciencias del Mar, Departamento de Geociencias Marinas, 36200 Vigo, Spain
[5] Departamento de Geodinámica i Geofísica, Universitat de Barcelona, 08028 Barcelona, Spain
[6] Facultad de Ciencias del Mar, Universidad de Cádiz, 11510 Puerto Real, Cádiz, Spain

Abstract. Multichannel and high resolution seismic profiles from the central Scotia Sea and northern Weddell Sea show a sequence of seismic units interpreted to be the result of high-energy bottom currents. The seismic character of the units is indicative of active bottom flows, which developed extensive drifts under the influence of the Weddell Sea Bottom Water (WSBW) and the Antarctic Circumpolar Current (ACC). The opening of the connection between Jane Basin and the Scotia Sea is marked by a major regional unconformity that recorded a reorganization of bottom flows. The uppermost deposits are characterized by intensified bottom currents, which may reflect increased production of WSBW.

Introduction

The Scotia Sea is a small (some 700 × 1 900 km^2) oceanic basin, located eastward of the Antarctic Peninsula and South America (Fig. 8.5-1). The Weddell Sea is a larger oceanic basin southward of the Scotia Sea. The boundary between the two seas is a complex area with ridges, banks and basins that are a direct consequence of the processes of seafloor spreading and subduction that were active in the region (Lawver et al. 1992; Livermore and Woollett 1993; Livermore et al. 1994; Maldonado et al. 1998; Lawver and Gahagan 1998).

The seafloor of the Scotia Sea is influenced by the Antarctic Circumpolar Current (ACC), which provides the only linkage between the three main Atlantic, Pacific and Indian oceans and facilitates the transport of heat, salt and nutrients around the globe (Nowlin and Klinck 1986; Duplessy et al. 1988). Another important water mass in the region is the Weddell Sea Bottom Water (WSBW), which flows clockwise following the Weddell Gyre, southward of the ACC. The Weddell Gyre is an important component of the global thermohaline circulation and it is the source for a majority of the cold bottom water of the oceans (Locarnini et al. 1993; Fahrbach et al. 1995; Orsi et al. 1999). The influence of the deep ocean circulation on marine sedimentation has been illustrated in several studies of the region (e.g., Pudsey and Howe 1998; Howe et al. 1998; Pudsey 2002; Maldonado et al. 2003, 2005).

The objectives of this study are to analyze the distribution of the different types of deposits above the igneous crust and their temporal evolution. The deposits show a significant evolution in time that reflects the bottom processes related to the dynamics of the main water masses in the region, one key aspect for paleoceanographic and environmental reconstructions.

Methods

We collected geophysical data along several profiles in the central Scotia Sea and northern Weddell Sea during the SCAN 97 and SCAN 2001 cruises (Fig. 8.5-2). The multichannel reflection seismic profiles were obtained with a tuned array of five BOLT air guns with a total volume of 22.14 l and a 96 channel streamer with a length of 2.4 km. The shot interval was 50 m. Data were recorded with a DFS V digital system and a sampling record interval of 2 ms and 10 s record lengths. The data were processed with a standard sequence, including migration using a DISCO/FOCUS system. High resolution subbottom profiles were obtained with a Topographic Parametric Sonar (TOPAS) Konsberg Simrad PS018. A near-field trace of the MCS profiles was also processed with the high resolution DELPH system. The swath bathymetric data were collected with a SIMRAD EM12 System and post-processed with the NEPTUNE software.

The stratigraphic analysis and the regional distribution of depositional units and discontinuities in the area were complemented with the data of additional MCS profiles acquired in previous cruises by Italian, Russian and Spanish institutions (Osservatorio Geofisico Sperimentales, OGS: IT90-91, IT95; Sevmorgeologia 1990, 1992; Instituto Andaluz de Ciencias de la Tierra, IACT: HESANT 92-93).

From: Fütterer DK, Damaske D, Kleinschmidt G, Miller H, Tessensohn F (eds) (2006) Antarctica: Contributions to global earth sciences. Springer-Verlag, Berlin Heidelberg New York, pp 441–446

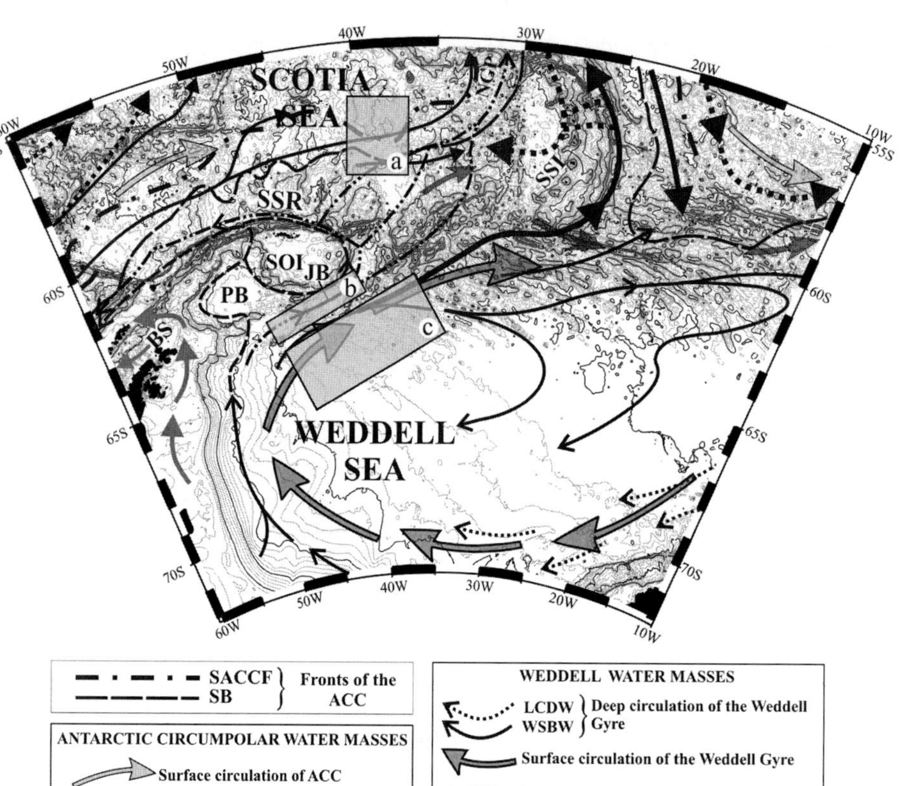

Fig. 8.5-1.
Regional oceanographic framework of the water masses from the Weddell Sea and the southern part of the Scotia Sea showing the pattern of the Antarctic Circumpolar Current (ACC) fronts. Study areas *shaded in grey* (modified from Maldonado et al. 2005). ACC fronts: *SACCF:* Southern ACC front and *SB:* Southern Boundary of the ACC. Physiographic reference points: *BS:* Bransfield Strait; *JB:* Jane Basin; *PB:* Powell Basin; *SOI:* South Orkney Islands; *SSI:* South Sandwich Islands; *SSR:* South Scotia Ridge; Legend of the water masses: *ACC:* Antarctic Circumpolar Current; *CDW:* Circumpolar Deep Water; *LCDW:* Lower Circumpolar Deep Water; *UCDW:* Upper Circumpolar Deep Water; *WSC:* Weddell Scotia Confluence; *WDW:* Warm Deep Water; *WSBW:* Weddell Sea Bottom Water

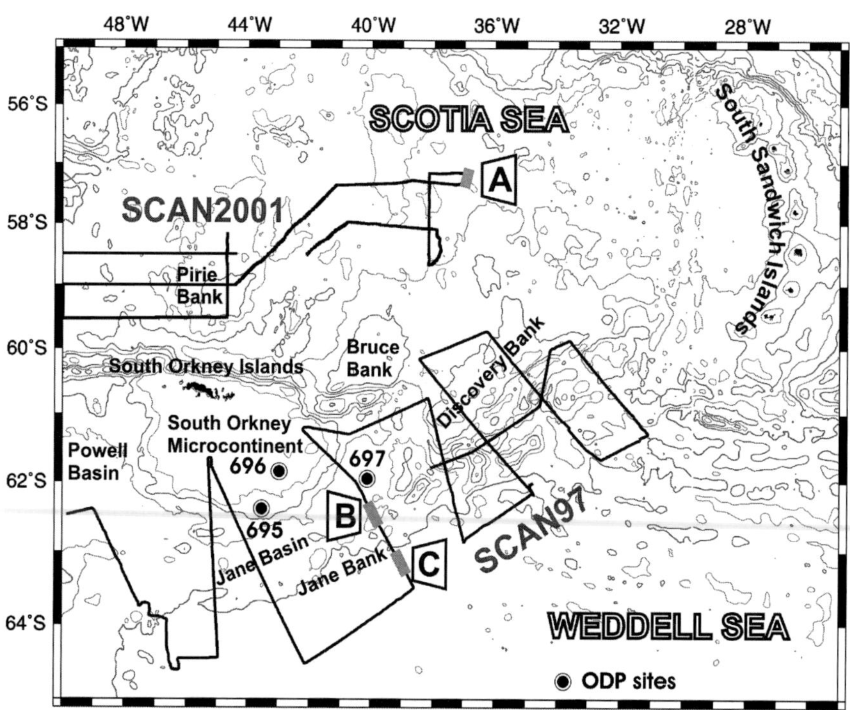

Fig. 8.5-2.
Location of the *SCAN 97* and *SCAN 2001* profiles collected in the study areas: *A:* Scotia Sea, *B:* Jane Basin and *C:* Weddell Sea

For details and additional MCS profile locations see Maldonado et al. (1998, their Fig. 8.5-2).

The age of the main seismic units was tentatively calculated on the basis of: *(a)* the age of the igneous basement provided by the magnetic anomalies, *(b)* the total thickness of the depositional sequence for selected stratigraphic sections, *(c)* the sedimentation rate of surface sediment cores, and *(d)* the results of an ODP borehole in Jane Basin (site 697), located close to a multichannel seismic profile (SM06; Fig. 8.5-2).

Seismic Stratigraphy

The basement in the MCS profiles shows a rough, high amplitude reflection at the top with discontinuous high-amplitude reflectors and irregular diffractions below, underlain by a thicker layer with sparse and weak reflectors. This basement represents the oceanic igneous crust in the area (Maldonado et al. 2000, 2003). The sedimentary cover above the basement is up to 1.4 s thick, but shows significant thickness variations and it is locally absent. Several major discontinuities characterized by high-amplitude continuous reflectors separate five seismic units (named 1 to 5, from top to bottom) in the Weddell Sea and Jane Basin, and six seismic units (named I to VI) in Scotia Sea that are regionally identified within each basin (Fig. 8.5-3 and 8.5-4). The unit boundaries are conformable or slightly erosional above the underlying deposits and are overlain by discontinuous reflectors, which are either conformable or exhibit downlap terminations.

Weddell Sea and Jane Basin

The lowermost unit 5 is an irregular deposit with poorly organized internal reflectors. This unit is tectonically deformed by faults and it fills the basement depressions. The overlying unit 4 is well stratified, with an aggradational configuration (Fig. 8.5-3b,c). The reflectors onlap the basement structural highs, whereas they downlap or become concordant with the deposits of unit 5 basinwards. In the

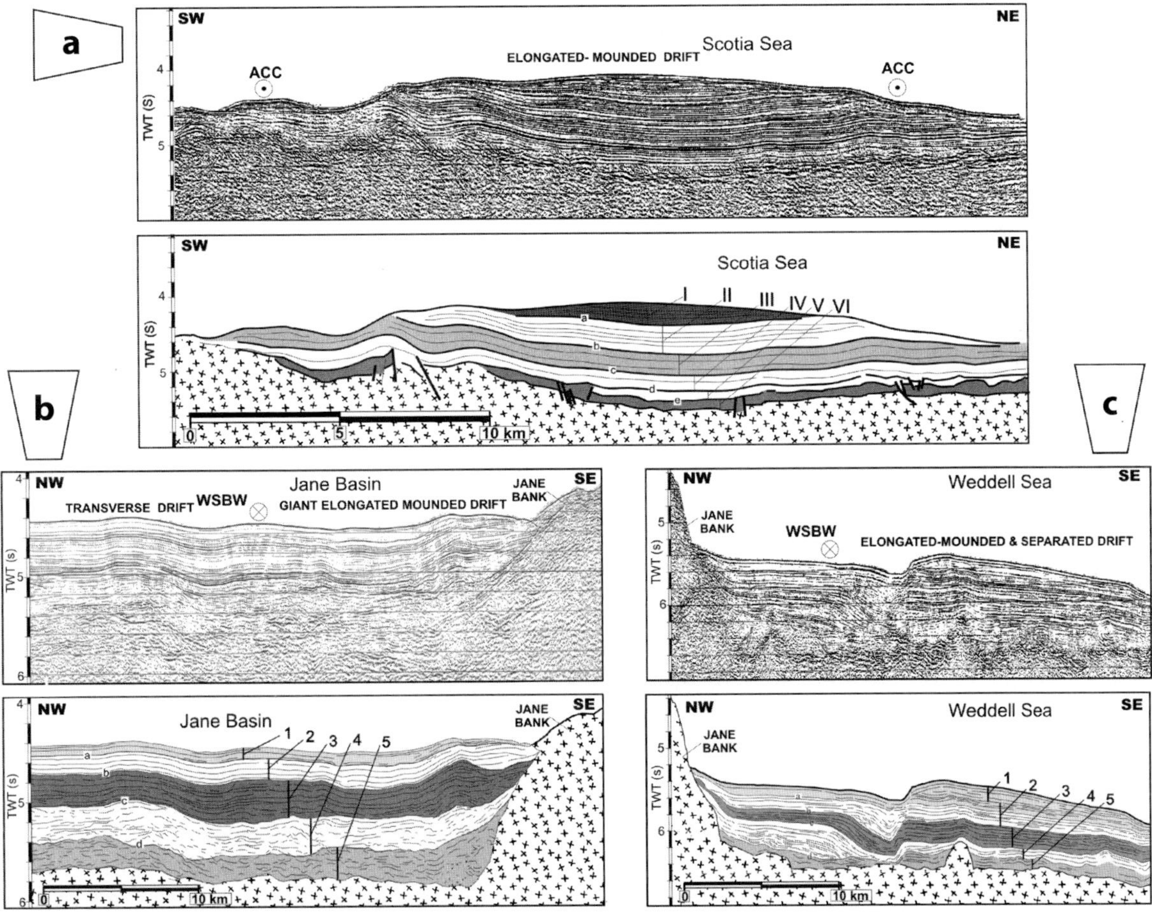

Fig. 8.5-3. Selected segments of MCS profiles and line drawing interpretations with the seismic stratigraphy units of **a** the Scotia Sea; **b** Jane Basin and **c** the Weddell Sea. Legend of the bottom current flow direction: *circle with cross:* flow away from the viewpoint; *circle with dot:* flow towards the viewpoint

Weddell Sea, the geometry of the unit is lenticular and bounded between basement structural highs (Fig. 8.5-3c). The top boundary (reflector c) is a high-amplitude regional reflector, with a clear erosional character and irregular relief that in Jane Basin is the most important regional unconformity (Fig. 8.5-3b). Unit 3 is characterized by mounded and channelized bodies, wavy reflectors and numerous internal discontinuities, all indicative of high-energy bottom current deposits. The basal reflectors of unit 3 show downlap terminations against reflector c (Fig. 8.5-3b,c). Unit 2 is sheeted and largely conformable over the underlying deposits in the basins, whereas it thins and shows onlap terminations over the morphological highs. The internal reflectors are wavy, channelized and locally contain chaotic facies. The geometry of the uppermost unit 1 is controlled by the occurrence of channels and morphological highs associated with Jane Bank. This unit is generally characterized by transparent and chaotic seismic facies (Fig. 8.5-3b,c and Fig. 8.5-4). Wavy, mounded, channelized and chaotic facies attributed to high-energy bottom current deposits are also observed throughout the area in most of the units.

Scotia Sea

The deposits of unit VI above the igneous oceanic crust fill in basement depressions (Fig. 8.5-3A and Fig. 8.5-4). The internal reflectors are irregular and laterally discontinuous. Seismic unit V shows an aggradational seismic pattern and a lens-like external shape, with well stratified internal reflectors.

Unit IV has an external lens-like geometry and is internally characterized by foreset reflectors in a progradational pattern over the area which laterally evolves to subparallel configurations (see Maldonado et al. 2003, Fig. 10). The progradation locally shows a distinct southwards trend, opposite to the orientation of the overlying units. The internal reflectors of unit III are sub-parallel and locally wavy, with onlap terminations over the lower boundary. Reflector c at the base is a major discontinuity that marks a change in the direction of progradation to northward trend. Unit II has a mounded morphology. The internal reflectors are sub-parallel, wavy and laterally continuous, with a cyclic pattern. Low-angle onlap and downlap terminations over the underlying high-amplitude, laterally continuous Reflector b are locally observed (see Maldonado et al. 2003, Fig. 10). This reflector is a major discontinuity in the area that marks the most important change in the depositional seismic facies sequence. The youngest unit I also has a mounded morphology. It is characterized by sub-parallel, gently undulating and wavy reflectors (Fig. 8.5-3a). A cyclic pattern of alternating low-amplitude and high-amplitude internal reflectors is observed in this unit.

Seismic Interpretation

We interpret that the seismic characteristics of the depositional units are related to bottom contour currents. These features include: *(a)* discontinuities that can be traced across the area, *(b)* lenticular, convex-upward geometries, *(c)* progradational and aggradational reflectors that converge towards the marginal zones of the depositional bodies and are truncated by numerous internal discontinuities, and *(d)* wavy reflectors (Fig. 8.5-3).

In the Weddell Sea and Jane Basin, unit 4 shows extensive deposits of contourite character, which were not identified in the underlying unit 5. The contourite deposits are better developed in the Weddell Sea, indicating the occurrence of permanent and well established sea-bottom flows that are persistent throughout the deposition of the entire succession until the present. Seismic facies indicate bottom flows with a northeastward orientation. Similar contourite deposits are extensively developed in the central Scotia Sea above the igneous oceanic crust (Maldonado et al. 2003). The contourites of the Scotia Sea, however, show two distinct styles of deposition. The two lowermost units have southward progradation patterns, whereas the uppermost units have either a northward progradational trend in the southern area or an east-southeastward direction in the central part of the Scotia Sea. All these deposits were developed as the results of the influence of the two most important flows in the area, the ACC and the WSBW on bottom processes.

Discussion and Conclusions

The main source of the Antarctic Bottom Water is considered to be located in the Weddell Sea, where the deep circulation is controlled by the clockwise flow of the Weddell Gyre (Fig. 8.5-1). The circumpolar eastward flow of the ACC is another important deep current system in the Southern Ocean which interacts with the sea-floor in many places (Nowlin and Klinck 1986; Foldvik and Gammelsrød 1988; Naveira-Garabato et al. 2002). The ACC was initiated after the opening of the Drake Passage, which was the final barrier to a complete circum-Antarctic deep-water flow (Lawver et al. 1992). The ridges and basins that were active during the early stages of development of the Scotia Sea controlled the development of the ACC and the circumpolar deep water flows (Barker 2001; Maldonado et al. 2003).

The deposits of the northern Weddell Sea, with the exception of the older unit 5, are formed by high-energy seismic facies indicative of significant bottom current activity from the Miocene to the present (Fig. 8.5-3 and 8.5-4). The deposits of units 4 to 1 exhibit, in addition, seismic characteristics that suggest an evolution in the bottom current flows, sediment supply and water mass distribution. Unit 4 (tentative age of Early to Middle Miocene) shows the devel-

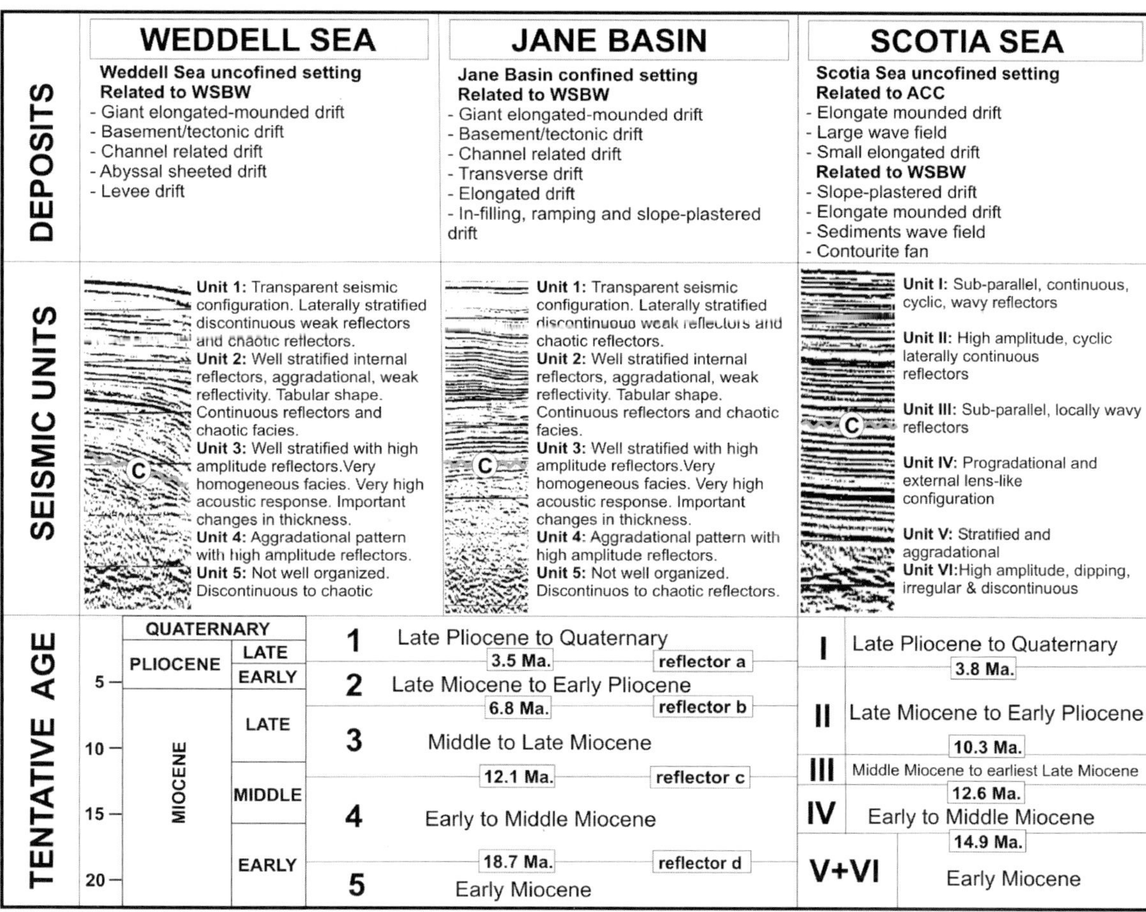

Fig. 8.5-4. Summary of the seismic units, types of deposits in the *Weddell Sea*, *Jane Basin* and *Scotia Sea*. The main characteristics and tentative age of each of the seismic units is presented. See text for discussion

opment of giant elongated mounded drifts in proximity to Jane Bank, together with a variety of contourite drift deposits that resulted from the northeastward flow of bottom currents. Unit 3 (Middle to Late Miocene) recorded a reorganization of bottom flows, probably due to a changing configuration of seaways that resulted from the end of spreading in Jane Basin and of the subduction of the northern Weddell Sea below Jane Bank (Bohoyo et al. 2002). The deposits of unit 2 (Late Miocene to Early Pliocene) in the northern Weddell Sea show an increase of high-energy, sheeted deposits which we believe reflect a higher production of WSBW and intensified bottom currents. Unit 1 (Late Pliocene to Quaternary) is characterized in the Weddell Sea and Jane Basin by the extensive development of chaotic, high-energy sheeted facies, in addition to a variety of contourite deposits. Units 1 and 2 also show a distinct cyclic pattern of deposition, more continuous wavy deposits and development of internal unconformities, all of which attest to increased bottom current energy (Fig. 8.5-4).

The older deposits of the Scotia Sea, units VI and V (tentative age of Early Miocene) recorded strong bottom current flows and high sediment supply. The units have a mounded geometry and wavy reflectors indicative of high-energy bottom current deposits. The internal reflectors of unit IV (Middle Miocene) show a southward migration trend, which is attributed to the influence of an active ACC flow into the central Scotia Sea. During these earlier stages, the bottom part of the ACC flow entering into the Scotia Sea was probably forced into a clockwise gyre because of the bottom topography, and developed the important southward prograding sediment drifts.

The initial incursions of the WSBW into the Scotia Sea are observed in the deposits of unit III (Middle Miocene), which influenced a northward progradational pattern similar to the overlying deposits. This evolution in the sequences was controlled by the opening of gaps in the South Scotia Ridge and the connection of the Weddell Sea and Scotia Sea through Jane Basin (Maldonado et al. 2003). By 15 Ma, however, due to waning proto-Antarctic Bottom Water (AABW) production, bottom current activity due to AABW was mainly restricted to water depths greater than ~5 500 m (Sykes et al. 1998). Later, waxing of proto-AABW during the Late Miocene permitted the reconnection of the deep basins (Sykes et al. 1998), and therefore we might expect the influence of WSBW to increase at this time. Evidence for development of a relatively strong thermohaline overturn at rates comparable to the present interglacial interval during the latest Miocene has also been reported recently by Billups (2002) in the deep-water circulation from the sub-Antarctic South Atlantic.

The deposits of the Late Miocene recorded intensified bottom currents, particularly in the northern Weddell Sea abyssal plain. This may reflect the increased production of WSBW as result of the expansion of the Antarctic ice sheets. The Early/Late Pliocene boundary is also marked by a major regional unconformity (Fig. 8.5-4). The deposits above this unconformity are characterized by extensive development of high energy bottom current facies, which suggest development of intensified deep water production.

In summary, a large variety of contourite deposits are observed throughout the depositional sequences in the northern Weddell Sea and Scotia Sea. The contourites resulted from the interplay between the northeastward flows of the WSBW and the complex bathymetry of the northern Weddell Sea, and from the influence of the ACC and the WSBW in the Scotia Sea. The opening of the deep passageway between Jane Basin and the Scotia Sea is marked by a major regional unconformity near the boundary between the Middle and Late Miocene, which is also recognized in the Pacific margin of the Antarctic Peninsula by the development of giant contourite drifts in the rise (Rebesco et al. 1997; Maldonado et al. 2003).

Acknowledgments

The Spanish "Comisión Interministerial de Ciencia y Tecnología (CYCIT)" supported this research through Project ANT94-0020, ANT99-08, REN2001-2143/ANT and CGL2004-05646/ANT.

References

Barker PF (2001) Scotia Sea regional tectonics evolution: implications for mantle flow and palaeocirculation. Earth-Sci Rev 55:1–39

Billups K (2002) Late Miocene through early Pliocene deep water circulation and climate change viewed from the sub-Antartic South Atlantic. Palaeogeogr Palaeoclimatol Palaeoecol 185:287–307

Bohoyo F, Galindo-Zaldívar J, Maldonado A, Schreider AA, Suriñach E (2002) Basin development subsequent to ridge-trench collision: the Jane Basin, Antarctica. Mar Geophys Res 23:413–421

Duplessy JC, Shackleton NJ, Fairbanks RG, Labeyrie LD, Oppo D, Kallel N (1988) Deep water source variations during the last climatic cycle and their impact on the global deep water circulation. Paleoceanography 3:343–360

Fahrbach E, Rohardt G, Scheele N, Strass V, Wisotzki A (1995) Formation and discharge of deep and bottom water in the northwestern Weddell Sea. J Mar Res 53:515–538

Foldvik A, Gammelsrød T (1988) Notes on Southern Ocean hydrography, sea-ice and bottom water formation. Palaeogeogr Palaeoclimatol Palaeoecol 67:3–17

Howe JA, Livermore RA, Maldonado A (1998) Mudwave activity and current-controlled sedimentation in the Powell Basin, northern Weddell Sea, Antarctica. Mar Geol 149:229–241

Lawver LA, Gahagan LM (1998) Opening of Drake Passage and its impact on Cenozoic ocean circulation. In: Crowley TJ, Burke KC (eds) Tectonic boundary conditions for climate reconstructions. Oxford University Press, Oxford, pp 212–223

Lawver LA, Gahagan LM, Coffin MF (1992) The development of paleoseaways around Antarctica. In: Kennett JP, Warnke DA (eds) The Antarctic paleoenvironment: a perspective on global change, Amer Geophys Union Antarctic Res Ser 56, pp 7–30

Livermore RA, Woollett RW (1993) Seafloor spreading in the Weddell Sea and southwest Atlantic since the Late Cretaceous. Earth Planet Sci Letters 117:475–495

Livermore R, McAdoo D, Marks K (1994) Scotia Sea tectonics from high-resolution satellite gravity. Earth Planet Sci Letters 123: 255–268

Locarnini RA, Whitworth III T, Nowlin WDJ (1993) The importance of the Scotia Sea on the outflow of Weddell Sea Deep Water. J Mar Res 51:135–153

Maldonado A, Zitellini N, Leitchenkov G, Balanyá JC, Coren F, Galindo-Zaldívar J, Lodolo E, Jabaloy A, Zanolla C, Rodríguez-Fernández J, Vinnikovskaya O (1998) Small ocean basin development along the Scotia-Antarctica plate boundary and in the northern Weddell Sea. Tectonophysics 296:371–402

Maldonado A, Balanyá JC, Barnolas A, Galindo-Zaldívar J, Hernández J, Jabaloy A, Livermore RA, Martínez JM, Rodríguez-Fernández J, Sanz de Galdeano C, Somoza L, Suriñach E, Viseras C (2000) Tectonics of an extinct ridge-transform intersection, Drake Passage (Antarctica). Mar Geophys Res 21:43–68

Maldonado A, Barnolas A, Bohoyo F, Galindo-Zaldívar J, Hernández-Molina J, Lobo F, Rodríguez-Fernández J, Somoza L, Vázquez JT (2003) Contourite deposits in the central Scotia Sea: the importance of the Antarctic Circumpolar Current and the Weddell Gyre flows. Palaeogeogr Palaeoclimatol Palaeoecol 198:187–221

Maldonado A, Barnolas A, Bohoyo F, Escutia C, Galindo-Zaldívar J, Hernández-Molina J, Jabaloy A, Lobo F, Nelson H, Rodríguez-Fernández J, Somoza L, Vázquez JT (2005) Miocene to recent contourite drift development in the northern Weddell Sea (Antarctica). Global Planet Change 45:99–129

Naveira-Garabato AC, Heywood KJ, Stevens DP (2002) Modification and pathways of Sourthern Ocean Deep Waters in the Scotia Sea. Deep-Sea Res I 49:681–705

Nowlin WD Jr, Klinck JM (1986) The physics of the Antarctic Circumpolar Current. Rev Geophys 24:469–491

Orsi AH, Johnson GC, Bullister JL (1999) Circulation, mixing and production of Antarctic Bottom Water. Prog Oceanogr 43:55–109

Pudsey CJ (2002) The Weddell Sea: contourites and hemipelagites at the northern margin of Weddell Gyre. In: Stow DAV, Pudsey CJ, Howe JA, Faugères JC, Viana AR (eds) Deep-water countourite systems: modern drifts and ancient series, seismic and sedimentary characteristics. Geol Soc Mem 22, pp 289–303

Pudsey CJ, Howe JA (1998) Quaternary history of the Antarctic Circumpolar Current: evidence from the Scotia Sea. Mar Geol 148:83–112

Rebesco M, Larter RD, Barker PF, Camerlenghi A, Vanneste LE (1997) The history of sedimentation on the continental rise west of the Antarctic Peninsula. In: Barker PF, Cooper AK (eds) Geology and seismic stratigraphy of the Antarctic margin, part 2. Amer Geophys Union Antarctic Res Ser 71, pp 29–49

Sykes TJS, RamsayTS, Kidd RB (1998) Southern hemisphere Miocene bottom-water circulation: a paleobathymetric analysis. In: Cramp A, MacLeod CJ, Lee SV, Jones EJW (eds) Geological evolution of ocean basin: results from the Ocean Drilling Program. Geol Soc London Spec Publ 131, pp 43–54

Limnology and Sedimentary Record of Radok Lake, Amery Oasis, East Antarctica

Bernd Wagner[1] · Holger Cremer[2]

[1] University of Leipzig, Institute for Geophysics and Geology, Talstraße 35, 04103 Leipzig, Germany, <wagnerb@rz.uni-leipzig.de>

[2] Utrecht University, Department of Palaeoecology, Laboratory of Palaeobotany and Palynology, Budapestlaan 4, 3584 CD Utrecht, The Netherlands
Present address: Netherlands Organization of Applied Scientific Research TNO, Core area Built Environment and Geosciences, Geological Survey of the Netherlands, Princetonlaan 6, 3584 CB Utrecht, The Netherlands, <holger.cremer@tno.nl>

Abstract. Radok Lake in Amery Oasis, East Antarctica, has a water depth of ca. 360 m, making it the deepest non-subglacial lake in Antarctica. Limnological analyses revealed that the lake had, despite a 3 m thick ice cover, a completely mixed water column during austral summer 2001/2002. High oxygen contents, low ion concentrations, and lack of planktonic diatoms throughout the water column indicate that Radok Lake is ultra-oligotrophic today. The late glacial and postglacial lake history is documented in a succession of glacial, glaciolimnic, and limnic sediments at different locations in the lake basin. The sediments record regional differences and past changes in allochthonous sediment supply and lake productivity. However, the lack of age control on these changes, due to extensive sediment redeposition and the lack of applicable dating methods, excluded Radok Lake sediments for advanced paleoenvironmental reconstructions.

Introduction

Freshwater lake environments in Antarctica react quickly to climatic and environmental changes (Quayle et al. 2002; Spaulding and McKnight 1999). Due to their ability to respond rapidly, lake sediments have been used successfully to reconstruct climatic and environmental changes in the Antarctic ice free regions, the so-called oases (e.g., Doran et al. 1994; Melles et al. 1997; Hendy 2000; Roberts et al. 2001). In East Antarctica, Amery Oasis is one of the largest ice-free regions that hosts lakes of different size and depth. Its remote location, however, has hampered detailed limnological and sedimentological investigations so far.

Radok Lake (70°52' S, 067°58' E), located in Amery Oasis (Fig. 8.6-1), is known to be the deepest non-subglacial lake in Antarctica. Measurements of the water depth by leadline revealed a minimum water depth of ca. 360 m (Wand et al. 1987; Wagner 2003). The lake is located at an elevation of 7 m a.s.l. in an overdeepened basin whose cliffs and steep slopes rise up to 400 m above the lake. Radok Lake is ca. 10 km long and up to 3 km wide. The northern part of the lake is split into two branches (Fig. 8.6-1). At the southwestern corner Battye Glacier enters Radok Lake forming a floating ice tongue. Meltwater supply from the glacier, the melt stream from Glossopteris Gully (Adamson et al. 1997) and to a lesser degree from the surrounding slopes feeds the lake during summer (Wand et al. 1987). The lake outlet is formed by the deeply incised Pagodroma Gorge that leads into Beaver Lake, an epishelf lake that is hydrologically connected to the ocean underneath Amery Ice Shelf (Fig. 8.6-1).

Radok Lake Basin was formed by Late Cenozoic glacial erosion, i.e., repeated excavation by the paleo-Battye Glacier (Adamson et al. 1997). However, a widespread ice coverage of Amery Oasis seems to have not occurred during the Last Glacial Maximum. This is suggested by the lack of geomorphological evidence for isostatic uplift during the latter half of the Holocene (Adamson et al. 1997). The overdeepening of Radok Lake likely occurred along a tectonically weak zone, the Amery Fault, which runs along the long axis of the lake. The Amery Fault separates Precambrian metamorphic and intrusive rocks to the west from Permian Radok Conglomerate to the south and Permo-Triassic sandstones to the east of the lake (McLoughlin and Drinnan 1997; Mikhalsky et al. 2001).

Despite the extraordinary setting and status of Radok Lake, little is known of its limnological and sedimentological characteristics. The study of its limnology would contribute significantly to a better understanding of the

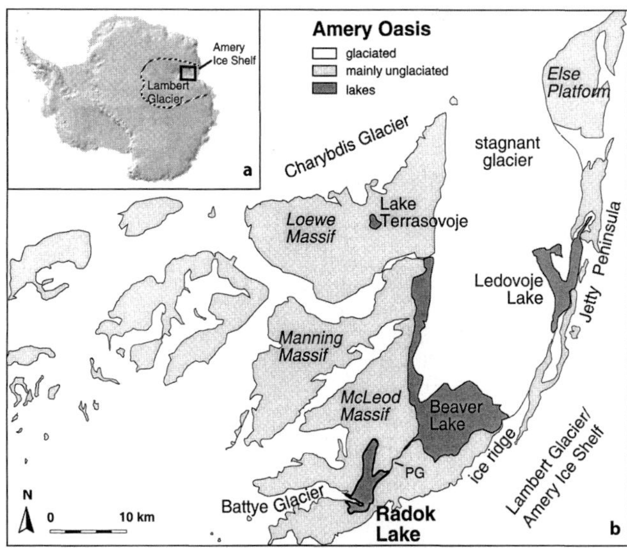

Fig. 8.6-1. a Map of Antarctica showing the location of Amery Oasis at the margin of the Lambert Glacier-Amery Ice Shelf drainage system (*dashed line*). **b** Location of Radok Lake in Amery Oasis

From: Fütterer DK, Damaske D, Kleinschmidt G, Miller H, Tessensohn F (eds) (2006) Antarctica: Contributions to global earth sciences. Springer-Verlag, Berlin Heidelberg New York, pp 447–454

hydrologic conditions in deep, high polar lakes, of which only a few exist. The investigation of sediment sequences from Radok Lake will indicate, if these records can be used for the reconstruction of past climatic and environmental changes in Amery Oasis. Water samples and sediment sequences were collected from Radok Lake during an expedition in austral summer 2001/2002. Water samples were analysed and the sediment sequences described and biogeochemically characterised.

Materials and Methods

Subsequent to bathymetrical measurements, water sampling and analyses were carried out at locations Lz1007 and Lz1012 (Fig. 8.6-2). At both locations a water sampler (UWITEC Co., Austria) was used to sample the water column along a vertical profile. Immediately after the recovery of a water sample, conductivity (WTW LF 197 probe), pH (WTW pH 197 probe), oxygen content, and temperature (WTW Oxi 196 probe) were measured. Aliquots of 60 ml were filtered through a 0.45 µm glass fibre filter for laboratory analyses of major cations (ICP-OES, Perkin-Elmer Co.) and anions (ion chromatograph, Dionex Co.). 1 l of each water sample was filtered through a 0.45 µm cellulose acetate filter, dried, and stored in a plastic container for identification of diatom communities.

Sediment coring was carried out using gravity and piston corers (UWITEC Co.; Melles et al. 1994; Wagner 2003) at sites Lz1007, Lz1010, and Lz1012 (Fig. 8.6-2) through holes in the lake ice-cover. After core opening and documentation in the laboratory, subsamples were taken at 1–2 cm intervals, freeze-dried, and water contents (%) determined. Subsamples were ground and homogenized to measure total carbon (*TC*), total nitrogen (*TN*), and total sulphur (*TS*) using an Elementar III (VARIO Co.) analyser.

Five samples from cores Lz1007 (2 cm and 151 cm) and Lz1010 (2 cm, 101 cm, and 302 cm) were used for diatom analyses. Preparation of diatom slides followed the procedure described in Cremer et al. (2001) for identification and enumeration at 1 000× magnification on an Olympus BX51 microscope with differential interference contrast. A Nikon Coolpix 990 digital camera was used for photographic documentation of diatom taxa.

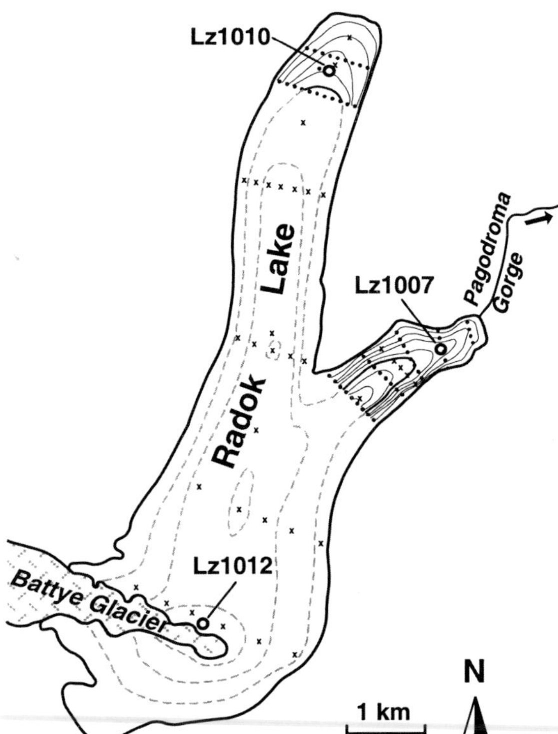

Fig. 8.6-2. Bathymetry of Radok Lake, based on single spot measurements carried out in 2001/2002 (*dots*) and during an earlier expedition (*crosses*; Adamson et al. 1997). *Continuous contour lines* are given in 20 m intervals, *dashed lines* in 100 m intervals. *Open circles* indicate the location of water profiles and sedimentary records obtained during the expedition 2001/2002

Results and Discussion

Limnology

In early January 2002, Radok Lake was covered by ca. 3 m of ice and up to 1 m of snow. In late January 2002, the lake ice-cover was observed to be melting at the tip of the northeastern branch. The melt seemed to be of minor importance compared to earlier years (Adamson et al. 1997), due to thick snow accumulation and consequent increased albedo during early summer 2001/2002. Snow cover was noticably reduced in early February 2002 and lake ice became exposed, particularly in the central areas of the lake.

Water sample analyses from Lz1007 (74 m water depth) and Lz1012 (357 m water depth) performed on January 14 and 27 2002 revealed only minor variations throughout the water column (Table 8.6-1). Notably low ion concentrations, conductivity, and pH measured at 4 m depth from location Lz1007 can be explained by the influence of melted snow penetrating through the ice hole into the water column (Table 8.6-1). Apart from this sample, the water columns from both the northeastern branch and the front of Battye Glacier are characterised by low ion concentrations, low conductivities (ca. 160 µS cm^{-1}), and a pH between 7.5 and 7.9 (see also Wand et al. 1987). Only the bottom water sample at 357 m depth has a pH of 8.3 (Table 8.6-1). A slight increase of temperature with increasing water depth is observed at both locations. A maximum temperature of 1.6 °C was measured from the bottom water of the northeastern branch. Variations in oxygen content (11.4 to 14.2 mg l^{-1}) can be attributed to temperature changes, however, oxygen saturation amounts to more than 85% throughout the water columns at both locations.

Table 8.6-1. Hydrological and geochemical data from water profiles at locations Lz1007 (74 m water depth) and Lz1012 (357 m water depth) in Radok Lake (Fig. 8.6-2). Major cation and anion concentrations were only measured at location Lz1007

Lz1007												Lz1012				
Depth (m)	Temp. (°C)	O_2 (mg l^{-1})	Cond. (µS cm^{-1})	pH	Ca (mg l^{-1})	K (mg l^{-1})	Mg (mg l^{-1})	Na (mg l^{-1})	Si (mg l^{-1})	Cl (mg l^{-1})	SO_4 (mg l^{-1})	Depth (m)	Temp. (°C)	O_2 (mg l^{-1})	Cond. (µS cm^{-1})	pH
4	0.7	12.6	15.0	5.4	0.1	0.1	0.1	0.9	<0.1011	1.1	0.7	5	0.4	13.5	161.5	7.8
7	1.0	13.3	157.3	7.5	4.4	0.9	2.5	11.9	0.5	8.0	4.6	10	0.4	13.7	164.3	7.8
10	1.1	12.7	159.2	7.8	3.5	0.9	2.5	12.2	1.2	8.0	4.7	30	0.4	13.9	162.8	7.8
15	1.1	12.9	158.0	7.9	3.8	0.9	2.5	12.2	1.5	8.1	4.7	60	0.4	13.9	164.5	7.7
20	1.1	12.6	160.4	7.9	4.0	0.9	2.5	12.0	1.3	8.1	4.4	100	0.4	13.3	164.7	7.7
25	1.1	12.3	159.7	7.8	3.3	0.9	2.5	12.0	1.4	7.9	4.6	150	0.5	13.6	162.8	7.8
30	1.1	12.0	156.5	7.8	3.7	0.9	2.5	11.9	1.2	8.0	4.5	200	0.6	13.9	162.0	7.8
35	1.1	12.0	157.6	7.8	4.4	0.8	2.5	11.8	0.9	7.7	4.3	250	0.8	14.1	161.0	7.8
40	1.4	12.0	156.0	7.7	3.2	0.9	2.5	12.1	1.2	8.0	4.3	300	0.8	13.9	161.0	7.8
45	1.4	11.9	157.0	7.7	3.1	0.9	2.5	12.1	1.4	8.1	4.5	330	0.8	13.7	161.3	7.8
50	1.4	11.9	157.7	7.6	3.7	0.9	2.5	12.1	1.3	8.0	4.6	350	0.9	13.5	162.5	7.7
55	1.5	11.8	157.9	7.6	4.1	0.9	2.5	12.2	1.5	8.1	4.7	357	0.4	14.2	164.0	8.3
60	1.4	11.8	158.2	7.7	3.8	0.9	2.5	12.1	1.5	7.9	4.5					
65	1.5	11.4	157.0	7.6	3.9	0.9	2.6	12.2	1.5	8.0	4.5					
70	1.5	11.4	156.5	7.6	4.2	0.9	2.6	12.2	1.4	8.1	4.5					
73	1.6	11.5	158.0	7.5	3.4	0.9	2.6	12.5	0.4	8.0	4.7					

Minor variations in the vertical hydrological profiles reveal that Radok Lake was completely mixed in January 2002 and confirm earlier observations (Wand et al. 1987; Adamson et al. 1997). Low conductivities and ion concentrations are caused by the supply of ion depleted meltwater from Battye Glacier and the surrounding slopes. Ion depleted waters in general are poor in nutrients and thus hamper lake primary productivity. Productivity in Antarctic lakes is additionally constrained by ice and snow cover, particularly if the ice and snow cover persist during summer. Although photosynthesis can take place when ice and snow cover permit less than 1% of incident light to penetrate through (McMinn et al. 2000; Hawes et al. 2001), reduced light penetration into the water column confines the occurrence of photo-autotrophic algae and microbial organisms. They form the major part of primary production in Antarctic lakes. The short growing season in combination with low nutrient availability may explain the planktonic diatom absence in Radok Lake. This observation agrees with other studies of Antarctic lakes (e.g., Hodgson et al. 2001; Laybourn-Parry et al. 2001; Sabbe et al. 2003; Verleyen et al. 2003).

The low productivity in Radok Lake was also reflected in the oxygen gradients throughout the water column. Neither a horizon of enhanced primary productivity indicated by oxygen supersaturation nor a horizon of oxygen depleted bottom waters, caused by enhanced bacterial decomposition of organic matter, was observed in the northeastern branch or in front of Battye Glacier. Continued oxygen saturation throughout the water column even after at least 10–12 months of ice coverage confirmed that Radok Lake is ultra-oligotrophic today and has reduced photo-autotrophic primary production.

Sedimentary Records

The sediment records obtained from the three locations Lz1007, Lz1010, and Lz1012 in Radok Lake (Fig. 8.6-2) indicate distinct local and stratigraphic differences. However, interpretation of the latter is hampered by lack of chronological data. Radiocarbon dating of the sediments was excluded due to the presence of finely disseminated coal fragments, which were mainly derived from coal-bearing Permo-Triassic sandstones to the east of the lake and would have produced erroneous ages. Dating via optically stimulated luminescence (OSL/IRSL) was rejected, due to the strong probability of incomplete bleaching of feldspars and quartz grains due to the short travel distances of lateral meltwater inflows or subglacial sediment supply from Battye Glacier (see also Krause et al. 1997). Finally, dating of the surface sediments via the ^{210}Pb method was not applicable, because surface sediments at Beaver Lake nearby (Fig. 8.6-1) revealed that recent atmospheric input of ^{210}Pb into the lake is not sufficient to obtain reliable ages.

Lz1007

The 202 cm long sediment sequence Lz1007 from a water depth of 74 m in the northeastern branch of Radok Lake exhibits decreasing grain sizes from the bottom to the top (Fig. 8.6-3). The core base is overconsolidated, has a stiff consistency, is coarse sand dominated, and contains high amounts of gravel and pebbles, yet this occurrence is distinctly reduced between 125 and 58 cm depth. Above that

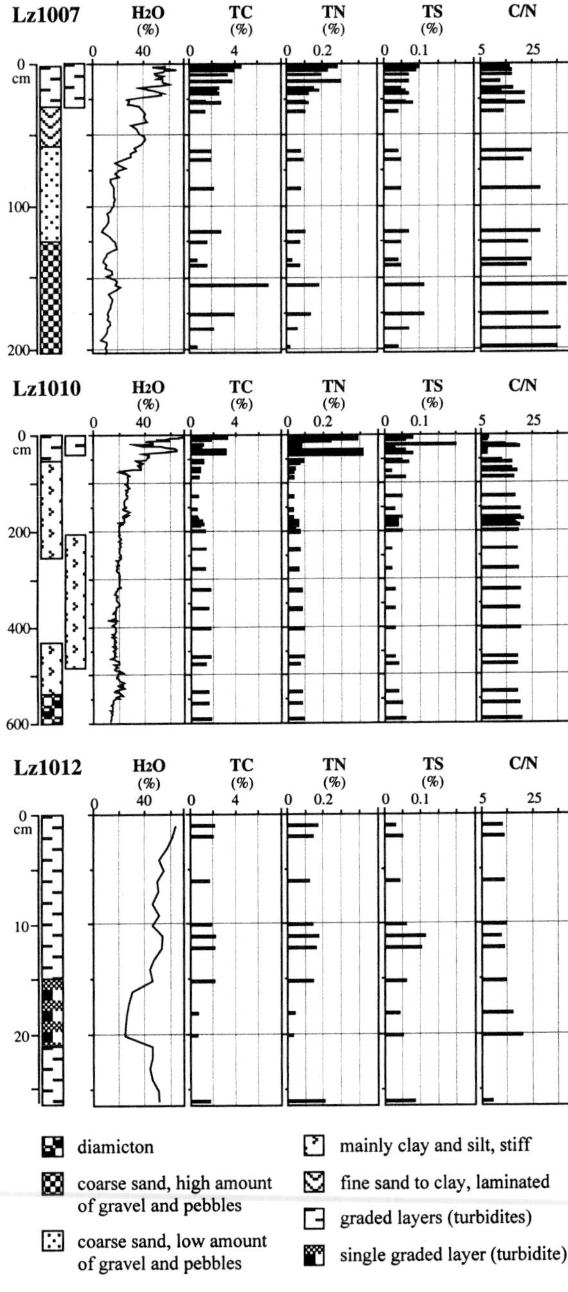

Fig. 8.6-3. Lithology, water content, and biogeochemical parameters of the sediment sequences obtained at coring locations *Lz1007*, *Lz1010*, and *Lz1012* in Radok Lake. Note different scales of *y*-axis

Table 8.6-2. Relative abundance (%) of diatom taxa in surface sediments (2 cm depth) of Radok Lake. Detailed morphological and morphometric information on the taxa is available in Cremer et al. (2004)

Diatom taxa	Lz1007	Lz1010
Amphora veneta Kützing	3.0	9.0
Craticula cf. *molesta* (Krasske) Lange-Bertalot et Willmann	1.3	1.0
Diadesmis spp.[a]	2.0	23.6
Muelleria peraustralis (West et West) Spaulding et Stoermer[c]	43.0	38.0
Psammothidium spp.[b]	7.2	6.7
Stauroneis anceps Ehrenberg	43.5	21.7

[a] Includes *Diadesmis* cf. *ingeae* Van de Vijver, *Diadesmis* cf. *perpusilla* (Grunow) Mann, *Diadesmis* sp. 1.
[b] Includes *P. metakryophilum* (Lange-Bertalot et Schmidt) Sabbe[c], *P. stauroneioides* (Manguin) Bukhtiyarova[c].
[c] Endemic to Antarctic inland waters.

depth, a laminated fine mud occurs, with thin turbidites in the topmost 30 cm. The general fining upward is correlated with an increase in water content and a decline of sediment consolidation. Coal particles are common throughout the sequence, particularly in the unlaminated sediments below 58 cm depth. Other organic remains were not found, except for the sediment surface, where a 1–2 cm thick layer of water mosses was growing. Diatoms are absent at 151 cm depth, however, a benthic diatom assemblage, mainly composed of *Muelleria peraustralis*, *Stauroneis anceps*, and *Psammothidium* spp., is preserved at 2 cm depth (Table 8.6-2; Fig. 8.6-4). Relatively high amounts of TC between 0.7 and 7.0% and low amounts of TN and TS throughout the sediment sequence are mainly due to the coal particles in the sediment, likely originating from the coal bearing sandstones to the east of the lake.

Sediment sequence Lz1007 documents the deglaciation of the northeastern branch of Radok Lake. The deposition of overconsolidated coarse and stiff sediments at the basis can be related to the latest stage of glaciation or to the early deglaciation. The fining upward reflects gradual glacier retreat. The onset of deglaciation is uncertain, but probably simultaneous with local ice retreat in Amery Oasis prior to 12 ka B.P. (D. Gore, pers. communication; Wagner et al. in press). Relatively high sedimentation rates during the deglaciation may explain lack of diatoms at 151 cm depth. Above 58 cm the glacier had retreated too far to have a direct impact on sediment deposition at the coring site. However, lamination and turbidite layers suggest disturbed sedimentation, with sporadic embanking of sand layers from ice related transport or mass movement processes from the steep slopes in the vicinity of the coring location. A decrease of the C/N ratio from ca. 30 at the base to ca. 15 at the sediment surface is likely to be due to higher deposition of autochthonous organic matter in the younger lake history (Fig. 8.6-3). Whereas single coal particles have a C/N ratio between 28 and 41, corresponding with the deeper sediments in core Lz1007, autochthonous matter commonly has a C/N ratio between 7 and 10 (Meyers and Ishiwatari 1995). The autochthonous organic matter can be formed *in situ*, e.g., from water mosses. However, it can also originate from shallower parts of the lake and be supplied to the coring site by sediment redeposition. Sediment redeposition is confirmed by the occurrence of benthic diatoms (1.6 valves $\times 10^6$ g^{-1} dry sediment) at 2 cm depth in core Lz1007. Most of these diatoms prefer relatively shallow waters (Sabbe et al. 2003) and thus likely originate from lateral parts of the lake. The occurrence of benthic diatoms in contrast to the lacking planktonic diatoms probably is promoted by a direct turnover of the few nutrients, which are released from the sediment surface by bacterial decomposition.

Lz1010

Sediment sequence Lz1010 is 593 cm in length (Fig. 8.6-3). The base is formed from a stiff diamicton of brown to dark grey colour with low water content and a lack of stratification. Between 530 and 60 cm follows a thick horizon of mainly silt and clay with sporadic occurrences of coarse sand grains and gravel. The base of this horizon is slightly laminated. The occurrence of all grain sizes, the stiff consistency, and the low water content throughout this sequence suggest deposition as a till or, alternatively, glaciolimnic sediments with ice rafted debris (IRD) supply and without overconsolidation by grounded ice. Its stiff consistency could then be due to a very old age and former overburden by sediments that have become eroded by mass movement. The top 60 cm of sediment sequence Lz1010 are characterised by mud with fine laminated turbidites, relatively high water content, and dark grey brown to olive brown colour. Coal particles were common throughout the complete sequence, particularly below 60 cm depth. Other organic remains were not found, except in the surface sediments. Whilst diatoms are absent at depths of 302 and 101 cm, a benthic diatom assemblage (6.0 valves $\times 10^6$ g^{-1} dry sediment) mainly composed of *Muelleria peraustralis*, *Diadesmis* spp., *Stauroneis anceps*, *Amphora veneta*, and *Psammothidium* spp. was identified at a depth of 2 cm (Table 8.6-2; Fig. 8.6-4). The sediment surface was covered by growing water mosses.

Sediment sequence Lz1010 documents the deglaciation of northern Radok Lake by the up-core decrease in grain-size distribution and consistency. TC and TN values in the glacial or glaciolimnic facies indicate incorporation of coal particles from the sedimentary rocks to the east of the coring location, although to a minor content than recorded at location Lz1007. A relatively higher sediment supply from the Precambrian metamorphic and intrusive

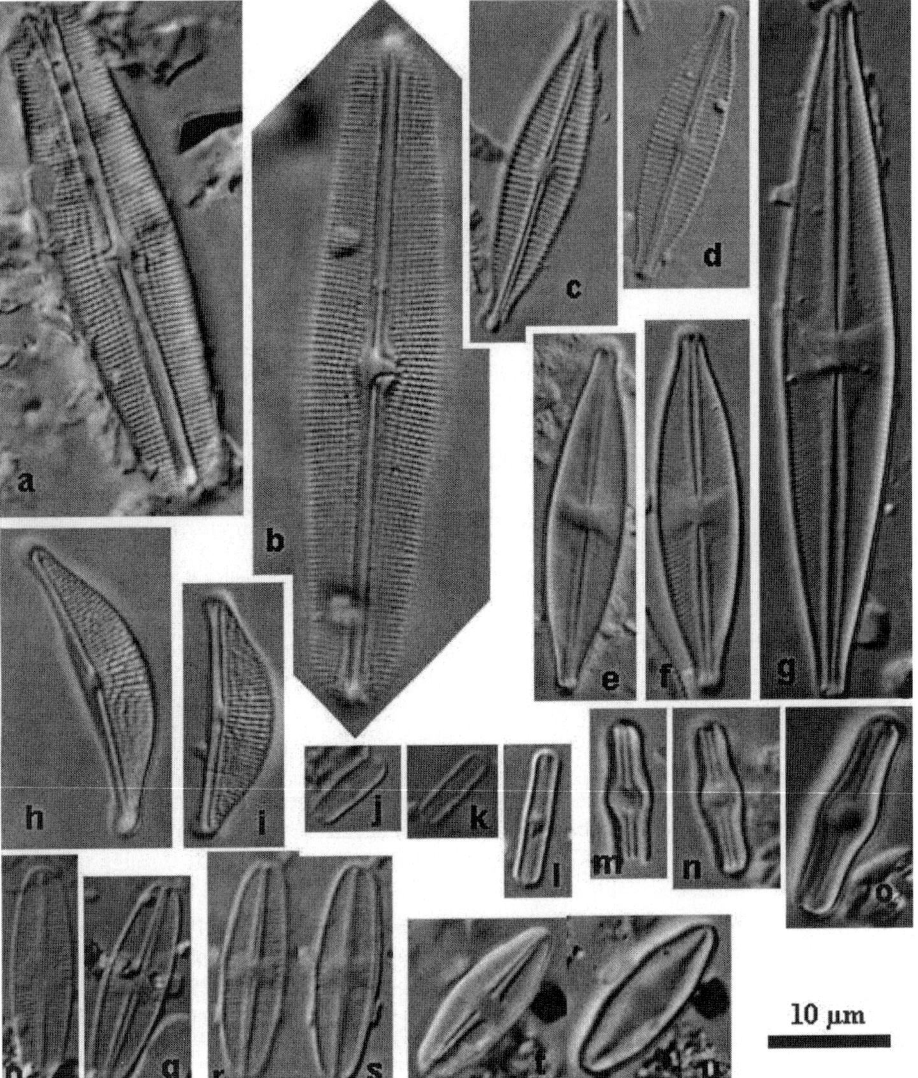

Fig. 8.6-4. Light micrographs of diatoms identified in surface sediments from Radok Lake. All photos in valve view. **a, b** *Muelleria peraustralis* (West et West) Spaulding et Stoermer; **c, d** *Craticula* cf. *molesta* (Krasske) Lange-Bertalot et Willmann; **e–g** *Stauroneis anceps* Ehrenberg; **h, i** *Amphora veneta* Kützing; **j, k** *Diadesmis* sp. 1; **l** *Diadesmis* cf. *ingeae* Van de Vijver; **m–o** *Diadesmis* cf. *perpusilla* (Grunow) Mann; **p–s** *Psammothidium metakryophilum* (Lange-Bertalot et Schmidt) Sabbe; **t, u** *Psammothidium stauroneioides* (Manguin) Bukhtiyarova

rocks to the west of the lake is also recorded in the C/N ratio, which is in general lower in core Lz1010. C/N ratios of less than 10 and the occurrence of the benthic diatom assemblage in the surface sediments imply that autochthonous organic matter was particularly contributed in recent times, probably by redeposition of lake edge sediments. This is supported by the occurrence of turbidites in the top 60 cm of core Lz1010. The diatom assemblage in core Lz1010 is more or less conformable with core Lz1007 except a total diatom abundance four times higher and differences in the relative abundance of *Diadesmis spp.* and *Stauroneis anceps* (Table 8.6-2). These differences might be attributable to different geological and bathymetrical situations or different amounts of sediment redeposition at both sites. Overall, the benthic diatom flora in surface sediment samples from Radok Lake consists of 14 taxa, of which three taxa are endemic to Antarctic inland waters (Table 8.6-2, Fig. 8.6-4). Diatom diversity is relatively low compared, for example, to Larsemann Hills and Rauer Islands (Sabbe et al. 2003) or from lake environments of the maritime and sub-Antarctic (e.g., Jones 1996, and references therein). However, recently relatively broad and unprecise taxonomic concepts might lead to underestimation of the number of diatom taxa and the grade of endemism in Antarctic inland waters (Jones 1996; Sabbe et al. 2003).

Lz1012

The 25 cm long sediment sequence Lz1012 is made up of alternating horizons of sand, silt, and clay of different colours (Fig. 8.6-3). Single layers show a gradual upward fining of grain size, which is typical for turbidites. The thickest of these turbidites occurs between 21 and 15 cm depth. The overall consistency of the sediment is soft and the water content is similar to surface sediments at the other coring locations. Coal particles and other organic remains were not observed macroscopically, although TC, TN, and TS abundances indicate their presence (Fig. 8.6-3).

Because coring location Lz1012 is in front of Battye Glacier and in the deepest part of the lake, sediment supply from the glacier and from the surrounding subaquatic slopes can be assumed. Dropstones were not observed, indicating a relatively stable glacier front in recent times, low sediment supply from the glacier, or both. Sediment lamination is similar to the surface sediments from the northern and northeastern branches. The turbidites indicate unstable sedimentation conditions with accumulation of sediments from the slopes above. This is confirmed by C/N ratios, which encompass a similar range as the surface sediments at the other two locations.

Conclusions

By the study of the recent hydrological conditions and the sedimentary records of Radok Lake the following conclusions can be drawn concerning the recent and historical conditions in the lake.

High oxygen saturation, low ion concentrations, and the lack of planktonic diatoms throughout the water column indicate that Radok Lake is ultra-oligotrophic today and completely mixed even after at least 10–12 months of ice coverage.

The sedimentary records of Radok Lake document glaciation and deglaciation of the basin in historical times and the transition to the recent postglacial limnic setting. Geological differences in the catchment and their influences on sediment supply are reflected in the C/N ratio, which also indicates a recent increase in lake primary productivity.

The lithology of the uppermost sediments and the occurrence of benthic diatom assemblages in relatively deep waters suggest unstable subaquatic slopes and sediment redeposition. This, along with the lack of applicable dating methods make sediment records from Radok Lake unsuitable for detailed paleoenvironmental reconstructions.

Acknowledgments

The project was initiated by Martin Melles (University Leipzig, Germany) and Damian Gore (Macquarie University, Sydney). It was funded by the Deutsche Forschungsgemeinschaft (Grant Me 1169/4) and the Australian Antarctic Science Advisory Committee (ASAC Grant 1071). Special thanks are due to Martin Klug, Gerald Müller, Andy Cianchi, Margie Jenkin and Rob Ferguson for their assistance in the field. The data of the water contents and ion concentrations were provided by Nadja Hultzsch (Alfred Wegener Institute, Potsdam, Germany). The manuscript benefitted from comments of Damian Gore, Martin Melles, and an anonymous reviewer.

References

Adamson DA, Mabin MCG, Luly JG (1997) Holocene isostasy and late Cenozoic development of landforms including Beaver and Radok Lake basins in the Amery Oasis, Prince Charles Mountains, Antarctica. Antarct Sci 9:299–306

Cremer H, Wagner B, Melles M, Hubberten H-W (2001) The Holocene environmental development of Raffles Sø, East Greenland: inferences from a 10 000 year diatom record. J Paleolim 26:67–87

Cremer H, Gore D, Hultzsch N, Melles M, Wagner B (2004) The diatom flora and limnology of lakes in the Amery Oasis, East Antarctica. Polar Biol 27:513–531

Doran PT, Wharton Jr RA, Lyons WB (1994) Paleolimnology of the McMurdo Dry Valleys, Antarctica. J Paleolim 10:85–114

Hawes I, Moorhead D, Sutherland D, Schmeling J, Schwarz AM (2001) Benthic primary production in two perennially ice-covered Antarctic lakes: patterns of biomass accumulation with a model of community metabolism. Antarct Sci 13:18–27

Hendy CH (2000) Late Quaternary Lakes in the McMurdo Sound Region of Antarctica. Geogr Ann 82:411–432

Hodgson DA, Noon PE, Vyverman W, Bryant CL, Gore DB, Appleby P, Gilmour M, Verleyen E, Sabbe K, Ellis-Evans JC, Wood PB (2001) Were the Larsemann Hills ice-free through the Last Glacial Maximum? Antarct Sci 13:440–454

Jones VJ (1996) The diversity, distribution and ecology of diatoms from Antarctic inland waters. Biodiv Conserv 5:1433–1449

Krause WE, Krbetschek MR, Stolz W (1997) Dating of Quatarnary lake sediments from the Schirmacher Oasis (East Antarctica) by infra-red stimulated luminiscence (IRSL) detected at the wavelength of 560 nm. Quat Sci Rev 16:387–392

Laybourn-Parry J, Quayle WC, Henshaw T, Ruddell A, Marchnat HJ (2001) Life on the edge: the plankton and chemistry of Beaver Lake, an ultra-oligotrophic epishelf lake, Antarctica. Freshw Biol 46:1205–1217

McLoughlin S, Drinnan AN (1997) Fluvial sedimentology and revised stratigraphy of the Triassic Flagstone Bench Formation, northern Prince Charles Mountains, East Antarctica. Geol Mag 134:781–806

McMinn A, Bleakley N, Steinburner K, Roberts D, Trenerry L (2000) Effect of permanent sea ice cover and different nutrient regimes on the phytoplankton succession of fjords of the Vestfold Hills Oasis, eastern Antarctica. J Plankt Res 22:287–303

Melles M, Kulbe T, Overduin PP, Verkulich S (1994) The expedition Bunger Oasis 1993/94 of the AWI Research Unit Potsdam. In: Melles M (ed) The expeditions Norilsk/Taymyr 1993 and Bunger Oasis 1993/94 of the AWI Research Unit Potsdam, Reports Polar Res 148. Alfred Wegener Institute for Polar and Marine Research, Bremerhaven, pp 27–80

Melles M, Kulbe T, Verkulich SR, Pushina ZV, Hubberten H-W (1997) Late Pleistocene and Holocene environmental history of Bunger Hills, East Antarctica, as revealed by fresh-water and epishelf lake sediments. In: Ricci CA (ed) The Antarctic region: geological evolution and processes. Terra Antartica Publication, Siena, pp 809–820

Meyers PA, Ishiwatari R (1995) Organic matter accumulation records in lake sediments. In: Lerman A, Imboden D, Gat J (eds) Physics and Chemistry of Lakes. Springer, Berlin Heidelberg New York, pp 279–328

Mikhalsky EV, Sheraton J, Laiba AA, et al. (2001) Geology of the Prince Charles Mountains, Antarctica. AGSO, Geoscience Australia, Canberra

Quayle WC, Peck LS, Paet H, Ellis-Evans JC, Harrigan PR (2002) Extreme responses to climate change in Antarctic lakes. Science 295:645–650

Roberts D, van Ommen TD, McMinn A, Morgan V, Roberts JL (2001) Late-Holocene East Antarctic climate trends from ice-core and lake-sediment proxies. The Holocene 11:117–120

Sabbe K, Verleyen E, Hodgson DA, Vanhoutte K, Vyermann W (2003) Benthic diatom flora of freshwater and saline lakes in the Larsemann Hills and Rauer Islands, East Antarctica. Antarct Sci 15:227–248

Spaulding SA, McKnight DM (1999) Diatoms as indicators of environmental change in Antarctic freshwaters. In: Stoermer EF, Smol JP (eds) The diatoms: applications for the environmental and earth sciences. Cambridge University Press, Cambridge, pp 245–263

Verleyen E, Hodgson DA, Vyvermann W, Roberts D, McMinn A, Vanhoutte K, Sabbe K (2003) Modelling diatom responses to climate induced fluctuations in the moisture balance in continental Antarctic lakes. J Paleolim 30:195–215

Wagner B (2003) The expeditions Amery Oasis, East Antarctica, 2001/02 and Taylor Valley, southern Victoria Land, 2002. Reports on Polar and Marine Research 460, Bremerhaven

Wagner B, Cremer H, Hultzsch N, Gore DB, Melles M (in press) Late Pleistocene and Holocene history of Lake Terrasovoje, Amery Oasis, East Antarctica, and its climatic and environmental implications. J Paleolim

Wand U, Hermichen W-D, Höfling R, Mühle K, Klokov VD, Ufimcev AV (1987) Stable isotope and hydrogeochemical studies of Beaver Lake and Lake Radok, MacRobertson Land, East Antarctica. In: Wand U, Strauch G (eds): Proc 4th Working Meeting, Isotopes in Nature, Leipzig, pp 647–659

New Data Related to Holocene Landform Development and Climatic Change from James Ross Island, Antarctic Peninsula

Jorge A. Strelin[1] · Toshio Sone[2] · Junko Mori[3] · Cesar A. Torielli[4] · Toshio Nakamura[5]

[1] Instituto Antártico Argentino, Centro Austral de Investigaciones Científicas, and Universidad de Córdoba, Argentina
Present address: Departamento de Geología Básica, Universidad Nacional de Córdoba, Ciudad Universitaria, Avda. Vélez Sársfield 1611, 5000 Córdoba, Argentina
[2] Institute of Low Temperature Science, Hokkaido University, Sapporo 060-0918, Japan
[3] Graduate School of Engineering, Hokkaido University, Sapporo 060-8628, Japan
[4] Departamento de Geología Básica, Universidad Nacional de Córdoba, Ciudad Universitaria, Avda. Vélez Sársfield 1611, 5000 Córdoba, Argentina
[5] Dating and Materials Research Center, Nagoya University, Furo-cho, Chikusa-ku Nagoya 464-8602, Japan

Abstract. A survey of some relict and active landforms in the Lachman and Rink Crags areas of NW James Ross Island on the Antarctic Peninsula has yielded new evidence that provides a better understanding of the Holocene morphoclimatic evolution of this currently deglaciated sector of James Ross Island. Six Holocene (mainly land-grounded) glacial advances were delimited by these morphological and stratigraphic studies, dated at 6 700–6 400, 4 900–4 400, and shortly after 3 900 ^{14}C yr B.P., with three more recent advances, dated by regional correlations, that occurred between the Holocene regional climatic optimum (3 900–3 000 ^{14}C yr B.P.) and the Little Ice Age (ca. 300 ^{14}C yr B.P.). In some cases, ice-cored rock glacier formation followed the younger glacier advances. In recent decades, the significant climatic warming recorded in the NE region of James Ross Island has produced a number of changes in the periglacial and glacial landforms. Stone-banked terraces and lobes have developed in the Rink Crags, and protalus rampart formation has ceased in favor of protalus lobe development in the Cerro Triple area. Conical depressions filled with water have also increased in area over the surface of the Lachman II ice-cored rock glacier, threatening to destroy this landform.

Introduction

Since 2001, the Instituto Antártico Argentino Research Group (Criología) in cooperation with Japanese scientists of the Institute of Low-Temperature Science (ILTS), University of Hokkaido, and Italian scientists from the Tersa Universidad di Roma has been working on a project entitled "Morphogenesis of Southern South America, Scotia Arc, and Antarctic Peninsula". Geomorphological transects along climatic gradients represent one of the primary keys to understanding past climate change, and data from the longest Southern Hemisphere emerged transect have been used in conjunction with the interpretation of landforms resulting from the three principal cryosphere variables (snow, glacier-ice and permafrost) to reconstruct Holocene environmental conditions in this region. Here, the results of studies on James Ross Island (Fig. 8.7-1) are presented. This area is located in a region of inner shelf deglaciation and fjord formation since approximately 10 000 ^{14}C yr before present (B.C.).

All dates presented in this paper are uncalibrated radiocarbon dates. Following Gordon and Harkness (1992), and in an attempt to simplify the regional time correlations, the radiocarbon data for marine living organisms (marked with asterisks) are corrected by subtracting 1 000 years as a reservoir effect.

Deglaciation

Based on ^{14}C-dated moss from Cape Lachman (Fig. 8.7-2), initial deglaciation of James Ross Island occurred 9 500 yr B.P. (Ingolfsson et al. 1992), which is roughly consistent with the date of 8 700 yr B.P. obtained for the Fildes Península, South Shetland Islands, by Matthies et al. (1990). Proximal glaciomarine deposits, including shell fragments from 7 600 yr B.P.*, overlie Last Glaciation (possibly late glacial) deformation till or deforming bed and lodgment till on the NE coast of Croft Bay, James Ross Island (Ingolfsson et al. 1992; Hjort et al. 1997). No later till deposits are represented in this sequence. The top of the sequence exhibits prograding beach-face sediments,

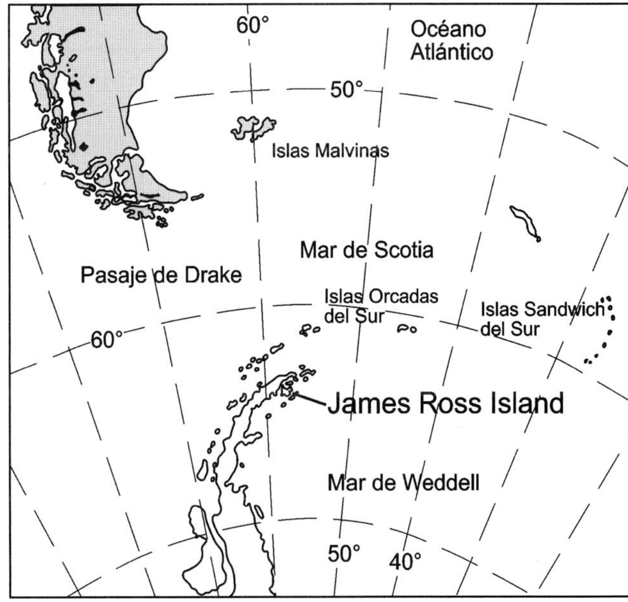

Fig. 8.7-1. Location map of the Antarctic Peninsule region and James Ross Island

From: Fütterer DK, Damaske D, Kleinschmidt G, Miller H, Tessensohn F (eds) (2006) Antarctica: Contributions to global earth sciences. Springer-Verlag, Berlin Heidelberg New York, pp 455–460

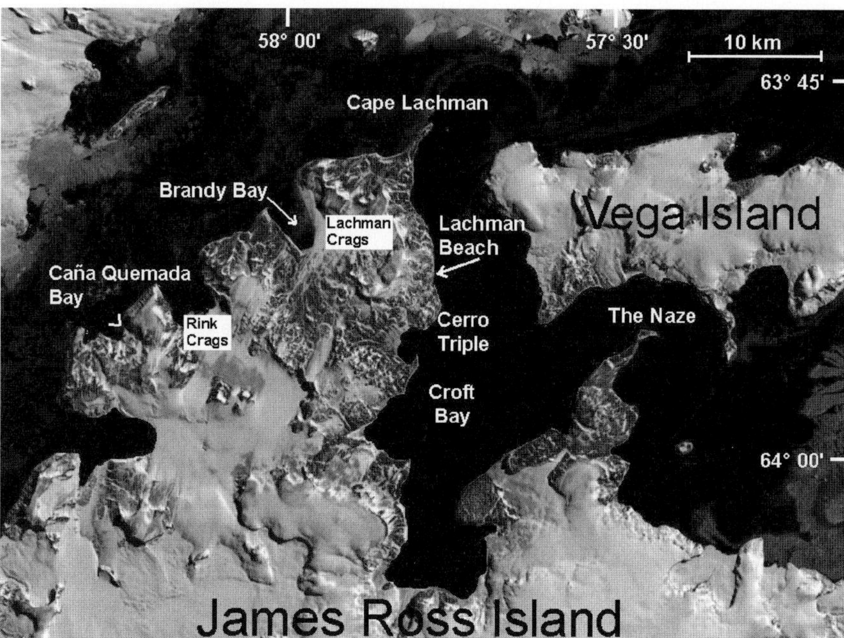

Fig. 8.7-2. Location map of investigation sites in the northwest sector of James Ross Island

including offshore mud to nearshore coarse-grained spit facies, deposited 6 700–6 600 yr B.P.*, forming a terrace at 15–20 m above sea level (a.s.l.). Hjort et al. (1997) noted that, due to contamination, the ^{14}C dates based on mosses in Antarctica often appear much older. Consequently, Hjort et al. (1997) considered the James Ross Island deglaciation date of 9 500 yr B.P. (Ingolfsson et al. 1992) to be too old as a minimum age and proposed that the onset of deglaciation occurred shortly before 7 600 yr B.P.*, during the highest (30 m a.s.l.) Holocene marine terraces formation.

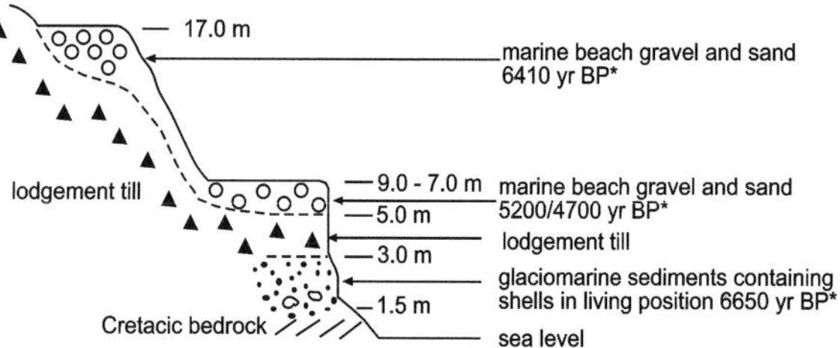

Fig. 8.7-3. Profile of Lachman Beach, James Ross Island

Based on these results, the existing assumption that the first glacier advance occurred on this island 4 900 to 4 400 yr B.P.* (Hjort et al. 1997) should be revised.

Neoglaciations

Lachman Beach Area

In the Lachman Beach area, NW coast of Croft Bay (Fig. 8.7-2), detailed fieldwork lead to the identification of a left tributary-valley lateral moraine hosting two marine terraces (upper and middle). At the base of the ~8 m high marine cliff, glaciomarine sediments containing shells dated at 6 650 yr B.P.* are exposed (Fig. 8.7-3, Table 8.7-1), covered by a lodgment till layer. Considering the proximity of the moraine ridge to the till outcrop and the orientation of striations of blocks in the till, this till layer appears to be associated with the lateral moraine. Shell remains from 6 400 yr B.P.* were recovered from a pit dug into the upper marine terrace (18 m a.s.l.), providing a possible minimum age for the glacier event that deposited the tributary valley lateral moraine. Shells collected in marine nearshore sediments of the middle terrace (5–9 m a.s.l.) have been dated at 5 200 to 4 700 yr B.P.*.

Brandy Bay

The glacial morphology and profiles in Brandy Bay, NW James Ross Island (Fig. 8.7-2), have been described previously (Rabassa 1987; Hjort et al. 1997). Proximal glaciomarine inter/upper sediments and glaciotectonically deformed underflow sediments containing shells dated at 6 400 to 4 200 yr B.P.* outcrop in this area, covered by a deformation/lodgment till. Hjort et al. (1997) interpreted the glacier advance that deposited these glaciogenic units to have occurred between 4 900 and 4 400 yr B.P., although this interpretation does not consider Rabassa's (1987) shells dated at 6 400 and 4 200 yr B.P.* from the glaciotectonically deformed marine sediments. Taking into account Rabassa's dates and the discussion of Lachman Beach above, it is possible to identify two Middle Holocene glaciations: the first around 6 400 yr B.P.*, and the second between 4 900 and 4 400 yr B.P.*. Palynological studies in

Table 8.7-1. New radiocarbon dates obtained in Lachman Beach and Caña Quemada Bay

Sample	Lab. number	Dated shells	^{14}C yr BP	Age error (±yr)	Reservoir effect (yr)	Corrected ^{14}C yr BP*
97M01	NUTA 6567	*Laternula* sp.	7650[a]	100	1000	6650
M000224	Beta 155912	*Laternula* sp.	7410[a]	60	1000	6410
10M02	NUTA 3908	*Laternula* sp.	5800[b]	60	1000	4800
97M18	NUTA 6568	*Laternula* sp.	5670[a]	100	1000	4670
05R01	Beta 155911	*Yoldia* sp.	4990[a]	50	1000	3990
CQ3	NUTA 6581	*Yoldia* sp.	4930[a]	100	1000	3930

[a] Conventional date. [b] Measured date.

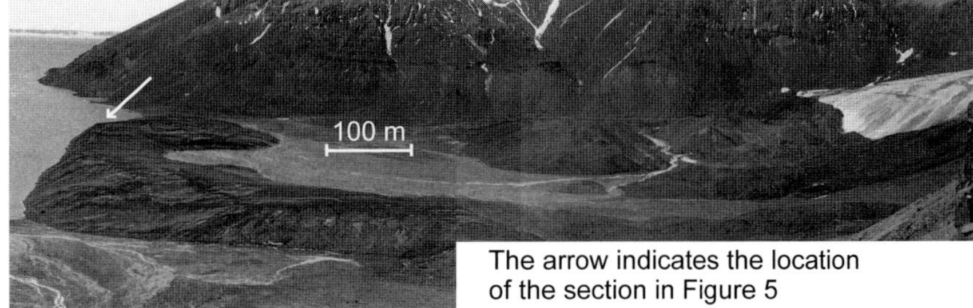

Fig. 8.7-4. Arial view of moraines at Caña Quemada Bay

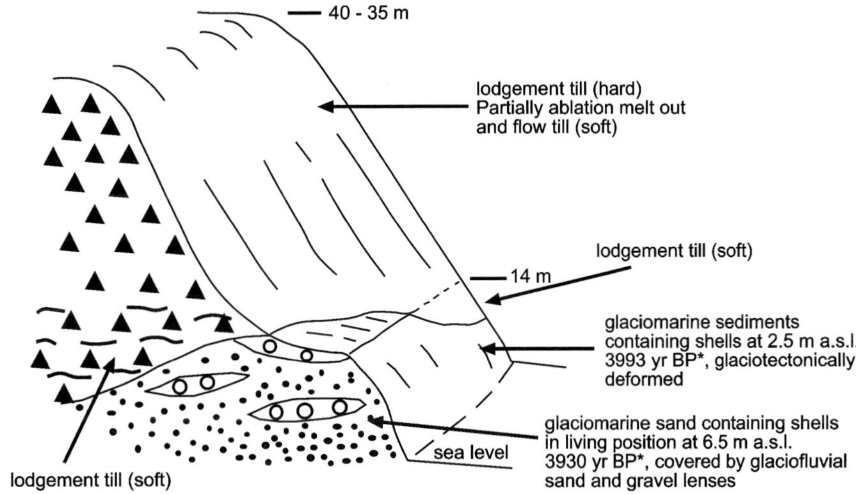

Fig. 8.7-5. Outcrop of a frontal moraine on the beach of Caña Quemada Bay; for location see Fig. 8.7-4

the South Georgia Islands (Clapperton and Sugden 1988) revealed a warm period between 5600 and 4800 yr B.P., reinforcing the possible existence of two regional middle Holocene glacial advances.

Caña Quemada Bay

In Caña Quemada Bay, NW James Ross Island (Fig. 8.7-2), the oldest lodgment till deposit was recognized on a marine cliff cut on the left side of the bay, probably associated with a lateral moraine. These deposits were preliminarily assigned to one or both of the oldest glacier advances (ca. 6500 to 4400 yr B.P.*). The next glaciogenic feature is a frontal moraine (Fig. 8.7-4) facing the internal bay shore. One outcrop (Fig. 8.7-5) on the beach also exhibits a lower glaciomarine unit associated with glaciotectonically deformed near-shore sediments containing shells (paired *Yoldia eightsi*) dated at ca. 4000 yr B.P.* covered by glaciofluvial sediments and soft lodgment till (deformable bed till or deformation till). This lower facies correlates laterally with a terrace of glaciofluvial or glaciomarine origin. The lower glaciomarine deposits of this terrace contain shells (paired *Yoldia eightsi*) dated at ca. 3900 yr B.P.*, suggesting that the glacier that deposited this frontal moraine overrode the glaciofluvial or glaciomarine terrace at or shortly after 3900 yr B.P.*

Upstream, a less-extended, well-formed frontal moraine belt was recognized (Fig. 8.7-4), evidencing a glacial advance that occurred after the regional Holocene Climate Optimum (3 900–3 000 yr B.P.*; Björck et al. 1996). This belt can be tentatively correlated with the main neoglacial advance of the South Georgia Islands (2 700–2 000 yr B.P.; Clapperton and Sugden 1988) and/or with the glacial advance detected in lake core obtained from Hidden Lake on James Ross Island (ca. 1 400–1 100 yr B.P.; Zale and Karlen 1989). The South Georgian glacial advance correlates with a cooling event in the region of the West Antarctic Peninsula fjords and channels (Shevenell et al. 1996; Domack et al. 2001).

The early and late Little Ice Age (LIA) is inferred to be responsible for the last frontal moraine deposit in Caña Quemada Bay, and also for the formation of many ice-cored moraines and rock glaciers recognized in other localities of James Ross Island. After the late-LIA glacial readvance, depending on local climatic conditions, the glaciers either retreated rapidly, leaving back-ice-cored frontal moraines, or receded slowly, inducing ice-cored rock glacier formation following the glacier ice models proposed by Potter (1972) and Whalley and Martin (1992). The ice-cored rock glaciers are considered to have principally developed by melt-out and ablation till accumulation and basal till extrusion.

Periglacial Features As Morphoclimatic Indicators

Large Gelifluction Lobes and Terraces

The fieldwork in Rink Crags revealed large gelifluction lobes and terraces classified as stone-banked terraces and lobes (Mori et al. 2006). The oldest features are up to 40 m long and 30 m wide, and reach up to 5 m high on frontal slopes. Considering the weathering of the clastic components, as well as lichen, moss cover, and the flow rate over the last 10 years, the lobes appear to be presently inactive, probably due to a shortage of water supply. These features are considered to have formed during the strong ablation of a nearby ice cap, probably during the Holocene regional Climatic Optimum (3 900–3 000 yr B.P.*). While most of the stone-banked terraces are inactive due to low water supply, stone-banked lobes, close to the ice-cap, are still active.

Protalus Lobes and Protalus Ramparts

In the Cerro Triple area near the Lachman Crags (Fig. 8.7-2), two landforms indicative of different climatic conditions were identified: protalus lobes and protalus ramparts (Strelin and Sone 1998). The protalus lobes (Whally and Martin 1992) are currently active and are related to the present, more arid (warmer and drier) climatic conditions. The protalus ramparts require strong creep activity and the presence of a permanent snow patch (nivation hollow) for formation. As the climate is becoming warmer and drier, the snow tends to melt out, halting the formation of this landform, turning it into an inactive and later fossil landform. The ramparts are likely to have been active during the LIA, suggesting a transition between these two landforms as a consequence of this climate change.

Lachman II Rock Glacier

Close to the Cerro Triple area, along the NE foot of the Lachman Crags (Fig. 8.7-2), several ice-cored rock glaciers

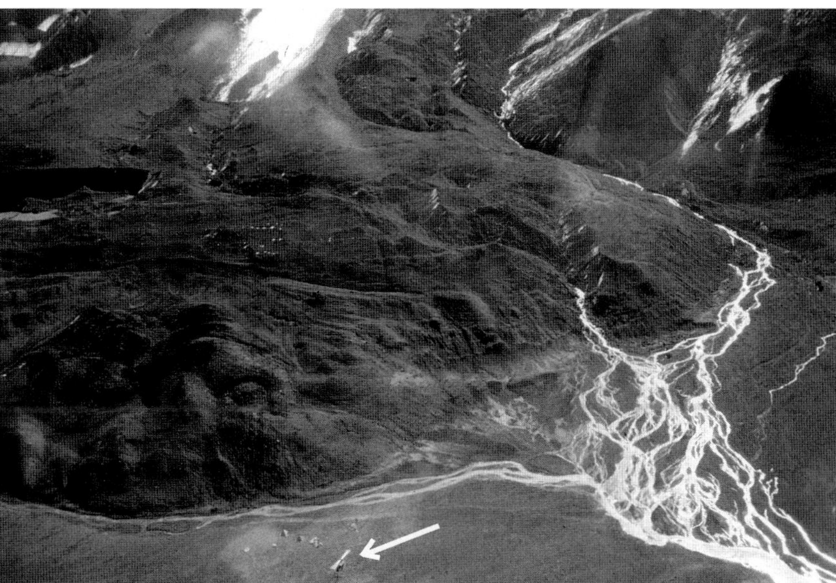

Fig. 8.7-6. Arial view of Lachman II rock glacier (see *helicopter* for scale)

can be recognized, related to glacial advances during the late LIA. In the case of the Lachman II rock glacier, a multi-temporal detailed morphometric study was conducted to determine the dynamic behavior of the rock glacier (Strelin and Sone 1998). As a result, the rock glacier was found to be part of a composite geomorphic system integrated by landforms with distinctly different morphodynamic behavior: glacier, ice-cored moraine, ice-cored rock glacier, and old rock glacier. Each of these landforms is related to specific paleoclimatic conditions: the retreating glacier tongue corresponds to warming over the last 50 years, with a marked increase in prominence over the last 15 years; the ice-cored moraine reflects an early 20th Century glacial advance; the main ice-cored rock glaciers would have developed in response to a late-LIA glacial advance and subsequent slow ablation (melt-out till, ablation till, and shear till formation); and the old rock glacier reflects the recession of the early LIA or a previous glaciation event.

Large conical holes, partially filled with water, were also recognized in the rock glacier. The increase in the diameter of these holes is clearly related to the air temperature trend over the last 15 years, evidence that climatic warming threatens to destroy this landform.

Conclusion

A survey of some relict and active landforms in the Lachman and Rink Crags areas of NW James Ross Island on the Antarctic Peninsula has yielded new evidence that provides a better understanding of the Holocene morphoclimatic evolution of this currently deglaciated sector of James Ross Island. In chronological order, this study has identified the following sequence of events:

a Deglaciation of the inner shelf and main fjords around James Ross Island occurred between 9 400 and 7 600 yr B.P.
b The first Neoglacial advance took place ca. 6 500 yr B.P. and was clearly detected and dated on Lachman Beach.
c The next glacier advance occurred 4 900 to 4 400 yr B.P.
d The third glacial advance took place ca. 3 900 yr B.P. and was clearly detected and dated in Caña Quemada Bay.
e Several minor glacier advances occurred after the regional Holocene Climatic Optimum.
f Ice-cored moraines and rock glaciers identified close to present glacial fronts in various localities of James Ross Island represent early/late LIA features and reflect more recent climatic fluctuations.
g Deglaciation initiated ca. 50 years ago, and there is evidence of strong glacial recession and wasting of periglacial forms over the last 15 years.

Acknowledgments

We thank the Instituto Antártico Argentino (Buenos Aires), the Centro Austral de Investigaciones Científicas (Ushuaia), Argentina, and the Institute of Low Temperature Science, Hokkaido University (Sapporo), Japan, for the logistical and technical support. This work was partly funded by the Agencia Nacional de Promoción Científica y Tecnológica and Dirección Nacional del Antártico, grant PICTO 2002 N 07-11573, Argentina.

References

Bjoerk S, Olsson S, Ellis-Evans C, Hakansson H, Humlum O, Lirio J (1996) Late Holocene palaeoclimatic records from lake sediments on James Ross Island, Antarctica. Palaeogeogr Palaeoclimatol Paleoecol 113:195–220

Clapperton C, Sugden D (1988) Holocene glacier fluctuation in South America and Antarctica. Quat Sci Rev 7:185–198

Domack E, Levanter A, Dunbar R, Taylor F, Brachfield S, Sjunneskog C, ODP Leg 178 Scientific Party (2001) Chronology of the Palmer Deep site, Antarctic Peninsula: a Holocene palaeoenvironmental reference for the circum-Antarctic. The Holocene 11:1–9

Gordon J, Harkness D (1992) Magnitude and geographic variation of the radiocarbon content in Antarctic marine life: implications for reservoir corrections in radiocarbon dating. Quat Sci Rev 711:697–708

Hjort C, Ingolfsson O, Moeller P, Lirio J (1997) Holocene glacial history and sea-level change on James Ross Island, Antarctic Peninsula. J Quat Sci 4:259–273

Ingolfsson O, Hjort C, Bjoerck S, Smith R (1992) Late Pleistocene and Holocene glacial history of James Ross Island, Antarctic Peninsula. Boreas 21:209–222

Matthies D, Mäusbacher R, Storzer D (1990) Deception Island tephra: a stratigraphical marker for limnic and marine sediments in Bransfield Strait area, Antarctica. Zentralbl Geol Palaeontol 153–165

Mori J, Sone T, Strelin J, Torielli C (2006) Surface movement of stone-banked lobes and terraces on Rink Crags, James Ross Island, Antarctic Peninsula. In: Fütterer DK, Damaske D, Kleinschmidt G, Miller H, Tessensohn F (eds) Antarctica – Contributions to global earth sciences. Springer, Berlin Heidelberg New York, pp 461–466

Potter N (1972) Ice-cored rock glacier, Galena Creek, Northern Absaroka Mountains, Wyoming. Geol Soc Amer Bull 83:3025–3058

Rabassa J (1987) Drumlins and drumlinoid forms in northern James Ross Island, Antarctic Peninsula. In: Menzies J, Rose J (eds) Drumlin Symposium. Balkema, Rotterdam, pp 267–288

Shevenell A, Domack E, Kernan G (1996) Record of Holocene palaeoclimate change along the Antarctic Peninsula: evidence from glacial marine sediments, Lallemand Fjord. Paper Proc Royal Soc Tasmania 130:55–64

Strelin J, Sone T (1998) Rock Glaciers on James Ross Is-land, Antarctica. Proc 7th Internat Permafrost Conf pp 1027–1033

Whalley B, Martin E (1992) Rock glaciers: II models and mechanisms. Progr Phys Geogr 16:127–186

Zale R, Karlen W (1989) Lake sediment cores from the Antarctic Peninsula and surrounding islands. Geograf Annaler 71(A):211–220

Surface Movement of Stone-Banked Lobes and Terraces on Rink Crags Plateau, James Ross Island, Antarctic Peninsula

Junko Mori[1] · Toshio Sone[2] · Jorge A. Strelin[3,4] · Cesar A. Torielli[5]

[1] Graduate School of Engineering, Hokkaido University, N13 W8, Kita-ku, Sapporo 060-8628, Japan, <jmori@eng.hokudai.ac.jp>

[2] Institute of Low Temperature Science, Hokkaido University, N19 W8, Kita-ku, Sapporo 060-0918, Japan

[3] Instituto Antártico Argentino and Centro Austral de Investigaciones Científicas, Av. Malvinas Argentinas S/N°, 9410 Ushuaia, Tierra del Fuego, Argentina

[4] *Present address:* Universidad Nacional de Córdoba, Ciudad Universitaria, Avda. Vélez Sársfield 1611, 5000 Córdoba, Argentina

[5] Departamento de Geología Básica, Universidad Nacional de Córdoba, Vélez Sársfield 299, 5000 Córdoba, Argentina

Abstract. Rink Crags Plateau is located on the northwestern part of James Ross Island, northeast of Antarctic Peninsula. Stone-banked lobes and terraces of various sizes are developed in a 1 km^2 area of average 4° slope upon the plateau. The largest lobes rise up to 5 m above the surrounding surface. Monitoring of painted markers upon the stone-banked lobes reveal that superficial movement of the tread surface was rapid on the central upper part of the lobes and terraces, whereas no movement was recorded near the risers. The solifluction velocity on the tread surface is up to 7.4 cm yr^{-1}. The lobes and terraces have a distinct frontal ridge. Three lobes were excavated and their internal structures were described. Coarse material accumulates on the riser slope and bank raise. The risers contain two layers of gravels of contrasting fabric. Wedge structures occur at the boundary between the frontal ridge and the tread. We propose that the formation of high risers upon gentle slopes results from rapid gelifluction movement. Differences in the internals structures described in this and previous studies suggest the process of formation is different to those previously proposed.

Introduction

James Ross Island is situated to the east of the Antarctic Peninsula (Fig. 8.8-1). The dry climate results in large ice-free areas on the northwestern part of the island. Periglacial features such as rock glaciers, protalus ramparts, patterned ground and solifluction features are well developed (e.g., Strelin and Malagnino 1992; Sone and Strelin 1997, 1998; Strelin and Sone 1998).

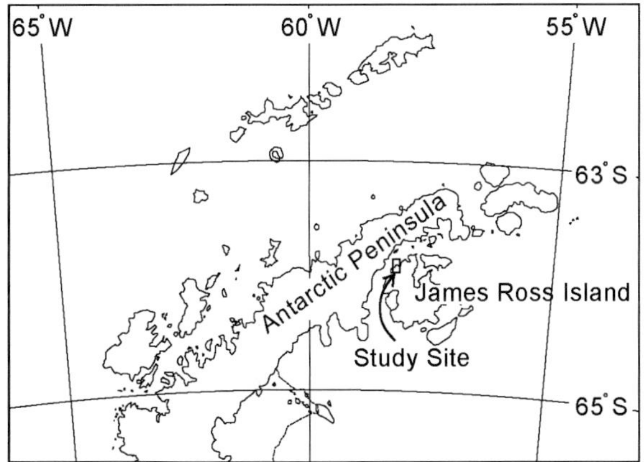

Fig. 8.8-1. Location map of the study area on James Ross Island

Rink Crags Plateau (Fig. 8.8-2), on northwest James Ross Island, contains large-scale stone-banked lobes and terraces developed over a 1 km^2 area of average 4° slope. Previous studies generally describe risers of >2 m height from slopes steeper than 10° (Matsuoka 2001; Grab 2000). Matsuoka (2001) concluded that the minimum frontal height of solifluction features is dependant upon the maximum depth of movement.

The development of stone-banked lobes and high risers is not fully understood. The internal structure of the high risers may provide clues to processes of their formation. But there are limited descriptions of the internal structure of stone-banked lobes and terraces (e.g., Francou 1990; Boelhouwers 1994; Grab 2000; Holness 2003).

In this paper we present the results of ongoing investigations, initiated in 1995, into the formation of stone-banked lobes and terraces on Rink Crags Plateau. We de-

Fig. 8.8-2. Aerial photo of the study area of northwestern part of Rink Crags Plateau. *Squares* indicate location of lobes analyzed for movement trends

From: Fütterer DK, Damaske D, Kleinschmidt G, Miller H, Tessensohn F (eds) (2006) Antarctica: Contributions to global earth sciences. Springer-Verlag, Berlin Heidelberg New York, pp 461–466

scribe the morphology, internal structure and dynamics of stone-banked lobes and terraces. Differences between our observations and those of previous studies are discussed. The processes of lobe formation will be considered in a subsequent paper.

Study Site

Rink Crags Plateau is situated on the northwestern part of James Ross Island (63°55' S, 58°10' W; Fig. 8.8-1) and has a mean altitude of 450 m a.s.l. The geology of the plateau consists of Cenozoic lavas and pyroclastites underlain by friable Mesozoic sedimentary rocks (Strelin and Malagnino 1992). The upper part of the plateau is covered with a small ice cap. Rock glaciers and protalus ramparts develop on the talus slope of the Crags.

Solifluction features are widely developed on the plateau (Fig. 8.8-2). These lobes and terraces have stony frontal risers and are classified as either stone-banked terraces or lobes, depending upon the ratio of width and length (Benedict 1970). As there is no difference between them except for the ratio of width and length, we treat them all as lobes hereafter.

Methods

Surface movement of the lobes was determined by annual monitoring of marker lines, of 2 cm width, painted during 1995 and 1998 on lobes E11, E15 and E18 (Fig. 8.8-2, 8.8-3). Two or three marker lines were painted across the lobes. Each line was painted between two benchmarks situated away from the lobe. A theodolite was used to measure the positions of the benchmarks relative to a datum point on bedrock in the study area.

The frontal banks of three lobes were excavated in January 2000 to reveal the internal structure of the lobes. The excavated lobes are different to those used for the monitoring of movement, and are located near lobe E18. Automated air temperature measurements have been conducted since 1995 on the tread of a lobe. At the site of temperature measurement, active layer soil temperatures have been measured at depths of 35, 20, 10, and 5 cm on the tread surface of the lobe since 1997. The depth of the deepest sensor was increased to 40 cm in 1999. From 1995 to 1999 the temperature was measured with DATAMARK of Hakusan Corporation with thermocouple sensors, from 1999, Ondotori Jr. of T and D Corporation with thermistor sensors.

Maximum snow accumulation was determined at 18 points on the plateau using snow meters that consist of vertical poles with horizontal wires fixed every 20 cm. Movement of the ice cap margin has been monitored by annual theodolite surveys since 1995.

Climatic Conditions

The mean annual air temperature at Rink Crags Plateau during the period 1996–2000 was –6.3 °C. The depth of the active layer, as estimated from the thermal profile of the active layer soil temperature was 40–70 cm. During the period 1995–2000, the ice cap retreated at an average rate of 10 m yr^{-1}. Annual precipitation occurs mostly as snow and is around 200 mm water equivalent as estimated at Lachman Crag, about 20 km east of the study site (Strelin and Sone 1998). Snow accumulation monitoring indicates that snow depth is greatest on the riser slope of lobes, reaching 140 cm depth. Nivation hollows occur on the foot of the frontal riser slopes of some of the lobes. Snow depth on the tread of the lobes is generally less than 20 cm.

The amount of frost heaving has not been measured, but adfreezing frost heave of steel stakes inserted to a depth of 35 cm indicates a maximum of 6.7 cm yr^{-1}. Soil temperatures at 5 cm depth recorded 61 freeze-thaw cycles between February 2000 and January 2001.

Description of the Stone-Banked Lobes

Shape of the Lobes

A sketch of typical lobe features observed on Rink Crags Plateau is shown in Fig. 8.8-3. The height of risers ranges from 0.23 m to 5 m and the inclination of riser slopes ranges from 7° to 27°. The widths of the lobes are up to 30 m. A raised ridge is typically developed on the margin of tread surface. The tread surface is covered with platy, angular to subangular and pebble to boulder clasts which are horizontally bedded. Sorted patterned ground is developed on the ground surface. Patterned ground within the upslope part of the tread consists of elongate shapes

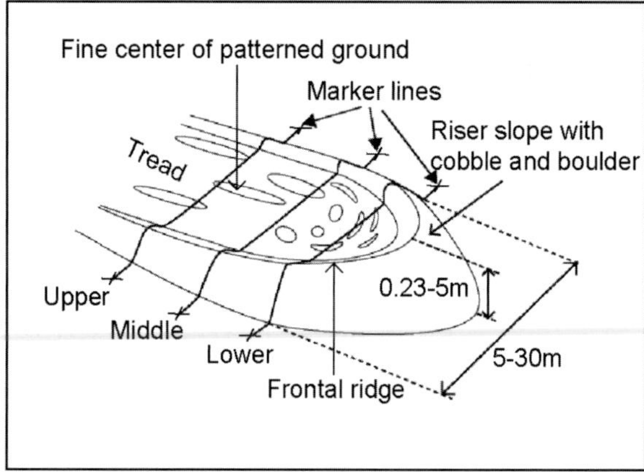

Fig. 8.8-3. Generalised sketch of a typical stone-banked lobe, Rink Crags Plateau

parallel to the slope direction, while on the area close to the frontal bank they are deformed to crescent shapes parallel to the frontal ridge (Fig. 8.8-3). Platy shaped stones near the frontal ridge are orientated vertically. Risers increase in height toward the edge of the plateau (Fig. 8.8-4).

Internal Structure of the Lobes

A cross section of one of the excavated lobes is shown in Fig. 8.8-5. The upper 0.2 m of the tread consists of silt and sand. The coarse part of the sorted patterned ground is less than 0.1 m in depth. Sediments deeper than 0.2 m are composed of matrix-supported pebble or cobble. Coarse clasts are platy-shaped and angular to subangular.

At the boundary between the tread and frontal ridge, clast-supported gravel forms wedge structures in the ground. Wedges within the lower slope are narrower and deeper than those further upslope. A layer of silty material occurs below the wedges but is absent from the riser slope.

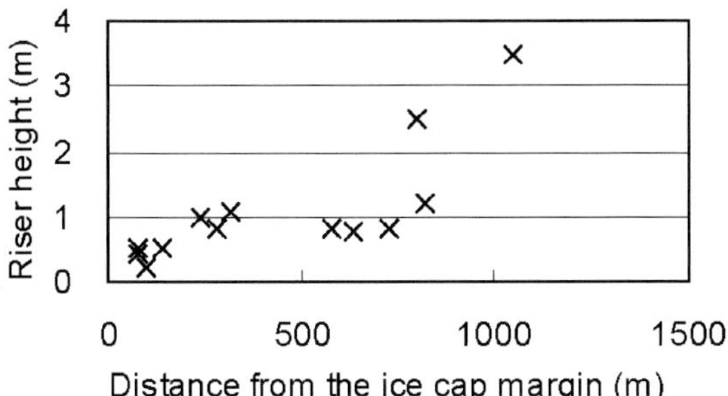

Fig. 8.8-4. Relationship between riser height and distance from the ice cap margin

Below the frontal ridge, the upper 0.35–0.4 m consists of platy pebble aligned parallel to the riser surface. The underlying layer consists of pebble with silty sand matrix. In the upper parts of this lower layer, sand and silt only cover the surface of the pebble. The deeper parts of this layer contain more matrix but remain clast-supported. Clasts are imbricated in this layer.

Within the excavated pit, the thawing depth was 0.4 m below the tread surface. Sediments are water-saturated on the frozen table. The thawing depth is deeper below the wedge structures, while below the riser slope it was deeper than 0.6 m. All three excavations revealed similar internal structure.

Surface Movement of Lobes

The annual displacement of three marker lines painted across lobe E18 is shown in Fig. 8.8-6. The upper, middle and lower lines are positioned 16 m, 8.75 m and 2.75 m respectively from the frontal ridge of the lobe. The upper marker line moved more rapidly than the middle line, while no movement was recorded upon the lower line. Measured rates of movement are variable from year to year, but the maximum recorded movement is always near the centre of the lobe.

Measured surface movements are similar for lobes E15 and E18. While downslope movement occurred on the tread surfaces of lobes E15 and E18, no movement was recorded across the ridges of these lobes (Fig. 8.8-6). The mean annual maximum movement of the upper marker lines is presented in Table 8.8-1, and was calculated as the average of ten data points about the maximum recorded value.

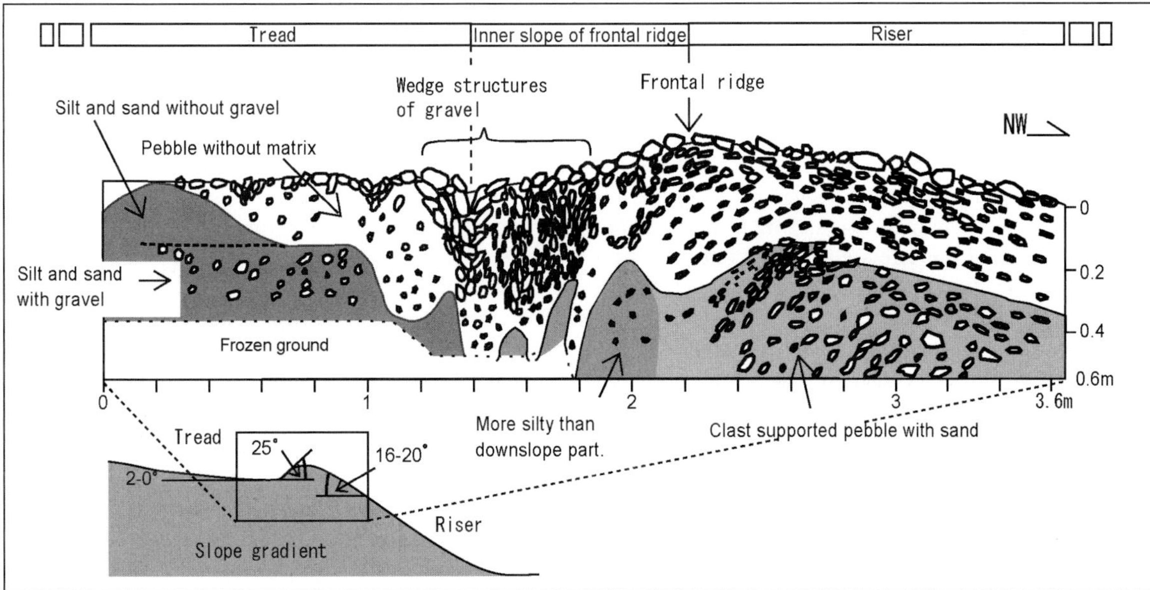

Fig. 8.8-5. Internal structure of an excavated stone-banked lobe

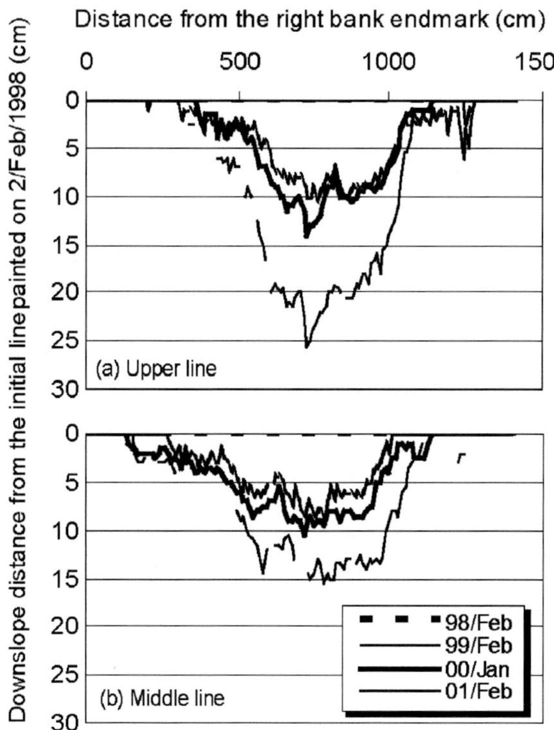

Fig. 8.8-6. Movement of marker lines across lobe E18, Feb. 1998 to Feb. 2001. No movement was recorded across the lower marker line during this period

Table 8.8-1. Distance from ice cap margin and maximum movement of the painted lines

	E15	E18	E11
Distance from ice cap margin (m)	300	600	850
Mean annual maximum movement of upper lines (cm yr^{-1})	4.78	7.37	<0.2

Discussion

The stone-banked lobes on the distal part of the Rink Crags Plateau have relatively high risers of >2 m in elevation. Until now, risers of this size are generally reported from surfaces with a gradient of >10° (Grab 2000; Matsuoka 2001).

The surface movement pattern described in this study is similar to that documented in previous studies (e.g., Benedict 1970). Matsuoka (2001) reviewed the solifluction movement velocity V_s (cm yr^{-1}) in permafrost areas as:

$$V_s = 100\tan\theta$$

with θ = slope gradient (°).

This formula predicts a velocity of solifluction for the study site of <6.3 cm yr^{-1}, which is less than the maximum recorded of 7.37 cm yr^{-1}.

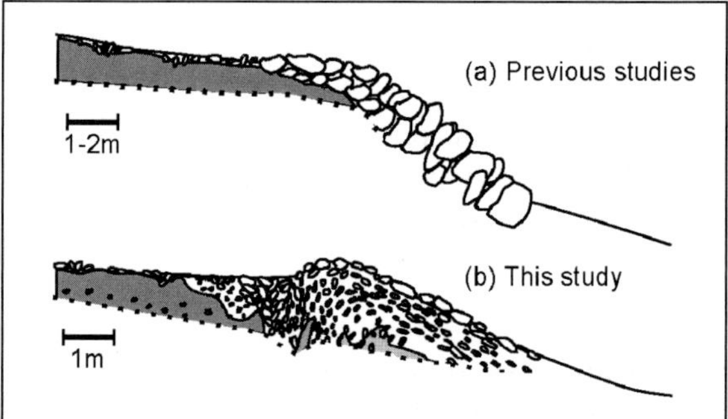

Fig. 8.8-7. a Schematic cross section of a typical stone-banked lobe/terrace as described in previous studies (Boelhouwers 1994; Van Steijin et al. 1995; Holness 2003). **b** Schematic cross section of a typical stone-banked lobe/terrace as described in this study

A comparison of the internal structure of stone-banked lobes described in this and previous studies (Holness 2003; Van Steijin et al. 1995; Boelhouwers 1994) is shown in Fig. 8.8-7. The lobes described in this study have a distinct frontal ridge and gravel wedges. These wedges penetrate deeper into the lobe than the wedge shapes within the patterned ground on the tread surface (Fig. 8.8-5). Patterned ground near the frontal ridge consists of crescent shapes oriented parallel to the frontal ridge (Fig. 8.8-3). Risers described in previous studies contain imbricated clasts, whereas the risers analyzed in this study contain imbricated clasts only at lower levels. Clasts within the upper part of the riser are parallel to the riser surface (Fig. 8.8-7b). These observations suggest that the stone-banked features on Rink Crags Plateau formed by processes different to those proposed in previously studies.

The rate of frontal advance of solifluction lobes has been calculated to be between 0.6 and 3.5 mm yr^{-1} (Harris 1981). During 5 years of monitoring during the current study, no frontal advance was detected despite the rapid movement of the tread surface. The increase in riser height probably results from the difference in velocity between solifluction movement on the tread and the riser. Caterpillar-like advance (Matsuoka 2001) is not considered a likely mechanism of riser growth in the studied lobes.

Lobe risers near the ice cap are about 50 cm high while those further from the ice cap are up to 5 m in height. We interpret this to indicate an increase in lobe height with increasing age of the riser.

We propose that movement of the lobes occurs dominantly by gelifluction. This interpretation is consistent with the gentle inclination of the plateau and the small amount of adfreezing frost heave recorded at the study site. The process of gelifluction requires a large supply of water. Water supply from the melting ice cap to the plateau sediments would be expected to result in faster surface movement in areas closer to the ice cap. Water is also supplied from snow patches and nivation hollows at the foot of high risers. These supplies of water would also contribute to the high rates of solifluction movement recorded on the plateau.

Conclusion

Rink Crags Plateau contains numerous stone-banked lobes. These solifluction features developed risers of up to 5 m in height despite the gentle gradient of the slope and an active layer of <50 cm thickness.

Monitoring of marker lines painted across three lobes indicates that lobe movement is rapid on the central and upper part of the tread and decreases down-slope. No movement of the frontal and lateral raise was detected. Movement of material on the tread surface is more rapid than that described in previous studies of lobes located on similarly gentle slopes.

The lobes have a distinct frontal ridge. The internal structure of the lobes reveals wedges of gravel developed between the ridge and the tread surface, and two layers of gravel within the risers, each of contrasting fabric.

Rapid gelifluction movement associated with adequate water supply is supposed to be a key process in the formation of high risers on a gentle slope such as that of the Rink Crags Plateau. Differences in the internal structure of lobes described in this study compared with previous studies suggest that the process of lobe formation is different to that described in previous models.

Acknowledgments

We thank the Instituto Antártico Argentino for logistical and technical support. We are grateful to Professor Masami Fukuda for helpful comments on the manuscript, to Dr. Kunio Watanabe and Dr. Kotaro Fukui for their assistance in the field.

References

Benedict JB (1970) Downslope soil movement in a Colorado Alpine Region: rates, processes, and climatic significance. Arctic Alpine Res 2:165–226

Boelhouwers JC (1994) Periglacial landforms at Giant's Castle, Natal Drakensberg, South Africa. Permafrost Perigl Processes 5: 129–136

Francou B (1990) Stratification mechanisms in slope deposits in high subequatorial mountains. Permafrost Perigl Processes 1: 249–263

Grab S (2000) Stone-banked lobes and environmental implications, High Drakensberg, Southern Africa. Permafrost Perigl Processes 11:177–187

Harris C (1981) Periglacial mass-wasting: a review of research. BGRG Research Monograph 4, Geo Abstracts, Norwich, pp 204

Holness SD (2003) The periglacial record of Holocene environmental change, subantarctic Marion Island. Permafros Perigl Processes 14:69–74

Matsuoka N (2001) Solifluction rates, processes and landforms: a global review. Earth Sci Rev 55:107–134

Sone T, Strelin JA (1997) Solifluction lobes in James Ross Island, Antarctic Peninsula. Proc General Meeting Association Japanese Geogr 53:368–369 (in Japanese)

Sone T, Strelin JA (1999) Air temperature conditions and climatic geomorphological characteristics of James Ross Island, Antarctic Peninsula region. Cuartas Jornadas Comun Investig Antárticas 2:272–377

Strelin JA, Malagnino EC (1992) Geomorfologia de la Isla James Ross. In: Rinaldi CA (ed) Geología de la Isla James Ross, Antártida. Inst Antártico Argentino, Buenos Aires, pp 7–36

Strelin JA, Sone T (1998) Rock glaciers on James Ross Island, Antarctica. Proc 7th Internat Permafrost Conf, 1027–1033

Van Steijn H, Bertran P, Francou B, Hétu B, Texier JP (1995) Models for the genetic and environmental interpretation of stratifeid slope deposits: review. Permafrost Perigl Processes 6:125–146

Soil Characteristics along a Transect on Raised Marine Surfaces on Byers Peninsula, Livingston Island, South Shetland Islands

Ana Navas[1] · Jerónimo López-Martínez[2] · José Casas[2,3] · Javier Machín[1] · Juan José Durán[4] · Enrique Serrano[5] · José-Antonio Cuchi[6]

[1] Estación Experimental de Aula Dei, CSIC, Apartado 202, 50080 Zaragoza, Spain, <anavas@eead.csic.es>

[2] Departamento de Química Agrícola, Geología y Geoquímica, Facultad de Ciencias, Universidad Autónoma de Madrid, 28049 Madrid, Spain, <jeronimo.lopez@uam.es>

[3] Centro de Ciencias Medioambientales, CSIC, Serrano 115, 28006 Madrid, Spain, <j.casassainzdeaja@uam.es>

[4] Instituto Geológico y Minero de España, Ríos Rosas 23, 28003 Madrid, Spain, <jj.duran@igme.es>

[5] Departamento de Geografía, Universidad de Valladolid, 47011, Valladolid, Spain, <serranoe@fyl.uva.es>

[6] Escuela Politécnica Superior de Huesca, Universidad de Zaragoza, 22071 Huesca, Spain, <cuchi@unizar.es>

Abstract. This paper presents the results of a soil survey carried out in Byers Peninsula to characterize some of the main features of soils and processes involved in soil development in the largest ice-free area in the South Shetland Islands. Soils were sampled in four different sites at altitudes between 24 and 88 m a.s.l., along a transect of about 1.5 km in the southwestern sector of the peninsula. The sampling sites are located on different geomorphological units from the Holocene raised beaches to the pre-Holocene upper marine platforms. It was found that general soil properties and elemental composition differ among the distinctive edaphic environments that conform the soils on the platforms and on the raised beaches. From the platform to the beach, pH and electrical conductivity values decrease, as well as the clay and silt percentages. Increases in carbonate and sand contents were observed along the transect. The variability in some elements (K, Fe, Al, Ca, Sr) appears to be related to mineralogy and parent materials. In the studied soils, the main factors affecting soil development are related to cryogenic processes. Lixiviation and other weathering processes also play a role in soil evolution although their influence is restricted because water circulation is limited to the summer period.

Introduction and Investigation Site

Soils provide a record of past and present conditions, and are a source of information on environmental changes. In cold regions the presence of underground ice produces chemical and physical effects that affect the distribution of components in water and in the solid phase of soils. Until now the soil forming processes under extreme climate are not completely understood. Studies of the influence of ground ice on soil formation are relatively rare in Antarctica, in part due to the scarcity of ice and snow-free areas (less than 3% of the total surface). A first survey on soils of the South Shetland Islands and adjacent parts of the Antarctic Peninsula, including observations on Byers Peninsula and other places of Livingston Island, was carried out by Everett (1976). Other studies on soils and landscapes of coastal areas have been carried out in maritime Antarctica more recently (e.g., Blume et al. 2001).

The present work aims to increase the knowledge on the soil characteristics and soil forming processes by studying some representative soils of ice free areas in the maritime Antarctic that comprises part of the Antarctic Peninsula and the surrounding islands. The climatic regime of this region is conducive to an ice-melting period during the summer, as well as the occurrence of snow free areas where soil development is favoured. In contrast, other Antarctic regions have more extreme climatic conditions and the soil surface is almost permanently covered by ice.

The Byers Peninsula, in the maritime Antarctic, is located at the western end of Livingston Island, South Shetland Islands (62°37' S, 61°06' W) and has the largest ice-free area in the archipelago that occupies about 60 km^2 (Fig. 8.9-1). The rock outcrops in the peninsula are mainly mudstones, sandstones, conglomerates and volcanic and volcaniclastic rocks (Upper Jurassic to Lower Cretaceous) intruded by different igneous bodies (e.g., Smellie et al. 1984; Crame et al. 1993). The study of lake sediments in Byers Peninsula indicates that the last major deglaciation of the peninsula began around 5–4 Ka B.P. (Björck et al. 1991, 1996; Björck and Zale 1996). The geomorphology of Byers Peninsula is characterized by the presence of extensive raised marine platforms and beaches, jointly with erosive and depositional features of glacial origin and a well developed drainage system, with many streams active in summer and more than 60 lakes. All these features, including the landforms and surface deposits of different origins have been represented in a geomorphological map of the Peninsula at 1:25 000 scale (López-Martínez et al. 1995) and the knowledge on both its geomorphology and Quaternary evolution has been compiled in López-Martínez et al. (1996).

Research carried out in Byers Peninsula identified the presence of wet permafrost in certain areas and an active layer of about 30 cm, but the thickness of permafrost still remains unknown. Studies on the distribution of permafrost (e.g., Bockheim 1995) and on the characteristics and fragility of Antarctic gelisols (e.g., Beyer et al. 1999) outline the existence of differences according to latitude and proximity to the coast. Thus an increase on soil water content and accumulation of organic matter occur in some areas of the Antarctic Peninsula region.

The Byers Peninsula belongs to the warmest and moistest area within the Antarctic region, that includes a part of Antarctic Peninsula and a series of islands. The

From: Fütterer DK, Damaske D, Kleinschmidt G, Miller H, Tessensohn F (eds) (2006) Antarctica: Contributions to global earth sciences. Springer-Verlag, Berlin Heidelberg New York, pp 467–474

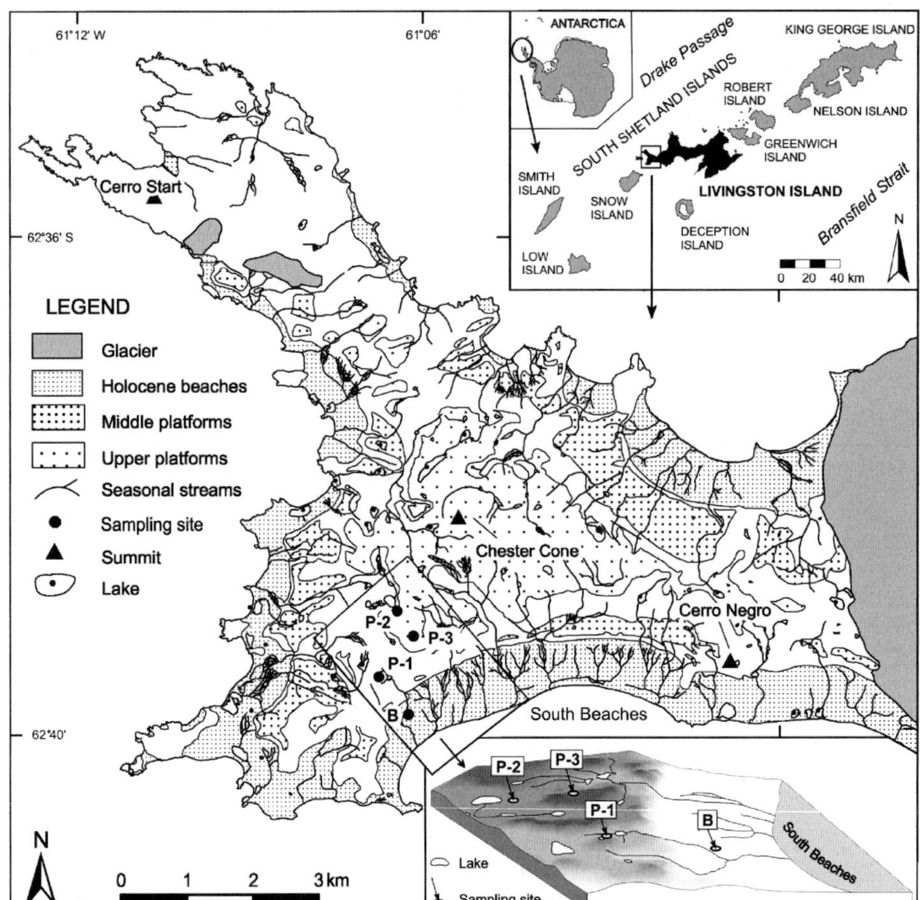

Fig. 8.9-1. Location and geomorphological setting of Byers Peninsula. The positions of sampling sites *B*, *P-1*, *P-2*, *P-3* are indicated

climate during summer is characterized by frequent precipitations and presence of clouds, mean day temperatures above 0 °C and moderate thermal amplitude. According to data recorded in Fildes Peninsula and Admiralty Bay, King George Island, and in Hurd Peninsula, the mean annual temperature in Livingston Island is about –2 °C at sea level and annual precipitation reaches 800 mm (Blümel and Eitel 1989; Rakusa-Suszczewski 1993; Bañón 2001). These climatic conditions are responsible for the soil differences that have been found between maritime Antarctica and other Antarctic areas (Allen and Heal 1970; Everett 1976; Campbell and Claridge 1987).

As part of an ongoing research on the surface formations in maritime Antarctica, several samples of representative soils from Byers Peninsula were collected along a transect of different geomorphological surfaces. In this study the general soil properties, as well as some characteristics on the geochemistry and mineralogy of the sampled soils are described. The aim of this work is to contribute to the knowledge of the soils and to identify the processes involved in soil formation under the presence of permafrost in the context of the specific climatic conditions existing in maritime Antarctica.

Methods

Soil samples were collected during the 2002/2003 Antarctic summer season and were stored at 4 °C until their analysis in laboratory. Soil around the sampling sites was excavated in order to extract samples in cylindrical containers thus maintaining their original volume and position. Subsamples were obtained after sectioning the soil samples according to soil texture and colour. Samples were air-dried, grinded, homogenized and quartered, to pass through a 2 mm sieve.

The general soil properties studied were: pH, electrical conductivity (*EC*), carbonate content, bulk density, granulometric fractions (>2 mm) and soil texture (<2 mm), organic matter, carbon, nitrogen, extractable phosphorous and potassium (CSIC 1976). Granulometric analysis of the clay, silt and sand fractions were performed using a laser equipment. To eliminate the organic matter, samples were chemically disaggregated with 10% H_2O_2 heated at 80 °C, then stirred and treated ultrasonically to facilitate particle dispersion. pH (1:2.5 soil:water) was measured using a pH-meter. Organic matter was determined by the Sanerlandt method (Guitian and Carballas 1976) using a titrimeter with selective electrode. Carbonate was measured using a pressure calcimeter (CSIC 1976).

Analysis of the total elemental composition was carried out after total acid digestion with HF (48%) in microwave. Samples were analysed for the following 17 elements: Li, K, Na (alkaline), Mg, Ca, Sr, Ba (light metals) and Cr, Cu, Mn, Fe, Al, Zn, Ni, Co, Cd and Pb (heavy metals). Analysis were performed by atomic emission spectrometry using an inductively coupled plasma ICP-OES (solid state detector). Concentrations, obtained after three measurements per element, are expressed in mg kg^{-1}.

The mineralogical study was made by means of X-ray diffraction using a difractometer equiped with Si-Li detector using CuKα radiation on random powder of bulk sample. The <2 µm fraction was studied on oriented samples after standard treatments. The reflecting powers of Schultz (1964) for bulk sample were used for the semiquantitative estimation of the identified minerals.

Studied Material and Sampling Sites

The studied soil samples are representative of different lithological and altitudinal contexts. In order to provide comparable results corresponding to the upper part of the soils, the depth of sampling was 13 cm in all sites. This sampling depth was selected because it was intended to describe the characteristics of the upper part of the soils where processes are supposed to be most active. The presence of permafrost limits both the depth of water circulation and its timing to just the summer season after ice melting. Besides, the common presence of large stones below this depth was an additional sampling difficulty. Therefore, the use of automatically driven coring equipment would have been required to extract undisturbed soil samples at deeper depths.

In Byers Peninsula, the high stone content is one of the main characteristic of very poorly developed soils showing almost no horizon differentiation. Snow covers the soil for most of the year and climatic conditions determine the existence of a very scarce vegetation cover in the region. However, Byers Peninsula and other ice-free areas of Livingston Island show a relatively high biodiversity (Sancho et al. 1999; Soechting et al. 2004). In the soils on the platforms there are lichens on rock outcrops and stony soils. Lichens are usually located on areas with positive relief where wind favours snow-free surfaces, whereas mosses develop in small depressions on finer grained soils. On the raised beaches the vegetation is also scarce and is mainly composed of grasses.

As can be seen in Fig. 8.9-1, the soil samples were collected along a transect of around 1.5 km located in the south-western sector of the peninsula. The selected four sites lie at altitudes of 24 m (B), 72 m (P-1), 70 m (P-2) and 88 m a.s.l. (P-3). According to the geomorphological units distinguished in the study area (López-Martínez et al. 1995), site B is on the upper part of the raised Holocene beaches. The other three sites are located in a glacially over-deepened catchment on marine platforms. The lower set of raised beaches below aprox 22 m a.s.l. are Holocene and the eroded upper marine surfaces are pre-Holocene (López-Martínez et al. 1996). Site P-1 is close to a relatively small lake and the other two sites are located in the catchment of another lake. Site P-2 is 2 m above the water surface of the lake and P-3 is in the upper part of the slope.

The rocks in the surrounding area of the sampling sites are mainly Upper Jurassic-Lower Cretaceous marine sandstones and conglomerates and Lower Cretaceous volcanoclastic materials. The surface detritic deposits in the area include microconglomerates, sandstones, lutites and fragments of andesitic basalts, cherts and calcitic veins. The sampling sites are on surfaces where the last main deglaciation started around 5–4 Ka B.P. (Björck et al. 1996). The geomorphological and hydrogeological processes in the area reflect the presence of permafrost and of an active layer (López-Martínez et al. 1996; Cuchi et al. 2004).

Analytical Results

In Byers Peninsula, the soils are very stony, mainly in the upper part of the transect at about 88 m a.s.l. (Cryic Leptosols) (ISSS – ISRIC-FAO 1998). Although the studied soils present a variety of textures, silty loams are predominant, specially in the pre-Holocene raised marine platforms, and sandy loam on the Holocene raised beaches. The coarsest fractions (>2 mm) are most abun-

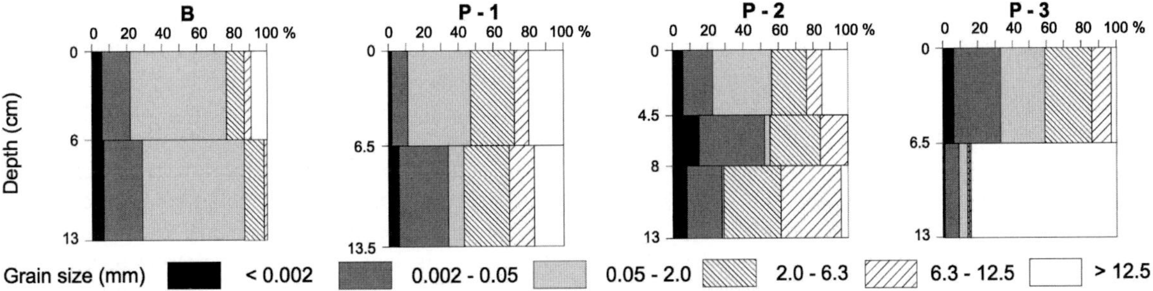

Fig. 8.9-2. Distribution of grain size fractions in the studied soils. For location of sampling sites see Fig. 8.9-1

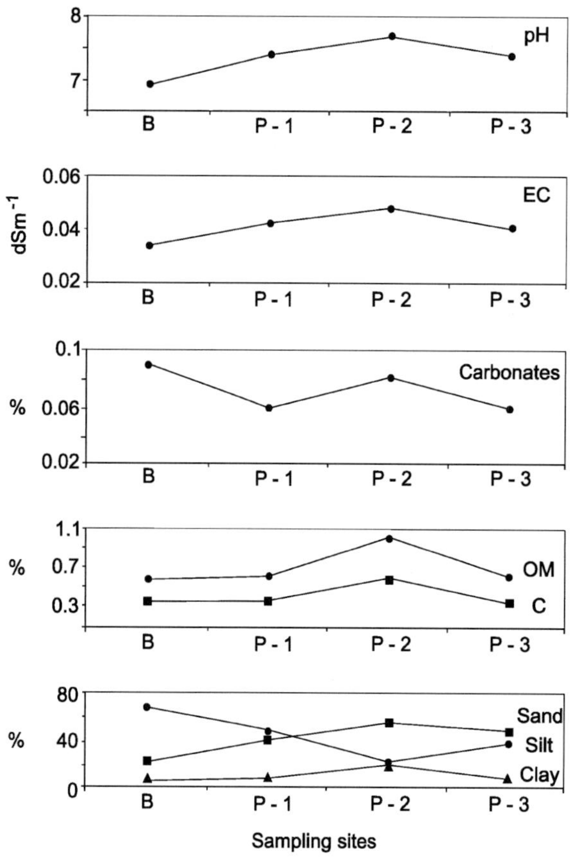

Fig. 8.9-3. Values of pH, electrical conductivity (*EC*), carbonates, carbon and organic matter, and grain size in soils along the transect. For location of sampling sites see Fig. 8.9-1

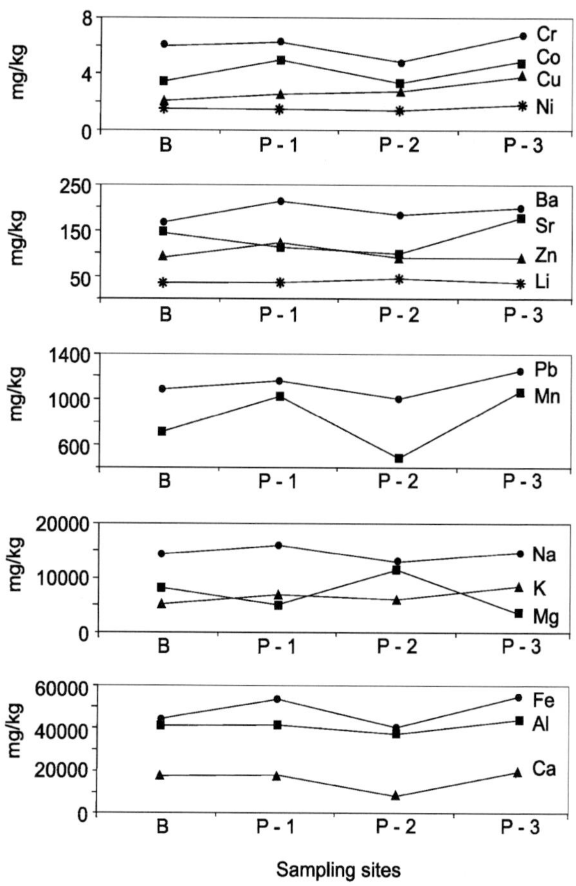

Fig. 8.9-4. Average elemental composition in the soils along the transect. For location of sampling sites see Fig. 8.9-1

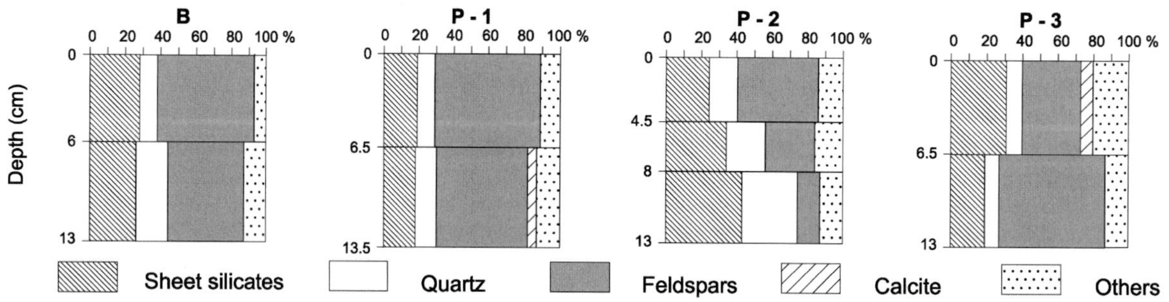

Fig. 8.9-5. Mineralogy of the studied soils. For location of sampling sites see Fig. 8.9-1

dant in the soils of the marine platforms (Fig. 8.9-2). The abundance of coarse material is related to cryogenic processes and in general they predominate at upper slope positions. Although coarse materials can be transported downslope by solifluction, the normal pattern of soil redistribution as a result of the processes of grain size classification, determines the abundance of finer materials at bottom slope positions (Navas et al. 2005).

Considering the grain-size fraction <2 mm, sand (4–78%) and silt (18–68%) are the most abundant, whereas clay contents are the lowest (4–28%). The grain size distribution varies along the studied transect. The highest silt and clay percentages and lowest sand content are found in site P-2 on the marine platform, whereas the opposite is registered in the raised beach site (B) (Fig. 8.9-3).

Concerning the general soil properties, the studied soils have low organic matter contents (0.40–1.13%) and carbonate contents are also very low (0.06–0.12%). Other soil parameters: carbon (0.29–0.66%), nitrogen (<0.1%) and extractable P (P_2O_5, 1.2–16.7 g 100 g^{-1}) and K (48–71 mg kg^{-1}) are also present in low ranges.

Table 8.9-1. Basic statistics of the elemental composition in all the studied soils

Elements ($mg\,kg^{-1}$)	Average	Standard error	Min	Max
Mg	6288	454	4417	8894
K	7397	1271	3702	14300
Na	13675	76	9371	16761
Pb	1114	37	924	1313
Ba	193	12	143	257
Zn	101	6	87	148
Sr	132	13	82	187
Li	43	2	37	52
Mn	788	101	356	1253
Co	4.5	0.4	2.7	6.7
Ni	1.7	0.1	1.5	2.2
Cu	2.9	0.2	2.2	4.0
Cr	6.0	0.3	5.0	7.3
Fe	46661	2504	34229	56032
Al	40476	1086	36580	47636
Ca	15330	1806	6482	22648

Salinity does not exceed 0.056 dS m^{-1} and pH ranges between 6.59 and 8.07. Soil bulk density values are between 1.3 and 2.1 g cm^{-3} (Fig. 8.9-3).

Regarding the elemental composition of the soils analysed by ICP-OES (Table 8.9-1, Fig. 8.9-4), major elements were Fe and Al, with average contents of 4.7 and 4.1% respectively. Ca was less abundant, around 1.5%. Mean percentages of Na, K, Mg were 1.4%, 0.7% and 0.6% respectively. Pb and Mn were around 0.1%. Barium, Sr, Zn and Li were below 0.02% and the remaining elements (Co, Ni, Cu and Cr) occured as traces, whereas Cd was not detected. Therefore, Fe and Al total almost 66% of the elements analysed and together with Ca reach as much as 88%. Although little is known on the geochemistry of soils in Antarctic ice free areas, it appears that the contents on major and trace elements fall within the range for soils developed on similar parent materials (Kabata-Pendias and Pendias 2001).

The studied soil samples are clastic, with feldspars and rock fragments composed of feldspars and quartz grains. Radial fibre zeolites were detected infilling cavities. The mineralogical composition of the studied powdered whole samples is shown in Table 8.9-2 and in Fig. 8.9-5. The fraction <2 µm located among the grains identified by XRD as a randomly interestratified chlorite-smectite being smectite the dominant term. Small amounts of illite or kaolinite were occasionally detected.

Table 8.9-2. Mineralogical composition of the studied soils. Other minerals are cristobalite and zeolites, mainly thomsonite and laumontite

Site	Depth (m)	Sheet silicate (%)	Qtz (%)	Fsp (%)	Calc. (%)	Other (%)
B	0 – 6	28	10	55	0	7
	6 – 13	26	18	43	0	13
P-1	0 – 6.5	19	10	60	0	11
	6.5 – 13.5	18	12	52	5	13
P-2	0 – 4.5	24	17	45	0	14
	4.5 – 8	34	22	28	0	16
	8 – 13	43	31	13	0	13
P-3	0 – 6.5	31	9	33	7	20
	6.5 – 13	19	8	60	0	13
Mean values						
B	0 – 13	27	14	49	0	10
P-1	0 – 13.5	18.5	11	56	2.5	12
P-2	0 – 13	33.7	23.3	28.7	0	14.3
P-3	0 – 13	25	8.5	46.5	3.5	16.5

Discussion and Conclusions

Along the studied transect pH and EC values slightly decrease from the upper platform to the Holocene raised beach and this could be related to the different composition of the parent materials. On the contrary, apart from site P-2, the carbonate content increases slightly at the raised beach. The highest organic matter and carbon contents are found at P-2, whereas the remaining sites have very similar values. This could be due to the location of this site at the footslope of an overdeepened catchment that could favour the accumulation of organic matter associated with a larger content in finer soil fractions.

The sand content increases from the platform towards the raised beach whereas silt and clay show an inverse behaviour. The difference in soil textures is caused by the different parent materials on which soils are developed. On the marine platforms, the predominance of silt in the soils is related to the outcropping lithologies of Cretacic age that are composed of marine mudstones and sandstones together with some volcanic and intrusive rocks. Whereas in the Holocene raised beaches materials are mainly composed by gravels and sand.

The analysis of the relationships between the elements and the general soil properties showed that in general, most of the elements have an inverse relationship with the clay fraction. This was not the case for Li and K as their direct relationship with the clay fraction could be due to the association of these two elements with feldspars and clay minerals. The high levels of Fe seem to be related with Fe oxides. The organic matter and soil nutrients (C and P) are also significantly correlated with the clay fraction ($r = 0.9$ and $r = 0.7$, respectively).

Concerning the relationships among the elements, apart from Ba and Zn that do not show any relationship with the rest of the elements, most of the elements (Mg, Mn, Sr, Al, Fe, Ca, Pb) are directly related between them with significant values of the Pearson correlation coefficient ($r = 0.6$). A highest correlation is found between the trace elements that are directly related between them ($r = 0.9$).

The mineralogy of the samples is clearly related with the distribution of grain size, specially concerning the contents of sheet silicates. Furthermore, these minerals are more abundant in the lower part of the profile P-2. It appears that the geochemical variability of these soils is in close relation with the parent materials and their mineralogical composition. Thus, site P-2 on the marine platform, has the highest K content and this could be related with the highest percentage of the finer fraction as well as with the abundance of sheet silicates in this site. Besides, the lowest content in Ca is related to the total absence of calcite in this site. Also, in comparison with the remaining sites, site P-2 has the lowest Mn content, whereas the highest Sr content is found in site P-3.

These results indicate that there are some differences in the general soil properties as well as in the elemental composition among the distinctive morphoedaphic environments that conform the soils on the platforms and on the beaches in Byers Peninsula. Variations in some soil properties may be due to water movement. It has been pointed out the importance of permafrost and the active layer on the underground water recharge and flow in Byers Peninsula (Cuchi et al. 2004). However in this environment, geomorphologic (cryogenic) and physical processes, as soil movement, are of greater importance as observed along the slope between P-3 and P-2, where the content of finer fractions increase at the bottom slope.

In general, the soils in Byers Peninsula are influenced by limited water availability during summer. Cryoturbation occurs only during approximately three months without snow coverage during the summer period. As consequence of the climatic and physiographic characteristics in this environment, the soils are scarcely developed and show very poor horizon differentiation. In agreement with other studies in King George Island (Jie et al. 2000), poorly developed soils are expected in areas with patterned ground and freeze-thaw cycles, this is generally accepted and has also been observed in Fildes Peninsula. The role of freeze-thaw, wetting and drying and other weathering mechanisms has also been studied by several authors (Hall 1993; Serrano et al. 1996).

In these environments freeze-taw weathering is generally recognized to be the most important process causing rock disintegration. Hall (1992) indicates that in Byers Peninsula wetting and drying, salt weathering and chemical weathering could be more important than freeze-thaw. Nevertheless, from the first results of this research it appears that in the studied transect, the main processes affecting soil development and evolution are related to cryogenic processes resulting from freeze-thaw cycles that lead to rock disintegration and supply the material for soil development. Lixiviation and other weathering processes also play a role in soil evolution, particularly in relation with water movement down the soil profile. However, these processes seem to be of more limited effect because of the short period of water circulation that only occurs during the summer.

Acknowledgements

This paper is a contribution to the project REN2001-0643 of the DGCYT, Spanish Ministry of Science and Technology. The comments by two anonymous referees have improved the manuscript. The authors thank the cooperation of A. Quesada, E. Rico, A. Fernández-Valiente and the other members of the project REN2000-0435. The logistic support provided by A. Simón and the personnel of vessel Las Palmas and Juan Carlos I Spanish station is thanked.

References

Allen SE, Heal OW (1970) Soils of the maritime Antarctic Zone. In: Holdgate MW (ed) Antarctic ecology. Acad Press, London, pp 693–696

Bañón M (2001) Observaciones meteorológicas en la Base Antártica Española Juan Carlos I. Publ A-151, Ser Monograf Inst Nacional Meteorología, Madrid

Beyer L, Bockheim JG, Campbell IB, Claridge GGC (1999) Genesis, properties and sensitivity of Antarctic Gelisols. Antarctic Sci 11(4):387–398

Björck S, Zale R (1996) Late Holocene tephrochronology and palaeoclimate, based on lake sediment studies. In: López-Martínez J, Thomson MRA, Thomson JW (eds) Geomorphological map of Byers Peninsula, Livingston Island. BAS Geomap Series 5-A. British Ant Surv, Cambridge, pp 43–48

Björck S, Hakansson H, Zale R, Karlen W, Jönsson BL (1991) A late Holocene lake sediment sequence from Livingston Island, South Shetland Islands, with palaeoclimatic implications. Antarctic Sci 3:61–72

Björck S, Hjort C, Ingólfsson O, Zale R, Ising J (1996) Holocene deglaciation chronology from lake sediments. In: López-Martínez J, Thomson MRA, Thomson JW (eds) Geomorphological map of Byers Peninsula, Livingston Island. BAS Geomap Series 5-A. British Ant Surv, Cambridge, pp 49–51

Blume HP, Kuhn D, Bölter M (2001) Soils and soilscapes. In: Beyer L, Bölter M (eds) Geoecology of Antarctic ice-free coastal landscapes. Springer, Berlin Heidelberg New York, pp 92–113

Blümel WD, Eitel B (1989) Geoecological aspects of maritime-climatic and continental periglacial regions in Antarctica (South Shetlands, Antarctic Peninsula and Victoria Land). Geoökodynamik 10:201–214

Bockheim JG (1995) Permafrost distribution in the southern circumpolar region and its relation to the environment: a review and recommendations for further research. Permafrost Periglac Processes 6:27–45

Campbell IB, Claridge CGC (1987) Antarctic: soils, weathering processes and environment. Elsevier, Amsterdam, pp 330–334

Crame JA, Pirrie, D, Crampton JS, Duane AM (1993) Stratigraphy and regional significance of the Upper Jurassic-Lower Cretaceous Byers Group, Livingston Island, Antarctica. J Geol Soc London 150:1075–1087

CSIC (1976). Comisión de métodos analíticos. Anales de Edafología y Agrobiología 35:813–814

Cuchi JA, Durán JJ, Alfaro P, Serrano E, López-Martínez J (2004) Discriminación mediante parámetros fisicoquímicos de diferentes tipos de agua presentes en un área con permafrost (península Byers, isla Livingston, Antártida occidental). Bol R Soc Esp Hist Nat 99(1–4):75–82

Everett KR (1976) A survey of the soils in the region of the South Shetland Islands and adjacent parts of the Antarctic Peninsula. Inst Polar Stud Rep 58, Ohio State Univ, Columbus

Guitian F, Carballas T (1976) Técnicas de análisis de suelos. Ed Pico Sacro. Santiago de Compostela

Hall KJ (1992) Mechanical weathering on Livingston Island, South Shetland Islands, Antarctica. In: Yoshida Y, Kaminuma K, Shiraishi K (eds) Recent progress in Antarctic earth science. Terra Scientific Publishing, Tokyo, pp 756–762

Hall KJ (1993) Enhanced bedrock weathering in association with late-lying snowpatches: evidence from Livingston Island, Antarctica. Earth Surface Process Landforms 18:121–129

ISSS-ISRIC-FAO (1998). World Reference Base for Soil Resources

Jie Ch, Zitong G, Blume HP (2000) Soils of Fildes Peninsula, King George Island, the maritime Antarctica, Part I, Formation processes and pedogenic particularities. Chinese J Polar Sci 11(1): 25–38

Kabata-Pendias A, Pendias H (2001) Trace elements in soils and plants (3rd ed). CRC Press, Boca Raton

López-Martínez J, Martínez de Pisón E, Serrano E, Arche A (1995) Geomorphological map of Byers Peninsula, Livingston Island, E. 1:25 000. BAS Geomap Series 5-A. British Ant Surv, Cambridge

López-Martínez J, Thomson MRA, Thomson JW (eds) (1996) Geomorphological map of Byers Peninsula, Livingston Island. BAS Geomap Series 5-A. British Ant Surv, Cambridge

Navas A, Soto J, López-Martínez J (2005) Radionuclides in soils of Byers Peninsula, South Shetland Islands, Western Antarctica. Appl Radiat Isotopes 62(5):809–816

Rakusa-Suszczewski S (ed) (1993) The maritime Antarctic coastal ecosystem of Admiralty Bay. Polish Academy of Science, Warsaw

Sancho LG, Schulz F, Schroeter B, Kappen L (1999) Bryophyte and lichen flora of South Bay (Livingston Island, South Shetland Islands, Antarctica). Nova Hedwigia 68:301–337

Schultz LG (1964) Quantitative interpretation of mineralogical composition from X-ray and chemical data of the Pierre Shale. U.S. Geol Surv Prof Paper 391C

Serrano E, Martínez de Pisón E, López-Martínez J (1996) Periglacial and nival landforms and deposits. In: López-Martínez J, Thomson MRA, Thomson JW (eds) Geomorphological map of Byers Peninsula, Livingston Island. BAS Geomap Series 5-A. British Ant Surv, Cambridge, pp 15–19

Smellie JL, Pankhurst RJ, Thomson MRA, Davies RES (1984) The geology of the South Shetland Islands: VI. Stratigraphy, geochemistry and evolution. British Ant Surv Sci Rep 87, Cambridge

Soechting U, Oevstedal DG, Sancho LG (2004) The lichens of Hurd Peninsula, Livingston Island, South Shetlands, Antarctica. Bibliotheca Lichenologica 88:607–658

Index

The numbers refer to the first page of each chapter concerning the index entry; in brackets: numbers of chapters concerned.

A

A-type granites 45 (2.4)
Actinium-227 (^{227}Ac) 409 (8.1)
Adare Basin 319 (6.5)
ADMAP 109 (3.1)
aeromagnetics 83, 95, 117 (2.8, 2.9, 3.2)
age determination 69, 191, 205, 217, 249 (2.7, 4.4, 4.6, 5.1, 5.5)
Allan Hills 291 (6.1)
alteration 55 (2.5)
Amery Ice Shelf 83 (2.8)
Amery Oasis 447 (8.6)
Amundsen Bay 15 (2.1)
Antarctic Circumpolar Current 433, 441 (8.4, 8.5)
Antarctic Peninsula 217, 229, 383, 455, 461 (5.1, 5.2, 7.6, 8.7, 8.8)
Antarctic Plate 351 (7.1)
Argentina Islands 383 (7.6)

B

Balleny Islands 351 (7.1)
banded iron formation 15 (2.1)
basin(s) 117, 309, 327 (3.2, 6.4, 6.6)
Beacon Supergroup 291 (6.1)
Beardmore Glacier 181 (4.3)
"Beardmore Orogeny" 181 (4.3)
Bellingshausen Sea 351 (7.1)
Bowers Terrane 205 (4.6)
Bransfield Basin 243 (5.4)
Bransfield Strait 229, 243 (5.2, 5.4)
Byrd Glacier 181, 191 (4.3, 4.4)
Byrd Group 181, 191 (4.3, 4.4)

C

Cambrian 181, 191 (4.3, 4.4)
Cape Roberts Project 309 (6.4)
Casey Bay 63 (2.6)
climate change 423, 447, 455 (8.3, 8.6, 8.7)
continental margin 327 (6.6)
continental slope 341, 423 (6.7, 8.3)
cores 309, 415, 423 (6.4, 8.2, 8.3)

D

data base 109, 351, 397, 403 (3.1, 7.1, 7.8, 7.9)
Deception Island 283, 391, 397 (5.10, 7.7, 7.8)
DECVOL 283 (5.10)
Dee Island 255 (5.6)
deformation 175, 195, 255, 271, 277, 357 (4.2, 4.5, 5.6, 5.8, 5.9, 7.2)
diatoms 447 (8.6)
Discovery Bank 237 (5.3)
Drake Passage 243, 433 (5.4, 8.4)
Dronning Maud Land 37, 45, 55, 357 (2.3, 2.4, 2.5, 7.2)
Drygalski 7 (1.2)

E

earthquakes 351, 361 (7.1, 7.3)
East Antarctic Craton 165 (4.1)
East Antarctica 341 (6.7)
Elephant Island 271, 277 (5.8, 5.9)
EMP dating 63 (2.6)
Enderby Land 15, 63, 147, 361 (2.1, 2.6, 3.7, 7.3)

F

faults 175, 195, 255, 271 (4.2, 4.5, 5.6, 5.8)
Ferrar Group 291 (6.1)
Ferrar Large Igneous Province 165, 255 (4.1, 5.6)
Filchnerfjella 55 (2.5)
fluid infiltration 55 (2.5)
fracture zones 175, 229, 271, 277 (4.2, 5.2, 5.8, 5.9)

G

Gaussberg 7 (1.2)
Gawler Craton 165 (4.1)

geochemistry 45, 423 (2.4, 8.3)
GEODEC 283 (5.10)
geoid 369, 383, 391 (7.4, 7.6, 7.7)
geomagnetics 377 (7.5)
geomorphology 141, 277, 455 (3.6, 5.9, 8.7)
German South Polar Expedition 1901–1903 7 (1.2)
GIS 397 (7.8)
Gondwana 37, 83, 109, 165, 175, 195, 217 (2.3, 2.8, 3.1, 4.1, 4.2, 4.5, 5.1)
GPS 357, 383 (7.2, 7.6)
granulites 23, 37, 63 (2.2, 2.3, 2.6)
gravity 117, 129, 327 (3.2, 3.4, 6.6)
Greenwich Island 255 (5.6)
Grenvillian 95 (2.9)
Grove Mountains 95 (2.9)

H

Hero Fracture Zone 229 (5.2)
history 3, 7 (1.1, 1.2)
Hurd Peninsula 217, 249 (5.1, 5.5)

I

Indian Ocean 415 (8.2)
International Symposia on Antarctic Earth Science (ISAES) 3 (1.1)

J

James Ross Island 455, 461 (8.7, 8.8)

K

K-Ar ages 205 (4.6)
King George Island 261 (5.7)

L

Lake Concordia 129 (3.4)
Lake Vostok 129, 135, 141 (3.4, 3.5, 3.6)
Lambert Glacier 69, 83 (2.7, 2.8)
Larsemann Hills 83 2.8
Livingston Island 217, 249, 467 (5.1, 5.5, 8.9)
Lützow-Holm Bay 23, 147, 361 (2.2, 3.7, 7.3)

M

MacRobertson Land 83, 447 (2.8, 8.6)
magnetics 15, 83, 95, 109, 117, 165, 319, 327, 377, 415 (2.1, 2.8, 2.9, 3.1, 3.2, 4.1, 6.5, 6.6, 7.5, 8.2)
Marie Byrd Land 123, 299 (3.3, 6.2)
Matusevich Glacier 175 (4.2)
Mawson Continent/Craton/block 165 (4.1)
Mawson Escarpment 69, 83 (2.7, 2.8)
Miers Bluff Formation 249 (5.5)
Moho 147, 155, 229 (3.7, 3.8, 5.2)
Mühlig-Hofmannfjella 45, 55 (2.4, 2.5)

N

Northern Basin 319 (6.5)

O

Oates Coast/Land 165, 175, 205 (4.1, 4.2, 4.6)

P

paleo-climate 433 (8.4)
paleontology 249 (5.5)
paleo-Pacific margin 165, 195, 217 (4.1, 4.5, 5.1)
Pan-African 45, 55, 63, 69, 83, 95 (2.4, 2.5, 2.6, 2.7, 2.8, 2.9)
PANGEA information system 403 (7.9)
Patagonia 217 (5.1)
Pb-Pb ages 69 (2.7)
permafrost 467 (8.9)
post-glacial rebound 123 (3.3)
post-rift sediments 341 (6.7)
Prince Charles Mountains 69, 83 (2.7, 2.8)
Princess Elizabeth Land 83, 95 (2.8, 2.9)
Prydz Bay 83, 95 (2.8, 2.9)

Q

Quaternary 277, 423, 455, 467 (5.9, 8.3, 8.7, 8.9)

R

radio echo sounding 135, 141 (3.5, 3.6)
radionucleides as tracers 409 (8.1)
Radium-228 (^{228}Ra) 409 (8.1)
Radok Lake 447 (8.6)
Rauer Islands 83 (2.8)
Rayner Complex 63 (2.6)
Rennick Glacier 175, 195 (4.2, 4.5)
rift 299, 304, 309, 327 (6.2, 6.3, 6.4, 6.6)
Robertson Bay Terrane 205 (4.6)
Rodinia 37, 69, 165 (2.3, 2.7, 4.1)

Ross Orogen(y) 165, 175, 181, 195, 205 (4.1, 4.2, 4.3, 4.5, 4.6)
Ross Sea 117, 123, 155, 303, 309, 319, 423 (3.2, 3.3, 3.8, 6.3, 6.4, 6.5, 8.3)

S

sapphirine granulite 37 (2.3)
SCAR 3 (1.1)
Schirmacher Hills/Oasis 37 (2.3)
Scotia Arc 369 (7.4)
Scotia Plate 237 (5.3)
Scotia Sea 217, 237, 351, 369, 433, 441 (5.1, 5.3, 7.1, 7.4, 8.4, 8.5)
sea-level changes 123, 361 (3.3, 7.3)
secular variation 377 (7.5)
seismicity 155, 351, 361 (3.8, 7.1, 7.3)
seismics 141, 147, 229, 237, 243, 309, 327, 341, 361, 441 (3.6, 3.7, 5.2, 5.3, 5.4, 6.4, 6.6, 6.7, 7.3, 8.5)
Shackleton Fracture Zone 271, 277 (5.8, 5.9)
SHRIMP U-Pb ages 69, 217 (2.7, 5.1)
SKS splitting 155 (3.8)
Sm-Nd model ages 69 (2.7)
soil 467 (8.9)
South Sandwich Islands 369 (7.4)
South Scotia Ridge 237, 271 (5.3, 5.8)
South Shetland Block 271, 277 (5.8, 5.9)
South Shetland Islands 249, 255, 261, 397, 467 (5.5, 5.6, 5.7, 7.8, 8.9)
Southern Ocean 341, 409 (6.7, 8.1)
stone-banked lobes 455, 461 (8.7, 8.8)
structural geology 175, 195, 255, 271, 277 (4.2, 4.5, 5.6, 5.8, 5.9)
subglacial lakes 129 (3.4)
subsidence 123 (3.3)

T

Tasmania 165 (4.1)
Terra Nova Bay 155, 205 (3.8, 4.6)
Terra Adélie 327, 351 (6.6, 7.1)
terraces 123, 455, 461 (3.3, 8.7, 8.8)
Thorium-234 (^{234}Th) 409 (8.1)
Transantarctic Mountains 155, 181, 191 (3.8, 4.3, 4.4)
Trinity Peninsula Group 217 (5.1)

U

uplift 205, 271, 277, 299, 351, 357, 361 (4.6, 5.8, 5.9, 6.2, 7.1, 7.2, 7.3)

V

Vestfold Hills 83 (2.8)
Victoria Land 155, 165, 195, 205, 291, 351 (3.8, 4.1, 4.5, 4.6, 6.1, 7.1)
Victoria Land Basin 303, 309 (6.3, 6.4)
volcanic activity 283, 391 (5.10, 7.7)
volcanism 261, 299 (5.7, 6.2)

W

Weddell Sea 217, 237, 441 (5.1, 5.3, 8.5)
West Antarctic Ice Sheet 117, 123 (3.2, 3.3)
West Antarctic Rift (System) 117, 123, 319 (3.2, 3.3, 6.5)
West Antarctica 123 (3.3)
Wilkes Land 165, 327, 351, 415 (4.1, 6.6, 7.1, 8.2)
Wilson Terrane 195, 205 (4.5, 4.6)